PARASITES AND DISEASES
OF WILD BIRDS IN FLORIDA

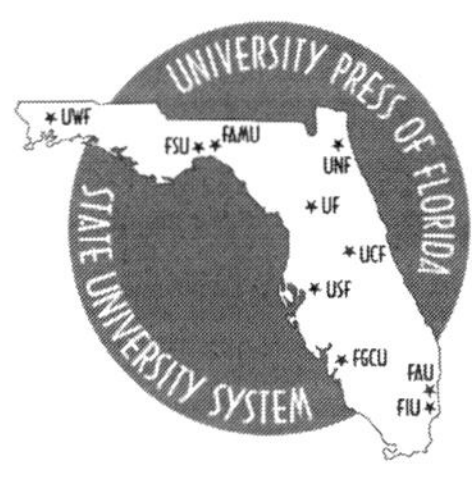

Parasites and Diseases of Wild Birds in Florida

Donald J. Forrester and Marilyn G. Spalding

Wildlife Drawings by David S. Maehr

UNIVERSITY PRESS OF FLORIDA

Gainesville / Tallahassee / Tampa / Boca Raton / Pensacola / Orlando / Miami / Jacksonville / Ft. Myers

Library of Congress Cataloging-in-Publication Data

Forrester, Donald J., 1937–
    Parasites and diseases of wild birds in Florida / Donald J. Forrester and Marilyn G. Spalding ;
    wildlife drawings by David S. Maehr.
        p.   cm.
    Includes bibliographical references.
    ISBN 0-8130-2560-5
    1. Birds—Parasites—Florida.   2. Birds—Diseases—Florida.   3. Wildlife diseases—Florida.
    I. Spalding, Marilyn, 1947–   .   II. Title.
    SF995.6P35 F67 2002
    571.9'18'09759—dc21   2002035994

The University Press of Florida is the scholarly publishing agency for the State University System of
Florida, comprising Florida A&M University, Florida Atlantic University, Florida Gulf Coast University,
Florida International University, Florida State University, University of Central Florida, University of
Florida, University of North Florida, University of South Florida, and University of West Florida.

University Press of Florida
15 Northwest 15th Street
Gainesville, FL 32611-2079
http://www.upf.com

Frontispiece photo courtesy of Stephen A. Nesbitt.

We dedicate this book to our good colleague and friend

Dr. Lovett E. Williams, Jr.

who recognized the importance of the study of parasites and diseases of free-ranging wildlife and while a member of the Division of Research of the Florida Game and Fresh Water Fish Commission (now called the Florida Fish and Wildlife Conservation Commission) fostered and encouraged in numerous ways our cooperative research program on wild birds in Florida. His willingness to share his knowledge, his friendship, and his insights and enthusiasm for high-quality field research were very important to us and formed a foundation for much of what is reported in this book.

Contents

Foreword

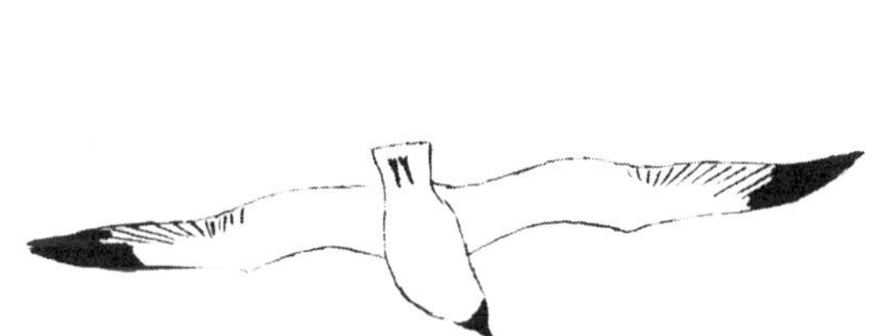

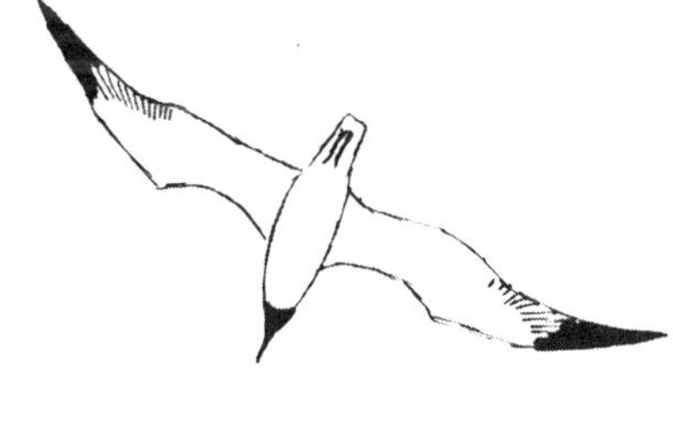

This book represents a wonderful collaboration of two scientists who have dedicated much of their careers to assisting our understanding of the various maladies affecting wildlife species in Florida. *Parasites and Diseases of Wild Birds in Florida* clearly is a valuable contribution to our scientific knowledge of avian diseases, but its significance may be fully appreciated only when one realizes that Dr. Forrester preceded this work with a book of similar proportion, his *Parasites and Diseases of Wild Mammals of Florida.*

The authors describe the many diseases and other mortality factors of wild birds they have researched during three decades of biomedical consultation for the Florida Fish and Wildlife Conservation Commission. Although managers must sometimes treat a disease of an individual animal, the free-roaming nature of wildlife species typically dictates their health be managed at population levels, to preclude compromising otherwise normal dynamics of host populations by diseases. This is especially true in recovery efforts for threatened and endangered species, and it is this ecological role of diseases in the dynamics of wild populations that is of paramount importance to those charged with management and recovery responsibilities for wildlife species.

The biomedical research of Drs. Forrester and Spalding has always been guided by a clear appreciation for the management implications of their findings, and it is for this reason that this book is of such significance. I congratulate them both and thank them as well, for their valuable contribution to our improved management of Florida wildlife through a better understanding of the diseases affecting them.

Tom H. Logan
Endangered Species Coordinator
Florida Fish and Wildlife Conservation Commission

Preface

Florida has a rich and diverse avian fauna, partly because of its climate and geography. Of the 457 species of free-ranging birds that occur in the state or its offshore waters, 446 are native and 11 are established exotics (Robertson and Woolfenden 1992). Among Florida's birds are a number of unique species or subspecies, such as Whooping Cranes (a recently re-established population), Florida Sandhill Cranes, Great White Herons, Fulvous Whistling-Ducks, Snail Kites, Crested Caracaras, White-crowned Pigeons, White-winged Doves, Florida Scrub-Jays, Florida Grasshopper Sparrows, and several subspecies of seaside sparrows, to name a few. Others are important as game birds, including Wild Turkeys, Northern Bobwhites, Mourning Doves, and several species of waterfowl.

In order to properly manage and conserve these birds, an understanding of their health and diseases is necessary. The purpose of this book is to provide a reference on the various parasites, diseases, and other factors that cause or may cause morbidity and mortality in Florida's wild birds, with emphasis on the distribution, prevalence, and significance of those factors. Information is included on 311 species of Florida birds, although the amount and quality of such information is variable; we have no information at all on the other 146 species. Much of the information we report was found scattered throughout the published literature, in the so-called gray literature, and in personal files, notes, and reports of numerous individuals and organizations. A large amount of unpublished data has been included through the cooperation and generosity of many colleagues. Much of this information has come from our own research and that of our technicians, graduate students, and postdoctoral associates during the past thirty-two years, as well as from many colleagues at the University of Florida, other universities, and state, federal, and private resource agencies.

For some birds (e.g., cranes, wading birds, Wild Turkeys, Northern Bobwhites, Mourning Doves, and Bald Eagles) the coverage is fairly extensive, whereas for others it is more limited. This was not by choice, but because many birds in Florida have not been well studied and little is known about their parasites and diseases. We hope that this book will help wildlife biologists, ornithologists, conservationists, resource administrators, educators, students, veterinarians, and others in the health professions to understand what is known and to stimulate investigations on those birds that have been poorly studied or neglected.

This book is not a diagnostic manual, although it may prove useful to some degree in that regard. For help in the recognition of the common diseases and parasites of many of the birds of Florida, the reader is referred to the recent publications by Davidson and Nettles (1997) and Friend and Franson (1999). These manuals provide excellent color plates and descriptions of clinical signs and lesions for many of the diseases of birds that occur in Florida and should be used as companion volumes to the present book. In addition, there are two recent books edited by Beyer et al. (1996) and Fairbrother et al. (1996) that will help the reader understand environmental contamination problems and interpret the meaning of tissue concentrations of various toxins.

The information is presented on a host basis, that is, the chapters are organized by species or groups of related species, rather than by disease agents. Our hope is that this will facilitate the use of this book by ornithologists, wildlife ecologists, and others who are "host oriented" and want to know the health/disease status of a certain bird or group of birds. We have followed Robertson and Woolfenden (1992) and American Ornithologists' Union (1998) for common and scientific names of species and American Ornithologists' Union (1998) for organization by orders and families.

Each chapter is preceded by an introduction in which the population and survival status of the bird or group of birds in question are discussed. In addition, appropriate reviews and bibliographies are mentioned in the introduction, with references to hematology, serum chemistry, nutrition, and various physiological topics, if such are available. Following the introduction, there are sections on the various morbidity and mortality factors, disease agents, and conditions, including the distribution, prevalence, and intensity (in some cases) of each. This is followed by statements on the significance of these diseases to populations as well as considerations of public health importance. Each chapter ends with a summary and conclusions section and a list of references cited. It is our hope that this book will demonstrate the need for further study, stimulate future disease/health research, and contribute to the database and concepts needed for the management and conservation of the valuable and unique avian fauna of Florida.

Although we have both worked on each chapter to some degree, we divided the major responsibilities of preparing each chapter: chapters 1, 2, 8–17, and 22–26 were done by DJF and chapters 3–7 and 18–21 by MGS.

Literature cited

American Ornithologists' Union. 1998. *Check-list of North American birds.* 7th ed. American Ornithologists' Union, Washington, D.C. 829 pp.

Beyer, W.N., G.H. Heinz, and A.W. Redmon-Norwood (eds.). 1996. *Environmental contaminants in wildlife. Interpreting tissue concentrations.* CRC Press, Boca Raton, Fla. 494 pp.

Davidson, W.R., and V.F. Nettles. 1997. *Field manual of wildlife diseases in the southeastern United States.* 2d ed. Southeastern Cooperative Wildlife Disease Study, Univ. of Georgia, Athens. 417 pp.

Friend, M., and J.C. Franson (eds.). 1999. *Field manual of wildlife diseases. General field procedures and diseases of birds.* U.S. Dept. of the Interior, U.S. Geol. Surv., Biol. Res. Div., Inf. Tech. Rep. 1999–001. Washington, D.C. 425 pp.

Fairbrother, A., L.N. Locke, and G.L. Hoff (eds.). 1996. *Noninfectious diseases of wildlife.* 2d ed. Iowa State Univ. Press, Ames. 219 pp.

Robertson, W.B., Jr., and G.E. Woolfenden. 1992. *Florida bird species: an annotated list.* Spec. Publ. 6, Florida Ornithological Society, Gainesville. 260 pp.

Acknowledgments

Funding for much of the original research that is reported in this book and for the preparation of the manuscript was provided by the Florida Fish and Wildlife Conservation Commission (formerly the Florida Game and Fresh Water Fish Commission) and the federal Pittman-Robertson Wildlife Restoration Program. J.R. Brady was particularly supportive, as were T.H. Logan, L.E. Williams, Jr., F.W. Stanberry, J.A. Powell, T.C. Hines, S.L. Sharpley-Evans, and B.A. Millsap.

A number of organizations and people were generous in contributing information, much of it unpublished. We are especially indebted to the personnel of the Southeastern Cooperative Wildlife Disease Study (SCWDS, University of Georgia, Athens) and the U.S. National Wildlife Health Center (Madison, Wisconsin) for their cooperation and help in providing data on Florida birds. Data were also obtained from the U.S. National Parasite Collection (Beltsville, Maryland), the files of the Pathology Service of the Department of Pathobiology and the Veterinary Medical Teaching Hospital (College of Veterinary Medicine, University of Florida, Gainesville), the Florida State Collection of Arthropods (Gainesville), Florida Audubon Society's Center for Birds of Prey (Maitland), and Suncoast Seabird Sanctuary (St. Petersburg, Florida). Personnel from these and other organizations who contributed included C.L. Abercrombie,

N. Abou-Madi, B.L. Akey, H.A. Albers, K. Allander, J.L. Allen, W.T. Atyeo, M.L. Avery, W.W. Baker, G.F. Bennett, R.E. Bennetts, W.J. Bigler, J. Birney, R. Bjork, B.H.D. Bolon, R.E. Brannian, M.B. Brown, K.E. Brugger, C.D. Buergelt, M.B. Calderwood-Mays, R. Caligiuri, C.H. Calisher, B.G. Campbell, R.E. Carleton, S. Citino, V. Clyde, R.A. Cole, R. Collins, M.W. Collopy, B.U. Constantin, J.A. Conti, K.A. Converse, T.E. Cornish, C.E. Couvillion, J. Cox, D.F. Coyner, L.E. Creekmore, A.M. Crowley, W.R. Davidson, J.F. Day, T.F. Dean, S.L. Deem, M.F. Delaney, A.A. Dhondt, D.E. Docherty, G.L. Doster, P.J. Douglass, P.R. Douglass, V. Dreitz, R.M. Duncan, L.A. Durden, R.J. Dusek, B.A. Dusenbury, N.F. Eichholz, K.C. Emerson, B. Fenwick, J.R. Fischer, G.W. Foster, J. C. Franson, P. C. Frederick, M. Friend, E.J. Galbreath, R.R. Gerrish, M.C. Garvin, M.L. Gay, J.R. Gaydos, C. Gilliland, P.E. Ginn, N. Gourlie, E.C. Greiner, T.L. Gross, J.E. Hanley, S.B. Haseltine, R.T. Heath, Jr., C.M. Herman, T.C. Hines, V. Hinshaw, G.L. Hoff, B.L. Homer, L.T. Hon, H. Hoogstraal, J.A. Hovis, E.W. Howerth, P.K. Humphlett, P.P. Humphrey, R. Isaza, J.M. Johnson, H.W. Kale, J.E. Keirans, F.E. Kellogg, S.M. Kerr, J.B. Kethley, A.R. Kiehl, J.M. Kinsella, R.M. Kocan, L.H. Kooistra, A.J. Krynitsky, R.E. Lange, J.N. Layne, A.W.C.V. Layton, S.L. Lin, S.B. Linda, S.E. Little, L.N. Locke, R.D. Lord, N.P. Lung, G.T. Mahnke, M.B. Main, H.A. McAllister, R. McCracken, G.S. McLaughlin, R.G. McLean, J.W. Mertins, C.U. Meteyer, K.D. Meyer, M.B. Mihalik, K.E. Miller, E.G. Milstrey, J.L. Morrison, B.M. Mulhern, J.K. Nayar, S.A. Nesbitt, V.F. Nettles, A.P. Nol, K. Papineau, D.B. Pence, J.A. Popp, B. Pranty, R.D. Price, R.M. Prouty, B.I. Purich, L. Quinn, C.F. Quist, T.W. Regan, W.L. Reichel, M.K. Reinhard, J.A. Richardson, L.G. Rickard, M.W. Riggs, B. Roberts, W.B. Robertson, Jr., M. Robson, T.E. Rocke, J.A. Rodgers, Jr., T.J. Roffe, C.M. Romagosa, L. Roth, S.W. Russell, B. Sasse, S.K. Schmeling, L.F. Schorr, J.C. Schwartz, M.S. Sepulveda, B.G. Short, L.M. iegfried, L. Sileo, G.L. Slater, K.E. Smith, B. Snyder, C.J. Stafford, L.M. Stark, H. Steinberg, E.R. Stetzer, J.A. Stevenson, L. Straub, R.K. Stroud, S.F. Sundlof, B.J. Suto, P.W. Sykes, Jr., K.A. Tarvin, S.R. Telford, Jr., S.A. Temple, S.P. Terrell, N.J. Thomas, N.P. Thompson, J.E. Thul, K.S. Todd, Jr., B.N. Tuggle, E.W. Uhl, M.P. Wallace, F.J. Ware, T.A. Webber, E.E. Wehr, M.R. Wells, F.M. Wellings, R.A. Westhouse, F.H. White, J.H. White, N.A. Wilson, J. Wolf, D.A. Wood, P.B. Wood, J.C. Woodard, K.L. Woods, G.E. Woolfenden, E.J. Wright, S.D. Wright, T.J. Wright, and Q.Y. Zeng.

We thank the following people who reviewed drafts of various chapters or parts of chapters and provided helpful suggestions: C.T. Atkinson, P.L. Barrows, R.E. Bennetts, R.A. Cole, R. Collins, J.A. Conti, D.F. Coyner, W.R. Davidson, S.L. Deem, G.L. Doster, R.J. Dusek, M.J. Folk, P.C. Frederick, M. Friend, M.C. Garvin, E.C. Greiner, L.F. Kiff, L.N. Locke, M.P. Luttrell, K.D. Meyer, B.A. Millsap, J.L. Morrison, S.A. Nesbitt, L.G. Rickard, W.B. Robertson, Jr., B.J. Suto, N.J. Thomas, J.E. Thul, C. Van Riper III, L.E. Williams, Jr., and M.D. Young. We want to give special thanks to J.M. Kinsella and J.W. Mertins for checking all of our helminth and arthropod sections, respectively, in each chapter for taxonomic accuracy.

Those who donated original photographs and figures for us to include in the book included R.E. Bennetts, J.A. Conti, D.F. Coyner, W.R. Davidson, J.R. Fischer, G.W. Foster, E.C. Greiner, B.L. Homer, P.P. Humphrey, M.B. Mihalik, J.K. Nayar, S.A. Nesbitt, R.T. Paul, L. Quinn, C.F. Quist, L.G. Rickard, L. Sileo, K.E. Smith, N.F.R. Snyder, S.R. Telford, Jr., S.P. Terrell, J.E. Thul, E.W. Uhl, J. Whidden, L.E. Williams, Jr., P.B. Wood, and S.D. Wright. The wildlife drawings were done by D.S. Maehr, to whom we are especially grateful.

We owe special gratitude to G.W. Foster and E.A. Dusenbury for their many hours of dedicated work with word processing of the chapter drafts and especially the tables. Without their assistance on this and many other tasks we would have not been able to complete the work for this book. Copy editor L. Treadwell gave the text, tables, and illustrations a thorough scrutiny; her corrections and queries helped us put the finishing touches on the manuscript, as did project editor J.K. Brown of the University Press of Florida. Others who assisted in various ways over the years included R.M. Anderson, R. Bireline, T.A. Cames, D.K. Couch, T.L. Creel, T.E. DeLaFuente, T.A. Gibbs, and H.C. McGill.

We also give special thanks to the following individuals who provided help with the identification of parasites: G.F. Bennett and E.C. Greiner (blood protozoans), J.M. Kinsella (helminths), J.W. Mertins and W.T. Atyeo (mites), J.E. Keirans and J.W. Mertins (ticks), R.D. Price, K.C. Emerson, and J.W. Mertins (chewing lice), and N.A. Wilson (louse flies).

Finally we wish to thank our families, especially our spouses, Gabriele and Peter, for their understanding and encouragement during the long process of preparing this book.

We are grateful to the editors and publishers of the following journals and books for permission to use figures from their publications:

The Auk for Figure 24.6

Bulletin of the Society of Vector Ecology for figures 17.9, 17.14, 17.15, 17.18, 17.20, 17.22, 17.30, 17.43, 17.53, 17.56, 17.58, and 17.79

Canadian Journal of Zoology for figure 24.14

Florida Entomologist for figure 17.37

Florida Field Naturalist for figures 24.4 and 24.5

Florida Game and Fresh Water Fish Commission Technical Bulletins for figure 17.1

Iowa State University Press for figure 17.72

Journal of Medical Entomology for figures 17.27, 17.28, and 24.16

Journal of Parasitology for figures 19.13, 19.20, 22.29, 22.31, 22.32, and 23.4

Journal of Protozoology for figure 17.41

Journal of Raptor Research for figure 15.5

Journal of Wildlife Diseases for figures 2.10, 2.11, 6.11, 6.15, 6.16, 10.4, 16.7, 16.8, 16.9, 16.11, 16.12, 17.33, 17.54, 17.55, 17.57, 20.1, 22.8, 23.1, 23.2, and 23.3

Journal of Wildlife Management for figure 17.3

Journal of Zoo and Wildlife Medicine for figure 19.17

Oxford University Press for figures 17.38, 17.52, and 24.10

Proceedings of the Helminthological Society of Washington for figures 16.6, 16.10, 17.64, 17.66, 17.67, 17.69, 17.71, 17.74, 17.76, 22.6, and 22.7

Proceedings of the 15th International Congress of Ornithology for figure 11.6

Science for figure 24.7

Stackpole Books for figures 17.80, 17.81, 22.10, and 22.11

University Press of Florida for figures 17.77 and 17.78

University of California Press for figures 8.1, 8.2, and 15.2

Veterinary Pathology for figure 20.2, and 20.3.

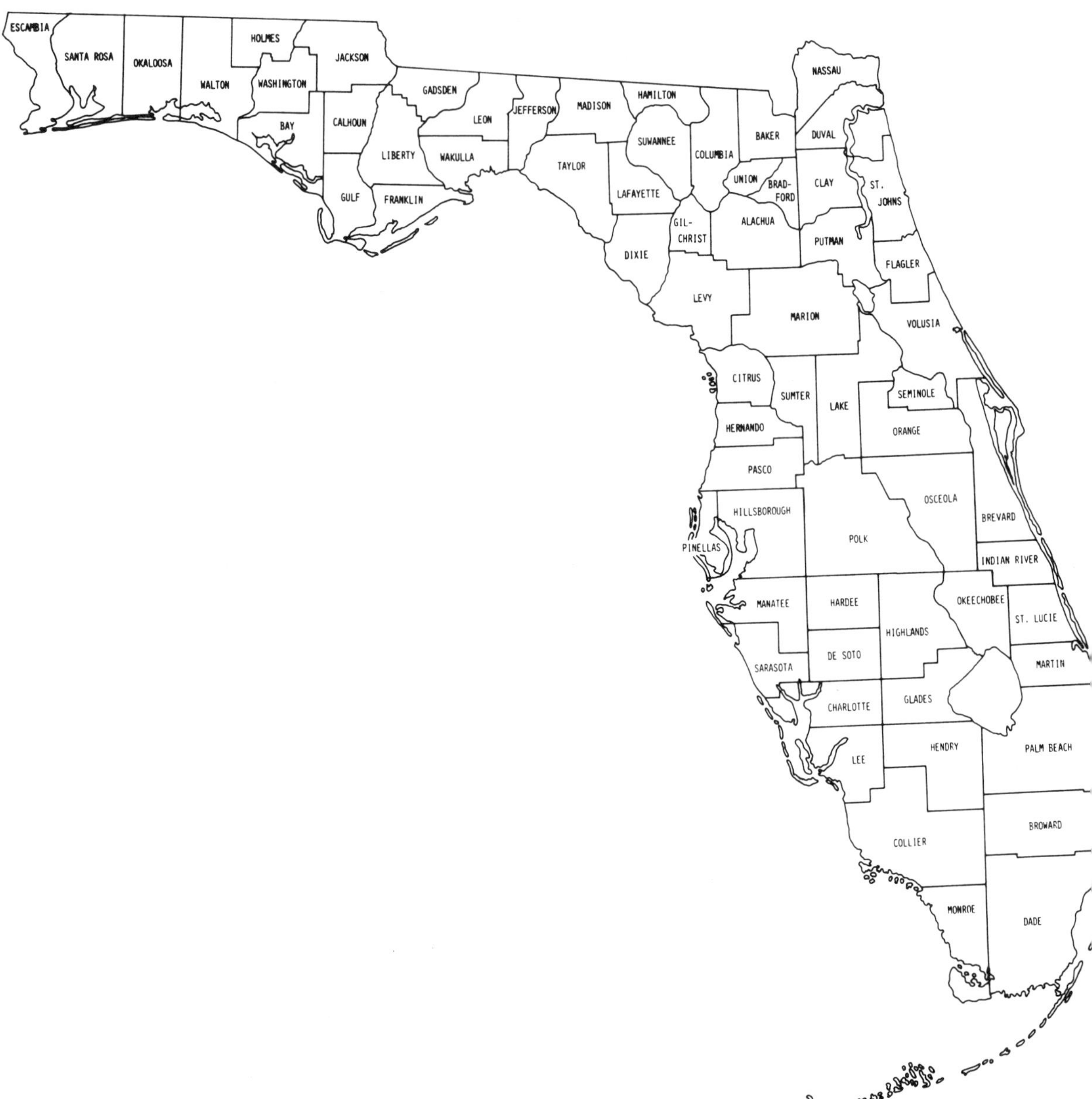

ESCAMBIA
SANTA ROSA
OKALOOSA
WALTON
HOLMES
WASHINGTON
BAY
JACKSON
CALHOUN
GADSDEN
LIBERTY
WAKULLA
LEON
JEFFERSON
MADISON
HAMILTON
TAYLOR
SUWANNEE
COLUMBIA
BAKER
NASSAU
DUVAL
GULF
FRANKLIN
LAFAYETTE
GIL-CHRIST
DIXIE
UNION
BRAD-FORD
ALACHUA
CLAY
ST. JOHNS
PUTNAM
FLAGLER
LEVY
MARION
VOLUSIA
CITRUS
SUMTER
LAKE
SEMINOLE
ORANGE
HERNANDO
PASCO
HILLSBOROUGH
OSCEOLA
BREVARD
PINELLAS
POLK
INDIAN RIVER
MANATEE
HARDEE
OKEECHOBEE
ST. LUCIE
HIGHLANDS
DE SOTO
MARTIN
SARASOTA
CHARLOTTE
GLADES
LEE
HENDRY
PALM BEACH
COLLIER
BROWARD
MONROE
DADE

CHAPTER ONE

Introduction

In its more advanced stages, game management is in effect the art of maintaining a population which is vigorous and healthy in spite of its density.
—Aldo Leopold, *Game Management*

The term *disease* has been defined in many different ways. For the purposes of this book a broad definition is employed as presented by Wobeser (1997): disease is "any impairment that interferes with or modifies the performance of normal functions." This would include diseases caused by infectious agents (prions, viruses, bacteria, fungi) and parasites (protozoans, trematodes, cestodes, nematodes, acanthocephalans, pentastomes, ticks, mites, fleas, lice, biting flies) as well as other factors such as environmental contaminants, trauma, inclement weather, neoplasia, anomalies, and nutritional deficiencies. The causes of morbidity and mortality of birds can be grouped into 13 categories (table 1.1), which include the items listed above and others.

A clear distinction must be made between the terms infection and disease. They are not the same. A bird can have an infection, i.e., it can be a host to bacteria, viruses, or parasites and experience little or no effects, or it can have a disease, in which case the infectious agent may overpower the resistance of the animal and cause considerable harm, possibly resulting in death. Often the suffix *-iasis* is used for a word to indicate an infection and *-osis* for disease. For example, small numbers of the proventricular nematode *Dispharynx nasuta* infecting Wild Turkeys result in dispharynxiasis, but if large numbers of the nematode are involved and disease is a consequence, it is called dispharynxosis. These two concepts

Table 1.1. The fundamental categories of morbidity and mortality in wild birds[a]

Category	Comments
1. Anomalies	Usually rare; confined to an occasional bird
2. Stress	Varies depending on the nature of the stress or stressors. Usually reflects other conditions such as chronic malnutrition or extreme parasitism.
3. Trauma	Includes predation, roadkills, tower kills, intraspecific conflicts, effects of inclement weather, and electrocution.
4. Suffocation	Can be caused by drowning or any kind of obstruction of air passages.
5. Neoplasms	Includes both benign and malignant tumors; most cause only occasional mortality.
6. Toxicoses	May cause considerable mortality in a localized area. Most caused by environmental contaminants and due to human activities.
7. Nutritional diseases	Malnutrition due to deficient intake of energy and protein. Often associated with infectious diseases and most severe in the young and the very old.
8. Viral diseases	Can cause widespread and rapid mortality.
9. Bacterial diseases	Can result in considerable mortality especially in dense populations, but also can be of an insidious nature.
10. Mycotic diseases	Usually are secondary invaders along with or following other diseases.
11. Parasitic diseases	Most are insidious and cause morbidity and mortality over a long period of time. Commonly associated with malnutrition.
12. Senility	Not usually an important mortality factor in wild birds.
13. Brood parasitism	Important as a mortality factor in some passeriforms.

a. Adapted and modified from Hayes and Prestwood (1969) and Kellogg (1981).

should be kept in mind when reading this book in which both conditions are presented, sometimes without clear distinction. A given bird may have a long list of infectious agents, parasites, and other conditions, but all may not cause disease, or if they do, the effects may not be serious. Wild birds often serve as reservoir hosts or carrier hosts in which case they maintain infections with little or no apparent harm to themselves, but serve as a source of infection for other wild birds, domestic animals, or man. In many cases the mechanisms whereby infections become diseases are poorly understood or not known at all.

Another concept that must be emphasized is that disease agents and conditions often interact within a host and result in synergistic effects. This means that separately various infectious and parasitic agents may cause little harm to an animal, but concurrent infections can be more harmful than would be expected from a mere additive effect. In addition various factors such as nutrition, environmental contaminants, stress due to overcrowding, inclement weather, etc. may interact with infectious and parasitic agents and result in serious disease. This complex phenomenon is only partially understood for most wildlife diseases.

Basic to an understanding of wildlife diseases is information on distribution, prevalence, and intensity. The distribution of diseases in time and space is of great relevance. Some diseases are widespread geographically, occur at all times of the year, and infect many different species of birds. Others are more restricted in distribution and occur only at certain times of the year; still others are found in only one host or in a few closely related species. In addition the prevalence of diseases or disease agents is of great interest. The term prevalence refers to the number or percentage of birds in a sample or population that are infected with a given disease agent or parasite. It is helpful to understand prevalence in terms of the time span during which the data were collected. Intensity is also of value in assessing the significance of infections and relates to the number of disease agents (i.e., helminths, protozoans, arthropods, etc.) per infected bird. Uninfected animals are not taken into consideration by the concept of intensity. It is not always possible to determine the intensity of infections, but where this can be done the information is of great value relative to understanding the severity of a given disease.

Diseases that cause large-scale mortality (epizootics) often receive considerable attention since the results are dramatic. However, the effects of more subtle or chronic problems can be important as well, or even more important. An attempt is made in each chapter to assess the importance of the various diseases to the host in question and to relate this to three aspects: (1) effects on the populations of wild birds, (2) effects on and interactions with domesticated animals, and (3) public health implications. Much of the information on these topics is incomplete, sketchy, or lacking entirely. An attempt is made in the summary and conclusion (chapter 26) to bring the available information for Florida's birds into focus, but the reader should understand that the discipline of wildlife disease investigation is still in its infancy compared to human and veterinary medicine. There is still much to be done, but the first steps are to assemble and examine what is known about the various diseases and their distribution and prevalence among native populations. This information should form the basis for future research as well as provide a foundation for management and conservation plans for the valuable and unique avian fauna of Florida.

Literature cited

Friend, M. 1985. Wildlife health implications of sewage disposed in wetlands. In: *Ecological considerations in wetlands. Treatment of municipal wastewaters*. P.J. Godfrey, E.R. Kaynur, S. Pelczaisliu, and J. Benforado (eds.). Van Nostrand Reinhold, New York. pp. 262–269.

Hayes, F.A., and A.K. Prestwood. 1969. Some considerations for diseases and parasites of white-tailed deer in the southeastern United States. *Proceedings, white-tailed deer in the southern forest habitat symposium*. USDA, Forest Service, Southern Forest Experiment Station, Nacogdoches, Texas. pp. 32–36.

Kellogg, F.E. 1981. Field considerations. In: *Diseases and parasites of white-tailed deer*. W.R. Davidson, F.A. Hayes, V.F. Nettles, and F.E. Kellogg (eds.). Misc. Publ. no. 7, Tall Timbers Research Station, Tallahassee, Fla. pp. 1–5.

Wobeser, G.A. 1997. *Diseases of wild waterfowl*. 2d ed. Plenum Press, New York. 324 pp.

Loons and Grebes

I. Introduction

Three species of loons (Family Gaviidae) and 5 species of grebes (Family Podicipedidae) occur in Florida (Robertson and Woolfenden 1992). All are winter visitors except for the Pied-billed Grebe, which breeds throughout the mainland and in the Keys (table 2.1). None of these species is threatened or endangered (Logan 1997).

Except for a small amount of information on Horned Grebes and Pied-billed Grebes, most of what is known about these two families of birds in Florida has been obtained from Common Loons that overwinter in coastal waters in fairly large numbers. Franson and Cliplef (1993) have published an overview of the causes of mortality of Common Loons in United States. Their database consisted of 222 loons obtained from 18 states, 1976–91, and included 36 birds from Florida. Alexander (1991) published an overview of the patterns of winter mortality among Common Loons in the northeastern Gulf of Mexico, but gave no data on the causes of this annual event. Forrester et al. (1997) analyzed the extent and causes of winter mortality in Common Loons overwintering in Florida waters from 1970 to 1994.

In her book on Common Loons, McIntyre (1988) included a short discussion of several parasites and diseases and provided a list of parasites reported from this host. Chafel and Pokras (1993) reported on the parasitic helminths of 20 Common Loons from 4 New

Table 2.1. Occurrence, distribution, and abundance of loons and grebes in Florida[a]

Species of loon or grebe		Range	Seasonal occurrence	Relative abundance
Common Loon	*Gavia immer*	Statewide, esp. coasts & offshore	Winter visitor	Uncommon to common
		Statewide	Summer visitor	Many reports
Red-throated Loon	*Gavia stellata*	Panhandle & northern peninsula	Winter visitor	Very rare to uncommon
Pacific Loon	*Gavia pacifica*	Coasts	Winter visitor	Very rare
Least Grebe	*Tachybaptus dominicus*	Southern peninsula & Keys	Winter visitor	Very rare (5 reports)
Pied-billed Grebe	*Podilymbus podiceps*	Mainland	Resident	Fairly common
Horned Grebe	*Podiceps auritus*	Mainland	Winter visitor	Fairly common to very common
Eared Grebe	*Podiceps nigricollis*	Mainland only	Winter visitor	Rare (80 reports)
Western Grebe	*Aechmophorus occidentalis*	Northern & central coasts	Winter visitor	Very rare

a. Modified from Robertson and Woolfenden (1992).

England states and Kinsella and Forrester (1999) reported on the helminths of 127 Common Loons overwintering in Florida waters. A brief summary of the parasites of Least Grebes was published by Storer (1992). A book titled *Proceedings of the 1992 Conference on the Loon and its Ecosystem: Status, Management, and Environmental Concerns* was published in 1993 and, in addition to the papers mentioned above by Franson and Cliplef (1993) and Chafel and Pokras (1993), contains a number of treatises on environmental contaminants and the causes of loon morbidity and mortality in various regions of the northern United States and Canada (Morse et al. 1993). The monograph by McIntyre and Barr (1997) contains some information on parasites and diseases of Common Loons.

Table 2.2 Primary diagnostic findings for 434 Common Loons from Florida, 1970–94

Primary diagnostic finding (% prevalence)	Number of loons from each area					
	Gulf Coast	Atlantic Coast	Inland lakes[a]	Keys[b]	Unknown	Totals
Emaciation syndrome (66)	219	58	2	0	7	286
Oiling (18)	5	71	0	0	0	76
Aspergillosis (7)	14	13	1	0	3	31
Trauma (5)	9	10	1	1	0	21
Miscellaneous (1)[c]	2	2	1	0	0	5
Undetermined (3)	8	7	0	0	0	15
Totals	257	161	5	1	10	434

Source: Forrester et al. (1997).
a. Includes freshwater lakes in Alachua, Clay, and Marion counties in north central Florida.
b. Big Pine Key in the Florida Keys (Monroe County).
c. Includes single cases of tuberculosis, intestinal coccidia, cardiac failure (associated with bacterial infection), encephalitis, and pneumonia.

II. Trauma

Trauma was the cause of death in 5% (21 of 434) of Common Loons found dead in Florida from 1970 to 1994 (table 2.2). One of these was due to gunshot wounds (Forrester et al. 1997), a 3.06 kg male found washed up on a beach in Palm Beach County in December of 1971 (Forrester 1971). In his report on seabird mortality on the Atlantic and Gulf coasts over an 8-year period (1975–83), Simons (1985) listed 2 Common Loons that had been shot and 1 entangled in a gill net. The exact localities of these mortalities were not given. Over the 8-year period of 1988 to 1995, 150 of 460 Common Loons (33%) that were recovered from beaches in the Tampa Bay area and treated at a wildlife rehabilitation center in Pinellas County were victims of trauma (Suto 1996). One of these had been shot, 18 had been hooked by fishhooks or tangled in fishing line, 9 had signs of impact trauma, and 122 died of unknown trauma. Longstreet (1953) reported a wounded Common Loon in May 1936 and an injured Red-throated Loon in February 1942, both in Volusia County. No details were given concerning the nature or cause of the trauma. Langridge (1991) reported that 36 Common Loons landed on the asphalt runways of the airport at Lake City during a heavy rainstorm in April of 1991, apparently mistaking the slick runways for a pond. The loons were unable to take off again and suffered bloody feet from struggling on the asphalt. They were captured subsequently and released into a lake by Florida Game and Fresh Water Fish Commission personnel.

Crawford (1981) reported that 65 Pied-billed Grebes and 1 Horned Grebe were found dead at the base of a 1,010-foot TV tower near Tall Timbers Research Station in Leon County. These numbers were based on his examination of records on more than 42,000 birds over a 25-year period. Taylor and Anderson (1973) reported the killing of 4 Pied-billed Grebes due to striking a 1,484-foot TV tower in Orange County. This latter study was based on records of 7,782 birds killed during autumn migrations

of 1969, 1970, and 1971. In a third study, Taylor and Kershner (1986) examined more than 5,000 birds killed by flying into the Vehicle Assembly Building at the John F. Kennedy Space Center (Brevard County) during 1970–81 and reported 1 Pied-billed Grebe. This building is one of the largest in the world, occupying more than 32,000 m^2 in ground area and standing 160 m tall. Grebes have not been found in other studies of such mortality involving tall structures in Florida (Kale 1971; Taylor and Anderson 1974; Maehr et al. 1983; Maehr and Smith 1988). Weston (1966) reported a Pied-billed Grebe killed by striking cables associated with the 3-mile Pensacola Bay Bridge (Santa Rosa and Escambia counties) in 1940. This was the only grebe he found over a 12-year period (1938–49), when he tabulated 740 specimens of dead birds representing 75 species, most of them passeriforms.

In a 4-year survey of roadkills in Florida state parks and recreation areas from 1990 to 1993, only 1 grebe, a Pied-billed Grebe, was encountered, in 1992 in St. Andrews State Recreational Area in Bay County (Stevenson 1994; Snyder 1994).

In October 1997, an estimated 11 Pied-billed Grebes died of trauma near Ft. Lauderdale (Broward County). The specific cause of the trauma was not determined (Meteyer 1997).

III. Predation

The causes of mortality of 20 of 434 Common Loons found dead on Florida beaches between 1970 and 1994 were attributed to sharks and other fish (Forrester et al. 1997). In his report on seabird mortality on the Atlantic and Gulf coasts over an 8-year period (1975–83), Simons (1985) listed 2 loons that were killed by sharks or some other fish. The exact localities of these mortalities were not given.

The remains of 2 Pied-billed Grebes were identified from a collection of 88 pellets from a family of Great Horned Owls over a 4-week period in January 1949 in Putnam County (Burns 1952). Fragments of Pied-billed Grebes were also found in the nest of these owls. Bent (1937) reported that Bald Eagles in Florida preyed on Pied-billed Grebes, but gave no specific data. McEwan and Hirth (1980) examined 16 active Bald Eagle nests in a 4-county area (Alachua, Marion, Putnam, and Volusia) during 1975 and 1976 and found the remains of 3 Pied-billed Grebes. It is not clear, however, whether these were cases of predation or scavenging by the Bald Eagles. Stolen (2001) observed a Great Blue Heron capture and eat a Pied-billed Grebe in a marsh in the Merritt Island National Wildlife Refuge (Brevard County) in 2001. Stevenson and Anderson (1994) reported an observation of an American alligator capturing a Pied-billed Grebe in Florida, but gave no details on the specific locality. Delany (1986, 1996) and Delany and Abercrombie (1986) found the remains of a Pied-billed Grebe in the stomach of 1 of 465 American alligators from Orange, Lochloosa, and Newnans Lakes (Alachua County) examined during 1981–85. They concluded that birds make up only a small unimportant component of the diet of alligators in north central Florida.

In the previous section on trauma, reference was made to an incident reported by Langridge (1991) near Lake City in which a number of Common Loons landed on a wet airport runway during a rainstorm, probably mistaking the runway for a body of water. Because their feet are located posteriorly, loons are unable to take off from dry land and must be in water to do so. Over the years we have become aware of a number of incidents in north central Florida in which loons have landed on roads (usually at night during rainstorms) and are unable to regain flight. In several cases it was suspected that the birds were injured or sick when they were found floundering on a road or at the side of a road. In reality they were perfectly healthy and when released into a lake were able to regain flight normally.

Two Pied-billed Grebes were identified as casualties of Hurricane Hilda in Brevard County during October 6–8, 1964 (Case et al. 1965). These were among 4,707 birds killed during the storm, most of which were passeriforms. No other data are available on the effects of inclement weather on grebes in Florida.

IV. Inclement weather

Winter storms undoubtedly cause mortality among Common Loons overwintering in Florida coastal waters, but the extent of this is virtually unknown. Simons (1985) reported mortality among Common Loons along the Florida Gulf coast during the severe winter of 1976–77, but gave no specific data. He also reported heavy mortality of loons on the Gulf Coast during the following winter, which was abnormally cold, but not as cold as the previous winter.

A severe die-off of Common Loons during the winter of 1983 was believed to be linked to an emaciation syndrome, discussed in section XVIII. Part of the complex series of events that led to that die-off was linked to weather, including a drought that caused changes in salinity of estuarine habitats where loons were feeding and a series of cold fronts that put additional thermoregulation stress on the birds.

V. Organochlorines

Between 1972 and 1994 residues of 11 organochlorines were detected in tissues of 46 Common Loons from Florida (table 2.3). DDE and PCBs were the most common residues. DDE was found in 45 of the 46 loons in concentrations ranging from 0.16 to 80 ppm (wet weight) and PCBs (Aroclor 1260) were present in 40 of the 46 loons in concentrations from 0.72 to 120 ppm. The 9 other organochlorine residues were at low prevalences and concentrations.

The concentrations of DDE and PCB in brain tissues were considerably below those reported as being lethal for birds (Stickel et al. 1970; Stickel 1975), but the sublethal effects of these contaminants on Common Loons are unknown. The concentrations for DDE and PCB are comparable to those reported in Common Loons in several northern states (Fay and

Table 2.3. Organochlorine residues in tissues of Common Loons[a] from Florida

Tissue County (or area)	Year	No. examined	No. with residue / Geometric mean (range) ppm wet weight[b]			Data source
			DDE	DDD	Dieldrin	
Brain						
Bay	1983	1	1 / 15	0 / ND	0 / ND	A
Broward	1994	3[c]	3 / 2.3	0 / ND	0 / ND	B
Duval	1983	1	1 / 0.86	0 / ND	0 / ND	C
Franklin	1983	7	7 / 7.4(5.2–20)	0 / ND	7 / 0.66(0.25–1.0)	D
	1983	2	2 / 8.14(6.9–9.6)	0 / ND	0 / ND	A
Wakulla	1983	3	3 / 6.79(3.1–13)	0 / ND	0 / ND	A
Carcass[d]						
Duval	1983	1	1 / 1.4	0 / ND	0 / ND	C
Franklin	1983	7	7 / 1.7(0.98–3.0)	0 / ND	5 / 0.12(ND–0.24)	D
Fat						
NG	1974	2	2 / 0.39(0.23–0.65)	NG	NG	E
Liver						
Bay	1983	1	1 / 13	0 / ND	0 / ND	A
Duval	1973	1	1 / 2.2	0 / ND	1 / 0.26	F
	1974	2	2 / 0.88(0.46–1.7)	0 / ND	2 / 0.18	F
Franklin	1983	2	2 / 5.4(4.4–6.7)	0 / ND	0 / ND	A
	1983	7	7 / 9.3(3.7–25)	0 / ND	7 / 0.76(0.11–1.7)	D
Indian River	1973	2	2 / 2.7(0.52–14)	1 / 0.31	1 / 0.21(ND–0.89)	F
Nassau	1972	1	1 / 3.8	0 / ND	1 / 0.21	F
Palm Beach	1987	9	9 / 0.78(0.22–9.18)	0 / ND	0 / ND	G
Pinellas	1973	2	1 / 0.09(ND–0.16)	0 / ND	0 / ND	F
	1974	5	5 / 4.0(0.68–80)	1 / 0.08(ND–0.59)	5 / 0.33(0.14–3.3)	F
Sarasota	1973	1	1 / 12	1 / 0.13	1 / 0.45	F
St. Johns	1972	1	1 / 26	1 / 0.33	1 / 0.73	F
	1974	3	3 / 1.1(0.31–11)	2 / 0.18(ND–0.92)	2 / 0.21(ND–1.1)	F
Wakulla	1983	3	3 / 7.3(2.7–16)	0 / ND	0 / ND	A
Uropygial gland						
NG	1974	2	2 / 0.39(0.23–0.65)	NG	0 / ND	E

Tissue County (or area)	Year	No. examined	Heptachlor epoxide	Oxychlordane	*cis*-chlordane	Data source
Brain						
Bay	1983	1	0 / ND	NA	0 / ND	A
Broward	1994	3[c]	0 / ND	0 / ND	0 / ND	B
Duval	1983	1	0 / ND	0 / ND	0 / ND	C
Franklin	1983	7	6 / 0.14(ND–0.21)	6 / 0.22(ND–0.36)	1 / 0.06(ND–0.15)	D
	1983	2	0 / ND	0 / ND	0 / ND	A
Wakulla	1983	3	0 / ND	0 / ND	0 / ND	A
Carcass[d]						
Duval	1983	1	0 / ND	0 / ND	0 / ND	C
Franklin	1983	7	0 / ND	0 / ND	0 / ND	D
Fat						
NG	1974	2	NG	NG	NG	E
Liver						
Bay	1983	1	0 / ND	0 / ND	0 / ND	A
Duval	1973	1	0 / ND	0 / ND	0 / ND	F
	1974	2	0 / ND	0 / ND	0 / ND	F
Franklin	1983	2	0 / ND	0 / ND	0 / ND	A
	1983	7	6 / 0.22(ND–0.4)	6 / 0.25(ND–0.51)	3 / 0.09(ND–0.33)	D
Indian River	1973	2	1 / 0.16(ND–0.51)	0 / ND	0 / ND	F
Nassau	1972	1	0 / ND	0 / ND	0 / ND	F
Palm Beach	1987	9	0 / ND	0 / ND	0 / ND	G
Pinellas	1973	2	0 / ND	0 / ND	0 / ND	F

(continued)

Table 2.3 *(continued)*

Tissue County (or area)	Year	No. exam.	Heptachlor epoxide	Oxychlordane	*cis*-chlordane	Data source
	1974	5	1 / 0.08(ND–0.59)	2 / 0.12(ND–0.70)	0 / ND	F
Sarasota	1973	1	1 / 0.10	1 / 0.45	0 / ND	F
St. Johns	1972	1	0 / ND	1 / 0.22	0 / ND	F
	1974	3	1 / 0.09(ND–0.28)	1 / 0.07(ND–0.11)	0 / ND	F
Wakulla	1983	3	0 / ND	0 / ND	0 / ND	A
Uropygial gland						
NG	1974	2	NG	NG	NG	E

Tissue County (or area)	Year	No. exam.	*trans*-nonachlor	*cis*-nonachlor	PCB (Aroclor 1260)	Data source
Brain						
Bay	1983	1	NA	NA	1 / 37	A
Broward	1994	3[c]	0 / ND	0 / ND	3 / 5.7	B
Duval	1983	1	0 / ND	0 / ND	1 / 3.1	C
Franklin	1983	7	6 / 0.18(ND–0.30)	0 / ND	7 / 20(14–51)	D
	1983	2	0 / ND	0 / ND	2 / 19(18–20)	A
Wakulla	1983	3	0 / ND	0 / ND	3 / 21(11–47)	A
Carcass[d]						
Duval	1983	1	0 / ND	0 / ND	1 / 5.3	C
Franklin	1983	7	0 / ND	0 / ND	7 / 6.1(2.5–32)	D
Fat						
NG	1974	2	NG	NG	NG	E
Liver						
Bay	1983	1	0 / ND	0 / ND	1 / 31	A
Duval	1973	1	1 / 0.16	0 / ND	1 / 2.5	F
	1974	2	1 / 0.07(ND–0.11)	0 / ND	2 / 1.3(0.78–2.2)	F
Franklin	1983	2	0 / ND	0 / ND	2 / 14(8.4–22)	A
	1983	7	6 / 0.26(ND–0.45)	0 / ND	7 / 26(9.3–66)	D
Indian River	1973	2	1 / 0.16(ND–0.52)	1 / 0.11(ND–0.23)	2 / 4.2(1.2–15)	F
Nassau	1972	1	1 / 0.16	0 / ND	1 / 9.9	F
Palm Beach	1987	9	0 / ND	0 / ND	9 / 2.1(0.64–24)	G
Pinellas	1973	2	0 / ND	0 / ND	0 / ND	F
	1974	5	4 / 0.23(ND–0.81)	2 / 0.08(ND–0.18)	5 / 6.1(0.87–120)	F
Sarasota	1973	1	1 / 0.42	1 / 0.13	1 / 18	F
St. Johns	1972	1	1 / 0.18	0 / ND	1 / 55	F
	1974	3	1 / 0.11(ND–0.55)	1 / 0.08(ND–0.20)	2 / 1.3(ND–15)	F
Wakulla	1983	3	0 / ND	0 / ND	3 / 27(9–59)	A
Uropygial gland						
NG	1974	2	NG	NG	NG	E

Sources: A = Davidson (1983), B = Smith (1995), C = Stafford and Haseltine (1983), D = Gay and Haseltine (1983), E = Johnston (1976), F = Gay and Haseltine (1986), G = Davidson and Hayes (1989).
ND = not detected, NG = not given, NA = not analyzed for.
a. A mixture of juveniles and adults; ages not determined.
b. Limits of detection for sources A, B, and G were 0.01 ppm for aldrin, BHC, DDD, DDE, DDT, heptachlor, and lindane; 0.02 ppm for dieldrin, endrin, and heptachlor epoxide; 0.05 ppm for methoxychlor and parathion; 0.10 ppm for carbophenolthin, chlordane, diazinon, ethon, malathion, methyl parathion, mirex, rabon, and toxaphene; and 0.20 ppm for PCB (Aroclor 1260). For sources C, D, and F they were 0.1 ppm for DDE, DDD, DDT, dieldrin, heptachlor epoxide, oxychlordane, *cis*-chlordane, *trans*-nonachlor, *cis*-nonachlor, endrin, and toxaphene; and 0.5 ppm for PCB (Aroclor 1260). For source E they were 0.01 ppm for PCB, DDE, DDT, and dieldrin. In addition to the residues listed above, source F detected 0.28 ppm of endrin and 0.35 ppm of toxaphene in a liver sample from 1 of 2 loons from Indian River County in 1973 and 0.18 ppm of endrin in 1 of 3 loons from St. Johns County in 1974; no residues were detected in other samples tested for endrin and toxaphene by sources A, B, C, D, F, and G. No residues of aldrin, BHC, DDT, heptachlor, lindane, carbophenalthion, chlordane, diazinon, ethion, malathion, methyl parathion, mirex, rabon, methoxychlor or parathion were detected by sources A, B, and G. No residues of DDT were detected by sources C, D, and F.
c. Pooled sample of brain tissues from three loons.
d. Values are based on the body minus the skin, feet, wings, liver, and gastrointestinal tract.

Table 2.4. Concentrations of lead in samples of liver and kidney from Common Loons[a] in Florida

County	Year	No. examined	No. with lead / Geom. mean(range) ppm wet wt.[b]		Data source
			Liver	Kidney	
Bay	1983	1	1 / 7.5	1 / 3.5	A
	1991	3	0 / ND	NA	B
Collier	1994	1	0 / ND	NA	C
Duval	1983	1	1 / 0.28	NA	D
	1991	1	0 / ND	NA	B
Franklin	1983	7	0 / ND	NA	E
	1983	2	2 / 3.3(3.2–3.4)	2 / 2.1(1.1–4.0)	A
Martin	1990	1	0 / ND	NA	B
	1991	1	0 / ND	NA	B
	1993	2	0 / ND	NA	B
Sarasota	1994	1	0 / ND	NA	F
Wakulla	1983	3	3 / 4.2(3.5–5.3)	3 / 2.1(0.89–4.3)	A
Totals		24	7 / 0.24(ND–7.5)	6 / 2.3(0.89–4.3)	

Sources: A = Davidson (1983), B = Franson (1991–93), C = Meteyer (1994), D = Stafford and Haseltine (1983), E = Prouty and Haseltine (1983), F = Fischer and Davidson (1994).
ND = not detected, NA = not analyzed for.
a. A mixture of juveniles and adults; ages not determined.
b. Limits of detection were 0.25 ppm for sources A, B, C, and F and 0.10 ppm for sources D and E.

Youatt 1967; Ream 1976; Belant and Anderson 1990), Canada (Frank et al. 1983), and Mississippi (Prouty et al. 1975). DDD, a metabolite of DDT, was determined to be responsible for the death of a Common Loon in Mississippi (Prouty et al. 1975).

No information is available on organochlorines in grebes from Florida.

VI. Lead

From 1983 to 1994 samples of liver and/or kidney from 24 Common Loons in Florida were examined for the presence of lead (table 2.4). Lead was detected in 7 of the loons (29%) and concentrations ranged from 0.28 to 7.5 ppm (wet weight). Concentrations in 2 of the loons (5.3 and 7.5 ppm) were in the range (>5 ppm) reported by Cook and Trainer (1966) to represent toxicity in waterfowl. Similar concentrations were reported by Pokras and Chafel (1992) in Common Loons that were diagnosed with lead poisoning in New England; however, in the 2 Florida birds, typical signs other than

emaciation were not observed at necropsy and lead poisoning was thereby ruled out as the final diagnosis. In addition to the New England report by Pokras and Chafel (1992), lead poisoning has been reported also from Common Loons in Minnesota, Wisconsin, New Hampshire, and Maine and in all cases was linked to ingestion of lead fishing sinkers (Ensor et al. 1992; Locke et al. 1981).

The overall effects of lead on the health of Common Loons overwintering in Florida coastal waters are not known. It is possible, however, that lead contamination may contribute to a general compromise in the health of these birds or render them more susceptible to infectious diseases such as aspergillosis, but there are no data to support this theory.

There are no data on lead contamination of grebes in Florida.

VII. Mercury and selenium

Between 1973 and 1998 samples of liver from 83 Common Loons from Florida were tested

Table 2.5. Concentrations of mercury and selenium in liver samples from Common Loons[a] in Florida

| County | Year | No. examined | No. positive / Geom. mean(range) ppm wet wt.[b] | | Data source |
			Mercury	Selenium	
Bay	1983	1	1 / 54	NA	A
	1991	7	7 / 8.4(4.6–15)	NA	B
Brevard	1993	2	2 / 4.8(4.1–5.6)	NA	C
Broward	1994	5	5 / 4.3(1.8–17)	NA	D
Duval	1983	1	1 / 20	1 / 39	E
	1991	1	1 / 3.9	NA	B
Franklin	1983	7	7 / 17(7.9–90)	7 / 12(5.2–41)	F
	1983	2	2 / 28(25–32)	NA	A
	1984	24	24 / 4.6(1.1–56)	24 / 5.8(2.9–20)	G
Martin	1990	1	1 / 1.3	NA	B
	1991	1	1 / 1.4	NA	B
Palm Beach	1987	9	8 / 2.2(ND–9.7)	NA	H
Pinellas	1973	1	1 / 11	1 / 10	G
	1974	10	10 / 15(4.5–40)	10 / 8.2(3.5–16)	G
	1975	1	NA	1 / 5.3	G
	1998	1	1 / 20	1 / 6.9	I
Sarasota	1994	1	1 / 3.4	NA	J
St. Johns	1974	6	6 / 5.2(2.3–9.9)	6 / 7.3(4.0–13)	G
Wakulla	1983	3	3 / 26(20–35)	NA	A

Sources: A = Davidson (1983), B = Franson (1991–93), C = Quist (1993), D = Smith (1995), E = Stafford and Haseltine (1983), F = Prouty and Haseltine (1983), G = Gay and Haseltine (1986), H = Davidson and Hayes (1989), I = Cornish (1998), J = Fischer and Davidson (1994).
NA = not analyzed, ND = not detected.
a. A mixture of juveniles and adults; ages were not determined.
b. Limits of detection: mercury = 0.02 ppm, sources B, E, F, G; 0.20 ppm, sources A, C, D, H, I, J; selenium = 0.10 ppm.

for mercury and 51 of these were tested also for selenium (table 2.5). Mercury was found in all of the samples and concentrations ranged from 0.15 to 90 ppm (wet weight). Selenium was found in all 51 of the samples that were tested for that metal and concentrations ranged from 2.9 to 41 ppm (wet weight). In table 2.6 concentrations of mercury and selenium are compared for 80 of these loons based on geographic origin in Florida (Atlantic versus Gulf coasts), condition of the birds (emaciated versus normal or oiled), and year. Concentrations of mercury and selenium were not different in males versus females. Concentrations of mercury were higher in loons from the Gulf Coast than in those from the Atlantic Coast. Concentrations of both mercury and selenium were higher in emaciated loons than in normal loons (Forrester et al. 1997). There was a linear correlation between concentrations of mercury and selenium (figure 2.1).

Published values for concentrations of mercury in Common Loons from several northern locations, including Minnesota (Ensor et al. 1992), Wisconsin (Belant and Anderson 1990), and Ontario (Fimreite 1974; Frank et al. 1983; Barr 1986), are comparable to the values determined from loons in Florida and range from 0.08 to 91 ppm (wet weight). By examination of chicks it has been determined that mercury is acquired by loons on their nesting grounds (Barr 1986). From this information it is tempting to conclude that loons are contaminated with mercury (and probably also selenium) before migrating to their wintering areas in Florida, but this has not been verified and remains only a conjecture.

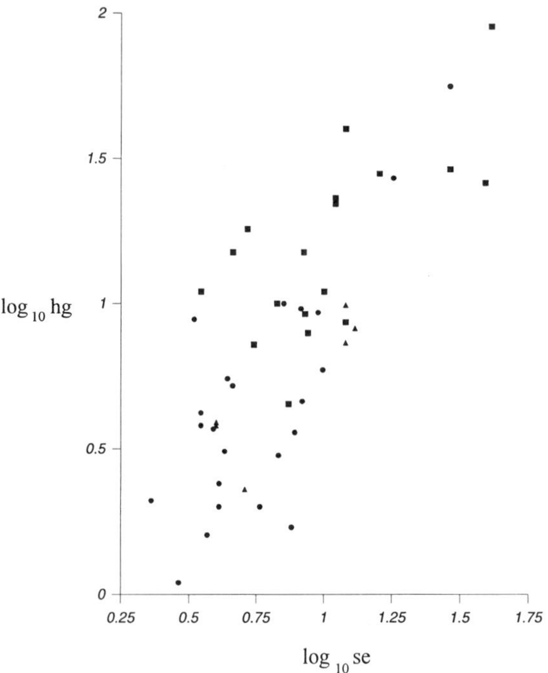

FIGURE 2.1. The relationship between concentrations of mercury and selenium in samples of liver from emaciated and normal or oiled loons in Florida, 1973–94. *Solid circles* = normal loons, *solid squares* = emaciated loons, *solid triangles* = oiled loons.

The effects of mercury on the health of Common Loons are unknown; there have been no experimental studies conducted on this species and conclusions must at the present time be based on information learned from other species. Eisler (1987) concluded that the effects of mercury contamination on birds vary according to a number of factors including "the form of the element, dose, route of administration, species, sex, age, and physiological condition." In addition, the situation is complicated further by the possible interactions between mercury, selenium, cadmium, and a number of other contaminants such as DDE and PCBs (Mullins et al. 1977; Eisler 1987). As an example of this, it has been shown from experimental studies with Japanese Quail (*Coturnix c. japonica*) that there is a mutual protective effect of mercury and selenium (Stoewsand et al. 1974; El-Begearmi et al. 1977). Such an effect

may be operational in loons, but this has not been investigated.

There is no information on mercury and selenium contamination in grebes in Florida.

VIII. Other metals

Concentrations of cadmium in samples of kidney and/or liver from 15 Common Loons from Florida in 1983 and 1993 are presented in table 2.7. Concentrations were detected in every loon and varied from 1.5 to 12 ppm (wet weight). The significance of cadmium contamination in Common Loons has not been determined. According to Eisler (1985) birds in general are believed to be resistant to harmful effects of this metal. White and Finley (1978) observed no mortality when they fed cadmium to Mallards for a 90-day period. In

Table 2.6. Concentrations (ppm, wet wt.)[a] of mercury and selenium in liver samples from Common Loons from Atlantic and Gulf Coast waters in Florida

| Coast | | | Mercury | | | Selenium | | |
Condition	Year(s)	No. examined	Mean[b]	Range	No. examined	Mean[b]	Range
Atlantic Coast							
Emaciated	1987	9	2.3	0.15–9.7	0	—	—
	1990–91	2	1.4	1.3–1.4	0	—	—
	1993	2	4.8	4.1–5.6	0	—	—
	1994	5	4.3	1.8–17	0	—	—
Oiled	1974	6	5.2	2.3–9.9	6	7.3	4.0–13
Gulf Coast							
Emaciated	1973–74	5	17	9.2–29	5	12	8.5–29
	1974–75	6	13	4.5–40	7	6.8	3.5–16
	1983	13	22	7.9–90	6	10	5.2–12
	1984	1	7.9	—	1	8.7	—
	1991	8	7.7	3.9–15	0	—	—
Normal[c]	1984	23	4.5	1.1–56	23	5.8	2.3–29

Source: Forrester et al. (1997).
a. Limits of detection: mercury = 0.02 ppm or 0.20 ppm; selenium = 0.10 ppm.
b. Geometric mean.
c. Collected by shooting.

some of their ducks tissue concentrations were >100 ppm.

Arsenic was detected in the liver of an emaciated adult female Common Loon found dead after a storm in Martin County during February of 1993 (Franson 1993). The concentration was 39 ppm (dry weight) and was considered to be above normal background values. Most concentrations found in livers of birds examined by various authors have been <1 ppm (wet weight) (Eisler 1988). An exception to this was the finding of 17 ppm (wet weight) in an Osprey from Maryland (Wiemeyer et al. 1980). In this case the authors felt

Table 2.7. Concentrations of cadmium in samples of liver and kidney from Common Loons[a] in Florida

| County | Year | No. examined | No. positive / Geom. mean(range) ppm wet wt.[b] | | Data source |
			Liver	Kidney	
Bay	1983	1	1 / 1.8	1 / 3.1	A
Duval	1983	1	NA	1 / 2.5	B
Franklin	1983	7	NA	7 / 5.9(3.5–10)	C
	1983	2	2 / 2.4(1.5–3.8)	2 / 10(9.0–12)	A
Martin	1993	1	1 / 5.1[c]	NA	D
Wakulla	1983	3	3 / 2.3(1.5–4.4)	3 / 4.8(4.2–5.4)	A
Totals		15	7 / 2.5(1.5–5.1)	14 / 5.5(2.5–12)	

Sources: A = Davidson (1983), B = Stafford and Haseltine (1983), C = Prouty and Haseltine (1983), D = Franson (1993).
NA = not analyzed for.
a. A mixture of juveniles and adults; ages were not determined.
b. Limits of detection = 0.03 ppm, source A; 0.10 ppm, sources B, C; limits for source D were not given.
c. Dry weight value.

Table 2.8. Oil spills in Florida waters affecting Common Loons and Horned Grebes

Year	County or site	No. birds oiled	Data source
1925	Volusia	"dozens of loons"	Howell (1932)
1940s	Volusia	NG	Longstreet (1953)
NG	"Florida"	NG	Sprunt (1954)
1968	Duval	16 loons	Locke (1968)
1970	Tampa Bay	~500 loons and ~200 grebes	Stevenson (1970), Sims (1970) Clark (1973)
1974	Duval	~225 loons	Stevenson (1974), White et al. (1976)

NG = not given.

that the arsenic contributed to the death of the bird; however, the role of arsenic in the death of the loon from Martin County remains unknown.

There are no data on cadmium or arsenic for grebes in Florida.

IX. Oiling

There are several accounts in the literature concerning mortality of loons and grebes in Florida due to oil spills (table 2.8). The numbers involved varied from a few birds to several hundred. The largest spill occurred in February of 1970 in Tampa Bay when approximately 9,000 seabirds and waterfowl were affected, including at least 500 Common Loons and 200 Horned Grebes (Clark 1973; Sims 1970). This event, in which 10,000 gallons of Bunker C oil were spilled and formed a 100-square-mile slick, was linked to a significant reduction in the local wintering population of Common Loons in that area for several subsequent years (Woolfenden 1973; Edscorn 1974; Stevenson 1974, 1977; Clapp et al. 1982).

In his 8-year study of beached birds along the Atlantic and Gulf coasts from 1975 to 1983, Simons (1985) reported 140 (18%) oiled Common Loons of a total of 789 loons examined. However, oiling along the Florida Gulf Coast was almost nonexistent. In contrast to Simons's findings, Clapp et al. (1982) stated that although Common Loons are very vulnerable to oiling, they are not usually found

in large numbers during beached bird surveys. Reasons for this include (1) populations are small and dispersed, (2) when oiled, loons may not seek the shore very quickly, and (3) compared with other seabirds, loons are less buoyant and oiled loons are more likely to sink when they die. Forrester et al. (1997) reported that 18% of a sample of 434 Common Loons examined at necropsy from 1970 through 1994 were oiled, but that most of these were from one spill in the Jacksonville Beach area.

Common Loons are sometimes found contaminated with oil when no major oil spill can be associated with the occurrence. These events are probably due to small amounts of oil being discharged into the ocean by ships. This oil subsequently comes into contact with seabirds such as loons. Four of 460 Common Loons from Gulf Coast beaches near Redington Beach (Pinellas County) from 1988 through 1995 contained oil (Suto 1996). Two were found in 1988, 1 in 1991, and 1 in 1993 that had been processed at the Suncoast Seabird Sanctuary, a wildlife rehabilitation center.

X. Biotoxins

Both Type C and Type E botulism have been reported in Florida waterbirds (Forrester et al. 1980; Franson and Duncan 1991). Since there have been epizootics among Common Loons on Lake Michigan due to Type E toxin produced by *Clostridium botulinum* (Brand et al. 1988) and at least 1 instance of mortality due

Table 2.9. Common Loons from Florida tested[a] for Types C and E toxins of *Clostridium botulinum*

Year	County	No. loons tested		Data source
		Type C	Type E	
1983	Franklin	6	3	Duncan (1983)
1987	Martin	1	0	Duncan (1987)
1991	Bay	7	6	Duncan (1991)
	Duval/St. Johns	1	1	Ibid.
	Martin	1	1	Ibid.
1994	Collier	1	1	Duncan (1994)

a. Mouse protection test; all results were negative.

to Type C (Franson and Cliplef 1993), these biotoxins were included in the list of differentials for investigation into causes of mortality of loons in Florida. Between 1983 and 1994 17 Common Loons found dead in Florida coastal areas were tested for botulinum toxin Type C and 11 for Type E (table 2.9). All results were negative (Duncan 1983–94).

Mortality of Common Loons has not been associated with red tide outbreaks, in which other species of waterbirds have been involved (Forrester et al. 1977), maybe because most red tides occur during the warmer months of the year, when loons are absent from Florida waters for the most part. However, Forrester et al. (1997) pointed out that although there was not a red tide at the time of the 1983 epizootic, a red tide did occur offshore several months prior to the die-off and resulted in extensive fish kills. This fish mortality may have reduced fish populations inshore as well and contributed to the emaciation observed in these loons.

One Pied-billed Grebe died during an outbreak of necrotizing enteritis or enterotoxemia in 1971 in Fisheating Creek Bay on the western edge of Lake Okeechobee (Jasmin et al. 1972). No other information is available on the effects of biotoxins on grebes in Florida.

XI. Neoplasia

There is only one known case of neoplasia in a Common Loon from Florida. A pulmonary tumor was diagnosed in an adult female found dead after a storm in Martin County during 1993 (Meteyer and Franson 1993). There were multiple tumor masses with areas of necrosis and hemorrhage and evidence of invasion of surrounding lung tissue. The tumor was judged to be a primary pulmonary tumor, possibly an adenocarcinoma, but was not studied further or characterized. Nothing is known about neoplasms in grebes in Florida.

Tumors in free-ranging wild birds are extremely rare (Siegfried 1983), but there is a report of a squamous cell carcinoma on the toe of a juvenile Common Loon from Minnesota in 1982 (Hier et al. 1986). No other records are known.

XII. Viruses

Only 1 virus has been isolated from Common Loons in Florida. During the 1983 epizootic a lentogenic strain (nonpathogenic) of Newcastle disease virus was isolated from 1 loon from the Gulf Coast (Hinshaw 1983). This was probably an incidental finding and considered to be of little or no significance to loon populations. Tissues from several additional Common Loons (exact number unknown) found dead during the 1983 epizootic were tested using various virus isolation techniques, but no viruses were identified (Davidson 1983; Lange 1983). No information is available on viruses of grebes in Florida.

Table 2.10. Bacteria identified from Common Loons[a] in Florida

County	Year	Organ or tissue	Primary diagnostic finding	Data source
Acinetobacter calcoaceticus				
Franklin	1988	Heart, air sacs	Aspergillosis	Wright (1988)
Aeromonas hydrophila				
Franklin	1983	Intestines	Emaciation syndrome	Davidson (1983)
Pinellas	1998	Liver, lungs, intestines	Emaciation syndrome	Cornish (1998)
Aeromonas sp.				
St. Johns	1998	Intestines	Emaciation syndrome	Spalding & Forrester (1998)
Campylobacter sp.				
Franklin	1983	Intestines[b]	Emaciation syndrome	Duncan (1983)
Citrobacter freundii				
St. Johns	1974	Large intestine	Oiling	White & Forrester (1996)
Citrobacter sp.				
St. Johns	1974	Large intestine	Oiling	Ibid.
Clostridium paraperfringens				
Levy	1987	Intestines	Emaciation syndrome	Duncan (1987)
Clostridium sp.				
Levy	1987	Intestines	Emaciation syndrome	Ibid.
Edwardsiella tarda				
Palm Beach	1972	Intestines	Aspergillosis	White et al. (1973), White & Forrester (1996)
Enterobacter sp.				
Broward	1976	Large intestine	Emaciation syndrome	White & Forrester (1996)
Duval	1974	Liver, intestines	Oiling	Ibid.
Indian River	1973	Large intestine	Aspergillosis	Ibid.
Pinellas	1974	Large intestine	Aspergillosis	Ibid.
	1974	Large intestine	Emaciation syndrome	Ibid.
St. Johns	1974	Large intestine[c]	Oiling	Ibid.
	1974	Large intestine	Aspergillosis	Ibid.
Escherichia coli				
Broward	1976	Large intestine[d]	Emaciation syndrome	Ibid.
Duval	1974	Large intestine[d]	Emaciation syndrome	Ibid.
	1974	Large intestine[e]	Oiling	Ibid.
	1974	Large intestine	Aspergillosis	Ibid.
Franklin	1983	Liver, intestines	Emaciation syndrome	Duncan (1983)
	1983	Kidney	Emaciation syndrome	Buergelt & Short (1983)
	1988	Heart, air sacs	Aspergillosis	Wright (1988)
Hillsborough	1976	Liver	Trauma	White & Forrester (1996)
Indian River	1973	Large intestine	Trauma	Ibid.
Levy	1987	Lung	Emaciation	Duncan (1987)
Martin	1987	Kidney	Unknown	Ibid.
Palm Beach	1972	Intestines[b]	Emaciation syndrome	White & Forrester (1996)
	1972	Intestines	Aspergillosis	Ibid.
Pinellas	1973	Liver	Oiling	Ibid.
	1974	Large intestine[f]	Emaciation syndrome	Ibid.
	1974	Large intestine[f]	Aspergillosis	Ibid.
	1974	Large intestine[f]	Trauma	Ibid.
	1975	Large intestine[d]	Emaciation syndrome	Ibid.
	1975	Large intestine[d]	Oiling	Ibid.
St. Johns	1972	Cecum	Emaciation syndrome	Ibid.
	1974	Liver	Oiling[c]	Ibid.

(continued)

Table 2.10. *(continued)*

County	Year	Organ or tissue	Primary diagnostic finding	Data source
	1974	Large intestine	Oiling[g]	Ibid.
	1974	Large intestine	Aspergillosis	Ibid.
	1974	Large intestine	Emaciation syndrome	Ibid.
St. Lucie	1972	Large intestine	Emaciation syndrome	Ibid.
Volusia	1972	Intestines	Emaciation syndrome	Ibid.
	1974	Liver	Oiling	Ibid.
	1974	Large intestine	Oiling	Ibid.
	1988	Lungs[h]	Aspergillosis + TB	Wright (1988)
Klebsiella pneumoniae				
St. Johns	1974	Large intestine[d]	Oiling	White & Forrester (1996)
Micrococcus sp.				
St. Johns	1974	Large intestine	Oiling	Ibid.
Mycobacterium sp.				
Volusia	1988	Lungs[h]	Aspergillosis + TB	Wright (1988)
Pasteurella multocida				
Franklin	1983	Intestines	Emaciation syndrome	Buergelt & Short (1983)
	1983	Brain	Emaciation syndrome	Ibid.
Peptastreptococcus sp.				
Levy	1987	Intestines	Emaciation syndrome	Duncan (1987)
Proteus vulgaris				
Pinellas	1974	Large intestine[d]	Emaciation syndrome	White & Forrester (1996)
St. Johns	1974	Large intestine[d]	Oiling	Ibid.
Proteus sp.				
Broward	1976	Large intestine[d]	Emaciation syndrome	Ibid.
Flagler	1974	Large intestine[d]	Oiling	Ibid.
Franklin	1983	Lungs	Emaciation syndrome	Duncan (1983)
	1988	Heart, air sacs	Aspergillosis	Wright (1988)
Indian River	1973	Large intestine	Aspergillosis	White & Forrester (1996)
Levy	1987	Abdominal fluid	Emaciation	Duncan (1987)
Palm Beach	1972	Intestines	Normal	White & Forrester (1996)
Pinellas	1974	Large intestine[b]	Aspergillosis	Ibid.
	1974	Large intestine[b]	Emaciation syndrome	Ibid.
Sarasota	1972	Large intestine	Emaciation syndrome	Ibid.
St. Johns	1974	Large intestine	Oiling	Ibid.
Volusia	1974	Large intestine[b]	Oiling	Ibid.
Pseudomonas putrifaciens				
Levy	1987	Intestines	Emaciation syndrome	Duncan (1987)
Pinellas	1975	Large intestine	Emaciation syndrome	White & Forrester (1996)
Pseudomonas sp.				
Levy	1987	Liver	Emaciation syndrome	Duncan (1987)
St. Johns	1974	Liver	Oiling	White & Forrester (1996)
Salmonella agona				
Pinellas	1974	Large intestine[c]	Emaciation syndrome	White et al. (1976), White & Forrester (1996)
Salmonella blockley				
Pinellas	1974	Large intestine	Emaciation syndrome	Ibid.
Salmonella infantis				
Pinellas	1973	Large intestine	Oiling	Ibid.
	1974	Large intestine[d]	Oiling	Ibid.
St. Johns	1974	Large intestine[d]	Oiling	Ibid.
Volusia	1974	Large intestine[b]	Oiling	Ibid.

(continued)

Table 2.10. *(continued)*

County	Year	Organ or tissue	Primary diagnostic finding	Data source
Salmonella montevideo				
St. Johns	1974	Large intestine[e]	Oiling	Ibid.
Salmonella muenchen				
St. Johns	1974	Large intestine[e]	Oiling	Ibid.
Volusia	1974	Large intestine[e]	Oiling	Ibid.
Salmonella newport				
St. Johns	1974	Large intestine[e]	Oiling	Ibid.
Volusia	1974	Large intestine[e]	Aspergillosis	Ibid.
Salmonella saint-paul				
Flagler	1974	Large intestine[e]	Oiling	Ibid.
St. Johns	1974	Large intestine[e]	Oiling	Ibid.
Salmonella typhimurium				
Duval	1974	Large intestine[e]	Emaciation syndrome	Ibid.
Serratia marcescens				
Levy	1987	Kidney, liver	Emaciation syndrome	Duncan (1987)
Shawanella putrifaciens				
Pinellas	1998	Liver, lungs, intestines	Emaciation syndrome	Cornish (1998)
Staphylococcus epidermidis				
Pinellas	1974	Liver	Emaciation syndrome	White & Forrester (1996)
	1974	Large intestine	Emaciation syndrome	Ibid.
Staphylococcus sp.				
St. Johns	1974	Liver	Oiling	Ibid.
Streptococcus sp.				
Franklin	1988	Heart, air sacs	Aspergillosis	Wright (1988)
Levy	1987	Abdominal fluid	Emaciation syndrome	Duncan (1987)
Pinellas	1974	Liver	Aspergillosis	White & Forrester (1996)
St. Johns	1974	Liver	Oiling	Ibid.
Volusia	1988	Lungs[b]	Aspergillosis + TB	Wright (1988)
Vibrio alginolyticus				
Levy	1987	Intestines, lungs[d]	Emaciation syndrome	Duncan (1987)
Vibrio damsela				
Collier	1994	Liver, spleen	Emaciation syndrome	Duncan (1994)
Vibrio parahemolyticus				
Levy	1994	Intestines	Emaciation syndrome	Duncan (1987)
	1994	Lungs	Emaciation syndrome	Ibid.
Vibrio sp.				
Volusia	1972	Intestines	Emaciation syndrome	White et al. (1973), White & Forrester (1996)
Weeksella virosa[i]				
Monroe	1987	Intestines	Trauma	Duncan (1987)

a. Isolations made from 1 loon, unless otherwise indicated.
b. Isolations made from 3 different loons.
c. Isolations made from 5 different loons.
d. Isolations made from 2 different loons.
e. Isolations made from 4 different loons.
f. Isolations made from 14 different loons.
g. Isolations made from 9 different loons.
h. Had been in captivity for 3 months prior to its death and subsequent examination for bacteria.
i. Formerly known as CDC Group IIf (see Holmes et al. 1986).

Table 2.11. *Aspergillus* infections in lungs and air sacs of Common Loons from Florida

County	Year	No. cases	Data source
Alachua	1997	1[a]	Homer & Papineau (1997)
Broward	1976	5	Forrester (1996)
	1994	1	Smith (1995)
Collier	1946	1	Hartman (1946)
Duval	1974	1	Forrester (1996)
Franklin	1983	2	Gross & Reinhard (1983)
	1988	1	Wright (1988)
Indian River	1973	2	Forrester (1996)
Levy	1993	1	Woodard & Westhouse (1993)
Marion	1983	1	Stetzer & Reinhard (1983)
Martin	1977	1	Forrester (1996)
Palm Beach	1971	1	Locke (1971)
	1971	1	Forrester (1996)
	1972	1	Ibid.
Pinellas	1973	2	Ibid.
	1974	8	Ibid.
	1975	3	Ibid.
Sarasota	1973	1	Ibid.
St. Johns	1974	2	Ibid.
St. Lucie	1977	1	Forrester & Kale (1977)
Volusia	1974	3	Forrester (1996)
	1988	1[b]	Wright (1988)
Unknown	1991	1	Roth (1991)

a. Had been in captivity for 1 month.
b. Had been in captivity for 3 months.

XIII. Bacteria

Thirty-six species of bacteria were isolated from Common Loons in Florida between 1972 and 1998 (table 2.10). The significance of these bacteria to the health of loons is not known. In none of these cases was the cause of death attributed to bacterial infection. Most of these bacteria are probably secondary in nature and some are most likely postmortem contaminants. Some of those found in the large intestines such as *Escherichia coli* and *Proteus* sp. should probably be considered normal enteric bacteria and are of little concern under normal circumstances. *Escherichia coli*, however, has been reported as causing a fatal septicemia in an adult male Common Loon from Minnesota that also had a broken wing (Pichner and Wolff 1993).

Others such as *Edwardsiella tarda*, *Weeksella virosa*, the various serotypes of *Salmonella*, and the species of *Vibrio* are of possible zoonotic significance. White et al. (1973) isolated *E. tarda* from 1 of 11 Common Loons found dead on Florida beaches and examined in 1971 and 1972. Eight serotypes of *Salmonella* were identified in Common Loons that were found dead in 1974 (White et al. 1976). Most of the positive loons were victims of an oil spill on the eastern coast of Florida ($n = 16$); 5 others were emaciated and 1 died of aspergillosis (table 2.10). Five of these 8 serotypes (namely *S. agona*, *S. infantis*, *S. newport*, *S. saint-paul*, and *S. typhimurium*) were among the 7 serotypes most commonly isolated from humans in 1974 (Center for Disease Control 1975). All were isolated from contents of the large intestines of the loons and probably represent the carrier state. Further studies on the antibiotic resistance of

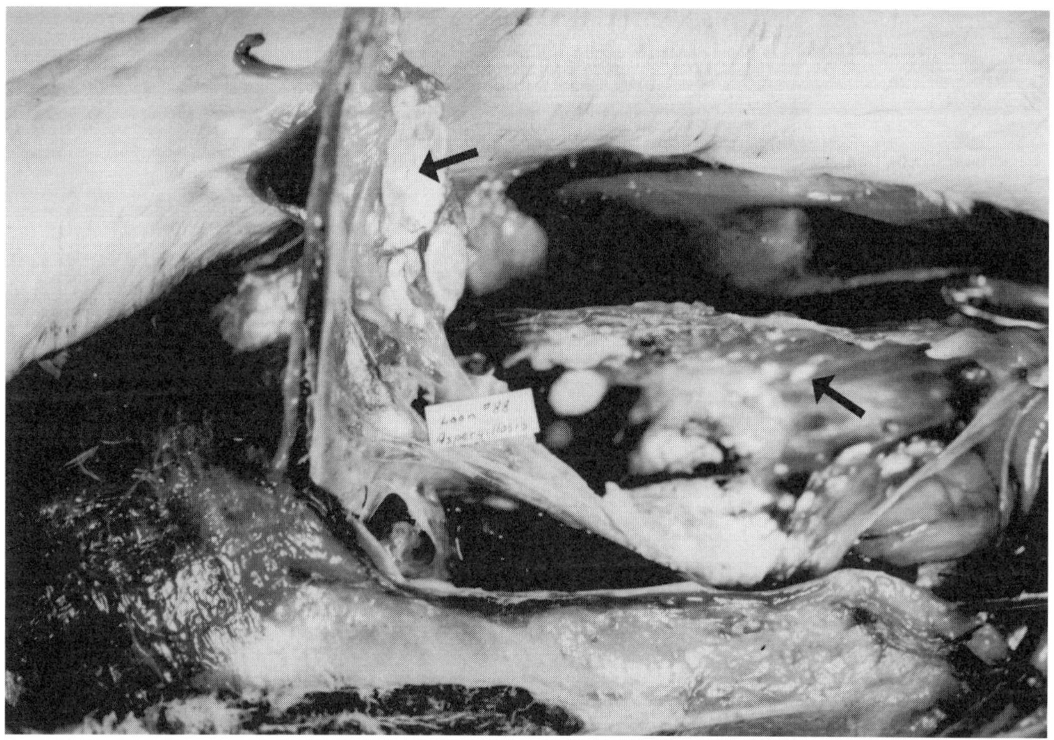

FIGURE 2.2. Aspergillosis in a Common Loon found in moribund condition in Pinellas County, January 1, 1974. In this view the loon is lying on its back and the sternum reflected back so the internal organs can be seen. Note the numerous large and small whitish lesions on the air sacs *(arrows)*. It was estimated that 90% of the air sac surface was covered by these plaques.

some of these serotypes of *Salmonella* were conducted (White and Forrester 1979). Three of the serotypes (*S. agona, S. infantis,* and *S. saintpaul*) were resistant to 2 or more antibiotics. It was postulated that these resistant serotypes from loons originated from treated infections of man or domesticated animals. It was pointed out that many of these infected loons (and also Double-crested Cormorants with similar resistant serotypes of *Salmonella*) were from the Tampa Bay area, where there were considerable sources of pollution from nearby sewage plants, hospitals, feed mills, domestic animal slaughter houses, etc.

Very little is known about bacterial infections in grebes from Florida, although 5 species of bacteria have been identified. *Pasteurella multocida* was identified in 1 of 2 Pied-billed Grebes found dead during an outbreak of avian cholera in Everglades National Park in the winter of 1967–68 (Klukas and Locke 1970). The outbreak involved mainly several thousand American Coots and a smaller number of waterfowl and is the only recorded epizootic of avian cholera known to have occurred in Florida. A *Streptococcus* sp. was isolated from the other grebe, although the site and extent of the infection were not given (Locke 1968). In 1997 *Klebsiella pneumoniae* and *Chromobacterium* sp. were isolated from the liver of 1 Pied-billed Grebe and *Aeromonas* sp. from the liver of another Pied-billed Grebe. Both grebes died of an undetermined type of trauma in Broward County (Meteyer 1997).

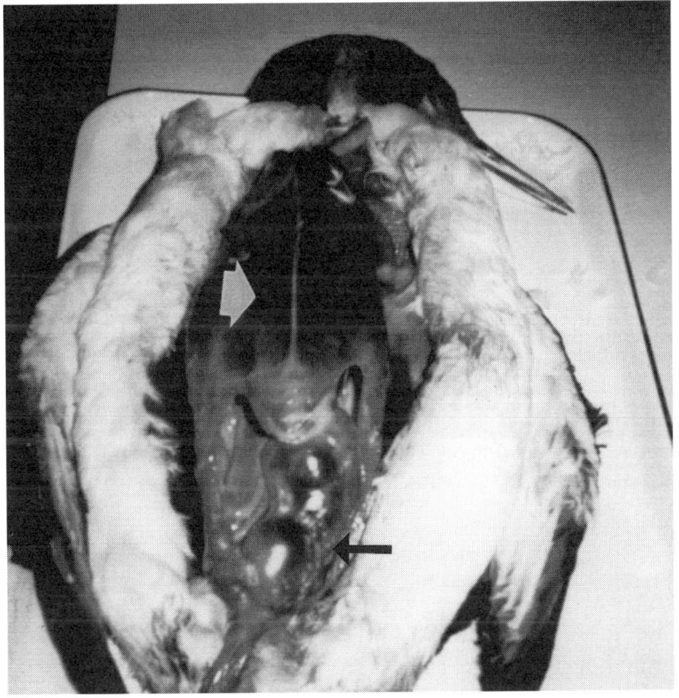

FIGURE 2.3. An emaciated Common Loon found on a Florida beach in the early 1970s with a mild case of aspergillosis. Note the sunken pectoral muscles *(white arrowhead)*, sunken keel and whitish plaques on the air sacs *(black arrow)*.

XIV. Fungi

Between 1946 and 1997 42 cases of *Aspergillus* infections (figure 2.2) were reported from Common Loons in Florida (table 2.11). These were most likely all *Aspergillus fumigatus,* but tests to determine definitive identifications were not accomplished on each case. The prevalence of *Aspergillus* infections in Common Loons in Florida has been estimated to be between 7 and 18% (White et al. 1976; Forrester et al. 1997). In other studies the prevalence was 7% in Minnesota (Ensor et al. 1992) and 2% in Ontario (Frank et al. 1983). Franson and Cliplef (1993) reported aspergillosis in 13 of 222 (6%) Common Loons examined at necropsy between 1976 and 1991. Their specimens came from 18 states, including 36 from Florida.

The significance of aspergillosis as a disease in loons is not clear. In some cases it has not always been possible to determine if the fungal infections were responsible for the death of the loon or if the infection resulted as a sequela to other stress factors such as oil spills or emaciation. Locke and Young (1967) described a case of aspergillosis that developed rapidly in a captive Common Loon. They suggested that loons were "highly susceptible to this mycotic infection." In waterfowl, aspergillosis has been recognized as a sequel to captivity, malnutrition, oiling, or other concurrent diseases (Wobeser 1981). However, White et al. (1976) found that aspergillosis was more prevalent in non-oiled loons than in oiled loons, suggesting that this stress factor may not be important. Almost all of the birds listed in table 2.11 were emaciated (figure 2.3). The mean weight for 35 of these cases (for which values were available) was 2.05

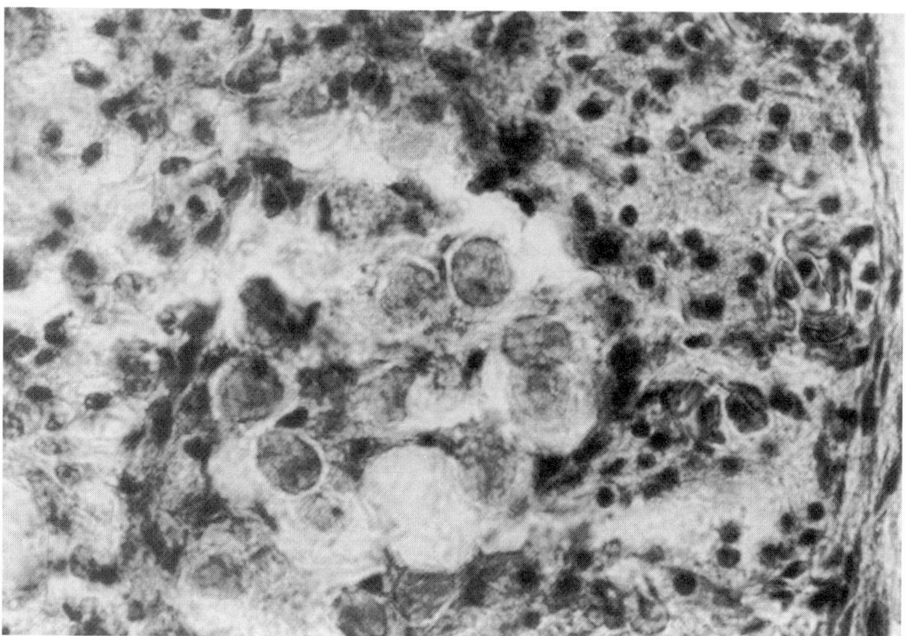

FIGURE 2.4. Photomicrograph of the kidney of a Common Loon from Alachua County (1985), showing coccidial organisms in the epithelium of the collecting ducts. H&E, X1,000.

kg, whereas normal Common Loons overwintering in Florida weighed more than 4 kg (Forrester et al. 1997). Ultimately, it is difficult to know whether emaciation leads to aspergillosis or if aspergillosis leads to emaciation. More research on this aspect is needed.

We have no data on mycotic infections in grebes from Florida.

XV. Protozoans

Five species of protozoans are known from Common Loons in Florida. Two of these are coccidians, 1 from the kidneys and 1 from the intestines, 2 are blood protozoans, and 1 is an intestinal flagellate.

Fourteen cases of renal coccidial infections

Table 2.12. Infections of renal coccidia in Common Loons from Florida

County/site	Year	No. loons infected[a]	Data source
Alachua	1977	1	Popp & Forrester (1978)
	1985	1	Fenwick & Crowley (1986)
Broward	1994	1	Smith (1995)
"Florida"	1991	1	Roth (1991)
Franklin	1983	1	Buergelt & Short (1983)
	1983	1	Stetzer & Short (1983)
	1983	6	Stroud (1983)
Levy	1993	1	Woodard & Westhouse (1993)
St. Johns	1998	1	Spalding & Forrester (1998)

a. Diagnosis was by examination of histologic sections of kidney tissue.

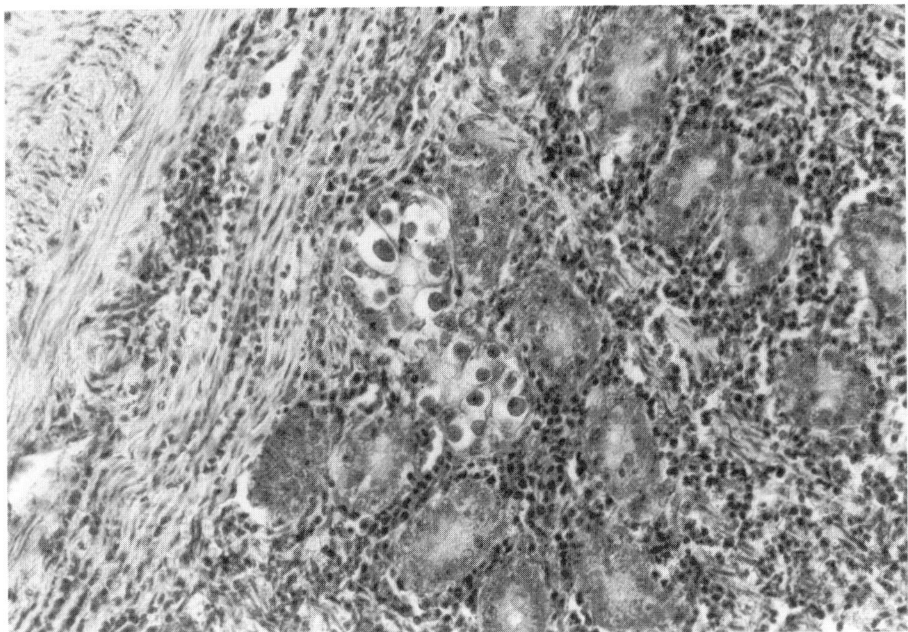

FIGURE 2.5. Photomicrograph showing coccidial organisms in the caecum of a Common Loon from Alachua County (1985). H&E, X400.

were found in Common Loons from 5 localities in Florida between 1977 and 1998 (table 2.12). The species involved is probably *Eimeria gaviae*, which was described originally from a Common Loon in Maryland by Montgomery et al. (1978). These infections resulted in mild to moderate inflammation in the kidneys. In figure 2.4 coccidial organisms can be seen in the epithelium of the collecting ducts in the kidney of a Common Loon from Alachua County.

The loon mentioned above from Alachua County was emaciated and also had moderate intestinal coccidiosis involving the ileum, jejunum, and cecum (figure 2.5) and an unidentified flagellate in the duodenum (figure 2.6). The species of coccidia involved was not determined. The combined renal and intestinal coccidial infections were judged to be contributing factors leading to the death of this one loon (Fenwick and Crowley 1986). The overall importance of coccidiosis to wintering populations of loons in Florida is not known.

Between 1972 and 1997 blood smears from 128 Common Loons were examined for protozoan infections (table 2.13). All but 1 were negative. In April of 1997 a moribund Common Loon was found in a watermelon field in Newberry (Alachua County) and submitted to the Wildlife Clinic at the Veterinary Medical Teaching Hospital (VMTH) at the University of Florida. It was diagnosed as having aspergillosis and eye problems perhaps due to trauma and was treated for these conditions. The bird then was given to a local wildlife rehabilitator who attempted to nurse it back to health prior to release into the wild. After about 1 month in captivity it was still not doing well and was returned to the VMTH for further observation and treatment, when a blood smear was made and the bird found to have a dual infection of *Plasmodium polare* and *Leucocytozoon* sp. (Greiner 1997). A week later it was found in a distressed condition and was then euthanized and examined at necropsy (Homer and Pap-

Table 2.13. Common Loons from Florida waters examined by thin blood smear for protozoans

County	Year(s)	No. loons Examined	Positive
Alachua	1974	1	0
	1997	1	1[a]
Duval	1974	5	0
Flagler	1974	3	0
Franklin	1983	1	0
	1984	12	0
Indian River	1973	3	0
Lee	1974	3	0
Martin	1972	1	0
	1977	1	0
Palm Beach	1972	1	0
Pinellas	1973	2	0
	1974	27	0
	1975	1	0
Sarasota	1972	3	0
St. Johns	1974	53	0
Volusia	1972	1	0
	1974	9	0
Totals	1972–97	128	1

Source: Forrester and Bennett (1996), unless otherwise specified.
Note: In their checklist, Bishop and Bennett (1992) refer to the same slides from which the above dataset was derived and record an "unknown parasite." In reexamination of the slides, no such parasite was located.
a. This loon was found injured and had been in captivity for about 1 month when the blood film was taken and found to be positive for *Plasmodium polare* and *Leucocytozoon* sp. (Greiner 1997) (see text for more details).

ineau 1997). This is the first known case of blood protozoans in Common Loons and may have resulted because of some abnormal circumstances. Both hematozoans most likely were acquired in captivity from vectors that had fed on other birds. Debilitation associated with concurrent aspergillosis and other problems may have led to an immunocompromised host, rendering it more susceptible to blood parasites that normally do not infect loons. A similar situation has been reported in Sandhill Cranes in Florida (Telford et al. 1994).

Thirteen Common Loons from various localities in Florida were examined serologically by the IHA test for antibodies to *Toxoplasma gondii* (Burridge et al. 1979). All were negative; however, as pointed out elsewhere, the IHA test may not be as sensitive in birds as in mammals and may result in false-negative results (Frenkel 1981). There are no records of toxoplasmosis in loons.

There are no data on protozoan infections in grebes in Florida.

XVI. Helminths

The helminths of Common Loons overwintering in Florida coastal waters include 31 species of trematodes (tables 2.14 and 2.15), 5 cestodes (table 2.16), 11 nematodes (table 2.17), and 3 acanthocephalans (table 2.16). No information on parasitic helminths is available on other species of loons that occur in Florida.

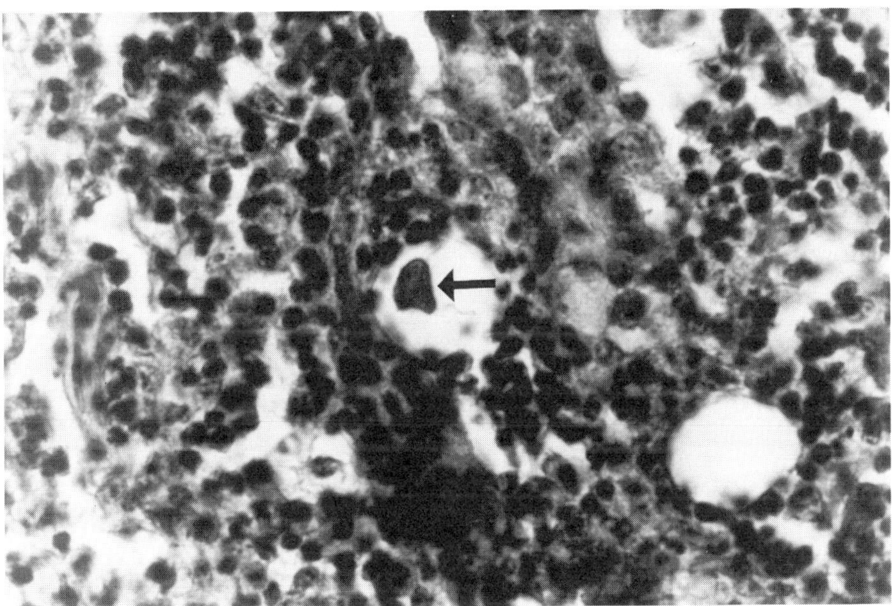

FIGURE 2.6. Unidentified flagellate in the duodenum of a Common Loon from Alachua County (1985). Note the anterior and posterior flagellae extending from the organism *(arrow)* in the vacuole in the center of the photograph. H&E, X1,000.

Little is known about the pathologic significance of these helminths to Common Loons. There has been speculation that the microphallid trematodes found at times in large numbers in the small intestines of Common Loons overwintering in Florida could be pathogenic (Alexander 1985; McIntyre 1988; Spitzer 1995), but Forrester et al. (1997) discounted this because of the absence of histopathologic evidence that this was occurring. During the epizootic of 1983 (see section XVIII, Emaciation syndrome) emaciated loons had high prevalences and intensities of intestinal trematodes, with total numbers averaging 7,665 per loon. The great majority of these consisted of 1 species, *Microphallus forresteri* (Kinsella and Forrester 1996, 1999; Kinsella and Deblock 1997). In several loons, numbers exceeded 10,000 and 1 loon had more than 82,000 trematodes (Forrester et al. 1997). These numbers were 600 times those found in a sample of normal loons from the same general area examined in the following year (1984), a nonepizootic year (table 2.18). It is tempting to attach primary patho-

logic significance to these observations, but, as mentioned above, such evidence is lacking. The elevated numbers observed in 1983 were attributed to changes brought on by a number of environmental influences (drought leading to increased salinity of estuaries, increases in numbers of macroinvertebrates that serve as intermediate hosts of the trematodes, depletion of preferred prey fish, and subsequent changes in food habits) as discussed further in section XVIII, Emaciation syndrome, and represented diagrammatically in figure 2.12.

The pancreatic trematode *Erschiovorchis lintoni* is not common in Common Loons, but has been associated with some remarkable tissue changes. Smith (1995) described one such case from Broward County in 1994. The pancreas in this bird was 3–5 times normal size (figure 2.7) and "yellowish-tan with greenish brown flecks scattered along the surface" (Smith 1995). The trematodes were found in small fluid-filled structures surrounded by thin fibrous capsules that occurred in approximately 2/3 of the pancreas. There was little or

Table 2.14. Trematode infections in the small intestines of Common Loons from Florida

Trematode County	Year(s)	No. loons Examined	Infected	%	Intensity Mean	Range
Apophallus brevis						
Broward	1972	1	1	—	2	—
Duval	1972–74	10	3	30	124	10–277
Indian River	1971–73	5	4	—	69	1–225
Lee	1972–74	3	2	—	566	2–1,130
Martin	1972	2	1	—	6	—
Nassau	1972	1	1	—	1,633	—
Palm Beach	1972	2	2	—	25	7–42
Pinellas	1972–74	18	4	22	404	5–815
Sarasota	1972–73	7	3	—	4	1–6
St. Johns	1972–74	30	4	13	89	7–165
Volusia	1972–74	10	2	20	31	26–35
Cotylurus erraticus						
Nassau	1972	1	1	—	2	—
Cotylurus platycephalus						
Pinellas	1972–74	18	1	6	1	—
St. Johns	1972–74	7	1	—	1	—
Diplostomum gavium						
Alachua	1977	1	1	—	7	—
Broward	1972	1	1	—	2	—
Lee	1972–74	3	1	—	2	—
Nassau	1972	1	1	—	2	—
Pinellas	1972–74	18	2	11	8	5–11
St. Johns	1972–74	30	1	3	1	—
Diplostomum immer						
Duval	1972–74	10	1	10	3	—
Flagler	1974	2	1	—	2	—
Indian River	1971–73	5	1	—	78	—
Lee	1972–74	3	1	—	10	—
Nassau	1972	1	1	—	82	—
Palm Beach	1972	2	1	—	1	—
Pinellas	1972–74	18	4	22	17	3–37
St. Johns	1972–74	30	9	30	17	1–61
Echinochasmus skrjabini						
Nassau	1972	1	1	—	24	—
Palm Beach	1972	2	1	—	3	—
Pinellas	1972–74	18	3	17	10	1–28
St. Johns	1972–74	30	6	20	9	1–27
Volusia	1972–74	10	3	30	5	1–14
Himasthla alincia						
Indian River	1971–73	5	1	—	4	—
Maritrema spp.[a]						
Alachua	1974	1	1	—	5	—
Duval	1972–74	10	1	10	1	—
Franklin	1984	23	1	4	3	—

(continued)

Table 2.14. *(continued)*

Trematode County	Year(s)	No. loons			Intensity	
		Examined	Infected	%	Mean	Range
Pinellas	1972–74	18	2	11	407	4–810
Sarasota	1972–73	7	2	—	102	86–117
Mesorchis denticulatus						
Duval	1972–74	10	1	10	5	—
Franklin	1984	23	2	9	1	—
Indian River	1971–73	5	2	—	58	23–93
Nassau	1972	1	1	—	2	—
Pinellas	1972–74	18	1	6	5	—
Sarasota	1972–73	7	1	—	1	—
St. Johns	1972–74	30	1	3	45	—
Volusia	1972–74	10	2	20	4	1–6
Mesostephanus appendiculatoides[b]						
Broward	1972	1	1	—	1	—
Hillsborough	1974	7	3	—	40	10–90
Indian River	1971–73	5	1	—	3	—
Lee	1972–74	3	2	—	16	6–25
Palm Beach	1972	2	1	—	1	—
Pinellas	1972–74	18	5	28	14	4–23
Sarasota	1972–73	7	1	—	1	—
Microphallus spp.[c]						
Alachua	1974	1	1	—	25	—
	1977	1	1	—	20	—
Duval	1972–74	10	6	60	155	2–642
Franklin	1984	23	6	26	12	1–32
Hillsborough	1974	7	1	—	29	—
Indian River	1971–73	5	3	—	397	4–1,126
Lee	1972–74	3	1	—	40	—
Nassau	1972	1	1	—	3,752	—
Palm Beach	1972	2	1	—	5	—
Pinellas	1972–74	18	5	28	23	1–90
Sarasota	1972–73	7	3	—	10	5–15
St. Johns	1972–74	30	8	27	2,356	4–10,325
Volusia	1972–74	10	5	50	1,543	1–4,411
Parvatrema sp.						
Volusia	1972–74	10	1	10	3	—
Phagicola longa						
Broward	1972	1	1	—	3	—
Hillsborough	1974	7	2	—	6	4–7
Indian River	1971–73	5	1	—	32	—
Palm Beach	1972	2	1	—	95	—
Pinellas	1972–74	18	3	17	87	1–230
Sarasota	1972–73	7	1	—	5	—
St. Johns	1972–74	30	1	3	32	9–54
Volusia	1972–74	10	1	10	2	—

(continued)

Table 2.14. *(continued)*

Trematode County	Year(s)	No. loons			Intensity	
		Examined	Infected	%	Mean	Range
Posthodiplostomum minimum						
Alachua	1977	1	1	—	4	—
Posthodiplostomum sp.						
St. Johns	1972–74	30	1	3	20	—
Stictodora lariformicola						
Hillsborough	1974	7	2	—	7	2–12

Sources: Kinsella and Forrester (1996, 1999).

Note: Our database on helminths of Common Loons in Florida is based on 134 loons examined at necropsy, with the exception of *Splendidofilaria fallinsensis*, which is a special case explained in a footnote in table 2.17. Otherwise the data are presented only for samples in which at least 1 loon was positive for a given helminth. Other samples in which no helminths of a given species were found are not presented. The total numbers of loons sampled are based on the following values for each county and in the years shown in parentheses: Alachua = 1 (1974) and 1 (1977), Broward = 1 (1972), Clay = 1 (1989), Duval = 10 (1972–74), Flagler = 2 (1974), Franklin = 1 (1983) and 24 (1984), Hillsborough = 7 (1974), Indian River = 5 (1971–73), Lee = 3 (1972–74), Nassau = 1 (1972), Palm Beach = 2 (1972) and 2 (1993), Pinellas = 18 (1972–74) and 5 (1983), Sarasota = 7 (1972–73), St. Johns = 30 (1972–74), Martin = 2 (1972), Volusia = 10 (1972–74), Unknown county = 1 (1971).

a. A composite of 2 undescribed species; also found in Pinellas County in 1983, but there are no data on the prevalence and intensity.

b. Also found in Pinellas County in 1983, but there are no data on the prevalence and intensity.

c. A composite of 3 species, *M. nicolli*, *M. forresteri*, and 1 undescribed species; also found in Pinellas County in 1983, but there are no data on the prevalence and intensity.

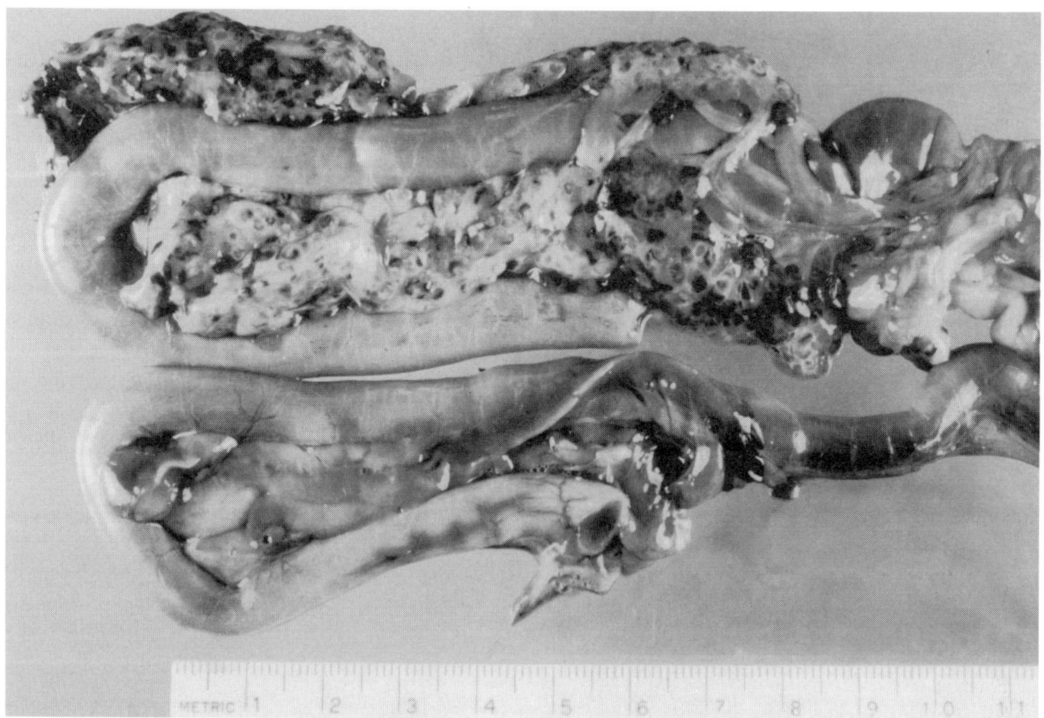

FIGURE 2.7. The pancreas of a Common Loon from Broward County (1994) infected with the trematode *Erschoviorchis lintoni (top)*, compared with a normal pancreas from an uninfected loon *(bottom)*. Courtesy of Kirk E. Smith.

Table 2.15. Trematode infections in Common Loons from Florida in sites other than the small intestine

Trematode (Site)[a] County	Year(s)	No. loons			Intensity	
		Examined	Infected	%	Mean	Range
Amphimerus arcticus[b] (LV)						
Duval	1972–74	10	4	40	3	1–9
Flagler	1974	2	1	—	12	—
Franklin	1983–84	25	6	24	2	1–3
Hillsborough	1974	7	2	—	2	1–3
Indian River	1971–73	5	2	—	4	1–7
Lee	1972–74	3	1	—	1	—
Nassau	1972	1	1	—	4	—
Pinellas	1972–74	18	5	28	8	1–29
Sarasota	1972–73	7	1	—	2	—
St. Johns	1972–74	30	3	10	3	1–8
Austrobilharzia terrigalensis[b] (BV)						
Franklin	1983–84	25	2	8	3	2–3
Indian River	1971–73	5	2	—	2	1–2
Pinellas	1972–74	18	1	6	1	—
Sarasota	1972–73	7	1	—	1	—
Dendritobilharzia pulverulenta (HT)						
Flagler	1974	2	1	—	1	—
Franklin	1983–84	25	1	4	1	—
Erschoviorchis lintoni (PA)						
Broward[c]	1994	5	1	—	10	—
Indian River	1971–73	5	1	—	10	—
Pinellas	1972–74	18	1	6	3	—
Microparyphium facetum (CL)						
Franklin	1983–84	25	1	4	1	—
Odhneria odhneri[d] (CE)						
Duval	1972–74	10	1	10	9	—
Franklin	1983–84	25	7	28	3	2–4
Lee	1972–74	3	1	—	103	—
Martin	1972	2	1	—	44	—
Pinellas	1972–74	18	1	6	5	—
	1983	5	3	—	4	1–7
Sarasota	1972–73	7	3	—	36	2–103
Volusia	1972–74	10	1	10	1	—
Parorchis acanthus (CL)						
Franklin	1983–84	25	1	4	1	—
Sarasota	1972–73	7	1	—	3	—
Plotnikovia fodiens (PA, LV)						
Duval	1972–74	10	1	10	3	—
Flagler	1974	2	1	—	27	—
Franklin	1983–84	25	2	8	3	2–3
Hillsborough	1974	7	1	—	10	—
Indian River	1971–73	5	2	—	3	2–3
Lee	1972–74	3	2	—	5	1–9
Palm Beach	1972	2	1	—	6	—

(continued)

Table 2.15. *(continued)*

| Trematode (Site)[a] | | No. loons | | | Intensity | |
County	Year(s)	Examined	Infected	%	Mean	Range
Pinellas	1972–74	18	7	39	13	1–49
	1983	5	2	—	31	8–53
Sarasota	1972–73	7	2	—	12	7–17
St. Johns	1972–74	30	5	17	9	2–33
Volusia	1972–74	10	2	20	3	1–5
Prosthogonimus ovatus (CL)						
Duval	1972–74	10	1	10	16	—
Flagler	1974	2	1	—	2	—
Lee	1972–74	3	1	—	55	—
Martin	1972	2	1	—	2	—
Pinellas	1972–74	18	1	6	3	—
Sarasota	1972–73	7	1	—	12	—
St. Johns	1972–74	30	2	7	1	1
Volusia	1972–74	10	2	20	2	1–2
Renicola pollaris (KD)						
Duval	1972–74	10	1	10	31	—
Franklin	1983–84	25	9	36	5	1–16
Indian River	1971–73	5	3	—	7	3–16
Lee	1972–74	3	2	—	6	1–10
Pinellas	1972–74	18	3	17	4	1–8
	1983	5	3	—	7	4–11
St. Johns	1972–74	30	1	3	2	—
Unknown	1971	1	1	—	25	—
Ribeiroia ondatrae (PR)						
Franklin	1983–84	25	1	4	4	—
Pinellas	1972–74	18	1	6	1	—
Sarasota	1972–73	7	1	—	9	—
Tanasia fedtschenkoi (KD)						
Hillsborough	1974	7	1	—	1	—

Sources: Kinsella and Forrester (1996, 1999) unless otherwise noted. Also, see Note in Table 2.14.
a. Infection sites: BV = blood vessels, CE = cecum, CL = cloaca, HT = heart, KD = kidney, LV = liver, PA = pancreas, PR = proventriculus.
b. Also found in Okaloosa and Walton counties in 1991, but no data are available on prevalence and intensity (Kinsella and Davidson 1996).
c. Data from Smith (1995).
d. Also found in Okaloosa and Walton counties in 1991 and Wakulla County in 1983, but no data are available on prevalence and intensity (Kinsella and Davidson 1996).

no inflammatory reaction to the parasites. The loon in question was emaciated (Smith 1995) and the trematode infection may have played a secondary role in the cause of death of the bird, but that is speculative.

The nematode *Eustrongylides tubifex* was found in 3 loons in low numbers and, although infections were associated with lesions, may not be a significant morbidity or mortality factor. In 1 loon from Broward County (1994), 3 worms had penetrated the proventriculus and were associated with a mild inflammatory reaction (Smith 1995). It was observed histologically that the worms were dead and were degenerating. One end of each worm protruded into the lumen of the proventriculus and the other end was

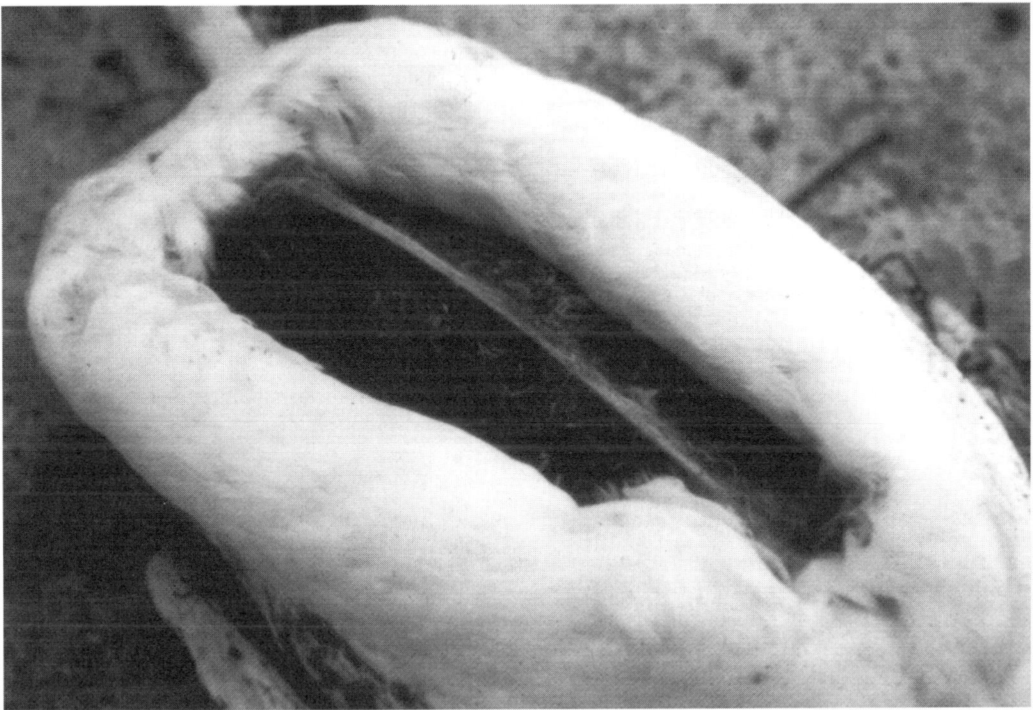

FIGURE 2.8. An emaciated Common Loon found dead on the beach in the panhandle region of Florida during the 1983 epizootic. The loon is lying on its back and the skin and feathers have been reflected to show the sunken pectoral muscles and prominent keel. Courtesy of William R. Davidson.

"coiled in a thin layer of tan to gray granular exudate on the serosal surface of the proventriculus." The mucosal surface of the proventriculus was coated with a layer of mucus. The proventricular tumors elicited by this worm in fish-eating birds in eastern Canada have been described by Measures (1988). She reported *E. tubifex* in 1 of 25 Common Loons from Ontario and New Brunswick; Chafel and Pokras (1993) found 2 of 20 Common Loons from New England infected with a species of *Eustrongylides* that was probably *E. tubifex*. Measures (1988) also studied the development and pathogenesis of experimental infections of *E. tubifex* in a number of piscivorous birds (excluding loons) and found that the nematodes matured rapidly (eggs being produced within 10–17 days of infection), after which the worms degenerated. From this it appears that *E. tubifex* is most likely acquired by loons on their northern breeding grounds who then bring infections with them during migration to the south; however, the worms probably do not survive through the winter in loons overwintering in Florida.

Grebes have not been well investigated in Florida and information is available only on the Pied-billed Grebe, in which 9 species have been identified: 3 trematodes, 3 cestodes, and 3 nematodes (table 2.19). MacInnis (1959) stated that he found cestodes in 1 of 3, and nematodes in 2 of 3 Pied-billed Grebes collected from Alligator Point (Franklin County) (date not given), but gave no details on specific identification, prevalence, or intensity. Premvati (1968) reported that he found 25 specimens of the trematode *Echinochasmus donaldsoni* in the intestines of 2 Pied-billed Grebes from Leon County (no collection dates given). He also described a new species of echinostome trematode (*Petasiger floridanus*) presumably from the same grebe.

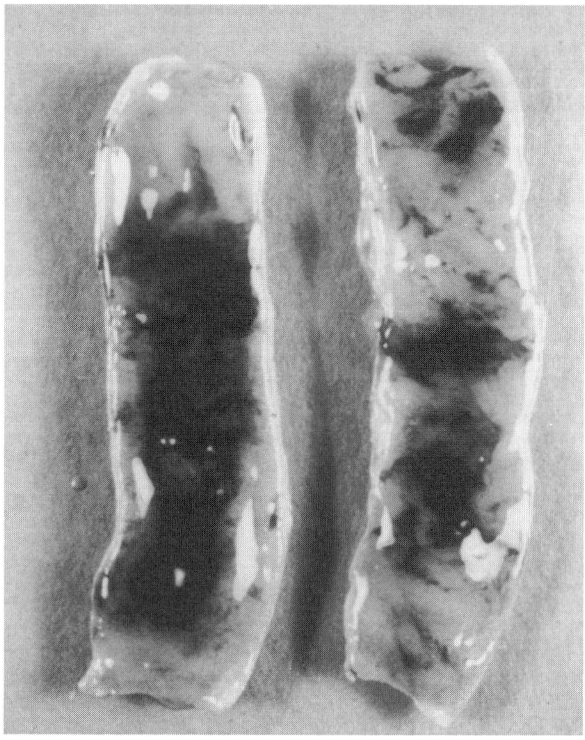

FIGURE 2.9. Parts of the upper small intestine of a Common Loon found dead on the beach at St. Marks National Wildlife Refuge during the 1983 epizootic. Note areas of hemorrhage *(dark areas)*. Courtesy of William R. Davidson.

His description was based on 4 specimens from the intestine of 1 bird from Leon County. He gave no date of collection in his paper, but the specimens he deposited in the U.S. National Parasite Collection are labeled with the date "July, 1966." Nothing is known about the pathogenic significance of these helminths.

XVII. Arthropods

Three species of ectoparasites (2 mites and 1 chewing louse) have been found on Common Loons, 2 species (both chewing lice) on Horned Grebes, and 2 species (1 mite and 1 chewing louse) on Pied-billed Grebes in Florida (table 2.20). Apparently parasitic arthropods are not common on Common Loons. Only 3 of 75 loons examined carefully for ectoparasites from 1972 to 1993 were positive, and in each of those cases, only 1 arthropod was found on each bird (Forrester 1996).

XVIII. Emaciation syndrome

The principal diagnostic finding related to mortality of Common Loons overwintering in Florida coastal waters is an emaciation syndrome (table 2.2). From 1970 to 1994, 286 of 434 (66%) dead or moribund loons examined at necropsy were emaciated, some very severely (Forrester et al. 1997). Data from Suncoast Seabird Sanctuary, a wildlife rehabilitation center in Pinellas County, were similar. From 1988 to 1995, 290 of 460 Common Loons (63%)

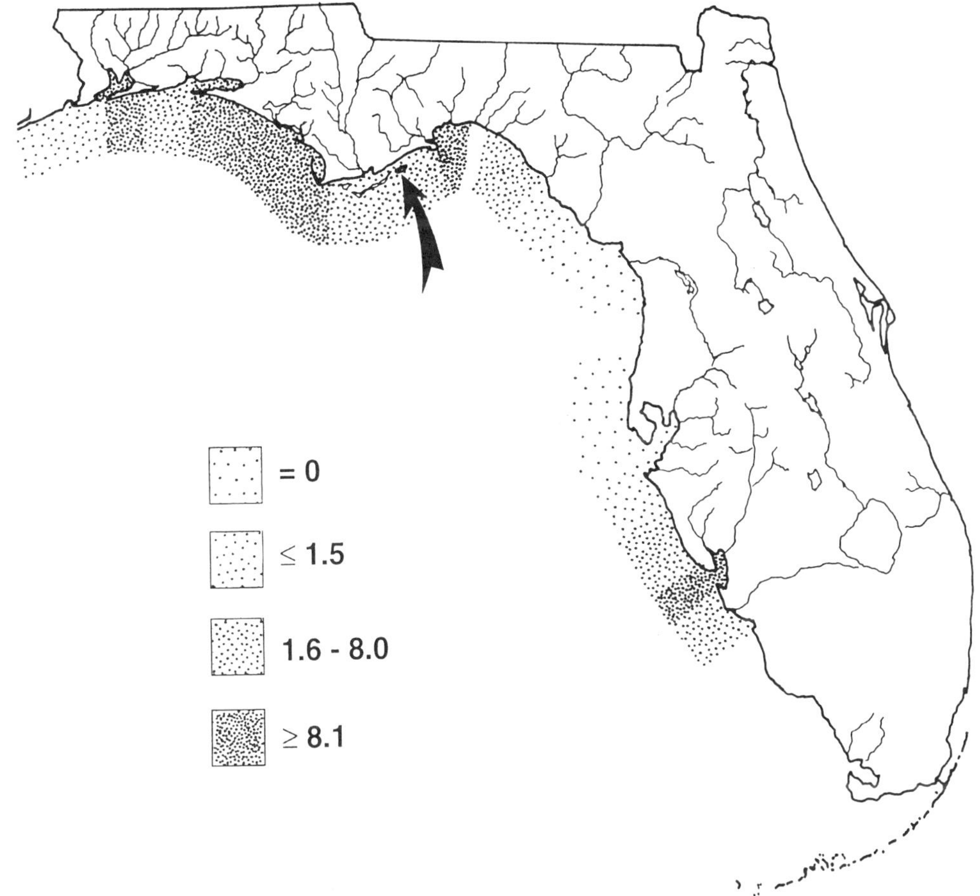

FIGURE 2.10. Distribution and magnitude of mortality of Common Loons in the Gulf Coast of Florida as determined by beach surveys, March 18–25, 1983. From Forrester et al. 1997; by permission of *Journal of Wildlife Diseases*.

found on beaches in the Tampa Bay area and submitted for treatment were emaciated (Suto 1996). This syndrome is characterized by severe atrophy of pectoral muscles (figure 2.8), loss of body fat, and hemorrhagic enteritis (figure 2.9). Total body weights of these emaciated loons averaged 2.00 kg compared with 4.19 for normal birds (Forrester et al. 1997). Each winter since at least the early 1970s dead Common Loons on beaches of the Atlantic and Gulf coasts of Florida have been recorded in varying numbers (Stevenson 1970, 1971, 1972, 1974, 1977; Ogden 1987; Simons 1985; Alexander 1991). Most of these loons were emaciated. Cata-

strophic mortality of Common Loons occurred in the winter of 1983 in Florida coastal waters and almost every bird examined was severely emaciated. This event has been discussed in a number of publications (Hoffman 1983; Imhof 1983; Kale 1983; Hamel 1983; Graham 1984; Alexander 1985, 1991; Klein 1985; McIntyre 1988; Stevenson and Anderson 1994), and many have provided suggestions as to the cause of the die-off; most of these were based on incomplete data and many focused on mercury contamination as the problem. More than 13,000 loons were estimated to have died between January and March of 1983 (Forrester et

Table 2.16. Cestode and acanthocephalan infections in the small intestines of Common Loons from Florida

Helminth County	Year(s)	No. of loons Examined	infected	%	Intensity Mean	Range
Cestoda						
Armadoskrjabinia rostellata[a]						
Alachua	1974	1	1	—	2	—
	1977	1	1	—	1	—
Broward	1972	1	1	—	29	—
Clay	1989	1	1	—	5	—
Duval	1972–74	10	6	60	136	3–606
Franklin	1984	23	12	52	10	1–63
Hillsborough	1974	7	4	—	17	9–30
Indian River	1971–73	5	3	—	18	1–49
Lee	1972–74	3	2	—	35	10–60
Martin	1972	2	1	—	1	—
Nassau	1972	1	1	—	375	—
Palm Beach	1972	2	1	—	728	—
	1993	2	1	—	17	—
Pinellas	1972–74	18	12	67	231	1–1,196
Sarasota	1972–73	7	5	—	113	2–540
St. Johns	1972–74	30	18	60	56	1–436
Volusia	1972–74	10	6	60	65	5–157
Unknown	1971	1	1	—	109	—
Cyclustera ibisae						
Broward	1972	1	1	—	1	—
Duval	1972–74	10	1	10	2	—
Hillsborough	1974	7	1	—	4	—
Indian River	1971–73	5	1	—	1	—
St. Johns	1972–74	30	5	17	6	1–12
Microsomacanthus pseudorostellatuas[a]						
Broward	1972	1	1	—	—	—
Duval	1972–74	10	1	10	—	—
Franklin	1984	23	4	17	—	—
Hillsborough	1974	7	2	—	—	—
Nassau	1972	1	1	—	—	—
Pinellas	1972–74	18	6	33	—	—
St. Johns	1972–74	30	5	17	—	—
Volusia	1972–74	10	2	20	—	—
Neovalipora parvispinae[a]						
Alachua	1974	1	1	—	17	—
Duval	1972–74	10	5	50	11	5–18
Franklin	1984	23	20	87	20	1–116
Hillsborough	1974	7	5	—	23	1–78
Indian River	1971–73	5	1	—	1	—
Lee	1972–74	3	3	—	28	4–54
Palm Beach	1972	2	1	—	93	—
Pinellas	1972–74	18	11	61	63	1–559
Sarasota	1972–73	7	1	—	2	—

(continued)

Table 2.16. *(continued)*

Helminth County	Year(s)	No. of loons			Intensity	
		Examined	infected	%	Mean	Range
St. Johns	1972–74	30	10	33	53	3–184
Volusia	1972–74	10	3	30	136	11–367
Unknown	1971	1	1	—	1	—
Tetrabothrius macrocephalus[a]						
Alachua	1974	1	1	—	2	—
	1977	1	1	—	46	—
Clay	1989	1	1	—	11	—
Duval	1972–74	10	3	30	8	2–15
Franklin	1984	23	6	26	2	1–5
Hillsborough	1974	7	1	—	1	—
Indian River	1971–73	5	3	—	3	1–6
Martin	1972	2	2	—	5	1–9
Palm Beach	1972	2	1	—	10	—
Pinellas	1972–74	18	4	22	6	1–12
Sarasota	1972–73	7	3	—	2	1–3
St. Johns	1972–74	30	3	10	2	1–3
Volusia	1972–74	10	5	50	11	1–24
Acanthocephala						
Andracantha gravida						
Alachua	1977	1	1	—	1	—
Franklin	1984	23	3	13	1	1–2
Pinellas	1972–74	18	4	22	2	1–3
Polymorphus brevis						
Franklin	1984	23	1	4	1	—
Southwellina hispida						
Franklin	1984	23	1	4	1	—

Sources: Kinsella and Forrester (1996, 1999). Also see Note in Table 2.14.
a. Also found in Pinellas County in 1983, but there are no data on prevalence and intensity. Intensities of *A. rostellata* and *M. pseudorostellatus* are combined since some scoleces lacked hooks and could not be differentiated.

al. 1997). Mortality occurred along both the Atlantic and Gulf coasts, but was most severe on the Gulf Coast, particularly in the area from Pensacola to Naples (figure 2.10). An epizootic curve illustrating the timing of the mortality was constructed for loons in the vicinity of Dog Island in Franklin County (figure 2.11). The events leading up to this mortality event are believed to be complex and have been postulated to include a number of synergistic factors, including the stress of migration, molting of flight feathers, changes in food resources and subse-

quent salt-loading, intense intestinal parasitism, environmental contaminants, and inclement weather (Forrester et al. 1997). These interrelated factors are presented in a schematic diagram in figure 2.12.

We have no data on emaciation in grebes from Florida. Although the Eared Grebe is rare in Florida, and there is no mortality information on it, large numbers (>25,000) of this species were reported to have died of an emaciation syndrome during the winter of 1982–83 along the coast of southern California and Baja

Table 2.17. Nematode infections in Common Loons from Florida

| Nematode (Site)[a] | | No. of loons | | | Intensity | |
County	Year(s)	Examined	Infected	%	Mean	Range
Capillaria mergi (SI)						
Alachua	1974	1	1	—	30	—
Flagler	1974	2	1	—	2	—
Franklin	1984	23	1	4	1	—
Hillsborough	1974	7	1	—	5	—
Indian River	1971–73	5	1	—	2	—
Palm Beach	1972	2	1	—	2	—
Pinellas	1972–74	18	2	11	1	—
St. Johns	1972–74	30	4	13	1	—
Volusia	1972–74	10	1	10	1	—
Contracaecum sp. (larvae)[b] (PR)						
Clay	1989	1	1	—	1	—
Flagler	1974	2	1	—	2	—
Hillsborough	1974	7	3	—	3	1–5
Indian River	1971–73	5	1	—	1	—
Pinellas	1972–74	18	3	17	11	1–18
St. Johns	1972–74	30	10	33	3	1–12
Cosmocephalus obvelatus (ES)						
Alachua	1974	1	1	—	3	—
Duval	1972–74	10	2	20	2	—
Flagler	1974	2	1	—	1	—
Hillsborough	1974	7	3	—	2	1–4
Indian River	1971–73	5	1	—	14	—
Lee	1972–74	3	1	—	1	—
Martin	1972	2	1	—	1	—
Palm Beach	1993	2	1	—	1	—
Pinellas	1972–74	18	5	28	3	1–5
Sarasota	1972–73	7	5	—	1	1–2
St. Johns	1972–74	30	9	30	2	1–4
Volusia	1972–74	10	5	50	1	1–3
Cyathostoma phenisci (TR)						
Franklin	1984	23	2	9	1	—
Palm Beach	1993	2	1	—	5	—
Volusia	1972–74	10	1	10	1	—
Eustrongylides tubifex (PR)						
Broward[c]	1994	5	1	—	3	—
St. Johns	1972–74	30	1	3	12	—
Volusia	1972–74	10	1	10	1	—
Paracuaria adunca (PR,ES)						
Alachua	1974	1	1	—	10	—
	1977	1	1	—	1	—
Duval	1972–74	10	3	30	2	1–3
Flagler	1974	2	1	—	2	—
Hillsborough	1974	7	4	—	2	1–4
Indian River	1971–73	5	2	—	3	1–5
Lee	1972–74	3	1	—	1	—
Palm Beach	1972	2	1	—	2	—
Pinellas	1972–74	18	3	17	1	—
Sarasota	1972–73	7	2	—	3	1–4

(continued)

Table 2.17. *(continued)*

| Nematode (Site)[a] | | No. of loons | | | Intensity | |
County	Year(s)	Examined	Infected	%	Mean	Range
St. Johns	1972–74	30	11	37	3	1–13
Volusia	1972–74	10	3	30	1	—
Sciadiocara rugosa (ES)						
St. Johns	1972–74	30	1	3	1	—
Splendidofilaria fallisensis[d] (SQ,BV)						
Alachua	1974	1	0	—	—	—
Duval	1974	5	0	—	—	—
Flagler	1974	3	2	—	ND	—
Franklin	1983–84	14	0	—	—	—
Indian River	1973	3	0	—	—	—
Lee	1974	3	0	—	—	—
Martin	1972	1	1	—	ND	—
	1977	1	0	—	—	—
Palm Beach	1972	2	1	—	ND	—
Pinellas	1973–75	29	3	10	ND	—
Sarasota	1972	2	0	—	—	—
St. Johns	1974	53	5	9	ND	—
Volusia	1972–74	10	1	10	ND	—
Stegophorus diomedeae (VE)						
St. Johns	1972–74	30	1	3	1	—
Streptocara crassicauda longispiculatus[e] (VE)						
Broward	1972	1	1	—	1	—
Duval	1972–74	10	2	20	6	3–9
Franklin	1984	23	1	4	1	—
Hillsborough	1974	7	1	—	2	—
Indian River	1971–73	5	4	—	2	1–3
Lee	1972–74	3	2	—	4	2–5
Martin	1972	2	1	—	1	—
Palm Beach	1972	2	1	—	6	—
Pinellas	1972–74	18	6	33	13	1–57
Sarasota	1972–73	7	2	—	5	2–7
St. Johns	1972–74	30	6	20	2	1–3
Volusia	1972–74	10	4	40	4	1–7
Streptocara formosensis (VE)						
Pinellas	1972–74	18	1	5	1	—
St. Johns	1972–74	30	1	3	1	—

Sources: Kinsella and Forrester (1996, 1999) unless otherwise noted. Also see *Note* in Table 2.14.

a. Infection sites: BV = blood vessels, ES = esophagus, PR = proventriculus, SI = small intestine, SQ = subcutaneous, TR = trachea, VE = ventriculus.

b. Also found in Pinellas County in 1983, but no data are available on prevalence and intensity.

c. Data from Smith (1995).

d. Based on examinations of blood films for the presence of microfilariae typical of this species and therefore only prevalence data are available (Forrester and Bennett 1996). The 127 loons on which these data are based are not all the same as in other parts of the table, although there is some overlap. Adults of *S. fallisensis* were found in the fascia of the legs and pectoral muscles of 3 loons, 1 from Alachua County in 1974 (4 worms), 1 from Martin County in 1972 (9 worms), and 1 from Volusia County in 1974 (1 worm) (Anderson and Forrester 1974; Kinsella and Forrester 1996). There may be a second species of filariid in Common Loons, judging from the data on the microfilariae found in Geimsa-stained blood films (Forrester, 1996). Two populations of microfilariae were noted, based on total lengths. Measurements of 20 microfilariae from 9 Common Loons were as follows: 3 loons had microfilariae measuring 96μ (n = 3) in total length and all 9 loons had microfilariae ranging from 120 to 158μ (mean = 143μ, n = 17). The shorter microfilariae fall in the range of measurements given by Anderson (1954) for *S. fallisensis,* but the larger ones may be from another species, the adults of which have not been collected or identified from Common Loons.

e. Also found in Pinellas County in 1983, but no data are available on prevalence and intensity.

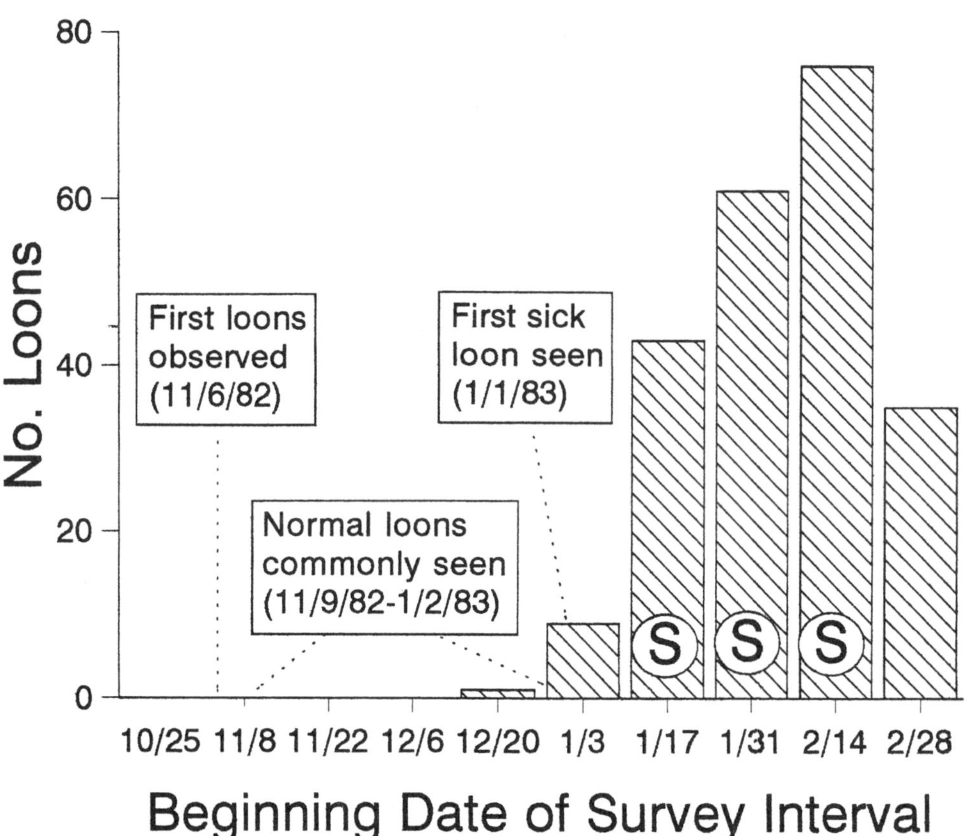

FIGURE 2.11. Epizootic curve illustrating the numbers of sick and dead Common Loons observed per two-week interval, December 1982–March 13, 1983, near Dog Island (Franklin County). An *S* within a circle indicates storms during that time period. From Forrester et al. 1997; by permission of *Journal of Wildlife Diseases*.

Table 2.18. Trematode[a] infections in the small intestines of emaciated and normal Common Loons from the Gulf Coast of Florida

Year	Status	No. loons examined	Prevalence (%)	Intensity		Abundance
				Mean	Range	
1983	Emaciated	23	96	7,665	1–82,625	7,358
1984	Normal[b]	24	46	23	1–85	11

Source: Forrester et al. (1997).
a. Represents 18 species of trematodes, the majority microphallids (total numbers of specimens).
b. Collected by shooting.

Table 2.19. Helminth infections in Pied-billed Grebes in Florida

| Helminth (Site)[a] | | No. grebes | | Intensity | | |
County	Year	Examined	Infected	Mean	Range	Data source
Trematoda						
Echinochasmus donaldsoni (SI)						
Leon	NG	2	NG	NG	—	Premvati (1968)
Microparyphium facetum (CL)						
Collier	1997	1	1	1	—	Kinsella & Foster (1998)
Petasiger floridanus (SI)						
Collier	1997	1	1	10	—	Ibid.
Leon	1966	1	1	4	—	Premvati (1968)
Cestoda						
Fimbriaria sp. (SI)						
Leon	NG	NG	—	NG	—	Loftin (1961)
Hymenolepis sp. (SI)						
Leon	NG	NG	—	NG	—	Ibid.
Wardium lobulatus (SI)						
Collier	1995	1	1	1	—	Kinsella & Foster (1998)
	1997	1	1	4	—	Ibid.
Nematoda						
Capillaria podicipitis (CL)						
Collier	1995	1	1	2	—	Ibid.
	1997	1	1	10	—	Ibid.
Capillaria sp. (ES)						
Collier	1995	1	1	1	—	Ibid.
Tetrameres gubanovi (PR)						
Collier	1997	1	1	5	—	Ibid.

NG = not given by authors.
a. Infection sites: CL = cloaca, ES = esophagus, PR = proventriculus, SI = small intestine.

(Jehl and Bond 1983). This event in California had a number of the same characteristics as the emaciation problem in Common Loons in Florida, including food shortages and inclement weather.

XIX. Summary and Conclusions

The most important cause of morbidity and mortality in Common Loons overwintering in Florida is an emaciation syndrome that accounted for 66% of 434 loons found dead on beaches along Florida coasts from 1970 to 1994. The cause of this syndrome appears to be multifaceted and related to a number of synergistic factors, including the stress of migration, molting of flight feathers, changes in food resources and subsequent salt-loading, intense intestinal parasitism, environmental contaminants, and inclement weather. Other significant factors for loons include oiling, aspergillosis, and trauma. One case of neoplasia has been recorded. One hundred and nine disease agents and parasites have been found in Common Loons in Florida. These include organochlorines (11), metals (4), viruses (1), bacteria (35), fungi (1), protozoans (5) trematodes (31), cestodes (5), nematodes (11), acanthocephalans (3), mites (2), and chewing lice (1). Almost

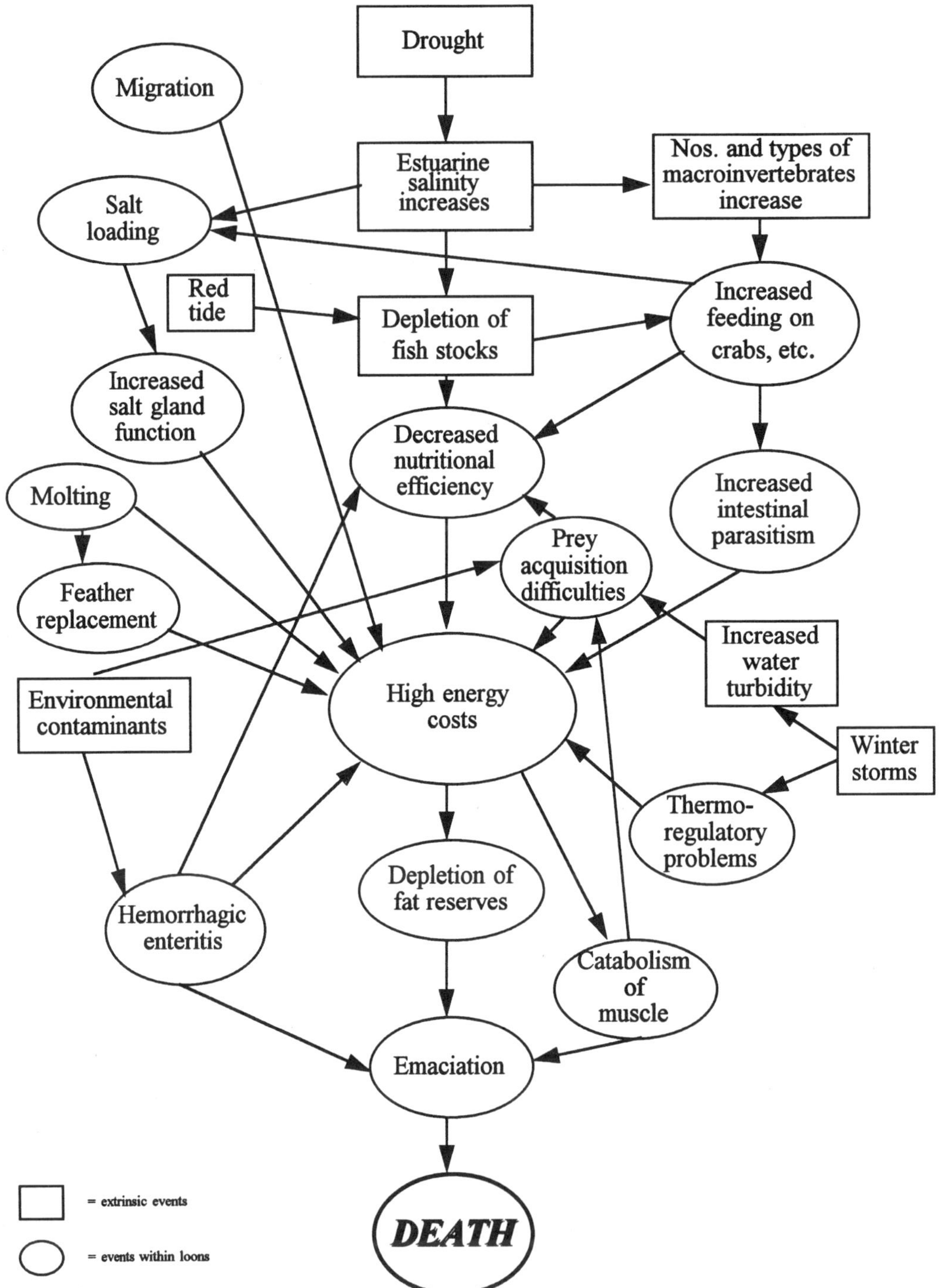

FIGURE 2.12. Schematic diagram of the many interrelated events and factors postulated to have led to the catastrophic mortality of Common Loons in the Gulf Coast of Florida in the winter of 1983.

Table 2.20. Arthropod infestations in loons and grebes from Florida

Host Species of arthropod	County	Year	Data source
Common Loons			
Feather mites			
Alloptes sp.[a]	Sarasota	1972	Forrester & Mertins (1995)
Avenzoariidae[b]	Pinellas	1983	Ibid.
Chewing lice			
Craspedonirmus immer	Pinellas	1973	Forrester et al. (1995)
Horned Grebes			
Chewing lice			
Aquanirmus bucomfishi	Dade	1918	Ibid.
Pseudomenopon dolium	Levy	1957	FSCA[c], Mertins (1999)[d]
Pied-billed Grebes			
Feather mites			
Schizurolichus elegans	Collier	1995	Foster & Mertins (1999)
	Collier	1997	Ibid.
Chewing lice			
Aquanirmus podilymbus	Collier	1997	Ibid.

a. An undescribed species.
b. An undescribed genus, close to *Scutomegninia*.
c. Florida State Collection of Arthropods, Division of Plant Industry, Florida Department of Agriculture and Consumer Services, Gainesville.
d. The host of this specimen in the Florida State Collection of Arthropods is given as *Colymbus auritus*, an old scientific name for the Horned Grebe. The currently accepted scientific name is *Podiceps auritus*, according to the American Ornithologists' Union (1983). The specimen was examined by Mertins (1999), who determined it to be *Pseudomenopon dolium*.

nothing is known about the pathologic significance of these agents for Common Loons.

Information on diseases and parasites of grebes in Florida is limited. One instance of oiling of Horned Grebes was recorded. A small amount of data exists on biotoxins (1), bacteria (5), trematodes (3), cestodes (3), and nematodes (3) from Pied-billed Grebes. Two species of chewing lice are recorded from the Horned Grebe; 1 species of feather mite and 1 chewing louse are known from Pied-billed Grebes. The significance of these is not known. Grebes should be studied further as opportunities become available.

XX. Literature cited

Alexander, L.L. 1985. Trouble with loons. *Living Bird Q.* 4:10–13.

———. 1991. Patterns of mortality among common loons wintering in the northeastern Gulf of Mexico. *Fla. Field Nat.* 19:73–79.

American Ornithologists' Union. 1983. *Check-list of North American birds.* 6th ed. American Ornithologists' Union, Washington, D.C. 877 pp.

Anderson, R.C. 1954. *Ornithofilaria fallisensis* n. sp. (Nematoda: Filarioidea) from the domestic duck with descriptions of microfilariae in waterfowl. *Can. J. Zool.* 32:125–137.

Anderson, R.C., and D. J. Forrester. 1974. *Splendidofilaria fallisensis* (Nematoda) in the common loon, *Gavia immer* (Brünnich). *Can. J. Zool.* 52:547–548.

Barr, J.R. 1986. Population dynamics of the common loon (*Gavia immer*) associated with mercury-contaminated waters in northwestern Ontario. Occas. Paper 56, Canadian Wildlife Service, Ottawa, Ontario. 25 pp.

Belant, J.L., and R. K. Anderson. 1990. Environmental contaminants in common loons from northern Wisconsin. *Passenger Pigeon* 52:307–310.

Bent, A.C. 1937. *Life histories of North American birds of prey.* Part 1. U.S. Natl. Mus. Bull. 167. 409 pp.

Bishop, M.A., and G.F. Bennett. 1992. *Host-parasite catalogue of the avian Haematozoa.* Suppl. 1. Meml. Univ. Nfld. Occas. Pap. Biol. 15. 210 pp.

Brand, C.J., S.M. Schmitt, R.M. Duncan, and T.M. Cooley. 1988. An outbreak of type E botulism among common loons (*Gavia immer*) in Michigan's Upper Peninsula. *J. Wildl. Dis.* 24:471–476.

Buergelt, C.D., and B.G. Short. 1983. Unpublished data. University of Florida, Gainesville.

Burns, B.J. 1952. Food of a family of great horned owls, *Bubo virginianus*, in Florida. *Auk* 69:86–87.

Burridge, M.J., W.J. Bigler, D.J. Forrester, and J.M. Hennemann. 1979. Serologic survey for *Toxoplasma gondii* in wild animals in Florida. *J. Am. Vet. Med. Assoc.* 175:964–967.

Case, L.D., H. Cruickshank, A.E. Ellis, and W.F. White. 1965. Weather causes heavy bird mortality. *Fla. Nat.* 38:29–30.

Center for Disease Control. 1975. *Salmonella* surveillance annual summary for 1974. Atlanta, Ga.

Chafel, R.M., and M. Pokras. 1993. Helminths of the common loon, *Gavia immer,* in New England. In: *Proceedings of the 1992 conference on the loon and its ecosystem: status, management, and environmental concerns.* L. Morse et al. (eds.). U.S. Fish and Wildlife Service, Concord, N.H. pp. 2–12.

Clapp, R.B., R.C. Banks, D. Morgan-Jacobs, and W.A. Hoffmann. 1982. *Marine birds of the southeastern United States and Gulf of Mexico. Part 1, Gaviiformes through Pelecaniformes.* U.S. Fish and Wildlife Service, Office of Biological Services, Washington, D.C. 637 pp.

Clark, R.B. 1973. Impact of acute and chronic oil pollution on sea birds. In: *Background papers for a workshop on inputs, fates, and effects of petroleum in the marine environment II.* Oceanic Affairs Board, National Academy of Sciences, Washington, D.C. pp. 619–634.

Cook, R.S., and D.O. Trainer. 1966. Experimental lead poisoning of Canada geese. *J. Wildl. Manag.* 30:1–8.

Cornish, T. 1998. Unpublished data. Southeastern Cooperative Wildlife Disease Study, University of Georgia, Athens.

Crawford, R.L. 1981. Bird casualties at a Leon County, Florida TV Tower: A 25-year migration study. *Bull. Tall Timbers Res. Stn.* 22:1–30.

Davidson, W.R. 1983. Unpublished data. Southeastern Cooperative Wildlife Disease Study, University of Georgia, Athens.

Davidson, W.R., and L.E. Hayes. 1989. Unpublished data. Southeastern Cooperative Wildlife Disease Study, University of Georgia, Athens.

Delany, M.F. 1986. Bird bands recovered from American alligator stomachs in Florida. *N. Am. Bird Bander* 11:92–94.

———. 1996. Unpublished data. Florida Game and Fresh Water Fish Commission, Gainesville.

Delany, M.F., and C.L. Abercrombie. 1986. American alligator food habits in north central Florida. *J. Wildl. Manag.* 50:348–353.

Duncan, R.M. 1978–94. Unpublished data. National Wildlife Health Center, Madison, Wis.

Edscorn, J.B. 1974. The fall migration: Florida region. *Am. Birds* 28:40–44.

Eisler, R. 1985. Cadmium hazards to fish, wildlife, and invertebrates: A synoptic review. U.S. Fish and Wildlife Service Biological Report 85 (1.2). 46 pp.

———. 1987. Mercury hazards to fish, wildlife, and invertebrates: A synoptic review. U.S. Fish and Wildlife Service Biological Report 85 (1.10). 90 pp.

———. 1988. Arsenic hazards to fish, wildlife, and invertebrates: A synoptic review. U.S. Fish and Wildlife Service Biological Report 85 (1.12). 92 pp.

El-Begearmi, M.M., M.L. Sunde, and H.E. Ganther. 1977. A mutual protective effect of mercury and selenium in Japanese quail. *Poult. Sci.* 56:313–322.

Ensor, K.L., D.D. Helwig, and L.C. Wemmer. 1992. Mercury and lead in Minnesota common loons (*Gavia immer*). Minnesota Pollution Control Agency, St. Paul. 32 pp.

Fay, L.D., and W.G. Youatt. 1967. Residues of chlorinated hydrocarbon insecticides in loons, grebes, a gull, and a sample of alewives from Lake Michigan. Michigan Department of Conservation Research and Development Report 109. 7 pp.

Fenwick, B., and A.M. Crowley. 1986. Unpublished data. University of Florida, Gainesville.

Fimreite, N. 1974. Mercury contamination of aquatic birds in northwestern Ontario. *J. Wildl. Manag.* 38:120–131.

Fischer, J.R., and W. R. Davidson. 1994. Unpublished data. Southeastern Cooperative Wildlife Disease Study, University of Georgia, Athens.

Forrester, D.J. 1971. Unpublished data. University of Florida, Gainesville.

———. 1996. Unpublished data. University of Florida, Gainesville.

Forrester, D.J., and G.F. Bennett. 1996. Unpublished data. University of Florida, Gainesville.

Forrester, D.J., W.R. Davidson, R.E. Lange, Jr., R.K. Stroud, L.L. Alexander, J.C. Franson, S.D. Haseltine, R.C. Littell, and S.A. Nesbitt. 1997. Winter mortality of common loons in Florida coastal waters. *J. Wildl. Dis.* 33:833–847.

Forrester, D.J., J.M. Gaskin, F.H. White, N.P. Thompson, J.A. Quick, Jr., G.E. Henderson, J.C. Woodard, and W.D. Robertson. 1977. An epizootic of waterfowl associated with a red tide episode in Florida. *J. Wildl. Dis.* 13:160–167.

Forrester, D.J., and H.W. Kale II. 1977. Unpublished data. University of Florida, Gainesville.

Forrester, D.J., H.W. Kale II, R.D. Price, K.C. Emerson, and G.W. Foster. 1995. Chewing lice (Mallophaga) from birds in Florida: A listing by host. *Bull. Fla. Mus. Nat. Hist.* 39:1–44.

Forrester, D.J., and J.W. Mertins. 1995. Unpublished data. University of Florida, Gainesville.

Forrester, D.J., K.C. Wenner, F.H. White, E.C. Greiner, W.R. Marion, J.E. Thul, and G.A. Berkhoff. 1980. An epizootic of avian botulism in a phosphate mine settling pond in northern Florida. *J. Wildl. Dis.* 16:323–327.

Foster, G.W., and J.W. Mertins. 1999. Unpublished data. University of Florida, Gainesville.

Frank, R., H. Lunsden, J.F. Barr, and H.E. Braun. 1983. Residues of organochlorine insecticides, industrial chemicals, and mercury in eggs and in tissues taken from healthy and emaciated common loons, Ontario, Canada, 1968–1980. *Arch. Environ. Contam. Toxicol.* 12:641–654.

Franson, J.C. 1991–93. Unpublished data. National Wildlife Health Center, Madison, Wis.

Franson, J.C., and D.J. Cliplef. 1993. Causes of mortality in common loons. In: *Proceedings of the 1992 conference on the loon and its ecosystem: status, management, and environmental concerns.* L. Morse et al. (eds.). U.S. Fish and Wildlife Service, Concord, N.H. pp. 2–12.

Franson, J.C., and R.M. Duncan. 1991. Unpublished data. National Wildlife Health Center, Madison, Wis.

Frenkel, J.K. 1981. False-negative serologic tests for *Toxoplasma* in birds. *J. Parasitol.* 67:952–953.

Gay, M.L., and S.B. Haseltine. 1983–86. Unpublished data. Patuxent Wildlife Research Center, Laurel, Md.

Graham, F., Jr. 1984. Mystery at Dog Island. *Audubon* 86 (March):30–33.

Greiner, E.C. 1997. Unpublished data. University of Florida, Gainesville.

Gross, T.L., and M.K. Reinhard. 1983. Unpublished data. University of Florida, Gainesville.

Hamel, P.B. 1983. The changing seasons. *Am. Birds* 37:840–843.

Hartman, F.A. 1946. Notes on the pathology of a loon and a pelican. *Auk* 63:588–589.

Hier, R.H., P.S. Perry, and M. Sperry. 1986. Foot tumor found on juvenile common loon. *Loon* 58:41–42.

Hinshaw, V. 1983. Unpublished data. St. Jude's Children's Research Hospital, Memphis, Tenn.

Hoffman, W. 1983. The winter season: Florida region. *Am. Birds* 37:293–296.

Holmes, B., A.G. Steigerwalt, R.E. Weaver, and D.J. Brenner. 1986. *Weeksella virosa* gen. nov., sp. nov. (formerly group IIf), found in human clinical specimens. *Syst. Appl. Microbiol.* 8:185–190.

Homer, B.L., and K. Papineau. 1997. Unpublished data. University of Florida, Gainesville.

Howell, A.H. 1932. *Florida bird life.* Coward-McCann, New York. 579 pp.

Imhof, T.A. 1983. The spring migration: central southern region. *Am. Birds* 37:878–882.

Jasmin, A.M., D.E. Cooperrider, C.P. Powell, and J.N. Baucom. 1972. Enterotoxemia of wildfowl due to *Cl. perfringens* type C. *J. Wildl. Dis.* 8:79–84.

Jehl, J.R., and S.I. Bond. 1983. Mortality of eared grebes in winter of 1982–83. *Am. Birds* 37:832–835.

Johnston, D.W. 1976. Organochlorine pesticide residues in uropygial glands and adipose tissue of wild birds. *Bull. Environ. Contam. Toxicol.* 16:149–155.

Kale, H.W. II. 1971. Regional reports: Florida region. *Am. Birds* 25:723–725, 730–735.

———. 1983. The spring migration: Florida region. *Am. Birds* 37:860–863.

Kinsella, J.M., and W.R. Davidson. 1996. Unpublished data. University of Florida, Gainesville.

Kinsella, J.M., and S. Deblock. 1997. Contribution à l'étude des Microphallidae (Trematoda). LI. De cinq espèces du plongeon imbrin *Gavia immer* (Aves) des Etats-Unis, dont *Microphallus forresteri* n. sp. Pluralité vraisemblable de l'espèce *Microphallus nicolli* (Cable & Hunninen, 1938). *Syst. Parasitol.* 37:139–148.

Kinsella, J.M., and D.J. Forrester. 1996. Unpublished data. University of Florida, Gainesville.

———. 1999. Parasitic helminths of the common loon, *Gavia immer,* on its wintering grounds in Florida. *J. Helminthol. Soc. Wash.* 66:1–6.

Kinsella, J.M., and G.W. Foster. 1998. Unpublished data. University of Florida, Gainesville.

Klein, T.J. 1985. *Loon magic.* Paper Birch Press. Ashland, Wis. 130 pp.

Klukas, R.W., and L.N. Locke. 1970. An outbreak of fowl cholera in Everglades National Park. *J. Wildl. Dis.* 6:77–79.

Lange, R.E., Jr. 1983. Unpublished data. National Wildlife Health Center, Madison, Wis.

Langridge, H.P. 1991. The spring migration: Florida region. *Am. Birds* 45:436–438.

Locke, L.N. 1968–71. Unpublished data. National Wildlife Health Center, Madison, Wis.

Locke, L.N., S.M. Kerr, and D. Zoromski. 1981. Lead poisoning in common loons (*Gavia immer*). *Avian Dis.* 26:392–396.

Locke, L.N., and L.T. Young. 1967. Aspergillosis in a common loon (*Gavia immer*). *Bull. Wildl. Dis. Assoc.* 3:34.

Loftin, H. 1961. An annotated check-list of trematodes and cestodes and their vertebrate hosts from northwest Florida. *Q. J. Fla. Acad. Sci.* 23:302–314.

Logan, T.H. 1997. Florida's endangered species, threatened species and species of special concern. Official lists. Florida Game and Fresh Water Fish Commission, Tallahassee. 14 pp.

Longstreet, R.J. 1953. Ornithology of the Mosquitoes. *Fla. Nat.* 26:103–114.

MacInnis, A.J. 1959. Some helminth parasites, mostly from marine birds from the northwest Gulf Coast of Florida. M.S. thesis, Florida State University, Tallahassee. 37 pp.

Maehr, D.S., and J.Q. Smith. 1988. Bird casualties at a central Florida power plant: 1982–1986. *Fla. Field Nat.* 16:57–80.

Maehr, D.S., A.G. Spratt, and D.K. Voigts. 1983. Bird casualties at a central Florida power plant. *Fla. Field Nat.* 11:45–49.

McEwan, L.C., and D.H. Hirth. 1980. Food habits of the bald eagle in north-central Florida. *Condor* 82:229–231.

McIntyre, J.W. 1988. *The common loon: spirit of northern lakes.* University of Minnesota Press, Minneapolis. 229 pp.

McIntyre, J.W., and J.F. Barr. 1997. Common loon. *Birds N. Am.* 313:1–31.

Measures, L.N. 1988. The development and pathogenesis of *Eustrongylides tubifex* (Nematoda: Dioctophymatoidea) in piscivorous birds. *Can. J. Zool.* 66:2223–2232.

Mertins, J.W. 1999. Unpublished data. USDA, National Veterinary Services Laboratories, Ames, Iowa.

Meteyer, C.U. 1994–97. Unpublished data. National Wildlife Health Center, Madison, Wis.

Meteyer, C.U., and J.C. Franson. 1993. Unpublished data. National Wildlife Health Center, Madison, Wis.

Montgomery, R.D., M.N. Novilla, and R.B. Shillinger. 1978. Renal coccidiosis caused by *Eimeria gaviae* n. sp. in a common loon (*Gavia immer*). *Avian Dis.* 22:809–814.

Morse, L., S. Stockwell, and M. Pokras (eds.).

1993. *Proceedings of the 1992 conference on the loon and its ecosystem: status, management, and environmental concerns.* U.S. Fish and Wildlife Service, Concord, N.H. 247 pp.

Mullins, W.H., E.G. Bizeau, and W.W. Benson. 1977. Effects of phenyl mercury on captive game farm pheasants. *J. Wildl. Manag.* 41:302–308.

Ogden, J.C. 1987. The winter season: Florida region. *Am. Birds* 41:272–274.

Pichner, J., and P.L. Wolff. 1993. Causes of common loon, *Gavia immer,* morbidity and mortality: a background, history and update of the Minnesota Zoo program. In: *Proceedings of the 1992 conference on the loon and its ecosystem: status, management, and environmental concerns.* L. Morse et al. (eds.). U.S. Fish and Wildlife Service, Concord, N.H. pp. 70–72.

Pokras, M.A., and R. Chafel. 1992. Lead toxicosis from ingested fishing sinkers in adult common loons (*Gavia immer*) in New England. *J. Zoo Wildl. Med.* 23:92–97.

Popp, J.A., and D.J. Forrester. 1978. Unpublished data. University of Florida, Gainesville.

Premvati, G. 1968. Echinostome trematodes from Florida birds. *Proc. Helminthol. Soc. Wash.* 35:197–200.

Prouty, R.M., and S.D. Haseltine. 1983. Unpublished data. Patuxent Wildlife Research Center, Laurel, Md.

Prouty, R.M., J.E. Peterson, L.N. Locke, and B.M. Mulhern. 1975. DDD poisoning in a loon and the identification of the hydroxylated form of DDD. *Bull. Environ. Contam. Toxicol.* 14:385–388.

Quist, C.F. 1993. Unpublished data. Southeastern Cooperative Wildlife Disease Study, University of Georgia, Athens.

Ream, C.H. 1976. Loon productivity, human disturbance, and pesticide residues in northern Minnesota. *Wilson Bull.* 88:427–432.

Robertson, W.B., and G.E. Woolfenden. 1992. *Florida bird species: an annotated list.* Spec. Publ. 6, Florida Ornithological Society, Gainesville. 260 pp.

Roth, L. 1991. Unpublished data. University of Florida, Gainesville.

Siegfried, L.M. 1983. Neoplasms identified in free-flying birds. *Avian Dis.* 27:86–99.

Simons, M.M., Jr. 1985. Beached bird survey project on the Atlantic and Gulf coasts. *Am. Birds* 39:358–362.

Sims, H.W., Jr. 1970. Operation bird wash. *Fla. Nat.* 43:43–45.

Smith, K.E. 1995. Unpublished data. Southeastern Cooperative Wildlife Disease Study, University of Georgia, Athens.

Snyder, B. 1994. Unpublished data. Florida Department of Environmental Protection, Tallahassee.

Spalding, M.G., and D.J. Forrester. 1998. Unpublished data. University of Florida, Gainesville.

Spitzer, P.R. 1995. Common loon mortality in marine habitats. *Environ. Rev.* 3:223–229.

Sprunt, A., Jr. 1954. *Florida bird life.* Coward-McCann, New York. 527 pp.

Stafford, C.J., and S.B. Haseltine. 1983–86. Unpublished data. Patuxent Wildlife Research Center, Laurel, Md.

Stetzer, E.R., and M.K. Reinhard. 1983. Unpublished data. University of Florida, Gainesville.

Stetzer, E.R., and B.G. Short. 1983. Unpublished data. University of Florida, Gainesville.

Stevenson, H.M. 1970. The winter season: Florida region. *Aud. Field Notes* 24:493–497.

———. 1971. Regional reports: Florida region. *Am. Birds* 25:567–570.

———. 1972. The winter season: Florida region. *Am. Birds* 26:562–596.

———. 1974. The winter season: Florida region. *Am. Birds* 28:628–632.

———. 1977. The winter season: Florida region. *Am. Birds* 31:322–325.

Stevenson, H.M., and B.H. Anderson. 1994. *The birdlife of Florida.* University Press of Florida, Gainesville. 892 pp.

Stevenson, J.A. 1994. Unpublished data. Florida Department of Environmental Protection, Tallahassee.

Stickel, W.H. 1975. Some effects of pollutants in terrestrial ecosystems. In: *Ecological toxicology research.* A. D. McIntyre and C. F. Mills (eds.). Plenum Publishing, New York. pp. 25–74.

Stickel, W.H., L.F. Stickel, and F.B. Coon. 1970. DDE and DDD residues correlated with mortality of experimental birds. In: *Inter-American conference on toxicology and occupational medicine, pesticide symposia*. W. P. Deichmann (ed.). Halos and Associates, Miami, Fla. pp. 287–294.

Stoewsand, G.S., C.A. Bache, and D.J. Lisk. 1974. Dietary selenium protection of methylmercury intoxication of Japanese quail. *Bull. Environ. Contam. Toxicol.* 11:152–156.

Stolen, E.D. 2001. Great blue heron eating a pied-billed grebe. *Fla. Field Nat.* 29:87.

Storer, R.W. 1992. Least grebe. *Birds N. Am.* 74:1–12.

Stroud, R.K. 1983. Unpublished data. National Wildlife Health Center, Madison, Wis.

Suto, B.J. 1996. Unpublished data. Suncoast Seabird Sanctuary, Redington Beach, Fla.

Taylor, W.K., and B.H. Anderson. 1973. Nocturnal migrants killed at a Central Florida TV tower, autumns 1969–71. *Wilson Bull.* 85:42–51.

———. 1974. Nocturnal migrants killed at a Central Florida TV tower, autumn 1972. *Fla. Field Nat.* 2:40–43.

Taylor, W.K., and M.A. Kershner. 1986. Migrant birds killed at the Vehicle Assembly Building (VAB), John F. Kennedy Space Center. *J. Field Ornithol.* 57:142–154.

Telford, S.R., Jr., S.A. Nesbitt, M.G. Spalding, and D.J. Forrester. 1994. A species of *Plasmodium* from sandhill cranes in Florida. *J. Parasitol.* 80:497–499.

Weston, F.M. 1966. Bird casualties on the Pensacola Bay bridge (1938–1949). *Fla. Nat.* 39:53–55.

White, D.H., and M.T. Finley. 1978. Uptake and retention of dietary cadmium in mallard ducks. *Environ. Res.* 17:53–59.

White, F.H., and D.J. Forrester. 1979. Antimicrobial resistant *Salmonella* spp. isolated from double-crested cormorants (*Phalacrocorax auritus*) and common loons (*Gavia immer*) in Florida. *J. Wildl. Dis.* 15:235–237.

———. 1996. Unpublished data. University of Florida, Gainesville.

White, F.H., D.J. Forrester, and S.A. Nesbitt. 1976. *Salmonella* and *Aspergillus* infections in common loons overwintering in Florida. *J. Am. Vet. Med. Assoc.* 169:936–937.

White, F.H., C.F. Simpson, and L.E. Williams, Jr. 1973. Isolation of *Edwardsiella tarda* from aquatic animal species and surface waters in Florida. *J. Wildl. Dis.* 9:204–208.

Wiemeyer, S.N., T.G. Lamont, and L.N. Locke. 1980. Residues of environmental pollutants and necropsy data for eastern United States ospreys, 1964–1973. *Estuaries* 3:155–167.

Wobeser, G.A. 1981. *Diseases of wild waterfowl*. Plenum Press, New York. 300 pp.

Woodard, J.C., and R.A. Westhouse. 1993. Unpublished data. University of Florida, Gainesville.

Woolfenden, G.E. 1973. The winter season: Florida region. *Am. Birds* 27:603–607.

Wright, S.D. 1988. Unpublished data. University of Florida, Gainesville.

Oceanic Birds

I. Introduction

Although a total of 24 species of oceanic birds has been recorded in Florida (table 3.1), the majority of them are only seen offshore except when blown near shore or inland by storms. Exceptions to this are Magnificent Frigatebirds and Masked Boobies that nest and roost on islands in Florida, and jaegers, gannets, and alcids that are seasonally found along the coasts. The Magnificent Frigatebird is listed as threatened by the Florida Committee on Rare and Endangered Plants and Animals (Rodgers et al. 1996) but is not listed by state or federal agencies.

Life history and population information has been reported for Leach's Storm-Petrel (Huntington et al. 1996), Masked Boobies (Anderson 1993), and Magnificent Frigatebirds (Robertson and Wilmers 1996). Clapp et al. (1982) provided considerable life history and observational records for oceanic birds in Florida. A review of normal weights, hematology, and serum chemistries for Pacific Brown Boobies and other oceanic birds was reported by Work (1996, 1999). Some general health information can be found in Friend and Franson (1999).

II. Trauma and human disturbance

Compared with other groups of birds, few oceanic birds collide with tall manmade structures on land. Crawford (1981) recorded birds

Table 3.1. Species of oceanic birds that occur on Florida coastal waters and beaches

	Species	Range	Seasonal occurrence	Relative abundance
Albatrosses				
Yellow-nosed Albatross	*Thalassarche chlororhynchos*	Wakulla, Brevard, Monroe counties	July	3 records
Shearwaters and Petrels				
Black-capped Petrel	*Pterodroma hasitata*	Atlantic offshore	Year-round	Common
Cory's Shearwater	*Calonectris diomedea*	Esp. Atlantic offshore	Summer	Common
Greater Shearwater	*Puffinus gravis*	Esp. Atlantic offshore	Year-round	Common
Sooty Shearwater	*Puffinus griseus*	Esp. Atlantic offshore	Year-round	Rare
Manx Shearwater	*Puffinus puffinus*	Esp. Atlantic offshore	Winter	Rare
Audubon's Shearwater	*Puffinus lherminieri*	Esp. Atlantic offshore	Summer	Common
Storm-Petrels				
Wilson's Storm-Petrel	*Oceanites oceanicus*	Both coasts	Summer	Common
Leach's Storm-Petrel	*Oceanodroma leucorhoa*	Throughout	Summer	Rare
Band-rumped Storm-Petrel	*Oceanodroma castro*	Both coasts	Summer & Fall	Rare
Tropicbirds				
White-tailed Tropicbird	*Phaethon lepturus*	Esp. Atlantic coast	Summer	Rare
Red-billed Tropicbird	*Phaethon aethereus*	NE Atlantic coast	Fall	Very rare
Boobies and Gannets				
Masked Booby[a]	*Sula dactylatra*	Dry Tortugas, esp. Gulf & S. Atlantic offshore	Year-round	Rare to common
Brown Booby	*Sula leucogaster*	Both coasts	Year-round	Rare
Red-footed Booby	*Sula sula*	Both coasts	Summer–Fall	Very rare
Northern Gannet	*Morus bassanus*	Both coasts	Winter-Spring-Summer	Common
Frigatebirds				
Magnificent Frigatebird[a]	*Fregata magnificens*	Both coasts	Year-round	Common
Jaegers				
Pomarine Jaeger	*Stercorarius pomarinus*	Both coasts	Winter	Rare to abundant
Parasitic Jaeger	*Stercorarius parasiticus*	Esp. Atlantic coast	Year-round	Rare to common
Long-tailed Jaeger	*Stercorarius longicaudus*	Both coasts	Winter-Spring	Rare
Auks, Murres and Puffins				
Dovekie	*Alle alle*	Atlantic coast	Winter	Rare
Razorbill	*Alca torda*	Coasts	Winter	Rare
Marbled Murrelet	*Brachyramphus marmoratus*	Pinellas County	Winter	1 record
Atlantic Puffin	*Fratercula arctica*	Martin County	Winter	1 record
Thick-billed Murre[b]	*Uria lomvia*	Palm Beach County	Winter	1 record

Source: Robertson & Woolfenden (1992), except where otherwise indicated.
a. Nests in Florida.
b. *Source:* Langridge & Woolfenden (1998).

found dead below a TV tower near Tall Timbers Research station in Leon County between 1955 and 1980. During this period 42,384 birds of 189 species were found. Of the oceanic birds, only 1 Black-capped Petrel (during Hurricane Dora) was among those listed. Oceanic birds have not been found in other studies of mortality involving tall structures in Florida (Kale 1971; Maehr et al. 1983; Maehr and Smith 1988; Taylor and Anderson 1973;

Table 3.2. Reports of mortality of Northern Gannets in Florida

Year	Month	County	Mortalities	No. examined	Findings	Data Source
1983	January	Palm Beach	3	3	Aspergillosis	A
1984	NG	St. Johns	1	1	Trauma	B
1987	Dec.–Jan.	Martin	2	2	Nephrosis, bacteria	C
1991[a]	April–May	Duval, St. Johns	10	4	Emaciation, trauma, and aspergillosis	D
1992–93[b]	Nov.–Feb. (esp. Jan. & Feb.)	St. Lucie to Broward	>100	2	Septicemia and muscle necrosis	E
1993[c]	Feb.–Mar.	Brevard	21	2	Emaciation	F
1993	March	Broward	40–50	8	Emaciation, trauma	G
1993	July	Gulf	1	1	Aspergillosis	H
1993	August	Franklin	2	2	Emaciation	H
1993	August	Duval	1	1	Emaciation	G
1993	Aug.–Sept.	Gulf	3	3	Emaciation	G
1997	November	St. Johns	75	1	Salmonellosis	I
1997	December	Volusia	40–50	2	Aspergillosis, emaciation	I
1998–99	December	St. Johns	>10	2	Aspergillosis, sarcocystosis	I,J

Sources: A = Richardson (1983), B = Jenkins (1984), C = Roffe & McAllister (1988), D = Franson (1992), E = Spalding & Robson (1993), F = Quist (1993), G = Franson (1993), H = Locke (1993), I = Spalding & Terrell (1998), J = Spalding at al. (2002). NG = not given.
a. One Loon died also.
b. More than 50 Common Loons also died.
c. This die-off also included 40 Common Loons, 5 Brown Pelicans, 5 Ring-billed Gulls, 3 Blue-footed Boobies, 3 Double-crested Cormorants, and 2 Royal Terns.

Taylor and Anderson 1974; Taylor and Kershner 1986).

Ingestion of foreign objects, especially floating plastic, may be a significant problem for oceanic birds (Sileo et al. 1990). Many pieces of plastic were found in the stomachs of 6 of 7 Greater Shearwaters examined during a die-off in 1993, Brevard County. The stomachs also contained cephalopod beaks, but neither appeared to cause lesions on the stomach wall (Spalding and Robson 1993). A Northern Gannet found in St. Johns County that died 24 hours after capture had a stainless steel welding rod, 44.8 cm long and 1 cm wide, in the esophagus (Jenkins 1984). The rod caused an erosion of the esophageal mucosa and distended the stomach enough to cause partial obstruction of the cloaca and fecal distention of the bursa of Fabricius.

Chronic traumatic wounds have been observed on oceanic birds collected for various reasons. A Northern Gannet picked up during a die-off in St. Johns County in 1991 with leg paralysis had peritonitis from a chronic wound on the abdominal wall (Franson 1992). Another from a large die-off in Broward County in 1993 had a laceration in the pectoral muscle (Franson 1993). These wounds may have been inflicted by large fish or sharks.

Boat use and aircraft tours are likely the reason for Magnificent Frigatebirds gradually abandoning the Marquesas Keys nesting site by 1989 and moving to the Dry Tortugas, Monroe County (Wilmers *in* Robertson and Wilmers 1996).

III. Inclement weather

Inclement weather is the most common cause for sightings of oceanic birds and for finding them dead or sick on the mainland of Florida. Stevenson and Anderson (1994) reported museum records of a Red-footed Booby found in Franklin County after a hurricane in October

Table 3.3. Organochlorine residues found in the brain of a Northern Gannet from Franklin County, 1993[a]

Contaminant	ppm wet weight
α-BHC	0.002
o,p'-DDD	0.48
p,p'-DDD	0.003
p,p'-DDE	2.2
o,p'-DDT	0.019
Dieldrin	0.41
Endrin	0.028
Heptachlor epoxide	0.06
Nonachlor	0.26
PCB	4.2

Source: Locke (1993).

a. In addition to the compounds listed, Locke (1993) did not detect aldrin, β-BHC, p,p'-DDT, heptachlor, or lindane.

1985 and 10 Leach's Storm-Petrels that were found on beaches of Nassau and Duval Counties, May 1991, after a period of stormy weather. Unusual findings of dead or dying birds in Volusia County included a Sooty Shearwater in June 1929, Audubon's Shearwaters all in August of 1925, 1928, 1929, and 1939, Leach's Petrel in May 1944, and Wilson's Petrel in August 1911 (Longstreet 1953–55). Drowning, which might be a consequence of foul weather, has not been documented in Florida; however, this type of death is difficult to document unless carcasses are very fresh.

Dovekies were observed in large numbers along the Atlantic coast from Daytona Beach southward during the winters of 1932 and 1936–37 (Sprunt 1938). These were the first sightings of this species in Florida since 1871 (Maynard *in* Longstreet 1953–55). Longstreet (1953–55) described finding dead (16) and living (16) Dovekies on the Atlantic shore of Volusia County in November–December 1932 after 4 days of a strong NE wind, but no weather event was associated with the 1936–37 sightings when he found 16 living and 24 dead birds in December (Sprunt 1938). There was no speculation made about the cause of death.

Because they nest on the leeward (west) side of islands, Magnificent Frigatebirds in southern Florida are more sensitive to winter cold fronts. For example, a March storm in 1993 caused the failure of 50 nests (Robertson and Wilmers 1996).

During the winter of 1992–93 more than 100 Northern Gannets and 100 Common Loons and small numbers of other oceanic and shore birds were found moribund and dead on the east coast of Florida from St. Lucie to Broward Counties (Spalding and Robson 1993; Quist 1993; Franson 1993). Gannets and loons also died as far north as New Jersey (Spitzer 1995). Peak mortality occurred in late January and early February and appeared to be associated with a strong cold front and heavy surf. Birds were consistently emaciated; a wide range of other findings are mentioned in other sections and in table 3.2. It was concluded that birds were unable to forage in the rough seas and succumbed to starvation with secondary traumatic and infectious causes of death (Spalding and Robson 1993). Robson noted that a similar event occurred in 1988. Although an infectious disease or exposure to contaminants was not ruled out, it appears likely that weather and/or limited food resources may have been the cause of the higher than usual mortality in that year. Common Loon mortality was also increased in 1993 and Spitzer (1995) proposed that storms and limited food resources were responsible for their mortality (see also chapter 2, Loons). Frederick (1993) reported that in southern Florida there were extreme high water (fresh water) conditions throughout the season (Dec.–Jun.) in 1993 and that much windier conditions than normal prevailed. The frequent association of loon and gannet mortality may be due to the fact that they both depend upon resources that become unavailable because of rough weather and/or increased rainfall.

Following the winter die-off, in June of 1993 as many as 80 Greater Shearwaters, a Cory's Shearwater, and a Leach's Storm Petrel were found washed up on beaches of Brevard County. Birds were emaciated, anemic, and had

Table 3.4. Concentrations of lead and mercury in liver tissues of Northern Gannets from Florida

County (or area)	Year	No. examined	Lead Mean[a]	(Range)	Mercury Mean[a]	(Range)	Data source
Brevard	1993	3	NA	—	3.3	(2.5–4.0)	A
Broward	1993	1	ND	—	NA	—	B
Duval	1991	1	NA	—	3.7	—	B
	1993	1	ND	—	NA	—	B
Franklin	1993	1	ND	—	NA	—	C
Gulf	1993	1	NA	—	5.6	—	C
	1993	3	ND	—	NA	—	D
St. Johns	1991	1	ND	—	2.9	—	B

Sources: A = Quist (1993), B = Franson (1992), C = Locke (1993), D = Franson (1993).
NA = not analyzed, ND = not detected
a. Geometric mean, ppm wet weight.

a hemorrhagic gastroenteritis (Spalding and Robson 1993). Again, this die-off was associated with an unseasonable storm. The change in species composition of this die-off was probably due to the change in species foraging offshore during the summer months. Brichetti et al. (2000) reported a negative correlation between the survival of Cory's Shearwaters in a colony in Italy with the number of intense hurricane days.

A detailed discussion about the effects of weather on seabirds, especially effects of El Niño-Southern Oscillation events, can be found in Schreiber (2001).

IV. Environmental contaminants

Residues of organochlorines were detected in the brain of 1 of 2 emaciated immature Northern Gannets collected in Franklin County in August 1993 (table 3.3). Concentrations were judged to be too low to be of significance in the death of these birds (Locke 1993). In 7 cases Northern Gannets have been tested for cholinesterase inhibition, and no evidence of organophosphate or carbamate poisoning has been detected (Franson 1991, 1993; Locke 1993).

Metals have been studied only occasionally in oceanic birds from Florida (table 3.4). None of the metals detected was considered to be of

concern. Mercury is generally high in long-lived oceanic birds, and they seem to tolerate the higher concentrations more than nonmarine birds (Thompson 1996).

V. Oiling

Clapp et al. (1982) reviewed the susceptibility to oiling for all seabirds in the southeastern United States. Procellariiformes appear to be less affected by oil than other oceanic birds; however, this may be an artifact of the pelagic nature of the birds and the reduced chance of finding dead birds. A single Cory's Shearwater was found oiled and dead at Ormond Beach, Volusia County, in November 1962 (Hudson 1963). An oiled Leach's Storm-petrel was found entangled in seaweed on Anna Maria Island, Manatee County, in May 1965 (Rohwer and Woolfenden 1968). A slightly oiled Red-footed Booby that was found moribund on the shore of Pinellas County in September 1963, following a hurricane, died shortly after being taken into captivity (Woolfenden 1965). Exposure to oil can cause direct mortality and reduced reproductive success (Jessup and Leighton 1996). Oil-dosed adult Leach's Storm-petrels in Maine had reduced reproductive success, in terms of lower chick survival and growth (Trivelpiece et al. 1984).

VI. Neoplasia

We found no reports of neoplastic diseases for this group of birds.

VII. Biotoxins

Tissues from Northern Gannets from Florida were tested occasionally for botulism toxins and found to be negative on 3 occasions (Roffe and McAllister 1988; Franson 1992; Franson 1993). We could find no records of oceanic birds dying from exposure to red tide. This is remarkable considering the frequency of red tides especially on the west coast of Florida.

VIII. Viruses

We found no reports of viral infections in oceanic birds from Florida. Wingate et al. (1980) noted poxvirus lesions on the legs and feet of fledgling White-tailed Tropicbirds in Bermuda that were considered to be severe enough to prevent flight.

IX. Bacteria

Bacteria and fungi isolated from seabird tissues are listed in table 3.5. Diphtheroid bacteria were isolated from liver, lung, and kidney of 3 of 4 Northern Gannets and a Royal Tern from a multispecies die-off on Hobe Sound National Wildlife Refuge in Martin County, December 1987. These were associated with a severe esophagitis in 2 of the gannets and the tern. There was no evidence of septicemia. Two of the gannets also had degenerative changes in the kidney. Although it was felt that the diphtheroids might have been a significant pathogen, it could not be proven. *Pasteurella/Actinobacillus* was cultured from the kidney of one of the gannets (Roffe and McAllister 1988).

Salmonella Group B (4,5,12:i-monophasic) was isolated from the kidney of an emaciated Northern Gannet from St. Augustine, St. Johns County. Lesions consistent with salmonellosis included very severe enterocolitis, myositis, nephritis, and septicemia (figure 3.1) (Spalding and Terrell 1998). It was the only bird examined from a large die-off along the Atlantic coast from St. Augustine to New Smyrna Beach in November of 1997. More than 75 were found in St. Johns County alone. There were also reports of dead gannets from Duval and Volusia Counties. Those that were alive were lethargic, weak, and had tremors. Most appeared to be in fair nutritional condition.

X. Fungi

Aspergillus sp., frequently *A. fumigatus*, has been isolated from Northern Gannets in Florida 8 times but has not been observed in other oceanic birds in Florida (table 3.5). Although *Aspergillus* sp. is usually considered to be a secondary invader in birds stressed for some other reason, and is commonly found in captive birds (Redig 1993), reports of aspergillosis in wild Northern Gannets seem to be relatively common, especially in young birds (tables 3.2 and 3.5).

XI. Protozoans

Oceanic birds from Florida have not been studied systematically for protozoan parasites, but protozoal infections have been noted occasionally. An adult Northern Gannet from St. Johns County in 1998 that was taken to a rehabilitation center, died the next day with neurologic signs. It had severe meningoencephalitis due to infection with *Sarcocystis* sp. (figure 3.2) (Spalding et al. 2002). Sarcocysts were also found in the heart and striated muscle. This species of *Sarcocystis* was closely related, if not identical to sarcocysts shed by Virginia opossums (*Didelphis virginiana*) collected on the east coast of Florida. Unidentified protozoal cysts were an incidental finding

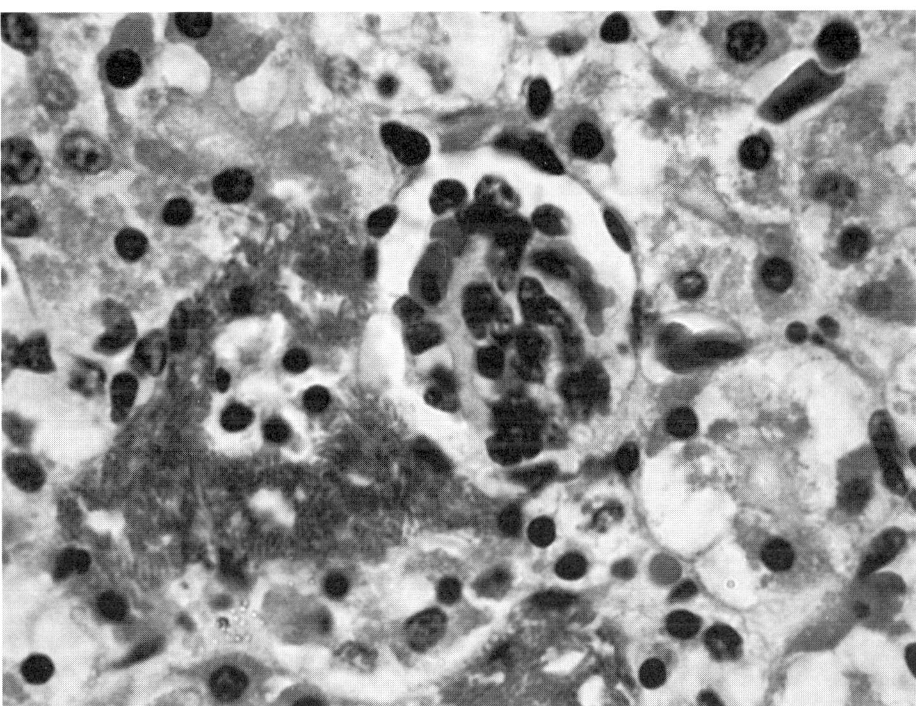

FIGURE 3.1. *Salmonella* sp. clustered to the left of a glomerulus *(center)* in the kidney of a Northern Gannet with salmonellosis.

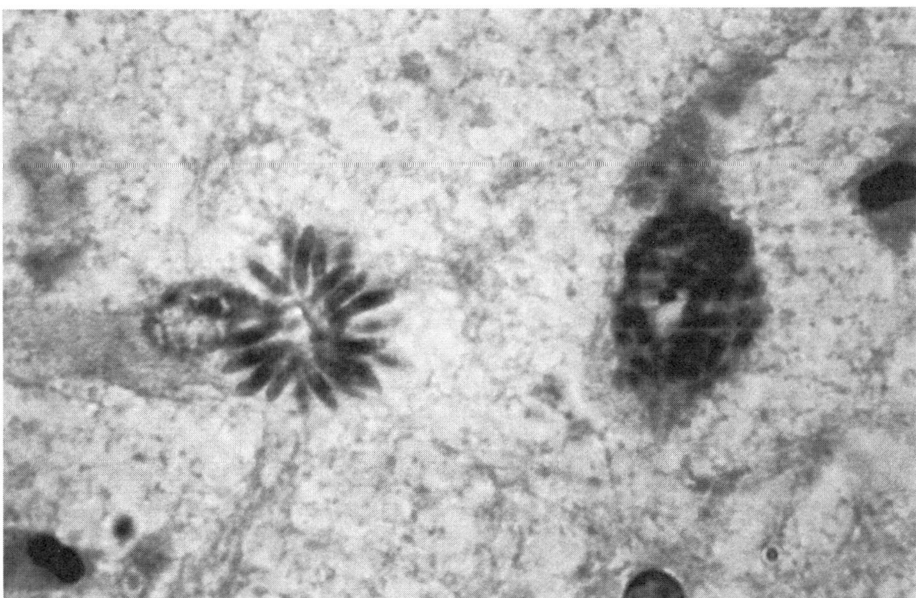

FIGURE 3.2. Photomicrograph of the cerebellum of a Northern Gannet with *Sarcocystis* sp. meningioencephalitis from St. Johns County in 1998. Two developmental stages of schizionts are illustrated.

Table 3.5. Bacteria and fungi isolated from oceanic birds in Florida

Host species	Organism	County	Year	Organ or tissue	Data source
Greater Shearwater					
	Bacteria				
	Escherichia coli	Brevard	1975	Intestine, liver	A
	Staphylococcus, coagulase positive	Brevard	1975	Liver, heart	A
	Streptococcus, beta hemolytic	Brevard	1975	Heart	A
Northern Gannet					
	Bacteria				
	Acinetobacter calcoaceticus	Broward	1993	Liver, lung	B
	Aeromonas hydrophila	Broward	1993	Liver, lung	B
	Diphtheroids	Martin	1987	Liver, lung, kidney	C
			1988	Brain, esophagus, kidney, intestine, trachea, liver, lung	C
	Escherichia coli	Broward	1993	Lung, intestine	B
	Moraxella sp.	Martin	1988	Trachea	C
	Pasteurella/Actinobacillus	Martin	1988	Kidney	C
	Proteus sp.	Broward	1993	Lung, liver, intestine	B
	Pseudomonas sp.	Broward	1993	Lung	B
		Duval	1993	Liver	B
	Salmonella sp.[b]	St. Johns	1997	Kidney[a]	D
	Serratia marcescens	Franklin	1993	Liver	E
	Fungi				
	Aspergillus fumigatus	Broward	1993	Lung, intestine[a]	B
		Duval	1991	Air sac	B
		Gulf	1993	Air sac[a], esophagus	E
		Martin	1987	Air sac[a]	C
		Palm Beach	1983	Air sac, lung	F
		St. Johns	1991	Air sac	G
			1998	Air sac[a]	D
		Volusia	1997	Air sac[a]	D
	Rhizopus sp.	St. Johns	1998	Air sac[a]	D

Sources: A = Schwartz (1975), B = Franson (1993), C = Roffe & McAllister (1988), D = Spalding & Terrell (1998), E = Locke, F = Richardson (1983), G. = Franson (1992).
a. Believed to be a significant pathogen.
b. The *Salmonella* sp. isolated was Group B (4,5,12:i-monophasic).

in heart and skeletal muscle of 1 of the Northern Gannets that was part of the die-off in the winter of 1993 (Spalding and Robson 1993).

An unidentified protozoan parasite (figure 3.3) was found in the kidney of a Northern Gannet that died from salmonellosis (see Bacteria, section IX), St. Johns County, 1997. The protozoans, probably coccidians, occupied a significant proportion, about 10%, of the kidney tissue, but there was no evidence of inflammation associated with these organisms and they were not found in feces (Spalding and Terrell 1998).

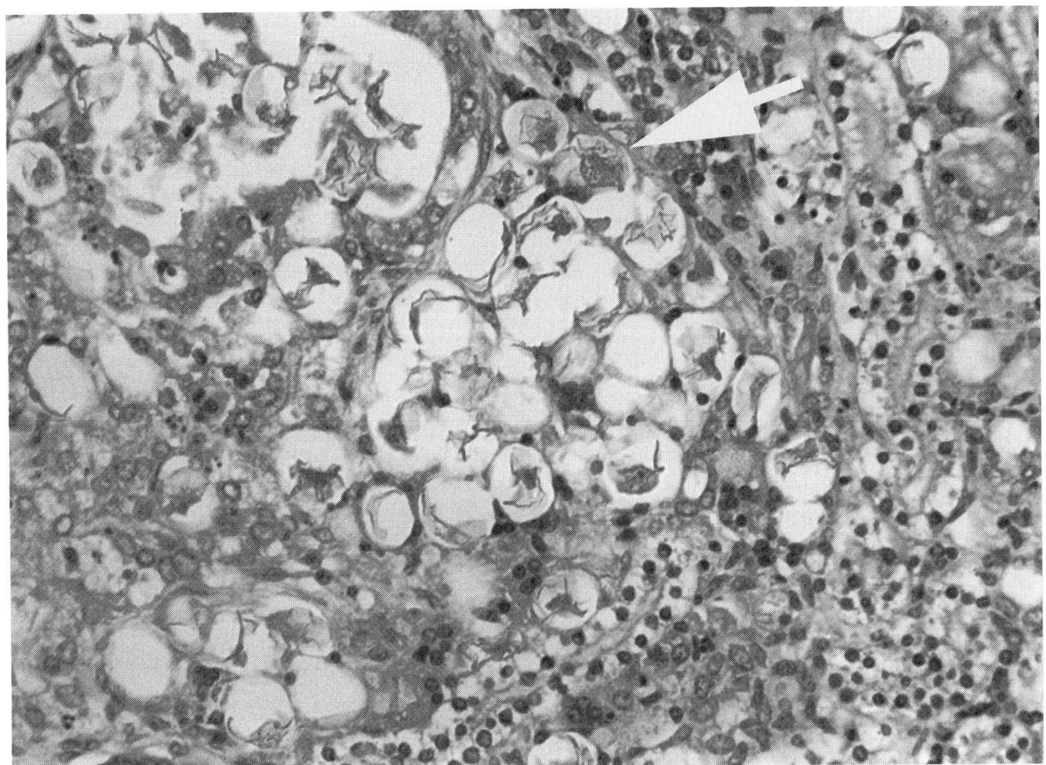

FIGURE 3.3. Degenerating protozoan, probably *Eimeria* sp. *(arrow)*, in the kidney of a Northern Gannet.

XII. Helminths

The helminths collected from oceanic birds in Florida are listed in table 3.6. Most were incidental findings at necropsy during investigations of die-off events. Only one parasitological survey was conducted and that was of Greater Shearwaters from the 1993 die-off (Foster et al. 1996). In no case were significant lesions attributed to helminths found.

In 2 cases, trematode eggs were observed in the lumen of tubules in the kidneys of Northern Gannets. They were not identified (Locke 1993). Schistosome eggs were observed histologically in the intestine and liver of a Northern Gannet that was part of a die-off in Brevard County, 1993 (Quist 1993), but were not identified to species.

XIII. Arthropods

In tables 3.7 and 3.8 the ectoparasites collected from oceanic birds in Florida are listed. The most detailed information comes from the Greater Shearwaters examined during a die-off in 1993 (Foster et al. 1996). The nymphal tick *Ixodes auritulus* found on one of the Greater Shearwaters has not been reported previously from eastern North America. Wanless et al. (1997) found a negative correlation between hematocrit levels and infestations by *I. uriae* ticks for 4 species of oceanic birds in the North Atlantic.

Subcutaneous mites, *Neottialges bassani*, were found on a Northern Gannet from Gulf County (Cole and Pence 1993) and on a Pomarine Jaegar necropsied in 1993 (a collection

Table 3.6. Characteristics of helminth infections in shearwaters, gannets, and frigatebirds collected in Florida

Host Class Species (Location)[a]	County (or area)	Year	No. birds			Intensity		Data source
			Exam.	Pos.	%	Mean	Range	
Cory's Shearwater								
Nematoda								
Seuratia shipleyi (ST)	Duval	1976	1	1	1	3	—	A
Greater Shearwater								
Cestoda								
Tetrabothrius diomedea (SI)	Brevard, Martin	1993	15	8	53	1	1–2	B
Tetrabothrius filiformis (SI)	Brevard, Martin	1993	15	15	100	99	6–305	B
Tetrabothrius laccocephalus (SI)	Brevard, Martin	1993	15	14	93	7	1–22	B
Tetrabothrius minor (SI)	Brevard, Martin	1993	15	1	7	3	—	B
Nematoda								
Contracaecum sp. (imm) (SW)	Brevard, Martin	1993	15	12	80	9	1–67	B
Seuratia shipleyi (ST)	Brevard, Martin	1993	15	9	60	4	1–8	B
Stegophorus diomedeae (GI,GL)	Brevard, Martin	1993	15	14	93	4	1–12	B
Stegophorus stellae-polaris (GL)	Brevard, Martin	1993	15	3	20	1	1–2	B
Larval spiruids (ST,GL)	Brevard, Martin	1993	15	3	20	5	1–13	B
Northern Gannet								
Trematoda								
Echinostoma sp. (IN)	St. Johns	1991	1	1	—	NG	—	C
	Duval	1991	2	2	—	NG	NG	C
Galactosomum cochleariformum[b] (IN)	Duval	1991	1	1	—	NG	—	C
Galactosomum sp. (IN)	St. Johns	1991	1	1	—	NG	—	C
	Duval	1991	2	2	—	NG	NG	C
	Franklin	1993	2	2	—	NG	NG	D
Mesostephanus sp. (IN)	Duval	1991	1	1	—	NG	—	C
Phagicola sp. (IN)	St. Johns	1991	1	1	—	NG	—	C
	Duval	1991	1	1	—	NG	—	C
	Franklin	1993	2	2	—	NG	NG	D
Stephanoprora sp. (IN)	Franklin	1993	2	2	—	NG	NG	D
Nematoda								
Capillaria sp. (IN)	Franklin	1993	1	1	—	NG	—	D
Contracaecum sp. (imm) (SW)	St. Johns	1991	1	1	—	NG	—	C
Contracaecum sp. (IN)	Duval	1991	1	1	—	NG	—	C
Cosmocephalus sp. (IN)	Duval	1991	1	1	—	NG	—	C
Magnificent Frigatebird								
Trematoda								
Galactosomum cochleariformum[b,c] (IN)	Charlotte	1955–62	NG	NG	—	NG	NG	E
	Monroe	1907	NG	NG	—	NG	NG	F
	Monroe	1910	NG	NG	—	NG	NG	G
		NG	1	1	—	NG[d]	—	H
	Monroe	1955–62	NG	NG	—	NG	NG	E

(continued)

Table 3.6. *(continued)*

Host Class Species (Location)[a]	County (or area)	Year	No. birds Exam.	Pos.	%	Intensity Mean	Range	Data source
Galactosomum *fregatae*[e] (IN)	"Florida"	1955–62	NG	NG	—	NG	NG	E
Mesostephanus sp. (IN)	Charlotte	1955–62	NG	NG	—	NG	NG	E
Cestoda								
Tetrabothrius *sulcatus* (SI)	Monroe	1907	NG	NG	—	NG	NG	I
Nematoda								
Contracaecum *rudolphi*[f] (SI)	Charlotte	1955–62	NG	NG	—	NG	NG	E

Source: A = Spalding & Kinsella (1997), B = Foster et al. (1996), C = Franson (1992), D = Locke (1993), E = Hutton (1964), F = Linton (1928), G = Pratt (1911), H = Manter (1929–30), I = Linton (1927).

NG = not given.

a. Location in host: GI = gizzard, GL = gizzard lining, IN = intestine, SI = small intestine, ST = stomach, SW = stomach wall.

b. Described as *G. cochleariforme* by Pratt (1911), Linton (1928), Manter (1929–30), and Hutton & Sogandares-Bernal (1960) and later synonymized to *G. cochleariformum* by Pearson (1973).

c. Pratt (1911) described this from a frigatebird (*Frigata aquila*); however, it was probably from a Magnificent Frigatebird (*F. minor*).

d. Reported as "large numbers."

e. Described as *Galactosomum* sp. by Hutton & Sogandares-Bernal (1960) and as *G. fregatae* by Pearson (1973).

f. *C. spiculigerum* is a synonym of *C. rudolphii.*

date was not given, Pence and Cole 1995). The epidermoptid mite that causes mange in some oceanic species (Gilardi et al. 2001), *Myialges nudus,* has not been found on oceanic birds in Florida.

XIV. Emaciation

Emaciation with no evident cause is a common finding in seabirds that are involved in die-offs and is frequently associated with inclement weather (see section III, Inclement weather). Emaciation, and emaciation with aspergillosis, were particularly common findings in Northern Gannets especially during 1991–93 (table 3.2). Emaciation was the only finding in 3 Northern Gannets found dead in August and September 1993 in Gulf County (Franson 1993).

XV. Summary and Conclusions

Most oceanic birds probably die unobserved at sea and thus our detection of disease is greatly biased toward those that die on or near shore. Northern Gannets and Greater Shearwaters fall into this category. Weather events are clearly the most common cause for the finding of sick and dead oceanic birds on the mainland in Florida, probably due to difficult foraging conditions, changes in prey, and being dislocated by storms. Several infectious diseases have been reported; however, they have not been clearly demonstrated to cause primary disease. Salmonellosis and aspergillosis often co-occur with stress and emaciation associated with storms. Twelve bacteria and a single species of fungus have been isolated. Infections with 2 protozoans were noted. No significant helminth diseases have been reported, but parasitic helminths have been examined closely only in Greater Shearwaters. Five cestodes, 5 nematodes, 5 trematodes, and no acanthocephalans were found in this group of birds. A diverse group of arthropod parasites (43) has been collected from many oceanic birds, but their significance relative to health is not known. Exposure to oil poses a significant threat that is difficult to document. Environmental contaminants have not been investigated adequately. Lead has not been found in livers of the few birds tested, although relatively high concentrations of mercury have. Only 1 gannet was tested for exposure to organochlorine pesticides.

Table 3.7. Arthropod infestations of shearwaters, petrels, and tropicbirds in Florida

Host	Species of parasite	County	Year(s)
Audubon's Shearwater			
	Chewing lice		
	Austromenopon paululum	Dade	1980
	Halipeurus spadix	Brevard	1980
		Lee	1983
	Naubates harrisoni	Dade	1980
	Trabeculus mirabilis	Dade	1980
Black-capped Petrel			
	Chewing lice		
	Austromenopon popellus	Brevard	1972
	Halipeurus theresae	Brevard	1972
		Volusia	1977
	Saemundssonia jamaicensis	Brevard	1972
Greater Shearwater			
	Ticks		
	Ixodes auritulus	Brevard	1993
	Mites		
	Alloptoides sp.	Brevard	1993
	Brephosceles puffini	Brevard, Martin	1993
	Dermation (Neodermation) sp.	Martin	1993
	Harpyrhynchus sp.	Martin	1993
	Ingrassia sp.	Brevard, Martin	1993
	Mesostigmata: Rhinonyssidae	Brevard	1993
	Microspalax sp.	Martin	1993
	Opetiopoda sp.	Brevard	1993
	Zachvatkinia puffini	Brevard, Martin	1993
	Chewing lice		
	Ancistrona vagelli	Brevard, Martin	1993
	Austromenopon paululum	Brevard	1972, 1993
		Indian River	1972, 1978–79
		Martin	1993
		St. Lucie	1979
	Halipeurus gravis	Brevard	1972, 1993

(continued)

Table 3.7. *(continued)*

Host	Species of parasite	County	Year(s)
		Indian River	1978–79
		Martin	1993
		St. Lucie	1979
	Naubates harrisoni	Brevard	1972, 1993
		Indian River	1978–79
		Martin	1993
	Saemundssonia peusi	Brevard	1972
		Martin	1978, 1993
	Trabeculus hexakon	Brevard	1972, 1993
		Indian River	1978–79
		Martin	1978, 1993
		St. Lucie	1979
Leach's Storm-Petrel			
	Mites		
	Brephosceles decapus	Martin	1993
	Brephosceles lanceolatus	Martin	1993
	Brephosceles sp.	Martin	1993
	Dermation (Neodermation) sp.	Martin	1993
	Ingrassia oceanodromae	Martin	1993
	Zachvatkinia zygoloba	Martin	1993
	Zachvatkinia sp.	Martin	1993
	Chewing lice		
	Halipeurus pelagicus	Martin	1993
Red-billed Tropicbird			
	Chewing lice		
	Saemundssonia phaetona	Martin	1979
White-tailed Tropicbird			
	Chewing lice		
	Saemundssonia uppalensis	Martin	1983

Sources: Foster et al. (1996) for Greater Shearwaters, Foster et al. (1994) for Leach's Storm Petrel, and Forrester et al. (1995) for all others.

Table 3.8. Arthropod infestations of boobies, gannets, frigatebirds, jaegers, and a dovekie in Florida

Host	Species of parasite	County	Year(s)
Brown Booby			
	Chewing lice		
	Pectinopygus garbei	St. Lucie	1981
Dovekie			
	Chewing lice		
	Austromenopon merguli	St. Lucie	1966
	Quadraceps klatti	St. Lucie	1966
	Saemundssonia merguli	St. Lucie	1966
Magnificent Frigatebird			
	Chewing lice		
	Colpocephalum spineum	Levy	1956
		Monroe	1977
	Fregatiella aurifasciata	Levy	1956
	Pectinopygus frigatiphagus	Levy	1956
		Monroe	1968
	Louse flies		
	Olfersia spinifera	Franklin	1967[a]
Masked Booby			
	Chewing lice		
	Eidmanniella albescens	Franklin	1983
	Pectinopygus annulatus	Franklin	1983
Northern Gannet			
	Subcutaneous mites		
	Neottialges bassani	Liberty	1993[b]
	Chewing lice		
	Eidmanniella pustulosa	Dade	1957, 1981
		Lee	1981
		Martin	1978, 1993
		Palm Beach	1988
		Volusia	1927
	Pectinopygus bassani	Dade	1957
		Indian River	1969, 1971, 1974–76, 1978–79, 1981
		Lee	1981
		Martin	1978, 1993
		Palm Beach	1988, 1993
		Volusia	1927
Pomarine Jaeger			
	Hypoderid mites		
	Thalassornectes aukletae	Pinellas	1993[c]

Source: Forrester et al. (1995), unless otherwise noted
a. Wilson et al. (1990)
b. Pence et al. (1997)
c. Pence & Cole (1995)

XVI. Literature cited

Anderson, D.J. 1993. Masked booby. *Birds N. Am.* 73:1–16.

Brichetti, P., U.F. Foschi, and G. Boano. 2000. Does El Niño affect survival rate of Mediterranean populations of Cory's shearwater? *Waterbirds* 23:147–154.

Clapp, R.B., R.C. Banks, D. Morgan-Jacobs, and W.A. Hoffman. 1982. *Marine birds of the southeastern United States and Gulf of Mexico*. Part I: *Gaviiformes through Pelecaniformes*. U.S. Fish and Wildlife Service, Office of Biological Services. 637 pp.

Cole, R., and D. Pence. 1993. Unpublished data. National Wildlife Health Center, Madison, Wis.

Crawford, R.L. 1981. Bird casualties at a Leon County, Florida TV tower: a 25-year migration study. *Bulletin Tall Timbers Research Station* 22:1–30.

Forrester, D.J., H.W. Kale II, R.D. Price, K.C. Emerson, and G.W. Foster. 1995. Chewing lice (Mallophaga) from birds in Florida: a listing by host. *Bull. Fla. Mus. Nat. Hist.* 39:1–44.

Foster, G.W., J.M. Kinsella, R.D. Price, J.W. Mertins, and D.J. Forrester. 1996. Parasitic helminths and arthropods of greater shearwaters (*Puffinus gravis*) from Florida. *J. Helminthol. Soc. Wash.* 63:83–88.

Foster, G.W., J.W. Mertins, and R.D. Price. 1994. Unpublished data. University of Florida, Gainesville.

Franson, J.C. 1990–93. Unpublished data. National Wildlife Health Center, Madison, Wis.

Frederick, P.C. 1993. Wading bird nesting success studies in the Water Conservation Areas of the Everglades, 1992 and 1993. Final report to South Florida Water Management District, West Palm Beach. 43 pp.

Friend, M., and J.C. Franson. 1999. *Field manual of wildlife diseases. General field procedures and diseases of birds.* U.S. Department of the Interior, U.S. Geological Survey, Biological Research Division, Information and Technology Report 1999-001. Washington, D.C. 425 pp.

Gilardi, K.V.K., J.D. Gilardi, A. Frank, M.L. Goff, and W.M. Boyce. 2001. Epidermotid mange in Laysan albatross fledglings in Hawaii. *J. Wildl. Dis.* 37: 185–188.

Hudson, R.D. 1963. Cory's shearwater at Ormond Beach. *Fla. Nat.* 36:27.

Huntington, C.E., R.G. Butler, and R.A. Mauck. 1996. Leach's storm-petrel. *Birds N. Am.* 233:1–32.

Hutton, R.F. 1964. A second list of parasites from marine and coastal animals of Florida. *Trans. Am. Microsc. Soc.* 83:439–447.

Hutton, R.F., and F. Sogandares-Bernal. 1960. Studies on helminth parasites from the coast of Florida. II. Digenetic trematodes from shore birds of the west coast of Florida. *Bull. Mar. Sci. Gulf Caribb.* 10:40–54.

Jenkins, R.L. 1984. Marine birds injured by welding rods. *Fla. Field Nat.* 12:13–15.

Jessup, D.A., and F.A. Leighton. 1996. Oil pollution and petroleum toxicity to wildlife. In: *Noninfectious diseases of wildlife.* 2d ed. A. Fairbrother, L.N. Locke, and G.L. Hoff (eds.). Iowa State University Press, Ames. pp. 141–156.

Kale, H.W. II. 1971. Florida region. *Am. Birds* 25:723–733.

Langridge, H.P., and G.E. Woolfenden. 1998. First record of thick-billed murre from Florida. *Fla. Field Nat.* 26:139–144.

Linton, E. 1927. Notes on cestode parasites of birds. *Proc. U.S. Natl. Mus.* 70:1–73.

———. 1928. Notes on trematode parasites of birds. *Proc. U.S. Natl. Mus.* 73:1–36.

Locke, L.N. 1993. Unpublished data. National Wildlife Health Research Center, Madison, Wis.

Longstreet, R.J. 1953–55. Ornithology of the Mosquitoes. *Fla. Naturalist* 26:103–114, 27: 175–188, 28:9–20.

Maehr, D.S., and J.Q. Smith. 1988. Bird casualties at a central Florida power plant: 1982–86. *Fla. Field Nat.* 16:57–80.

Maehr, D.S., A.G. Spratt, and D.K. Voigts. 1983. Bird casualties at a central Florida power plant. *Fla. Field Nat.* 11:45–49.

Manter, H.W. 1929–30. Studies on the trematodes of Tortugas fishes. *Carnegie Inst. Wash. Year Book* 29:338–340.

Pearson, J.C. 1973. A revision of the subfamily Haplorchinae Looss, 1899 (Trematoda: Heterophyidae) II. Genus *Galactosomum*. *Philos. Trans. R. Soc. Lond. B Biol. Sci.* 266:341–447.

Pence, D.B., and R.A. Cole. 1995. First record of an hypopus (Acari: Hypoderatidae) from a jaeger (Aves: Charadriiformes: Stercorariidae). *J. Med. Entomol.* 32:394–396.

Pence, D.B., M.G. Spalding, J.F. Bergan, R.A. Cole, S. Newman, and P.N. Gray. 1997. New records of subcutaneous mites (Acari: Hypoderatidae) in birds with examples of potential host colonization events. *J. Med. Entomol.* 34:411–416.

Pratt, H.S. 1911. On *Galactosomum cochleariforme* Rudolphi. *Zool. Anz.* 38:143–148.

Quist, C. 1993. Unpublished data. Southeastern Cooperative Wildlife Disease Study, University of Georgia, Athens.

Redig, P.T. 1993. Avian aspergillosis. In: *Zoo and wild animal medicine: current therapy.* 3d ed. M.E. Fowler (ed.). W.B. Saunders, Philadelphia. pp. 178–181.

Richardson, J.A. 1983. Unpublished data. University of Florida, Gainesville.

Robertson, W.B., Jr., and T. Wilmers. 1996. Magnificent Frigatebird. In: *Rare and endangered biota of Florida.* Vol. 5, *Birds.* J.A. Rodgers, Jr., H.W. Kale II, and H.T. Smith (eds.). University Press of Florida, Gainesville. pp. 156–169.

Robertson, W.B., and G.E. Woolfenden. 1992. *Florida bird species: an annotated list.* Spec. Publ. 6, Florida Ornithological Society, Gainesville. 260 pp.

Rodgers, J.A., Jr., H.W. Kale II, and H.T. Smith (eds.). 1996. *Rare and endangered biota of Florida.* Vol. 5, *Birds.* University Press of Florida, Gainesville. 688 pp.

Roffe, T.J., and H.A. McAllister. 1988. Unpublished data. National Wildlife Health Center, Madison, Wis.

Rohwer, S.A., and G.E. Woolfenden. 1968. A specimen of Leach's petrel from the Gulf coast of Florida. *Auk* 85:319.

Schreiber, E.A. 2001. Climate and weather effects on seabirds. In: *Biology of marine birds.* E.A. Schreiber and J. Burger (eds.). CRC Press, Boca Raton, Fla. pp. 179–215.

Schwartz, J.C. 1975. Unpublished data. Kissimmee Diagnostic Laboratory, Kissimmee, Fla.

Sileo, L., P.R. Stevert, and M.D. Samuel. 1990. Causes of mortality of albatross chicks at Midway Atoll. *J. Wildl. Dis.* 26:329–338.

Spalding, M.G., and J.M. Kinsella. 1997. Unpublished data. University of Florida, Gainesville.

Spalding, M.G., and M. Robson. 1993. Unpublished data. University of Florida, Gainesville, and Florida Game and Fresh Water Fish Commission, West Palm Beach.

Spalding, M.G., and S.P. Terrell. 1998. Unpublished data. University of Florida, Gainesville.

Spalding, M.G., C.A. Yowell, D.S. Lindsay, E.C. Greiner, and J.B. Dame. 2002. *Sarcocystis* meningoencephalitis in a northern gannet (*Morus bassanus*). *J. Wildl. Dis.* 38: 432–437.

Spitzer, P.R. 1995. Common loon mortality in marine habitats. *Environ. Rev.* 3:223–229.

Sprunt, A., Jr. 1938. The southern dovekie flight of 1936. *Auk* 55:85–88.

Stevenson, H.M., and B.H. Anderson. 1994. *The birdlife of Florida.* University Press of Florida, Gainesville. 892 pp.

Taylor, W.K., and B.H. Anderson. 1973. Nocturnal migrants killed at a central Florida TV tower, autumns 1969–71. *Wilson Bull.* 85:42–51.

———. 1974. Nocturnal migrants killed at a central Florida TV tower, autumn 1972. *Fla. Field Nat.* 2:40–43.

Taylor, W.K., and M.A. Kershner. 1986. Migrant birds killed at the Vehicle Assembly Building (VAB), John F. Kennedy Space Center. *J. Field Ornithol.* 57:142–154.

Thompson, D. 1996. Mercury in birds and terrestrial mammals. In: *Interpreting environmental contaminants in animal tissues.* W.N. Beyer, G.H. Heinz, and A. Redmon (eds.). Lewis, Boca Raton, Fla. pp. 341–356.

Trivelpiece, W., R.G. Butler, D.S. Miller, and D.B. Peakall. 1984. Reduced survival of chicks of oil-dosed adult Leach's storm-petrel. *Condor* 86:81–82.

Wanless, S., T.S. Barton, and M.P. Harris. 1997. Blood hematocrit measurements of 4 species of North Atlantic seabirds in relation to levels of infestation by the tick *Ixodes uriae*. *Colon. Waterbirds* 20:540–544.

Wilson, N.A., H.W. Kale II, and W.W. Baker.

1990. Unpublished data. University of Northern Iowa, Cedar Falls.

Wingate, D.B., I.K. Barker, and N.W. King. 1980. Poxvirus infection of the white-tailed tropicbird (*Phaethon lepturus*) in Bermuda. *J. Wildl. Dis.* 16:619–621.

Woolfenden, G.E. 1965. A specimen of the red-footed booby from Florida. *Auk* 82:102–103.

Work, T.M. 1996. Weights, hematology and serum chemistry of seven species of free-ranging tropical pelagic seabirds. *J. Wildl. Dis.* 32:643–647.

———. 1999. Weights, hematology, and serum chemistry of free-ranging brown boobies (*Sula leucogaster*) in Johnston Atoll, Central Pacific. *J. Zoo Wildl. Med.* 30:81–84.

Pelicans

I. Introduction

Two species of pelicans occur in Florida, the Brown Pelican (*Pelecanus occidentalis*), which breeds in Florida and is resident throughout the year, and the American White Pelican (*Pelecanus erythrorhynchos*), which is found primarily during the winter months and breeds in the western United States, Texas, and Canada (Robertson and Woolfenden 1992). Recently larger numbers of White Pelicans are remaining during the summer months. The Brown Pelican in Florida was listed federally as an endangered species from 1970 to 1985, and remains listed in Louisiana, in Texas, and on the Pacific coast. It is now listed as a species of special concern by the Florida Game and Fresh Water Fish Commission (Logan 1997) and as threatened by the

Florida Committee on Rare and Endangered Plants and Animals (Rodgers et al. 1996).

The nesting status of Brown Pelicans in Florida is summarized by Williams and Martin (1969, 1970), Nesbitt et al. (1977), Wilkinson et al. (1994), and Nesbitt (1996). Some additional information on eggs and breeding characteristics is found in Anderson and Hickey (1970). An excellent summary of serum chemistry values of young and old, captive and wild Brown Pelicans, sex differences, and seasonal effects has been published (Wolf et al. 1985).

Life history information and a summary of disease information from other locations can be found for the American White Pelican in Evans and Knopf (1993), and Clapp et al. (1982) summarize life history information in the southeastern United States. Some general health information on pelicans can be found in Fairbrother et al. (1996) and Friend and Franson (1999).

II. Trauma

As with most birds in Florida, human-caused changes in the environment present physical hazards to pelicans. Brown Pelicans are a small proportion of the dead birds found along roadsides in Florida. Road-kill surveys conducted on roads in state parks during 1990–93 included the following: 1 Brown Pelican from Duval County (1991), 11 from Indian River County (1992), 5 from Dade County (1992), 1 from Gulf County (1992), and 1 from Flagler County (1993)—a total of only 19 pelicans from a grand total of 1,922 birds of all species found dead (Stevenson 1994; Snyder 1994). Two percent of 4,604 Brown Pelicans taken to a rehabilitation center in the Tampa Bay area during 1994–98 had injuries associated with roads, including vehicular and power line collisions (Suto 2000). An adult White Pelican from Lee County in 1993 with evidence of trauma appeared to have collided with a power line or bridge (Spalding 1998). There have been 2 reports of Brown Pelicans being electrocuted, 1 in Tampa Bay in 1978 (Stroud 1980), the other in Dade County in 1980 (Schmeling 1980). Pelicans have not been listed in studies of mortality involving tall structures in Florida (Kale 1971; Crawford 1981; Maehr et al. 1983; Maehr and Smith 1988; Taylor and Anderson 1973, 1974; Taylor and Kershner 1986). Rodgers (1986) described a bony lesion on the maxilla of a Brown Pelican from Indian River County (1985) that suggested the bill had fractured and healed. Tiemeier (1941) found that 13.3% of specimens in Kansas museums of birds of the family Pelecanidae had evidence of repaired bone injuries.

Ingestion of, and entanglement in, foreign objects is a more significant mortality factor for Brown Pelicans than collisions with manmade structures. An immature Brown Pelican was found with a 0.3 x 35 cm wire rod (the core from a 3/16 welding rod) that had penetrated the bird's breast (Jenkins 1984). The bird recovered and was released. Jenkins also noted that welding rods have been reported as causing death in other birds in Florida (see chapter 3, Oceanic Birds) and elsewhere. Schreiber (1978) estimated that approximately 500 Brown Pelicans a year were killed from entanglement or ingestion of fishing gear in Florida. In Pinellas County 62% of 4,604 Brown Pelicans taken to a rehabilitation center over a 5-year period, 1994–98, had injuries associated with fishing tackle (Suto 2000). Of 269 pelicans, most collected from the wild, examined at the Laboratory of Wildlife Disease Research, University of Florida, between 1969 and 1998, 10 (3%) had fishing tackle wounds or had ingested foreign material (Spalding and Forrester 1998). In many cases fishhooks had perforated the esophagus or stomach and were surrounded by inflammatory tissue, but in a few cases hooks were retained within the stomach. Hansen et al. (1998) examined radiographs of more than 250 Brown Pelicans brought into rehabilitation centers in Florida. Some type of fishing tackle (mostly hooks) was visible on nearly 50% of the radiographs. About 3% of the birds had ingested fishing sinkers.

Human disturbance of colony sites can cause reproductive failure (Schreiber 1978). Egg predation by Fish Crows, associated with human disturbance, was the most common cause of egg loss on Tarpon Key, Tampa Bay in 1969–70 (Schreiber and Risebrough 1972). Rodgers and Smith (1997) found Brown Pelicans to be more sensitive to disturbance than any other bird they looked at and suggested a buffer distance of 126 meters for motorboats. On occasion, Brown Pelicans have been killed by fishermen because they compete for the same resources. Miller (1950) reported the killing of 400 at Pelican Island, Indian River County, in 1918 by fishermen. More than 1,000 Brown Pelican chicks (estimated at 80% of chicks in the colonies) were killed in colonies in Brevard County in 1924 (Longstreet 1953–55; Sprunt 1954). More recently, an estimated 30 Brown Pelicans were killed by fisher-

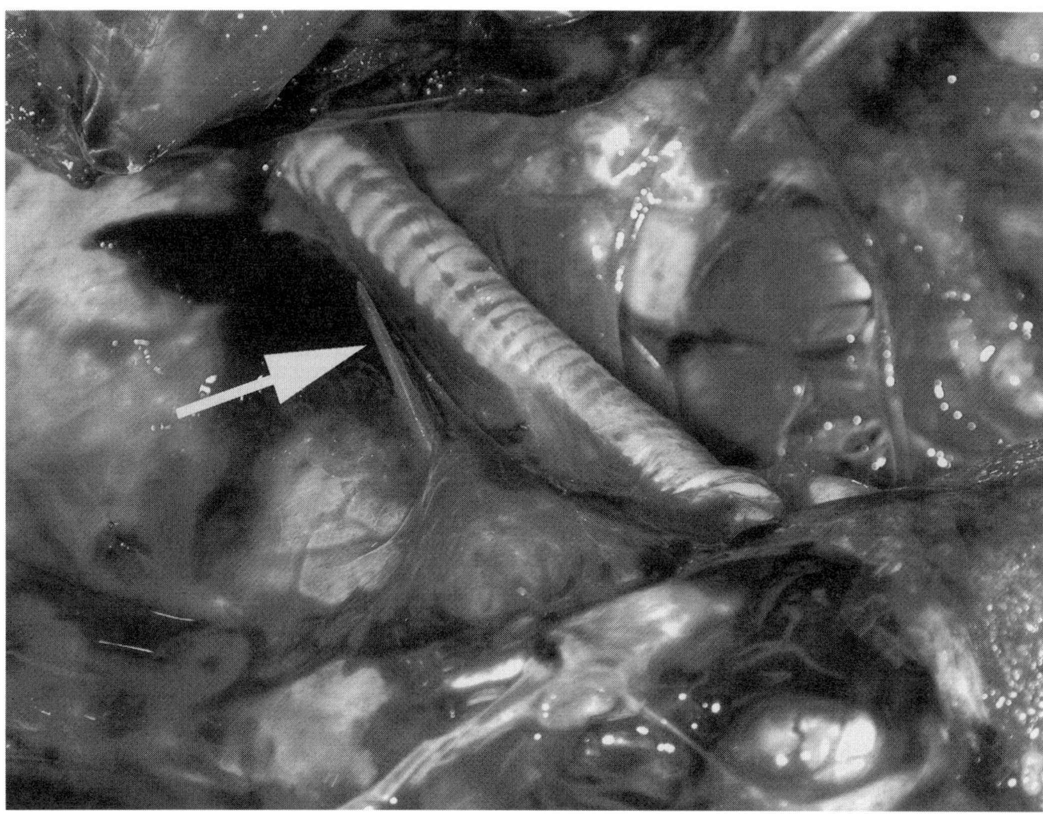

FIGURE 4.1. A catfish spine *(arrow)* that perforated the esophagus near the heart and caused massive hemorrhage and death in a White Pelican from Lake County in 2001.

men in Volusia County in 1993 (Spalding and Hovis 1993) (table 4.1).

Natural causes of injury also occur. In 2 cases, Pinellas County in 1973 and Duval County in 1987, Brown Pelicans had fish spines that had perforated the esophagus or stomach, causing airsacculitis (Forrester 1987; Spalding 1993). Forty-four (1%) of 4,604 Brown Pelicans taken to a rehabilitation center in Pinellas County in 1994–98 were captured because they had swallowed unusually large fish (Suto 2000). In Lake County in 1999 an adult White Pelican, attracted to large numbers of surfacing catfish, died from a laceration of the jugular vein caused by a catfish spine (Spalding 1999). In the same area in 2001 another White Pelican died when 2 catfish spines perforated its esophagus (figure 4.1) and stomach, resulting in massive hemorrhage (Spalding 2001). A hatch-year Brown Pel-

ican was found dead with a Double-crested Cormorant wedged in its pouch and the pouch was torn. It was speculated that both birds had drowned following the capture by the pelican in Bay County in 1980 (Francis 1981). Predation on adults by "wildcats," presumably bobcats, was recorded by Bogart (1953) in Dade County. Wood Storks will occasionally take small Brown Pelican chicks (Kale 1972). Alligator attacks on White Pelicans were suspected in 2 cases by Spalding (1990–2000). One pelican was emaciated and had an injured leg in Orange County in 1990, the second had a broken tail and ruptured intestine in Brevard County in 2000.

III. Inclement weather

Brown Pelicans almost disappeared from the panhandle area following destruction of a large

Table 4.1. Brown Pelican die-off events in Florida

Year	Month	County or area	No. of Pelicans involved	Other species involved	No. of Pelicans examined	Cause of death	Other findings	Data source
1974	July	Collier	Unknown	?	12	Undetermined	Bacteria, parasites, aspergillosis, fish hook	A
1980	January	?	?	?	5	Suspect oiling	Parasites, coccidia	B
1982	April	Brevard	10	Double-crested Cormorants, Ring-billed Gulls	1	Undetermined	Botulism in a gull	C
1983	Jan.–June	Pinellas	400	Double-crested Cormorants	40	Bacterial enteritis?	Emaciation	S, T, U
1985	February	Citrus	2	Coots	0	Undetermined	*Vibro* & *Aeromonas* in coots	D
		Wakulla	50–100	None	3	Emaciation, Frostbite	—	E
1986	May	Brevard	57–100	Gulls, Blue Crabs	7	Undetermined	Bacteria	F
1987	April	Brevard	30–40	Laughing Gulls, Loons	3	Undetermined	Gunshot	G
		Duval	?	?	2	Undetermined	—	H
1988	October	Volusia	150–200	Snowy Egrets, Little Blue Herons	9	Aerial spraying?	—	I
1989	Mar.–Apr.	Brevard	>100	?	2	Toxin suspected	Necrotic enteritis	J, K
1991	March	Bay	4	Common Loons	3	Botulism C & E	—	L
1992	February	Duval	500	None	2	*Erysipelothrix rhusiopathiae* septicemia	—	M
1993	February	Volusia	30	None	6	Trauma (fishermen)	—	N
	Feb.–Mar.	Brevard	5	Common Loons, Northern Gannets, Ring-billed Gulls, Blue-footed Boobies, Double-crested Cormorants, Royal Terns	0	Emaciation	—	O
	March	Duval	?	None	2	Undetermined	Bone perforation	P
	March	Volusia	7	None	3	Undetermined	Emaciation	M
1998	Jan.–Mar.	Duval	200	Loons, gulls, gannets	2	Undetermined	Septicemia	Q
	March	Brevard	50	None	4	Undetermined	Thyroid hyperplasia and bronchitis	R

Sources: A = Forrester (1974), B = Stroud (1980), C = Stroud (1982), D = Thomas (1985), E = Stroud (1985), F = Locke (1986), G = Roffe (1987), H = Forrester (1987), I = Spalding et al. (1988), J = Langenberg (1989), K = Locke (1989), L = Franson (1991), M = Spalding (1994), N = Spalding & Hovis (1993), O = Quist (1993), P = Spalding (1993), Q = Brannian (1998), R = Spalding (1998), S = Stroud (1983), T = Ankerberg (1983), U = Forrester (1983). *Note:* ? = Unknown.

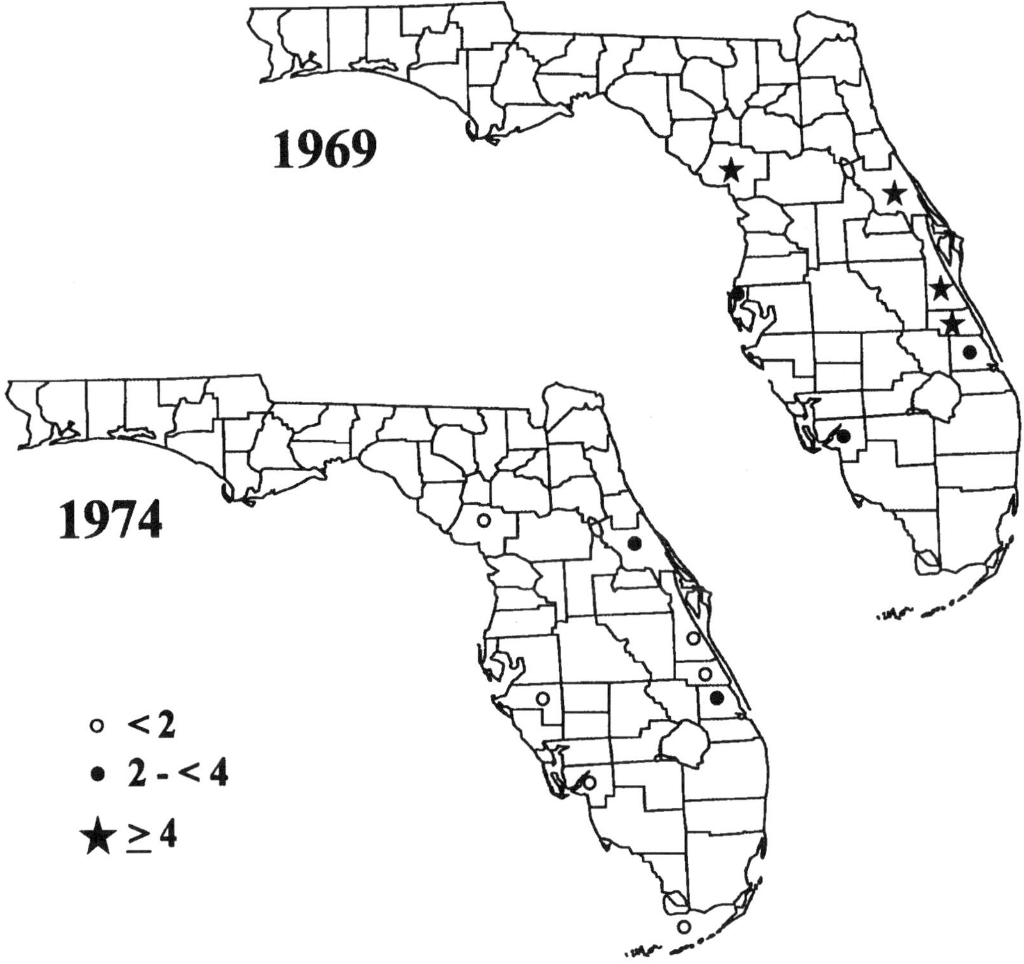

FIGURE 4.2. Comparison of Brown Pelican eggs tested for DDE residues in 1969 and 1974 in Florida (data from Blus et al. 1974b, 1979). Symbols are used to represent the highest concentration of DDE (ppm wet weight) from each collection.

Louisiana breeding colony by Hurricane Audrey in 1957 (Stevenson 1994). Hurricanes are potential sources of mortality, but such mortality is rarely reported. After Hurricane Donna in September 1960, 12 Brown Pelicans were found dead in Florida Bay and around Flamingo (Robertson and Paulson 1961). Eight Wood Storks, 3 White Ibises, 2 Ring-billed Gulls and 1 Brown Pelican were found dead along the beach in South Carolina following Hurricane Hugo in 1989 (Post 1992). No pelican mortality was noted following Hurricane Andrew in August 1992 (FGFWFC 1992). Following the "storm of the century" in March 1993, with hurricane-strength winds followed by unseasonably cold weather, it was estimated that 60% of the Brown Pelican nests at Alafia Bank in Hillsborough County were destroyed (Langridge 1993).

Cold weather can cause freezing of gular pouches and webbing on feet (Nesbitt 1996), and this has been reported in Franklin County (Stevenson 1994). About 50–100 Brown Pelicans died during a cold spell in February 1985 in Wakulla County. Three of those birds exam-

Table 4.2. DDE, DDD, and DDT residues in Brown Pelican eggs collected in Florida

Tissue County or area	Year	No. exam.	DDE			DDD			DDT			Data source
			No.[a]	Mean[b]	(Range)	No.[a]	Mean[b]	(Range)	No.[a]	Mean[b]	(Range)	
Whole eggs												
"Atlantic Coast"	1979	37	NG	0.41	NG	—	—	—	—	—	—	E
Brevard	1969	5	5	2.5	(0.92–4.0)	5	0.66	(0.18–1.2)	5	0.66	(0.42–0.90)	A
	1974	15	15	1.1	(0.44–1.8)	15	0.31	(0.13–0.78)	NA	—	—	C
"Charlotte Harbor"	1972	1	1	2.5[c]	—	NA	—	—	NA	—	—	D
"Florida Bay"	1970	20	20	1.0	(0.40–2.9)	18	0.15	(ND–0.35)	10	0.09	(ND–0.35)	A
	1972	8	8	0.14	(0.08–0.65)	8	0.055	(0.016–0.11)	8	0.05	(0.016–0.11)	F
	1974	11	11	0.31	(0.11–1.1)	2	NC	(0.10–0.14)	NA	—	—	C
	1979	18	NG	0.10	NG	—	—	—	—	—	—	E
"Gulf Coast"	1979	37	NG	0.42	NG	—	—	—	—	—	—	E
Indian River	1969	5	5	2.1	(1.1–6.0)	5	1.0	(0.19–2.0)	5	0.47	(0.22–0.99)	A
	1974	14	14	1.1	(0.49–1.8)	13	0.32	(ND–0.62)	NA	—	—	C
Lee	1969	15	15	1.2	(0.46–3.0)	15	0.37	(0.12–1.5)	13	0.19	(ND–0.71)	A
	1974	30	27	0.28	(ND–1.0)	12	0.08	(ND–0.41)	NA	—	—	C
Levy	1969	7	7	1.4	(0.84–5.5)	7	0.37	(0.2–0.96)	7	0.24	(0.14–0.39)	A
	1974	15	11	0.29	(ND–0.64)	4	NC	(ND–0.31)	NA	—	—	C
Manatee	1974	15	13	0.45	(ND–1.5)	13	0.12	(ND–0.22)	NA	—	—	C
Pinellas	1969	5	5	1.9	(1.2–3.8)	5	0.87	(0.62–2.0)	4	0.32	(ND–0.57)	A
St. Lucie	1969	7	7	2.4	(1.5–3.4)	7	0.94	(0.72–1.2)	7	0.54	(0.3–0.67)	A
	1974	8	8	1.2	(0.6–2.2)	8	0.31	(0.12–0.65)	NA	—	—	C

(continued)

Table 4.2. (*continued*)

Tissue County or area	Year	No. exam.	DDE			DDD			DDT			Data source
			No.[a]	Mean[b]	(Range)	No.[a]	Mean[b]	(Range)	No.[a]	Mean[b]	(Range)	
Volusia	1969	5	5	1.5	(0.16–5.9)	5	0.75	(0.24–1.7)	4	0.14	(ND–0.45)	A
	1974	14	14	1.3	(0.45–2.6)	14	0.33	(0.1–0.74)	NA	—	—	C
Egg yolks												
Brevard	1970	22	NG	28[d]	NG	NG	2.9[d]	NG	NG	1.4[d]	NG	B
	1971	29	NG	0.90[e]	NG	NG	0.26[e]	NG	NG	0.17[e]	NG	G
"Charlotte Harbor"	1970	20	NG	18d	NG	NG	4.5[d]	NG	NG	1.1[d]	NG	B
Gulf and Lee[f]	1971	14[f]	NG	0.84[e]	NG	NG	0.29[e]	NG	NG	0.07[e]	NG	G
Indian River	1970	10	NG	21[d]	NG	NG	4.0[d]	NG	NG	1.0[d]	NG	B
"Tampa Bay"	1969	14	NG	37[d]	NG	NG	13[d]	NG	NG	6.0[d]	NG	B
	1970	21	NG	26[d]	NG	NG	6.6[d]	NG	NG	3.2[d]	NG	B

Sources: A = Blus et al. (1974b), B = Schreiber & Risebrough (1972), C = Blus et al. (1979), D = Lincer & Salkind (1973), E = Blus et al. (1997), F = Ogden et al. (1974), G = Thompson et al. (1977b).

NA = not analyzed, NC = not calculated, ND = not detected, NG = not given.

a. Number with residue.

b. Geometric mean, ppm wet weight unless otherwise indicated.

c. Dry weight. Lincer & Salkind (1973) also give extractable fat weights.

d. Values given for egg yolks by Schreiber & Risebrough (1972) are lipid weights for all but PCBs. They also report total DDT lipid and wet weights and PCB wet weights not presented in this table.

e. Values given for egg yolks by Thompson et al. (1977b) are arithmetic means.

f. 11 eggs from Lee County and 3 eggs from Gulf County.

Table 4.3. Other pesticide residues in Brown Pelican eggs collected in Florida

Tissue County or area	Year	No. exam.	Dieldrin			PCBs			Heptachlor epoxide			Data source
			No.[a]	Mean[b]	(Range)	No.[a]	Mean[b]	(Range)	No.[a]	Mean[b]	(Range)	
Whole eggs												
Brevard	1969	5	5	0.63	(0.29–1.5)	5	3.8	(1.3–7.5)	2	NC	(ND–0.1)	A
	1974	15	15	0.32	(0.14–0.78)	15	4.9	(2.5–9.2)	0	ND	—	C
"Charlotte Harbor"	1972	1	1	0.10[c]	—	1	1.3[c]	—	NA	—	—	D
"Florida Bay"	1970	20	5	0.07	(ND–0.33)	20	0.8	2 (<1.0–2.7)	NA	—	—	A
	1972	8	8	0.032	(0.012–0.23)	8	0.5	8 (0.25–3.2)	NA	—	—	E
	1974	11	4	NC	(0.10–0.14)	7	1.1	(0.41–2.4)	0	ND	—	C
Indian River	1969	5	5	0.41	(0.10–1.1)	5	2.3	(1.3–5.5)	3	NC	(ND–0.29)	A
	1974	14	14	0.31	(0.13–0.60)	14	5.5	(2.0–9.7)	1	NC	(ND–0.11)	C
Lee	1969	15	15	0.08	(<0.10–0.22)	15	0.7	1 (<1.0–2.3)	1	<0.10	(ND–<0.1)	A
	1974	30	20	0.11	(ND–0.65)	30	1.1	(0.15–4.1)	1	NC	(ND–0.13)	C
Levy	1969	7	7	0.08	(<0.1–0.15)	7	0.8	9 (<1.0–4.2)	0	ND	—	A
	1974	15	7	NC	(ND–0.58)	15	0.8	0 (0.36–1.6)	0	ND	—	C
Manatee	1974	15	15	0.16	(ND–0.38)	15	1.7	(0.75–10)	0	ND	—	C
Pinellas	1969	5	5	0.18	(<0.1–0.40)	5	1.7	(1.1–2.9)	4	NC	(ND–<0.1)	A
St. Lucie	1969	7	7	0.32	(0.20–0.57)	7	2.0	(1.0–3.9)	0	ND	—	A
	1974	8	8	0.31	(0.19–0.42)	8	7.8	(4.0–13)	1	NG	(ND–0.12)	C
Volusia	1969	5	3	0.22	(ND–0.75)	5	1.7	(1.7–7.1)	0	ND	—	A
	1974	14	14	0.40	(0.18–0.83)	14	7.5	(4.2–12)	1	NG	(ND–0.10)	C
Egg yolks												
Brevard	1970	22	NA	—	—	NG	2.9	NG	NA	—	—	B
	1971	29	NG	0.21[d]	NG	NG	2.2[d,e]	NG	NA	—	—	F
"Charlotte Harbor"	1970	20	NA	—	—	NG	2.2	NG	NA	—	—	B
Gulf and Lee[f]	1971	14[f]	NG	0.15[d]	NG	NG	1.0[d,e]	NG	NA	—	—	F
Indian River	1970	10	NA	—	—	NG	4.2	NG	NA	—	—	B
"Tampa Bay"	1969	14	NA	—	—	NG	6.2	NG	NA	—	—	B
	1970	21	NA	—	—	NG	3.7	NG	NA	—	—	B

(continued)

Table 4.3. *(continued)*

Tissue County or area	Year	No. exam.	cis-chlordane			trans-nonachlor			cis-nonachlor			Data source
			No.[a]	Mean[b]	(Range)	No.[a]	Mean[b]	(Range)	No.[a]	Mean[b]	(Range)	
Whole Eggs												
Brevard	1974	15	10	0.11	(ND–0.30)	8	0.07	(ND–0.18)	10	0.10	(ND–0.23)	C
"Florida Bay"	1974	11	0	ND	—	0	ND	—	0	ND	—	C
Indian River	1974	14	10	0.12	(ND–0.35)	10	0.11	(ND–0.30)	8	0.09	(ND–0.25)	C
Lee	1974	30	13	NC	(ND–0.83)	8	NC	(ND–0.48)	5	NC	(ND–0.44)	C
Levy	1974	15	2	NC	(ND–0.14)	1	NC	(ND–0.12)	1	NC	(ND–0.10)	C
Manatee	1974	15	11	0.14	(ND–0.33)	9	0.14	(ND–1.0)	7	0.06	(ND–0.18)	C
St. Lucie	1974	8	6	0.13	(ND–0.22)	3	NC	(ND–0.18)	6	0.10	(ND–0.20)	C
Volusia	1974	14	9	0.14	(ND–0.53)	10	0.12	(ND–0.34)	8	0.09	(ND–0.33)	C

Tissue County or area	Year	No. exam.	Mirex			Toxaphene			Data source
			No.[a]	Mean[b]	(Range)	No.[a]	Mean[b]	(Range)	
Whole Eggs									
Brevard	1974	15	0	ND	—	10	0.15	(ND–0.46)	C
"Florida Bay"	1974	11	0	ND	—	2	NC	(ND–0.14)	C
Indian River	1974	14	0	ND	—	6	NC	(ND–0.30)	C
Lee	1974	30	4	NC	(ND–0.31)	3	NC	(ND–0.60)	C
Levy	1974	15	0	ND	—	0	ND	—	C
Manatee	1974	15	0	ND	—	0	ND	—	C
St. Lucie	1974	8	0	ND	—	1	NC	(ND–0.20)	C
Volusia	1974	14	0	ND	—	11	0.19	(ND–1.53)	

Sources: A = Blus et al. (1974b), B = Schreiber & Risebrough (1972), C = Blus et al. (1979), D = Lincer & Salkind (1973), E = Ogden et al. (1974), F = Thompson et al. (1977b).
Notes: NA = not analyzed, NC = not calculated, ND = not detected, NG = not given.
a. Number with residue.
b. Geometric mean, ppm wet weight unless otherwise indicated.
c. Dry weight. Lincer & Salkind (1973) also give extractable fat weights.
d. Values given for egg yolks by Thompson et al. (1977b) are arithmetic means.
e. Aroclor 1254.
f. 11 eggs from Lee County and 3 eggs from Gulf County.

ined had necrotic pododermatitis due to frostbite (Stroud 1985). Nestlings have been killed by cold weather: 600–700 died in November 1906 and a second brood was killed in 1907 by both cold weather and a storm (Miller 1950), and nestlings in Brevard County were killed by cold weather in 1978–79 (Schreiber 1980). Most Brown Pelican nestlings were killed by a storm in Brevard County in December 1925 (Sprunt 1954). Freezing temperatures also result in loss of black mangrove trees used for nesting, as occurred on the Atlantic coast from Brevard County northward and on the Gulf coast north of Pinellas County in the 1980s (Nesbitt 1996).

Five Brown Pelicans were found as part of a large die-off of at least 80 seabirds in Brevard County in February and March of 1993 (Quist 1993). No pelicans were examined, but all other species examined were emaciated; it was believed that unusual weather and reduced food availability contributed to this die-off (see chapter 3, Oceanic Birds, for more information).

IV. Organochlorines

The most extensive examination of organochlorine pesticide residues in birds in Florida was of Brown Pelican eggs in the 1960s and 1970s (tables 4.2 and 4.3). The highest concentrations of DDE (>2 ppm) were reported from Brevard, Indian River, and St. Lucie counties on the east coast of Florida in 1969. One egg from Indian River County contained 6 ppm DDE. Blus et al. (1979) reported mean concentrations of DDE+DDD+DDT in 1969 in eggs from the Atlantic coast (3.7 ppm) that were greater than those from the Gulf coast (2.3 ppm), with lower values in Florida Bay eggs (1.2 ppm). By 1974 these concentrations had declined to 1.5, 0.4, and 0.4 ppm, respectively (figure 4.2). Thompson et al. (1977b) also documented generally higher residues in eggs on the east coast, especially for PCBs. Blus (1982) reported that 3 ppm of DDE in eggs was associated with impaired reproduc-

tive success and 4 ppm with total reproductive failure, making the Brown Pelican the most sensitive species to this compound. Endrin, on the other hand, which was not measured in Florida pelicans, had effects on reproduction at 0.5 ppm in Louisiana, and likely played a role in the extirpation of pelicans in that state. The researchers were not able to establish a similar concentration at which dieldrin produced effects, but suggested that it would be greater than 1 ppm (Blus et al. 1979). Sales of DDT ceased in the United States in 1972 (Dunlap 1981). The most recent information we could find for pelican eggs in Florida was from 1979 (table 4.2). At that time, eggshells averaged less than 6% thinner than pre-exposure measurements. Thompson et al. (1977b) measured organochlorine residues in fish on both coasts in the early 1970s and found them to be relatively low, and not significantly different relative to location.

Much of the effect of organochlorines in reduced reproduction appears to be associated with its effect on calcium availability for the deposition of eggshells (Peakall 1970). Blus et al. (1971, 1972) reported a logarithmic correlation between DDE (but not PCB) residues and eggshell thinning in Brown Pelicans, with no effect on reproduction at <2.5 ppm in eggs (Blus et al. 1974a). This effect resulted in population declines in California and South Carolina (Blus et al. 1979) and extirpation in Louisiana (Blus et al. 1975). Anderson and Hickey (1972) found that most species that had sustained a 20% or greater decrease in shell thickness had also experienced population declines. Thinning was as high as 47% in California Brown Pelicans (Jehl 1973). Eggshell thinning in Florida rarely exceeded 15% (table 4.4) and a population effect could not be demonstrated. Blus et al. (1997) could find no effect of DDE on other egg measurements, including eggshell size, mass, or shape. When compared to pre-1947 Brown Pelican eggshells collected in Florida and Georgia, eggshells collected in Florida in 1969 were both thinner (8%) and lighter (17%) (Anderson and Hickey 1970). Atlantic coast eggshells were lighter and thinner than Gulf coast

Table 4.4. Eggshell weight and thickness of Brown Pelican eggs collected in Florida

| County/area | Year(s) | Shell weight (g) | | | Shell thickness (mm) | | | | Data source |
		No. examined	Mean[a]	% Change	No. examined	Mean[a]	(Range)	% Change[c]	
Brevard	1969	10	8.40 ± 0.20	—	10	0.497 ± 0.011	—	−11	C
	1970	NA	—	—	22	0.501 ± 0.013	(0.46–0.56)	−10	F
	1970	NA	—	—	10	0.482 ± 0.019	—	−13	H
	1974	NA	—	—	15	0.499 ± 0.010	—	−10	H
"Florida Bay"									
Buchanan Key	1969	3	8.64 ± 0.61	—	3	0.530 ± 0.015	—	−5	C
	1970	NA	—	—	10	0.545 ± 0.013	—	−2	H
Fanny Key	1970	NA	—	—	7	0.523 ± 0.019	—	−6	H
Frank Key	1972	NA	—	—	4	0.575	(0.545–0.621)	+3	D
Marquesas	1970	NA	—	—	10	0.541 ± 0.012	—	−3	H
Nest Key	1970	NA	—	—	10	0.532 ± 0.012	—	−4	H
Palm Key	1972	NA	—	—	4	0.566	(0.536–0.640)	+2	D
Indian River	1969	10	8.11 ± 0.21	—	10	0.499 ± 0.012	—	−10	C
	1970	NA	—	—	10	0.502 ± 0.018	(0.46–0.53)	−10	F
	1970	NA	—	—	9	0.498 ± 0.017	—	−11	H
	1974	NA	—	—	8	0.508 ± 0.011	—	−9	H
Lee									
Boca Grande Pass	1969	10	9.04 ± 0.27	—	10	0.559 ± 0.014	—	0	C
(Bird Key)	1970	NA	—	—	10	0.517 ± 0.014	—	−7	H
	1974	NA	—	—	15	0.549 ± 0.013	—	−1	H
Hemp Island	1969	10	8.42 ± 0.20	—	10	0.516 ± 0.012	—	−7	C
	1970	NA	—	—	20	0.518 ± 0.240	(0.43–0.61)	−7	F
	1970	NA	—	—	10	0.519 ± 0.015	—	−7	H
	1974	NA	—	—	15	0.549 ± 0.012	—	−1	H
Matlacha Pass	1969	9	8.64 ± 0.55	—	9	0.522 ± 0.023	—	−6	C
	1970	NA	—	—	10	0.504 ± 0.019	—	−10	H
Levy	1969	6	8.88 ± 0.36	—	6	0.530 ± 0.015	—	−5	C
	1970	NA	—	—	10	0.531 ± 0.016	—	−5	H
	1974	NA	—	—	15	0.547 ± 0.009	—	−2	H
Manatee	1970	NA	—	—	10	0.502 ± 0.012	—	−10	H
	1974	NA	—	—	15	0.534 ± 0.010	—	−4	H
Pinellas	1969	7	8.39 ± 0.29	—	8	0.509 ± 0.015	—	−9	C
	1970	NA	—	—	10	0.487 ± 0.015	—	−13	H
St.Lucie	1969	6	8.45 ± 0.23	—	6	0.513 ± 0.012	—	−8	C

	1970	NA	—	—	9	0.504 ± 0.009	—	−10	H
	1974	NA	—	—	8	0.508 ± 0.011	—	−9	H
"Tampa Bay"	1969	NA	—	—	14	0.506 ± 0.022	(0.42–0.55)	−9	F
	1970	NA	—	—	21	0.509 ± 0.024	(0.39–0.58)	−9	F
	1972	NA	—	—	32	0.529 ± 0.053	(0.40–0.62)	−5	E
	1973	NA	—	—	15	0.487 ± 0.065	(0.39–0.58)	−12	E
	1974	NA	—	—	6	0.524 ± 0.061	(0.44–0.58)	−6	E
	1975	NA	—	—	31	0.533 ± 0.053	(0.39–0.64)	−4	E
	1976	NA	—	—	31	0.545 ± 0.037	(0.47–0.62)	−2	E
	1972–75	NA	—	—	6[d]	0.403 ± 0.012	(0.390–0.422)	−28	E
	1969–76	NA	—	—	11[e]	0.459 ± 0.019	(0.417–0.481)	−18	E
Volusia									
Crane Island	1970	NA	—	—	10	0.491 ± 0.009	—	−12	H
(Bird Island)									
Port Orange	1969	9	8.43 ± 0.26	—	9	0.488 ± 0.012	—	−12	C
	1970	NA	—	—	9	0.487 ± 0.009	—	−13	H
	1974	NA	—	—	14	0.476 ± 0.013	—	−15	H
Summary									
"Florida/Georgia"	pre-1947	208	9.78 ± 0.12	—	172	0.557 ± 0.004	—	—	B
"Florida"	1950–53	9	8.10 ± 0.14	−17	9	—	—	−15	B
	1969	80	8.51 ± 0.07	−13[b]	81	0.515 ± 0.005	—	−7[b]	C
	1970	NA	—	—	87	0.508 ± 0.009	(0.39–0.61)	−9	F
"Atlantic Coast"	pre-1947	68	9.41 ± 0.11	—	55	0.547 ± 0.005	—	—	A
	1969	35	8.33 ± 0.11	−11[b]	35	0.498 ± 0.006	—	−9[b]	C
	1970	NA	—	—	47	0.494 ± 0.006	—	−10	H
	1974	NA	—	—	50	0.494	—	−11	G
	1979	NA	—	—	37	0.537	—	−4	G
"Florida Bay"	1970	NA	—	—	20	0.535	—	−4	G
	1974	NA	—	—	11	0.521	—	−6	G
	1979	NA	—	—	18	0.544	—	−2	G
"Gulf Coast"	pre-1947	115	9.93 ± 0.08	—	98	0.560 ± 0.004	—	+1	A
	1969	42	8.65 ± 0.16	−13[b]	43	0.527 ± 0.007	—	−6[b]	C
	1970	NA	—	—	60	0.510 ± 0.006	—	−9	H
	1974	NA	—	—	60	0.545	—	−2	G
	1979	NA	—	—	37	0.527	—	−5	G

Sources: A = Anderson & Hickey (1969), B = Anderson & Hickey (1970), C = Blus (1970), D = Ogden et al. (1974), E = Schreiber (1977), F = Schreiber & Risebrough (1972), G = Blus et al. (1997), and H = Blus et al. (1979).

NA = not analyzed.

a. ±Standard error for Anderson & Hickey (1969) and Blus (1970); ±95% confidence limit for Anderson & Hickey (1970), Schreiber & Risebrough (1972), and Schreiber (1977).

b. Significant at P<0.01: means compared by Student's t-test.

c. Compared with a pre-1947 thickness of 0.557 mm for Florida and Georgia eggs in Anderson & Hickey (1970), except where more site-specific data are available.

d. Schreiber (1977) found 6 crushed eggshells: one in 1972 (0.400 mm), 4 in 1973 (0.390, 0.402, 0.415, 0.422 mm) and one in 1975 (0.393 mm).

e. Eleven eggs that hatched.

shells. Schreiber (1977) analyzed eggshell thickness at a colony in Tampa Bay over an 8-year period. Shell thickness appeared to have increased since 1975. He also measured the thickness of eggs that were crushed and found them to be thinner overall than those that hatched, but there was considerable overlap. A slow trend to recovery can be seen in the most recent data in 1979, but we could find no more recent data to confirm this trend.

Nesbitt et al. (1981) found higher concentrations of DDT and its metabolites, dieldrin, and PCBs in tissues of sick Brown Pelicans than in those that appeared to be healthy (tables 4.5 and 4.6), but they could not demonstrate a population-wide effect. Mean concentrations were below those considered lethal in normal birds. In some cases birds found sick or dead had concentrations of contaminants that were considered to be lethal. Nesbitt et al. (1981) reported concentrations as high as 91 ppm DDT, 13 ppm DDD, and 24 ppm DDT in brains of sick or dead pelicans. Stickel et al. (1966) reported a minimum lethal concentration in brain tissue of 30 ppm wet weight of DDT+DDD in a passerine species and Greichus and Hannon (1973) estimated that 30 ppm DDD was evidence of toxicity in Double-crested Cormorants.

Thompson et al. (1977a) experimentally investigated the interaction between starvation and p,p'-DDE naturally acquired by nestling Brown Pelicans collected from Vero Beach, Indian River County. They found significantly higher residues in fat of starved birds than in nonstarved control birds. They did not find similar differences in liver, muscle, or brain.

A large scale die-off of about 1,000 White Pelicans occurred during the winter and spring of 1999–2000 in Orange County. Organochlorine toxicosis was the presumed cause of death of many of the pelicans. Tissues from these birds were tested but data were not available for use in this publication. Some other information from this die-off has been included in other sections. No other information about organochlorine contaminants in White Pelicans from Florida was available.

V. Organophosphates and carbamates

No evidence for poisoning with these compounds was found. Birds were infrequently tested. A few Brown Pelicans were tested for cholinesterase activity from die-off events listed in table 4.1: Brevard County in 1989 and Duval County in 1998.

VI. Metals

Mercury concentrations in eggs in the 1960s and 1970s were high (table 4.7) compared with other piscivorous species (see for example chapter 5, Cormorants). Blus et al. (1971) found a significant negative correlation between mercury concentrations in eggs and eggshell thickness for Brown Pelican eggs collected in 10 sites in Florida and 2 sites in South Carolina. There was no significant correlation between mercury concentrations and eggshell weight or shell thickness index (= 10 x weight in grams/length x breadth in cm: Anderson and Hickey 1970). Mercury concentrations measured in livers of Brown Pelicans, although as high as 36 ppm (table 4.8), were never high enough to be considered to be the cause of death. The effect of mercury on reproduction of pelicans has not been investigated.

Bone lead concentrations averaging 23 ppm were measured in Brown Pelicans collected from Tampa Bay (table 4.8) (Connors et al. 1972). No indication of their health status was given. Concentrations greater than 20 ppm are considered to indicate severe clinical poisoning in waterfowl (Pain 1996); however, bone concentrations are less meaningful in the interpretation of acute toxicity than are other tissues such as liver and kidney, which were not tested in these cases.

We could find metal analysis of White Pelican tissues in only one case. Liver from 2 birds that were part of a die-off in Orange County in 1990 had >.05, 0.06 ppm lead, 0.2, 0.14 ppm cadmium, and 2.7 and 6 ppm wet weight mercury respectively (Spalding et al. 1996).

VII. Oiling

Although pelicans frequently are involved in oiling events, there is very little published documentation of oiling in Florida for Brown or White Pelicans (Clapp et al. 1982). An oil spill in Tampa Bay in the winter of 1970 that killed several thousand birds also killed smaller numbers of Brown Pelicans. A spill also occurred in Jacksonville, Duval County, in 1970; however, a list of birds killed was not reported (Stevenson 1970). Three oiled Brown Pelicans were examined from Tampa Bay in 1978 (Stroud 1980). A cause of death was not determined. A diesel oil spill in Duval County preceded a die-off of Brown Pelicans in 1992 from bacterial infection (see below), but a direct connection with the oil spill was never established (Spalding 1992). Birds that were shivering and had wettable feathers were noted also in the following year in the same area; however, a specific spill was not identified (Spalding 1993). There is evidence that pelicans rehabilitated from oil spills have lower survival and reproduction rates than control birds during the first 6 months after release (Anderson et al. 1996).

VIII. Neoplasia

Only 2 cases of neoplasia have been reported in pelicans from Florida. A pulmonary adenocarcinoma of alveolar cell origin with possible metastases was diagnosed in the lung of a mature Brown Pelican from Monroe County in 1993 (Woodard 1993). A benign fibroma was removed from the bill of an adult Brown Pelican in Monroe County in 1995 (figure 4.3) (Spalding and Quinn 1995).

IX. Anomalies

A large fluid-filled cyst, possibly an outpouching of the atrium, was found attached to the right atrial wall within the pericardium of an adult Brown Pelican from Brevard County in 1998 (Spalding 1998). It was an incidental finding and did not appear to cause a problem for the pelican.

X. Biotoxins

Botulinum toxin has caused mortality in Brown Pelicans on occasion. More detailed information about this toxin can be found in chapter 10 (Ducks). One Brown Pelican tested negative for botulinum type C from a multi-species die-off from which a Ring-billed Gull tested positive in Brevard County in 1982. Ten pelicans were found dead in that die-off (Stroud 1982). Tissues from a Brown Pelican submitted by a rehabilitation facility on Sanibel Island, Lee County, in 1988 tested positive for botulinum toxin type C (Franson 1988). It could not be determined if this was a wild pelican. Three Brown Pelicans and 10 Common Loons were collected from the beach in Bay County, March 1991. Botulism type E and type C were each diagnosed in one pelican but none of the loons. This was the first report of botulism type E in Florida and in pelicans (Franson 1991). Type C botulinum testing was negative for a number of the die-off events listed in table 4.1 (1985 Wakulla County, 1983 Pinellas County, 1989 Brevard). Approximately 25 sick and dead White Pelicans and 1 Snowy Egret were found in Lake, Marion, and Orange counties during the months of May to July 2001. Botulism type C (and not E) was diagnosed using the mouse protection test from 2 of the White Pelicans tested from Marion County (Spalding 2001).

Mortality from domoic acid poisoning has been reported in Brown Pelicans and Brant's Cormorants (*Phalacrocorax penicillatus*) in California. It is believed that they acquired the poison from eating northern anchovies that contained the toxin produced by a planktonic diatom (Work et al. 1993). No reports of this type of toxicosis were found in records from Florida, nor did we find any reports of birds being tested for this toxin. Although red tide epizootics are common along the west coast of

Table 4.5. Residues of DDE, DDD, and DDT in Brown Pelican tissues collected in Florida

Age Tissue	County/ area	Physical condition	Year(s)	No. exam.	DDE			DDD			DDT			Data Source
					No.[a]	Mean[b]	(Range)	No.[a]	Mean[b]	(Range)	No.[a]	Mean[b]	(Range)	
Nestlings														
Brain	"Florida"	normal	1971–72	13	NG	0.05[c]	(0.01–0.14)	NG	0.02[c]	(0.01–0.08)	NG	0.03[c]	(0.002–0.08)	A
Carcass	"East Coast"	normal	1970	3	3	0.37	(0.10–1.4)	2	0.25	(ND–1.2)	1	0.068	(ND–0.51)	E
	"Florida Bay"	normal	1970	2	2	0.31	(0.29–0.34)	0	ND	—	0	ND	—	E
	"West Coast"	normal	1970	3	3	0.32	(0.13–1.0)	2	0.14	(ND–1.1)	2	0.15	(ND–0.78)	E
Fat	"Florida"	normal	1971–72	3	NG	13[c]	(1.4–36)	NG	3.2[c]	(0.53–8.3)	NG	7.2[c]	(0.11–21)	A
Liver	"Florida"	normal	1971–72	13	NG	0.22[c]	(0.18–1.2)	NG	0.09[c]	(0.01–0.42)	NG	0.036[c]	(0.008–0.21)	A
Muscle	"Florida"	normal	1971–72	13	NG	0.20[c]	(0.008–0.97)	NG	0.08[c]	(0.005–0.54)	NG	0.10[c]	(0.006–0.67)	A
Adults and fledged juveniles														
Brain	Collier	die-off	1974	5	5	0.47	(0.12–0.51)	5	0.046	(0.009–0.514)	5	0.12	(0.024–1.3)	D
	"Florida"	normal	1971–72	12	NG	0.03[c]	(0.002–0.02)	NG	0.006[c]	(0.001–0.09)	NG	0.004[c]	(0.001–0.18)	A
		sick	1971–72	5	NG	1.4[c]	(0.008–5.6)	NG	0.27[c]	(0.01–0.82)	NG	0.73[c]	(0.07–2.5)	A
		unknown	1971–72	1	1	0.03[c]	—	1	0.001[c]	—	1	1.1[c]	—	A
	"Florida Bay"	dead	1972	4	4	0.3	(0.1–0.9)	0	ND	—	0	ND	—	B
	Martin	dead	1972	5	5	2.2	(0.4–5.3)	4	0.5	(0.3–0.8)	4	0.3	(0.2–0.5)	B
	Monroe	sick[d]	1970	1	1	11	—	1	2.2	—	1	1.3	—	E
Carcass[e]	"East Coast"	normal	1970	2	2	11	(8.0–15)	2	1.2	(0.76–1.9)	1	0.13	(ND–0.69)	E
	"Florida Bay"	normal	1970	2	2	2.1	(1.2–3.6)	2	0.52	(0.38–0.70)	1	0.97	(ND–0.38)	E
		dead	1972	4	4	1.3	(0.9–1.7)	2	0.3	(0.2–0.6)	1	0.2	—	B
	Martin	dead	1972	5	5	7.9	(5.8–10)	5	1.7	(1.3–2.0)	5	0.7	(0.5–0.9)	B
	Monroe	sick[d]	1970	1	1	11	—	1	2.8	—	1	0.31	—	E

	"West Coast"	normal	1970	2	2	17	(9.6–31)	2	1.1	(0.63–1.9)	1	0.17	(ND–1.2)	E
		sick	1970	1	1	4.6	—	1	0.84	—	0	ND	—	E
Fat	"Florida"	normal	1971–72	10	NG	5.6[c]	(0.04–60)	NG	3.3[c]	(0.001–11)	NG	1.9[c]	(2.4–8.0)	A
		sick	1971–72	1	1	17[c]	—	1	8.0[c]	—	1	0.94[c]	—	A
		unknown	1971–72	1	1	21[c]	—	1	4.3[c]	—	1	10[c]	—	A
Liver	Collier	die-off	1974	4	4	0.59	(0.21–2.8)	4	0.056	(0.019–0.28)	4	0.13	(0.047–0.75)	D
	"Florida"	normal	1971–72	12	NG	0.40[c]	(0.001–1.5)	NG	0.17[c]	(0.001–0.60)	NG	0.12[c]	(0.01–0.80)	A
		sick	1971–72	5	NG	2.5[c]	(0.01–10)	NG	1.9[c]	(0.04–4.7)	NG	0.68[c]	(0.02–1.9)	A
		unknown	1971–72	3	NG	0.11[c]	(0.11)	NG	0.07[c]	(0.04–0.09)	NG	5.1[c]	(0.009–10)	A
Muscle	Collier	die-off	1974	5	5	1.0	(0.37–2.8)	5	0.13	(0.032–0.85)	5	0.30	(0.095–2.0)	D
	"Florida"	normal	1971–72	12	NG	1.4[c]	(0.002–4.8)	NG	0.58[c]	(0.01–0.68)	NG	5.8[c]	(0.01–2.4)	A
		sick	1971–72	5	NG	3.3[c]	(0.26–12)	NG	2.0[c]	(0.60–4.2)	NG	1.3[c]	(0.09–3.7)	A
		unknown	1971–72	3	NG	0.63[c]	(0.43–0.83)	NG	0.09[c,f]	(0.10–0.70)[f]	NG	0.21[c]	(0.17–0.26)	A
Unspecified age														
Brain	Dade	dead	1980	1	1	1.3	—	1	0.16	—	ND	—	—	C
	"Florida"	sick	1971–72	22	NG	4.9[c]	(0.001–91)	NG	1.1[c]	(0.001–13)	NG	0.38[c]	(0.009–24)	A
Carcass	Dade	dead	1980	1	1	0.48	—	1	ND	—	1	ND	—	C
Fat	"Florida"	sick	1971–72	3	NG	7.5[c]	(0.28–20)	NG	4.8[c]	(0.29–13)	NG	2.1[c]	(0.55–5.0)	A
Liver	"Florida"	sick	1971–72	22	NG	2.0[c]	(0.001–14)	NG	0.80[c]	(0.002–4.0)	NG	0.66[c]	(0.01–3.3)	A
Muscle	"Florida"	sick	1971–72	22	NG	4.4[c]	(0.10–19)	NG	2.2[c]	(0.09–8.5)	NG	1.4[c]	(0.004–5.0)	A

Sources: A = Nesbitt et al. (1981), B = Blus et al. (1977), C = Schmeling (1980), D = Forrester & Thompson (1974), E = Blus et al. (1974b).

ND = not detected, NG = not given.

a. Number with residue.

b. Geometric mean, ppm wet weight unless otherwise indicated.

c. Data presented from Nesbitt et al. (1981) are arithmetic means.

d. This bird died in tremors.

e. Brain, skin, feet, wings, liver, kidney, and gastrointestinal tract removed.

f. Mean given is smaller than lowest value of range given. This is assumed to be a typographical error and the data are represented as given in source.

Table 4.6. Other pesticide residues in Brown Pelican tissues collected in Florida[f]

Age Tissue	County/ area	Physical condition	Year(s)	No. exam.	Dieldrin No.[a]	Dieldrin Mean[b]	Dieldrin (Range)	PCBs No.[a]	PCBs Mean[b]	PCBs (Range)	Heptachlor epoxide No.[a]	Heptachlor epoxide Mean[b]	Heptachlor epoxide (Range)	Data Source
Nestlings														
Brain	"Florida"	normal	1971–72	13	NG	0.02[c]	(0.002–0.04)	NG	0.56[c]	(0.04–1.5)	NA	—	—	A
Carcass	"East Coast"	normal	1970	3	2	0.16	(ND–0.44)	3	1.1	(TR–7.5)	NA	—	—	E
	"Florida Bay"	normal	1970	2	1	0.043	(ND–0.13)	2	TR	—	NA	—	—	E
	"West Coast"	normal	1970	3	1	0.057	(ND–0.29)	3	0.59	(TR–3.5)	NA	—	—	E
Fat	"Florida"	normal	1971–72	3	NG	1.4[c]	(0.38–3.2)	NG	53[c]	(4.6–145)	NA	—	—	A
Liver	"Florida"	normal	1971–72	13	NG	0.3[c]	(0.002–0.08)	NG	0.67[c]	(0.09–1.8)	NA	—	—	A
Muscle	"Florida"	normal	1971–72	13	NG	0.02[c]	(0.001–0.10)	NG	1.2[c]	(0.09–9.4)	NA	—	—	A
Adults and fledged juveniles														
Brain	Collier	die-off	1974	5	5	0.093	(0.019–3.4)	5	1.3	(0.31–5.5)	NA	—	—	D
	"Florida"	normal	1971–72	12	NG	0.008[c]	(0.001–0.01)	NG	0.32[c]	(0.01–2.5)	NA	—	—	A
		sick	1971–72	5	NG	0.37[c]	(0.07–2.5)	NG	3.8[c]	(0.45–11)	NA	—	—	A
		unknown	1971–72	1	1	0.01[c]	—	1	0.001[c]	—	NA	—	—	A
	"Florida Bay"	dead	1972	4	2	0.2	(0.1–0.3)	4	0.4	(0.3–0.9)	0	ND	—	B
	Martin	dead	1972	5	4	0.7	(0.5–1.0)	5	3.8	(0.5–9.4)	2	0.12	(0.12–0.13)	B
	Monroe	sick	1970	1	1	1.8	—	1	10	—	1	TR	—	E
Carcass[d]	"East Coast"	normal	1970	2	2	0.99	(0.61–1.6)	2	19	(12–29)	NA	—	—	E
	"Florida Bay"	normal	1970	3	2	0.11	(ND–0.47)	3	2.4	(1.0–4.4)	NA	—	—	E
		dead	1972	4	4	0.2	(0.2–0.4)	4	1.4	(0.9–1.8)	1	0.2	—	B
	Martin	dead	1972	5	5	1.1	(0.9–1.2)	5	9.9	(5.0–25)	3	0.1	(0.1–0.2)	B
	Monroe	sick	1970	1	1	1.2	—	1	20	—	1	TR	—	E
	"West Coast"	normal	1970	2	2	1.2	(0.74–1.8)	2	25	(20–34)	NA	—	—	E
		sick	1970	1	1	0.60	—	1	19	—	NA	—	—	E
Fat	"Florida"	normal	1971–72	10	NG	0.87[c]	(0.008–2.6)	NG	37[c]	(0.14–123)	NA	—	—	A
		sick	1971–72	1	1	11[c]	—	1	31[c]	—	NA	—	—	A
		unknown	1971–72	1	1	0.99[c]	—	1	61[c]	—	NA	—	—	A
Liver	Collier	die-off	1974	4	4	0.12	(0.023–3.7)	4	1.4	(0.44–4.3)	NA	—	—	D
	"Florida"	normal	1971–72	12	NG	0.06[c]	(0.10–0.24)	NG	1.4[c]	(0.29–6.4)	NA	—	—	A
		sick	1971–72	5	NG	0.43[c]	(0.10–1.6)	NG	18[c]	(0.75–62)	NA	—	—	A
		unknown	1971–72	3	NG	0.55[c]	(0.03–0.99)	NG	1.1[c]	(0.99–1.2)	NA	—	—	A
Muscle	Collier	dieoff	1974	5	5	0.14	(0.056–0.33)	5	2.6	(0.20–22)	NA	—	—	D
	"Florida"	normal	1971–72	12	NG	0.29[c]	(0.008–1.7)	NG	6.3[c]	(0.40–17)	NA	—	—	A
		sick	1971–72	5	NG	1.3[c]	(0.04–4.4)	NG	23[c]	(3.3–70)	NA	—	—	A
		unknown	1971–72	3	NG	0.02[c]	(0.001–0.04)	NG	2.7[c]	(1.6–3.7)	NA	—	—	A
Unspecifed age														
Brain	"Florida"	sick	1971–72	22	NG	0.68[c]	(0.002–9.0)	NG	9.3[c]	(0.10–120)	NA	—	—	A
	Dade	—	1980	1	1	0.11	—	1	4.2	—	1	ND	—	C

Tissue	Location	Condition	Year	n	(a)	(b)	Range	(a)	(b)	Range	(a)	(b)	Range	Source
Carcass	Dade	—	1980	1	1	ND	—	1	1.8	—	1	ND	—	C
Fat	"Florida"	sick	1971–72	3	NG	1.5[c]	(0.30–2.0)	NG	53[c]	(10–125)	NA	—	—	A
Liver	"Florida"	sick	1971–72	22	NG	0.33[c]	(0.008–2.3)	NG	23[c]	(0.67–215)	NA	—	—	A
Muscle	"Florida"	sick	1971–72	22	NG	0.81[c]	(0.009–4.0)	NG	29[c]	(2.0–129)	NA	—	—	A
					cis-chlordane			*trans*-nonachlor			*cis*-nonachlor			
Adults and fledged juveniles														
Carcass[d]	"Florida Bay"	dead	1972	4	1	0.2	—	NA	—	—	0	ND	—	B
	Martin	dead	1972	5	0	ND	—	NA	—	—	0[e]	ND	—	B
	"Florida Bay"	dead	1972	4	3	0.3	(0.1–1.3)	NA	—	—	2	0.1	(0.1–0.2)	B
	Martin	dead	1972	5	1	0.1	—	NA	—	—	0[e]	ND	—	B
Unspecified age														
Brain	Dade	—	1980	1	1	ND	—	1	0.26	—	1	0.14	—	C
Carcass	Dade	—	1980	1	1	ND	—	1	ND	—	1	ND	—	C
					Mirex			Toxaphene						
Nestlings														
Carcass	"East Coast"	normal	1970	3	0	ND	—	NA	—	—	—	—	—	E
	"Florida Bay"	normal	1970	2	0	ND	—	NA	—	—	—	—	—	E
	"West Coast"	normal	1970	3	0	ND	—	NA	—	—	—	—	—	E
Adults and fledged juveniles														
Brain	"Florida Bay"	dead	1972	4	0	ND	—	1	0.3	—	—	—	—	B
	Martin	dead	1972	5	1	0.1	—	NA	—	—	—	—	—	B
Carcass[d]	"East Coast"	normal	1970	2	2	0.20	(0.20–0.20)	NA	—	—	—	—	—	E
	"Florida Bay"	normal	1970	3	0	ND	—	NA	—	—	—	—	—	E
		dead	1972	4	1	0.1	—	2	0.2	(0.2–0.3)	—	—	—	B
	Martin	dead	1972	5	3	0.3	(0.2–0.4)	NA	—	—	—	—	—	B
	"West Coast"	normal	1970	2	2	0.66	(0.30–1.4)	NA	—	—	—	—	—	E
		sick	1970	1	1	0.31	—	NA	—	—	—	—	—	E
Unspecified age														
Brain	Dade	—	1980	1	1	ND	—	1	ND	—	—	—	—	C
Carcass	Dade	—	1980	1	1	ND	—	1	ND	—	—	—	—	C

Sources: A = Nesbitt et al. (1981), B = Blus et al. (1977), C = Schmeling (1980), D = Forrester & Thompson (1974), E = Blus et al. (1974b).

NA = not analyzed, ND = not detected, NG = not given, TR = trace.

a. Number with residue.

b. Geometric mean, ppm wet weight unless otherwise indicated.

c. Data presented from Nesbitt et al. (1981) are arithmetic means.

d. Brain, skin, feet, wings, liver, kidney, and gastrointestinal tract removed.

e. Sample size = 1.

f. In addition to the compounds listed, Blus et al. (1977) did not detect oxychlordane in any of the samples. Schmeling (1980) did not detect oxychlordane, endrin or HCB.

Table 4.7. Metal concentrations in eggs of Brown Pelicans collected in Florida

County or area	Year(s)	No. exam.	Lead Mean[a]	(Range)	Mercury Mean[a]	(Range)	Selenium Mean[a]	(Range)	Data Source
Brevard	1969	10	TR	—	0.38	(0.20–1.5)	NA	—	A
"Florida"	1969–70	6	0.03	(0.01–0.05)	0.39	(0.20–0.84)	0.28	(0.19–0.38)	A
"Florida Bay"	1970	20	TR	—	0.48	(0.11–0.98)	NA	—	A
	1972	8	<0.02	—	0.34	(0.12–0.65)	NA	—	B
Lee	1969	15	TR	—	0.55	(0.16–1.5)	NA	—	A
Levy	1969	7	TR	—	0.32	(0.16–0.49)	NA	—	A
Pinellas	1969	5	TR	—	0.47	(0.17–1.2)	NA	—	A
St. Lucie	1969	7	TR	—	0.46	(0.16–0.70)	NA	—	A
Volusia	1969	5	TR	—	0.28	(0.12–0.48)	NA	—	A

County or area	Year(s)	No. exam.	Cadmium Mean	(Range)	Copper Mean	(Range)	Chromium Mean	(Range)	Data Source
"Florida"	1969–70	6	ND	(ND–0.01)	0.97	(0.78–1.1)	0.01	(ND–0.07)	C
"Florida Bay"	1972	8	ND	<0.05	0.97	(0.72–1.1)	NA	—	B

County or area	Year(s)	No. exam.	Arsenic Mean	(Range)	Zinc Mean	(Range)	Magnesium Mean	(Range)	Data Source
"Florida"	1969–70	6	0.1	(0.07–0.18)	6.4	(4.3–8.3)	74	(42–107)	C
"Florida Bay"	1972	8	<0.1	—	6.7	(4.7–9.3)	NA	—	B

County or area	Year(s)	No. exam.	Nickel Mean	(Range)	Data Source
"Florida"	1969–70	6	0.02	(0.01–0.04)	C

Sources: A = Blus et al. (1974b), B = Ogden et al. (1974), C = Blus et al. (1977).
NA = not analyzed, ND = not detected, TR = trace (<0.1 ppm).
a. Geometric mean, ppm wet weight.

Florida (Quick and Henderson 1975), we found no reports of pelican mortality associated with these events.

Hartman (1946) captured a paralyzed White Pelican on Lake Okeechobee in April 1945. The stomach contained numerous parasites and he proposed that toxins from the parasites might have caused the paralysis. We have found no further reports of intestinal parasites producing a toxin.

XI. Viruses

Viral diseases are poorly studied in pelicans. The only data available are for mosquito-transmitted arboviruses. Four of 42 (10%) adult and juvenile Brown Pelicans at a rehabilitation center in Lee County in 1978 were seropositive for St. Louis encephalitis virus (Spalding et al. 1994). In a study to identify birds bitten by mosquitoes, pelicaniform blood was identified in <0.1% of blood meals collected from *Culiseta melanura* in Indian River County (Edman et al. 1972). Virus isolation attempts were negative for 2 Brown Pelicans from Brevard County in 1989 (Locke 1989), and for 2 from Collier County in 1974 (Forrester 1974).

The only testing for virus in White Pelicans was by Johnson and Spalding (1999), who, in a survey of Double-crested Cormorants for Newcastle Disease virus, included 6 White Pelicans: 5 from Orange County and 1 from Brevard County. All but 1 of the birds from Orange County had low positive hemagglutination inhibition titers ranging from 1:5 to 1:20 using La Sota virus antigen. No virus was isolated and immunohistochemistry (both Q24 and anti-NP

Table 4.8. Metal concentrations in tissues of Brown Pelicans collected in Florida

Tissue	County or area	Year(s)	No. exam.	Lead Mean[a]	(Range)	Mercury Mean[a]	(Range)	Selenium Mean[a]	(Range)	Data Source
Bone	"Tampa Bay"	1969	5	23[b,c]	(20–26)[c]	0.28[b]	(0.15–0.41)	NA	—	A
Brain	Dade	1973	1	NA	—	5.6	—	NA	—	B
	Pinellas	1975	1	NA	—	15[i]	—	NA	—	B
Liver	Bay	1991	3	NA	—	1.9	(1.5–3.1)	NA	—	C
	Brevard	1989	1	0.09	—	NA	—	NA	—	I
		1998	2	ND	—	6.9	(2.7–18)	NA	—	D
	Dade	1973	1	NA	—	36	—	NA	—	B
	"Florida"	1970	1[d]	0.10	—	4.1	—	3.4	—	E
		1970	1[e]	0.10	—	1.7	—	4.4	—	E
	"Florida Bay"	1973	1[f]	0.21	—	6.2	—	4.0	—	E
		1984	1	0.45	—	NA	—	NA	—	F
	Martin	1972	1[g]	0.10	—	6.3	—	2.8	—	E
	Pinellas	1973	1	NA	—	13	—	NA	—	B
		1975	1	NA	—	2.4[i]	—	NA	—	B
		1983	6	0.48	(0.35–0.74)	1.5	(ND–5.8)	1.8	(1.2–2.3)	G
	"Tampa Bay"	1969	5[h]	NA	—	9.7[b]	(5.1–17)	NA	—	A
	Wakulla	1985	1	0.45	—	NA	—	NA	—	H
Muscle	Dade	1973	1	NA	—	5.2	—	NA	—	B
	Pinellas	1975	1	NA	—	0.09[i]	—	NA	—	B
	"Tampa Bay"	1969	5	NA	—	1.8[b]	(1.0–2.3)	NA	—	A

Tissue	County or area	Year(s)	No. exam.	Cadmium Mean	(Range)	Copper Mean	(Range)	Chromium Mean	(Range)	Data Source
Bone	"Tampa Bay"	1969	5	1.7[b,c]	(1.4–1.9)[c]	2.8[b,c]	(0.30–6.2)[c]	15[b,c]	(8.8–23)[c]	A
Liver	Brevard	1998	2	NA	—	NA	—	0.99	(0.63–1.6)	D
	"Florida"	1970	1[d]	1.06	—	6.7	—	0.11	—	E
		1970	1[e]	0.21	—	7.5	—	0.065	—	E
	"Florida Bay"	1973	1[f]	0.23	—	9.0	—	0.05	—	E
	Martin	1972	1[g]	0.73	—	5.4	—	0.07	—	E
	Pinellas	1983	3[j]	0.50	(0.34–0.62)	6.0	(3.1–10)	0.40	(0.11–1.3)	G
	"Tampa Bay"	1969	5	1.8[b,c]	(1.3–2.4)[c]	26[b,c]	(19–48)[c]	0.92[b,c]	(0.8–1.2)[c]	A
Muscle	"Tampa Bay"	1969	5	0.28[b,c]	(0.25–0.32)[c]	17[b,c]	(14–23)[c]	4.0[b,c]	(2.6–4.7)[c]	A

Tissue	County or area	Year(s)	No. exam.	Arsenic Mean	(Range)	Zinc Mean	(Range)	Magnesium Mean	(Range)	Data Source
Bone	"Tampa Bay"	1969	5	NA	—	160[b,c]	(140–170)[c]	NA	—	A
Liver	Brevard	1998	2	NA	—	57	(38–87)[c]	NA	—	D
	"Florida"	1970	1[d]	0.63	—	55	—	220	—	E
		1970	1[e]	0.89	—	32	—	230	—	E
	"Florida Bay"	1973	1[f]	0.54	—	50	—	180	—	E
	Martin	1972	4[g]	0.47	—	41	—	210	(180–230)	E
	"Tampa Bay"	1969	5	NA	—	120[b,c]	(110–140)[c]	NA	—	A
Muscle	"Tampa Bay"	1969	5	NA	—	58[b,c]	(47–70)[c]	NA	—	A

Tissue	County or area	Year(s)	No. exam.	Nickel Mean	(Range)	Silver Mean	(Range)	Cobalt Mean	(Range)	Data Source
Bone	"Tampa Bay"	1969	5	20[b,c]	(15–26)[c]	2.5[b,c]	(2.0–3.1)[c]	6.1[b,c]	(4.8–7.3)[c]	A
Liver	"Florida"	1970	1[d]	0.058	—	NA	—	NA	—	E
		1970	1[e]	0.048	—	NA	—	NA	—	E

(continued)

Table 4.8. *(continued)*

Tissue	County or area	Year(s)	No. exam.	Nickel Mean[a]	(Range)	Silver Mean[a]	(Range)	Cobalt Mean[a]	(Range)	Data Source
	"Florida Bay"	1973	1[f]	0.05	—	NA	—	NA	—	E
	Martin	1972	1[g]	0.02	—	NA	—	NA	—	E
	"Tampa Bay"	1969	5	<2.0	—	NA	—	NA	—	A
Muscle	"Tampa Bay"	1969	5	4.4[b,c]	(3.7–5.2)[c]	NA	—	NA	—	A

Sources: A = Connors et al. (1972), B = Gourlie (1984), C = Franson (1991), D = Spalding (1998), E = Blus et al. (1977), F = Stroud (1984), G = Stroud (1983), H = Stroud (1985), I = Langenberg (1989).
NA = not analyzed, ND = not detected.
a. Geometric mean, ppm wet weight unless otherwise indicated.
b. Arithmetic mean.
c. Dry weight.
d. Adult female bird that was shot.
e. Immature male bird that was shot.
f. Immature male that died from gunshot.
g. Includes one adult female that died from enteritis.
h. Found dead.
i. The value reported for mercury in brain of this individual is exceptionally high and it is suspected that it was exchanged with the muscle value (0.09) or liver value (2.4) in error. The bird was immature.
j. In addition to the compounds listed, Stroud (1983) also measured thallium (ND, 0.70, and 0.67) in the livers of these 3 pelicans from Pinellas County.

antibodies) was negative for the presence of virus (Brown 2000). See chapter 5, Cormorants and Anhingas, for additional information about Newcastle Disease virus.

Aransas Bay virus, an RNA virus from the Upolu serogroup, was isolated from ticks (*Ornithodoros capensis*) collected from Brown Pelican nests in Texas (Yunker et al. 1979). We found no reports of testing for this virus in Florida. See also chapter 6, Herons, Egrets, and Bitterns, for additional information.

XII. Bacteria

Bacteria isolated from Brown Pelican tissues are listed in table 4.9. Most of these data came from the investigation of a die-off that occurred in Pinellas County during January–June 1983 (Ankerberg 1983; Stroud 1983; Forrester 1983). The species involved included Brown Pelicans, Double-crested Cormorants, various unspecified gulls, Great Blue Herons, Great Egrets, and Cattle Egrets. Approximately 400 birds died. A municipal sewage lift station had released raw sewage into Boca Ciega Bay and this was suspected to be the cause of the die-off. Many of the bacterial organisms cultured from water and sediment in the area were the same as those cultured from the pelicans; however, a definitive link to the cause of death was not made. Several of the fresher pelicans examined had enteritis that may have been due to bacterial infection, but most were too autolyzed or had been frozen, obscuring the lesions.

White et al. (1973) examined 42 Brown Pelicans from Florida in 1971–72, and isolated *Edwardsiella tarda* from the intestines of 3 with hemorrhagic enteritis, and from the lungs and liver of another that was too autolyzed to determine the cause of death. *Salmonella litchfield* was also cultured from 1 of the Brown Pelicans with hemorrhagic enteritis. *Edwardsiella tarda* was cultured from sick fish and from fresh surface water in north central Florida (White et al. 1973). In 1974, *E. tarda* was isolated from the large intestine of both normal and sick birds associated with a die-off in Collier County (Forrester and White 1974). Cell-associated hemolysins produced by *E. tarda* that were isolated from a

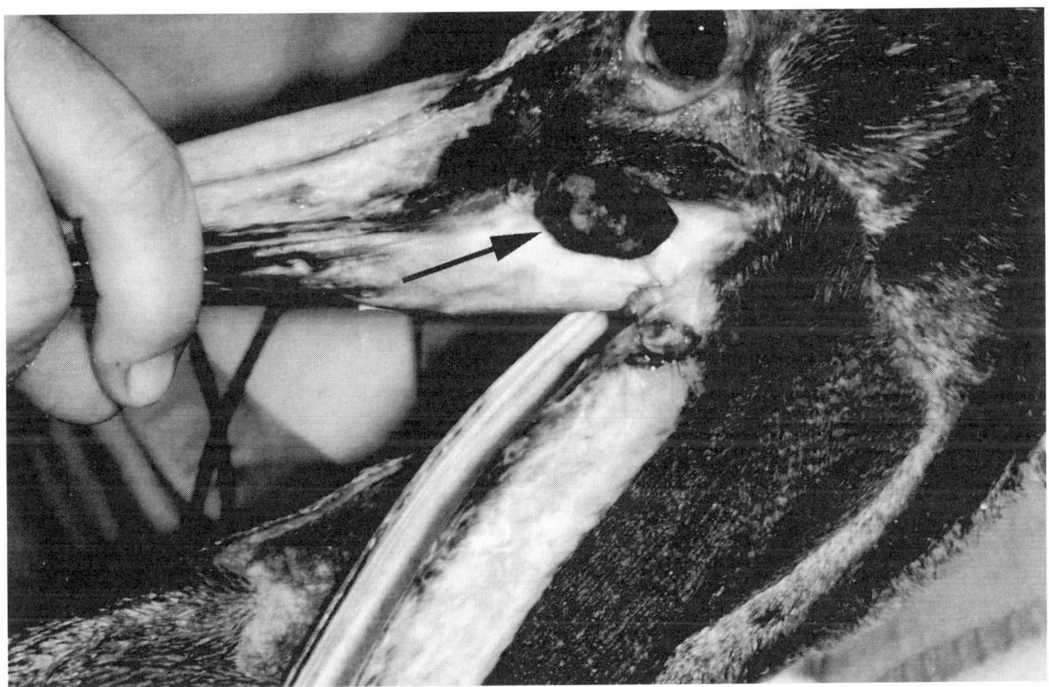

FIGURE 4.3. Fibroma on the base of the bill of a Brown Pelican. Courtesy of Laura Quinn.

Brown Pelican in Florida were characterized by Watson and White (1979).

Erysipelothrix rusiopathiae was determined to be the cause of morbidity and mortality in 500 Brown Pelicans from Duval County in February 1992 (Spalding 1992). The pelicans were weak, reluctant to enter the water, and easily caught. A diesel spill had occurred 1 week earlier, and local use of a fire retardant was suspected. Although feathers appeared to be more wettable than normal, the relevance of that spill to the die-off could not be documented. At necropsy examination, there was evidence of septicemia, petechial hemorrhages in many tissues, especially the adrenal gland (figure 4.4), and thrombosis. Birds in this area frequently had fish scraps available to them. Exposed bones from fish scraps probably cause injuries to the pouch and esophagus with subsequent bacterial infection. In 1998 a similar die-off occurred at the same location (Brannian 1998). Again, fire retardant was suspected. Bacteria associated with splenic necrosis were noted; however, tissues were not cultured. In California in 1988 a similar erysipelas die-off involved 400 Brown Pelicans and was associated also with the ingestion of fish scraps (Hunter 1988). Fish in the diet was also suspected to cause septicemia with *E. rusiopathiae* in a captive Bald Eagle (Franson et al. 1994).

Five species of *Salmonella* have been identified from Brown Pelicans in Florida. In one case from Florida Bay in 1972, necrotic enteritis was observed; however, *Edwardsiella tarda* was also isolated. Interestingly, none of the species identified overlap with the species found in Double-crested Cormorants (see chapter 5, Cormorants and Anhingas).

We could find only 1 White Pelican that was tested for bacteria. *Edwardsiella tarda* and nonenteric gram-negative rods were cultured from the serosal surface of the intestine of a White Pelican with a ruptured intestine and acute peritonitis (Spalding 2000). It was suspected that the pelican had been attacked by an alligator since it also had a broken tail.

Table 4.9. Bacteria and fungi isolated from Brown Pelicans in Florida

Class	Species	County or area	Year	Organ or tissue	Data source
Bacteria					
	Acinetobacter calcaoaceticus var. *anitratum*	Pinellas	1983	Liver	A
	Acinetobacter sp.	Pinellas	1983	Heart, liver, lungs, mesentery	A,B
	Aeromonas hydrophila	Brevard	1998	Liver[a]	C
		Pinellas	1983	Liver, intestine[c]	A
	Bacillus cereus	Pinellas	1983	Heart, blood	B
	Bacillus sp.	Pinellas	1983	Liver	B
	Citrobacter sp.	Pinellas	1983	Liver, stomach	B
	Clostridium beijerinckii	Pinellas	1983	Intestine, stomach	B
	Clostridium hastiforme	Pinellas	1983	Feces	B
	Clostridium histolyticum	Pinellas	1983	Feces	B
	Clostridium limosum	Pinellas	1983	Feces	B
	Clostridium perfringens	Martin	1972	Intestine[a]	M
		Pinellas	1983	Intestine[c], stomach, feces, mesentery	A,B
	Clostridium ramosum	Pinellas	1983	Spleen, stomach	B
	Clostridium spp.	Pinellas	1983	Feces, stomach	B
	Clostridium subterminale	Pinellas	1983	Feces	B
	Diphtheroid	Volusia	1988	Conjunctiva[a]	D
	Edwardsiella sp.	Pinellas	1983	Mesentery	A
	Edwardsiella tarda	Brevard	1971	Liver, lung	F
			1986	Intestine	G
		Charlotte	1972	Large intestine	E
		Collier	1972	Liver[a,b]	E
			1974	Large intestine, liver	E
		Duval	1972	Intestine[a]	F,E
		"Florida Bay"	1972	Intestine[a,b]	F,E
		Lee	1973	Large intestine	E
		Martin	1972	Large intestine	K
		Pinellas	1973	Intestine	E
			1983	Granulomas in muscle[a], intestine[c], liver	A
	Enterobacter cloacae	Pinellas	1983	Liver	A,B
	Enterobacter sp.	Pinellas	1983	Liver, heart, lungs, mesentery	A,B
	Enterococcus Group D	Pinellas	1983	Mesentery, stomach, blood	A,B
	Erysipelothrix rusiopathiae	Duval	1992	Liver[a], kidney[a] (septicemia)	H
	Escherichia coli	Pinellas	1983	Heart, blood, liver, feces, stomach, lungs, mesentery, intestine[c]	A,B
		Volusia	1988	Conjunctiva[a]	D
	Escherichia coli (hemolytic)	Pinellas	1983	Heart, stomach, liver	B
	Gaffkya anaerobia	Pinellas	1983	Liver	A
	Klebsiella pneumoniae	Brevard	1987	Lung	I
	Klebsiella sp.	Pinellas	1983	Heart, liver, lungs	B
	Listonella damsela	Brevard	1998	Liver[a]	C
	Peptostreptococcus	Pinellas	1983	Spleen, stomach	B
	Plesiomonas shigalloides	Pinellas	1983	Mesentery, intestine[c]	A,B
	Proteus mirabilis	Pinellas	1983	Feces	B
	Proteus sp.	Pinellas	1983	Stomach, liver	B
		Volusia	1988	Conjunctiva[a]	D
	Providencia sp.	Pinellas	1983	Heart, mesentery, stomach	B

(continued)

Table 4.9. *(continued)*

Class	Species	County or area	Year	Organ or tissue	Data source
	Pseudomonas florescens	Pinellas	1983	Liver	A
	Pseudomonas putrificiens	Pinellas	1983	Liver	A
	Pseudomonas sp.	Pinellas	1983	Heart, blood, liver, lungs	A,B
		Volusia	1988	Conjunctiva[a]	D
	Salmonella bareilly	Duval	1972	Large intestine	E
	Salmonella hartford	Indian River	1972	Cloaca	E
	Salmonella litchfield	"Florida Bay"	1972	Large & small intestine[a,b]	F,E
	Salmonella miami	Collier	1974	Feces	E
	Salmonella poona	Lee	1973	Large intestine	E
	Salmonella sp. (type B)	Pinellas	1983	Heart, liver, lungs	B
	Salmonella sp.	Collier	1974	Large intestine	E
		Pinellas	1975	Intestine	E
			1983	Stomach	B
	Staphylococcus aureus	Brevard	1998	Spleen (septicemia and endocarditis)	C
	Staphylococcus sp.	Pinellas	1983	Intestine, liver, spleen, heart, lungs	B
			1983	Granulomas in muscle	A
	Streptococcus fecalis	Pinellas	1983	Granulomas in muscle	A
	Streptococcus intermedius	Pinellas	1983	Mesentery, intestine[c]	A
	α *Streptococcus*	Pinellas	1983	Stomach, liver, heart, lungs	B
	α *Streptococcus* (not group D)	Pinellas	1983	Mesentery, feces	B
	α hemolytic *Streptococcus* sp.	Volusia	1988	Conjunctiva[a]	D
	Streptococcus sp.	Pinellas	1983	Granulomas in muscle	A
	Vibrio alginoltyicus	Pinellas	1983	Mesentery, intestine	A,B
	Vibrio damsela	Brevard	1987	Lung	I
	Vibrio fluvialis	Brevard	1987	Lung	I
	Vibrio parahemolyticus	Pinellas	1983	Mesentery, intestine[c]	A,B
	Vibrio spp.	Pinellas	1983	Intestine, heart, liver, lungs	B
	Yersinia enterocolitica	Brevard	1987	Lung	I
		Pinellas	1983	Intestine	A
Fungi					
	Candida tropicalis	Pinellas	1983	Pouch[a]	A
	Aspergillus niger	Pinellas	1983	Granulomas in muscle	A
	Aspergillus sp.	Collier	1974	Not specified	L
		"Florida Bay"	1973	Air sac[a]	J
		St. Johns	1998	Air sac[a]	C

Sources: A = Stroud (1983), B = Ankerberg (1983), C = Spalding (1998), D = Riggs (1988), E = Forrester & White (1974), F = White et al. (1973), G = Duncan (1986), H = Spalding (1992), I = Roffe (1987), J = Blus et al. (1977), K = Forrester & White (1972), L = Forrester (1974), M = Locke (1972).

a. Associated with changes in tissue.

b. Both *S. litchfield* and *E. tarda* were isolated from the intestine of this bird.

c. Of 4 pelicans examined, one had severe bacterial/parasitic enteritis and hepatitis and another had moderate bacterial enteritis. No specific etiologic agent was determined. *Candida tropicalis* was cultured from the pouch, which had proliferative lesions. Poxvirus was suspected but not confirmed.

XIII. Fungi

Three species of fungi have been identified from pelicans in Florida (table 4.9). The most frequently reported is *Aspergillus* sp., commonly presumed to be *A. fumigatus,* but generally the diagnosis is not pursued past the gross or histologic observation. A diagnosis of aspergillosis is frequently made in emaciated Brown Pelicans in rehabilitation centers; how-

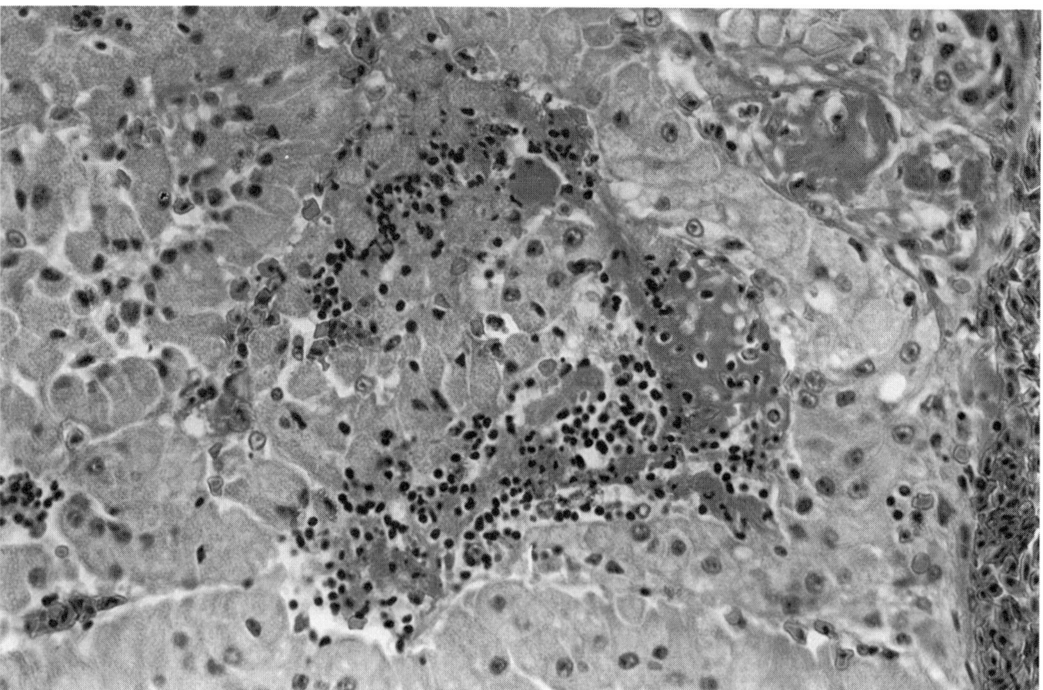

FIGURE 4.4. Hemorrhage and necrosis in the adrenal gland of a Brown Pelican with erysipilas.

ever, the listed cases are only those reported from birds that died in the wild, or shortly after being taken into captivity. All cases involve individual birds that were part of a die-off but in no case was aspergillosis a consistent finding for the die-off. In addition to the usual finding of fungal colonies growing in air sacs or forming granulomas within lungs, Stroud (1983) reported granulomas in muscle in the scapular region in one Brown Pelican. *Aspergillus niger* was cultured from these lesions. Severe aspergillosis involved the air sacs and lungs of 1 of 7 White Pelicans that were examined from a die-off in Orange County (Spalding et al. 1996). *Candida tropicalis* was isolated from proliferative pouch and esophageal lesions of a Brown Pelican from Pinellas County (Stroud 1983). In that case poxvirus was a differential but was not ruled out. Pouch lesions are common in Brown Pelicans and appear to be caused by fish spines; however, they are rarely cultured.

XIV. Protozoans

Evidence of protozoal infections in pelicans is relatively rare compared with other species. Courtney and Ernst (1975) described *Eimeria pelecani* from 4 of 30 nestling Brown Pelicans collected in Lee County in 1972. Unidentified eimerian oocysts were observed in the intestine of a Brown Pelican from Tampa Bay in 1978 (Stroud 1980). Oocysts, probably of this same species, were found in the kidney and feces of an adult pelican from Brevard County in 1998 (Spalding 1998).

Sarcocystis sp. cysts were present in the pectoral muscle of a White Pelican from Marion County in 1999 (Spalding 1999). *Toxoplasma gondii* antibodies were not detected in sera from 16 Brown Pelicans from Florida (dates and locations not given) by Burridge et al. (1979). These were tested by hemagglutination inhibition, which may not be as sensitive in birds as in mammals, resulting in false negatives (Frenkel 1981).

One hundred and six blood smears were examined from Brown Pelicans from Indian River (40), Volusia (19), Duval (1), Brevard (2), Monroe (1), Charlotte (30), Collier (9), Lee (3), and Alachua (1) counties, 1972–1981, and 1 from a juvenile White Pelican from Clay County in 1982. All were negative for blood parasites (Forrester and Bennett 1985).

XV. Helminths

Brown Pelicans from Florida were infected with 14 species of trematodes, 3 cestodes, 3 acanthocephalans, and 12 nematodes (tables 4.10–4.11). The most prevalent helminths were *Phagicola longus, Mesostephanus appendiculatoides,* and several species of *Contracaecum,* all parasites that can be acquired from eating mullet (*Mugil* sp.). Humphrey et al. (1978) investigated the ecology of Brown Pelican helminths reported by Courtney and Forrester (1974). They found low overall helminth diversity and lower diversity in younger birds and in birds translocated to Louisiana. The low overall helminth diversity was due to very large numbers of *M. appendiculatoides* in young birds and large numbers of *P. longus* in older birds. The increasing diversity with age could be due to increasing diversity of food items as birds grow, and the heavy feeding on mullet when birds fledge. Brown Pelican populations on the Gulf coast had more balanced numbers of *M. appendiculatoides* and *P. longus* than populations on the Atlantic coast. The researchers interpreted the very high populations of *P. longus* in translocated Louisiana birds as evidence of a lack of interspecific competition by *M. appendiculatoides.* The decline of *P. longus* populations in adult birds may be due to immunity.

Greve et al. (1985) described lesions caused by *Phagicola longus, Mesostephanus appendiculatoides,* and *Contracaecum* spp. in experimentally infected Brown Pelicans. *Phagicola longus* occasionally penetrated the lamina propria of villi of the small intestine. *Mesostephanus appendiculatoides* was usually attached to the tips of villi and sometimes penetrated the epithelium.

Small ulcers were found at the attachment sites of *Contracaecum* spp. in the distal esophagus and proventriculus. Infected pelicans lost weight compared with control birds; however, the diet fed to the control birds, sardines, was preferred, and control birds had increased food intake that might have confounded the results. They concluded that these parasites were characterized by low virulence; however, in combination with other factors they may play a role in population fluctuations. In 2 cases, renal trematodes were associated with moderate nephrosis in Brown Pelicans that were collected from a die-off in Pinellas County in 1983 (Stroud 1983; Gross 1983). *Renicola thapari* was very common (93%) in the kidneys of Brown Pelicans from the west coast (table 4.10), and almost always found in kidneys examined histologically, but these trematodes were rarely associated with severe disease. Anthelmentic therapy for helminths in Brown Pelicans was investigated by Courtney et al. (1977).

Contracaecum spp. are commonly present in moderate to large numbers in pelican stomachs. Oglesby (1960) suggested that 1,100 *Contracaecum* sp. found in the stomach of an emaciated White Pelican may have caused or contributed to its death. Owre (1962) suggested that the presence of large numbers of these ascaridoid nematodes in the stomach may contribute to the mechanical breakdown of large food masses like fish, and become a pathogen only when ingesta are no longer available, such as during starvation.

Although larvae were not found, lesions that resembled chronic eustrongylidosis, as seen in wading birds, were found in 3 of 4 Brown Pelicans that were part of a die-off in Brevard County in 1998 (Spalding 1998). Since this parasite is found rarely in birds foraging in marine habitats (Spalding et al. 1993), its presence may indicate that these birds fed in fresh water. A description of a large red nematode within fibrinous coils on the serosal surface of the proventriculus of a Brown Pelican from Tampa Bay in 1980 (Stroud 1980) was undoubtedly *Eustrongylides* sp. Unfortunately the nematode was not saved.

Table 4.10. Trematodes collected from Brown Pelicans in Florida

Species (location)[a]	County/area	Year(s)	No. of pelicans			Intensity		Data source
			Examined	Positive	%	Mean	Range	
Ascocotyle sp. (SI)	"East coast"	1971–73	39	1	3	80	—	A
Astrobilharzia sp. (BV)	"East coast"	1971–73	39	21	54	3	1–12	A
Carneophallus turgidus (SI,CE)	"East coast"	1971–73	39	4	10	178	1–650	A
Echinochasmus sp. (SI)	"East coast"	1971–73	39	4	10	12	1–25	A
Galactosomum sp. (SI)	Charlotte	1955–62	NG	NG	—	NG	—	B
Galactosomum darbyi (SI,CE,LI,CL)	"East coast"	1971–73	39[b]	8	20	174	NG	A
	"West coast"	1971–73	14	12	86	56	NG	A
Galactosomum fregatae (SI,CL)	"East coast"	1971–73	39	2	5	2	1–8[c]	A
	"West coast"	1971–73	14	4	29	4	1–8[c]	A
Mesorchis denticulatus[f] (SI,CE,LI,CL)	"East coast"	1971–73	39	28	72	43	NG	A
	"West coast"	1971–73	14	6	43	2	NG	A
Mesostephanus appendiculatoides (SI,CE,LI,CL)	Charlotte & Pinellas	1958	3	3	—	NG	"hundreds"	C
	Pinellas	1983	NG	NG	—	NG	—	D
	"East coast"	1971–73	39	31	79	785	NG	A
	"West coast"	1971–73	14	14	100	1,224	NG	A
Mesostephanus microbursa (SI)	"Florida"	1971–73	99	few	—	NG	"small numbers"	A
Mesostephanus yedeae[d] (SI)	Charlotte		NG	—	—	—	—	E
Micropharyphium facetum (SI)	"East coast"	1971–73	39	1	3	1	—	A
Phagicola longa[e] (SI,CE,LI,CL)	Charlotte	1955–62	NG	NG	—	NG	—	B,C
	"East coast"	1971–73	39	39	100	9,058	NG	A
	Pinellas	1983	NG	NG	—	NG	—	D
	"West coast"	1971–73	14	14	100	3,677	NG	A
Phagicola sp. (SI)	"Florida Bay"	1984	1	1	—	NG	—	F
Pholeter sp. (SI)	"East coast"	1971–73	NG	1	1	16	—	A
Renicola thapari (KD)	"East coast"	1971–73	39	13	33	20	1–1,500[c]	A
	"West coast"	1971–73	14	13	93	201	1–1,500[c]	A

Sources: A = Courtney & Forrester (1974), B = Hutton (1964), C = Hutton & Sogandares-Bernal (1960), D = Ankerberg (1983), E = Dennis (1967), F = Stroud (1984). NG = not given.

a. Location in host: BV = blood vessels, CE = cecum, CL = cloaca, KD = kidney, LI = large intestine, SI = small intestine.
b. These specimens discussed by Pearson (1973) and Pearson and Courtney (1977).
c. Range for east and west coasts combined.
d. The gastropod host *Cerithium literatum* was collected in Sarasota County (Dennis 1967)
e. A complex of two species: *P. longus* and *Phagicola* sp. like *minutus*.
f. *Stephanophora denticulata* is a synonym of *Mesorchis denticulatus*.

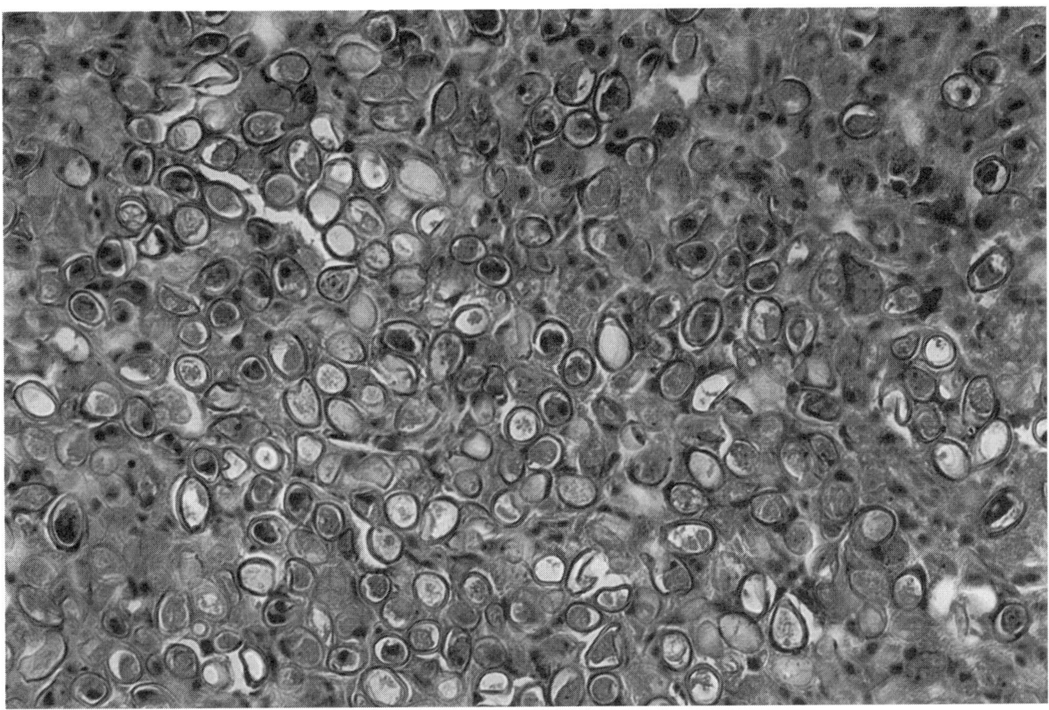

FIGURE 4.5. Trematode eggs, probably *Pholeter* sp., within a granulomatous cyst on the small intestine of a White Pelican. The eggs measure about 12 × 22 μm.

Helminth parasites from White Pelicans in Florida have been collected from only a few individuals (table 4.12). Eighteen species of trematodes, 4 cestodes, and 6 nematodes were identified. Lesions associated with these infections were noted only for the nematode, *Eustrongylides*, and the trematode, *Pholeter* sp. The White Pelicans from Orange County in 1990 were emaciated, and one had larval *Eustrongylides* (probably *E. ignotus*) that recently had perforated the stomach and entered the coelomic cavity. It was not the cause of death. The birds were foraging in a water reclamation facility and probably ingested infected fish there (Spalding et al. 1996). *Pholeter* causes nodules on the surface of the intestine that contain coiled trematodes and eggs (figure 4.5) (Forrester 1973; Spalding et al. 1996). These parasites were probably not responsible for the death of these birds, but may have contributed to the debilitated condition. Schistosome eggs were frequently found in the lamina propria of

villi in the small intestine of White Pelicans from Orange County in 1999, and caused mild enteritis. In some of these cases the schistosomes could also be found within vessels in the muscular layers of the intestine (figure 4.6) (Spalding, 1999). In 2001 schistosomes were observed in vessels in the brain of a White Pelican from Orange County. At one location granulomatous inflammation may have been associated with one of these parasites (Meteyer 2001).

XVI. Arthropods

Three species of chewing lice, 1 of them a pouch louse, 3 species of subcutaneous mites, 1 feather mite, 1 tick, 2 hippoboscid flies and 1 flesh fly have been found on Brown Pelicans in Florida (table 4.13). Subcutaneous mites were common and have been identified from the subcutis and air sacs near the trachea from Florida Brown Pelicans by Pence and Courtney (1973). They were

Table 4.11. Cestodes, nematodes, and acanthocephalans collected from Brown Pelicans in Florida

Class	Species (location)[a]	County/area	Year(s)	No. of pelicans			Intensity		Data Source
				Examined	Positive	%	Mean	Range	
Cestoda									
	Cyclustera ibisae[b] (SI,LI)	"East coast"	1971–73	39	14	36	7	NG	A
		"West coast"	1971–73	14	5	36	12	NG	A
	Glossocercus caribaensis[c] (SI)	Lee	1971–73	>30	1	<1	2	—	A
	Tetrabothrius sp. (SI,CL)	"East coast"	1971–73	39	24	61	3	NG	A
		"West coast"	1971–73	14	11	79	5	NG	A
Nematoda									
	Capillaria sp.(like *contorta*) (ES,PR)	"East coast"	1971–73	39	8	20	1	NG	A
	(like *mergi*) (SI,CE,CL)	"East coast"	1971–73	39	11	28	806	1–8,850	A
	Contracaecum microcephalum (ST)	Brevard	1982	1	1	—	NG	F	
		Dade	1980	1	1	—	NG	—	G
	Contracaecum multipapillatum (ST)	Dade	1980	1	1	—	NG	many	G
	Contracaecum rudolphi (ST)	"Tampa Bay"	1978	5	2	—	NG	many	H
		"Florida"	NG	NG	NG	—	NG	—	I
		Pinellas & Charlotte	1955–62	NG	NG	—	NG	—	B
		Lee	1966	1	1	—	100	100	J
	Contracaecum spp.[d] (ES,PR,SI)	Brevard	1989	1	1	—	NG	—	K
		Collier	1974	9	9	—	1,021	49–8,342	L
		Monroe	1984	1	1	—	—	"numerous"	C
		Pinellas	1983	5	NG	—	NG	—	E
		"East coast"	1971–73	39	37	95	22	NG	A
		"West coast"	1971–73	14	14	100	66	NG	A

Cosmocephalus obvelatus (ES,PR)	"East coast"	1971–73	39	28	72	3	NG	A
	"West coast"	1971–73	14	5	36	8	NG	A
Cyathostoma phenisci (TR,LU)	"East coast"	1971–73	39	10	26	2	1–4	A
Eustrongylides sp.[e] (PR)	"Tampa Bay"	1978	5	1	—	1	—	H
Paracuaria tridentata (ES,PR)	"East coast"	1971–73	39	13	33	10	NG	A
	"West coast"	1971–73	14	3	21	10	NG	A
Physaloptera sp. (SI)	"East coast"	1971–73	39	1	3	1	NG	A
Schistorophid larvae (ES,PR)	"West coast"	1971–73	14	2	14	20	10–30	A
Synhimantus sp.	Pinellas	1983	5	NG	—	NG	—	E
Tetrameres inerme (PR)	"West coast"	1971–73	14	1	7	1	NG	A
Acanthocephala								
Andracantha gravida (SI)	Pinellas	1970	NG	NG	—	NG	—	D
Corynosoma sp. (SI)	"West coast"	1971–73	14	1	7	2	—	A
Southwellina hispida (SI,CE,LI)	"East coast"	1971–73	39	5	13	3	NG	A
	"West coast"	1971–73	14	6	44	86	NG	A
Southwellina sp. (SI)	Pinellas	1983	1	1	—	NG	—	E

Sources: A = Courtney & Forrester (1974), B = Hutton (1964), C = Stroud (1984), D = Schmidt (1975), E = Forrester (1983), F = Stroud (1982), G = Tuggle (1980), H = Stroud (1980), I = Walton (1927), J = Huizinga (1971), K = Langenberg (1989), L = Forrester (1974).

NG = not given.

a. Location in host: CE = cecum, CL = cloaca, ES = esophagus, LI = large intestine, LU = lung, PR = proventriculus, SI = small intestine, ST = stomach, TR = trachea.

b. Published as *Parvitaenia ibisae.*

c. Not sexually mature (the Brown Pelican may not be a suitable host). This parasite is described by Schmidt and Courtney (1973) as *Parvitaenia heardi.*

d. A complex of 2 species, *C. spiculigerum* and *C. multipapillatum.*

e. The presence of this parasite was described grossly and not confirmed by identification. It was found within the fibrinous tissue on the surface of the proventriculus with the ends protruding into the lumen.

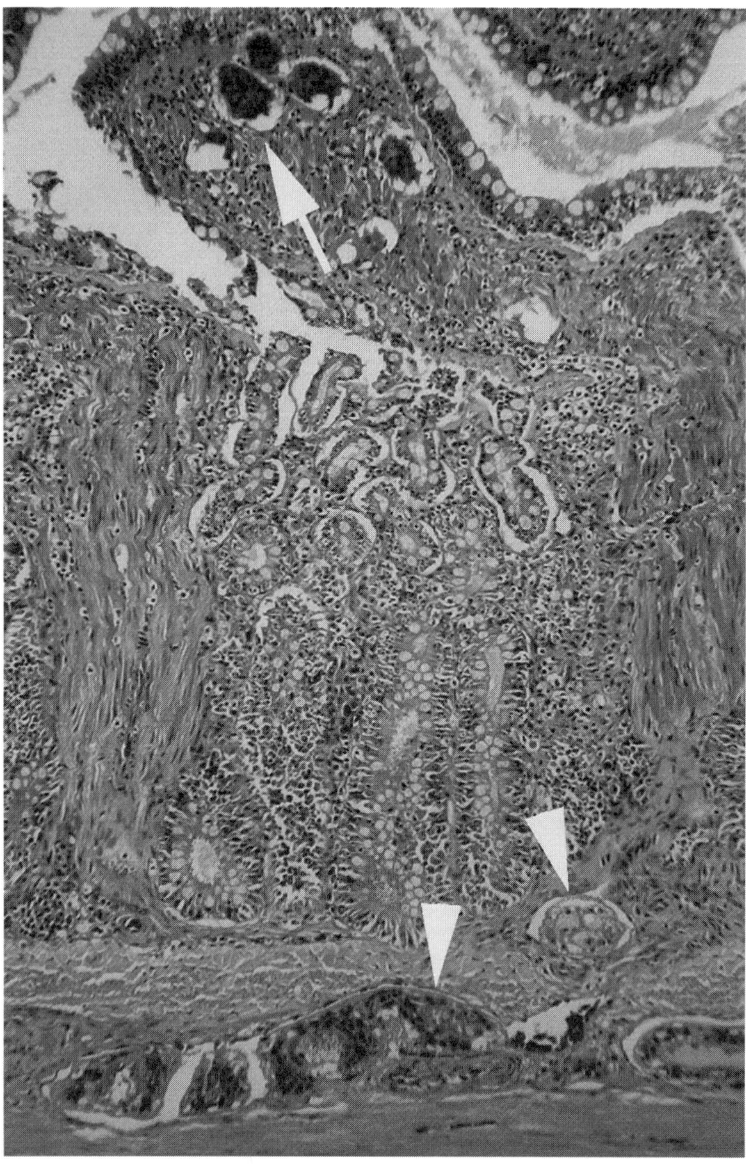

FIGURE 4.6. Schistosome eggs in the lamina propria *(arrow)* and trematodes in the vessels of the tunica muscularis *(arrowheads)* of the small intestine of a White Pelican from Orange County in 1999.

not found in the axilla and inguinal regions, as most subcutaneous mite species usually are. We found only 1 record of ticks (*Argas radiatus*) on Brown Pelicans in Florida. Ticks have been associated with nesting failure of Brown Pelicans in Texas (King et al. 1977) and with transmission of a Soldado (and Soldado-like) virus and

Aransas Bay virus (arbovirus, Upola serogroup) (King et al. 1977; Yunker et al. 1979). Mosquitoes were believed to cause the abandonment of Brown Pelican nests in June 1911 in Brevard County (Schreiber 1980, summary table).

Two species of chewing lice, 1 species of pouch louse, and 2 species of mites were found

Table 4.12. Helminths collected from White Pelicans in Florida

| Class | | | No. of birds | | Intensity | | |
Species (location)[a]	County	Year	Examined	Positive	Mean	Range	Data source
Trematoda							
Ascocotyle gemina (SI,CE)	Brevard	1999	1	1	5	—	G
	Clay	1982	1	1	10	—	B
Ascocotyle leighia (CE)	Brevard	1999	1	1	38	—	G
	Orange	1999	5	1	2	—	G
Ascocotyle sp. (SI)	Clay	1982	1	1	1	—	B
Bolbophorus confusus (SI)	Brevard	1999	1	1	1	—	G
	Orange	1999	5	1	1	—	G
Bursacetabulus macrobursas (IN)	Brevard	1999	1	1	119	—	G
	Orange	1999	5	5	45	9–82	G
Bursacetabulus pelecanus (IN)	Brevard	1999	1	1	464	—	G
	Orange	1999	5	5	227	12–624	G
Clinostomum complanatum (TR)	Brevard	1999	1	1	4	—	G
	Orange	1999	5	1	9	—	G
Dendritobilharzia pulverulenta (BV)	Clay	1982	1	1	4	—	B
Echinochasmus dietzevi (SI)	Clay	1982	1	1	43	—	B
	Orange	1999	5	3	1	1–2	G
Gigantobilharzia sp. (BV)	Lake	1999	3	2	NG	NG	G
Mesorchis denticulatus (SI)	Brevard	1999	1	1	1	—	G
	Orange	1999	5	2	1	—	G
Mesostephanus appendiculatoides (SI)	Clay	1982	1	1	10	—	B
Mesostephanus microbursa (SI)	Brevard	1999	1	1	2	—	G
	Orange	1999	5	1	1	—	G
Phagicola longa (IN)	Brevard	1999	1	1	1,300,240[f]	—	G
	Clay	1982	1	1	330	—	B
	Orange	1999	5	5	472	25–1130	G
Phagicola nana (SI)	Clay	1982	1	1	20	—	B
	Orange	1999	5	2	414	50–777	G
Posthodiplostomum minimum (SI)	Orange	1999	5	1	1	—	G
Renicola thapari (KD)	Brevard	1999	1	1	4	—	G
	Clay	1982	1	1	4	—	B
	Orange	1990	1	1	90	—	C
		1999	5	3	10	6–15	G
Renicola sp. (KD)	Orange	1999	5	1	10	—	G
Ribeiroia ondatrae (PR,ST)	Brevard	1999	1	1	2	—	G
	Clay	1982	1	1	4	—	B
	Orange	1999	5	5	132	1–620	G
Pholeter sp.[b] (SI)	Orange	1973	1	1	NG	—	A
Cestoda							
Cyclustera ibisae (SI)	Brevard	1999	1	1	1	—	G
	Orange	1999	5	1	5	—	G
Paradilepis caballeroi (SI)	Clay	1982	1	1	2	—	B
Paradilepis longivaginosus (SI)	Brevard	1999	1	1	107	—	G
	Orange	1999	5	3	13	1–20	G
Spirometra mansonoides (spargana) (SQ)	Orange	1999	5	1	5	—	G
Nematoda							
Capillaria mergi (CE,LI,SI)	Clay	1982	1	1	40	—	B
	Orange	1990	1	1	20	—	C
Capillaria sp. (SI)	Brevard	1999	1	1	4	—	G
	Orange	1999	5	4	13	4–24	G

(continued)

Table 4.12. *(continued)*

| Class | | | No. of birds | | Intensity | | |
Species (location)[a]	County	Year	Examined	Positive	Mean	Range	Data source
Contracaecum rudolphi[e]	Florida	NG	NG	—	—	—	F
Contracaecum spp.[c] (ES,SI,ST)	Brevard	1999	1	1	1413	—	G
	Clay	1982	1	1	1,065	—	B
	Orange	1990	1	1	675	—	C
		1999	5	5	363	103–900	G
Contracaecum sp.	Dade	1957	2	2	"many"	—	E
	Wakulla	1959	1	1	1,100	—	D
Eustrongylides sp. larva[d] (PR)	Orange	1990	1	1	1	—	C
Microtetrameres pelecani (PR)	Orange	1999	5	5	18	10–29	G
Strongyloides sp. (SI)	Orange	1999	5	1	1	—	G
Tetrameres sp. (PR)	Clay	1982	1	1	7	—	B
	Orange	1990	1	1	1	—	C

Sources: A = Forrester (1973), B = Forrester & Kinsella (1996), C = Spalding et al. (1996), D = Oglesby (1960), E = Owre (1962), F = Walton (1927), G = Spalding & Kinsella (1999).
NG = not given.
a. Location in host: BV = blood vessels, CE = ceca, ES = esophagus, IN = intestine, KD = kidney, LI = large intestine, PR = proventriculus, SI = small intestine, ST = stomach, SQ = subcutaneous, TR = trachea.
b. Associated with cystic cavities and chronic enteritis.
c. Many are larval, adults of *C. multipapillatum*, *C. micropapillatum*, and *C. rudolphii* are present.
d. Probably *Eustrongylides ignotus*.
e. *Contracaecum spiculigerum* is a synonym of *C. rudolphii* (Hartwich 1964).
f. This number is a combination of *P. longa* and *P. nana*.

on White Pelicans in Florida. *Piagetiella peralis*, a pouch louse, has been observed to severely infest nestlings in Saskatchewan (Wobeser et al. 1974; Samuel et al. 1982). Mild infestations have been found on juvenile and adult White Pelicans in Florida. However, because these birds do not nest in Florida, such severe infestations have not been observed here. No subcutaneous mites have been identified from Florida White Pelicans; however, *Pelecanectes apunctatus* has been reported elsewhere (Tuggle 1983).

XVII. Unresolved die-offs

The cause of die-off events of Brown Pelicans in Florida more often than not goes unresolved (table 4.1).

In April 1982, there was a die-off that included 10 Brown Pelicans, 47 Double-crested Cormorants and 1 Ring-billed Gull at Patrick Air Force Base in Brevard County. Two Brown Pelicans were examined. A Ring-billed Gull tested positive for botulism type C; however, 1 of the pelicans examined did not. It was suspected that pesticides were involved in this dieoff, but this was not confirmed by appropriate tests (Stroud 1982).

A large-scale die-off of Brown Pelicans took place in Pinellas County in 1983 (see section XII, Bacteria, for more details). Birds were tested for exposure to floridone, the active ingredient of SONAR, which had been used in the area as a herbicide for cattails, but the results were negative (Forrester 1983). A conclusive cause for the die-off was never found.

Two Brown Pelicans were found as part of a die-off of more than 40 Coots in Citrus County in February 1985 (Thomas 1985). A cause for the die-off was not determined (see chapter 18, Rails, Coots, Gallinules, Moorhens, and Limpkins).

In April 1987 30–40 Brown Pelicans, a loon, and 4–5 gulls were found dead in Brevard County near a dump site. Pelicans had wobbly heads and loss of balance (Roffe 1987). There was no consistent finding at necropsy examination. A toxin was suspected, but not identified.

Table 4.13. Ectoparasites collected from pelicans in Florida

Host	Species of parasite	County	Year(s)	Data source
American White Pelican				
Chewing Lice				
	Colpocephalum unciferum	Monroe	1957	A
		Lee	1981	A
		Clay	1982	A
		Orange	1972, 1990	A
	Pectinopygus tordoffi	Clay	1982	A
		Orange	1972, 1990	A
	Piagetiella peralis	Lee	1993	A
		Orange	1972, 1990	A
Mites				
	Alloptes sp.	Orange	1990	B
	Scutomegninia sp.[b]	Orange	1990	B
Brown pelican				
Chewing Lice				
	Colpocephalum occidentalis	Brevard	1972	A
		Indian River	1976	A
		Lee	1980–85	A
		Pinellas	1930	C
			1973	A
	Pectinopygus occidentalis	Brevard	1972	A
		Indian River	1976, 1981	A
		Lee	1980–85	A
		Palm Beach	1973	A
		Pinellas	1930, 1973	A
	Piagetiella bursaepelecani	Pinellas	1930	A
Mites				
	Neottialges apunctatus[c]	Brevard	1977	D
		Dade	1971	D
		Lee	1972	D
	Phalacrodectes pelecani[c]	Indian River	1971	D
		Lee	1972	D
		Sarasota	1972	D
	Phalacrodectes punctatissimus	Brevard	1982	E
		Lee	1972	D
		Sarasota	1972	D
	Scutomegninia sp.	St. Johns	1998	F
Ticks				
	Argas radiatus	Lee	1982	G
Diptera				
	Olfersia sordida	Brevard	NG	H
		Charlotte	NG	H
		Monroe	1970	I
	Olfersia spinifera	Indian River	NG	H
	Wohlfahrtia vigil larvae[a]	Volusia	1988	J

Sources: A = Forrester et al. (1995), B = Spalding & Atyeo (1992), C = Price (1967), D = Pence & Courtney (1973), E = Stroud (1982), F = Spalding & Mertins (1999), G = Keirans (1982), H = Bequaert (1957), I = Wilson et al. (1990), J = Riggs (1988).

NG = not given.

a. Maggots of this flesh fly (Sarcophagidae) infested ocular and skin lesions on the heads of the birds associated with die-off. The cause of the lesions was not determined.

b. Very similar to *Scutomegninia* in wading birds in Florida and may represent a new genus.

c. Identified as *Pelecanectes apunctatus* and *Phalacrodectes pelecani* and later reclassified by Pence & Duncan (1995).

A large-scale die-off of Brown Pelicans (145), Snowy Egrets (3) and a single Little Blue Heron occurred late in October of 1988 on an island off Port Orange in the Halifax River, Volusia County (Spalding et al. 1988). When first discovered, many pelicans, mostly immature, had maggot-infested eyes and heads, and weakened birds were wandering on the island. The maggots were from flesh flies, *Wohlfahrtia virgil*, secondary to an undetermined insult to the skin. A cause of mortality was not found from necropsies of 5 Brown Pelicans, a Snowy Egret, and a Little Blue Heron. The pelicans had an acute bacterial conjunctivitis, and mild hepatitis and gastroenteritis that were probably due to mild helminth infections. *Clostridium botulinum* toxin and viral isolation tests were negative. The possibility of a toxin applied from the air, such as fuel or pesticides, was considered but could not be confirmed. Birds on adjacent islands were not affected.

All 4 of the Brown Pelicans examined from a die-off of about 50 in Brevard, Martin, and St. Lucie counties in March 1998 had thyroid follicular atrophy and mild bronchitis (Spalding 1998). Both adult and juvenile birds were found begging aggressively for food and were observed to have difficulty swallowing fish. This followed a period of very heavy rains with considerable agricultural runoff and reduced bait fish availability. Not all of the birds were emaciated. A biotoxin and/or pesticide exposure was suspected, but not confirmed.

A cause was not determined for a die-off of 200 Brown Pelicans in Duval County, January–March of 1998 (Brannian 1998). Exposure to oil or another contaminant was suspected; however, the finding of bacteria histologically supports the possibility that it was due to *Erysipelothrix rusiopathiae* or other bacterial septicemia, as occurred in that location 2 years earlier.

At least 9 juvenile White Pelicans died in Orange County during the late summer of 1990 at a sewage water reclamation site. All birds examined were emaciated. Other inconsistent findings in individuals were aspergillosis and eustrongylidosis. Lead, cadmium, and mercury concentrations in liver were relatively low (Spalding 1990).

XVIII. Emaciation

For pelicans found dead in Florida, emaciation is probably the most consistent finding. Whether this indicates that food resources are limiting is debatable. The foraging technique used by Brown Pelicans, plunge diving, is probably energetically costly and successful only when both food resources are good and the bird is in excellent health. Minor health perturbations, therefore, may result in a vicious cycle whereby health continues to decline because of poor nutrition. Additionally, as with many other species, hunting skills increase with experience, and young birds are less efficient foragers (Orians 1969).

Thompson et al. (1977a) investigated the interaction between starvation and p,p'-DDE naturally acquired by nestling Brown Pelicans collected from Vero Beach, Indian River County. They found significantly higher residues in fat in starved birds than in nonstarved control birds. They did not find similar differences in brain, liver, and muscle.

XIX. Summary and conclusions

Trauma, more commonly associated with fishing gear than with vehicles and power lines, was the most significant cause of injury and death for Brown Pelicans. After this, severe emaciation was commonly observed, but more often than not, the cause of emaciation was not clear. Pesticides, particularly DDE, were well monitored in eggs of Brown Pelicans in the past. Although eggshell thinning was well documented, it did not appear to be severe enough to result in population declines in Florida as it did elsewhere. Very little information is available about contaminants in pelican eggs recently. Mercury concentrations are not as high as in other fish-eating birds; however, they have not been well studied recently and no information is available about the effect of mercury on reproduction. Compared with other species, bacteria have been well studied in Brown Pelicans, particularly associated with a sewage spill. For many bacteria a relationship to illness and death has not been established. One bacterial pathogen associated with the feeding of fish scraps, *Erysipelothrix rusiopathiae,* has caused mortality in large numbers of Brown Pelicans in Florida and elsewhere. Protozoa are rare in both White and Brown Pelicans; only 2 were found. Viruses are very poorly studied: antigen to only 2 viruses has been detected. Thirty-one helminth parasites were identified in Brown Pelicans. Helminth parasites of White Pelicans are poorly studied. Very large numbers of *Contracaecum* spp. nematodes were associated with starvation-emaciation; however, their role in causing emaciation remains undetermined. Although a diverse group of arthropods has been identified on pelicans, investigations of prevalence, intensity, and significance of these are lacking.

XX. Literature cited

Anderson, D.W., F. Gress, and D.M. Fry. 1996. Survival and dispersal of oiled brown pelicans after rehabilitation release. *Mar. Pollut. Bull.* 32:711–718.

Anderson, D.W., and J.J. Hickey. 1969. Some breeding characteristics of brown pelicans from oological data. Progress Report to U.S. Fish and Wildlife Service, March 1, 1969, contract 14-06-D-6182. 23 pp.

———. 1970. Oological data on egg and breeding characteristics of brown pelicans. *Wilson Bull.* 82:14–28.

———. 1972. Eggshell changes in certain North American birds. *Proc. Int. Ornithol. Congr.* 15:514–540.

Ankerberg, C.W., Jr. 1983. An investigation of the death of *Pelecanus occidentalis* (brown pelican) in the Boca Ciega Bay. Suncoast Seabird Sanctuary, Redington Beach, Florida.

Bequaert, J. 1957. The Hippoboscidae or louse-flies (Diptera) of mammals and birds. II. Taxonomy, evolution and revision of American genera and species. *Entomol. Am.* 36:417–611.

Blus, L.J. 1970. Measurements of brown pelican eggshells from Florida and South Carolina. *BioScience* 20:867–869.

———. 1982. Further interpretation of the relation of organochlorine residues in brown pelican eggs to reproductive success. *Environ. Pollut. Ser. A Ecol. Biol.* 28:15–33.

Blus, L.J., A.A. Belisle, and R.M. Prouty. 1974b. Relations of the brown pelican to certain environmental pollutants. *Pestic. Monit. J.* 7:181–194.

Blus, L.J., C.D. Gish, A.A. Belisle, and R.M. Prouty. 1972. Logarithmic relationship of DDE residues to eggshell thinning. *Nature* 235:376–377.

Blus, L.J., R.G. Heath, C.D. Gish, A.A. Belisle, and R.M. Prouty. 1971. Eggshell thinning in the brown pelican: implications of DDE. *BioScience* 21:1213–1215.

Blus, L.J., T. Joanen, A.A. Belisle, and R.M. Prouty. 1975. The brown pelican and certain environmental pollutants in Louisiana. *Bull. Environ. Contam. Toxicol.* 13:646–655.

Blus, L.J., T.G. Lamont, and B.S. Neely, Jr. 1979. Effects of organochlorine residues on eggshell thickness, reproduction, and population status of brown pelican (*Pelecanus occidentalis*) in South Carolina and Florida, 1969–76. *Pestic. Monit. J.* 12:172–184.

Blus, L.J., B.S. Neely, Jr., A.A. Belisle, and R.M. Prouty. 1974a. Organochlorine residues in brown pelican eggs: relation to reproductive success. *Environ. Pollut.* 7:81–91.

Blus, L.J., B.S. Neely, T.G. Lamont, and B. Mulhern. 1977. Residues of organochlorines and heavy metals in tissues and eggs of brown pelicans, 1969–73. *Pestic. Monit. J.* 11:40–53.

Blus, L.J., S.N. Wiemeyer, and C.M. Bunck. 1997. Clarification of effects of DDE on shell thickness, size, mass, and shape of avian eggs. *Environ. Pollut.* 95:67–74.

Bogart, D.E. 1953. Brown pelican falls prey to wildcat. *Everglades Nat. Hist.* 1:76–77.

Brannian, R.E. 1998. Unpublished data. National Wildlife Health Center, Madison, Wis.

Brown, C. 2000. Unpublished data. University of Georgia, Athens.

Burridge, M.J., W.J. Bigler, D.J. Forrester, and

J.M. Hennemann. 1979. Serologic survey for *Toxoplasma gondii* in wild animals in Florida. *J. Am. Vet. Med. Assoc.* 175:964–967.

Clapp, R.B., R.C. Banks, D. Morgan-Jacobs, and W.A. Hoffman. 1982. *Marine birds of the southeastern United States and Gulf of Mexico. Part I, Gaviiformes through Pelecaniformes.* U.S. Fish and Wildlife Service, Biological Services Program FWS-OBS-82/01. 637 pp.

Connors, P.G., V.C. Anderlini, R.W. Risebrough, J.H. Martin, R.W. Schreiber, and D.W. Anderson. 1972. Heavy metal concentrations in brown pelicans from Florida and California. *Cal-Neva Wildl.* 1972:56–64.

Courtney, C.H., and J.V. Ernst. 1975. *Eimeria pelecani* sp. n. from the brown pelican, *Pelecanus occidentalis carolinensis,* from Florida. *J. Parasitol.* 61:1081–1082.

Courtney, C.H., and D.J. Forrester. 1974. Helminth parasites of the brown pelican in Florida and Louisiana. *Proc. Helminthol. Soc. Wash.* 41:89–93.

Courtney, C.H., D.J. Forrester, and F.H. White. 1977. Anthelmintic treatment of brown pelicans. *J. Am. Vet. Med. Assoc.* 171:991–992.

Crawford, R.L. 1981. Bird casualties at a Leon County, Florida TV Tower: a 25-year migration study. *Bull. Tall Timbers Res. Stn.* 22:1–30.

Dennis, E.A. 1967. Biological studies on the life history of *Mesostephanus yedeae* sp. n. (Trematoda: Cyathocotylidae). Ph.D. diss., University of Connecticut, Storrs. 94 pp.

Duncan, R.M. 1986. Unpublished data. National Wildlife Health Center, Madison, Wis.

Dunlap, T.R. 1981. *DDT: scientists, citizens, and public policy.* Princeton University Press, Princeton, N.J.

Edman, J.D., L.A. Webber, and H.W. Kale II. 1972. Host-feeding patterns of Florida mosquitoes II. Culiseta. *J. Med. Entomol.* 9:429–434.

Evans, R.M., and F.L. Knopf. 1993. American white pelican (*Pelecanus erythrorhynchos*). *Birds N. Am.* 57:1–24.

Fairbrother, A., L.N. Locke, and G.L. Hoff (eds.). 1996. *Noninfectious diseases of wildlife.* 2d ed. Iowa State University Press, Ames. 219 pp.

[FGFWFC] Florida Game and Fresh Water Fish Commission. 1992. Effects of Hurricane An-

drew on fish and wildlife of southern Florida, a preliminary assessment. Report, September 8, 1992, Tallahassee. 19 pp.

Forrester, D.J. 1973–87. Unpublished data. University of Florida, Gainesville.

Forrester, D.J., and G.F. Bennett. 1985. Unpublished data. University of Florida, Gainesville.

Forrester, D.J., H.W. Kale II, R.D. Price, K.C. Emerson, and G.W. Foster. 1995. Chewing lice (Mallophaga) from birds in Florida: a listing by host. *Bull. Fla. Mus. Nat. Hist.* 39:1–44.

Forrester, D.J., and J.M. Kinsella. 1996. Unpublished data. University of Florida, Gainesville.

Forrester, D.J., and N.P. Thompson. 1974. Unpublished data. University of Florida, Gainesville.

Forrester, D.J., and F.H. White. 1972–1974. Unpublished data. University of Florida, Gainesville.

Francis, T.L. 1981. Brown pelican found dead with adult double-crested cormorant in pouch. *Fla. Field Nat.* 9:62–63.

Franson, J.C. 1988–93. Unpublished data. National Wildlife Health Center, Madison, Wis.

Franson, J.C., E.J. Galbreath, S.N. Wiemeyer, and J.M. Abell. 1994. *Erysipelothrix rhusiopathiae* infection in a captive bald eagle (*Haliaeetus leucocephalus*). *J. Zoo Wildl. Med.* 25:446–448.

Frenkel, J.K. 1981. False-negative serologic tests for *Toxoplasma* in birds. *J. Parasitol.* 67:952–953.

Friend, M., and J.C. Franson. 1999. *Field manual of wildlife diseases. General field procedures and diseases of birds.* U.S. Department of the Interior, U.S. Geological Survey, Biological Research Division, Information and Technology Report 1999-001. Washington, D. C. 425 pp.

Gourlie, N. 1984. Unpublished data. Department of Environmental Regulation, Tallahassee, Florida.

Greichus, Y.A., and M.R. Hannon. 1973. Distribution and biochemical effects of DDT, DDD and DDE in penned double-crested cormorants. *Toxicol. Appl. Pharmacol.* 26:483–494.

Greve, J.H., H.F. Albers, B. Suto, and J. Grimes. 1985. Pathology of gastrointestinal helminthiasis in the brown pelican (*Pelecanus occidentalis*). *Avian Dis.* 30:482–487.

Gross, T.L. 1983. Unpublished data. University of Florida, Gainesville.

Hansen, S.P., J.C. Franson, and T.E. Creekmore.

1998. Unpublished data. National Wildlife Health Center, Madison, Wis.

Hartman, F. 1946. Notes on the pathology of a loon and a pelican. *Auk* 63:588–589.

Hartwich, G. 1964. Revision der vogelparasitischen Nematoden Mitteleuropas-II. Die Gattung *Contracaecum* Railliet & Henry, 1912 (Ascaridoidea). *Mitt. Zool. Mus. Berl.* 40:15–53.

Huizinga, H.W. 1971. Contracaeciasis in pelicaniform birds. *J. Wildl. Dis.* 7:198–202.

Humphrey, S.R., C.H. Courtney, and D.J. Forrester. 1978. Community ecology of the helminth parasites of the brown pelican. *Wilson Bull.* 90:587–598.

Hunter, D.L. 1988. *Ersipelothrix rusiopathiae* infection in brown pelicans (*Pelecanus occidentalis californicus*) along California coast. Abstract. *Proc. Wildl. Dis. Assoc. Annu. Conf.* 37:33.

Hutton, R.F. 1964. A second list of parasites from marine and coastal animals of Florida. *Trans. Am. Microsc. Soc.* 83:439–447.

Hutton, R.F., and F. Sogandares-Bernal. 1960. Studies on helminth parasites from the coast of Florida. II. Digenetic trematodes from shore birds of the west coast of Florida. *Bull. Mar. Sci. Gulf Caribb.* 10:40–54.

Jehl, J.R., Jr. 1973. Studies of a declining population of brown pelicans in northwestern Baja California. *Condor* 75:69–79.

Jenkins, R.L. 1984. Marine birds injured by welding rods. *Fla. Field Nat.* 12:13–15.

Johnson, J.M., and M.G. Spalding. 1999. Unpublished data. University of Florida, Gainesville.

Kale, H.W. II. 1971. Florida region. *Am. Birds* 25:723–733.

———. 1972. The spring migration: Florida region. *Am. Birds* 26:751–774.

Keirans, J.E. 1982. Unpublished data. U.S. National Tick Collection, Georgia Southern University, Statesboro.

King, K.A., D.R. Blankinship, R.T. Paul, and R.C.A. Rice. 1977. Ticks as a factor in the 1975 nesting failure of Texas brown pelicans. *Wilson Bull.* 89:157–158.

Langenberg, J.A. 1989. Unpublished data. National Wildlife Health Center, Madison, Wis.

Langridge, H.P. 1993. Florida region. *Am. Birds* 47:406–408.

Lincer, J.L., and D. Salkind. 1973. A preliminary note on organochlorine residues in the eggs of fish-eating birds of the west coast of Florida. *Fla. Field Nat.* 1:19–22.

Locke, L.N. 1972. Unpublished data. Patuxent Wildlife Research Center, Laurel, Md.

———. 1986–89. Unpublished data. National Wildlife Health Research Center, Madison, Wis.

Logan, T.H. 1997. Florida's endangered species, threatened species, and species of special concern: official lists. Florida Game and Fresh Water Fish Commission, Tallahassee. 14 pp.

Longstreet, R.J. 1953–55. Ornithology of the mosquitoes. *Fla. Nat.* 26:103–114, 27:175–188, 28:9–20.

Maehr, D.S., and J.Q. Smith. 1988. Bird casualties at a central Florida power plant: 1982–86. *Fla. Field Nat.* 16:57–80.

Maehr, D.S., A.G. Spratt, and D.K. Voigts. 1983. Bird casualties at a central Florida power plant. *Fla. Field Nat.* 11:45–49.

Meteyer, C.U. 2001. Unpublished data. National Wildlife Health Research Center, Madison, Wis.

Miller, H.A. 1950. Ol' Bill pelican. *Fla. Wildl.* 4:10.

Nesbitt, S.A. 1996. Eastern brown pelican. In: *Rare and endangered biota of Florida.* Vol. 5, *Birds.* J.A. Rodgers, Jr., H.W. Kale II, and H.T. Smith (eds.). University Press of Florida, Gainesville. pp. 144–155.

Nesbitt, S.A., P.E. Cowan, P.W. Rankin, N.P. Thompson, and L.E. Williams, Jr. 1981. Chlorinated hydrocarbon residues in Florida brown pelicans. *Colon. Waterbirds* 4:77–84.

Nesbitt, S.A., M.J. Fogarty, and L.E. Williams, Jr. 1977. Status of Florida nesting brown pelicans, 1971–1976. *Bird-Banding* 48:138–144.

Ogden, J.C., W.B. Robertson, G.E. Davis, and T.W. Schmidt. 1974. Pesticides, polychlorinated biphenyls and heavy metals in upper food chain levels, Everglades National Park and vicinity. National Technical Information Service, Department of the Interior, Atlanta. 27 pp.

Oglesby, L.C. 1960. Heavy nematode infestation of white pelican. *Auk* 77:354.

Orians, G.H. 1969. Age and hunting success in the brown pelican *Pelecanus occidentalis. Anim. Behav.* 17:316–318.

Owre, O.T. 1962. Nematodes in birds of the order Pelicaniformes. *Auk* 79:114.

Pain, D.J. 1996. Lead in waterfowl. In: *Interpreting environmental contaminants in animal tissues.* W.N. Beyer, G.H. Heinz, and A.W. Redmon-Norwood (eds.). CRC Press, Boca Raton, Fla. pp. 251–264.

Peakall, D.B. 1970. Pesticides and the reproduction of birds. *Sci. Am.* April: 73–78.

Pearson, J.Ç. 1973. A revision of the subfamily Haplorchinae Looss, 1899 (Trematoda: Heterophyidae) II. Genus *Galactosomum. Philos. Trans. R. Soc. Lond. B Biol. Sci.* 266:341–447.

Pearson, J.C., and C.H. Courtney. 1977. *Pholeter anterouterus* Fischtal and Nasir, 1974 (Digenea: Opisthorchiidae) redescribed, together with remarks on the genera *Pholeter* Odhner, 1914 and *Phocitrema* Goto and Ozaki, 1930 and their relationship to the centrocestine heterophyids. *J. Parasitol.* 74:255–271.

Pence, D.B., and C.H. Courtney. 1973. The hypopi (Acarina: Hypoderidae) from the subcutaneous tissues of the brown pelican *Pelecanus occidentalis carolinensis* Gmelin. *J. Parasitol.* 59:711–719.

Pence, D.B., and M. Duncan. 1995. Hypopi (Acari: Hypoderatidae) from subcutaneous tissues of the African spoonbill (Aves: Ciconiiformes: Threskiornithidae). *J. Med. Entomol.* 32:166–173.

Post, W. 1992. Wood stork mortality from Hurricane Hugo. *Fla. Field Nat.* 20:107.

Price, R.D. 1967. The *Colpocephalum* (Mallophaga: Menoponidae) of the Pelecaniformes. *Can. Entomol.* 99:273–280.

Quick, J.A., Jr., and G.E. Henderson. 1975. Effects of *Gymnodinium breve* red tide on fishes and birds: a preliminary report on behavior, anatomy, hematology, and histopathology. *Proc. Gulf Coast Reg. Symp. Dis. Aquat. Anim.* :85–113.

Quist, C. 1993. Unpublished data. Southeastern Cooperative Wildlife Disease Study, University of Georgia, Athens.

Riggs, M.W. 1988. Unpublished data. University of Florida, Gainesville.

Robertson, W.B., Jr., and D.R. Paulson. 1961. Region reports: Florida region. *Audubon Field Notes* 15:26–35.

Robertson, W.B., and G.E. Woolfenden. 1992. *Florida bird species: an annotated list.* Spec. Publ. 6, Florida Ornithological Society, Gainesville. 260 pp.

Rodgers, J.A., Jr. 1986. Repaired bill injury in the brown pelican. *Fla. Field Nat.* 14:46–47.

Rodgers, J.A., Jr., H.W. Kale II, and H.T. Smith (eds.). 1996. *Rare and endangered biota of Florida.* Vol. 5, *Birds.* University Press of Florida, Gainesville. 688 pp.

Rodgers, J.A., and H.T. Smith. 1997. Buffer zone distances to protect foraging and loafing waterbirds from human disturbance in Florida. *Wildl. Soc. Bull.* 25:139–145.

Roffe, T.J. 1987. Unpublished data. National Wildlife Health Center, Madison, Wis.

Samuel, W.M., E.S. Williams, and A.B. Rippin. 1982. Infestations of *Piagetiella peralis* (Mallophaga: Menoponidae) on juvenile white pelicans. *Can. J. Zool.* 60:951–953.

Schmeling, S.K. 1980. Unpublished data. National Wildlife Health Center, Madison, Wis.

Schmidt, G.D. 1975. *Andracantha,* a new genus of Acanthocephala (Polymorphidae) from fish-eating birds, with descriptions of three species. *J. Parasitol.* 61:615–620.

Schmidt, G.D., and C.H. Courtney. 1973. *Parvitaenia heardi* sp. n. (Cestoidea: Dilepididae), from the great blue heron, *Ardea herodias,* in South Carolina. *J. Parasitol.* 59:821–823.

Schreiber, R.W. 1977. Shell thickness in brown pelican eggs from Tampa Bay, Florida. *Fla. Field Nat.* 5:31–34.

———. 1978. Eastern brown pelican. In: *Rare and endangered biota of Florida.* Vol. 2. H.W. Kale II (ed.). University Press of Florida, Gainesville. pp. 23–25.

———. 1980. Nesting chronology of the eastern brown pelican. *Auk* 97:491–508.

Schreiber R.W., and R.W. Risebrough. 1972. Studies of the brown pelican. *Wilson Bull.* 84:118–135.

Snyder, B. 1994. Unpublished data. Florida Department of Environmental Protection, Tallahassee.

Spalding, M.G. 1990–2001. Unpublished data. University of Florida, Gainesville.

Spalding, M.G., and W.T. Atyeo. 1992. Unpublished data. University of Florida, Gainesville.

Spalding, M.G., G.T. Bancroft, and D.J. Forrester. 1993. The epizootiology of eustrongylidosis in wading birds (Ciconiiformes) in Florida. *J. Wildl. Dis.* 29:237–249.

Spalding, M.G., and D.J. Forrester. 1998. Unpublished data. University of Florida, Gainesville.

Spalding, M.G., and J. Hovis. 1993. Unpublished data. University of Florida, Gainesville.

Spalding, M.G., and J.M. Kinsella. 1999. Unpublished data. University of Florida, Gainesville.

Spalding, M.G., J.M. Kinsella, and J. Birney. 1996. Unpublished data. University of Florida, Gainesville.

Spalding, M.G., R.G. McLean, J.H. Burgess, and L.J. Kirk. 1994. Arboviruses in water birds (Ciconiiformes, Pelecaniformes) from Florida. *J. Wildl. Dis.* 30:216–221.

Spalding, M.G., and J.W. Mertins. 1999–2001. Unpublished data. University of Florida, Gainesville.

Spalding, M.G., and L. Quinn. 1995. Unpublished data. University of Florida, Gainesville.

Spalding, M.G., S. Wright, V. Clyde, M.W. Riggs, and J.A. Hovis. 1988. Unpublished data. University of Florida, Gainesville.

Sprunt, A., Jr. 1954. *Florida bird life.* Coward-McCann, New York. 527 pp.

Stevenson, H.M. 1970. The winter season: Florida region. *Audubon Field Notes* 24:493–497.

Stevenson, J.A. 1994. Unpublished data. Florida Department of Environmental Protection, Tallahassee.

Stickel, L.F., W.H. Stickel, and R. Christensen. 1966. Residues of DDT in brains and bodies of birds that died on dosage and in survivors. *Science* 151:1549–1551.

Stroud, R.K. 1980–1987. Unpublished data. National Wildlife Health Center, Madison, Wis.

Suto, B. 2000. Unpublished data. Suncoast Seabird Sanctuary, Redington Beach, Fla.

Taylor, W.K., and B.H. Anderson. 1973. Nocturnal migrants killed at a central Florida TV tower, autumns 1969–71. *Wilson Bull.* 85:42–51.

———. 1974. Nocturnal migrants killed at a central Florida TV tower, autumn 1972. *Fla. Field Nat.* 2:40–43.

Taylor, W.K., and M.A. Kershner. 1986. Migrant birds killed at the Vehicle Assembly Building (VAB), John F. Kennedy Space Center. *J. Field Ornithol.* 57:142–154.

Thomas, N. 1985. Unpublished data. National Wildlife Health Center, Madison, Wis.

Thompson, N.P., C.H. Courtney, D.J. Forrester, and F.H. White. 1977a. Starvation-pesticide interactions in juvenile brown pelicans. *Bull. Environ. Contam. Toxicol.* 17:485–490.

Thompson, N.P., P.W. Rankin, P.E. Cowan, L.E. Williams, Jr., and S.A. Nesbitt. 1977b. Chlorinated hydrocarbon residues in the diet and eggs of the Florida brown pelican. *Bull. Environ. Contam. Toxicol.* 18:331–339.

Tiemeier, O.W. 1941. Repaired bone injuries in birds. *Auk* 58:350–359.

Tuggle, B. 1980. Unpublished data. National Wildlife Health Center, Madison, Wis.

———. 1983. The white pelican, *Pelecanus erythrorhynchos,* as a host of *Pelecanectes apunctatus* (Acarina: Hypoderidae). *J. Parasitol.* 69:1083.

Walton, A.C. 1927. A revision of the nematodes of the Leidy collections. *Proc. Acad. Nat. Sci. Phila.* 79:49–163.

Watson, J.J., and F.H. White. 1979. Hemolysins of *Edwardsiella tarda. Can. J. Comp. Med.* 43:78–83.

White, F.H., C.F. Simpson, and L.E. Williams, Jr. 1973. Isolation of *Edwardsiella tarda* from aquatic animal species and surface waters in Florida. *J. Wildl. Dis.* 9:204–208.

Wilkinson, P.M., S.A. Nesbitt, and J.F. Parnell. 1994. Recent history and status of the eastern brown pelican. *Wildl. Soc. Bull.* 22:420–240.

Williams, L.E., Jr., and L. Martin. 1969. Nesting status of the brown pelican in Florida in 1968. *Q. J. Fla. Acad. Sci.* 31:130–140.

———. 1970. Nesting populations of brown pelicans in Florida. *Proc. Southeast Assoc. Game Fish Comm.* 24:154–169.

Wilson, N.A., H.W. Kale II, and W.W. Baker.

1990. Unpublished data. University of Northern Iowa, Cedar Falls.

Wobeser, G., G.R. Johnson, and G. Acompanado. 1974. Stomatitis in a juvenile white pelican due to *Piagetiella peralis* (Mallophaga: Menoponidae). *J. Wildl. Dis.* 10:135–138.

Wolf, S.H., R.W. Schrieber, L. Kahana, and J.J. Torres. 1985. Seasonal, sexual, and age-related variation in the blood parameters of the brown pelican (*Pelecanus occidentalis*). *Comp. Biochem. Physiol.* 82A:837–946.

Woodard, J.C. 1993. Unpublished data. University of Florida, Gainesville.

Work, T.M., B. Barr, A.M. Beale, L. Fritz, M.A. Quilliam, and J.L.C. Wright. 1993. Epidemiology of domoic acid poisoning in brown pelicans (*Pelecanus occidentalis*) and Brant's cormorants (*Phalacrocorax penicillatus*) in California. *J. Zoo Wildl. Med.* 24: 54–62.

Yunker, C.E., C.M. Clifford, J.E. Keirans, L.A. Thomas, and R.C.A. Rice. 1979. Aransas Bay virus, a new arbovirus of the Upola serogroup from *Ornithodoros capensis* (Acari: Argasidae) in coastal Texas. *J. Med. Entomol.* 16:453–460.

Cormorants and Anhingas

I. Introduction

Two species of cormorants and one species of darter occur in Florida (Robertson and Woolfenden 1992). Double-crested Cormorants (*Phalacrocorax auritus*) are common residents and fall–winter visitors on both coasts and inland. Nesting occurs throughout the year in southern Florida. The Great Cormorant (*Phalacrocorax carbo*) is a rare winter visitor along both coasts. Anhingas (*Anhinga anhinga*) are common residents throughout Florida except for the western panhandle, where they are uncommon. They are rare migrant visitors in the Keys during fall and winter.

Double-crested Cormorant populations in North America reached a low point in about 1970, and have increased since because of pro-

tection, increased food resources, and decreased pesticide use (Weseloh et al. 1995). The Florida subspecies, *P. a. floridanus,* is poorly known and may be declining (Hatch 1995). Because they nest year-round (Kushlan and McEwan 1982; Nesbitt et al. 1982), population numbers are difficult to estimate. Runde (1991) felt confident that within Florida there had been a shift in colony locations in the 1980s away from coastal areas to inland sites and an increase in more northern colonies when compared with locations in the 1970s. Coastal development and the creation of new habitat at phosphate mines might account for some of these changes. The population of Double-crested Cormorants that migrates to

Florida in the winter nests in the Great Lakes area (Dolbeer 1991).

O'Meara et al. (1986) presented some information on diets of Double-crested Cormorants on inland lakes in central and northern Florida. Clapp et al. (1982) published life history reviews of Great and Double-crested Cormorants. Kuiken et al. (1999) reported causes of mortality in a nesting colony of Double-crested Cormorants in Saskatchewan. Life history information on Anhingas can be found in Frederick and Siegel-Causey (2000). General information about health of these species can be found in Friend and Franson (1999).

II. Trauma

Injured and sick Double-crested Cormorants are presented commonly to rehabilitators in Florida. At one rehabilitation center in the Florida Keys, Monroe County, 12% of the 1,098 birds they received in 1993 were Double-crested Cormorants and only 10% of these recovered sufficiently to release back into the wild (Quinn 1994). Of 3 of these submitted for cause-of-death determination in 1992, 1 was emaciated, 1 had a broken wing due to entanglement in monofilament, and a third had a broken leg (Franson 1993). Two additional moribund birds were killed humanely and submitted to the Southeastern Cooperative Wildlife Disease Study in Athens, Georgia. One of these had been shot with steel shot (Smith and Nettles 1995) and a cause for emaciation was not found for the other (Little 1995).

Of 117 Double-crested Cormorants submitted for cause-of-death determination between 1973 and 1994 (most from a rehabilitation center in Pinellas County), fishhooks were found in the stomach of 2, a fish bone perforated the stomach of one, 2 collided with a power line or vehicle, 1 had a broken bill, 1 had a broken wing, 1 had a broken leg, and 5 were believed to have been attacked by sharks (Spalding and Forrester 1994). Two of 6 Double-crested Cormorants examined from a 1976–77 die-off in Lee County (see section XX, Unresolved die-

offs) had bony spines that penetrated the stomach, abdominal wall, and femoral musculature in 1 case, and small intestine in another (Kerr and Reichel 1977). A fishhook with attached beaded leader had perforated the esophagus of a Double-crested Cormorant examined from a multiple species die-off in Brevard County during the winter of 1992–93 (Quist 1993).

Road-kills are relatively uncommon for this group of birds. Snyder (1994) conducted a survey of roadkills from 1990 to 1993 from in and around state parks, and of a total of 1,562 birds found dead, 9 were Double-crested Cormorants and 1 was an Anhinga. Spalding and Forrester (1994) found that 2 of 117 cormorants submitted for cause-of-death determination had collided with a power line or vehicle.

Crawford (1981) recorded birds found dead below a TV tower on Tall Timbers Research station in Leon County between 1955 and 1980. It is presumed that most of these birds were killed during migration. Of the 42,384 birds of 189 species collected during this period only a single Double-crested Cormorant and no Anhingas were tallied.

III. Predation

Delany (1986) reported Anhinga remains in less than 1% of the stomachs of 350 American alligators from Alachua County, 1981–85. Ellis (1980) found a Double-crested Cormorant within the stomach of an exotic caiman (*Caiman crocodilus*) in Dade County. A juvenile pelican was found dead with a Double-crested Cormorant wedged in its pouch and the pouch was torn. It was speculated that both birds had drowned following capture by the pelican in Bay County (Francis 1981).

Predators of eggs and small nestlings include Fish Crows, vultures, and probably rat snakes (Stevenson and Anderson 1994).

IV. Human disturbance

Schreiber and Schreiber (1978) stated that nesting Double-crested Cormorants were very susceptible to disturbance and eggs were lost when birds fled nests or they were taken by Fish Crows. Investigators entering colonies have caused nest abandonment and egg predation in other locations (Ellison and Cleary 1978). Rodgers and Smith (1997) discussed buffer distances for this group of birds.

V. Inclement weather

Nestling Double-crested Cormorants and Anhingas were all killed by a hard freeze in a colony in Polk County, December 1962 (Funderburg 1963). After Hurricane Donna in September 1960 5 Double-crested Cormorants were found dead in Florida Bay and around Flamingo (Robertson and Paulson 1961).

VI. Organochlorines

Double-crested Cormorants were not included in the extensive egg collections used to document DDT contamination in the 1970s, but Anhingas were (see also chapter 4, Pelicans, and chapter 6, Herons, Egrets, and Bitterns). Only 5 cormorant eggs were tested and they contained relatively low concentrations of DDE when compared with those of Anhingas (tables 5.1 and 5.2). Twenty-eight Anhinga eggs collected from 3 locations in Florida during 1972–73 were tested. DDE concentrations were lower than those measured from Louisiana and Mississippi; however, PCB concentrations, especially at Merritt Island, were higher than in those measured elsewhere in the southeast. Residues occurred more often at inland sites than at coastal sites in the southeast (Ohlendorf et al. 1979). Residues were reported also for Anhinga eggs collected at a landfill site in Palm Beach County in the 1980s to 1990s (Rumbold et al. 1997) and no decline in contamination since the 1970s was noted.

Eggshell thinning has been associated with organochlorine contamination and is discussed in greater detail in chapter 4, Pelicans. Eggshell thickness as measured by a thickness index de-

clined by 7% when Double-crested Cormorant eggs collected in 1960 were compared with those collected before 1947 (prior to DDT use); however, this was based on a small sample size (table 5.3). Thinning on the order of 20% occurred in Wisconsin where high concentrations of DDE were found (Anderson et al. 1969) and was associated with a population decline (Anderson and Hamerstrom 1967). No decrease in thickness was found in Anhinga eggshells during this period by Ohlendorf et al. (1979). Rumbold et al. (1996) found an increase in thickness in Anhinga eggshells collected near a landfill in the 1980s and 1990s when compared with pre-1946 eggs.

Concentrations of chlorinated hydrocarbons measured in various tissues of Double-crested Cormorants and Anhingas in Florida are presented in tables 5.1–5.2. Lethal concentrations were never recorded. The highest DDE concentration in brain was 14 ppm from a Tampa Bay cormorant. Lethal concentrations for DDT are believed to begin at 200–300 ppm (Stickel et al. 1966; Henny and Meeker 1981). One bird tested from Monroe County had significantly elevated fat DDE (54 ppm) and fat PCBs (130); however, brain concentrations were not measured in this individual. Greichus and Hannon (1973) reported that DDD was the best indicator of DDT toxicity and that 30 ppm DDD wet weight, in brain tissue, was diagnostic of DDT toxicity in captive Double-crested Cormorants experimentally exposed to DDT. The highest DDD concentration reported for Florida cormorants was 0.56 ppm for brain.

Although no lethal concentrations were recorded, sublethal effects may have occurred in cormorants. Forrester and Thompson (1975) collected Double-crested Cormorants both by shooting and from rehabilitation centers on the west coast of Florida in the 1970s. In figure 5.1 these 2 groups are compared by using concentrations of various organochlorines in brain tissue. Significantly higher concentrations of DDE, DDT, DDD, dieldrin, and PCBs were found in birds submitted to rehabilitation centers. This result would be expected if sublethal concentrations caused birds to be-

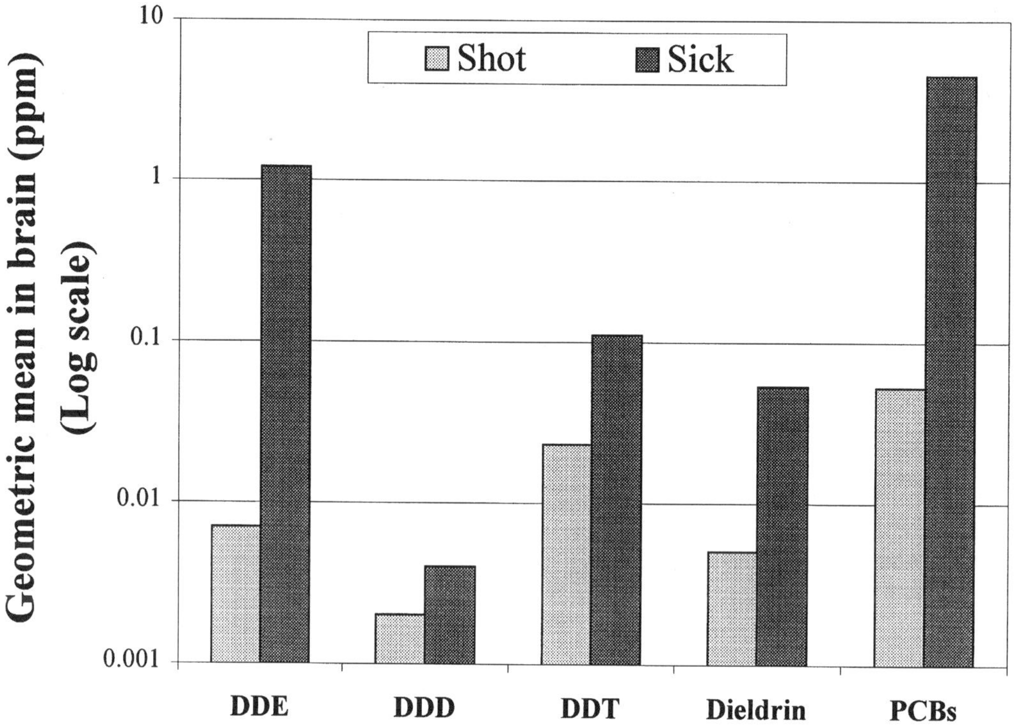

FIGURE 5.1. Geometric mean organochlorine residues in brains from Double-crested Cormorants that were collected by gunshot (*n* = 10) compared with those that died in rehabilitation centers (*n* = 19).

come ill, or caused birds to be more susceptible to traumatic hazards.

The organochlorine residues in brains and carcasses of 6 sick Double-crested Cormorants that were collected while still alive from the Gulf beach of Sanibel Island, Lee County, Florida, during the winter of 1976–77 are listed in table 5.1 (see also section XX, Unresolved die-offs). All displayed an inability to coordinate movements and loss of equilibrium. Residues were considered to be below those that would cause death or such signs (Kerr and Reichel, 1977).

Carcass concentrations of DDE in Double-crested Cormorants from Lee County in the 1970s (mean = 1.2 ppm) were somewhat higher than those reported by King (1989) and King et al. (1987) in Galveston Bay, Texas, in Olivaceous Cormorants, 1980 (mean = 0.8 ppm wet weight), 1982 (mean = 0.66 ppm), and 1983 (mean = 0.93 ppm).

Johnston (1976) reported the concentrations of contaminants in adipose tissue and uropygial glands of cormorants and Anhingas (tables 5.1–5.2). For all of the species that he tested, DDT and dieldrin in the adipose tissue were higher than in the uropygial gland (2.2:1 and 2.6:1, respectively), and for 1 species concentrations were higher in the expressed oil than in the gland itself (13% lipid). The uropygial gland can thus serve to excrete DDT. Concentrations were considerably higher in the Anhinga tested than in the 2 Double-crested Cormorants. He found no PCBs in fat or uropygial gland. PCB concentrations were measured in 5 Double-crested Cormorant eggs and 41 Anhinga eggs, and brains of 35 cormorants. Concentrations in eggs were well below those considered to be associated with reduced hatching and deformities (8 ppm), and in brains below lethality concentration (75 ppm) (Hoffman et al. 1996). The highest concentration in cormorants (130 ppm) was in the fat of a bird from Tampa Bay in 1973–75. The significance of sublethal concentrations is not well studied.

Wheeler et al. (1977) measured concentrations of mirex in tissues of a wide variety of species on a ranch in Duval and St. Johns counties over a 2-year period beginning in 1972. Mirex was applied as a 0.1% bait from the air at a rate of 1 pound per acre. Concentrations peaked at 6 months in fat of Anhingas and then declined to pretreatment concentrations after 2 years (table 5.4). The highest concentrations were found in insectivorous mammals and birds. Low concentrations were found in fish at 9 months; the greatest concentrations (0.33 ppm wet weight) occurred before treatment and none was detected after 9 months.

Dioxins and furans were measured in Anhinga nestlings and eggs in Palm Beach County in 1989–94 as part of a study of the effects of a landfill operation (Rumbold et al. 1997) (table 5.5). The colony tested was close to the landfill facility. Generally, concentrations were higher in eggs than in nestling tissues. They found concentrations of 2,3,7,8-TCDD in eggs (up to 9.5 ppb) that were higher than the experimental LD50 concentrations reported for pheasant embryos of 1.4 ppb (egg albumin injection) (Nosek et al. 1993). Rumbold et al. (1997) did not report embryo mortality or nestling deformity at this colony. Anhinga eggs contained higher concentrations of 2,3,7,8-TCDD than did eggs of White Ibises at the same colony and eggs of Double-crested Cormorants and Caspian Terns from the Great Lakes area where embryo mortality and deformities were noted (Yamashita et al. 1993). This is contradictory to another study where Anhingas in Florida had lower concentrations than those reported for both contaminated and control or "unimpacted" tern colonies in the Great Lakes (Stalling et al. 1985). It is likely that direct correlations between these contaminants and deformations and mortality are confused by the presence of multiple contaminants or other factors.

VII. Organophosphates and carbamates

No evidence of cholinesterase inhibition was found in a Double-crested Cormorant found dead with other seabirds in Lee County in 1987 (Stroud 1987), in 4 tested from die-offs in Monroe County in 1994–95 (Franson 1995; Brannian 1995), or in 1 that was emaciated and depressed in Monroe County, 1995 (Little 1995). These were the only reports we could find in which cormorants and Anhingas from Florida were tested for exposure to these compounds.

VIII. Metals

Metal concentrations for individual tissues and eggs of Double-crested Cormorants and Anhingas are listed in tables 5.6–5.7. Most of the information available comes from a study of mercury and selenium in Double-crested Cormorants submitted to a rehabilitation center in Florida Bay (Sepulveda et al. 1998). Mercury and selenium concentrations were correlated with each other, and both were significantly elevated (as high as 250 ppm mercury and 110 ppm selenium in liver). The concurrent elevation of these elements has been found in many fish-eating birds, and is thought to be due to the protective effect of selenium in birds exposed to mercury (Heinz and Hoffman 1998). It could not be determined whether the elevated mercury and/or selenium was responsible for illness in these birds. Healthy birds have not been sampled.

Considerable information on various trace metals in Double-crested Cormorants was available from a study of birds using phosphate settling ponds in Florida (O'Meara et al. 1986). They found moderately elevated concentrations of selenium in Double-crested Cormorants using the settling ponds when compared with those using natural ponds; these concentrations were below that which would be of concern for health. They did not test for mercury concentrations.

Rumbold et al. (1997) reported on metal

Table 5.1. Organochlorine residues in eggs and tissues of Double-crested Cormorants collected in Florida[f,g]

Tissue / County or area	Year(s)	No. exam.	DDE No.[a]	DDE Mean[b]	DDE (Range)	DDD No.[a]	DDD Mean[b]	DDD (Range)	DDT No.[a]	DDT Mean[b]	DDT (Range)	Data source
Eggs												
Monroe	1992	5	5	0.041	(0.031–0.53)	1	0.002	(ND–0.031)	0	ND	—	B
Brain												
Charlotte/Lee	1974	5[d]	5	1.2	(0.53–4.5)	5	0.089	(0.04–0.56)	5	0.17	(0.10–0.53)	A
Citrus/Levy	1975	10[c]	6	0.006	(ND–0.12)	2	0.002	(ND–0.054)	7	0.023	(ND–0.34)	A
"Tampa Bay"	1973–74	14[c]	14	1.1	(0.035–14)	1	0.001	(ND–0.02)	12	0.090	(ND–1.5)	A
Lee	1976–77	6[f]	6	2.9	(1.4–5.2)	0	ND	—	0	ND	—	D
Carcass												
Lee	1976–77	6[f]	6	1.2	(0.7–1.9)	0	ND	—	0	ND	—	D
Fat												
Citrus/Levy	1975	7[c]	7	0.54	(0.23–2.2)	5	0.022	(ND–5.6)	6	0.13	(ND–5.2)	A
Monroe	1973	1[d]	1	54	—	0	ND	—	1	19	—	A
"Tampa Bay"	1973–74	1[d]	1	4.0	—	0	ND	—	1	0.64	—	A
"Florida"	1974	2[f]	2	1.2	(0.70–2.2)	NA	—	—	NA	—	—	C
Liver												
Charlotte/Lee	1974	2[d]	2	3.6	(1.8–7.1)	2	0.39	(0.30–0.50)	1	0.060	(ND–3.6)	A
Citrus/Levy	1975	9[c]	9	0.040	(0.013–0.094)	2	0.003	(ND–0.15)	7	0.027	(ND–0.39)	A
Monroe	1973	1[d]	1	2.1	—	0	ND	—	1	0.88	—	A
"Tampa Bay"	1973–74	13[c]	13	2.0	(0.12–31)	0	ND	—	13	0.21	(ND–14)	A
Muscle												
Charlotte/Lee	1974	1[d]	1	0.55	—	1	0.87	—	0	ND	—	A
Citrus/Levy	1975	10[c]	9	0.045	(ND–0.31)	0	ND	—	4	0.006	(ND–0.13)	A
Monroe	1973	1[d]	1	2.9	—	0	ND	—	1	0.81	—	A
"Tampa Bay"	1973–75	12[d]	13	2.5	(0.21–11)	0	ND	—	11	0.60	(ND–4.2)	A
Uropygial gland												
"Florida"	1974	2[f]	2	1.0	(0.73–1.4)	NA	—	—	NA	—	—	C

Tissue / County or area	Year(s)	No. exam.	Dieldrin No.[a]	Dieldrin Mean[b]	Dieldrin (Range)	PCBs No.[a]	PCBs Mean[b]	PCBs (Range)	Mirex No.[a]	Mirex Mean[b]	Mirex (Range)	Data source
Eggs												
Monroe	1972	5	0	ND	—	1	0.003	(0.25)	NA	—	—	B

	Year	n										Source
Brain												
Charlotte/Lee	1974	5[d]	4	0.066	(ND–0.95)	5	7.4	(2.9–20)	NA	—	—	A
Citrus/Levy	1975	10[c]	5	0.006	(ND–0.20)	6	0.021	(ND–0.77)	NA	—	—	A
"Tampa Bay"	1973–74	14[d]	13[e]	0.049	(ND–1.9)	14	3.8	(0.13–63)	NA	—	—	A
Lee	1976–77	6[f]	1	0.1	—	6	7.1	(2.6–20)	1	0.14	—	D
Carcass												
Lee	1976–77	6[f]	0	ND	—	6	2.9	(1.5–7.6)	1	0.13	—	D
Fat												
Citrus/Levy	1975	7[c]	4	0.010	(ND–0.16)	5	0.15	(ND–7.2)	NA	—	—	A
Monroe	1973	1[d]	1	0.059	—	1	130	—	NA	—	—	A
"Tampa Bay"	1973–74	1[d]	1	1.9	—	1	11	—	NA	—	—	A
"Florida"	1974	2[f]	0	ND	—	0	ND	—	NA	—	—	C
Liver												
Charlotte/Lee	1974	2[d]	1	0.036	(ND–1.3)	2	13	(5–34)	NA	—	—	A
Citrus/Levy	1975	9[c]	9	0.006	(ND–0.10)	6	0.38	(ND–1.1)	NA	—	—	A
Monroe	1973	1[d]	0	ND	—	1	6.3	—	NA	—	—	A
"Tampa Bay"	1973–74	13[d]	13	0.17	(0.009–3.2)	13	7.0	(0.32–110)	NA	—	—	A
Muscle												
Charlotte/Lee	1974	1[d]	0	ND	—	1	1.7	—	NA	—	—	A
Citrus/Levy	1975	10[c]	2	0.002	(ND–0.030)	9	0.94	(ND–1.5)	NA	—	—	A
Monroe	1973	1[d]	1	0.041	—	1	8.1	—	NA	—	—	A
"Tampa Bay"	1973–74	12[d]	13	0.24	(0.006–1.0)	13	14	(0.23–61)	NA	—	—	A
Uropygial gland												
"Florida"	1974	2[f]	1	0.34	—	0	ND	—	NA	—	—	C

Sources: A = Forrester & Thompson (1975), B = Ogden et al. (1974), C = Johnston (1976), D = Kerr & Reichel (1977).

NA = not analyzed, ND = not detected.

a. Number with residue.

b. Geometric mean, ppm wet weight.

c. Birds collected by Forrester & Thompson (1975) from Citrus/Levy counties were collected by gun shot.

d. Sick and injured birds from rehabilitation centers (Forrester & Thompson 1975).

e. Number sampled = 13.

f. In addition to the compounds listed, Johnston (1976) also measured 1.5 (0.79–3.0) ppm total DDTs in fat and 1.4 (1.2–1.7) in uropygial gland in 2 cormorants from Florida. Kerr & Reichel (1977) measured 0.14 (0.11–0.19) ppm oxychlordane in 5 of 6 brains and none of 6 carcasses; 0.22 (0.13–0.35) ppm cis-chlordane in 4 of 6 brains and none of 6 carcasses; and they did not detect heptachlor epoxide, trans-nonachlor, cis-nonachlor, endrin, toxaphene, or HCB in any of the brains or carcasses.

g. In addition to the cases listed, no compounds were detected in liver and brain of 4 cormorants examined from a 1995 die-off in Monroe County (Franson 1995). Brannian (1995) reported trimethoprim, o,p'-DDE and PCBs in tissues from 2 cormorants collected from Monroe County in 1994, however no quantitative data were obtained; endosulfan was not detected.

Table 5.2. Organochlorine residues in eggs and tissues of Anhingas collected in Florida[f]

Age / County or area)	Year(s)	No. exam.	DDE No.[a]	DDE Mean[b]	DDE (Range)	DDD No.[a]	DDD Mean[b]	DDD (Range)	DDT No.[a]	DDT Mean[b]	DDT (Range)	Data source
Eggs												
Alachua	1972–73	7	7	0.76	(0.36–1.6)	0	ND	—	6	0.58	(ND–7.2)	B
Brevard	1972–73	10	9	0.39	(ND–0.93)	0	ND	—	7	1.1	(ND–6.4)	B
"Lake Okeechobee"	1992–73	9	9	0.44	(0.08–1.30)	0	ND	—	0	ND	—	C
Lee	1972–73	11	11	0.41	(0.12–2.5)	1	0.02	(ND–0.18)	6	0.21	(ND–0.72)	B
Osceola	1992–73	4	4	0.14	(0.09–0.18)	0	ND	—	0	ND	—	C
Palm Beach	1989	5[c]	5	0.71	(0.12–2.1)	0	ND	—	NA[d]	—	—	D
	1990	5[c]	5	0.43	(0.11–2.0)	4	0.020	(ND–0.20)	NA[d]	—	—	D
	1991	5[c]	5	0.29	(0.077–0.88)	3	0.006	(ND–0.23)	NA[d]	—	—	D
Nestlings												
Carcass												
Palm Beach	1989	5[c]	5	0.075	(0.017–0.25)	0	ND	—	NA[d]	—	—	D
	1990	5[c]	5	0.049	(0.021–0.36)	3	0.003	(0.003–0.015)	NA[d]	—	—	D
	1991	5[c]	5	0.052	(0.021–0.28)	3	0.008	(0.017–0.077)	NA[d]	—	—	D
Adult												
Adipose												
"Florida"	1972	1[h]	1	20	—	0	ND	—	0	ND	—	A
Uropygial gland												
"Florida"	1972	1[h]	1	23	—	0	ND	—	0	ND	—	A

Age / County or area)	Year(s)	No. exam.	Heptachlor No.[a]	Heptachlor Mean[b]	Heptachlor (Range)	Oxychlordane No.[a]	Oxychlordane Mean[b]	Oxychlordane (Range)	Mirex No.[a]	Mirex Mean[b]	Mirex (Range)	Data source
Eggs												
Alachua	1972–73	7	NA	—	—	1	0.1	(ND–0.08)	2	0.05	(ND–0.24)	B
Brevard	1972–73	10	NA	—	—	0	ND	—	0	ND	—	B
"Lake Okeechobee"	1992–93	9	0	ND	—	NA	—	—	NA	—	—	C
Lee	1972–73	11	NA	—	—	1	0.01	(ND–0.14)	0	ND	—	B

Osceola	1992–93	4	0	ND	—	NA	—	—	NA	—	—	C
Palm Beach	1989	5[c]	0	ND	—	NA	—	—	NA	—	—	D
	1990	5[c]	0	ND	—	NA	—	—	NA	—	—	D
	1991	5[c]	1	NC	(ND–0.003)	NA	—	—	NA	—	—	D
Nestlings												
Carcass												
Palm Beach	1989	5[c]	0	ND	—	NA	—	—	NA	—	—	D
	1990	5[c]	0	ND	—	NA	—	—	NA	—	—	D
	1991	5[c]	0	ND	—	NA	—	—	NA	—	—	D
Age Not Stated												
Brain												
Duval/St. Johns	1972	1[e]	NA	—	—	NA	—	—	1	0.03	(NG)	E
Fat												
Duval/St. Johns	1972	1[e]	NA	—	—	NA	—	—	1	0.06	(NG)	E
Liver												
Duval/St. Johns	1972	1[e]	NA	—	—	NA	—	—	1	0.06	(NG)	E
Muscle												
Duval/St. Johns	1972	1[e]	NA	—	—	NA	—	—	1	0.02	(NG)	E
Stomach contents												
Duval/St. Johns	1972	1[e]	NA	—	—	NA	—	—	1	0.01	(NG)	E

Sources: A = Johnston (1976), B = Ohlendorf et al. (1979), C = Rodgers (1997), D = Rumbold et al. (1996), E = Wheeler et al. (1977).

NA = not analyzed, ND = Not detected, NC = not calculated, NG = not given.

a. Number with residue.

b. Geometric mean, ppm wet weight.

c. Ten nestlings and 10 eggs were sampled and paired to produce 5 samples of each. Bills, gastrointestinal tract, legs, and feathers were removed from nestlings.

d. Specific dioxin and furan data listed in Table 5.5.

e. A pooled sample of tissues from 3 birds. These are pretreatment concentrations (see Table 5.3).

f. In addition to the compounds listed, Ohlendorf et al. (1979) did not detect DDT, DDD, cis-chlordane, cis-nonachlor, HCB, or toxaphene in 17 eggs from Alachua (7) and Osceola (10) counties. Rumbold et al. (1996) detected heptachlor epoxide (ND–0.012 ppm) in 2 of 5 pooled egg samples and endosulfan II in 1 of 5 pooled egg samples (ND - 0.12 ppm dry weight) from Palm Beach County (geometric means were not calculated). He did not detect DDT, BHC, endosulfan I, endrin, endosulfan sulfate, methoxychlor, chlordane, or toxaphene in any of the pooled egg and pooled nestling-carcass samples in 1989 and 1990. Rodgers (1997) detected DDT in Osceola County only (he did not detect DDT at Lake Okeechobee) and reported a range of (ND–0.12) ppm; he did not detect DDD, chlordane, endrin, toxaphene, or trans-nonachlor in any of the 13 eggs from either Osceola County (4) or Lake Okeechobee. Johnston (1976) also measured total DDTs for the single nestling from Florida (26 ppm for adipose tissue and 24 ppm for uropygial gland).

Table 5.3. Eggshell weight and thickness of Double-crested Cormorant and Anhinga eggs from Florida

County/area	Year(s)	No. examined	Shell weight ± SD (g)	Percent change[a]	No. examined	Shell thickness ± SD or (range)(mm)	Percent change[a]	Thickness index[b]	Percent change[a]	Data source
Double-crested Cormorants										
"Florida"	pre-1947	194	4.45 ± 0.07	—	126	0.398 ± 0.004	—	1.96 ± 0.02	—	A
	1947	7	4.88 ± 0.39	NS	NA	—	—	2.09 ± 0.11	NS	A
	1953	8	4.53 ± 0.26	NS	NA	—	—	1.89 ± 0.05	NS	A
	1960	6	4.51 ± 0.33	+1	NA	—	—	1.82 ± 0.05	-7	A
Monroe	1972	NA	—	—	5	0.436 (0.041–0.49)	—	NA	—	B
Anhingas										
"Florida"	pre-1946	NA	—	—	104	0.343	—	NA	—	C
	1947–73	NA	—	—	45	0.345	NS	NA	—	C
Palm Beach	1989–91	NA	—	—	31	0.37 ± 0.0	+9	NA	—	D

Sources: A = Anderson & Hickey (1972), B = Ogden et al. (1974), C = Ohlendorf et al. (1979), D = Rumbold et al. (1996).
NA = not analyzed, NS = not significantly different from pre-1947 measurement.
a. Compared with pre-1947 data.
b. Thickness index = weight of shell (mg) divided by product of length x breadth (mm) (Ratcliffe 1967).

Table 5.4. Mirex residues in tissues of Anhingas after a single treatment of 10-5 bait, on a ranch in Duval and St. Johns counties, 1972–1974

	Amount of residue[a] (no. pooled samples)[b]							
		Months after treatment						
Tissue	Pretreatment	1	3	6	9	12	18	24
Brain	0.03 (3)	ND (3)	ND	0.03 (3)	0.06	0.88	ND	0.01 (2)
Fat	0.06 (3)	ND (3)	0.47	4.98 (2)	1.90	2.25	0.03	ND (2)
Liver	0.06 (3)	ND (3)	0.11	0.18 (3)	0.23	0.55	ND	0.01 (2)
Muscle	0.02 (3)	ND (3)	ND	0.15 (3)	0.19	0.75	ND	0.03 (2)
Stomach contents	0.01 (3)	ND (3)	NA	ND (3)	NA	NA	NA	NA

Source: Wheeler et al. (1977).
NA = not analyzed, ND = not detected (at about 0.01 ppm).
a. Values are ppm wet weight.
b. Sample size = 1 if not indicated.

Table 5.5. Dioxin and furan residues in eggs and tissues of nestling Anhingas collected in Florida

Tissue	County	Year	No. examined	Total TCDD			2,3,7,8-TCDD			Total TCDF			2,3,7,8-TCDF		
				No.[a]	Mean[b]	(Range)	No.[a]	Mean[b]	(Range)	No.[a]	Mean[b]	(Range)	No.[a]	Mean[b]	(Range)
Eggs															
	Palm Beach	1989	5[c]	4	1.6	(ND–2.6)	4	1.6	(ND–2.6)	4	0.2	(ND–0.4)	4	0.2	(ND–0.1)
		1990	5[c]	5	3.6	(NG–9.5)	5	3.6	(NG–9.5)	5	0.3	(NG–0.5)	5	0.2	(NG–0.4)
		1991	5[c]	4	1.4	(ND–2.3)	4	1.3	(ND–2.1)	3	0.3	(ND–0.8)	3	0.2	(ND–0.3)
		1994	5[c]	4	0.3	(ND–0.4)	4	0.3	(ND–0.4)	3	0.2	(ND–0.3)	2	NC	(ND–0.2)
Nestling carcasses															
	Palm Beach	1989	5[c]	1	NC	(ND–1.3)	1	NC	(ND–1.3)	2	NC	(ND–4.6)	2	NC	(ND–0.4)
		1990	5[c]	1	NC	(ND–0.5)	1	NC	(ND–0.5)	1	NC	(ND–0.2)	1	NC	(ND–0.2)
		1991	5[c]	1	NC	(ND–0.2)	1	NC	(ND–0.2)	4	0.2	(ND–0.3)	4	0.2	(ND–0.3)
		1994	5[c]	5	0.1	(NG–0.3)	1	NC	(ND–0.2)	5	0.3	(NG–0.3)	5	0.2	(NG–0.2)

Source: Rumbold et al. (1997).
NC = not calculated (<50% of samples contained the residue), ND = not detected, NG = not given.
a. Number with residue.
b. Geometric mean, ppm wet weight.
c. Ten nestlings and 10 eggs were sampled and paired to produce 5 samples of each. Bills, gastrointestinal tract, legs, and feathers were removed from nestlings.

Table 5.6. Metal concentrations in eggs and tissues of Double-crested Cormorants from Florida

Tissue	County/area	Year(s)	No. examined	Lead Mean[a]	Lead (Range or ±SE[b])	Mercury Mean[a]	Mercury (Range or ±SE[b])	Cadmium Mean[a]	Cadmium (Range or ±SE[b])	Data source
Egg										
	Monroe	1972–75	5	0.2	(0.20–0.40)	0.39	(0.31–0.44)	0.05	(0.05–0.05)	A
Bone										
	Alachua	1981–82	18	3.3[c]	(±1.2)	NA	—	NA	—	B
	Hamilton[d]	1981–82	19	1.9[c]	(±0.5)	NA	—	NA	—	B
	Osceola	1981–82	19	2.8[c]	(±0.7)	NA	—	NA	—	B
	Polk[d]	1981–82	17	5.3[c]	(±2.0)	NA	—	NA	—	B
Brain										
	"Florida Bay"	1994–97	29[e]	NA	—	1.6[f]	(0.05–8.4)	NA	—	C
Kidney										
	"Florida Bay"	1994–97	64[e]	NA	—	7.1[f]	(0.25–77)	NA	—	C
		1995	1	ND	—	0.97	—	NA	—	J
Liver										
	Brevard	1977	1[g]	NA	—	2.3	—	NA	—	D
		1993	1	NA	—	9.8	—	NA	—	E
	Pinellas	1975	1	NA	—	6.9	—	NA	—	D
	Sarasota	1975	2	NA	—	19	(18–20)	NA	—	D
	Lee	1990	1	ND	—	NA	—	NA	—	F
	Monroe	1993	1	ND	—	NA	—	NA	—	G
	"Florida Bay"	1994	2[e]	ND	—	NA	—	NA	—	I
		1994–97	84	NA	—	24	(2.2–200)	NA	—	C
		1995	4	ND	—	5.3	(1.1–11)	NA	—	H
		1995	1	ND	—	2.2	—	NA	—	J

Tissue	County/area	Year(s)	No. examined	Selenium Mean[a]	Selenium (Range or ±SE[b])	Copper Mean[a]	Copper (Range or ±SE[b])	Zinc Mean[a]	Zinc (Range or ±SE[b])	Data source
Egg										
	Monroe	1972	5	NA	—	0.9	(0.66–1.25)	7.7	(7.2–8.4)	A
Bone										
	Alachua	1981–82	18	NA	—	5.0[c]	(±0.50)	220[c]	(±19)	B
	Hamilton[d]	1981–82	19	NA	—	4.3[c]	(±0.40)	220[c]	(±16)	B

Tissue	Location	Year	N							
	Osceola	1981–82	19	NA	—	4.1[c]	(±0.50)	200[c]	(±14)	B
	Polk[d]	1981–82	17	NA	—	5.8[c]	(±0.80)	220[c]	(±26)	B
Brain	"Florida Bay"	1994–97	33[e]	0.78[f]	(0.18–11)	NA	—	NA	—	C
Kidney	Alachua	1981–82	17	1.3[h]	(±0.18)	2.4[h]	(±0.15)	17[h]	(±0.66)	B
	"Florida Bay"	1994–97	30[e]	6.3[f]	(1.1–43)	NA	—	NA	—	C
		1995	1	1.1	—	NA	—	NA	—	J
	Hamilton[d]	1981–82	20	2.8[h]	(±0.36)	2.8[h]	(±0.14)	20[h]	(±0.84)	B
	Osceola	1981–82	19	1.6[h]	(±0.14)	3.1[h]	(±0.09)	20[h]	(±0.46)	B
	Polk[d]	1981–82	13	3.0[h]	(±0.57)	2.9[h]	(±0.09)	20[h]	(±0.66)	B
Liver	Alachua	1981–82	17	2.2[h]	(±0.67)	4.2[h]	(±0.35)	28[h]	(±2.3)	B
	"Florida Bay"	1994–97	65[e]	11[f]	(0.78–110)	NA	—	NA	—	C
		1995	1	0.94	—	NA	—	NA	—	J
	Hamilton[d]	1981–82	20	5.8[h]	(±1.1)	4.2[h]	(±0.21)	27[h]	(±1.6)	B
	Osceola	1981–82	19	2.1[h]	(±0.31)	5.5[h]	(±0.28)	30[h]	(±1.1)	B
	Polk[d]	1981–82	17	5.0[h]	(±1.4)	5.3[h]	(±0.60)	32[h]	(±1.6)	B

Tissue	Location	Year	N	Bromine		Manganese		Rubidium		
Bone	Alachua	1981–82	18	24[c]	(±2.6)	5.7[c]	(±0.50)	21[c]	(±3.0)	B
	Hamilton[d]	1981–82	19	15[c]	(±1.4)	5.4[c]	(±0.40)	18[c]	(±2.6)	B
	Osceola	1981–82	19	17[c]	(±1.5)	5.2[c]	(±0.50)	12[c]	(±1.7)	B
	Polk[d]	1981–82	17	22[c]	(±2.5)	6.8[c]	(±1.0)	22[c]	(±3.5)	B
Kidney	Alachua	1981–82	17	12[h]	(±0.88)	1.4[h]	(±0.11)	11[h]	(±0.53)	B
	Hamilton[d]	1981–82	20	9.6[h]	(±0.48)	1.5[h]	(±0.12)	13[h]	(±0.58)	B
	Osceola	1981–82	19	13[h]	(±1.2)	1.9[h]	(±0.09)	6.4[h]	(±0.28)	B
	Polk[d]	1981–82	18	13[h]	(±0.44h)	1.6[h]	(±0.11)	14[h]	(±0.64)	B
Liver	Alachua	1981–82	17	11[h]	(±0.96)	2.9[h]	(±0.26)	12[h]	(±0.55)	B
	Hamilton[d]	1981–82	20	9.9[h]	(±0.94)	2.6[h]	(±0.16)	12[h]	(±0.73)	B
	Osceola	1981–82	19	13[h]	(±0.95)	4.6[h]	(±0.14)	8.7[h]	(±0.98)	B
	Polk[d]	1981–82	17	12[h]	(±0.83)	3.0[h]	(±0.26)	14[h]	(±0.75)	B

(continued)

Table 5.6. (continued)

Tissue	County/area	Year(s)	No. examined	Aluminum		Vanadium		Arsenic		Data source
				Mean[a]	(Range or ±SE[b])	Mean[a]	(Range or ±SE[b])	Mean[a]	(Range or ±SE[b])	
Egg					—		—		—	
	Monroe	1972–75	5	NA		NA		ND		A
Bone										
	Alachua	1981–82	16	130[c]	(±9.0)	0.36[c]	(±0.07)	NA	—	B
	Hamilton[d]	1981–82	19	160[c]	(±8.0)	0.22[c]	(±0.04)	NA	—	B
	Osceola	1981–82	16	140[c]	(±5.0)	0.33[c]	(±0.06)	NA	—	B
	Polk[d]	1981–82	18	160[c]	(±15)	0.37[c]	(±0.09)	NA	—	B
Kidney										
	Alachua	1981–82	10	2.2[h]	(±1.1)	0.05[h]	(±0.02)	NA	—	B
	Hamilton[d]	1981–82	14	2.4[h]	(±0.48)	0.04[h]	(±0.007)	NA	—	B
	Osceola	1981–82	17	12[h]	(±8.7)	0.08[h]	(±0.02)	NA	—	B
	Polk[d]	1981–82	12	2.0[h]	(±0.44)	0.05[h]	(±0.01)	NA	—	B
Liver										
	Alachua	1981–82	15	4.8[h]	(±2.0)	0.07[h]	(±0.03)	NA	—	B
	Hamilton[d]	1981–82	17	8.6[h]	(±2.2)	0.02[h]	(±0.003)	NA	—	B
	Osceola	1981–82	19	3.1[h]	(±0.34)	0.03[h]	(±0.01)	NA	—	B
	Polk[d]	1981–82	8	7.1[h]	(±4.7)	0.03[h]	(±0.01)	NA	—	B

Sources: A = Ogden et al. (1974), B = O'Meara et al. (1986), C = Sepulveda et al. (1998), D = Gourlie (1984), E = Quist (1993), F = Franson (1990), G = Franson (1993), H = Franson (1995), I = Brannian (1995), J = Little (1995).
NA = not analyzed, ND = not detected.
a. Geometric mean, ppm wet weight unless otherwise indicated.
b. Standard error.
c. O'Meara et al. (1986) presented bone values as ash weight.
d. Collection sites in Hamilton and Polk counties were phosphate-mine settling ponds.
e. Samples from adults and fledged juveniles.
f. Arithmetic mean.
g. Chick.
h. Kidney and liver means and standard errors in O'Meara et al. (1986) have been converted from dry weight to wet weight values using conversion values presented in the report.

concentrations in Anhinga eggs and nestlings from a colony located near a landfill. Mercury concentrations in eggs (<1 ppm) were similar to those reported for wading birds in the Everglades (see chapter 6, Herons, Egrets, and Bitterns), but lower than those reported for ibises in the same study. It is interesting to note that Anhinga nestling carcasses contained more mercury than those of nestling White Ibises in the same study, suggesting that the Anhinga food had higher concentrations than the ibis food. The Anhingas did not forage on the landfill itself, while the ibises did.

IX. Radionuclides

Radium-226 concentrations were 1.6–1.8 times higher in bones of Double-crested Cormorants collected at phosphate settling ponds when compared with natural lakes in Florida. They were similar to concentrations in ducks, but below concentrations considered hazardous in humans (table 5.8) (Myers et al. 1989; O'Meara et al. 1986). Researchers did not find a correlation between dietary contamination and tissue contamination, probably because of foraging by the cormorants at other

locations. See chapter 10, Ducks, for additional discussion on this topic.

X. Oiling

An oil spill in Tampa Bay during the winter of 1970 killed large numbers of birds, including Double-crested Cormorants (Sims 1970; Stevenson 1970). A spill also occurred in Jacksonville; however, a list of birds killed was not reported (Stevenson 1970). An adult female Double-crested Cormorant found sick on Madeira Beach, Johns Pass, Pinellas County, in 1974 was suspected to have gotten into diesel oil (Forrester 1978). Clapp et al. (1982) felt that Double-crested Cormorants were probably more susceptible to population losses from oil development than any other species in the southeast United States.

XI. Neoplasia and anomalies

We found no records of neoplasia in Double-crested Cormorants or Anhingas in Florida. Johnson (1999) found a single Double-crested Cormorant embryo with a malformed bill (figure 5.2) in Lake County in 1999. Developmen-

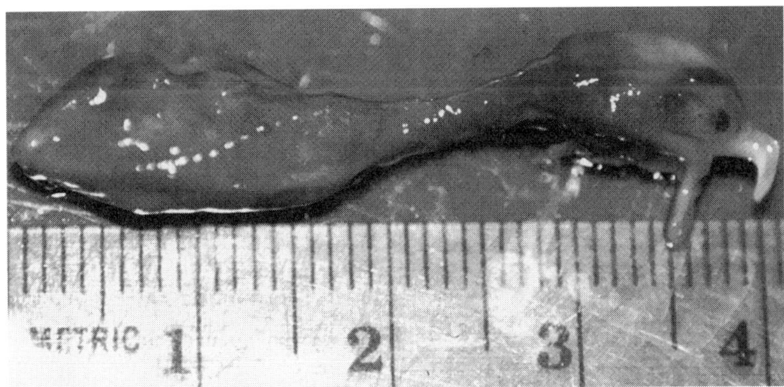

FIGURE 5.2. Double-crested Cormorant embryo with a malformed maxilla. Courtesy of Jacqueline M. Farley.

Table 5.7. Metal concentrations in eggs and tissues of Anhingas from Florida[d]

Tissue	County/area	Year	No. examined	Lead		Mercury		Selenium		Data source
				Mean[a]	(Range)	Mean[a]	(Range)	Mean[a]	(Range)	
Eggs										
	"Lake Okeechobee"	1992–93	9	NC	(ND–0.15)	0.13	(0.04–0.85)	NA	—	A
	Osceola	1992–93	4	0	ND	0.05	(0.03–0.13)	NA	—	A
	Palm Beach	1989	5[b]	NC	(ND–0.32)[c]	0.22[c]	(0.12–0.41)	0.58[c]	(0.49–0.68)	B
		1990	5[b]	ND	—	0.20[c]	(0.09–0.34)	0.61[c]	(0.46–0.71)	B
		1991	5[b]	NC	(ND–0.41)[c]	0.22[c]	(0.10–0.32)	0.56[c]	(0.44–0.73)	B
		1994	5[b]	ND	—	0.05[c]	(NG–0.12)	0.41[c]	(NG–0.49)	C
Nestlings										
	Palm Beach	1989	5[b]	NC	(ND–0.71)[c]	0.17[c]	(0.08–0.27)	0.44[c]	(0.23–0.67)	B
		1990	5[b]	ND	—	0.11[c]	(0.04–0.21)	0.38[c]	(0.27–0.57)	B
		1991	5[b]	NC	(ND–0.36)[c]	0.21[c]	(0.15–0.27)	0.36[c]	(0.25–0.59)	B
		1994	5[b]	NC	(NG–0.30)[c]	0.17[c]	(NG–0.27)	0.23[c]	(NG–0.32)	C

Tissue	County/area	Year	No. examined	Nickel		Cadmium		Beryllium		Data source
				Mean[a]	(Range)	Mean[a]	(Range)	Mean[a]	(Range)	
Eggs										
	Palm Beach	1989	5[b]	NC	(ND–0.26)[c]	ND	—	NC	(ND–0.05)[c]	B
		1990	5[b]	NC	(ND–1.76)[c]	ND	—	ND	—	B
		1991	5[b]	ND	—	ND	—	ND	—	B
		1994	5[b]	ND	—	ND	—	ND	—	C
Nestlings										
	Palm Beach	1989	5[b]	0.42[c]	(ND–0.82)	ND	—	ND	—	B
		1990	5[b]	ND	—	ND	—	ND	—	B
		1991	5[b]	ND	—	NC	(ND–0.06)[c]	ND	—	B
		1994	5[b]	ND	—	ND	—	ND	—	C

Sources: A = Rodgers (1997), B = Rumbold (1992), C = Rumbold et al. (1997)
NA = not analyzed, NC = not calculated (<50% of samples contained the residue), ND = not detected.
a. Geometric mean, ppm wet weight unless otherwise indicated.
b. Ten nestlings and 10 eggs were sampled and paired to produce 5 samples of each. Bills, gastrointestinal tract, legs, and feathers were removed from nestlings.
c. Wet weight values have been calculated using the formula given in Rumbold (1992).
d. In addition to the metals listed, Rumbold (1992) and Rumbold et al. (1997) did not detect arsenic in any of the samples.

Table 5.8. Concentrations of radium-226 in tissues of Double-crested Cormorants in Florida, 1981–82

Tissue	County	No. examined (No. pooled samples)	Mean[a]	(Range)	±SE[b]	Data source
Bone						
	Alachua	10(2)	3.0	(2.6–3.3)	NG	A
	Hamilton[c]	15(3)	4.6	(4.4–4.8)	NG	A
	Osceola	20(2)	3.7	(3.0–4.4)	NG	A
	Polk[c]	16(3)	4.4	(5.2–7.4)	NG	A
Kidney						
	Alachua	20(2)	0.090[d]	(NG)	±0.14[d]	B
	Hamilton[c]	9(1)	0.041[d]	(NG)	±0.053[d,e]	B
	Osceola	21(2)	0.120[d]	(NG)	±0.051[d]	B
	Polk[c]	20(2)	0.024[d]	(NG)	±0.040[d]	B
Liver						
	Alachua	15(3)	0.032[d]	(NG)	NG	B
	Hamilton[c]	10(2)	0.018[d]	(NG)	±0.018[d]	B
	Osceola	15(3)	0.062[d]	(NG)	NG	B
	Polk[c]	20(4)	0.057[d]	(NG)	NG	B

Sources: A = Myers at al. (1989), B = O'Meara et al. (1986).
NG = not given.
a. Values are Becquerel per kg wet weight unless otherwise indicated.
b. Standard error.
c. Collection sites in Hamilton and Polk counties were phosphate-mine settling ponds.
d. Dry weight values presented in O'Meara et al. (1986) have been converted to wet weight values using conversion ratios presented in the report.
e. For results based on a single sample, value is the estimated error of the analytical determination.

tal anomalies have been reported from cormorants in the Great Lakes area; although suspected to be due to environmental contaminants, the specific cause remains uncertain (Larson et al. 1996).

XII. Biotoxins

Mortality of Double-crested Cormorants from red tide exposure probably occurs frequently, but had never been documented with certainty and remains poorly documented in terms of dates and numbers. Blooms of the dinoflagellate *Karenia brevis* (= *Gymnodinium breve*) and fish kills occur on the west coast of Florida every 3–5 years (Morton and Burkley 1969) and appear to be increasing in frequency. From October 1973 to May 1974 a red tide occurred in Boca Ciega and Tampa Bay, Pinellas County. Double-crested Cormorants, Red-breasted Mergansers and Lesser Scaup (see chapter 10, Ducks)

were found moribund and dead (Quick and Henderson 1975). Experimental feeding of *Karenia brevis* toxins, the organism associated with this red tide, to White Pekin ducklings produced behavioral effects and death similar to those seen in the Lesser Scaup (Forrester et al. 1977). Mortality of Double-crested Cormorants (number not specified) declined early in the epizootic and many of the cormorants and mergansers responded to antibiotic therapy, suggesting that these species might have been ill from some other cause (Heath and Albers 1974). See chapter 10, Ducks, for more details about this epizootic. Cormorants in England have died from exposure to dinoflagellate toxins (Coulson et al. 1968; Armstrong et al. 1978).

Mortality from domoic acid poisoning has been reported in Brown Pelicans and Brant's Cormorants (*Phalacrocorax penicillatus*) in California that probably acquired it from eating northern anchovies that contained the poison (Work et al. 1993). No reports of this poison were found in records from Florida, nor could we find any reports of birds being tested for this toxin.

Botulism has never been documented for Double-crested Cormorants or Anhingas in Florida. Six Double-crested Cormorants from a die-off during the 1976–77 winter on Sanibel Island tested negative for types C and E botulism (see section XX, Unresolved die-offs) (Kerr and Reichel 1977). Two of 3 Double-crested Cormorants submitted to a rehabilitation center in Monroe County, 1992, also tested negative for botulism types C and D (Franson 1993). One of 4 was tested for botulism type C and 1 for both C and E from a die-off of 25 birds in Monroe County in 1995. Tests were negative (Franson 1995).

XIII. Viruses

Newcastle Disease was diagnosed by egg inoculation from a group of 5 Double-crested Cormorants that were found dying, some with convulsions, on a beach in Pinellas County in 1980. This apparently was not a particularly virulent strain, since it took 7 days to cause mortality in chicken eggs. Hemorrhagic gastroenteritis and inflammatory infiltrates in the liver, intestine, and lungs were not further characterized (Forrester and Hanley 1980). Another report of Newcastle disease virus (NDV) was from an Anhinga from an animal attraction in the Orlando area. Seal et al. (1995) listed this isolate as a mesogenic strain that was very similar to the isolate from cormorants during the 1992 epizootic in Canada (Wobeser et al. 1993) and Michigan (Banerjee et al. 1994). No other information was available about the health of that Anhinga. Farley et al. (2001; see also Johnson 1999) surveyed Double-crested Cormorant eggs and nestlings in Florida in 1998–99 and found that 78% of eggs (*n* = 126) from 11 colonies in Florida (figure 5.3) had antibodies to NDV (using hemagglutination inhibition with B1-LaSota strain). Since they could not isolate any virus from

these, it is likely these antibodies were maternally derived. They could find no antibodies (*n* = 45) in blood, and no virus (*n* = 51) in cloacal swabs from chicks in these same colonies. Newcastle disease does cause mortality in nesting Double-crested Cormorants in Canada and northern United States (Wobeser et al. 1993; Glaser et al. 1999; Kuiken et al. 1999).

One of 4 Anhingas but not a single Double-crested Cormorant from central Florida during the spring and early summer of 1958 was positive (neutralized at least 1.5 logs) by the serum neutralization test for EEE virus antibody (Favorite 1960). No virus was isolated. Two of 13 (15%) nestling Double-crested Cormorants and all of 3 Anhinga nestlings from Lee County in 1978 were seropositive for SLE virus (Spalding et al. 1994).

No virus was isolated from the brains of 2 emaciated Double-crested Cormorants from Lee County in 1990 (Franson 1990) or from various tissues of cormorants from Seminole, Lee, and Collier counties in 1994 (Meteyer 1994).

XIV. Bacteria

Bacteria have been cultured from a large number of Double-crested Cormorants (*n* = 164) and a few Anhingas (*n* = 4) (table 5.9), but no clear association with lesions or illness has been documented. Seven *Salmonella agona* isolates from the large intestines of 104 "sick" Double-crested Cormorants from the Tampa Bay region were resistant to various antimicrobial products (White and Forrester 1979). No lesions were described in association with these isolates. This organism was first recorded in the United States from Peruvian fish meal shipped to the southeastern states in 1970 and has since been associated with infections in man (Clark et al. 1973). *Salmonella typhimerium* and *S. infantis* isolated in the same study did not show antibiotic resistance.

Salmonella group B was isolated from 2 of 5 Double-crested Cormorants in Pinellas County during an unresolved die-off in 1980. Three of these birds, including 1 of those from which the

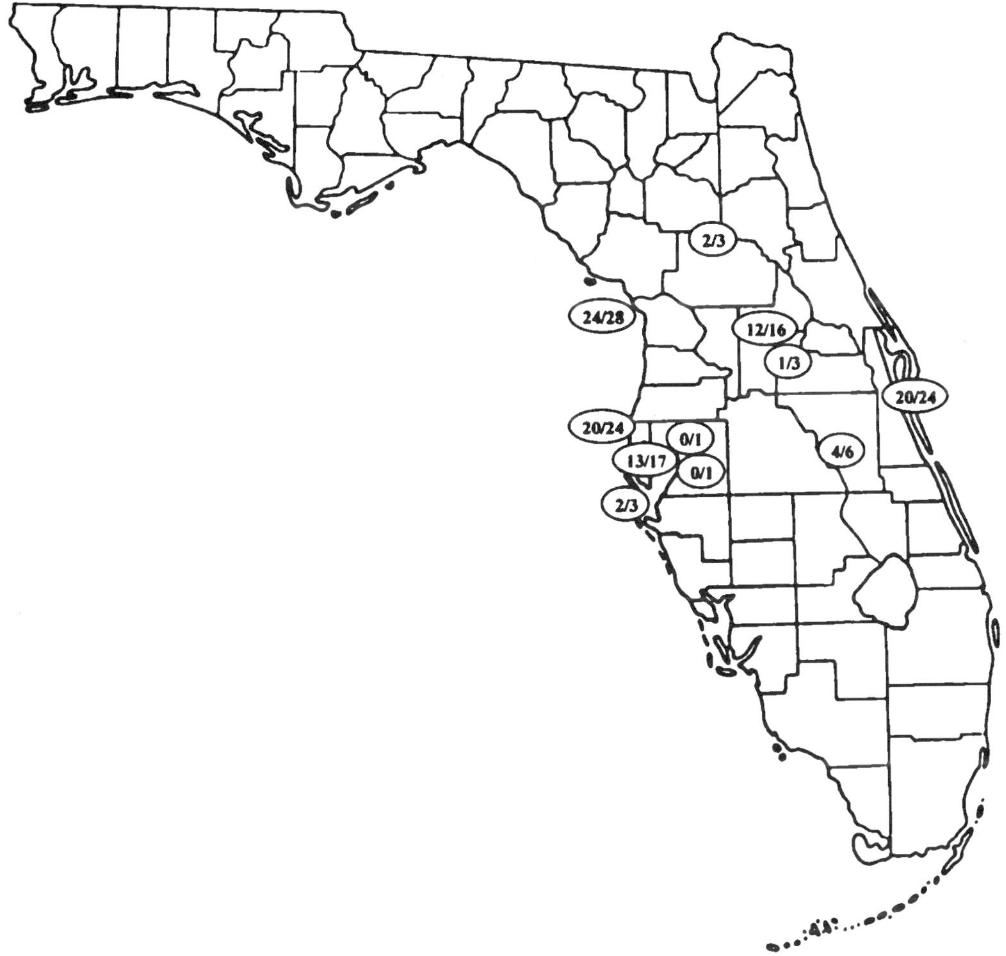

FIGURE 5.3. Distribution of eggs and chicks tested for Newcastle disease virus antibodies. Number of eggs positive for antibodies/number of eggs tested. Courtesy of Jacqueline M. Farley.

Salmonella was isolated, had hemorrhagic gastroenteritis (Forrester and Hanley 1980).

Edwardsiella tarda was isolated from the large intestine of 1 Double-crested Cormorant and the liver of another from Pinellas County in 1974. Both of these birds were found "sick" and died shortly thereafter (Forrester and White 1980).

XV. Fungi

Aspergillosis has occasionally been observed in both captive and wild Double-crested Cormorants in Florida. Aspergillosis was diagnosed in the lung and air sacs of 6 of 7 Double-crested Cormorants examined from Lee County in 1978 (Forrester 1978), and was cultured from the lungs, liver, and air sacs of 1 of 5 Double-crested Cormorants examined from the 1980 unresolved die-off in Pinellas County (Forrester and Hanley 1980). A fungal infection of muscle tissue (location not specified) was diagnosed by histopathology in a Double-crested Cormorant from Monroe County in 1992 (Homer 1992). The fungus appeared similar to *Aspergillus* sp.

Table 5.9. Bacteria and fungi isolated from Double-crested Cormorants and Anhingas in Florida

Host	Class	Species	County or area	Year(s)	Organ or tissue[a]	Prevalence[b]	Data Source
Double-crested Cormorants							
	Bacteria						
		Bacillus sp.	Pinellas	1975	LV	1%	A[c]
		Clostridium perfringens	"Florida Bay"	1995	KD	—	H
		Corynebacterium sp.	Brevard	1977	LV	3%	A[c]
		Edwardsiella tarda	Lee	1994	AS	—	B
			"Tampa Bay"	1974	LI	3%	A[c]
		Enterococcus sp.	Lee	1987	IN	—	C
			Seminole	1994	IN	—	B
		Escherichia coli	Brevard	1977	LI	30%	A[c]
			Charlotte	1973–75	LI	—	A[c]
			Duval	1974	LI	—	A[c]
			"Florida Bay"	1995	KD	—	H
			Lee	1978	IN	—	A[c]
			Levy	1975	LI	83%	A[c]
			"Tampa Bay"	1973–75	LI	43%	A[c]
				1974	LV	1%	A[c]
			Sarasota	1974–75	LI	—	A[c]
		Klebsiella oxytoca	Monroe	1992	LV	—	D
		Pasteurella multocida	Collier	1994	SP[d]	—	B
		Proteus sp.	Levy	1975	LI	6%	A[c]
			"Tampa Bay"	1974	LI	8%	A[c]
		Salmonella agona	"Tampa Bay"	1974	LI	7%	E
		Salmonella infantis	Pinellas	1973	LI	1%	E
		Salmonella typhimerium	"Tampa Bay"	1974	LI	4%	E
		Salmonella group B	Pinellas	1980	NG	—	F
		Salmonella sp.	Lee	1974	LI	—	A[c]
		Staphylococcus sp.	Lee	1994	AS,LV	—	B
			Seminole	1994	LV,IN	—	B
		Staphylococcus aureus	Lee	1987	LV	—	C
	Fungi						
		Aspergillus sp.	Lee	1978	LG,AS	—	G
			Pinellas	1980	LG,LV,AS	—	F
		Aspergillus fumigatus	Lee	1994	AS[e]	—	B
Anhingas							
	Bacteria						
		Citrobacter sp.	Sarasota	1975	LI	—	A[c]
		Edwardsiella tarda	Brevard	1972	LI	—	A[c]
		Escherichia coli	Brevard	1972	LI	—	A[c]

Sources: A = Forrester & White (1980), B = Meteyer (1994), C = Stroud (1987), D = Franson (1993), E = White & Forrester (1979), F = Forrester & Hanley (1980), G = Forrester (1978), H = Little (1995).
a. AS = air sac, IN = intestine, KD = kidney, LG = lung, LV = liver, LI = large intestine, NG = not given, SP = spleen.
b. Prevalences calculated for samples of 10 or more cormorants.
c. Forrester & White (1980) cultured liver and intestine of 156 Double-crested Cormorants from Brevard (32), Duval (1), Levy (18), Lee/Charlotte/Sarasota (8), Volusia (1), and Tampa Bay (104), from 1973 to 1977; and 4 Anhingas from Sarasota (2), Pinellas (1), and Brevard (1) counties.
d. Adult female in good nutritional condition with some abdominal trauma suggesting a predator might have attacked it. It was not felt that death was due to avian cholera.
e. Associated with a mild multifocal aspergillosis in this emaciated subadult, but not felt to be the cause of death.

XVI. Protozoans

Toxoplasma gondii antibodies were not detected in sera from a single Double-crested Cormorant from Florida (date and location not given) by Burridge et al. (1979). These were tested by IHA, which may not be as sensitive in birds as in mammals, resulting in false negatives (Frenkel 1981).

Blood parasites have never been found in Double-crested Cormorants or Anhingas from Florida or elsewhere (Greiner et al. 1975). Five adult and 5 juvenile Double-crested Cormorants from Columbia County, and 4 adult and 4 juvenile Double-crested Cormorants from Alachua County in 1981 were negative for blood parasites. One Anhinga from Brevard County in 1972 and 1 from Glades County were negative (Forrester and Bennett 1985). Ten nestling Anhingas collected in Palm Beach County in 1990 were negative for blood parasites (Spalding 1994).

Coccidian oocysts were found by fecal flotation and on small intestine scrapings from a sick Anhinga found in Melbourne, Brevard County, in 1972 (Forrester 1978). The species was not identified.

XVII. Helminths

A survey of helminth parasites of Double-crested Cormorants was undertaken by Threlfall (1982a) on 42 adult birds from Citrus to Lee counties on the west coast, mostly from rehabilitation facilities, and on 17 nestling and 17 adult cormorants collected from the east coast by gunshot, Brevard and Indian River counties, 1973–77 (tables 5.10–5.12). He found that west coast birds were most frequently infected with a greater number of species and higher intensities than east coast birds. In comparing nestlings with adults from the east coast, he found nestlings had fewer species present.

Ascaroid nematodes (*Contracaecum* sp.) appear to be the most prevalent and numerous of the helminths in Double-crested Cormorants and Anhingas (figure 5.4). Owre (1962) suggested that the presence of large numbers of nematodes in the stomach may contribute to the digestion of large food masses like fish and become pathogens only when ingesta are no longer available, such as during starvation. He found large numbers of ascaroid nematodes in 15 of 24 Double-crested Cormorants from south Florida and more than 100 nematodes each in the stomachs of 2 of them. Huizinga (1971) reported that *Contracaecum* sp. commonly perforates the mucosa of the proventriculus, causing hemorrhages and ulcerations, and also suggested that they may aid in digestion by breaking up fish. He disagreed with Owre, however, and felt that *Contracaecum* sp. would attach and cause lesions in healthy birds, even when fish were present in the stomach.

A nematode, *Serratospiculum helicinum*, found between the pia mater and the arachnoid of the meningies over the cerebellum (figure 5.5) of Anhingas in the 1800s (Wyman 1869) was not reported again until 2000 in Alachua County (Spalding and Kinsella 2000). This may simply be due to the fact that Anhingas are rarely examined.

Tetrameres sp. found histologically in Double-crested Cormorant proventricular glands (figure 5.6, table 5.11) from Alachua County in 1988 has not been identified to species (Uhl 1988). This same individual also had a perforated stomach and verminous peritonitis from infection with *Eustrongylides* sp. (figure 5.4), which was found only once in a cormorant and once in an Anhinga from the Everglades (Measures 1988).

Renal trematodes, possibly *Renicola* spp., are commonly observed histologically in kidney, but they have not been identified to species. They have been reported from Double-crested Cormorants in Pinellas County in 1980 (Forrester and Hanley 1980), Martin County in 1987 (Roffe and McAllister 1988), Lee County in 1987 and 1994 (Stroud 1987; Meteyer 1994), from 3 cormorants from Monroe County in 1992 (Spalding 1994; Homer 1992), and from 1 cormorant from Monroe County in 1995 (Little 1995).

A schistosome-like fluke was found in the

Table 5.10. Trematodes collected from Double-crested Cormorants in Florida

Species (Location)[a]	County or area	Year(s)	Number of cormorants			Intensity		Data source
			Examined	Positive	%	Mean	Range	
Ascocotyle sp. (SI)	Pinellas	1955–62	—	—	—	—	—	A
Clinostomum marginatum (TR,LU)	"West coast"[b]	1973–76	42[c]	4	9	3	1–7	B
Clinostomum attenuatum	Hillsborough	1958	4	1	—	—	—	C
Drepanocephalus spathans (SI)	Brevard, Indian River	1977	34	1	3	1	1	B
	"West coast"[b]	1973–76	42[c]	6	14	263	6–956	B
Drepanocephalus sp. (SI)	Lee	1990	2	2	—	NG	NG	D
Hysteromorpha triloba (SI)	Pinellas	1958	6	1	—	—	hundreds	C
	"West coast"[b]	1973–76	42[c]	7	17	4,169	1–19,727	B
Hysteromorpha sp. (SI)	Lee	1990	2	2	—	NG	NG	D
Mesostephanus appendiculatoides (SI)	Brevard, Indian River	1977	34	22	65	69	1–624	B
	"West coast"[b]	1973–76	42[c]	7	17	50	1–249	B
Mesostephanus sp. (SI)	Pinellas	1955–62	—	—	—	—	—	A
Ornithobilharzia sp. (BV)	"West coast"[b]	1973–76	42[c]	2	5	1	1–2	B
Parorchis acanthus (CL)	"West coast"[b]	1973–76	42[c]	1	2	1	1	B
Phagicola diminuta[d] (SI)	Pinellas	1958	6	1	—	—	hundreds	C
Phagicola longa (SI)	Brevard, Indian River	1977	34	19	56	106	1–400	B
	"West coast"[b]	1973–76	42[c]	16	38	54	1–488	B
	"numerous localities"							A
Phagicola sp.	Lee	1987	1	1	—	NG	NG	E
	Monroe	1994	4	NG	—	NG	—	F
Renicola sp.(*thapari*?) (KD)	"West coast"[b]	1973–76	42[c]	1	2	NG	many	B
Ribeiroia sp. (SI)	Lee	1990	2	1	—	—	NG	D

Sources: A = Hutton (1964), B = Threlfall (1982a), C = Hutton & Sogandares-Bernal (1960), D = Campbell (1990), E = Stroud (1987), F = Brannian (1995).
NG = not given.

a. Location in host: BV = blood vessels, CL = cloaca, KD = kidney, LU = lung, SI = small intestine, and TR = trachea.
b. "West Coast" = Citrus to Lee counties on the west coast.
c. Adult birds.
d. *Parascocotyle* is a synonym of *Phagicola*.

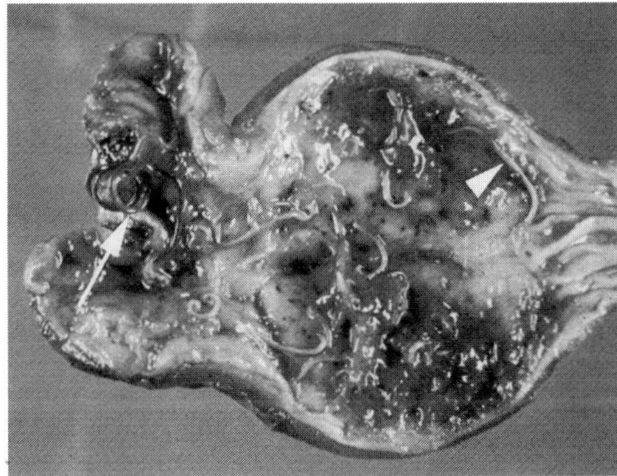

FIGURE 5.4. *Contracaecum* sp. (*arrowhead*) and *Eustrongylides* sp. (*arrow*) in the stomach of a Double-crested Cormorant. Courtesy of Elizabeth W. Uhl.

liver of an emaciated Double-crested Cormorant from Lee County in 1994 (Meteyer 1994). A trematode, *Amphimerus elongatus*, causing lesions in the liver and bile ducts has been reported in Louisiana (Pence and Childs 1972) but not from Florida.

XVIII. Arthropods

Threlfall (1982b) examined the ectoparasites of Double-crested Cormorants from both the east ($n = 33$) and west ($n = 44$) coasts of Florida (table 5.14). In comparing adult birds with nestlings, he found that the chewing louse *Pectinopygus farallonii* was absent from the crown and the auricular and gular regions of the chicks. Ticks, *Argus radiatus,* were found on the nestlings only, with the greatest numbers on nestlings that were only a few days old. *Michaelia* sp. mites were found only on the wings, whereas *Ornithonyssus* mites were found on the bodies of both nestlings and adults and on the legs of nestlings and jugulum of the adults.

Red imported fire ants (*Solenopsis invicta*) have been reported to cause significant reduction in waterbird reproduction (92%, including Olivaceus Cormorants) on Texas spoil islands (Drees 1994). We found no such reports for cormorants and Anhingas in Florida.

XIX. Emaciation

The weight listed for normal Double-crested Cormorants in del Hoyo et al. (1992) is 1.67–2.10 kg. Twenty-four (21%) of 117 Double-crested Cormorants submitted for cause-of-death determination between 1973 and 1994 (most from a rehabilitation center in Pinellas County) weighed less than 1 kg (Spalding and Forrester 1994). Also most Double-crested Cormorants presented to a rehabilitation center in Monroe County were severely emaciated (Quinn 1994). It appears that emaciation—whether due to starvation or secondary to disease—is relatively common in cormorants that are easily captured or are found dead.

All of 3 Double-crested Cormorants collected in Lee County, 1990, were emaciated and 2 of them were infected with nematodes, trematodes, and cestodes (Franson 1990). One of 3 Double-crested Cormorants collected in Monroe County in 1992 was emaciated (Franson 1993). A Double-crested Cormorant collected during a multiple species die-off during the winter of 1992–93 was emaciated; however, a fishhook perforating the esophagus may have been the primary cause of death (Quist 1993). Emaciation was the only significant

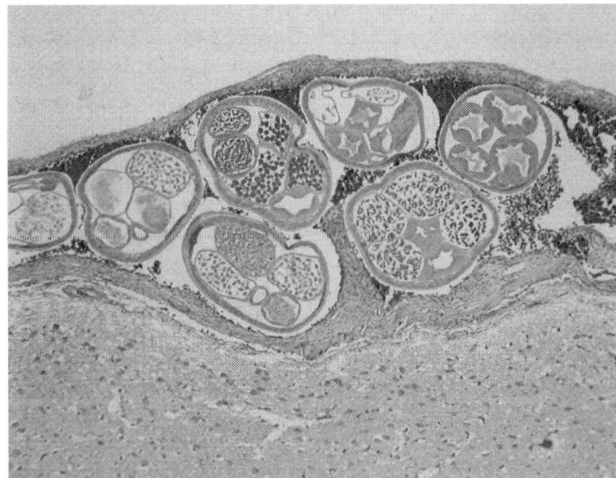

FIGURE 5.5. Photomicrograph of a nematode, *Serratospiculum helicinum,* found between the pia mater and the arachnoid of the meningies over the cerebellum of an Anhinga, from Alachua County.

Table 5.11. Nematodes collected from Double-crested Cormorants in Florida

Species (Location)[a]	County or area	Year(s)	Number of cormorants			Intensity		Data Source
			Examined	Positive	%	Mean	Range	
Capillaria carbonis (SI)	Brevard, Indian River	1977	34	7	21	13	1–50	A
	"West coast"[b]	1973–76	42[c]	11	26	44	1–227	A
Capillaria contorta (ES)	Brevard, Indian River	1977	34	2	6	5	2–8	A
	"West coast"[b]	1973–76	42[c]	6	14	7	1–26	A
Contracaecum multipapillatum (ST)	Indian River	1977		see *Contracaecum* spp.				A
	"Tampa Bay"	1974		see *Contracaecum* spp.				A
Contracaecum rudolphi[d] (ES,ST)	"Florida"	NG	NG	NG	—	NG	—	B
	Indian River	1977		see *Contracaecum* spp.				A
	Lee	1966	1	1	—	50	50	E
	"Long Island"	1913	NG	NG	—	NG	—	C[e]
	Pinellas	1955–62	NG	NG	—	NG	—	D
	Santa Rosa	1929	NG	—	—	NG	—	C[f]
	Taylor	1926	NG	—	—	NG	—	C[g]
Contracaecum spp.[h] (ES,ST)	Brevard, Indian River	1977	34	32	94	129	3–550	A
	"West coast"[b]	1973–76	42[c]	41	98	240	2–1,178	A
Contracaecum sp. (ES,ST)	Brevard	1993	1	1	—	NG	numerous	H
	Lee	1987	1	1	—	NG	NG	F
		1990	2	2	—	NG	—	G
	Monroe	1992	1	1	—	NG	—	J
		1995	4	NG	—	NG	—	K
	"South Florida"	NG	NG	—	—	NG	lg. numbers	I

Desmidoercella incognita (TR,LU,AS)	Brevard, Indian River	1977	34	7	21	7	1–15	A
	"West coast"[b]	1973–76	42[c]	17	40	49	1–100	A
Desmidoercella sp. (LU,TR)	Monroe	1992	1	1	—	7	—	J
Eustrongylides sp.	Alachua	1988	1	1	—	NG	—	M
Skrjabinocara squamatum (ES,ST)	Brevard, Indian River	1977	34	1	3	1	1	A
	"West coast"[b]	1973–76	42[c]	14	33	2	1–4	A
Syngamus trachea (TR)	Collier	1995	3	1	—	1	—	L
Synhimantus sp.	Monroe	1994	4	NG	—	NG	—	K
Tetrameres microspinosa (ES,ST)	Brevard, Indian River	1977	34	6	18	23	4–54	A
	"West coast"[b]	1973–76	42[c]	30	71	58	1–445	A
Tetrameres sp. (PR)	Alachua	1988	1	1	—	NG	—	M
	Lee	1990	2	1	—	NG	—	G
	Monroe	1994	4	NG	—	NG	—	K

Sources: A = Threlfall (1982a), B = Walton (1927), C = U.S. National Parasite Collection, D = Hutton (1964), E = Huizinga (1966), F = Stroud (1987), G = Campbell (1990), H = Quist (1993), I = Owre (1962), J = Spalding & Kinsella (1994), K = Brannian (1995), L = Sileo (1996), M = Uhl (1988).

NG = not given.

a. Location in host: AS = air sacs, ES = esophagus, LU = lung, FR = proventriculus, SI = small intestine, ST = stomach, TR = trachea.

b. "West Coast" = Citrus to Lee counties on the west coast.

c. Adult birds.

d. *C. spiculigerum* is a synonym of *C. rudolphii*.

e. Case number 030585.

f. Case number 028952.

g. Case number 026834.

h. A complex of 2 species (ratio 10:1): *C. spiculigerum* (*C. rudolphii*) and *C. multipapillatum.*

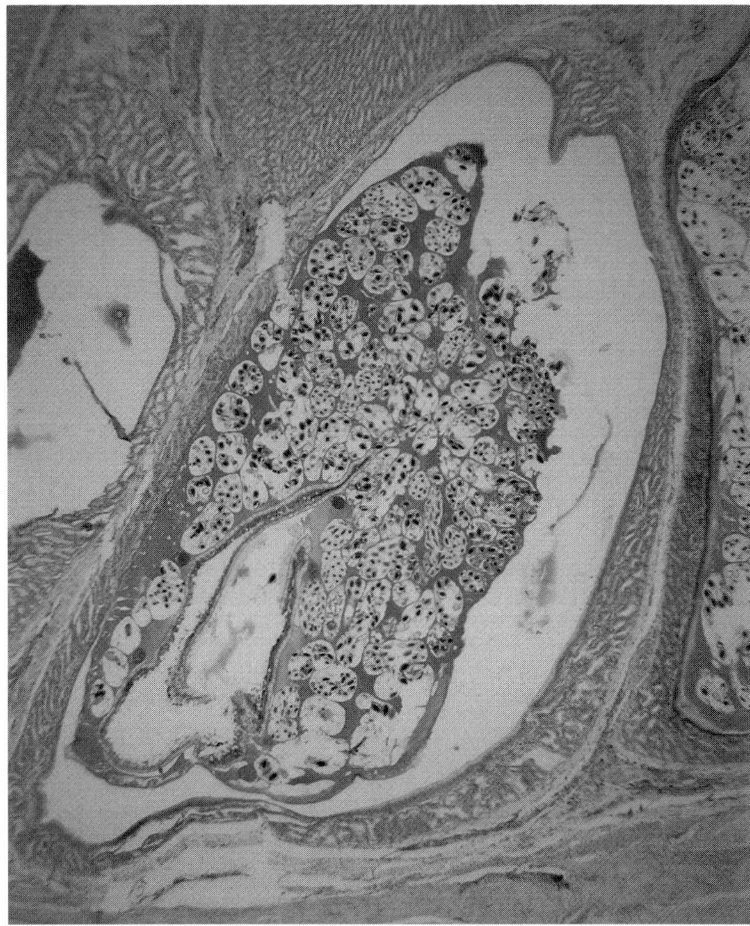

FIGURE 5.6. *Tetrameres* sp. in the proventricular gland of a Double-crested Cormorant. Case material courtesy of Elizabeth W. Uhl.

finding for 2 of 3 Double-crested Cormorants that died in central Florida in 1994 with neurologic signs (Meteyer 1994). Emaciation with no other significant findings was observed in 4 Double-crested Cormorants examined from a die-off of 25 in the Florida Keys, Monroe County, in February–March 1995 (Franson 1995). Emaciation with oral plaques of sloughed cells (inanition plaques) was observed in 3 Double-crested Cormorants examined from a die-off of 12 on Marco Island, Collier County, in October–November 1995 (Sileo 1996). Other findings were unremarkable.

XX. Unresolved die-offs

Six adult female Double-crested Cormorants were examined from a die-off during the 1976–77 winter on Sanibel Island, Lee County. All birds were living, emaciated, and had loss of equilibrium and inability to coordinate movements. Several birds had evidence of feeding accidents (see section II, Trauma). Types C and E botulism (see section XII, Biotoxins), organochlorine pesticides (see section VI, Organochlorines), and lead were ruled out as causes. Cultures of livers, intestines and brains

Table 5.12. Cestodes and acanthocephalans collected from Double-crested Cormorants in Florida

| Class | County/ | | Number of cormorants | | | Intensity | | Data |
Species (Location)[a]	area	Year(s)	Examined	Positive	%	Mean	Range	Source
Cestoda								
Cyclustera sp.[b] (SI)	Brevard,							
	Indian River	1977	34	4	12	27	1–100	A
	"West coast"[c]	1973–76	42[d]	9	21	48	1–334	A
	Lee	1990	2	2	NG	—		B
Acanthocephala								
Andracantha gravida (SI)	"Florida"	NG	NG	—	—	—	—	C
Andracantha spp.[e] (SI)	Brevard,	1977	34	11	32	5	1–13	A
	Indian River							
	Monroe	1994	4	NG	—	NG	—	D
	"West coast"[c]	1973–76	42[d]	31	74	24	1–129	A
Corynosoma sp. (LI)	Hillsborough	1955–62	NG	—	—	—	—	E
Polymorphus	"West coast"	1973–76	42[d]	13	31	2	1–2	A
obtusus (LI)								

Sources: A = Threlfall (1982a), B = Campbell (1990), C = Schmidt (1975), D = Brannian (1995), E = Hutton (1964).
NG = not given.
a. Location in host: LI = large intestine, SI = small intestine.
b. Threlfall (1982a) and Campbell (1990) identified these as *Parvitaenia* sp. Threlfall suggested they were probably immature *P. eudocimi*. The current name for this parasite is *Cyclustera ibisae* (Bona 1994).
c. "West Coast" = Citrus to Lee counties on the west coast.
d. Adult birds.
e. A complex of 2 species.

were negative for known pathogens (Kerr and Reichel 1977).

An unspecified number of Double-crested Cormorants was found convulsing and dying on the beaches of Pinellas County in 1980 (see sections XIII, Viruses, XIV, Bacteria, and XV, Fungi) (Forrester and Hanley 1980). A cause was not determined.

A die-off including 10 Brown Pelicans, 47 Double-crested Cormorants and 1 Ring-billed Gull was reported at Patrick Air Force Base, Brevard County, April 1982. The Ring-billed Gull tested positive for botulism type C, the 1 pelican examined did not. No cormorants were examined. It was suspected that pesticides were involved in this die-off, but diagnostic assays were not conducted (Stroud 1982).

The birds examined from a die-off of gulls, terns, and cormorants in Lee County, 1987, included a Double-crested Cormorant. All birds were emaciated, and most had parasites. No evidence of cholinesterase inhibition was found

and no unifying diagnosis was made (Stroud 1987).

A pelican die-off in Boca Ciega Bay, Pinellas County, during January–June 1983, was reported by Ankerberg (1983) (see chapter 4, Pelicans, for more details). Species involved also included the Double-crested Cormorant. Approximately 400 birds were estimated to have died. A municipal sewage lift station releasing raw sewage into Boca Ciega Bay was implicated. Many of the bacterial organisms cultured from water and sediment in the area were the same as those cultured from the pelicans. No health data were collected from Double-crested Cormorants.

Multiple species (40 Common Loons, 21 Northern Gannets, 5 Brown Pelicans, 5 Ring-billed Gulls, 3 Blue-Footed Boobies, 3 Double-crested Cormorants, and 2 Royal Terns) were found dying on Merritt Island National Wildlife Refuge, Brevard County, during the winter of 1992–93 (Quist 1993). Weather may have been a factor in their mortality, as no other common

Table 5.13. Helminths collected from Anhingas in Florida

| Class | Species (Location)[a] | County/area | Year(s) | Number of anhingas | | | Intensity | | Data source |
				Examined	Positive	%	Mean	Range	
Trematoda									
	Ascocotyle sp.[b] (SI)	Palm Beach	1990	10[c]	2	20	230	60–400	C
	Diplostomum (*Tylodelphys*) sp. (SI)	Palm Beach	1990	10[c]	1	10	1	—	C
	Echinochasmus dietzevi (SI)	Palm Beach	1990	10[c]	4	40	135	60– 220	C
	Euparyphium anhingae (IN)	Leon	1966	1	1	—	12	—	H
	Euparyphium capitaneum (DU,SI)	Brevard	1972	1	1	—	7	—	B
		Charlotte	1993	1	1	—	72	—	K
		Palm Beach	1990	10[c]	4	40	140	2–240	C
	Macrobilharzia macrobilharzia[d] (BV)	"Florida"	1930	NG	—	—	NG[e]	—	E
		Palm Beach	1990	10[c]	3	30	250	100–400	C
	Posthodiplostomum minimum (DU, SI)	Charlotte	1993	1	1	—	45	—	K
		Palm Beach	1990	10[c]	1	10	1	—	C
	Posthodiplostomum nanum (SI)	Charlotte	1993	1	1	—	6	—	K
		Palm Beach	1990	10[c]	4	40	70	2–100	C
Nematoda									
	Capillaria sp. (SI)	Brevard	1972	1	1	—	4	—	B
	Contracaecum caballeroi (CL, TN)	Alachua	1956	NG	—	—	—	—	L[f]
	Contracaecum microcephalum	Charlotte	1993	1	1	—	122[g]	—	K
	Contracaecum multipapillatum (ST)	Brevard	1972	1	1	—	12	—	B
		Charlotte	1993	1	1	—	122[g]	—	K
		Jefferson	1924	NG	—	—	NG	—	L[h]
	Contracaecum quincuspis (ST,SI)	Brevard	1972	1	1	—	7	—	B
		Charlotte	1993	1	1	—	122[g]	—	K
		Jefferson	1924	NG	—	—	NG	—	D
		Palm Beach	1990	10[c]	1	10	1	—	C
	Contracaecum rudolphi[i]	"Florida"	NG	NG	—	—	NG	—	J
	Contracaecum tricuspe	Jefferson	1924	NG	—	—	NG	—	L[j]
	Contracaecum sp. (ES,PR,SI,ST)	"South Florida"	NG	2	2	—	>100	NG	G
			NG	24	15	62	NG	"lg. numbers"	G

Contracaecum spp.[k] (OR,ES,ST)	Palm Beach	1990	10[c]	10	100	19	2–58	C
Eustrongylides ignotus (IN,BC)	"Everglades"	1930–31	NG	1	—	NG	—	F
Serratospiculum helicinum[l] (BR)	Alachua	2000	1	1	—	2	—	O
	"St. John's River"	1861, 1867	NG	—	—	NG	—	J, M
	"St. John's FL"	1890	NG	—	—	NG	—	L[m]
Tetrameres spiculigerum (PR)	Jefferson	1924	NG	—	—	NG	—	L[n]
Tetrameres sp.[o] (PR)	Charlotte	1993	1	1	—	23	—	K
	Brevard	1972	1	1	—	4	—	B
Acanthocephala								
Echinorhynchus hystrix	Dade	NG	NG	—	—	NG	—	L[p]
Polymorphus brevis (SI)	Charlotte	1993	1	1	—	1	—	K
Polymorphus obtusus (IN)	"Florida"	1898	NG	—	—	NG	—	I, L[q]

Sources: A = Huizinga (1971), B = Kinsella (1997), C = Spalding & Kinsella (1994), D = Lucker (1941), E = McIntosh (1933), F = Measures (1988), G = Owre (1962), H = Premvati (1968), I = Van Cleave (1918), J = Walton (1927), K = Kinsella & Spalding (1997), L = U.S. National Parasite Collection, M = Wyman (1869), O = Spalding & Kinsella (2000). NG = not given.

a. Location in host: BC = body (coelomic) cavity, BR = meninges over cerebellum, BV = blood vessels, CL = cloaca, DU = duodenum, GI = gizzard/ventriculus, IN = intestine, PR = proventriculus, SI = small intestine, ST = stomach, TN = tongue.

b. *Ascocotyle* sp. near *filippei,* but cannot be identified without oral spines.

c. Nestlings.

d. *Macrobilharzia macrobilharzia* identified by Price (1931) and listed by Kohn (1964).

e. Females in large numbers in small mesenteric vessels leading to viscera, males in large hepatic portal vein in small numbers, usually 1–3 per bird.

f. Case number 38069.

g. Numbers combined for 3 species of *Contracaecum* present: *C. quincuspis, C. multipapillatum,* and *C. microcephalum.* Adult *C. quincuspis* mostly in SI.

h. Case number 44556

i. *C. spiculigerum* is a synonym of *C. rudolphii.*

j. Case number 30591.

k. Most of the *Contracaecum* appear to be *C. multipapillatum,* a few are *C. microcephalum.*

l. Previously identified as *Filaria anhingae* by Wyman (1869) and as *Filaria helicina* by Leidy (1886).

m. Case number 6588.

n. Case number 34385.

o. Probably *Tetrameres elyi* of Mollhagen (1976), which is unpublished and thus has no taxonomic validity.

p. Case number 5211.

q. Case number 2967.

Table 5.14. Ectoparasites collected from Double-crested Cormorants and Anhingas in Florida

Host	Species of parasite	County/area	Year(s)	Data source
Double-crested Cormorant				
Ticks				
	Argas radiatus[b]	Brevard	1977	A
Mites				
	Dinalloptes chelionatus	Charlotte	1960	B
	Ornithonyssus sp.[b]	Brevard	1977	A
	Scutomegninia sp.	Monroe	1992	C
	Stibarokris phalacrus	Monroe	1967	D
	Michaelia sp.	Monroe	1992	C
	Michaelia sp. (near *urile*)[b]	Brevard	1977	A
Chewing lice				
	Eidmanniella pellucida	Brevard	1977	E[d]
		Indian River	1974	F
		Lee	1980	F
			1981	F
		Monroe	1992	F
		"West coast"[a]	1973–76	A
	Pectinopygus farallonii[b]	Brevard	1977	A
			1977	F
		Lee	1979, 1981	F
			1978, 1980	F
			1983–84	F
		Monroe	1989	F
			1992	F
		Pinellas	1929, 1931	F
		"West coast"[a]	1973–76	A
	Piagetiella incomposita[c]	Brevard	1977	E[e]
		"Florida"	NG	G
		"Florida Bay"	1995	H
		Lee	1984	F
		Monroe	1992	F
		"West coast"[a]	1973–76	A
Anhinga				
Chewing Lice				
	Ciconiphilus decimfasciatus	Indian River	1969	F
	Pectinopygus anhingae	Indian River	1969	F
		Lee	1982–83	F

Sources: A = Threlfall (1982b), B = Atyeo & Peterson (1966), C = Forrester & Atyeo (1992), D = Kethley (1970), E = U.S. National Parasite Collection[d], F = Forrester et al. (1995), G = Price (1970), H = Little (1995).
NG = not given.
a. "West Coast" = Citrus to Lee counties on the west coast, fledged birds only.
b. Threlfall (1982b) examined 33 birds from Brevard County (17 nestlings, 8 adult females, and 8 adult males). *P. farallonii* was found on 88% of birds, mean = 30, range = 1–91. *Ornithonyssus* sp. and *Michaelia* sp. (near *urile*) were each found on 3% of birds. *Argas radiatus* was found only on nestlings (n = 17, prevalence on nestlings = 71%, mean = 27, range = 1–207).
c. Pouch louse.
d. Case number 075900.
e. Case number 075899.

factor was found (see chapter 2, Loons and Grebes, and chapter 21, Gulls, Terns, and Skimmers, for additional details, and sections II, Trauma; V, Inclement weather; VIII, Metals; XVII, Helminths; and XIX, Emaciation).

XXI. Cormorants in conflict with humans

The cormorant as a competitor with fishermen and a depredator of fish farms is treated in great detail in a special issue of *Colonial Waterbirds* (vol. 18, Spec. Publ. 1, 1995). Within that issue, Brugger (1995) discusses the problem in Florida. She reports little conflict between cormorants and the food-fish and game-fish industries. There have been occasional problems, however, at exotic aquaculture facilities. Other fish-eating birds, such as wading birds, appear to be more of a problem than cormorants in Florida, but little information has been gathered. Cormorants foraging in fishponds may transmit parasitic diseases to fish, affecting their survival, growth, or suitability as ornamental fish. Avian diseases, such as Newcastle disease and avian influenza, may be transmitted between domestic and wild birds at such locations.

XXII. Summary and conclusions

Double-crested Cormorants are often the most common birds found in rehabilitation facilities in Florida and thus carcasses have been made available for various tests and surveys. A comprehensive quantification of the cause of morbidity in these birds has never been attempted. As is the case with most large birds, trauma associated with human-created hazards appears to be the most common and most observed cause of mortality of Double-crested Cormorants. Unexplained emaciation is a very common finding and the cause of this, whether poor food availability, weather, or secondary to disease or contamination, bears study. We found documentation of 10 organochlorine

pesticides, 12 metals, 1 radionuclide, 2 viruses, 14 bacteria, 18 trematodes, 1 cestode, 12 nematodes, 5 acanthocephalans, and 11 arthropods in cormorants and Anhingas in Florida. The helminth and arthropod parasites of cormorants have been reasonably well studied, as has pesticide contamination in the 1970s. Probably the most neglected area of study has been the effects of various pathogens, such as bacteria, viruses, and biotoxins, on the health of cormorants. Except for studies on exposure to dioxins and furans, the health of Anhingas has been almost entirely neglected.

XXIII. Literature cited

Anderson, D.W., and F. Hamerstrom. 1967. The recent status of Wisconsin cormorants. *Passenger Pigeon* 29:3–15.

Anderson, D.W., and J.J. Hickey. 1972. Eggshell changes in certain North American birds. *Proc. Intl. Ornithol. Congr.* 15:514–540.

Anderson, D.W., J.J. Hickey, R.W. Risebrough, D.F. Hughes, and R.E. Christensen. 1969. Significance of chlorinated hydrocarbon residues to breeding pelicans and cormorants. *Can. Field Nat.* 83:91–112.

Ankerberg, C.W., Jr. 1983. An investigation of the death of *Pelecanus occidentalis* (brown pelican) in the Boca Cicga Bay. Suncoast Scabird Sanctuary, Redington Beach, Fla.

Armstrong, I.H., J.C. Coulson, P. Haskey, and M.J. Hudson. 1978. Further mass seabird deaths from paralytic shellfish poisoning. *Br. Birds* 71:58–68.

Atyeo, W.T., and P.C. Peterson. 1966. The feather mite genus *Dinalloptes* (Acarina, Proctophyllodidae). *Acaralogia* (Paris) 8:470–474.

Banerjee, M., W.M. Reed, S.D. Fitzgerald, and B. Panigrahy. 1994. Neurotrophic velogenic Newcastle disease in cormorants in Michigan: pathology and virus characterization. *Avian Dis.* 38:873–878.

Bona, V. 1994. Family Dilepididae. In: *Keys to the cestode parasites of vertebrates.* L.F. Khalil, A. Jones, and R.A. Bray (eds.). Cambridge University Press, Cambridge. pp. 443–554.

Brannian, R.E. 1995–96. Unpublished data. National Wildlife Health Center, Madison, Wis.

Brugger, K.E. 1995. Double-crested cormorants and fisheries in Florida. *Colon. Waterbirds* 18, Spec. Publ. 1:8–24.

Burridge, M.J., W.J. Bigler, D.J. Forrester, and J.M. Hennemann. 1979. Serologic survey for *Toxoplasma gondii* in wild animals in Florida. *J. Am. Vet. Med. Assoc.* 175:964–967.

Campbell, B.G. 1990. Unpublished data. National Wildlife Health Center, Madison, Wis.

Clapp, R.B., R.C. Banks, D. Morgan-Jacobs, and W.A. Hoffman. 1982. *Marine birds of the southeastern United States and Gulf of Mexico. Part I, Gaviiformes through Pelecaniformes.* U.S. Fish and Wildlife Service, Biological Services Program FWS-OBS-82/01. 637 pp.

Clark, G.M., A.F. Kaufmann, E.J. Gangarosa, and M.A. Thompson. 1973. Epidemiology of an international outbreak of *Salmonella agona*. *Lancet* 2:490–493.

Coulson, J.C., G.R. Potts, I.R. Deans, and S.M. Fraser. 1968. Exceptional mortality of shags and other seabirds caused by paralytic shellfish poison. *Br. Birds* 61:381–404.

Crawford, R.L. 1981. Bird casualties at a Leon County, Florida TV tower: a 25-year migration study. *Bull. Tall Timbers Res. Stn.* 22:1–30.

del Hoyo, J., A. Elliott, and J. Sargatal. 1992. *Handbook of the birds of the world.* Vol. 1. Lynx Edicions, Barcelona. 696 pp.

Delany, M.F. 1986. Bird bands recovered from American alligator stomachs in Florida. *N. Am. Bird Bander* 11:92–94.

Dolbeer, R.A. 1991. Migration patterns of double-crested cormorants east of the Rocky Mountains. *J. Field Ornithol.* 62:83–93.

Drees, B.M. 1994. Red imported fire ant predation on nestlings of colonial waterbirds. *Southwest. Entomol.* 19:355–359.

Ellis, T.M. 1980. *Caiman crocodilus:* an established exotic in south Florida. *Copeia* 1980:152–154.

Ellison, L.N., and L. Cleary. 1978. Effects of human disturbance on breeding of double-crested cormorants. *Auk* 95:510–517.

Farley, J.M., C.H. Romero, M.G. Spalding, M.L. Avery, and D.J. Forrester. 2001. Newcastle disease virus in double-crested cormorants in Alabama, Florida, and Mississippi. *J. Wildl. Dis.* 37:808–812.

Favorite, F.G. 1960. Some evidence of local origin of EEE virus in Florida. *Mosq. News* 20:87–92.

Forrester, D.J. 1978. Unpublished data. University of Florida, Gainesville.

Forrester, D.J., and W.T. Atyeo. 1992. Unpublished data. University of Florida, Gainesville.

Forrester, D.J., and G.F. Bennett. 1985. Unpublished data. University of Florida, Gainesville.

Forrester, D.J., J.M. Gaskin, F.H. White, N.P. Thompson, J.A. Quick, Jr., G.E. Henderson, J.C. Woodard, and W.D. Robertson. 1977. An epizootic of waterfowl associated with a red tide episode in Florida. *J. Wildl. Dis.* 13:160–167.

Forrester, D.J., and J.E. Hanley. 1980. Unpublished data. University of Florida, Gainesville, and Live Oak Diagnostic Laboratory, Live Oak, Fla.

Forrester, D.J., H.W. Kale II, R.D. Price, K.C. Emerson, and G.W. Foster. 1995. Chewing lice (Mallophaga) from birds in Florida: a listing by host. *Bull. Fla. Mus. Nat. Hist.* 39:1–44.

Forrester, D.J., and N.P. Thompson. 1975. Unpublished data. University of Florida, Gainesville.

Forrester, D.J., and F.H. White. 1980. Unpublished data. University of Florida, Gainesville.

Francis, T.L. 1981. Brown pelican found dead with adult double-crested cormorant in pouch. *Fla. Field Nat.* 9:62–63.

Franson, J.C. 1990–1995. Unpublished data. National Wildlife Health Center, Madison, Wis.

Frederick, P.C., and D. Siegel-Causey. 2000. Anhinga. *Birds N. Am.* 522:1–24.

Frenkel, J.K. 1981. False-negative serologic tests for *Toxoplasma* in birds. *J. Parasitol.* 67:952–953.

Friend, M., and J.C. Franson. 1999. *Field manual of wildlife diseases. General field procedures and diseases of birds.* U.S. Department of the Interior, U.S. Geological Survey, Biological Research Division, Information and Technology Report 1999-001. Washington, D.C. 425 pp.

Funderburg, J.B., Jr. 1963. Report on research activities, 1962–63. *Lake Reg. Nat.* 4:2–5.

Glaser, L.C., I.K. Barker, D.V.C. Weseloh, J. Ludwig, R.M. Windingstad, D.W. Key, and

T.K. Bollinger. 1999. The 1992 epizootic of Newcastle disease in double-crested cormorants in North America. *J. Wildl. Dis.* 35:319–330.

Gourlie, N. 1984. Unpublished data. Florida Department of Environmental Regulation, Tallahassee.

Greichus, Y.A., and M.R. Hannon. 1973. Distribution and biochemical effects of DDT, DDD and DDE in penned double-crested cormorants. *Toxicol. Appl. Pharmacol.* 26:483–494.

Greiner, E.C., G.F. Bennett, E.M. White, and R.F. Coombs. 1975. Distribution of the avian hematozoa of North America. *Can. J. Zool.* 53:1762–1787.

Hatch, J.J. 1995. Changing populations of double-crested cormorants. *Colon. Waterbirds* 18, Spec. Publ. 1:8–24.

Heath, R.T., Jr., and H.A. Albers. 1974. Unpublished data. Suncoast Seabird Sanctuary, St. Petersburg, Fla.

Heinz, G.H., and D.J. Hoffman. 1998. Methylmercury chloride and selenomethionine interactions on health and reproduction in mallards. *Environ. Toxicol. Chem.* 17:139–145.

Henny, C.J., and D.L. Meeker. 1981. An evaluation of blood plasma for monitoring DDE in birds of prey. *Environ. Pollut.* 25A:291–304.

Hoffman, D.J., C.P. Rice, and T.J. Kubiak. 1996. PCBs and dioxins in birds. In: *Interpreting environmental contaminants in animal tissues.* W.N. Beyer, G.H. Heinz, and A. Redmon (eds.). Lewis, Boca Raton, Fla. 494 pp.

Homer, B.L. 1992. Unpublished data. University of Florida, Gainesville.

Huizinga, H.W. 1966. Studies on the life cycle and development of *Contracaecum spiculingerum* (Rudolph, 1809) (Ascaroidea; Heterocheilidae) from marine piscivorous birds. *J. Elisha Mitchell Sci. Soc.* 82:181–195.

———. 1971. Contracaeciasis in pelicaniform birds. *J. Wildl. Dis.* 7:198–202.

Hutton, R.F. 1964. A second list of parasites from marine and coastal animals of Florida. *Trans. Am. Microsc. Soc.* 83:439–447.

Hutton, R.F., and F. Sogandares-Bernal. 1960. Studies on helminth parasites from the coast of Florida. II. Digenetic trematodes from shore birds of the west coast of Florida. *Bull. Mar. Sci. Gulf Caribb.* 10:40–54.

Johnson, J.M. 1999. The prevalence and distribution of Newcastle disease virus in double-crested cormorants in Alabama, Mississippi, and Florida, 1998–1999. M.S. thesis, University of Florida, Gainesville. 77 pp.

Johnston, D.W. 1976. Organochlorine pesticide residues in uropygial glands and adipose tissue of wild birds. *Bull. Environ. Contam. Toxicol.* 16:149–155.

Kerr, S.M., and W.L. Reichel. 1977. Unpublished data. National Wildlife Health Center, Madison, Wis.

Kethley, J.B. 1970. A revision of the family Syringophilidae (Prostigmata: Acarina). *Contrib. Am. Entomol. Inst.* 5:1–76.

King, K.A. 1989. Food habits and organochlorine contaminants in the diet of olivaceous cormorants in Galveston Bay, Texas. *Southwest. Nat.* 34:338–343.

King, K.A., C.J. Stafford, B.W. Cain, A.J. Mueller, and H.D. Hall. 1987. Industrial, agricultural, and petroleum contaminants in cormorants wintering near the Houston Ship Channel, Texas, USA. *Colon. Waterbirds* 10:93–99.

Kinsella, J.M. 1997. Unpublished data. University of Florida, Gainesville.

Kinsella, J.M., and M.G. Spalding. 1997. Unpublished data. University of Florida, Gainesville.

Kohn, A. 1964. Sobre o Genero *Macrobilharzia* Travassos, 1922 (Trematoda, Schistosomatoidea). *Mem. Inst. Oswaldo Cruz* 62:1–6.

Kuiken, T., F.A. Leighton, G. Wobeser, and B. Wagner. 1999. Causes of morbidity and mortality and their effect on reproductive success in double-crested cormorants from Saskatchewan. *J. Wildl. Dis.* 35:331–346.

Kushlan, J.A., and L.C. McEwan. 1982. Nesting phenology of the double-crested cormorant. *Wilson Bull.* 94:201–206.

Larson, J.M., W.H. Karasov, L. Sileo, K.L. Stromborg, B.A. Hanbidge, J.P. Giesy, P.D. Jones, D.E. Tillitt, and D.A. Verbrugge. 1996. Reproductive success, developmental anomalies, and environmental contaminants in double-crested cormorants (*Phalacrocorax auritus*). *Environ. Toxicol. Chem.* 15:553–559.

Leidy, J. 1886. Notices of nematoid worms. *Pest Articles and News Summaries* 38:308–313.

Little, S.E. 1995. Unpublished data. Southeastern Cooperative Wildlife Disease Study, University of Georgia, Athens.

Lucker, J.T. 1941. *Contracaecum quincuspis,* a new species of nematode from the American waterturkey. *J. Wash. Acad. Sci.* 31:33–37.

McIntosh, A. 1933. Polyandry and polygamy in parasitic worms. *J. Parasitol.* 20:141.

Measures, L.N. 1988. Revision of the genus *Eustrongylides* Jagerskiold, 1909 (Nematoda: Dioctophymatoidea) of piscivorous birds. *Can. J. Zool.* 66:885–895.

Meteyer, C. 1994. Unpublished data. National Wildlife Health Center, Madison, Wis.

Mollhagen, T.R. 1976. A study of the systematics and hosts of the parasitic nematode genus *Tetrameres* (Habronematoidea: Tetrameridae). Ph.D. diss., Texas Tech University, Lubbock. 546 pp.

Morton, R.A., and M.A. Burkley. 1969. Florida shellfish toxicity following blooms of the dinoflagellate, *Gymnodium breve. Fla. Dep. Nat. Resour. Mar. Res. Lab. Tech. Ser.* 60:1–26.

Myers, O.B., W.R. Marion, T.E. O'Meara, and C.E. Roessler. 1989. Radium-226 in wetland birds from Florida phosphate mines. *J. Wildl. Manag.* 53:1110–1116.

Nesbitt, S.A., J.C. Ogden, H.W. Kale II, B.W. Patty, and L.A. Rouse. 1982. *Florida atlas of breeding sites for herons and their allies: 1976–78.* U.S. Fish and Wildlife Service, Biological Services Program FWS-OBS 81/49.

Nosek, J.A., J.R. Sullivan, S.R. Craven, A. Gendron-Fitzpatrick, and R.E. Peterson. 1993. Embryotoxicity of 2,3,7,8-tetrachlorodibenzo-p-dioxin in the ring-necked pheasant. *Environ. Toxicol. Chem.* 12:1215–1222.

Ogden, J.C., W.B. Robertson, G.E. Davis, and T.W. Schmidt. 1974. Pesticides, polychlorinated biphenyls and heavy metals in upper food chain levels, Everglades National Park and vicinity. Department of the Interior, National Technical Information Service, Atlanta. 27 pp.

Ohlendorf, H.M., E.E. Klaas, and T.E. Kaiser. 1979. Environmental pollutants and eggshell thickness: anhingas and wading birds in the eastern United States. U.S. Fish and Wildlife Service, Special Scientific Report—Wildlife 216. 94 pp.

O'Meara, T.E., W.R. Marion, C.E. Roessler, G.S. Roessler, H.A. Van Rinsvelt, and O.B. Meyers. 1986. Environmental contaminants in birds: phosphate-mine and natural wetlands. Publ. 05-003-045. Florida Institute of Phosphate Research, Bartow. 77 pp.

Owre, O.T. 1962. Nematodes in birds of the order Pelicaniformes. *Auk* 79:114.

Pence, D.B., and G.E. Childs. 1972. Pathology of *Amphimerus elongatus* (Digenea: Opisthorchiidae) in the liver of the double-crested cormorant. *J. Wildl. Dis.* 8:221–224.

Premvati, G. 1968. Echinostome trematodes from Florida birds. *Proc. Helminthol. Soc. Wash.* 35:197–200.

Price, E.W. 1931. Note on *Macrobilharzia* Travassos. *J. Parasitol.* 17:230–231.

Price, R.C. 1970. The Piagetiella (Mallophaga: Menoponidae) of the Pelecaniformes. *Can. Entomol.* 102:389–404.

Quick, J.A., Jr., and G.E. Henderson. 1975. Effects of *Gymnodinium breve* red tide on fishes and birds: a preliminary report on behavior, anatomy, hematology, and histopathology. *Proc. Gulf Coast Reg. Symp. Dis. Aquat. Anim.:* 85–113.

Quinn, L. 1994. Unpublished data. Florida Keys Wild Bird Center, Tavernier.

Quist, C. 1993. Unpublished data. Southeastern Cooperative Wildlife Disease Study, University of Georgia, Athens.

Ratcliffe, D.A. 1967. Decrease in eggshell weight in certain birds of prey. *Nature* 215:208–210.

Robertson, W.B., Jr., and D.R. Paulson. 1961. Region reports: Florida region. *Audubon Field Notes* 15:26–35.

Robertson, W.B., and G.E. Woolfenden. 1992. *Florida bird species: an annotated list.* Spec. Publ. 6, Florida Ornithological Society, Gainesville. 260 pp.

Rodgers, J.A. 1997. Pesticide and heavy metal levels of waterbirds in the Everglades agricultural area of south Florida. *Fla. Field Nat.* 25:33–84.

Rodgers, J.A., and H.T. Smith. 1997. Buffer zone distances to protect foraging and loafing waterbirds from human disturbance in Florida. *Wildl. Soc. Bull.* 25:139–145.

Roffe, T.J., and H.A. McAllister. 1988. Unpublished data. National Wildlife Health Center, Madison, Wis.

Rumbold, D.G. 1992. Ecotoxicological surveillance of chemical residues in wildlife near the North County Resource Recovery Facility: a baseline study. Final Report to Solid Waste Authority of Palm Beach County, West Palm Beach, Fla. 57 pp.

Rumbold, D.G., M.C. Bruner, M.B. Mihalik, E.A. Marti, and L.L. White. 1996. Organochlorine pesticides in Anhingas, White Ibises and apple snails collected in Florida, 1989–1991. *Arch. Environ. Contam. Tox.* 30:379–383.

Rumbold, D.G., M.C. Bruner, M.B. Mihalik, and E.A. Marti. 1997. Biomonitoring environmental contaminants near a municipal solid-waste combustor. *Environ. Pollut.* 96: 99–105.

Runde, D.E. 1991. Trends in wading bird nesting populations in Florida 1976–78 and 1986–89. Nongame Wildlife Section, Florida Game and Fresh Water Fish Commission, Tallahassee. 105 pp.

Schmidt, G.D. 1975. *Andracantha,* a new genus of Acanthocephala (Polymorphidae) from fish-eating birds, with descriptions of three species. *J. Parasitol.* 61:615–620.

Schreiber, R.W., and E.A. Schreiber. 1978. *Colonial bird use and plant succession on dredged material islands in Florida.* Vol. 1, *Sea and wading bird colonies.* Dredged Material Research Program, Tech. Rep. D-78-14. Environmental Laboratory, U.S. Army Engineer Waterways Experiment Station, Vicksburg, Mo. Contract DACW 39-76-0161. 63 pp.

Seal, B.S., D.J. King, and J.D. Bennett. 1995. Characterization of Newcastle disease virus isolated by reverse transcription PCR coupled to direct nucleotide sequencing and development of sequence database for pathotype prediction and molecular epidemiological analysis. *J. Clin. Microbiol.* 33:2624–2630.

Sepulveda, M.S., R.H. Poppenga, J.J. Arrecis, and L.B. Quinn. 1998. Concentrations of mercury and selenium in tissues of double-crested cormorants (*Phalacrocorax auritus*) from southern Florida. *Colon. Waterbirds* 21:35–42.

Sileo, L. 1996. Unpublished data. National Wildlife Health Center, Madison, Wis.

Sims, H.W., Jr. 1970. Operation bird wash. *Fla. Nat.* 43:43–45.

Smith, K., and V.F. Nettles. 1995. Unpublished data. Southeastern Cooperative Wildlife Disease Study, University of Georgia, Athens.

Snyder, B. 1994. Unpublished data. Florida Department of Environmental Protection, Tallahassee.

Spalding, M.G. 1994–2000. Unpublished data. University of Florida, Gainesville.

Spalding, M.G., and D. J. Forrester. 1994. Unpublished data. University of Florida, Gainesville.

Spalding, M.G., and J.M. Kinsella. 1994–2000. Unpublished data. University of Florida, Gainesville.

Spalding, M.G., R.G. McLean, J.H. Burgess, and L.J. Kirk. 1994. Arboviruses in water birds (Ciconiiformes, Pelecaniformes) from Florida. *J. Wildl. Dis.* 30:216–221.

Stalling, D.L., R.J. Norstrom, L.M. Smith, and M. Simin. 1985. Patterns of PCDD, PCDF, and PCB contamination in Great Lakes fish and birds and their characterization by principal components analysis. *Chemosphere* 14:627–643.

Stevenson, H.M. 1970. The winter season: Florida region. *Audubon Field Notes* 24:493–497.

Stevenson, H.M., and B.H. Anderson. 1994. *The birdlife of Florida.* University Press of Florida, Gainesville. 892 pp.

Stickel, L.F., W.H. Stickel, and R. Christensen. 1966. Residues of DDT in brains and bodies of birds that died on dosage and in survivors. *Science* 151:1549–1551.

Stroud, R.K. 1982–87. Unpublished data. National Wildlife Health Center, Madison, Wis.

Threlfall, W. 1982a. Endoparasites of the double-crested cormorant (*Phalacrocorax auritus*) in Florida. *Proc. Helminthol. Soc. Wash.* 49:103–108.

———. 1982b. Ectoparasites (Mallophaga, Acarina) from the double-crested cormorant (*Phalacrocorax auritus*) in Florida. *Proc. Entomol. Soc. Wash.* 84:369–375.

Uhl, E.W. 1988. Unpublished data. University of Florida, Gainesville.

Van Cleave, H.J. 1918. The Acanthocephala of North American birds. *Trans. Am. Microsc. Soc.* 37:19–47.

Walton, A.C. 1927. A revision of the nematodes of the Leidy collections. *Proc. Acad. Nat. Sci. Phila.* 79:49–163.

Weseloh, D.V., P.J. Ewins, J. Struger, P. Mineau, C.A. Bishop, S. Postupalsky, and J.P. Ludwig. 1995. Double-crested cormorants of the Great Lakes: changes in population size, breeding distribution and reproductive output between 1913 and 1991. *Colon. Waterbirds* 18, Spec. Publ. 1:48–59.

Wheeler, W.B., D.P. Jouvenaz, D.P. Wojcik, W.A. Banks, C.H. VanMiddelem, C.S. Lofgren, S. Nesbitt, L. Williams, and R. Brown. 1977. Mirex residues in nontarget organisms after application of 10-5 bait for fire ant control, northeast Florida—1972–1974. *Pestic. Monit. J.* 11:146–156.

White, F.H., and D.J. Forrester. 1979. Antimicrobial resistant *Salmonella* spp. isolated from double-crested cormorants (*Phalacrocorax auritus*) and common loons (*Gavia immer*) in Florida. *J. Wildl. Dis.* 15:235–238.

Wobeser, G., F.A. Leighton, R. Norman, D.J. Myers, D. Onderla, M.J. Pybus, J.L. Neufeld, G.A. Fox, and D.J. Alexander. 1993. Newcastle disease in wild water birds in western Canada, 1990. *Can. Vet. J.* 34:353–359.

Work, T.M., B. Barr, A.M. Beale, L. Fritz, M.A. Quilliam, and J.L.C. Wright. 1993. Epidemiology of domoic acid poisoning in brown pelicans (*Pelecanus occidentalis*) and Brant's cormorants (*Phalacrocorax penicillatus*) in California. *J. Zoo Wildl. Med.* 24:54–62.

Wyman, J. 1869. On a threadworm (*Filaria anhingae*) infesting the brain of the snake-bird (*Plotus anhinga* Linn.). *Proc. Boston Soc. Nat. Hist.* 12:100–104.

Yamashita, N., S. Tanabe, J.P. Ludwig, H. Kurita, M.E. Ludwig, and R. Tatsukawa. 1993. Embryonic abnormalities and organochlorine contamination in double-crested cormorants (*Phalacrocorax auritus*) and Caspian terns (*Hydroprogne caspia*) from the upper Great Lakes. *Environ. Pollut.* 79:163–173.

Herons, Egrets, and Bitterns

I. Introduction

Twelve species of herons, egrets, and bitterns (family Ardeidae) are found in Florida, and all but the American Bittern regularly nest in Florida (table 6.1). The Little Blue Heron, Snowy Egret, Tricolored Heron, and Reddish Egret are listed as species of special concern by the Florida Game and Fresh Water Fish Commission (Logan 1997). The Reddish Egret is a candidate for federal listing and listed as rare by the Florida Committee on Rare and Endangered Plants and Animals (Rodgers et al. 1996). The Least Bittern, Great White Heron, Great Egret, Little Blue Heron, Snowy Egret, Tricolored Heron, Black-crowned Night-Heron, and Yellow-crowned Night-Heron are listed as species of special concern by the

Florida Committee on Rare and Endangered Plants and Animals (Rodgers et al. 1996). The Cattle Egret is a recent invader of Florida; it was first sighted in 1941 (Arendt 1988).

Life history information and disease summaries are given for Great Blue Herons (Butler 1997), Great White Herons (Powell and Bjork 1996), Great Egrets (Runde 1996), Reddish Egrets (Paul 1996), Little Blue Herons (Rodgers and Smith 1995; Rodgers 1996), Cattle Egrets (Browder 1973b; Telfair, 1994), Snowy Egrets (Ogden 1996a), Tricolored Herons (Ogden 1996b; Frederick 1997), Green Herons (Davis and Kushlan 1994), Black-crowned Night-Herons (Davis 1993; Below 1996), Yellow-crowned Night-Herons (Watts

Table 6.1. Species of herons and bitterns that occur in Florida

Species	Distribution	Seasonal occurrence	Relative abundance
American Bittern (*Botaurus lentiginosus*)	Throughout	Fall to spring, nests rarely	Uncommon
Least Bittern (*Ixobrychus exilis*)	Throughout	Year-round	Locally common
Great Blue Heron			
Blue morph (*Ardea herodias*	Throughout	Year-round	Common
White morph (*A. h. oxidentalis*)	South Florida, disperses throughout	Year-round	Locally common (south Florida)
Great Egret (*Ardea alba*)	Throughout	Year-round	Common
Snowy Egret (*Egretta thula*)	Throughout	Year-round	Locally common
Little Blue Heron (*Egretta caerulea*)	Throughout	Year-round	Locally common
Tricolored Heron (*Egretta tricolor*)	Throughout	Year-round	Locally common
Reddish Egret (*Egretta rufescens*)	South Florida	Year-round	Locally common
Cattle Egret (*Bubulcus ibis*)	Throughout	Year-round	Common
Green Heron (*Butorides virescens*)	Throughout	Year-round	Common
Black-crowned Night-Heron (*Nycticorax nycticorax*)	Throughout	Year-round	Common
Yellow-crowned Night-Heron (*Nyctanassa violacea*)	Throughout	Year-round	Common

Source: Robertson & Woolfenden (1992).

1995; Bancroft and Strong 1996), American Bitterns (Gibbs et al. 1992a) and Least Bitterns (Gibbs et al. 1992b; Frederick 1996). McVaugh (1972) presents information on nestling development in 4 species. Other general sources of information about the health of ardeids can be found in Fowler (1986), Friend and Franson (1999), Fairbrother et al. (1996), and Davidson and Nettles (1997).

Florida historically had the largest documented numbers of breeding ardeids in North America in the 1930s (Ogden 1978). Breeding populations of most of these species have declined in southern Florida by more than 90% since 1940 (Robertson and Kushlan 1974; Ogden 1978, 1994; Kushlan and White 1977; Frederick and Collopy 1989b; Runde 1991; Frederick and Spalding 1994) following their partial recovery from plume hunting during the late 1800s and early 1900s. The causes for the decline are not completely understood but it has been attributed, particularly in south Florida, to a number of factors, including water management practices (Browder 1978; Frederick and Spalding 1994) and emigration

to other regions (Sprunt 1954; Ogden 1978; Frederick and Collopy 1987; Frederick et al. 1996a). Emigration, however, may be a consequence of poor foraging conditions in southern Florida (Walters et al. 1992). Although biologists studying reproductive success have noted high mortality of nestlings, the effects of mortality on recruitment are not well understood because of the migratory nature of these birds. Spalding and Forrester (1991) documented the causes of mortality that occurred in ciconiiforms in Florida, especially southern Florida during the years 1987–91. A total of 2,088 ardeid birds (12 species), which included 1,835 nestlings, was examined from 61 colonies. That study is the basis for much of the work presented in this chapter.

II. Trauma

Trauma, especially that associated with tall structures, power lines, and vehicular traffic, was the most apparent cause of mortality of fledged birds in a study by Spalding and For-

Table 6.2. Causes of mortality for ardeids[a] collected in southern Florida

Cause of mortality	Birds per age category (%)		
	Nestlings	Juveniles	Adults
	(n = 715)	(n = 78)	(n = 174)
Unknown	17	9	5
Hatch failure	<1	0	0
Emaciation	34	10	9
Parasitic disease	19	12	18
Infectious disease	11	8	13
Toxicosis	<1	1	2
Nutritional disease	1	0	0
Neoplasia	0	0	1
Trauma			
Unknown	6	6	25
Predation	8	19	5
Human-related	2	31	18
Aggression	2	1	2
Feeding accident	<1	3	1

Source: Modified from Table 4 in Spalding & Forrester (1991).
a. A few White Ibises are included in this table; however, most are ardeids.

rester (1991) (table 6.2). The larger species, Great Blue and White Herons and Great Egrets, were the most commonly killed by collisions with power lines. Because birds dying along roads are more visible, their collection likely magnifies the perception of this problem. However, the finding of road trauma in 5 of 31 juvenile, and 1 of 8 adult Great White Herons with radio transmitters (birds found regardless of the location of death) supports the suggestion that this is a very real and significant cause of mortality (Spalding and Forrester 1991). Characteristic lesions included rupture of the liver at the base of the sternum, fracture of the acetabular region of the pelvis and synsacrum, and occasionally singed feathers, legs, or bill (figure 6.1) if the bird was electrocuted. In some cases there was evidence of both power line contact (i.e., singed feathers) and vehicular collision. Spalding and Forrester (1991) concluded that vehicular collision was much less frequent than power line collision, and that most mortality that appeared to be due to vehicular traffic was actually attributable to

power line collisions and subsequent falling or wandering onto roads. Birds with evidence of electrocution and no evidence of trauma probably landed on power poles, spanning wires.

During 1990–91, 165 birds were found dead under power lines in the Everglades, Palm Beach County. Of these, 66 were Great Blue Herons, 13 were Great Egrets, and 3 were night-herons (FPL and FGFWFC 1991). Seven

FIGURE 6.1. A Great White Heron electrocuted by coming into contact with a power line.

Great Blue Herons examined as part of that study all had evidence of trauma with internal hemorrhage; none had clear evidence of electrocution (Locke 1990; Spalding 1991), supporting the hypothesis that injury is most commonly caused by collision with wires rather than by contact between wires. Great Blue Herons were the most commonly found species under a new power line erected in Hendry County. In that study, 21 dead birds (7 Great Blue Herons, 3 Black-crowned Night-Herons, 2 Glossy Ibises, 2 Tricolored Herons, 2 Yellow-crowned Night-Herons, and 1 each of Anhinga, Green Heron, Red-shouldered Hawk, Snowy Egret, and Turkey Vulture) were found on 3 different days in July 1999 (Sasse 1999).

Weston (1966) surveyed a bridge with an overhead cable on Pensacola Bay from 1938 to 1949. Of the 740 birds he found dead, 1 was a Yellow-crowned Night-Heron and 1 a Least Bittern. Mortality ceased when a new bridge without an overhead cable was built. In a survey of road kills from 1990 to 1993 in and around state parks, a total of 1,562 birds was found dead, including 18 Green Herons, 9 Cattle Egrets, 3 Great Egrets, 2 Least Bitterns, 2 Black-crowned Night-Herons, 2 Snowy Egrets, 1 Great White Heron, 9 Great Blue Herons, and 1 Little Blue Heron (Stevenson 1994; Snyder 1994).

Although it seems that the larger ardeids more commonly hit power lines or may be electrocuted, the smaller species also are at risk of hitting tall structures, probably during migration. Crawford (1981) recorded birds found dead below a TV tower on Tall Timbers Research station in Leon County between 1955 and 1980. Of 42,384 birds of 189 species found, 102 were ardeids. The total numbers of each species were: Green Heron 34, Great Egret 1, Little Blue Heron 4, Cattle Egret 10, Snowy Egret 1, Tricolored Heron 2, Yellow-crowned Night-Heron 5, Least Bittern 18, American Bittern 27. Kale (1971) reported the death of 2 Least Bitterns in a total of 2,195 birds found dead at the base of a lunar vehicle assembly building in Brevard County, April 22–May 2, 1971. These were the only ardeids recorded between 1971 and 1981, of 5,046 individuals found dead at this site (Taylor and Kershner 1986). Taylor and Anderson (1973) reported that 13 Green Herons, 1 Cattle Egret, 1 Yellow-crowned Night-Heron, and 2 American Bitterns were among the 7,782 individuals found dead below a recently erected TV tower in Orange County during the autumns of 1969 and 1971.

Least Bitterns are struck occasionally and killed by airboats. Frederick et al. (1990) found that less than 3% of those flushed by airboats were killed. About 10 Cattle Egrets foraging in fields at an airport were killed by planes in Alachua County in 1991 (Spalding 1991).

Within nesting colonies, intraspecific aggression (brood reduction or attacks by siblings) was suspected when traumatic lesions were found on nestlings (Mock et al. 1987). These types of lesions were seen more commonly in emaciated and parasitized chicks than in otherwise healthy nestlings, probably because starving and diseased chicks wandered from their nests onto the nests and territories of other birds (Spalding and Forrester 1991).

Feeding accidents resulting in death were uncommon. A nestling Tricolored Heron swallowed a twig that perforated the stomach (Spalding and Forrester 1991). Deaths related to food items in adults and juveniles (2%, table 6.2) all involved fish that were too large to pass through the thoracic inlet of Great Egrets and had perforated the esophagus (Spalding and Forrester 1991). Two Great Blue Herons that had their feet entangled in blue-green filamentous algae (*Lyngbya* spp.) on Newnans Lake, Alachua County, were captured and released. The birds appeared to be otherwise healthy but probably would have died if not released (Delany and Woodward 1988).

Ardeids are not as prone to injury by fishing gear as Brown Pelicans (see chapter 4, Pelicans); however, these species, especially the Great Blue Heron, occasionally become entangled in monofilament, are injured by fish hooks, or ingest lead sinkers (Quinn 1994; Suto 2000). Esophageal perforation resulting in hemorrhage and death occurred in a Great Blue Heron in Alachua County in 1990, and in a Great White

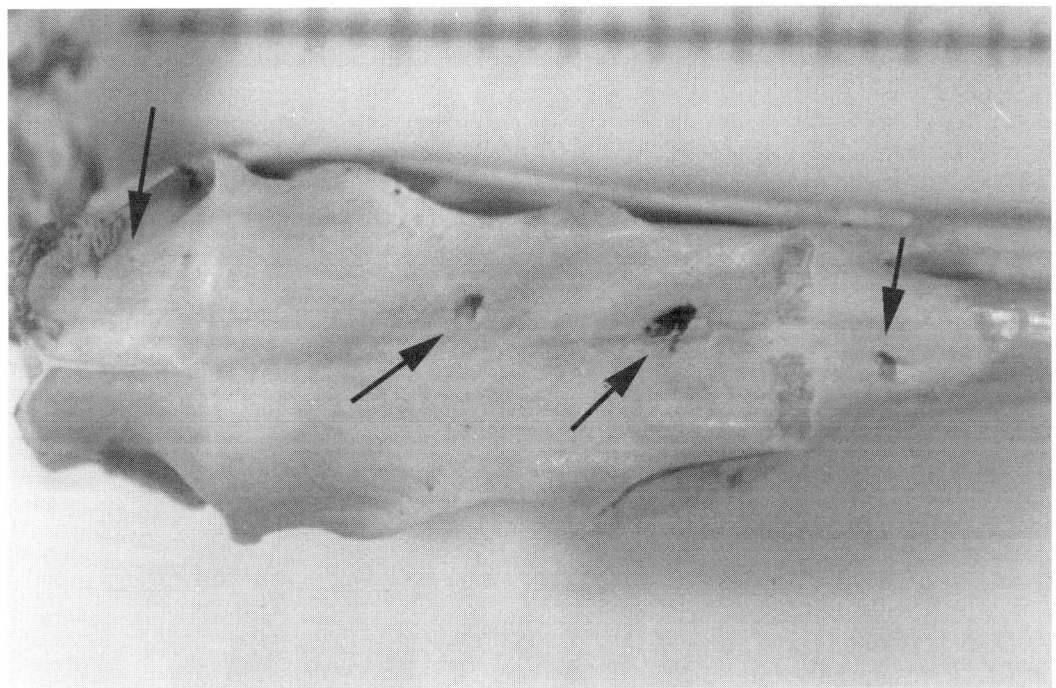

FIGURE 6.2. The skull of a Great White Heron killed by a Bald Eagle in Monroe County. Note talon perforations *(arrows)* in the skull.

Heron in Monroe County in 1990 (Spalding 1990).

III. Predation

Predation of nest contents is a common cause of nest failure in ardeids, especially when human disturbance keeps adults away from the nest. Predation by Fish and American Crows on Great White Heron nests, Fish Crows on Great Blue Heron nests, and Black Vultures, Fish Crows, Boat-tailed Grackles, and large snakes on Great Egret nests have been observed (Jenni 1969; Frederick and Collopy 1989c; Spalding and Forrester 1991; Stevenson and Anderson 1994). Egg mortality is most often attributed to abandonment of the nest by the adults or predation on the nest contents (Frederick and Collopy 1989c). By using a tracking medium, researchers found that snakes took 23% of eggs and mammals took

20%. Failure of entire nests was most common. Vultures were suspected predators of 10 Tricolored Heron nestlings from Florida Bay. The legs and unopened stomachs were all that remained in the nest (Spalding and Forrester 1991). Both Black and Turkey Vultures have been reported to take live nestlings (Sprunt 1949, 1969). Paul (1996) lists raccoons, Bald Eagles, Fish Crows, American Crows, and dogs as predators of Reddish Egrets, and in one colony in southwest Florida Bald Eagles took most of the Reddish Egret young produced over a 2-year period.

Postfledging mortality is even more difficult to assess than nestling mortality and is reported rarely. Predation by Bald Eagles was responsible for the death of 39% of nestling Great White Herons, and of 13–25% of juveniles during the first year after fledging in Florida Bay (Spalding and Forrester 1991). Lesions consisted of hemorrhage and talon perforations on the head and neck (figure 6.2). Observations of predation on

Cattle Egrets by Bald Eagles, Crested Caracaras, Red-tailed Hawks, bobcats, and red foxes were summarized in Layne et al. (1977). They reported that avian predation on Cattle Egrets is especially common in south central Florida, where habitat alteration to drained pastureland has increased numbers of Cattle Egrets, eagles, and caracaras. Other authors have observed also that Cattle Egrets are frequently the target of Bald Eagle attacks (Knight 1976; McEwan and Hirth 1980; Jennings and Jennings 1982), and McEwan and Hirth (1980) noted that they were 2% of food items found in Bald Eagle nests in north central Florida. Folk (1992) found that 11 of 13 prey items taken by cooperatively foraging Bald Eagles in Alachua County were Cattle Egrets. Nicholson (in Bent 1937) found carcasses of Little Blue Herons and Snowy Egrets in Bald Eagle nests. Callahan and Carey (1979) reported predation by a Great Horned Owl (*Bubo virginianus*) on a Snowy Egret in Pasco County, 1977. The neck was lacerated and broken. A tethered Great Horned Owl in Welaka Game Reserve, Putnam County, in 1949 was studied for food items brought to it and pellets contained remains of a Least Bittern (Burns 1952).

Predation on wading birds by alligators is probably common, but rarely reported. Spalding and Forrester (1991) observed an alligator run after a nestling Great Egret and chase it partway up a tree. Since most nesting occurs over water, alligators undoubtedly eat nestlings that fall from trees. Delany et al. (1988) reported Cattle Egret carcasses in 1.8% of 113 nuisance alligator stomachs from north central Florida in 1977.

IV. Habitat alteration and human disturbance

Historically, the most significant mortality factor reported for wading bird populations in Florida was the extensive killing of ardeids for their beautiful breeding plumes, especially the "aigrettes" of the Great Egret. Since most of the collections were made from mixed-species breeding colonies, even species that were not hunted were impacted adversely. Scott (1887, 1889, 1890) gave a graphic description of the effects of hunting in these colonies. Powell et al. (1989) summarized the near extinction of Great White Herons and Reddish Egrets in Florida Bay from the 1800s to 1930s due to the effects of plume hunting and harvesting for food. Current disturbance of colonies includes development close to colony sites and human recreational or research visitation. Frederick and Collopy (1989a) measured no effect of researchers on nesting success of Tricolored Herons in the Everglades, but cautioned that these results might not extend to colonies with more predators. Spalding and Forrester (1991) occasionally found nestlings wedged in branches and wondered if the occurrence of these accidents was increased by researcher visitations, when nestlings would move rapidly through the treetops.

Loss of intertidal flats due to dredge and fill operations, and boat use, particularly personal watercraft that can use shallow water, are particularly detrimental to the Reddish Egret on the southwest coast of Florida, where significant human population growth is occurring (Paul 1996). Rodgers and Smith (1997) suggest buffer distances for this group of birds. Because wading birds are attracted to and hunt in commercial fish farms they are subject to both legal and illegal hunting (Rodgers 1996).

A number of reports attempt to link reproductive failure or reduction to human-caused environmental changes that affect food resources. Powell (1983) showed that adult Great White Herons supplemented with food were able to fledge more young.

V. Inclement weather

Violent weather is among the potential sources of mortality, but such mortality is rarely reported. The large numbers of wading birds that migrate into Florida have mostly dispersed out of the state before the onset of the hurricane season (June–November), nevertheless there are some records of hurricane-related mortality. The particularly powerful hurricane of 1935 took a large toll on the already low numbers of Great White Herons; 211 birds were seen on a survey before the storm, but only 146 after the storm (Powell et al. 1989; Sprunt 1954). After Hurricane Donna in September 1960, 78 Great Egrets, 53 Great White Herons, 31 Great Blue Herons, 12 Snowy Egrets, 4 Yellow-crowned Night-Herons, 2 Tricolored Herons, and 2 Black-crowned Night-Herons were found dead in Florida Bay (Robertson and Paulson 1961). Counts of Great White Herons before (898) and after (516) Hurricane Donna in 1960 led to an estimate of 39% mortality (Robertson and Paulson 1961). Dead Cattle Egrets were found in the Homestead area following Hurricane Andrew in August 1992, and vegetative damage was severe in 14 wading bird rookeries and moderate in 57 rookeries (FGFWFC 1992). Little damage was noticed in marshes; however, effects upon food resources were not known. Winds and elevated water levels during storms can interfere with foraging ability (see the example of tropical storm Marco in section XX, Emaciation, below).

Jenni (1969) reported losses of chicks and nests during windstorms in Alachua County. Several days after a severe storm, approximately 65 herons and egrets and a rail (species not specified) washed ashore in Collier County in April 1997. Trauma or drowning was suspected to be the cause of death, but the 3 birds that were examined were too autolyzed to confirm this (Sileo 1997).

Eleven dead Cattle Egrets were found among many other dead birds during the second week of May 1962 on the Dry Tortugas, Monroe County. All of these birds were severely emaciated and an extreme spring drought was suspected to have contributed to the mortality (Paulson and Stevenson 1962).

VI. Organochlorines

Information on organochlorine contamination of ardeids in Florida consists of a large statewide sampling of eggs in the early 1970s, followed by miscellaneous tissue sampling from adults and 2 surveys of nestlings in southern Florida. There have been a few cases of acute poisoning; however, the effects of sublethal poisoning are poorly understood.

Large numbers of ardeid eggs of all species but Reddish Egrets and American Bitterns were tested for organochlorine contaminants in 1972–73, and fewer numbers were tested in 1992–93 (tables 6.3–6.13). Although Peakall (1975) found ciconiiforms to be highly sensitive to DDE-induced eggshell thinning, pesticide concentrations in eggs were found to be lower in the southern states when compared with northern states in the early 1970s (Ohlendorf et al. 1978). Custer et al. (1983) found a negative effect on hatching success of Black-crowned Night-Herons in New England when DDE concentrations were greater than 4 ppm, but little effect on overall reproductive success. DDE concentrations higher than Custer's threshold of 4 ppm were detected in eggs of 7 ardeid species tested in Florida during 1972–73. Locations of egg collections and of species with DDE concentrations >4 ppm are illustrated in figure 6.3. Embryonic mortality and congenital defects did not appear to be a severe problem in southern Florida during 1987–91, except possibly in the Lake Okeechobee region, where 3 malformed nestlings were found (see section XI, Anomalies and Malformations, below). One of these, a Little Blue Heron with a deformed bill, contained no measurable organochlorine contaminants (Spalding et al. 1997). Several authors reported that at least 80% and usually more than 90% of all eggs present in active nests at the expected time of hatch did hatch (Girard and Taylor 1979; Rodgers 1980a,b, 1987; Black et al. 1984; Frederick and Collopy 1987). Unfortunately most of these studies were conducted well after the contaminant testing in the early 1970s, so the effects of organochlorine uses on hatching success during the 1970s may never truly be known. More recent egg sampling by Rodgers (1997) resulted in the finding of somewhat lower concentrations of these persistent contaminants.

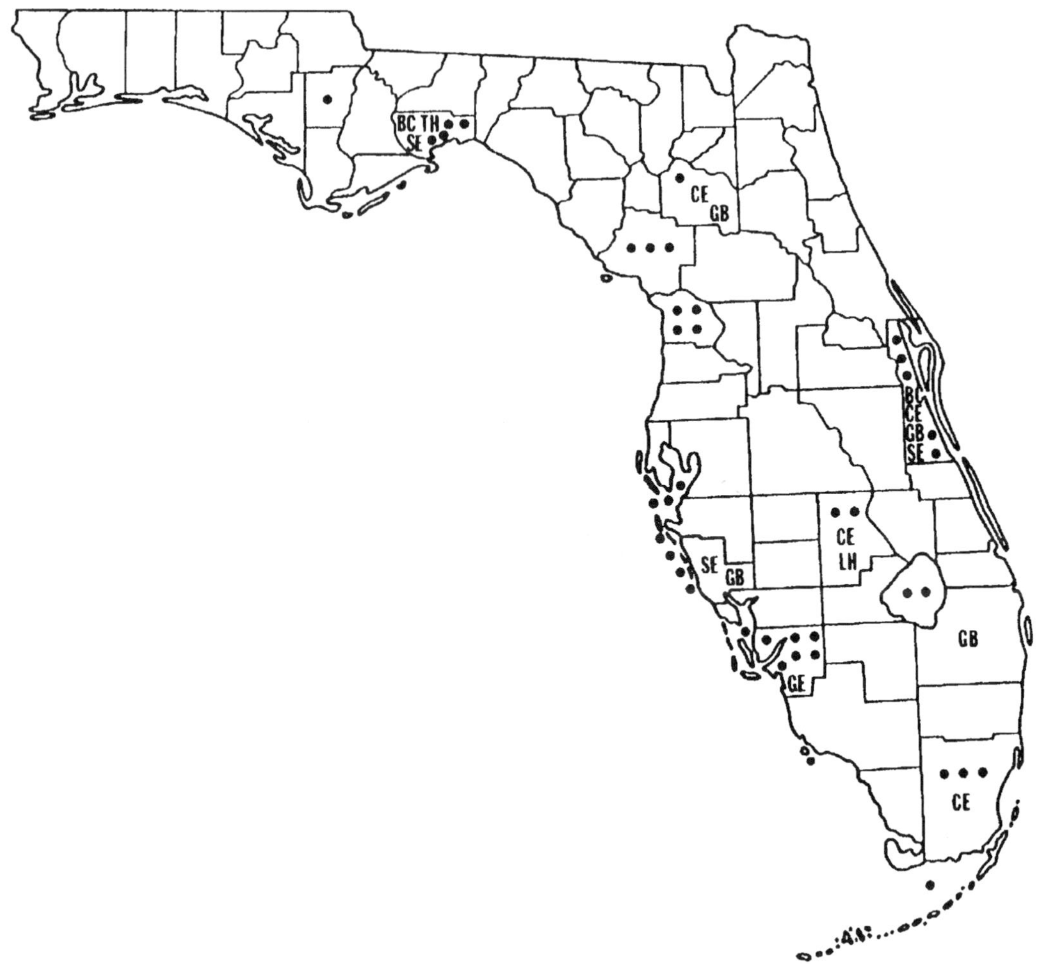

FIGURE 6.3. Locations *(dots)* of egg collections tested for DDE during 1972–73 by Ohlendorf et al. (1979). Species for which individual eggs had >4 ppm DDE: TH = Tricolored Heron, SE = Snowy Egret, BC = Black-crowned Night-Heron, CE = Cattle Egret, GB = Great Blue Heron, GE = Great Egret, LH = Little Blue Heron.

The eggshell thinning effects of organochlorine contaminants on ardeids were not as dramatic as in other species such as the Brown Pelican, and were less well documented. Anderson and Hickey (1972) found up to 18% eggshell thinning and a 22% decrease in eggshell weight in Black-crowned Night-Heron eggs collected from the eastern coast of the United States when compared with pre-1947 eggs. In Florida, however, Ohlendorf et al. (1979) found only a 5% decrease for Black-crowned Night-Herons and Great Blue Herons, and no significant thinning in other ardeid species from Florida (table 6.14). See chapter 4, Pelicans, and chapter 11, Eagles, for more discussion of eggshell thinning.

Of other tissues sampled from ardeids, brain and entire carcasses were the most common (tables 6.3–6.13). A wide variety of organochlorine compounds was found in most samples tested. Only one documented case of acute organochlorine toxicosis was reported in

Table 6.3. Organochlorine residues in tissues of American Bitterns and eggs of Least Bitterns collected in Florida[c]

Tissue County/area	Year(s)	No. examined	DDE			Dieldrin			PCBs			Data Source
			No.[a]	Mean[b]	(Range)	No.[a]	Mean[b]	(Range)	No.[a]	Mean[b]	(Range)	
American Bitterns												
Adipose												
"Florida"	1973	2	2	5.4	(2.4–12)	0	ND	—	0	ND	—	A
Uropygial gland												
"Florida"	1973	2	2	3.0	(1.8–5.1)	1	0.02	(ND–0.46)	0	ND	—	A
Least Bitterns												
Eggs												
Alachua	1972–73	2	2	0.67	(0.21–1.3)	0	ND	—	0	ND	—	B
Brevard	1972–73	8	8	0.29	(0.080–1.0)	0	ND	—	3	0.17	(ND–0.88)	B
"Lake Okeechobee"	1972–73	9	9	1.0	(0.73–2.1)	0	ND	—	0	ND	—	B
Wakulla	1972–73	2	2	1.1	(0.67–1.6)	1	1.1	(ND–3.4)	1	0.48	(ND–1.2)	B

Sources: A = Johnston (1976), B = Ohlendorf et al. (1979).

ND = not detected.

a. Number with residue.

b. Geometric mean, ppm wet weight.

c. In addition to the compounds listed, Johnston (1976) measured 6.3 (3.2–13) ppm total DDTs in adipose tissue and 4.1 (1.8–9.2) in uropygial gland in 2 American Bitterns from Florida. Ohlendorf et al. (1979) detected 0.12 ppm DDE and 0.10 ppm DDT in eggs from Wakulla County and 0.10 ppm *cis*-chlordane in 1 egg from Lake Okeechobee. They did not detect mirex, *cis*-nonachlor, *trans*-nonachlor, HCB, toxaphene, heptachlor epoxide, or endrin in any of the eggs.

Table 6.4. Organochlorine residues in eggs of Night-Herons collected in Florida[c]

Species County	Year(s)	No. examined	DDE			DDT			Dieldrin		
			No.[a]	Mean[b]	(Range)	No.[a]	Mean[b]	(Range)	No.[a]	Mean[b]	(Range)
Black-crowned Night-Herons											
Brevard	1972–73	9	9	1.01	(0.32–6.2)	0	ND	—	4	0.23	(ND–2.8)
Citrus	1972–73	5	5	0.30	(ND–1.2)	1	0.024	(ND–0.13)	0	ND	—
Lee	1972–73	5	4	0.33	(ND–1.59)	0	ND	—	0	ND	—
Wakulla	1972–73	11	11	1.50	(0.45–11)	1	0.018	(ND–0.2 2)	0	ND	—
Yellow-crowned Night-Herons											
Calhoun	1972–73	10	10	0.35	(0.20–0.89)	1	0.009	(ND–0 .096)	0	ND	—
Wakulla	1972–73	1	0	ND	—	0	ND	—	0	ND	—
			Mirex			cis-chlordane			PCBs		
Black-crowned Night-Herons											
Brevard	1972–73	9	1	0.018	(ND–0.18)	2	0.035	(ND–0.24)	9	1.80	(0.25–6.2)
Citrus	1972–73	5	0	ND	—	0	ND	—	3	0.14	(ND– 0.25)
Lee	1972–73	5	0	ND	—	0	ND	—	4	0.20	(ND– 0.25)
Wakulla	1972–73	11	1	0.030	(ND–0.42)	1	0.025	(ND–0.3 1)	10	0.68	(ND– 4.1)
Yellow-crowned Night-Herons											
Calhoun	1972–73	10	1	0.040	(ND–0.49)	0	ND	—	0	ND	—
Wakulla	1972–73	1	0	ND	—	0	ND	—	0	ND	—

Sources: Ohlendorf et al. (1978, 1979).
ND = not detected.
a. Number with residue.
b. Geometric mean, ppm wet weight.
c. In addition to the compounds listed, Ohlendorf et al. (1978) detected 0.01 ppm oxychlordane in 1 Black-crowned Night-Heron egg from Wakulla County. They did not detect DDD, *cis*-nonachlor, *trans*-nonachlor, HCB, toxaphene, heptachlor epoxide, or endrin in any of the eggs.

Table 6.5. Organochlorine residues in eggs of Great Blue Herons collected in Florida

County	Year(s)	No. examined	DDE			DDD			DDT			Data source
			No.[a]	Mean[b]	(Range)	No.[a]	Mean[b]	(Range)	No.[a]	Mean[b]	(Range)	
Alachua	1972–73	3	3	2.2	(1.2–4.0)	0	ND	—	0	ND	—	A
Brevard	1972–73	10	10	2.1	(0.28–24)	3	0.059	(ND–0.42)	2	0.064	(ND–0.66)	A
Citrus	1972–73	3	3	0.46	(0.25–0.75)	0	ND	—	0	ND	—	A
Lee	1972–73	8	3	1.4	(0.44–3.1)	0	ND	—	0	ND	—	A
Palm Beach	1972–73	4	4	12	(5.4–33)	1	0.026	(ND–0.11)	0	ND	—	A
Sarasota	1972	1	1	20[c]	—	NA	—	—	NA	—	—	B
			Mirex			Heptachlor epoxide			Dieldrin			
Alachua	1972–73	3	2	0.45	(ND–1.5)	0	ND	—	3	0.36	(0.12–0.97)	A
Brevard	1972–73	10	4	0.35	(ND–5.8)	0	ND	—	1	0.028	(ND–0.31)	A
Citrus	1972–73	3	0	ND	—	0	ND	—	0	ND	—	A
Lee	1972–73	8	1	0.099	(ND–0.73)	1	0.10	(ND–1.2)	2	0.047	(ND–0.31)	A
Palm Beach	1972–73	4	0	ND	—	0	ND	—	2	0.094	(ND–0.31)	A
Sarasota	1972	1	NA	—	—	NA	—	—	1	2.2[c]	—	B
			Oxychlordane			cis-chlordane			cis-nonachlor			
Alachua	1972–73	3	3	0.20	(0.087–0.41)	3	0.49	(0.24–1.1)	1	0.042	(ND–0.13)	A
Brevard	1972–73	10	2	0.021	(ND–0.11)	3	0.047	(ND–0.21)	1	0.009	(ND–0.087)	A
Citrus	1972–73	3	0	ND	—	1	0.028	(ND–0.087)	0	ND	—	A
Lee	1972–73	8	0	ND	—	7	0.22	(ND–0.75)	1	0.021	(ND–0.19)	A
Palm Beach	1972–73	4	0	ND	—	2	0.086	(ND–0.22)	0	ND	—	A
			t-nonachlor			Toxaphene			PCBs			
Alachua	1972–73	3	0	ND	—	0	ND	—	3	6.5	—	A
Brevard	1972–73	10	0	ND	—	2	0.040	(ND–0.30)	7	2.4	(ND–7.4)	A
Citrus	1972–73	3	0	ND	—	0	ND	—	2	0.43	(ND–0.87)	A
Lee	1972–73	8	0	ND	—	0	ND	—	8	2.6	(0.81–20)	A
Palm Beach	1972–73	4	0	ND	—	2	0.054	(ND–0.13)	4	1.5	(0.40–4.0)	A
Sarasota	1972	1	NA	—	—	NA	—	—	1	29	—	B

Sources: A = Ohlendorf et al. (1979), B = Lincer & Salkind (1973).
ND = not detected, NA = not analyzed.
a. Number with residue.
b. Geometric mean, ppm wet weight unless otherwise indicated.
c. Data presented are dry weights. Lincer & Salkind (1973) also give extractable fat weights.

Table 6.6. Organochlorine residues found in tissues of Great Blue Herons collected in Florida[e]

Age / Tissue	County/area	Year(s)	No. examined	DDE No.[a]	DDE Mean[b]	DDD No.[a]	DDD Mean[b]	DDT No.[a]	DDT Mean[b]	Data source
Nestlings										
Brain	"Lake Okeechobee"	1987–91	1[c]	1	0.38	0	ND	0	ND	C
	"Florida Bay"	1987–91	1[d]	1	0.02	0	ND	0	ND	C
Liver	"Lake Okeechobee"	1992–93	1	1	1.6	0	ND	0	ND	B
Adults										
Brain	Dade	1973	1	1	8.0	1	2.0	1	0.15	A
	Monroe	1973	1	1	2.6	0	ND	0	ND	A
	Monroe	1973	1	1	28	1	1.5	1	0.21	A
Carcass	Dade	1973	1	1	20	1	5.5	0	ND	A
	Monroe	1973	1	1	4.3	0	ND	0	ND	A
	Monroe	1973	1	1	12	1	1.1	1	0.11	A

Age / Tissue	County/area	Year(s)	No. examined	Mirex No.[a]	Mirex Mean[b]	Heptachlor epoxide No.[a]	Heptachlor epoxide Mean[b]	Dieldrin No.[a]	Dieldrin Mean[b]	Data source
Nestlings										
Brain	"Lake Okeechobee"	1987–91	1[c]	0	ND	1	0.03	0	ND	C
	"Florida Bay"	1987–91	1[d]	1	0.02	0	ND	0	ND	C
Liver	"Lake Okeechobee"	1992–93	1	NA	—	NA	—	0	ND	B
Adults										
Brain	Dade	1973	1	0	ND	0	ND	1	3.9	A
	Monroe	1973	1	0	ND	0	ND	0	ND	A
	Monroe	1973	1	1	0.17	1	0.26	1	6.6	A
Carcass	Dade	1973	1	0	ND	1	0.30	1	6.5	A
	Monroe	1973	1	0	ND	0	ND	0	ND	A
	Monroe	1973	1	0	ND	1	0.13	1	4.5	A

Age / Tissue	County/area	Year(s)	No. examined	Oxychlordane No.[a]	Oxychlordane Mean[b]	cis-chlordane No.[a]	cis-chlordane Mean[b]	cis-nonachlor No.[a]	cis-nonachlor Mean[b]	Data source
Nestlings										
Brain	"Lake Okeechobee"	1987–91	1[c]	1	0.03	NA	—	0	ND	C
	"Florida Bay"	1987–91	1[d]	0	ND	NA	—	0	ND	C

				[a]							
Adults											
Brain	Dade	1973	1		1	0.17	1	0.75	1	0.34	A
	Monroe	1973	1		0	ND	0	ND	0	ND	A
	Monroe	1973	1		1	0.20	1	0.75	1	0.38	A
Carcass	Dade	1973	1		1	0.19	1	1.7	1	0.40	A
	Monroe	1973	1		0	ND	0	ND	0	ND	A
	Monroe	1973	1		0	ND	1	0.35	1	0.19	A

					trans-nonachlor		Toxaphene		PCBs		
Nestlings											
Brain	"Lake Okeechobee"	1937–91	1[c]		1	0.05	1	0.03	NA	—	C
	"Florida Bay"	1937–91	1[d]		0	ND	0	ND	0	ND	C
Liver	"Lake Okeechobee"	1992–93	1		1	0.07	0	ND	0	ND	B
Adults											
Brain	Dade	1973	1		0	ND	1	0.17	1	0.75	A
	Monroe	1973	1		0	ND	0	ND	0	ND	A
	Monroe	1973	1		1	0.17	1	0.20	1	0.76	A
Carcass	Dade	1973	1		0	ND	1	0.19	1	1.7	A
	Monroe	1973	1		0	ND	0	ND	0	ND	A
	Monroe	1973	1		0	ND	0	ND	1	0.35	A

Sources: A = Ohlendorf et al. (1981), B = Rodgers (1997), C = Spalding et al. (1997).

NA = not analyzed, ND = not detected.

a. Number with residue.

b. Geometric mean, ppm wet weight.

c. Pooled sample of 5 brains.

d. Pooled sample of 10 brains of the white morph, Great White Heron.

e. In addition to the compounds listed, Spalding et al. (1997) detected 0.01 ppm HCB in the pooled brain sample of 5 nestlings from Lake Okeechobee. They did not detect α-BHC, Γ-BHC, β-BHC, δ-BHC, Γ-chlordane, α-chlordane, endrin, *cis*-nonachlor, endosulfan I, endosulfan II, or HCB from any of the pooled nestling brain samples (percent lipid range = 5.10–9.10; percent moisture range = 78.9–85.0; and lower level of detection = 0.01 ppm, 0.05 ppm for toxaphene and PCBs). Ohlendorf et al. (1981) detected 0.20 ppm endrin in 1 of the adult brains from Monroe County. They did not detect HCB in any of the samples (3 adult brains and carcasses) from Dade (1) and Monroe (2) counties. Rodgers (1997) did not detect chlordane, endrin, or heptachlor in the nestling brain from Lake Okeechobee.

Table 6.7. Organochlorine residues in eggs and tissues of Great Egrets collected in Florida[g]

Age Tissue	County/area	Year(s)	No. examined	DDE			DDD			DDT			Data source
				No.[a]	Mean[b]	(Range)	No.[a]	Mean[b]	(Range)	No.[a]	Mean[b]	(Range)	
Eggs	Brevard	1972–73	13	13	0.66	(0.23–1.8)	0	ND	—	0	ND	—	C
	"Charlotte Harbor"	1972	1	1	10[c]	—	NA	—	—	NA	—	—	A
	Citrus	1972–73	7	7	0.42	(0.15–1.0)	0	ND	—	0	ND	—	C
	Dade	1972–73	5	5	0.56	(0.29–0.89)	5	0.027	(0.022–0.041)	0	ND	—	B
	"Florida Bay"	1972	5	5	0.24	(0.097–0.94)	3	0.006	(ND–0.031)	5	0.028	(0.016–0.094)	B
	Palm Beach	1993	6	NG	0.08	(ND–7.0)	NG	0.04	(ND–7.0)	NG	NC	(ND–0.17)	E
	Lee	1972–73	10	10	0.98	(0.38–9.0)	1	0.028	(ND–0.3 1)	0	ND	—	C
	Levy	1972–73	1	1	1.0	—	0	ND	—	0	ND	—	C
	"Sarasota Bay"	1972–73	1	1	0.28	—	0	ND	—	0	ND	—	C
Nestlings													
Brain	Dade	1989	1	1	1.3	—	0	ND	—	0	ND	—	D
	"Florida Bay"	1989	1[d]	0	ND	—	0	ND	—	0	ND	—	D
	"Lake Okeechobee"	1989–90	1[e]	1	0.80	—	0	ND	—	0	ND	—	D
	Monroe	1987–88	1[f]	1	0.11	—	0	ND	—	0	ND	—	D

Age Tissue	County/area	Year(s)	No. examined	Dieldrin			Mirex			Oxychlordane			Data source
				No.[a]	Mean[b]	(Range)	No.[a]	Mean[b]	(Range)	No.[a]	Mean[b]	(Range)	
Eggs	Brevard	1972–73	13	5	0.10	(ND–0.64)	0	ND	—	1	0.018	(ND–0.26)	C
	"Charlotte Harbor"	1972	1	1	0.06[c]	—	NA	—	—	NA	—	—	A
	Citrus	1972–73	7	1	0.014	(ND–0.10)	1	0.091	(ND–0.85)	1	0.014	(ND–0.10)	C
	Dade	1972	5	3	0.006	(ND–0.075)	NA	—	—	NA	—	—	B
	"Florida Bay"	1972	5	3	0.007	(ND–0.03)	NA	—	—	NA	—	—	B
	Palm Beach	1993	6	NG	0.10	(ND–1.60)	NA	—	—	NA	—	—	E
	Lee	1972–73	10	1	0.091	(ND–1.4)	0	ND	—	1	0.014	(ND–0.14)	C
	Levy	1972–73	1	0	ND	—	0	ND	—	0	ND	—	C
	"Sarasota Bay"	1972–73	1	1	0.14	—	1	0.23	—	0	ND	—	C
Nestlings													
Brain	Dade	1989	1	0	ND	—	0	ND	—	0	ND	—	D
	"Florida Bay"	1989	1[d]	0	ND	—	0	ND	—	0	ND	—	D
	"Lake Okeechobee"	1989–90	1[e]	0	ND	—	0	ND	—	0	ND	—	D
	Monroe	1987–88	1[f]	0	ND	—	0	ND	—	0	ND	—	D

					cis-chlordane			PCBs			Heptachlor epoxide			
Eggs	Brevard	1972–73	13	0	0.042	(ND–0.26)	12	1.5	(ND–12)	0	ND	—		C
	"Charlotte Harbor"	1972	1	NA	—	—	1	7.5[c]	—	NA	—	—		A
	Citrus	1972–73	7	1	0.038	(ND–0.30)	2	0.13	(ND–0.6 2)	0	ND	—		C
	Dade	1972	5	NA	—	—	5	0.30	(0.19–0.56)	NA	—	—		B
	"Florida Bay"	1972	5	NA	—	—	3	0.15	(ND–5.0)	NA	—	—		B
	Palm Beach	1993	6	NA	—	—	0	ND	—	NA	—	—		E
	Lee	1972–73	10	2	0.023	(ND–0.14)	9	0.65	(ND–1.7)	0	ND	—		C
	Levy	1972–73	1	0	ND	—	0	ND	—	1	0.091	—		C
	"Sarasota Bay"	1972–73	1	0	ND	—	1	0.54	—	0	ND	—		C
Nestlings														
Brain	Dade	1989	1	0	ND	—	0	ND	—	1	0.07	—		D
	"Florida Bay"	1989	1[d]	0	ND	—	0	ND	—	0	ND	—		D
	"Lake Okechobee"	1989–90	1[e]	0	ND	—	0	ND	—	1	0.03	—		D
	Monroe	1987–88	1[f]	0	ND	—	0	ND	—	1	0. 02	—		D

					Heptachlor			trans-nonachlor			
Eggs	Palm Beach	1993	6	NG	0.02	(ND–0.27)	NG	0.14	(ND–0.42)		E

Sources: A = Lincer & Salkind (1973), B = Ogden et al. (1974), C = Ohlendorf et al. (1979), D = Spalding et al. (1997), E = Rodgers (1997).

NA = not analyzed, ND = not detected, NG = not given, NC = not calculated (<50% of samples contained the residue).

a. Number with residue.

b. Geometric mean, ppm wet weight unless otherwise indicated.

c. Data presented is dry weight. Lincer & Salkind (1973) also give extractable fat weights.

d. Pooled sample of 2 brains.

e. Pooled sample of 10 brains.

f. Pooled sample of 5 brains.

g. In addition to the compounds listed, Spalding et al. (1997) also detected 0.04 ppm *trans*-nonachlor in the pooled sample of 10 nestling brains from Lake Okeechobee, 0.05 ppm from the single nestling brain from Dade County, and 0.02 ppm from the pooled sample of 5 nestling brains from Monroe County. They did not detect α-BHC, Γ-BHC, β-BHC, δ-BHC, Γ-chlordane, α-chlordane, endrin, *cis*-nonachlor, toxaphene, endosulfan I, or endosulfan II in any of the nestling brain samples (percent lipid range = 5.10–9.10; percent moisture range = 78.9–85.0; and lower level of detection = 0.01 ppm, 0.05 ppm for toxaphene and PCBs). Ohlendorf et al. (1979) did not detect *cis*-nonachlor, *trans*-nonachlor, toxaphene, HCB or endrin in any of their 32 egg samples examined. Rodgers (1997) did not detect chlordane, endrin, or toxaphene in any of the 6 eggs listed.

Table 6.8. Organochlorine residues in eggs of Snowy Egrets collected in Florida[d]

County/area	Year(s)	No. examined	DDE			DDD			DDT			Data source
			No.[a]	Mean[b]	(Range)	No.[a]	Mean[b]	(Range)	No.[a]	Mean[b]	(Range)	
Brevard	1972–73	10	10	0.54	(0.090–4.2)	0	ND	—	1	0.014	(ND–0.16)	B
Dade	1972–73	2	2	0.19	(0.096–0.29)	0	ND	—	0	ND	—	B
Highlands	1972–73	5	5	0.96	(0.37–1.7)	1	0.066	(ND–0.3 8)	0	ND	—	B
Lee	1972–73	7	7	0.59	(0.23–1.2)	1	0.026	(ND–0.20)	0	ND	—	B
Levy	1972–73	10	10	0.37	(0.098–0.93)	0	ND	—	2	0.01 8	(ND–0.096)	B
Osceola	1992–93	5	0	ND	—	0	ND	—	0	ND	—	C
Sarasota	1972	1	1	21[c]	—	NA	—	—	NA	—	—	A
"Sarasota Bay"	1972–73	2	2	0.64	(0.48–0.81)	1	0.042	(ND–0.085)	0	ND	—	B
Wakulla	1972–73	10	10	1.9	(0.33–5.2)	0	ND	—	0	ND	—	B

County/area	Year(s)	No. examined	Dieldrin			Mirex			Oxychlordane			Data source
			No.[a]	Mean[b]	(Range)	No.[a]	Mean[b]	(Range)	No.[a]	Mean[b]	(Range)	
Brevard	1972–73	10	2	0.042	(ND–0.35)	0	ND	—	0	ND	—	B
Dade	1972–73	2	0	ND	—	0	ND	—	0	ND	—	B
Highlands	1972–73	5	0	ND	—	0	ND	—	1	0.054	(ND–0.31)	B
Lee	1972–73	7	0	ND	—	0	ND	—	1	0.062	(ND–0.53)	B
Levy	1972–73	10	1	0.012	(ND–0.12)	0	ND	—	0	ND	—	B
Osceola	1992–93	5	0	ND	—	NA	—	—	NA	—	—	C
Sarasota	1972	1	1	0.90[c]	—	NA	—	—	NA	—	—	A
"Sarasota Bay"	1972–73	2	1	0.11	(ND–0.23)	1	0.18	(ND–0.40)	2	0.21	(0.16–0.26)	B
Wakulla	1972–73	10	2	0.050	(ND–0.36)	1	0.030	(ND–0.36)	2	0.030	(ND–0.23)	B

			cis-chlordane			PCBs			Heptachlor epoxide			
Brevard	1972–73	10	0	ND	—	4	0.53	(ND– 2.8)	0	ND	—	B
Dade	1972–73	2	0	ND	—	0	ND	—	0	ND	—	B
Highlands	1972–73	5	1	0.091	(ND–0.55)	4	0.46	(ND– 1.3)	0	ND	—	B
"Lake Okeechobee"	1992–93	5	NA	—	—	0	ND	—	NA	—	—	D
Lee	1972–73	7	1	0.062	(ND–0.51)	4	0.29	(ND–0.76)	0	ND	—	B
Levy	1972–73	10	0	ND	—	6	0.24	(ND–0.59)	0	ND	—	B
Osceola	1992–93	5	NA	—	—	0	ND	—	NA	—	—	C
Sarasota	1972	1	NA	—	—	1	161[c]	—	NA	—	—	A
"Sarasota Bay"	1972–73	2	2	0.30	(0.21–0.40)	2	1.5	(1.2–1.8)	2	0.21	(0.16–0.26)	B
Wakulla	1972–73	10	4	0.050	(ND–0.16)	10	1.9	(0.25–11)	1	0.021	(ND–0.23)	B

Sources: A = Lincer & Salkind (1973), B = Ohlendorf et al. (1979), C = Rodgers (1997).
NA = not analyzed, ND = not detected.
a. Number with residue.
b. Geometric mean, ppm wet weight unless otherwise indicated.
c. Data presented is dry weight. Lincer & Salkind (1973) also give extractable fat weights.
d. In addition to the compounds listed, Rodgers (1997) did not detect chlordane, endrin, toxaphene, heptachlor, or *trans*-nonachlor in the 5 eggs from Osceola County. Ohlendorf et al. (1979) did not detect *cis*-nonachlor, *trans*-nonachlor, toxaphene, HCB, or endrin in any of their 46 egg samples.

Table 6.9. Organochlorine residues in tissues of Snowy Egrets collected in Florida[f]

Age Tissue	County/area	Year(s)	No. examined	DDE			DDD			DDT			Data source
				No.[a]	Mean[b]	(Range)	No.[a]	Mean[b]	(Range)	No.[a]	Mean[b]	(Range)	
Nestlings													
Brain	Dade	1989	1[c]	1	0.06	—	0	ND	—	0	ND	—	A
	"Lake Okechobee"	1989–90	1[d]	1	0.95	—	0	ND	—	0	ND	—	A
	Monroe	1988–90	1[e]	1	0.05	—	0	ND	—	0	ND	—	A
Liver	"Lake Okeechobee"	1992–93	5	NG	0.57	(ND–4.6)	NG	NC	(ND–0.15)	NG	NC	(ND–0.15)	B
Age Unspecified													
Brain	Indian River	1966	5	5	0.82	(0.16–5.1)	4	0.099	(ND–0.43)	4	0.0 72	(ND–0.24)	C
Liver	Indian River	1966	5	5	0.83	(3.6–0.38)	5	0.096	(0.027–0.78)	3	0.019	(ND–0.045)	C

Age Tissue	County/area	Year(s)	No. examined	Dieldrin			Mirex			Oxychlordane			Data source
				No.[a]	Mean[b]	(Range)	No.[a]	Mean[b]	(Range)	No.[a]	Mean[b]	(Range)	
Nestlings													
Brain	Dade	1989	1[c]	0	ND	—	0	ND	—	0	ND	—	A
	"Lake Okeechobee"	1989–90	1[d]	0	ND	—	0	ND	—	1	0.04	—	A
	Monroe	1988–90	1[e]	0	ND	—	0	ND	—	0	ND	—	A
Liver	"Lake Okeechobee"	1992–93	5	NG	NC	(ND–0.14)	NA	—	—	NA	—	—	B
Age Unspecified													
Brain	Indian River	1966	5	5	0.09	(0.032–0.73)	NA	—	—	NA	—	—	C
Liver	Indian River	1966	5	5	0.14	(0.054–0.61)	NA	—	—	NA	—	—	C

				cis-chlordane			PCBs			Heptachlor epoxide			
Nestlings														
Brain	Dade	1989	1[c]	0	ND	—	0	ND	—	0	ND	—	A
	"Lake Okechobee"	1989–90	1[d]	0	ND	—	0	ND	—	1	0.04	—	A
	Monroe	1988–90	1[e]	0	ND	—	0	ND	—	0	ND	—	A
Liver	"Lake Okeechobee"	1992–93	5	NA	—	—	0	ND	—	NA	—	—	B
Age Unspecified														
Brain	Indian River	1966	5	NA	—	—	NA	—	—	1	0.068	(ND–0.23)	C
Liver	Indian River	1966	5	NA	—	—	NA	—	—	3	0.080	(ND–0.39)	C

Sources: A = Spalding et al. (1997), B = Rodgers (1997), C = Locke (1967).

NA = not analyzed, ND = not detected, NG = not given, NC = not calculated (<50% of samples contained the residue).

a. Number with residue.
b. Geometric mean, ppm wet weight.
c. Pooled sample of 2 brains.
d. Pooled sample of 4 brains.
e. Pooled sample of 2 brains.
f. In addition to the compounds listed, Rodgers (1997) also reported a range of ND–0.32 ppm of *trans*-nonachlor, and did not detect chlordane, endrin, toxaphene, or heptachlor in the 5 liver samples from Lake Okeechobee. Spalding et al. (1997) detected 0.03 ppm *trans*-nonachlor, and did not detect α-BHC, Γ-BHC, β-BHC, δ-BHC, Γ-chlordane, α-chlordane, *cis*-nonachlor, endrin, HCB, toxaphene, endosulfan I, or endosulfan II in the pooled nestling brain sample from Lake Okeechobee (percent lipid range = 5.10–9.10; percent moisture range = 78.9–85.0; lower level of detection = 0.01 ppm, 0.05 ppm for toxaphene and PCBs).

Table 6.10. Organochlorine residues in eggs and tissues of Little Blue Herons and Reddish Egrets collected in Florida[d]

Age Tissue	County/area	Year(s)	No. examined	DDE No.[a]	DDE Mean[b]	DDE (Range)	Dieldrin No.[a]	Dieldrin Mean[b]	Dieldrin (Range)	Mirex No.[a]	Mirex Mean[b]	Mirex (Range)	Data source
Little Blue Herons													
Eggs	Brevard	1972–73	10	10	0.41	(0.090–1.5)	2	0.028	(ND–0.20)	1	0.01	(ND–0.18)	A
	Highlands	1972–73	2	2	1.5	(0.13–4.6)	1	1.0	(ND–3.1)	0	ND	—	A
	Palm Beach	1993	6	6	0.34	(0.05–1.7)	6	NC	(ND–0.20)	NA	—	—	C
	Lee	1972–73	10	10	0.14	(ND–0.26)	0	ND	—	0	ND	—	A
	Osceola	1993	5	0	ND	—	NG	NC	(ND–0.20)	NA	—	—	C
	"Tampa Bay"	1972–73	1	1	0.21	—	0	ND	—	1	0.91	—	A
	Wakulla	1972–73	10	10	0.87	(0.046–3.3)	2	0.021	(ND[n]0.11)	0	ND	—	A
Malformed nestling	"L. Okeechobee"	1990	1	0	ND	—	0	ND	—	0	ND	—	B
Reddish Egrets Nestlings													
Brain	"Florida Bay"	1988–90	1[c]	1	0.02	—	0	ND	—	0	ND	—	B

Age Tissue	County/area	Year(s)	No. examined	Oxychlordane No.[a]	Oxychlordane Mean[b]	Oxychlordane (Range)	cis-chlordane No.[a]	cis-chlordane Mean[b]	cis-chlordane (Range)	PCBs No.[a]	PCBs Mean[b]	PCBs (Range)	Data source
Little Blue Herons													
Eggs	Brevard	1972–73	10	1	0.009	(ND–0.097)	1	0.01 4	(ND–0.16)	5	0.54	(ND–4.3)	A
	Highlands	1972–73	2	1	0.042	(ND–0.084)	1	0.045	(ND–0.090)	1	8.4	(ND–88)	A
	Palm Beach	1993	6	NA	—	—	NA	—	—	0	ND	—	C
	Lee	1972–73	10	10	0.064	(ND–0.14)	0	ND	—	0	ND	—	A
	Osceola	1993	5	NA	—	—	NA	—	—	0	ND	—	C
	"Tampa Bay"	1972–73	1	0	ND	—	0	ND	—	1	3.2	—	A
	Wakulla	1972–73	10	0	ND	—	0	ND	—	0	ND	—	A

							trans-nonachlor		Toxaphene						
Malformed nestling	"L. Okeechobee"	1990	1	0	ND	—			0	ND	—		0	ND	—	B
Reddish Egrets Nestlings Brain	"Florida Bay"	1988–90	1ᶜ	0	ND	—			0	ND	—		0	ND	—	B
Little Blue Herons Eggs	Palm Beach	1993	6	NG	NC	(ND–0.26)	0	ND	—							C
	Highlands	1972–73	2	0	ND	—	1	0.088	—							A
	Osceola	1992–93	5	NG	NC	(ND–0.26)	0	ND	—							C
Malformed nestling	"L. Okeechobee"	1990	1	0	ND	—	0	ND	—							B

Sources: A = Ohlendorf et al. (1979), B = Spalding et al. (1997), C = Rodgers (1997).

NA = not analyzed, ND = not detected, NC = not calculated.

a. Number with residue.

b. Geometric mean, ppm wet weight.

c. Pooled sample from 2 nestlings.

d. In addition to the compounds listed, Ohlendorf et al. (1979) did not detect DDD, DDT, *cis*-nonachlor, *trans*-nonachlor, HCB, heptachlor epoxide, or endrin in any of the 33 Little Blue Heron egg samples reported. Spalding et al. (1997) did not detect DDD, DDT, α-BHC, Γ-BHC, β-BHC, δ-BHC, Γ-chlordane, α-chlordane, *cis*-nonachlor, heptachlor epoxide, endrin, endosulfan I, or endosulfan II in either the malformed Great Blue Heron nestling from Lake Okeechobee or the pooled brain sample of 2 Reddish Egret nestlings from Florida Bay (percent lipid range = 5.10 – 9.10; percent moisture range = 78.9–85.0; and lower level of detection = 0.01 ppm, 0.05 ppm for toxaphene and PCBs). Rodgers (1997) did not detect DDT, DDD, chlordane, endrin, or heptachlor in the 11 eggs reported from Palm Beach and Osceola counties.

Table 6.11. Organochlorine residues in eggs and tissues of Tricolored Herons collected in Florida[e]

Age Tissue	County/area	Year(s)	No. examined	DDE No.[a]	DDE Mean[b]	DDE (Range)	DDD No.[a]	DDD Mean[b]	DDD (Range)	DDT No.[a]	DDT Mean[b]	DDT (Range)	Data source
Eggs	Brevard	1972–73	9	9	0.49	(0.18–1.0)	0	ND	—	3	0.15	(ND–0.65)	A
	Citrus	1972–73	5	4	0.15	(ND–0.35)	0	ND	—	0	ND	—	A
	Dade	1972–73	10	10	0.16	(0.097–0.38)	0	ND	—	0	ND	—	A
	Highlands	1972–73	5	5	0.50	(0.16–1.8)	0	ND	—	0	ND	—	A
	Lee	1972–73	10	9	0.41	(ND–1.2)	0	ND	—	0	ND	—	A
	Levy	1972–73	2	2	0.21	(0.20–0.23)	0	ND	—	0	ND	—	A
	"Sarasota Bay"	1972–73	1	1	1.10	—	0	ND	—	0	ND	—	A
	"Tampa Bay"	1972–73	1	1	0.38	—	0	ND	—	0	ND	—	A
	Wakulla	1972–73	11	11	1.30	(0.15–5.9)	2	0.072	(ND[n]0.86)	2	0.081	(ND–1.1)	A
Nestlings													
Brain	Dade	1989	1[c]	1	0.07	—	0	ND	—	0	ND	—	B
	"Florida Bay"	1989–90	1[d]	1	0.01	—	0	ND	—	0	ND	—	B

Age Tissue	County/area	Year(s)	No. examined	Dieldrin No.[a]	Dieldrin Mean[b]	Dieldrin (Range)	Mirex No.[a]	Mirex Mean[b]	Mirex (Range)	Oxychlordane No.[a]	Oxychlordane Mean[b]	Oxychlordane (Range)	Data source
Eggs	Brevard	1972–73	9	6	0.17	(ND–0.49)	1	0.026	(ND–0.26)	0	ND	—	A
	Citrus	1972–73	5	0	ND	—	0	ND	—	0	ND	—	A
	Dade	1972–73	10	0	ND	—	0	ND	—	0	ND	—	A
	Highlands	1972–73	5	0	ND	—	0	ND	—	0	ND	—	A
	Lee	1972–73	10	1	0.014	(ND–0.14)	0	ND	—	0	ND	—	A
	Levy	1972–73	2	0	ND	—	0	ND	—	0	ND	—	A
	"Sarasota Bay"	1972–73	1	1	0.086	—	1	0.27	—	1	0.13	—	A
	"Tampa Bay"	1972–73	1	0	ND	—	0	ND	—	0	ND	—	A
	Wakulla	1972–73	11	1	0.084	(ND–1.4)	0	ND	—	1	0 .012	(ND–0.14)	A

				cis-chlordane			cis-nonachlor			PCBs			
Nestlings													
Brain	Dade	1989	1[c]	0	ND	—	0	ND	—	0	ND	—	B
	"Florida Bay"	1989–90	1[d]	0	ND	—	0	ND	—	0	ND	—	B
Eggs	Brevard	1972–73	9	0	ND	—	0	ND	—	8	0.81	(ND–1.6)	A
	Citrus	1972–73	5	0	ND	—	0	ND	—	1	0.20	(ND–1.5)	A
	Dade	1972–73	10	0	ND	—	0	ND	—	0	ND	—	A
	Highlands	1972–73	5	0	ND	—	0	ND	—	1	0.19	(ND–1.4)	A
	Lee	1972–73	10	2	0.028	(ND–0.18)	1	0.009	(ND–0.08 6)	5	0.13	(ND–0.53)	A
	Levy	1972–73	2	0	ND	—	0	ND	—	1	0.12	(ND–0.2 5)	A
	"Sarasota Bay"	1972–73	1	0	ND	—	0	ND	—	1	2.0	—	A
	"Tampa Bay"	1972–73	1	0	ND	—	0	ND	—	1	1.0	—	A
	Wakulla	1972–73	11	1	0.042	(ND–0.56)	1	0.012	(ND– 0.14)	9	1.4	(ND–12)	A
Nestlings													
Brain	Dade	1989	1[c]	0	ND	—	0	ND	—	0	ND	—	B
	"Florida Bay"	1989–90	1[d]	0	ND	—	0	ND	—	0	ND	—	B

Sources: A = Ohlendorf et al. (1979), B = Spalding et al. (1997).

ND = not detected.

a. Number with residue.

b. Geometric mean, ppm wet weight.

c. Pooled sample of 2 brains.

d. Pooled sample of 5 brains.

e. In addition to the compounds listed, Spalding et al. (1997) did not detect α-BHC, Γ-BHC, β-BHC, δ-BHC, Γ-chlordane, α-chlordane, endrin, HCB, heptachlor epoxide, toxaphene, endosulfan I, or endosulfan II in any of the samples reported (percent lipid range = 5.10–9.10; percent moisture range = 78.9–85.0; lower level of detection = 0.01 ppm, 0.05 ppm for toxaphene and PCBs). Ohlendorf et al. (1979) detected 0.089 ppm toxaphene in 1 of the 5 eggs from Highlands County and 0.007 ppm heptachlor epoxide in 1 of the 11 eggs from Wakulla County. They did not detect *trans*-nonachlor, HCB, or endrin in any of the samples.

Table 6.12. Organochlorine residues in eggs and tissues of Cattle Egrets collected in Florida[e]

Age Tissue	County/area	Year(s)	No. examined	DDE No.[a]	DDE Mean[b]	DDE (Range)	DDD No.[a]	DDD Mean[b]	DDD (Range)	DDT No.[a]	DDT Mean[b]	DDT (Range)	Data source
Eggs													
	Alachua	1972–73	10	10	0.66	(0.12–17)	0	ND	—	0	ND	—	C
	Brevard	1972–73	10	8	0.93	(ND–8.3)	0	ND	—	2	0.04	(ND–0.37)	C
	Dade	1972	5	5	0.52	(0.12–5.5)	5	0.03	(0.008–0.11)	5	0.065	(0.035–0.21)	B
	Highlands	1972–73	10	8	0.58	(ND–5.4)	0	ND	—	2	0.20	(ND–2.5)	C
	"Sarasota Bay"	1972–73	4	4	0.38	(0.092–1.2)	0	ND	—	1	0.052	(ND–0.23)	C
	"Tampa Bay"	1972–73	3	3	0.53	(0.15–1.1)	0	ND	—	0	ND	—	C
	Wakulla	1972–73	1	1	0.24	—	0	ND	—	0	ND	—	C
Nestlings													
Liver													
	"Lake Okeechobee"	1992–93	8	8	NC	(ND–1.3)	NG	NC	(ND–1.0)	0	ND	—	D
Adults													
Adipose													
	"Florida"	1973–74	2	2	4.7	(0.60–37)[c]	NA[c]	—	—	NA[c]	—	—	A
Uropygial gland													
	"Florida"	1973–74	2	2	0.11	(ND–2.61)[c]	NA[c]	—	—	NA[c]	—	—	A

Age Tissue	County/area	Year(s)	No. examined	Dieldrin No.[a]	Dieldrin Mean[b]	Dieldrin (Range)	Mirex No.[a]	Mirex Mean[b]	Mirex (Range)	Oxychlordane No.[a]	Oxychlordane Mean[b]	Oxychlordane (Range)	Data source
Eggs													
	Alachua	1972–73	10	2	0.030	(ND–0.22)	2	0.038	(ND–0.18)	2	0.030	(ND–0.22)	C
	Brevard	1972–73	10	4	0.23	(ND–3.3)	2	0.021	(ND–0.13)	1	0.012	(ND–0.12)	C
	Dade	1972	5	4	0.049	(ND–0.66)	NA	—	—	2[d]	0.055	(0.035–0.087)	B
	Highlands	1972–73	10	2	0.20	(ND–3.1)	1	0.012	(ND–0.11)	0	ND	—	C
	"Sarasota Bay"	1972–73	4	1	0.20	(ND–1.1)	4	0.37	(0.11–1.3)	1	0.10	(ND–0.49)	C
	"Tampa Bay"	1972–73	3	0	ND	—	2	0.24	(ND–0.77)	1	0.79	(ND–0.26)	C
	Wakulla	1972–73	1	0	ND	—	0	ND	—	0	ND	—	C

	Year	n		*cis*-chlordane			PCBs			Heptachlor epoxide		Source
Nestlings												
Liver												
"Lake Okeechobee"	1992–93	8	NG	NC	(ND–0.31)	NA	—	—	NA	—	—	D
Adults												
Adipose												
"Florida"	1973–74	2	1	0.028	(ND–0.16)	NA	—	—	NA	—	—	A
Uropygial gland												
"Florida"	1973–74	2	0	ND	—	NA	—	—	NA	—	—	A
Eggs												
Alachua	1972–73	10	0	ND	—	5	0.12	(ND–0.25)	0	N D	—	C
Brevard	1972–73	10	1	0.012	(ND–0.11)	8	0.53	(ND–3.5)	0	ND	—	C
Dade	1972	5	NA	—	—	1	0.02	(ND–0.35)	NA	—	—	B
Highlands	1972–73	10	0	ND	—	0	ND	—	0	ND	—	C
"Sarasota Bay"	1972–73	4	1	0.16	(ND–0.84)	0	ND	—	1	0.052	(ND–0.22)	C
"Tampa Bay"	1972–73	3	1	0.11	(ND–0.38)	0	ND	—	0	ND	—	C
Wakulla	1972–73	1	0	ND	—	0	ND	—	0	ND	—	C
Nestlings												
Liver												
"Lake Okeechobee"	1992–93	8	NA	—	—	0	ND	—	NA	—	—	D
Adults												
Adipose												
"Florida"	1973–74	2	NA	—	—	0	ND	—	NA	—	—	A
Uropygial gland												
"Florida"	1973–74	2	NA	—	—	0	ND	—	NA	—	—	A

Sources: A = Johnston (1976), B = Ogden et al. (1974), C = Ohlendorf et al.(1979), D = Rodgers (1997).

ND = not detected, NA = not analyzed, NC = not calculated.

a. Number with residue.

b. Geometric mean, ppm wet weight.

c. Johnston (1976) reported 6.6 (0.86–51) ppm total DDTs in adipose and 0.11 (ND–2.6) ppm total DDTs in uropygial glands.

d. Two eggs tested.

e. In addition to the compounds listed, Ohlendorf et al. (1979) detected 0.098 ppm *cis*-nonachlor in 1 egg from Sarasota Bay. They did not detect *trans*-nonachlor, toxaphene, HCB, or endrin in any of the samples. Rodgers (1997) did not detect chlordane, endrin, toxaphene, heptachlor, or *trans*-nonachlor.

Table 6.13. Organochlorine residues in eggs and tissues of Green Herons collected in Florida[d]

Tissue	County/area	Year(s)	No. examined	DDE			Oxychlordane			PCBs			Data source
				No.[a]	Mean[b]	(Range)	No.[a]	Mean[b]	(Range)	No.[a]	Mean[b]	(Range)	
Eggs													
	Brevard	1972–73	10	9	0.49	(ND–1.5)	0	ND	—	5	0.44	(ND–3.7)	A
	"Lake Okeechobee"	1972–73	3	3	0.56	(0.34–0.99)	1	0.057	(ND–0.18)	0	ND	—	A
	Lee	1972–73	1	1	0.28	—	0	ND	—	0	ND	—	A
	Wakulla	1972–73	8	7	0.40	(ND–1.3)	0	ND	—	5	0.15	(ND–0.25)	A
Post fledging juvenile													
Whole body													
	North Liberty	1978	7	7	1.0[c]	(0.55–1.6)	7	0.01[c]	(0.01–0.02)	7	1.7[c]	(1.2–2.3)	B
	South Liberty	1978	3	3	0.26[c]	(0.11–0.34)	3	0.02[c]	(0.02–0.03)	3	0.73[c]	(0.50–1.0)	B

Tissue	County/area	Year(s)	No. examined	DDD			Total DDT			PCB (Aroclor 1254)			Data source
Eggs													
	Brevard	1972–73	10	0	ND	—	0	ND	—	0	ND	—	A
	"Lake Okeechobee"	1972–73	3	0	ND	—	0	ND	—	0	ND	—	A
	Lee	1972–73	1	0	ND	—	0	ND	—	0	ND	—	A
	Wakulla	1972–73	8	0	ND	—	0	ND	—	0	ND	—	A
Post fledging juvenile													
Whole body													
	North Liberty	1978	7	7	0.11[c]	(0.07–0.27)	7	1.1[c]	(0.67–1.9)	7	0.41[c]	(0.30–0.70)	B
	South Liberty	1978	3	3	0.02[c]	(0.01–0.03)	3	0.29[c]	(0.13–0.39)	3	0.26[c]	(0.10–0.40)	B

Tissue	County/area	Year(s)	No. examined	PCB (Aroclor 1260)			Toxaphene			Dieldrin			Data source
Eggs													
	Brevard	1972–73	10	0	ND	—	0	ND	—	0	ND	—	A
	"Lake Okeechobee"	1972–73	3	0	ND	—	0	ND	—	0	ND	—	A
	Lee	1972–73	1	0	ND	—	0	ND	—	0	ND	—	A
	Wakulla	1972–73	8	0	ND	—	0	ND	—	0	ND	—	A

Category	Location	Year	N[a]	n	Mean[b]	Range	n	Mean[b]	Range	n	Mean[b]	Range	Source
Post fledging juvenile													
Whole body	North Liberty	1978	7	7	1.2[c]	(0.80–1.7)	7	1.3[c]	(0.40–2.6)	7	0.03[c]	(0.02–0.06)	B
	South Liberty	1978	3	3	0.46[c]	(0.40–0.60)	2	0.10[c]	(ND–0.20)	3	0.08[c]	(0.01–0.21)	B

Category	Location	Year	N[a]	Endrin			Heptachlor epoxide			cis-chlordane			Source
				n	Mean[b]	Range	n	Mean[b]	Range	n	Mean[b]	Range	
Eggs	Brevard	1972–73	10	0	ND	—	0	ND	—	0	ND	—	A
	"Lake Okeechobee"	1972–73	3	0	ND	—	0	ND	—	1	0.12	—	A
	Lee	1972–73	1	0	ND	—	0	ND	—	0	ND	—	A
	Wakulla	1972–73	8	0	ND	—	0	ND	—	0	ND	—	A
Post fledging juvenile													
Whole body	North Liberty	1978	7	7	0.01[c]	(0.01–0.02)	7	0.01[c]	(0.01–0.02)	NG	0.02[c]	(ND–0.05)	B
	South Liberty	1978	3	0	ND[c]	—	3	0.02[c]	(0.01– 0.04)	0	ND[c]	—	B

Category	Location	Year	N[a]	cis-nonachlor			trans-nonachlor			Source
				n	Mean[b]	Range	n	Mean[b]	Range	
Eggs	Brevard	1972–73	10	0	ND	—	0	ND	—	A
	"Lake Okeechobee"	1972–73	3	0	ND	—	0	ND	—	A
	Lee	1972–73	1	0	ND	—	0	ND	—	A
	Wakulla	1972–73	8	0	ND	—	0	ND	—	A
Post fledging juvenile										
Whole body	North Liberty	1978	7	7	0.06[c]	(0.05–0.10)	7	0.07[c]	(0.06–0.10)	B
	South Liberty	1978	3	3	0.02[c]	(0.02–0.03)	3	0.03[c]	(0.02–0.04)	B

Sources: A = Ohlendorf et al.(1979), B = Winger et al. (1984).

ND = not detected, NG = not given.

a. Number with residue.

b. Geometric mean, ppm wet weight unless otherwise indicated.

c. Means reported by Winger et al. (1984) are arithmetic means.

d. In addition to the compounds listed, Ohlendorf et al. (1979) did not detect mirex, HCB, or toxaphene in any of the samples. Winger et al. (1984) did not detect α-BHC or *trans*-chlordane.

the literature for Florida, although others undoubtedly occurred. Dieldrin (6.6 ppm) and DDE (28 ppm) were detected in the brain of an adult Great Blue Heron from Monroe County in 1973 that may have died from dieldrin poisoning (Ohlendorf et al. 1981). Call et al. (1976) and Ohlendorf et al. (1981) have reported poisonings of Great Blue Herons outside of Florida. Johnston (1976) measured the concentrations of contaminants in adipose tissue and uropygial glands of American Bitterns and Cattle Egrets from Florida (tables 6.3 and 6.12). For all of the species that he tested, including birds from other groups, DDTs and dieldrin in the adipose tissue were higher than in the uropygial gland (2.2:1 and 2.6:1 respectively). The uropygial gland may serve to excrete DDTs. He found no PCBs in fat or uropygial glands.

Spalding et al. (1997) and Rodgers (1997) did more recent testing of nestling ardeids in southern Florida in the 1980s and 1990s. Both detected concentrations of organochlorines in herons and egrets in the Lake Okeechobee region; these were sublethal but higher than expected in light of restrictions against their use. Great Egrets had higher concentrations and compounds were more commonly detected than in the other species examined. DDE was found in all samples, with the exception of Great Egrets in Florida Bay.

VII. Organophosphates and carbamates

Chlorpyrifos, an insecticide, was suspected to be the cause of death (0.34 ppm in stomach contents) in a Cattle Egret collected from a die-off of 20 birds found dead in a horse pasture in Alachua County in 1992 (Spalding and Brady 1992). All of the birds examined had large numbers of mole crickets in their stomachs. Although there was no history of chlorpyrifos use on the farm, it was situated in an agricultural area. Outside of Florida, the organophosphate fenthion, used for mosquito control, was suspected to cause cholinesterase suppression and death in Great Blue Herons,

Snowy Egrets, and Great Egrets in California (Zinkl et al. 1981).

We could find no documentation of cholinesterase suppression in ardeids in Florida. Cholinesterase activity that has been measured in 4 Great Blue and 2 Great White Herons is listed in table 6.15.

VIII. Metals

As a byproduct of the organochlorine testing in the 1970s and concern for mercury accumulation in the Everglades, a lot of data are available about metal contaminants in Florida ardeids (tables 6.16–6.18). Little testing has been conducted using eggs. The highest mercury concentration in eggs (0.19–0.91 ppm, geometric mean = 0.31) was in Great Egret eggs from Dade County in 1972 (Ogden et al. 1974).

Mercury concentrations in tissues of nestling ardeids in the Everglades are quite high compared with other species and ecosystems. Mean blood and feather concentrations measured in Great Egret nestlings in southern Florida by Sepulveda et al. (1999a) were higher than for any other species previously reported. The highest liver concentration reported from a nestling was from a Great Egret (19 ppm) in the Everglades (Sundlof et al. 1994). A juvenile Great Blue Heron, also from the Everglades, had 75 ppm mercury in liver. Both of these birds were moribund when found. Sundlof et al. (1994) reported higher concentrations of mercury in livers of nestling ardeids in south Florida that ate larger fish, were older, were emaciated, and were found in the central Everglades. Fledgling Great White Herons that died from acute causes (such as trauma and poisoning) had lower concentrations of mercury in livers than those that died from chronic, often multiple, diseases, suggesting an effect of mercury on the immune system (Spalding et al. 1994b). Three fledgling Great White Herons with liver mercury greater than 25 ppm had evidence of kidney disease. Frederick et al. (1999) estimated dietary exposure of Great Egrets to be 0.41 ppm mercury in food consumed in the Everglades between 1994 and

Table 6.14. Eggshell thickness of wading bird eggs collected in Florida

Species Area	Year(s)	No. exam.	Shell Thickness	(SE or range)	% change	Data source
Great Blue Heron						
"Gulf Coastal Plain"	pre-1947	81	0.400	(±0.005)	—	A[a]
"Florida Keys"	pre-1947	34	0.402		—	B[b]
"Florida, Tennessee"	pre-1947	114	0.400		—	B
"Florida, Tennessee"	1947–73	33	0.379		-5.2[c]	B
Great Egret						
"Florida"	pre-1947	83	0.293		—	B
	1947–73	89	0.295		+0.9	B
"Florida Bay"	1972	5	0.336	(0.287–0.351)		C
"Everglades"	1972	5	0.346	(0.335–0.371)		C
Snowy Egret						
"Florida"	pre-1947	82	0.228		—	B
"Florida"	1947–73	58	0.224		-1.7	B
Louisiana Heron						
"North Florida"	pre-1947	82	0.228		—	B
	1947–73	58	0.231		+1.3	B
"South Florida"	pre-1947	43	0.232		—	B
	1947–73	27	0.226		-2.9	B
Black-Crowned Night Heron						
"Fla., Ga., S.C."	pre-1947	21	0.287		—	B
	1947–73	36	0.274		-4.6[b]	B
Yellow-Crowned Night Heron						
"Fla., Ga., Ala., N.C.,	pre-1947	49	0.280		—	B
S.C., Va."	1947–73	23	0.280		+2.8	B
Least Bittern						
"Florida"	pre-1947	62	0.143		—	B
	1947–73	25	0.140		-2.1	B
Green Heron						
"Florida"	pre-1947	58	0.183		—	B
	1947–73	41	0.184		+0.3	B
Little Blue Heron						
"Florida"	pre-1947	157	0.241		—	B
"Florida"	1947–73	39	0.242		+1.7	B
Cattle Egret						
"Florida"	1947–73	67	0.228		—	B

Sources: A = Anderson & Hickey (1972), B = Ohlendorf et al. (1979), C = Ogden et al. (1974).
a. Anderson & Hickey (1972) found no significant difference in shell thickness or weight when comparing pre-1947 with post-1947 data. They measured shell weight and calculated shell thickness index for Florida 1950–60 but provided no shell thickness data for those years.
b. Ohlendorf et al. (1979) used the mean shell thickness for the clutch.
c. Significantly different from pre-1947 ($P<0.01$).

1996. Spalding et al. (2000a,b) and Bouton et al. (1999) tested captive Great Egrets fed a concentration of methylmercury similar to that found in Everglades fish. They found sublethal effects, including: depressed appetite, decreased weight, dingy feathers, fewer active behaviors, decreased sun tolerance, decreased motivation to hunt, lower packed cell volume, and changed appearance of the lung and hematopoetic tissues including bursa, thymus, and bone marrow.

Lead concentrations in ardeid nestling livers in Florida were all below 1 ppm except for 3 cases (table 6.16). These were 2 Tricolored Heron nestlings from 2 colonies in the central Everglades (6.3 and 3.1 ppm), and a Snowy Egret from Lake Okeechobee (2.4 ppm). In experimentally dosed American Kestrels, Hoffman et al. (1985) suggested that 2 ppm in liver was associated with impaired growth, and 5 ppm with mortality. Thus the diagnositic criterion for lead poisoning was met in 1 of the nestling Tricolored Herons, if these species can be compared. Cattle Egrets could be at risk for lead toxicosis based on a study by Udevitz et al. (1980) in which high concentrations of lead were found in insects collected along highways where Cattle Egrets commonly forage.

The highest cadmium concentration found in ardeid livers was 0.68 ppm wet weight in the liver of a Tricolored Heron from a central Everglades colony; this same Tricolored Heron also had elevated lead (Spalding et al. 1997) (table 6.18). The highest cadmium concentration found in eggs collected in Florida was 0.05 ppm wet weight. The suggested upper limit for human food is 0.075 ppm wet weight; residues in liver that exceed 10 ppm wet weight are likely to have cadmium contamination (Eisler 1985).

Custer and Mulhern (1983) measured copper in livers of prefledgling Black-crowned Night-Herons in 3 colonies on the northeastern coast of the United States. They found that values varied significantly between colonies, ranging from 23 to 381 ppm, dry weight. Copper dry weight values for nestlings in Florida ranged between 6 and 890 ppm (Spalding et al. 1997). The significance of these values is unknown;

copper is an essential element. Frank and Borg (1979) found >1000 ppm, wet weight, copper in the livers of Mute Swans (*Cygnus olar*).

Arsenic concentrations were low (<1.0 ppm, wet weight) in ardeids from Florida. The significance of arsenic in bird tissues is poorly understood (Eisler 1988). Zinc and chromium have rarely been measured in tissues and eggs and the significance of the data for these metals is unknown.

IX. Oil spills

Ardeids are rarely involved in oiling in Florida. An oil spill in Tampa Bay in February 1970 was estimated to have killed as many as 9,000 birds, including "herons" (Sims 1970; Stevenson 1970), but the numbers and species of herons involved were not given.

X. Neoplasia

Neoplasia has been reported only once in ardeids in Florida. Spalding and Woodard (1992) investigated a chondrosarcomatous mass that formed in the nictitating membrane of a Great White Heron from Florida Bay in 1990. The mass interfered with vision, resulting in severe emaciation and death of the bird.

XI. Anomalies and malformations

Congenital malformations were observed in 4 of more than 2,000 ardeids examined by Spalding and Forrester (1991) in southern Florida during 1988–91. Three of these were from Lake Okeechobee. One Little Blue Heron less than 1 week of age had a significantly shortened maxilla (figure 6.4) and deformed ribs. These malformations were probably the cause of death (no organochlorine residues were detected, see table 6.10). A Snowy Egret nestling 1–2 weeks old had a cardiovascular malformation and no glottis was present. Death appeared to be related to the heart malformation and bacterial infection. A Great Egret fledgling also had a cardiovascular malformation. A fledgling Great White Heron from Florida Bay, which failed to fledge because of severe malformations of the bill, carpi (bilateral), humerus, and sternum, was killed by a Bald Eagle. The cause of these malformations was not determined.

XII. Biotoxins

Botulism, resulting in mortality due to a toxin produced by the bacterium *Clostridium botulinum* type C, is relatively rare in Florida when compared with the western United States. Seven mortality events were reported in Florida between 1970 and 2001 (see chapter 4, Pelicans, chapter 10, Ducks, and table 10.22). Although botulism is primarily a disease of waterfowl, 1 Snowy Egret was found in the die-off in Hamilton County in 1979 (Forrester et al. 1980). It is believed that the conditions produced in the shallow phosphate mining operation in conjunction with high temperatures were responsible for these outbreaks. *Clostridium botulinum* type C was identified in the substrate of these ponds between April and October (Marion et al. 1983).

A die-off during June–July 1971 in a shallowly inundated area on the west shore of Lake Okeechobee, Glades County, involved 398 birds of 19 species including ducks (see chapter 10, Ducks), ibises, rails, spoonbills, skimmers, terns,

FIGURE 6.4. Little Blue Heron chick with malformed maxilla.

grebes, shorebirds, and wading birds (Jasmin et al. 1972). Six Snowy Egrets, a Great Egret, a Great Blue Heron, and a Tricolored Heron were identified among those that died. Birds were weak and had severe hemorrhagic enteritis and acute hepatitis. It was concluded that *Clostridium perfringens* type C, causing necrotizing enteritis, was the most likely cause of mortality based on culture and mouse protection tests.

Biotoxicosis, probably from a naturally occurring organism, was considered to be the cause of death of 6 Great White Herons in 1988. Six birds (5 fledglings, 1 adult) in Florida Bay died suddenly and unexpectedly between January 28 and February 17 (Spalding 1988). Many of these birds were being monitored daily with radio transmitters. One bird was ataxic the evening before it was found dead. All

Table 6.15. Cholinesterase activity measured in brains of fledgling and juvenile Great White and Blue Herons

Species	Cause of death	County/ area	Year	Original test[a]	After 18 hrs (35°C)[a]	Data source
Great White Heron	Sudden death[b]	"Florida Bay"	1988	18.07 ± 0.28	20.15 ± 0.37	A
		"Florida Bay"	1988	9.59 ± 0.05	11.60 ± 0.05	A
		"Florida Bay"	1988	10.66 ± 0.18	12.09 ± 0.28	A
		"Florida Bay"	1988	14.76 ± 0.09	17.16 ± 0.63	A
Great Blue Heron	Eagle predation	"Florida Bay"	1988	13.16 ± 0.05	14.46 ± 0.23	A
	Powerline trauma	Palm Beach	1990	13.30 ± 0.0	—	B

Sources: A = Spalding (1988), B = Locke (1990).
a. minutes/gram of original tissue ± SE.
b. A naturally occurring algal toxin was suspected; see Biotoxin section.

Table 6.16. Metal concentrations in eggs of ardeids from Florida

Species	County or area	Year	No. examined	Lead Mean[a]	(Range)	Mercury Mean[a]	(Range)	Zinc Mean[a]	(Range)	Data source
Black-crowned Night Heron										
	Brevard	1972	9	0.18[b]	(0.088–0.34)	0.07[b]	(0.018–0.14)	13[b]	(7.0–31)	A
Cattle Egret										
	Dade	1972	5	NA	—	0.05	(0.05–0.05)	NA	—	B
Great Egret										
	Dade	1972	5	ND	<0.4	0.31	(0.19–0.91)	7.5	(6.3–9.8)	B
	"Florida Bay"	1972	5	ND	<0.2	0.16	(0.13–0.19)	7.8	(5.9–8.5)	B
	Palm Beach	1993	6	ND	—	0.08	(0.04–0.18)	NA	—	C
Little Blue Heron										
	"Lake Okeechobee"	1993	6	ND	—	0.09	(0.04–0.30)	NA	—	C
	Osceola	1992–93	5	ND	—	0.16	(0.03–0.13)	NA	—	C
Snowy Egret										
	Osceola	1992–93	5	ND	—	0.22	(0.16– 0.27)	NA	—	C

Species	County or area	Year	No. examined	Arsenic Mean[a]	(Range)	Cadmium Mean[a]	(Range)	Copper Mean[a]	(Range)	Data source
Black-crowned Night Heron										
	Brevard	1972	9	0.007[b]	(0.002–0.01)	0.013[b]	(0.009–0.017)	1.1[b]	(0.77–1.4)	A
Cattle Egret										
	Dade	1972	5	0.11	(<0.07–0.3)	NA	—	NA	—	B
Great Egret										
	Dade	1972	5	ND	<0.10	ND	<0.05	0.93	(0.78–1.3)	B
	"Florida Bay"	1972	5	ND	<0.10	ND	<0.05	1.1	(0.94–1.2)	B

Species	County or area	Year	No. examined	Chromium Mean[a]	(Range)	Data source
Black-crowned Night Heron						
	Brevard	1972	9	0.18[b]	(0.004–0.64)	A

Sources: A = Ohlendorf et al. (1978), B = Ogden et al. (1974), C = Rodgers (1997).
NA = not analyzed, ND = not detected.
a. Geometric mean, ppm wet weight unless otherwise indicated.
b. Arithmetic mean.

of the birds were in excellent nutritional condition. The only consistent lesions noted were congestion of subcutaneous vessels and most organs, acute granulocytic enteritis, and empty stomachs. Bacterial pathogens, mercury toxicosis, organophosphate and carbamate toxicity, and botulism were all ruled out as causes of death. This mortality occurred during a red tide episode that moved from Naples, around the southern tip of Florida, to the coast of North Carolina (Red tide 1988). A seventh juvenile Great White Heron died with similar lesions in October 1988. This bird was the nestling of a marked adult "panhandler" that was fed fish by humans in residential yards or fish cleaning houses (Powell 1983; Powell and Powell 1986); it was being fed frozen baitfish at the time and it is possible that toxin in the stomach of baitfish could have caused toxicosis. Algae collected from the marine flats in Florida Bay at the time of the die-off were examined for toxin-producing organisms (Roberts 1988). *Gambierbiscus toxicus,* the organism responsible for ciquetera poisoning, and *Prorocentrum* sp. and *Ostreopis* sp., toxin-producing dinoflagellates, were found.

Anabaena sp., a blue-green algae, was identified in the water of a small freshwater pond in Polk County in June 1999. A total of 20 birds (including Great Blue Herons, Black-crowned Night-Herons, and White Ibises), 15 soft shelled turtles, an American alligator and numerous fish died in this pond (Fischer 1999). Although a definitive diagnosis of algal toxicosis was not made, other possible causes of mortality could not be found for these animals.

XIII. Viruses

Avian pox lesions are uncommon on ardeids. When they occur, they are more common on nestlings and are usually found on the head. There seems to be no evident taxonomic or geographic predisposition. Locke (1972) noted a pox lesion near the nares in a Great Blue Heron from Indian River County in 1972. Death due to avian poxvirus was reported in a juvenile Reddish Egret from Florida Bay in 1978 with lesions in front of both eyes (figure 6.5) (Conti et al. 1986). Spalding and Forrester (1991) reported minor pox lesions in 4 nestlings (Great White Heron, Great Egret, Snowy Egret), 3 juvenile birds (Great White and Blue Herons) and 1 adult (Snowy Egret) of 2,088 ardeids examined from southern Florida, 1988–91.

Titers to eastern equine encephalitis (EEE) virus were found in 4 of 85 Cattle Egrets collected in Florida (Bigler et al. 1967). Favorite (1960) reported that 2 of 4 Snowy Egrets, 2 of 8 Great Egrets, and 1 of 2 Little Blue Herons collected from central Florida in 1958 were seropositive for EEE virus by serum neutralization test. In addition, 1 of 2 Little Blue Herons was also seropositive for Highlands J virus. No viruses were isolated from any of these birds. Jennings et al. (1969) collected 5 Cattle Egrets and 2 Little Blue Herons from a roost in the Tampa Bay area in 1962, during a St. Louis encephalitis (SLE) epidemic. All had hemagglutination-inhibition antibody for Group B arbovirus. It was assumed that positive results indicated exposure to SLE virus; however, this could not be confirmed by serum neutralization. Serum neutralizing antibody to EEE virus was found in 2% and to SLE virus in 6% of 245 nestling ardeids from south Florida (figure 6.6) (Spalding et al. 1994d). The highest prevalence of EEE virus was in Great Blue Herons (3%), and of SLE virus in Green Herons (18%). These titers may have been from local transmission, or from maternal antibodies, since no virus isolations were made. Edman et al. (1972) found that wading bird blood (14% of samples), especially Green Heron and Black-crowned Night-Heron blood, was more frequently included in blood meals of the mosquito *Culiseta melanura,* in Indian River County, than would be expected when the relative abundance of birds present is taken into consideration. See the Arthropod section below regarding antimosquito behaviors of ardeids.

West Nile virus, recently introduced into North America, was first detected by using PCR and virus isolation techniques in an ardeid in Florida, a Green Heron from Clay County, in

Table 6.17. Lead and mercury concentrations in tissues of ardeids from Florida

Age						Pb		Hg		
Tissue	Species	County/area	Year(s)	No. examined	Mean[a]	(Range)	Mean[a]	(Range)		Data source
Nestlings										
Blood	Great Egret	Dade, Broward	1994–95	393	NA	—	1.70	(0.07–3.9)[b]		H
Feathers	Great Blue Heron	"South Florida"[f]	1987–90	7	NA	—	3.50	(1.8–7.7)[b]		G
		Dade, Broward[c]	1987–90	9	NA	—	7.10	(1.6–15)		G
		"South Florida"[c]	1994–95	9	NA	—	7.10	(1.6–15)		G
	Great White Heron	"South Florida"[c]	1987–90	10	NA	—	4.70	(1.0–9.1)		G
Liver	Cattle Egret	"Lake Okeechobee"	1993	8	0.01	(ND–0.62)	0.02	(ND–0.18)		D
	Great Blue Heron	"Florida Bay"	1989–90	4	NA	—	1	(0.21–2.7)		A
		"Lake Okeechobee"	1987–91	6	0.28	(0.20–0.44)	0.42[d]	(0.32 –1.1)		A,B
			1992–93	1	0.08	—	12	—		D
	Great Egret	Collier	1987–91	5	0.18	(0.09–0.32)	NA	—		B
		Dade	1989	1	0.16	—	19	—		A,B
		"Lake Okeechobee"	1987–91	6	0.26	(0.18–0.33)	0.44[d]	(0.18 –1.6)		A,B
		Palm Beach	1992	2	NA	—	0.79	(0.63–0.98)		A
	Great White Heron	"Florida Bay"	1987–89	5	0.32	(0.19–0.51)	1.40[e]	(0.22–7.3)		A,B
	Little Blue Heron	"Florida Bay"	1988	1	NA	—	0.41	—		A
		"Lake Okeechobee"	1990	4	0.28	(0.19–0.52)	0.38	(0.29–0.73)		A,B
	Snowy Egret	Dade	1989	1	NA	—	5.38	—		A
		"Florida Bay"	1988	1	NA	—	0.75	—		A
		"Lake Okeechobee"	1989–90	3	NA	—	0.37	(0.29–0.42)		A
		″	1992–93	5	0.01	(ND–2.4)	0.47	(0.18–1.8)		D
		Monroe	1988–90	5	0.26	(0.16–0.41)	0.34[f]	(0.15–1.4)		A,B
	Tricolored Heron	Dade, Broward, Palm Beach	1987–90	5	NA	—	2.70	(1.7–5.0)		A
		"Everglades"	1987–91	4	1.4	(0.27–6.3)	0.57	(0.20–4.96)		A,B
		"Florida Bay"	1989–90	5	NA	—	0.55	(0.30–2.7)		A
		"Lake Okeechobee"	1990	3	NA	—	0.28	(0.20–0.50)		A
		Monroe	1987–88	2	0.20	(0.16–0.24)	0.24	(0.12–0.50)		A,B

Adults & Juveniles	Species	Location	Year	n					Source
Feathers	Great White Heron	"South Florida"[c]	1987–90	19	NA	—	7	(2.7–15)	G
Liver	American Bittern	"Florida Bay"	1989	1	NA	—	7	—	A
	Black-crowned Night Heron	Collier	1990	1	NA	—	0.62	—	A
	Cattle Egret	Palm Beach	1989	2	0.77	(0.75–0.79)	NA	—	A
	Great Blue Heron	Alachua	1990	2	NA	—	11	(6.4–18)	A
		Collier	1989–91	9	NA	—	6.00	(0.77–18)	A
		Dade/Broward	1987	3	NA	—	37	(13–75)	A
		Levy	1991	1	NA	—	4.10	—	A
		Marion	1988	1	NA	—	4.50	—	A
		Palm Beach	1989	3	0.067	(ND–0.12)	NA	—	F
			1990–91	2	NA	—	1.80	(0.70–5.8)	A
	Great Egret	Collier	1990	9	NA	—	1.40	(0.44–4.3)	A
	Great White Heron	Collier	1988	2	NA	—	4.90	(2.9–8.4)	A
		Dade/Broward	1987–88	3	NA	—	11	(2.9–59)	A
		"Florida Bay"	1987–89	15	NA	—	4.40	(0.87–29)	A
	Little Blue Heron	"Everglades"	1973	1	NA	—	2.00	—	C
		"Florida Bay"	1989	2	NA	—	0.24	(0.23–0.26)	A
		Pinellas	1975	1	NA	—	2.00	—	C
	Tricolored Heron	"Florida Bay"	1990	1	NA	—	0.67	—	A
	Yellow-crowned Night Heron	Pinellas	1975	1	NA	—	2.10	—	C
		Pinellas	1975	1	NA	—	3.40	—	C
Kidney	Cattle Egret	Palm Beach	1989	2	1.5	(0.78–2.9)	NA	—	A
Whole carcass		North Liberty	1978	7	NA	—	0.46[g]	(0.21–0.75)	E
	Green Heron	South Liberty	1978	3	NA	—	0.67[g]	(0.55–0.74)	E

Sources: A = Sundlof et al. (1994), B = Spalding et al. (1997), C = Gourlie (1984), D = Rodgers (1997), E = Winger et al. (1984), F = Locke (1990), G = Beyer et al. (1997), H = Sepulveda et al. (1999a).

NA = not analyzed, ND = not detected.

a. Geometric mean, ppm wet weight unless otherwise indicated.

b. Range is for first-hatch chicks.

c. Most of the samples collected by Beyer et al. (1997) were from Florida Bay.

d. No. examined = 7

e. No. examined = 14

f. No. examined = 6

g. Arithmetic mean.

FIGURE 6.5. Poxvirus lesion on the eye of a Reddish Egret from Florida Bay, 1978. Courtesy of Richard T. Paul.

July of 2001 (Conti et al. 2002). It was assumed that this heron died of the West Nile virus infection, but this was not proved at necropsy. See chapter 24, Perching Birds, for more details.

In other locations, arboviruses, such as Lake Clarendon virus in Cattle Egrets in Australia (McKilligan 1987), and Aransas Bay virus (Texas, wading bird or Roseate Spoonbill nests, Yunker et al. 1979) have been found in association with tick infestations.

XIV. Bacteria

Bacterial diseases are relatively well documented for ardeids. Bacteria isolated from birds in this family are listed in table 6.19.

Salmonellosis, or paratyphoid, has been reported in ardeids frequently enough that it should be considered in the differential diagnoses for sick birds in this family. In a study of diseases of ardeids in southern Florida, *Salmonella* group B, 4,5,12:i-monophasic, which is similar to *S. typhimerium,* was isolated from 2 nestlings (Great Egret and Snowy Egret) and 1 adult (Great White Heron) (Spalding and Forrester 1991). Lesions suspected to be due to *Salmonella* were seen in 6 others, including an adult Great White Heron, 2 nestling Great Egrets, a nestling Reddish Egret, a nestling Snowy Egret, and a nestling Great White Heron, from southern Florida. This disease was characterized by multifocal, necrotizing hepatitis and splenitis, hepatomegaly (figure 6.7), occasionally grossly visible necrotic foci in muscle (figure 6.8), airsacculitis, and, in severe cases, valvular endocarditis and thrombosis. Enteritis was not a prominent feature of any of the cases from which *Salmonella* was isolated. A serological study of 76 Cattle Egrets from 15 counties in Florida resulted in 2 birds with "suspicious" titers to *S. typhimerium* and 2 to *S. pullorum*

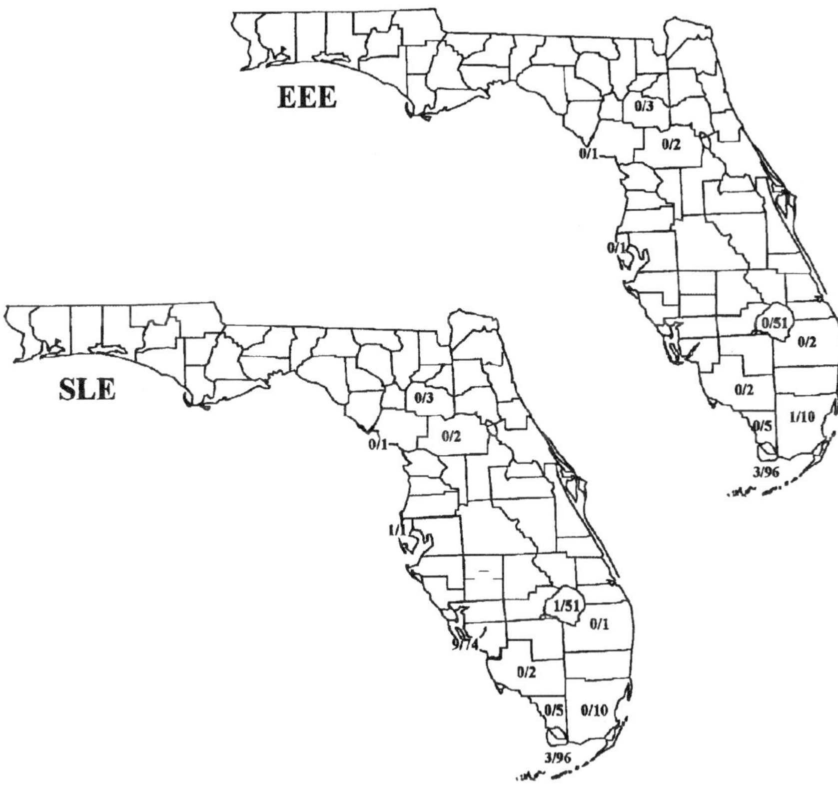

FIGURE 6.6. Distribution of ardeids tested for antibodies to eastern equine encephalitis (EEE) virus and St. Louis encephalitis (SLE) virus. Number positive/number tested (serum neutralization, >80% plaque reduction). Modified from Spalding et al. 1994d; by permission of *Journal of Wildlife Diseases*.

(Bigler et al. 1967). No pathogenic bacteria were cultured from the liver and large intestine of a fledgling Reddish Egret from Florida Bay, but *S. typhimerium* was cultured from the large intestine of an adult in Texas (Conti et al. 1986). *Salmonella* sp. was isolated from 2 of 3 Great Egrets collected from a eustrongylidosis epizootic in Louisiana (Roffe 1988), but not from birds dying in an epizootic in Delaware (Weise et al. 1977). Locke et al. (1974) diagnosed salmonellosis (*S. typhimerium*) in 9 individuals of 6 species of ardeids that were in captivity. The lesions were necrotic and granulomatous hepatitis and splenitis, and in some cases caseonecrotic enteritis.

Edwardsiella tarda was likely a pathogen in a Great White Heron nestling from Florida Bay in 1988 with hepatic necrosis, septicemia, and gout. It was isolated from both the heart blood and large intestine (*Clostridium perfringens* and *Escherichia coli* were also isolated from the large intestine). Mesenteric vessels were congested; however, autolysis was too severe to allow for evaluation of the intestinal mucosa (Spalding and Forrester 1991). *Edwardsiella tarda* was also isolated from the small intestine of 2 Great Blue Herons with enterocolitis (*C. perfringens* was also present). *Edwardsiella tarda* was isolated from the liver, pericardium, and coelomic cavity of Tricolored Heron nestlings experimentally infected with *Eustrongylides ignotus* (Spalding and Forrester 1993). *Edwardsiella* sp. was also isolated from the abdominal cavities of Great

Table 6.18. Other metal concentrations in tissues of ardeids from Florida

Age / Tissue	Species	County/area	Year(s)	No. examined	Arsenic Mean[a]	(Range)	Cadmium Mean[a]	(Range)	Copper Mean[a]	(Range)	Data source
Nestlings											
Liver	Great Blue Heron	"Lake Okeechobee"	1987–91	6	NA	—	0.05	(0.02–0.10)	5.5	(1.5–13)	A
	Great Egret	Collier	1987–91	5	NA	—	0.07	(0.06–0.09)	1.7[b]	(2.4–15)	A
		Dade	1989	1	NA	—	0.07	—	NA	—	A
		"Lake Okeechobee"	1987–91	6	NA	—	0.07	(0.05–0.11)	43	(22–220)	A
	Great White Heron	"Florida Bay"	1987–89	5	NA	—	0.10	(0.04–0.21)	20	(5.3–57)	A
	Little Blue Heron	"Lake Okeechobee"	1987–91	4	NA	—	0.06	(0.04–0.09)	23	(4.1–44)	A
	Snowy Egret	Monroe	1988–90	5	NA	—	0.08	(0.03–0.14)	21	(11–58)	A
	Tricolored Heron	"Everglades"	1987–91	4	NA	—	0.22	(0.08–0.68)	23	(2.7–170)	A
		Monroe	1987–88	2	NA	—	0.15	(0.13–0.17)	12	(3.6–37)	A
Adults and Juveniles											
Liver	Cattle Egret	Palm Beach	1989	2	0.033	(ND–0.11)	NA	—	NA	—	B
Kidney	Cattle Egret	Palm Beach	1989	2	0.11	(0.08–0.16)	NA	—	NA	—	B
Whole	Green Heron	North Liberty	1978	7	0.07	(0.05–0.22)	0.01	(0.01–0.01)	NA	—	C
carcass		South Liberty	1978	3	0.05[c]	(0.05–0.05)	0.07[c]	(0.03–0.17)	NA	—	C

Age / Tissue	Species	County/area	Year(s)	No. examined	Selenium Mean[a]	(Range)	Zinc Mean[a]	(Range)			Data source
Adults and Juveniles											
Kidney	Cattle Egret	Palm Beach	1989	2	NA	—	10	(8–13)	—	—	B
Liver	Cattle Egret	Palm Beach	1989	2	NA	—	12	(12–12)	—	—	B
Whole	Green Heron	North Liberty	1978	7	0.29[c]	(0.15–0.45)	NA	—	—	—	C
carcass		South Liberty	1978	3	0.32[c]	(0.28–0.36)	NA	—	—	—	C

Sources: A = Spalding et al. (1997), B = Spalding & Forrester (1991), C = Winger et al. (1984).
NA = not analyzed, ND = not detected.
a. Geometric mean, ppm wet weight unless otherwise indicated.
b. No. examined = 4.
c. Arithmetic mean.

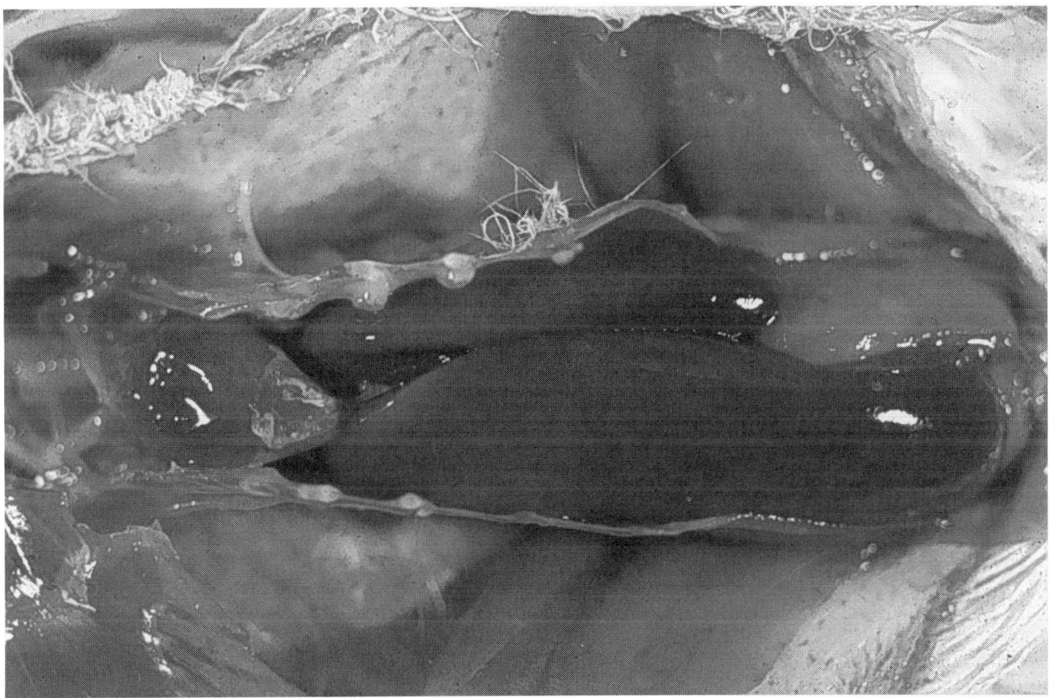

FIGURE 6.7. Hepatomegaly in a Snowy Egret nestling with salmonellosis.

Egrets from Louisiana with eustrongylidosis (Roffe 1988). White et al. (1973) reported isolating *E. tarda* from the large intestine of 1 of 2 Great Blue Herons cultured, though no associated lesions were observed. Taylor (1992) was able to detect *E. ictaluri* (a significant pathogen of catfish) in intestines of several species of ardeids in Mississippi that were foraging in catfish ponds. This organism has not been detected in Florida, to our knowledge.

Aeromonas hydrophila was isolated from several nestling ardeids and an adult Great Egret with enteritis (Spalding and Forrester 1991). Its role in association with the enteritis could not be determined since 2 species of *Clostridium* were also cultured from the same birds. *Aeromonas hydrophila* is commonly associated with peritonitis secondary to infection with *Eustrongylides* sp. It was isolated from the coelomic cavity of nestling Tricolored Herons experimentally infected with *E. ignotus* (Spalding and Forrester 1993). *Aeromonas* sp. was isolated from the abdominal cavities of Great

Egrets from Louisiana (Roffe 1988), and from Snowy Egrets from Delaware (Weise et al. 1977) with eustrongylidosis.

Pasteurella multocida, the organism responsible for avian cholera, has been isolated on several occasions from birds in Florida, but only once from an ardeid. Avian cholera was reported in the Everglades in 1967–68 when

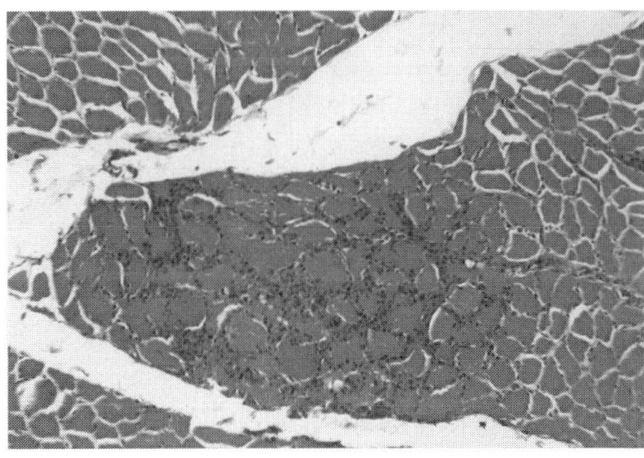

FIGURE 6.8. Focal necrotizing myositis in a Great White Heron with salmonellosis.

Table 6.19. Bacteria cultured from tissues of wading birds from Florida

Host species Organism isolated	Age[a]	County/area	Year	Tissue	Lesion/disease	Data source
Great Blue Heron						
Aeromonas hydrophila	NS	"Florida Bay"	1989	Small intestine	Enteritis	A
Bacillus sp.	AD	Palm Beach	1989	Cloacal swab	None	B
Clostridium perfringens	NS	"Florida Bay"	1989	Small intestine	Enteritis	A
Clostridium sordelli	NS	"Florida Bay"	1989	Small intestine	Enteritis	A
Diphtheroids	NS	Broward	1988	Foot	Pododermatitis	A
Edwardsiella tarda	NS	Alachua	1970–71	Large intestine	None	D
	NS	"Florida Bay"	1989	Small intestine	Enteritis	A
Escherichia coli	NS	"Florida Bay"	1989	Small intestine	Enteritis	A
Pasteurella multocida	JU	Orange	1999	Liver	Septicemia[b]	C
Pseudomonas putrificians	NS	Broward	1988	Foot	Pododermatitis	A
Anaerobic *Streptococcus*	NS	"Florida Bay"	1988	Duodenum	Biotoxicosis?	A
Streptococcus group D	AD	Palm Beach	1989	Cloacal swab	None	B
Alpha-hemolytic *Streptococcus*	NS	Broward	1988	Foot	Pododermatitis	A
Non-enteric Gram-negative rods	NS	Broward	1988	Foot	Pododermatitis	A
Great Egret						
Aeromonas hydrophila	AD	"Florida Bay"	1989	Small intestine	Enteritis	A
Clostridium perfringens	AD	"Florida Bay"	1989	Small intestine	Enteritis	A
Clostridium sordelli	AD	"Florida Bay"	1989	Small intestine	Enteritis	A
Escherichia coli	NS	Monroe	1988	Air sac	Peritonitis[c]	A
	AD	"Florida Bay"	1989	Small intestine	Enteritis	A
Klebsiella pneumoniae	AD	"Florida Bay"	1989	Small intestine	Enteritis	A
Salmonella sp.[e]	NS	Monroe	1988	Liver	Peritonitis[c], hepatitis[b]	A
Staphylococcus sp.	NS	Monroe	1988	Air sac	Peritonitis[c]	A
Non-hemolytic *Streptococcus*	NS	Monroe	1988	Liver	Peritonitis[c]	A
Great White Heron						
Clostridium perfringens	NS	"Florida Bay"	1988	Large intestine	Hepatic necrosis, septicemia	A
Edwardsiella tarda	NS	"Florida Bay"	1988	Heart blood, large intestine	Hepatic necrosis, septicemia, gout	A
Escherichia coli	NS	"Florida Bay"	1988	Large intestine	Hepatic necrosis, septicemia	A

Enterobacter cloacae	JU	"Florida Bay"	1988	Lung	Enterocolitis, septicemia	A
Proteus sp.	JU	"Florida Bay"	1988	Liver	Biotoxicosis	A
Salmonella sp.[e]	AD	"Florida Bay"	1988	Muscle, liver	Myositis[b], hepatitis[b]	A
Staphylococcus sp.	NS	"Florida Bay"	1988	Liver	Biotoxicosis	A
Vibrio alginolyticus	JU	"Florida Bay"	1988	Liver, lung	Enterocolitis, septicemia	A
Snowy Egret						
Salmonella sp.[e]	NS	"Lake Okeechobee"	1989	Liver	Hepatitis	A
Staphylococcus sp.	NS	"Lake Okeechobee"	1989	Liver	Hepatitis	A
Tricolored Heron						
Aeromonas calcoaceticus	NS	Captive	1990	Liver	Peritonitis[d]	A
Aeromonas hydrophila	NS	Captive	1990	Coelomic cavity	Peritonitis[d]	A
Citrobacter freundii	NS	Captive	1990	Coelomic cavity	Peritonitis[d]	A
Clostridium perfringens	NS	Captive	1990	Liver, coelomic cavity	Peritonitis[d]	A
Edwardsiella tarda	NS	Captive	1990	Coelomic cavity, pericardium, liver	Peritonitis[d]	A
Eschericia coli	NS	Captive	1990	Liver	Peritonitis[d]	A
Klebsiella pneumoniae	NS	Captive	1990	Liver	Peritonitis[d]	A
Plesiomonas shigelloides	NS	Captive	1990	Coelomic cavity	Peritonitis[d]	A
Proteus mirabilis	NS	Captive	1990	Coelomic cavity, liver	Peritonitis[d]	A
Pseudomonas aeruginosa	NS	Captive	1990	Coelomic cavity, pericardium, liver	Peritonitis[d]	A
Pseudomonas/*Alcaligenes* sp.	NS	Captive	1990	Pericardium	Peritonitis[d]	A
Staphylococcus-coagulase neg.	NS	Captive	1990	Liver	Peritonitis[d]	A
Gamma *Streptococcus* group D	NS	Captive	1990	Pericardium	Peritonitis[d]	A
Vibrio cholerae (presumptive)	NS	Captive	1990	Liver	Peritonitis[d]	A
Yeast	NS	Captive	1990	Liver	Peritonitis[d]	A

Sources: A = Spalding & Forrester (1991), B = Locke (1990), C = Spalding (1999), D = White et al. (1973).

a. Age: AD = adult, JU = juvenile, NS = nestling.
b. Bacteria felt to be the cause of the lesion.
c. Peritonitis associated with eustrongylidosis.
d. Nestlings experimentally infected with *Eustrongylides ignotus* (Spalding & Forrester 1993).
e. *Salmonella* sp. group B, serotype 4,5,12:i-monophasic.

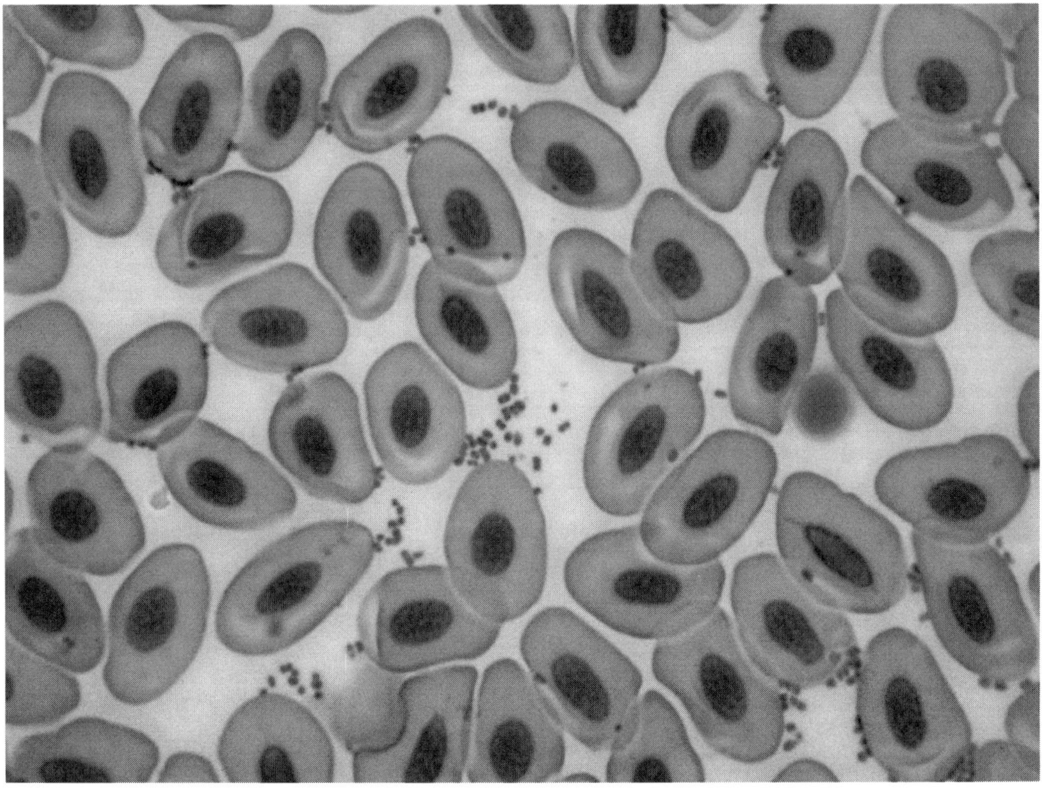

FIGURE 6.9. *Pasteurella multocida* in a blood smear of a Great Blue Heron with avian cholera. Note numerous short bipolar rods.

up to 6,000 American coots (see chapter 18, Rails) were estimated to have died. Four other species of waterfowl were involved, although no wading birds were found dead (Klukas and Locke 1970). A single juvenile Great Blue Heron died from avian cholera in Orange County in 1999 during an organochlorine die-off (see chapter 4, Pelicans) (Spalding 1999). In addition to septicemia (figure 6.9), hepatitis, and spleenomegaly (figure 6.10), this bird also had traumatic lesions and may have survived an attack by a predator with subsequent infection. Since *P. multocida* is a common oral resident in raccoons, dogs, and cats, this was a possible source. No other birds from the die-off had similar bacterial infections. Avian cholera has been reported to kill ardeids in the western United States (Rosen and Bischoff 1949; Oddo et al. 1978; Brand and Duncan 1983; Brogden and Rhoades 1983; Mensik and Botzler 1989; Friend 1999). Generally, avian cholera is a common disease of migratory waterfowl of central and western states and only rarely occurs in the southeastern United States. The conditions necessary for mortality from cholera to occur are poorly understood (Brand 1984; Brand et al. 1988; Botzler 1991).

There have been no published accounts of chlamydiosis from Florida ardeids, but there have been reports outside of Florida (Rubin et al. 1951; Pollard 1947) and Locke (1987) and Franson (1999) list this disease as common for this family. Chlamydiosis, caused by the rick-

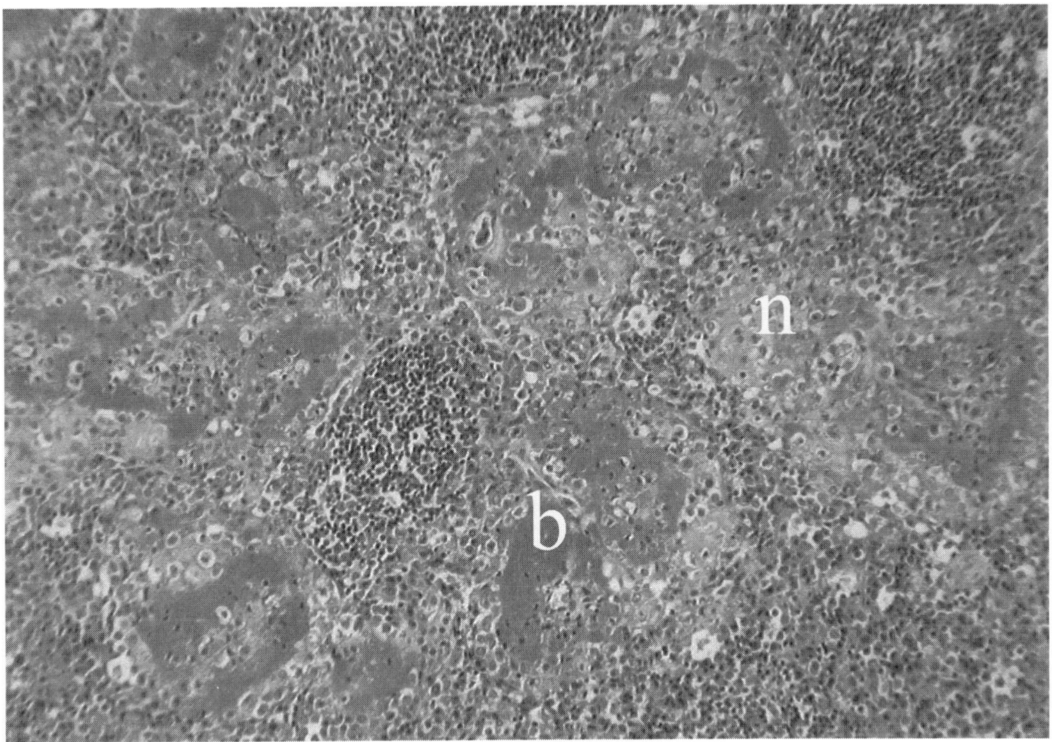

FIGURE 6.10. Colonies of *Pasteurella multocida* in the spleen of a Great Blue Heron with spleenomegaly. Note areas of necrosis (*n*) and bacterial colonies (*b*).

ettsia-like organism *Chlamydia psittaci*, was suspected in 5 nestling ardeids, based on gross and histologic lesions (Spalding and Forrester 1991). However, isolation or other definitive proof was not accomplished. A single Snowy Egret from Lake Okeechobee in 1990 tested negative by an antigen test.

Mortality can also be caused by the toxins produced by bacteria, especially of the genus *Clostridium*. *Clostridium perfringens* was commonly cultured from the intestines of normal ardeids as well as from those with enteritis (Spalding and Forrester 1991). This organism can be a common nonpathogenic resident of bird intestines, consequently diagnosis of *Clostridium*-caused enteritis is difficult. *Clostridium perfringens* was found commonly outside of the intestinal tract in nestlings experimentally infected with *Eustrongylides* sp. that perforated the gut, but it also was isolated from the liver of 1 control bird. The identification of *C. perfringens* type C, the causative agent of necrotic enteritis or enterotoxemia, can be made only by serologic analysis of the toxin produced (see Biotoxins section above).

Three of 76 Cattle Egrets collected in Florida (sites not specified) had titers to *Mycoplasma* sp. (Bigler et al. 1967). Lesions suggestive of mycoplasmosis have not been described in Florida ardeids, nor are we aware of such reports elsewhere. Spalding and Forrester (1991) found no evidence for mycobacterial infections in wading birds in Florida; however, only sporadic cultures were attempted.

Table 6.20. Protozoal parasites found in ardeids in Florida[e,f]

Host species / Hemoparasite	Age[a]	County/area	Year(s)	No. of birds Examined	Positive	Prevalence	Data source
Great Blue Heron							
Plasmodium relictum	N	"Florida Bay"	1987–91	11	1	9	A
Plasmodium elongatum	J	"North Central Florida"	1987–91	6	1	—	A
Plasmodium sp.	N	"Florida Bay"	1987–91	11	1	9	A
	N	"Lake Okeechobee"	1987–91	13	1	8	A
Haemoproteus herodiadis	N	"Florida Bay"	1987–91	11	1	9	A
	N	"Lake Okeechobee"	1987–91	13	2	15	A
Haemoproteus sp.	N	"Lake Okeechobee"	1987–91	13	1	8	A
Leucocytozoon ardeae	J	Alachua	1973	4	1	—	B
	J	Alachua	1979	1	1	—	B
	J	"North Central Florida"	1987–91	6	3	—	A
	A	"North Central Florida"	1987–91	6	2	—	A
Sarcocystis spaldingae[b]	A	Alachua, Collier, "Florida Bay," Hendry, Marion, Pinellas	1987–2000	82	8	20[d]	D, E
Trypanosoma avium	A	"North Central Florida"	1987–91	6	1	—	A
Great White Heron[e]							
Plasmodium relictum	A	"Florida Bay"	1987–91	27	1	4	A
Plasmodium sp.	N	"Florida Bay"	1987–91	66	2	3	A
	A	"Florida Bay"	1987–91	27	1	4	A
Haemoproteus herodiadis	N	"Florida Bay"	1987–91	66	1	2	A
Haemoproteus sp.	N	"Florida Bay"	1987–91	66	1	2	A
Leucocytozoon ardeae	A	"Florida Bay"	1987–91	27	1	4	A
Sarcocystis spaldingae[b,g]	A	"Florida Bay"	1987–91	—	—	—	D
Trypanosoma avium	A	"Florida Bay"	1987–91	27	2	7	A
Tricolored Heron							
Haemoproteus sp.	N	"Florida Bay"	1987–91	12	1	8	A
Trypanosoma avium	N	"Florida Bay"	1987–91	12	1	8	A
Great Egret							
Sarcocystis spaldingae[b]	A	Broward, "Lake Okeechobee," Hillsborough, "Florida Bay"[c]	1989–91	43[b]	5	20[d]	D

(continued)

XV. Fungi

Aspergillosis is reported to be infrequent in wading birds (Locke 1987; Friend 1999). Aspergillosis (*Aspergillus fumigatus*) was identified in an adult Great White Heron taken into captivity with a chronic head wound from Florida Bay in 1989 and in an adult Great Egret with eustrongylidosis from Lake Okeechobee in 1991 (Spalding and Forrester 1991). Yeast infections, probably *Candida albicans*, were seen in 2 nestling Great Blue Herons from Florida Bay in 1988 and Lake Okeechobee in 1990 and in a nestling Tricolored Heron from Florida Bay in 1990. In all cases, yeast was found in the peritoneum, along with bacteria. It was never identified as a pathogen by itself.

XVI. Protozoans

Blood smears were examined from 424 ardeids in 4 studies (table 6.20). Conti et al. (1986) found no blood parasites in a single fledgling Reddish Egret from Florida Bay in 1978.

Table 6.20. *(continued)*

| Host species | | | | No. of birds | | | Data |
Hemoparasite	Age[a]	County/area	Year(s)	Examined	Positive	Prevalence	source
Snowy Egret							
Sarcocystis spaldingae[b]	A	Volusia[c]	1988	1[b]	1	—	C
Little Blue Heron							
Sarcocystis spaldingae[b]	A	"Florida Bay"[c]	1989	6[b]	3	29[d]	D
Green Heron							
Sarcocystis spaldingae[b]	A	Broward, "Florida Bay"[c]	1989–90	15[b]	2	29[d]	D
Yellow-crowned Night Heron							
Sarcocystis spaldingae[b]	A	"Florida Bay"[c]	1989	5[b]	1	29[d]	D

Sources: A = Telford et al. (1992), B = Forrester & Bennett (1985), C = Riggs (1988), D = Spalding et al. (1994a), E = Spalding & Cheadle (2000).

a. A = adult, J = juvenile, N = nestling.

b. Reported by Spalding et al. (1994a) and later named by Odening (1997); very similar to *S. falcatula* in a heron from Alachua County in 2000.

c. Adult ardeids were examined from peninsular Florida, and only those counties and dates for which infected birds were found are listed. For the total number and distribution of birds examined, see Telford et al. (1992).

d. The prevalence was calculated using information from both gross and histologic examination. Spalding et al. (1994a) also examined 234 ardeid nestlings and found no sarcocysts.

e. In addition to the hemoparasites listed, see Telford et al. (1992), who also found 3 infections of unidentified trophozoites in nestling Great White Herons from Florida Bay.

f. In addition to the birds reported in this table, Telford et al. (1992) found no blood parasites in the following: 4 nestling Black-crowned Night Herons from Florida Bay; 1 adult Yellow-crowned Night Heron from Florida Bay; 7 juvenile Great White Herons from Florida Bay; 76 nestling Great Egrets from North-central Florida (1), South Florida (14), Florida Bay (14), and Lake Okeechobee (47), and 3 adult Great Egrets from Florida Bay (2) and Lake Okeechobee (1); 14 nestling Little Blue Herons from Lake Okeechobee (9) and Florida Bay (5) and 3 adult Little Blue Herons from Florida Bay; 3 nestling Reddish Egrets from Florida Bay; 15 nestling Snowy Egrets from Lake Okeechobee (7) and South Florida (8), and 4 adult Snowy Egrets from South Florida; 4 adult Tricolored Herons from Florida Bay (1) and South Florida (3); and 1 nestling and 1 adult Cattle Egret from Palm Beach County. Conti et al. (1986) found no blood parasites in a fledgling Reddish Egret from Florida Bay in 1978. Bradley and Kirkham (1966–67) found no blood parasites in 105 Cattle Egrets from Glades, Hendry, Marion, and Sumter counties. Forrester & Bennett (1985) found no blood parasites in 21 Cattle Egrets from Duval, Levy, Glades, and Alachua counties (1970–73); 2 juvenile Great Blue Herons from Alachua County (1974, 1976); birds of unspecified age from Polk (1974), Brevard (2 in 1977), Glades (1977) and Marion (1978) counties; a Least Bittern from Alachua County (1975); 2 Snowy Egrets from Glades County (1971–72); 6 juvenile Reddish Egrets from Florida Bay (1978); or a Great Egret from Alachua County (1975).

g. Data for *Sarcocystis* for Great White Herons are combined with Great Blue Herons.

Bradley and Kirkham (1966–67) found no blood parasites in blood smears from 105 Cattle Egrets in central Florida. Blood smears from 16 ardeids in northern and central Florida contained 2 that were positive for *Leucocytozoon ardeae* in Great Blue Herons (Forrester and Bennett 1985). Telford et al. (1992) examined blood smears from 302 ardeids in Florida. They found 3 species of *Plasmodium*, 2 species of *Haemoproteus, Leucocytozoon ardeae,* and *Trypanosoma avium* in Great Blue and White Herons and Tricolored Herons.

The most prevalent hemoparasite was *Leu-cocytozoon ardeae* in Great Blue Herons (Telford et al. 1992). *Leucocytozoon ardeae* was found in 45.5% of the Great Blue Herons north of Lake Okeechobee, in none of the Great Blue Herons south of the lake, and in only 1 (1%) of the Great White Herons from Florida Bay. Two *Plasmodium* species were present in the Great Blue and White Heron, *P. relictum* in both hosts, and *P. elongatum* in the Great Blue Heron. A third was not identified to species. The youngest bird positive for an infection by a *Plasmodium* sp. was about 19 days old, and a mixed infection of *Trypanosoma avium* and

FIGURE 6.11. Elongate sarcocysts in the pectoral muscle of a Great White Heron. From Spalding et al. 1994a; by permission of *Journal of Wildlife Diseases*.

Haemoproteus sp. was found in a 10-day-old Tricolored Heron.

Two distinct species of *Haemoproteus* were found in Great White and Blue Herons and Tricolored Herons: *H. herodiadis* and an unidentified species (Telford et al. 1992). *Trypanosoma avium* was found in a nestling Tricolored Heron. Single trypanosomes were found in blood films from a Great Blue Heron and a Great White Heron and in a smear of bone marrow from a Great White Heron. These were not present in adequate numbers for identification.

Sarcocysts (*Sarcocystis spaldingae*) in ciconiiformes are filamentous and very long (figure 6.11). They were first described from herons in Florida (Spalding et al. 1994a; Odening 1997). Adult Great Blue Herons, Great White Herons, Great Egrets, Green Herons, Little Blue Herons, Snowy Egrets, and Yellow-crowned Night-Herons were infected, and for those species for which more than ten individuals were examined, prevalence ranged from 20 to 29% (table 6.20). None was found in examination of 235

nestlings and smaller numbers of adult Tricolored Herons, Reddish Egrets, and Black-crowned Night-Herons. Birds with cysts were found throughout the state with no detectable effect of region (figure 6.12). In 1999 sarcocysts in a Great Blue Heron from Alachua County were determined to be very similar to *Sarcocystis falcatula* (Spalding and Cheadle 2000).

Toxoplasma gondii antibodies were detected in serum from 1 of 40 Cattle Egrets (1:128) but were not detected in 12 Tricolored Herons, 6 Snowy Egrets, 2 Great Blue Herons, 9 Green Herons, and 2 Black-crowned Night-Herons from Florida (dates and locations not given) by Burridge et al. (1979). These were tested by inhibition hemagglutination, which may not be as sensitive in birds as in mammals; thus some may have been false-negatives (Frenkel 1981).

Giardia ardeae has been found in several ardeid species, including Great Blue Herons in New York (Hegner 1925; Georgi et al. 1986; Erlandsen et al. 1990; Kulda and Nohynkova 1995); however, there have been no reports from Florida.

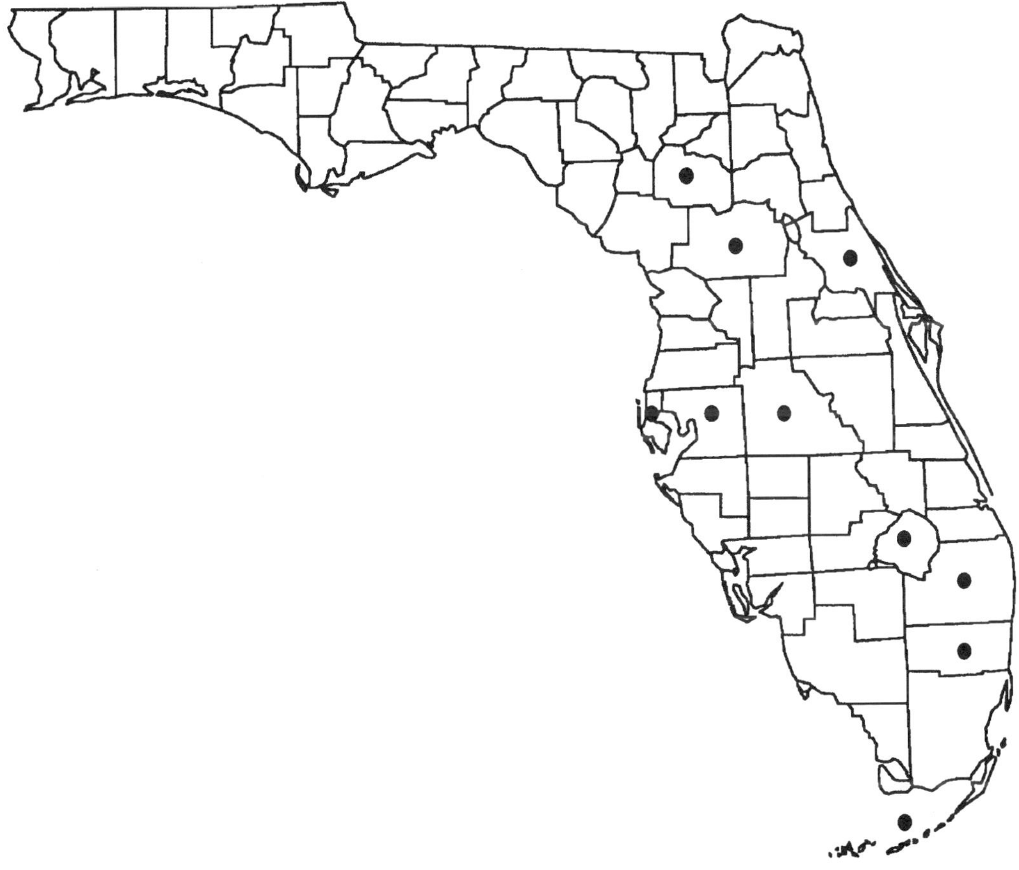

FIGURE 6.12. Distribution of ardeids examined for sarcocysts. *Solid circles* = counties with positive cases.

XVII. Helminths

Parasitic helminths of ardeids in Florida are listed in tables 6.21–6.34. The most complete helminth surveys have been made of Great Egrets (Sepulveda et al. 1999b), Great Blue/White Herons (Spalding and Kinsella 1999), Little Blue Herons (Sepulveda et al. 1996), and Cattle Egrets (Stuart et al. 1972). Fewer than ten of the other ardeid species have been examined for all parasites. Most of the information available stems from a study of dead and moribund birds, mostly nestlings from southern Florida (Spalding and Forrester 1991). Many of the helminths were shared by more than 1 host species (21 trematodes, 4 cestodes, 11 ne-

matodes, and 2 acanthocephalans), a situation likely due to overlapping of prey species and shared foraging habitats. The host sharing the fewest helminth parasites was the Cattle Egret, which rarely eats fish.

Eustrongylidosis, caused by infection with a large red nematode, *Eustrongylides ignotus,* is clearly the most significant infectious disease of ardeids in Florida (Spalding 1990; Spalding et al. 1993, 1994e; Spalding and Forrester 1993). This disease has been linked with nutrient pollution of foraging sites, resulting in increased numbers of infected intermediate host fish (Spalding et al. 1993; Coyner 1998). In Florida, the prevalence of sites with infected fish declined from the northern part of the peninsula

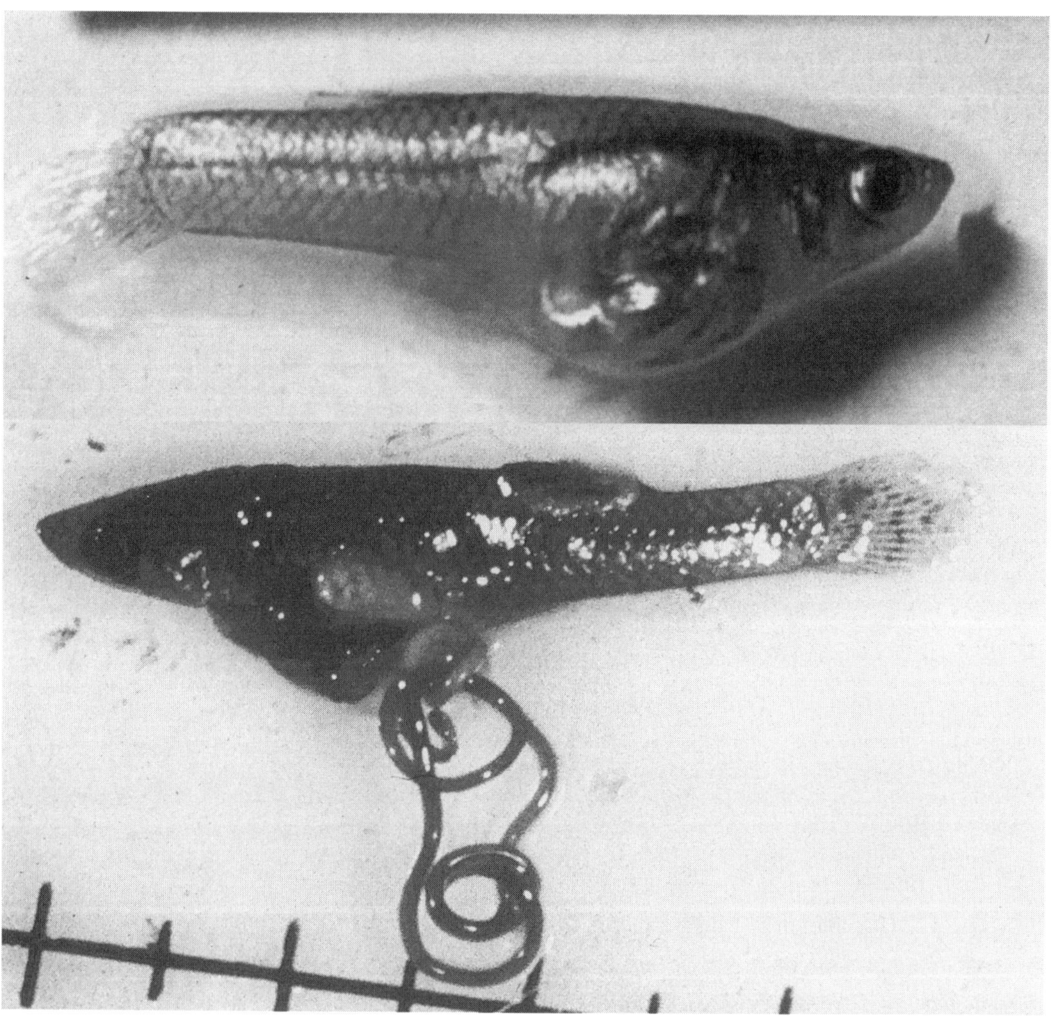

FIGURE 6.13. Intact *(above)* and excised mosquitofish (*Gambusia affinis*) *(below)* infected with larval *Eustrongylides ignotus*. Note the appearance of the coiled nematode on the surface of the abdomen of the top fish.

(39%) to the central part (15%) to the southern part (10%) (Coyner 1998). Infected fish (figure 6.13) were found only at sites with a history of anthropogenic nutrient effluent and human disturbance (Spalding et al. 1993; Coyner 1998). Frederick et al. (1996b) examined museum specimens of mosquitofish (*Gambusia holbrooki*) and documented an increase in the prevalence of infected fish with time, especially in north central Florida. Coyner (1998) found that the water at sites with infected fish was characterized by lower dissolved oxygen and higher total nitrogen, total phosphorus, and chlorophyll-a. The sediment had higher soil oxygen demand, total phosphorus, and mean grain size. The life cycle of this parasite has been further defined by Coyner (1998) to include direct transmission from eggs to fish, and is illustrated in figure 6.14.

Infection with *E. ignotus* results in a firm serpentine peritonitis that can be seen and palpated on the ventral surface of the stomach in

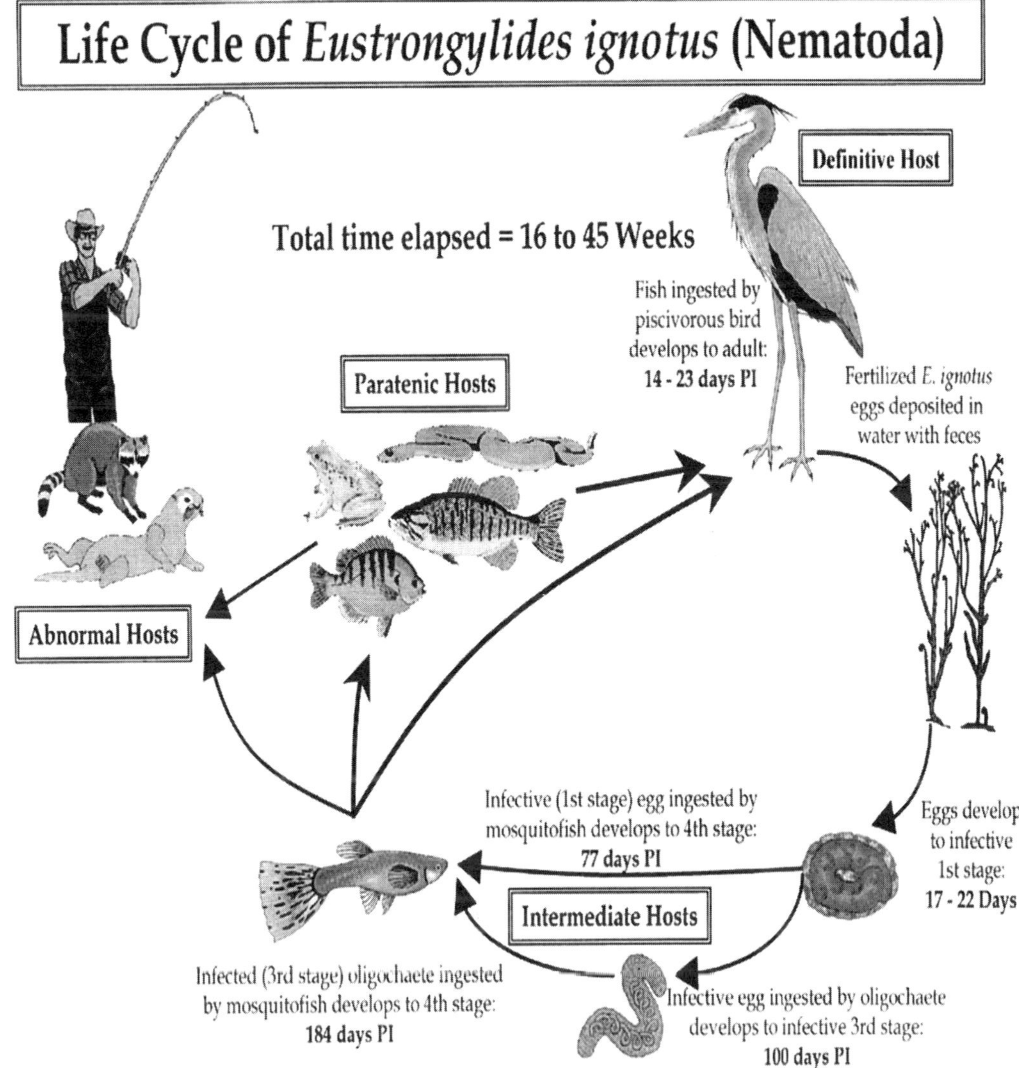

FIGURE 6.14. Life cycle of *Eustrongylides ignotus*. Courtesy of Donald F. Coyner.

chronic and severe cases (figure 6.15) (Spalding 1990). The nematode has characteristic features when examined histologically (figure 6.16). Lesions were more severe and caused more mortality in nestlings than in adults (Spalding and Forrester 1993), and resulted in mortality as high as 80% in some colonies (Spalding et al.1993). The impact on the adult population is more difficult to estimate. Infec-

tions and scars from resolved infections are very common (>30% prevalence), indicating that many adults survive infection. *Eustrongylides* larvae in nestlings were more prevalent in mainland colonies (17%) than in marine colonies (3%) (table 6.21). Adult and fledged Great Egrets (32%) and Great Blue Herons (47%) had the highest prevalence (Spalding and Forrester 1991). Larger preda-

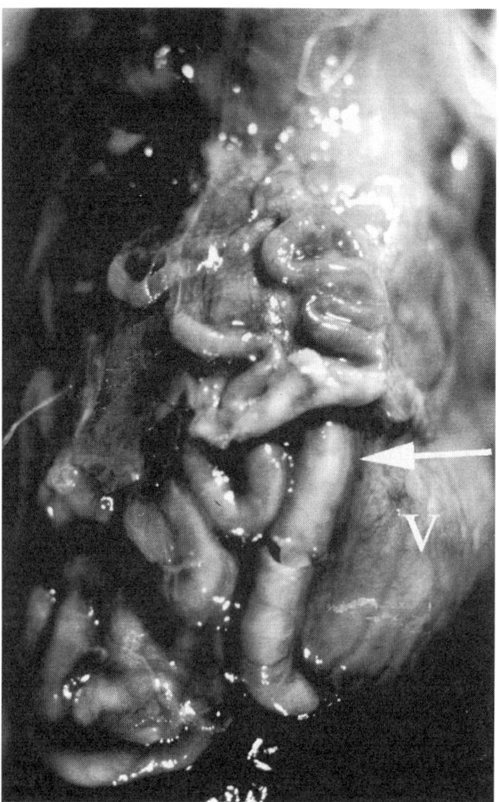

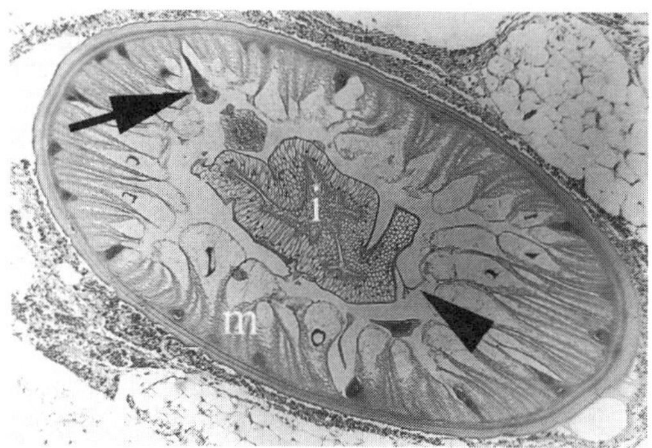

Figure 6.16. Cross section of fourth-stage *Eustrongylides ignotus* larvae in adipose tissue below the serosa of the ventriculus. Note the large ventral cord *(arrow)*, coelomyarian musculature *(m)*, pseudocoelomic membrane *(arrowhead)*, and intestine *(i)*. From Spalding and Forrester 1993; by permission of *Journal of Wildlife Diseases*.

Figure 6.15. Chronic eustrongylidosis in a Snowy Egret. Arrow points to a fibrino-fibrous tubule containing a nematode on the surface of the ventricular portion of the stomach *(v)*. From Spalding and Forrester 1993; by permission of *Journal of Wildlife Diseases*.

tory fish may act as transport hosts (Coyner et al. 2001), which may explain the higher prevalence in larger ardeid species. Mortality of ardeids with eustrongylidosis has also been reported across North America, including California, Delaware, Louisiana, and Rhode Island (Weise et al. 1977; Roffe 1988; Franson and Custer 1994).

Most of the other helminth parasites recorded from ardeids are not associated with significant pathology, such as a nematode in the heart of a Great White Heron (figure 6.17). A nematode, probably *Avioserpens* sp., coiled in the submucosa of the throat was visible near the base of the tongue (figure 6.18). In a few cases their presence caused a bulge in the throat that could be seen externally and rarely hemorrhage was associated with the nematodes. These lesions were observed grossly in 7 Snowy Egrets, 3 Tricolored Herons, 1 Great Egret, 1 Little Blue Heron, 1 Great Blue Heron, and 1 Cattle Egret in Florida (Spalding 2001). Trematodes are very common in the intestine (figure 6.19) and rarely cause more than mild localized inflammation.

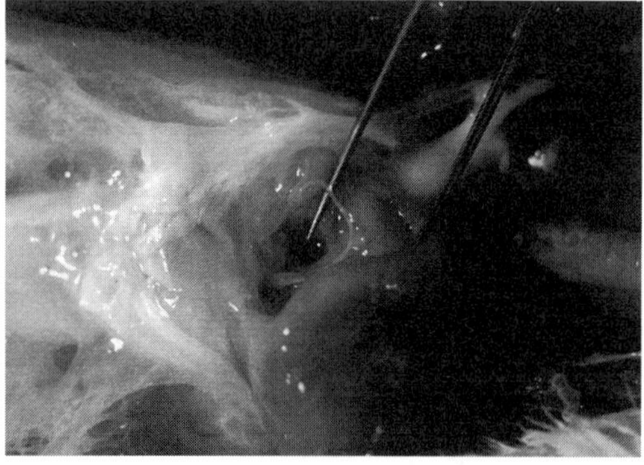

Figure 6.17. Nematode, probably *Paroncocerca tonkinensis*, in the heart of a Great White Heron.

XVIII. Arthropods

Ectoparasites reported from ardeid birds in Florida are listed in tables 6.35–6.38. No clear evidence of host morbidity or mortality related to ectoparasitism has been recorded for Florida. Subcutaneous mites (table 6.36) were found in 9.8% of all ardeid carcasses examined by Spalding and Forrester (1991), including Black-crowned Night-Herons, Cattle Egrets, Great Blue and Great White Herons, Great Egrets, Snowy Egrets, and Little Blue Herons. Infested nestlings (all ardeid species combined) were found in all regions except for the northern Everglades. Pence et al. (1997) found that some mite species might infest atypical host species, possibly because of the sharing of nest material.

Cattle Egrets can serve in the dissemination of the tropical bont tick, *Amblyomma variegatum*, in the Caribbean (Corn et al. 1993). This is of great concern because of the potential for this tick to transmit heartwater disease (*Cowdria ruminantium*) to mainland North America. McKilligan (1987) found increased mortality and lower body weights in Cattle Egret chicks that were infested with ticks (*Argus robertsi*) in Australia. Yunker et al. (1979) collected *Ornithodoros capensis* ticks from wading bird or

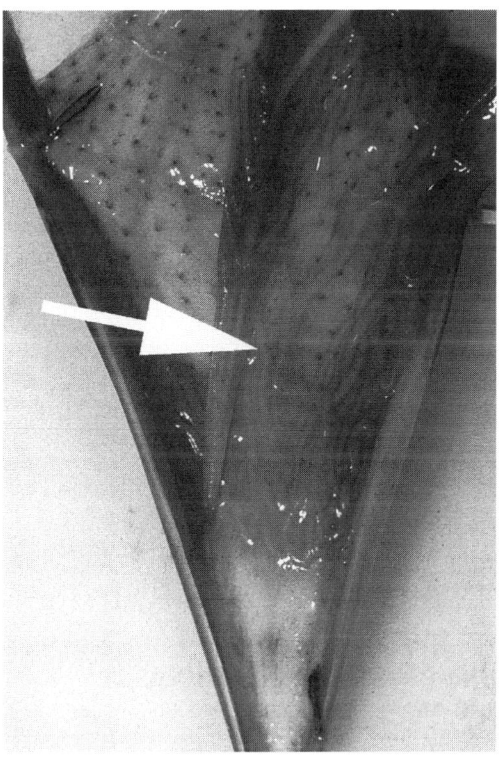

FIGURE 6.18. Nematode, probably *Avioserpens* sp., *(arrow)* in a coiled mass within the submucosa of the oral cavity to the right of the base of the tongue *(under arrow)* of a Snowy Egret from Monroe County in 1988.

Table 6.21. Prevalence of *Eustrongylides ignotus* by location for selected ardeid nestlings in Florida (combining data from both postmortem examinations and palpations[a])

Location	Year(s)	Great Blue Heron[b] n	Great Blue Heron[b] %	Great Egret n	Great Egret %	Snowy Egret n	Snowy Egret %	Tricolored Heron n	Tricolored Heron %	Little Blue Heron n	Little Blue Heron %	All n	All %
Mainland Colonies													
Monroe	1987–90	0	—	35	49	164	20	113	12	4	75	316	20
Polk	1990	0	—	0	—	3	66	1	100	8	75	12	75
Palm Beach	1990	0	—	53	25	40	25	69	23	108	12	270	19
Dade/Broward	1987–89	3	33	6	17	21	19	17	12	13	0	60	13
Lake Okeechobee	1989–90	31	3	246	20	183	18	162	6	103	3	725	13
All Mainland Colonies	1987–90	34	6	340	24	411	20	362	12	235	10	1,383	17
Florida Bay Colonies	1987–90	57	0	53	4	17	0	104	5	17	0	248	3

Source: Spalding & Forrester (1991).
a. Spalding (1990).
b. Includes both the blue and white color phases.

Table 6.22. Helminths recorded from Bitterns in Florida

Host Class	Species (location)[a]	County/ area	Year	No. of birds Examined	No. of birds Positive	Intensity Mean	Intensity Range	Data source
American Bittern								
Trematoda								
	Apharyngostrigea pipientis (SI)	Glades	1970	1	1	15	—	A
	Pegosomum asperum[b] (LV)	Levy	1971	1	1	2	—	A
	Posthodiplostomum boydae (SI)	Glades	1970	1	1	235	—	A
		Monroe	1989	1	1	320	—	B
	Posthodiplostomum grande (SI)	Glades	1970	1	1	3	—	A
	Posthodiplostomum opisthosicya (SI)	Glades	1970	1	1	10	—	A
Nematoda								
	Capillaria sp. (SI)	Monroe	1989	1	1	1	—	B
	Chandleronema longigutturata (ST)	Monroe	1989	1	1	12	—	B
	Contracaecum microcephalum (ST)	"Florida"	NG	—	—	—	—	C
	Contracaecum sp. (larval) (ST)	Glades	1970	1	1	1	—	A
		Monroe	1989	1	1	1	—	B
	Desportesius trianuchae (ST)	Collier	1998	1	1	4	—	D
		Glades	1970	1	1	5	—	A
		Levy	1971	1	1	3	—	A
	Larval spirurids	Levy	1971	1	1	2	—	A
		Monroe	1989	1	1	3	—	B
	Paronchocerca tonkinensis (CC)	Levy	1971	1	1	3	—	A
	Tetrameres microspinosa (PR)	Collier	1998	1	1	9	—	D
	Tetrameres sp. 1 (PR)	Glades	1970	1	1	1	—	A
	Tetrameres sp. 2 (PR)	Glades	1970	1	1	2	—	A
	Tetrameres sp. 3[c] (PR)	Collier	1998	1	1	9	—	D
Acanthocephala								
	Polymorphus brevis (SI)	Collier	1998	1	1	1	—	D
Least Bittern								
Trematoda								
	Echinochasmus sp. (SI)	Dade	1994	1	1	1	—	D
	Posthodiplostomum sp. (SI,LI)	Dade	1994	1	1	72	—	D
	Posthodiplostomum boydae (SI,LI)	Dade	1994	1	1	227	—	D

Sources: A = Kinsella (1971), B = Spalding & Kinsella (1989), C = Walton (1927), D = Spalding & Kinsella (1998).
NG = not given.
a. Location in host: CC = coelomic cavity, LI = large intestine, LV = liver, PR = proventriculus, SI = small intestine, ST = stomach.
b. First record for North America.
c. The undescribed species mentioned in Mollhagen (1976: 309).

Roseate Spoonbill nests from Texas and isolated Aransas Bay virus.

Nest-dwelling larvae of scavenging dermestid beetles commonly chew holes in dead and sometimes living ardeid nestlings in Florida. The species of beetle has not been identified in southern Florida, but in northern and central Florida it is *Dermestes nidum* (Rodgers et al. 1993). Great Blue and Great White Heron carcasses most frequently had dermestids or evidence of dermestids present (27%), followed by Tricolored Herons (16%), Great Egrets (11%), Little Blue Heron (9%), Snowy Egrets (5%), and a single (of 6) Black-crowned Night-Heron (Spalding and Forrester 1991). Dermestid beetle larvae also caused le-

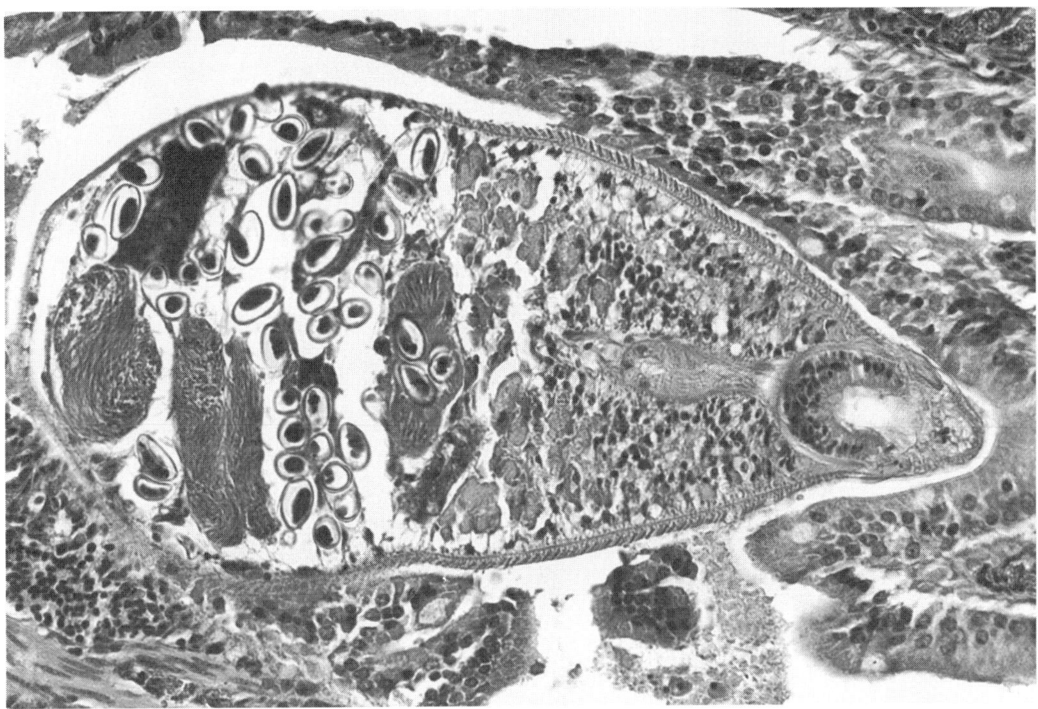

FIGURE 6.19. Photomicrograph of a trematode of the *Ascocotyle-Phagicola* complex (family Heterophyidae) in the intestinal mucosa of a Great White Heron.

sions on 5 living nestlings in southern Florida, including 1 Great Blue Heron, 1 Great White Heron, 2 Tricolored Herons, and 1 Great Egret (Spalding and Forrester 1991). See chapter 7 (Ibises, Storks, and Spoonbills) for additional information and illustration of a lesion. Other anecdotal reports of dermestid infestations of ardeids concern a Great Egret in Florida (Snyder et al. 1984) and Snowy Egrets, Tricolored Herons, and Cattle Egret nests in central Florida (Black et al. 1984).

Red imported fire ants (*Solenopsis invicta*) reportedly caused significant (92%) reduction in waterbird reproduction (including Great Egrets, Great Blue Herons, Snowy Egrets, and Tricolored Herons) on Texas spoil islands (Drees 1994). This fire ant is present in Florida; however, we have no data regarding its impact on nestling ardeids in Florida.

Edman et al. (1984), and Maxwell and Kale (1977) recorded anti-insect behaviors in several ardeid species due to biting mosquitoes and *Culicoides* sp.

XIX. Nutritional diseases

Nutritional diseases are rare in free-ranging wildlife. A condition called pansteatitis (inflammation of fat) has been reported in Great Blue Herons from the Chesapeake Bay region (Nichols et al. 1986). This condition is suspected to be due to either a deficiency in Vitamin E and selenium, or to consumption of rancid fish or fish with high concentrations of polyunsaturated fatty acids. Fat necrosis is a common incidental finding in ardeids in Florida (Spalding 1999); however, its cause in wild birds is unclear and it has not been reported as a cause of death in Florida.

Table 6.23. Trematodes recorded from Great Blue and Great White Herons in Florida

Species (location)[a]	County/area	Year(s)	No. of birds			Intensity		Data source
			Examined	Positive	%	Mean	Range	
Apharyngostrigea cornu (LS)	Alachua	1973–94	6	1	—	10	—	A
Apharyngostrigea simplex (SI)	Alachua	1973–94	6	4	—	27	2–85	A
	Collier	1989	1	1	—	3	—	B
	Dade	1987–88	4	2	—	6	2–9	B
	"Florida Bay"	1987–95	16	1	6	1	—	B
Apharyngostrigea sp. (SI)	Glades	1977	1	1	—	1	—	C
	"L. Okeechobee"	1981–90	5	1	—	2	—	B
Ascocotyle gemina (LI)	Alachua	1973–94	6	1	—	3	—	A
	Florida Bay	1987–95	16	1	6	3	—	B
	Hendry	1988	1	1	—	11	—	B
	"L. Okeechobee"	1981–90	5	1	—	60	—	B
Ascocotyle tenuicollis (IN)	Alachua	1973–94	6	1	—	248	—	A
	Dade	1987–88	4	2	—	38	30–45	B
	"Florida Bay"	1987–95	16	4	25	12	5–25	B
	Hendry	1988	1	1	—	100	—	B
	"L. Okeechobee"	1981–90	5	4	—	97	60–181	B
	Lee	1988	1	1	—	NG	—	B
Clinostomum attenuatum (OR)	Alachua	1973–94	6	2	—	8	1–14	A
	Hendry	1988	1	1	—	3	—	B
Clinostomum complanatum (OR)	Alachua	1973–94	6	3	—	11	6–18	A
	Dade	1987–88	4	2	—	20	14–26	B
	"L. Okeechobee"	1981–90	5	1	—	5	—	B
	Orange	1977–81	2	1	—	3	—	C
Clinostomum heluans (OR)	Hendry	1988	1	1	—	1	—	B
Diasiella diasi (PA)	Alachua	1973–94	6	1	—	2	—	A
Diplostomum ardeae (SI)	Alachua	1973–94	6	1	—	15	—	A
	"Florida Bay"	1987–95	16	1	6	4	—	B
	"L. Okeechobee"	1981–90	5	1	—	1	—	B
Echinochasmus dietzevi (SI)	Dade	1987–88	4	1	—	1	—	B
	"Florida Bay"	1987–95	16	1	6	8	—	B
	Hendry	1988	1	1	—	48	—	B
	"L. Okeechobee"	1981–90	5	3	—	90	1–180	B
Ignavia venusta (KD)	Osceola	1993	1	1	—	1	—	B
Mesorchis denticulatus (SI)	"L. Okeechobee"	1981–90	5	2	—	197	93–300	B
Mesostephanus appendiculatoides (SI)	"Florida Bay"	1987–95	16	2	13	6	3–9	B

Parasite (location in host)[a]	County	Years						Source
Microparyphium facetum (CL)	Alachua	1973–94	6	2	—	2	—	A
	"Florida Bay"	1987–95	16	1	6	16	—	B
	Hendry	1988	1	1	—	2	—	B
Parastrigea brasiliana (SI)	Alachua	1973–94	6	2	—	87	1–173	A
	Collier	1989	1	1	—	3	—	B
	Dade	1987–88	4	1	—	7	—	B
	Hendry	1988	1	1	—	11	—	B
Phagicola longa (SI)	"Florida Bay"	1987–95	16	13	81	765	2–3810	B
	"L. Okeechobee"	1981–90	5	2	—	207	55–358	B
Phagicola sp. (SI)	"Florida Bay"	1987–95	16	1	6	2176	—	B
	Hendry	1988	1	1	—	920	—	B
Philophthalmus hegeneri (OC)	"Florida Bay"	1987–95	16	1	6	2	—	B
Posthodiplostomum boydae (DU)	Alachua	1973–94	6	1	—	5	—	A
Posthodiplostomum macrocotyle (SI)	Alachua	1973–94	6	3	—	NG	3–225	A
	Glades	1977	1	1	—	51	—	C
	Hendry	1988	1	1	—	20	—	B
Posthodiplostomum minimum (SI)	Alachua	1973–94	6	4	—	1386	10–5225	A
	Dade	1987–88	4	2	—	606	21–1190	B
	Hendry	1988	1	1	—	65	—	B
	"L. Okeechobee"	1981–90	5	5	—	159	20–400	B
	Orange	1977–81	2	1	—	26	—	C
Posthodiplostomum nanum (SI)	Alachua	1973–94	6	4	—	NG	1–125	A
	Dade	1987–88	4	1	—	NG	—	B
Posthodiplostomum opisthosicya (SI)	Alachua	1973–94	6	2	—	4	3–5	A
	Dade	1987–88	4	1	—	NG	—	B
	"L. Okeechobee"	1981–90	5	3	—	NG	—	B
Posthodiplostomum sp. (SI)	Lee	1988	1	1	—	NG	—	B
Prosthogonimus ovatus (CL)	"Florida Bay"	1987–95	16	1	6	2	—	B
Pygiodopsis pindoramensis (SI)	Alachua	1973–94	6	1	—	5	—	A
	Dade	1987–88	4	2	—	7	1–13	B
	"L. Okeechobee"	1981–90	5	4	—	191	2–400	B
Renicola sp. (KD)	"Florida Bay"	1987–95	16	4	25	17	1–49	B
	Orange	1977–81	2	1	—	132	—	A
Ribeiroia ondatrae (PR, SI)	Hendry	1988	1	1	—	3	—	B
	Lee	1988	1	1	—	1600	—	B

Sources: A = Spalding et al. (1991), B = Spalding & Kinsella (1999), C = Forrester & Kinsella (1999).
NG = not given.
a. Location in host: CL = cloaca, DU = duodenum, KD = kidney, IN = intestine, LI = large intestine, LS = lower small intestine, SI = small intestine, OC = under nictitating membrane of eye, OR = oral cavity, PA = pancreas, PR = proventriculus.

Table 6.24. Nematodes recorded from Great Blue and Great White Herons in Florida

Species (location)[a]	Country/area	Year(s)	No. of birds			Intensity		Data source
			Examined	Positive	%	Mean	Range	
Acuaria multispinosa (PR)	Alachua	1973–94	6	1	—	4	—	A
	Dade	1987–88	4	1	—	3	—	B
	"Florida Bay"	1987–95	16	2	13	2	1–2	B
Capillaria herodiae (IN)	Alachua	1973–94	6	2	—	11	1–21	A
	Dade	1987–88	4	3	—	10	5–20	B
	"Florida Bay"	1987–95	16	2	13	3	1–5	B
	Glades	1977	1	1	—	1	—	C
	Hendry	1988	1	1	—	32	—	B
	"L. Okeechobee"	1981–90	5	1	—	1	—	B
	Lee	1988	1	1	—	2	—	B
Capillaria obsignata (ES)	Alachua	1973–94	6	1	—	1	—	A
Capillaria sp. (ES,SI)	"L. Okeechobee"	1981–90	5	1	—	2	—	B
	Orange	1977–81	2	1	—	1	—	C
Chandleronema longigutturata (PR)	Alachua	1973–94	6	1	—	7	—	A
Contracaecum microcephalum (ST)	Osceola	1993	1	1	—	10	—	B
	"Florida"	NG	NG	NG	—	NG	—	D
Contracaecum micropapillatum (ST)	"Florida Bay"	1987–95	16	1	6	2	—	B
Contracaecum multipapillatum (ST)	Alachua	1973–94	6	1	—	4	—	A
	Dade	1987–88	4	1	—	3	—	B
	"Florida Bay"	1987–95	16	1	6	5	—	B
	"L. Okeechobee"	1981–90	5	2	—	8	—	B
	Lee	1988	1	1	—	11	—	B
	Orange	1977–81	2	1	—	1	—	C
Contracaecum spp. (ES,ST)	Alachua	1973–94	6	3	—	35	2–99	A
	Dade	1987–88	4	3	—	16	4–25	B
	"Florida Bay"	1987–95	16	4	25	217	1–757	B
	Hendry	1988	1	1	—	159	—	B
	"L. Okeechobee"	1981–90	5	3	—	15	5–22	B
	Orange	1977–81	2	1	—	3	—	C
	Osceola	1993	1	1	—	20	—	B
Cosmocephalus obvelatus (PR)	Orange	1977–81	2	1	—	1	—	C
Desmidocercella numidica (BC)	Alachua	1973–94	6	5	—	19	3–45	A
	Collier	1989	1	1	—	1	—	B
	Dade	1987–88	4	1	—	15	—	B
	Orange	1977–81	2	1	—	4	—	C

Desportesius invaginatus (ST)	Alachua	1973–94	6	2	—	9	1–16	A
	Dade	1987–88	4	1	—	1	—	B
	"Florida Bay"	1987–88	16	2	13	4	1–6	B
	Glades	1977	1	1	—	1	—	C
	Hendry	1988	1	1	—	1	—	B
	"L.Okeechobee"	1981–90	5	1	—	1	—	B
Eustrongylides ignotus[b] (ST,CO)	Alachua	1973–2000	34	10	30	3	1–22	A, G
	Collier	1989	1	1	—	1	—	B
	Duval	1977	1	1	—	14	—	C
	"Florida Bay"	1990–94	7	1	—	1	1–12	G
	Hendry	1988	1	1	—	9	—	B
	Hillsborough	1989–90	8	1	—	1	1–17	G
	Lee	1988	1	1	—	2	—	B
	Orange	1977–81	2	1	—	5	—	C
	Pinellas	1989–90	16	3	19	3	1–24	G
	Polk	1999	1	1	—	1	—	F
Paronchocerca tonkinensis (ES,LU,KD,LV)	Alachua	1973–94	6	1	—	4	—	A
	"Florida Bay"	1987–95	16	1	6	9	—	B
Porrocaecum reticulatum (ST,SI)	Alachua	1973–94	6	2	—	10	4–16	A
	Collier	1989	1	1	—	5	—	B
	Florida Bay	1987–95	16	1	6	4	—	B
	Lee	1988	1	1	—	1	—	B
	Orange	1977–81	2	1	—	2	—	C
spirurid larvae (PR)	Dade	1987–88	4	1	—	18	—	B
	"Florida Bay"	1987–95	16	1	6	5	—	B
	Hendry	1988	1	1	—	3	—	B
	Orange	1977–81	4	2	—	6	3–9	C
Strongyloides sp. (SI)	Alachua	1973–94	6	2	—	58	55–60	A
	Dade	1987–88	4	1	—	4	—	B
	"Florida Bay"	1987–95	16	2	13	44	2–86	B
	Glades	1977	1	1	—	17	—	C
	"L. Okeechobee"	1981–90	5	1	—	5	—	B
	Orange	1977–81	2	1	—	28	—	C
Tetrameres ardamericanus (PR)	Alachua	1973–94	6	5	—	11	5–30	A
	Dade	1987–88	4	3	—	43	6–105	B
	"Florida Bay"	1987–95	16	5	31	18	3–59	B
	"L.Okeechobee"	1981–90	5	2	—	5	4–6	B
Tetrameres microspinosa (PR)	Bradford	1971	1	1	—	NG	—	E
	Pinellas	1971	1	1	—	NG	—	E

Sources: A = Spalding et al. (1991), B = Spalding & Kinsella (1959), C = Forrester & Kinsella (1999), D = Walton (1927), E = Mollhagen (1976), F = Fischer (1999), G = Spalding (2000). NG = not given.

a. Location in host: BC = body cavity, CO = coelum, ES = esophagus, IN = intestine, KD = kidney, LU = lung, LV = liver, PR = proventriculus, SI = small intestine, ST = stomach.

b. Data for nestlings is presented in Table 6.21.

Table 6.25. Cestodes and acanthocephalans recorded from Great Blue and Great White Herons in Florida

Class			No. of birds			Intensity		Data
Species (location)[a]	County/area	Year(s)	Examined	Positive	%	Mean	Range	source
Cestoda								
Dendrouterina ardeae (SI)	Alachua	1973–94	6	3	—	7	1–15	A
	"Florida Bay"	1987–95	16	1	6	1	—	B
	Hendry	1988	1	1	—	1	—	B
Glossocercus caribaensis (SI)	Dade	1987–88	4	1	—	6	—	B
	"Florida Bay"	1987–95	16	6	38	14	3–60	B
	Hendry	1988	1	1	—	22	—	B
Neogryporhynchus cheilancristrotus (SI)	Alachua	1973–94	6	1	—	15	—	A
Valipora mutabile (SI)	Alachua	1973–94	6	1	—	5	—	A
	Dade	1987–88	4	1	—	70	—	B
	"Florida Bay"	1987–95	16	2	13	2	—	B
Valipora sp. (SI)	Dade	1987–88	4	1	—	NG	—	B
	"Florida Bay"	1987–95	16	1	6	3	—	B
Acanthocephala								
Echinorhynchus striatus	"Florida"	1885	—	—	—	—	—	C[b]
Plagiorhynchus sp. (ST)	Dade	1987–88	4	1	—	7	—	B
Polymorphis brevis (SI)	Alachua	1973–94	6	3	—	14	5–24	A
	Collier	1989	1	1	—	5	—	B
	Dade	1987–88	4	3	—	9	1–17	B
	"Florida Bay"	1987–95	16	3	19	3	1–5	B
	Hendry	1988	1	1	—	4	—	B
	"L. Okeechobee"	1981–90	5	1	—	9	—	B
Southwellina dimorpha (SI)	Alachua	1973–94	6	1	—	24	—	A
	Hendry	1988	1	1	—	3	—	B
Unidentified acanthocephalan (SI)	Dade	1987–88	4	1	—	1	—	B

Sources: A = Spalding et al. (1991), B = Spalding & Kinsella (1999), C = U.S. National Parasite Collection.
NG = not given.
a. Location in host: SI = small intestine, ST = stomach.
b. Case number 5205.

XX. Emaciation

Emaciation was a common finding (54%) in nestling ardeids, especially for the smallest nestling in a nest, and was most likely the result of competition for food among asynchronously hatched nestlings (Spalding and Forrester 1991). A clear example of emaciation resulting from reduced food delivered to nests was found during a study at the Rodgers River Bay colony, which was visited following the passage of tropical storm Marco in 1990 (Spalding and Forrester 1991). Most nestling Snowy Egrets were found in their nests dead or cold and moribund. There was no evidence of infectious disease at postmortem exam. Nestling Tricolored Herons in the same colony were warm, in excellent nutritional condition, and had been fed recently. A likely explanation is that Tricolored Herons, foraging nearby in tidal mangrove streams, did not have difficulty finding food for their nestlings. Adult Snowy Egrets, however, may have been prohibited from reaching their more distant foraging sites because of weather conditions, or foraging sites may have been flooded, making food more difficult to find.

Williams (1997) was able to demonstrate reduced appetite and weight in Great Egret nestlings given methylmercury in their diet at concentrations within the realm of natural exposure in certain parts of the Florida Everglades. In his study the controls were also exposed to elevated mercury concentrations, so the true magnitude of this effect could not be demonstrated.

Newly fledged birds are less efficient in the capture of prey (Recher and Recher 1969), and thus starvation likely contributes significantly to mortality in recently fledged birds. It is also likely that food stress that might not otherwise lead to death in older birds may predispose younger birds to predation or disease.

Migrating Cattle Egrets landing on the Dry Tortugas, March through June 1968, were weak and emaciated and had an unusually high mortality rate. Several factors have been suggested for this mortality, including erratic wind patterns (Browder 1973a) and inability to find food (Harrington and Dinsmore 1975).

XXI. Unexplained Die-offs

A large-scale die-off of Brown Pelicans (145), Snowy Egrets (3) and a single Little Blue Heron occurred late in October of 1988 on an island off Port Orange in the Halifax River, Volusia County (Spalding et al. 1988). When first discovered, many pelicans, mostly immature, had maggot-infested eyes and heads, and weakened birds were wandering on the island. The herons and egrets did not have the eye and skin lesions. The possibility of a toxin applied from the air, such as fuel or pesticides, was considered but could not be confirmed. Birds on adjacent islands were not affected (see chapter 4, Pelicans, for additional details).

XXII. Summary and Conclusions

A drastic decline in the numbers of ardeids in Florida during the latter 1900s has been well documented. Although the cause of that decline is poorly understood, a number of factors appear to have contributed. Habitat loss has been significant and degradation of food resources and habitat quality are important contributors. As with other groups of birds, trauma from man-made objects is a commonly observed cause of mortality, but that mortality is unlikely to have significant demographic consequences. Except for the large numbers of birds killed for their plumes, large-scale die-off events have been rare for this group of birds.

The most significant infectious disease is eustrongylidosis, which appears to be strongly tied to nutrient pollution of wetland sites. There is a good possibility that this disease does have effects at the population level, but because it kills mostly nestlings confined to colonies, much of its effect goes unnoticed.

Recently investigators have begun to study the possibility that sublethal concentrations of contaminants are affecting reproduction. This is

Table 6.26. Helminths recorded from Great Egrets in Florida

Class / Species (location)[a]	County/area	Year(s)	No. of birds			Intensity		Data source
			Examined	Positive	%	Mean	Range	
Trematoda								
Apharyngostrigea pipientis (CL,LI,SI)	"South Florida"[f]	1987–97	70	29	41	26	1–240	E
Ascocotyle tenuicollis (LI)	"South Florida"[f]	1987–97	68	37	54	112	1–1,140	E
Ascocotyle gemina (LI)	"South Florida"[f]	1987–97	68	18	26	44	1–260	E
Ascocotyle mcintoshi (SI)	"South Florida"[f]	1987–97	70	4	6	7	1–9	E
Ascocotyle sp. (LI,SI)	Pinellas	NG	NG	—	—	NG	—	A
Clinostomum complanatum (OR,ES)	"South Florida"[f]	1987–97	73	22	30	4	1–35	E
Clinostomum attenuatum (OR,ES)	"South Florida"[f]	1987–97	73	13	18	6	1–36	E
Diasiella diasi (PA)	"South Florida"[f]	1987–97	24	1	4	5		E
Diplostomum ardeae (SI)	"South Florida"[f]	1987–97	70	11	16	27	1–240	E
Echinochasmus dietzevi (SI)	"South Florida"[f]	1987–97	70	21	30	107	1–520	E
Ignavia venusta (KD)	"South Florida"[f]	1987–97	47	3	6	3	1–4	E
Mesorchis denticulatus (SI)	"South Florida"[f]	1987–97	70	9	13	4	1–10	E
Mesostephanus appendiculatoides[b]	experimental		NG	NG	NG	NG	NG	A, B
Microparyphium facetum (CL)	"South Florida"[f]	1987–97	68	18	26	7	1–47	E
Phagicola diminuta[c] (SI)	"South Florida"[f]	1987–97	70	23	33	64	1–3,580	E
Phagicola longa[b]	Pinellas	1958	12	8	67	3	1	C
	experimental		NG	NG	NG	NG	NG	A, B
Phagicola nana[c] (SI)	"South Florida"[f]	1987–97	70	28	40	171	2–1,360	E
Pholeter anterouterus (SI)	"South Florida"[f]	1987–97	70	6	9	13	1–40	E
Posthodiplostomum sp. (SI)	"South Florida"[f]	1987–97	70	46	66	556	1–16,000	E
Posthodiplostomum opisthosicya (SI)	"South Florida"[f]	1987–97	70	30	43	318	1–1,260	E
Posthodiplostomum boydae (SI)	"South Florida"[f]	1987–97	70	7	10	79	1–270	E
Renicola sp. (KD)	"South Florida"[f]	1987–97	47	1	2	24		E
Ribeiroia ondatrae (PR,VE)	"South Florida"[f]	1987–97	103	9	9	7	1–17	E
Cestoda								
Cyclustera ibisae (SI)	"South Florida"[f]	1987–97	70	1	1	1		E
Dendrouterina ardeae (SI)	"South Florida"[f]	1987–97	70	5	7	2	1–3	E

Glossocercus caribaensis (SI)	"South Florida"[f]	1987–97	70	5	7	7	1–26	E
Plerocercoid (SQ)	"South Florida"[f]	1987–97	NG	—	—	NG	—	E
Nematoda								
Acuaria multispinosa (PR,VE)	"South Florida"[f]	1987–97	103	2	2	4	1–6	E
Avioserpens galliardi (PR,VE)	"South Florida"[f]	1987–97	63	3	4	1	1–2	E
Capillaria herodiae (SI)	"South Florida"[f]	1987–97	70	15	21	7	1–21	E
Chandleronema	"South Florida"[f]	1987–97	103	1	1	2	—	E
longigutterata (PR,VE)								
Contracaecum microcephalum	"Florida"	NG	NG	—	—	NG	—	D
Contracaecum multipapillatum	"South Florida"[f]	1987–97	103	79	77	37	1–203	E
(ES,PR,OR,VE)								
Cosmocephalus obvelatus (OR,ES)	"South Florida"[f]	1987–97	73	1	1	2		E
Desportesius invaginatus (GI,GL,PR,VE)	"South Florida"[f]	1987–97	103	6	6	2	1–5	E
Desportesius trianuchae (PR,VE)	"South Florida"[f]	1987–97	103	9	9	26	1–131	E
Desportesius larvae (PR,VE)	"South Florida"[f]	1987–97	103	9	9	1	1–60	E
Desmidocercella numidica	"South Florida"[f]	1987–97	70	15	21	74	1–660	E
(CL,GI,HT,SI,CC)								
Eustrongylides ignotus[d] (PR,VE)	"South Florida"[f]	1987–97	103	39	38	3	1–10	E
Filaria dentata	"Florida"	NG	NG	—	—	NG	—	D
Tetrameres microspina and sp.[e] (PR,VE)	"South Florida"[f]	1987–97	103	39	38	70	1–871	E
Acanthocephala								
Arhythmorhynchus pumiliorostris (SI)	"South Florida"[f]	1987–97	70	8	11	5	1–13	E
Polymorphus brevis (SI)	"South Florida"[f]	1987–97	70	27	39	22	1–180	E

Sources: A = Hutton (1964), B = Hutton & Sogandares-Bernal (1959), C = Hutton & Sogandares-Bernal (1960), D = Walton (1927), E = Sepulveda et al. (1999b).
NG = not given.
a. Location in host: CC = coelomic cavity, CL = cloaca, ES = esophagus, GI = gizzard, GL = gizzard lining, HT = heart, KD = kidney, LI = large intestine, OR = oral cavity, PA = pancreas, PR = proventriculus, SI = small intestine, SQ = subcutaneous, VE = ventriculus.
b. Hutton and Sogandares-Bernal (1959) experimentally infected 2 Great Egrets with *Phagicola longa* and *Mesostephanus appendiculatoides*.
c. *Ascocotyle diminuta* and *A. nana* are synonyms of *Phagicola diminuta* and *P. nana* respectively.
d. See also Table 6.21 for prevalence in nestlings.
e. A complex of 2 species, the second not identified
f. Sepulveda et al. (1999b) examined 106 nestling, fledgling, and adult Great Egrets from Okeechobee, Monroe, Collier, Dade, Pinellas, Broward, Lee, Palm Beach, and Hillsborough counties. Only the counties where each parasite was found were used for calculations.

Table 6.27. Helminths recorded from Snowy Egrets in Florida

Class Species (location)[a]	County/area	Year(s)	No. of birds			Intensity		Data source
			Examined	Positive	%	Mean	Range	
Trematoda								
Apatemon gracilis	"Florida"	1967–71	NG	NG	—	NG	NG	A
Ascocotyle gemina (CL)	Dade	1997	1	1	—	43	—	B
Ascocotyle tenuicollis (LI)	Dade	1997	1	1	—	90	—	B
Clinostomum complanatum (OR)	Dade	1997	1	1	—	3	—	B
Clinostomum attenuatum (OR,ES)	Dade	1997	1	1	—	3	—	B
Diplostomum ardeae (SI)	Dade	1997	1	1	—	1	—	B
Echinochasmus dietzevi (SI)	Dade	1997	1	1	—	30	—	B
Echinochasmus sp. (SI)	Monroe	1990	1	1	—	2	—	B
Mesorchis denticulatus (SI)	Dade	1991	1	1	—	5	—	B
Microparyphium facetum (CL)	Dade	1997	1	1	—	2	—	B
Phagicola nana (SI)	Dade	1997	1	1	—	5	—	B
Posthodiplostomum macrocotyle (SI)	Dade	1997	1	1	—	30	—	B
Posthodiplostomum opisthosicya (SI)	Dade	1997	1	1	—	60	—	B
Cestoda								
Valipora mutabile (SI)	Monroe	1990	1	1	—	10	—	B
Nematoda								
Capillaria sp. (CL)	Dade	1997	1	1	—	1	—	B
Contracaecum sp. (ES)	Collier	1997	2	2	—	2	—	B
Contracaecum spp.[c] (OR,VE,SI,ES)	Dade	1997	1	1	—	78	—	B
Eustrongylides ignotus[b] (VE, ME)	Dade	1997	1	1	—	2	—	B
Tetrameres sp. (PR)	Glades	1971	1	1	—	1	1	C

Sources: A = Palmieri (1973), B = Spalding & Kinsella (1997), C = Mollhagen (1976).
NG = not given
a. Location in host: PR = proventriculus, SI = small intestine, CL = cloaca, LI = large intestine, OR = oral cavity, ES = esophagus, ME = mesenteries, VE = ventriculus.
b. See Table 6.21 for prevalence in nestlings. Only this 1 adult was positive, of the 20 adult Snowy Egrets examined from all of Florida.
c. Larvae and adults only; males present are C. *multipapillatum*.

Table 6.28. Helminths recorded from Little Blue Herons in Florida

| Class | | | | No. of birds | | | Intensity | | Data |
Species (location)[a]	County/area	Year(s)	Examined	Positive	%	Mean	Range		source
Trematoda									
Apharyngostrigea multiovata (PR/VE,SI)	"South Florica"[c]	1970–92	17	4	24	4	2–9		A
Apharyngostrigea simplex (SI)	"South Florica"[c]	1970–92	17	1	6	1	—		A
Ascocotyle sp. (CL)	"South Florida"[c]	1970–92	13	1	8	14	—		A
Ascocotyle gemina (PR/VE,SI,LI,CL)	"South Florida"[c]	1970–92	13	3	23	7	1–18		A
Ascocotyle tenuicollis (SI)	Dade	NG	NG	NG	—	NG	NG		B
Clinostomum complanatum (OR/ES)	"South Florida"[c]	1970–92	19	5	26	2	1–2		A
Echinochasmus donalsoni (SI,LI,CL)	"South Florida"[c]	1970–92	13	4	31	456	9–1770		A
Microphallus turgidus (SI)	"South Florida"[c]	1970–92	17	2	12	15	10–19		A
Phagicola diminuta (SI)	"South Florida"[c]	1970–92	17	3	18	38	5–100		A
Phagicola nana[b] (SI,LI,CL)	"South Florida"[c]	1970–92	13	4	31	49	22–90		A
Pholeter anterouterus (SI)	"South Florida"[c]	1970–92	17	2	12	2	1–2		A
Posthodiplostomum macrocotyle (PR/VE,SI,LI,CL)	"South Florida"[c]	1970–92	13	1	8	120	2–503		A
Prosthogonimus ovatus (SI)	"South Florida"[c]	1970–92	17	1	6	1	—		A
Ribeiroia ondatrae (PR/VE)	"South Florida"[c]	1970–92	35	1	3	3	—		A
Cestoda									
Taenia sp.	"Florida"	NG	NG	NG	—	NG	—		C
Unidentified protoglottids (SI)	"South Florida"[c]	1970–92	17	4	24	2	1–2		A

(continued)

Table 6.28. *(continued)*

Class Species (location)[a]	County/area	Year(s)	No. of birds			Intensity		Data source
			Examined	Positive	%	Mean	Range	
Nematoda								
Acuaria multispinosa (PR/VE)	"South Florida"[e]	1970–92	35	1	3	1	—	A
Capillaria mergi (SI)	"South Florida"[e]	1970–92	17	1	6	3	—	A
Chandleronema longigutturata (PR/VE)	"South Florida"[e]	1970–92	35	1	3	30	—	A
Contracaecum spp.[c] (PR/VE)	"South Florida"[e]	1970–92	35	24	69	25	1–236	A
Contracaecum microcephalum	"Florida"	NG	NG	NG	—	NG	—	D
Eustrongylides ignotus[d] (PR/VE)	"South Florida"[e]	1970–92	35	7	20	4	1–10	A
Syncuaria sp. (OR/ES)	"South Florida"[e]	1970–92	19	1	5	1	—	A
Tetrameres sp. (OR/ES,PR/VE)	"South Florida"[e]	1970–92	20	7	35	9	1–42	A
Acanthocephala								
Neoechinorhynchus sp. (PR/VE)	"South Florida"[e]	1970–92	35	2	6	1	—	A
Southwellina sp. (SI/LI)	"South Florida"[e]	1970–92	17	1	6	2	—	A

Sources: A = Sepulveda et al. (1996), B = Leigh (1956), C = Stiles & Hassall (1894), D = Walton (1927).
NG = not given.
a. Location in host: CL = cloaca, LI = large intestine, OR/ES = oral cavity/esophagus, PR/VE = proventriculus/ventriculus, SI = small intestine.
b. Metacercariae of *P. nana* have been described in sunfish from Seminole County by Font et al. (1984).
c. A complex of larvae and adults of 2 species, C. *multipapillatum* and C. *microcephalum*.
d. See also Table 6.21 for prevalence in nestlings.
e. Sepulveda et al. (1996) examined 4 Little Blue Herons from the Everglades in 1970 and 31 (April 1987–August 1992) from Polk (7), Okeechobee (4), Palm Beach (1), Collier (1), Dade (6), and Monroe (12) counties. These data are taken from the same birds reported in Spalding et al. (1993).

Table 6.29. Helminths recorded from Tricolored Herons in Florida

| Class | | | No. of birds | | | Intensity | | Data |
Species (location)[a]	County/area	Year(s)	Examined	Positive	%	Mean	Range	source
Trematoda								
Ascocotyle tenuicollis	Dade	NG	NG	—	—	NG	—	A
Ascocotyle sp.	Pinellas	NG	NG	—	—	NG	—	B
Phagicola diminuta[b]	Pinellas	1958	2	1	—	100s	—	C
Posthodiplostomum sp. (IN)	Wakulla	NG	NG	—	—	NG	—	D
Nematoda								
Contracaecum microcephalum (ST)	"Florida"	NG	NG	—	—	NG	—	E
Eustrongylides ignotus (CC,ST)[c]	Monroe	1990	exp.	NG	—	NG	—	G
	Polk	1991	1	1	—	11	—	F
	Broward	1990	1	1	—	1	—	G

Sources: A = Leigh (1956), B = Hutton (1964), C = Hutton & Sogandares-Bernal (1960), D = Loftin (1961), E = Walton (1927), F = Spalding (1988), G = Spalding (1990).
NG = not given
a. Location in host: CC = coelomic cavity, IN = intestine, ST = stomach.
b. *Parascotyle diminuta* is a synonym of *Phagicola diminuta*.
c. See also Table 6.21 for prevalence in nestlings. Spalding & Forrester (1993) experimentally infected Tricolored Herons.

Table 6.30. Helminths recorded from a single Reddish Egret from Monroe County, Florida, in 1978

Class	Species (location)[a]	Intensity
Trematoda	*Apharyngostrigea multiovata* (SI)	3
	Bolbophorus confusus (SI)	5
	Mesoophorodiplostomum pricei (SI)	21
Cestoda	*Glossocercus caribaensis*[b] (SI)	98
Nematoda	*Capillaria* sp.(SI)	3
	Contracaecum multipapillatum (ES,GI,PR)	44
	Contracaecum rudolphi[c] (ES,GI,PR)	22
	Contracaecum sp. (imm.) (ES,GI,PR)	52
	Desportesius invaginatus[d] (PR)	19
	Tetrameres sp.(PR)	22
Acanthocephala	*Arhythmorhynchus pumiliorostris* (SI)	2

Source: Conti et al. (1986).
a. Location in host: ES = esophagus, GI = gizzard, PR = proventriculus, SI = small intestine.
b. According to Schmidt (1986), *Parvitaenia heardi* is a synonym of *G. caribaensis*.
c. *C. spiculigerum* is a synonym of *C. rudolphii*.
d. *Synhimantus invaginatus* is a synonym of *D. invaginatus*.

Table 6.31. Helminths recorded from Cattle Egrets in Florida, Georgia, and Alabama[c]

Class Species (location)[a]	County	Year(s)	No. of birds			Intensity	
			Examined	Positive	%	Mean	Range
Trematoda							
Apatemon gracilis (SI)	Clay	NG	30	6	20	3	NG
Clinostomum complanatum (OR)	Clay	NG	30	1	3	1	NG
Nephrostomum ramosum[d] (SI)	Jackson, Clay	NG	30	12	40	4	NG
Nematoda							
Amplicaecum sp. (ST)	Montgomery	NG	30	1	3	3	NG
Desportesius invaginatus[b,d] (PR)	Jackson, Clay, Montgomery	NG	30	14	47	6	NG
Habronema sp. (ST)	Montgomery	NG	30	1	3	1	NG
Hadjelia sp.[d] (ST)	Jackson	NG	30	1	3	1	NG
Microtetrameres spiralis[d] (PR)	Jackson, Clay, Houston, Montgomery	NG	30	22	73	200+	NG
Physaloptera sp.[d] (ST)	Jackson, Clay, Montgomery	NG	30	9	30	2	NG
Tetrameres cochleariae (PR)	Clay	NG	30	1	3	13	NG

Source: Stuart et al. (1972).

NG = not given.

a. Location in host: OR = oral cavity, PR = proventriculus, SI = small intestine, ST = stomach.

b. *Synhimantus invaginatus* is a synonym of *D. invaginatus*.

c. Thirty birds were collected from Jackson County, Florida; Montgomery and Houston counties, Alabama; and Clay County, Georgia. Because of their close proximity, all of the data are presented.

d. Collected in Florida.

Table 6.32. Helminths recorded from Green Herons in Florida

Class / Species (location)[a]	County/area	Year(s)	No. of birds		Intensity		Data source
			Examined	Positive	Mean	Range	
Trematoda							
Ascocotyle leighia (SI)	Alachua	1971	2	1	10	—	A
Posthodiplostomum macrocotyle (SI)	Alachua	1971	2	1	8	—	A
Cestoda							
Dilepis unilateralis[b] (SI)	"Florida"	NG	NG	1	—	—	B
Valipora mutabile (SI)	Pinellas	1997	1	1	17	—	C
Nematoda							
Acuaria multispinosa (PR)	Alachua	1971	2	1	2	—	A
Capillaria sp. (SI)	Alachua	1971	2	2	1	1–2	A
Cardiofilaria pavlovskyi (LS)	Pinellas	1997	1	1	1	—	C
Chandleronema longigutturata (PR)	Alachua	1971	1	1	2	—	A
Contracaecum microcephalum (ST)	"Florida"	NG	NG	—	—	—	D
Contracaecum micropapillatum (ST)	Alachua	1971	2	2	14	14–15	A
Desportesius invaginatus (PR)	Alachua	1971	2	1	2	—	A
Eustrongylides ignotus (ST)	Hillsborough	1989	9	1	2	—	E
Physaloptera sp. (PR)	Pinellas	1997	1	1	2	1	C
Tetrameres sp.[c] (PR)	Alachua	1971	1	1	4	—	A
	"Florida"	NG	NG	—	—	—	F
larval spiurids (SI)	"Florida"	1971	2	1	1	—	A

Sources: A = Kinsella (1971), B = Stiles & Hassall (1894), C = Spalding & Kinsella (1997), D = Walton (1927), E = Spalding et al. (1993), F = Mollhagen (1976).
NG = not given.

a. Location in host: LS = lower small intestine, PR = proventriculus, SI = small intestine, ST = stomach.
b. *Taenia unilateralis* is a synonym of *D. unilateralis*.
c. Described by Mollhagen (1976) as *T. butorides* but is not a valid species since it was not published.

Table 6.33. Helminths recorded from Black-crowned Night Herons in Florida

Class Species (location)[a]	County/area	Year(s)	No. of birds			Intensity		Data source
			Examined	Positive	%	Mean	Range	
Trematoda								
Ascocotyle sp. (LI)	Pinellas	NG	NG	—	—	NF	—	A
	Collier	1994	1	1	—	1	—	B
Bolbophorus confusus (SI)	Pinellas	NG	NG	—	—	NG	—	A
Mesostephanus appendiculatoides (SI)[b]	experimental							C
Mesostephanus sp. (SI)	Pinellas	NG	NG	—	—	NG	—	A
Phagicola longa (SI)	Pinellas	NG	NG	—	—	NG	—	A
	experimental							C
Posthodiplostomum nanum (LI,SI)	Collier	1994	1	1	—	9	—	B
Pseudascocotyle mollienesicola (CE,SI)	Collier	1994	1	1	—	74	—	B
Pygiodopsis pindoramensis (SI)	Collier	1994	1	1	—	419	—	B
Nematoda								
Capillaria contorta (ES)	Collier	1997	1	1	—	1	—	E
Contracaecum microcephalum	"Florida"	NG	NG	—	—	NG	—	D
Contracaecum	Collier	1994	1	1	—	15	—	B
multipapillatum (ES,PR)	Collier	1997	1	1	—	114	—	E
Contracaecum rudolphi (ES,GI,PR)[c]	Pinellas	NG	NG	—	—	NG	—	A
Contracaecum sp. (ST)	Dade	1918	NG	—	—	NG	—	F
Eustrongylides ignotus (ST)	Collier	1994	1	1	—	1	—	B
	Collier	1997	1	1	—	1	—	E
	Monroe	1991	3	1	1	—	—	G
	Orange	1999	1	1	1	—	—	H
Larval spiruids (ST)	Collier	1997	1	1	—	10	—	E
Tetrameres sp. (PR)	Collier	1994	1	1	—	3	—	B
Acanthocephala								
Polymorphus brevis (ES,SI,ST)	Collier	1994	1	1	—	31	—	B
	Collier	1997	1	1	—	291	—	E

Sources: A = Hutton (1964), B = Spalding & Kinsella (1995), C = Hutton & Sogandares-Bernal (1959), D = Walton (1927), E = Spalding & Kinsella (1997), F = U.S. National Parasite Collection, number 26146, G = Spalding et al. (1993), H = Spalding (1999).

NG = not given.

a. Location in host: CE = cecum, ES = esophagus, GI = gizzard, LI = large intestine, PR = proventriculus, SI = small intestine, ST = stomach.

b. Hutton & Sogandares-Bernal (1959) experimentally infected a Black-crowned Night Heron with *Phagicola longa* and *Mesostephanus appendiculatoides*.

c. *C. spiculigerum* is a synonym of *C. rudolphii*.

Table 6.34. Helminths recorded from Yellow-crowned Night Herons in Florida

Class Species (location)[a]	County	Year(s)	No. of birds		Mean intensity	Data source
			Examined	Positive		
Trematoda						
Ascocotyle gemina (SI,LI)	Pinellas	1989	1	1	868	B
Brachylaima mcintoshi (ST)	Pinellas	1975	2	1	8	A
Clinostomum complanatum (OR)	Pinellas	1989	1	1	17	B
Microphallus turgidus (SI)	Lee	1989	1	1	500	B
Philophthalmus hegeneri[b] (CJ)	Manatee to Charlotte	NG	NG	—	—	C
Posthodiplostomum nanum (SI)	Pinellas	1989	1	1	870	B
Pygiodopsis pindoramensis (SI)	Pinellas	1989	1	1	190	B
Cestoda						
Glossocercus caribaensis[c] (SI,LI)	Pinellas	1975	2	1	69	A
	Pinellas	1989	1	1	234	B
Unidentified cestode (SI)	Lee	1989	1	1	1	B
Nematoda						
Acuaria microspinosa (PR)	Pinellas	1989	1	1	3	B
Acuaria multispinosa (PR)	Pinellas	1975	2	1	2	B
Capillaria sp. (SI)	Lee	1989	1	1	1	B
Contracaecum larvae (SI,GI)	Pinellas	1989	1	1	31	B
Contracaecum micropapillatum (GI)	Pinellas	1975	2	1	31	A
	Pinellas	1989	1	1	47	B

Desportesius invaginatus (PR)	Pinellas	1989	1	1	4	B
Eustrongylides ignotus (ST)	Pinellas	1989	1	1	1	B
	"South Florida"	1987–91	6	1	1	D
Strongyloides sp. (SI,LI)	Lee	1989	1	1	234	B
	Pinellas	1975	2	1	1	A
Synhimantus	Lee	1989	1	1	165	B
magnipapillatus (ES,PR,SI)	Pinellas	1989	1	1	6	B
Tetrameres micropenis (PR)	Lee	1989	1	1	5	B
Tetrameres microspinosa (PR)	Lee	1989	1	1	1	B
	Pinellas	1989	1	1	6	B
Uncinaria sp.[d] (SI)	Lee	1989	1	1	27	B
Acanthocephala						
Southwellina hispida (SI)	Pinellas	1975	2	1	190	A
	Pinellas	1989	1	1	24	B

Sources: A = Forrester & Kinsella (1996), B = Spalding & Kinsella (1996), C = Penner & Fried (1963), D = Spalding et al. (1993).
NG = not given.
a. Location in host: CJ = conjunctiva, ES = esophagus, GI = gizzard, LI = large intestine, OR = oral cavity, PR = proventriculus, SI = small intestine, ST = stomach.
b. Penner & Fried (1963) described this species from adults reared from cercariae in the marine snail *Batillaria minima* from the southwest coast of Florida.
c. According to Schmidt (1986), *Parvitaenia heardi* is a synonym of *G. caribaensis*.
d. This record and one from Clapper Rails are the only records of hookworms in birds.

Table 6.35. Mites reported from ardeids in Florida

Host species	Ectoparasite species	County	Year(s)	Data source
American Bittern	*Ardeacarus ardeae*[a]	Collier	1998	A
Black-crowned Night-Heron	*Megniniella* sp.[b]	Collier	1994	A
	Metanalges sp.[b]	Collier	1994	A
	Neomyssus sp.	Pinellas	1955–62	B
	Speleognathus sp.	Pinellas	1955–62	B
	Taeniosikya sp.[b]	Collier	1994	A
Cattle Egret	*Neomyssus* sp.	Lafayette	1975	C
	Uropodidae deutonymphs	Duval	1972	C
Great Blue Heron	*Ardeacarus ardeae*[a]	Monroe	1995	D
		Alachua	1994	D
		Palm Beach	1994	D
	Ardeialges sp.	Alachua	1980	E
		Monroe	1995	D
		Alachua	1994	D
	Dermation (*Neodermation*) sp.	Alachua	1981	E
	Scutomegninia sp.	Monroe	1995	D
Great Egret	*Ardeacarus ardeae*[a]	Collier	1994	D
		Dade	1994	D
		Monroe	1994	D
	Ardeialges sp.	Collier	1994	D
		Dade	1994	D
		Palm Beach	1994	D
	Dermation (*Neodermation*) sp.	Collier	1994	D
	Scutomegninia sp.	Collier	1995	D
Great White Heron	*Ardeacarus ardeae*[a]	Monroe	1994	D
			1995	D
	Ardeialges sp.	Monroe	1995	D
	Scutomegninia sp.	Monroe	1995	D
Least Bittern	*Ardeacarus ardeae*[a]	Dade	1994	A
	Herodialges pentimantus	Dade	1994	D
	Pteralloptes stellaris	Dade	1994	D
	Pteroclichidae	Dade	1994	D
Snowy Egret	*Ardeialges* sp.	Dade	1994	D

Sources: A = Spalding & Mertins (1999), B = Hutton (1964), C = Forrester & Atyeo (1992), D = Sepulveda & Mertins (1995), E = Forrester & Mertins (1995).

a. This genus contains only 1 described species of feather mite found on more than a half-dozen genera of ardeid birds worldwide. There are subtle differences among the mites, and they may be different species.

b. Usually found on storks and rails and may be contaminants.

Table 6.36. Subcutaneous mites collected from ardeids in Florida

Host species / Parasite species	County/area	Year(s)	No. of birds Examined	Positive	Data source
Black-crowned Night Heron					
Hypodectes nycticoracis	Volusia	1987	1	1	A
Great Blue Heron					
Hypodectes propus bubulci	Hillsborough	1989–90	2	2	A
	Levy	1991	1	1	A
	Pinellas	1990	2	2	A
Neottialges montagui	Pinellas	1990	2	1	B
Great Egret					
Hypodectes propus bubulci	Broward	1992	1	1	A
	Collier	1988	3	3	A
	Dade	1989	1	1	A
	"Florida Bay"	1988	1	1	A
	Hillsborough	1989	4	4	A
	"Lake Okeechobee"	1989	1	1	A
	Lee	1989	1	1	A
	Pinellas	1989–90	3	3	A
Neottialges evansi	Lee	1989	1	1	B
Neottialges kutzeri	Glades	1989	1	1	B
Great White Heron					
Hypodectes propus bubulci	Collier	1989	1	1	A
	Dade	1987–88	2	2	A
	"Florida Bay"	1987–90	9	9	A
	Hendry	1988	1	1	A
Greenbacked Heron					
Hypodectes propus bubulci	"Florida Bay"	1988	1	1	A
Little Blue Heron					
Hypodectes propus bubulci	"Florida Bay"	1989	5	4	B
	Pinellas	1989	1	1	A
Neottialges ajajae	"Florida Bay"	1989	5	1	A
Snowy Egret					
Hypodectes propus bubulci	Broward	1990	1	1	A
	Collier	1988	1	1	A
	Lee	1989	1	1	A
	Pinellas	1989	1	1	A
	Polk	1988	1	1	A
Yellow-crowned Night Heron					
Hypodectes nycticoracis	Hillsborough	1989	1	1	B
	Lee	1989	1	1	A
Hypodectes propus bubulci	Lee	1989	1	1	A

Sources: A = Spalding & Pence (1996), B = Pence et al. (1997).

Table 6.37. Chewing lice reported from ardeids in Florida

Host species Ectoparasite species	County	Year(s)	Data source
Black-crowned Night-Heron			
Ardeicola goisagi	Collier	1994	A
	Lee	1989	B
Ciconiphilus decimfasciatus	Collier	1994	A
	St. Lucie	1966	B
Eulaemobothrion atrum[a]	Collier	1994	A
Pseudomenopon sp.[a]	Collier	1994	A
Cattle Egret			
Ardeicola expallidus	Palm Beach	1989	B
Ardeiphilus sp.	Indian River	1972	B
Ciconiphilus decimfasciatus	Indian River	1969, 1972	B
	Palm Beach	1989	B
Great Blue Heron			
Ardeicola cruscula	Collier	1994	B
	Glades	1976	B
	Monroe	1995	B
Ciconiphilus decimfasciatus	Alachua	1994	B
	Brevard	1969	B
	Collier	1993–94	B
	Lee	1985	B
	Monroe	1966, 1988–89, 1995	B
	Okeechobee	1981	B
	Osceola	1993	B
Great Egret			
Ardeicola expallidus	Collier	1995	B
	Dade	1993–94	B
	Monroe	1987–89, 1994	B
	Okeechobee	1990	B
Ciconiphilus decimfasciatus	Collier	1995	B
	Dade	1987, 1989, 1993–94	B
	Monroe	1987–89	B
	Okeechobee	1990	B
	Palm Beach	1990	B
Great White Heron			
Ardeicola cruscula	Collier	1994	B
	Monroe	1995	B
Ardeicola sp.	Collier, Dade,		
	Monroe, Palm Beach	1988	B
Ciconiphilus decimfasciatus	Collier, Dade	1988	B
	Monroe	1987–89, 1994–95	B
Green Heron			
Ciconiphilus butoridiphagus	Indian River	1973	B
Little Blue Heron			
Ardeicola florida	Monroe	1987–89	B
	Palm Beach	1990	B
	Lee	1983	B

(continued)

Table 6.37. *(continued)*

Host species Ectoparasite species	County	Year(s)	Data source
Ardeiphilus sp.	Indian River	1972	B
Ciconiphilus decimfasciatus	Collier	1992	B
	Dade	1994	B
	Indian River	1966, 1972, 1974	B
	Lee	1979	B
	Monroe	1988	B
	St. Lucie	1966	B
Reddish Egret			
Ardeicola florida nigra	Monroe	1978, 1987–89	B,C
Ciconiphilus decimfasciatus	Monroe	1978	C
Snowy Egret			
Ardeicola expallidus	Collier	1993	B
	Dade	1989	B
	Monroe	1987–88, 1990	B
	Okeechobee	1989–90	B
Ciconiphilus decimfasciatus	Collier	1993	B
	Glades	1990	B
	Indian River	1967–69, 1972	B
	Lee	1979	B
	Monroe	1987–88	B
	Okeechobee	1989–90	B
	St. Lucie	1966	B
Tricolored Heron			
Ardeicola florida nigra	Dade	1989	B
	Glades	1990	B
	Indian River	1966	B
	Lee	1983	B
	Monroe	1987–90	B
Ciconiphilus decimfasciatus	Indian River	1968, 1972	B
	Lee	1982–83	B
	Monroe	1970, 1987–90	B
	St. Lucie	1970	B

Sources: A = Spalding & Mertins (1999), B = Forrester et al. (1995), C = Conti et al. (1986).
a. Usually found on cranes and coots and suspected to be contaminants.

Table 6.38. Dipterans found on ardeids in Florida

Host species	Ectoparasite species	County or area	Year(s)	Data source
Cattle Egret	*Icosta albipennis*[a]	Indian River	1973	A
		Palm Beach	1964	B
	Ornithoica confluenta	"Florida"	1976	C
		Alachua	1992	D
		"Central Florida"	NG	B
		Duval	1972	A
		Glades	NG	E
		Hendry	NG	E
		Marion	1966	E
		Palm Beach	1972	A
		Sumter	NG	E
Great Blue Heron	*Icosta albipennis*[a]	Dade	1987	F
		Levy, Hendry	1991	F
		Indian River	1971	A
		Alachua, Glades	1976	C
		"Florida"	NG	G
Great Egret	*Icosta albipennis*[a]	Dade	1989	F
		Collier	1987	F
	Ornithoica confluenta	Monroe, Dade	1989	F
		St. Lucie	1971	A
		"Florida"	1863	H
Great White Heron	*Icosta albipennis*[a]	Monroe	1987–88	F
		"Bowlegs Key"	NG	H
		St. Johns	1887	I,H
	Ornithoica confluenta	Monroe	1988	F
Green Heron	*Icosta albipennis*[a]	St. Lucie	1972	A
Little Blue Heron	*Icosta albipennis*[a]	Indian River	1973	A
		St. Lucie	1966	A
		Sarasota	NG	H
	Ornithoica confluenta	Indian River	1970	A
		St. Lucie	1966	A
Reddish Egret	*Icosta albipennis*[a]	"Florida Bay"	1978	J
Snowy Egret	*Icosta albipennis*[a]	Collier	1988	F
		Indian River	1968	A
		St. Lucie	1967	A
	Ornithoica confluenta	Collier	1988	F
		Indian River	1974	A
		St. Lucie	1966	A
Yellow-crowned Night-Heron	*Ornithoica confluenta*	Hillsborough	1989	F
Host not specified	*Icosta ardeae botaurinorum*[b]	Dade	1962	A
		"Florida"	NG	G

Sources: A = Wilson et al. (1990), B = Funderburg et al. (1968), C = Forrester & Wilson (1990), D = Forrester & Wilson (1992), E = Bradley & Kirkham (1966–67), F = Spalding & Wilson (1992), G = Maa (1969), H = Bequaert (1954), I = Johnson (1913), J = Conti et al. (1986).
NG = not given.
a. *Icosta albipennis* = *Lynchia albipennis* = *Olfersia albipennis*.
b. Normal host = American Bittern.

difficult to demonstrate, especially since there are so many types of contaminants, the source is widespread for migratory adults, and effects on adults are difficult to measure. Most of the work to date has focused on methylmecury, which is found in particularly high concentrations in the aquatic food web in the Everglades. There has clearly been a decline in the concentrations of organochlorines in eggs since the use of these chemicals has been mostly banned, but these persistent compounds continue to be detected and cause mortality in specific sites.

Ardeids have been relatively well studied in Florida. We have the best information for Great Blue Herons and Great Egrets, good information for Snowy and Cattle Egrets, Little Blue and Tricolored Herons, and little information for Reddish Egrets, Green Herons, bitterns, and night-herons. We documented a large array of organochlorine and metal contaminants, only 1 case of organophosphate poisoning, 3 viruses, 3 biotoxins, 20 bacteria, 6 protozoa, 50 trematodes, 11 cestodes, 42 nematodes, 8 acanthocephalans, 19 mites, 10 chewing lice, 2 louse flies, and 1 beetle.

XXIII. Literature cited

Anderson, D.W., and J.J. Hickey. 1972. Eggshell changes in certain North American birds. *Proc. Int. Ornithol. Congr.* 15:514–540.

Arendt, W.J. 1988. Range expansion of the cattle egret (*Bubulcus ibis*) in the greater Caribbean basin. *Colon. Waterbirds* 11:252–262.

Bancroft, G.T., and A.M. Strong. 1996. Yellow-crowned night heron. In: *Rare and endangered biota of Florida*. Vol. 5, *Birds*. J.A. Rodgers, Jr., H.W. Kale II, and H.T. Smith (eds.). University Press of Florida, Gainesville. pp. 450–456.

Below, T.H. 1996. Black-crowned night heron. In: *Rare and endangered biota of Florida*. Vol. 5, *Birds*. J.A. Rodgers, Jr., H.W. Kale II, and H.T. Smith (eds.). University Press of Florida, Gainesville. pp. 442–449.

Bent, A.C. 1937. *Life histories of North American birds of prey*. Part 1. U.S. Natl. Mus. Bull. 167. 409 pp.

Bequaert, J. 1954. The Hippoboscidae or louse-flies (Diptera) of mammals and birds. Part II. Taxonomy, evolution and revision of American genera and species. *Entomol. Am.* 34:1–232.

Beyer, W.N., M. Spalding, and D. Morrison. 1997. Mercury concentrations in feathers of wading birds from Florida. *Ambio* 26:97–100.

Bigler, W.J., J.B. Nichols, and T.L. Landers. 1967. A serological survey of the old world cattle egret. *Southwest. Vet.* 20:99–100.

Black, B.B., M.W. Collopy, H.F. Percival, A.A. Tiller, and P.G. Bohall. 1984. Effects of low level training flights on wading bird colonies in Florida. Florida Coop. Fish and Wildl. Res. Unit, Tech. Rep. 7. 189 pp.

Botzler, R.G. 1991. Epizootiology of avian cholera in wildfowl. *J. Wildl. Dis.* 27:367–395.

Bouton, S.N., P.C. Frederick, M.G. Spalding, and H. McGill. 1999. Effects of chronic, low concentrations of dietary methylmercury on the behavior of juvenile great egrets. *Environ. Toxicol. Chem.* 18:1934–1939.

Bradley, R.E., and W.W. Kirkham. 1966–1967. The possible role of the cattle egret (*Bubulcus ibis*) as a reservoir of animal diseases in Florida. *Annu. Rep. Fla. Agric. Exp. Stn.*: 206.

Brand, C.J. 1984. Avian cholera in the Central and Mississippi Flyways during 1979–80. *J. Wildl. Manag.* 48:399–406.

Brand, C.J., and R.M. Duncan. 1983. Avian cholera in an American flamingo, *Phoenicopterus ruber*: a new host record. *Calif. Fish Game* 69:190–191.

Brand, C.J., R.M. Windingstad, L.M. Siegfried, R.M. Duncan, and R.M. Cook. 1988. Avian morbidity and mortality from botulism, aspergillosis, and salmonellosis at Jamaica Bay Wildlife Refuge, New York, U.S.A. *Colon. Waterbirds* 11:284–292.

Brogden, K.A., and K.R. Rhoades. 1983. Prevalence of serologic types of *Pasteurella multocida* from 57 species of birds and mammals in the United States. *J. Wildl. Dis.* 19:315–320.

Browder, J.A. 1973a. Long-distance movements of cattle egrets. *Bird-Banding* 44:158–170.

———. 1973b. Studies on the feeding ecology and morphological variation of the cattle egret *Bubulcus ibis* (Linnaeus) (Aves: Ardeidae).

Master's thesis, University of Miami, Coral Gables, Fla.

———. 1978. A modeling study of water, wetlands and wood storks. In: *Wading Birds.* A. Sprunt, J.C. Ogden, S. Winkler (eds.). Research Report no. 7, National Audubon Society, New York. pp. 325–347.

Burns, B.J. 1952. Food of a family of great horned owls, *Bubo virginianus,* in Florida. *Auk* 69:86–87.

Burridge, M.J., W.J. Bigler, D.J. Forrester, and J.M. Hennemann. 1979. Serologic survey for *Toxoplasma gondii* in wild animals in Florida. *J. Am. Vet. Med. Assoc.* 175:964–967.

Butler, R.W. 1997. *The great blue heron.* University of British Columbia Press, Vancouver. 167 pp.

Call, D.J., H.J. Shave, H.C. Binger, M.E. Bergeland, B.D. Ammann, and J.J. Worman. 1976. DDE poisoning in the wild great blue heron. *Bull. Environ. Contam. Toxicol.* 16:310–313.

Callahan, R., and W. Carey. 1979. Great horned owl suspected of preying upon snowy egret. *Fla. Field Nat.* 7:29.

Conti, J.A., D.J. Forrester, and R.T. Paul. 1986. Parasites and diseases of reddish egrets (*Egretta rufescens*) from Texas and Florida. *Trans. Am. Microsc. Soc.* 105:79–82.

Conti, L., R. Oliveri, and C. Blackmore. 2002. Unpublished data. Florida Department of Health, Tallahassee.

Corn, J.L., N. Barre, B. Thiebot, T.E. Creekmore, G.I. Garris, and V. Nettles. 1993. Potential role of cattle egrets, *Bubulcus ibis* (Ciconiformes: Ardeidae), in the dissemination of *Amblyomma variegatum* (Acari: Ixodidae) in the eastern Caribbean. *J. Med. Entomol.* 30:1029–1037.

Coyner, D.F. 1998. The epizootiology and transmission of *Eustrongylides ignotus* (Dioctophymatoidea) in intermediate hosts in Florida. Ph.D. diss., University of Florida, Gainesville. 245 pp.

Coyner, D.F., S.R. Schaack, M.G. Spalding, and D.J. Forrester. 2001. Altered predation susceptibility of mosquitofish infected with *Eustrongylides ignotus. J. Wildl. Dis.* 3:556–560.

Crawford, R.L. 1981. Bird casualties at a Leon County, Florida TV tower: a 25-year migration study. *Bull. Tall Timbers Res. Stn.* 22:1–30.

Custer, T.W., and B.L. Mulhern. 1983. Heavy metal residues in prefledgling black-crowned night-herons from three Atlantic coast colonies. *Bull. Environ. Contam. Toxicol.* 30:178–185.

Custer, T.W., G.L. Hensler, and T.E. Kaiser. 1983. Clutch size, reproductive success, and organochlorine contaminants in Atlantic coast black-crowned night-herons. *Auk* 100:699–710.

Davidson, W.R., and V.F. Nettles. 1997. *Field manual of wildlife diseases in the southeastern United States.* 2d ed. Southeastern Cooperative Wildlife Disease Study, University of Georgia, Athens. 417 pp.

Davis, W.E., Jr. 1993. Black-crowned night-heron (*Nycticorax nycticorax*). *Birds N. Am.* 74:1–20.

Davis, W.E., Jr., and J.A. Kushlan. 1994. Green heron. *Birds N. Am.* 129:1–24.

Delany, M.F., and A.R. Woodward. 1988. Great blue herons trapped in algae. *Fla. Field Nat.* 16:96.

Delany, M.F., A.R. Woodward, and I.H. Kochel. 1988. Nuisance alligator food habits in Florida. *Fla. Field Nat.* 16:90–96.

Drees, B.M. 1994. Red imported fire ant predation on nestlings of colonial waterbirds. *Southwest. Entomol.* 19:355–359.

Edman, J.D., L.A. Webber, and H.W. Kale II. 1972. Host-feeding patterns of Florida mosquitoes II. Culiseta. *J. Med. Entomol.* 9:429–434.

Edman, J.D., J.F. Day, and E.D. Walker. 1984. Field confirmation of laboratory observations on the differential antimosquito behavior of herons. *Condor* 86:91–92.

Eisler, R. 1985. Cadmium hazards to fish, wildlife, and invertebrates: a synoptic review. U.S. Fish and Wildlife Service Biological Report 85(1.2). 46 pp.

———. 1988. Arsenic hazards to fish, wildlife, and invertebrates: a synoptic review. U.S. Fish and Wildlife Service Biological Report 85(1.12). 92 pp.

Erlandsen, S.L., W.J. Bemrick, C.L. Wells, D.E. Feely, L. Knudson, S.R. Campbell, H. Van Keulen, and E.L. Jarroll. 1990. Axenic culture and characterization of *Giardia ardeae* from the great blue heron (*Ardea herodias*). *J. Parasitol.* 76: 717–724.

Fairbrother, A., L.N. Locke, and G.L. Hoff (eds.). 1996. *Noninfectious diseases of wildlife*. Iowa State University Press, Ames. 219 pp.

Favorite, F.G. 1960. Some evidence of local origin of EEE virus in Florida. *Mosq. News* 20:87–92.

[FGFWFC] Florida Game and Fresh Water Fish Commission. 1992. Effects of Hurricane Andrew on fish and wildlife of southern Florida, a preliminary assessment. Report, September 8, 1992. Tallahassee. 19 pp.

Fischer, J. 1999. Unpublished data. Southeastern Cooperative Wildlife Disease Study, University of Georgia, Athens.

Folk, M.J. 1992. Cooperative hunting of avian prey by a pair of bald eagles. *Fla. Field Nat.* 20:110–112.

Font, W.R., R.M. Overstreet, and R.W. Heard. 1984. Taxonomy and biology of *Phagicola nana* Digenea Heterophyidae. *Trans. Am. Microsc. Soc.* 103:408–422.

Forrester, D.J., and W.T. Atyeo. 1992. Unpublished data. University of Florida, Gainesville.

Forrester, D.J., and G.F. Bennett. 1985. Unpublished data. University of Florida, Gainesville.

Forrester, D.J., H.W. Kale II, R.D. Price, K.C. Emerson, and G.W. Foster. 1995. Chewing lice (Mallophaga) from birds in Florida: a listing by host. *Bull. Fla. Mus. Nat. Hist.* 39:1–44.

Forrester, D.J., and J.M. Kinsella. 1996–99. Unpublished data. University of Florida, Gainesville.

Forrester, D.J., and J.W. Mertins. 1995. Unpublished data. University of Florida, Gainesville.

Forrester, D.J., and N. Wilson. 1974–1992. Unpublished data. University of Florida, Gainesville.

Forrester, D.J., K.C. Wenner, F.H. White, E.C. Greiner, W.R. Marion, J.E. Thul, and G.A. Berkhoff. 1980. An epizootic of avian botulism in a phosphate mine settling pond in northern Florida. *J. Wildl. Dis.* 16:323–327.

Fowler, M. 1986. Storks and flamingos (Ciconiiformes and Phoenicopteriformes). In: *Zoo and wild animal medicine*. 2d ed. M. Fowler (ed.). W.B. Saunders, Philadelphia. pp. 326–331.

[FPL and FGFWFC] Florida Power and Light Company and Florida Game and Fresh Water Fish Commission. 1991. Unpublished correspondence. Lake Harbor bird mortality project: daily inspections of test framing designs.

Frank, A., and K. Borg. 1979. Heavy metals in tissues of the mute swan (*Cygnus olor*). *Acta Vet. Scand.* 20:447.

Franson, J.C. 1999. Chlamydiosis. In: *Field manual of wildlife diseases. General field procedures and diseases of birds*. M. Friend and J. C. Franson (eds.). U.S. Department of the Interior, U.S. Geological Survey, Biological Research Division, Information and Technology Report 1999-001. Washington, D.C. pp. 111–114.

Franson, J.C., and T.W. Custer. 1994. Prevalence of eustrongylidosis in wading birds from colonies in California, Texas, and Rhode Island, USA. *Colon. Waterbirds* 17:168–172.

Frederick, P.C. 1996. Least bittern. In: *Rare and endangered biota of Florida*. Vol. 5, *Birds*. J.A. Rodgers, Jr., H.W. Kale II, and H.T. Smith (eds.). University Press of Florida, Gainesville. pp. 381–387.

———. 1997. Tricolored heron (*Egretta tricolor*). *Birds N. Am.* 306:1–28.

Frederick, P.C., and M.W. Collopy. 1987. Chronic tidally-induced nest failure in a colony of white ibises. *Condor* 89:413–419.

———. 1989a. Researcher disturbance in colonies of wading birds: effects of frequency of visit and egg-marking on reproductive parameters. *Colon. Waterbirds* 12:152–157.

———. 1989b. Nesting success of five ciconiiform species in relation to water conditions in the Florida Everglades. *Auk* 106:625–634.

———. 1989c. The role of predation in determining reproductive success of colonially nesting wading birds in the Florida Everglades. *Condor* 91:860–867.

Frederick, P.C., and M.G. Spalding. 1994. Factors affecting reproductive success by wading birds (Ciconiiformes) in the Everglades ecosystem. In: *The Everglades: the ecosystem and its restoration*. S. Davis and J. Ogden (eds.). St. Lucie Press, Delray Beach, Fla. pp. 659–691.

Frederick, P.C., K.L. Bildstein, B. Fleury and J.C. Ogden. 1996a. Conservation of nomadic populations of white ibis (*Eudocimus albus*) in the United States. *Conserv. Biol.* 10:203–216.

Frederick, P.C., N. Dwyer, S. Fitzgerald, and R.E. Bennetts. 1990. Relative abundance and habitat preferences of least bitterns (*Ixobrychus exilis*) in the Everglades. *Fla. Field Nat.* 18:1–9.

Frederick, P.C., S.M. McGehee, and M.G. Spalding. 1996b. Prevalence of *Eustrongylides ignotus* in mosquito fish (*Gambusia holbrooki*) in Florida: historical and regional comparisons. *J. Wildl. Dis.* 32:552–555.

Frederick, P.C., M.G. Spalding, M.S. Sepulveda, G.E. Williams, L. Nico, and R. Robins. 1999. Exposure of great egret (*Ardea albus*) nestlings to mercury through diet in the Everglades ecosystem. *Environ. Toxicol. Chem.* 18:1940–1947.

Frenkel, J.K. 1981. False-negative serologic tests for *Toxoplasma* in birds. *J. Parasitol.* 67:952–953.

Friend, M. 1999. Aspergillosis. In: *Field manual of wildlife diseases. General field procedures and diseases of birds.* M. Friend and J.C. Franson (eds.). U.S. Department of the Interior, U.S. Geological Survey, Biological Research Division, Information and Technology Report 1999-001. Washington, D.C. pp.129–133.

Friend, M., and J.C. Franson (eds.). 1999. *Field manual of wildlife diseases. General field procedures and diseases of birds.* U.S. Department of the Interior, U.S. Geological Survey, Biological Research Division, Information and Technology Report 1999-001. Washington, D.C. 425 pp.

Funderburg, J.B., M.L. Gilbert, and E.L. Bostelman. 1968(1969). Hippoboscid flies from cattle egrets in central Florida. *Q. J. Fla. Acad. Sci.* 31:141–142.

Georgi, M.E., M.S. Carlisle, and L.E. Smiley. 1986. Giardiasis in a great blue heron (*Ardea herodias*) in New York State: another potential source of waterborne giardiasis. *Am. J. Epidemiol.* 123:916–917.

Gibbs, J.P., S. Melvin, and F.A. Reid. 1992a. American bittern. *Birds N. Am.* 18:1–12.

Gibbs, J.P., F.A. Reid, and S. Melvin. 1992b. Least bittern. *Birds N. Am.* 17:1–12.

Girard, G.T., and W.K. Taylor. 1979. Reproductive parameters for nine avian species at Moore Creek, Merritt Island National Wildlife Refuge, Florida. *Fla. Sci.* 42:94–102.

Gourlie, N. 1984. Unpublished data. Florida Department of Environmental Regulation, Tallahassee.

Harrington, B.A., and J.J. Dinsmore. 1975. Mortality of transient cattle egrets at Dry Tortugas, Florida. *Bird-Banding* 46:7–14.

Hegner, R.W. 1925. *Giardia felis* n. sp. from the domestic cat and giardias from birds. *Am. J. Hyg.* 5:258–273.

Hoffman, D.J., J.C. Franson, O.H. Pattee, C.M. Bunck, and A. Anderson. 1985. Survival, growth, and accumulation of ingested lead in nestling American kestrels (*Falco sparverius*). *Arch. Environ. Contam. Toxicol.* 14:89–94.

Hutton, R.F. 1964. A second list of parasites from marine and coastal animals of Florida. *Trans. Am. Microsc. Soc.* 83:439–447.

Hutton, R.F., and F. Sogandares-Bernal. 1959. Studies of the trematode parasites encysted in Florida mullets. Spec. Sci. Rep. 1, Florida State Board of Conservation Marine Laboratory no. 59-12.

———. 1960. Studies on helminth parasites from the coast of Florida. II. Digenetic trematodes from shore birds of the west coast of Florida. *Bull. Mar. Sci. Gulf Caribb.* 10:40–54.

Jasmin, A.M., D.E. Cooperrider, C.P. Powell, and J.N. Baucom. 1972. Enterotoxemia of wildfowl due to *Cl. perfringens* type C. *J. Wildl. Dis.* 8:79–84.

Jenni, D.A. 1969. A study of the ecology of four species of herons during the breeding season at Lake Alice, Alachua County, Florida. *Ecol. Monogr.* 39:245–269.

Jennings, M.L., and D.P. Jennings. 1982. Bald eagle preys on cattle egret. *Fla. Field Nat.* 10:60.

Jennings, W.L., W.G. Winkler, D.D. Stamm, P.H. Coleman, and A.L. Lewis. 1969. Serologic studies of possible avian or mammalian reservoirs of St. Louis encephalitis virus in Florida. *Fla. State Board Health Monogr. Ser.* 12:118–125.

Johnson, C.W. 1913. Insects of Florida. I. Diptera. *Bull. Am. Mus. Nat. Hist.* 32:37–90.

Johnston, D.W. 1976. Organochlorine pesticide residues in uropygial glands and adipose tissue of wild birds. *Bull. Environ. Contam. Toxicol.* 16:149–155.

Kale, H.W. II. 1971. Florida region. *Am. Birds* 25:723–733.

Kinsella, J.M. 1971. Unpublished data. University of Florida, Gainesville.

Klukas, R.W., and L.N. Locke. 1970. An outbreak of fowl cholera in Everglades National Park. *J. Wildl. Dis.* 6:77–79.

Knight, C. 1976. Observation of a bald eagle capturing a cattle egret in flight. *Fla. Field Nat.* 4:14.

Kulda, J., and E. Nohynkova. 1995. Giardia in humans and animals. In: *Parasitic Protozoa.* Vol. 10. J. P. Kreier (ed.). Academic Press, San Diego. pp. 225–422.

Kushlan, J.A., and D.A. White. 1977. Nesting wading bird populations in southern Florida. *Fla. Sci.* 40:61–64.

Layne, J.N., F.E. Lohrer, and C.E. Winegarner. 1977. Bird and mammal predators on the cattle egret in Florida. *Fla. Field Nat.* 5:1–4.

Leigh, W.H. 1956. Observations on life-histories of members of the genus *Ascocotyle* Looss (Heterophyidae). *J. Parasitol.* 42:39.

Lincer, J.L., and D. Salkind. 1973. A preliminary note on organochlorine residues in the eggs of fish-eating birds of the west coast of Florida. *Fla. Field Nat.* 1:19–22.

Locke, L.N. 1967–1990. Unpublished data. National Wildlife Health Research Center, Madison, Wis.

———. 1987. Chlamydiosis. In: *Field guide to wildlife diseases.* Vol. 1, *General field procedures and diseases of migratory birds.* M. Friend (ed.). U.S. Fish and Wildlife Service Resource Bulletin 167. pp. 107–113.

Locke, L.N., H.M. Ohlendorf, R.B. Shillinger, and T. Jareed. 1974. Salmonellosis in a captive heron colony. *J. Wildl. Dis.* 10:143–145.

Loftin, H. 1961. An annotated check-list of trematodes and cestodes and their vertebrate hosts from northwest Florida. *J. Fla. Acad. Sci.* 23:302–314.

Logan, T.H. 1997. Florida's endangered species, threatened species, and species of special concern: official lists. Florida Game and Fresh Water Fish Commission, Tallahassee. 14 pp.

Maa, T.C. 1969. Studies in Hippoboscidae (Diptera). Part 2. *Pacific Insects Monogr.* 20:1–312.

Marion, W.R., T.E. O'Meara, G.D. Riddle, and H.A. Berkhoff. 1983. Prevalence of *Clostridium botulinum* type C in substrates of the phosphate-mine settling ponds and implications for epizootics of avian botulism. *J. Wildl. Dis.* 19:302–307.

Maxwell, G.R.I., and H.W. Kale II. 1977. Maintenance and anti-insect behavior of six species of ciconiiform birds in south Florida. *Condor* 79:51–55.

McEwan, L.C., and D.H. Hirth. 1980. Food habits of the bald eagle in north-central Florida. *Condor* 82:229–231.

McKilligan, N. 1987. Causes of nesting losses in the cattle egret *Ardeola ibis* in eastern Australia with special reference to the pathogenicity of the tick *Argas* (*Persicargas*) *robertsi* to nestlings. *Aust. J. Ecol.* 12:9–16.

McVaugh, W., Jr. 1972. The development of four North American herons. *Living Bird* 11:155–183.

Mensik, J.G., and R.G. Botzler. 1989. Epizootiological features of avian cholera on the north coast of California. *J. Wildl. Dis.* 25:240–245.

Mock, D.W., T.C. Lamey, and B.J. Ploger. 1987. Proximate and ultimate roles of food amount in regulating egret sibling aggression. *Ecology* 68:1760–1772.

Mollhagen, T.R. 1976. A study of the systematics and hosts of the parasitic nematode genus *Tetrameres* (Habronematoidea: Tetrameridae). Ph.D. diss., Texas Tech University, Lubbock. 546 pp.

Nichols, D.K., V.L. Campbell, and R.J. Montali. 1986. Pansteatitis in great blue herons. *J. Am. Vet. Med. Assoc.* 189:1110–1112.

Oddo, A.F., R.D. Pagan, L. Worden, and R.C. Botzler. 1978. The January 1977 avian cholera epornitic in northwest California. *J. Wildl. Dis.* 14:317–321.

Odening, V.K. 1997. Die Sarcocystis-infektion: Wechselbeziehungen zwischen freilebenden Wildtieren, Haustieren und Zootieren. *Zool. Gart.* 67:317–340.

Ogden, J.C. 1978. Recent population trends of colonial wading birds on the Atlantic and Gulf coastal plains. In: *Wading Birds.* A. Sprunt, J.C. Ogden, S. Winkler (eds.). Research Report 7. National Audubon Society, New York. pp. 137–153.

———. 1994. A comparison of wading bird nesting dynamics, 1931–1946 and 1974–1989, as an indication of changes in ecosystem conditions in the southern Everglades, Florida. In: *The Everglades: the ecosystem and its restoration.* S. Davis and J. C. Ogden (eds.). St. Lucie Press, Delray Beach, Fla. pp. 533–570.

———. 1996a. Snowy egret. In: *Rare and endangered biota of Florida.* Vol. 5, *Birds.* J.A. Rodgers, Jr., H. W. Kale II, and H.T. Smith (eds.). University Press of Florida, Gainesville. pp. 420–431.

———. 1996b. Tricolored heron. In: *Rare and endangered biota of Florida.* Vol. 5, *Birds.* J.A. Rodgers, Jr., H. W. Kale II, and H.T. Smith (eds.). University Press of Florida, Gainesville. pp. 432–441.

Ogden, J.C., W.B. Robertson, G.E. Davis, and T.W. Schmidt. 1974. Pesticides, polychlorinated biphenyls and heavy metals in upper food chain levels, Everglades National Park and vicinity. Deparment of the Interior, National Technical Information Service, Atlanta. 27 pp.

Ohlendorf, H.M., D.M. Swineford, and L.N. Locke. 1981. Organochlorine residues and mortality of herons. *Pestic. Monit. J.* 14:125–135.

Ohlendorf, H.M., E.E. Klaas, and T.E. Kaiser. 1978. Environmental pollutants and eggshell thinning in the black-crowned night heron. In: *Wading birds.* A. Sprunt IV, J.C. Ogden, and S. Winckler (eds.) Research Report 7. National Audubon Society, New York. pp. 63–82.

———. 1979. Environmental pollutants and eggshell thickness: anhingas and wading birds in the eastern United States. U.S. Fish and Wildlife Service Special Scientific Report— Wildlife 216. 94 pp.

Palmieri, J.R. 1973. New definitive and intermediate hosts and host localities for *Apatemon gracilis. J. Parasitol.* 59:1063.

Paul, R.T. 1996. Reddish egret. In: *Rare and endangered biota of Florida.* Vol. 5, *Birds.* J.A. Rodgers, Jr., H. W. Kale II, and H. T. Smith (eds.). University Press of Florida, Gainesville. pp. 281–294.

Paulson, D.R., and H.M. Stevenson. 1962. Florida region. *Audubon Field Notes* 16:398–404.

Peakall, D.B. 1975. Physiological effects of chlorinated hydrocarbons on avian species. In: *Environmental Dynamics of Pesticides.* R. Hague and V. H. Freed (eds.). Plenum, New York. pp. 343–360

Pence, D.B., M.G. Spalding, J.F. Bergan, R.A. Cole, S. Newman, and P.N. Gray. 1997. New records of subcutaneous mites (Acari: Hypoderatidae) in birds with examples of potential host colonization events. *J. Med. Entomol.* 34:411–416.

Penner, L.R., and B. Fried. 1963. *Philophthalmus hegeneri* sp. n., an ocular trematode from birds. *J. Parasitol.* 49:974–977.

Pollard, M. 1947. Ornithosis in seashore birds. *Proc. Soc. Exp. Biol. Med.* 64:200.

Powell, G.V.N. 1983. Food availability and reproduction by great white herons, *Ardea herodias:* a food addition study. *Colon. Waterbirds* 6:139–147.

Powell, G.V.N., and R.B. Bjork. 1996. Great white heron. In: *Rare and endangered biota of Florida.* Vol. 5, *Birds.* J.A. Rodgers, Jr., H.W. Kale II, and H.T. Smith (eds.). University Press of Florida, Gainesville. pp. 388–403.

Powell, G.V.N., and A.H. Powell. 1986. Reproduction by great white herons *Ardea herodias* in Florida Bay as an indicator of habitat quality. *Biol. Conserv.* 36:101–113.

Powell, G.V.N., R.B. Bjork, J.C. Ogden, R.T. Paul, and A.H. Powell. 1989. Population trends in some Florida Bay wading birds. *Wilson Bull.* 101:436–457.

Quinn, L. 1994. Unpublished data. Florida Keys Wild Bird Center, Tavernier.

Recher, H.F., and J.A. Recher. 1969. Comparative foraging efficiency of adult and immature little blue herons (*Florida caerulea*). *Anim. Behav.* 17:320–322.

Red tide leaves Carolina coast. 1988. *IAAAM News* 19:1.

Riggs, M.W. 1988. Unpublished data. University of Florida, Gainesville.

Roberts, B. 1988. Unpublished data. Florida Department of Natural Resources, St. Petersburg.

Robertson, W.B., Jr., and J.A. Kushlan. 1974. The southern Florida avifauna. *Miami Geol. Soc. Mem.* 2:414–452.

Robertson, W.B., Jr., and D.R. Paulson. 1961. Region reports: Florida region. *Audubon Field Notes* 15:26–35.

Robertson, W.B., Jr., and G.E. Woolfenden. 1992. *Florida bird species: an annotated list.* Spec.

Publ. 6, Florida Ornithological Society, Gainesville. 260 pp.

Rodgers, J.A. 1980a. Breeding ecology of the little blue heron on the west coast of Florida. *Condor* 82:164–169.

———. 1980b. Reproductive success of three heron species on the west coast of Florida. *Fla. Field Nat.* 8:37–56.

———. 1987. Breeding chronology and reproductive success of cattle egrets and little blue herons on the west coast of Florida, USA. *Colon. Waterbirds* 10:38–44.

———. 1996. Little blue heron. In: *Rare and endangered biota of Florida*. Vol. 5, *Birds*. J.A. Rodgers, Jr., H.W. Kale II, and H.T. Smith (eds.). University Press of Florida, Gainesville. pp. 413–419.

———. 1997. Pesticide and heavy metal levels of waterbirds in the Everglades agricultural area of south Florida. *Fla. Field Nat.* 25:33–84.

Rodgers, J.A., and H.T. Smith. 1995. Little blue heron. *Birds N. Am.* 145:1–12.

———. 1997. Buffer zone distances to protect foraging and loafing waterbirds from human disturbance in Florida. *Wildl. Soc. Bull.* 25:139–145.

Rodgers, J.A., H.W. Kale II, and H.T. Smith (eds.). 1996. *Rare and endangered biota of Florida*. Vol. 5, *Birds*. University Press of Florida, Gainesville. 688 pp.

Rodgers, J.A., S.T. Schwikert, and A.S. Wenner. 1993. The prevalence of abdominal lesions on wood stork nestlings in north and central Florida. *Condor* 95:473–475.

Roffe, T.J. 1988. *Eustrongylides* sp. epizootic in young common egrets (*Casmerodius albus*). *Avian Dis.* 32:143–147.

Rosen, M.N., and A.I. Bischoff. 1949. The 1948–49 outbreak of fowl cholera in birds in the San Francisco Bay area and surrounding counties. *Calif. Fish Game* 35:185.

Rubin, H., R.E. Kessling, R.W. Chamberlain, and M.E. Eidson. 1951. Isolation of a psittacosis-like agent from the blood of snowy egrets. *Proc. Soc. Exp. Biol. Med.* 78:696.

Runde, D.E. 1991. Trends in wading bird nesting population in Florida 1976–78 and 1986–89. Florida Game and Fresh Water Fish Commission, Nongame Wildlife Section. 105 pp.

———. 1996. Great egret. In: *Rare and endangered biota of Florida*. Vol. 5, *Birds*. J.A. Rodgers, Jr., H.W. Kale II, and H.T. Smith (eds.). University Press of Florida, Gainesville. pp. 404–412.

Sasse, B. 1999. Unpublished data. Florida Game and Fresh Water Fish Commission, Tallahassee.

Schmidt, G.D. 1986. *Handbook of tapeworm identification*. CRC, Boca Raton, Fla. 675 pp.

Scott, W.E.D. 1887. The present condition of some of the bird rookeries of the gulf coast of Florida. *Auk* 4:135–144, 213–222, 273–284.

———. 1889. A summary of observations on the birds of the gulf coast of Florida. *Auk* 6:13–18.

———. 1890. A summary of observations on the birds of the gulf coast of Florida. *Auk* 7:14–23.

Sepulveda, M.S., and J.W. Mertins. 1994–95. Unpublished data, University of Florida, Gainesville.

Sepulveda, M.S., P.C. Frederick, M.G. Spalding, and G.E. Williams, Jr. 1999a. Mercury contamination in free-ranging great egret (*Ardea albus*) nestlings from southern Florida. *Environ. Toxicol. Chem.* 18:985–992.

Sepulveda, M.S., M.G. Spalding, J.M. Kinsella, and D.J. Forrester. 1996. Parasitic helminths of the little blue heron, *Egretta caerulea*, in southern Florida. *J. Helminthol. Soc. Wash* 63:136–140.

———. 1999b. Parasites of the great egret (*Ardea albus*) in Florida and a review of the helminths reported for the species. *J. Helminthol. Soc. Wash.* 66:7–13.

Sileo, L. 1997. Unpublished data. National Wildlife Health Research Center, Madison, Wis.

Sims, H.W., Jr. 1970. Operation bird wash. *Fla. Nat.* 43:43–45.

Snyder, B. 1994. Unpublished data. Florida Department of Environmental Protection, Tallahassee.

Snyder, N.F.R., J.C. Ogden, J.D. Bittner, and G.A. Grau. 1984. Larval dermestid beetles feeding on nestling snail kites, wood storks, and great blue herons. *Condor* 86:170–174.

Spalding, M.G. 1988–2001. Unpublished data. University of Florida, Gainesville.

———. 1990. Antemortem diagnosis of eustrongylidosis in wading birds (Ciconiiformes). *Colon. Waterbirds* 13:75–77.

Spalding, M.G., C.T. Atkinson, and R.E. Carleton. 1994a. *Sarcocystis* sp. in wading birds (Ciconiiformes) from Florida. *J. Wildl. Dis.* 30:29–35.

Spalding, M.G., G.T. Bancroft, and D.J. Forrester. 1993. The epizootiology of eustrongylidosis in wading birds (Ciconiiformes) in Florida. *J. Wildl. Dis.* 29:237–249.

Spalding, M.G., R.D. Bjork, G.V.N. Powell, and S.F. Sundlof. 1994b. Mercury and cause of death in great white herons. *J. Wildl. Manag.* 58:735–739.

Spalding, M.G., and J. Brady. 1992. Unpublished data. University of Florida, Gainesville.

Spalding, M.G., and M. Cheadle. 2000. Unpublished data. University of Florida, Gainesville.

Spalding, M.G., and D.J. Forrester. 1991. Effects of parasitism and disease on the reproductive success of colonial wading birds (Ciconiiformes) in southern Florida. Final report. Florida Game and Fresh Water Fish Commission, Tallahassee. 121 pp.

———. 1993. Pathogenesis of *Eustrongylides ignotus* (Nematoda: Dioctophymatoidea) in Ciconiiformes. *J. Wildl. Dis.* 29:250–260.

Spalding, M.G., D.J. Forrester, and J.M. Kinsella. 1991. Unpublished data. University of Florida, Gainesville.

Spalding, M.G., P.C. Frederick, H.C. McGill, S.N. Bouton, and L.R. McDowell. 2000a. Methylmercury accumulation in tissues and effects on growth and appetite in captive great egrets. *J. Wildl. Dis.* 36:411–422.

Spalding, M.G., P.C. Frederick, H.C. McGill, S.N. Bouton, L.J. Richey, I.M. Schumacher, C.G.M. Blackmore, and J. Harrison. 2000b. Histologic, neurologic and immunologic effects of methylmercury in captive great egrets. *J. Wildl. Dis.* 36:423–435.

Spalding, M.G., and J.M. Kinsella. 1989–1999. Unpublished data. University of Florida, Gainesville.

Spalding, M.G., R.G. McLean, J.H. Burgess, and L.J. Kirk. 1994d. Arboviruses in water birds (Ciconiiformes, Pelecaniformes) from Florida. *J. Wildl. Dis.* 30:216–221.

Spalding, M.G., and J.W. Mertins. 1994–99. Unpublished data. University of Florida, Gainesville.

Spalding, M.G., and D.B. Pence. 1996. Unpublished data. University of Florida, Gainesville.

Spalding, M.G., J.P. Smith, and D.J. Forrester. 1994e. Natural and experimental infections of *Eustrongylides ignotus:* effect on growth and survival of nestling wading birds. *Auk* 111:328–336.

Spalding, M.G., C.K. Steible, S.F. Sundlof, and D.J. Forrester. 1997. Metal and organochlorine contaminants in tissues of nestling wading birds (Ciconiiformes) from southern Florida. *Fla. Field Nat.* 25:42–50.

Spalding, M.G., and N.A. Wilson. 1992. Unpublished data. University of Florida, Gainesville.

Spalding, M.G., and J.C. Woodard. 1992. Chondrosarcoma in a wild great white heron from southern Florida. *J. Wildl. Dis.* 28:151–153.

Spalding, M.G., S. Wright, V. Clyde, M.W. Riggs, and J.A. Hovis. 1988. Unpublished data. University of Florida, Gainesville.

Sprunt, A., Jr. 1954. *Florida bird life.* Coward-McCann, New York. 527 pp.

Sprunt, A.J. 1949. Predation on living prey by the black vulture. *Auk* 63:260–261.

Sprunt, S.A. 1969. A case of turkey vulture piracy on great blue herons. *Wilson Bull.* 77:257–263.

Stevenson, H.M. 1970. The winter season: Florida region. *Audubon Field Notes* 24:493–497.

Stevenson, H.M., and B.H. Anderson. 1994. *The birdlife of Florida.* University Press of Florida, Gainesville. 892 pp.

Stevenson, J.A. 1994. Unpublished data. Florida Department of Environmental Protection, Tallahassee.

Stiles, C.W., and A. Hassall. 1894. A preliminary catalogue of the parasites contained in the collections of the United States Bureau of Animal Industry, United States Army Medical Museum, Biological Department of the University of Pennsylvania (Coll. Leidy) and in Coll. Stiles and Coll. Hassall. *Vet. Mag.* 1:245–253; 331–354.

Stuart, J.J., J.F. Dismukes, and C.F. Dixon. 1972. Endoparasites of the cattle egret (*Bubulcus ibis*) in Alabama. *J. Parasitol.* 58:518.

Sundlof, S.F., M.G. Spalding, J.D. Wentworth, and C.K. Steible. 1994. Mercury in livers of wading birds (Ciconiiformes) in southern Florida. *Arch. Environ. Contam. Toxicol.* 27:299–305.

Suto, B. 2000. Unpublished data. Suncoast Seabird Sanctuary, Redington Beach, Fla.

Taylor, P.W. 1992. Fish-eating birds as potential vectors of *Edwardsiella ictaluri*. *J. Agric. Anim. Health* 4:240–243.

Taylor, W.K., and B.H. Anderson. 1973. Nocturnal migrants killed at a central Florida TV tower, autumns 1969–71. *Wilson Bull.* 85:42–51.

Taylor, W.K., and M.A. Kershner. 1986. Migrant birds killed at the Vehicle Assembly Building (VAB), John F. Kennedy Space Center. *J. Field Ornithol.* 57:142–154.

Telfair, R.C. II. 1994. Little blue heron. In: *Birds N. Am.* 113:1–32.

Telford, S.R., Jr., M.G. Spalding, and D.J. Forrester. 1992. Hemoparasites of wading birds (Ciconiiformes) in Florida. *Can. J. Zool.* 70:1397–1408.

Udevitz, M.S., C.A. Howard, R.J. Robel, and B. Carnutte, Jr. 1980. Lead contamination in insects and birds near an interstate highway, Kansas. *Environ. Entomol.* 9:35–36.

Walters, C., L. Gunderson, and C.S. Holling. 1992. Experimental policies for water management in the Everglades. *Ecol. Appl.* 2:189–202.

Walton, A.C. 1927. A revision of the nematodes of the Leidy collections. *Proc. Acad. Nat. Sci. Phila.* 79:49–163.

Watts, B.D. 1995. Yellow-crowned night-heron. *Birds N. Am.* 161:1–24.

Weise, J.H., W.R. Davidson, and V.F. Nettles. 1977. Large scale mortality of nestling ardeids caused by nematode infection. *J. Wildl. Dis.* 13:376–382.

Weston, F.M. 1966. Bird casualties on the Pensacola Bay bridge (1938–1949). *Fla. Nat.* 39:53–55.

White, F.H., C.F. Simpson, and L.E. Williams, Jr. 1973. Isolation of *Edwardsiella tarda* from aquatic animal species and surface waters in Florida. *J. Wildl. Dis.* 9:204–208.

Williams, G.E., Jr. 1997. The effects of methylmercury on the growth and food consumption of great egret nestlings in the central Everglades. Master's thesis, University of Florida, Gainesville. 105 pp.

Wilson, N.A., H.W. Kale II, and W.W. Baker. 1990. Unpublished data. University of Northern Iowa, Cedar Falls.

Winger, P.V., C. Sieckman, T.W. May, and W. Johnson. 1984. Residues of organochlorine insecticides, polychlorinated biphenyls and heavy metals in biota from Apalachicola River, Florida. *J. Assoc. Off. Anal. Chem.* 67:325–333.

Yunker, C.E., C.M. Clifford, J.E. Keirans, L.A. Thomas, and R.C.A. Rice. 1979. Aransas Bay virus, a new arbovirus of the Upola serogroup from *Ornithodoros capensis* (Acari: Argasidae) in coastal Texas. *J. Med. Entomol.* 16:453–460.

Zinkl, J.G., D.A. Jessup, A.I. Bischoff, T.E. Lew, and F.B. Wheeldon. 1981. Fenthion poisoning of wading birds. *J. Wildl. Dis.* 17:117–119.

CHAPTER SEVEN

Ibises, Spoonbills, Flamingos, and Storks

I. Introduction

Four species of ibis, 1 spoonbill, 1 stork, and 1 flamingo occur in Florida (table 7.1). Some consider the White Ibis and Scarlet Ibis to be the same species (Hancock et al. 1992). The Wood Stork is listed as endangered by the Florida Game and Fresh Water Fish Commission (Logan 1997), the Florida Committee on Rare and Endangered Plants and Animals (Rodgers et al. 1996a), and the U.S. Fish and Wildlife Service. The Roseate Spoonbill and White Ibis are listed as species of special concern by the Florida Game and Fresh Water Fish Commission (Logan 1997). In addition, the Roseate Spoonbill is listed as rare, and both the Glossy and White Ibis are listed as species of special concern by the Florida Committee on Rare and Endangered Plants and Animals (Rodgers et al. 1996a).

Life history and population information have been reported for Wood Storks (Kahl 1964; Kushlan and Frohring 1986; Stangel et al. 1990; Ogden 1994, 1996a; Rodgers et al. 1996b; Rodgers and Schwikert 1997; Frederick and Ogden 1997), Roseate Spoonbills (Allen 1942; Bjork and Powell 1996; Dumas 2000), Glossy Ibis (Ogden 1996b), White Ibis (Sprunt 1944; Nesbitt et al. 1974; Kushlan and Kushlan 1975; Kushlan and Bildstein 1992; Bildstein 1993; Frederick 1996; Frederick and Ogden 1997), White-faced Ibis (Ryder and Manry 1994), and several other species (Sprunt 1954; Runde 1991). Frederick and Spalding (1994)

Table 7.1. Species of ibises, spoonbills, storks, and flamingos that occur in Florida

	Species	Distribution	Seasonal occurrence	Relative abundance
White Ibis	*Eudocimus albus*	Throughout	Year-round[a]	Common
Scarlet Ibis	*Eudocimus ruber*	Throughout	Year-round[a]	Rare
Glossy Ibis	*Plegadis falcinellus*	Throughout	Year-round[a]	Locally common
White-faced Ibis	*Plegadis chihi*	Throughout	Year-round[a]	Few records
Roseate Spoonbill	*Ajaia ajaja*	Throughout	Year-round[a]	Locally common
Wood Stork	*Mycteria americana*	Throughout	Year-round[a]	Common
Greater Flamingo	*Phoenicopterus ruber*	South Fla.	Fall/Winter	Uncommon

Source: Robertson & Woolfenden (1992).
a. Breeds in Florida.

discussed factors affecting reproduction. Much of the information in this chapter comes from a survey of the causes of mortality of ciconiiforms, primarily of nestlings, by Spalding and Forrester (1991). In addition, general health information can be found in Fowler (1986), Friend and Franson (1999), Fairbrother et al. (1996), and Davidson and Nettles (1997).

II. Trauma

Roadside trauma, consisting of collision with a power line or a vehicle or a combination of the two, was the most common cause (22/48) of mortality in Wood Storks examined by Spalding and Forrester (1991) and by Spalding (1999) and by the National Wildlife Health Center (Thomas 1988–94). Characteristic lesions of power line collision included rupture of the liver at the base of the sternum, fracture of the sternum, and fracture of the acetabular region of the pelvis. If the bird was electrocuted, additional findings were singed feathers, legs, or bill. The sample was undoubtedly biased because of the visibility of dead storks along roadsides. It was concluded that vehicular collision was much less important than power line collision, and that most fatalities that appeared to be due to vehicular traffic were actually caused by falling or wandering onto roads after colliding with power lines. In some cases storks appeared to be sick and were observed near the road for several days prior to being found dead with ev-

idence of trauma. For example, one of these birds had peritonitis secondary to a stomach impacted with plastic fishing worms and a hook, and another was severely emaciated, with many helminth parasites (Spalding 1987).

A total of 165 birds, including 5 Wood Storks, was found during a survey of dead birds under power lines across the Everglades in Palm Beach County during 1990–91 (FPL and FGFWFC 1991). Most of the storks examined had evidence of trauma, but a few had singed feathers. In a 1990–93 survey of road kills in state parks, 2 White Ibis and 1 Wood Stork were found among a total of 1,562 birds (Stevenson 1994; Snyder 1994). Deng (1998) noted that birds foraging at night, such as Wood Storks, might be more likely to strike power lines but also that birds flying at night tended to fly at higher altitudes. Evidence from roadside collections and submissions to rehabilitators indicates that roadside trauma appears to be less of a problem for ibis and spoonbills (Spalding and Forrester 1991). A Roseate Spoonbill radio-tagged as a fledgling was found entangled in a power line and electrocuted (Spalding and Bjork 1990).

Collision with tall objects, however, appears to occur less often among this group. Crawford (1981) recorded birds found dead below a TV tower on Tall Timbers Research station in Leon County between 1955 and 1980. During this period 42,384 birds of 189 species were found dead, presumably killed during migration, with only a single White Ibis among them.

Ingestion of foreign objects is less common in this group of birds than in pelicans. As mentioned, an immature Wood Stork that was hit by a car near a popular fishing location in Dade County in 1987 had a stomach that was severely distended and impacted with hardened plastic fishing worms and a fish hook (Spalding 1987). The hook had perforated the stomach wall, causing peritonitis.

The importance of natural causes of accidental injury is poorly understood. Wood Storks nest in very tall trees in some colonies. A nestling Wood Stork that died after it fell from its nest in Collier County in 1974 appeared to have been in good health otherwise (Spalding 1995).

III. Predation

In the Everglades in 1988–89, nest predation by birds, mammals, and snakes was relatively uncommon (Frederick and Collopy 1989c). In Brevard County in 1975, however, Girard and Taylor (1979) reported frequent egg predation by birds, probably Fish Crows and Boat-tailed Grackles, especially on Wood Stork eggs. Sprunt and Kahl (1960) reported that groups of unpaired Wood Storks would chase off parent birds and destroy eggs and kill nestlings. Raccoon predation was suspected in the destruction of eggs in a small colony of Roseate Spoonbills in Dade County (Baker 1940) and one in Monroe County (Allen 1942). Human disturbance of nesting colonies, discussed later in this chapter, can increase nest predation.

Twelve Roseate Spoonbill nestlings from Florida Bay were suspected to have been killed by an avian predator. Chick parts, consisting of stomachs that were full of ingesta, and legs and wings with the flesh removed, were found in or below nests. The chicks were believed to have been healthy because there was abundant fat on the stomachs and chicks in adjacent nests appeared healthy. Vultures were observed (from a blind) scavenging on dead chicks in the nest; however, they were never observed to kill live nestlings (Spalding and Forrester 1991; Bjork and Powell 1994). Both Black and Turkey Vultures have been reported to take live nestlings (Sprunt 1949, 1969). Roseate Spoonbill nestling carcasses have been collected in a Bald Eagle nest on Sandy Key, Monroe County (Spalding and Forrester 1991). Adult Bald Eagles hunting cooperatively were able to capture an adult White Ibis that they forced to the ground in Alachua County (Folk 1992). Bald Eagle talons were the most likely cause of perforations and hemorrhage into the thoracic and abdominal cavities of an adult Roseate Spoonbill found floating in the water in Florida Bay (Spalding 1988).

Mammalian predation was suspected to be the cause of death of an adult Wood Stork in Everglades National Park, Dade County, in 1993 (Thomas 1993). Frederick and Collopy (1988) reported that White Ibis nests were predated by raccoons in Dade County, and they suggested that nesting over water that contained alligators might be a strategy to avoid mammalian predation.

IV. Human disturbance

Powell et al. (1989) summarized the near extinction from the 1800s to 1930s and recovery of Roseate Spoonbills and other wading birds from Florida Bay due to the effects of plume hunting and harvesting for food. More recently, despite protection, 5 of 48 Wood Storks examined at necropsy had been shot, resulting in death of 2 of these (Thomas 1991; Forrester 1971). Two adult Wood Storks from Collier County in 1993 died with encapsulated lead pellets in their air sacs. One was emaciated, the other died from aspiration pneumonia (Thomas 1993). An immature female Wood Stork that may have been emaciated secondary to a chronic gunshot wound in the wing was found dead in Everglades National Park in Dade County in 1993 (Thomas 1993). A Greater Flamingo that died in Everglades National Park from septicemia had 2 pellets in the wing and neck that did not appear to be the cause of death (Stroud 1985).

There have been cases in which researcher visitation to colonies was believed to be the cause of death of nestlings; however, this is difficult to document. Frederick and Spalding (1989) noted extensive egg predation by Fish Crows on Seahorse Key (Levy County) in 1989 after they had checked nests, and they had to abandon their work in the colony. In one case, the combination of overheating and an extraordinarily distended stomach likely caused death of a nestling spoonbill. In another case, a Roseate Spoonbill nestling died with its foot wedged between branches, possibly in an attempt to escape (Spalding and Forrester 1991). The extent of mortality caused directly by human disturbance was not thought to be a significant factor (<2% of nestlings) (Spalding and Forrester 1991). Marking of eggs by researchers may result in unnatural egg orientation. Marks on White Ibis eggs were more frequently found oriented downward than in other positions (Frederick and Collopy 1989a). It was not known if this affected hatchability. Bjork and Powell (1994) found that nesting of Roseate Spoonbills increased when boating activities were restricted by a 100-foot no-access zone around nesting colonies. Rodgers and Smith (1997) suggested a buffer distance for Wood Storks of 77 meters for boats. Human disturbance of wading bird colonies can result in massive mortality of young White Ibis; the adults will readily abandon nests (Frederick 1996).

V. Inclement weather

In the Everglades, large-scale abandonment of nests, especially those of White Ibis, in response to heavy rainfall and rising surface water is common (Frederick and Collopy 1989b). In the case of Wood Storks, late initiation of nesting since the 1960s (Feb.–March, rather than Nov.–Jan.) has frequently resulted in abandonment at the start of the rainy season in late spring (Ogden 1994). Nestlings typically starve to death or are eaten by predators following these abandonments. Though coastal tidal flooding can be a major source of nest failure where ibis nest on the ground (Frederick 1987), most coastal White Ibis nests in Florida are in trees.

Cold weather and high winds have occasionally caused destruction of Wood Stork nests in Collier County (3,000 to 4,000 in February 1963, Stevenson 1963) and mortality of nestlings (from 4,000 nests in January 1966, Cunningham 1966). All 4 Wood Stork colonies in Everglades National Park were abandoned because of cold weather in January 1969, but some renested in February (Stevenson 1969). Roseate Spoonbills abandoned their nests in Florida Bay in unseasonably cold weather in January 1973 (Ogden 1978). Cold weather can also make prey less available to storks, ibises, and spoonbills, since fish will burrow in the mud and move to deeper water (Frederick and Loftus 1993).

Hurricanes are potential sources of mortality, but such mortality is reported rarely. The large numbers of wading birds that migrate into Florida are mostly gone during the hurricane season (June–November). Robert Porter Allen (1942) mentions mortality of eggs and nestlings of Roseate Spoonbills in Florida Bay in 1940 associated with the passage of a hurricane. After Hurricane Donna in September 1960, 64 White Ibis and 6 Roseate Spoonbills were found dead in Florida Bay and around Flamingo (Robertson and Paulson 1961). No measurable loss of these species was noticed following Hurricane Andrew in August 1992; however, severe vegetation damage was sustained in 14 wading bird rookeries and moderate damage in 57 rookeries (FGFWFC 1992). No effects on nesting were detected in subsequent years. Eight Wood Storks and 3 White Ibis were found dead along the beach in South Carolina following Hurricane Hugo in 1989 (Post 1992), and a large colony of White Ibis failed to nest in the years following the storm, probably because of salinity changes in adjacent marshes (Bildstein 1993).

VI. Reproductive failure

Because of their colonial nesting habits there is considerably more information about nesting in this group than for some other birds. Hatching failure or death of nestlings can be caused by a number of factors, but for these species, human disturbance, inclement weather, or inadequate food resources, all leading to abandonment of the nests by the adults, are among the most important causes. Girard and Taylor (1979) reported that most Wood Stork nestling mortality (25%) occurred in the first 2 weeks after hatching in Brevard County, 1975. Rodgers (1996) found addled eggs in 3% (1% of eggs) of Wood Stork nests in north and central Florida during 1981–85 and concluded that this loss was not a significant mortality factor. This did not include abandoned nests, however, where he estimated mortality of 11–55% prior to hatching. Rodgers and Schwikert (1997) found a 96% hatching success for Wood Stork nests that successfully hatched at least 1 young, and an overall 88–99% hatching success for all colony-years. Brood reduction, or loss of the youngest nestlings, to insure the survival of the remaining chicks is a common phenomenon in ciconiiformes (Mock et al. 1987) and is discussed as a strategy for Wood Storks in Florida by Rodgers and Schwikert (1997). Also see comments in sections on inclement weather and human disturbance, this chapter, for further information.

VII. Organochlorines

Organochlorine residues measured in eggs of ibis, spoonbills, and storks are listed in tables 7.2–7.5. DDE concentrations in White Ibis eggs ranged as high as 1.6 ppm wet weight and 8.7 ppm dry weight in the 1970s, and DDE was still detectable in more recent samples collected near a landfill in Palm Beach County (table 7.2). No significant eggshell thinning was measured for White Ibis, Glossy Ibis, or Roseate Spoonbills when compared with pre-1946 eggs; in fact, Rumbold et al. (1997) reported an increase in shell thickness in White Ibis feeding at a landfill (table 7.6). DDE concentrations in Wood Stork eggs were higher (up to 19 ppm) than in those of 12 other species of fish-eating birds tested in Brevard County in 1973 (geometric mean = 4.0 ppm ww), suggesting that they were selecting food items more likely to be contaminated (Ohlendorf et al. 1979) (table 7.5). Fleming et al. (1984) examined Wood Stork eggs from 8 colonies in northern and central Florida in 1982. They found detectable amounts of DDE in all eggs (geometric mean = 1.1 ppm), PCBs in 63% of eggs, and other organochlorines in less than 30% of eggs, at concentrations lower than in 1973. Although eggshell thinning in 1982 (4%) was still correlated with DDE concentrations, it was less severe than during 1967–73 (9%) (table 7.6). Higher DDE concentrations occurred in nests with less than 100% hatching success; however, no correlation existed with fledging success. Anderson and Hickey (1972) reported that eggshell thinning of 15–20% or more was associated with population declines. See chapters 4, Pelicans, and 11, Eagles, for more discussion about eggshell thinning and organochlorine contaminants in Florida. PCBs in Wood Stork eggs declined from an average of 1.2 ppm in eggs from Brevard County in 1973 (Ohlendorf et al. 1978) to 0.53 ppm in 1982 (Fleming et al. 1984). The number of eggs with PCBs also declined from 100% to 62%. Eleven other organochlorine contaminants were found inconsistently and at low concentrations in Wood Stork, ibis, and spoonbill eggs (tables 7.2–7.5).

Concentrations of organochlorines measured in tissues of ibis and spoonbills were consistently below 2 ppm (tables 7.3, 7.4, 7.7). Low concentrations of DDE and *trans*-nonachlor were the only organochlorine contaminants found in the brains of ibis and spoonbill nestlings tested from south Florida in 1989–91 (Spalding et al. 1997). Dioxin and furan concentrations measured from White Ibis nestlings near a landfill were low (table 7.3). We could

Table 7.2. Organochlorine residues in eggs of White Ibises collected in Florida[f]

County or area	Year(s)	No. examined	DDE			DDD			DDT			Data source
			No.[a]	Mean[b]	(Range)	No.[a]	Mean[b]	(Range)	No.[a]	Mean[b]	(Range)	
Brevard	1972–73	10	9	0.27	(ND–0.84)	0	ND	—	1	0.014	(ND–0.15)	A
"Charlotte Harbor"	1972	1	1	8.7[c]	—	NA	—	—	NA	—	—	B
Dade	1972	5	5	0.24	(0.062–0.80)	4	0.011	(ND–0.031)	4	0.012	(ND–0.037)	D
"Everglades National Park"	1972–73	10	6	0.080	(ND–0.25)	0	ND	—	0	ND	—	A
"Florida Bay"	1972	4	4	0.080	(0.031–0.16)	4	0.022	(0.016–0.031)	3	0.009	(ND–0.039)	D
Highlands	1972–73	11	7	0.14	(ND–0.52)	0	ND	—	1	0.014	(ND–0.16)	A
Lee	1972–73	1	1	0.11	—	0	ND	—	0	ND	—	A
Levy	1972–73	10	9	0.12	(ND–0.19)	0	ND	—	0	ND	—	A
Palm Beach	1989	5[d]	5	0.11	(0.04–0.65)	0	ND	—	2	NC	(ND–0.073)	C
	1990	5[d]	5	0.13	(0.03–1.6)	2	NC	(ND–0.073)	2	0.027	(0.006–1.2)	C
	1991	5[d]	5	0.080	(0.039–0.25)	0	ND	—	4	0.030	(ND–0.11)	C
"Sarasota Bay"	1972–73	2	2	0.29	(0.19–0.40)	0	ND	—	0	ND	—	A
"Tampa Bay"	1972–73	3	3	0.67	(0.29–1.6)	0	ND	—	1	0.030	(ND–0.097)	A

County or area	Year(s)	No. examined	Dieldrin			Heptachlor epoxide			PCBs			Data source
			No.[a]	Mean[b]	(Range)	No.[a]	Mean[b]	(Range)	No.[a]	Mean[b]	(Range)	
Brevard	1972–73	10	1	0.012	(ND–0.13)	0	ND	—	5	0.21	(ND–1.4)	A
"Charlotte Harbor"	1972	1	1	0.09[c]	—	NA	—	—	1	9.8[c]	—	B
Dade	1972	5	4	0.007	(ND–0.020)	5	0.17	(0.002–0.057)	0	ND	—	D
"Everglades National Park"	1972–73	10	1	0.012	(ND–0.13)	0	ND	—	0	ND	—	A
"Florida Bay"	1972	4	3	0.008	(ND–0.039)	NA	—	—	1	0.004	(ND–0.19)	D
Highlands	1972–73	11	1	0.033	(ND–0.41)	0	ND	—	0	ND	—	A
Lee	1972–73	1	0	ND	—	0	ND	—	0	ND	—	A
Levy	1972–73	10	0	ND	—	0	ND	—	5	0.12	(ND–0.25)	A

(continued)

Table 7.2. (*continued*)

County or area	Year(s)	No. examined	Dieldrin			Heptachlor epoxide			PCBs			Data source
			No.[a]	Mean[b]	(Range)	No.[a]	Mean[b]	(Range)	No.[a]	Mean[b]	(Range)	
Palm Beach	1989	5[d]	0	ND	—	0	ND	—	NA[e]	—	—	C
	1990	5[d]	4	0.056	(ND–0.61)	2	NC	(ND–0.008)	NA[e]	—	—	C
	1991	5[d]	5	0.056	(0.022–0.22)	2	NC	(ND–0.015)	NA[e]	—	—	C
"Sarasota Bay"	1972–73	2	0	ND	—	0	ND	—	0	ND	—	A
"Tampa Bay"	1972–73	3	1	0.028	(ND–0.087)	1	0.14	(ND–0.48)	1	0.25	(ND–0.93)	A

County or area	Year(s)	No. examined	Oxychlordane			Heptachlor			Mirex			Data source
			No.[a]	Mean[b]	(Range)	No.[a]	Mean[b]	(Range)	No.[a]	Mean[b]	(Range)	
Brevard	1972–73	10	0	ND	—	NA	—	—	1	0.014	(ND–0.16)	A
"Everglades National Park"	1972–73	10	NA	—	—	NA	—	—	0	ND	—	A
Highlands	1972–73	11	0	ND	—	NA	—	—	0	ND	—	A
Lee	1972–73	1	0	ND	—	NA	—	—	0	ND	—	A
Levy	1972–73	10	0	ND	—	NA	—	—	1	0.014	(ND–0.15)	A
Palm Beach	1989	5[d]	NA	—	—	0	ND	—	NA	—	—	C
	1990	5[d]	NA	—	—	0	ND	—	NA	—	—	C
	1991	5[d]	NA	—	—	2	NC	(ND–0.013)	NA	—	—	C
"Sarasota Bay"	1972–73	2	0	ND	—	NA	—	—	0	ND	—	A
"Tampa Bay"	1972–73	3	1	0.033	(ND–0.10)	NA	—	—	0	ND	—	A

Sources: A = Ohlendorf et al. (1979), B = Lincer & Salkind (1973), C = Rumbold et al. (1996), D = Ogden et al. (1974).

ND = not detected, NA = not analyzed, NC = not calculated (<50% of samples contained the residue).

a. Number with residue.

b. Geometric mean, ppm wet weight unless otherwise indicated.

c. Values expressed in dry weight. Lincer & Salkind (1973) also give extractable fat weights.

d. Ten eggs were sampled and paired to produce 5 samples.

e. See Table 7.3.

f. In addition to the compounds listed above, Ohlendorf et al. (1979) did not detect *cis*-chlordane, *trans*-nonachlor, *cis*-nonachlor, HCB, or toxaphene. Rumbold et al. (1996) did not detect BHC, endosulfan I, endosulfan II, endrin, endosulfan sulfate, methoxychlor, chlordane, or toxaphene in 1989 or 1990.

Table 7.3. Dioxin and furan residues in eggs and tissues of White Ibises collected in Florida

Tissue County	Year(s)	No. examined	Total TCDD			2,3,7,8-TCDD			Total TCDF			2,3,7,8-TCDF		
			No.[a]	Mean[b]	(Range)	No.[a]	Mean[b]	(Range)	No.[a]	Mean[b]	(Range)	No.[a]	Mean[b]	(Range)
Eggs														
West Palm	1989	5[c]	3	0.4	(NG–1.0)	3	0.4	(NG–1.0)	3	0.1	(NG–0.3)	3	0.1	(NG–0.1)
	1990	5[c]	4	2.6	(NG–7.2)	4	2.6	(NG–7.2)	4	0.2	(NG–0.4)	3	0.1	(NG–0.4)
	1991	5[c]	4	0.2	(NG–0.4)	4	0.2	(NG–0.4)	0	ND	—	0	ND	—
	1994	5[c]	5	0.6	(NG–2.3)	5	0.3	(NG–0.4)	3	0.1	(NG–0.2)	3	0.1	(NG–0.2)
Nestlings														
Carcass														
West Palm	1989	5[c,d]	0	ND	—	0	ND	—	0	ND	—	0	ND	—
	1990	5[c,d]	1	NC	(NG–0.5)	0	ND	—	1	NC	(NG–0.8)	0	ND	—
	1991	5[c,d]	4	0.1	(NG–0.2)	0	0.1	(NG–0.2)	1	NC	(NG–0.2)	1	NC	(NG–0.2)
	1994	5[c,d]	5	0.2	(NG–0.3)	0	0.1	(NG–0.1)	4	0.2	(NG–0.4)	4	0.1	(NG–0.2)
Adults														
Combined organs and muscle														
West Palm	1989	1[e]	0	<0.4	—	0	<0.14	—	0	<0.03	—	0	ND	—

Source: Rumbold et al. (1997).
TCDD = tetrachlorodibenzo-*p*-dioxin, TCDF = tetrachlorodibenzofuran.
ND = not detected, NG = not given, NC = not calculated (<50% of samples contained the residue).
a. Number with residue.
b. Geometric mean, ppb wet weight.
c. Ten nestlings and 10 eggs were paired to produce 5 samples each.
d. Bills, gastrointestinal tract, legs, and feathers were removed from nestlings.
e. Mixed organs and muscle from 2 adults.

Table 7.4. Organochlorine residues in eggs and tissues of Glossy Ibises and Roseate Spoonbills from Florida[d,e]

Species Tissue(s)	County/area	Year(s)	No. examined	DDE			DDT			cis-chlordane			Data source
				No.[a]	Mean[b]	(Range)	No.[a]	Mean[b]	(Range)	No.[a]	Mean[b]	(Range)	
Glossy Ibis													
Eggs	Brevard	1972–73	5[d]	5	0.34	(0.13–0.92)	1	0.045	(ND–0.24)	1	0.021	(ND–0.11)	A
Roseate Spoonbill													
Nestlings													
(brain)	"Florida Bay"	1989–91	2[c,e]	1	0.01	(ND–0.02)	0	ND	—	NA	—	—	B

Sources: A = Ohlendorf et al. (1979), B = Spalding et al. (1997).
NA = not analyzed, and ND = not detected.
a. Number with residue.
b. Geometric mean, ppm wet weight.
c. Each sample consisted of 10 brains pooled.
d. In addition to the compounds listed, Ohlendorf et al. (1979) did not detect DDD, dieldrin, mirex, oxychlordane, cis-nonachlor, trans-nonachlor, HCB, toxaphene, PCBs, or heptachlor epoxide.
e. In addition to the compounds listed, Spalding et al. (1997) did not detect α-BHC, Γ-BHC, β-BHC, δ-BHC, Γ-chlordane, toxaphene, PCB, o,p'-DDE, α-chlordane, dieldrin, o,p'-DDD, o,p'-DDT, p,p'-DDD, p,p'-DDT, HCB, endrin, cis-nonachlor, trans-nonachlor, oxychlordane, heptachlor epoxide, mirex, endosulfan I, endosulfan II, or aldrin in any of the samples (percent lipid range = 5.10–9.10; percent moisture range = 78.9–85.0; lower level of detection = 0.01 ppm, 0.05 ppm for toxaphene and PCBs).

Table 7.5. Organochlorine residues in eggs of Wood Storks from Florida[d]

County/area	Year(s)	No. examined	DDE			DDD			DDT			Data source
			No.[a]	Mean[b]	(Range)	No.[a]	Mean[b]	(Range)	No.[a]	Mean[b]	(Range)	
Brevard	1973	10	10	4.0	(1.2–19)	2	0.03	(ND– 0.13)	3	0.24	(ND–1.9)	A
"North Florida"[c]	1982	40	40	1.1	(0.16–9.4)	NA	—	—	1	NG	(ND–0.19)	B
			Dieldrin			Mirex			Oxychlordane			
Brevard	1973	10	2	0.053	(ND–0.50)	7	0.30	(ND–1.4)	2	0.030	(ND–0.20)	A
"North Florida"[c]	1982	40	NA	—	—	NA	—	—	11	NG	(ND–0.51)	B
			cis-chlordane			cis-nonachlor			trans-nonachlor			
Brevard	1973	10	2	0.032	(ND–0.24)	1	0.063	(ND–0.84)	0	ND	—	A
"North Florida"[c]	1982	40	3	NG	(ND–0.29)	5	NG	(ND–0.60)	2	NG	(ND–0.44)	B
			HCB			Toxaphene			PCBs			
Brevard	1973	10	1	0.004	(ND–0.047)	3	0.059	(ND–0.41)	10	1.2	(0.43–3.3)	A
"North Florida"[c]	1982	40	NA	—	—	5	NG	(ND–0.24)	25	0.53	(ND–3.5)	B

Sources: A = Ohlendorf et al. (1979), and B = Fleming et al. (1984).
NA = Not analyzed, ND = not detected, NG = not given by author.
a. Number with residue.
b. Geometric mean, ppm wet weight.
c. Fleming et al. (1984) collected 5 eggs from each of 8 colonies in Leon, Lake, Polk, Duval, Brevard, and Indian River counties.
d. In addition to the compounds listed, Fleming et al. (1984) detected heptachlor epoxide in 2 of 40 eggs (<0.1 ppm) from North Florida in 1982. Ohlendorf et al. (1979) did not detect heptachlor epoxide from any of the samples.

Table 7.6. Eggshell thickness for stork, ibis and spoonbill eggs collected in Florida

Species	Area	Year(s)	No. exam.	Mean shell thickness (mm)	% change	Data source
White Ibis	"Florida"	pre-1947	297	0.349	—	A
		1947–73	123	0.350	+0.3	A
	Palm Beach	1989–90	20	0.380	+9	B
Glossy Ibis	"Florida"	pre-1947	29	0.322	—	A
		1947–73	50	0.318	-1	A
Roseate Spoonbill	"Florida"	pre-1947	32	0.425	—	A
		1947–73	7	0.419	-1	A
Wood Stork	"Florida"	1865–1946	93	0.530	—	A
		1947–73	20	0.483	-9[b]	A
	"North Florida"[c]	1982	40	0.508	-4	C

Sources: A = Ohlendorf et al. (1979), B = Rumbold et al. (1996), C = Fleming et al. (1984).
a. Compared with pre-1947 data.
b. Significant difference ($P < 0.001$).
c. Fleming et al. (1984) collected 5 eggs from each of 8 colonies in Leon, Lake, Polk, Duval, Brevard, and Indian River counties.

find no record of organochlorine contaminants measured in Wood Stork tissues. The significance of these low concentrations on health is not known.

VIII. Organophosphates and carbamates

There were only a few records of poisoning from other classes of pesticides. A chlorpyrifos spill was suspected as the cause of death of a wide variety of birds, fish, and reptiles in Highlands County in 1988 (Wentworth 1988). Results of testing for organophosphate residues and brain cholinesterase activity in a White Ibis were negative for organophosphate poisoning. Brain cholinesterase activity measured for 6 Wood Storks is listed in table 7.8. Mild cholinesterase inhibition was suspected in 1 adult bird; however, the number of controls was inadequate for comparison, and the diagnosis was never confirmed by analysis of stomach contents (Thomas 1994).

IX. Metals

Tissues of White Ibis, Wood Storks, and Roseate Spoonbills in Florida have been tested for metals (tables 7.9–7.12). Eggs have been tested only from White Ibis and Wood Storks (tables 7.9, 7.12). Mercury concentrations were much higher in Wood Stork (geometric mean = 0.22 ppm wet weight) than in White Ibis eggs (geometric mean = <0.08 ppm wet weight). The significance of these or most other metals in eggs is not known. Heinz and Locke (1976) observed effects on eggs and nestlings of Mallard hens fed methylmercury when eggs contained up to 5 ppm. Egg mercury concentrations were generally similar to those reported in White-faced Ibis eggs at a contaminated lake in Nevada (Henny and Herron 1989, highest mean = 1.09 ppm dry weight) where no effect upon reproduction was detected.

The concentrations of 2 metals, mercury and lead, may be high enough in tissues to be of concern in this group of birds in Florida. Sundlof et al. (1994) found significantly lower concentrations of mercury in the livers of Roseate Spoonbills and White Ibis when compared with wading birds that are more piscivorous (see chapter 6, Herons, Egrets, and Bitterns, for more details about mercury). Unfortunately that study did not include Wood Storks, which consume larger fish. They also found strong effects of location within southern Florida, with higher concentrations found in birds in the central Everglades and eastern

Florida Bay. The highest concentration of mercury in livers (19 ppm) was reported in fledged Wood Storks from Dade County (table 7.12). The importance of the concentrations of selenium in livers (as high as 8.5 ppm) is poorly understood, but a direct correlation with mercury contamination, as well as a protective effect, has been noted in many studies (Cuvin-Aralar and Furness 1991).

Lead concentrations in the liver of 1 Roseate Spoonbill nestling from eastern Florida Bay (2.8 ppm, Spalding et al. 1997) and from White Ibis nestlings in Palm Beach County (as high as 3.1 ppm, Rumbold et al. 1997) may have been high enough to cause or contribute to death. The highest lead concentration (9 ppm) measured was from the muscle of an adult White Ibis collected in Collier County in 1971 (Ogden et al. 1974).

Other metals were in low concentrations, and little is known about the significance of their presence. The highest cadmium concentration (9.0 ppm) found in adult birds was from a White Ibis collected in Collier County in 1971, the same bird with high lead concentrations (see above). The suggested upper limit for human food is 0.075 ppm wet weight; residues in liver that exceed 10 ppm wet weight are considered to have cadmium contamination (Eisler 1985). In only 1 case, a White Ibis from Collier County in 1971 (Ogden et al. 1974), were concentrations of more than 1.0 ppm measured in Florida ciconiiformes. The significance of arsenic concentrations in bird tissues is poorly understood (Eisler 1988). Arsenic concentrations were not detected in nestlings.

No toxicologic information was available for Greater Flamingos in Florida.

X. Neoplasia

We found no reports of neoplasia in this group of birds. Bililary hyperplasia, a preneoplastic condition or response to toxin exposure, was noted in an emaciated adult female Wood Stork found dead in Lee County in 1989 (Thomas 1991).

XI. Biotoxins

A large scale die-off during June–July 1971 in a shallow inundated area on the west shore of Lake Okeechobee, Glades County, involved 398 birds of 19 species including ducks (see chapter 10, Ducks), ibis, rails, spoonbills, skimmers, terns, grebes, shorebirds, and wading birds (Jasmin et al. 1972). After Mallards, Glossy Ibis (58) were the most commonly found, followed by Roseate Spoonbills (26) and White-faced Ibis (11). It is possible that the White-faced Ibis were misidentified, since there are only a few records of this species in Florida (Robertson and Woolfenden 1992). Birds were weak and had severe hemorrhagic enteritis and acute hepatitis. It was concluded that *Clostridium perfringens* type C, causing necrotizing enteritis, was the most likely cause of mortality, based on culture and mouse protection tests.

Botulism, resulting in mortality due to a toxin produced by the bacteria *Clostridium botulinum*, has not been reported in these species in Florida. A Wood Stork found dead with a large number of birds, including cormorants, in the northern end of the Florida Keys, 1993, was tested for botulism and found to be negative (Meteyer 1993). A single Glossy Ibis was diagnosed with avian botulism in a two-year-long die-off in New York that involved 8 ciconiiforms of 4 species (Brand et al. 1988).

XII. Viruses

Viral diseases of wading birds have been poorly studied. Many viruses can be transmitted by mosquitoes (including poxviruses, Forrester 1991). Poxvirus lesions were found on a nestling Roseate Spoonbill in Florida Bay in 1990 (Spalding and Forrester 1991). Several arboviruses, some of them zoonotic, have been identified in wild birds in Florida (bird species not specified, Chamberlain et al. 1969). Jennings et al. (1969) reported that 1 Wood Stork collected in a roost in the Tampa Bay area in 1962, during a St. Louis encephalitis (SLE) epidemic, had hemagglutination-inhibition antibody for

Table 7.7. Organochlorine residues in tissues of White Ibises collected in Florida[g]

Age Tissue(s)	County/area	Year(s)	No. examined	DDE No.[a]	DDE Mean[b]	DDE (Range)	DDD No.[a]	DDD Mean[b]	DDD (Range)	DDT No.[a]	DDT Mean[b]	DDT (Range)	Data source
Nestlings													
Brain	"Lake Okeechobee"	1990	1[c]	1	0.40	—	0	ND	—	0	ND	—	B
Carcass	Palm Beach	1989	5[d]	5	0.024	(0.010–0.11)	0	ND	—	0	ND	—	A
		1990	5[d]	5	0.012	(0.004–0.067)	0	ND	—	3	0.003	(ND–0.007)	A
		1991	5[d]	5	0.026	(0.012–0.079)	1	NC	(ND–0.008)	1	NC	(0.008)	A
Liver	"Lake Okeechobee"	1992–93	3	3	1.9	(1.40–2.3)	0	ND	—	0	ND	—	D
Adults													
Brain	Collier	1971	3	3	0.27	(0.092–0.69)	1	0.013	(ND–0.092)	0	ND	—	C
Brain & muscle	Dade	1972	1	1	0.010	—	1	0.005	—	1	0.005	—	C
Organs & muscle	Palm Beach	1989	1[e]	1	0.053	—	0	ND	—	1	0.047	—	A
Muscle	Collier	1971	3	3	0.11	(0.081–0.16)	0	ND	—	0	ND	—	C

Age Tissue(s)	County/area	Year(s)	No. examined	Dieldrin No.[a]	Dieldrin Mean[b]	Dieldrin (Range)	Heptachlor epoxide No.[a]	Heptachlor epoxide Mean[b]	Heptachlor epoxide (Range)	PCBs No.[a]	PCBs Mean[b]	PCBs (Range)	Data source
Nestlings													
Brain	"Lake Okeechobee"	1990	1[c]	0	ND	—	0	ND	—	0	ND	—	B
Carcass	Palm Beach	1989	5[d]	0	ND	—	0	ND	—	NA[f]	—	—	A
		1990	5[d]	5	0.008	(0.004–0.014)	0	ND	—	NA[f]	—	—	A
		1991	5[d]	5	0.013	(0.008–0.035)	1	NC	(ND–0.008)	NA[f]	—	—	A
Liver	"Lake Okeechobee"	1992–93	3	NG	NC	(ND–0.33)	NA	—	—	NG	NC	(ND–0.63)	D
Adults													
Brain	Collier	1971	3	0	ND	—	NA	—	—	1	0.15	(ND–1.3)	C
Brain & muscle	Dade	1972	1	1	0.025	—	NA	—	—	1	0.050	—	C

				Oxychlordane			Heptachlor			*trans*-nonachlor			
Organs & muscle	Palm Beach	1989	1[e]	0	ND	—	0	ND	—	NA	—	—	A
Muscle	Collier	1971	3	1	0.006	(ND–0.010)	NA	—	—	2	0.007	(ND–0.37)	C
Nestlings													
Brain	"Lake Okeechobee"	1990	1[c]	0	ND	—	NA	—	—	1	0.01	—	B
Carcass	Palm Beach	1989	5[d]	0	ND	—	0	ND	—	NA	—	—	A
		1990	5[d]	5	0.008	(0.004–0.014)	0	ND	—	NA	—	—	A
		1991	5[d]	5	0.013	(0.008–0.035)	4	0.004	(ND–0.009)	NA	—	—	A
Liver	"Lake Okeechobee"	1992–93	3	NA	—	—	NG	NC	(ND–0.09)	0	ND	—	D
Adults													
Organs & muscle	Palm Beach	1989	1[e]	0	ND	—	1	0.02	—	NA	—	—	A

Sources: A = Rumbold et al. (1996), B = Spalding et al. (1997), C = Ogden et al. (1974), D = Rodgers (1997).

NA = not analyzed, NC = not calculated, ND = not detected, NG = not given.

a. Number with residue.

b. Geometric mean, ppm wet weight.

c. The brains from 2 White Ibises were pooled to make 1 sample.

d. Ten nestlings and 10 eggs were sampled and paired to produce 5 samples. Bills, gastrointestinal tract, legs, and feathers were removed from nestlings.

e. Mixed organs and muscle from 2 adults pooled into 1 sample.

f. See Table 7.3.

g. In addition to the compounds listed, Rumbold et al. (1996) detected 0.16 ppm PCB (Aroclor 1016) in the pooled sample of 2 nestling carcasses from Palm Beach County in 1990, and 0.01 ppm aldrin in pooled mixed organs and muscle from the 2 adults from Palm Beach County in 1989. They did not detect BHC, endosulfan I, endosulfan II, endrin, endosulfan sulfate, methoxychlor, chlordane, or toxaphene in 1989 or 1990. Spalding et al. (1997) did not detect α-BHC, Γ-BHC, β-BHC, δ-BHC, Γ-chlordane, toxaphene, PCB, α-chlordane, aldrin, endrin, *cis*-nonachlor, endosulfan I, endosulfan II, mirex, or HCB in the pooled sample of 2 nestling brains from Lake Okeechobee (percent lipid range = 5.10–9.10; percent moisture range = 78.9–85.0; lower level of detection = 0.01 ppm, 0.05 ppm for toxaphene and PCBs). Rodgers (1997) did not detect chlordane, endrin, or toxaphene in the 3 nestling livers from Lake Okeechobee.

Table 7.8. Brain cholinesterase activity for Wood Storks collected in 1993 in Florida

Age County	Cholinesterase activity		Findings	Data Source
	Routine	18 hrs. at 37°C		
Juveniles				
Collier	20.9(2)+0.3	—	Roadside trauma	Meteyer (1993)
Indian River	15.5(2)+0.1	—	Emaciation	Thomas (1993)
Monroe	19.8(2)+0.3	—	Emaciation	Meteyer (1993)
Adults				
Collier	15.3(2)+0.5	—	Emaciation	Thomas (1993)
	19.2(2)+0.5	24.1(2)+0.1	Emaciation, chronic gunshot, increased liver Hg	Thomas (1994)
	10.4(2)+0.2[a]	11.9(2)+0.2[a]	Trauma, sublethal pesticide toxicosis	Thomas (1994)

a. Possible cholinesterase inhibition.

Group B arbovirus. It was assumed that positive results indicated exposure to SLE virus; however, this was not confirmed by serum neutralization. Neutralizing antibody to eastern equine encephalitis (EEE) virus was detected in 1 of 44 Roseate Spoonbills from Florida Bay and in none of 12 White Ibis and 1 Wood Stork tested from southern Florida, 1988–1990 (Spalding et al. 1994b). Antibodies to SLE virus were detected in 1 White Ibis from Palm Beach County in 1990 from this same group of birds.

A single Wood Stork tested for antibodies to Newcastle disease virus from Orange County in 1999 was negative (Johnson and Spalding 1999). Herpesviruses and paramyxoviruses have been isolated from storks in Europe with no clinical signs (Kaleta and Kummerfeld 1983). Aransas Bay virus, Upolu serogroup, was isolated from ticks (*Ornithodoros capensis*) collected from wading bird or Roseate Spoonbill nests in Texas (Yunker et al. 1979).

XIII. Bacteria

Aeromonas hydrophila was isolated from several nestling Roseate Spoonbills and an adult Greater Flamingo with septicemia (table 7.13). Its association with enteritis could not be confirmed (Stroud 1985; Spalding and Forrester 1991). A Roseate Spoonbill nestling from Florida Bay in 1989 died from severe airsacculitis and septicemia with intralesional gram-negative rods. The organisms were suspected to be *Salmonella* sp. or *Aeromonas* sp. but were not cultured (Spalding and Forrester 1991).

Pasteurella multocida, the causative agent of avian cholera, was cultured from a Greater Flamingo found dead in Everglades National Park in 1985; however, the lesions were not typical of avian cholera, and other bacteria were present (Stroud 1985).

Clostridium perfringens was cultured frequently from the intestines of ciconiiforms in Florida (see also chapter 6, Herons, Egrets, and Bitterns). This organism can be a common non-pathogenic inhabitant of bird intestines, so the diagnosis of *Clostridium*-caused enteritis is difficult to make. The identification of *C. perfringens* type C, the causative agent of necrotic enteritis or enterotoxemia, can be made only by serologic analysis of the toxin produced. *Clostridium perfringens* and *Aeromonas hydrophilla* were cultured from both the bone marrow and the small intestine of a nestling Roseate Spoonbill with focal fibrinonecrotic enteritis and valvular endocarditis. The causative agent could not be determined.

Table 7.9. Metal concentrations in eggs of White Ibises from Florida

County	Year	No. examined	Lead		Mercury		Selenium		Data source
			Mean[a]	(Range)	Mean[a]	(Range)	Mean[a]	(Range)	
Dade	1972	5	0.17	(ND–0.4)	0.04	(ND–0.09)	NA	—	A
Monroe	1972	4	ND	—	0.07	(0.05–0.09)	NA	—	A
Palm Beach	1989	5[b]	NC	(ND–0.52)[d]	0.04[c,d]	(0.02–0.07)[d]	0.38[c,d]	(0.32–0.43)[d]	B
	1990	5[b]	ND	—	0.07[c,d]	(ND–0.11)[d]	0.29[c,d]	(0.23– 2.5)[d]	B
	1991	5[b]	0.25[c,d]	(ND–0.40)[d]	NC	(ND–0.07)[d]	0.31[c,d]	(0.25–0.36)[d]	B
	1994	5[b]	ND	—	ND	—	0.31[c,d]	(NG–0.34)[d]	B

County	Year	No. examined	Arsenic		Cadmium		Zinc		Data source
			Mean[a]	(Range)	Mean[a]	(Range)	Mean[a]	(Range)	
Dade	1972	5	ND	—	0.02	(ND–0.05)	8.4	(6.6–9.6)	A
Monroe	1972	4	0.07	(ND–0.24)	ND	—	7.8	(6.6–9.6)	A
Palm Beach	1989	5[b]	ND	—	ND	—	NA	—	B
	1990	5[b]	ND	—	ND	—	NA	—	B
	1991	5[b]	ND	—	ND	—	NA	—	B
	1994	5[b]	ND	—	ND	—	NA	—	B

County	Year	No. examined	Beryllium		Copper		Nickel		Data source
			Mean[a]	(Range)	Mean[a]	(Range)	Mean[a]	(Range)	
Dade	1972	5	NA	—	1.1	(0.64–1.6)	NA	—	A
Monroe	1972	4	NA	—	1.1	(0.64–1.6)	NA	—	A
Palm Beach	1989	5[b]	ND	—	NA	—	ND	—	B
	1990	5[b]	ND	—	NA	—	ND	—	B
	1991	5[b]	NC	(ND–0.05)[d]	NA	—	NC	(ND–0.29)[d]	B
	1994	5[b]	ND	—	NA	—	0.49[c,d]	(NG–1.3) [d]	B

Sources: A = Ogden et al. (1974), B = Rumbold et al. (1997).
NA = not analyzed, ND = not detected, NC = not calculated (<50% of the samples were positive).
a. Geometric mean, ppm wet weight unless otherwise indicated.
b. Ten eggs were paired to produce 5 samples.
c. Arithmetic means.
d. Wet weight values have been calculated using the formula given in Rumbold (1992).

Table 7.10. Metal concentrations in tissues of White Ibises from Florida

Age Tissue	County/ area	Year	No. examined	Lead Mean[a]	(Range)	Mercury Mean[a]	(Range)	Selenium Mean[a]	(Range)	Data source
Nestlings										
Carcass	Palm Beach	1989	5[b]	0.34[c,d]	(ND–0.62)[d]	0.02[c,d]	(0.01–0.03)[d]	0.31[c,d]	(0.25–0.36)[d]	A
		1990	5[b]	1.5[c,d]	(0.64–3.1)[d]	0.03[c,d]	(0.006–0.07)[d]	0.31[c,d]	(0.25–0.42)[d]	A
		1991	5[b]	NC	(ND–2.7)[d]	ND	—	0.28[c,d]	(0.20–0.39)[d]	A
		1994	5[b]	1.4[c,d]	(NG–1.6)[d]	ND	—	0.22[c,d]	(NG–0.34)[d]	B
Liver	Broward	1987	1	NA	—	0.77	—	NA	—	C
	"Lake Okeechobee"	1989–90	4	0.25[e]	(0.24–0.27)[e]	0.35	(0.24–0.66)	NA	—	C
		1992–93	3	0.22	(0.05–0.71)	0.72	(0.35–3.0)	NA	—	D
	Palm Beach	1990	6	0.23[f]	(0.18–0.28)[f]	0.50	(0.23–1.2)	NA	—	C
Adults										
Feather	Monroe	1989	1	NA	—	4.5[g]	—	NA	—	E
Kidney	Palm Beach	1989	2	1.2	(0.54–2.6)	NA	—	NA	—	C
Liver	"Florida Bay"	1989	1	NA	—	1.3	—	NA	—	C
	"Lake Okeechobee"	1990	1	NA	—	0.47	—	NA	—	C
	Palm Beach	1989	2	1.3	(1.2–1.4)	0.09	(ND–0.15)	NA	—	C
Muscle	Collier	1971	3	0.28	(ND–9.0)	0.50	(0.50–2.6)	NA	—	F
Muscle & brain	Dade	1972	1	0.2	—	1.4	—	NA	—	F
Muscle & organs	Palm Beach	1989	2	2.9[g]	(NG)	0.70[g]	(NG)	1.8g	—	A

Age Tissue	County/ area	Year	No. examined	Arsenic Mean[a]	(Range)	Cadmium Mean[a]	(Range)	Zinc Mean[a]	(Range)	Data source
Nestlings										
Carcass	Palm Beach	1989	5[b]	ND	—	ND	—	NA	—	A
		1990	5[b]	ND	—	NC	(ND–0.14)[d]	NA	—	A
		1991	5[b]	ND	—	ND	—	NA	—	A
		1994	5[b]	ND	—	ND	—	NA	—	B

Tissue	Location	Year	n							Source
Liver	Glades	1990	2	NA	—	0.03	(0.03–0.04)	NA	—	C
	Palm Beach	1990	4	NA	—	0.07	(0.02–0.21)	NA	—	
Adults										
Kidney	Palm Beach	1989	2	0.10	(ND–0.10)	NA	—	13	(12–14)	C
Liver	Palm Beach	1989	2	0.09	(0.05–0.16)	NA	—	2.2	(0.30–16)	C
Muscle	Collier	1971	2	0.30	(ND–1.8)	0.60	(ND–9.0)	0.10	(ND–0.20)	F
Muscle & brain	Dade	1972	1	0.07	—	0.05	—	10	—	F
Muscle & organs	Palm Beach	1989	2	ND	—	0.60[e]	—	NA	—	B

Tissue	Location	Year	n	Beryllium		Copper		Nickel		Source
Nestlings										
Carcass	Palm Beach	1989	5[b]	NC	(ND–0.03)[d]	NA	—	0.59[c,d]	(0.42–0.76)[d]	A
		1990	5[b]	NC	(ND–0.14)[d]	NA	—	0.36[c,d]	(ND–0.92)[d]	A
		1991	5[b]	NC	(ND–0.03)[d]	NA	—	ND	—	A
		1994	5[b]	ND	—	NA	—	0.56[c,d]	(NG–0.95)[d]	B
Liver	Glades	1990	2	NA	—	20	(8.8–45)	NA	—	C
	Palm Beach	1990	4	NA	—	27	(18–55)	NA	—	C
Adults										
Muscle	Collier	1971	2	NA	—	0.41	(ND–3.4)	NA	—	F
Muscle & brain	Dade	1972	1	NA	—	7.2	—	NA	—	F
Muscle & organs	Palm Beach	1989	2	ND	—	NA	—	ND	—	B

Sources: A = Rumbold (1992), B = Rumbold et al. (1997), C = Spalding & Forrester (1991), D = Rodgers (1997), E = Beyer et al. (1997), F = Ogden et al. (1974).
NA = not analyzed, ND = not detected, NC = not calculated (<50% of the samples were positive).
a. Geometric mean, ppm wet weight unless otherwise indicated.
b. Ten nestlings and 10 eggs were sampled and paired to produce 5 samples. Bills, gastrointestinal tract, legs, and feathers were removed from nestlings.
c. Arithmetic mean.
d. Wet weight values have been calculated using the formula given in Rumbold (1992).
e. $n = 2$.
f. $n = 4$.
g. Values in ppm dry weight.

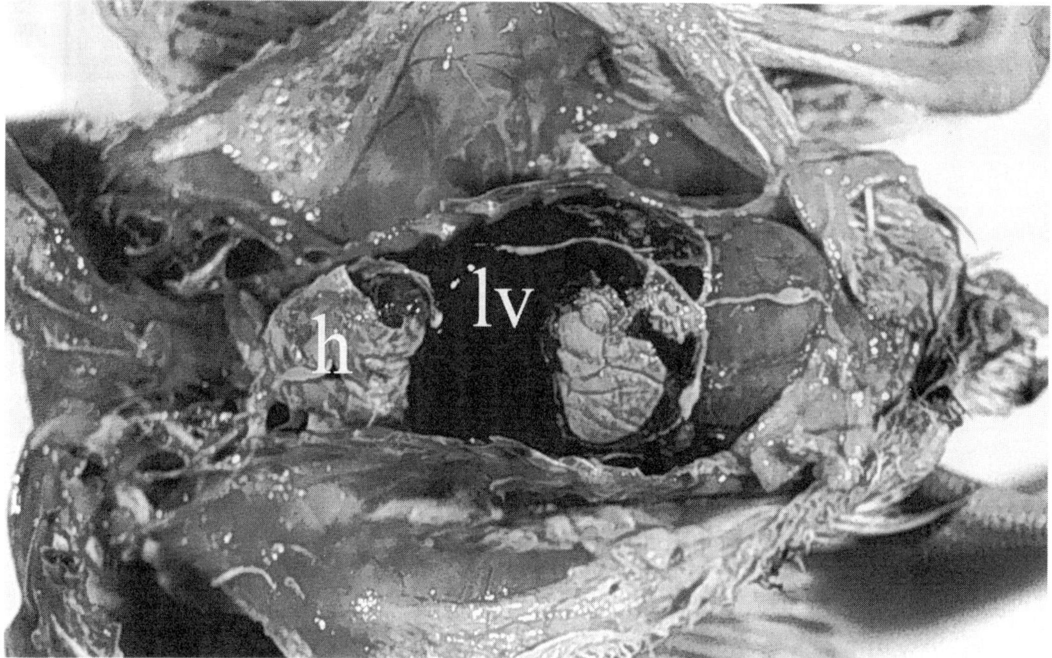

FIGURE 7.1. Roseate Spoonbill nestling with chlamydiosis. Note the fibrinous pericarditis (*h*), fibrin on the surface of the liver (*lv*), and peritonitis.

Chlamydiosis, infection with *Chlamydia psittaci,* was diagnosed histologically from a moribund nestling Roseate Spoonbill from Florida Bay in 1989 (Spalding and Forrester 1991) with fibrinous pericarditis and peritonitis (figure 7.1). Chlamydiosis has not been reported from Florida ciconiiform birds before (see also chapter 6, Herons, Egrets, and Bitterns), but it has been reported elsewhere in North America in these species in Texas and Louisiana at a time when diagnostic tests were questionable (Rubin et al. 1951; Pollard 1947). Franson (1999) lists chlamydiosis as a frequent disease in herons, egrets, and ibis but specific species and locations are not given.

XIV. Fungi

Aspergillosis was suspected as the cause of mild pulmonary lesions in an emaciated immature Wood Stork unable to stand in Pinellas County in 1993 (Thomas 1993). Aspergillosis is reported to be infrequent in wading birds (Friend 1999).

XV. Protozoans

Toxoplasma gondii antibodies were not detected in sera from 1 White Ibis, 1 Glossy Ibis, and 1 Wood Stork from Florida (dates and locations not given) by Burridge et al. (1979). However, these were tested by indirect hemagglutination, which may not be as sensitive in birds as in mammals, resulting in false negatives (Frenkel 1981).

Sarcocysts, the elongate form, were observed in the skeletal muscle tissue of 2 of 5 adult White Ibis (40%) examined histologically but not in 37 examined grossly, from southern Florida (table 7.14). Sarcocysts were not found in 67 Roseate Spoonbills (30 examined histologically) and 3 Wood Storks (one examined histologically) (Spalding et al. 1994a). Immature sarcocysts were observed in skeletal muscle

Table 7.11. Lead, copper, cadmium, and mercury concentrations in tissues of Roseate Spoonbills collected in Florida

| Tissue | | | | Pb | | Cu | | Cd | | Hg | | |
Age	County	Year(s)	No. examined	Mean[a]	(Range)	Mean[a]	(Range)	Mean[a]	(Range)	Mean[a]	(Range)	Data source
Feathers												
	"South Florida"	1987–90	32	NA	—	NA	—	NA	—	2.0[b]	(0.4–5.7)	B
Liver												
Nestling	Monroe	1989–90	6	0.55	(0.18–2.8)	17	(6.4–46)	0.08	(0.02–0.40)	NA	—	A
		1989–90	13	NA	—	NA	—	NA	—	0.53	(0.16–5.4)	A
Adult	Dade	1990	1	NA	—	NA	—	NA	—	0. 84	—	A

Sources: A = Spalding & Forrester (1991), B = Beyer et al. (1997).
NA = not analyzed.
a. Geometric mean, ppm wet weight unless otherwise indicated.
b. Arithmetic mean.

Table 7.12. Metal concentrations in eggs, tissues, and feathers of Wood Storks from Florida

Age	Tissue	County/area	Year(s)	No. examined	Mercury		Cadmium		Lead		Data source
					Mean[a]	(Range)	Mean[a]	(Range)	Mean[a]	(Range)	
Eggs	Eggs	Duval	1982	5	0.09	NG	NA	—	ND		
		"North Florida"	1982	40	0.22	(0.04–0.73)	NA	—	NA	—	D
Nestlings	Blood	Dade	1995	18	0.43	(0.25–0.89)	NA	—	NA	—	G
	Feather[b]	Dade	1995	29	7.4	(5.6–9.3)	G				
		Indian River	1993	1	3.8	—	NA	—	NA	—	A
	Liver	Indian River	1993	1	3.2	—	NA	—	NA	—	A
Adults & Juveniles	Feather[b]	Collier	1993	1	5.1	—	NA	—	NA	—	A
		Dade	1993	2	3.1[c]	(<2.2–8.7)	NA	—	NA	—	A
		Duval	1991	15	1.6	NG	0.12	NG	3.5	NG	B
		Indian River[d]	1993	7	2.2[c]	(1.1–4.3)	NA	—	NA	—	A
		Manatee	1991	1	9.9	—	NA	—	NA	—	C
	Liver	Brevard	1973	1	3.5	—	NA	—	NA	—	E
			1990	1	NA	—	NA	—	ND	—	J
		Collier	1974	1	0.44	—	NA	—	NA	—	H
			1989	1	3.7	—	NA	—	NA	—	H
			1993	2	11	(7.1–18)	NA	—	0.05	(ND–0.25)	A,L
			1993	2	NA	—	NA	—	ND	—	F,K
		Dade	1993	2	11	(6.3–19)	NA	—	NA	—	A,K

					Selenium		Chromium		Manganese		
		Duval	1975	1	11	—	NA	—	NA	—	E
		Indian River	1993	1	3.1	—	NA	—	<0.25	—	K
		Lee	1989	1	NA	—	NA	—	0.05	—	I
		Manatee	1991	1	10	—	NA	—	NA	—	C
		Monroe	1993	1	NA	—	NA	—	ND	—	F
		Pasco	1976	1	8.8	—	NA	—	NA	—	E
Adults & juveniles	Feather[b]	Duval	1991	15	1.6	NG	8.5	NG	3.5	NG	B
	Liver	Collier	1993	2	4.5	(3.0–6.9)	NA	—	NA	—	A,L
		Dade	1993	2	2.7	(0.87–8.5)	NA	—	NA	—	A,K
		Indian River	1993	1	0.91	—	NA	—	NA	—	K

Sources: A = Beyer et al. (1997), B = Burger et al. (1993), C = Facemire & Chlebowski (1991), D = Fleming et al. (1984), E = Gourlie (1984), F = Meteyer (1993), G = Spalding (1995), H = Spalding & Forrester (1991), I = Thomas (1991), J = Thomas (1992), K = Thomas (1993), L = Thomas (1994).
NA = not analyzed, ND = not detected, NG = not given.
a. Geometric mean, ppm wet weight unless otherwise indicated.
b. Feather values are given in ppm dry weight.
c. Arithmetic means.
d. Feathers collected at nests.

Table 7.13. Bacteria cultured from tissues of Roseate Spoonbills, Wood Storks, and Greater Flamingos from Florida

Host species Organism isolated	County/area	Host Age[a]	Year	Tissue	Lesion/disease	Data source
Roseate Spoonbill						
Aeromonas hydrophila	"Florida Bay"	Nestling	1989	Small intestine	Enteritis/septicemia	A
	"Florida Bay"	Nestling	1989	Bone marrow	Enteritis/septicemia	A
Clostridium perfringens	"Florida Bay"	Nestling	1989	Small intestine	Enteritis/septicemia	A
	"Florida Bay"	Nestling	1989	Bone marrow	Enteritis/septicemia	A
Plesiomonas shigelloides	"Florida Bay"	Nestling	1989	Small intestine	Enteritis/endocarditis	A
	"Florida Bay"	Nestling	1989	Bone marrow	Enteritis/endocarditis	A
	"Florida Bay"	Nestling	1989	Liver	Enteritis/endocarditis	A
Gamma *Streptococcus* group D	"Florida Bay"	Nestling	1989	Heart	Enteritis/septicemia	A
	"Florida Bay"	Nestling	1989	Bone marrow	Enteritis/septicemia	A
	"Florida Bay"	Nestling	1989	Liver	Enteritis/septicemia	A
Wood Stork						
Acinetobacter calcoaceticus var *lwoffi*	Collier	Adult	1993	Lung	None	C
Alcaligenes sp.	Brevard	Adult	1990	Liver	Hock cellulitis	D
Enterococcus sp.	Collier	Adult	1993	Liver	None	E
Escherichia coli	Osceola	Adult	1988	Intestine	None	F
Hafnia alveii	Osceola	Adult	1988	Intestine	None	F
Klebsiella oxytoca	Brevard	Adult	1990	Liver	Hock cellulitis	D
Proteus sp.	Brevard	Adult	1990	Hock	Cellulitis	D
	Brevard	Adult	1973	Large intestine	Emaciated	G
Pseudomonas aeruginosa	Collier	Adult	1993	Liver	None	E
Pseudomonas fluorescens	Brevard	Adult	1990	Liver	Hock cellulitis	D
Pseudomonas paucimobilis	Osceola	Adult	1988	Brain	Contaminant type growth	F
Serratia sp.	Collier	Adult	1993	Liver	None	C
Greater Flamingo						
Aeromonas hydrophila	Monroe	Adult	1985	Liver, lung	Septicemia	B
Escherichia coli	Monroe	Adult	1985	Liver, lung	Septicemia	B
Pasteurella multocida	Monroe	Adult	1985	Liver, lung	Septicemia	B

Sources: A = Spalding & Forrester (1991), B = Stroud (1985), C = Thomas (1994), D = Thomas (1992), E = Thomas (1993), F = Thomas (1988), G = Forrester & White (1975).

Table 7.14. Protozoans reported from storks, ibises, and spoonbills from Florida[d]

Host species / Protozoan	County/area	Year(s)	Age[a]	No. Birds Exam.	Pos.	%	Data source
White Ibis							
Haemoproteus plataleae	Alachua	1970–72	AD	6	6	—	A
		1972	JU	20	0	0	A
	"Everglades"	1974	UN	4	1	—	A
	"Florida"	1975	UN	3	3	—	A
	Glades	1972–74	UN	12	11	92	A
		1973	AD	8	6	—	A
	Manatee	1972	AD	15	13	87	A
	Lake Okeechobee	1971	UN	1	1	—	A
	Levy	1972	AD	10	10	100	A
		1972	JU	20	4	20	A
	"Florida Bay"	1987–91	AD	1	1	—	B
	"Lake Okeechobee"	1987–91	AD	1	1	—	B
Sarcocystis sp.	Florida	1987–92	AD	5[b]	2	—	C
Wood Stork							
Haemoproteus crumenium[c]	Brevard	1981	UN	1	1	—	D
	Dade	1995	NS	31	0	0	E
	Duval	1982–83	AD	2	1	—	D
	Glades	1973	UN	2	2	—	F
		1981	JU	1	0	—	D
	Polk	1981	UN	1	0	—	D
Sarcocystis sp.	Monroe	1993	JU	1	1	—	G

Sources: A = Forrester (1980), B = Telford et al. (1992), C = Spalding et al. (1994a), D = Forrester & Bennett (1985), E = Spalding (1995), F = Forrester et al. (1977), G = Meteyer (1993).

a. AD = adult, JU = immature, NS = nestling, UN = unknown age.

b. Six birds were examined grossly and 5 of these were examined histologically for sarcocysts.

c. Forrester et al. (1977) described this parasite as *H. brodkorbi* a few months after the description of *H. crumenium* (Peirce & Cooper 1977) was published. *H. brodkorbi* is a synonym of *H. crumenium*.

d. In addition to the birds listed, Spalding et al. (1994a) found no sarcocysts in 3 nestling and juvenile Wood Storks; 64 nestling and juvenile and 3 adult Roseate Spoonbills; and 31 nestling and juvenile White Ibis examined from Florida, 1987–92. Telford et al. (1992) found no blood parasites in 1 adult and 1 juvenile Wood Stork from South Florida, 51 nestling Roseate Spoonbills from Florida Bay, and 45 White Ibis nestlings from Florida Bay (1), Lake Okeechobee (10), Levy County (34), and South Florida (10). Tilmant (1980) found no blood parasites in 60 blood smears from nestling Wood Storks in Florida Bay (10), Indian River County (40), and Monroe County (10) in 1979.

of an emaciated immature female Wood Stork, Monroe County, in 1993 (Meteyer 1993). The species of sarcocysts in ibises and storks have not been identified.

Telford et al. (1992) found infections of *Haemoproteus plataleae* in each of 2 adult White Ibis from Lake Okeechobee and Florida Bay. They found no blood parasites in 45 nestling White Ibis, 51 nestling Roseate Spoonbills, and 1 adult and 1 juvenile Wood Stork ex-amined from southern Florida. Forrester (1980), on the other hand, reported *Haemoproteus plataleae* in 54 of 98 (55%) White Ibis collected from Alachua, Levy, Manatee, and Glades counties between 1970 and 1975. Of those for which age was determined, 35 of 39 (90%) adults and 4 of 40 (10%) juvenile birds were infected. In June, 4 of 20 (20%) nestlings from Levy County contained immature forms, indicating that the birds had become infected before leaving their

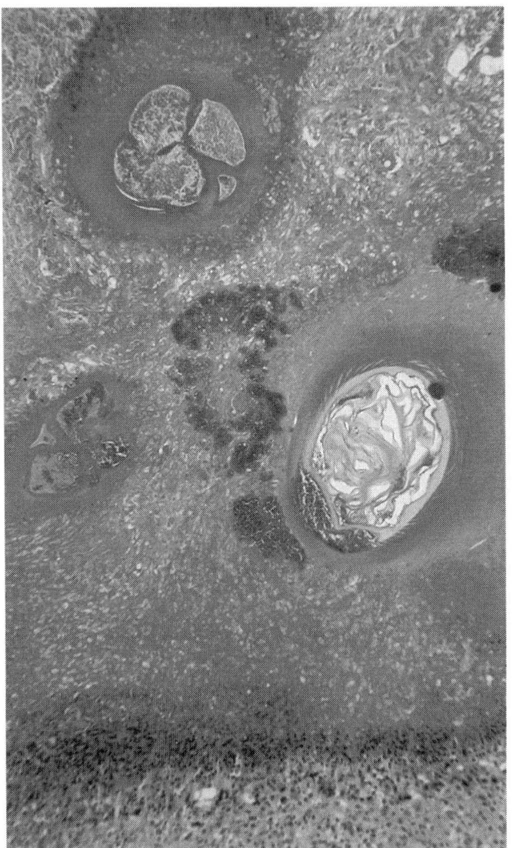

FIGURE 7.2. Cross section of *Contracaecum* sp. mouthparts (trilobed) and degenerate nematode (round) embedded in fibrinonecrotic debris containing bacterial colonies *(center)* in the esophagus of a Roseate Spoonbill with severe obstructive esophagitis.

nesting colony. None of 20 nestlings collected in July and August had detectable parasitemias. *Haemoproteus crumenium* was described from 2 Wood Storks collected from Glades County in 1973 (Forrester et al. 1977). Since then, it has been observed in Duval and Brevard counties. Tilmant (1980) found no *Haemoproteus* infections among 10 nestling Wood Storks from Madeira Rookery, Dade County, 10 from Lane Rookery, Monroe County, or 40 from Pelican Island, Brevard County, in 1979. These nestlings were between 20 and 40 days of age.

Giardia spp. have not been reported from this group of birds in Florida (Kulda and Nohynkova 1995), although they have been documented from storks in Europe (Franssen et al. 2000) and ibis in Australia (McRoberts et al. 1996).

XVI. Helminths

Helminth parasites of this group of birds are listed in tables 7.15–7.20. In addition, a *Tetrameres* sp., collected from a Greater Flamingo in Pasco County in 1959, resides in the National Parasite Collection (number 055826) and is the only parasite collected from flamingos in Florida. It could not be determined if it came from a wild or a captive flamingo.

The helminth parasites of White Ibis have been well studied in central and northern Florida. Three trematodes, 2 found in the cloaca, *Parorchis acanthus,* and *Stomylotrema vicarum,* and 1 in the small intestine, *Parastrigea diovadena,* were present in more than 50% of the birds examined. Bush and Forrester (1976) found a greater diversity of trematodes in ibis foraging in saltwater than in those foraging in freshwater sites. Lesions were not noted in this study of ibis that were collected by gunshot or from nests.

One Roseate Spoonbill and 5 White Ibis nestlings collected from south Florida were infected with *Eustrongylides* larvae. The Roseate Spoonbill nestling was the only one of this species with eustrongylidosis. Among the White Ibis nestlings (from Palm Beach County in 1990) the lesions were severe and appeared to be the cause of death. These nestlings were representative of high mortality in this colony. It is likely that a proportion of the adults from this colony were foraging at an intensively infected site, since this disease was not identified in any other White Ibis nestlings. Since mature *Eustrongylides* spp. were not found in White Ibis, Roseate Spoonbills, or Wood Storks, it is unlikely these species serve as definitive hosts for this parasite (Spalding and Forrester 1993). See

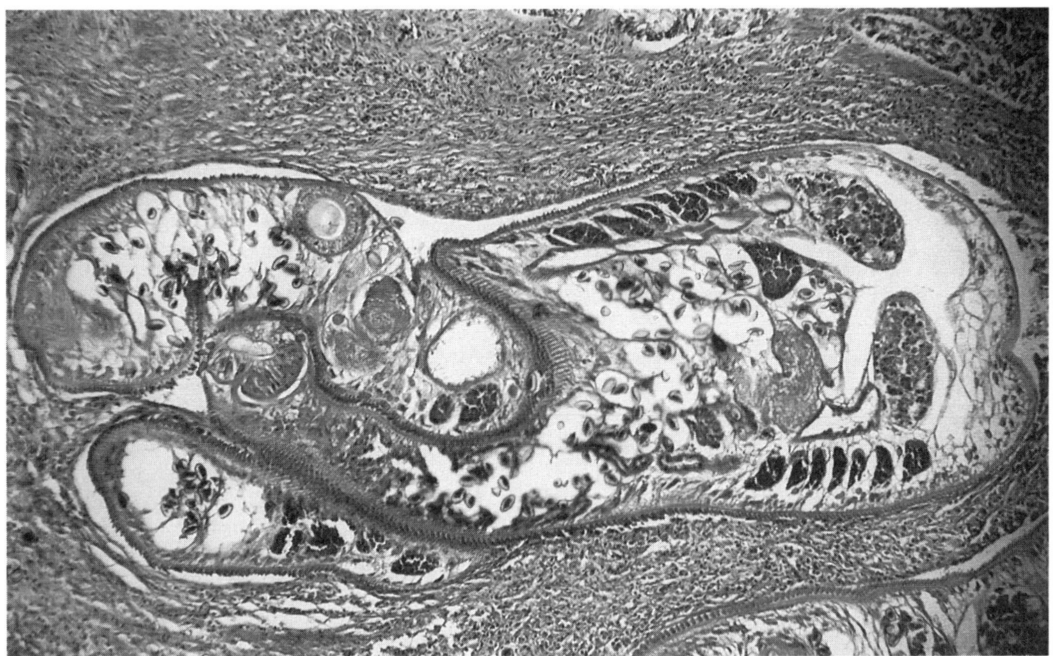

FIGURE 7.3. Trematodes (likely *Pholeter* sp.) encysted within nodular lesions in the muscular wall of the small intestine of a Roseate Spoonbill.

chapter 6, Herons, Egrets, and Bitterns, for more information about this disease.

Roseate Spoonbill helminths were examined by Sepulveda et al. (1994) and found to be very similar to those of the White Ibis in Florida, possibly indicating similar food habits. Only 1 parasite, *Contracaecum multipapillatum*, was found in more than 50% of the birds examined. Sepulveda et al. (1994) noted an inverse relationship between *Cosmocephalus obvelatus* and amount of body fat. Two other helminths were associated with lesions in spoonbills (Spalding 1991). In most cases there was a small focus of inflammation associated with the attachment site of ascaroid nematodes such as *Contracaecum multipapillatum*. In a number of nestling Roseate Spoonbills from Florida Bay in 1991, however, very large nodules and fibrosis caused occlusion of the esophagus and resulted in death (figure 7.2). This was an unusual presentation for these very common parasites. Unfortunately the individ-

ual nematodes involved were not isolated to determine whether they might be a different species. Some of the spoonbill nestlings also had nodular lesions in the wall of the small intestines that were caused by *Pholeter* sp. migrating into the muscular layer of the intestine (figure 7.3).

Several helminths were associated with lesions in Wood Storks. Severe verminous hepatitis, caused by an infection with *Erschoviorchis lintoni* (figure 7.4), and a fractured leg caused the mortality of a Wood Stork in Alachua County in 1999 (Spalding 1999). This was only the third record for this parasite in Florida, which was previously reported in the pancreas of a Common Loon in Florida (see chapter 2, Loons and Grebes). Mild to moderately severe schistosomosis (figure 7.5), caused by *Gigantobilharzia* sp., has been observed in juvenile Wood Storks from Lake County in 1999 (Spalding 1999) and Collier County in 1993 (Meteyer 1993). Microfilarid larvae were

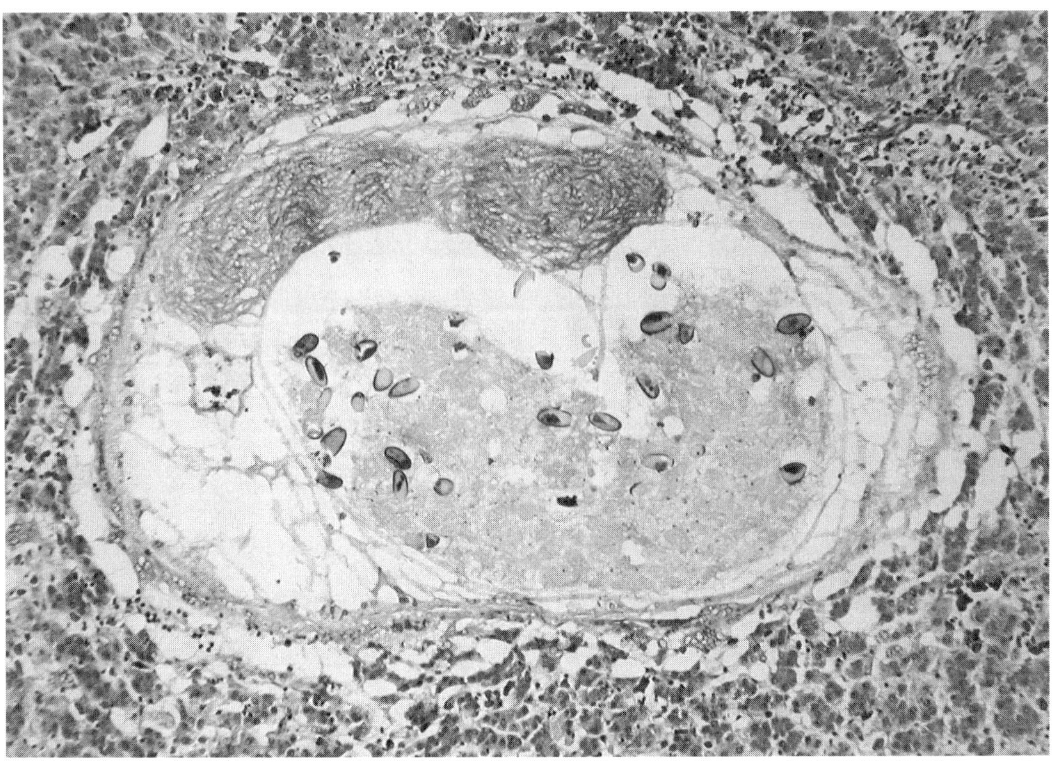

Figure 7.4. A trematode, *Erschoviorchis lintoni,* in the liver of a Wood Stork with verminous peritonitis.

seen in blood smears of 1 of 75 Wood Storks from Georgia (Fedynich et al. 1998) but have not been reported in Florida.

XVII. Arthropods

Subcutaneous mites (hypoderatid deutonymphs) have been well documented in this group of birds from Florida, especially from Roseate Spoonbills in Florida Bay (table 7.21). Five species were found in spoonbills, although only 2 were common. The others may represent cases of "host capture" from other wading birds, since spoonbills commonly share the same nesting colonies, and sometimes the same nests with other species (Pence and Spalding 1996; Pence et al. 1997). In some cases, the numbers of mites seemed sufficient to cause health problems, especially in nestling Roseate Spoonbills, but a cause and effect relationship was not established (Spalding and Forrester 1991). Subcutaneous mites were also found in all of the other species of birds in this group, except for the flamingo.

Other ectoparasites collected from ibis, spoonbills, storks, and flamingos are listed in tables 7.22–7.24. We found no reports of ticks on birds in this group in Florida. Kohls et al. (1965) lists a Roseate Spoonbill from Texas as a host for *Ornithodoros capensis* (*Carios capensis*). Yunker et al. (1979) collected *O. capensis* ticks from wading bird or Roseate Spoonbill nests from Texas and isolated Aransas Bay virus from them.

Nest-dwelling larvae of scavenging dermestid beetles cause lesions and sometimes death in nestling Snail Kites and Wood Storks in Florida (Snyder et al. 1984) (table 7.25). Spalding and Forrester (1991) found postmortem evidence (holes chewed in skin) of dermestid beetle larvae infestation in 18% of Roseate Spoonbill nests in Florida Bay. Occasionally, dermestids cause le-

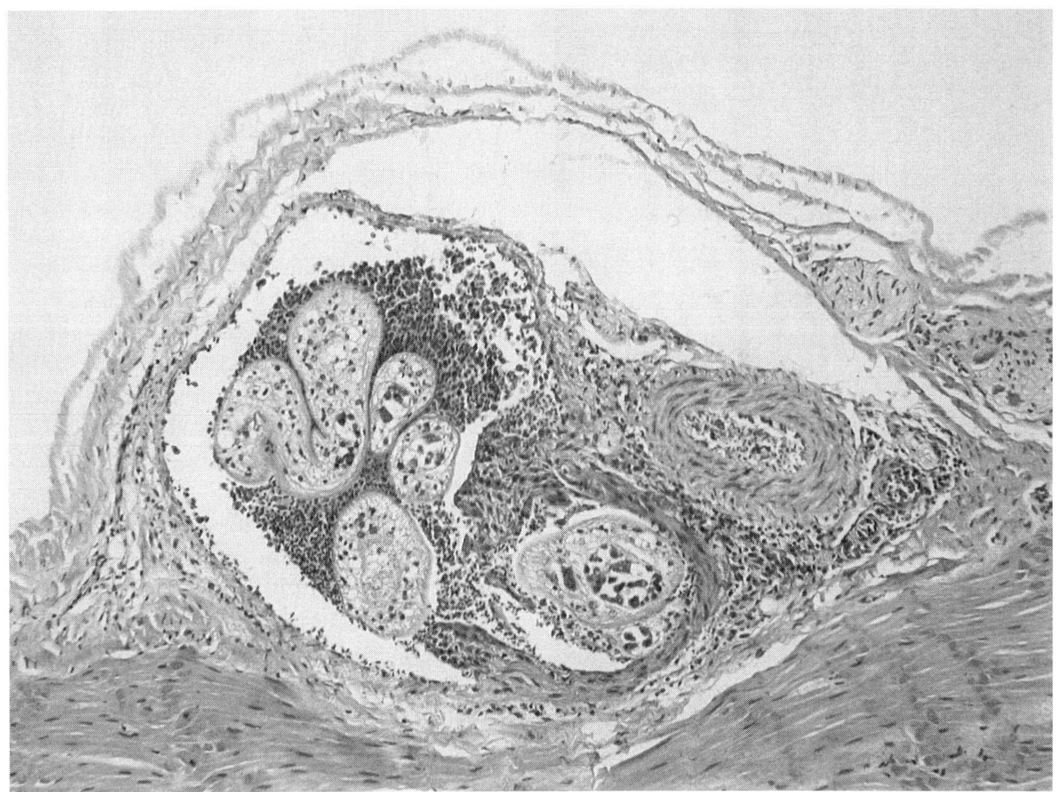

FIGURE 7.5. Schistosomes, probably *Gigantobilharzia* sp., within a serosal vessel of the small intestine of a Wood Stork.

sions on living chicks, as we observed in 4 Roseate Spoonbills (Florida Bay) and 4 White Ibis nestlings (Florida Bay and Levy County). In all cases except 1 spoonbill, the nestlings were emaciated and weak. The White Ibis nestling from Levy County (figure 7.6) had lost significant amounts of blood and had a packed cell volume of 10%. Rodgers et al. (1993) reported skin lesions caused by dermestid larvae (*Dermestes nidum*) on Wood Stork nestlings from 11 colonies in northern and central Florida during 1981–1985. Skin lesions were observed on 0.4–6.5% of the nestlings in these colonies. In only 1 case did it appear that the lesions might have been severe enough to cause death. In the same study, soldier fly larvae (*Hermetia illucens*) were found in all nests examined. These flies generally feed on decaying organic material. Rodgers et al. (1988) felt that green nest material de-

posited early in the nesting cycle might help to isolate chicks from these larvae. Lesions caused by dermestid larvae also were documented in Wood Stork nestlings in Georgia (Jewell 1987).

Red imported fire ants (*Solenopsis invicta*) have been reported to cause significant (92%) reduction in waterbird reproduction (including that of Roseate Spoonbills) on Texas spoil islands (Drees 1994), but we have no such data from Florida. In Texas, ants were observed entering a pipped Roseate Spoonbill egg; however, that nestling successfully hatched and appeared healthy (Ramsey 1968).

XVIII. Nutritional diseases

Nutritional diseases, other than malnutrition or starvation, are rare in wild birds. Nutri-

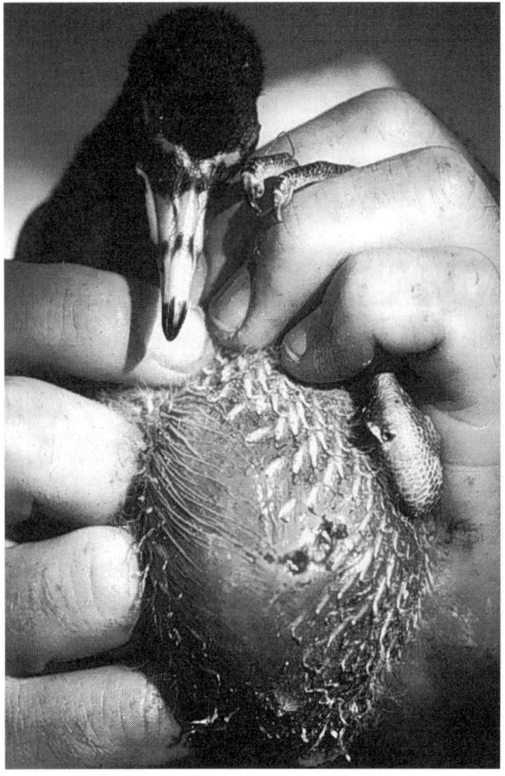

FIGURE 7.6. Bleeding lesion on the abdomen of a nestling White Ibis with severe anemia caused by dermestid larvae.

tional deficiency was suspected to be a possible contributor to leg bone malformation in 7 nestling Roseate Spoonbills on Sandy Key in Florida Bay in 1989 (Spalding and Forrester 1991). Six nestlings and 1 juvenile Roseate Spoonbill had bone malformations of the proximal tarsometatarsus (figure 7.7). In most cases, this malformation was not severe enough to be the cause of death. In a few cases the deformity prevented walking and may have interfered with flight. Others apparently fledged; 1 of these juveniles was found dead, hanging from a power line.

Habitat change, whether the result of human activity or weather changes, can cause modifications in the diet. If this change results in salty prey (isohaline invertebrates) delivered to nestlings that have immature salt glands, a decline in growth rate and death can occur. Although not reported in Florida, such events have been documented elsewhere (Bildstein 1993). In both Trinidad, where freshwater was diverted, and South Carolina following Hurricane Hugo, where salty flood water was trapped in marshes, the net result was abandonment of nesting colonies of White and Scarlet Ibises. Changes in salinity are also further discussed in chapter 2, Loons and Grebes.

Emaciation with no other cause of death was relatively common (16 of 44) in Wood Storks, especially in juvenile birds presented for cause of death determinations (Spalding and Forrester 1991; Spalding 1998; Thomas 1988–94). Abandoned nestlings and fledglings that fail to learn to find sufficient food frequently die from starvation. Wood Storks in south Florida that initiate nests late in the season frequently fail when late spring rains inundate foraging sites, making it more difficult to catch fish. This frequently results in abandonment of nestlings.

XIX. As pests and public health hazards

Although rarely reported in birds of this group within Florida, chlamydiosis can be a serious health hazard to persons handling infected species (Wobeser and Brand 1982; Brand 1989; Franson 1999).

XX. Summary and Conclusions

The white plumage and size of many of these species bring attention to them, especially when ill or when large numbers die. This probably explains why more health-related information is available, especially for White Ibis, Wood Storks, and Roseate Spoonbills, than for other groups of birds. Additionally, colonial nesting birds are included commonly in research studies because of the large numbers of reproducing birds in one location. For this reason, we probably know more about diseases associated with nestlings for these groups, than

FIGURE 7.7. Legs of a juvenile Roseate Spoonbill that had malformations of both tarsometatarsi. Note the bony swelling *(arrow)* of the proximal tarsometatarsus, rotation of the ligament, and lateral deviation of the distal leg and foot.

for the other groups of birds. Storks, ibis, and spoonbills were less commonly involved in large die-off events, when compared with pelicans and gulls. Trauma, especially due to roadside causes such as power line collisions and vehicle collisions, was particularly common especially for Wood Storks. Eagles and nest predators are probably the most common natural causes of death. Historically, collection of eggs and of adults for plumes caused severe population declines. Currently, loss or degradation of nesting and foraging habitat is a significant, but poorly documented, population limiter. Inclement weather probably has more effects on reproductive success than on survival. Pesticides and metal contaminants have been frequently measured, but clear effects on health are poorly understood. The Wood Stork is especially at risk since it consumes larger fish; however, it is less well studied because of its endangered status. Storks, ibis, and spoonbills were found dead occasionally in suspected

biotoxin die-offs but never as the primary species. Death due to viral disease has not been documented in ciconiiforms in Florida, but lesions and titers to 4 viruses have been reported. Sixteen different bacteria have been isolated from stork, spoonbill, and flamingo tissues. Apparently ibis have never been tested for bacteria in Florida. No bacterial associated die-offs have been reported. Only 3 protozoans, 2 species of *Hemoproteus* and 1 *Sarcocystis* sp., have been reported from this group. Helminth populations have been well documented. Thirty-nine species of trematodes, 3 cestodes, 21 nematodes, and 7 acanthocephalans have been identified from storks, ibis, and spoonbills in Florida. Only a few, however, are associated with adverse health effects. Subcutaneous mites and chewing lice have frequently been identified, but little is known about their significance. Dermestid larvae have caused lesions in chicks that might be debilitated for other reasons.

Table 7.15. Trematodes recorded from White Ibises in Florida

Species (location)[a]	County/area[b]	Year	No. of birds Exam.	Pos.	%	Intensity Mean	(Range)	Data source
Acanthoparyphium sp. (SI)	Manatee	1972	20	2	10	5	(3–6)	B
Ascocotyle ampullacea (SI)	Levy	1972	30	4	13	363	(5–1,324)	B
Ascocotyle mcintoshi (SI)	Monroe	1929	NG	NG	—	NG	—	A
	Manatee, Levy	1972	50	22	44	174	(1–3,276)	B
Carneophallus turgidus (SI)	Manatee, Levy	1972	50	6	12	590	(2–1,078)	B
Clinostomum marginatum (OR)	Levy	1972	30	1	3	1	—	B
Gynaecotyla adunca (SI)	Manatee, Levy	1972	50	8	16	258	(1–899)	B
Levinseniella sp. (SI)	Manatee, Levy	1972	50	2	4	605	(31–1,170)	B
Lyperosomum sinuosum (PA)	Alachua, Glades, Manatee	1972	110	12	11	5	—	B
Maritrema sp. (SI)	Manatee	1972	20	1	5	11	—	B
Mesostephanus sp.	"Florida"	NG	NG	—	—	NG	—	C
Microparyphium facetum (CL)	Levy	1972	30	1	3	1	—	B
Ophthalmophagus sp. (ES)	Glades	1972	35	1	3	1	—	B
Ornithobilharzia sp. (BV)	Manatee	1972	20	1	5	1	—	B
Parastrigea diovadena (SI)	Alachua, Glades, Manatee, Levy	1972	140	85	61	123	(1–2,900)	B
Parorchis acanthus (CL)	Manatee	1972	20	15	75	3	(1–8)	B
Parvatrema sp. (SI)	Manatee	1972	20	2	10	8	(2–14)	B
Patagifer vioscai (SI, GL)	Alachua, Glades, Manatee, Levy	1972	140	51	36	5	(1–40)	B
Polycyclorchis eudocimi (TR)	Alachua, Glades, Manatee, Levy	1972	140	9	6	1	(1–3)	B, E
Posthodiplostomum minimum (SI)	Alachua, Glades, Levy	1972	120	36	30	162	(1–1,014)	B
Probolocoryphe glandulosa (SI)	Manatee	1972	20	5	25	2230	(1–6,200)	B
Stephanoprora sp. (SI)	Levy	1972	30	1	3	2	—	B
Stomylotrema vicarium (CL)	Alachua, Glades	1972	90	64	71	36	NG	B
	Manatee, Levy	1972	50	31	62	84	NG	B
Stomylotrema sp. (CL)	Lee	NG	NG	NG	—	NG	—	D
Tanaisia fedtschenkoi (KD)	Alachua, Glades, Manatee, Levy	1972	140	10	7	7	(4–12)	B

Sources: A = Price (1936), B = Bush & Forrester (1976), C = Hutton & Sogandares-Bernal (1959), D = Hutton (1964), E = Pence & Bush (1973).

NG = not given.

a. Location in host: BV = blood vessels, CL = cloaca, ES = esophagus, GL = gizzard lining, KD = kidney, OR = oral cavity, PA = pancreas, SI = small intestine, TR = trachea.

b. Bush & Forrester (1976) examined 140 birds from Alachua ($n = 55$), Glades ($n = 35$), Levy ($n = 30$), and Manatee ($n = 20$) counties. Only counties where each parasite was found are listed.

Table 7.16. Cestodes and acanthocephalans recorded from White Ibises in Florida

Class Species (location)[a]	County[b]	Year	No. of birds Exam.	Pos.	%	Intensity Mean	(Range)	Data source
Cestoda								
Microsomacanthus sp. (SI)	Alachua, Glades, Manatee, Levy	1972	140	25	18	8	(1–35)	A
Cyclustera ibisae[c] (SI)	Alachua, Glades, Manatee, Levy	1972	140	105	75	49	(1–436)	A,B
Acanthocephala								
Corynosoma sp. (SI)	Charlotte	NG	NG	—	—	NG	—	C
Southwellina dimorpha[d] (SI)	Alachua, Glades, Manatee, Levy	1972	140	97	69	10	(1–35)	A
	Indian River	1964	NG	—	—	NG	—	D[e]

Sources: A = Bush & Forrester (1976), B = Schmidt & Bush (1972), C = Hutton (1964), and D = U.S. National Parasite Collection. NG = not given.
a. Location in host: SI = small intestine.
b. Bush & Forrester (1976) examined 140 birds from Alachua (*n* = 55), Glades (*n* = 35), Levy (*n* = 30), and Manatee (*n* = 20) counties. Only counties where each parasite was found are listed.
c. Identified as *Parvitaenia ibisae*. The name was changed by Bona (1994).
d. These are also mentioned in Schmidt (1973).
e. Case number 47697.

Table 7.17. Nematodes recorded from White Ibises in Florida

Species (location)[a]	County/area[b]	Year(s)	No. of birds Exam.	Pos.	%	Intensity Mean	(Range)	Data source
Ancyracanthopsis coronata	Manatee	1972	20	6	30	6	(2–15)	B
Capillaria contorta (ES)	Alachua, Glades, Manatee, Levy	1972	140	29	21	3	(1–8)	B
Capillaria sp. (SI)	Glades	1972	35	1	3	1	—	B
Cyathostoma sp. 1 (TR,LU)	Alachua	1972	55	6	11	6	(1–26)	B
Cyathostoma sp. 2 (TR,LU)	Alachua, Glades	1972	90	1	1	1	—	B
Eustrongylides ignotus (CC)	"South Florida"	1987–91	161[c]	5	3	2	(1–4)	C
Gnathostoma procyonis (SI)	Glades	1972	35	1	3	1	—	B
Sciadiocara chabaudi (GL)	Alachua	1972	55	1	2	1	—	B
Sciadiocara umbellifera (GL)	Manatee	1972	20	10	50	11	(1–26)	B
Skrjabinoclava thapari (PR)	Alachua, Glades, Manatee, Levy	1972	140	7	5	2	(2–7)	B
Strongyloides sp. (SI)	Alachua, Glades	1972	90	18	20	17	(1–111)	B
Syncuaria sp. (GL)	Manatee, Levy	1972	50	2	4	3	(1–5)	B
Syngamus trachea (TR)	Alachua	1972	55	1	2	2	—	B
Synhimantus sp. (ES)	Alachua, Levy	1972	85	2	2	2	(1–2)	B
Tetrameres williamsi (PR)	Alachua, Glades, Manatee, Levy	1972	140	74	53	14	(1–101)	A,B
Viktorocara sp. (GL)	Alachua	1972	55	1	2	1	—	B
Larval spirurids (GL)	Alachua	1972	55	12	22	2	(1–10)	B

Sources: A = Bush et al. (1973), B = Bush & Forrester (1976), C = Spalding et al. (1993).
a. Location in host: CC = coelomic cavity, ES = esophagus, GL = gizzard lining, LU = lung, PR = proventriculus, SI = small intestine, TR = trachea.
b. Bush & Forrester (1976) examined 140 birds from Alachua (55), Glades (35), Levy (30), and Manatee (20) counties. Only counties where each parasite was found are listed.
c. Nine of these were fledged juveniles and adults, all others were nestlings.

Table 7.18. Helminths of Roseate Spoonbills from Florida

| Class | No. of birds[b] | | Intensity | | |
Species (location)[a]	Positive	%	Mean	(Range)	Data source
Trematoda					
Ascocotyle chandleri (SI)	61	45	385	(1–3986)	A
Ascocotyle mcintoshi (SI)	39	29	84	(1–963)	A
Apharyngostrigea multiovata (LI, SI)	5	4	6	(1–10)	A
Clinostomum complanatum (OC/ES)	15	11	2	(1–3)	A
Cotylotretus grandis (BC)	NG	—	NG	—	B
Dendritobilharzia pulverulenta[c] (HT)	1	<1	1	—	A
Echinochasmus dietzevi (SI)	58	43	518	(1–3140)	A
Levinseniella sp.[c] (LI, SI)	1	<1	6	—	A
Mesoophorodiplostomum anterovarium (SI)	4	3	46	(1–140)	A
Mesorchis denticulatus[d] (SI)	14	10	54	(2–248)	A
Microparyphium facetum[c] (CL)	1	<1	1	—	A
Microphallus turgidus (LI, SI)	45	33	112	(1–1386)	A
Phagicola longus (SI)	52	38	1942	(1–47,449)	A
Pholeter sp.[i] (SI)	3	NG	NG	—	C
Posthodiplostomum minimum (LI, SI)	14	10	34	(1–202)	A
Pygidiopsis pindoramensis (LI, SI)	9	7	104	(1–782)	A
Renicola ralli (CL, BC, KD, LI)	16	12	22	(1–140)	A
Cestoda					
Cyclustera capito (SI)	58	43	39	(1–240)	A
Cyclustera ibisae (SI)	23	17	26	(1–307)	A
Microsomacanthus sp. (SI)	1	<1	25	—	A
Nematoda					
Capillaria mergi (LI, ST, SI)	11	8	5	(1–9)	A
Contracaecum multipapillatum[e] (OC/ES, ST, SI)	86	63	13	(1–152)	A
Cosmocephalus obvelatus[e] (OC/ES, ST)	54	40	8	(1–68)	A
Eustrongylides sp.[f] (ST)	1	<1	1	—	A
Syncuaria diacantha (LI, ST)	8	6	2	(1–3)	A
Synhimantus magnipapillatus[g] (ST)	1	<1	107	—	A
Tetrameres micropenis (PR)	3	2	2	—	A
Acanthocephala					
Arhythmorhynchus sp. (SI)	1	<1	1	—	A
Leptorhynchoides sp.[h] (ST)	1	<1	1	—	A
Southwellina hispida (SI)	33	24	11	(1–135)	A

Sources: A = Sepulveda et al. (1994), B = Caballero (1939), C = Spalding (2000).
NG = not given.
a. Location in host: BC = body cavity, CL = cloaca, HT = heart, KD = kidneys, LI = large intestine, OC/ES = oral cavity/esophagus, PR = proventriculus, SI = small intestine, ST = stomach.
b. Sepulveda et al. (1994) examined 136 birds from Monroe, Dade, and Collier counties in 1987–91, and 1 adult male from Lake Okeechobee in 1971 (indicated by note c).
c. Found in an adult male, Lake Okeechobee, 1971.
d. *Stephanophora denticulata* is a synonym of *M. denticulatus*.
e. A combination of adults and larvae.
f. Immature specimen, probably *Eustrongylides ignotus*.
g. Listed as *Synhimantus* sp. by Sepulveda et al. (1994) and later identified as *S. magnipapillatus* by J.M. Kinsella.
h. Probably represents an undescribed species.
i. Only observed histologically within nodules on the small intestine.

Table 7.19. Trematodes and cestodes collected from Wood Storks in Florida

Class Species (location)[a]	County	Year(s)	No. of birds			No. of parasites		Data source
			Examined	Positive	%	Mean	Range	
Trematoda								
Clinostomum attenuatum (OR,ES)	Collier	1974–95	10	3	33	3	(1–6)	A
	Lee	1989	1	1	—	NG	—	C
Diplostomum americanum (IN)	Alachua	1992	1	1	—	19	—	A
	Brevard	1973	1	1	—	1,232	—	A
	Broward	1971–96	2	1	—	8,600	—	A
	Collier	1974–95	10	6	—	397	(9–1500)	A
		1991–95	5	3	—	NG	—	C, E, G
	Dade	1979–93	5	2	—	7	(5–9)	A
		1993	2	1	—	NG	—	E
	Glades	1981	1	1	—	410	—	A
	Indian River	1993	1	1	—	NG	—	E
	Lee	1989	1	1	—	NG	—	C
	Orange	1994–99	2	2	—	1,515	(10–3020)	A
	Osceola	1988	1	1	—	NG	—	B
	Pasco	1975	1	1	—	106	—	A
Echinochasmus dietzevi (SI)	Broward	1971–96	2	1	—	5	—	A
	Collier	1974–95	10	3	33	89	(2–260)	A
		1991–95	5	2	—	NG	—	E, F
	Dade	1979–93	5	3	—	35	(5–95)	A
		1993	2	1	—	NG	—	E
	Pasco	1975	1	1	—	2	—	A
Erschoviorchis lintoni (LV)	Alachua	1999	1	1	—	NG	—	A
Gigantobilharzia sp. (BV)	Collier	1991–95	5	1	—	NG	—	G
		1974–99	10	2	20	1	(1–1)	A
	Orange	1999	1	1	—	4	—	A
Ignavia venusta (KD)	Collier	1974–95	10	3	33	3	(1–2)	A
	Dade	1979–93	5	1	—	1	—	A
Mesorchis denticulatus (LI,CL)	Alachua	1992	1	1	—	29	—	A
	Brevard	1973	1	1	—	352	—	A
Microparyphium facetum (LI,SI)	Collier	1974–95	10	1	10	5	—	A
	Dade	1979–93	5	3	—	5	(1–11)	A
	Glades	1981	1	1	—	1	—	A
	Orange	1994	1	1	—	16	—	A
	Pasco	1975	1	1	—	1	—	A

(continued)

Table 7.19. *(continued)*

Class / Species (location)[a]	County	Year(s)	No. of birds			No. of parasites		Data source
			Examined	Positive	%	Mean	Range	
Phagicola byrdi (SI)	Dade	1979–93	5	1	—	10	—	A
Phagicola longa (IN)	Brevard	1990	1	1	—	5	—	D
Phagicola macrostomus (SI)	Alachua	1992	1	1	—	2	—	A
	Brevard	1973	1	1	—	184	—	A
		1990	1	1	—	NG	—	D
	Broward	1971–96	2	1	—	100	—	A
	Collier	1974–95	10	5	50	49	(1–220)	A
		1991–95	5	3	—	NG	—	E, F, G
	Dade	1979–93	5	3	—	424	(50–980)	A
		1993	2	1	—	NG	—	E
	Glades	1981	1	1	—	44	—	A
	Indian River	1993	1	1	—	NG	—	E
	Lee	1989	1	1	—	NG	—	C
	Levy	1989	1	1	—	NG	—	C
	Monroe	1993	1	1	—	NG	—	G
	Orange	1994–99	2	2	—	2,385	(10–4760)	A
	Osceola	1988	1	1	—	NG	—	B
Phagicola nana (SI)	Dade	1979–93	5	1	—	1	—	A
Phagicola sp. (IN)	Brevard	1990	1	1	—	NG	—	D
	Collier	1974–95	10	1	10	1	—	A
		1991–95	5	1	—	NG	—	E
	Dade	1979–93	5	2	—	6	(6–7)	A
	Indian River	1993	1	1	—	NG	—	E
	Lee	1984	1	1	—	30	—	A
	Orange	1994	1	1	—	290	—	A
	Pasco	1975	1	1	—	1	—	A
Philophthalmus gralli (OC)	Collier	1974–95	10	1	10	1	—	A
Posthodiplostomum sp. (IN)	Collier	1974–95	10	1	10	1	—	A
		1991–95	5	1	—	NG	—	F, G
Cestoda								
Microstomacanthus sp. (SI)	Orange	1994–99	2	1	—	3	—	A

Sources: A = Spalding et al. (1999), B = Thomas (1988), C = Thomas (1991), D = Thomas (1992), E = Thomas (1993), F = Thomas (1994), G = Meteyer (1993).
NG = not given.
a. Location in host: BC = body cavity, BV = blood vessels, CL = cloaca, ES = esophagus, GL = gizzard lining, IN = intestine, KD = kidney, LI = large intestine, OR = oral cavity, SI = small intestine, ST = stomach, LV = liver, OC = ocular.

Table 7.20. Nematodes and acanthocephalans collected from Wood Storks in Florida

Class	Species (location)[a]	County	Year(s)	No. of birds			No. of parasites		Data source
				Examined	Positive	%	Mean	Range	
Nematoda									
	Capillaria avellari (ES)	Alachua	1992	1	1	—	44	—	A
		Brevard	1973	1	1	—	34	—	A
		Broward	1971–96	2	1	—	30	—	A
		Calhoun	1991	1	1	—	2	—	B
		Collier	1991–95	5	2	—	NG	—	E,F
			1974–98	11	5	45	17	(2–38)	A
		Dade	1979–93	5	4	—	9	(1–18)	A
			1993	2	1	—	NG	—	D
		Glades	1981	1	1	—	6	—	A
		Indian River	1993	1	1	—	NG	—	D
		Lafayette	1984	1	1	—	28	—	A
		Lee	1984	1	1	—	10	—	A
			1989	1	1	—	NG	—	B
		Monroe	1993	1	1	—	NG	—	F
		Orange	1994–99	2	2	—	18	(15–21)	A
		Pasco	1975	1	1	—	2	—	A
	Capillaria mergi (SI)	Alachua	1992	1	1	—	7	—	A
		Brevard	1973	1	1	—	2	—	A
		Collier	1974–95	10	3	33	9	(3–21)	A
			1991–95	5	1	—	NG	—	B
		Glades	1981	1	1	—	7	—	A
		Lafayette	1984	1	1	—	5	—	A
		Lee	1984	1	1	—	3	—	A
	Capillaria sp. (IN)	Broward	1971–96	2	1	50	1	—	A
		Lafayette	1984	1	1	—	2	—	A
	Chandleronema longigutturata (IN)	Alachua	1992	1	1	—	4	—	A
	Contracaecum multipapillatum (ST,IN)	Alachua	1992	1	1	—	2	—	A
		Broward	1971–96	2	1	—	2	—	A
		Collier	1959–62	NG	NG	—	NG	—	G[b]
			1974–95	10	8	80	45	(1–130)	A
			1991–95	5	1	—	NG	—	E
		Dade	1979–93	5	2	—	15	(8–22)	A
		Lee	1984	1	1	—	25	—	A
			1989	1	1	—	NG	—	B

(continued)

Table 7.20. *(continued)*

Class	Species (location)[a]	County	Year(s)	No. of birds			No. of parasites		Data source
				Examined	Positive	%	Mean	Range	
		Monroe	1993	1	1	—	NG	—	F
		Orange	1994	1	1	—	24	—	A
		Pinellas	1993	1	1	—	NG	—	D
	Contracaecum sp. (ST)	Broward	1971–96	2	1	—	54	—	A
		Collier	1974–98	11	3	27	21	(2–26)	A
		Dade	1979–93	5	1	—	1	—	A
		Glades	1981	1	1	—	3	—	A
	Cosmocephalus obvalatus (ES)	Lee	1984	1	1	—	1	—	A
		Monroe	1993	1	1	—	NG	—	F
	Eustrongylides ignotus (BC)	Collier	1974–95	10	1	10	1	—	A
		Dade	1979–93	5	2	—	1	(1–2)	A
	Strongyloides sp. (SI)	Glades	1981	1	1	—	16	—	A
		Pasco	1975	1	1	—	1	—	A
	Syncuaria sp. (ST,GL)	Brevard	1973	1	1	—	4	—	A
		Broward	1971–96	2	1	—	3	—	A
		Collier	1974–98	11	6	54	6	(1–24)	A
			1991–95	5	1	—	NG	—	E
		Dade	1979–93	5	1	20	2	—	A
		Indian River	1993	1	1	—	18	—	D
		Lafayette	1984	1	1	—	1	—	A
		Lee	1984	1	1	—	6	—	A
		Monroe	1993	1	1	—	NG	—	F
		Pinellas	1993	1	1	—	NG	—	D
Acanthocephala									
	Larval acanthocephalan (IN)	Collier	1974–95	10	1	10	1	—	A
	Polymorphus crassus (IN)	"Florida"	1856	NG	—	—	NG	—	H
	Southwellina dimorpha (IN)	Alachua	1992	1	1	—	1	—	A
		Brevard	1990	1	1	—	2	—	C
		Collier	1974–95	10	3	30	1	(1–2)	A
		Dade	1979–93	5	1	—	1	—	A
		Lee	1984	2	2	—	1	—	A
			1989	2	1	—	2	—	B

Sources: A = Spalding et al. (1999), B = Thomas (1991), C = Thomas (1992), D = Thomas (1993), E = Thomas (1994), F = Meteyer (1993), G = U.S. National Parasite Collection, H = Van Cleave (1924).

NG = not given.

a. Location in host: BC = body cavity, ES = esophagus, GL = gizzard lining, IN = intestine, LI = large intestine, SI = small intestine, ST = stomach.

b. Case numbers 056352 and 056643.

Table 7.21. Subcutaneous mites of spoonbills, ibises, and storks in Florida

Host Species	County or area	Year(s)	No. of birds Examined	Positive	%[a]	Data source
Roseate Spoonbill						
Hypodectes propus bubulci[b]	"Florida Bay"	1990	29	1	3	A,C
Neottialges ajajae	"Florida Bay"	1989–90	29	23	79	A,C
	Dade	1990	1	1	—	A,C
Neottialges montagui	"Florida Bay"	1989	29	8	28	A,C
Phalacrodectes parvus[c]	"Florida Bay"	1990	29	1	3	A,C
Phalacrodectes whartoni[d]	"Florida Bay"	1989–90	29	2	7	A,C
White Ibis						
Hypodectes propus bubulci	Palm Beach	1990	1	1	—	B,C
Neottialges eudocimae	Collier	1987	2	1	—	C
	"Lake Okeechobee"	1990	2	1	—	C
Phalacrodectes whartoni	Collier	1987	2	1	—	C
	"Lake Okeechobee"	1990	2	1	—	C
	Monroe	1990	1	1	—	A,C
Glossy Ibis						
Neottialges plegadicola	Okeechobee	1994	4	4	—	B
Wood Stork						
Hypoderatidae	Lafayette	1984	1	1	—	D
Neottialges eudocimae	Orange	1994	1	1	—	B
Neottialges kutzeri	Brevard	1990	1	1	—	E
	Brevard	1990	1	1	—	E
	Broward	1971	1	1	—	F
	Collier	1989	1	1	—	C
		1993	2	2	—	E
	Dade	1993	2	1	—	E
	Monroe	1989	1	1	—	C
		1993	1	1	—	E
Neottialges mycteriae	Brevard	1990	1	1	—	E
	Brevard	1990	1	1	—	E
	Broward	1971	1	1	—	F
	Collier	1989	1	1	—	C
		1993	2	2	—	E
	Dade	1993	2	1	—	E
	Monroe	1989	1	1	—	C
		1993	1	1	—	E
Neottialges sp.	Indian River	1993	1	1	—	G
Pelargolichus sp.(females only)	Alachua	1982				D
Phalacrodectes mycteria	Collier	1993	2	2	—	E
	Dade	1993	2	1	—	E

Sources: A = Pence & Spalding (1996), B = Pence et al. (1997), C = Spalding & Pence (1996), D = Forrester & Atyeo (1992), E = Pence & Thomas (1995), F = Pence (1973).

a. Percentages represent the proportion of infected birds that were positive for each species of subcutaneous mites, not the entire set of birds examined.

b. Usually found on ardeids; the finding of this species on spoonbills may represent a case of "host capture" since they share nesting sites with ardeids (Pence & Spalding 1996).

c. This species was described from these spoonbills.

d. Since *P. whartoni* is uncommon on Roseate Spoonbills, Pence & Spalding (1996) suggest that the White Ibis is the normal definitive host.

Table 7.22. Ectoparasites reported from ibises in Florida

Host species Class	Species	County	Year(s)	Data source
White Ibis				
Chewing lice	*Ardeicola elongata*	Broward	1978	A
		Glades	1989	A
		Indian River	1972	A
		Lee	1981, 1983	A
		Levy	1988	A
		Monroe	1987	A
		Okeechobee	1990	A
	Ardeicola robusta	Alachua	1971	A,B
		Broward	1978	A
		Glades	1989	A
		Indian River	1972	A
		Lee	1981, 1983	A
		Levy	1988	A
		Monroe	1987	A
		Okeechobee	1990	A
		Palm Beach	1989	A
	Colpocephalum fusconigrum	Alachua	1971	A
		Broward	1978	A
		Glades	1989	A
		Indian River	1972	A
		Lee	1981, 1983	A
		Levy	1988	A
		Monroe	1987, 1989	A
		Okeechobee	1990	A
		Palm Beach	1989	A
	Ibidoecus bimaculatus	Alachua	1971	A
		Broward	1978	A
		Indian River	1969, 1972, 1976	A
		Lee	1981, 1983	A
		Levy	1988	A
		Monroe	1987, 1989	A
		Okeechobee	1990	A
		Palm Beach	1989	A
		Pinellas	1932	A
	Plegadiphilus eudocimus	Alachua	1971–72	A
		Glades	1989	A
		Indian River	1972	A
		Lee	1981, 1983	A
		Levy	1988	A
		Monroe	1987–88	A
		Okeechobee	1990	A
		Palm Beach	1989	A
Diptera	*Ornithoica confluenta*	Glades	1989	C
Glossy Ibis				
Feather mites	Ingrassinae[a]	Glades	1971	D
Chewing lice	*Ardeicola rhaphidius*	Glades	1971	A
	Colpocephalum leptopygos	Glades	1971	A
	Ibidoecus bisignatus	Glades	1971	A
	Plegadiphilus plegadis	Glades	1971	A

Sources: A = Forrester et al. (1995), B = Forrester (1980), C = Spalding & Wilson (1992), D = Forrester & Atyeo (1992).
a. Xolalgidae, could not be identified further, only females present.

Table 7.23. Chewing lice reported from Roseate Spoonbills in Florida

Species	County/area	Year(s)	Data source
Ardeicola ajajae	Monroe	1918, 1987–92	A
	"Florida"	NG	B
Colpocephalum ajajae	Lee	1982	A
	Monroe	1987–88, 1990	A
Eucolpocephalum femorale	Dade	1990	A
	Monroe	1989–92	A
Ibidoecus iberoamericanus	Dade	1990	A
	Glades	1971	A
	Lee	1982–83	A
	Monroe	1918	A
		1987–90	A

Sources: A = Forrester et al. (1995), B = Hajela & Tandan (1970).
a. NG = not given.

Table 7.24. Ectoparasites reported from storks and flamingos in Florida

Host Species Class	Species	County or area	Year(s)	Data source
Wood Stork				
Feather mites	*Analloptes* sp.	Collier	1993–94	A,B
		Dade	1993	A
	Ingrassia sp.	Lafayette	1984	H
	Mycteralges mesomorphus	Collier	1993	A
		Dade	1992–93	A,B
		Orange	1994	B
	Mycteralges sp.	Collier	1993–94	A,B
		Dade	1993	A
		Orange	1994	B
	Taeniosikya sp.	Collier	1992–93	A
		Dade	1993	A
		Orange	1994	B
Skin mites	*Ornithonyssus bursa*	Collier	1992	A
	Unidentified chigger	Levy	1989	C
Chewing lice	*Ardeicola loculator*	Alachua	1982	D
		Broward	1963	D
		Collier	1993–94	D
		Highlands	1960	E
		Indian River	1971	D
		Lafayette	1984	D
		Lee	1984–89	D
		Monroe	1993	D
		Orange	1984, 1994	D
		Pasco	1975	D
		St. Lucie	1978	D
		"Florida"	NG	G

(continued)

Table 7.24. *(continued)*

Host Species Class	Species	County or area	Year(s)	Data source
	Ciconiphilus quadripustulatus	Collier	1992–94	D
		Orange	1984, 1994	D
		Pasco	1975	D
	Colpocephalum mycteriae	Indian River	1971	D
		Lee	1985	D
	Colpocephalum scalariforme	Bradford	1981	D
		Broward	1963	F
		Collier	1979, 1989,	D
			1993–94	D
		Highlands	1960	F
		Lee	1984–85	D
		Orange	1984, 1994	D
		Pinellas	1929	F
	Neophilopterus heteropygus	Bradford	1981	D
		Collier	1989, 1993–94	D
		Dade	1979, 1992–93	D
		Indian River	1971	D
		Lafayette	1984	D
		Lee	1984–89	D
		Levy	1989	D
		Liberty	1991	D
		Monroe	1993	D
		Orange	1976, 1984, 1994	D
		Pasco	1975	D
		St. Lucie	1978	D
Greater Flamingo				
Chewing lice	*Anaticola candidus*	Dade	1956	D
	Colpocephalum heterosoma	"Florida"	NG	F

Sources: A = Spalding & Mertins (1994), B = Spalding & Mertins (1995), C = Thomas (1991), D = Forrester et al. (1995), E = Tuff (1967), F = Price & Beer (1965), G = Kumar & Tandan (1971), H = Forrester & Atyeo (1992). NG = not given.

Table 7.25. Stork, ibis, and spoonbill chicks examined for dermestid lesions in Florida

| Species | County or area | Year(s) | No. of birds | | | Data source |
			Examined	Positive	%[a]	
Roseate spoonbill	"South Florida"	1987–91	140	3	2[a]	A
White ibis	"South Florida"	1987–91	193	4	2[a]	A
Wood stork	Brevard	1979	36	7	32	B
		1981–85	1,127	8	>1	C
	Dade	1979	55	0	0	B
		1995	31	2	6	D
	Duval	1981–85	1,242	35	3	C
	Hardee	1979	21	5	24	B
	Indian River	1979	270	10	4	B
	Lake	1981–85	902	12	1	C
	Leon	1981–85	2,194	20	>1	C
	Monroe	1979	45	0	0	B
	Pasco	1983–85	1,402	11	>1	C
	Polk	1981–82	329	7	2	C
	St. Johns	1981–85	1,024	15	1	C

Sources: A = Spalding & Forrester (1991), B = Snyder et al. (1984), C = Rodgers et al. (1993), D = Spalding (1995).
a. In addition to antemortem lesions, postmortem holes in skin were observed in 18% of dead Roseate Spoonbill nestlings and 5% of White Ibis nestlings.

XXI. Literature cited

Allen, R.P. 1942. *The roseate spoonbill*. Dover, New York. 142 pp.

Anderson, D.W., and J.J. Hickey. 1972. Eggshell changes in certain North American birds. *Proc. Int. Ornithol. Congr.* 15:514–540.

Baker, J.H. 1940. The director reports to you. *Bird Lore* 42:183–192.

Beyer, W.N., M. Spalding, and D. Morrison. 1997. Mercury concentrations in feathers of wading birds from Florida. *Ambio* 26:97–100.

Bildstein, K.L. 1993. *White ibis: wetland wanderer*. Smithsonian Institution Press, Washington, D.C. 242 pp.

Bjork, R.D., and G.V.N. Powell. 1994. Relationships between hydrologic conditions and quality and quantity of foraging habitat for roseate spoonbills and other wading birds in the C-111 basin. Final Report to the South Florida Research Center, Everglades National Park, Homestead.

———. 1996. Roseate spoonbill. In: *Rare and endangered biota of Florida*. Vol. 5, *Birds*. J.A. Rodgers, H.W. Kale II, and H.T. Smith (eds.). University Press of Florida, Gainesville. pp. 295–308.

Bona, V. 1994. Family Dilepididae. In: *Keys to the cestode parasites of vertebrates*. L.F. Khalil, A. Jones, and R.A. Bray (eds.). Cambridge University Press, Cambridge. pp. 443–554.

Brand, C.J. 1989. Chlamydial infections in free-living birds. *J. Am. Vet. Med. Assoc.* 195:1531–1535.

Brand, C.J., R.M. Windingstad, L.M. Siegfried, R.M. Duncan, and R.M. Cook. 1988. Avian morbidity and mortality from botulism, aspergillosis, and salmonellosis at Jamaica Bay Wildlife Refuge, New York, U.S.A. *Colon. Waterbirds* 11:284–292.

Burger, J., J.A. Rodgers, and M. Gochfeld. 1993. Heavy metal and selenium levels in endangered wood storks *Mycteria americana* from nesting colonies in Florida and Costa Rica. *Arch. Environ. Contam. Toxicol.* 24:417–420.

Burridge, M.J., W.J. Bigler, D.J. Forrester, and J.M. Hennemann. 1979. Serologic survey for *Toxoplasma gondii* in wild animals in Florida. *J. Am. Vet. Med. Assoc.* 175:964–967.

Bush, A.O., and D.J. Forrester. 1976. Helminths of the white ibis in Florida. *Proc. Helminthol. Soc. Wash.* 43:17–23.

Bush, A.O., D.B. Pence, and D.J. Forrester. 1973. *Tetrameres (Gynaecophila) williamsi* sp. n. (Nematoda: Tetrameridae) from the white ibis, *Eudocimus albus*, with notes on *Tetrameres (Tetrameres) grusi* Shumakovich from the sandhill crane, *Grus canadensis. J. Parasitol.* 59:788–792.

Caballero, E. 1939. Sobre la presencia de *Cotylotretus grandis* (Rud., 1819) Odhner, 1910, en las aves de Mexico. *Ann. Inst. Biol.* 10:65–72.

Chamberlain, R.W., W.D. Sudia, T.H. Work, P.H. Coleman, V.F. Newhouse, and J.G. Johnston, Jr. 1969. Arbovirus studies in south Florida, with emphasis on Venezuelan equine encephalomyelitis virus. *Am. J. Epidemiol.* 89:197–210.

Crawford, R.L. 1981. Bird casualties at a Leon County, Florida TV tower: a 25-year migration study. *Bull. Tall Timbers Res. Stn.* 22:1–30.

Cunningham, R.L. 1966. Florida region. *Audubon Field Notes* 20:412–413.

Cuvin-Aralar, M.L., and R.W. Furness. 1991. Mercury and selenium interaction: a review. *Ecotoxicol. Environ. Saf.* 21:348–364.

Davidson, W.R., and V.F. Nettles. 1997. *Field manual of wildlife diseases in the southeastern United States.* 2d ed. Southeastern Cooperative Wildlife Disease Study, University of Georgia, Athens. 417 pp.

Deng, J. 1998. Bird-strike mortality of wetland birds on a 550 kv high-voltage powerline in the Everglades in Florida. M.S. thesis, University of Florida, Gainesville.

Drees, B.M. 1994. Red imported fire ant predation on nestlings of colonial waterbirds. *Southwest. Entomol.* 19:355–359.

Dumas, J.V. 2000. Roseate spoonbill. *Birds N. Am.* 490:1–32.

Eisler, R. 1985. Cadmium hazards to fish, wildlife, and invertebrates: a synoptic review. U.S. Fish and Wildlife Service, Biological Report 85(1.2). 46 pp.

———. 1988. Arsenic hazards to fish, wildlife, and invertebrates: a synoptic review. U.S.

Fish and Wildlife Service, Biological Report 85(1.12). 92 pp.

Facemire, C.F., and L. Chlebowski. 1991. Mercury contamination in a wood stork (*Mycteria americana*) from west-central Florida. U.S. Fish and Wildlife Service Publication VBFO-91-CO3. Vero Beach, Fla. 6 pp.

Fairbrother, A., L.N. Locke, and G.L. Hoff (eds.). 1996. *Noninfectious diseases of wildlife.* Iowa State University Press, Ames. 219 pp.

Fedynich, A.M., L.A. Bryan, Jr., and M.J. Harris. 1998. Hematozoa in the endangered wood stork from Georgia. *J. Wildl. Dis.* 34:165–167.

[FGFWFC] Florida Game and Fresh Water Fish Commission. 1992. Effects of Hurricane Andrew on fish and wildlife of southern Florida, a preliminary assessment. Report, September 8, 1992. Tallahassee. 19 pp.

Fleming, J.W., J.A. Rodgers, and C.J. Stafford. 1984. Contaminants in wood stork eggs and their effects on reproduction, Florida, 1982. *Colon. Waterbirds* 7:88–93.

Folk, M.J. 1992. Cooperative hunting of avian prey by a pair of bald eagles. *Fla. Field Nat.* 20:110–112.

Forrester, D.J. 1971. Unpublished data. University of Florida, Gainesville.

———. 1980. Hematozoa and Mallophaga from the white ibis *Eudocimus albus* in Florida U.S.A. *J. Parasitol.* 66:58.

———. 1991. The ecology and epizootiology of avian pox and malaria in wild turkeys. *Bull. Soc. Vector Ecol.* 16:127–148.

Forrester, D.J., and W.T. Atyeo. 1992. Unpublished data. University of Florida, Gainesville.

Forrester, D.J., and G.F. Bennett. 1985. Unpublished data. University of Florida, Gainesville.

Forrester, D.J., E.C. Greiner, G.F. Bennett, and M.K. Kigaye. 1977. Avian Haemoproteidae. 7. A review of the haemoproteids of the family Ciconiidae (storks) and descriptions of *Haemoproteus brodkorbi* sp. nov. and *H. peircei* sp. nov. *Can. J. Zool.* 55:1268–1274.

Forrester, D.J., H.W. Kale II, R.D. Price, K.C. Emerson, and G.W. Foster. 1995. Chewing lice (Mallophaga) from birds in Florida: a listing by host. *Bull. Fla. Mus. Nat. Hist.* 39:1–44.

Forrester, D.J., and F.H. White. 1975. Unpublished data. University of Florida, Gainesville.

Fowler, M. 1986. Storks and flamingos (Ciconiiformes and Phoenicopteriformes). In: *Zoo and Wild Animal Medicine*. 2d ed. M. Fowler (ed.). W. B. Saunders, Philadelphia. pp. 326–331.

[FPL and FGFWFC] Florida Power and Light Company and Florida Game and Fresh Water Fish Commission. 1991. Unpublished correspondence. Lake Harbor bird mortality project—daily inspections of test framing designs.

Franson, J.C. 1999. Chlamydiosis. In: *Field manual of wildlife diseases. General field procedures and diseases of birds*. M. Friend and J.C. Franson (eds.). U.S. Department of the Interior, U.S. Geological Survey, Biological Research Division, Information and Technology Report 1999-001. Washington, D.C. pp. 111–114.

Franssen, F.F.J., J. Hooimeijer, B. Blankenstein, and D.J. Houwers. 2000. Giardiasis in a white stork in The Netherlands. *J. Wildl. Dis.* 36:764–766.

Frederick, P.C. 1987. Chronic tidally-induced nest failure in a colony of white ibises. *Condor* 89:413–419.

———. 1996. White ibis. In: *Rare and endangered biota of Florida*. Vol. 5, *Birds*. J.A. Rodgers, H.W. Kale II, and H.T. Smith (eds.). University Press of Florida, Gainesville. pp. 466–474.

Frederick, P.C., and M.W. Collopy. 1988. Reproductive ecology of wading birds in relation to water conditions in the Florida Everglades. Florida Cooperative Fish and Wildlife Research Unit, Technical Report 30, University of Florida. 259 pp.

———. 1989a. Researcher disturbance in colonies of wading birds: Effects of frequency of visit and egg-marking on reproductive parameters. *Colon. Waterbirds* 12:152–157.

———. 1989b. Nesting success of five ciconiiform species in relation to water conditions in the Florida Everglades. *Auk* 106:625–634.

———. 1989c. The role of predation in determining reproductive success of colonially nesting wading birds in the Florida Everglades. *Condor* 91:860–867.

Frederick, P.C., and W.F. Loftus. 1993. Responses of marsh fishes and breeding wading birds to low temperatures: a possible behavioral link between predator and prey. *Estuaries* 16:216–222.

Frederick, P.C., and J.C. Ogden. 1997. Philopatry and nomadism: contrasting long-term movement behavior and population dynamics of white ibises and wood storks. *Colon. Waterbirds* 20:316–323.

Frederick, P.C., and M.G. Spalding. 1989. Unpublished data. University of Florida, Gainesville.

———. 1994. Factors affecting reproductive success by wading birds (Ciconiiformes) in the Everglades ecosystem. In: *The Everglades: the ecosystem and its restoration*. S. Davis and J. Ogden, (eds.). St. Lucie Press, Delray Beach, Fla. pp. 659–691.

Frenkel, J.K. 1981. False-negative serologic tests for *Toxoplasma* in birds. *J. Parasitol.* 67:952–953.

Friend, M. 1999. Aspergillosis. In: *Field manual of wildlife diseases. General field procedures and diseases of birds*. M. Friend and J.C. Franson (eds.). U.S. Department of the Interior, U.S. Geological Survey, Biological Research Division, Information and Technology Report 1999-001. Washington, D.C. pp.129–133.

Friend, M., and J.C. Franson (eds.). 1999. *Field manual of wildlife diseases. General field procedures and diseases of birds*. U.S. Department of the Interior, U.S. Geological Survey, Biological Research Division, Information and Technology Report 1999-001. Washington, D.C. 425 pp.

Girard, G.T., and W.K. Taylor. 1979. Reproductive parameters for nine avian species at Moore Creek, Merritt Island National Wildlife Refuge, Florida. *Fla. Sci.* 42:94–102.

Gourlie, N. 1984. Unpublished data. Florida Department of Environmental Regulation, Tallahassee.

Hajela, K.P., and B.K. Tandan. 1970. Species of *Ardeicola* (Insecta: Mallophaga) parasitic on birds of the family Threskiornithidae. *Zool. J. Linn. Soc.* 49:309–334.

Hancock, J.A., J.A. Kushlan, and M.P. Kahl. 1992. *Storks, ibises and spoonbills of the world*. Academic, London. 385 pp.

Heinz, G.H., and L.N. Locke. 1976. Brain lesions in mallard ducklings from parents fed methylmercury. *Avian Dis.* 20:9–17.

Henny, C.J., and G.B. Herron. 1989. DDE, selenium, mercury, and white-faced ibis reproduction at Carson Lake, Nevada. *J. Wildl. Manag.* 53:1032–1045.

Hutton, R.F. 1964. A second list of parasites from marine and coastal animals of Florida. *Trans. Am. Microsc. Soc.* 83:439–447.

Hutton, R.F., and F. Sogandares-Bernal. 1959. Studies of the trematode parasites encysted in Florida mullets. Special Scientific Report 1, Florida State Board of Conservation Marine Laboratory no. 59-12.

Jasmin, A.M., D.E. Cooperrider, C.P. Powell, and J.N. Baucom. 1972. Enterotoxemia of wildfowl due to *Cl. perfringens* type C. *J. Wildl. Dis.* 8:79–84.

Jennings, W.L., W.G. Winkler, D.D. Stamm, P.H. Coleman, and A.L. Lewis. 1969. Serologic studies of possible avian or mammalian reservoirs of St. Louis encephalitis virus in Florida. *Fla. State Board Health Monogr. Ser.* 12:118–125.

Jewell, S.D. 1987. Larval dermestid infestation of nestling wood storks (*Mycteria americana*) in Georgia. *Oriole* 52:11–13.

Johnson, J.M., and M.G. Spalding. 1999. Unpublished data. University of Florida, Gainesville.

Kahl, M.P., Jr. 1964. Food ecology of the wood stork (*Mycteria americana*). *Ecol. Monogr.* 34:97–117.

Kaleta, E.F., and N. Kummerfeld. 1983. Herpes viruses and Newcastle disease viruses in white storks (*Ciconia ciconia*). *Avian Pathol.* 12:347–352.

Kohls, G.M., D.E. Sonenshine, and C.M. Clifford. 1965. The systematics of the subfamily Ornithodorinae (Acarina: Argasidae). II. Identification of the larvae of the western hemisphere and descriptions of three new species. *Ann. Entomol. Soc. Am.* 58:331–364.

Kulda, J., and E. Nohynkova. 1995. *Giardia* in humans and animals. In: *Parasitic protozoa.* Vol. 10. J.P. Kreier (ed.). Academic Press, San Diego. pp. 225–422.

Kumar, P., and B.K. Tandan. 1971. The species of *Adeicola* (Phthiraptera: Ischnocera) parasitic on the Ciconiidae. *Bull. Br. Mus. (Nat. Hist.) Entomol.* 26:119–158.

Kushlan, J.A., and K.L. Bildstein. 1992. White ibis. *Birds N. Am.* 9:1–20.

Kushlan, J.A., and P.C. Frohring. 1986. The history of the southern Florida wood stork population. *Wilson Bull.* 98:368–386.

Kushlan, J.A., and M.S. Kushlan. 1975. Food of the white ibis in southern Florida. *Fla. Field Nat.* 3:31–38.

Lincer, J.L., and D. Salkind. 1973. A preliminary note on organochlorine residues in the eggs of fish-eating birds of the west coast of Florida. *Fla. Field Nat.* 1:19–22.

Locke, L.N. 1987. Chlamydiosis. In: *Field guide to wildlife diseases.* Vol. 1, *General field procedures and diseases of migratory birds.* M. Friend (ed.). U.S. Fish and Wildlife Service, Resource Bulletin 167. pp. 107–113.

Logan, T.H. 1997. Florida's endangered species, threatened species, and species of special concern: official lists. Florida Game and Fresh Water Fish Commission, Tallahassee. 14 pp.

McRoberts, K.M., B.P. Meloni, U.M. Morgan, R. Marano, N. Binz, S.L. Erlandsen, S.A. Halse, and R.C.A. Thompson. 1996. Morphological and molecular characterization of *Giardia* isolated from the straw-necked ibis (*Threskiornis spinicollis*) in western Australia. *J. Parasitol.* 82:711–718.

Meteyer, C. 1993. Unpublished data. National Wildlife Health Center, Madison, Wis.

Mock, D.W., T.C. Lamey, and B.J. Ploger. 1987. Proximate and ultimate roles of food amount in regulating egret sibling aggression. *Ecology* 68:1760–1772.

Nesbitt, S.A., W.M. Hetrick, and L.E. Williams, Jr. 1974. Foods of white ibis from seven collection sites in Florida. In: *Proc. Annu. Conf. Southeast. Assoc. Game Fish Comm.* 28:517–532.

Ogden, J.C. 1978. Roseate spoonbill. In: *Rare and endangered biota of Florida.* Vol. 2, *Birds.* H.W. Kale II (ed.). University Press of Florida, Gainesville. pp. 52–54.

———. 1994. A comparison of wading bird nesting dynamics, 1931–1946 and 1974–1989, as an indication of changes in ecosystem conditions in the southern Everglades, Florida. In: *The Everglades: the ecosystem and its restoration.* S. Davis and J.C. Ogden (eds.). St. Lucie Press, Delray Beach, Fla. pp. 533–570.

———. 1996a. Wood stork. In: *Rare and endangered biota of Florida.* Vol. 5, *Birds.* J.A. Rodgers, H.W. Kale II, and H.T. Smith (eds.). University Press of Florida, Gainesville. pp. 31–41.

———. 1996b. Glossy ibis. In: *Rare and endan-*

gered biota of Florida. Vol. 5, *Birds.* J.A. Rodgers, H.W. Kale II, and H.T. Smith (eds.). University Press of Florida, Gainesville. pp. 457–465.

Ogden, J.C., W.B. Robertson, G.E. Davis, and T.W. Schmidt. 1974. Pesticides, polychlorinated biphenyls and heavy metals in upper food chain levels, Everglades National Park and vicinity. National Technical Information Service, Department of the Interior, Atlanta. 27 pp.

Ohlendorf, H.M., E.E. Klaas, and T.E. Kaiser. 1978. Organochlorine residues and eggshell thinning in wood storks and anhingas. *Wilson Bull.* 90:608–618.

———. 1979. Environmental pollutants and eggshell thickness: anhingas and wading birds in the eastern United States. U.S. Fish and Wildlife Service, Special Scientific Report 216. 94 pp.

Peirce, M.A., and J.E. Cooper. 1977. Haematozoa of East African Birds. IV. Redescription of *Haemoproteus crumenium,* a parasite of the Marabou stork. *East Afr. Wildl. J.* 15:169–172.

Pence, D.B. 1973. Hypopi (Acarina: Hypoderidae) from the subcutaneous tissues of the wood ibis, *Mycteria americana* L. *J. Med. Entomol.* 10:240–243.

Pence, D.B., and A.O. Bush. 1973. *Polycyclorchis eudocimi* gen. et sp. n. (Trematoda: Cyclocoelidae) from the trachea of the white ibis *Eudocimus albus* L. *J. Parasitol.* 59:85–89.

Pence, D.B., and M.G. Spalding. 1996. Hypopi (Acari: Hypoderatidae) from the roseate spoonbill (Aves: Ciconiiformes; Threskiornithidae). *J. Med. Entomol.* 33:244–249.

Pence, D.B., M.G. Spalding, J.F. Bergan, R.A. Cole, S. Newman, and P.N. Gray. 1997. New records of subcutaneous mites (Acari: Hypoderatidae) in birds with examples of potential host colonization events. *J. Med. Entomol.* 34:411–416.

Pence, D.B., and N.J. Thomas. 1995. Hypopi (Acari: Hypoderatidae) of the wood stork (Aves: Ciconiiformes; Ciconiidae). *J. Med. Entomol.* 32:895–899.

Pollard, M. 1947. Ornithosis in seashore birds. *Proc. Soc. Exp. Biol. Med.* 64:200.

Post, W. 1992. Wood stork mortality from Hurricane Hugo. *Fla. Field Nat.* 20:107.

Powell, G.V.N., R.D. Bjork, J.C. Ogden, R.T. Paul, and A.H. Powell. 1989. Population

trends in some Florida Bay wading birds. *Wilson Bull.* 101:436–457.

Price, E.W. 1936. A new heterophyid trematode of the genus *Ascocotyle* (Centrocestinae). *Proc. Helminthol. Soc. Wash.* 3:31–34.

Price, R.D., and J.R. Beer. 1965. The *Colpocephalum* (Mallophaga: Menoponidae) of the Ciconiiformes. *Ann. Entomol. Soc. Am.* 58:111–131.

Ramsey, J.J. 1968. Roseate spoonbill chick attacked by ants. *Auk* 85:325.

Robertson, W.B., Jr., and D.R. Paulson. 1961. Region reports: Florida region. *Audubon Field Notes* 15:26–35.

Robertson, W.B., Jr., and G.E. Woolfenden. 1992. *Florida bird species: an annotated list.* Spec. Publ. 6, Florida Ornithological Society, Gainesville. 260 pp.

Rodgers, J.A. 1997. Pesticide and heavy metal levels of waterbirds in the Everglades agricultural area of south Florida. *Fla. Field Nat.* 25:33–84.

———. 1996. Frequency of addled eggs of nesting wood storks in north and central Florida. *Fla. Field Nat.* 24:38–40.

Rodgers, J.A., Jr., H.W. Kale II, and H.T. Smith (eds.). 1996a. *Rare and endangered biota of Florida.* Vol. 5, *Birds.* University Press of Florida, Gainesville. 688 pp.

Rodgers, J.A., Jr., and S.T. Schwikert. 1997. Breeding success and chronology of wood storks *Mycteria americana* in northern and central Florida, U.S.A. *Ibis* 139:76–91.

Rodgers, J.A., Jr., S.T. Schwikert, and A. Shapiro-Wenner. 1996b. Nesting habits of wood storks in north and central Florida, USA. *Colon. Waterbirds* 19:1–169.

Rodgers, J.A., Jr., S.T. Schwikert, and A.S. Wenner. 1993. The prevalence of abdominal lesions on wood stork nestlings in north and central Florida. *Condor* 95:473–475.

Rodgers, J.A., Jr., and H.T. Smith. 1997. Buffer zone distances to protect foraging and loafing waterbirds from human disturbance in Florida. *Wildl. Soc. Bull.* 25:139–145.

Rodgers, J.A., A.S. Wenner, and S.T. Schwikert. 1988. The use and function of green material in wood stork nests. *Wilson Bull.* 100:411–423.

Rubin, H., R.E. Kessling, R.W. Chamberlain, and M.E. Eidson. 1951. Isolation of a psittacosis-like agent from the blood of snowy egrets. *Proc. Soc. Exp. Biol. Med.* 78:696.

Rumbold, D.G. 1992. Ecotoxicological surveillance of chemical residues in wildlife near the North County Resource Recovery Facility: a baseline study. Final Report to Solid Waste Authority of Palm Beach County, West Palm Beach, Fla. 57 pp.

Rumbold, D.G., M.C. Bruner, M.B. Mihalik, and E.A. Marti. 1997. Biomonitoring environmental contaminants near a municipal solid-waste combustor. *Environ. Pollut.* 96: 99–105.

Rumbold, D.G., M.C. Bruner, M.B. Mihalik, E.A. Marti, and L.L. White. 1996. Organochlorine pesticides in anhingas, white ibises and apple snails collected in Florida, 1989–1991. *Arch. Environ. Contam. Toxicol.* 30:379–383.

Runde, D.E. 1991. Trends in wading bird nesting population in Florida 1976–78 and 1986–89. Florida Game and Fresh Water Fish Commission, Nongame Wildlife Section, Tallahassee. 105 pp.

Ryder, R.A., and D.E. Manry. 1994. White-faced ibis. *Birds N. Am.* 130:1–24.

Schmidt, G.D. 1973. Resurrection of *Southwellina* Witenberg, 1932, with a description of *Southwellina dimorpha* sp. n. and a key to the genera in Polymorphidae (Acanthocephala). *J. Parasitol.* 59:299–305.

Schmidt, G.D., and A.O. Bush. 1972. *Parvitaenia ibisae* sp. n. (Cestoidea: Dilepididae), from birds in Florida. *J. Parasitol.* 58:1095–1097.

Sepulveda, M.S., M.G. Spalding, J.M. Kinsella, R.E. Bjork, and G.S. McLaughlin. 1994. Helminths of the roseate spoonbill, *Ajaia ajaja*, in southern Florida. *J. Helminthol. Soc. Wash.* 61:179–189.

Snyder, B. 1994. Unpublished data. Florida Department of Environmental Protection, Tallahassee.

Snyder, N.F.R., J.C. Ogden, J.D. Bittner, and G.A. Grau. 1984. Larval dermestid beetles feeding on nestling snail kites, wood storks, and great blue herons. *Condor* 86:170–174.

Spalding, M.G. 1987–2001. Unpublished data. University of Florida, Gainesville.

Spalding, M.G., C.T. Atkinson, and R.E. Carleton. 1994a. *Sarcocystis* sp. in wading birds (Ciconiiformes) from Florida. *J. Wildl. Dis.* 30:29–35.

Spalding, M.G., G.T. Bancroft, and D.J. Forrester. 1993. The epizootiology of eustrongylidosis in wading birds (Ciconiiformes) in Florida. *J. Wildl. Dis.* 29:237–249.

Spalding, M.G., and R. Bjork. 1990. Unpublished data. University of Florida, Gainesville.

Spalding, M.G., and D.J. Forrester. 1991. Effects of parasitism and disease on the reproductive success of colonial wading birds (Ciconiiformes) in southern Florida. Final Report. Florida Game and Fresh Water Fish Commission, Tallahassee. 130 pp.

———. 1993. Pathogenesis of *Eustrongylides ignotus* (Nematoda: Dioctophymatoidea) in Ciconiiformes. *J. Wildl. Dis.* 29:250–260.

Spalding, M.G., D.J. Forrester, and J.M. Kinsella. 1999. Unpublished data. University of Florida, Gainesville.

Spalding, M.G., R.G. McLean, J.H. Burgess, and L.J. Kirk. 1994b. Arboviruses in water birds (Ciconiiformes, Pelecaniformes) from Florida. *J. Wildl. Dis.* 30:216–221.

Spalding, M.G., and J.W. Mertins. 1994–1999. Unpublished data. University of Florida, Gainesville.

Spalding, M.G., and D.B. Pence. 1996. Unpublished data. University of Florida, Gainesville.

Spalding, M.G., C.K. Steible, S.F. Sundlof, and D.J. Forrester. 1997. Metal and organochlorine contaminants in tissues of nestling wading birds (Ciconiiformes) from southern Florida. *Fla. Field Nat.* 25:42–50.

Spalding, M.G., and N.A. Wilson. 1992. Unpublished data. University of Florida, Gainesville.

Sprunt, A., Jr. 1954. *Florida bird life*. Coward-McCann, New York. 527 pp.

Sprunt, A.I., and M.P. Kahl, Jr. 1960. Mysterious *Mycteria* our American stork. *Audubon Mag.* 62:206–209, 234, 252.

Sprunt, A.J. 1944. Northward extension of the breeding range of the white ibis. *Auk* 61:144–145.

———. 1949. Predation on living prey by the black vulture. *Auk* 63:260–261.

Sprunt, S.A. 1969. A case of turkey vulture piracy on great blue herons. *Wilson Bull.* 77:257–263.

Stangel, P.W., J.A. Rodgers, and A.L. Bryan. 1990. Genetic variation and population structure of the Florida wood stork. *Auk* 107:614–619.

Stevenson, H.M. 1963. Region reports: Florida region. *Audubon Field Notes* 17:319–323.

———. 1969. The winter season: Florida region. *Audubon Field Notes* 23:468–473.

Stevenson, J.A. 1994. Unpublished data. Florida Department of Environmental Protection, Tallahassee.

Stroud, R.K. 1985. Unpublished data. National Wildlife Health Center, Madison, Wis.

Sundlof, S.F., M.G. Spalding, J.D. Wentworth, and C.K. Steible. 1994. Mercury in livers of wading birds (Ciconiiformes) in southern Florida. *Arch. Environ. Contam. Toxicol.* 27:299–305.

Telford, S.R., Jr., M.G. Spalding, and D.J. Forrester. 1992. Hemoparasites of wading birds (Ciconiiformes) in Florida. *Can. J. Zool.* 70:1397–1408.

Thomas, N.J. 1988–1994. Unpublished data. National Wildlife Health Center, Madison, Wis.

Tilmant, L. 1980. Haemoproteid blood parasites in natural populations of wood storks, *Mycteria americana* Linne. Unpublished manuscript. University of Florida, Gainesville.

Tuff, D.W. 1967. A review of North American *Ardeicola* (Mallophaga: Philopteridae). *J. Kans. Entomol. Soc.* 40:241–263.

Van Cleave, H.J. 1924. A critical study of the Acanthocephala described and identified by Joseph Leidy. *Proc. Acad. Nat. Sci. Phila.* 76:279–334.

Wentworth, E.J. 1988. Unpublished data. Southeastern Cooperative Wildlife Disease Study, University of Georgia, Athens.

Wobeser, G., and C.J. Brand. 1982. Chlamydiosis in two biologists investigating disease occurrence in wild waterfowl. *Wildl. Soc. Bull.* 10:170–172.

Yunker, C.E., C.M. Clifford, J.E. Keirans, L.A. Thomas, and R.C.A. Rice. 1979. Aransas Bay virus, a new arbovirus of the Upola serogroup from *Ornithodoros capensis* (Acari: Argasidae) in coastal Texas. *J. Med. Entomol.* 16:453–460.

Vultures

I. Introduction

Two species of New World vultures (family Cathartidae) occur in Florida, the Black Vulture (*Coragyps atratus*) and the Turkey Vulture (*Cathartes aura*). Both breed in Florida and their populations are augmented by northern migrants in the fall and winter. The Black Vulture is uncommon to locally common throughout the mainland and the Turkey Vulture is fairly common to common throughout the state, including the Keys (Robertson and Woolfenden 1992). In a recent analysis of data from breeding bird surveys, Florida had the highest breeding season densities of both Black and Turkey Vultures among the 12 southeastern states (Coleman and Fraser 1990). The same authors also found from Christmas bird count data that winter densities in 1986 and 1987 were highest in Florida for Turkey Vultures (86 birds per 100 miles) and second highest for Black Vultures (20 birds per 100 miles) for the 12 southeastern states. They also found that the numbers of Turkey Vultures during the breeding season had decreased significantly in Florida over the 22-year period 1966–87, while numbers of Black Vultures remained the same. On the other hand there was an increase in numbers of both species in Florida during the winter months for the same period of time.

For additional information on distribution and population status and for general information on biology and management of vultures the reader is referred to Bent (1937), Wilbur and Jackson (1983), Rabenold (1985), Jackson (1988), and Coleman and Fraser (1990).

II. Trauma

Because both species of vultures habitually feed on road-killed carcasses of various animals, collisions with vehicles undoubtedly result in significant morbidity and mortality of both in Florida. In a 4-year survey conducted from 1990 to 1993 in state parks and recreation areas in Florida, 55 vultures were found dead on highways (table 8.1). Black Vultures outnumbered Turkey Vultures slightly in that study, but more Turkey Vultures were injured or killed on highways than Black Vultures, according to the records of birds submitted to 3 wildlife rehabilitation centers in Florida between 1988 and 1995 (table 8.2). These cases of vehicle strike trauma were from counties close to the rehabilitation centers where they were submitted, which indicates a possible locality bias in understanding geographical locations of this type of morbidity and mortality. For example, the cases reported by Deem and Terrell (1996) from the records of the Veterinary Medical Teaching Hospital in Gainesville (Alachua County), for which locality data were known, came from Alachua (n = 17 vultures), Marion (n = 2), and Levy (n = 1) counties. Likewise, the cases from the Center for Birds of Prey in Maitland (Orange County) came from Orange (n = 8 vultures), Seminole (n = 5), Volusia (n = 4), Osceola (n = 4), Polk (n = 2), Brevard (n = 1), and Sumter (n = 1) counties (Collins and Gilliland 1995). Undoubtedly the numbers killed throughout the state are considerably higher, particularly on highways with higher speed limits. In all probability most of the 32 vultures listed in table 8.2 as "unknown trauma" were vehicle strike cases.

Mortality due to collisions with power lines, radio and TV towers, and other tall structures probably also occurs, but extensive data throughout the state are lacking. Collins and Gilliland (1995) have records of 1 Turkey Vulture being fatally injured by striking a power line (1989, county unknown) and another by striking a TV tower (1994, Orange County). Between October of 1989 and July of 1991, 15 vultures were found dead near power lines and poles in the vicinity of the Miami Canal across from Lake Harbor in Palm Beach County (Regan 1996). Only one, a Turkey Vulture, was identified to species. These vultures died either because of trauma inflicted when they struck a power line or by electrocution, or a combination of the two factors. Necropsies were not performed on these birds, however, to verify this assumption. Crawford (1981) reported that 16 Turkey Vultures and 4 Black Vultures were found dead at the base of a 1,010-foot TV tower at Tall Timbers Research Station in Leon County. These numbers were based on examination of more than 42,000 birds over a 25-year period. Vultures have not been found in other studies of such mortality involving tall structures in Florida (Kale 1971; Taylor and Anderson 1973; Taylor and Anderson 1974; Maehr et al. 1983; Taylor and Kershner 1986; Maehr and Smith 1988).

Table 8.1. Numbers of vultures killed on highways in Florida state parks and recreation areas between 1990 and 1993

Species of vulture	No. vultures reported by year				
	1990	1991	1992	1993	Totals
Black Vulture	5	7	7	11	30
Turkey Vulture	6	5	2	9	22
Unidentified Vultures	0	0	0	3	3
No. parks reporting	45	64	64	61	—

Sources: Snyder (1994), Stevenson (1994).

Table 8.2. Types of trauma causing morbidity and mortality in vultures submitted to three wildlife rehabilitation centers in Florida, 1988–95

Species of vulture	Vehicle strike	Gunshot	Power line strike	TV tower strike	Unknown	Total
Black Vulture	5	1	0	0	17	23
Turkey Vulture	22	3	1	1	15	42
Totals	27	4	1	1	32	65

Sources: Collins and Gilliland (1995), 29 cases; Deem and Terrell (1996), 24 cases; Suto (1996), 12 cases.

Three Turkey Vultures and 1 Black Vulture were victims of gunshot (table 8.2). The Black Vulture came from Alachua County and the 3 Turkey Vultures from Alachua, Levy, and Sumter counties (Deem and Terrell 1996; Collins and Gilliland 1995).

III. Predation

Predation has been linked with mortality of vultures, especially of eggs and nestlings, although we have no specific data for Florida. In other areas there is evidence implicating foxes, opossums, dogs, and crows (Jackson 1983).

IV. Electrocution

Electrocution of vultures due to perching on power lines and poles probably occurs from time to time. We know of only one such account, that of Hallinan (1922), who reported observations of Turkey Vultures being electrocuted on high-tension electric transmission lines in South Jacksonville (Duval County) in the summer of 1921. He gave no details as to how many birds were involved, however. The significance of this type of mortality to vulture populations in Florida is unknown.

stumps, logs, and trees, and in thickets where they are less protected from sudden flood waters (Jackson 1983, 1988). Baynard (1909) reported Black Vulture nest losses to flooding in Florida (exact locality not given), stating that hundreds of nestlings and many clutches of eggs were lost during the flood that occurred in February of either 1908 or 1909.

Mote (1969) recorded an interesting occurrence of Turkey Vultures landing on a ship at sea. The incident occurred as a 57-foot vessel was crossing Florida Bay in November of 1968. The sea was calm and smooth and there was a low-lying fog that reduced visibility to about a half mile. About 55 exhausted vultures tried to land on the moving ship, but were unable to do so until the ship slowed down. Several birds fell into the sea and drowned. The others remained on the vessel and eventually flew off when the vessel came close to land. The author stated that the vultures "were so exhausted that they would let us come up and actually touch them. Some of them after landing regurgitated partly digested mice and other objects on the deck. The birds showed no interest whatsoever in the passengers, among whom needless to say they caused some apprehension. The vessel certainly did look like a 'ship of death' going through the fog with its load of vultures" (Mote 1969: 766).

V. Inclement weather

Although vultures often build their nests up off the ground (in tree cavities for instance), a significant number are constructed in hollow

VI. Organochlorines

Carcasses of both species of vultures were obtained in the early 1970s from north central Florida (except for 4 Turkey Vultures from Leon

Table 8.3. Concentrations of chlorinated hydrocarbons in various tissues of Black Vultures from Florida

County Year	Age[b]	Sex[c]	Organ/ tissue[d]	Residue (ppm wet weight)[a]				Data source
				DDE	ΣDDT	Dieldrin	PCB[e]	
Alachua								
1972	AD	M	AD	6.4	11	1.2	NT	Johnston (1978)
1972	UN	UN	AD[f]	3.0	4.5	0.71	3.8[h]	Sundlof et al. (1986)
			MU[g]	0.13	0.16	0.02	0.29[h]	Ibid.
1973	AD	F	AD	1.3	1.5	0.04	NT	Johnston (1978)
			BR	0.10	0.11	0.00	NT	Ibid.
1973	AD	M	AD	3.1	7.7	0.66	NT	Ibid.
			UP	5.4	6.8	0.24	NT	Ibid.
1974	AD	M	AD	11	25	2.0	NT	Ibid.
			UP	9.3	14	0.69	NT	Ibid.
Marion								
1973	AD	F	AD	3.8	6.9	0.22	NT	Ibid.
			UP	10	15	0.50	NT	Ibid.

a. Lowest limits of detection were 0.01 ppm.
b. AD = adult, UN = unknown.
c. M = male, F = female, UN = unknown.
d. AD = adipose tissue, BR = brain, UP = uropygial gland, MU = muscle.
e. NT = not tested for this residue.
f. Additional residues determined were 0.98 ppm of DDD and 0.52 ppm of DDT.
g. Additional residues determined were 0.03 ppm of DDD and 0 ppm of DDT.
h. Aroclor 1260.

County) and tested for various organochlorine residues (Johnston 1976; Johnston 1978; Sundlof et al. 1986). Residues of DDE, DDD, DDT, and dieldrin were found in tissues of all 6 Black Vultures sampled from Florida and PCBs were found in the 1 Black Vulture that was tested for this residue (table 8.3). DDE and DDT residues were found in all 11 Turkey Vultures that were sampled and dieldrin was present in 10 of them (table 8.4). In general, values for Black Vultures were higher than those for Turkey Vultures. All of the concentrations are well below those associated with acute mortality (Stickel et al. 1969; Stickel et al. 1970; Stickel et al. 1984).

Samples of eggs of Black and Turkey Vultures from Florida collected before 1947 (the pre-DDT era) were measured and compared with others obtained from 1947 to 1968 in order to investigate the effects of organochlorines on eggshell thinning (Wilbur 1978; Kiff et al. 1983). Shell thicknesses decreased by 10–12% for Turkey Vultures and by almost 16% for Black Vultures (table 8.5). These results are presented graphically in figures 8.1 and 8.2. The mean concentration of DDE in Black Vultures from the north central Florida area was higher (4.85 ppm) than that in Turkey Vultures (2.32 ppm) from the same area and same time period. It is not known if the eggshell thinning noted in these retrospective studies resulted in egg breakage and subsequent harmful effects on vulture reproduction. There are no records of such breakage, but Kiff et al. (1983) pointed out that 33% of the sample of Black Vulture eggs and 14% of the Turkey Vulture eggs from Florida had 20% decreases in shell thickness, a decrease that has been associated with reproductive failure and population decline in other birds. They also stated that there was a high prevalence of hairline cracks and other damage in the thinner eggshells that were examined and they felt that there were egg losses during the time represented by the samples. There were notations

Table 8.4. Concentrations of chlorinated hydrocarbons in various tissues of Turkey Vultures from Florida

County Year	Age[b]	Sex[c]	Organ or tissue[d]	Residue (ppm wet weight)[a]		
				DDE	ΣDDT	Dieldrin
Alachua						
1971	UN	UN	AD	3.4	3.5	0.00
1973	AD	F	AD	0.78	1.3	0.46
			BR	0.11	0.11	0.00
1973	AD	M	AD	4.0	5.0	0.16
			BR	0.69	0.87	0.02
1974	AD	F	AD	3.0	3.7	0.74
			UP	0.64	1.0	0.00
1974	AD	M	AD	1.2	1.4	0.05
			UP	0.00	0.00	0.00
1974	AD	F	AD	9.0	13	0.30
			UP	6.4	6.4	0.00
1974	AD	F	AD	7.9	9.7	1.1
			UP	5.0	5.0	0.16
1973	AD	M	AD	1.3	3.0	0.07
			BR	0.17	0.17	0.00
1973	AD	M	AD	1.6	2.3	0.07
			BR	0.01	0.01	0.00
1973	AD	M	AD	2.9	4.2	0.19
			BR	0.04	0.04	0.00
Not specified						
1974	AD	F	AD	4.7	4.9	0.17
			UP	1.9	1.9	0.00

Sources: 1971–73 based on Johnston (1978); 1974 based on Johnston (1976).
a. Lowest limits of detection were 0.01 ppm.
b. AD = adult, UN = unknown.
c. M = male, F = female, UN = unknown.
d. AD = adipose tissue, BR = brain, UP = uropygial gland.

by some of the egg collectors that 1 of a set of 2 eggs was broken during preparation, "apparently because of its extreme fragility" (Kiff et al. 1983: 451). Values for later years (from the late 1970s to present) are not available, but would be important in determining if the thickness of eggshells of vultures has returned to those observed in the pre-DDT era. Even though the use of DDT and other organochlorines has been severely reduced in the United States, this is not the case in some areas in Latin America where North American vultures overwinter (Wilbur 1983). The DDT threat would seem to be still pertinent to vultures in Florida.

VII. Organophosphates and carbamates

There is little information available on organophosphates and carbamates in vultures in Florida. Mortality of approximately 100 Black Vultures occurred at a roost site in Hendry County during March of 1985. Organophosphate poisoning was suspected since there was a history of use of organophosphate pesticides in that area, specifically methomyl on melon crops. Laboratory tests, however, did not support that diagnosis (Regan and Locke 1986).

Organophosphates pose a particular threat as a secondary poisoning problem since vultures often scavenge livestock carcasses to

Table 8.5. Measurements of eggshells of Black and Turkey Vultures from Florida, pre-1947 and post-1947

| | Pre-1947 | | 1947–68 | | | |
Species of vulture	No. eggs sampled	Mean thickness index[a]	No. eggs sampled	Mean thickness index[a]	% change	Data source
Black Vultures	27	2.37	55	2.00	-15.6	Kiff et al. (1983)
Turkey Vultures	20	2.09	22	1.84	-12.0	Wilbur (1978)[b]
	17	2.08	14	1.87	-10.1	Kiff et al. (1983)[b]

a. Thickness index = weight of eggshell (mg) divided by the product of length x breadth (mm) (Ratcliffe 1967).
b. The data given by Wilbur (1978) for Turkey Vultures may also be included in the data given by Kiff et al. (1983). This inference is not clear from the papers.

which organophosphates have been applied prior to death (Henny et al. 1987). Research should be conducted on this neglected aspect of the health of vultures in Florida.

VIII. Metals

Platt et al. (1999) reported a case of peripheral neuropathy in a Turkey Vulture due to lead poisoning. The bird was found recumbent and in poor body condition in Florida (county unknown) in 1997. It was unable to fly, but could stand and was able to walk a bit. It was maintained in captivity for 8 days during which time it was reluctant to eat and eventually died. Lead concentrations in the blood were 277 µg/dl, more than 10 times normal values. There was no evidence of lead in the gastrointestinal tract, and even though multiple gunshot pellets

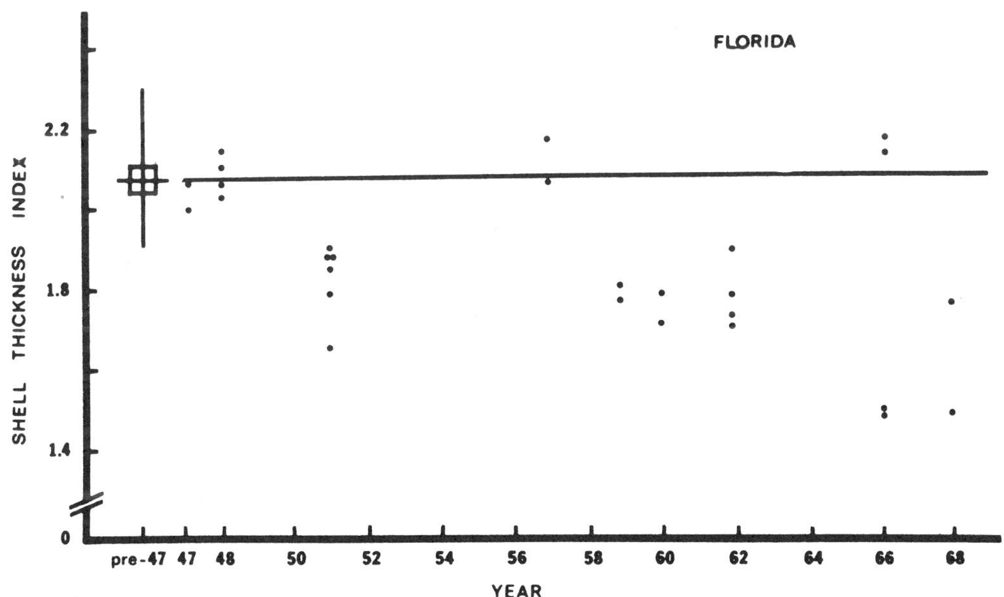

FIGURE 8.1. Eggshell thickness indexes of Turkey Vulture eggs from Florida compared by years of collection. All pre-1947 values fell within the ranges indicated by the box at the left. Horizontal line = mean thickness index for eggshells obtained before 1947. From Kiff et al. 1983; by permission of Lloyd F. Kiff.

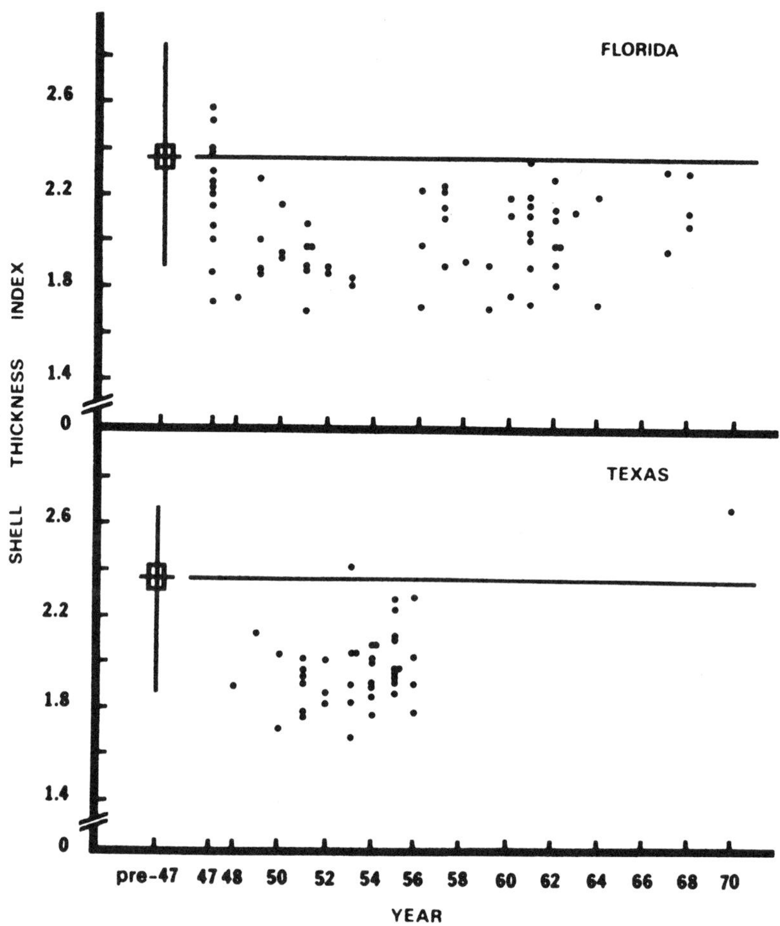

Figure 8.2. Eggshell thickness indexes of Black Vulture eggs from Florida and Texas compared by years of collection. All pre-1947 values fell within the ranges indicated by the box at the left. Horizontal lines = mean thickness indexes for eggshells obtained before 1947. From Kiff et al. 1983; by permission of Lloyd F. Kiff.

and metallic fragments were identified on whole-body radiographs, the cause and source of the lead poisoning was not determined. In 1995 another Turkey Vulture from Alachua County was suspected to have lead poisoning. It was examined at necropsy and had acute tubular necrosis of the kidney and multifocal degeneration of the brain stem and the sciatic nerve (Uhl and Wolf 1995). The lead concentration in blood was 3.3 ppm. The significance of lead poisoning as a mortality factor in populations of vultures in Florida is not known. No other information on metals such as mercury or cadmium is available for vultures in Florida.

IX. Pentobarbital poisoning

In September of 1989 3 Turkey Vultures from Osceola County were submitted to the Center for Birds of Prey in Maitland. All 3 came from the same area and were diagnosed as having barbituate poisoning (Collins and Gilliland 1995). The birds were treated and later released. It was assumed that they had fed on the carcass of a cow that was in the same area and been euthanized with pentobarbital.

X. Neoplasia

No information is available on this topic.

XI. Biotoxins

No information is available on this topic.

XII. Viruses

The diphtheritic form of avian pox was diagnosed in 3 of 20 nestling Turkey Vultures examined in Highlands and Glades counties in 1978 (Wallace and Temple 1994). The diagnosis was based on the observation of typical pox lesions and inclusion bodies in the cytoplasm of chorioallantoic membranes of inoculated chicken eggs. Thirty-four nestling Black Vultures from the same area were negative for avian pox. All 3 of the infected Turkey Vultures

died within 6 days of being removed from their nests in the wild.

Three arboviruses (EEE virus, SLE virus, and Everglades virus) have been identified from vultures in Florida (Lord et al. 1973; Greiner and McLean 1993). Both EEE and SLE viruses are fairly common (especially SLE virus) in Black Vultures (table 8.6). Vultures may play some role in the maintenance and spread of these arboviruses of public health significance, but the exact nature of that role remains to be determined.

Albers (1984) tested 5 Black Vultures and 16 Turkey Vultures for evidence of rabies virus infection. The vultures came from a wildlife rehabilitation facility in Pinellas County in 1982–83. One of the Turkey Vultures was seropositive, but did not have signs of rabies. The significance of this finding is not known. The infection was probably acquired from feeding on prey or carrion.

XIII. Bacteria

In 1978 cloacal swabs from 9 apparently normal nestling Turkey Vultures and 2 nestling

Table 8.6. Arbovirus infections in vultures from Florida

Species of vulture			No. of vultures[a]			
Arbovirus	County or site	Year(s)	Exam.	Inf.	%	Data source
Black Vultures						
Everglades	South Florida[b]	1965–67	19	0	—	Lord et al. (1973), Calisher & Lord (1994)
EEE[c]	Citrus	1990	(24)	(5)	21	Greiner & McLean (1993)
	Orange	1990	(44)	(16)	36	Ibid.
	Osceola	1991	(10)	(5)	50	Ibid.
SLE[d]	Orange	1990	(44)	(33)	75	Greiner & McLean (1993)
Turkey Vultures						
Everglades	South Florida[b]	1965–67	28	4	14	Lord et al. (1973), Calisher & Lord (1994)
	Osceola	1992–93	(3)	(0)	—	Spalding & McLean (1994)

a. Results are from hemagglutination-inhibition tests; parentheses indicate plaque-reduction neutralization test in Vero cell cultures.
b. Everglades National Park and Big Cypress Swamp.
c. EEE = Eastern equine encephalitis.
d. SLE = St. Louis encephalitis.

Table 8.7. Prevalences of bacteria isolated from fecal swabs taken from 44 adult Black Vultures in Orange County, Florida, 1990

Species of bacteria	No. vultures	
	Positive	%
Bacillus sp.	37	84.1
Bacteroides sp.	3	6.8
Corynebacterium sp.	17	38.6
Enterococci	28[a]	63.6
Escherichia coli	31[b]	70.5
Flavobacterium sp.	2[c]	4.5
Klebsiella-Enterobacter sp.	11	25.0
Lactobacillus sp.	6	13.6
Micrococcus sp.	5	11.4
Plesiomonas shigilloides	23	52.3
Proteus mirabilis	1	2.3
Proteus sp.	31	70.5
Providencia sp.	1	2.3
Salmonella Group C1	2	4.5
Salmonella Group D	1	2.3
Staphylococcus sp.	18	40.9
Streptococcus sp. (non-hemolytic)	14	31.8
Streptococcus sp. (alpha-hemolytic)	29	65.9
Streptococcus sp. (anaerobic)[d]	23	52.3
Streptococcus sp.	1	2.3

Source: Greiner and Purich (1993).
a. Includes 22 presumptive identifications.
b. Includes 23 presumptive identifications.
c. Includes 1 presumptive identification.
d. Presumptive identifications.

Black Vultures from Highlands County were cultured specifically for enteric bacteria (Duncan and Wright 1978). None of the Black Vultures was positive, but 2 of the Turkey Vultures were positive for *Salmonella anatum.*

Cloacal swabs from 44 apparently normal adult Black Vultures from Orange County were cultured for bacteria in 1990 (Greiner and Purich 1993). Twenty species of bacteria were isolated and identified (table 8.7). The 5 most prevalent isolants were *Bacillus* sp. (84.1%), *Escherichia coli* (70.5%), *Proteus* sp. (70.5%), alpha-hemolytic *Streptococcus* (65.9%), and Enterococci (63.6%); some of these may be normal components of the bacterial gut flora.

The wide range of species isolated from vultures in Florida is in general agreement with findings from other studies on the gut flora of raptors (see review by Needham [1981]); however, the significance of these bacteria to the health of vultures in Florida is not known. The *Bacillus* sp. is probably not of pathologic significance, but the other 3 most common species have been associated with diseases of various birds of prey (Needham 1981) and may be important. Several of the other bacteria, particularly *Corynebacterium* sp., *Plesiomonas shigelloides,* and the species of *Salmonella* and *Streptococcus* may be of public health significance.

The Turkey Vulture has been shown to be naturally resistant to the harmful effects of the toxins of *Clostridium botulinum,* especially type C, which is the cause of massive mortality in other birds (see chapter 10, Ducks). Kalmbach (1939) found that Turkey Vultures could withstand doses of type C toxin 100,000 times the amount that would kill a domestic pigeon. This explains how vultures can feed on carcasses of other birds that have died during botulism outbreaks and experience no harm.

XIV. Fungi

There are reports of 2 immature male Turkey Vultures from Alachua County, 1 in 1992 and 1 in 1998, that had air sac lesions resembling those of aspergillosis, although this diagnosis was not verified by culture techniques (Woodard 1992; Terrell 1998). Both birds had been in captivity for only a day or two before they died of other causes.

Candida sp. was isolated from the crop of a nestling Turkey Vulture in south central Florida in 1978 (Wallace and Temple 1994). Two unidentified species of yeasts and 1 unidentified species of mold have been identified from cloacal swabs taken from adult Black Vultures in Orange County in 1990 (Greiner and Purich 1993). One of the yeast species occurred in 20 of 44 (45.5%) of the vultures and the other in 2 of 44 vultures (4.5%). The 2 vultures with the second species of yeast were positive also for the first species. The unidentified mold was found in 6 of the 44 (13.6%) vultures, 2 of which were also infected concurrently with the first type of yeast. The significance of these infections to the health of vultures is not known.

XV. Protozoans

Protozoan infections are not common in vultures in Florida. Both Black and Turkey Vultures from various parts of the state have been tested for several types of protozoan infections, including blood parasites and coccidia.

A number of Black Vultures (n = 211) and Turkey Vultures (n = 11) from 7 counties were examined for blood protozoan infections from 1989 to 1998 (table 8.8). Five of these same vultures (all negative) were included in a publi-

cation on blood parasites of raptors in Florida (Forrester et al. 1994). All except 2 from Alachua County were negative. One was a Black Vulture with a concurrent infection of *Plasmodium elongatum* and *Leucocytozoon toddi* in 1995 and the other a Turkey Vulture with an infection of *Plasmodium elongatum* in 1994. The significance of these infections in vultures in Florida is not known, but since they were so low in prevalence, they are probably of little or no concern.

According to Burridge et al. (1979) 4 Black Vultures (dates and localities not given) from Florida were seronegative for antibodies to *Toxoplasma gondii*, but, as mentioned in other chapters, these were tested via the IHA test, which may not be as sensitive in birds as in mammals and may produce false-negative results (Frenkel 1981). One of 6 Turkey Vultures from California was seropositive for *T. gondii* (Franti et al. 1975; Franti et al. 1976). Recently Lindsay et al. (1993) examined tissues from 4 Turkey Vultures and 2 Black Vultures in Alabama for encysted forms of *T. gondii*, but found none.

Uhl and Wolf (1995) observed an unidentified protozoan cyst in heart muscle of an emaciated vulture (species not given) from Alachua County in 1995 (figure 8.3). Carleton (1994) obtained fecal samples from 13 Black Vultures from Orange County (1990), 4 Turkey Vultures from Broward County (1989), and 2 Turkey

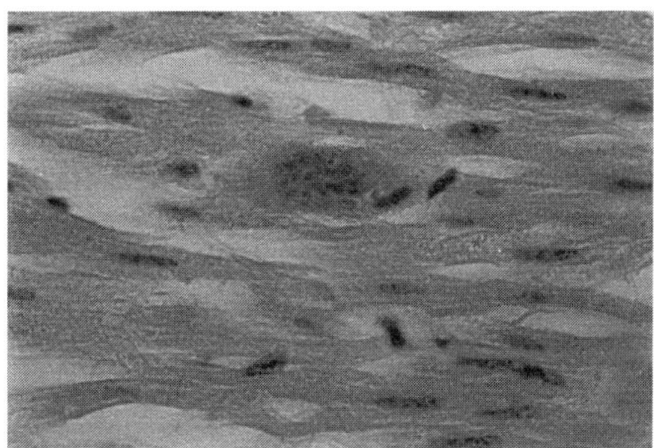

FIGURE 8.3. Unidentified protozoan cyst in heart muscle of an emaciated vulture (species unknown) from Alachua County (1995). Courtesy of Elizabeth W. Uhl.

Table 8.8. Blood protozoan infections in Black and Turkey Vultures from Florida

| Species of vulture | County | Year | No. vultures | | Data source |
			Exam.[a]	Inf.	
Black Vultures	Alachua	1989	2	0	Forrester et al. (1993)
		1995	2	1[b]	Telford et al. (1996)
		1996	1	0	Ibid.
		1997	9	0	Foster et al. (1997)
		1997	18	0	Foster & Avery (1997)
	Citrus	1991	28	0	Greiner (1993)
	Marion	1989	1	0	Forrester et al. (1993)
	Orange	1990	51	0	Greiner (1993)
		1998	44	0	Foster & Avery (1998)
	Osceola	1990	19	0	Greiner (1993)
		1991	18	0	Ibid.
		1992	7	0	Ibid.
	Sumter	1997	10	0	Foster & Avery (1998)
	Unknown	1994	1	0	Telford et al. (1996)
Turkey Vultures	Alachua	1989	1	0	Forrester et al. (1993)
		1990	1	0	Ibid.
		1994	1	1[c]	Telford et al. (1996)
		1995	1	0	Ibid.
	Martin	1995	1	0	Ibid.
	Sumter	1997	6	0	Foster & Avery (1998)

a. Examinations were by thin blood smear techniques.
b. Infected with both *Plasmodium elongatum* and *Leucocytozoon toddi*.
c. Infected with *Plasmodium elongatum*.

Vultures from Dade County (1990); all were examined via the flotation test and were negative for oocysts of *Sarcocystis* spp. While there are no published reports of *Sarcocystis* spp. in vultures from North America, Cape Vultures (*Gyps coprotheres*) and White-backed Vultures (*Gyps africanus*) have been shown experimentally to be definitive hosts of at least 1 species of *Sarcocystis* that infects impala (*Aepyceros melampus*) in Zimbabwe (Markus et al. 1985).

Coccidial oocysts were observed in the intestines of a nestling Turkey Vulture obtained in south central Florida in 1978 (Wallace and Temple 1994). These were not identified to species and their significance is unknown.

XVI. Helminths

Very little is known about the helminths of vultures in Florida. Greiner (1993) examined 1 Black Vulture from Orange County in 1990 and found no helminths. Another Black Vulture from Alachua County in 1972 had 3 specimens of *Strigea* sp. in its small intestine and a Turkey Vulture from Alachua County in 1994 had 1 specimen of *Mesostephanus appendiculatoides* and 13 specimens of *Phagicola longa* in its small intestine (Kinsella and Forrester 1994). Further research should be conducted on vultures in Florida in order to determine the significance of helminth parasites.

XVII. Arthropods

Four species of chewing lice from Black Vultures and 4 species of mites and 4 chewing lice from Turkey Vultures have been reported in Florida (table 8.9). Three of the species of chewing lice were found on both Black and Turkey Vultures. Nothing is known about the impor-

Table 8.9. Parasitic arthropod infestations on vultures from Florida

Host Species of arthropod	County or site	Year[a]	Data source
Black Vultures			
Chewing lice			
Colpocephalum foetens	Alachua	1953	Forrester et al. (1995)
	Leon	1936	Ibid.
Cuculiphilus alternatus	Alachua	1953	Ibid.
	Leon	1936	Ibid.
	Pinellas	1929	Ibid.
	"Florida"	NG	Peters (1936)
Falcolipeurus marginalis	Alachua	1953, 1972	Forrester et al. (1995)
	Leon	1936	Ibid.
	Pasco	1965	Ibid.
Laemobothrion glutinans	Alachua	1973	Ibid.
	Pinellas	1929	Ibid.
	"Florida"	NG	Peters (1936)
Turkey Vultures			
Ticks[b]			
Feather mites			
Ancyralges sp.	Alachua	1995	Foster & Mertins (1995)
Glaucalges sp.	Alachua	1995	Foster & Mertins (1995)
Hieracolichus sp.	Alachua	1995	Ibid.
Tropical fowl mites			
Ornithonyssus bursa	"Florida"	NG	Peters (1936)
Chewing lice			
Colpocephalum kelloggi	Alachua	1953	Forrester et al. (1995)
	Alachua	1994	Foster & Price (1994)
	Brevard	1936	Forrester et al. (1995)
	Collier	1937	Ibid.
	Dade	1917, 1980	Ibid.
Cuculiphilus alternatus	Alachua	1953	Ibid.
	Alachua	1994	Foster & Price (1994)
	Dade	1980	Forrester et al. (1995)
Falcolipeurus marginalis	Alachua	1953	Ibid.
	Alachua	1994	Foster & Price (1994)
	Dade	1980	Forrester et al. (1995)
	"Florida"	NG	Peters (1936)
Laemobothrion glutinans	Alachua	1953	Forrester et al. (1995)
	Dade	1980	Ibid.
	Monroe	1937	Ibid.
	Osceola	1933–34	Ibid.
	"Florida"	NG	Peters (1936)

a. NG = not given.
b. Travis (1941) examined 2 Turkey Vultures from Orange, Osceola, or Collier county in 1936–37; neither was infested with ticks.

tance of these ectoparasites to the health of vultures in Florida.

XVIII. Vultures as pests

Claims were made in the early 1900s that Black and Turkey Vultures were responsible for contaminating water and spreading livestock diseases such as anthrax and hog cholera (Bent 1937; Jackson 1988; Coleman and Fraser 1990). This, along with the fact that vultures are known to attack newborn livestock and cattle that are in labor or have just given birth, led to widespread trapping and killing in many states including Florida (Parmalee 1954; Jackson 1988). It was later reported that vultures could feed on the flesh of animals infected with hog cholera and anthrax but that the etiologic agents were actually inactivated, thereby clearing the vulture as a suspected disseminator of these diseases (Howell 1932; Redington and McAtee 1932). Subsequently it has been found from experimental studies on Black and Turkey Vultures in Chile and Whiteback Griffon Vultures (*Gyps africanus*) in Kenya that the vegetative stages of the anthrax bacillus do not survive passage through the digestive tract, whereas the resistant spores pass through in viable form (Bullock 1956; Houston and Cooper 1975). It was also found that spores could survive and be transported on the feet and feathers of vultures; thus the spread of anthrax could be facilitated by birds flying from an infected area to an uninfected area. Further research on this topic should be conducted on North American vultures.

Nevertheless, the vulture remains a threat to livestock as a predator and is a pest in other ways, such as congregating in large numbers in recreational facilities, like Disney World of Florida, where they cause aesthetic and sanitation problems. They are known also to cause damage to power and communications equipment (Coleman and Fraser 1990).

XIX. Summary and conclusions

Excluding trauma, electrocution, and inclement weather, 52 different morbidity/mortality factors and disease agents are known from vultures in Florida. Most of the information is on Black Vultures, from which 36 agents have been identified, including 4 organochlorines, 2 viruses, 20 bacteria, 3 fungi, 2 protozoans, 1 helminth, and 4 chewing lice. Twenty-three agents have been found in Turkey Vultures: 3 organochlorines, 1 metal, 1 barbiturate, 3 viruses, 1 bacterium, 2 fungi, 2 protozoans, 2 helminths, 4 chewing lice, and 4 mites.

Almost nothing is known about the population significance of the disease agents and parasites that have been identified from vultures in Florida. The organochlorines (DDE especially) are of historical interest and significance; their use may have resulted in harmful effects on reproduction because of eggshell thinning, although conclusive data on this are not available. This could continue to be a problem, even though the use of DDT in the United States and Canada has been curtailed significantly, since some vultures from Florida may migrate into Latin American countries (although this migration has not been well documented; see Jackson 1988) where these pesticides are still used. Although vultures have been more or less vindicated from being involved in the dissemination of such diseases as anthrax and hog cholera, they may be of some significance in the epizootiology of salmonellosis and several arboviruses (EEE virus, SLE virus, and Everglades virus) of human health significance.

XX. Literature cited

Albers, H.F. 1984. Naturally occurring immunity to rabies in raptors. *Proc. Annu. Natl. Wildl. Rehabil. Symp.* 2:171–174.

Baynard, O.E. 1909. Notes from Florida on *Catharista urubu. Oölogist* 26:191–193.

Bent, A.C. 1937. *Life histories of North American*

birds of prey. Part 1. U.S. Natl. Mus. Bull. 167. 409 pp.

Bullock, D.S. 1956. Vultures as disseminators of anthrax. *Auk* 73:283–284.

Burridge, M.J., W.J. Bigler, D.J. Forrester, and J.M. Hennemann. 1979. Serologic survey for *Toxoplasma gondii* in wild animals in Florida. *J. Am. Vet. Med. Assoc.* 175:964–967.

Calisher, C.H., and R.D. Lord. 1994. Unpublished data. Colorado State University, Fort Collins.

Carleton, R.E. 1994. Unpublished data. University of Florida, Gainesville.

Coleman, J.S., and J.D. Fraser. 1990. Black and turkey vultures. *Proc. Southeast Raptor Manag. Symp. Workshop.* Natl. Wildl. Fed. Sci. Tech. Ser. 14:78–88.

Collins, R., and C. Gilliland. 1995. Unpublished data. Florida Audubon Society, Center for Birds of Prey, Maitland.

Crawford, R.L. 1981. Bird casualties at a Leon County, Florida, TV tower: a 25-year migration study. *Bull. Tall Timbers Res. Stn.* 22:1–30.

Deem, S.L., and S.P. Terrell. 1996. Unpublished data. University of Florida, Gainesville.

Duncan, R.M., and E.J. Wright. 1978. Unpublished data. National Wildlife Health Center, Madison, Wis.

Forrester, D.J., H.W. Kale II, R.D. Price, K.C. Emerson, and G.W. Foster. 1995. Chewing lice (Mallophaga) from birds in Florida: a listing by host. *Bull. Fla. Mus. Nat. Hist.* 39:1–44.

Forrester, D.J., S.R. Telford, Jr., G.W. Foster, and G.F. Bennett. 1993. Unpublished data. University of Florida, Gainesville.

———. 1994. Blood parasites of raptors in Florida. *J. Raptor Res.* 28:226–231.

Foster, G.W., and M.L. Avery. 1997–98. Unpublished data. University of Florida, Gainesville.

Foster, G.W., and J.W. Mertins. 1995. Unpublished data. University of Florida, Gainesville.

Foster, G.W., and R.D. Price. 1994. Unpublished data. University of Florida, Gainesville.

Foster, G.W., M.G. Spalding, and E.C. Greiner. 1997. Unpublished data. University of Florida, Gainesville.

Franti, C.E., G.E. Connolly, H.P. Riemann, D.E. Behymer, R. Ruppanner, C.M. Willadsen, and W. Longhurst. 1975. A survey for *Toxoplasma gondii* antibodies in deer and other wildlife on a sheep range. *J. Am. Vet. Med. Assoc.* 167:565–568.

Franti, C.E., H.P. Riemann, D.E. Behymer, D. Suther, J.A. Howarth, and R. Ruppanner. 1976. Prevalence of *Toxoplasma gondii* antibodies in wild and domestic animals in northern California. *J. Am. Vet. Med. Assoc.* 169:901–906.

Frenkel, J.K. 1981. False-negative serologic tests for *Toxoplasma* in birds. *J. Parasitol.* 67:952–953.

Greiner, E.C. 1993. Unpublished data. University of Florida, Gainesville.

Greiner, E.C., and R.G. McLean. 1993. Unpublished data. University of Florida, Gainesville.

Greiner, E.C., and B.I. Purich. 1993. Unpublished data. University of Florida, Gainesville.

Hallinan, T. 1922. Bird interference on high tension electric transmission lines. *Auk* 39:573.

Henny, C.J., E.J. Kolbe, E.F. Hill, and L.J. Blus. 1987. Case histories of bald eagles and other raptors killed by organophosphorus pesticides topically applied to livestock. *J. Wildl. Dis.* 23:292–295.

Houston, D.C., and J.E. Cooper. 1975. The digestive tract of the whiteback griffon vulture and its role in disease transmission among wild ungulates. *J. Wildl. Dis.* 11:306–313.

Howell, A.H. 1932. *Florida bird life.* Coward-McCann, New York. 579 pp.

Jackson, J.A. 1983. Nesting phenology, nest site selection, and reproductive success of Black and Turkey Vultures. In: *Vulture biology and management.* S.R. Wilbur, and J.A. Jackson (eds.). University of California Press, Berkeley. pp. 245–270.

———. 1988. American black vulture and turkey vulture. In: *Handbook of North American birds.* Vol. 4. R.S. Palmer (ed.). Yale University Press, New Haven, Conn. pp. 11–42.

Johnston, D.W. 1976. Organochlorine pesticide residues in uropygial glands and adipose tissue of wild birds. *Bull. Environ. Contam. Toxicol.* 16: 149–155.

———. 1978. Organochlorine pesticide residues in Florida birds of prey, 1969–76. *Pestic. Monit. J.* 12:8–15.

Kale, H.W. II. 1971. Regional reports: Florida region. *Am. Birds* 25:723–733.

Kalmbach, E.R. 1939. American vultures and the toxin of *Clostridium botulinum*. *J. Am. Vet. Med. Assoc.* 47:187–191.

Kiff, L.F., D.B. Peakall, M.L. Morrison, and S.R. Wilbur. 1983. Eggshell thickness and DDE residue levels in vulture eggs. In: *Vulture biology and management*. S.R. Wilbur and J.A. Jackson (eds.). University of California Press, Berkeley. pp. 440–458.

Kinsella, J.M., and D.J. Forrester. 1994. Unpublished data. University of Florida, Gainesville.

Lindsay, D.S., P.C. Smith, F.J. Hoerr, and B.L. Blagburn. 1993. Prevalence of encysted *Toxoplasma gondii* in raptors from Alabama. *J. Parasitol.* 79:870–873.

Lord, R.D., C.H. Calisher, W.D. Sudia, and T.H. Work. 1973. Ecological investigations of vertebrate hosts of Venezuelan equine encephalomyelitis virus in south Florida. *Am. J. Trop. Med. Hyg.* 22:116–123.

Maehr, D.S., and J.Q. Smith. 1988. Bird casualties at a central Florida power plant: 1982–1986. *Fla. Field Nat.* 16:57–80.

Maehr, D.S., A.G. Spratt, and D.K. Voigts. 1983. Bird casualties at a central Florida power plant. *Fla. Field Nat.* 11:45–49.

Markus, M.B., P.J. Mundy, and T.J.M. Daly. 1985. Vultures *Gyps* spp. as final hosts of *Sarcocystis* of the impala *Aepyceros melampus*. *S. Afr. J. Sci.* 81:43.

Mote, W.R. 1969. Turkey vultures land on vessel in fog. *Auk* 86:766–767.

Needham, J.R. 1981. Bacterial flora of birds of prey. In: *Recent advances in the study of raptor diseases. Proceedings of the International Symposium on Diseases of Birds of Prey.* J.E. Cooper and A.G. Greenwood (eds.). Chiron, West Yorkshire, England. pp. 3–9.

Parmalee, P.W. 1954. The vultures: their movements, economic status, and control in Texas. *Auk* 71:443–453.

Peters, H.S. 1936. A list of external parasites from birds of the eastern part of the United States. *Bird-Banding* 7:9–27.

Platt, S.R., K.E. Helmick, J. Graham, R.A. Bennett, L. Phillips, C.L. Chrisman, and P.E. Ginn.

1999. Peripheral neuropathy in a turkey vulture with lead toxicosis. *J. Am. Vet. Med. Assoc.* 214:1218–1220.

Rabenold, P.P. 1985. Identifying causes of population decline in black and turkey vultures in the U.S.A. *Vulture News* 14:16–18.

Ratcliffe, D.A. 1967. Decrease in eggshell weight in certain birds of prey. *Nature* 215:208–210.

Redington, P.G., and W.L. McAtee 1932. Policies of the Bureau of Biological Survey relative to the control of injurious birds. USDA Miscellaneous Publications 145. pp. 1–8.

Regan, T.W. 1996. Unpublished data. Florida Game and Fresh Water Fish Commission, West Palm Beach.

Regan, T.W., and L.N. Locke. 1986. Unpublished data. Florida Game and Fresh Water Fish Commission, West Palm Beach.

Robertson, W.B., Jr., and G.E. Woolfenden. 1992. *Florida bird species: an annotated list.* Spec. Publ. 6, Florida Ornithological Society, Gainesville. 260 pp.

Snyder, B. 1994. Unpublished data. Florida Deptartment of Environmental Protection, Tallahassee.

Spalding, M.G., and R.G. McLean. 1994. Unpublished data. University of Florida, Gainesville.

Stevenson, J.A. 1994. Unpublished data. Florida Department of Environmental Protection, Tallahassee.

Stickel, W.H., L.F. Stickel, and F.B. Coon. 1970. DDE and DDD residues correlated with mortality of experimental birds. In: *Inter-American conference on toxicology and occupational medicine, pesticide symposia*. W.P. Deichmann (ed.). Halos and Associates, Miami. pp. 287–294.

Stickel, W.H., L.F. Stickel, R.A. Dyrland, and D.L. Hughes. 1984. Aroclor 1254 residues in birds: lethal levels and loss rates. *Arch. Environ. Contam. Toxicol.* 13:7–13.

Stickel, W.H., L.F. Stickel, and J.W. Spann. 1969. Tissue residues of dieldrin in relation to mortality in birds and mammals. In: *Chemical fallout*. M.W. Miller and G.G. Berg (eds.). Charles C. Thomas, Springfield, Ill. pp. 174–204.

Sundlof, S.F., D.J. Forrester, N.P. Thompson, and M.W. Collopy. 1986. Residues of chlorinated hydrocarbons in tissues of raptors in Florida. *J. Wildl. Dis.* 22:71–82.

Suto, B.J. 1996. Unpublished data. Suncoast Seabird Sanctuary, Redington Beach, Fla.

Taylor, W.K., and B.H. Anderson. 1973. Nocturnal migrants killed at a central Florida TV tower, autumns 1969–1971. *Wilson Bull.* 85:42–51.

———. 1974. Nocturnal migrants killed at a central Florida TV tower, Autumn 1972. *Fla. Field Nat.* 2:40–43.

Taylor, W.K., and M.A. Kershner. 1986. Migrant birds killed at the Vehicle Assembly Building (VAB), John F. Kennedy Space Center. *J. Field Ornithol.* 57:142–154.

Telford, S.R., Jr., G.W. Foster, J.K. Nayar, S.P. Terrell, and D.J. Forrester. 1996. Unpublished data. University of Florida, Gainesville.

Terrell, S.P. 1998. Unpublished data. University of Florida, Gainesville.

Travis, B.V. 1941. Examinations of wild animals for the cattle tick *Boophilus annulatus microplus* (Can.) in Florida. *J. Parasitol.* 27:465–467.

Uhl, E.W., and J. Wolf. 1995. Unpublished data. University of Florida, Gainesville.

Wallace, M.P., and S.A. Temple. 1994. Unpublished data. University of Wisconsin, Madison.

Wilbur, S.R. 1978. Turkey vulture eggshell thinning in California, Florida, and Texas. *Wilson Bull.* 90:642–643.

———. 1983. The status of vultures in the Western Hemisphere. In: *Vulture biology and management.* S.R. Wilbur and J.A. Jackson (eds.). University of California Press, Berkeley. pp. 113–123.

Wilbur, S.R., and J.A. Jackson (eds.). 1983. *Vulture biology and management.* University of California Press, Berkeley. 497 pp.

Woodard, J.C., and R.A. Westerhouse. 1992. Unpublished data. University of Florida, Gainesville.

CHAPTER NINE

Whistling-Ducks, Swans, and Geese

I. Introduction

The family Anatidae is divided into two subfamilies, Anserinae and Anatinae. In the Anserinae are included whistling-ducks, swans, and geese and in the Anatinae are the ducks. Eight species of Anserinae occur in Florida, including 2 whistling-ducks, 1 swan, and 5 geese (table 9.1). None is classified as threatened or endangered (Logan 1997). Only 3 of these, the Fulvous Whistling-Duck, the Black-bellied Whistling-Duck, and the Canada Goose, breed in Florida. The others are uncommon winter visitors.

Disease information is available only for Fulvous Whistling-Ducks and Canada Geese in Florida and relates mainly to toxicoses and parasitic infections. We have no data on the other 6 species. The reader is referred to Wobeser (1997) and Friend and Franson (1999) for additional information on the diseases and parasites of free-ranging waterfowl.

II. Trauma

Traumatic injuries are fairly common in anatids. Tiemeier (1941) evaluated 256 museum specimens of various species of Anatidae and found evidence of injuries in 33 birds. The species and origin of the anatids were not given, nor was an explanation of this finding provided by the author. Wobeser (1997) stated that collisions during flight may result in considerable mortality, especially in the case of large species of waterfowl. Most of these collisions are associated with power lines, telephone wires, and

Table 9.1. Distribution, occurrence, and abundance of whistling-ducks, swans, and geese of the subfamily Anserinae in Florida[a]

Species	Range	Seasonal occurrence	Relative abundance
Fulvous Whistling-Duck *Dendrocygna bicolor*	Peninsula & Keys	Resident	~6,000
Black-bellied Whistling-Duck *Dendrocygna autumnalis*	Peninsula	Resident	Rare
Tundra Swan *Cygnus columbianus*	Panhandle & Northern Peninsula	Winter transient	Rare to uncommon
Greater White-fronted Goose *Anser albifrons*	Panhandle, Northern Peninsula, & Collier, Dade, & Sarasota counties	Winter transient	Rare to fairly common
Snow Goose *Chen caerulescens*	Statewide	Winter transient	Rare to locally common
Ross's Goose *Chen rossii*	Northern Florida	Winter transient	Very rare
Brant *Branta bernicla*	Panhandle & Northern Peninsula	Winter transient	Rare
Canada Goose *Branta canadensis*	Statewide Mainland	Winter transient Resident[b]	Rare ?

a. Modified from Robertson and Woolfenden (1992) and American Ornithologists' Union (1998).
b. Feral, nonmigratory, free-flying population.

fences. Whistling-ducks, geese, and swans have not been found in studies of trauma-related mortality involving collisions with vehicles (Stevenson 1994; Snyder 1994) or with tall structures in Florida (Crawford 1981; Kale 1971; Maehr et al. 1983; Maehr and Smith 1988; Taylor and Anderson 1973, 1974; Taylor and Kershner 1986).

A small number of Black-bellied Whistling-Ducks are killed by hunters each year in Florida, but during the ten-year period of 1981–90, the average annual total harvest for Fulvous Whistling-Ducks was 1,042 (Martin 1996). These were killed in 15 counties in the state, with the majority from Glades, Okeechobee, and Palm Beach counties.

III. Predation

Two juvenile Canada Geese from Kanapaha Prairie (Alachua County) were found in 1972 with multiple wounds, probably caused by a predator (Forrester 1996). One had a broken wing. Domestic dogs were suspected as the predators. Native fire ants (*Solenopsis xyloni*) have been reported as predators of pipping eggs of Black-bellied Whistling-Ducks in Texas (Delnicki and Bolen 1977), but we have no such information for Florida.

IV. Inclement weather

No information is available on this topic.

V. Pesticide residues

Information on pesticides is available for only 1 species, the Fulvous Whistling-Duck. Turnbull et al. (1989) reported low concentrations of residues of 13 organochlorines and 4 organophosphates in a sample of 30 Fulvous Whistling-Ducks collected in the Everglades Agricultural Area (Palm Beach County) in 1984–85 (tables

Table 9.2. Organochlorine residues in tissues of Fulvous Whistling-Ducks from Palm Beach County, Florida

Age of duck[a] Tissue	Year	No. examined	No. with residue[b]	ppb wet wt. Geometric mean (highest concentration)[c]	
			Aldrin	Dieldrin	Heptachlor
Ducklings[d]					
Liver	1984	9	5 / 2.1(180)	8 / 4.4(16)	7 / 0.8(3.5)
HY/Adults					
Liver	1985	21	19 / 9.4(181)	20 / 5.5(71)	8 / NG(2.5)
Pectoral muscle	1985	15	4 / NG(3.7)	12 / 0.7(23)	1 / NG(NG)
			Heptachlor epoxide	DDE	*trans*-nonachlor
Ducklings					
Liver	1984	9	8 / 1.5(7.6)	1 / NG(5.3)	2 / NG(12)
HY/Adults					
Liver	1985	21	18 / 2.3(35)	13 / 1.9(32)	8 / NG(19)
Pectoral muscle	1985	15	10 / 0.1(6.7)	13 / 2.6(110)	5 / NG(9.5)
			Oxychlordane	Mirex	Lindane
Ducklings					
Liver	1984	9	2 / NG(6.2)	3 / NG(12)	2 / NG(2.6)
HY/Adults					
Liver	1985	21	5 / NG(5.7)	4 / NG(7.7)	4 / NG(3.2)
Pectoral muscle	1985	15	2 / NG(5.1)	ND / —	ND —
			Hexachlorobenzene	DDT[e]	Endosulfan
Ducklings					
Liver	1984	9	2 / NG(1.8)	ND / —	ND / —
HY/Adults					
Liver	1985	21	2 / NG(1.6)	ND / —	ND / —
Pectoral muscle	1985	15	1 / NG(NG)	4 / NG(96)	4 / NG(NG)

Source: Turnbull et al. (1989).
a. HY = hatch year birds.
b. The authors stated that the "detection limit for most compounds was 5 ppb."
c. NG = not given by authors, ND = not detected.
d. Ducklings that had not yet reached flight stage.
e. DDD was detected in liver tissue of 1 duck collected in 1985; the concentration was not given by the authors.

9.2 and 9.3). Contaminants were found in all 30 livers examined and in 14 of 15 samples of pectoral muscle. The most common residues were dieldrin, heptachlor epoxide, DDE, aldrin, heptachlor, and DDT. There were no differences in concentrations of contaminants between sexes of ducks and no interactions between age and sex. Liver samples from adult ducks had higher concentrations of DDE than did samples from ducklings and hatch-year birds. The authors concluded that these low concentrations would not be associated with mortality, but stated that the sublethal effects are unknown. The pesticides of most concern are aldrin and dieldrin. The Fulvous Whistling-Duck has been found to be highly susceptible to aldrin-dieldrin poisoning compared with other species of ducks (Flickinger and King 1972), but the concentra-

Table 9.3. Organophosphate residues in tissues of Fulvous Whistling-Ducks from Palm Beach County, Florida

| Age of duck | | | ppb wet wt. | | |
| | | | No. with residue[a] / Geometric mean (highest concentration)[b] | | |
Tissue	Year	No. examined	Diazinon	Ethyl parathion[c]	Chlorpyrifos
Ducklings[d]					
Liver	1984	9	2 / NG(10)	1 / NG(0.4)	1 / NG(15)
HY[e]/Adults					
Liver	1985	21	9 / NG(160)	4 / NG(34)	2 / NG(3.9)
Pectoral muscle	1985	15	ND / —	3 / NG(2.2)	2 / NG(9.7)

Source: Turnbull et al. (1989).
a. The authors stated that "the detection limit for most compounds was 5 ppb."
b. NG = not given by authors, ND = not detected.
c. Methyl parathion was detected in pectoral muscle from 1 duck collected in 1985; the concentration was not given by the authors.
d. Ducklings that had not yet reached flight stage.
e. HY = hatch year birds.

tions found in Turnbull's study in Florida were considerably lower than those known to cause acute mortality.

VI. Lead poisoning

Lead poisoning or plumbism in waterfowl is a toxic disease caused by the absorption of harmful concentrations of lead into body tissues (Friend 1999). The disease is more often a chronic one with low levels of mortality that usually go undetected, although die-offs of birds have been recorded as well (Wobeser 1997). This topic has been discussed in a number of publications and the reader is referred to Trainer (1982), Locke and Thomas (1996), Pain (1996), Wobeser (1997), and Friend (1999) for overviews. The typical source of lead, especially for waterfowl, is pellets from shotgun shells that are ingested from bottom sediments of marshes and lakes along with food items. Lead shot imbedded in tissue is not believed to contribute to lead poisoning. It has been estimated that more than 1,300 metric tons of lead shot were deposited by hunters each year in waterfowl habitat in the United States (U.S. Fish and Wildlife Service 1976) before the use of lead shot for hunting waterfowl was banned in 1991 (Morehouse 1992). Other less common sources include bullets, paint, mine wastes, and lead fishing sinkers (Friend 1999).

Field signs of lead poisoning in waterfowl include reluctance to fly, inability to maintain flight, no attempt to escape in the presence of man, holding wings in a "roof-shaped" position, fluid discharge from the bill, and bile-staining of feces and in the vent area (Friend 1999). At necropsy carcasses often are emaciated because of severe wasting of the breast muscles, contain little or no visceral fat, have

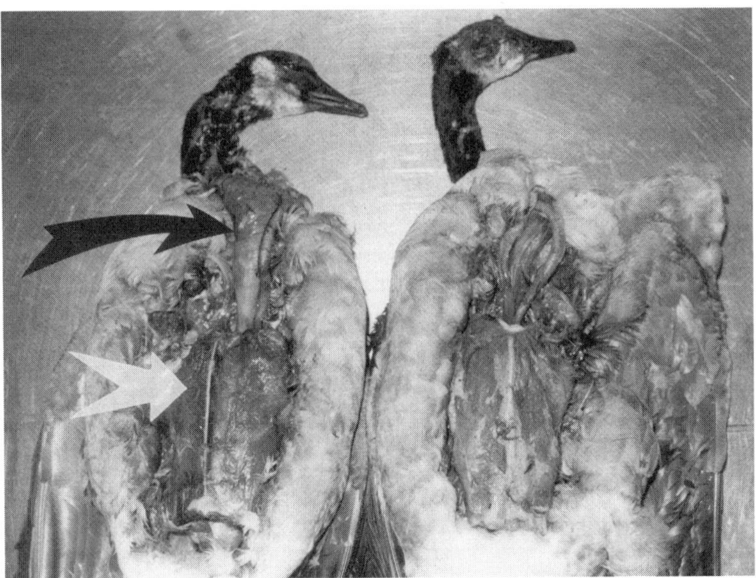

FIGURE 9.1. Canada Goose that died of lead poisoning *(left)* compared with a normal goose *(right)* that died of trauma (trapping mortality) in Leon County, 1982. The goose on the left has an atrophied pectoral muscle *(white arrow)* and an impacted esophagus *(black arrow)*.

esophagi and proventriculi impacted with food, and have swollen gall bladders and dark-green stained gizzard linings. Lead pellets may or may not be present in the gizzards or proventriculi and their presence is not a requirement for a diagnosis of lead poisoning. The field signs and lesions described above are suggestive of lead poisoning, but a definitive diagnosis must include the application of appropriate laboratory tests and the finding of concentrations of lead in livers of at least 6 ppm (wet weight) or 20 ppm (dry weight) (Friend 1999).

Lead poisoning was diagnosed in nonmigratory Canada Geese from Leon County in January of 1982 (Forrester et al. 1987). The exact extent of the mortality in this event was not determined, but it was reported that up to 15 carcasses were counted. The population at risk was not known, but at the time there was a flock of 300 to 400 geese in Leon and Jefferson counties (Forrester et al. 1987) that according to Stevenson (1976) were descendants of captive Giant Canada Geese (*B.c. maxima*). These geese had been feeding on agricultural land that had been shot over by dove hunters for many years. Only 1 of these geese was examined and it had many of the signs and lesions of lead poisoning including emaciation (atrophied pectoral muscles and depletion of subcutaneous fat) and an esophagus impacted with food (figures 9.1 and 9.2). There were 4 lead pellets in the gizzard and the concentration of lead in the liver was 31 ppm (wet weight). Two months later 3 additional geese from the same area were killed during trapping operations and were examined at necropsy. One of these was emaciated and had 27 ppm (wet weight) of lead in its liver. The other 2 birds were more normal and had lower residues of lead (table 9.4).

In October of 1989 a Graylag Goose (*Anser anser*), an established exotic domestic goose, was diagnosed with lead poisoning in Leon County (Mahnke and Hayes 1989). This bird was 1 of 2 that died in a flock of Domestic

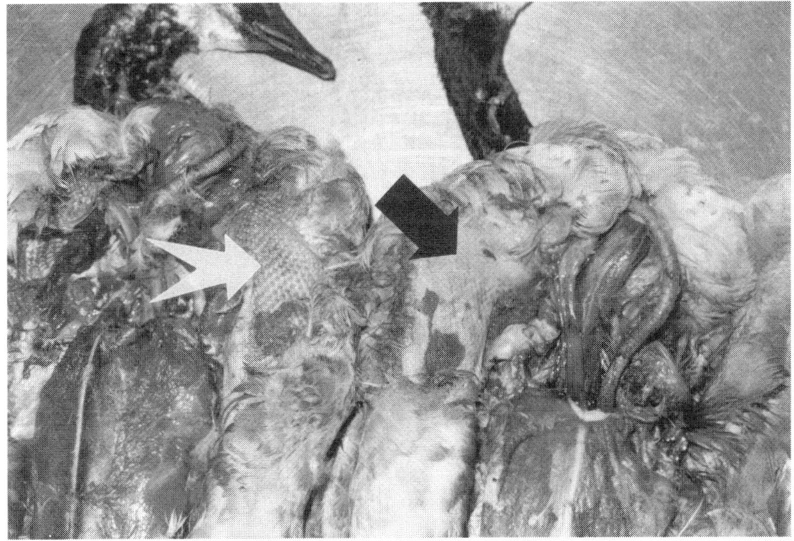

FIGURE 9.2. Canada Goose that died of lead poisoning *(left)* compared with a normal goose *(right)* that died of trauma (trapping mortality) in Leon County, 1982. The goose on the left is depleted of subcutaneous fat *(white arrow)*, while the goose on the right has normal amounts of subcutaneous fat *(black arrow)*.

Geese, Mallards, and Muscovies that used a 2-acre pond on private land. The water in the pond had been very low throughout the summer, but had risen 1 to 2 feet during the 2 months prior to the incident. At necropsy the bird had moderate atrophy of pectoral muscles, a distended gall bladder, bile-stained gizzard linings, and 25 pieces of lead shot in the upper gastrointestinal tract, many flattened and partially eroded. No free-ranging waterfowl were involved, although the pond was used by a few Wood Ducks.

There is no information on the prevalence of lead shot in the gizzards of geese in Florida, but

Table 9.4. Relationship of lead concentrations and physical condition of 4 nonmigratory Canada Geese from Leon County, 1982

Goose no.	Wt. (kg)	No. lead shot in gizzard	Lead concentrations[a] Liver	Kidney	Humerus	Index of emaciation[b]	Impaction of proventriculus and esophagus
1[c]	2.56	4	31	47	140	++	Yes
2	3.02	0	7	13	180	+	No
3	3.02	0	<1	2	18	0	No
4[c]	1.95	0	27	45	48	+++	No

Source: Forrester et al. (1987).
a. ppm wet weight.
b. 0 = no emaciation, + = slight emaciation, ++ = moderate emaciation, +++ = considerable emaciation.
c. Diagnosed as lead poisoning.

there are some data on Fulvous Whistling-Ducks. Baker and Thompson (1978) examined gizzards from 10 hunter-killed ducks from Merritt Island National Wildlife Refuge. Two of these were collected in 1973–74 and 8 in 1975–76. None contained lead shot. Similar negative findings were encountered by Hines (1980) and Thul (1983), who examined 14 ducks from Broward and St. Johns counties in 1979–80 and 1982–83. These findings are consistent with Friend's (1999) statement that lead shot ingestion rates for whistling-ducks are low and therefore lead poisoning is rare in this duck.

Additional information on lead poisoning in waterfowl is found in chapter 10, Ducks.

VII. Avian vacuolar myelinopathy

This disease is not known to occur in Florida, but has been diagnosed in waterfowl in nearby states (see chapter 11, Eagles).

VIII. Neoplasia

No information is available on this topic.

IX. Biotoxins

In February and March of 1954 a mortality event involving more than 100 Canada Geese was reported in Leon County in a small shallow freshwater pond (Grice et al. 1956). Botulism and infectious diseases were suspected but ruled out after appropriate laboratory tests were performed; there was no mention of tests to detect environmental contaminants such as organophosphates or heavy metals. The pond contained a bloom of the dinoflagellate *Peridinium volzi,* absent from 4 other ponds and lakes in the area, 1 of which had geese on it that did not experience mortality. The concentration of *P. volzi* was 2,738 per cc of water on February 25 and declined to 12 per cc on April 19, at which time there was no further evidence of waterfowl mortality. There

are no reports of this dinoflagellate being toxic (Andersen 1996), but apparently this aspect has not been studied.

X. Viruses

No information is available on this topic.

XI. Bacteria

No information is available on this topic.

XII. Fungi

No information is available on fungal infections in free-ranging whistling-ducks, geese, or swans; however, an outbreak of rhinosporidiosis in captive Mute Swans (*Cygnus olor*) and Black Swans (*C. atratus*) on a lake in central Florida (exact locality and date not given) was reported by Kennedy et al. (1995). Forty-one swans were involved in the outbreak. The etiologic agent was not identified, but was suspected to be *Rhinosporidium seeberi*. Rhinosporidiosis (nasal polyp) was reported in 1 of 234 free-ranging Wood Ducks examined from Louisiana (Davidson and Nettles 1977), but this is the only such report from Florida.

XIII. Protozoans

No parasitic protozoans are known from whistling-ducks, geese, or swans from Florida. A sample of blood films from 52 nonmigratory Canada Geese (32 from Leon County and 19 from Jefferson County in 1978 and 1 from Leon County in 1982) were examined for blood parasites (Thul et al. 1980; Forrester 1996). All were negative. Blood films taken from 21 Fulvous Whistling-Ducks from Palm Beach County in 1985 were negative also (Forrester et al. 1994). These findings are in

Table 9.5. Helminth infections in non-migratory Canada Geese from Florida

Helminth Species (site)[a] County	Year(s)	No. geese Exam.	No. geese Inf.	Intensity Mean	Intensity Range	Data source
Trematoda						
Hypoderaeum conoideum (LI)						
Leon	1982	4	1	1	—	Forrester et al. (1987)
Cestoda						
Unidentified fragments (SI)[b]						
Leon	1982	4	2	—	—	Ibid.
Nematoda						
Amidostomum anseris (GI,PR)						
Leon	1982	4	2	7	6–8	Ibid.
Wakulla	1932	NG	NG	NG	NG	Wehr (1932)
	1948–53	5	4	NG	NG	Herman & Wehr (1954)
Capillaria sp. (SI)						
Leon	1982	4	1	1	—	Forrester et al. (1987)
Contracaecum sp. (SI)						
Leon	1982	4	1	3	—	Ibid.
Epomidiostomum crami (GI,PR)						
Leon	1982	4	3	5	3–7	Ibid.
Tetrameres sp. (PR)						
Leon	1982	4	3	6	3–11	Ibid.
Trichostrongylus cramae (CE,LI)						
Leon	1982	4	4	29	2–48	Ibid.

a. Infection sites: CE = cecum, GI = gizzard, LI = large intestine, PR = proventriculus, SI = small intestine.
b. Specimens in poor condition because of freezing and recovery techniques.

agreement with data obtained from other species of resident waterfowl in Florida (i.e., Mottled Ducks and Wood Ducks) showing blood protozoans to be absent (Thul et al. 1980). Although other species of waterfowl migrate to Florida during the winter months and bring blood parasites with them, these parasites cannot be transmitted because of the absence of appropriate vectors or perhaps because peak populations of vectors do not occur at the same time of year the migratory birds are present.

XIV. Helminths

A small amount of information is available on the helminths of Canada Geese in Florida, but the data are limited and based on examinations of only a few birds. Eight species have been identified and include 1 trematode, 1 cestode, and 6 nematodes (table 9.5). Only 3 of these helminths (*Amidostomum anseris*, *Epomidiostomum crami*, and *Trichostrongylus cramae* (= *T. tenuis*) have been reported from Canada Geese previously (McDonald 1969) and only 1 species (*Hypoderaeum conoideum*) has been collected from resident waterfowl in Florida, i.e., from Mottled Ducks (Kinsella and Forrester 1972). Although some of these helminths can cause harm if present in large numbers, probably the most important is the gizzard worm *A. anseris*. This trichostrongyle is very common in Canada Geese in many parts of its range and has a direct life cycle. Adults are located under the epithelial lining of the gizzard;

Table 9.6. Helminth infections in 30 Fulvous Whistling-Ducks from Palm Beach County, Florida, 1984–85

Helminth	Infection site	No. ducks Infected	No. ducks %	Intensity Mean	Intensity Range
Trematoda					
Apatemon gracilis	Small intestine	1	3	1	1
Cotylurus gallinulae	Small intestine	12	40	3	1–7
Echinoparyphium sp.	Small intestine	11	37	3	1–10
Echinostoma trivolvis	Large intestine	20	67	5	1–31
Philophthalmus gralli	Eye	3	10	3	1–5
Prosthogonimus ovatus	Cloaca	7	23	3	1–10
Tanasia fedtschenkoi	Kidney	3	10	8	1–22
Typhlocoelum cucumerinum	Trachea/air sacs	19	63	10	1–31
Cestoda					
Cloacotaenia megalops	Cloaca	9	30	2	1–4
Diorchis sp. I	Small intestine	15	50	3	1–20
Diorchis sp. II	Small intestine	15	50	5	1–20
Microsomacanthus hopkinsi	Ceca	4	13	4	1–6
Microsomacanthus teresoides	Small intestine	10	33	3	1–15
Sobolevicanthus sp.	Small intestine	7	23	4	1–15
Nematoda					
Strongyloides sp.	Small intestine	3	10	2	1–4

Source: Forrester et al. (1994).

in cases where numbers exceed 150, erosions of the gizzard lining occur (Herman and Wehr 1954). In infections of high intensity these erosions can be severe and result in emaciation and weight losses. Herman and Wehr (1954) felt that the presence of the parasites alone would not normally result in death, but along with additional adverse conditions such as the presence of other parasites and diseases, competition for food by other species of waterfowl, and various negative environmental factors, such mortality could occur. Wobeser (1997) pointed out that because of its direct cycle this nematode is of concern as a pathogen in areas where there are dense populations of geese associated with mild weather conditions. Other than the fact that nonmigratory Canada Geese in Florida are infected with low numbers of *A. anseris*, nothing else is known about the status of this nematode.

Fulvous Whistling-Ducks in Florida have been found infected with 15 species of helminths including 8 trematodes, 6 cestodes, and 1 nematode (table 9.6). In comparison with helminths infecting 2 other resident waterfowl species in Florida (the Wood Duck and the Mottled Duck), those of the Fulvous Whistling-Duck have less species richness and lower prevalences and intensities (Forrester et al. 1994). The most striking feature of the helminth community of the Fulvous Whistling-Duck in Florida is the virtual absence of nematodes; there is only 1 species in the Fulvous Whistling-Duck, whereas there are 10 in the Wood Duck and 14 in the Mottled Duck. This is most likely due to the habitat and diet of the Fulvous Whistling-Duck in the rice fields of Palm Beach County. In addition, the rice field monoculture practices, including the use of pesticides, may have resulted in the elimination of the intermediate hosts required by some species of nematodes in their life cycles.

Table 9.7. Parasitic arthropods from 30 Fulvous Whistling-Ducks from Palm Beach County, Florida, 1984–85

Species	No. ducks		Total no. specimens collected
	Infested	%	
Feather mites			
Alloptoides sp.	5	17	8
Brephosceles sp. I	12	40	74
Brephosceles sp. II	6	20	11
Ingrassia (*Vingrassia*) sp.	29	97	1,288
Rectijanua striata	4	13	8
Sctomegninia sp.	1	3	1
Zygochelifer edentulus	11	37	31
Quill mites			
Paralges sp.	1	3	1
Skin mites			
Dermation (*Neodermation*) sp.	13	43	43
Chewing lice			
Acidoproctus rostratus	23	77	54
Anatoecus icterodes	19	63	64
Holomenopon leucoxanthum	29	97	248
Trinoton aculeatum	1	3	1

Source: Forrester et al. (1994, 1995).

The pathological significance of these helminths to Fulvous Whistling-Ducks in Florida is not known. Three of the trematodes (*Typhlocoelum cucumerinum*, *Cotylurus gallinulae*, and *Echinoparyphium* sp.) and 1 of the cestodes (*Cloacotaenia megalops*) may be pathogens, judging from known effects of these or closely related species in other species of ducks (Wobeser 1997), but this remains to be established for Fulvous Whistling-Ducks.

There are no additional data on the helminths of other species of whistling-ducks, geese, or swans in Florida.

XV. Arthropods

Thirteen species of parasitic arthropods have been reported from Fulvous Whistling-Ducks in Florida and include 7 species of feather mites, 1 skin mite, 1 quill mite, and 4 chewing lice (table 9.7). Two of the arthropods (the feather mite *Ingrassia* sp. and the chewing louse *Holomenopon leucoxanthum*) were especially common and occurred on 97% of the ducks sampled (Forrester et al. 1994). *Ingrassia* sp. was the most numerous species. The significance of these ectoparasites to Fulvous Whistling-Duck populations in Florida is unknown.

We have no information concerning ectoparasites infesting other species of whistling-ducks, geese or swans.

XVI. Summary and conclusions

Excluding trauma, 54 different morbidity/mortality factors and disease agents are known for whistling-ducks and geese in Florida.

Most of the information is on Fulvous Whistling-Ducks, from which 45 agents have been identified, including 13 organochlorines, 4 organophosphates, 8 trematodes, 6 cestodes, 1 nematode, 7 feather mites, 1 skin mite, 1 quill mite, and 4 chewing lice. Ten agents have been found in nonmigratory Canada Geese and include 1 metal, 1 biotoxin, 1 trematode, 1 cestode, and 6 nematodes. One episode of lead poisoning was documented in nonmigratory Canada Geese in Leon County in 1982.

Nothing is known about the population significance of these disease agents and parasites. In addition, no viruses, bacteria, or protozoans have been identified from whistling-ducks or geese in Florida; rather than an indication of their absence, this is probably a reflection of the small amount of work done on these species in Florida. Nothing is known about Tundra Swans, Black-bellied Whistling-Ducks, and the other species of geese that occur occasionally in Florida in low numbers.

XVII. Literature cited

American Ornithologists' Union. 1998. *Checklist of North American birds.* 7th ed. American Ornithologists' Union, Washington, D.C. 829 pp.

Andersen, P. 1996. *Design and implementation of some harmful algal monitoring systems.* Intergovernmental Oceanic Commission, UNESCO. Tech. Ser. Publ. 44. 102 pp.

Baker, J.L., and R.L. Thompson. 1978. Shot ingestion by waterfowl on national wildlife refuges in Florida. *Proc. Annu. Conf. Southeast. Assoc. Fish Wildl. Agencies* 32:256–262.

Crawford, R.L. 1981. Bird casualties at a Leon County, Florida TV tower: a 25-year migration study. *Bull. Tall Timbers Res. Stn.* 22:1–30.

Davidson, W.R., and V.F. Nettles. 1977. Rhinosporidiosis in a wood duck. *J. Am. Vet. Med. Assoc.* 171:989–990.

Delnicki, D.E., and E.G. Bolen. 1977. Use of black-bellied whistling duck nest sites by other species. *Southwest. Nat.* 22:275–277.

Flickinger, E.L., and K.A. King. 1972. Some effects of aldrin-treated rice on Gulf Coast wildlife. *J. Wildl. Manag.* 36:706–727.

Forrester, D.J. 1996. Unpublished data. University of Florida, Gainesville.

Forrester, D.J., J.A. Conti, and J.U. Bell. 1987. Lead poisoning and parasitism of non-migratory Canada geese in Florida. *Fla. Field Nat.* 15:71–76.

Forrester, D.J., H.W. Kale II, R.D. Price, K.C. Emerson, and G.W. Foster. 1995. Chewing lice (Mallophaga) from birds in Florida: a listing by host. *Bull. Fla. Mus. Nat. Hist.* 39:1–44.

Forrester, D.J., J.M. Kinsella, J.W. Mertins, R.D. Price, and R.E. Turnbull. 1994. Parasitic helminths and arthropods of fulvous whistling-ducks (*Dendrocygna bicolor*) in southern Florida. *J. Helminthol. Soc. Wash.* 61:84–88.

Friend, M. 1999. Lead. In: *Field manual of wildlife diseases. General field procedures and diseases of birds.* M. Friend and J.C. Franson (eds.). U.S. Department of the Interior, U.S. Geological Survey, Biological Research Division, Information and Technology Report 1999-001. Washington, D.C. pp. 317–334.

Friend, M., and J.C. Franson (eds.). 1999. *Field manual of wildlife diseases. General field procedures and diseases of birds.* U.S. Department of the Interior, U.S. Geological Survey, Biological Research Division, Information and Technology Report 1999-001. Washington, D.C. 426 pp.

Grice, G.D., Jr., E.L. Tyson, and E.B. Chamberlain, Jr. 1956. An unexplained mortality of Canada geese in north Florida. *J. Wildl. Manag.* 20:330–331.

Herman, C.M., and E.E. Wehr. 1954. The occurrence of gizzard worms in Canada geese. *J. Wildl. Manag.* 18:509–513.

Hines, T.C. 1980. Unpublished data. Florida Game and Fresh Water Fish Commission, Gainesville.

Kale, H.W. II. 1971. Regional reports: Florida region. *Am. Birds* 25:723–725, 730–735.

Kennedy, F.A., R.R. Buggage, and L. Ajello. 1995. Rhinosporidiosis: a description of an unprecendented outbreak in captive swans (*Cygnus* spp.) and a proposal for revision of the ontogenic nomenclature of *Rhinosporidium seeberi*. *J. Med. Vet. Mycol.* 33:157–165.

Kinsella, J.M., and D.J. Forrester. 1972. Helminths of the Florida duck, *Anas platyrhynchos fulvigula*. *Proc. Helminthol. Soc. Wash.* 39:173–176.

Locke, L.N., and N.J. Thomas. 1996. Lead poisoning of waterfowl and raptors. In: *Noninfectious diseases of wildlife*. 2d ed. A. Fairbrother, L.N. Locke, and G.L. Hoff (eds.). Iowa State University Press, Ames. pp. 108–117.

Logan, T.H. 1997. Florida's endangered species, threatened species and species of special concern. Official lists. Florida Game and Fresh Water Fish Commission, Tallahassee. 14 pp.

Maehr, D.S., and J.Q. Smith. 1988. Bird casualties at a central Florida power plant: 1982–1986. *Fla. Field Nat.* 16:57–80.

Maehr, D.S., A.G. Spratt, and D.K. Voigts. 1983. Bird casualties at a central Florida power plant. *Fla. Field Nat.* 11:45–49.

Mahnke, G.T., and L.E. Hayes. 1989. Unpublished data. Southeastern Cooperative Wildlife Disease Study, University of Georgia, Athens.

Martin, E.M. 1996. Distribution in states and counties of waterfowl species harvested during the 1981 to 1990 hunting seasons. Unpublished report. Office of Migratory Bird Management, U.S. Fish and Wildlife Service, Laurel, Md.

McDonald, M.E. 1969. *Catalogue of helminths of waterfowl (Anatidae)*. U.S. Department of the Interior, Bureau of Sport Fisheries and Wildlife, Special Scientific Research Publication—Wildlife 126. Washington, D.C. 692 pp.

Morehouse, K.A. 1992. Lead poisoning of migratory birds: the U.S. Fish and Wildlife Service position. In: *Lead poisoning in waterfowl*. D.J. Pain (ed.). Proc. IWRB Workshop, Brussels, Belgium, 13–15 June 1991. IWRB Spec. Publ. 16. International Waterfowl and Wetlands Resource Bureau, Slimbridge, Gloucester, United Kingdom. pp. 51–55.

Pain, D.J. 1996. Lead in waterfowl. In: *Environmental contaminants in wildlife: interpreting tissue concentrations*. W.N. Beyer, G.H. Heinz, and A.W. Redmon-Norwood (eds.). CRC, Boca Raton, Florida. pp. 251–264.

Robertson, W.B., Jr., and G.E. Woolfenden. 1992. *Florida bird species: an annotated list*. Spec. Publ. 6, Florida Ornithological Society, Gainesville. 260 pp.

Snyder, B. 1994. Unpublished data. Florida Department of Environmental Protection, Tallahassee.

Stevenson, H.M. 1976. Vertebrates of Florida: identification and distribution. University Press of Florida, Gainesville. 607 pp.

Stevenson, J.A. 1994. Unpublished data. Florida Department of Environmental Protection, Tallahassee.

Taylor, W.K., and B.H. Anderson. 1973. Nocturnal migrants killed at a central Florida TV tower; autumns 1969–1971. *Wilson Bull.* 85:42–51.

———. 1974. Nocturnal migrants killed at a central Florida TV tower, autumn 1972. *Fla. Field Nat.* 2:40–43.

Taylor, W.K., and M.A. Kershner. 1986. Migrant birds killed at the Vehicle Assembly Building (VAB), John F. Kennedy Space Center. *J. Field Ornithol.* 57:142–154.

Thul, J.E. 1983. Unpublished data. Florida Game and Fresh Water Fish Commission, Gainesville.

Thul, J.E., D.J. Forrester, and E.C. Greiner. 1980. Hematozoa of wood ducks (*Aix sponsa*) in the Atlantic flyway. *J. Wildl. Dis.* 16:383–390.

Tiemeier, O.W. 1941. Repaired bone injuries in birds. *Auk* 58:350–359.

Trainer, D.O. 1982. Lead poisoning of waterfowl. In: *Noninfectious diseases of wildlife*. G.L. Hoff and J.W. Davis (eds.). Iowa State University Press, Ames. pp. 24–30.

Turnbull, R.E., F.A. Johnson, M.A. Hernandez, W.B. Wheeler, and J.P. Toth. 1989. Pesticide residues in fulvous whistling-ducks from south Florida. *J. Wildl. Manag.* 53:1052–1057.

U.S. Fish and Wildlife Service. 1976. *Final environmental statement on proposed use of steel shot for hunting waterfowl in the United States.* U.S. Government Printing Office, Washington, D.C. 276 pp.

Wehr, E.E. 1932. Unpublished data. U.S. National Parasite Collection, Beltsville, Md. Cat. no. 32276.

Wobeser, G.A. 1997. *Diseases of wild waterfowl.* 2d ed. Plenum, New York. 324 pp.

Ducks

I. Introduction

The subfamily Anatinae of the family Anatidae contains a large group of waterfowl subdivided into 5 tribes (Bellrose 1976). In Florida there are 33 species, including 2 perching ducks, 12 dabbling ducks, 14 diving ducks, 3 mergansers, and 2 stiff-tailed ducks (table 10.1). None is classified as threatened or endangered (Logan 1997). Three of these species (Muscovy Ducks, Wood Ducks, and Mottled Ducks) breed in Florida and range in numbers from rare to locally abundant. The others are primarily winter visitors and occur in small numbers except for Green-winged Teal, Blue-winged Teal, Northern Pintails, Northern Shovelers, American Wigeons, Redheads, Ring-necked Ducks, and Lesser Scaup, which are often abundant. The latter is

the "most numerous species of waterfowl in Florida, at times occurring in rafts of tens of thousands" (Robertson and Woolfenden 1992).

The reader is referred to the treatises by Wobeser (1997), Fairbrother et al. (1996), Beyer et al. (1996), and Friend and Franson (1999) for general information on the diseases, parasites, and environmental contaminants of free-ranging waterfowl. In addition, the book by Bellrose and Holm (1994) contains a section on natural mortality and a fairly detailed chapter on parasites and diseases of Wood Ducks (including some information on Wood Ducks from Florida). Host-parasite lists for waterfowl have been published by Lapage (1961) and McDonald (1969a). McDonald has published an

Table 10.1. Distribution, occurrence, and abundance of ducks of the subfamily Anatinae in Florida[a]

	Species	Range	Seasonal occurrence	Relative abundance
Perching ducks				
Muscovy Duck	*Cairina moschata*	Statewide	Resident[b]	Uncommon to very common
Wood Duck	*Aix sponsa*	Statewide	Resident	Rare to common
		Keys	Winter	Rare
Dabbling ducks				
Green-winged Teal	*Anas crecca*	Peninsula	Winter	Uncommon to abundant
		Keys	Winter	Rare
		Western panhandle	Fall & Spring	Common
American Black Duck	*Anas rubripes*	Northern Florida	Winter	Rare to fairly common
		Southern Florida	Winter	Rare to irregular
Mottled Duck	*Anas fulvigula*	Peninsula	Resident	Fairly common to locally abundant
Mallard	*Anas platyrhynchos*	Northern Florida	Winter	Uncommon to fairly common
		Southern Florida	Winter	Rare to irregular
White-cheeked Pintail	*Anas bahamensis*	Peninsula & Keys	Spring & Winter	Rare to irregular
Northern Pintail	*Anas acuta*	Peninsula	Winter	Abundant
		Western panhandle & Keys	Winter	Rare to uncommon
Blue-winged Teal	*Anas discors*	Statewide	Transient & Winter	Fairly common to abundant
		Western panhandle	Winter	Rare
Cinnamon Teal	*Anas cyanoptera*	Statewide	Winter	Rare (40 reports)
Northern Shoveler	*Anas clypeata*	Statewide	Transient & Winter	Uncommon to locally abundant
		Keys	Winter	Rare
Gadwall	*Anas strepera*	Northern peninsula	Winter	Rare to fairly common
		Panhandle	Winter	Rare to fairly common
		Southern Florida & Keys	Winter	Rare
Eurasian Wigeon	*Anas penelope*	Eastern panhandle & peninsula	Winter	Rare
American Wigeon	*Anas americana*	Statewide	Winter	Common to abundant
		Western panhandle & Keys	Winter	Rare to uncommon

Common name	Scientific name	Location	Season	Status
Diving ducks				
Canvasback	*Aythya valisineria*	Panhandle & northern peninsula	Winter	Rare to uncommon
		Southern peninsula (except Keys)	Winter	Rare
Redhead	*Aythya americana*	Panhandle & northern peninsula	Winter	Locally common to abundant
		Southern Florida	Winter	Rare to uncommon
Ring-necked Duck	*Aythya collaris*	Statewide	Transient & Winter	Fairly common to abundant
Greater Scaup	*Aythya marila*	Panhandle & northern peninsula	Winter	Uncommon to fairly common
		Southern Florida (except Keys)	Winter	Very rare
Lesser Scaup	*Aythya affinis*	Statewide	Winter	Common to abundant
		Keys	Winter	Rare to uncommon
Common Eider	*Somateria mollissima*	Atlantic coast	Winter	Very rare (10 reports)
King Eider	*Somateria spectabilis*	Coasts	Winter	Occasional (6 reports)
Harlequin Duck	*Histronicus histronicus*	Coasts	Winter	Rare (20 reports)
Oldsquaw	*Clangula hyemalis*	Panhandle & northern peninsula	Winter	Rare to locally uncommon
		Southern Florida	Winter	Very rare
Black Scoter	*Melanitta nigra*	Panhandle & Atlantic coast	Winter	Rare to fairly common
Surf Scoter	*Melanitta perspicillata*	Panhandle & northern peninsula	Winter	Rare to uncommon
		Southern Florida	Winter	Rare
White-winged Scoter	*Melanitta fusca*	Panhandle & N. Atlantic coasts	Winter	Rare
Common Goldeneye	*Bucephala clangula*	Panhandle & northern Florida	Winter	Rare to uncommon
		Southern peninsula	Winter	Sporadic to occasional
Bufflehead	*Bucephala albeola*	Panhandle & northern Florida	Winter	Fairly common to very common
		Southern Florida (except Keys)	Winter	Rare
Mergansers				
Hooded Merganser	*Lophodytes cucullatus*	Statewide	Winter	Rare to locally abundant
		Keys	Transient & Winter	Very rare
Common Merganser	*Mergus merganser*	Statewide (coasts)	Winter	Rare
Red-breasted Merganser	*Mergus serrator*	Statewide (coasts)	Winter	Common to abundant
Stiff-tailed ducks				
Ruddy Duck	*Oxyura jamaicensis*	Statewide	Winter	Fairly common to locally abundant
Masked Duck	*Nomonyx dominicus*	Southern peninsula & Keys	Winter	Rare

a. Modified from Robertson and Woolfenden (1992) and American Ornithologists' Union (1998).

b. An established exotic.

Table 10.2. Numbers of ducks killed by colliding with a 1,010-foot TV tower at Tall Timbers Research Station, Leon County, Florida, 1955–1980

Group	Species	Number found dead
Perching-ducks	Wood Duck	7
Dabbling ducks	Green-winged Teal	5
	American Black Duck	1
	Blue-winged Teal	13
	Northern Shoveler	1
	Gadwall	2
	American Wigeon	2
Diving ducks	Redhead	2
	Ring-necked Duck	61
	Lesser Scaup	10
Mergansers	Hooded Merganser	5
	Red-breasted Merganser	2
Stiff-tailed ducks	Ruddy Duck	1
	Total	112

Source: Crawford (1981).

annotated bibliography of the helminths of waterfowl (McDonald 1969b) and separate keys to the nematodes (McDonald 1974), trematodes (McDonald 1981), and acanthocephalans (McDonald 1988).

II. Trauma

As pointed out in chapter 9, Whistling-Ducks, Swans, and Geese, traumatic injuries are fairly common in anatids (Tiemeier 1941; Wobeser 1997). Crawford (1981) reported that 112 ducks representing 13 species were found dead at the base of a 1,010-foot TV tower near Tall Timbers Research Station in Leon County (table 10.2), the result of striking the tower during nocturnal migrations. The numbers were based on his examination of records on more than 42,000 birds over a 25-year period. More than half of these 112 ducks were Ring-necked Ducks, possibly a reflection of the large numbers of this species that migrate into Florida each winter. Ducks have not been found in other studies of such mortality involving tall structures in Florida (Kale 1971; Maehr

et al., 1983; Maehr and Smith 1988; Taylor and Anderson 1973, 1974; Taylor and Kershner 1986).

In a 4-year survey of roadkills in Florida state parks and recreation areas from 1990 to 1993, 3 ducks were found. These included 2 Wood Ducks (1 from Fakahatchee Strand State Preserve, Collier County, in 1992 and 1 from Paynes Prairie State Preserve, Alachua County, in 1993) and 1 Mallard from Lake Griffin State Recreation Area, Lake County, in 1993 (Stevenson and Snyder 1994).

Weston (1966) reported a Ruddy Duck killed by striking cables associated with the 3-mile Pensacola Bay Bridge (Santa Rosa and Escambia counties) in 1949. This was the only species of waterfowl that he found over a 12-year period (1938–49), during which he tabulated 740 specimens of dead birds representing 75 species, most of which were passeriforms. A Mottled Duck was among about 50 aquatic and wading birds that were killed by flying into a power line near the Miami Canal across from Lake Harbor in Palm Beach County in 1989. The bird had a broken neck, broken clavicle, and massive hemorrhage at the thoracic inlet (Locke 1990). During 1990–91 3 Green-winged Teal, 2 Blue-winged Teal, and 3 other unidentified ducks were found dead near this same area and were assumed to have died because of striking a power line, although this was not verified by examinations at necropsy (Regan 1996). This type of mortality is probably of some significance in the case of ducks, but few data are available concerning such incidents.

The average annual total harvest of ducks by hunters in Florida for the 10-year period 1981–90 was 179,887 and included 85,314 diving ducks, 69,402 dabblers, 21,329 perching ducks, 2,867 mergansers, and 975 stiff-tailed ducks (Martin 1996). The most heavily harvested species were Ring-necked Ducks (64,165/yr), Blue-winged and Cinnamon Teal (30,811/yr), Wood Ducks (21,319/yr), and Lesser Scaup (17,447/yr). In 1997–98 an estimated 248,232 ducks (all legally hunted species) were harvested by hunters in Florida (Linda 1998). Trauma due to crippling by hunters is undoubtedly a factor of some significance, but we have no relevant data for waterfowl in Florida.

III. Predation

Remains of 2 Greater Scaup and 5 Lesser Scaup were identified from a collection of 88 pellets from a family of Great Horned Owls over a 4-week period in January 1949 in Putnam County (Burns 1952). Likewise, the remains of 9 Lesser Scaup were identified in 1,098 pellets collected at a roosting site of a Northern Harrier in Leon County during the winter of 1925–26 (Stoddard 1931). Fargo (1926) found bones of mergansers (probably Red-breasted Mergansers) under Bald Eagle nests in Pinellas County during 1923–26. Bent (1937) listed Mottled Ducks and Lesser Scaup as prey items of Bald Eagles in Florida, but gave no specific dates or localities. McEwan and Hirth (1980) examined 16 active Bald Eagle nests in a 4-county area (Alachua, Marion, Putnam, and Volusia) during 1975–76 and found the re-

mains of 14 Ruddy Ducks, 7 Lesser Scaup, 2 Mottled Ducks, and 1 Blue-winged Teal. It is not clear, however, whether these were cases of predation or of scavenging by the Bald Eagles. Delany (1986) found a band of a Northern Pintail in the stomach of an American alligator from Orange Lake. This pintail had been banded 1 year previously in Northwest Territories, Canada. He also reported finding the remains of a Wood Duck in the stomach contents of another alligator (from either Orange, Lochloosa, or Newnans Lake) but, based on examination of 350 alligator stomachs, concluded that alligator predation on waterfowl was not significant. Predation of Wood Duck ducklings and pipped eggs in nest boxes by the red imported fire ant (*Solenopsis invicta*) has been reported in Texas (Ridlehuber 1982), but this has not been seen in Florida.

IV. Inclement weather

No information is available on this topic.

V. Vagrancy

An unusual influx of Black Scoters was reported as far south as Fort Lauderdale on the Atlantic coast and Naples on the Gulf coast from November 1981 through May 1982 (Bancroft and Hoffman 1985). Normally only small numbers of Black Scoters overwinter in Florida and these are found primarily in the upper Atlantic Coast. Several hundred scoters were seen during this time and many were found sick and dead on beaches. Complete necropsies were not performed, but the birds appeared to be "under physiological stress and in poor condition." The weights of these birds were only a little above half of what is normal for this species; in addition they were in delayed and protracted molt. The authors concluded that "To the extent that birds' ranges are the product of stabilizing selection for conservative migratory and dispersive behavior, low survivorship and

Table 10.3. Organochlorine residues in wings[a] of Mallards[b] killed by hunters in Florida[c]

Years	No. pools[d]	Mean (range) ppm wet weight[e]		Data source
		DDE	**DDD**	
1965–66	4	0.53 (0.35–0.70)	0.05 (ND–0.10)	A
	5[f]	0.66 (0.11–1.8)	ND —	A
1969–70	2	0.82 (0.49–1.3)	ND —	B
1972–73	3	0.54 (0.18–0.97)	0.02 (0.009–0.03)	C
1976–77	3	0.22 (0.20–0.25)	0.01 (ND–0.01)	D
1979–80	1	0.02 —	ND —	E
1981–82	1	0.13 —	ND —	F
		DDT	**Dieldrin**	
1965–66	4	0.09 (ND-0.15)	ND —	A
	5[f]	0.11 (ND-0.29)	ND —	A
1969–70	2	0.06 (0.05–0.06)	0.02 (ND-0.02)	B
1972–73	3	0.07 (0.04–0.11)	0.03 (0.006–0.07)	C
1976–77	3	0.05 (ND-0.05)	0.04 (0.01–0.09)	D
1979–80	1	0.02 —	ND —	E
1981–82	1	ND —	ND —	F
		PCBs		
1965–66	4	NG —		A
	5[f]	NG —		A
1969–70	2	0.97 (0.25–1.7)		B
1972–73	3	0.48 (0.37–0.62)		C
1976–77	3	1.4 (0.51–2.8)		D
1979–80	1	0.09 —		E
1981–82	1	0.14 —		F

Sources: A = Heath (1969), B = Heath and Hill (1974), C = White and Heath (1976), D = White (1979), E = Cain (1981), F = Prouty and Bunck (1986).

ND = not detected, NG = not given.

a. Consists of wings (trimmed of flight feathers) that were chopped and blended in homogenates with a food cutter.

b. All birds are adults, unless otherwise noted.

c. Data for 1965–66, 1969–70, 1972–73, and 1976–77 are combined Florida and Georgia birds.

d. Each pool has wings from 25 birds, except for 1979–80 and 1981–82 data, which were based on <25 birds per pool (exact numbers not given).

e. Type of mean not stated, except for 1981–82 data, which were given as geometric means. Limits of detection: Reference A = 0.05 ppm for all residues; Reference B = 0.03 ppm for PCBs and 0.02 ppm for other residues; Reference C = 0.01 ppm for PCBs and 0.005 ppm for other residues; References D and E = 0.01 ppm for all residues; Reference F = 0.10 ppm for PCBs and 0.04 ppm for DDE.

f. Immature birds.

Table 10.4. Organochlorine residues in wings[a] of adult American Black Ducks killed by hunters in Florida[b]

| Years | Mean (ppm wet weight[c]) | | Data source |
	DDE	PCBs	
1979–80	0.11	0.45	Cain (1981)
1981–82	0.30	0.18	Prouty & Bunck (1986)

a. A pool of approximately 25 wings, (trimmed of flight feathers) that were chopped and blended in homogenates with a food cutter.
b. Ducks from Florida and Georgia were combined into 1 sample.
c. Type of means for the 1979–80 data was not described; 1981–82 means were geometric means. Limits of detection: for the 1979–80 data = 0.01 ppm for both residues; for the 1981–82 data = 0.10 ppm for PCBs and 0.04 ppm for DDE.

physiological abnormalities (in this case molt) might be expected."

VI. Organochlorines

Residues of 11 different organochlorines have been identified in Mallards (table 10.3), American Black Ducks (table 10.4), Red-breasted Mergansers (table 10.5), Lesser Scaup (table 10.6), and Northern Shovelers (table 10.7) from Florida.

The largest dataset is on Mallards and American Black Ducks and was obtained by the U.S. Fish and Wildlife Service as part of its National Biomonitoring Contaminant Program, designed to assess the body concentrations of organochlorine compounds in migratory birds. Researchers collected wings from several thousand hunter-killed ducks during the hunting season at 2- or 3-year intervals and analyzed them for various residues (Heath and Prouty 1967). The results of these surveys have been published in a series of papers (Heath 1969; Heath and Hill 1974; White and Heath 1976; White 1979; Cain 1981; Prouty and Bunck 1986) and each contains some data on ducks from Florida. It is difficult to interpret the significance of these findings since the values represent collective residues in bone, muscle, and feathers. Indeed, the monitoring study was designed to follow changes from year to year and to compare residues in various flyways rather than to provide information directly related to the health of waterfowl. It is interesting to note that there was a general decrease in the concentrations of DDE and other metabolites of DDT in the wings of Mallards from 1976 onward (table 10.3); this might be a reflection of the banning of the use of DDT in the United States in 1972 and subsequent decline in its use as a pesticide. However, it must be remembered that, because of the migratory nature of waterfowl, concentrations of organochlorines in tissues of Mallards in Florida are at least a partial indication of exposure along other points of their migration route (Jacknow et al. 1986). This complicates the picture considerably. The concentrations of PCBs in Mallards from Florida decreased after 1976–77, but this was true for Mallards in the nation as a whole, where PCB contamination remained widespread, despite the restrictions in the use of this compound by industry (Jacknow et al. 1986).

The residues of various organochlorines found in Red-breasted Mergansers (table 10.5), Lesser Scaup (table 10.6), and Northern Shovelers (table 10.7) from Florida were low and probably of little or no health significance. However, sample sizes for these 3 species were too small to allow definitive conclusions to be drawn.

Lehner et al. (1967) tested 4 moribund Muscovy Ducks that were found in Dade County 4–20 days after the area was sprayed to control the mosquito *Aedes aegypti* in July of 1965.

Table 10.5. Organochlorine residues in tissues of Red-breasted Mergansers from Florida

| Tissue County or site | Year(s) | Sex | Residues, ppm wet weight[a] | | | Data source |
			α-BHC	DDD	DDE	
Brain						
Sarasota	1994	M	ND	0.011	0.54	Locke (1995)
Sarasota	1994	F	0.002	0.008	0.029	Ibid.
Sarasota	1994	1M,4F[b]	ND	ND	0.5	Fischer (1994)
Fat						
"Florida"	1969–74	M	NT	NT	10.4	Johnston (1976)
"Florida"	1969–74	F	NT	NT	ND	Ibid.
Uropygial gland						
"Florida"	1969–74	M	NT	NT	7.0	Ibid.
"Florida"	1969–74	F	NT	NT	0.1	Ibid.
			DDT	Dieldrin	Endrin	
Brain						
Sarasota	1994	M	0.019	0.16	0.008	Locke (1995)
Sarasota	1994	F	ND	0.031	ND	Ibid.
Fat						
"Florida"	1969–74	M	NG	ND	NT	Johnston (1976)
"Florida"	1969–74	F	NG	ND	NT	Ibid.
Uropygial gland						
"Florida"	1969–74	M	NG	ND	NT	Ibid.
"Florida"	1969–74	F	NG	ND	NT	Ibid.
			Heptachlor epoxide	Nonachlor	PCBs	
Brain						
Sarasota	1994	M	0.04	0.04	3.9	Locke (1995)
Sarasota	1994	F	0.007	0.006	0.37	Ibid.
Sarasota	1994	1M,4F[b]	ND	ND	1.8	Fischer (1994)

a. Limits of detection = unknown for the Locke (1995) and Fischer (1994) data, 0.01 ppm for the Johnston (1976) data. Locke (1995) also tested for residues of aldrin, beta-BHC, heptachlor, and lindane, but none were detected. ND = not detected, NG = not given by authors, NT = not tested for.
b. A pool of brain tissues from 1 male and 4 females.

Concentrations of DDT, DDD, p,p'-DDE, and o,p'-DDE in brain tissue were all <1 ppm (wet weight) or below the levels of detection. Organochlorines were ruled out as the cause of sickness in these ducks.

Even though there is no evidence of direct lethality due to organochlorines in ducks in Florida, sublethal effects of low concentrations could be important. There is experimental evidence that DDT (Friend and Trainer 1974a), dieldrin (Friend and Trainer 1974b), and PCBs (Friend and Trainer 1970) can interact to cause increased mortality in Mallards infected with duck hepatitis virus, compared with ducks infected with the virus but not contaminated with the respective organochlorines. Even though the use of many organochlorines has been reduced in the United States, migratory ducks can become contaminated in other countries where such restrictions are not in effect. Thus the threat of these compounds is still present in populations of North American waterfowl, some of which migrate into Florida.

Table 10.6. Organochlorine residues (ppm wet wt.)[a] in tissues of three Lesser Scaup
from Florida in February 1974

Tissue County	Sex	DDE	DDD	DDT	PCB(Aroclor #)	Dieldrin
Brain						
Charlotte	F	0.14	0.015	0.06	0.2(1242) 0.4(1254)	0.015
Hillsborough	M	0.015	0.005	0.015	0.21(1248) 0.25(1254)	ND
	M	0.04	ND	0.05	0.5(1242) 0.6(1254)	0.015
Fat						
Hillsborough	M	1.8	0.13	2.1	15(1254)	ND
	M	1.1	0.11	0.053	0.68(1254)	0.11
Liver						
Charlotte	F	0.47	0.03	0.13	1.2(1254)	ND
Hillsborough	M	0.24	0.019	0.093	1.1(1254)	ND
	M	0.14	ND	0.15	1.8(1248) 0.95(1254)	0.016
Muscle						
Charlotte	F	0.16	0.016	0.063	0.25(1254)	ND
Hillsborough	M	0.12	0.011	0.055	0.51(1254) 0.38(1248)	0.02
	M	0.10	0.02	0.09	0.60(1248) 0.22(1254)	0.007

Source: Thompson and Forrester (1974).
ND = not detected.
a. Limits of detection not determined.

VII. Organophosphates

Diazinon was reported as the cause of death of several Mallards from Lake Talquin (Leon County) in May 1992 (Quist 1992). In the 1 bird examined, there was 58% inhibition of brain cholinesterase activity and the contents of the gizzard contained 0.5 ppm diazinon. There are numerous reports of this organophosphate as the cause of mortality in waterfowl in various parts of the country. Many of these involve Canada Geese on golf courses, where the compound is used commonly (Smith 1993), although we have no such data for Florida. As mentioned in chapter 9, Whistling-Ducks, Swans, and Geese, residues of diazinon have been reported from tissues of Fulvous Whistling-Ducks in Palm Beach County (Turnbull et al. 1989), but no mortality has been associated with this pesticide and these ducks.

VIII. Lead poisoning

Very few cases of lead poisoning in waterfowl have been documented in Florida. In addition to the accounts of this disease in Canada Geese and Graylag Geese discussed in chapter 9, Whistling-Ducks, Swans, and Geese, only 3 such reports are known from ducks. Locke (1968) diagnosed lead poisoning in 1 of 5 Lesser Scaup examined from Everglades National Park in January 1969,

Table 10.7. Organochlorine residues[a] in uropygial glands and their oil
from 2 Northern Shovelers from Florida[b]

Sex of duck	Uropygial gland		Oil	
	DDE	Dieldrin	DDE	Dieldrin
Male	0.17	0.07	0.68	1.03
Female	0.14	0.07	0.81	0.75

Source: Johnston (1976).
a. ppm wet weight; limits of detection = 0.01 ppm.
b. Dates and localities not given by author.

several weeks after the occurrence of an outbreak of avian cholera (Klukas and Locke 1970). In February 1998 a Lesser Scaup from Lake Jackson in Leon County died of lead poisoning (Terrell and Forrester 1998). Lead concentrations in its liver were 53 ppm (wet weight). Liver concentrations >6 ppm are considered toxic for waterfowl (Friend 1999b). Feierabend and Myers (1984) reported lead-poisoned waterfowl from Brevard, Leon, and Volusia counties, but gave no details on dates, species, and numbers of waterfowl involved. However, the actual prevalence of lead poisoning in waterfowl in Florida is most likely higher than indicated by available reports and publications. Most morbidity and mortality of waterfowl due to lead poisoning is chronic and low level and as such usually goes undetected (Wobeser 1997). Thul (1986a) pointed out that because of "Florida's generally lush plant cover, abundant predators, and relatively mild winters, carcasses are probably difficult to detect and disappear quickly."

There have been several investigations of the concentrations of lead in various tissues of ducks in Florida. Data from such studies on Wood Ducks, Mottled Ducks, and Ring-necked Ducks are presented in tables 10.8, 10.9, and 10.10, respectively. In addition, liver samples from 8 Red-breasted Mergansers obtained from Sarasota County in 1994 were tested for lead (Fischer 1994; Locke 1995). Only 1 of the

Table 10.8. Concentrations of lead in tissues of 5 Wood Ducks[a]
from Alachua County and 5[a] from Hamilton County, Florida,
1981–82

Tissue	County	Concentrations (ppm dry wt.)[b]	
		Mean[c]	SE[d]
Liver	Alachua	6.4	2.4
	Hamilton	4.6	1.8
Kidney	Alachua	2.3	1.8
	Hamilton	0.9	0.1
Muscle	Alachua	1.0	0.1
	Hamilton	0.9	0.0
Bone	Alachua	52	21
	Hamilton	30	13

Source: O'Meara et al. (1986).
a. A mixture of adults and subadults.
b. Limits of detection not given by authors.
c. Type of mean not given.
d. SE = standard error.

Table 10.9. Concentrations of lead in tissues of Mottled Ducks from Florida

| Tissue County or site | Year(s) | Age[a] | No. exam. | Concentrations (ppm dry wt.)[b] | | Data source |
				Mean[c]	SE[d]	
Liver						
Osceola	1981–82	AD&SA	13	20	8.0	O'Meara et al. (1986)
Polk	1981–82	AD&SA	17	7.6	2.0	Ibid.
Kidney						
Osceola	1981–82	AD&SA	14	23.0	9.0	Ibid.
Polk	1981–82	AD&SA	17	2.8	1.6	Ibid.
Muscle						
Glades	1980	AD	10	0.082	0.009	Montalbano et al. (1983)
Osceola	1981–82	AD&SA	14	1.1	0.2	O'Meara et al. (1986)
Polk	1980	AD&IM	20	0.25	0.15	Montalbano et al. (1983)
Polk	1981–82	AD&SA	17	0.9	0.1	O'Meara et al. (1986)
Bone						
Osceola	1981–82	AD&SA	14	220	116	O'Meara et al. (1986)
Polk	1981–82	AD&SA	17	38	15	Ibid.
"Florida"	1973	IM	60	37	9.2	Stendell et al. (1979)[e]
"Florida"	1973	AD	22	46	15	Ibid.

a. AD = adult, SA = subadult, IM = immature.
b. Limits of detection not given by authors.
c. Type of mean not given by authors.
d. SE = standard error.
e. Blus et al. (1977) were probably referring to the same dataset as published by Stendell et al. (1979) when they reported that Mottled Ducks from Florida, Louisiana, and Texas had the highest concentrations of any species from any area.

8 was positive and contained 2.9 ppm (wet weight). Care must be exercised in comparing these data as some were given on a wet-weight and some on a dry-weight basis. Pain (1996) stated that "For comparative purposes, 1 ppm wet weight = 3 to 4 ppm dry weight." Keeping these points in mind, and following the suggested interpretations of the significance of concentrations of lead in tissues of waterfowl given by Pain (1996), it is evident that 1 of the Red-breasted Mergansers and many of the Wood Ducks, Mottled Ducks, and Ring-necked Ducks sampled had lead (particularly in livers) in concentrations that would be considered elevated and some would be high enough to make one suspect subclinical or clinical poisoning.

Data from 2 Florida studies, involving experimental and field approaches, are helpful in understanding the effects of lead ingestion on Ring-necked Ducks. Mautino and Bell (1986) found that 4 of 27 ducks, each given 1 No. 4 lead shot by gastric intubation, were extremely lethargic, ataxic, and anorexic and died within 2–3 weeks after dosing. The other 23 ducks had signs of impaired gastrointestinal tract function (including green-stained vents and impaction) and had lower body weights compared with undosed controls. These signs decreased with time and by week 7 of the experiment, all birds appeared normal. Concentrations of lead in livers, kidneys, and bones of the dosed birds were significantly higher than in normal birds at the end of the 7-week experiment (table 10.11). In a field study, Thul (1986b) examined the correlation of several indexes of physical condition with the presence of lead shot in gizzards and lead in various tissues in 621 Ring-necked Ducks collected from 1979 to 1982. He found that >98% of the ducks sampled had either lead shot in their gizzards, high concentrations of lead in tissues, or lesions

Table 10.10. Concentrations of lead in tissues of Ring-necked Ducks from Florida

Tissue Site	Year(s)	Age[a]	Sex	No. exam.	Concentrations (ppm dry wt.)[b] Mean[c]	SE[d]	Data source
Blood							
"Florida"	1979–82	AD	M	NG	1.8	7.3	Thul (1986b)
	1979–82	AD	F	NG	0.92	2.6	Ibid.
	1979–82	JU	M	NG	2.6	5.9	Ibid.
	1979–82	JU	F	NG	1.5	3.5	Ibid.
Liver							
"Florida"	1979–82	AD	M	NG	3.4	8.7	Ibid.
	1979–82	AD	F	NG	1.7	5.6	Ibid.
	1979–82	JU	M	NG	6.4	13	Ibid.
	1979–82	JU	F	NG	3.8	8.0	Ibid.
"No. Fla."	NG	NG	NG	10	0.24	0.03	Mautino & Bell (1986)
Kidney							
"Florida"	1979–82	AD	M	NG	5.8	14	Thul (1986b)
	1979–82	AD	F	NG	6.0	25	Ibid.
	1979–82	JU	M	NG	10	28	Ibid.
	1979–82	JU	F	NG	10	64	Ibid.
"No. Fla."	NG	NG	NG	9	0.44	0.05	Mautino & Bell (1986)
Bone							
"Florida"	NG	AD	M	NG	47	54	Thul (1986b)
	NG	AD	F	NG	33	38	Ibid.
	NG	JU	M	NG	47	58	Ibid.
	NG	Ju	F	NG	53	109	Ibid.
"No. Fla."	NG	NG	NG	10	18	4.0	Mautino & Bell (1986)

NG = not given by authors.
a. AD = adult, JU = juvenile.
b. Limits of detection not given by authors.
c. Type of mean not given by authors.
d. SE = standard error.

due to lead poisoning. He concluded that almost the entire population of Ring-necked Ducks wintering in Florida during that time span was compromised to some extent by lead contamination. For example, fat index values for ducks with liver lead concentrations of 2 ppm or less increased during the winter, but such was not the case for ducks with >2 ppm.

In addition to locating and examining carcasses and tissues of waterfowl, researchers can assess the extent of lead poisoning as a problem by examining gizzards of hunter-killed birds for the presence of lead shot. Several such studies have been conducted in Florida on dabbling ducks (table 10.12), diving ducks (table 10.13), and Ruddy Ducks and mergansers (table 10.14). Data in these 3 tables were taken from several sources, including 2 large-scale surveys of the occurrence of lead shot in gizzards of hunter-killed ducks from Florida. Baker and Thompson (1978) analyzed 9,632 gizzards collected from 15 species of waterfowl that were killed on or near 4 National Wildlife Refuges (Merritt Island, Chassahowitzka, Loxahatchee, and Ding Darling) over a 5-year period (1973–78). In addition, Thul (1986a) examined 10,231 gizzards from 21 species of ducks killed by hunters in 37 counties (scattered throughout the state) over an 8-year period (1976–84). The mean prevalence of ingested shot for all samples was 17% for the

Table 10.11. Concentrations (ppm wet wt.[a]) of lead in tissues of Ring-necked Ducks 7 weeks after being dosed with one no. 4 lead shot

	Dosed			Controls[b]		
Tissue	Number tested	Mean[c]	SE[d]	Number tested	Mean[c]	SE
Liver	22	0.87	0.15	10	0.24	0.03
Kidney	20	1.5	0.28	9	0.44	0.05
Bone	20	32	3.9	10	18	4

Source: Mautino & Bell (1986).
a. Limits of detection not given by authors.
b. These birds were not dosed with lead shot during the experiment, but since they were wild-caught ducks, it is not unexpected that they have some lead in their tissues.
c. Type of mean not given by authors.
d. SE = standard error.

Baker-Thompson study and 12% for the Thul study. Both of these percentages are higher than the nationwide prevalence (10%) of lead in gizzards of comparable species of waterfowl determined by Bellrose (1959) for 1938–54. In both the Baker-Thompson and Thul studies, diving ducks had higher prevalences than dabbling ducks, 23 vs. 11% in the first study and 14 vs. 4% in the second. Ruddy Ducks also had high prevalences (between 10 and 55%). Mergansers were free of lead shot, which is not surprising since they are mainly fish-eaters. It is difficult and perhaps unwise to make many more comparisons between the 2 studies since the Baker-Thompson study was based on lead and steel shot combined, whereas data on the 2 types of shot were separated in the Thul study. The great majority of shot reported by Baker and Thompson, however, were most likely lead rather than steel; the exclusive use of steel shot was instituted on a small scale in certain areas beginning in 1975. Thul (1986a) found that the species with the highest prevalence of lead shot (>12%) were scaup, Mallards, Ring-necked Ducks, Mottled Ducks, and Northern Pintails. He also reported that the prevalence of lead shot was lowest in early fall, increased during the hunting season, and was the highest after the hunting season closed (figure 10.1). There also was an inverse relationship between water levels and ingestion of lead shot in ducks from central and southern Florida, i.e., from Citrus, Sumpter, Lake, and Volusia counties south to Dade and Monroe counties (figure 10.2). Of special management interest were his findings in relation to the implementation of steel shot as a replacement for lead shot. His data on the decline of the prevalence of lead shot in waterfowl gizzards and the increase in the prevalence of steel shot over the 8-year period 1976–77 to 1983–84 on a statewide basis are shown in figure 10.3.

Hopefully, with the total ban on the use of lead shot for waterfowl hunting that was instituted in 1991 (Morehouse 1992), we can expect reduced prevalences of ingested lead shot and fewer cases of lead poisoning in waterfowl in Florida. Even though lead shot is most likely still present in Florida's marshes, having accumulated as a result of many years of shooting, most of it is out of reach of feeding waterfowl under normal conditions. The majority of the lead that waterfowl ingest has been shown to be shot that was newly deposited in marshes by hunters during the most recent hunting season (Bellrose 1959). Lead shot deposited in years past, therefore, may not be a serious threat to present and future waterfowl populations, unless severe storms or other ecological events cause exposure of old lead shot located deep in bottom deposits of certain lakes and marshes and thereby allow waterfowl to ingest such shot during feeding.

For additional information on lead poisoning in waterfowl the reader is referred to chapter 9 (Whistling-Ducks, Swans, and Geese) and to the accounts by Friend (1985, 1999b), Pain (1996), and Locke and Thomas (1996).

Table 10.12. Prevalences of lead shot in gizzards of dabbling ducks from Florida

Species County or area[a]	Year(s)	No. ducks			Data source
		Exam.	Pos.	%	
Green-winged Teal					
"Interior Fla."	1938–54	12	0	0	Bellrose (1959)
Merritt Isl. NWR	1964–65	5	0	—	Stieglitz (1967)
	1973–74	41	0	0	Baker & Thompson (1978)[b]
	1974–75	72	1	1	Ibid.
	1975–76	256	5	2	Ibid.
Chassahowitzka NWR	1973–78	22	0	0	Ibid.
Ding Darling NWR[c]	1973–78	56	0	0	Ibid.
Statewide	1976–84	279	8	3	Thul (1986a)
American Black Duck					
"Interior Fla."	1938–54	5	0	—	Bellrose (1959)
Merritt Isl. NWR	1965	2	0	—	Stieglitz (1967)
	1973–74	4	1	—	Baker & Thompson (1978)[b]
	1974–75	8	1	—	Ibid.
	1975–76	10	4	40	Ibid.
	1976–77	3	1	—	Ibid.
Chassahowitzka NWR	1973–78	8	1	—	Ibid.
Statewide	1976–84	9	0	—	Thul (1986a)
Mottled Duck					
Merritt Isl. NWR	1964–66	40	7	18	Stieglitz (1967)
	1973–74	28	6	21	Baker & Thompson (1978)[b]
	1974–75	109	23	22	Ibid.
	1975–76	171	63	37	Ibid.
	1976–77	48	17	35	Ibid.
	1977–78	63	16	25	Ibid.
Loxahatchee NWR	1973–78	15	3	21	Ibid.
Ding Darling NWR[c]	1973–78	18	2	11	Ibid.
Statewide	1976–84	202	26	13	Thul (1986a)
Mallard					
"Interior Fla."	1938–54	3	0	—	Bellrose (1959)
Merritt Isl. NWR	1964–65	2	0	—	Stieglitz (1967)
	1973–74	1	0	—	Baker & Thompson (1978)[b]
	1974–75	2	0	—	Ibid.
	1975–76	12	5	42	Ibid.
	1976–77	20	10	50	Ibid.
Chassahowitzka NWR	1973–78	14	1	7	Ibid.
Ding Darling NWR[c]	1973–78	1	0	—	Ibid.
Statewide	1976–84	91	14	15	Thul (1986a)
Northern Pintail					
"Interior Fla."	1938–54	15	3	20	Bellrose (1959)
Merritt Isl. NWR	1964–65	7	1	—	Stieglitz (1967)
	1973–74	34	14	41	Baker & Thompson (1978)[b]
	1974–75	167	30	18	Ibid.
	1975–76	423	131	31	Ibid.
	1976–77	207	66	32	Ibid.
	1977–78	209	33	16	Ibid.
Chassahowitzka NWR	1973–78	88	12	14	Ibid.
Loxahatchee NWR	1973–78	1	1	—	Ibid.
Ding Darling NWR[c]	1973–78	5	1	—	Ibid.
Statewide	1976–84	114	10	9	Thul (1986a)
Blue-winged Teal					
Merritt Isl. NWR	1964–65	14	0	0	Stieglitz (1967)
	1973–74	73	2	3	Baker & Thompson (1978)[b]
	1974–75	273	9	3	Ibid.

(continued)

Table 10.12. *(continued)*

Species County or area[a]	Year(s)	No. ducks			Data source
		Exam.	Pos.	%	
	1975–76	585	4	1	Ibid.
Chassahowitzka NWR	1973–78	3	0	—	Ibid.
Loxahatchee NWR	1973–78	3	0	—	Ibid.
Ding Darling NWR[c]	1973–78	88	2	2	Ibid.
Osceola County	1976–77	80	3	4	Thul (1986a)
	1977–78	35	3	9	Ibid.
	1978–79	1	0	—	Ibid.
	1979–80	35	4	11	Ibid.
	1981–82	46	4	9	Ibid.
Polk County	1976–77	28	0	0	Ibid.
	1977–78	17	0	0	Ibid.
	1978–79	18	1	6	Ibid.
	1979–80	43	0	0	Ibid.
	1981–82	121	4	3	Ibid.
	1982–83	97	4	4	Ibid.
	1983–84	291	6	2	Ibid.
Northern Shoveler					
"Interior Fla."	1938–54	5	0	—	Bellrose (1959)
Merritt Isl. NWR	1964	5	0	—	Stieglitz (1967)
	1973–74	7	0	—	Baker & Thompson (1978)[b]
	1974–75	68	1	2	Ibid.
	1975–76	222	0	0	Ibid.
Chassahowitzka NWR	1973–78	1	0	—	Ibid.
Loxahatchee NWR	1973–78	1	0	—	Ibid.
Ding Darling NWR[c]	1973–78	9	0	—	Ibid.
Polk County	1976–77	15	0	0	Thul (1986a)
	1977–78	1	0	—	Ibid.
	1978–79	7	1	—	Ibid.
	1979–80	26	1	4	Ibid.
	1981–82	4	0	—	Ibid.
	1982–83	82	1	1	Ibid.
	1983–84	224	2	1	Ibid.
Gadwall					
"Interior Fla."	1938–54	15	0	0	Bellrose (1959)
Merritt Isl. NWR	1964–65	5	0	—	Stieglitz (1967)
	1973–74	4	0	—	Baker & Thompson (1978)[b]
	1974–75	19	0	0	Ibid.
	1975–76	48	0	0	Ibid.
Chassahowitzka NWR	1973–78	7	1	14	Ibid.
Ding Darling NWR[c]	1973–78	9	0	—	Ibid.
Statewide	1976–84	68	1	2	Thul (1986a)
American Wigeon					
"Interior Fla."	1938–54	15	2	13	Bellrose (1959)
Merritt Isl. NWR	1964–65	4	0	—	Stieglitz (1967)
	1973–74	53	0	0	Baker & Thompson (1978)[b]
	1974–75	181	4	2	Ibid.
	1975–76	597	6	1	Ibid.
Chassahowitzka NWR	1973–78	83	7	8	Ibid.
Ding Darling NWR[c]	1973–78	54	0	0	Ibid.
Statewide	1976–84	171	7	4	Thul (1986a)

a. NWR = National Wildlife Refuge.
b. Data from Baker & Thompson (1978) reflect steel and lead shot combined.
c. Ducks were from areas around Ding Darling NWR, but not from the refuge itself.

Table 10.13. Prevalences of lead shot in gizzards of diving ducks from Florida

| Species
County or area[a] | Year(s) | No. ducks | | | Data source |
		Exam.	Pos.	%	
Canvasback					
"Interior Fla."	1938–54	2	0	—	Bellrose (1959)
Merritt Island NWR	1964–65	1	0	—	Stieglitz (1967)
	1973–74	4	3	75	Baker & Thompson (1978)[b]
	1974–75	16	12	75	Ibid.
	1975–76	2	2	—	Ibid.
	1976–77	165	119	72	Ibid
	1977–78	42	18	43	Ibid.
Chassahowitzka NWR	1973–78	5	0	—	Ibid.
Statewide	1976–84	33	3	9	Thul (1986a)
Redhead					
Merritt Island NWR	1964–65	2	0	—	Stieglitz (1967)
	1973–74	22	17	77	Baker & Thompson (1978)[b]
	1974–75	24	8	33	Ibid.
	1975–76	20	8	40	Ibid.
	1976–77	51	19	37	Ibid.
	1977–78	58	1	2	Ibid.
Chassahowitzka NWR	1973–78	5	0	—	Ibid.
Statewide	1976–84	26	3	12	Thul (1986a)
Ring-necked Duck					
"Interior Fla."	1938–54	25	1	4	Bellrose (1959)
Merritt Island NWR	1965	4	1	—	Stieglitz (1967)
	1973–74	203	110	54	Baker & Thompson (1978)[b]
	1974–75	976	303	31	Ibid.
	1975–76	632	209	33	Ibid.
	1976–77	217	78	36	Ibid.
	1977–78	116	32	28	Ibid.
Chassahowitzka NWR	1973–78	19	1	5	Ibid.
Loxahatchee NWR	1963	20	5	25	Valentine (1963)[d]
	1973–78	1292	103	8	Baker & Thompson (1978)[b]
Alachua County	1976–77	70	11	16	Thul (1986a)
	1977–78	119	29	24	Ibid.
	1979–80	66	19	29	Ibid.
	1980–81	132	33	25	Ibid.
	1981–82	179	39	22	Ibid.
	1982–83	100	21	21	Ibid.
	1983–84	37	2	5	Ibid.
Broward County	1976–77	82	8	10	Ibid.
	1977–78	197	29	15	Ibid.
	1978–79	67	3	5	Ibid.
	1979–80	407	37	9	Ibid.
	1980–81	4	1	—	Ibid.
	1981–82	89	17	19	Ibid.
	1982–83	18	0	0	Ibid.
Dade County	1976–77	66	9	14	Ibid.
	1977–78	271	30	11	Ibid.
	1978–79	246	15	6	Ibid.
	1979–80	285	26	9	Ibid.
	1981–82	319	67	21	Ibid.
Glades County	1976–77	85	7	8	Ibid.
	1977–78	234	35	15	Ibid.
	1978–79	56	5	9	Ibid.

(continued)

Table 10.13. *(continued)*

| Species
County or area[a] | Year(s) | No. ducks | | | Data source |
		Exam.	Pos.	%	
	1979–80	9	1	—	Ibid.
	1981–82	99	14	14	Ibid.
	1982–83	188	13	7	Ibid.
	1983–84	188	34	18	Ibid.
Leon County	1976–77	304	106	35	Ibid.
	1977–78	155	60	39	Ibid.
	1979–80	51	10	20	Ibid.
	1980–81	3	1	—	Ibid.
	1981–82	41	8	20	Ibid.
Polk County	1976–77	50	9	18	Ibid.
	1977–78	6	0	—	Ibid.
	1978–79	7	1	—	Ibid.
	1979–80	13	1	8	Ibid.
	1981–82	169	19	11	Ibid.
	1982–83	14	3	21	Ibid.
	1983–84	123	7	6	Ibid.
Putnam County	1976–77	43	5	12	Ibid.
	1977–78	4	1	—	Ibid.
	1978–79	33	0	0	Ibid.
	1979–80	9	1	—	Ibid.
	1980–81	6	0	—	Ibid.
	1981–82	11	4	36	Ibid.
	1982–83	30	1	3	Ibid.
Greater Scaup					
Merritt Island NWR	1964–65	3	0	—	Stieglitz (1967)
Lesser Scaup					
"Interior Fla."	1938–54	6	0	—	Bellrose (1959)
Merritt Island NWR	1964–65	9	0	—	Stieglitz (1967)
Scaup[e]					
Merritt Island NWR	1973–74	109	4	4	Baker & Thompson (1978)[b]
	1974–75	283	17	6	Ibid.
	1975–76	321	6	2	Ibid.
Chassahowitzka NWR	1973–78	264	50	19	Ibid.
Loxahatchee NWR	1973–78	3	0	—	Ibid.
Ding Darling NWR[c]	1973–78	98	5	5	Ibid.
Polk County	1976–77	9	3	33	Thul (1986a)
	1977–78	3	0	—	Ibid.
	1978–79	1	0	—	Ibid.
	1979–80	4	2	—	Ibid.
	1981–82	73	21	29	Ibid.
	1982–83	37	8	22	Ibid.
	1983–84	181	51	28	Ibid.
Common Goldeneye					
Statewide	1976–84	2	0	—	Ibid.
Bufflehead					
Statewide	1976–84	10	2	20	Ibid.

a. NWR = National Wildlife Refuge.
b. Data from Baker & Thompson (1978) reflect steel and lead shot combined.
c. Ducks were from areas around Ding Darling NWR but not from the refuge itself.
d. Baker and Thompson (1978) cited Valentine's unpublished data.
e. Combined data on Greater and Lesser Scaup.

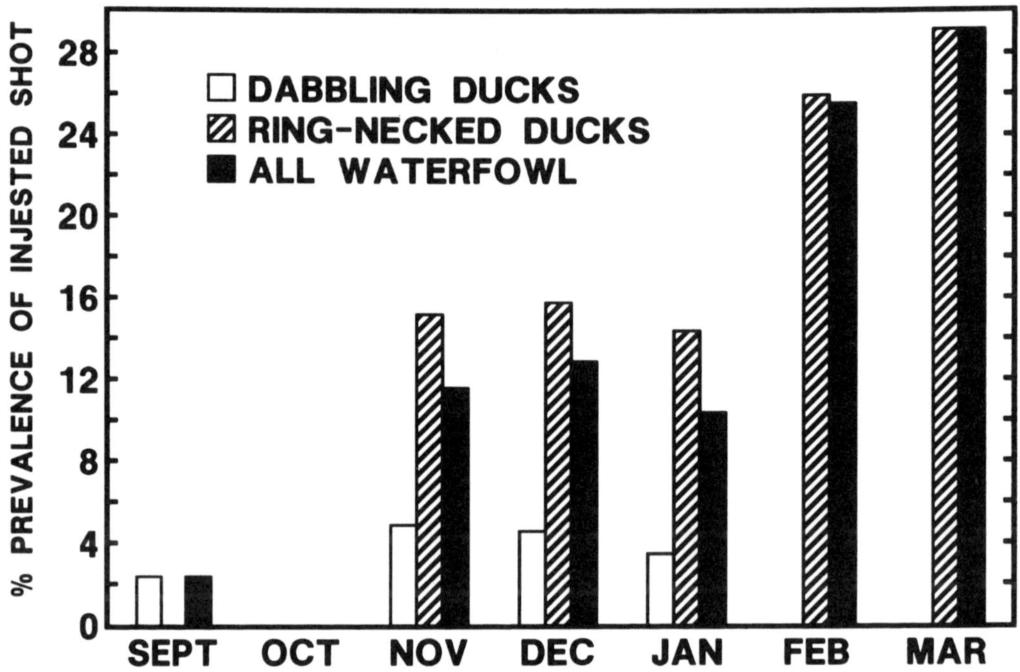

FIGURE 10.1. Monthly prevalences of ingested lead shot in gizzards of waterfowl wintering in Florida, based on data collected from 1976–77 through 1983–84. Courtesy of James E. Thul.

IX. Mercury and selenium

Concentrations of mercury have been determined for a sample of Mallards, Mottled Ducks, and Red-breasted Mergansers from Florida (table 10.15). Concentrations were low and not indicative of acute mercury poisoning (Eisler 1987; Thompson 1996). The concentrations from the mergansers were higher than from the Mallards and Mottled Ducks, which would be expected since the mergansers feed primarily on fish. On the other hand, the effects of low-level exposure to mercury may be a problem. For example, experimental studies have been conducted in which food contaminated with methylmercury (3 ppm dry weight) was given to Mallards over a 12-month period and resulted in decreased periods of egg output, decreased hatching success, and increased mortality in ducklings of the parents that had consumed contaminated feed (Heinz 1974). These aspects have not been studied in ducks from Florida.

When the toxic effects of mercury are evaluated, the presence of selenium must be determined and considered since it has been shown that there are interactions of mercury and selenium that can influence the toxicity of these 2 metals (Cuvin-Aralar and Furness 1991). For example, Heinz and Hoffman (1996) found experimentally that when selenium and mercury were both given in the feed of Mallards, there was a protective effect that resulted in no mortality or paralysis in males, whereas in other males fed only mercury there was some mortality and considerable paralysis. In contrast there was a decrease in hatching success and survival of ducklings produced by adult pairs of Mallards fed both mercury and selenium. Selenium by itself can cause detrimental effects on reproduction in Mallards. Heinz (1996a) evaluated the available experimental data and concluded that when concentrations of selenium in livers of egg-laying females are above 3 ppm (wet weight), reproductive impairment is a possibility. Concentrations

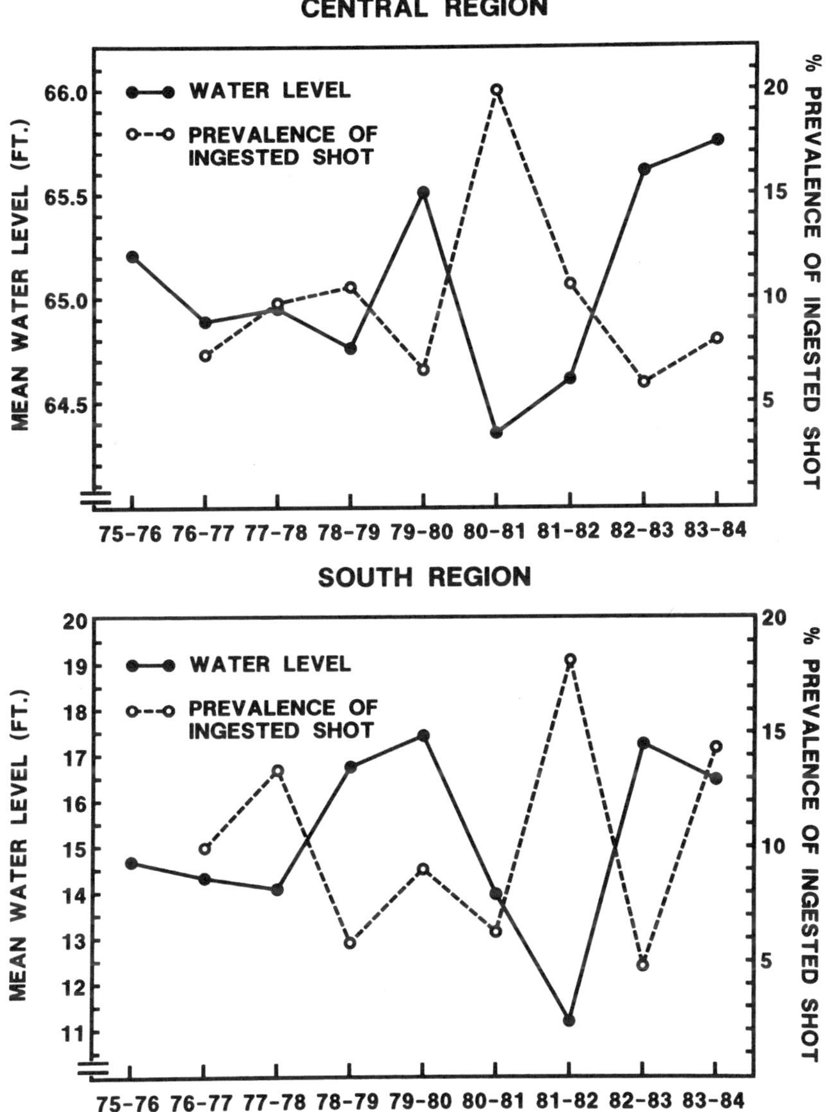

FIGURE 10.2. Prevalence of ingested shot in gizzards of wintering waterfowl compared with water levels in Florida from 1976–77 through 1983–84. Courtesy of James E. Thul.

of selenium were not determined for the tissues presented in table 10.15, but have been determined for samples of liver, kidney, and muscle from other Mottled Ducks (table 10.16) and also for a sample of Wood Ducks (table 10.17) from Florida. When dry weight values were converted to approximate wet weight values, none of the concentrations in liver samples was above 3 ppm (wet weight). Unfortunately, the concentrations of mercury in these tissues were not determined and cannot be factored into the equation in order to understand the total picture.

More information on mercury and selenium can be found in several recently published reviews by Heinz (1996a, 1996b), Ohlendorf (1996), and Thompson (1996).

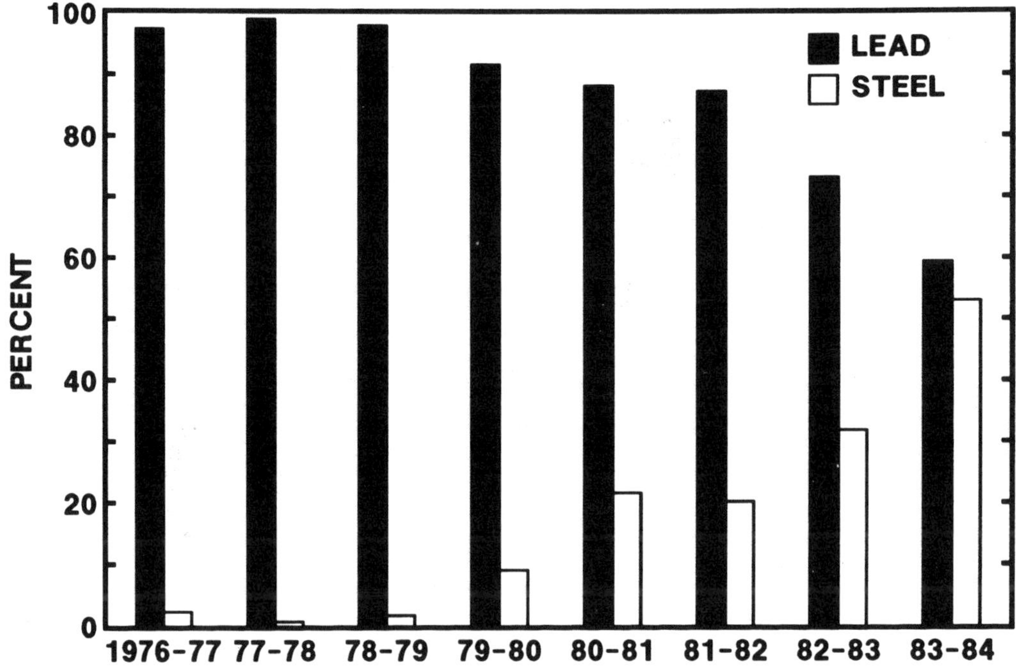

FIGURE 10.3. Relative proportions of lead and steel shot in gizzards of ducks shot by hunters in Florida each year from 1976–77 through 1983–84. Courtesy of James E. Thul.

X. Metals and trace elements

Phosphate mining in Florida is a significant industry and results in the disturbance of large tracts of land. Stowasser (1979) stated that more than 74,000 hectares have been so affected in Florida. In the process of removing phosphate ores, settling basins are utilized for the storage and settling of waste clays. These basins are in use over a 10–20-year period and are attractive to waterfowl as breeding and wintering areas (Myers et al. 1989; O'Meara et al. 1986). Another consequence of the phosphate mining operation is the redistribution of a number of trace elements to the surface where they are available to wildlife such as waterfowl (O'Meara et al. 1986). Some of these are essential elements while others can be toxic if present in high concentrations (Mertz 1981).

Tissues of Mottled Ducks and Wood Ducks from Florida were analyzed in the early 1980s for the presence of a number of trace elements (Montalbano et al. 1983; O'Meara et al. 1986). In addition to selenium (discussed in the previous section), 13 other trace elements were identified in various tissues of Mottled Ducks (table 10.18) and 11 elements from Wood Ducks (table 10.19). In both the Montalbano and O'Meara studies, comparisons were made of tissues from waterfowl obtained from phosphate mine settling basin areas and from nonmining areas. Montalbano et al. (1983) found no differences in concentrations of arsenic, fluoride, and barium in muscle samples of Mottled Ducks from a settling basin in Polk County and those from Mottled Ducks from a nonmining area on the northwest shore of Lake Okeechobee (Glades County). However, O'Meara et al. (1986) found statistically significant differences in concentrations of 6 elements (aluminum, copper, manganese, molybdenum, rubidium, and vanadium) in tissues of Mottled Ducks from a settling basin in Polk County compared with concentrations in Mottled Ducks from a non-

Table 10.14. Prevalences of lead shot in gizzards of Ruddy Ducks and mergansers from Florida

Species County or area[a]	Year(s)	No. ducks Exam.	Pos.	%	Data source
Ruddy Ducks					
Merritt Isl. NWR	1964–65	4	0	—	Stieglitz (1967)
	1973–74	22	17	77	Baker & Thompson (1978)[b]
	1974–75	56	26	46	Ibid.
	1975–76	9	5	—	Ibid.
	1976–77	14	8	57	Ibid.
	1977–78	6	3	—	Ibid.
Statewide	1976–84	94	9	10	Thul (1986a)
Mergansers[c]					
Statewide	1976–84	24	0	0	Ibid.

a. NWR = National Wildlife Refuge.
b. Data from Baker & Thompson (1978) reflect steel and lead shot combined.
c. Combined data on Hooded and Red-breasted Mergansers.

mining area in Osceola County. They also found differences in concentrations of the same 6 elements plus zinc in tissues of Wood Ducks from a settling basin in Hamilton County compared with concentrations in Wood Ducks from a non-mining area in Alachua County.

The effects of these concentrations of trace elements on the health of free-ranging waterfowl are uncertain (Eisler 1988, 1989, 1993). All concentrations with the exception of aluminum were within the ranges considered normal for other animals and therefore probably do not indicate toxicity to waterfowl (O'Meara et al. 1986). Even though some of the concentrations of aluminum were considered to be elevated, the significance of these observations is unknown since tissue concentrations have not been reported for wild birds. The authors of both studies concluded from comparisons of their data with available information in the literature that there were no significant human health risks connected with consuming muscle from these ducks.

XI. Radionuclides

Five radionuclides have been reported from tissues of Mottled Ducks and Wood Ducks in Florida (tables 10.20–10.21). Most of the available information is on concentrations of radium-226 (table 10.21). These data are the result of 2 studies performed in the early 1980s, one headed by the Florida Game and Fresh Water Fish Commission (Montalbano et al. 1983) and another by a multidisciplinary group at the University of Florida (Myers et al. 1989; O'Meara et al. 1986; Stabin 1983). Interest in this topic was generated by a desire to understand the effects of ionizing radiation on the health and reproduction of birds that use phosphate-mine settling ponds (as described in the previous section on toxic trace elements) and also because of the possible public health risks involved when waterfowl are consumed by humans. The settling ponds contain clay materials that have high concentrations of radium-226 and other radionuclides.

Montalbano et al. (1983) found that there were no differences in the concentrations of radium-226 in muscles of Mottled Ducks from a phosphate-mine settling pond in Polk County compared with Mottled Ducks from a natural marsh area in Glades County. They also found no correlation between radium-226 concentrations and several indexes of reproductive condition (testes length/body weight ratios and oviduct width/body weight ratios) and physical

Table 10.15. Concentrations of mercury in tissues of adult Mallards, Mottled Ducks, and Red-breasted Mergansers from Florida

Species of duck	Tissue	County or area	Year	No. examined	Concentrations[a] (ppm wet wt.)		Data source
					Mean[b]	Range	
Mallards	Wings[c]	"Ga. & Fla."[d]	1969	50	0.97	0.25–1.7	Heath & Hill (1974)
Mottled Ducks	Muscle	Palm Beach	1989	5	0.16	0.07–0.27	Ware et al. (1990), Ware (1994)
Red-breasted Mergansers	Liver	Monroe	1995	3	6.1	1.7–13	Sileo (1995)

a. Limits of detection for Heath and Hill (1974) = 0.05 ppm; limits not given by Ware et al. (1990), Ware (1994), and Sileo (1995).
b. Arithmetic means.
c. Wings (trimmed of flight feathers), chopped and blended in homogenate with a food cutter.
d. Samples from Georgia and Florida were combined into 2 pools of 25 wings each.

Table 10.16. Concentrations of selenium in tissues of adult and subadult Mottled Ducks from Florida, 1981–82

Tissue	County	No. exam.	Concentrations (ppm dry wt.)[b]	
			Mean[b]	SE[c]
Liver	Osceola	13	3.2	0.5
	Polk	17	9.9	1.4
Kidney	Osceola	14	4.6	0.4
	Polk	17	8.2	1.1
Muscle	Osceola	14	3.4	1.2
	Polk	17	3.6	0.7

Source: O'Meara et al. (1986).
a. Limits of detection not given by authors.
b. Type of mean not given by authors.
c. SE = standard error.

condition (weights, measurements, emaciation indexes, and fat deposits) in the ducks. They concluded that radionuclide contamination was not detrimental to waterfowl using phosphate-mine settling ponds and that the amounts of radium-226 in muscles of these birds did not pose a human health risk for people consuming these ducks.

The conclusions reached by investigators in the other study (Myers et al. 1989; O'Meara et al. 1986) were fairly similar. The latter study was larger in scope, involving both Mottled Ducks and Wood Ducks, and was conducted in both northern and central Florida. Radium-226 concentrations in soft tissues were low and variable, and differences between ducks from settling ponds and from natural areas were not statistically significant. However, they did find higher concentrations in bones of ducks from settling ponds than from natural areas, but concluded that it did "not appear that observed radium-226 levels in bones would constitute a health hazard to birds on settling ponds, given their short life span" (O'Meara et al. 1986). They agreed with Montalbano et al. (1983) that there was

Table 10.17. Concentrations of selenium in tissues of 5 Wood Ducks[a] from Alachua County and 5[a] from Hamilton County, Florida, 1981–82

Tissue	County	Concentrations (ppm dry wt.)[b]	
		Mean[c]	SE[d]
Liver	Alachua	0.8	0.3
	Hamilton	6.3	0.8
Kidney	Alachua	2.0	0.6
	Hamilton	7.5	0.9
Muscle	Alachua	1.1	0.1
	Hamilton	2.9	0.5

Source: O'Meara et al. (1986).
a. A mixture of adults and subadults.
b. Limits of detection not given by authors.
c. Type of mean not given by authors.
d. SE = standard error.

Table 10.18. Concentrations of metals and trace elements in tissues of Mottled Ducks from Florida

Organ or tissue County	Year(s)	Age[a]	No. exam.	Aluminum	No. exam.	Arsenic	No. exam.	Barium	Data source[c]
Liver									
Osceola	1981–82	AD&SA	8	11(29)	13	0.8(0)	NG	—	A
Polk	1981–82	AD&SA	14	26(9.6)	17	1.4(0.6)	NG	—	A
Kidney									
Osceola	1981–82	AD&SA	8	8.4(1.6)	NG	—	NG	—	A
Polk	1981–82	AD&SA	9	39(12)	NG	—	NG	—	A
Muscle									
Glades	1980	AD	NG	—	10	0.79(0.04)	10	17(6.5)	B
Osceola	1981–82	AD&SA	12	11(1.9)	14	0.5(0.1)	NG	—	A
Polk	1980	AD&IM	NG	—	20	0.71(0.02)	20	8.7(1.2)	B
	1981–82	AD&SA	15	37(8.0)	ND	—	NG	—	A
Bone									
Osceola	1981–82	AD&SA	13	150(6)	NG	—	NG	—	A
Polk	1981–82	AD&SA	16	180(9)	NG	—	NG	—	A

Organ or tissue County	Year(s)	Age[a]	No. exam.	Bromine	No. exam.	Cobalt	No. exam.	Copper	Data source[c]
Liver									
Osceola	1981–82	AD&SA	14	19(1.8)	14	5.3(0.9)	14	81(16)	A
Polk	1981–82	AD&SA	17	25(2.2)	17	6.5(1.1)	17	84(26)	A
Kidney									
Osceola	1981–82	AD&SA	14	33(2.6)	NG	—	14	24(4.1)	A
Polk	1981–82	AD&SA	17	33(3.0)	NG	—	17	16(1.4)	A
Muscle									
Osceola	1981–82	AD&SA	14	14(1.7)	14	0.8(0.1)	14	15(1.0)	A
Polk	1981–82	AD&SA	17	13(1.3)	17	0.8(0.1)	17	18(2.0)	A
Bone									
Osceola	1981–82	AD&SA	14	15(2.1)	NG	—	14	6.0(0.9)	A
Polk	1981–82	AD&SA	17	14(1.3)	NG	—	17	6.0(0.9)	A

Organ or tissue County	Year(s)	Age[a]	No. exam.	Fluoride	No. exam.	Manganese	No. exam.	Molybdenum	Data source[c]
Liver									
Osceola	1981–82	AD&SA	NG	—	13	7.4(0.9)	13	3.1(0.5)	A
Polk	1981–82	AD&SA	NG	—	17	4.7(0.6)	17	3.2(0.5)	A
Kidney									
Osceola	1981–82	AD&SA	NG	—	14	8.0(0.9)	14	2.0(0.3)	A

Mean (standard error) ppm dry wt.[b]

Tissue	Year	Age[a]	No. exam.	Nickel	No. exam.	Rubidium	No. exam.	Vanadium	[c]
Polk	1981–82	AD&SA	NG	—	17	6.0(0.5)	17	3.2(0.4)	A
Muscle									
Glades	1980	AD	10	0.026(0)	NG	—	NG	—	B
Osceola	1981–82	AD&SA	NG	—	14	1.9(0.3)	NG	—	A
Polk	1980	AD&IM	20	0.026(0.001)	NG	—	NG	—	B
	1981–82	AD&SA	NG	—	17	1.4(0.2)	NG	—	A
Bone									
Osceola	1981–82	AD&SA	NG	—	14	8.1(0.8)	NG	—	A
Polk	1981–82	AD&SA	NG	—	17	5.9(0.6)	NG	—	A
Liver									
Osceola	1981–82	AD&SA	13	0.08(0.04)	13	16(2.2)	8	0.09(0. 02)	A
Polk	1981–82	AD&SA	17	2.1(1.5)	17	31(2.1)	14	0.45(0.11)	A
Kidney									
Osceola	1981–82	AD&SA	NG	—	14	17(3.0)	8	0.08(0.002)	A
Polk	1981–82	AD&SA	NG	—	17	34(3.3)	9	0.58(0.08)	A
Muscle									
Osceola	1981–82	AD&SA	14	0.6(0.1)	14	20(2.9)	12	0.07(0.0 3)	A
Polk	1981–82	AD&SA	17	0.7(0)	17	29(3.3)	15	0.14(0.03)	A
Bone									
Osceola	1981–82	AD&SA	NG	—	14	12(2.7)	13	0.32(0.04)	A
Polk	1981–82	AD&SA	NG	—	17	18(1.6)	16	1.1(0.19)	A

Tissue	Year	Age[a]	No. exam.	Zinc	[c]
Liver					
Osceola	1981–82	AD&SA	13	130(8)	A
Polk	1981–82	AD&SA	17	160(13)	A
Kidney					
Osceola	1981–82	AD&SA	14	96(4)	A
Polk	1981–82	AD&SA	17	83(5.6)	A
Muscle					
Osceola	1981–82	AD&SA	14	45(2)	A
Polk	1981–82	AD&SA	17	41(3)	A
Bone					
Osceola	1981–82	AD&SA	14	350(10)	A
Polk	1981–82	AD&SA	17	370(16)	A

ND = not detected, NG = not given by authors.

a. AD = adult, SA = subadult, IM = immature.

b. Detection limits and type of mean not given by authors.

c. A = O'Meara et al. (1986), B = Montalbano et al. (1983).

Table 10.19. Concentrations of metals and trace elements in tissues of adult and subadult Wood Ducks from Florida, 1981–82

Organ or tissue County	Mean (standard deviation) ppm dry wt.[a]					
	No. exam.	Aluminum	No. exam.	Arsenic	No. exam.	Barium
Liver						
Alachua	5	50(40)	5	0.8(0)	5	5.5(0.9)
Hamilton	5	12(1.8)	5	0.8(0)	5	6.3(0.8)
Kidney						
Alachua	3	12(2.8)	NG	—	3	12(1.7)
Hamilton	3	15(2.2)	NG	—	3	15(2.7)
Muscle						
Alachua	4	17(5)	ND	—	4	4.3(0.4)
Hamilton	5	16(2.4)	ND	—	5	5.9(0.7)
Bone						
Alachua	5	150(9)	NG	—	5	12(2.3)
Hamilton	5	150(6)	NG	—	5	9.3(1.0)
	No. exam.	Cobalt	No. exam.	Copper	No. exam.	Manganese
Liver						
Alachua	5	4.3(1.1)	5	14(2.4)	5	6.2(0.9)
Hamilton	5	6.4(2.5)	5	59(25)	5	14(2.8)
Kidney						
Alachua	NG	—	3	14(0.6)	3	9.4(0.7)
Hamilton	NG	—	3	35(4.0)	3	11(1.9)
Muscle						
Alachua	4	0.7(0.2)	4	14(1.1)	4	1.1(0.2)
Hamilton	5	0.8(0.1)	5	16(1.7)	5	1.6(0.3)
Bone						
Alachua	NG	—	5	0.7(0.2)	5	11(4.6)
Hamilton	NG	—	5	8.5(0.2)	5	27(9.5)
	No. exam.	Molybdenum	No. exam.	Nickel	No. exam.	Rubidium
Liver						
Alachua	5	3.1(0.8)	5	0.4(0.1)	5	45(4.1)
Hamilton	5	3.2(0.7)	5	0.6(0.1)	5	28(5.5)
Kidney						
Alachua	3	1.6(0.8)	NG	—	3	60(7.1)
Hamilton	3	3.1(0.5)	NG	—	3	25(5.2)
Muscle						
Alachua	NG	—	4	0.6(0.1)	4	60(7.6)
Hamilton	NG	—	5	0.6(0.1)	5	19(2.1)
Bone						
Alachua	NG	—	NG	—	5	25(4.8)
Hamilton	NG	—	NG	—	5	8.0(1.3)
	No. exam.	Vanadium	No. exam.	Zinc		
Liver						
Alachua	5	0.33(0.23)	5	89(5)		
Hamilton	5	0.36(0.08)	5	160(20)		

(continued)

Table 10.19. *(continued)*

Organ or tissue County	No. exam.	Vanadium	No. exam.	Zinc
		Mean(standard deviation) ppm dry wt.[a]		
Kidney				
Alachua	3	0.12(0.03)	3	100(2.3)
Hamilton	3	1.3(0.6)	3	110(5.2)
Muscle				
Alachua	4	0.09(0.02)	4	31(2)
Hamilton	5	0.09(0.02)	5	31(1)
Bone				
Alachua	5	0.27(0.06)	5	310(14)
Hamilton	5	0.62(0.24)	5	320(17)

Source: O'Meara et al. (1986).
ND = not detected, NG = not given by authors.
a. Detection limits and type of mean not given by authors.

no human health risk involved in the consumption of these birds.

Additional information on radiation hazards to wildlife can be found in the monograph by Eisler (1994).

XII. Avian vacuolar myelinopathy

This disease is not known to occur in Florida, but has been diagnosed in waterfowl in nearby states (see chapter 11, Eagles).

XIII. Oiling

The relative susceptibility of waterfowl in the southeastern United States to oiling has been evaluated by Clapp et al. (1982). They concluded that sea ducks and diving ducks were the most vulnerable. Of these, they felt that Lesser Scaup have the highest susceptibility of all species, principally because they are very numerous during the winter and are found in large rafts offshore and in bays and estuaries where the likelihood of encountering an oil spill is high.

There are only 2 records of oiling of ducks in

Table 10.20. Concentrations of radionuclides (Bq/kg, dry weight) in composite samples[a] of muscle and bone from Mottled Ducks and Wood Ducks in central Florida, 1981–82

Radionuclide	Muscle		Bone	
	Osceola[b]	Polk[c]	Osceola[b]	Polk[c]
Uranium-238	0	0.32	0	9.3
Uranium-234	0	0.37	0	8.9
Thorium-230	0.30	0.52	0	0
Radium-226	0	0.19	10	42
Lead-210	0	0.78	34	56

Source: O'Meara et al. (1986).
a. Number of ducks in each composite sample not given by authors.
b. Lake Kissimmee (control).
c. Settling ponds of two phosphate mine areas.

Table 10.21. Concentrations of radium-226 in tissues of Mottled Ducks and Wood Ducks from Florida

| Organ or tissue | County | Date(s) | No. exam. | Becquerel/kg | | | Wt. type[c] | Data source |
				Mean[a]	Range	SE[b]		
Mottled Ducks								
Bone								
	Osceola[d]	1981–82	17	5.9	3.3–9.6	NG	WW	Myer s et al. (1989)
	Polk[e]	1981–82	16	63	32–130	NG	WW	Ibid.
Muscle								
	Glades[d]	1980	10	0.03	NG	0.01	WW	Montalbano et al. (1983)
	Osceola[d]	1981–82	17	0.48	NG	0.18	DW	O'Meara et al. (1986)
	Polk[e]	1980	20	0.11	NG	0.02	WW	Montalbano et al. (1983)
		1981–82	19	0.70	NG	0.18	DW	O'Meara et al. (1986)
Liver								
	Osceola[d]	1981–82	13	0.26	NG	NG	DW	Ibid.
	Polk[e]	1981–82	17	0.63	NG	0.11	DW	Ibid.
Wood Ducks								
Bone								
	Alachua[d]	1981–82	5	8.9	4.4–14	NG	WW	Myers et al. (1989)
	Hamilton[e]	1981–82	5	25	6.3–42	NG	WW	Ibid.
Muscle								
	Alachua[d]	1981–82	5	0.07	NG	0.19	DW	O'Meara et al. (1986)
	Hamilton[e]	1981–82	5	0.33	NG	0.37	DW	Ibid.
Liver								
	Alachua[d]	1981–82	5	0.26	NG	NG	DW	Ibid.
	Hamilton[e]	1981–82	3	0.85	NG	0.30	DW	Ibid.
Kidney								
	Hamilton[e]	1981–82	5	5.7	NG	0.74	DW	Ibid.

NG = not given by authors.
a. Type of mean not given by authors.
b. SE = standard error.
c. WW = wet weight, DW = dry weight.
d. Normal lake or marsh.
e. Settling ponds in phosphate mine area.

Florida waters. The first was by Robertson and Mason (1965), who reported 2 oiled Red-breasted Mergansers found at Long Key in the Dry Tortugas in January 1964. A second occurrence was in January 1970 when "hundreds" of Red-breasted Mergansers, "several thousand" Lesser Scaup and "smaller or undetermined numbers" of other species of ducks were killed following an oil spill in Tampa Bay (Stevenson 1970).

Oil contamination can lead to mortality because of feather matting, chilling, and starvation, and because of the harmful effects of ingestion of oil during preening (Clapp et al. 1982). In addition, ingestion of sublethal doses of oil by ducks can result in lowered resistance to infectious disease agents (Rocke et al. 1984). Additional information on oiling and its effects on birds can be found in chapter 2 (Loons and Grebes) and in the following publications: Wobeser (1997), Simons (1985), Hall and Coon (1988), Welte and Frink (1991), Jessup and Leighton (1996), and Rocke (1999).

XIV. Neoplasia

No information is available on this topic.

XV. Anomalies

An albino Mottled Duck from Everglades National Park was reported in 1953 by Stimson (1953). It was almost completely white except for the head, which was dusky, and a dark spot on each wing where the speculum occurs. It was accompanied by 5 other Mottled Ducks that were normal in feather coloration.

XVI. Biotoxins

Three types of biotoxins have been associated with waterfowl mortality events in Florida (table 10.22). These include type C botulinum neurotoxin produced by the anaerobic bacterium *Clostridium botulinum*, necrotizing enteritis toxin produced by the bacterium *C. perfringens*, and brevetoxin produced by the red tide marine dinoflagellate *Karenia brevis* (= *Gymnodinium breve*).

Botulism (also known as limberneck, western duck sickness, and duck disease) has been associated with numerous large-scale epizootics, especially in western North America (Wobeser 1997) and is "probably the most important disease of migratory birds" (Rocke and Friend 1999). Outbreaks are uncommon and smaller in the southeastern United States, but 6 such episodes have been documented in Florida since the 1970s (table 10.22) and undoubtedly other occurrences have gone unnoticed.

Botulism is a paralytic disease that is caused by the ingestion of toxin produced by the bacterium *Clostridium botulinum*. The most common type of toxin involved in outbreaks in waterfowl is type C. *Clostridium botulinum* is an obligate anaerobe and widespread in soil and mud, especially in freshwater marshes (Wobeser 1997). It persists in the environment as a spore form that is resistant to drying and to heat (Rocke and Friend 1999). The factors that lead to an outbreak are not fully understood, but several environmental conditions seem to be linked with the initiation of such an event. The optimum temperature for production of toxin is above 30–37°C (86–98°F) and

is dependent on the presence of certain bacteriophages (Wobeser 1997). In addition to warm temperatures, shallow and fluctuating water that is rich in organic matter and deficient in oxygen appear to be important. The richness in organic matter is often due to rotting vegetation and vertebrate and invertebrate carcasses (Rocke and Friend 1999). Once an outbreak is started, it is perpetuated often by a "carcass-maggot cycle." Toxin is produced in decaying bird carcasses, fly maggots in such carcasses concentrate the toxin, and these maggots in turn are ingested by other birds, which die and are associated with further toxin production and subsequent mortality (Rocke and Friend 1999). Field signs include carcasses in parallel rows at the edge of marshes and bodies of water (resulting from receding water levels); these are often in various stages of decomposition as the mortality usually continues over a period of time rather than all birds dying at more or less the same time. Living birds have muscular paralysis, the inability to fly and hold their heads up (hence the name "limberneck"), and paralysis of the inner eyelid (nictitating membrane). There are no lesions seen at necropsy that are characteristic of botulism (Rocke and Friend 1999). Afflicted birds are usually in good flesh. A definitive diagnosis is reached by observing the above signs and conditions, then confirming by demonstrating the presence of type C toxin via the mouse protection test. The latter is done by inoculating serum from sick ducks into groups of laboratory mice, some of which are also given type C antitoxin along with the serum from suspect birds. The mice receiving antitoxin will be protected and will remain healthy, whereas the mice receiving only serum will become sick or die with characteristic paralytic signs of botulism if the serum contains botulinum toxin. Additional details of this test are given by Haagsma (1987) and Wobeser (1997).

Even though there are still many unanswered questions about the epizootiology of avian botulism, this disease has been studied extensively and there are numerous publications available on it. These include 2 bibliographies (Allen and

Table 10.22. Mortality of waterfowl in Florida due to biotoxins

Cause of death / Year(s)	County (or area)	Species involved	No. ducks At risk	No. ducks Found dead	Data source
Botulism (Type C)[a]					
1970s	"Florida"	Mallards	ND	79	Harrison (1974)
1979 (May)	Hamilton	Wood Ducks	ND	26	Forrester et al. (1980)
		Northern Shovelers	ND	9	Ibid.
		Blue-winged Teal	ND	6	Ibid.
		Ring-necked Ducks	ND	3	Ibid.
1979 (December)	Hamilton	Northern Shovelers	ND	"few"	Ibid.
1981	Seminole	Mallards	~300	40	Stroud (1981)
1994	Hillsborough	Northern Shovelers	450	228	Smith (1994)
		Blue-winged Teal	ND	96	Ibid.
		Green-winged Teal	ND	22	Ibid.
		Mottled Ducks	ND	2	Ibid.
		Mallards	ND	"1 or more"	Ibid.
		Redheads	ND	"1 or more"	Ibid.
		Lesser Scaup	ND	"1 or more"	Ibid.
1998	Hamilton	Northern Shovelers	NG	NG[b]	Quist (1998)
		Ruddy Ducks	NG	NG	Ibid.
		Scaup[c]	NG	NG	Ibid.
		Gadwalls	NG	NG	Ibid.
Necrotizing enteritis					
1971	Glades	Mallards	NG	184	Jasmin et al. (1972)
		Mottled Ducks	NG	7	Forrester (1971)
		Ring-necked Ducks	NG	4	Jasmin et al. (1972)
		Blue-winged Teal	NG	2	Ibid.
Red tide intoxication					
1974	Hillsborough & Pinellas	Lesser Scaup	68,000	"several thousand"	Schreiber et al. (1975)
2002	Charlotte	Lesser Scaup	ND	200	Cunningham et al. (2002)

ND = not determined, NG = not given by authors.

a. All 6 botulism outbreaks were confirmed by the mouse-protection test.

b. Not given by author for each species; it was stated that 30–50 ducks were found dead or dying, November 21–30, 1998.

c. Species not given; probably were Lesser Scaup.

Wilson 1977; Wilson and Locke 1982), 2 books (Smith 1977; Eklund and Dowell 1987), and several excellent chapters in wildlife disease books (Rosen 1971; Smith 1982; Jessup 1986; Wobeser 1997; Rocke and Friend 1999). A book on cellular and molecular aspects of the pathogenesis of bacterial infections in animals contains a chapter on *Clostridium botulinum* by T.E. Rocke (1993).

All 6 of the botulism outbreaks that have been recorded in Florida (table 10.22) have been associated with unnatural ecological situations. Two of them (the outbreak in the "1970s" and the one in 1981) occurred on privately owned ponds that had been modified by human activity. Not much is known about the conditions surrounding the 1981 outbreak other than it occurred during July in a "local private pond" near Orlando in Seminole County (Stroud 1981). In the case of the "1970s" outbreak, however, the incident occurred in a half-acre shellrock pond located on a 5-acre game fowl farm. Apparently the pond was being used to dispose of dead game bird carcasses and spoiled eggs and had "several bloated, maggot-infested dead fish floating on the surface" at the time the die-off was investigated (Harrison 1974).

Two outbreaks occurred in 1979 in a phosphate mine settling pond in Hamilton County (Forrester et al. 1980) where a nesting study of Wood Ducks was being conducted. Artificial nesting boxes were used in the study area and this provided a unique opportunity to measure the effects of botulism on a local breeding population of ducks (Wenner and Marion 1981). The epizootic occurred over a 3-week period during May and June and involved 3 other species of waterfowl in addition to Wood Ducks (Northern Shovelers, Blue-winged Teal, and Ring-necked Ducks) and a variety of other aquatic birds (American Coots, Common Moorhens, Black-necked Stilts, Semipalmated Sandpipers, and Snowy Egrets). Three Wood Duck hens were found dead on eggs in nesting boxes. The outbreak caused a shortening of the nesting season of Wood Ducks to only 79 days (compared with 127 days for the 1978 season) and resulted in the termination of all nesting

activity; i.e., no nests were initiated or brooded after the peak of the outbreak in May. Nesting success dropped from 61% in 1978 to 30% in 1979, the year of the epizootic. Ecological conditions that accompanied this outbreak included (1) high ambient temperatures of up to 30°C (86°F), (2) shallow water (no greater than 0.8 m), (3) extensive dead and decaying vegetation, a result of previous drawdowns and reflooding initiated by the phosphate mining company as part of their process of stabilization of the waste colloidal slurry resulting from the phosphate mining operation, and (4) the presence of an abundant population of flies. An additional outbreak occurred in the same area 7 months later (December of 1979) when ambient temperatures were abnormally high for that time of year, but involved only a small number of American Coots and Northern Shovelers (Forrester et al. 1980). These 2 outbreaks of botulism stimulated further investigations into the prevalence of *C. botulinum* type C in substrates in 6 phosphate-mine settling ponds and 1 natural pond in Hamilton County (Marion et al. 1983). *Clostridium botulinum* Type C bacteria were found in 26 (5.6%) of 467 samples of sediment taken from settling ponds. Organisms were found in 4 of the 6 settling ponds, but not in the natural pond. There was a distinct seasonal pattern in the occurrence of the organisms that were identified only in the summer months (April–October), with the peak occurrence being in June (figure 10.4). A number of ecological conditions were measured for each sampling point during this study including water temperature, dissolved-oxygen concentration, slope of the pond bottom, turbidity, water movement, percent of cover due to detritus and vegetation, and major plant species present. *Clostridium botulinum* type C occurred independently of the ecological parameters that were measured, with the exception of dissolved-oxygen concentrations and percent vegetation coverage. In the case of dissolved oxygen, mean values were greater in water where positive substrate samples were found, which is the opposite of what might be expected since *C. botulinum* is an

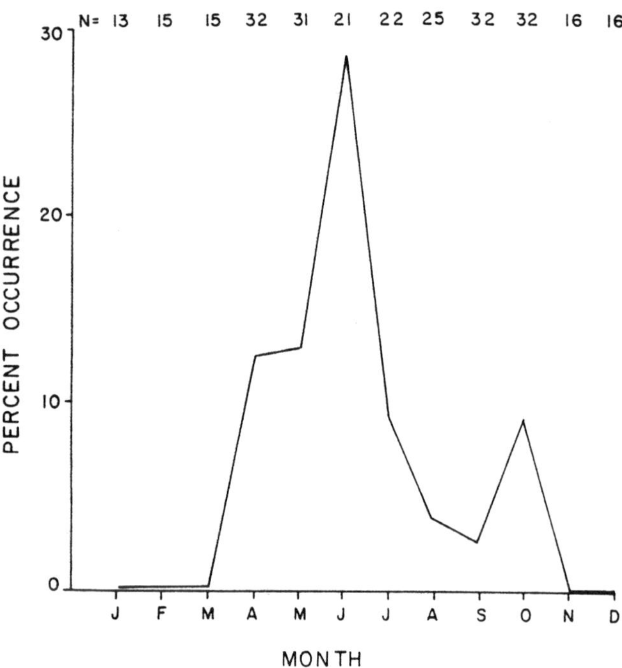

FIGURE 10.4. Frequency and percent frequency of *Clostridium botulinum* type C in samples of substrates from phosphate-mine settling ponds in Hamilton County, April 1981–March 1982. From Marion et al. 1983; by permission of *Journal of Wildlife Diseases*.

anaerobic bacterium. The authors pointed out that the dissolved-oxygen concentrations in the water column could be different from those found in the substrate where the bacteria are found. They concluded that their findings support the prediction that future outbreaks in northern Florida most likely would be during the summer months, although another small outbreak occurred in the same area in November of 1998 (Quist 1998).

The 1994 outbreak in Hillsborough County resulted in the highest mortality of the 6 botulism incidents, with 464 carcasses being counted (Smith 1994). This event occurred on Island 3D, a manmade dredge island in Hillsborough Bay, in November 1994 in 3 shallow settling ponds (2 small ponds and 1 large one) covering about 15 hectares. Most of the dead birds (91%) were found along the shoreline of one of the small ponds that also had an algal bloom. It was determined from further studies that the algae involved in the bloom were not blue-green algae known to cause animal toxicosis. Ambient temperatures at the time consisted of daytime highs of 27–32°C (80–90°F) and nighttime lows in the 16–21°C (60–70°F) range. At the time of the outbreak there were about 1,000 birds at risk, half waterfowl and half shorebirds. The numbers of dead waterfowl that were counted are given in table 10.22. In addition, small numbers of American Coots, American Avocets, Dunlins, Herring Gulls, Laughing Gulls, and Turkey Vultures were found dead, along with 1 Northern Harrier.

An outbreak of necrotizing enteritis or enterotoxemia caused by *Clostridium perfringens* type C occurred in June and July of 1971 in Fisheating Creek Bay on the western edge of Lake Okeechobee (Jasmin et al. 1972). A total of 398 dead aquatic birds representing 19 species was counted in a 12-hectare area of stagnant water and included 4 species of waterfowl: Mallards, Mottled Ducks, Ring-necked Ducks, and Blue-winged Teal (Forrester 1971; Jasmin et al. 1972). Botulism was suspected originally because of the nature of the ecological conditions and the signs in affected birds, but was ruled out after appropriate laboratory tests were performed, including the mouse protection test. For more details on this outbreak the reader is referred to earlier chapters on wading birds and shore birds. Leibovitz (1973) has reported individual cases in wild ducks (American Black Ducks and Mallards) and Canada Geese in New York, but this outbreak in Florida is the only record of a large-scale event in free-ranging waterfowl due to necrotizing enteritis.

An epizootic involving Lesser Scaup occurred in Tampa Bay during an 8-week period in February and March 1974 (Forrester et al. 1977; Quick and Henderson 1975). Several thousand birds died, of an estimated 68,000 at risk (Schreiber et al. 1975). It was suspected that this was due to brevetoxins produced by a bloom of the toxic red tide dinoflagellate *Karenia brevis*. Afflicted ducks were weak, dehydrated, and reluctant to fly, and had clear nasal discharges, viscous oral discharges, oil gland dysfunction, excessive lacrimation, chalky yellow diarrhea, dyspnea, tachypnea, tachycardia, decreased blood pressure, depressed body temperature, and diminished flexes (Quick and Henderson 1975). Much of the evidence for this was circumstantial and consisted of the following: (1) morbidity and mortality of scaup began 2 to 3 weeks after a red tide bloom entered Boca Ciega Bay and Tampa Bay, (2) many of the neurologic and hematological signs were similar to those seen in fishes affected by red tide toxins, (3) recovery of the ducks was rapid and often occurred within 1–6 hours after removal from the source of toxin, (4) mortality was higher in males than in females (which is a characteristic of other intoxication problems), (5) tests conducted to detect the presence of infectious diseases and other toxins were negative and parasites and pesticide residues, although present, were considered sublethal. A controlled feeding experiment was conducted in which clams

(*Mercenaria campechiensis*) and seawater containing brevetoxin were fed to 3-week-old White Pekin ducklings (Forrester et al. 1977). Exposed ducklings became ill and died with some of the signs seen in the scaup during the epizootic. Thus, although it was not possible to determine that brevetoxin was the cause of the scaup mortality, there was considerable circumstantial and experimental evidence to suggest that this was the case and that it was at least partially responsible. In March of 2002 a die-off of approximately 200 Lesser Scaup occurred in Dead Lake near Acline (Charlotte County). This mortality was believed to have been caused by brevetoxins produced by *K. brevis* since concentrations of this toxin detected by an ELISA test were extremely high in the tissues of the ducks (Cunningham et al. 2002). Red tide brevetoxin was implicated in mortality of Florida manatees, *Trichechus manatus latirostris*, in 1982 (O'Shea et al. 1991) and in 1996 (Bossart et al. 1998).

XVII. Viruses

There is little information on viruses and viral diseases of free-ranging ducks in Florida. One case of avian pox was diagnosed in a Wood Duck killed by a hunter on Lake Lafayette (Leon County) in September 1997 (Quist 1998). The duck had numerous 0.5–2.0-cm proliferative lesions at the base of the beak and side of the head but none on the legs and feet.

West Nile virus was identified by using PCR and virus isolation techniques in 4 American Black Ducks (1 each from Citrus, Duval, Marion, and Monroe counties), 1 Mallard (Gulf County), and 1 Wood Duck (Dixie County) in 2001 (Conti et al. 2002). These ducks were found dead, and it is assumed that they died of West Nile virus infection, but necropsies were not performed to prove this.

There are 2 published surveys of domestic or captive ducks examined serologically for evidence of infection with St. Louis encephalitis virus. Gainer et al. (1964) reported that 1 of 3 captive Wood Ducks from Orange County in

Table 10.23. Bacteria identified in ducks from Florida[a]

Host Bacteria	County	Year(s)	Organs cultured	Cause of death	Data source
Perching-ducks					
Muscovy Duck					
Escherichia coli	Orange	1980	Liver	Unknown	White & Forrester (1980)
Wood Duck					
Corynebacterium sp.	Gadsden[b]	1976	Liver	Unknown	White & Forrester (1976)
Enterobacter sp.	Gadsden[c]	1976	Large intestine, liver	Unknown	Ibid.
Streptococcus sp.	Gadsden[d]	1976	Liver	Unknown	Ibid.
Dabbling ducks					
Mallard					
Proteus sp.	Orange	1972	Large intestine	Unknown	White & Forrester (1972)
Blue-winged Teal					
Pasteurella multocida	Monroe	1967–68	Heart blood	Avian cholera	Klukas & Locke (1970)
Northern Shoveler					
Proteus sp.	Polk[e]	1978	Large intestine	Unknown	White & Forrester (1978)
Diving ducks					
Lesser Scaup					
Enterobacter sp.	Charlotte	1974	Large intestine	Emaciation	White & Forrester (1974)
Escherichia coli	Charlotte	1974	Large intestine	Emaciation	Ibid.
	Pinellas[f]	1974	Large intestine	Red tide	Ibid.
Pasteurella multocida	Monroe[g]	1967–68	Heart blood	Avian cholera	Klukas & Locke (1970)
Mergansers					
Red-breasted Merganser					
Escherichia coli	Sarasota	1994	Heart, liver	Emaciation	Duncan (1995)
Salmonella sp.	Pinellas	1973	Large intestine	Unknown	White & Forrester (1973)

a. Isolation was made from 1 duck, unless otherwise indicated.
b. Isolation was made from 1 of 12 Wood Ducks.
c. Isolations were made from 3 of 12 Wood Ducks.
d. Isolations were made from 3 of 12 Wood Ducks.
e. Isolation was made from 1 of 7 Northern Shovelers.
f. Isolation was made from 1 of 9 Lesser Scaup.
g. Isolations were made from 2 of 9 Lesser Scaup.

1962 was seropositive and Jennings et al. (1969) found that 64 of 115 Pekin Ducks and Muscovy Ducks (they did not separate their data for each species) from the Tampa Bay area were positive. The significance of these infections to ducks is unknown. Although wild waterfowl are known to be infected naturally, they do not seem to be important in the epidemiology of SLE virus (McLean and Bowen 1980). Other wild birds (mainly sparrows and pigeons) are known to maintain and amplify the virus, which infects mosquitoes and subsequently infections occur in aberrant hosts such as humans (Kemp 1981). A number of St. Louis encephalitis outbreaks and human cases have been recorded from Florida with the principal vector being the mosquito *Culex nigripalpus* (see review in Nayar 1982).

XVIII. Bacteria

Seven species of bacteria have been isolated from ducks in Florida between 1967 and 1994 (table 10.23). With the exception of *Pasteurella multocida* (the etiologic agent of avian cholera) and *Salmonella* sp. (which is of public health concern), the significance of these bacterial infections is unknown. Most are probably secondary or opportunistic invaders or perhaps contaminants.

The records of *Pasteurella multocida* infections were obtained during an outbreak of avian cholera at West Lake in Everglades National Park during December 1967 and January 1968 (Klukas and Locke 1970). This epizootic involved 5,000–6,000 American Coots and between 50 and 100 ducks, mainly Lesser Scaup and a few Blue-winged Teal. This is the only occurrence of avian cholera on record for Florida. In other parts of the United States, this disease occurs frequently and is one of the major causes of waterfowl mortality (Friend 1999a). For more details on this outbreak and on avian cholera in general the reader is referred to chapter 18, Rails, Coots, Gallinules, Moorhens, and Limpkins.

XIX. Protozoans

Eight species of parasitic protozoans have been identified from ducks in Florida (tables 10.24 and 10.25). Three of these are blood protozoans, 2 of which (*Haemoproteus nettionis* and *Leucocytozoon simondi*) are common parasites of North American waterfowl (Wobeser 1997).

Haemoproteus nettionis is the most common blood protozoan of waterfowl and has a widespread distribution in anatids in the Australian, Oriental, Ethiopian, Neotropical, and Nearctic regions (Williams and Bennett 1980). Fallis and Wood (1957) determined that the biting midge *Culicoides downesi* was the vector of *H. nettionis*. It is not believed to be pathogenic (Bennett et al. 1993). Julian and Galt (1980) reported that *H. nettionis* caused severe mortality in a flock of Muscovy Ducks in Ontario, but 5 years later the authors printed a retraction after it was discovered that the disease was actually caused by a species of *Bacillus* (Julian et al. 1985).

Leucocytozoon simondi is less common and has a more restricted distribution in Holarctic anatids. A number of ornithophilic black flies, including *Simulium ruggelsi*, *S. anatinum*, *S. innocens*, and *Cnephia ornithophilia* serve as vectors of *L. simondi* in North America (Bennett and Squires-Parsons 1992). This parasite is known to be lethal to domestic ducks, but does not cause mortality in free-ranging wild ducks such as Mallards and American Black Ducks (Trainer et al. 1962; Fallis and Bennett 1966).

Plasmodium vaughani is 1 of 15 species of *Plasmodium* that have been reported from waterfowl and is among the most commonly reported species, next to *P. circumflexum*. In addition to Blue-winged Teal it has been reported from 6 other species of waterfowl including Wood Ducks, Mallards, and American Black Ducks (Bennett et al. 1982; Bishop and Bennett 1992). Infections by *P. vaughani* may result in some degree of clinical disease in the early phase of infection, but are probably not fatal (Garnham 1966).

Transmission of blood protozoans of water-

Table 10.24. Blood protozoans reported from migratory ducks from Florida

Species of duck Protozoan	County or site	Year(s)	No. ducks		Data source
			Exam.[a]	Inf.	
Perching-ducks					
Wood Duck[b]					
Haemoproteus nettionis	Statewide	1976–79	2,074	257	Thul & O'Brien (1990)
Leucocytozoon simondi	Statewide	1976–79	2,074	48	Ibid.
Dabbling-ducks					
Blue-winged Teal					
Haemoproteus nettionis	Hamilton	1979	6	2	Forrester et al. (1980)
Plasmodium vaughani	Hamilton	1979	6	1	Ibid.
Northern Shoveler					
Haemoproteus nettionis	Hamilton	1979	7	1	Forrester (1979)
	Polk	1979	1	1	Ibid.
Leucocytozoon simondi	Hamilton	1979	7	1	Ibid.
Gadwall					
Leucocytozoon simondi	Hamilton	1979	1	1	Ibid.
Diving-ducks					
Ring-necked Duck					
Leucocytozoon simondi	Alachua	1979	1	1	Ibid.
		1979–82	125	10	Forrester et al. (2001)
	Broward	1979–82	73	1	Ibid.
	Hamilton	1979–82	1	0	Ibid.
	Jefferson	1979–82	20	0	Ibid.
	Leon	1979–82	52	14	Ibid.
	Putnam	1979–82	12	1	Ibid.
	Hamilton	1979	1	1	Forrester et al. (1980)
	Marion	1978–79	3	1	Forrester (1979)
Haemoproteus nettionis	Alachua	1979–82	125	7	Forrester et al. (2001)
	Broward	1979–82	73	0	Ibid.
	Hamilton	1979–82	1	1	Ibid.
	Jefferson	1979–82	20	1	Ibid.
	Leon	1979–82	52	5	Ibid.
	Putnam	1979–82	12	2	Ibid.
Lesser Scaup					
Haemoproteus nettionis	Hillsborough	1974	5	1	Forrester et al. (1977)

a. All data were obtained by examination of stained blood smears. Ducks found negative for blood protozoans: 1 Green-winged Teal (Alachua County, 1979), 1 Blue-winged Teal (Glades, 1972), 2 Lesser Scaup (Leon, 1979), 1 Lesser Scaup (Polk, 1978), 1 Red-breasted Merganser (Charlotte, 1974), and 1 Ruddy Duck (Columbia, 1981) (Forrester 1972, 1974, 1978–79, 1981).
b. Values for Wood Ducks were based on blood smears from ducks collected between September and February and represent a mixture of resident and migratory Wood Ducks (Thul & O'Brien 1990).

fowl does not occur in Florida. This conclusion is supported by 2 lines of evidence: (1) Resident populations of Wood Ducks and Mottled Ducks have been found to be free of blood protozoans (table 10.26). The same is true for res-

ident Canada Geese and Fulvous Whistling-Ducks (see chapter 9, Whistling-Ducks, Swans, and Geese). (2) Pekin Ducks used as sentinels failed to contract infections of blood protozoans. Such a study was conducted at the edge

Table 10.25. Protozoan infections (other than hematozoans) in ducks from Florida

Species of duck Protozoan	County or site	Year(s)	Method of diagnosis[a]	No. ducks Exam.	Inf.	Data source
Perching-ducks						
Wood Duck						
Sarcocystis sp.	Columbia	1988	N	1	1	Wright (1988)
Toxoplasma gondii	"Florida"	NG	S	16	1	Burridge et al. (1979)
Dabbling ducks						
Green-winged Teal						
Sarcocystis sp.	Alachua	1979	N	1	1	Forrester (1979)
Mottled Duck						
Cryptosporidium sp.	Alachua	1970	F	2	0	Forrester (1970)
	Glades	1970	F	20	2	Ibid.
Intestinal coccidia	Alachua	1970	F	2	0	Ibid.
(*Eimeria* sp.)	Glades	1970	F	20	6	Ibid.
Northern Pintail						
Sarcocystis sp.	Citrus/ Hernando	1976	N	1	1	Forrester (1976)
Blue-winged Teal						
Sarcocystis sp.	"Florida"	1981	N	1	1	Forrester (1981)
	Palm Beach	1997	N	1	1	Regan (1999)
American Wigeon						
Sarcocystis sp.	Citrus/ Hernando	1976	N	2	2	Forrester (1976)
Mergansers						
Red-breasted Merganser						
Renal coccidia	Sarasota	1994	H	3	1	Locke (1995)
(*Eimeria* sp.)	Sarasota	1994	H	3	0	Fischer (1994)

NG = not given.

a. F = fecal flotation, H = histopathology, N = necropsy, S = serology (IHA test).

of Paynes Prairie (Alachua County) from March 31 to November 16, 1980 (Thul and O'Brien 1990). In that study several groups of 2-week-old Pekin Ducks were housed in outdoor exposure pens for 5-week intervals to determine if they would acquire infections. In August of that year a pair of Wood Ducks infected with *H. nettionis* and *L. simondi* were obtained from Maine and placed in pens adjacent to the Pekin Ducks to serve as sources of the parasites for the Pekin Ducks. Blood smears were made from these birds on a weekly basis and examined for protozoan infections. None was found. It can be concluded from these data that even though blood protozoans are present in ducks that migrate to Florida in the fall each year, these infections (which were acquired in the north) are not transmitted to other ducks in Florida. This lack of transmission may be due to the absence of appropriate vectors or because peak populations of vectors do not occur at the same time of year when these migratory ducks are present. Thul and O'Brien (1990) used these data "to develop a model for estimating the proportion of northern migrants in the harvest of wood ducks in southern states." The technique was based on the unique geographic and temporal distribution of blood protozoans in Wood Ducks in the Atlantic Flyway. Thul (1990) applied this "biological tag" technique as a method of estimating the proportion of northern Wood Ducks in the annual harvests of southern Atlantic Flyway states during 1980–85. This was done so that adjustments in the times of the hunting seasons in various states could be made to insure that local populations were not over-harvested. This is an interesting example of how host-specific parasites such as blood protozoans can be used as a practical wildlife management tool.

Oocysts of an unidentified species of *Cryptosporidium* were found in fecal samples from 2 of 20 Mottled Ducks from Glades County (Forrester 1970). These ducks were all normal when examined at necropsy and apparently the infections were subclinical. Respiratory cryptosporidiosis has been reported in a commercial flock of Mallards in Australia (O'Donoghue et al. 1987), but no such reports exist for free-ranging waterfowl in North America (Dubey et al. 1990).

Six of the same 20 Mottled Ducks from Glades County had oocysts of an unidentified species of *Eimeria* in their feces (Forrester 1970). It was not determined whether these were intestinal or renal coccidia and their significance is unknown. A number of species of *Eimeria* have been reported from waterfowl (Todd and Hammond 1971; Pellerdy 1974) and some of these are pathogenic (Wobeser 1997). Renal coccidia were found in 1 of 6 Red-breasted Mergansers examined from Sarasota County in 1994. Developing coccidial stages within tubular epithelial cells were associated with necrosis. The species of *Eimeria* was not determined (Locke 1995). Renal coccidia have been reported from a number of species of ducks and infections appear to be widespread (Wobeser 1974; Nation and Wobeser 1977). Lesions in older birds are not believed to be of clinical significance, but infections in ducklings can be problematic (Wobeser 1997).

Macroscopic cysts of *Sarcocystis* sp. have been found in the pectoral muscles of 5 species of ducks in Florida (table 10.25). The species involved has not been determined but there may be more than one, including *S. rileyi*. Species of

this coccidian parasite are found commonly in waterfowl, especially dabbling ducks (Cornwell 1963; Drouin and Mahrt 1979), but are not known to cause morbidity or mortality (Wobeser 1997; Tuggle and Friend 1999a). Hunters often observe this conspicuous infection, which occurs as large whitish to yellowish cysts the size of rice grains arranged in parallel streaks. These are usually in breast muscles, but sometimes occur in other skeletal muscles or in the heart. The striped skunk (*Mephitis mephitis*) is the definitive host of *S. rileyi* and becomes infected by the ingestion of cysts when eating infected waterfowl; waterfowl in turn are infected when they ingest oocysts in water contaminated by skunk feces (Cawthorn et al. 1981; Wicht 1981). An unnamed species of *Sarcocystis* has been transmitted experimentally to ducks with oocysts from Virginia opossums (*Didelphis virginiana*) (Duszynski and Box 1978) and domestic cats (Golubkov 1979).

Sixteen Wood Ducks from Florida were examined serologically by the IHA test for antibodies to *Toxoplasma gondii* (Burridge et al. 1979). One was seropositive. However, as pointed out elsewhere, the IHA test may not be as sensitive in birds as in mammals and may result in false-negative results (Frenkel 1981). There are no records of toxoplasmosis in free-ranging ducks (Dubey and Beattie 1988; Wobeser 1997), but clinical toxoplasmosis has been reported from domestic ducks in Argentina, where 50% of a flock of 100 birds died (Boehringer et al. 1962).

XX. Helminths

Sixty species of parasitic helminths have been identified from migratory and nonmigratory ducks in Florida (tables 10.27–10.34). These include 21 species of trematodes, 13 cestodes, 24 nematodes, and 2 acanthocephalans. The Wood Duck (tables 10.28–10.31) and the Mottled Duck (tables 10.32–10.34) are the most extensively studied species and the only ducks for which there is adequate information on prevalence and intensity.

Table 10.26. Nonmigratory ducks from Florida examined by blood smear for protozoans

Species of duck	County or site	Year	No. ducks Examined	Positive	Data source[a]
Mottled Duck	Alachua	1970	4	0	A
		1980	1	0	B
	Glades	1970	29	0	A
		1971	33	0	A
	"Lake Okeechobee"[b]	1994	20	0	C
	Polk	1983	3	0	B
Mottled Duck/ Mallard cross	Alachua	1970	5	0	A
Wood Duck	Alachua	1976	3	0	D
	Gadsden	1976	10	0	D
	Glades	1976	1	0	D
	Hamilton	1976	4	0	D
	Hernando	1976	6	0	D
	Leon	1976	4	0	D
		1977	20	0	D
	Levy	1976	14	0	D

a. A = Thul et al. (1980) and Forrester (1997), B = Forrester (1983), C = Pence (1997), D = Thul et al. (1980).
b. Ducks were collected in an area to the northwest of Lake Okeechobee where Glades, Highlands, and Okeechobee counties come together.

Thul et al. (1985) examined ecological aspects of parasitic helminths of Wood Ducks throughout the Atlantic Flyway, including Florida. In that study ducks were collected during their postbreeding and premigratory period in order to compare the helminths of migratory northern birds with those of nonmigratory southern Wood Ducks before the northern ducks migrated into the range of the southern Wood Duck. There were some differences due to age and sex, but most of the variability in the helminths was linked with geographic factors. Three helminths (*Trichobilharzia* sp., *Amidostomum acutum,* and an undescribed species of *Tetrameres*) were very abundant throughout the Flyway. However, there were differences in the prevalences and abundances of 20 of the 24 most common species when northern ducks were compared with southern ducks. In another study in Florida, Thul (1986c) found that the helminth faunas of Wood Ducks collected in the summer were different from those collected in the winter (figures 10.5–10.13). This was true especially for the trematodes and cestodes. He concluded that these differences were due to the influx of migrant Wood Ducks from the north as well as environmental factors affecting both resident and migratory ducks. It was postulated also that some helminths acquired during the summer in the north were lost subsequently and others picked up on the wintering grounds in Florida.

The helminth community structure of Mottled Ducks in Florida was compared with that of Mottled Ducks in Texas by Fedynich et al. (1996). They found that there were fewer species of helminths in Mottled Ducks in Florida ($n = 28$) compared with those in Texas ($n = 37$). Twenty-three species occurred in ducks from both states. The authors also found that of the 23 common species of helminths in Mottled Ducks, 15 also occurred in Mallards and 14 in Blue-winged Teal. They concluded that waterfowl migration through or overwintering in Florida would result in the dispersal and exchange of helminths between migratory and nonmigratory ducks.

There are at least 960 species of parasitic helminths reported from waterfowl on a worldwide basis; of these, only 71 (7%) have

Table 10.27. Helminths reported from ducks in Florida

Species of duck Helminth	Site	County or site	Year(s)	Data source
Perching-ducks				
Wood Duck[a]				
Dabbling-ducks				
Mottled Duck[b]				
Mallard				
Trematoda				
Psilochasmus oxyurus	Small intestine	Leon	1966	Premvati (1969)
Blue-winged Teal				
Cestode[c]	NG[e]	"Florida"	NG	Stiles & Hassall (1894)
Northern Shoveler				
Acanthocephala				
Polymorphus sp.	NG	Glades	1949	USNPC[d] Accession #64984
Diving-ducks				
Ring-necked Duck				
Trematoda				
Apatemon gracilis	Small intestine	Alachua	1982	Kinsella & Thul (1997)
		Broward	1981–82	Ibid.
		Hamilton	1981	Ibid.
Cyclocoelum obscurum	Air sacs	Leon	NG	MacInnis (1959), Loftin (1960)
Echinoparyphium sp.	Small intestine	Alachua	1980	Kinsella & Thul (1997)
		Hamilton	1981	Ibid.
Prosthogonimus ovatus	Cloaca	Hamilton	1981	Ibid.
Zygocotyle lunata	Ceca	Alachua	1980–82	Ibid.
		Hamilton	1981	Ibid.
Cestoda				
Diploposthe laevis	Small intestine	Alachua	1980–82	Ibid.
		Hamilton	1981	Ibid.
Nematoda				
Amidostomum acutum	Gizzard	Alachua	1981	Ibid.
Capillaria contorta	Crop	Alachua	1980–82	Kinsella & Thul (1997)
		Broward	1981	Ibid.
		Collier	1981	Ibid.
		Hamilton	1981	Ibid.
		Jefferson	1981	Ibid.
		Osceola	1982	Ibid.
Capillaria spinulosa	Ceca	Alachua	1982	Ibid.
Echinuria uncinata	Proventriculus	Leon	1936	USNPC Accession #43220
		Hamilton	1979	Forrester et al. (1980)
Epomidiostomum crami[f]	Gizzard	Leon	1936	USNPC Accession #43221
		Jefferson	1932	USNPC Accession #46800
Splendidofilaria fallisensis[g]	Heart blood	Alachua	1979–82	Forrester et al. (2001)
		Broward	1979–82	Ibid.
		Hamilton	1979–82	Ibid.
		Jefferson	1979–82	Ibid.
		Leon	1979–82	Ibid.
		Putnam	1979–82	Ibid.
Streptocara crassicauda	Proventriculus	Alachua	1980	Kinsella & Thul (1997)

(continued)

Table 10.27. *(continued)*

Species of duck Helminth	Site	County or site	Year(s)	Data source
Tetrameres fissispina	Proventriculus	Alachua	1980–82	Ibid.
		Hamilton	1981	Ibid.
		Osceola	1982	Ibid.
Microfilaria Type I[h]	Heart blood	Alachua	1979–82	Forrester et al. (2001)
		Broward	1979–82	Ibid.
		Hamilton	1979–82	Ibid.
		Jefferson	1979–82	Ibid.
		Leon	1979–82	Ibid.
		Putnam	1979–82	Ibid.
Microfilaria Type II[i]	Heart blood	Alachua	1979–82	Ibid.
		Broward	1979–82	Ibid.
		Hamilton	1979–82	Ibid.
		Jefferson	1979–82	Ibid.
		Leon	1979–82	Ibid.
		Putnam	1979–82	Ibid.
Greater Scaup				
Trematoda				
Psilochasmus oxyurus	Small intestine	Leon	1966	Premvati (1969)
Lesser Scaup				
Trematoda				
Microphallus pygmaeus	NG	Pinellas	NG	Hutton (1964)
Microphallidae	Small intestine	Hillsborough	1974	Forrester et al. (1977)
Pachytrema sp.	NG	Pinellas	NG	Hutton (1964)
Spelotrema sp.	Intestine	Franklin	NG	Loftin (1960)
Zygocotyle sp.	Ceca	Hillsborough	1974	Forrester et al. (1977)
Cestoda				
Hymenolepidae	Small intestine	Hillsborough	1974	Ibid.
Nematoda				
Amidostomum sp.	Gizzard	Hillsborough	1974	Ibid.
Capillaria sp.	Ceca	Hillsborough	1974	Ibid.
Streptocara sp.	Gizzard	Hillsborough	1974	Ibid.
Tetrameres sp.	Proventriculus	Hillsborough	1974	Ibid.
"Nematode"	NG	NG	NG	MacInnis (1959)
Mergansers				
Red-breasted Merganser				
Nematoda				
Eustrongylides sp.	Proventriculus	Sarasota	1994	Fischer (1994), Cole (1995)

a. See Tables 10.28–31.
b. See Tables 10.32–34.
c. Listed as *Taenia* sp. by the authors.
d. USNPC = U.S. National Parasite Collection.
e. NG = not given by authors.
f. Not identified to species according to the USNPC paperwork, but were examined by J.M. Kinsella and determined to be *E. crami.*
g. Microfilariae identified on blood smears stained with Giemsa stain. Number infected (sample size in parentheses) for each county = Alachua 47 (125), Broward 31 (73), Hamilton 1 (1), Jefferson 1 (20), Leon 26 (52), Putnam 5 (12).
h. Unidentified microfilaria, Type I. Number infected (sample size in parentheses) for each county = Alachua 8 (125), Broward 6 (73), Hamilton 0 (1), Jefferson 0 (20), Leon 2 (52), Putnam 1 (12).
i. Unidentified microfilaria, Type II. Number infected (sample size in parentheses) for each county = Alachua 18 (125), Broward 16 (73), Hamilton 0 (1), Jefferson 0 (20), Leon 4 (52), Putnam 1 (12).

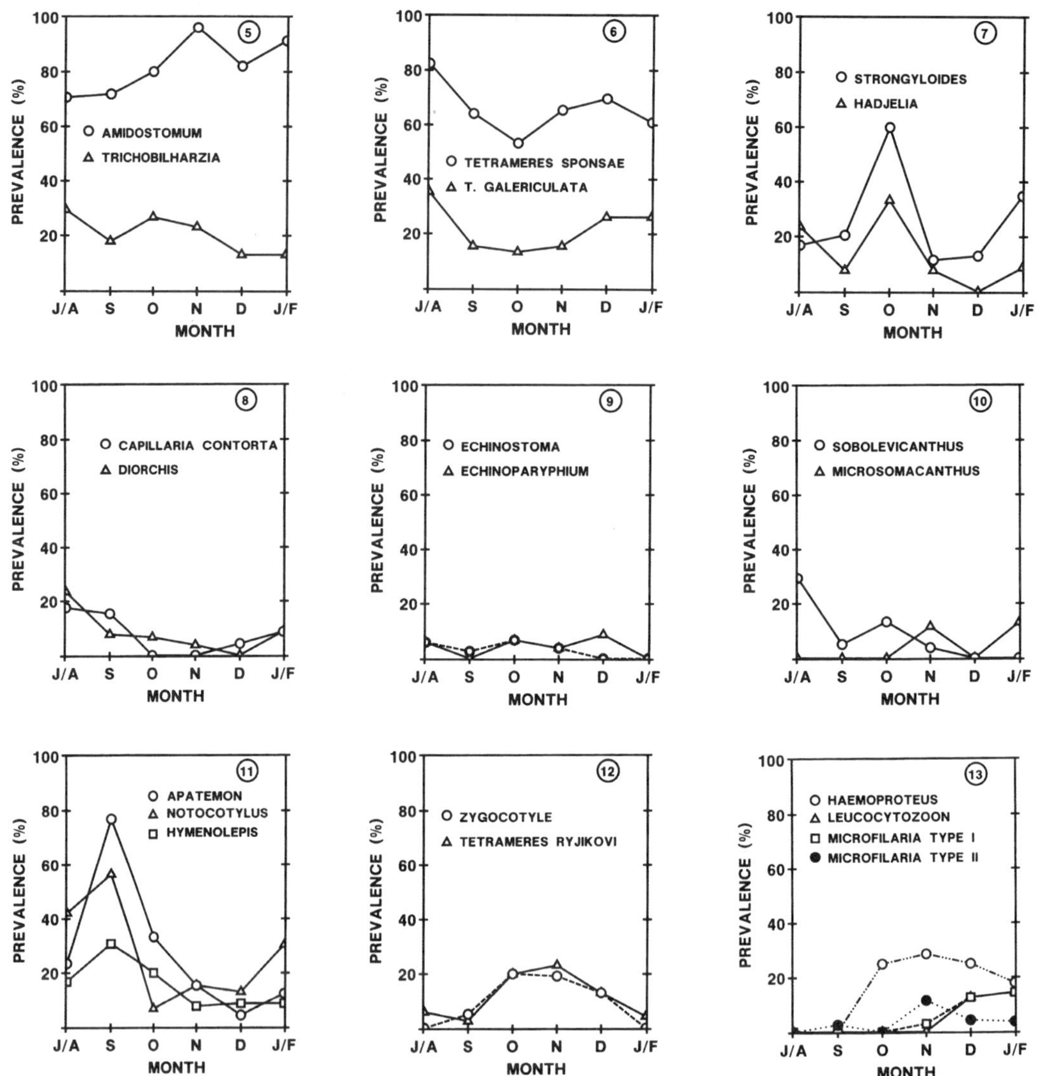

FIGURES 10.5–10.13. Seasonal variation in prevalences of helminths and hematozoans of Wood Ducks in Florida, 1976–79. (Note corrections: in Figure 10.6 *Tetrameres sponsae* should be *Tetrameres* sp., and in Figure 10.10 *Sobolevicanthus* should be *Calixolepis thuli*). Courtesy of James E. Thul.

been associated with disease (McDonald 1969a). Of the 60 species of helminths found in ducks from Florida, 16 (27%) are of possible pathologic significance. These include 3 trematodes (*Dendritobilharzia pulverulenta, Echinoparyphium recurvatum,* and *Typhlocoelum cucumerinum*), 12 nematodes (*Amidostomum acutum, Capillaria contorta, Echinuria uncinata, Epomidiostomum uncinatum, Eustrongylides*

sp., *Hystrichis tricolor, Streptocara crassicauda,* and 5 species of *Tetrameres*), and 1 acanthocephalan (*Polymorphus* sp.). No specific studies have been conducted in Florida to evaluate these parasites as pathogens of free-ranging ducks. However, based on fairly high prevalences in ducks from Florida and information from studies done elsewhere (Wobeser 1997), the most important of these helminths are probably

Table 10.28. Trematode infections in nonmigratory Wood Ducks from Florida[a]

Species (Site)[b] County	Year	Number examined	Number infected	Mean	Range
Apatemon gracilis (SI)					
Alachua	1976	3	2	30	28–31
Gadsden	1976	10	1	10	—
Glades	1976	1	1	1	—
Hamilton	1976	4	1	1	—
Hernando	1976	6	1	4	—
Leon	1976	15	8	4	1–11
	1977	20	18	8	1–22
Levy	1976	4	4	24	5–46
Dendritobilharzia pulverulenta (BV)					
Leon	1977	20	1	1	—
Echinostoma trivolvis (SI)					
Gadsden	1976	10	2	5	1–8
Hernando	1976	6	1	1	—
Leon	1976	15	1	9	—
Notocotylus urbanensis (CE)					
Alachua	1976	3	2	1	—
Gadsden	1976	10	4	21	7–33
Hamilton	1976	4	4	10	8–19
Hernando	1976	6	1	1	—
Leon	1976	15	11	4	1–16
	1977	20	6	4	1–8
Levy	1976	4	3	5	2–11
Prosthogonimus ovatus (BU,CL)					
Leon	1976	15	1	15	—
Levy	1976	4	1	1	—
Psilostomum sp. (SI)					
Gadsden	1976	10	1	1	—
Leon	1976	15	1	3	—
Ribeiroia ondatrae (PR)					
Leon	1976	15	1	3	—
Trichobilharzia sp. (BV)					
Alachua	1976	3	1	33	—
Gadsden	1976	10	3	22	3–37
Hernando	1976	6	2	3	1–4
Leon	1976	15	2	2	1–3
	1977	20	5	11	4–17
Levy	1976	4	2	8	1–15
Typhlocoelum cucumerinum sisowi (AS)					
Glades	1976	1	1	15	—
Zygocotyle lunata (CE)					
Leon	1977	20	2	1	—

Sources: Thul (1979a), Thul et al. (1985).

a. Sixty-three ducks were examined: 3 from Alachua County (1976), 10 from Gadsden County (1976), 1 from Glades County (1976), 4 from Hamilton County (1976), 6 from Hernando County (1976), 35 from Leon County (1976–77), and 4 from Levy County (1976). For each parasite, data are given for counties in which positive birds were found; negatives are not included in the table.

b. Infection sites: AS = air sacs, BU = bursa, BV = blood vessels, CE = cecum, CL = cloaca, PR = proventriculus, SI = small intestine.

Table 10.29. Cestode infections in nonmigratory Wood Ducks[a] from Florida

Species (Site)[b] County	Year	Number examined	Number infected	Mean	Range
Diorchis bulbodes (SI)					
Gadsden	1976	10	3	2	1–3
Hamilton	1976	4	3	2	1–3
Hernando	1976	6	1	1	—
Leon	1976	15	1	1	—
Microsomacanthus hopkinsi[c] (CE)					
Alachua	1976	3	1	10	—
Gadsden	1976	10	1	1	—
Hamilton	1976	4	1	2	—
Leon	1976	15	3	1	—
	1977	20	7	1	1–2
Levy	1976	4	1	4	—
Calixolepis thuli[d] (SI)					
Alachua	1976	3	1	1	—
Gadsden	1976	10	4	2	1–4
Hernando	1976	6	2	1	—
Leon	1976	15	1	1	—
Levy	1976	4	1	1	—

Sources: Thul (1979a), Thul et al. (1985).
a. See Table 10.28, note a, for details on sample sizes.
b. Infection sites: CE = cecum, SI = small intestine.
c. Identified as *Hymenolepis* sp. by Thul (1979a) and Thul et al. (1985), but upon examination by Kinsella (1998), determined to be *Microsomacanthus hopkinsi*.
d. Described by Macko and Hanzelová (1997) using material collected from Leon County by J.E. Thul; referred to as *Sobolevicanthus* sp. by Thul et al. (1985).

E. recurvatum, A. acutum, E. uncinatum, and the 5 species of *Tetrameres.*

The intestinal trematode *Echinoparyphium recurvatum* is a very common and characteristic helminth of waterfowl and is cosmopolitan in distribution. It has a wide host range; in addition to infecting numerous species of waterfowl, it is also a parasite of species in 9 other orders of birds as well as in rodents and carnivores (McDonald 1969a). The life cycle includes 2 intermediate hosts, first aquatic snails, then either mollusks or amphibians; ducks are infected by ingesting metacercariae encysted in snails or amphibians (Kingston 1978). The specifics of this cycle have not been worked out for Florida. Although there is no information on the effects of this trematode on ducks in Florida, it has caused mortality in Mute Swans, *Cygnus olor,* in England (Soulsby 1955).

Amidostomum acutum and *Epomidiosto-mum uncinatum* are both trichostrongylid nematodes with direct life cycles, ducks being infected by ingestion of infective larvae. Adults are located under the horny lining of the gizzard and are prevalent in both Wood Ducks and Mottled Ducks in Florida. Infections of high intensity result in ulceration and erosion of the grinding pads and necrosis of the gizzard muscle, which leads to debilitation and increased susceptibility to predation and other diseases (Tuggle and Friend 1999b).

Species in the genus *Tetrameres* are unusual in that males and females are markedly dimorphic; males are small and filiform, whereas females are larger and are globular and fusiform (Wehr 1971). Five of the 6 species of *Tetrameres* found in ducks from Florida have been described. One other from Wood Ducks is undescribed. Its life cycle has not been determined, nor have those of *Tetrameres galericulata* and *T.*

Table 10.30. Nematode infections in nonmigratory Wood Ducks[a] from Florida

Species (Site)[b] County	Year	Number examined	Number infected	Mean	Range
Amidostomum acutum (GL)					
Alachua	1976	3	2	4	1–7
Gadsden	1976	10	6	7	1–18
Glades	1976	1	1	19	—
Hamilton	1976	4	4	23	3–44
Hernando	1976	6	3	4	1–8
Leon	1976	15	10	3	1–9
	1977	20	14	4	1–12
Levy	1976	4	4	5	1–8
Aproctella stoddardi (BC)					
Leon	1976	15	2	4	1–6
Capillaria contorta (ES)					
Gadsden	1976	10	1	3	—
Leon	1976	15	4	2	1–2
	1977	20	4	1	—
Contracaecum rudolphi (SI)					
Leon	1977	20	1	1	—
Levy	1976	4	1	1	—
Hadjelia neglecta (GL,PR)					
Gadsden	1976	10	4	3	2–3
Hernando	1976	6	5	2	1–6
Leon	1976	15	1	1	—
	1977	20	1	1	—
Levy	1976	4	1	1	—
Hystrichis tricolor (GL)					
Hernando	1976	6	1	1	—
Strongyloides sp. (CE)					
Gadsden	1976	10	1	7	—
Glades	1976	1	1	4	—
Hernando	1976	6	3	13	7–21
Leon	1976	15	2	5	2–8
	1977	20	6	6	1–13
Tetrameres galericulata (PR)					
Alachua	1976	3	1	1	—
Gadsden	1976	10	4	5	1–12
Leon	1976	15	3	1	1–2
	1977	20	4	3	1–10
Levy	1976	4	1	64	—
Tetrameres ryjikovi (PR)					
Glades	1976	1	1	1	—
Leon	1976	15	1	1	—
Tetrameres sp.[c] (PR)					
Alachua	1976	3	1	5	—
Gadsden	1976	10	9	10	1–25
Hamilton	1976	4	1	6	—
Hernando	1976	6	1	23	—
Leon	1976	15	8	10	1–47
	1977	20	18	10	2–47
Levy	1976	4	3	9	2–20

Sources: Thul (1979a), Thul et al. (1985).

a. See Table 10.28, note a, for details on sample size.

b. Infection sites: BC = body cavity, CE = cecum, ES = esophagus, GL = gizzard lining, PR = proventriculus, SI = small intestine.

c. Identified erroneously as *Tetrameres sponsae* in Thul et al. (1985); described as *T. sponsae* n. sp. by Mollhagen (1976) in an unpublished Ph.D. dissertation and therefore became a *nomen nudum*. We refer to them here as *Tetrameres* sp.

ryjikovi, but may be similar to that of *T. crami*. The life cycle of *Tetrameres crami* is indirect, with freshwater crustaceans serving as intermediate hosts (Wehr 1971). As ducks feed, infective larvae are ingested with the intermediate host and eventually make their way to the proventriculus where they enter the crypts of Lieberkuehn. After further development and mating, adult males return to the lumen of the proventriculus, while females remain in the crypts where they feed on blood and can be seen as conspicuous dark spots in the glandular proventriculus. In infections of 100 or more female nematodes there is malfunction of the proventriculus and impaired digestion leading subsequently to poor body condition of the duck (Swales 1936). Wobeser (1997) stated that the buildup of *Tetrameres* larvae in crustacean populations might lead to harmful effects on ducklings. Nothing is known about these effects on ducks in Florida.

Larvae of *Eustrongylides* sp. were reported to cause mortality in Red-breasted Mergansers in Virginia in 1962 (Locke et al. 1964). Such mortality has not been recorded in Florida, but 3 of 10 Red-breasted Mergansers from Sarasota County were found infected with *Eustrongylides* sp. when examined at necropsy in 1994 (Fischer 1994, Cole 1995). These 10 birds were part of a die-off of Red-breasted Mergansers and Common Loons, the cause of which was not determined. Nematodes were found in cystic cavities in the muscular wall of the proventriculus (figure 10.14) and were surrounded by granulomatous inflammation; these were judged to be secondary or incidental and not the primary cause of death (Fischer 1994).

XXI. Arthropods

Fourteen species of parasitic arthropods have been reported from ducks in Florida. These include 3 species of feather mites, 1 skin mite, 1 nasal mite, 7 chewing lice, 1 flea, and 1 cimicid bug (tables 10.35 and 10.36). The Wood Duck is the only species of waterfowl in Florida for which we have data on prevalence (table 10.36),

thanks to the excellent work of Thul (1985a and 1985b). Thul's study included birds collected in the summer and early fall from Florida and 11 other states in the Atlantic Flyway. He found that 71% of his sample was infested with 1–5 species of ectoparasite (mean = 2.0), and mixed infestations of mites and chewing lice occurred in 28% of his birds. Ectoparasites in general were more prevalent on immature than on adult birds, but there was no difference in infestations of male versus female ducks.

The pathologic significance of the feather mites, the skin mite, and the chewing lice to ducks in Florida is unknown. Wobeser (1997) stated that the importance of ectoparasites to waterfowl is related to "debilitation of birds that are compromised in some other way." He claimed that populations of chewing lice could increase in a short period of time on sick birds. We have no data on this aspect for waterfowl in Florida.

The nasal mite (*Rhinonyssus rhinolethrum*) has been reported from a number of species of Anatidae, including Wood Ducks, in Texas and Louisiana and is probably restricted to this family of waterfowl (Mitchell and Rhodes 1960; Pence 1972). Details on the morphology of the various life stages of this mite have been reported by Mitchell (1963). It has not been reported to be pathogenic.

The sticktight flea (*Echidnophaga gallinacea*) was reported from a Mallard in Brevard County (no date given) (Layne 1971). This is primarily a parasite of domestic poultry in southern United States, but has been found also on a variety of wild birds and mammals (Loomis 1978). Although it is known to cause irritation and blood loss that can result in mortality in young chickens, nothing is known about its effects on waterfowl and it is probably an incidental finding.

Specimens of *Ornithocoris pallidus* were collected from manmade nesting boxes that had been used recently by Wood Ducks in Lake County (1990) and Leon County (1995 and 1996) (Eggeman et al. 1996). This is a true "bug" of the family Cimicidae and is related to the famous common bed bug (*Cimex lec-*

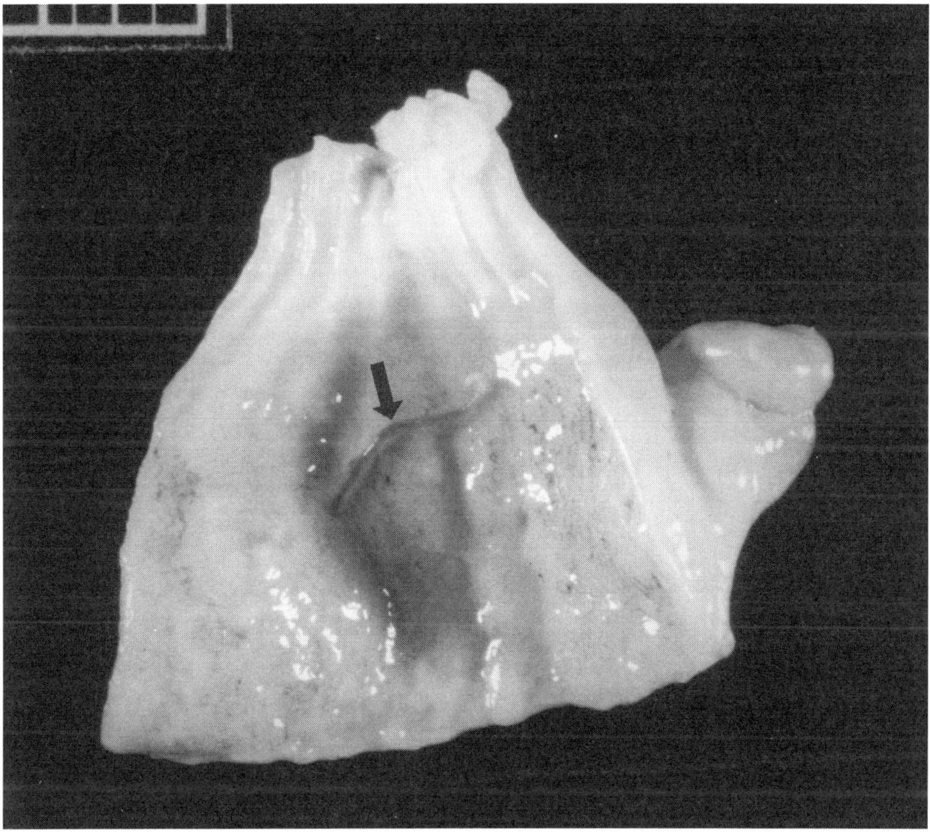

FIGURE 10.14. Proventriculus of a Red-breasted Merganser from Sarasota County in 1994 infected with *Eustrongylides* sp. The posterior end of a nematode *(arrow)* can be seen protruding from a cystic cavity. Courtesy of John R. Fischer.

turalis), a serious pest of humans in many parts of the world. *Ornithocoris pallidus* has been reported also from domestic chicken coops and Purple Martin nesting boxes in Florida and is believed to have been introduced originally from South America (Usinger 1966). The significance of this parasite to Wood Duck populations is unknown and should be investigated. There may be harmful effects on nestlings and on reproduction in general. This parasite is of public health importance since it feeds on humans; in fact the recent rediscovery of this parasite in Florida was precipitated by the experience that Eggeman and Ohme had when they were cleaning out Wood Duck boxes and were bitten numerous times by *O. pallidus* (Eggeman et al. 1996).

XXII. Summary and conclusions

Excluding trauma, 128 different morbidity/mortality factors and disease agents are known for ducks in Florida. These include 11 organochlorines, 1 organophosphate, 3 metals, 13 toxic trace elements, 5 radionuclides, 1 oil, 3 biotoxins, 2 viruses, 7 bacteria, 8 protozoans, 24 trematodes, 13 cestodes, 21 nematodes, 2 acanthocephalans, 4 feather mites, 1 nasal mite, 7 chewing lice, 1 flea, and 1 cimicid bug. Wood Ducks and Mottled Ducks are the best studied of the 33 species of ducks that occur in Florida. Small amounts of information are known for 21 other species; nothing is known about 10 additional species, all rare winter visitors.

Even though there is considerable information on some mortality factors and disease agents, little is known about their effects on duck populations in Florida. Mortality events ranging from a few birds to as many as several hundred have been attributed to trauma, diazinon poisoning, lead poisoning, avian botulism, avian cholera, and oiling. In addition, 1 die-off of several thousand Lesser Scaup was associated with a red tide bloom. Several agents are of some public health concern and include red tide brevitoxins, *Salmonella* infections, and cimicid bug infestations.

Table 10.31. Helminth infections in migratory Wood Ducks[a] from Florida, 1978–79

Species (Site)[b] County	Number examined	Number infected	Mean	Range
Trematoda				
Apatemon gracilis (SI)				
Levy	5	1	1	—
Echinoparyphium recurvatum (SI)				
Alachua	7	1	1	—
Eucotyle warreni (KD)				
Levy	5	1	1	—
Notocotylus urbanensis (CE)				
Alachua	7	1	1	—
Levy	5	2	3	2–3
Putnam	4	2	3	2–3
Prosthogonimus ovatus (CL)				
Leon	2	1	1	—
Psilostomum sp. (SI)				
Jefferson	1	1	1	—
Putnam	4	1	1	—
Trichobilharzia sp. (BV)				
Alachua	7	1	1	—
Levy	5	1	14	—
Zygocotyle lunata (CE)				
Alachua	7	1	1	—
Levy	5	1	1	—
Putnam	4	1	1	—
Cestoda				
Microsomacanthus hopkinsi[c] (CE)				
Jefferson	1	1	2	—
Leon	2	1	1	—
Putnam	4	1	1	—
Nematoda				
Alifilaria pseudolabiata (SQ)				
Alachua	7	1	1	—
Amidostomum acutum (GL)				
Alachua	7	7	3	1–8
Jefferson	1	1	6	—
Leon	2	2	7	2–12
Levy	5	5	7	3–12
Putnam	4	4	6	4–8

(continued)

Table 10.31. *(continued)*

Species (Site)[b] County	Number examined	Number infected	Mean	Range
Capillaria contorta (ES)				
Alachua	7	1	2	—
Capillaria spinulosa (CE)				
Alachua	7	1	3	—
Levy	5	1	2	—
Hystrichis tricolor (GL)				
Levy	5	1	5	—
Strongyloides sp. (CE)				
Levy	5	1	3	—
Tetrameres fissispina (PR)				
Putnam	4	1	1	—
Tetrameres galericulata (PR)				
Alachua	7	1	17	—
Levy	5	2	2	1–2
Tetrameres ryjikovi (PR)				
Alachua	7	1	1	—
Tetrameres sp.[d] (PR)				
Alachua	7	6	2	1–4
Jefferson	1	1	5	—
Leon	2	2	3	2–4
Levy	5	3	4	3–5
Putnam	4	2	6	4–8

Source: Thul (1979b).

a. Nineteen ducks were examined: 7 from Alachua County, 1 from Jefferson County, 2 from Leon County, 5 from Levy County, and 4 from Putnam County. For each parasite, data are given for counties in which positive birds were found; negatives are not included in the table.

b. Infection sites: BV = blood vessels, CE = cecum, CL = cloaca, ES = esophagus, GL = gizzard lining, KD = kidney, PR = proventriculus, SI = small intestine, SQ = subcutaneous.

c. Identified as *Hymenolepis* sp. by Thul (1979a) and Thul et al. (1985), but upon examination by Kinsella (1998), determined to be *Microsomacanthus hopkinsi*.

d. Identified erroneously as *Tetrameres sponsae* in Thul et al. (1985); described as *T. sponsae* n. sp. by Mollhagen (1976) in an unpublished Ph.D. dissertation and therefore became a *nomen nudum*. We refer to them here as *Tetrameres* sp.

Table 10.32. Trematode infections in Mottled Ducks[a] from Florida

Species (Site)[b] County or site	Year(s)	Number of ducks			
		Exam.	Inf.	Mean	Range
Apatemon gracilis (SI)					
Alachua	1970	4	3	8	1–21
Glades	1970–71	74	46	13	1–61
"Lake Okeechobee"[c]	1994	20	7	11	1–45
Dendritobilharzia pulverulenta (BV)					
Glades	1970–71	74	17	2	1–5
"Lake Okeechobee"	1994	20	2	1	—
Echinoparyphium recurvatum (LI,SI)					
Alachua	1970	4	4	21	3–48
Glades	1970–71	74	40	36	1–575
"Lake Okeechobee"	1994	20	11	12	1–86
Echinostoma trivolvis (CE,LI,SI)					
Alachua	1970	4	3	6	1–14
Glades	1970–71	74	25	9	1–53
"Lake Okeechobee"	1994	20	10	6	1–24
Eucotyle wehri (KD)					
Glades	1970–71	74	1	2	—
Gymnophallus sp. (NG)					
"Lake Okeechobee"	1994	20	1	2	—
Hypoderaeum conoideum (CL)					
Glades	1970–71	74	6	14	1–62
Levinseniella sp. (CE)					
Glades	1970–71	74	1	1	—
Prosthogonimus ovatus (CL)					
Glades	1970–71	74	5	2	1–5
Psilochasmus oxyurus (SI)					
Glades	1970–71	74	5	11	5–25
Trichobilharzia sp. (BV)					
Glades	1970–71	74	33	ND	ND
"Lake Okeechobee"	1994	20	13	6	1–22
Typhlocoelum cucumerinum sisowi (AS,LU,TR)					
Alachua	1970	4	2	5	2–7
Glades	1970–71	74	28	4	1–20
"Lake Okeechobee"	1994	20	9	2	1–5
Zygocotyle lunata (CE)					
Alachua	1970	4	2	1	—
Glades	1970–71	74	34	3	1–21
"Lake Okeechobee"	1994	20	3	1	1–2

Sources: 1994 data are from Fedynich et al. (1996); 1970–71 data are from Kinsella and Forrester (1972, 1997).
ND = not determined.
a. Ninety-eight ducks were examined: 4 from Alachua County (1970), 74 from Glades County (1970–71), and 20 from Lake Okeechobee (1994). For each parasite, data are given for counties in which parasites were found; negatives are not included in table.
b. Infection sites: AS = air sac, BV = blood vessel, CE = cecum, CL = cloaca, KD = kidney, LI = large intestine, LU = lung, NG = not given by authors, SI = small intestine, TR = trachea.
c. Ducks were collected in an area to the northwest of Lake Okeechobee where Glades, Highlands, and Okeechobee counties come together.

Table 10.33. Cestode infections in Mottled Ducks[a] from Florida

Species (Site)[b] County or site	Year(s)	Number of ducks			
		Exam.	Inf.	Mean	Range
Cloacotaenia megalops (CL,LI)					
Alachua	1970	4	2	3	1–5
Glades	1970–71	74	36	3	1–16
"Lake Okeechobee"[c]	1994	20	10	6	1–13
Dicranotaenia coronula (SI)					
Glades	1970–71	74	10	23	4–100
"Lake Okeechobee"	1994	20	3	6	4–8
Diorchis bulbodes (CE,LI,SI)					
Glades	1970–71	74	7	29	2–100
"Lake Okeechobee"	1994	20	1	1	—
Diploposthe laevis (SI)					
"Lake Okeechobee"	1994	20	1	1	—
Fimbriaria fasciolaris (SI)					
Glades	1970–71	74	8	6	1–30
"Lake Okeechobee"	1994	20	9	9	1–50
Hymenolepis sp. (SI)					
Alachua	1970	4	1	10	—
Glades	1970–71	74	20	4	1–15
Microsomacanthus compressa (SI)					
"Lake Okeechobee"	1994	20	1	8	—
Microsomacanthus hopkinsi (CE,LI,SI)					
Alachua	1970	4	3	15	1–42
Glades	1970–71	74	10	22	1–123
"Lake Okeechobee"	1994	20	8	10	1–35
Platyscolex ciliata (SI)					
"Lake Okeechobee"	1994	20	1	4	—
Sobolevicanthus filumferens (SI)					
Glades	1970–71	74	4	17	1–30
"Lake Okeechobee"	1994	20	5	3	1–6
Sobolevicanthus gracilis (SI)					
"Lake Okeechobee"	1994	20	6	3	1–7

Sources: 1994 data are from Fedynich et al. (1996); 1970–71 data are from Kinsella and Forrester (1972, 1997).
a. See Table 10.32, note a, for details on samples.
b. Infection sites: CE = cecum, CL = cloaca, LI = large intestine, SI = small intestine.
c. Ducks were collected in an area to the northwest of Lake Okeechobee where Glades, Highlands, and Okeechobee counties come together.

Table 10.34. Nematode and acanthocephalan infections in Mottled Ducks from Florida[a]

Helminth (Site)[b] County or site	Year(s)	Number of ducks			
		Exam.	Inf.	Mean	Range
Nematoda					
Amidostomum acutum (GI)					
Glades	1970–71	74	46	3	1–16
"Lake Okeechobee"[c]	1994	20	13	2	1–8
Capillaria contorta (BL, ES)					
Glades	1970–71	74	23	3	1–10
"Lake Okeechobee"	1994	20	16	10	1–37
Capillaria spinulosa[d] (CE)					
Alachua	1970	4	3	3	1–4
Glades	1970–71	74	44	3	1–9
"Lake Okeechobee"	1994	20	7	2	1–3
Echinuria uncinata (GI,PR)					
Glades	1970–71	74	11	3	1–106
"Lake Okeechobee"	1994	20	2	6	5–6
Epomidiostomum uncinatum (GI)					
Alachua	1970	4	1	1	—
Glades	1970–71	74	46	4	1–19
"Lake Okeechobee"	1994	20	14	5	1–13
Filaridae (unidentified) (LU)					
Glades	1970–71	74	2	1	—
Hadjelia neglecta (GI)					
Glades	1970–71	74	1	1	—
Porrocaecum crassum (GI, SI)					
Alachua	1970	4	1	4	—
Glades	1970–71	74	39	4	1–30
"Lake Okeechobee"	1994	20	7	4	1–8
Sciadiocara rugosa[e] (GI)					
Glades	1970–71	74	3	1	1–2
Spirurid larvae (ES,PR)					
Glades	1970–71	74	9	2	1–5

(continued)

Table 10.34. *(continued)*

| Helminth (Site)[b]
County or site | Year(s) | Number of ducks | | | |
		Exam.	Inf.	Mean	Range
Streptocara crassicauda (GI)					
Glades	1970–71	74	4	1	—
"Lake Okeechobee"	1994	20	6	3	1–6
Strongyloides sp. (CE, CL, LI, SI)					
Alachua	1970	4	2	5	3–6
Glades	1970–71	74	33	4	1–19
"Lake Okeechobee"	1994	20	6	12	1–52
Tetrameres crami (PR)					
Glades	1970–71	74	32	4	1–13
Tetrameres spp.[f] (PR)					
Alachua	1970	4	2	3	2–4
Glades	1970–71	74	23	3	1–19
"Lake Okeechobee"	1994	20	17	6	1–23
Acanthocephala					
Corynosoma peposacae (SI)					
"Lake Okeechobee"	1994	20	2	4	—
Corynosoma sp. (SI)					
Glades	1970–71	74	2	1	1–2

Sources: 1994 data are from Fedynich et al. (1996); 1970–71 data are from Kinsella and Forrester (1972, 1997).

a. See Table 10.32, note a, for details on samples.

b. Infection sites: BL = bill, CE = cecum, CL = cloaca, ES = esophagus, GI = gizzard, LI = large intestine, LU = lung, PR = proventriculus, SI = small intestine.

c. Ducks were collected in an area to the northwest of Lake Okeechobee where Glades, Highlands, and Okeechobee counties come together.

d. Kinsella (1997) examined specimens collected in 1971, referred to as *Capillaria* sp. by Kinsella and Forrester (1972), and in 1994, referred to as *C. anatis* by Fedynich et al. (1996), and determined that all were *C. spinulosa*.

e. Described by Schmidt and Kinsella (1972) from material collected by Kinsella and Forrester (1972).

f. Two species of *Tetrameres* are involved. One was referred to as "undescribed" by Kinsella and Forrester (1972); Bergan et al. (1994) later examined some of these specimens and identified them as *T. striata*. The other species was closest to *T. ryjikovi*, according to Kinsella and Forrester (1972).

Table 10.35. Infestations of parasitic arthropods on ducks (other than Wood Ducks) from Florida

Species of duck	Species of arthropod	County or site	Year
Dabbling-ducks			
American Black Duck			
Chewing lice	*Trinoton querquedulae*	Dade	1919
Mallard			
Chewing lice	*Anaticola crassicornis*	Dade	1919
Fleas	*Echidnophaga gallinacea*	Brevard	NG[a]
Mottled Duck			
Nasal mites	*Rhinonyssus rhinolethrum*	"Lake Okeechobee"	1994[b]
Chewing lice	*Anatoecus icterodes*	Glades	1971
	Trinoton querquedulae	Glades	1970
Northern Pintail			
Chewing lice	*Trinoton querquedulae*	Dade	1928
Northern Shoveler			
Chewing lice	*Anaticola crassicornis*	Hamilton	1995
	Anatoecus dentatus	Hamilton	1995
	Trinoton querquedulae	Hamilton	1995
Diving ducks			
Ring-necked Duck			
Chewing lice	*Anaticola crassicornis*	"Florida"	1983
	Anatoecus sp.	"Florida"	1983[c]
	Holomenopon clypeilargum	"Florida"	1983[c]
	Holomenopon leucoxanthum	Leon	1936
		"Florida"	1983
	Trinoton querquedulae	"Florida"	1983
Lesser Scaup			
Chewing lice	*Trinoton querquedulae*	Indian River	1973
Black Scoter			
Chewing lice	*Anaticola crassicornis*	Lee	1982
	Anatoecus icterodes	Lee	1982
	Holomenopon leucoxanthum	Lee	1982
Surf Scoter			
Chewing lice	*Anatoecus* sp.	Dade	1919
Bufflehead			
Chewing lice	*Anaticola crassicornis*	Dade	1919
Mergansers			
Red-breasted Merganser			
Chewing lice	*Anaticola crassicornis*	Dade	1919
		Lee	1982, 1986
	Anatoecus icterodes	Lee	1986
	Anatoecus sp.	Indian River	1973
	Trinoton querquedulae	Lee	1982, 1986
		Monroe	1953, 1974
		"Florida"	NG[d]
Hooded Merganser			
Feather mites	*Freyana anatina lophodytes*	"Florida"	1890[e]
Stiff-tailed ducks			
Ruddy Duck			
Chewing lice	*Anaticola crassicornis*	Lee	1982
	Holomenopon leucoxanthum	Lee	1982

Source: Forrester et al. (1995) except as noted.
a. From Layne (1971). NG = not given by author.
b. From Pence (1997). Ducks were collected in an area to the northwest of Lake Okeechobee where Glades, Highlands, and Okeechobee counties come together. Two of 20 ducks were infested; one had 1 mite, the other had 2.
c. From Thul and Price (1995).
d. From Peters (1936). No specific locality was given.
e. From Dubinin (1950). No specific locality was given.

Table 10.36. Parasitic arthropod infestations on Wood Ducks from Florida[a]

Arthropod County or site	Date	Adults			Immatures		
		No. exam.	No. inf.	Number of arthropods collected	No. exam.	No. inf.	Number of arthropods collected
Feather mites							
Bdellorhynchus sp.[b]							
Leon	1977	12	2	8	5	1	16
Freyana largifolia							
Gadsden	1976	6	1	5	3	0	—
Hernando	1976	2	1	1	4	0	—
Leon	1977	12	7	30	5	3	41
Ingrassia sp.[b]							
Alachua	1976	1	0	—	2	1	3
Gadsden	1976	6	0	—	3	1	1
Hamilton	1976	3	2	7	1	0	—
Hernando	1976	2	2	2	4	0	—
Leon	1976	11	0	—	3	1	41
	1977	12	6	42	5	3	9
Levy	1976	4	2	12	0	—	—
Nasal mites							
Rhinonyssus rhinolethrum							
Leon	1977	12	1	2	5	1	1
Chewing lice							
Anaticola crassicornis							
Alachua	1976	1	0	—	2	2	2
Hernando	1976	2	1	1	4	0	—
Leon	1977	12	3	4	5	3	4
Levy	1976	4	2	2	0	—	—
Anatoecus dentatus[c]							
Gadsden	1976	6	1	3	3	0	—
Glades	1976	1	1	1	0	—	—
Hamilton	1976	3	2	2	1	0	—
Hernando	1976	2	1	1	4	0	—
Leon	1977	12	1	1	5	2	4
Holomenopon clauseni[d]							
Alachua	1976	1	0	—	2	1	1
Gadsden	1976	6	1	6	3	1	1
Hamilton	1976	2	2	5	1	0	—
Hernando	1976	2	1	1	4	3	7
Leon	1976	11	1	1	3	1	1
	1977	12	3	4	5	3	4
Levy	1976	4	1	5	0	—	—
Trinoton querquedulae[e]							
Alachua	1976	1	0	—	2	1	1
Gadsden	1976	6	1	3	3	0	—
Hernando	1976	2	0	—	4	2	2
Leon	1976	11	0	—	3	1	1
	1977	12	0	—	5	1	1

(continued)

Table 10.36. *(continued)*

Arthropod County or site	Date	Adults			Immatures		
		No. exam.	No. inf.	Number of arthropods collected	No. exam.	No. inf.	Number of arthropods collected
Cimicid bugs							
Ornithicoris pallidus[f]							
Lake	1990	NG[g]	NG	NG	NG	NG	NG
Leon	1995–96	NG	NG	NG	NG	NG	NG

Sources: Thul (1985a,b), unless otherwise noted.

a. Forty ducks were examined: 1 from Alachua County (1976), 6 from Gadsden County (1976), 1 from Glades County (1976), 3 from Hamilton County (1976), 2 from Hernando County (1976), 23 from Leon County (1976–77), and 4 from Levy County (1976). For each parasite, data are given for counties in which positive birds were found; negatives are not included in the table.

b. Undescribed species.

c. Also reported from Dade County in 1927 (Forrester et al. 1995).

d. Also reported from Dade County in 1927 (Price 1971; Forrester et al. 1995).

e. Also reported from Jefferson County in 1926 (Forrester et al. 1995).

f. Data from Eggeman et al. (1996). These bugs were found in Wood Duck nesting boxes and not on Wood Ducks.

g. NG = not given by author.

XXIII. Literature cited

Allen, J.P., and S.S. Wilson. 1977. A bibliography of references to avian botulism. U.S. Fish and Wildlife Service Special Scientific Report—Wildlife 204. 6 pp.

American Ornithologists' Union. 1998. *Check-list of North American birds.* 7th ed. American Ornithologists' Union, Washington, D.C. 829 pp.

Baker, J.L., and R.L. Thompson. 1978. Shot ingestion by waterfowl on national wildlife refuges in Florida. *Proc. Annu. Conf. Southeast. Assoc. Fish Wildl. Agencies* 32:256–262.

Bancroft, G.T., and W. Hoffman. 1985. Vagrancy and associated mortality of black scoters in peninsular Florida. *Fla. Field Nat.* 13:14–17.

Bellrose, F.C. 1959. Lead poisoning as a mortality factor in waterfowl populations. *Ill. Nat. Hist. Surv. Bull.* 27:235–288.

———. 1976. *Ducks, geese and swans of North America.* Stackpole Books, Harrisburg, Pa. 543 pp.

Bellrose, F.C., and D.J. Holm. 1994. *Ecology and management of the wood duck.* Stackpole Books, Mechanicsburg, Pa. 588 pp.

Bennett, G.F., M.A. Peirce, and R.W. Ashford. 1993. Avian Haematozoa: mortality and pathogenicity. *J. Nat. Hist.* 27:993–1001.

Bennett, G.F., and D. Squires-Parsons. 1992. The leucocytozoids of the avian families Anatidae and Emberizidae s.l., with descriptions of three new *Leucocytozoon* species. *Can. J. Zool.* 70:2007–2014.

Bennett, G.F., M. Whiteway, and C. Woodworth-Lynas. 1982. *A host-parasite catalogue of the avian Haematozoa.* Meml. Univ. Nfld. Occas. Pap. Biol. 5. 243 pp.

Bent, A.C. 1937. *Life histories of North American birds of prey.* Part 1. U.S. Natl. Mus. Bull. 167. 409 pp.

Bergan, J.F., A.A. Radomski, D.B. Pence, and O.E. Rhodes, Jr. 1994. *Tetrameres (Petrowimeres) striata* in ducks. *J. Wildl. Dis.* 30:351–358.

Beyer, W.N., G.H. Heinz, and A.W. Redmon-Norwood (eds.). 1996. *Environmental contaminants in wildlife: interpreting tissue concentrations.* CRC, Boca Raton, Fla. 494 pp.

Bishop, M.A., and G.F. Bennett. 1992. *Host-parasite catalogue of the avian Haematozoa,* Suppl. 1. Meml. Univ. Nfld. Occas. Pap. Biol. 15. 210 pp.

Blus, L.J., S.N. Wiemeyer, J.A. Kerwin, R.C. Stendell, H.M. Ohlendorf, and L.F. Stickel. 1977. Impact of estuarine pollution on birds. In: *Estuarine pollution control and assessment, proceedings of a conference.* U.S. Environmental Protection Agency, Washington, D.C. pp. 57–71.

Boehringer, E.G., O.E. Fornari, and I.K. Boehringer. 1962. The first case of *Toxoplasma gondii* in domestic ducks in Argentina. *Avian Dis.* 6:391–396.

Bossart, G.D., D.G. Baden, R.Y. Ewing, B. Roberts, and S.C. Wright. 1998. Brevetoxicosis in manatees (*Trichechus manatus latirostris*) from the 1996 epizootic: gross, histologic, and immunohistological features. *Toxicol. Pathol.* 26:276–282.

Burns, B.J. 1952. Food of a family of great horned owls, *Bubo virginianus,* in Florida. *Auk* 69:86–87.

Burridge, M.J., W.J. Bigler, D.J. Forrester, and J.M. Hennemann. 1979. Serologic survey for *Toxoplasma gondii* in wild animals in Florida. *J. Am. Vet. Med. Assoc.* 175:964–967.

Cain, B.W. 1981. Nationwide residues of organochlorine compounds in wings of adult mallards and black ducks, 1979 80. *Pestic. Monit. J.* 15:128–134.

Cawthorn, R.J., D. Rainnie, and G. Wobeser. 1981. Experimental transmission of *Sarcocystis* sp. (Protozoa: Sarcocystidae) between the shoveler (*Anas clypeata*) duck and the striped skunk (*Mephitis mephitis*). *J. Wildl. Dis.* 17:389–394.

Clapp, R.B., D. Morgan-Jacobs, and R.C. Banks. 1982. *Marine birds of the southeastern United States and Gulf of Mexico.* Part II, *Anseriformes.* U.S. Fish and Wildlife Service, Office of Biological Services, Washington, D.C. 492 pp.

Cole, R.A. 1995. Unpublished data. Natl. Wildlife Health Research Center, Madison, Wis.

Conti, L., R. Oliveri, and C. Blackmore. 2002. Unpublished data. Florida Department of Health, Tallahassee.

Cornwell, G. 1963. New waterfowl host records for *Sarcocystis rileyi* and a review of sarcosporidiosis in birds. *Avian Dis.* 7:212–216.

Crawford, R.L. 1981. Bird casualties at a Leon County, Florida TV tower: a 25-year migration study. *Bull. Tall Timbers Res. Stn.* 22:1–30.

Cunningham, M.W., K. Ziober, N. Gottdenker, L. Flewelling, and J. Landsberg. 2002. Unpublished data. Florida Fish and Wildlife Conservation Commission, Gainesville.

Cuvin-Aralar, M.L.A., and R.W. Furness. 1991. Mercury and selenium interaction: a review. *Ecotoxicol. Environ. Saf.* 21:348–364.

Delany, M.F. 1986. Bird bands recovered from American alligator stomachs in Florida. *N. Am. Bird Bander* 11:92–94.

Drouin, T.E., and J.L. Mahrt. 1979. The prevalence of *Sarcocystis* Lankester, 1882, in some bird species in western Canada, with notes on its life cycle. *Can. J. Zool.* 57:1915–1921.

Dubey, J.P., and C.P. Beattie. 1988. *Toxoplasmosis of animals and man.* CRC, Boca Raton, Fla. 220 pp.

Dubey, J.P., C.A. Speer, and R. Fayer. 1990. *Cryptosporidiosis of man and animals.* CRC, Boca Raton, Fla. 199 pp.

Dubinin, V.B. 1950. [Systematic analysis of species of feather mites (Sarcoptiformes, Analgesoides) parasitizing Anserinae] (in Russian). *Parazitol. Sb.* 12:17–72.

Duncan, R.M. 1995. Unpublished data. National Wildlife Health Center, Madison, Wis.

Duszynski, D.W., and E.D. Box. 1978. The opossum (*Didelphis virginiana*) as a host for *Sarcocystis debonei* from cowbirds (*Molothrus ater*) and grackles (*Cassidix mexicanus, Quiscalus quiscula*). *J. Parasitol.* 64:326–329.

Eggeman, D.R., C.M. Ohme, and S. Halbert. 1996. Unpublished data. Florida Game and Fresh Water Fish Commission, Tallahassee.

Eisler, R. 1987. *Mercury hazards to fish, wildlife, and invertebrates: a synoptic review.* U.S. Fish and Wildlife Service, Biological Report 85 (1.10). 90 pp.

———. 1988. *Arsenic hazards to fish, wildlife,*

and invertebrates: a synoptic review. U.S. Fish and Wildlife Service, Biological Report 85 (1.12). 92 pp.

———. 1989. *Molybdenum hazards to fish, wildlife, and invertebrates: a synoptic review.* U.S. Fish and Wildlife Service, Biological Report 85(1.19). 61 pp.

———. 1993. *Zinc hazards to fish, wildlife, and invertebrates: a synoptic review.* U.S. Fish and Wildlife Service, Biological Report 10(26). 106 pp.

———. 1994. *Radiation hazards to fish, wildlife, and invertebrates: a synoptic review.* National Biological Service, Biological Report 26(29). 124 pp.

Eklund, M.W., and V.R. Dowell, Jr. (eds.). 1987. *Avian botulism: an international perspective.* Charles C. Thomas, Springfield, Ill. 405 pp.

Fairbrother, A., L.N. Locke, and G.L. Hoff (eds.). 1996. *Noninfectious diseases of wildlife.* 2d ed. Iowa State University Press, Ames. 219 pp.

Fallis, A.M., and G.F. Bennett. 1966. On the epizootiology of infections caused by *Leucocytozoon simondi* in Algonquin Park, Canada. *Can. J. Zool.* 144:101–112.

Fallis, A.M., and D.M. Wood. 1957. Biting midges (Diptera: Ceratopogonidae) as intermediate hosts for *Haemoproteus* of ducks. *Can. J. Zool.* 35:425–435.

Fargo, W.G. 1926. Notes on birds of Pinellas and Pasco counties, Florida. *Wilson Bull.* 38:140–155.

Fedynich, A.M., D.B. Pence, P.N. Gray, and J.F. Bergan. 1996. Helminth community structure and pattern in two allopatric populations of a nonmigratory waterfowl species (*Anas fulvigula*). *Can. J. Zool.* 74:1253–1259.

Feierabend, J.S., and O. Myers. 1984. A national summary of lead poisoning in bald eagles and waterfowl. Unpublished report. National Wildlife Federation, Washington, D.C. 87 pp.

Fischer, J.R. 1994. Unpublished data. Southeastern Cooperative Wildlife Disease Study, University of Georgia, Athens.

Forrester, D.J. 1970–97. Unpublished data. University of Florida, Gainesville.

Forrester, D.J., G.W. Foster, and J.E. Thul. 2001. Blood parasites of the ring-necked duck (*Aythya collaris*) on its wintering range in Florida, U.S.A. *Comp. Parasitol.* 68:173–176.

Forrester, D.J., J.M. Gaskin, F.H. White, N.P. Thompson, J.A. Quick, G.E. Henderson, J.C. Woodard, and W.D. Robertson. 1977. An epizootic of waterfowl associated with a red tide episode in Florida. *J. Wildl. Dis.* 13:160–167.

Forrester, D.J., H.W. Kale II, R.D. Price, K.C. Emerson, and G.W. Foster. 1995. Chewing lice (Mallophaga) from birds in Florida: a listing by host. *Bull. Fla. Mus. Nat. Hist.* 39:1–44.

Forrester, D.J., K.C. Wenner, F.H. White, E.C. Greiner, W.R. Marion, J.E. Thul, and G.A. Berkhoff. 1980. An epizootic of avian botulism in a phosphate mine settling pond in northern Florida. *J. Wildl. Dis.* 16:323–327.

Frenkel, J.K. 1981. False-negative serologic tests for *Toxoplasma* in birds. *J. Parasitol.* 67:952–953.

Friend, M. 1985. Interpretation of criteria commonly used to determine lead poisoning problem areas. U.S. Department of the Interior, Fish and Wildlife Leaflet no. 2, Washington, D.C. 4 pp.

———. 1999a. Avian cholera. In: *Field manual of wildlife diseases. General field procedures and diseases of birds.* M. Friend and J.C. Franson (eds.). U.S. Department of the Interior, U.S. Geological Survey, Biological Research Division, Information and Technology Report 1999-001. Washington, D.C. pp. 75–92.

———. 1999b. Lead. In: *Field manual of wildlife diseases. General field procedures and diseases of birds.* M. Friend and J.C. Franson (eds.). U.S. Department of the Interior, U.S. Geological Survey, Biological Research Division, Information and Technology Report 1999-001. Washington, D.C. pp. 317–334.

Friend, M., and J.C. Franson (eds.). 1999. *Field manual of wildlife diseases. General field procedures and diseases of birds.* U.S. Department of the Interior, U.S. Geological Survey, Biological Research Division, Information and Technology Report 1999-001. Washington, D.C. 426 pp.

Friend, M., and D.O. Trainer. 1970. Polychlorinated biphenyl: interaction with duck hepatitis virus. *Science* 170:1314–1316.

———. 1974a. Experimental DDT-duck hepatitis virus interaction studies. *J. Wildl. Manag.* 38:887–895.

———. 1974b. Experimental dieldrin-duck hepatitis virus interaction studies. *J. Wildl.* Manag. 38:896–902.

Gainer, J.H., W.G. Winkler, A.L. Lewis, W.L. Jennings, and P.H. Coleman. 1964. Isolations of St. Louis encephalitis virus from domestic pigeons, *Columba livia. Am. J. Trop. Med. Hyg.* 13:472–474.

Garnham, P.C.C. 1966. *Malaria parasites and other hemosporidia.* Blackwell Scientific, Oxford. 1,114 pp.

Golubkov, V.I. 1979. [On the infestation of dogs and cats with *Sarcocystis* from hens and ducks] (in Russian). *Veterinariya (Mosc.)* 1:55–56.

Haagsma, J. 1987. Laboratory investigation of botulism in wild birds. In: *Avian botulism: an international perspective.* M.W. Eklund and V.R. Dowell, Jr. (eds.). Charles C. Thomas, Springfield, Ill. pp. 283–293.

Hall, R.J., and N.C. Coon. 1988. Interpreting residues of petroleum hydrocarbons in wildlife tissues. U.S. Fish and Wildlife Service, Biological Report 88. 8 pp.

Harrison, G. 1974. *Clostridium botulinum* type C infection on a game fowl farm. *Proc. Annu. Meet. Am. Assoc. Zoo Vet.* pp. 221–224.

Heath, R.G. 1969. Nationwide residues of organochlorine pesticides in wings of mallard and black ducks. *Pestic. Monit. J.* 3:115–123.

Heath, R.G., and S.A. Hill. 1974. Nationwide organochlorine and mercury residues in wings of adult mallards and black ducks during the 1969–70 hunting season. *Pestic. Monit. J.* 7:153–164.

Heath, R.G., and R.M. Prouty. 1967. Trial monitoring of pesticides in wings of mallards and black ducks. *Bull. Environ. Contam. Toxicol.* 2:101–110.

Heinz, G.H. 1974. Effects of low dietary levels of methylmercury on mallard reproduction. *Bull. Environ. Contam. Toxicol.* 11:386–392.

———. 1996a. Selenium in birds. In: *Environmental contaminants in wildlife: interpreting tissue concentrations.* W.N. Beyer, G.H. Heinz, and A.W. Redmon-Norwood (eds.). CRC Press, Boca Raton, Fla. pp. 447–458.

———. 1996b. Mercury poisoning in wildlife. In: *Noninfectious diseases of wildlife.* 2d ed. A. Fairbrother, L.N. Locke, and G.L. Hoff (eds.). Iowa State University Press, Ames. pp. 118–127.

Heinz, G.H., and D.J. Hoffman. 1996. Unpublished data. Patuxent Environmental Science Center, National Biological Service, Laurel, Md.

Hutton, R.F. 1964. A second list of parasites from marine and coastal animals of Florida. *Trans. Am. Microsc. Soc.* 83:439–447.

Jacknow, J., J.L. Ludke, and N.C. Coon. 1986. Monitoring fish and wildlife for environmental contaminants: the National Contaminant Biomonitoring Program. U.S. Fish and Wildlife Service, Fish and Wildlife Leaflet 4. 15 pp.

Jasmin, A.M., D.E. Cooperrider, C.P. Powell, and J.N. Baucom. 1972. Enterotoxemia of wildfowl due to *Cl. perfringens* type C. *J. Wildl. Dis.* 8:79–84.

Jennings, W.L., W.G. Winkler, D.D. Stamm, P.H. Coleman, and A.L. Lewis. 1969. Serologic studies of possible avian or mammalian reservoirs of St. Louis encephalitis virus in Florida. In: *St. Louis encephalitis virus in Florida.* Fla. State Board Health Monogr. Ser. 12. Jacksonville. pp. 118–125.

Jessup, D.A. 1986. Botulism. In: *Zoo and wild animal medicine.* 2d ed. M.E. Fowler (ed.). W.B. Saunders, Philadelphia. pp. 349–352.

Jessup, D.A., and F.A. Leighton. 1996. Oil pollution and petroleum toxicity in wildlife. In: *Noninfectious diseases of wildlife.* 2d ed. A. Fairbrother, L.N. Locke, and G.L. Hoff (eds.). Iowa State University Press, Ames. pp. 141–156.

Johnston, D.W. 1976. Organochlorine pesticide residues in uropygial glands and adipose tissue of wild birds. *Bull. Environ. Contam. Toxicol.* 16:149–155.

Julian, R.J., and D.E. Galt. 1980. Mortality in muscovy ducks (*Cairina moschata*) caused by *Haemoproteus* infection. *J. Wildl. Dis.* 16:39–44.

Julian, R.J., T.J. Beveridge, and D.E. Galt. 1985. Muscovy duck mortality not caused by *Haemoproteus*. *J. Wildl. Dis.* 21:335–337.

Kale, H.W. II. 1971. Regional reports: Florida region. *Am. Birds* 25:723–725, 730–735.

Kemp, G.E. 1981. Saint Louis encephalitis (SLE). In: *CRC handbook series in zoonoses*. Vol. 1. G.B. Beran (ed.) Sect. B, *Viral zoonoses*. CRC Press, Boca Raton, Fla. pp. 71–83.

Kingston, N. 1978. Trematodes. In: *Diseases of poultry*. 7th ed. M.S. Hofstad, B.W. Calnek, C.F. Helmboldt, W.M. Reid, and H.W. Yoder, Jr. (eds.). Iowa State University Press, Ames. pp. 759–782.

Kinsella, J.M. 1997–98. Unpublished data. University of Florida, Gainesville.

Kinsella, J.M., and D.J. Forrester. 1972. Helminths of the Florida duck, *Anas platyrhynchos fulvigula*. *Proc. Helminthol. Soc. Wash.* 39:173–176.

———. 1997. Unpublished data. University of Florida, Gainesville.

Kinsella, J.M., and J.E. Thul. 1997. Unpublished data. University of Florida, Gainesville.

Klukas, R.W., and L.N. Locke. 1970. An outbreak of fowl cholera in Everglades National Park. *J. Wildl. Dis.* 6:77–79.

Lapage, G. 1961. A list of the parasitic protozoa, helminths and Arthropoda recorded from species of the family Anatidae (ducks, geese, swans). *Parasitology* 51:1–109.

Layne, J.N. 1971. Fleas (Siphonaptera) of Florida. *Fla. Entomol.* 54:35–51.

Lehner, P.N., T.O. Boswell, and F. Copeland. 1967. An evaluation of the effects of the *Aedes aegypti* eradication program on wildlife in south Florida. *Pestic. Monit. J.* 1:29–34.

Leibovitz, L. 1973. Necrotic enteritis of breeder ducks. *Am. J. Vet. Res.* 34:1053–1061.

Linda, S.B. 1998. Analysis of 1997–1998 statewide and wildlife management area hunter surveys: methods and results. Unpublished report. Florida Game and Fresh Water Fish Commission, Gainesville. 88 pp.

Locke, L.N. 1968–95. Unpublished data. National Wildlife Health Center, Madison, Wis.

Locke, L.N., J.B. DeWitt, C.M. Menzie, and J.A. Kerwin. 1964. A merganser die-off associated with larval *Eustrongylides*. *Avian Dis.* 8:420–427.

Locke, L.N., and N.J. Thomas. 1996. Lead poisoning of waterfowl and raptors. In: *Noninfectious diseases of wildlife*. 2d ed. A. Fairbrother, L.N. Locke, and G.L. Hoff (eds.). Iowa State University Press, Ames. pp. 108–117.

Loftin, H. 1961. An annotated check-list of trematodes and cestodes and their vertebrate hosts from northwest Florida. *Q. J. Fla. Acad. Sci.* 23:302–314.

Logan, T.H. 1997. Florida's endangered species, threatened species and species of special concern. Official lists. Florida Game and Fresh Water Fish Commission, Tallahassee. 14 pp.

Loomis, E.C. 1978. External parasites. In: *Diseases of poultry*. 7th ed. M.S. Hofstad, B.W. Calnek, C.F. Helmboldt, W.M. Reid, and H.W. Yoder, Jr. (eds.). Iowa State University Press, Ames. pp. 667–704.

MacInnis, A.J. 1959. Some helminth parasites, mostly from marine birds from the northwest Gulf Coast of Florida. M.S. thesis, Florida State University, Tallahassee. 37 pp.

Macko, J.K., and V. Hanzelová. 1997. *Calixolepis thuli* n.g., n.sp.(Cestoda: Hymenolepididae) from the wood duck *Aix sponsa* (Anatidae) in America. *System. Parasitol.* 38:137–145.

Maehr, D.S., and J.Q. Smith. 1988. Bird casualties at a central Florida power plant: 1982–1986. *Fla. Field Nat.* 16:57–80.

Maehr, D.S., A.G. Spratt, and D.K. Voigts. 1983. Bird casualties at a central Florida power plant. *Fla. Field Nat.* 11:45–49.

Marion, W.R., T.E. O'Meara, G.D. Riddle, and H.A. Berkhoff. 1983. Prevalence of *Clostridium botulinum* type C in substrates of phosphate-mine settling ponds and implications for epizootics of avian botulism. *J. Wildl. Dis.* 19:302–307.

Martin, E.M. 1996. Distribution in states and counties of waterfowl species harvested during the 1981 to 1990 hunting seasons. Unpublished data. Office of Migratory Bird Management, U.S. Fish and Wildlife Service, Laurel, Md.

Mautino, M., and J.U. Bell. 1986. Experimental lead toxicity in the ring-necked duck. *Environ. Res.* 41:538–545.

McDonald, M.E. 1969a. *Catalogue of helminths of waterfowl (Anatidae).* U.S. Department of the Interior, Bureau of Sport Fisheries and Wildlife, Special Scientific Research Publication—Wildlife 126. Washington, D.C. 692 pp.

———. 1969b. *Annotated bibliography of helminths of waterfowl (Anatidae).* U.S. Department of the Interior, Bureau of Sport Fisheries and Wildlife, Special Scientific Research Publication—Wildlife 125. Washington, D.C.

———. 1974. *Key to nematodes reported in waterfowl.* U.S. Department of the Interior, Bureau of Sport Fisheries and Wildlife, Research Publication 122. Washington, D.C. 44 pp.

———. 1981. *Key to trematodes reported in waterfowl.* U.S. Department of the Interior, Fish and Wildlife Service, Research Publication 142. Washington, D.C. 156 pp.

———. 1988. *Key to Acanthocephala reported in waterfowl.* U.S. Department of the Interior, Fish and Wildlife Service, Research Publication 173. Washington, D.C. 45 pp.

McEwan, L.C., and D.H. Hirth. 1980. Food habits of the bald eagle in north-central Florida. *Condor* 82:229–231.

McLean, R.G., and G.S. Bowen. 1980. Vertebrate hosts. In: *St. Louis Encephalitis.* T.P. Monath (ed.). American Public Health Association, Washington, D.C. pp. 381–450.

Mertz, W. 1981. The essential trace elements. *Science* 213:1332–1338.

Mitchell, R.W. 1963. Comparative morphology of the life stages of the nasal mite *Rhinonyssus rhinolethrum* (Mesostigmata: Rhinonyssidae). *J. Parasitol.* 49:506–515.

Mitchell, R.W., and W.L. Rhodes. 1960. New host records for the mesostigmatid mite *Rhinonyssus rhinolethrum* (Acarina: Rhinonyssidae). *Southwest. Nat.* 5:107–108.

Mollhagen, T.R. 1976. A study of the systematics and hosts of the parasitic nematode genus *Tetrameres* (Habronematoidea: Tetrameridae).

Ph.D. diss., Texas Tech University, Lubbock 546 pp.

Montalbano, F. III, J.E. Thul, and W.E. Bolch. 1983. Radium-226 and trace elements in mottled ducks. *J. Wildl. Manag.* 47:327–333.

Morehouse, K.A. 1992. Lead poisoning of migratory birds: The U.S. Fish and Wildlife Service position. In: *Lead poisoning in waterfowl.* D.J. Pain (ed.). Proc. IWRB Workshop, Brussels, Belgium, 13–15 June 1991. IWRB Spec. Publ. 16. International Waterfowl and Wetlands Research Bureau, Slimbridge, Gloucester, United Kingdom. pp. 51–55.

Myers, O.B., W.R. Marion, T.E. O'Meara, and C.E. Roessler. 1989. Radium-226 in wetland birds from Florida phosphate mines. *J. Wildl. Manag.* 53:1110–1116.

Nation, P.N., and G.A. Wobeser. 1977. Renal coccidiosis in wild ducks in Saskatchewan. *J. Wildl. Dis.* 13:370–375.

Nayar, J.K. 1982. *Bionomics and physiology of Culex nigripalpus (Diptera: Culicidae) of Florida: an important vector of diseases.* Florida Agricultural Experiment Stations Bulletin 827, Gainesville. 73 pp.

O'Donoghue, P.J., V.L. Tham, W.G. deSaram, K.L. Paull, and S. McDermott. 1987. *Cryptosporidium* infections in birds and mammals and attempted cross-transmission studies. *Vet. Parasitol.* 26:1–11.

Ohlendorf, H.M. 1996. Selenium. In: *Noninfectious diseases of wildlife.* 2d ed. A. Fairbrother, L.N. Locke, and G.L. Hoff (eds.). Iowa State University Press, Ames. pp. 128–140.

O'Meara, T.E., W.R. Marion, C.E. Roessler, G.S. Roessler, H.A. Van Rinsvelt, and O.B. Myers. 1986. *Environmental contaminants in birds: phosphate-mine and natural wetlands.* Florida Institute of Phosphate Research Publication 05-003-045. Bartow. 77 pp.

O'Shea, T.J., G.B. Rathbun, R.K. Bonde, C.D. Buergelt, and D.K. Odell. 1991. An epizootic of Florida manatees associated with a dinoflagellate bloom. *Mar. Mamm. Sci.* 7:118–146.

Pain, D.J. 1996. Lead in waterfowl. In: *Environmental contaminants in wildlife: interpreting tissue concentrations.* W.N. Beyer, G.H. Heinz, and A.W. Redmon-Norwood (eds.). CRC Press, Boca Raton, Fla. pp. 251–264.

Pellerdy, L.P. 1974. *Coccidia and coccidiosis.* 2d ed. Verlag Paul Parey, Berlin. 959 pp.

Pence, D.B. 1972. The nasal mites of birds from Louisiana. I. Dermanyssids (Rhinonyssinae) from shore and marsh birds. *J. Parasitol.* 58:153–168.

———. 1997. Unpublished data. Texas Tech University, Lubbock.

Peters, H.S. 1936. A list of external parasites from birds of the eastern part of the United States. *Bird-Banding* 7:9–27.

Premvati, G. 1969. Redescription of *Psilochasmus oxurus* (Creplin, 1825) Lühe, 1909 (Trematoda: Psilostomatidae), with proposal to synonymize three other species. *Parasitology* 59:493–496.

Price, R.D. 1971. A review of the genus *Holomenopon* (Mallophaga: Menoponidae) from the Anseriformes. *Ann. Entomol. Soc. Am.* 64:633–646.

Prouty, R.M., and C.M. Bunck. 1986. Organochlorine residues in adult mallard and black duck wings, 1981–1982. *Environ. Monit. Assess.* 6:49–57.

Quick, J.A., Jr., and G.E. Henderson. 1975. Effects of *Gymnodinium breve* red tide on fishes and birds: a preliminary report on behavior, anatomy, hematology, and histopathology. *Proc. Gulf. Coast Reg. Symp. Dis. Aquat. Anim.* pp. 85–115.

Quist, C.F. 1992. Unpublished data. Southeastern Cooperative Wildlife Disease Study, University of Georgia, Athens.

———. 1997–98. Unpublished data. Southeastern Cooperative Wildlife Disease Study, University of Georgia, Athens.

Regan, T.W. 1996–99. Unpublished data. Florida Game and Fresh Water Fish Commission, West Palm Beach.

Ridlehuber, K.T. 1982. Fire ant predation on wood duck ducklings and pipped eggs. *Southwest. Nat.* 27:222.

Robertson, W.B., Jr., and C.R. Mason. 1965. Additional bird records from the Dry Tortugas. *Fla. Nat.* 38:131–138.

Robertson, W.B., Jr., and G.E. Woolfenden. 1992. *Florida bird species. An annotated list.* Spec. Publ. 6, Florida Ornithological Society, Gainesville. 260 pp.

Rocke, T.E. 1993. *Clostridium botulinum.* In: *Pathogenesis of bacterial infections in animals.* 2d ed. C. L. Gyles and C.O. Thoen (eds.). Iowa State University Press, Ames. pp. 86–96.

———. 1999. Oil. In: *Field manual of wildlife diseases. General field procedures and diseases of birds.* M. Friend and J.C. Franson (eds.). U.S. Department of the Interior, U.S. Geological Survey, Biological Research Division, Information and Technology Report 1999-001. Washington, D.C. pp. 309–315.

Rocke, T.E., and M. Friend. 1999. Avian botulism. In: *Field manual of wildlife diseases. General field procedures and diseases of birds.* M. Friend and J.C. Franson (eds.). U.S. Department of the Interior, U.S. Geological Survey, Biological Research Division, Information and Technology Report 1999-001. Washington, D.C. pp. 271–281.

Rocke, T.E., T.M. Yuill, and R.D. Hinsdill. 1984. Oil and related toxicant effects on mallard immune defenses. *Environ. Res.* 33:343–352.

Rosen, M.N. 1971. Botulism. In: *Infectious and parasitic diseases of wild birds.* J.W. Davis, R.C. Anderson, L. Karstad, and D.O. Trainer (eds.). Iowa State University Press, Ames. pp. 100–117.

Schmidt, G.D., and J.M. Kinsella. 1972. Two new species of *Sciadiocara* Skrjabin, 1915 (Nematoda: Schistorophidae) from birds in Florida. *J. Parasitol.* 58:271–274.

Schreiber, R.W., F.M. Dunstan, and J.J. Dinsmore. 1975. Lesser scaup mortality in Tampa Bay, Florida, 1974. *Fla. Field Nat.* 3:13–15.

Sileo, L. 1995. Unpublished data. National Wildlife Health Center, Madison, Wis.

Simons, M.M., Jr. 1985. Beached bird survey project on the Atlantic and Gulf coasts. *Am. Birds* 39:358–362.

Smith, G.J. 1993. *Toxicology and pesticide use in relation to wildlife: organophosphorus and carbamate compounds.* C.K. Smoley, Boca Raton, Fla. 171 pp.

Smith, G.R. 1982. Botulism in waterfowl. In: *Animal diseases in relation to animal concentration*. Symp. Zool. Soc. London, no. 50. Academic, New York. pp. 97–119.

Smith, K.E. 1994. Unpublished data. Southeastern Cooperative Wildlife Disease Study, University of Georgia, Athens.

Smith, L.D. 1977. *Botulism: the organism, its toxins, the disease*. Charles C. Thomas, Springfield, Ill. 236 pp.

Soulsby, E.J.L. 1955. Deaths in swans associated with trematode infection. *Br. Vet. J.* 111:498–500.

Stabin, M.G. 1983. Radium-226 in waterfowl associated with Florida phosphate clay settling areas. M.S. thesis, University of Florida, Gainesville. 97 pp.

Stendell, R.C., R.I. Smith, K.P. Burnham, and R.E. Christensen. 1979. *Exposure of waterfowl to lead: A nationwide survey of residues in wing bones of seven species, 1972–73*. U.S. Fish and Wildlife Service, Special Scientific Report—Wildlife 223. pp. 1–12.

Stevenson, H.M. 1970. The winter season: Florida region. *Audubon Field Notes* 24:493–497.

Stevenson, J.A., and B. Snyder. 1994. Unpublished data. Florida Department of Environmental Protection, Tallahassee.

Stieglitz, W.O. 1967. Food habits of waterfowl on the Merritt Island National Wildlife Refuge. Unpublished report. U.S. Fish and Wildlife Service, Atlanta, Ga. 28 pp.

Stiles, C.W., and A. Hassall. 1894. A preliminary catalogue of the parasites contained in the collections of the U.S. Bureau of Animal Industry, U.S. Army Medical Museum, Biological Department of the University of Pennsylvania (Coll. Leidy) and in Coll. Stiles and Coll. Hassall. *Vet. Mag.* 1:245–253, 331–354.

Stimson, L.A. 1953. An albino Florida duck. *Everglades Nat. Hist.* 1:128.

Stoddard, H.L. 1931. *The bobwhite quail, its habits, preservation and increase*. Charles Scribner's Sons, New York. 559 pp.

Stowasser, W.F. 1979. *Phosphate-mineral commodity profiles*. U.S. Department of the Interior, Bureau of Mines, Washington, D.C. 19 pp.

Stroud, R.K. 1981. Unpublished data. National Wildlife Health Center, Madison, Wis.

Swales, W.E. 1936. *Tetrameres crami* Swales, 1933, a nematode parasite of ducks in Canada. Morphological and biological studies. *Can. J. Res. Sect. D Zool. Sci.* 14:151–164.

Taylor, W.K., and B.H. Anderson. 1973. Nocturnal migrants killed at a central Florida TV tower; autumns 1969–1971. *Wilson Bull.* 85:42–51.

———. 1974. Nocturnal migrants killed at a central Florida TV tower, Autumn 1972. *Fla. Field Nat.* 2:40–43.

Taylor, W.K., and M.A. Kershner. 1986. Migrant birds killed at the Vehicle Assembly Building (VAB), John F. Kennedy Space Center. *J. Field Ornithol.* 57:142–154.

Terrell, S.P., and D.J. Forrester. 1998. Unpublished data. University of Florida, Gainesville.

Thompson, D.R. 1996. Mercury in birds and terrestrial mammals. In: *Environmental contaminants in wildlife: interpreting tissue concentrations*. W.N. Beyer, G.H. Heinz, and A.W. Redmon-Norwood (eds.). CRC, Boca Raton, Fla. pp. 341–356.

Thompson, N.P., and D.J. Forrester. 1974. Unpublished data. University of Florida, Gainesville.

Thul, J.E. 1979a. A morphological and parasitological study of wood ducks in the Atlantic flyway. M.S. thesis, University of Florida, Gainesville. 279 pp.

———. 1979b. Unpublished data. Florida Game and Fresh Water Fish Commission, Gainesville.

———. 1985a. Parasitic arthropods of wood ducks, *Aix sponsa* L., in the Atlantic flyway. *J. Wildl. Dis.* 21:316–318.

———. 1985b. Unpublished data. Florida Game and Fresh Water Fish Commission, Gainesville.

———. 1986a. Waterfowl gizzard survey. Waterfowl Research, Final Performance Report (unpublished). Florida Game and Fresh Water Fish Commission, Gainesville. 38 pp.

———. 1986b. Ring-necked duck lead ingestion study. Wildlife Research, Final Performance Report (unpublished). Florida Game and Fresh Water Fish Commission, Gainesville. 42 pp.

———. 1986c. Wood Duck subpopulation study. Waterfowl Research, Final Performance Report (unpublished). Florida Game and Fresh Water Fish Commission, Gainesville. 51 pp.

———. 1990. Proportion of northern wood ducks in southern harvests: application of a biological tag. *Proc. 1988 N. Am. Wood Duck Symp.*, St. Louis. pp. 335–339.

Thul, J.E., D.J. Forrester, and C.L. Abercrombie. 1985. Ecology of parasitic helminths of wood ducks, *Aix sponsa,* in the Atlantic flyway. *Proc. Helminthol. Soc. Wash.* 52:297–310.

Thul, J.E., D.J. Forrester, and E.C. Greiner. 1980. Hematozoa of wood ducks (*Aix sponsa*) in the Atlantic flyway. *J. Wildl. Dis.* 16:383–390.

Thul, J.E., and T. O'Brien. 1990. Wood duck hematozoan parasites as biological tags: development of a population assessment model. *Proc. 1988 N. Am. Wood Duck Symp.*, St. Louis. pp. 323–334.

Thul, J.E., and R.D. Price. 1995. Unpublished data. Florida Game and Fresh Water Fish Commission, Gainesville.

Tiemeier, O.W. 1941. Repaired bone injuries in birds. *Auk* 58:350–359.

Todd, K.S., Jr., and D.M. Hammond. 1971. Coccidia of Anseriformes, Galliformes, and Passeriformes. In: *Infectious and parasitic diseases of wild birds.* J.W. Davis, R.C. Anderson, L. Karstad, and D.O. Trainer (eds.). Iowa State University Press, Ames. pp. 234–281.

Trainer, D.O., G.S. Schildt, R.A. Hunt, and L.R. John. 1962. Prevalence of *Leucocytozoon simondi* among some Wisconsin waterfowl. *J. Wildl. Manag.* 26:137–143.

Tuggle, B.N., and M. Friend 1999a. *Sarcocystis.* In: *Field manual of wildlife diseases. General field procedures and diseases of birds.* M. Friend and J.C. Franson (eds.). U.S. Department of the Interior, U.S. Geological Survey, Biological Research Division, Information and Technology Report 1999-001. Washington, D.C. pp. 219–222.

———. 1999b. Gizzard worms. In: *Field manual of wildlife diseases. General field procedures and diseases of birds.* M. Friend and J.C. Franson (eds.). U.S. Department of the Interior, U.S. Geological Survey, Biological Research Division, Information and Technology Report 1999-001. Washington, D.C. pp. 235–239.

Turnbull, R.E., F.A. Johnson, M.A. Hernandez, W.B. Wheeler, and J.P. Toth. 1989. Pesticide residues in fulvous whistling-ducks from south Florida. *J. Wildl. Manag.* 53:1052–1057.

Usinger, R.I. 1966. Monograph of Cimicidae (Hemiptera-Heteroptera). *Thomas Say Found.* 7:1–585.

Ware, F.J. 1994. Unpublished data. Florida Game and Fresh Water Fish Commission, Tallahassee.

Ware, F.J., H. Royals, and T. Lange. 1990. Mercury contamination in Florida largemouth bass. *Proc. Annu. Conf. Southeast. Assoc. Fish Wildl. Agencies* 44:5–12.

Wehr, E.E. 1971. Nematodes. In: *Infectious and parasitic diseases of wild birds.* J.W. Davis, R.C. Anderson, L. Karstad, and D.O. Trainer (eds.). Iowa State University Press, Ames. pp. 185–233.

Welte, S., and L. Frink. 1991. Rescue and rehabilitation of oiled birds. U.S. Fish and Wildlife Service, Fish and Wildlife Leaflet 13.2.8. 7 pp.

Wenner, K.C., and W.R. Marion. 1981. Wood duck production on a northern Florida phosphate mine. *J. Wildl. Manag.* 45:1037–1042.

Weston, F.M. 1966. Bird casualties on the Pensacola Bay bridge (1938–1949). *Fla. Nat.* 39:53–55.

White, D.H. 1979. Nationwide residues of organocholorine compounds in wings of adult mallards and black ducks, 1976–77. *Pestic. Monit. J.* 13:12–16.

White, D.H., and R.G. Heath. 1976. Nationwide residues of organochlorines in wings of adult mallards and black ducks, 1972–73. *Pestic. Monit. J.* 9:176–185.

White, F.H., and D.J. Forrester, 1972–80. Unpublished data. University of Florida, Gainesville.

Wicht, R.J. 1981. Transmission of *Sarcocystis rileyi* to the striped skunk (*Mephitis mephitis*). *J. Wildl. Dis.* 17:397–388.

Williams, N.A., and G.F. Bennett. 1980. Avian Haemoproteidae. 13. The haemoproteids of the ducks and geese (Anatidae). *Can. J. Zool.* 58:88–93.

Wilson, S.S., and L.N. Locke. 1982. *Bibliography of references to avian botulism: update.* U.S. Fish and Wildlife Service Special Scientific Report—Wildlife 243. 10 pp.

Wobeser, G.A. 1974. Renal coccidiosis in mallard and pintail ducks. *J. Wildl. Dis.* 10:249–255.

———. 1997. *Diseases of wild waterfowl.* 2d ed. Plenum, New York. 324 pp.

Wright, S.D. 1988. Unpublished data. University of Florida, Gainesville.

Eagles

I. Introduction

Two species of eagles occur in Florida, the Golden Eagle, *Aquila chrysaetos,* and the Bald Eagle, *Haliaeetus leucocephalus.* Golden Eagles are rare but regular visitors during the winter (October to April) in the panhandle and occur less frequently in the northern and central parts of peninsular Florida; there are about 55 reports. They do not breed in Florida. Bald eagles, however, are fairly common breeders in peninsular Florida and rare, but regular, breeders in the panhandle to the west of St. Marks National Wildlife Refuge (Robertson and Woolfenden 1992). During the summer months juveniles and immatures migrate northward as far as eastern Canada (Broley 1947); adults in southern Florida are permanent residents, but some adults from populations in north central Florida also migrate northward (Robertson and Woolfenden 1992). It is also reported that Bald Eagles from northern states migrate into northern Florida during the winter months (Postupalsky 1976).

Bald Eagles in Florida are classified currently as threatened by the U.S. Fish and Wildlife Service and the Florida Game and Fresh Water Fish Commission (Logan 1997) and the Florida Committee on Rare and Endangered Plants and Animals (Curnutt 1996). Florida has more nesting Bald Eagles than any other southeastern state (Nesbitt 1997). Between 1973 and 1995, 9,989 young were produced in Florida and in the 1996–97 nesting

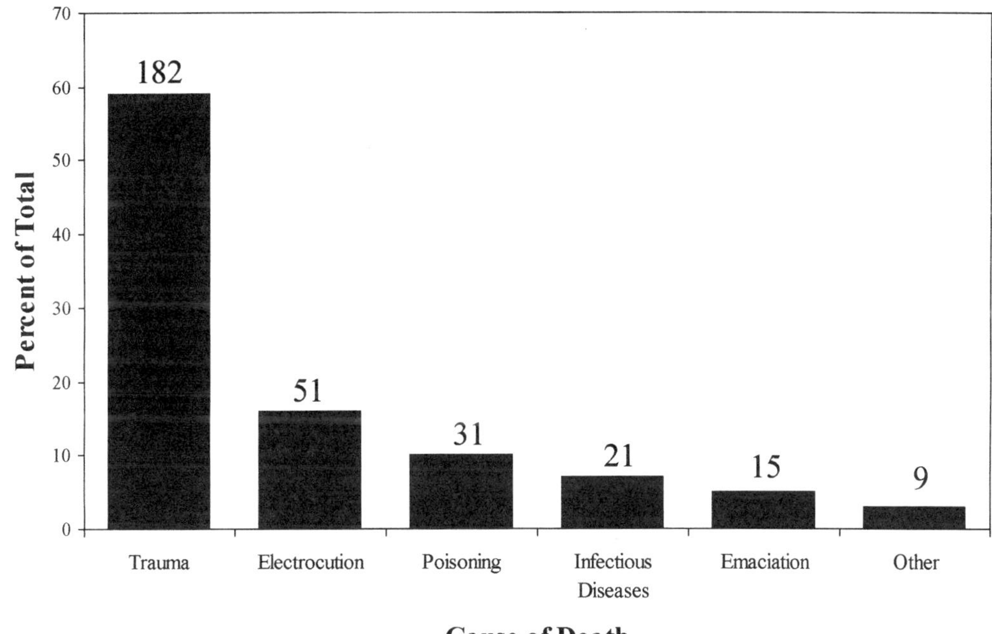

FIGURE 11.1. Causes of death for 309 Bald Eagles in Florida, 1963–94. Figures on the top of each bar indicate the number of eagles in that category. Not shown are an additional 24 eagles for which a cause of death could not be determined during the same time period. Data from National Wildlife Health Center records.

season 1,216 young were fledged. The population of Bald Eagles was estimated to be between 2,425 and 3,640 and it was concluded that the population in Florida was "healthy" (Nesbitt 1997).

Although some general information on diseases of eagles can be found in Cooper (1985), Fowler (1986, 1993), and Deem et al. (1998), the most pertinent specific information on mortality factors in Florida eagles is found in various files and in an unpublished report issued in 1985 by the National Wildlife Health Center (NWHC) in Madison, Wisconsin, in which the causes of mortality of 1,429 Bald Eagles in the United States, including 107 from Florida, during 1963–84 were summarized. The data obtained from various carcasses that were submitted for determination of cause of death must be interpreted with caution since these birds do not represent a random sample and there are numerous sources of bias, as listed in the NWHC report (1985). Nevertheless, such information does provide a useful database, even though there are limitations to its accuracy. Additional information can be found in a 1-page summary based on more than 4,300 Bald and Golden Eagles examined at the NWHC (Franson et al. 1995) and in a series of publications by personnel of the Patuxent Wildlife Research Center, U.S. Fish and Wildlife Service (Coon et al. 1970; Mulhern et al. 1970; Belisle et al. 1972; Prouty et al. 1977; Cromartie et al. 1975; Kaiser et al. 1980; Reichel et al. 1984). The files of the NWHC contain a wealth of information on eagles from Florida (mostly Bald Eagles, but also a few Golden Eagles) that have been examined at necropsy to determine the cause of death. Much of the information in this chapter was obtained from those files through the courtesy of Milton Friend, Nancy J. Thomas, Louis N. Locke, and other personnel of the Center.

Table 11.1. Numbers of Bald Eagles dying of trauma in Florida, 1963–94

Type of trauma	Number of cases	% of total
Struck by vehicle	81	44
Gunshot	18	10
Intraspecific aggression	12	7
Power line strike	8	4
Nest blown down in storm	5	3
Fell from nest	5	3
Predation by alligator	2	1
Penetration of esophagus by bone	2	1
Struck by airplane	1	<1
Caught in barbed-wire fence	1	<1
Unknown	47	26
Total	182	—

Source: Records of the National Wildlife Health Center, Madison, Wisconsin.

Other important sources of information were the files of Florida Audubon Society's Center for Birds of Prey in Maitland (provided by Resee Collins and Carol Gilliland) and the files of the Veterinary Medicine Teaching Hospital at the University of Florida (provided by Sharon L. Deem and Scott P. Terrell). These latter 2 data sets were composed of combined morbidity and mortality information on Bald Eagles submitted for rehabilitation purposes and are therefore of a different nature than the mortality data from the National Wildlife Health Center mentioned above. Most of the eagles involved in these rehabilitation datasets were not examined at necropsy and therefore the determinations of the primary diagnostic findings (i.e., causes of morbidity or mortality) are based on histories and physical examinations of the birds. Nevertheless these data when viewed along with information derived from birds examined at necropsy are of value in determining the major disease problems of free-ranging eagles in Florida.

II. Trauma

During the 30-year period between 1963 and 1994, carcasses of 333 Bald Eagles from Florida were examined at the National Wildlife Health Center. For 24 of these, the cause of death could not be determined. Of the other 309 eagles, 182 (59%) died because of trauma (figure 11.1), mostly that caused by collisions with vehicles on highways (46%), followed by gunshot (10%) and intraspecific aggression (7%) (table 11.1). Smaller numbers were associated with other traumatic events such as impacts with power lines or wires, entanglement in fences, predation by alligators, and nestlings falling from nests. The geographic distribution of mortality due to trauma correlates generally with areas where Bald Eagle populations are most dense (Nesbitt 1994). Gunshot and other types of trauma were found to be the leading causes of death in the nationwide study summarized by the NWHC (1985). Data on 107 Bald Eagles from Florida were included in that report and are part of the dataset discussed herein.

Data on morbidity and mortality from the Florida Audubon Society's Center for Birds of Prey in Maitland (Collins and Gilliland 1995) and the Veterinary Medical Teaching Hospital at the University of Florida in Gainesville (Deem and Terrell 1996) were collected from 1988 to 1994 and included information on 274 eagles, mostly from central and north central Florida (table 11.2). These eagles were submitted to the 2 rehabilitation centers for treatment and then were either rehabilitated and released,

Table 11.2. Primary reasons given for submission of Bald Eagles to 2 wildlife rehabilitation centers in Florida, 1988–94

Primary diagnostic finding	Number of cases per region			% of total cases
	Central[a]	North central[b]	Total	
Trauma	159	28	187	68
Poisoning	19	9	28	10
Inclement weather	20	0	20	7
Electrocution	15	0	15	6
Infectious diseases	12	2	14	5
Emaciation	8	1	9	3
Congenital cataract[c]	1	0	1	<1
Total	234	40	274	—

a. Records from Florida Audubon Society's Center for Birds of Prey, Maitland, 1989–94 (Collins and Gilliland 1995).
b. Records of Veterinary Medical Teaching Hospital, College of Veterinary Medicine, University of Florida, Gainesville, 1988–94 (Deem and Terrell 1996).
c. In the left eye of a 3-day-old nestling from Osceola County in 1991.

became unreleasable permanent cripples, or died. As in the other dataset from the NWHC, trauma was the most dominant category, involving 68% of the birds submitted. The most common types of trauma included vehicle strikes, intraspecific aggression, and gunshot (table 11.3).

Mortality due to collisions on highways is shown by county in figure 11.2. Such mortality occurred in 27 counties in peninsular Florida with concentrations in Alachua, Brevard, Orange, Osceola, Polk, and Volusia counties. None occurred in the panhandle region. It is possible that this pattern is an artifact of collection, since the counties where most road-killed eagles were reported are close geographically to active raptor rehabilitation centers, i.e., the Florida Audubon Society's Birds of Prey Center (Osceola County) and the University of Florida Veterinary Med-

Table 11.3. Types of trauma causing morbidity and mortality of Bald Eagles submitted to the Center for Birds of Prey, Maitland, Florida, 1989–94

Type of trauma	No. of cases	% of total cases
Struck by vehicle	34	21
Intraspecific aggression	31	19
Gunshot	12	8
Fell from nest	6	4
Struck power line	4	3
Trapped	2	1
Perforated esophagus	1	<1
Ruptured ovarian follicle	1	<1
Tangled in fishing line	1	<1
Nest destroyed	1	<1
Unknown	66	42
Total	159	—

Source: Collins and Gilliland (1995).

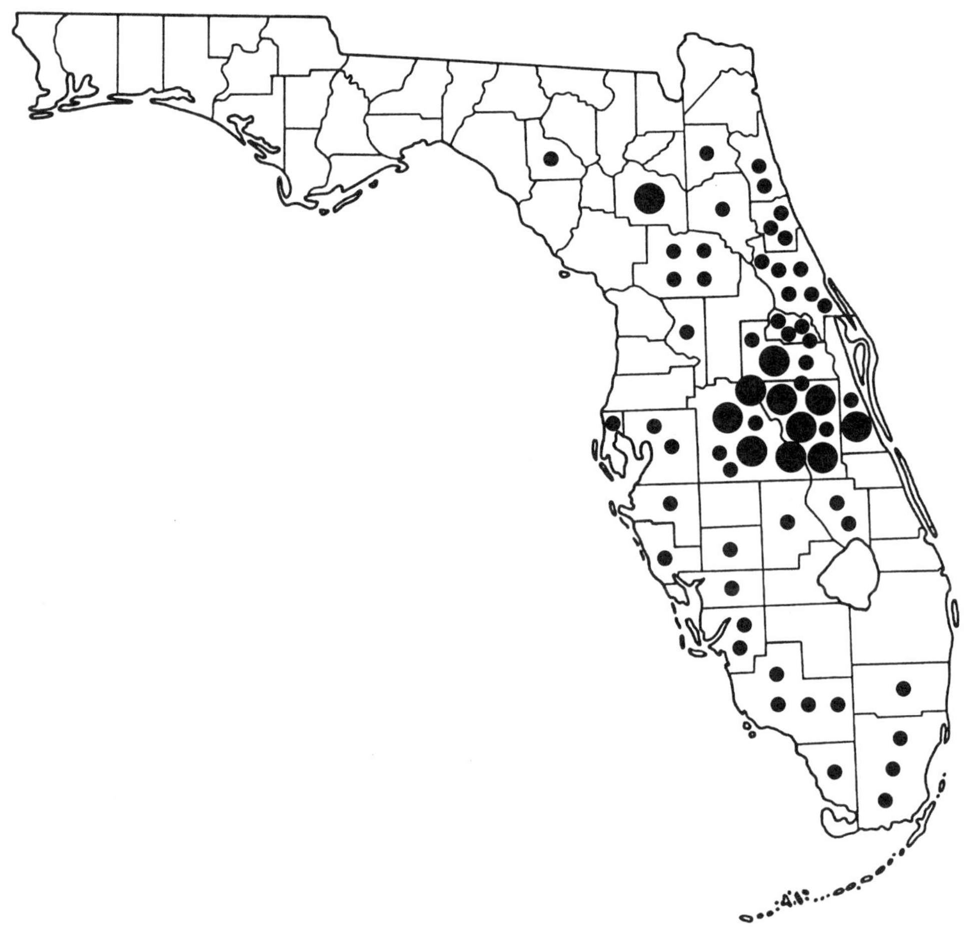

FIGURE 11.2. Distribution of mortality of Bald Eagles in Florida due to collisions with vehicles on highways, 1963–94. *Small circles* = 1 eagle; *large circles* = 5 eagles. Four additional eagles died on highways in Florida, but the county of origin was not given. Data from National Wildlife Health Center records and courtesy of Resee Collins and Carol Gilliland.

ical Teaching Hospital (Alachua County). The majority (71%) of the eagles examined by the NWHC died during November through March. Of the 81 road-killed eagles, 50 (62%) were adults and 31 (38%) were juveniles. Eagles feed on dead animals along roads and often are struck by passing vehicles. Such mortality related to carrion feeding can be expected to increase as the human population in Florida grows and the use of Florida's highways increases. As more vehicles use the roads the production of carrion will increase and so will

the probability of carrion-feeding Bald Eagles being injured or killed. It has been recommended that highway signs reading "Slow— Eagles Feeding" be posted along highways where eagles are known to frequently feed so that this mortality might be mitigated (NWHC 1985).

The distribution of mortality due to gunshot is shown in figure 11.3. This type of mortality occurred in 17 counties in peninsular Florida and 1 in the panhandle. Eighteen of these (60%) were from an 8-county cluster in central

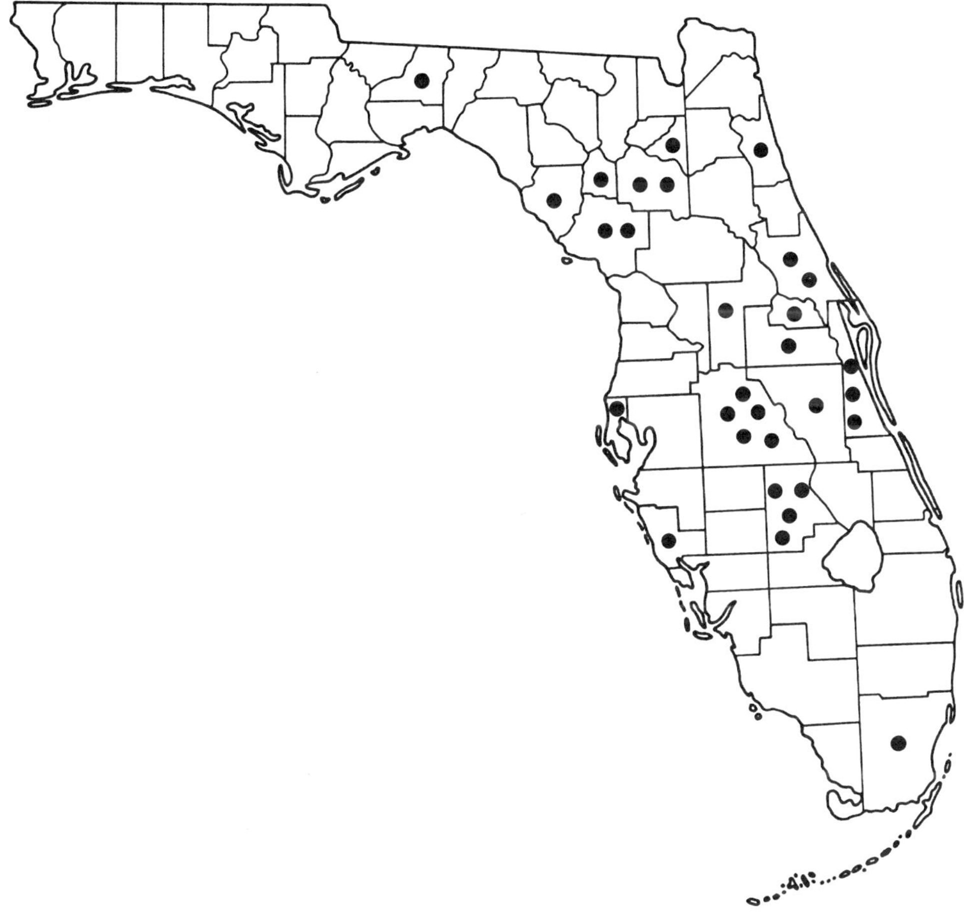

FIGURE 11.3. Distribution of mortality of Bald Eagles in Florida due to gunshot, 1963–94. *Circle* = 1 eagle. Data from National Wildlife Health Center records and courtesy of Resee Collins and Carol Gilliland.

Florida including Polk, Highlands, Osceola, Brevard, Orange, Seminole, Lake, and Volusia counties. Most (74%) of these gunshot incidents occurred between November and March. According to the NWHC dataset, 11 were adults, 6 were juveniles, and 1 was a nestling. This latter finding varies from the national dataset presented by the NWHC (1985), in which the majority of eagles dying of gunshot wounds were juveniles.

Intraspecific aggression was widespread throughout the state, as indicated by the occurrence of 43 such cases originating from 18 different counties. These diagnoses were determined by a combination of postmortem findings

and field observations. Twenty-six of the incidents occurred between October and February. Of the 12 eagles in the dataset from the NWHC in which age was known, 11 were adults and 1 was a juvenile.

In March 1982 a juvenile female Golden Eagle drowned after becoming tangled in a trotline in Lake Tohopekaliga in Osceola County (Siegfried 1982). Apparently an American Coot was caught in the line and the eagle attempted to take it and became entangled herself. Collins (1996) reported that an immature male Golden Eagle was found in a pasture in Pasco County in 1984; the eagle had a dislocated shoulder, the cause of which was not de-

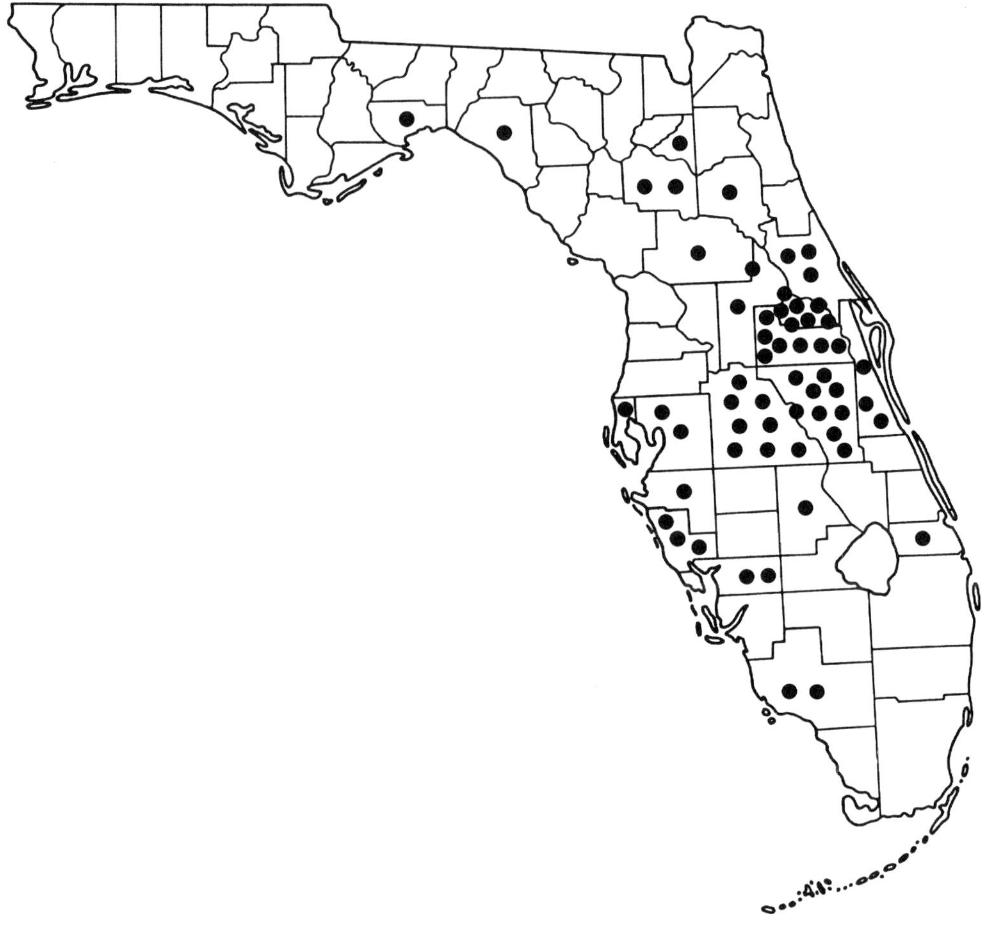

FIGURE 11.4. Distribution of mortality of Bald Eagles in Florida due to electrocution, 1963–94. *Circle* = 1 eagle. Only 47 of the 51 eagles that died of electrocution are plotted in this figure; another 4 birds came from Florida, but the specific localities were not given. Data from National Wildlife Health Center records and courtesy of Resee Collins and Carol Gilliland.

termined. No other information on trauma is available for Golden Eagles in Florida.

III. Electrocution

The primary cause of death for 16% of the Bald Eagles examined at necropsy by the NWHC from 1963 to 1994 was electrocution (figure 11.1). Of these, 55% were adults and 45% were juveniles, contrary to the nationwide dataset of the NWHC (1985) in which a lower frequency of electrocution was found

for adult eagles than for juveniles. Electrocutions of eagles occurred in 21 counties, with 39 (59%) of these eagles originating from 7 counties in central peninsular Florida (figure 11.4). They also occurred in every month of the year with most (74%) during October through April. Not shown on the map (figure 11.4) are 3 cases where electrocution was judged to be secondary to other causes of death, such as trauma and lead poisoning, for eagles from Indian River County (1979), Volusia County (1982), and Orange County (1988) (NWHC records).

Mortality due to electrocution probably occurs when eagles take off from or land on transmission poles carrying high-voltage wires. It has been suggested (Solt 1980) that young eagles are electrocuted because of their clumsiness around such poles and power lines, but this would not explain adult mortality. Zimmerman (1976) reported an incident that occurred in Florida (exact locality not given) in 1975 in which several people observed a young Bald Eagle pick up the carcass of a 36-inch kingfish that had been filleted and thrown over a seawall by some fishermen. The eagle flew with the carcass up onto a high-tension power line and was electrocuted when the fish connected across 2 of the wires. In most cases, however, the electrocution event is not observed; it is then necessary to look for other types of information to decide whether or not electrocution was involved. The classic signs of electrocution include singed feathers, burned flesh, and blood in the pericardial sac. This postmortem information, along with field observations such as the carcass being found beneath a power line, allow a diagnosis of electrocution to be made. In some cases where the signs are not clear-cut and field observations are scanty or lacking, a presumptive diagnosis has to be made.

Electrocution of eagles on power lines can be prevented by certain adaptations or modifications of the power line structures on which the birds tend to perch, and changes in the grounding setup can be made. Details on these practices have been given in a report published by the Raptor Research Institute and Edison Electric Institute (Miller et al. 1975). Apparently these changes have reduced the numbers of electrocutions of eagles, especially Golden Eagles in the western states (Miller et al. 1975), but specific data for eagles in Florida are not available.

Howell (1941) gave an account of a Bald Eagle being struck by lightning and killed while incubating eggs on its nest in a tall pine tree. The authors did not see the actual lightning strike occur, but the circumstantial evidence was compelling and included the recent removal of the bark on the limbs and trunk of the tree above the nest and damage to the truck below the nest. The event occurred on Merritt Island (Brevard County) in December of 1940.

IV. Inclement weather

Hurricanes are known to cause considerable damage to eagle nests and thereby to eggs and young. Heinzman (1961) reported significant losses in hurricanes that occurred in 1944 and 1960. Hurricane Donna in 1960 destroyed 90% of the Bald Eagle nests in Florida Bay (Everglades National Park); losses in central Florida were estimated to be at 30%. Hurricane Andrew destroyed 3 Bald Eagle nests in Everglades National Park in August 1992, although no eggs or young were likely affected since Bald Eagles do not nest at that time of year (FGFWFC 1992).

The "storm of the century" on March 13, 1993, resulted in considerable morbidity and mortality of Bald Eagles. Winds during this storm were measured as high as 118 miles per hour and blew some eagle nestlings out of their nests (Langridge 1993). An adult Bald Eagle and 2 nestlings were found dead at Marco Island and 8 nests were known to have been destroyed, 2 at Three Lakes Wildlife Management Area (Osceola County), 2 at Marco Island (Collier County), 2 in Charlotte County, 1 in Lee County, and 1 in Polk County (Wood 1993); other nests certainly were damaged or destroyed in areas where there was no surveillance. Shortly after the storm, 20 Bald Eagles, mostly nestlings, were brought to the Center for Birds of Prey in Maitland. Seven of these were dead on arrival, the other 13 were suffering from various degrees of trauma related to the storm (Collins and Gilliland 1995). These eagles originated from 8 different counties: Brevard (n = 2), Hillsborough (1), Lake (2), Orange (1), Polk (3), Seminole (7), St. Johns (1), and Volusia (3). White (1993) checked 333 active Bald Eagle nests in 18 counties after the storm and documented 23 dead nestlings and 4

Table 11.4. Residues of organochlorines (ppm wet wt.) in Bald Eagles from Florida, 1968–84[a]

Organochlorine compound	Carcass				Brain			
	n	% pos.	Median	Range	*n*	% pos.	Median	Range
p,p'-DDE	62	100	6.5	0.13–120	67	96	1.90	0.05–96
p,p'-DDD	62	74	0.45	0.05–5.9	64	50	0.23	0.05–3.8
p,p'-DDT	62	13	0.14	0.08–0.5	64	3	0.85	0.7–1.0
Dieldrin	62	86	0.35	0.06–7.0	66	70	0.19	0.07–8.7
Heptachlor epoxide	62	55	0.12	0.05–1.7	64	33	0.15	0.05–2.1
Oxychlordane	60	82	0.20	0.05–2.3	62	58	0.08	0.04–2.7
cis-chlordane	60	80	0.35	0.05–2.0	62	50	0.14	0.05–2.3
trans-nonachlor	60	88	0.70	0.06–4.0	61	69	0.21	0.06–4.8
cis-nonachlor	60	75	0.33	0.07–1.4	62	47	0.13	0.06–1.6
Endrin	61	10	0.08	0.07–0.13	63	6	0.13	0.07–0.14
Toxaphene	59	20	0.17	0.05–0.40	61	15	0.15	0.06–0.91
Hexachlorobenzene	45	4	0.09	0.08–0.1	49	2	0.16	—
Mirex	45	80	0.70	0.07–5.0	49	67	0.14	0.05–3.8
PCBs	60	97	0.45	0.25–200	64	92	3.30	0.19–83

Sources: Data modified from Reichel et al. (1969b) and from records of Patuxent Wildlife Research Center, Laurel, Md.
a. Reichel et al. (1969a) included data on 7 organochlorines in 1 Bald Eagle obtained from Florida in 1965, but residue values for this bird were combined with data from 27 other eagles from other states. The individual values for the Florida bird were not given.

juveniles. The dead eagles were from the following counties: Brevard (3), Highlands (2), Lake (1), Levy (2), Orange (1), Osceola (8), Polk (5), and Seminole (5). It was estimated that 12–15% of the Bald Eagle nests in Florida were affected by the storm; the decline of the mean number of young per active nest compared with the previous year (1.02 in 1993 vs. 1.12 in 1992) was attributed also to the storm (Nesbitt 1993). Fortunately eagles will rebuild destroyed nests. For example, according to C.L. Broley (Heinzman 1961), 64 of 65 nests that were destroyed in the hurricane of 1944 were rebuilt; however, in only 30% of these new nests were young successfully raised, compared with 70% nesting success the previous year. In 1960 2/3 of the destroyed nests were rebuilt in Florida Bay (Heinzman 1961).

V. Emaciation

Emaciation was given as the cause of death for 15 of 309 (5%) Bald Eagles from Florida examined at necropsy from 1963 to 1994 by the NWHC (figure 11.1). An additional 9 emaciated eagles were submitted to 2 wildlife rehabilitation centers in central and north central Florida between 1988 and 1994 (table 11.2). These eagles originated from 16 counties located throughout peninsular Florida (figure 11.5) with no obvious concentrations in any one region or during any particular season of the year. Of the 15 emaciated Bald Eagles examined at the NWHC, 9 were adults, 5 were juveniles, and 1 was a nestling. One emaciated juvenile Golden Eagle from Madison County was found dead in December of 1988 (Locke 1989).

Emaciation was diagnosed by NWHC personnel in eagles with depleted fat deposits, marked muscle and organ atrophy, no other apparent lesions, negative results from microbiological studies, and no significant tissue residues of likely environmental contaminants. The actual cause of emaciation (which is a condition rather than a disease) in these birds is unknown. On a national scale emaciation was found to account for 8% of total Bald Eagle mortality (NWHC 1985).

Table 11.5. Organochlorine residues in eggs of Bald Eagles from Florida

Organochlorine compound	Residues (ppm fresh wet wt.)						
	1965[a]	1968[b]	1969[c]	1975[d]	1976[e]	1977[f]	1979[g]
	No. of eggs tested						
	1	6	2	1	2	2	1
p,p'-DDE	13	11	18	13	10	3.3	2.0
p,p'-DDD	2.6	0.6	1.8	0.1	0.3	0.1	0
p,p'-DDT	0.5	0.2	0.2	0	0.05	0	0
Dieldrin	0.5	0.2	1.1	0.4	0.5	0.2	0.1
Heptachlor epoxide	NA	0.02	0.05	0	0.1	0.04	0
Oxychlordane	NA	NA	NA	0.1	0.2	0.1	0.1
cis-chlordane	NA	NA	NA	0.2	0.2	0.1	0
trans-nonachlor	NA	NA	NA	NA	0.5	0.5	0.3
cis-nonachlor	NA	NA	NA	0	0.1	0.1	0.1
Toxaphene	NA	NA	NA	0	0	0	0
Hexachlorobenzene	NA	NA	NA	0	0	0	0
Mirex	NA	NA	NA	0	1.5	0.4	0.3
Endrin	NA	NA	NA	NA	0[h]	0[h]	0[h]
PCBs	NA	NA	12	16	22	7.4	5.7

NA = not analyzed for this residue.
a. Specific locality not given; data from Stickel et al. (1966).
b. One egg from each of the following localities: Buoy Key, Cormorant Key, Palm Key, Deer Key, Crab Key, and Manatee Key (all within Everglades National Park); data from Krantz et al. (1970).
c. Both eggs from Lee County; data from Wiemeyer et al. (1972).
d. From Camp Key (Everglades National Park); data from Wiemeyer et al. (1984).
e. One egg each from Pine Island (Lee County) and Deer Key (Everglades National Park); data from Wiemeyer et al. (1984).
f. Both eggs from Pine Island (Lee County); data from Wiemeyer et al. (1984).
g. From Pine Island (Lee County); data from Wiemeyer et al. (1984).
h. Wiemeyer et al. (1984) stated that 5 eggs were tested for endrin residues 1975–79 but did not identify the years of collection or the specific localities; all were negative.

VI. Organochlorines

From 1968 to 1984, 69 Bald Eagles from Florida were analyzed at Patuxent Wildlife Research Center for residues of 14 organochlorines. In table 11.4 concentrations of these residues are given for brain and carcass. Values for carcasses are based on the body of the eagle minus the skin, feet, wings, liver, and gastrointestinal tract (Mulhern et al. 1970). DDE was present in all carcass samples and in 96% of the brains tested. PCBs were the next highest in terms of prevalence and were found in 97% of the carcasses and 92% of the brains. Other common residues included DDD, oxychlordane, cis-chlordane, trans-nonachlor, cis-nonachlor, and mirex. Data on some or all of the same 14 organochlorines are available for 15 eggs collected between 1965 and 1979 (table 11.5). Residues of 11 of the 14 organochlorines were detected. DDE, dieldrin, and PCBs were present in all eggs for which they were analyzed. DDE occurred in the highest concentrations and varied from 2.0 to 18.5 ppm.

Concentrations of dieldrin in brain tissue varying from 3.9 to 8.0 ppm have been shown to be lethal to Bald Eagles (Prouty et al. 1977; Coon et al. 1970). Three Bald Eagles from Florida have died of dieldrin poisoning, 2 in 1968 and 1 in 1976 (table 11.6, figure 11.5). The eagle that died in 1976 is of interest since it had a brain concentration of 8.7 ppm and this was 2 years after sales of dieldrin had been banned in the United States (Klaassen 1980).

Table 11.6. Data on Bald Eagles that died of dieldrin poisoning in Florida

County	Year	Age[a]	Sex	Brain[b]	Carcass	Sileo index[c]	Data source
				Dieldrin concentration (ppm)			
Brevard	1968	AD	M	5.5	1.5	1.8	Locke (1968), Mulhern et al. (1970)
Dade	1976	IM	M	8.7	7.0	1.9	Locke (1976), Reichel (1976), Kaiser et al. (1980)
Marion	1968	AD	F	7.0	6.5	2.4	Reichel et al. (1969b), Mulhern et al. (1970)

a. AD = adult, IM = immature.
b. Minimum lethal concentrations in brain have been reported to range from 3.9 ppm (Prouty et al. 1977) to 8 ppm (Coon et al. 1970).
c. After Sileo et al. (1977). Values of >1 are considered indicative of poisoning.

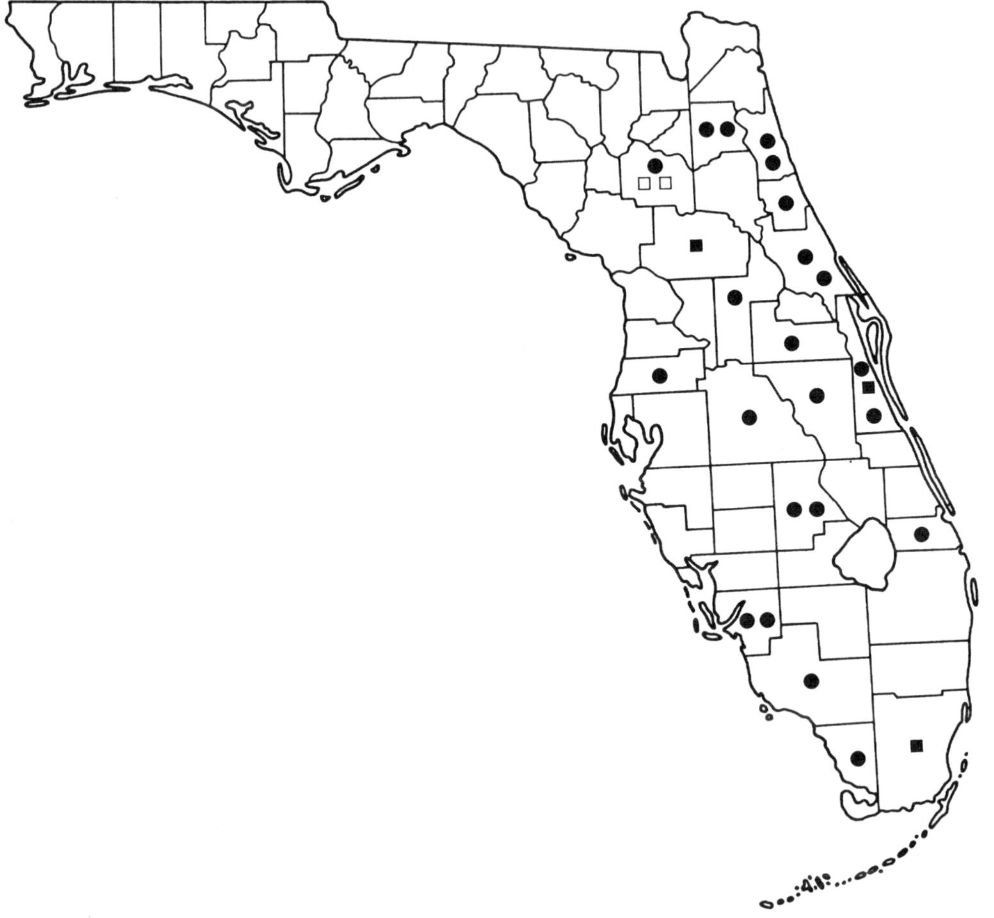

FIGURE 11.5. Distribution of mortality of Bald Eagles in Florida due to emaciation *(solid circles)*, organochlorine poisoning *(open boxes)*, and dieldrin poisoning *(solid boxes)*, 1963–88. Data from Patuxent Wildlife Research Center and National Wildlife Health Center records and courtesy of Resee Collins, Carol Gilliland, Sharon L. Deem, and Scott P. Terrell.

Table 11.7. Organochlorine residues (ppm, wet wt.) in brain tissue, from 2 adult male Bald Eagles that died of organochlorine poisoning in Alachua County, Florida

Organochlorine compound	1975	1977
p,p'-DDE	35	62
p,p'-DDD	1.90	1.10
p,p'-DDT	ND	ND
Dieldrin	2.60	3.70
Heptachlor epoxide	0.44	0.55
Oxychlordane	0.68	1.70
cis-Chlordane	1.20	0.66
trans-Chlordane	2.20	2.80
cis-Nonachlor	0.17	0.55
Endrin	0.12	0.14
Toxaphene	ND	0.38
Mirex	0.98	2.50
PCBs	68	45
Sileo index[a]	1.20	1.50

Source: Locke (1975, 1977) and Reichel (1975, 1977).
ND = none detected.
a. Values of >1 are considered indicative of organochlorine poisoning (see Sileo et al. 1977).

Two other Bald Eagles, 1 in 1975 and 1 in 1977, died of organochlorine poisoning (table 11.7, figure 11.5). In these 2 birds, none of the individual organochlorines occurred in concentrations high enough to be considered lethal, but the collective values resulted in Sileo Indexes greater than 1, which would indicate lethal effects (Sileo et al. 1977).

Even though organochlorines can be direct mortality factors as discussed above, their indirect effects on reproduction, including thinning of eggshells and subsequent destruction of eggs by incubating adults, have been more widespread and serious. Broley (1958) reported a drastic decline in Bald Eagles on the west coast of Florida, from Tampa to Fort Myers, during the late 1950s and speculated that DDT contamination was the cause of reproductive failure. Similarly, a 2/3 reduction in Bald Eagle numbers was reported for the east coast of Florida (Volusia and Brevard counties) between 1951 and 1961 (Howell 1963). Subsequent retrospective studies of Bald Eagle eggshells in museums and oological collections resulted in the recognition of 15–19% reductions in weights

and thicknesses of eggs of Florida Bald Eagles beginning around 1947 and extending into the early 1970s (table 11.8, figure 11.6) (Hickey and Anderson 1968; Anderson and Hickey 1972; Wiemeyer et al. 1972, 1984). Declining populations in Brevard and Osceola counties experienced 18–20% reductions in weights of their eggshells during 1947–62 (Hickey and Anderson, 1968). DDT and its metabolites (especially DDE) were found to be associated with this eggshell thinning in Bald Eagles (Newton 1979). Since the sale of DDT was banned in the United States in 1972, most populations of Bald Eagles have recovered (Grier 1982) and in Florida, populations appear to be increasing (Nesbitt 1997). Nevertheless, Wood et al. (1990) collected 87 eggshells from nests in Florida (specific localities not given) during 1984–87 and found that most shells were slightly thinner than pre-1947 values, and a few shells were 29% thinner. In 1 eagle nest there were 2 eggs, 1 broken and 1 infertile; both had thin shells. The authors concluded from these data that there might be reproductive problems in some areas in the state. PCBs also have been implicated in eggshell thin-

Table 11.8. Eggshell data from Bald Eagles in Florida showing changes in weights and thickness

County or site	Year(s)	Eggshell weights			Eggshell thickness			Data source[a]
		No. eggs examined	Mean wt. (g)	% change	No. eggs examined	Mean (mm)	% change	
Brevard	1886–1939	56	12.15	—	0	—	—	A
Osceola	1901–44	25	12.32	—	0	—	—	A
"Florida"	pre-1947	337	11.93	—	211	0.584	—	B
	1944–45	28	11.62	None[b]	0	—	—	B
	1947–48	6	10.09	-15.0[b]	0	—	—	B
	1949–50	9	9.86	-17.0[b]	3	0.487	-17.0[b]	B
	1958–59	4	9.89	-17.0[b]	0	—	—	B
	1961–62	8	9.67	-19.0[b]	0	—	—	B
	1984–87	87	NG	NG	87	NG	-29.0[e]	C
Brevard	1947–62	12	9.96	-18.0[c]	0	—	—	A
Osceola	1959–62	8	9.88	-19.8[d]	0	—	—	A
Lee	1968–70	2	9.65	-19.1[b]	2	0.518	-11.3[b]	D
Pine Island and Deer Key	1969–79	0	—	—	6	0.510	-13.0[b]	E

NG = not given.

a. A = Hickey and Anderson (1968), B = Anderson and Hickey (1972), C = Wood et al. (1990), D = Wiemeyer et al. (1972), E = Wiemeyer et al. (1984).

b. Compared with pre-1947 data (Anderson and Hickey 1972).

c. Compared with 1886–1939 data (Hickey and Anderson 1968).

d. Compared with 1901–44 data (Hickey and Anderson 1968).

e. Authors stated that the shells were "only slightly thinner, on average, than pre-1947 eggs," but a few were 29% thinner (Wood et al. 1990).

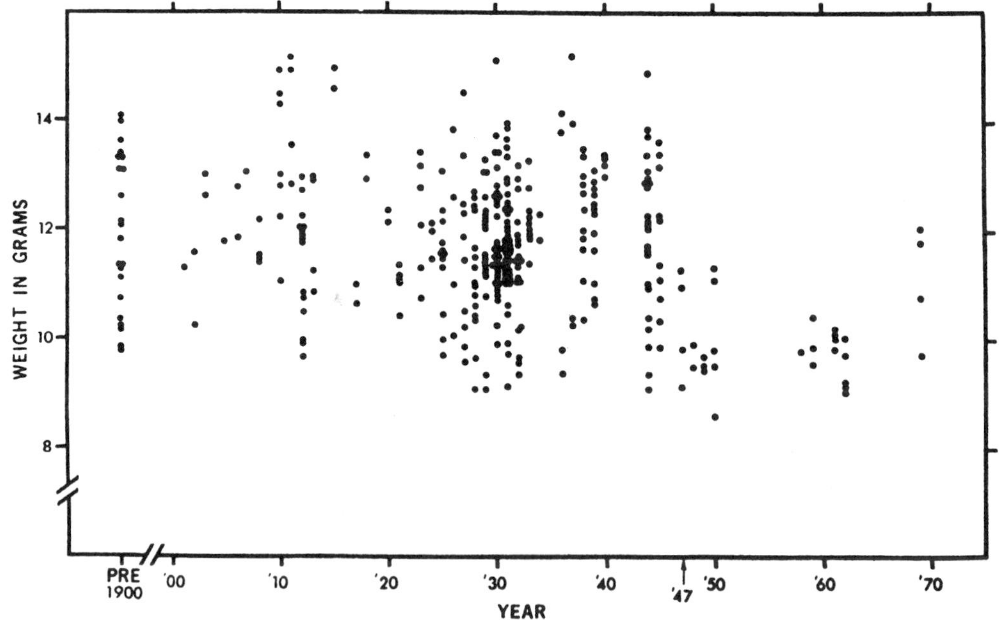

FIGURE 11.6. Eggshell weight versus time for Bald Eagles in Florida, 1947–70. After Anderson and Hickey 1972; courtesy of Daniel W. Anderson.

Table 11.9. Residues of dieldrin (ppm wet wt.) found in Bald Eagles from Florida[a]

Year	Carcass[b]				Brain			
	n	% positive	Median	Range	n	% positive	Median	Range
1968	2	100	4.0	1.5–6.5	2	100	6.3	5.5–7.0
1969	0	—	—	—	0	—	—	—
1970	0	—	—	—	0	—	—	—
1971	0	—	—	—	0	—	—	—
1972	0	—	—	—	0	—	—	—
1973	0	—	—	—	2	100	2.6	1.5–3.6
1974[c]	0	—	—	—	0	—	—	—
1975	5	80	1.9	0.4–2.3	5	100	0.2	0.11–2.6
1976	2	100	3.9	0.73–7.0	2	100	4. 5	0.2–8.7
1977	4	50	1.2	0.2–2.3	4	25	3.7	—
1978	11	91	0.3	0.08–1.1	12	58	0. 2	0.08–0.37
1979	11	82	0.5	0.06–1.0	12	58	0. 2	0.08–1.1
1980	2	100	0.5	0.16–0.8	2	10	0.2	0.09–0.23
1981	10	90	0.2	0.07–0.88	12	67	0.2	0.09–2.2
1982	6	100	0.5	0.07–2.4	5	80	0.3	0.08–4.2
1983	7	71	0.3	0.07–0.45	7	71	0.1	0.07–0.76
1984	2	100	0.2	0.09–0.25	1	100	0.1	—

a. Data calculated from the records of Patuxent Wildlife Research Center, Laurel, Md., except for 1 eagle in 1968, from Reichel et al. (1969b).
b. Values are based on the body minus the skin, feet, wings, liver, and gastrointestinal tract (Mulhern et al. 1970).
c. Sales of dieldrin were halted in the United States in 1974 (Klaassen 1980).

ning in Bald Eagles (Wiemeyer et al. 1984), but the evidence that they have been a major cause of this problem is not conclusive (Eisler 1986). Nevertheless, it is good news that in 1979 a ban was instituted in the United States on the manufacture, processing, distribution, and use of PCBs except in totally enclosed systems or unless exempted by the EPA (Bremer 1983).

The fact that there are limited data on dieldrin, DDE, and PCBs in eggs and carcasses of Bald Eagles in Florida might be interpreted to mean that there is a trend toward decreasing concentrations in recent years, subsequent to the banning of the use of these chemicals (tables 11.5, 11.9, 11.10, and 11.11). Since Bald Eagles in Florida migrate to the north, but not to the south (Broley 1947) where they might obtain some of these contaminants in countries where they are still used, the U.S. bans appear to have been appropriate and effective, judging by the favorable condition of these birds at the present time (Nesbitt 1997).

For overviews and additional details on organochlorines and their effects on Bald Eagles in North America, the reader is referred to Gerrard and Bortolotti (1988) and Nisbet (1989).

VII. Organophosphates and carbamates

According to Smith (1993:1) the "use of organophosphorus and carbamate pesticides has increased markedly during the past 2 decades. Currently, more than 100 different organophosphorus and carbamate chemicals are registered as the active ingredients in thousands of different pesticide products in the United States. More than 160 million acre-treatments of these pesticides are estimated to be applied to agricultural crops and forests each year."

Between 1991 and 1994 2 Bald Eagles from Florida (1 from Alachua County and 1 from Seminole County) were determined to have died

Table 11.10. Residues of PCBs (ppm wet wt.) found in Bald Eagles from Florida[a]

Year	Carcass[b]				Brain			
	n	% positive	Median	Range	n	% positive	Median	Range
1973	0	—	—	—	2	100	37	31–43
1974	0	—	—	—	0	—	—	—
1975	5	100	32	1–160	5	100	8	3–68
1976	2	100	130	53–200	2	100	10	9–12
1977	4	75	11	0.3–35	4	100	1	0.2–45
1978	11	100	7	2–28	12	92	1	0.5–8
1979[c]	11	100	7	1–65	12	92	1	0.3–83
1980	2	100	35	4–66	2	100	6	4–8
1981	10	100	13	1–24	12	92	4	1–25
1982	6	83	22	1–70	5	100	4	0.2–63
1983	7	100	5	0.4–25	7	100	1	1–76
1984	2	100	5	1–9	1	100	2	—

a. Data collected from the records of Patuxent Wildlife Research Center, Laurel, Md.
b. Values are based on the body minus the skin, feet, wings, liver, and gastrointestinal tract (Mulhern et al. 1970).
c. Use of PCBs was curtailed in the United States in 1979 (Bremer 1983).

Table 11.11. Residues of p,p′-DDE (ppm wet wt.) found in Bald Eagles from Florida[a]

Year	Carcass[b]				Brain			
	n	No. positive	Median	Range	n	No. positive	Median	Range
1968	2	2	31	29–32	2	2	61	27–96
1969	0	—	—	—	0	—	—	—
1970	0	—	—	—	0	—	—	—
1971	0	—	—	—	0	—	—	—
1972[c]	0	—	—	—	0	—	—	—
1973	0	—	—	—	2	2	25	23–26
1974	0	—	—	—	0	—	—	—
1975	5	5	23	0.5–54	5	5	3.4	2.5–35
1976	2	2	65	9.3–120	2	2	7.1	2.2–12
1977	4	4	4.5	0.2–8.5	4	4	1.0	0.1–62
1978	11	11	4.8	0.6–20	12	12	0.9	0.05–14
1979	11	11	4.3	0.23–14	12	11	1.8	0.09–21
1980	2	2	7.3	2.5–12	2	2	3.1	2.9–3.2
1981	10	10	6.0	0.38–12	12	12	1.4	0.06–21
1982	6	6	6.4	0.35–37	5	5	1.1	0.12–35
1983	7	7	3.5	0.13–11	7	5	1.7	0.78–23
1984	2	2	3.7	0.83–6.5	2	2	0.6	0.1–1.1

a. Data calculated from the records of Patuxent Wildlife Research Center, Laurel, Md., except for 1 eagle in 1968, from Reichel et al. (1969b).
b. Values are based on the body minus the skin, feet, wings, liver, and gastrointestinal tract (Mulhern et al. 1970).
c. Sales of DDT were halted in the United States in 1972 (Klaassen 1980).

Table 11.12. Data on Bald Eagles that died, or were suspected of dying, of organophosphorus or carbamate poisoning

County	Year	Age	Sex	% brain cholinesterase inhibition	Pesticide identified	Data source
Organophosphorus poisoning						
Alachua	1994	AD	F	88	Undetermined	Franson (1994)
Seminole	1991	AD	M	88	Parathion[a]	Thomas (1992)
Carbamate poisoning						
Alachua	1992	AD	M	54	Carbofuran	Thomas (1993)
Levy	1993	AD	M	49	Carbofuran	Franson (1993)
Orange	1992	AD	M	69	Undetermined[b]	Thomas (1993)

a. 24.4 ppm in the upper gastrointestinal tract contents.
b. Although there was a significant depression (69%) of brain cholinesterase activity in this eagle, and this was reversed with incubation of the sample, no carbamates were detected in the upper gastrointestinal tract contents. Carbamate poisoning was therefore suspected, but could not be confirmed.

of organophosphorus (OP) poisoning and 2 (1 from Alachua County and 1 from Levy County) of carbamate poisoning (table 11.12). One additional Bald Eagle was suspected to have died of carbamate poisoning, but tests did not confirm this. The definitive diagnosis of OP or carbamate poisoning is accomplished (1) by determining that there is a depression in the normal cholinesterase activity in brain tissue and (2) by the identification of the appropriate residues in the gastrointestinal tract of the bird (Hill and Fleming 1982). In 2 of the eagles listed in table 11.12 a definitive diagnosis was not possible, but in the other 3, the causative pesticides were identified as parathion or carbofuran (Franson 1993; Thomas 1992–93). These were probably cases of secondary poisonings that occurred when the eagles fed on carcasses of dead birds such as waterfowl that had encountered the pesticides by feeding on treated agricultural crops. There are numerous records of waterfowl die-offs in the United States due to parathion and carbofuran (Smith 1993). The full impact of OPs and carbamates on eagles in Florida is not known, but mortalities from secondary exposure to these compounds can be expected, since their use in certain agricultural operations in Florida is widespread and ongoing.

VIII. Pentobarbital

In October 1991 an adult Bald Eagle was found dead in Polk County and was determined to have died of pentobarbital toxicosis (Thomas 1991). Two immature Bald Eagles from Volusia County died of pentobarbital poisoning in November and December of 1993 (Meteyer 1993). In the latter 2 cases concentrations of pentobarbital in liver tissue were 2 and 14 ppm. Both immature eagles were found at a landfill where they most likely had been feeding on carcasses of animals that had been euthanized with pentobarbital, perhaps in a veterinary clinic or animal shelter. The above scenario was repeated in connection with the same landfill between October 1996 and February 1997 (4 Bald Eagles died: Converse 1997, and another was intoxicated, but survived: Fischer 1997). An additional Bald Eagle died of pentobarbital intoxication at the Orange County landfill in December 1997 (Cornish 1998). Stomach contents in this latter eagle contained 2400 ppm pentobarbital.

In addition to Florida, this type of secondary barbiturate poisoning has been recognized in several other areas in North America, including Alaska, Colorado, Montana, Washing-

Table 11.13. Concentrations of lead in livers of Bald Eagles from Florida

| Year | No. of eagles | | ppm (wet weight) | | | Data source[a] |
	Examined	Positive	Median	Mean	Range	
1973	1	0	—	—	—	NWHC
1975	5	4	1.6	7.5	0.6–26	NWHC
1977	1	1	1.4	1.4	—	PNWR
1978	7	7	1.8	12	0.4–38	PNWR
1979	10	10	0.6	0.8	0.1–1.7	PNWR
1980	4	3	0.7	0.9	0.4–1.5	PNWR
1981	13	8	0.2	1.0	0.1–4.5	PNWR
1982	6	2	0.2	0.2	0.1–0.4	PNWR
1983	8	4	0.9	9.8	0.4–37	PNWR & NWHC
1984	9	5	0.7	0.8	0.35–1.1	PNWC & NWHC
1985	13	8	0.5	0.6	0.3–1.2	NWHC
1986	14	2	1.8	1.8	0.6–3.1	NWHC
1987	23	5	0.4	0.8	0.3–2.1	NWHC
1988	38	8	13	18	0.2–50	NWHC
1989	19	1	0.4	0.4	—	NWHC
1990	12	3	0.6	11	0.3–33	NWHC
1991	23	4	7.3	7.9	1.0–16	NWHC
1992	24	3	31	22	0.6–35	NWHC
1993	16	3	14	15	12–20	NWHC
1994	6	5	2.6	8.7	0.2–24	NWHC
All years	252	86	0.9	6.4	0.1–50	—

a. NWHC = records of the National Wildlife Health Center, Madison, Wis. Limits of detection = 0.2 ppm. PWRC = records of the Patuxent Wildlife Research Center, Laurel, Md. Limits of detection = 0.1 ppm.

ton (Thomas 1992), and British Columbia (Langelier 1993). In some instances these poisonings may be intentional, but others can be prevented by the proper disposal of carcasses by burial or incineration.

IX. Lead

From 1973 to 1994, liver samples from 252 Bald Eagles in Florida were analyzed for the presence of lead. Concentrations were found in 34% of the samples and ranged from 0.1 to 50 ppm, median = 0.9 (table 11.13). These eagles originated from 29 counties scattered throughout the state, except for the western panhandle (figure 11.7). During that same period of time 19 eagles from 10 different counties, mostly in north central Florida, died of lead poisoning (table 11.14, figure 11.7). Of these, 9 were

males and 7 were females; 11 were adults and 5 were juveniles. Most (87%) occurred during the 5-month period of November through March. A diagnosis of lead poisoning in Bald Eagles is made when the concentration of lead in liver tissue is greater than 10 ppm (wet weight) and this finding is accompanied by 1 or more appropriate postmortem observations such as emaciation, enlarged gall bladder, or bile staining of the proventriculus (Pattee et al. 1981; Thomas 1994). A twentieth eagle, an adult female from Alachua County in 1975, had a high concentration of lead in its liver (26 ppm), but lacked lesions that usually accompany lead poisoning (NWHC records). This eagle was diagnosed as having died of trauma, but lead may have been a contributing factor.

In October 1997 an adult male Bald Eagle from Marion County died of lead poisoning (Fischer 1997). This bird had been banded and

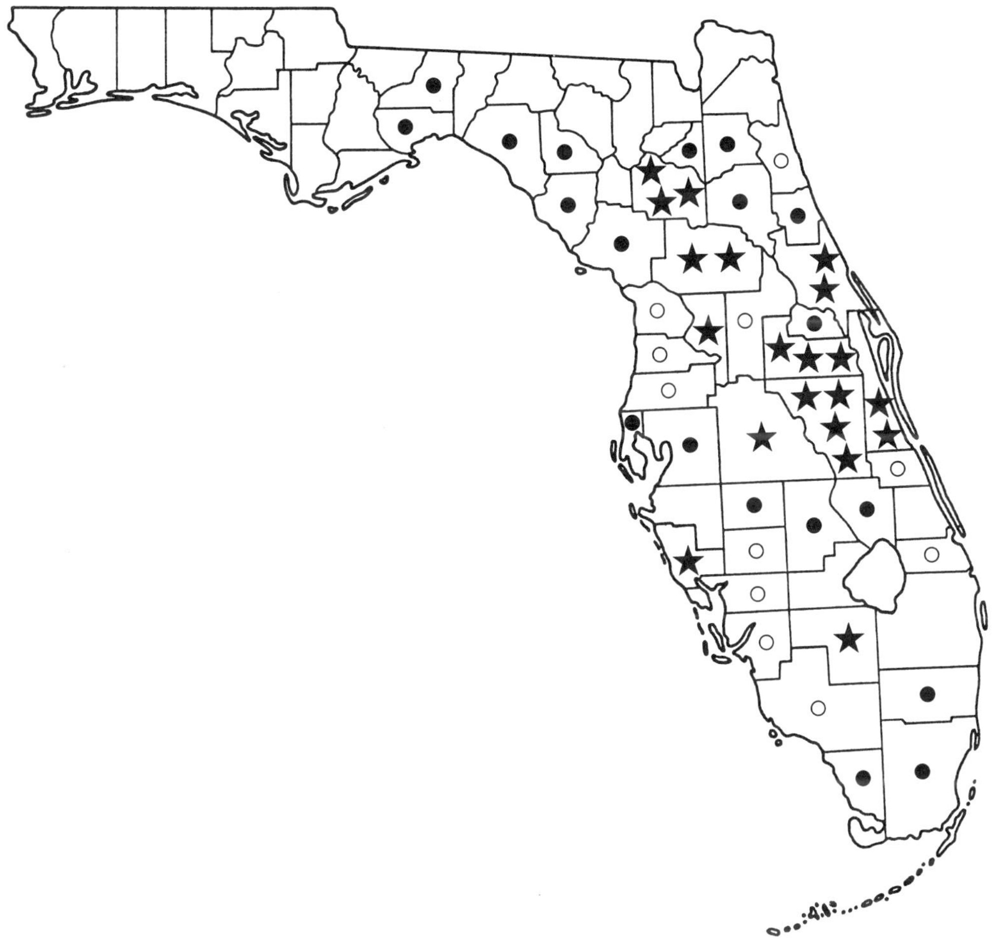

FIGURE 11.7. Distribution of mortality of Bald Eagles due to lead poisoning *(stars)* and counties in which at least 1 Bald Eagle (numbers varied from 1 to 18, mean = 2.9 eagles per county) has been found with lead in liver tissue *(solid circles)* in Florida, 1973–94. *Open circles* = counties in which at least 1 Bald Eagle (numbers varied from 1 to 12, mean = 3.6 eagles per county) has been tested and found to be negative for lead. Data from National Wildlife Health Center records.

equipped with a radio in 1989 on Lake Loch-loosa and had spent 2 summers on Chesapeake Bay. At necropsy fragments of lead and a .40 caliber bullet jacket were found within the lumen of the proventriculus. Concentrations of lead in tissues were 73 ppm in liver and 29 ppm in kidney. It was concluded that the eagle had ingested a spent pistol bullet, probably from the carcass of an animal that it had scavenged (Fischer 1997).

Most lead poisoning of Bald Eagles has been linked with the use of lead shot for waterfowl hunting (Feierabend and Myers 1984; NWHC 1985). Apparently eagles ingest lead shot when they feed on hunter-crippled waterfowl or waterfowl that have lead shot embedded in their tissues or present in their gizzards. Pattee and Hennes (1983) called attention to the associations of large concentrations of waterfowl and eagles between October and March, the use of lead shot for hunting, and the occurrence of lead poisoning in eagles. In 1987 the U.S. Fish

Table 11.14. Data on Bald Eagles from Florida that died of lead poisoning

Year	County	Age[a]	Sex	Concentrations of lead in liver (ppm wet wt.)[b]	Data source
1978	Alachua	JU	M	28	Kerr (1978), Reichel (1978), Reichel et al. (1984)
	Sarasota	AD	F	38[c]	Locke (1978), Reichel (1978), Reichel et al. (1984)
1983	Volusia	AD	F	37	Siegfried (1983), Mulhern (1983)
1988	Marion	AD	M	50	Franson (1990)
	Orange	JU	M	22	Franson (1992)
	Osceola	AD	M	16	Franson (1993)
	Osceola	AD	M	10	Ibid.
	Polk	JU	F	11	Franson (1994)
	Volusia	AD	F	36	Franson (1993)
1990	Orange	JU	F	33	Ibid.
1991	Brevard	JU	F	12	Thomas (1992)
	Orange	AD	M	16	Ibid.
1992	Brevard	AD	M	35	Franson (1992)
	Hendry	AD	F	31	Ibid.
1993	Osceola	AD	M	14	Thomas (1993)
	Osceola	AD	F	12	Franson (1993)
	Sumter	AD	M	20	Franson (1994)
1994	Alachua	AD	M	24	Thomas (1994)
	Marion	JU	F	17	Ibid.

a. JU = juvenile, AD = adult.
b. >10 ppm is indicative of lead poisoning in Bald Eagles (Pattee et al. 1981).
c. This eagle also had 13 ppm lead in its kidney.

and Wildlife Service banned the use of lead shot for hunting waterfowl in areas where eagles and waterfowl occur together in high numbers (Henny and Anthony 1989) and in 1991 it banned the use of lead shot completely for waterfowl hunting (Morehouse 1992). Hopefully, secondary lead poisoning of Bald Eagles will become a rare event in the future.

Six Bald Eagle eggs, 1 each from nests in Buoy Key, Cormorant Key, Palm Key, Deer Key, Crab Key, and Manatee Key, were collected in 1968 and examined for lead. No lead was detected (Krantz et al. 1970).

Golden Eagles rarely have problems with lead intoxication since they feed mainly on mammals, rather than birds such as waterfowl (Reichel et al. 1969a). Nevertheless, 2.5 ppm of lead was found in the liver of an emaciated immature female Golden Eagle from Madison County in 1988 (NWHC records). The significance of this finding is unknown. No Golden Eagles are known to have died of lead poisoning in Florida.

X. Mercury

From 1978 to 1993, liver samples from 59 Bald Eagles in Florida were analyzed for mercury. Concentrations were found in all of the samples and ranged from 0.1 to 27.0 ppm, median = 2.3 (table 11.15). These eagles originated from 22 counties scattered throughout the state (figure 11.8). Four eagles had concentrations above 10 ppm and included 2 from Osceola County (12 and 27 ppm) and 1 each from Brevard (11 ppm) and Seminole (13 ppm) counties. In another study conducted in Florida, Wood et al. (1993,

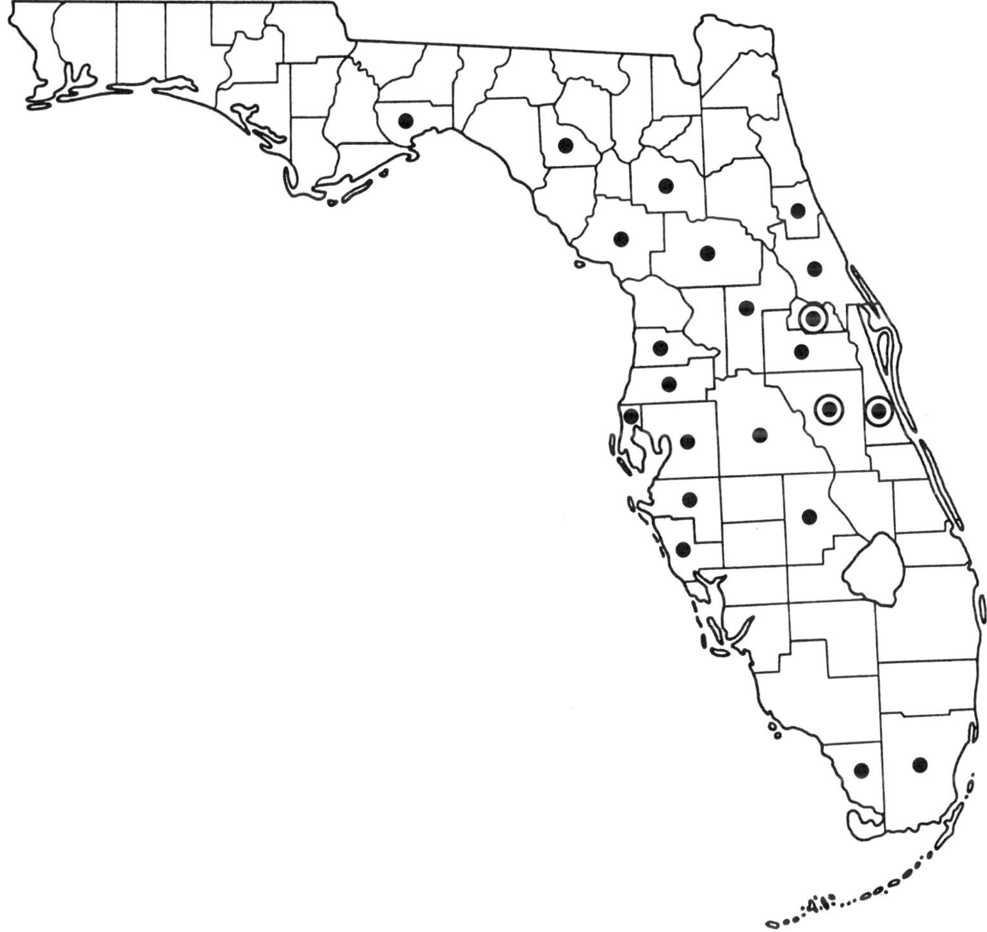

FIGURE 11.8. Distribution of counties *(solid circles)* in which at least 1 Bald Eagle (numbers varied from 1 to 9, mean = 2.7 eagles per county) has been found with mercury in liver tissue in Florida, 1978–93. *Solid circles within a circle* = counties in which 1 or more eagles were found with mercury concentrations >10 ppm. Data from National Wildlife Health Center records.

1996) examined 48 blood and 61 feather samples from nestling Bald Eagles at 42 nests and adult feathers from 20 nests during 1991–93. They also tested 32 liver, 10 feather, and 5 blood samples from 33 carcasses from throughout Florida during 1987–93. They concluded that the concentrations of mercury in Bald Eagles in Florida were above background amounts, but were similar to or lower than those in other eagle populations in the United States. They analyzed mercury in blood, liver,

and feathers of adults and nestlings and found that concentrations were highest in feathers and in adults (figure 11.9). They reported that the concentrations of mercury in feathers were higher in adults than in nestlings, but that the 2 were highly correlated. They suggested that collecting adult feathers at nest sites would be representative for both adults and nestlings and thereby could be used in a monitoring program.

Information on mercury in Golden Eagles in Florida is limited to 1 case. Kidney tissues from

Table 11.15. Concentrations of mercury in livers of Bald Eagles from Florida

| Year | No. of eagles | | ppm (wet weight)[a] | | | Data source |
	Examined	Positive	Median	Mean	Range	
1978	2	2	3.2	3.2	2.3–4.2	Reichel (1978)
1979	7	7	2.5	4.9	0.7–13	Reichel (1979)
1980	2	2	14	14	1.9–27	Reichel (1980)
1981	11	11	1.7	1.9	0.4–4.1	Reichel (1981)
1982	4	4	1.4	1.5	0.6–2.8	Reichel (1982)
1987	1	1	2.1	2.1	—	Wood et al. (1993)
1988	1	1	3.1	3.1	—	Ibid.
1990	2	2	1.8	1.8	—	Ibid.
1991	5	5	1.0	1.2	0.6–1.7	Ibid.
	1	1	4.0	4.0	—	Thomas (1992)
1992	9	9	2.0	2.1	0.9–4.4	Wood et al. (1993)
1993	14	14	3.1	3.4	0.1–12	Ibid.
All years	59	59	2.3	3.6	0.1–27	—

a. Limits of detection for the data from Reichel (1978–82) = 0.02 ppm (wet wt.); limits not given by Thomas (1992) or Wood et al. (1993).

an emaciated juvenile female Golden Eagle that died in Madison County in 1988 contained 0.5 ppm of mercury (Locke 1989).

The source of mercury contamination for eagles is most certainly linked with their food. In Florida fish make up a large part of the diet of Bald Eagles, but they also feed on wading birds, coots, waterfowl, and other birds, as well as various mammals, reptiles, and amphibians. Mercury has been found in 7 species of wading birds, and in Mottled Ducks, Mallards, largemouth bass (*Micropterus salmoides*), gar (*Lepisosteus* spp.), and bowfin (*Amia calva*) (Heath and Hill 1974; Sundlof et al. 1994; Ware et al. 1990), all of which are consumed by Florida Bald Eagles (Wood et al. 1993). The concentrations in Mottled Ducks were low, but high in wading birds and fish. No correlation was found by Wood et al. (1993) between the concentrations of mercury in largemouth bass and those in Bald Eagles from the same lakes or river systems, however.

Mercury was found in 8 eggs obtained from Bald Eagle nests in Florida between 1969 and 1979 (table 11.16). All concentrations were small, i.e., less than 1 ppm.

Mercury contamination may be of significance, since sublethal amounts of this metal are known to have adverse effects on repro-

duction in ducks and pheasants (see discussion in Wiemeyer et al. 1984). This has not been examined for Bald Eagles, however, nor have lethal concentrations of mercury been established (Belisle et al. 1972; Eisler 1987).

XI. Other metals

Concentrations of 8 other metals in livers of 25 Bald Eagles from Florida during 1978–82 are given in table 11.17. Also, in table 11.18 concentrations for copper, iron, and zinc are presented for 6 Bald Eagle eggs collected in 1968. The significance of these findings is not known.

XII. Avian vacuolar myelinopathy

In 1994 a fatal neurologic disease was recognized in Bald Eagles and American Coots in Arkansas. Since then the disease has been identified in several species of waterfowl (Mallards, Ring-necked Ducks, and Canada Geese), Great Horned Owls, and Killdeer. The range of the disease has been extended to North Carolina, South Carolina, Georgia, and Texas (Fischer 2000–01).

Mortality was high in eagles and low in coots

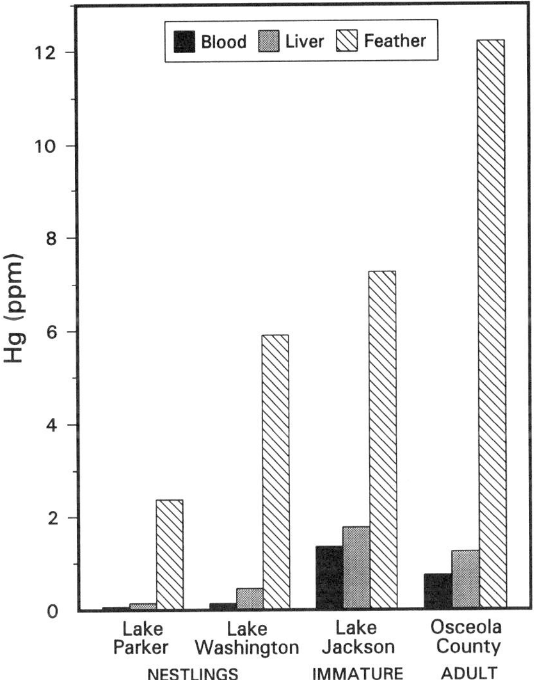

FIGURE 11.9. Comparison of concentrations of mercury in blood, liver, and feathers of 2 nestlings, 1 2-year-old, and 1 adult Bald Eagle from Brevard, Osceola, and Polk counties, 1990–93. Courtesy of Petra B. Wood.

and waterfowl. A description of the neuropathology has been provided by Thomas et al. (1998). Sick eagles were observed to overfly perches and collide with objects. Coots showed signs of incoordination and limb paralysis. The cause of this malady has not been determined, although toxins are suspected (Thomas et al. 1998). This disease has not been identified in Florida birds, but since it is in neighboring areas, we should be concerned about it and be on the lookout for it.

XIII. Neoplasia

No information is available on this topic.

XIV. Biotoxins

No information is available on this topic.

XV. Viruses

Two viral infections have been identified in Bald Eagles in Florida. These include 18 cases of pox (table 11.19) and 3 cases of an infection by an enveloped DNA virus (table 11.20).

Pox infections were found in 11 juveniles and 1 nestling; 7 were females, 3 were males. The genders of 8 and the ages of 6 were unknown. Such infections can be severe (figure 11.10) and if the lesions involve the eyes and mouth, mortality may result (Schmeling and Docherty 1982). In all of the juvenile eagles mentioned above, it was not known if mortality was due to the pox infections or to a combination of poxvirus and other complicating disease factors. The infection in the nestling was minor and of no importance to the health of the eagle at the time it was observed.

Table 11.16. Concentrations of mercury found in 8 eggs of Bald Eagles

Year	County/site	Lower limit of detection (ppm)	Amount in egg (ppm)	Data source
1969	Lee	0.05	0.30	Wiemeyer et al. (1972)
	Lee	0.05	0.70	Ibid.
1975	Camp Key	0.02	0.41	Wiemeyer et al. (1984)
1976	Pine Island	0.02	0.42	Ibid.
	Deer Key	0.02	0.25	Ibid.
1977	Pine Island	0.02	0.19	Ibid.
	Pine Island	0.02	0.18	Ibid.
1979	Pine Island	0.02	0.39	Ibid.

Table 11.17. Concentrations of heavy metals (other than lead and mercury) found in livers of Bald Eagles from Florida, 1978–82[a]

Metal	Lower limit of of detection (ppm)	No. eagles examined	% positive	Concentration (ppm wet wt.) Median	Range
Copper	0.10	25	100	2.9	0.99–75
Zinc	0.10	25	100	30	11–400
Nickel	0.10	25	64	0.32	0.10–1.20
Cadmium	0.10	25	16	0.10	0.10–0.26
Iron	1.00	25	100	420	150–6,200
Vanadium	0.20	18	0	—	—
Chromium	0.05	25	20	0.19	0.09–0.64
Arsenic	0.05	18	50	0.09	0.04–0.25
Selenium	0.10	25	100	0.44	0.04–3.50
Thallium	0.10	19	0	—	—

a. Data modified from Reichel (1978–82).

Table 11.18. Concentrations of copper, iron, and zinc found in 6 eggs from Bald Eagles in Florida, 1968

Egg No.	Site	Amount in egg (ppm)[a,b] Copper	Iron	Zinc
1	Buoy Key	0.7	10	4.6
2	Cormorant Key	1.0	17	5.1
3	Palm Key	1.0	13	6.5
4	Deer Key	0.7	15	7.1
5	Crab Key	0.9	18	8.1
6	Manatee Key	1.2	14	5.5

Source: Krantz et al. (1970).
a. Lower limits of detection were not given.
b. The 6 eggs were also tested for cadmium and nickel; results were negative.

Table 11.19. Characteristics of avian pox infections in Bald Eagles from Florida

Year	County	Age[a]	Sex[b]	Other conditions	Data source
1982	Okeechobee	JU	F	Acute pneumonia	Siegfried (1982)
1983	Collier	JU	M	Anemia, emaciation	Sileo (1983)
1987	Seminole	JU	M	None	Franson (1992)
1988	Volusia	NS[c]	U	None	Spalding & Wood (1988)
1989	NG	JU	F	Roadkill	Locke (1989)
1989	Orange	JU	F	Dermatitis, hepatitis, epidermitis[d], emaciation	Franson (1992)
1992	Osceola	JU	F	Emaciation	Ibid.
1992	Polk	JU	F	None	Ibid.
1992	Sarasota	JU	F	Emaciation	Collins & Gilliland (1995)
1993	Osceola	JU	M	Emaciation	Ibid.
1994	Brevard	JU	U	Emaciation	Ibid.
1995	Seminole	NG	NG	None	Collins (1999)
	Volusia	NG	NG	None	Ibid.
1996	Okeechobee	NG	NG	None	Ibid.
	Orange	NG	NG	None	Ibid.
	Volusia	JU	F	Pentobarbital poisoning	Brannian (1996)
1997	Alachua	NG	NG	None	Collins (1999)
1998	Brevard	NG	NG	None	Ibid.

NG = not given.
a. JU = juvenile, NS = nestling.
b. F = female, M = male, U = unknown.
c. Eight weeks old.
d. An enveloped DNA virus was isolated from the oral mucosa of this eagle.

The enveloped DNA virus (which has not been further characterized) was isolated from 3 Bald Eagles: an adult male, an adult female, and an immature female. In all 3 cases the eagles died of other complications and the virus was isolated from oral plaques. The significance of infections of this DNA virus to Bald Eagles is not known.

Albers (1984) tested 2 Bald Eagles in Florida for serologic evidence of rabies infection. These

FIGURE 11.10. Poxvirus infection in a juvenile male Bald Eagle from Collier County, April 20, 1983. Courtesy of Lou Sileo.

Table 11.20. Characteristics of DNA enveloped virus infections in Bald Eagles

Year	County	Age[a]	Sex	Other conditions	Data source
1988	Osceola	AD	M	Necrotic sinusitis and glossitis; high lead concentration (9.9 ppm)	Docherty (1988)
1989	Orange	JU	F	Avian pox and emaciation	Docherty (1993)
1993	Osceola	AD	F	Lead poisoning	Ibid.

a. AD = adult, JU = juvenile.

eagles came from a wildlife rehabilitation facility in Pinellas County in 1983. Both were seronegative.

XVI. Bacteria

Twenty-nine species of bacteria have been isolated from Bald Eagles in Florida (table 11.21). A number of these were found in eagles that had undergone trauma and the bacterial infections were probably secondary in nature. One eagle had an infection of *Staphylococcus aureus* that was associated with bumblefoot and this eagle and 2 others (1 of which was infected with both *S. aureus* and *S. simulans*) died of staphylococcal septicemia. The eagle with bumblefoot had been in captivity for some time until its death; bumblefoot is one of the most common clinical conditions of captive raptors (Cooper 1978) and is seldom seen in free-ranging birds. With the exception of 1 case, the enteric bacteria (*Salmonella* spp. and *Edwardsiella tarda*) were isolated from the intestinal contents of eagles and were not associated with lesions; although these pathogens can cause disease in raptors (Halliwell and Graham 1986), they are probably of more interest because of their zoonotic significance. The one exception was the isolation of *E. tarda* from the lungs of an eagle that died of lead poisoning, in which case the secondary bacterial infection was of pathologic importance. In addition to the above findings in Bald Eagles, *Streptococcus fecalis* and an unidentified species of *Enterococcus* were isolated from the pericardial sac of a juvenile female Golden Eagle that

became entangled in a trotline and drowned in Osceola County in 1982 (Duncan 1982).

The significance of most of these bacterial infections to eagles is not known. *Erysipelothrix rhusiopathiae*, while it has not been reported from eagles in Florida, has been documented as the cause of death of a captive Bald Eagle in Maryland (Franson et al. 1994). In his assessment of the bacterial diseases of birds of prey, Needham (1981) listed the following bacteria as important etiologic agents: *Staphylococcus aureus*, *Escherichia coli*, *Proteus* spp., *Mycobacterium* spp., *Pasteurella* spp., *Salmonella* spp., *Clostridium* spp., *Streptococcus* spp., and *Erysipelothrix* spp. Representatives of most of these (except for *Mycobacterium* spp., *Clostridium* spp., and *Erysipelothrix* spp.) have been found in eagles in Florida. Since some of these infections are often a secondary consequence of trauma, and since trauma accounts for a large proportion of morbidity and mortality in Bald Eagles in Florida (figure 11.1), these bacteria may play a more important role than is appreciated.

XVII. Fungi

Six (and possibly 7) species of fungi have been identified from Bald Eagles in Florida (table 11.22). Aspergillosis was determined as the primary cause of death for 12 Bald Eagles between 1978 and 1994. In 7 other cases *Aspergillus* infections were incidental or secondary findings. All cases involved lungs or air sacs and in 1 case the liver and kidneys were also involved. Overall, aspergillosis is seen infrequently in free-living raptors, but is often diagnosed in captive birds (Locke 1987). In many cases the histories of the eagles that were submitted for examination were incomplete, but in some it was known that the eagles had been in captivity for several weeks to several months and were submitted because of some other problem such as lead poisoning, emaciation, or injuries due to trauma. Two of the eagles were nestlings that had fallen out of their nests. Such events could have led to stress-related immunosuppression and subsequent increases in susceptibility to aspergillosis as explained by Redig (1993). Air sac infection by an unidentified species of fungus was the cause of death of an immature male Florida Bald Eagle (date and specific locality unknown). Incidental or secondary infections of 4 other fungi (1 species of *Alternaria* and 3 species of *Candida*) were identified in the esophagus, air sacs, or lungs of 5 Bald Eagles. *Candida parapsilosis* was cultured from the heart of an emaciated immature female Golden Eagle obtained in 1988 in Madison County (Duncan 1988). *Rhizopus* sp. was cultured from the trachea of a fledgling Bald Eagle with respiratory disease found in Brevard County in 1991 (Kiehl 1991).

The impact of these fungal infections on populations of eagles in Florida is unknown, but may not be significant unless eagles are subjected to some type of stress such as trauma or other concurrent infections or diseases.

XVIII. Protozoans

Nine species of parasitic protozoans have been identified from Bald Eagles in Florida (tables 11.23 and 11.24). Several of the blood protozoans are shown in figure 11.11. The mosquito vectors of *Plasmodium elongatum* from Bald Eagles in Florida were studied experimentally utilizing Pekin ducklings (Nayar et al. 1998). Three species of *Culex* (*C. nigripalpus*, *C. restuans*, and *C. salinarius*) were found to be capable of transmitting *P. elongatum* from duck to duck. It was concluded that these 3 species might be

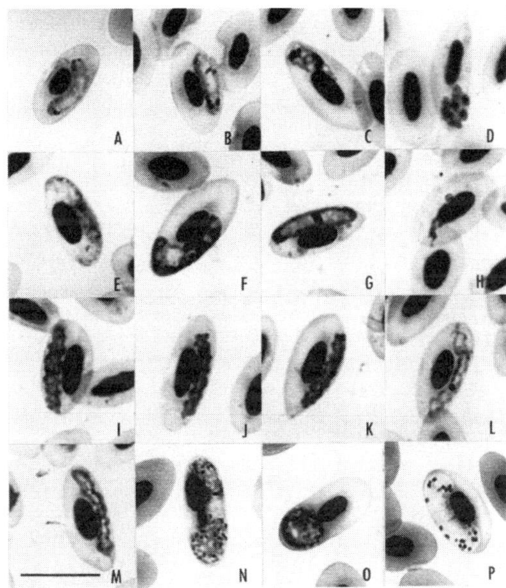

FIGURE 11.11. Blood parasites from Bald Eagles in Florida. *A–B*, macrogametocytes of *Plasmodium elongatum*; *C*, binucleate schizont of *P. polare*; *D*, schizont of *P. polare* with 10 nuclei; *E–F*, macrogametocytes of *P. polare*; *G*, microgametocyte of *P. polare*; *H*, schizont of *Plasmodium forresteri*; *I–M*, gametocytes of *Plasmodium forresteri*; *N–O*, macrogametocytes of *Haemoproteus elani*; *P*, microgametocyte of *H. elani*. Horizontal bar = 10 μm. Courtesy of Sam R. Telford, Jr.

natural vectors in Florida and because of the timing of their population peaks and their feeding habits could probably transmit *P. elongatum* to Bald Eagles throughout the year. The pathologic significance of these infections is not known, although the 3 species of *Plasmodium* may be of importance. Malaria has been implicated in the mortality of Bald Eagles that were transplanted from Alaska to North Carolina in 1987 (Sherrod et al. 1990), but this finding has not been verified.

Leucocytozoon toddi is known to occur in hawks in high intensities without being associated with mortality (Peirce 1981), but little is known about its effects on eagles. Stuht et al. (1999) described a nestling Bald Eagle from

Table 11.21. Bacteria identified from Bald Eagles in Florida

Bacterium County/site	Year	Organs cultured	Cause of death[a]	Data source
Achromobacter xylosoxidans				
Highlands	1991	Spleen	Ulcerative dermatitis	Duncan (1991)
Acinetobacter calcoaceticus				
"Florida"	1991	Liver	Open	Duncan (1991)
Actinobacillus sp.				
Osceola	1988	Sinus exudate	Lead poisoning	Duncan (1993)
Bacillus cereus				
Highlands	1991	Spleen	Ulcerative dermatitis	Duncan (1991)
"No. central Fla."	1986	Air sacs	Aspergillosis	Woodard (1986)
"Florida"	1991	Lung	Open	Duncan (1991)
Bacillus sp.				
Indian River	1986	Foot, tongue	Trauma	Duncan (1986)
Osceola	1988	Liver, lung	Aspergillosis	Duncan (1988)
Cephalosporium sp.				
Manatee	1993	Pericardial sac	Esophageal perforation	Duncan (1993)
Chlamydia sp.[b]				
Pinellas	2000	Pericardial sac	Unknown	Spalding (2001)
Edwardsiella tarda				
Lake / Orange	1971	Cloaca	"Sick"	White et al. (1973)
Osceola	1982	Intestines	Trauma	Duncan (1982)
	1988	Lungs	Lead poisoning	Duncan (1993)
Enterobacter cloacae				
Marion	1989	Brain	Congestion	Duncan (1989)
Orange	1989	Brain	Avian pox	Duncan (1992)
Enterococcus sp.				
Brevard	1989	Esophagus	Trauma	Duncan (1989)
Dade	1987	Liver	Gunshot	Duncan (1987)
Indian River	1986	Foot	Trauma	Duncan (1986)
Lee	1988	Air sacs	Trauma	Duncan (1988)
Osceola	1988	Liver	Aspergillosis	Ibid.
"No. central Fla."	1986	Femur	Aspergillosis	Woodard (1986)
"Florida"	1991	Lung	Open	Duncan (1991)
Escherichia coli				
Alachua	1987	Liver, kidney	Aspergillosis	Duncan (1987)
Brevard	1987	Liver, lung, pericardial sac, joints	Trauma	Ibid.
Orange	1992	Liver	Electrocution	Duncan (1992)
Polk	1987	Knee joint	Aspergillosis	Duncan (1987)
Seminole	1987	Pericardial sac, heart	Avian pox	Duncan (1992)
Sumter	1993	Lung, intestine	Lead poisoning	Duncan (1994)
"No. central Fla."	1986	Femur, liver, spleen	Aspergillosis	Woodard (1986)
"Florida"	1991	Pericardium	Open	Duncan (1991)
Haemophilus-like sp.				
Osceola	1988	Sinus exudate	Lead poisoning	Duncan (1993)
Klebsiella oxytoca				
Osceola	1988	Sinus exudate	Lead poisoning	Duncan (1989)
Klebsiella pneumoniae				
Alachua	1994	Liver	Metal poisoning	Duncan (1994)
Collier	1989	Liver	Trauma	Duncan (1989)
Pasteurella multocida				
"Florida"	1970	Liver	Gunshot	Belisle et al. (1972)
Pasteurella sp.				
Osceola	1988	Sinus exudate	Lead poisoning	Duncan (1993)

(continued)

Table 11.21. *(continued)*

Bacterium County/site	Year	Organs cultured	Cause of death[a]	Data source
Proteus sp.				
Brevard	1987	Humerus	Trauma	Duncan (1987)
Lake	1986	Intestine	Trauma, aspergillosis, encephalomyelitis	Duncan (1986)
Orange	1992	Spleen	Electrocution	Duncan (1992)
Osceola	1988	Liver	Aspergillosis	Duncan (1988)
	1992	Liver	Avian pox	Duncan (1992)
Seminole	1987	Pericardial sac, heart	Avian pox	Duncan (1992)
Pseudomonas aeruginosa				
Manatee	1993	Pericardial sac	Esophageal perforation	Duncan (1993)
Pseudomonas fluorescens				
Osceola	1985	Lung	Aspergillosis	Duncan (1992)
Pseudomonas sp.				
Brevard	1989	Esophagus	Trauma	Duncan (1989)
Dade	1987	Brain	Gunshot	Duncan (1993)
Salmonella kentucky				
Highlands	1992	Intestines	Open	Duncan (1992)
Salmonella typhimurium				
Osceola	1982	Intestines	Trauma	Duncan (1982)
Salmonella sp.				
Osceola	1982	Intestines	Trauma	Duncan (1982)
Serratia odorifera				
Osceola	1992	Liver	Avian pox	Duncan (1992)
Staphylococcus aureus				
Alachua	1994	Subcutaneous	Lead poisoning	Duncan (1994)
Dade	1978	Liver, heart	Bumblefoot and septicemia[c]	Duncan (1978)
	1987	Elbow joint, capsule & fluid	Gunshot	Duncan (1993)
Indian River	1986	Tongue	Trauma	Duncan (1986)
Leon	1984	Liver, spleen	Septicemia[c]	Duncan (1984)
Lee	1984	Leg	Septicemia[c]	Ibid.
Staphylococcus sciuri				
Dade	1987	Liver	Gunshot	Duncan (1993)
Staphylococcus simulans				
Leon	1984	Liver, spleen	Septicemia[c]	Duncan (1984)
Staphylococcus sp.				
Alachua	1989	Heart	Open	Duncan (1989)
Collier	1989	Heart	Trauma	Ibid.
Dade	1987	Liver	Gunshot	Duncan (1993)
Indian River	1986	Foot	Trauma	Duncan (1986)
Lee	1988	Air sacs	Trauma	Duncan (1988)
Streptococcus sp.				
Alachua	1989	Heart, liver	Open	Duncan (1989)
Brevard	1989	Esophagus	Trauma	Ibid.
Osceola	1992	Liver	Avian pox	Duncan (1992)
"No. central Fla."	1986	Liver, spleen	Aspergillosis	Woodard (1986)

a. From records of NWHC, Belisle et al. (1972), White et al. (1973).
b. Presumptive diagnosis; not verified by culture techniques. Histologic appearance and special stains (Machievelli) indicated that *Chlamydia* infection was likely. The eagle was a fledgling and its parents were observed feeding on psittacines from a nearby colony.
c. Septicemia resulting from staphylococcal infection.

Table 11.22. Fungi identified from Bald Eagles in Florida

Fungus County/site	Year	Organ/tissue cultured	Cause of death[a]	Data source
Aspergillus fumigatus				
Alachua	1987	Lungs, liver, kidney	Aspergillosis	Duncan (1993)
Charlotte	1987	Lungs, air sacs	Aspergillosis	Duncan (1993)
Lake	1986	Air sacs	Open[b]	Duncan (1986)
	1986	Air sacs	Trauma, encephalomyelitis, aspergillosis	Duncan (1992)
Lee	1985	Lungs	Emaciation and aspergillosis	Duncan (1985)
	1988	Air sacs	Aspergillosis	Duncan (1988)
Osceola	1985	Lungs	Aspergillosis	Duncan (1992)
	1988	Lungs	Aspergillosis	Ibid.
Polk	1981	Lungs	Aspergillosis	Duncan (1981)
	1984	Lungs	Aspergillosis	Duncan (1984)
	1987	Air sacs	Aspergillosis	Duncan (1992)
"No. central Fla."	1986	Lungs	Aspergillosis	Woodard (1986)
Unknown	Unknown	Air sacs	Aspergillosis	Duncan (1981)
Aspergillus sp.				
Alachua	1994	Lungs	Aspergillosis	Deem & Terrell (1996)
	1994	Lungs	Lead poisoning	Deem & Terrell (1996)
Hillsborough	1984	Air sacs	Trauma	Duncan (1984)
Levy	1993	Lungs	Trauma	Deem & Terrell (1996)
Seminole	1978	Lungs	Aspergillosis	Duncan (1978)
	1979	Lungs	Aspergillosis	Duncan (1979)
Sumter	1993	Lungs	Lead poisoning	Duncan (1994)
Alternaria sp.				
Hillsborough	1984	Air sacs	Trauma	Duncan (1984)
Candida albicans				
Brevard	1987	Lungs	Trauma	Duncan (1987)
Highlands	1992	Esophagus	Open	Duncan (1992)
Candida tropicalis				
Polk	1984	Lungs	Aspergillosis	Duncan (1984)
Candida zeylanoides				
Charlotte	1988	Esophagus	Trauma	Duncan (1992)
Rhizopus sp.				
Brevard	1991	Trachea	None (live bird)	Kiehl (1991)
Unidentified fungus				
"Florida"	Unknown	Air sacs	Mycotic airsacculitis	Duncan (1992)

a. From records of NWHC and Woodard (1986).
b. Cause of death undetermined.

Michigan that was infected with *L. toddi,* had signs of anemia, and was below normal in size for its age. However, they were not able to rule out other concurrent diseases or infections that may have contributed to the anemia. It has been suggested that blood protozoans may be important when they occur as infections concomitant with other diseases (Bennett et al. 1993), but this has not been studied in eagles.

The life cycles and pathogenicity of the 3

Table 11.23. Protozoans (other than blood parasites) identified from Bald Eagles

Protozoan County	Year(s)	Site in eagle	Data source
Sarcocystis sp. I[a]			
Alachua	1994	Heart & skeletal muscles	Meteyer (1994)
Charlotte	1987	Heart	Galbreath (1993)
Indian River	1991	Heart	Thomas (1992)
Marion	1968	Heart	Locke (1968)
Okeechobee	1982	Skeletal muscles	Siegfried (1982)
Orange	1988	Heart	Woods (1992)
Orange	1992	Heart	Steinberg (1993)
Osceola	1988	Heart	Woods (1992)
Polk	1988	Skeletal muscles	Ibid.
Volusia	1993	Heart	Meteyer (1994)
Sarcocystis sp. II[b]			
Brevard	1990–91	Feces	Carleton (1994)
Levy	1990–91	Feces	Ibid.
Marion	1990–91	Feces	Ibid.
Sarcocystis sp. III[c]			
Marion	1990–91	Feces	Ibid.
Intestinal coccidia[d]			
St. Johns	1973	Small intestine	Locke (1973)

a. Identified as sarcocysts in muscle tissue.
b. Identified as sporocysts in feces. Sporocysts measured 11.6 x 7.8 (range = 10.0–13.0 x 7.0–9.0)μm. Sporozoites measured 1.5 x 7.2 (range = 1.0–2.5 x 6.0–9.0)μm. Feces from 17 other eagles were negative for sporocysts and included birds from Alachua ($n = 8$), Brevard ($n = 1$), Charlotte ($n = 1$), Highlands ($n = 1$), Osceola ($n = 3$), Putnam ($n = 1$), Suwannee ($n = 1$), and Volusia ($n = 1$) counties.
c. Identified as sporocysts in feces. Sporocysts measured 11.2 x 8.0 (range = 10.4–12.0 x 7.0–9.0)μm. Sporozoites measured 3.5 x 2.5 (range = 3.4–3.5 x 2.5)μm.
d. This eagle had hemorrhagic enteritis. The diagnosis was based on seeing oocysts in intestinal tissue.

species of *Sarcocystis* and the intestinal coccidium are not known. According to Cawthorn (1993), sarcocysts in tissues are not harmful to raptors, whereas intestinal infections may be pathogenic. Crawley et al. (1982) found 2 morphologically distinct sarcocysts in cardiac and skeletal muscles of a Bald Eagle from Alabama, and Tuggle and Schmeling (1982) reported *Sarcocystis*-like sporocysts in feces of several Bald Eagles from Alaska, California, and Colorado.

XIX. Helminths

Twenty-three species of helminths have been identified from Bald Eagles in Florida, including 8 trematodes, 1 cestode, 3 acanthocephalans, and 11 nematodes (tables 11.25–11.27).

There is only 1 record of parasitism of Golden Eagles in Florida. Tuggle (1982) found the nematode *Porrocaecum angusticolle* in the proventriculus of an adult male Golden Eagle from Dade County in 1982. The number of nematodes was not determined.

Because prevalence and intensity data are limited, it is difficult to assess the pathological significance of these parasites on eagle populations. Kocan and Locke (1974) and Tuggle and Schmeling (1982) pointed out that pathogenicity has been reported for *Serratospiculoides amaculata* and *Capillaria contorta* in other raptors. Ackerman et al. (1992) described a case of acute respiratory distress in a Bald Eagle from Florida (exact locality and date not given) after it was treated for severe head trauma. At necropsy 2 adult filarids were found

Table 11.24. Protozoans identified in the peripheral blood of Bald Eagles[a]

County	Year(s)	No. examined	Number positive[b]				
			H.e.	L.t.	P.e.	P.p.[c]	P.f.[d]
Alachua	1975–90	7	1	0	0	1	1
Baker	1988	1	0	1	0	0	0
Bradford	1995	1	0	1	0	0	0
Brevard	1995–96	2	0	0	0	0	0
Citrus	1988	2	0	0	0	0	0
Collier	1990	1	0	0	0	0	0
Columbia	1988	1	0	0	1	0	0
Flagler	1996	1	0	0	1[e]	0	1[e]
Hernando	1996	1	0	0	0	0	0
Indian River	1996	1	0	0	0	0	0
Lafayette	1988	1	0	1	0	0	0
Levy	1991	1	0	0	1	0	0
Manatee	1996	2	0	0	0	0	0
Marion	1985–91	7	0	1	0	0	0
	1994–95	2	0	0	0	0	0
Orange	1996	1	0	0	0	0	0
Osceola	1995–96	5	0	0	0	0	0
Pinellas	1996	1	0	0	0	0	0
Putnam	1989	1	0	0	0	0	0
	1996	1	0	0	0	0	0
Sarasota	1995	1	0	0	0	0	0
Seminole	1995	2	0	0	0	0	0
Unknown	1985	1	1	1	0	0	0
	1994–96	2	0	0	0	0	0
All	1975–96	46	2	5	3	1	2

Source: 1975–91 data from Forrester et al. (1994); 1994–96 data from Telford et al. (1996).
a. During 1988–93, 40 nestling Bald Eagles (approximately 6–8 weeks old) were sampled and all were negative, from Alachua (*n* = 10), Marion (*n* = 4), Osceola (*n* = 16), Polk (*n* = 9), Putnam (*n* = 1), and Volusia (*n* = 4) counties.
b. H.e. = *Haemoproteus elani*, L.t. = *Leucocytozoon toddi*, P.e. = *Plasmodium elongatum*, P.p. = *P. polare*, P.f. = *P. forresteri*.
c. Greiner et al. (1981) reported a parasite resembling *Plasmodium polare* from a Bald Eagle from "north central Florida" in 1978 and pointed out that some of the characteristics of this organism varied from the known features of *P. polare*.
d. Referred to as an undescribed species by Telford and Forrester (1992) in a Barred Owl from Valdosta, Georgia, and by Forrester et al. (1994) in a Great Horned Owl from Alachua County, Florida. Described as *P. forresteri* by Telford et al. (1997).
e. Peripheral blood film was negative. Infection was identified via inoculation of blood from this eagle into a Pekin duckling (Nayar et al. 1998).

Table 11.25. Trematode infections in Bald Eagles from Florida[a]

Species (site)[b] County/site	Year	No. of trematodes found	Data source
Clinostomum complanatum (TR)			
Seminole	1980	1	Tuggle & Schmeling (1982), NWHC records
Volusia	1982	2	Ibid.
Mesostephanus appendiculatoides (SI)			
Alachua	1994	12	Kinsella & Foster (1995)
Hardee	1993	85	Ibid.
Lee	1994	70	Ibid.
Marion	1995	2	Ibid.
Osceola	1994	ND	Kinsella & Cole (1995)
Pasco	1993	20	Kinsella & Foster (1995)
Polk	1994	25	Ibid.
Putnam	1995	1	Ibid.
Mesostephanus sp. (SI)			
Alachua	1989	ND	Campbell (1990)
Microparyphium facetum (CL)			
Charlotte	1994	1	Kinsella & Foster (1995)
Marion	1995	1	Ibid.
Orange	1988	ND	Kinsella & Cole (1995)
	1994	2	Kinsella & Foster (1995)
Polk	1995	1	Ibid.
Volusia	1995	2	Ibid.
Neodiplostomum attenuatum (SI)			
Charlotte	1994	9	Kinsella & Foster (1995)
Osceola	1994	ND	Kinsella & Cole (1995)
Pasco	1993	3	Kinsella & Foster (1995)
"No. central Fla."	1986	1	Kinsella & Greiner (1993)
"Florida"	1993	3	Kinsella & Foster (1995)
Phagicola longa (SI)			
Charlotte	1994	312	Kinsella & Foster (1995)
Hillsborough	1994	77	Ibid.
Lee	1993	33	Ibid.
	1994	1,245	Ibid.
Marion	1992	67	Ibid.
Pinellas	NG	ND	Hutton (1964)
Polk	1994	10	Kinsella & Foster (1995)
Phagicola nana (SI)			
Orange	1993	ND	Kinsella & Foster (1995)
Osceola	1994	ND	Ibid.
Phagicola sp. (SI)			
Charlotte	1988	ND	Campbell (1990)
	1990	ND	Kinsella & Cole (1995)
	1994	25	Kinsella & Foster (1995)
Lake	1987	ND	Campbell (1990)
Polk	1987	ND	Thomas (1992)
	1988	ND	Kinsella & Cole (1995)
Volusia	1994	ND	Ibid.

(continued)

Table 11.25. *(continued)*

Species (site)[b] County/site	Year	No. of trematodes found	Data source
Posthodiplostomum minimum (KD)			
Charlotte	1994	1	Kinsella & Foster (1995)
Renicola thapari (KD)			
Charlotte	1994	9	Kinsella & Foster (1995)
Hillsborough	1994	1	Ibid.
Lee	1994	82	Ibid.
Marion	1992	8	Ibid.
Pasco	1993	4	Ibid.
Volusia	1995	2	Ibid.
Renicola sp. (KD)			
"Florida"	1986	ND	Kinsella & Cole (1995)
Strigea falconis (SI)			
Charlotte	1994	2	Kinsella & Foster (1995)
Pasco	1993	1	Ibid.
"Florida"	1992	2	Ibid.
Strigea sp. (SI)			
Charlotte	1988	ND	Campbell (1992)
	1990	ND	Kinsella & Cole (1995)
Orange	1994	ND	Ibid.
Volusia	1994	ND	Ibid.

ND = not determined, NG = not given.

a. Each entry represents 1 eagle. Thirty-eight eagles were examined, from the following counties: Alachua (*n* = 2), Charlotte (3), Clay (1), Hardee (1), Hillsborough (1), Lake (1), Lee (2), Marion (4), Orange (2), Osceola (2), Pasco (2), Pinellas (1), Polk (7), Putnam (1), Seminole (1), Volusia (3), unknown (4).

b. Infection sites: TR = trachea, SI = small intestine, KD = kidney, CL = cloaca.

in the air sacs and 2 others in the lungs. One of the nematodes had penetrated the lung and was partially in the coelomic cavity, resulting in the development of pneumocoelom, a condition the authors felt contributed significantly to the death of this eagle. Kinsella et al. (1998) examined the specimens that Tuggle and Schmeling (1982) and Ackerman et al. (1992) reported as *S. amaculata* and found that they were an undescribed species of *Hamatospiculum*. Specimens referred to as *S. amaculata* by Kocan and Locke (1974) were not available for examination, but these findings cast doubt on their report of this species from Bald Eagles.

Nematodes of the genus *Contracaecum* appear to be very common and widespread in Bald Eagles in Florida. Four species have been reported, but little is known about their significance, since good data on prevalence and

intensity are lacking. In 1 case, however, an adult male Bald Eagle that died of electrocution in Marion County in 1993 was found to have "an irregular, 5 x 10 mm crateriform ulcer" in its proventriculus in association with 3 nematodes identified as *Contracaecum* sp. (Cole 1993). The pathological significance of this ulcer was not discussed, but was probably minor.

XX. Arthropods

Fifteen species of parasitic arthropods have been reported from Bald Eagles in Florida and include 2 ticks, 3 mites, 2 louse flies, 2 blow flies, and 6 chewing lice (tables 11.28 and 11.29). Available data are limited to reports of the occurrence of these arthropods. There is no

Table 11.26. Cestode and acanthocephalan infections in Bald Eagles[a] from Florida

Species County/site	Year	Infection site[b]	No. of specimens found	Data source
Cestoda				
Cladotaenia banghami				
Polk	1971	NG	NG	Kocan & Locke (1974), NWHC records
Acanthocephala				
Polymorphus brevis				
Alachua	1994	SI	1	Kinsella & Foster (1995)
Charlotte	1994	SI	2	Ibid.
Hillsborough	1994	SI	2	Ibid.
Marion	1995	SI	10	Ibid.
Polk	1994	SI	1	Ibid.
"No. central Fla."	1986	LI	1	Kinsella & Greiner (1993)
"Florida"	1992	SI	7	Kinsella & Foster (1995)
Centrorhynchus kuntzi				
Lee	1994	SI	1	Ibid.
"Florida"	1992	SI	1	Ibid.
Centrorhynchus sp.				
Lake	1987	LI	NG	Campbell (1990)
Osceola	1988	LI	"several"	Ibid.
Polk	1971	LI	"few"	Kocan & Locke (1974), NWHC records
	1979	SI	"heavy infection"	Tuggle & Schmeling (1982), NWHC records
	1987	LI	NG	Thomas (1992)
Plagiorhynchus sp.				
Lee	NG	NG	6	Richardson & Cole (1997)
Polk	NG	NG	NG	Ibid.

NG = not given.
a. See Table 11.25, note a, for details on sample sizes and geographic locations.
b. LI = large intestine, SI = small intestine.

information on intensity and nothing is known about the significance of these infestations to eagle populations.

The finding of 3 adult specimens of *Ixodes affinis* on a Bald Eagle is unusual. This tick is usually found on mammals and in Florida has been reported from white-tailed deer, bobcats, Florida panthers, and gray foxes (Forrester 1992). The infestation was probably a result of the eagle's having fed on a road-killed mammal.

A case of myiasis, caused by the calliphorid fly *Lucilia illustris*, was reported from a nestling Bald Eagle from Weekiwachee Springs (Hernando County) in 1993 (Thomas and Cole 1993). This condition involved the wings and vent area and was judged to be secondary to soft tissue trauma. The nestling was originally found alive, but died 2 days later during surgery.

XXI. Summary and conclusions

The most common cause of death for Bald Eagles in Florida appears to be trauma due to collisions with motor vehicles, gunshot, intraspecific aggression, and flying into power lines, wires, etc. These factors accounted for the majority of the deaths of Bald Eagles from Florida

Table 11.27. Nematode infections in Bald Eagles[a] from Florida

Species[b] County/site	Year	Infection site[c]	No. of nematodes found	Data source
Capillaria contorta				
Hardee	1993	ES	1	Kinsella & Foster (1995)
Hillsborough	1994	ES	1	Ibid.
Marion	1992	ES	1	Ibid.
	1995	ES	2	Ibid.
Osceola	1987	ES	ND	Kinsella & Cole (1995)
Polk	1995	ES	5	Kinsella & Foster (1995)
"Florida"	1992	ES	18	Ibid.
	1993	ES	1	Ibid.
	1995	ES	17	Ibid.
Capillaria falconis				
Clay	1995	SI	1	Kinsella & Foster (1995)
Hardee	1993	SI	3	Ibid.
Lee	1994	SI	1	Ibid.
Marion	1992	SI	1	Ibid.
	1995	SI	1	Ibid.
Polk	1995	SI	14	Ibid.
Putnam	1995	SI	1	Ibid.
Capillaria sp.				
Alachua	1994	SI	1	Kinsella & Foster (1995)
Brevard	1987	ES	NG	Franson (1993)
Highlands	1970	OR	NG	Locke (1970)
Osceola	1987	ES	7	Schorr (1988)
Sumter	1973	SI	"several"	Locke (1973)
"No. central Fla."	1986	ES	1	Kinsella & Greiner (1993)
Contracaecum microcephalum				
Bradford	1985	NG	NG	Tuggle (1985)
Highlands	1981	PR	1	Tuggle & Schmeling (1982), NWHC records
Volusia	1979	NG	"several"	Tuggle (1979)
Contracaecum multipapillatum				
Hillsborough	1987	VE	ND	Kinsella & Cole (1995)
Indian River	1986	VE	ND	Ibid.
Pasco	1993	VE	1	Kinsella & Foster (1995)
Volusia	1990	VE	ND	Kinsella & Cole (1995)
"No. central Fla."	1986	VE	2	Kinsella & Greiner (1993)
Contracaecum rudolphi				
Marion	1994	VE	ND	Kinsella & Cole (1995)
Orange	1982	ES	17	Tuggle (1982)
	1990	VE	ND	Kinsella & Cole (1995)
Polk	1989	VE	ND	Ibid.
"No. central Fla."	1986	VE	ND	Kinsella & Greiner (1993)
	Unknown	VE	ND	Ibid.
Marion	1992	VE	1	Kinsella & Foster (1995)
Putnam	1995	VE	1	Ibid.
Volusia	1995	VE	1	Ibid.
Contracaecum sp.				
Alachua	1975	PR	1	Kerr (1975)
	1994	SI	1	Kinsella & Foster (1995)
DeSoto	1984	PR	1	Tuggle (1984)
Flagler	1978	PR	1	Tuggle (1978)

(continued)

Table 11.27. *(continued)*

Species[b] County/site	Year	Infection site[c]	No. of nematodes found	Data source
Hendry	1992	SI	NG	Campbell (1992)
Hillsborough	1987	PR	"several"	Thomas (1992)
Indian River	1986	PR	"approx. 20"	Franson (1992)
Lake	1987	PR	1	Campbell (1990)
Lee	1985	PR	4	Tuggle (1985)
	1988	PR	2	Schorr (1988)
Marion	1993	PR	3	Cole (1993)
Martin	1984	PR	NG	Sileo (1984)
Polk	1981	PR	2	Tuggle (1981)
	1987	PR	1	Locke (1992)
Putnam	1988	PR	1	Franson (1992)
Sarasota	1985	PR	"several"	Tuggle (1985)
Unknown	Unknown	PR	1	Tuggle (1981)
	Unknown	PR	>50 larvae	Campbell (1992)
Volusia	1988	PR	4	Locke (1989)
	1988	PR	1	Franson (1993)
Contracaecum sp. (larvae)				
Brevard	1994	VE	ND	Kinsella & Cole (1995)
Charlotte	1994	VE	1	Kinsella & Foster (1995)
Hendry	1992	VE	ND	Kinsella & Cole (1995)
Marion	1995	ES	2	Kinsella & Foster (1995)
Pasco	1993	VE	1	Ibid.
	1993	VE	75	Ibid.
"Florida"	1993	ES, VE	7	Ibid.
	1995	VE	3	Ibid.
Chandleronema longigutturata				
"Florida"	1992	VE	5	Ibid.
Desportesius invaginatus				
Hardee	1993	ES	11	Ibid.
"Florida"	1992	VE	13	Ibid.
Hamatospiculum sp.				
Alachua	1992	AS, LU	4	Ackerman et al. (1992)
Hendry	1992	AS	ND	Kinsella & Cole (1995)
Levy	1993	AS	ND	Ibid.
Pinellas	1994	AS	ND	Ibid.
Seminole	1979	AS, BC	3	Tuggle & Schmeling (1982); NWHC records
"No. central Fla."	1990	AS	NG	Kinsella & Greiner (1993)
Physaloptera sp. (larvae)				
Putnam	1995	VE	3	Kinsella & Foster (1995)
Spirurid larvae				
Alachua	1994	PR	2	Kinsella & Foster (1995)
Hardee	1993	VE	118	Ibid.
Lee	1993	BC	1	Ibid.
Polk	1994	CL	1	Ibid.
"Florida"	1992	VE	1	Ibid.

ND = not determined, NG = not given.
a. See table 11.25, note a, for details on sample sizes and geographic locations.
b. Tuggle and Schmeling (1982, table 1) listed *Paracuaria* sp. from a Bald Eagle in Florida. When the records of NWHC were checked, it was determined that there had been an error; the record was actually from an eagle from Massachusetts.
c. PR = proventriculus; ES = esophagus; AS = air sacs; BC = body cavity; CL = cloaca; LU = lung; OR = oral cavity; SI = small intestine; VE = ventriculus.

Table 11.28. Parasitic arthropod infestations (other than lice) of Bald Eagles in Florida

Arthropod County	Year	Infestation site	Number of arthropods found	Data source
Ticks				
Ixodes affinis				
Alachua	1975	Unknown	3	Forrester & Hoogstraal (1982)
Amblyomma sp.				
Volusia	1982	NG	1	Tuggle (1982)
Feather mites				
Aetacarus proctophyllus				
Polk	1985	Tips of primaries	"many"	Tuggle (1985)
Aetacarus sp.[a]				
Alachua	1994	Feathers	307	Foster & Mertins (1995)
Charlotte	1994	Feathers	10	Ibid.
Hardee	Unknown	Feathers	39	Ibid.
Hillsborough	1987	Feathers	59	Cole & Mertins (1996)
	1994	Feathers	2	Foster & Mertins (1995)
Lake	Unknown	Feathers	12	Ibid.
Lee	1993	Feathers	14	Ibid.
	1994	Feathers	3	Ibid.
Marion	1992	Feathers	6	Ibid.
	1992	Feathers	10	Ibid.
	1995	Feathers	26	Ibid.
Orange	1992	Feathers	971	Cole & Mertins (1996)
	1994	Feathers	43	Foster & Mertins (1995)
Osceola	1988	Feathers	82	Cole & Mertins (1996)
	1989	Feathers	813	Ibid.
	1993	Feathers	128	Ibid.
Pasco	1993	Feathers	14	Foster & Mertins (1995)
	1993	Feathers	8	Ibid.
Polk	1985	Feathers	538	Cole & Mertins (1996)
	1987	Feathers	468	Ibid.
	1988	Feathers	9	Ibid.
	1994	Feathers	17	Foster & Mertins (1995)
	1995	Feathers	11	Ibid.
	1995	Feathers	25	Ibid.
Putnam	1995	Feathers	5	Ibid.
Volusia	1988	Feathers	64	Cole & Mertins (1996)
	1995	Feathers	25	Foster & Mertins (1995)
Unknown	1987	Feathers	11	Cole & Mertins (1996)
	1993	Feathers	19	Ibid.
	1993	Feathers	8	Foster & Mertins (1995)
	1995	Feathers	37	Ibid.
	Unknown	Feathers	16	Ibid.
Pseudalloptinus aquilinus				
Alachua	1994	Feathers	74	Ibid.
Charlotte	1994	Feathers	22	Ibid.
Hardee	Unknown	Feathers	62	Ibid.
Hillsborough	1994	Feathers	18	Ibid.
Lake	Unknown	Feathers	9	Ibid.
Lee	1993	Feathers	7	Ibid.
	1994	Feathers	8	Ibid.
Marion	1992	Feathers	44	Ibid.
	1992	Feathers	13	Ibid.
	1995	Feathers	35	Ibid.

(continued)

Table 11.28. *(continued)*

Arthropod County	Year	Infestation site	Number of arthropods found	Data source
	1995	Feathers	19	Ibid.
Orange	1992	Feathers	58	Cole & Mertins (1996)
	1994	Feathers	9	Foster & Mertins (1995)
Osceola	1988	Feathers	1	Cole & Mertins (1996)
	1989	Feathers	4	Ibid.
	1993	Feathers	17	Ibid.
Pasco	1993	Feathers	54	Foster & Mertins (1995)
	1993	Feathers	14	Ibid.
Polk	1985	Feathers	2	Cole & Mertins (1996)
	1987	Feathers	21	Ibid.
	1994	Feathers	12	Foster & Mertins (1995)
	1995	Feathers	64	Ibid.
	1995	Feathers	44	Ibid.
Putnam	1995	Feathers	15	Ibid.
Volusia	1988	Feathers	3	Cole & Mertins (1996)
	1995	Feathers	15	Foster & Mertins (1995)
Unknown	1987	Feathers	23	Cole & Mertins (1996)
	1993	Feathers	2	Ibid.
	1993	Feathers	4	Foster & Mertins (1995)
	1995	Feathers	14	Ibid.
	Unknown	Feathers	28	Ibid.
Unidentified				
Lake	1987	Wing feathers	"numerous"	Campbell (1990)
Lee	1988	Wing feathers	NG	Schorr (1988)
Orange	1992	Wing feathers	NG	Thomas (1992)
Osceola	1987	Wing feathers	"few"	Schorr (1987)
	1988	Wing feathers	"numerous"	Campbell (1990)
	1988	Wing feathers	"numerous"	Thomas (1992)
	1992	Feathers	NG	Cole (1993)
	1993	Wing feathers	NG	Thomas (1993)
Polk	1987	Wing feathers	NG	Campbell (1992)
	1988	Wing feathers	"numerous"	Thomas (1988)
Putnam	1988	Wing feathers	NG	Franson (1992)
Seminole	1987	Wing feathers	NG	Campbell (1990)
Hemipterans (Psocidae)				
Unknown	Unknown	Lungs	NG	Campbell (1992)
Louse flies				
Icosta americana				
Alachua	NG	NG	NG	Bequaert (1955)
Olfersia fumipennis				
Alachua	1947	Unknown	1	Wilson et al. (1990)
	1972	Unknown	2	Wilson & Forrester (1990)
Wakulla	1973	Unknown	2	Wilson et al. (1990)
Blow flies				
Cochliomyia macellaria				
Hillsborough	1994	Feathers	3(eggs)	Foster & Mertins (1995)
Lucilia illustris				
Hernando	1993	Wing & vent areas	"numerous"	Cole (1993)

NG = not given.
a. An undescribed species.

Table 11.29. Infestations of chewing lice on Bald Eagles from Florida

Chewing louse County/site	Year(s)	Infestation site	No. of lice found	Data source
Colpocephalum flavescens				
Alachua	1972	Feathers	NG	Forrester et al. (1995)
Leon	1938	Feathers	NG	Ibid.
Polk	1985	Head and neck feathers	NG	Tuggle (1985)
Colpocephalum turbinatum				
Alachua	1994	Feathers	15	Forrester et al. (1995)
Dade	1917	Feathers	15	Ibid.
Lee	1980, 1985	Feathers	NG	Ibid.
Orange	1992	Feathers	2	Cole & Mertins (1996)
Colpocephalum sp.				
Charlotte	1988	Wing feathers	"rare"	Thomas (1991)
Lee	1988	Wing and body feathers	"numerous"	Schorr (1988)
Orange	1988	Feathers	NG	Thomas (1988)
Osceola	1988	Wing and body feathers	"numerous"	Schorr (1988)
Seminole	1987	Feathers	NG	Campbell (1990)
Craspedorrhynchus halieti				
Alachua	1994	Feathers	1	Forrester et al. (1995)
Dade	1917	Feathers	NG	Ibid.
	1973	Feathers	NG	Ibid.
Flagler	1978	Head feathers	"many"	Tuggle (1978)
Highlands	1981	Head feathers	NG	Tuggle (1981)
Hillsborough	1948	Feathers	NG	Forrester et al. (1995)
Leon	1938	Feathers	NG	Ibid.
Orange	1979	Head feathers	"many"	Tuggle (1979)
Polk	1977	Feathers	NG	Forrester et al. (1995)
	1985	Head feathers	NG	Tuggle (1985)
Seminole	1979	Head feathers	NG	Tuggle (1979)
Volusia	1982	Head feathers	NG	Tuggle (1982)
"Florida"	NG	Feathers	NG	Osborn (1896), Kellogg (1899)
Unknown	1991	Feathers	4	Cole & Mertins (1996)
Craspedorrhynchus sp.				
Brevard	1986	Head and neck feathers	"several"	Schorr (1986)
	1987	Head and neck feathers	"several"	Schorr (1987)
Osceola	1988	Feathers	"numerous"	Schorr (1988)
Polk	1986	Feathers	NG	Campbell (1990)
Degeeriella discocephalus				
Leon	1938	Feathers	NG	Forrester et al. (1995)
Degeeriella sp.				
Lee	1988	Wing and body feathers	"numerous"	Schorr (1988)
Kurodaia fulvofasciata				
Marion	1954	Feathers	NG	Forrester et al. (1995)

(continued)

Table 11.29. *(continued)*

Chewing louse County/site	Year(s)	Infestation site	No. of lice found	Data source
Laemobothrion vulturis				
Collier	1987	Feathers	NG	Ibid.
Dade	1917	Feathers	NG	Ibid.
Highlands	1991	Feathers	1	Cole & Mertins (1996)
Leon	1938	Feathers	1	Forrester et al. (1995)
Osceola	1993	Feathers	2	Ibid.
Laemobothrion sp.				
Highlands	1991	Leg feathers	1	Meteyer (1992)
Unidentified				
Orange	1992	Wing feathers	NG	Thomas (1993)
Osceola	1988	Wing feathers	"numerous"	Thomas (1988)

NG = not given.

examined over the 30-year period 1963–94. Vehicular collisions are of considerable importance and probably are related to the habitual feeding on carcasses of road-killed animals by Bald Eagles. Other significant factors are electrocution, poisoning, infectious diseases, and emaciation (of unknown causes). Inclement weather, especially hurricanes, is responsible for some mortality of eggs and young. One hundred and twelve disease agents and parasites have been found in Bald Eagles in Florida. These include organochlorines (14), organophosphates (1), carbamates (1), barbiturates (1), metals (10), viruses (2), bacteria (29), fungi (7), protozoans (9), trematodes (8), cestodes (1), acanthocephalans (3), nematodes (11), ticks (2), mites (3), dipterans (4), and chewing lice (6). Several of these disease agents are known to have caused direct mortality and include dieldrin, cumulative organochlorines, pentobarbital, lead, poxvirus, *Aspergillus fumigatus*, and an undescribed species of the nematode *Hamatospiculum*. In addition, DDT and its metabolites (especially DDE) have caused reproductive impairment due to eggshell thinning. There is evidence that the latter problem has diminished since the banning of DDT sales in the United States and currently Bald Eagle populations in Florida are increasing.

Little is known about mortality factors and diseases of Golden Eagles in Florida. One case each of emaciation (unknown cause), drowning, and trauma has been documented. In addition, small amounts of data are available from several birds on 1 nematode, 2 bacteria, 1 fungus, and lead. Golden Eagles should be further studied as opportunities become available.

XXII. Literature cited

Ackerman, N., R. Isaza, E. Greiner, and C.R. Berry. 1992. Pneumocoelom associated with *Serratospiculum amaculata* in a bald eagle. *Vet. Radiol. Ultrasound* 33:351–355.

Albers, H.F. 1984. Naturally occurring immunity to rabies in raptors. *Proc. Annu. Natl. Wildl. Rehabil. Symp.* 2:171–174.

Anderson, D.W., and J.J. Hickey. 1972. Eggshell changes in certain North American birds. *Proc. Int. Ornithol. Congr.* 15:514–540.

Belisle, A.A., W.L. Reichel, L.N. Locke, T.G. Lamont, B.M. Mulhern, R.M. Prouty, R.B. DeWolf, and E. Cromartie. 1972. Residues of organochlorine pesticides, polychlorinated biphenyls, and mercury and autopsy data for bald eagles, 1969 and 1970. *Pestic. Monit. J.* 6:133–138.

Bennett, G.F., M.A. Peirce, and R.W. Ashford.

1993. Avian Haematozoa: mortality and pathogenicity. *J. Nat. Hist.* 27:993–1001.

Bequaert, J.C. 1955. The Hippoboscidae or louseflies (Diptera) of mammals and birds. Part II. Taxonomy, evolution and revision of American genera and species. *Entomol. Am.* 35:233–416.

Brannian, R. 1996. Unpublished data. National Wildlife Health Center, Madison, Wis.

Bremer, K.E. 1983. Dealing with multi-agency and multi-state regulation of polychlorinated biphenyls. In: *PCBs: human and environmental hazards.* F.M. D'Itri and M.A. Kamrin (eds.). Butterworth, Woburn, Mass. pp. 367–373.

Broley, C.L. 1947. Migration and nesting of Florida bald eagles. *Wilson Bull.* 59:3–20.

———. 1958. The plight of the American bald eagle. *Audubon Mag.* 60:162–163, 171.

Campbell, B.G. 1990–92. Unpublished data. National Wildlife Health Center, Madison, Wis.

Carleton, R.E. 1994. Unpublished data. University of Florida, Gainesville.

Cawthorn, R.J. 1993. Cyst-forming coccidia of raptors: significant pathogens or not? In: *Raptor biomedicine.* P.T. Redig, J.E. Cooper, J.D. Remple, and D.B. Hunter (eds.). University of Minnesota Press, Minneapolis. pp. 14–20.

Cole, R.A. 1993. Unpublished data. National Wildlife Health Center, Madison, Wis.

Cole, R.A., and J.W. Mertins. 1996. Unpublished data. National Wildlife Health Center, Madison, Wis.

Collins, R. 1996–99. Unpublished data. Florida Audubon Society, Center for Birds of Prey, Maitland.

Collins, R., and C. Gilliland. 1995. Unpublished data. Florida Audubon Society, Center for Birds of Prey, Maitland.

Converse, K.A. 1997. Unpublished data. National Wildlife Health Center, Madison, Wis.

Coon, N.C., L.N. Locke, E. Cromartie, and W.L. Reichel. 1970. Causes of bald eagle mortality, 1960–1965. *J. Wildl. Dis.* 6:72–76.

Cooper, J.E. 1978. Preventive medicine in birds of prey. In: *Zoo and wild animal medicine.* M.E. Fowler (ed.). W.B. Saunders, Philadelphia. pp. 253–259.

———. 1985. *Veterinary aspects of captive birds of prey.* 2d ed. Standfast, Gloucestershire, England. 256+31 pp.

Cornish, T.E. 1998. Unpublished data. Southeastern Cooperative Wildlife Disease Study, University of Georgia, Athens.

Crawley, R.R., J.V. Ernst, and J.L. Milton. 1982. *Sarcocystis* in a bald eagle (*Haliaeetus leucocephalus*). *J. Wildl. Dis.* 18:253–255.

Cromartie, E., W.L. Reichel, L.N. Locke, A.A. Belisle, T.E. Kaiser, T.G. Lamont, B.M. Mulhern, R.M. Prouty, and D.M. Swineford. 1975. Residues of organochlorine pesticides and polychlorinated biphenyls and autopsy data for bald eagles, 1971–72. *Pestic. Monit. J.* 9:11–14.

Curnutt, J.L. 1996. Southern bald eagle. In: *Rare and endangered biota of Florida.* Vol. 5, *Birds.* J.A. Rodgers, Jr., H.W. Kale II, and H.T. Smith (eds.). University Press of Florida, Gainesville. pp. 179–187.

Deem, S.L., and S.P. Terrell. 1996. Unpublished data. University of Florida, Gainesville.

Deem, S.L., S.P. Terrell, and D.J. Forrester. 1998. A retrospective study of morbidity and mortality of raptors in Florida: 1988–1994. *J. Zoo Wildl. Med.* 29:160–164.

Docherty, D.E. 1988–93. Unpublished data. National Wildlife Health Center, Madison, Wis.

Duncan, R.M. 1978–94. Unpublished data. National Wildlife Health Center, Madison, Wis.

Eisler, R. 1986. *Polychlorinated biphenyl hazards to fish, wildlife, and invertebrates: a synoptic review.* U.S. Fish and Wildlife Service, Biological Report 85(1.7). 72 pp.

———. 1987. *Mercury hazards to fish, wildlife, and invertebrates: a synoptic review.* U.S. Fish and Wildlife Service, Biological Report 85(1.10). 90 pp.

Feierabend, J.S., and O. Myers. 1984. A national summary of lead poisoning in bald eagles and waterfowl. National Wildlife Federation Report, Washington, D.C. 89 pp.

[FGFWFC] Florida Game and Fresh Water Fish Commission. 1992. Effects of Hurricane Andrew on fish and wildlife in south Florida. Unpublished document. Tallahassee. 19 pp.

Fischer, J.R. 1997–2001. Unpublished data. Southeastern Cooperative Wildlife Disease Study, University of Georgia, Athens.

Forrester, D.J. 1992. *Parasites and diseases of wild mammals in Florida.* University Press of Florida, Gainesville. 459 pp.

Forrester, D.J., and H. Hoogstraal. 1982. Unpublished data. University of Florida, Gainesville.

Forrester, D.J., H.W. Kale II, R.D. Price, K.C. Emerson, and G.W. Foster. 1995. Chewing lice (Mallophaga) from birds in Florida: a listing by host. *Bull. Fla. Mus. Nat. Hist.* 39:1–44.

Forrester, D.J., S.R. Telford, Jr., G.W. Foster, and G.F. Bennett. 1994. Blood parasites of raptors from Florida. *J. Raptor Res.* 28:226–231.

Foster, G.W., and J.W. Mertins. 1995. Unpublished data. University of Florida, Gainesville.

Fowler, M.E. (ed.). 1986. *Zoo and wild animal medicine.* 2d ed. W.B. Saunders, Philadelphia. 1,127 pp.

———(ed.). 1993. *Zoo and wild animal medicine: current therapy.* 3d ed. W.B. Saunders, Philadelphia. 617 pp.

Franson, J.C. 1990–94. Unpublished data. National Wildlife Health Center, Madison, Wis.

Franson, J.C., E.J. Galbreath, S.N. Wiemeyer, and J.M. Abell. 1994. *Erysipelothrix rhusiopathiae* infection in a captive bald eagle (*Haliaeetus leucocephalus*). *J. Zoo Wildl. Med.* 25:446–448.

Franson, J.C., L. Sileo, and N.J. Thomas. 1995. Causes of eagle deaths. In: *Our living resources: a report to the nation on the distribution, abundance, and health of U.S. plants, animals, and ecosystems.* E.T. LaRoe, G.S. Farris, E.E. Puckett, and P.D. Doran (eds.). National Biological Service, Washington, D.C. p. 68.

Galbreath, E.J. 1993. Unpublished data. National Wildlife Health Center, Madison, Wis.

Gerrard, J.M., and G.R. Bortolotti. 1988. *The bald eagle: haunts of a wilderness monarch.* Smithsonian Institution Press, Washington, D.C. 177 pp.

Greiner, E.C., D.J. Black, and W.O. Iverson. 1981. *Plasmodium* in a bald eagle (*Haliaeetus leucocephalus*) in Florida. *J. Wildl. Dis.* 17:555–558.

Grier, J.W. 1982. Ban of DDT and subsequent recovery of reproduction in bald eagles. *Science* 218:1232–1235.

Halliwell, W.H., and D.L. Graham. 1986. Bacterial diseases of birds of prey. In: *Zoo and wild animal medicine.* 2d ed. M.E. Fowler (ed.). W.B. Saunders, Philadelphia. pp. 413–419.

Heath, R.G., and S.A. Hill. 1974. Nationwide organochlorine and mercury residues in wings of adult mallards and black ducks during the 1969–70 hunting season. *Pestic. Monit. J.* 7:153–164.

Heinzman, G. 1961. The American bald eagle. *Nat. Hist.* 70(6):18–21.

Henny, C.J., and R.G. Anthony. 1989. Bald eagle and osprey. *Natl. Wildl. Fed. Sci. Tech. Ser.* 12:66–82.

Hickey, J.J., and D.W. Anderson. 1968. Chlorinated hydrocarbons and eggshell changes in raptorial and fish-eating birds. *Science* 162: 271–273.

Hill, E.F., and W.J. Fleming. 1982. Anticholinesterase poisoning of birds: field monitoring and diagnosis of acute poisoning. *Environ. Toxicol. Chem.* 1:27–38.

Howell, J.C. 1941. Bald eagle killed by lightning while incubating its eggs. *Wilson Bull.* 53:42–43.

———. 1963. The 1961 status of some bald eagle nest sites in east-central Florida. *Auk* 79:716–718.

Hutton, R.F. 1964. A second list of parasites from marine and coastal animals of Florida. *Trans. Am. Microsc. Soc.* 83:439–447.

Kaiser, T.E., W.L. Reichel, L.N. Locke, E. Cromartie, A.J. Krynitsky, T.G. Lamont, B.M. Mulhern, R.M. Prouty, C.J. Stafford, and D.M. Swineford. 1980. Organochlorine pesticide, PCB, and PBB residues and necropsy data for bald eagles from 29 states—1975–77. *Pestic. Monit. J.* 13:145–149.

Kellogg, V.L. 1899. A list of the biting lice (Mallophaga) taken from birds and mammals of North America. *Proc. U.S. Natl. Mus.* 22:39–100.

Kerr, S.M. 1975–78. Unpublished data. National Wildlife Health Center, Madison, Wis.

Kiehl, A.R. 1991. Unpublished data. Doctor's and Physicians Laboratory, Leesburg, Fla.

Kinsella, J.M., and R.A. Cole. 1995. Unpublished data. University of Florida, Gainesville, and National Wildlife Health Center, Madison, Wis.

Kinsella, J.M., and G.W. Foster. 1995. Unpublished data. University of Florida, Gainesville.

Kinsella, J.M., G.W. Foster, R.C. Cole, and D.J. Forrester. 1998. Helminth parasites of the bald eagle, *Haliaeetus leucocephalus,* in Florida. *J. Helminthol. Soc. Wash.* 65:65–68.

Kinsella, J.M., and E.C. Greiner. 1993. Unpublished data. University of Florida, Gainesville.

Klaassen, C.D. 1980. Nonmetallic environmental toxicants: air pollutants, solvents and vapors, and pesticides. In: *Goodman and Gilman's The pharmacological basis of therapeutics*. 6th ed. A.G. Gilman, L.S. Goodman, and A. Gilman (eds.). Macmillan, New York. pp. 1649–1650.

Kocan, A.A., and L.N. Locke. 1974. Some helminth parasites of the American bald eagle. *J. Wildl. Dis.* 10:8–10.

Krantz, W.C., B.M. Mulhern, G.E. Bagley, A. Sprent IV, F.J. Ligas, and W.B. Robertson, Jr. 1970. Organochlorine and heavy metal residues in bald eagle eggs. *Pestic. Monit. J.* 3:136–140.

Langelier, K.M. 1993. Barbiturate poisoning in twenty-nine bald eagles. In: *Raptor biomedicine*. P.T. Redig, J.E. Cooper, J.D. Remple, and D.B. Hunter (eds.). University of Minnesota Press, Minneapolis. pp. 231–232.

Langridge, H.P. 1993. Florida region. *Am. Birds.* 47:406–408.

Locke, L.N. 1968. Unpublished data. Patuxent Wildlife Research Center, Laurel, Md.

———. 1970–92. Unpublished data. National Wildlife Health Center, Madison, Wis.

———. 1987. Aspergillosis. In: *Field guide to wildlife diseases*. Vol. 1, *General field procedures and diseases of migratory birds*. M. Friend (ed.). U.S. Fish and Wildlife Service, Resource Bulletin 167. Washington, D.C. pp. 145–150.

Logan, T.H. 1997. Florida's endangered species, threatened species and species of special concern. Official lists. Florida Game and Fresh Water Fish Commission, Tallahassee. 14 pp.

Meteyer, C.U. 1992–94. Unpublished data. National Wildlife Health Center, Madison, Wis.

Miller, D., E.L. Boeker, and R.S. Thorsell. 1975. Suggested practices for raptor protection on powerlines. Raptor Research Foundation, Provo, Utah. 19 pp.

Morehouse, K.A. 1992. Lead poisoning of migratory birds: the U.S. Fish and Wildlife Service position. In: *Lead poisoning in waterfowl*. D.J. Pain (ed.) Proc. IWRB Workshop, Brussels, Belgium, 13–15 June 1991. IWRB Spec. Publ. 16. International Waterfowl and Wetlands Research Bureau, Slimbridge, Gloucester, United Kingdom. pp. 51–55.

Mulhern, B.M. 1983. Unpublished data. Patuxent Wildlife Research Center, Laurel, Md.

Mulhern, B.M., W.L. Reichel, L.N. Locke, T.G. Lamont, A. Belisle, E. Cromartie, G.E. Bagley, and R.M. Prouty. 1970. Organochlorine residues and autopsy data from bald eagles, 1966–68. *Pestic. Monit. J.* 4:141–144.

Nayar, J.K., J.W. Knight, and S.R. Telford, Jr. 1998. Vector ability of mosquitoes for isolates of *Plasmodium elongatum* from raptors in Florida. *J. Parasitol.* 84:542–546.

Needham, J.R. 1981. Bacterial flora of birds of prey. In: *Recent advances in the study of raptor diseases. Proceedings of the International Symposium on Diseases of Birds of Prey.* J.E. Cooper and A.G. Greenwood (eds.). Chiron, West Yorkshire, England. pp. 3–9.

Nesbitt, S.A. 1993. Bald eagle population monitoring. Annual performance report, study no. II-L-1 7521. Florida Game and Fresh Water Fish Commission, Gainesville. 7 pp.

———. 1994. Unpublished data. Florida Game and Fresh Water Fish Commission, Gainesville.

———. 1997. Bald Eagle population monitoring. Annual performance report, study no. II-L-1 7521. Florida Game and Fresh Water Fish Commission, Gainesville. 8 pp.

Newton, I. 1979. *Population ecology of raptors*. Buteo Books, Vermillion, S. Dak. 399 pp.

Nisbet, I.C.T. 1989. Organochlorines, reproductive impairment and declines in bald eagle *Haliaeetus leucocephalus* populations: mechanisms and dose-response relationships. In: *Raptors in the modern world*. B.-U. Meyburg and R.D. Chancellor (eds.). World Working Group on Birds of Prey and Owls, London. pp. 483–489.

[NWHC] National Wildlife Health Research Center. 1985. *Bald eagle mortality from lead poisoning and other causes.* U.S. Fish and Wildlife Service, Madison, Wis. 48 pp.

Osborn, H. 1896. *Insects affecting domestic animals: an account of the species of importance in North America with mention of related forms occurring on other animals.* Bulletin no. 5 (New Series), Division of Entomology, USDA, Washington, D.C.

Pattee, O.H., and S.K. Hennes. 1983. Bald eagles

and waterfowl: the lead shot connection. *Trans. N. Am. Wildl. Nat. Resour. Conf.* 48:230–237.

Pattee, O.H., S.N. Wiemeyer, B.M. Mulhern, L. Sileo, and J.W. Carpenter. 1981. Experimental lead-shot poisoning in bald eagles. *J. Wildl. Manag.* 45:806–810.

Peirce, M.A. 1981. Current knowledge of the Haematozoa of raptors. In: *Recent advances in the study of raptor diseases. Proceedings of the International Symposium on Diseases of Birds of Prey.* J.E. Cooper and A.G. Greenwood (eds.). Chiron, West Yorkshire, England. pp. 15–19.

Postupalsky, S. 1976. Banded northern bald eagles in Florida and other southern states. *Auk* 93:835–836.

Prouty, R.M., W.L. Reichel, L.N. Locke, A.A. Belisle, E. Cromartie, T.E. Kaiser, T.G. Lamont, B.M. Mulhern, and D.M. Swineford. 1977. Residues of organochlorine pesticides and polychlorinated biphenyls and autopsy data for bald eagles, 1973–74. *Pestic. Monit. J.* 11:134–137.

Redig, P.T. 1993. Avian aspergillosis. In: *Zoo and wild animal medicine: current therapy.* 3d ed. M.E. Fowler (ed.). W.B. Saunders, Philadephia. pp. 178–181.

Reichel, W.L. 1975–1982. Unpublished data. Patuxent Wildlife Research Center, Laurel, Md.

Reichel, W.L., E. Cromartie, T.G. Lamont, B.M. Mulhern, and R.M. Prouty. 1969a. Pesticide residues in eagles. *Pestic. Monit. J.* 3:142–144.

Reichel, W.L., T.G. Lamont, E. Cromartie, and L.N. Locke. 1969b. Residues in two bald eagles suspected of pesticide poisoning. *Bull. Environ. Contam. Toxicol.* 4:24–30.

Reichel, W.L., S.K. Schmeling, E. Cromartie, T.E. Kaiser, A.J. Krynitsky, T.G. Lamont, B.M. Mulhern, R.M. Prouty, C.J. Stafford, and D.M. Swineford. 1984. Pesticide, PCB, and lead residues and necropsy data for bald eagles from 32 states—1978–81. *Environ. Monit. Assess.* 4:395–403.

Richardson, D.J., and R.A. Cole. 1997. Acanthocephala of the bald eagle (*Haliaeetus leucocephalus*) in North America. *J. Parasitol.* 83:540–541.

Robertson, W.B., Jr., and G.E. Woolfenden. 1992. *Florida bird species. An annotated list.* Spec. Publ. 6, Florida Ornithological Society, Gainesville. 260 pp.

Schmeling, S.K., and D.E. Docherty. 1982. Fatal avian pox in bald eagles from Alaska. In: *Raptor management and biology in Alaska and western Canada: proceedings of a symposium and workshop.* W.N. Ladd and P.F. Schempf (eds.). U.S. Fish and Wildlife Service, Alaska Regional Office, FWS/AK/Proc. 82. pp. 255–262.

Schorr, L.F. 1986–88. Unpublished data. National Wildlife Health Center, Madison, Wis.

Sherrod, S.K., M.A. Jenkins, and T.J. Cade. 1990. Avian reintroductions as a management tool. *Proc. Southeast Raptor Manag. Symp. Workshop.* Natl. Wildl. Fed. Sci. Tech. Ser. 14:150–157.

Siegfried, L.M. 1982–83. Unpublished data. National Wildlife Health Center, Madison, Wis.

Sileo, L. 1983–85. Unpublished data. National Wildlife Health Center, Madison, Wis.

Sileo, L., L. Karstad, R. Frank, M.V.H. Holdrinet, E. Addison, and H.E. Braun. 1977. Organochlorine poisoning of ring-billed gulls in southern Ontario. *J. Wildl. Dis.* 13:313–322.

Smith, G.J. 1993. *Toxicology and pesticide use in relation to wildlife: organophosphorus and carbamate compounds.* C.K. Smoley, Boca Raton, Fla. 171 pp.

Solt, V. 1980. Raptor mortalities from power lines studied. *Fish Wildl. News* Aug./Sept.:2, 16.

Spalding, M.G. 2001. Unpublished data. University of Florida, Gainesville.

Spalding, M.G., and P.B. Wood. 1988. Unpublished data. University of Florida, Gainesville.

Steinberg, H. 1993. Unpublished data. National Wildlife Health Center, Madison, Wis.

Stickel, L.F., N.J. Chura, P.A. Stewart, C.M. Menzie, R.M. Prouty, and W.L. Reichel. 1966. Bald eagle pesticide relations. *Trans. N. Am. Wildl. Conf.* 31:190–200.

Stuht, J.N., W.W. Bowerman, and D.A. Best. 1999. Leucocytozoonosis in nestling bald eagles in Michigan and Minnesota. *J. Wildl. Dis.* 35:608–612.

Sundlof, S.F., M.G. Spalding, J.D. Wentworth, and C.K. Steible. 1994. Mercury in livers of wading birds (Ciconiiformes) in southern Florida. *Arch. Environ. Contam. Toxicol.* 27:299–305.

Telford, S.R., Jr., and D.J. Forrester. 1992. Morphological comparisons of the *Plasmodium* (*Novyella*) species reported from North American birds, with comments on a species from the barred owl *Strix varia* Barton. *System. Parasitol.* 22:17–24.

Telford, S.R., Jr., J.K. Nayar, G.W. Foster, and J.W. Knight. 1997. *Plasmodium forresteri* n.sp., from raptors in Florida and southern Georgia: its distinction from *Plasmodium elongatum* morphologically within and among host species and by vector susceptibility. *J. Parasitol.* 83:932–937.

Telford, S.R., Jr., J.K. Nayar, and S.P. Terrell. 1996. Unpublished data. University of Florida, Gainesville.

Thomas, N.J. 1988–94. Unpublished data. National Wildlife Health Center, Madison, Wis.

Thomas, N.J., and R.A. Cole. 1993. Unpublished data. National Wildlife Health Center, Madison, Wis.

Thomas, N.J., C.U. Meteyer, and L. Sileo. 1998. Epizootic vacuolar myelinopathy of the central nervous system of bald eagles (*Haliaeetus leucocephalus*) and American coots (*Fulica americana*). *Vet. Pathol.* 35:479–487.

Tuggle, B.N. 1978–85. Unpublished data. National Wildlife Health Center, Madison, Wis.

Tuggle, B.N., and S.K. Schmeling. 1982. Parasites of the bald eagle (*Haliaeetus leucocephalus*) of North America. *J. Wildl. Dis.* 18:501–506.

Ware, F.J., H. Royals, and T. Lange. 1990. Mercury contamination in Florida fish and wildlife. *Proc. Annu. Conf. Southeast. Assoc. Fish Wildl. Agencies* 44:5–12.

White, F.W., C.F. Simpson, and L.E. Williams, Jr. 1973. Isolation of *Edwardsiella tarda* from aquatic animal species and surface waters in Florida. *J. Wildl. Dis.* 9:204–208.

White, J.H. 1993. Unpublished data. Florida Game and Fresh Water Fish Commission, Eustis.

Wiemeyer, S.N., T.G. Lamont, C.M. Bunck, C.R. Sindelar, F.J. Gramlich, J.D. Fraser, and M.A. Byrd. 1984. Organochlorine pesticide, polychlorobiphenyl, and mercury residues in bald eagle eggs—1969–79—and their relationships to shell thinning and reproduction. *Arch. Environ. Contam. Toxicol.* 13:529–549.

Wiemeyer, S.N., B.M. Mulhern, F.J. Ligas, R.J. Hensel, J.E. Mathisen, F.C. Robards, and S. Postupalsky. 1972. Residues of organochlorine pesticides, polychlorinated biphenyls, and mercury in bald eagle eggs and changes in shell thickness—1969 and 1970. *Pestic. Monit. J.* 6:50–55.

Wilson, N.A., and D.J. Forrester. 1990. Unpublished data. University of Florida, Gainesville.

Wilson, N.A., H.W. Kale II, and W.W. Baker. 1990. Unpublished data. University of Northern Iowa, Cedar Falls.

Wood, D.A. 1993. Unpublished data. Florida Game and Fresh Water Fish Commission, Tallahassee.

Wood, P.B., D.A. Buehler, and M.A. Byrd. 1990. Bald eagle. In: *Proc. Southeast Raptor Manag. Symp. Workshop.* Natl. Wildl. Fed. Sci. Tech. Ser. 14:13–21.

Wood, P.B., J.H. White, A. Steffer, J.M. Wood, C.F. Facemire, and H.F. Percival. 1996. Mercury concentrations in tissues of Florida Bald Eagles. *J. Wildl. Manag.* 60:178–185.

Wood, P.B., J.H. White, A. Steffer, J.M. Wood, and H.F. Percival. 1993. Mercury concentrations in tissues of Florida bald eagles. Final project report. West Virginia Cooperative Fish and Wildlife Research Unit, West Virginia University, Morgantown. 41 pp.

Woodard, J.C. 1986. Unpublished data. Pathology report no. N86-0255, CVM-TH, University of Florida, Gainesville.

Woods, K.L. 1992. Unpublished data. National Wildlife Health Center, Madison, Wis.

Zimmerman, D.R. 1976. The bald eagle bicentennial blues. *Nat. Hist.* 85(1):8–16.

Ospreys

I. Introduction

The Osprey, *Pandion haliaetus,* known also as the Fish Hawk or Fishing Eagle, actually is neither a true hawk nor an eagle (Henny 1988), although it does feed almost exclusively on fish (Ogden 1996). This intriguing raptor is cosmopolitan, being found throughout the world in each continent except Antarctica (Poole 1989). Roger Tory Peterson (1969) called this bird "a world citizen." In Florida, there are resident breeding populations and also transients that come from northern populations and pass through Florida during the spring and fall on their way to and from tropical wintering areas (Ogden 1996). They are found primarily along lakeshores and coastal areas throughout the state except for the Dry Tortugas (Robertson and Woolfenden 1992). Currently they are listed as a species of special concern in Monroe County by the Florida Game and Fresh Water Fish Commission (Logan 1997) and as threatened by the Florida Committee on Rare and Endangered Plants and Animals (Ogden 1996). According to Robertson and Woolfenden (1992), they are especially numerous in phosphate mine pits in central Florida, but their numbers decline in the northern peninsula and in the panhandle during the winter. Although the current numbers of breeding pairs are unknown (Ogden 1996) there were an estimated 1,500–2,000 for the entire state of Florida in 1988 (Westall 1990).

Several general reviews of the life history of

Table 12.1. Primary reasons given for submission of Ospreys to 3 wildlife rehabilitation centers in Florida, 1988–95

| | Number of cases per region | | | | |
Primary diagnostic finding	Central[a]	North central[b]	West central[c]	Total	% of total cases
Trauma	129	9	52	190	67
Emaciation	15	3	23	41	14
Electrocution	24	0	3	27	9
Inclement weather	10	0	0	10	4
Infectious diseases	8	2	0	10	4
Poisoning	3	1	2	6	2
Total	189	15	80	284	—

a. Records from Florida Audubon Society's Center for Birds of Prey, Maitland, 1989–94 (Collins and Gilliland 1995).
b. Records from Veterinary Medical Teaching Hospital, College of Veterinary Medicine, University of Florida, Gainesville, 1988–94 (Deem and Terrell 1996).
c. Records from Suncoast Seabird Sanctuary, St. Petersburg, Fla., 1988–95 (Suto 1996).

Ospreys have included sections on morbidity and mortality factors with special emphasis on environmental contaminants (Henny 1988; Henny and Anthony 1989; Poole 1989; Ogden 1996). The mortality chapter in Poole's monograph is especially thorough and includes information on predation, trauma, and nest robbing inflicted by humans, effects of land clearing and habitat degradation, organochlorines, mercury and other pollutants, and the significance of acid rain and snow. Two other publications contain information gathered by personnel of the U.S. Fish and Wildlife Service on residues of environmental pollutants and include results of necropsy findings (with limited data on infectious diseases) on Ospreys from several states in the eastern United States from 1964 to 1982 (Wiemeyer et al. 1980, 1987). Concentrations of organochlorines and mercury were reported for eggs of Ospreys in Massachusetts, Maryland, and Virginia by Audet et al. (1992). Some general aspects of morbidity and mortality of Ospreys in Florida, particularly the north central part of the state, have been published (Deem et al. 1998).

Important sources of specific information on Ospreys in Florida used in the preparation of this chapter were the files of the National Wildlife Health Center in Madison, Wisconsin (provided through the courtesy of Milton Friend), the files of Florida Audubon Society's Center for Birds of Prey in Maitland (provided by Resee Collins and Carol Gilliland), the files of Suncoast Seabird Sanctuary in St. Petersburg (provided by Barbara Suto), and the files of the Veterinary Medical Teaching Hospital at the University of Florida (provided by Sharon L. Deem and Scott P. Terrell). These latter 3 datasets were composed of combined morbidity and mortality information on Ospreys submitted for rehabilitation purposes.

II. Trauma

During the 8-year period between 1988 and 1995, trauma was the primary diagnostic finding for 190 of 284 (67%) Ospreys submitted to 3 wildlife rehabilitation centers in Florida (table 12.1). Some of these Ospreys were dead on arrival, some died shortly after arrival, some were euthanized, and others were rehabilitated and eventually released. The specific cause of the trauma in the majority of these cases was not determined, but based on the findings of the large subset of data from the Center for Birds of Prey and several other published and unpublished records, a considerable amount was due to young birds falling from their nests and adults being struck by vehicles or shot illegally (table

Table 12.2. Mortality in Ospreys from Florida attributed to trauma, 1910–94

Types of trauma	No. of cases	% of total cases
Fell from nest (nestlings)	17	12
Struck by vehicle	12	8
Gunshot	10	7
Power line strike	6	4
Interspecific conflict (eagles)	5	4
Entanglement in fishing line	2	1
Alligator predation	2	1
Struck by boat	1	<1
Caught in trap	1	<1
Collision with building	1	<1
Unknown	85	60
Total	142	—

Sources: Longstreet (1910), Wiemeyer et al. (1980), Delany (1986), Stroud (1986), Franson (1990), Locke (1992), Forrester (1993), Collins and Gilliland (1995).

12.2). Other less numerous causes of trauma include power line strikes, interspecific conflicts, entanglement in fishing lines, collisions with boats and buildings, and being caught in traps.

Ogden (1977) reported that nestlings sometimes fall from nests and if they are less than 4 weeks of age, even if they are not injured by the fall, are ignored by their parents and starve to death. Older nestlings that fall from nests apparently are cared for and fed on the ground by their parents until they fledge. Poole (1979, 1982) observed intraspecific aggression among sibling nestlings during his 1978–79 study of Ospreys in Florida Bay. Submissive nestlings often had signs of sustained aggression such as areas on the head and back that were stained with blood and had been pecked clean of feathers. The injuries to these nestlings were not judged to be so serious that movement or the ability to beg for food from the parent was impaired. Poole (1982) also observed that submissive siblings were driven to the edge of the nest and he postulated that some probably were forced out of the nest by this behavior, although he had no direct observations to support this idea. This type of sibling aggression was seen also in 2 colonies in New York, but to a lesser degree than in Florida Bay where food

delivery rates were reduced and the Ospreys were stressed for food.

Ogden (1996) stated that "Ospreys are regularly injured or killed by striking power lines and by becoming entwined in fishing monofilament," but gave no specific data. We have data on 6 incidents relating to power line strikes and 2 cases of Ospreys becoming entangled in fishing line in Florida (table 12.2).

Trauma due to interspecific aggression (probably from Bald Eagles) was observed in an Osprey (age not given) from Big Pine Key, Monroe County, in 1985 (Stroud 1986). Four other such reports were given by Collins and Gilliland (1995) for Ospreys in Orange and Volusia counties during 1993–94 (table 12.2).

III. Predation

Information on predation of Ospreys is limited. Poole (1989) listed raccoons and Great Horned Owls as predators on eggs and nestlings, respectively, but we have no quantitative data for Florida. Ogden (1977) reported that American Crows destroyed 2 nests in Florida Bay (Monroe County), but felt that this was an infrequent phenomenon. He also suspected that

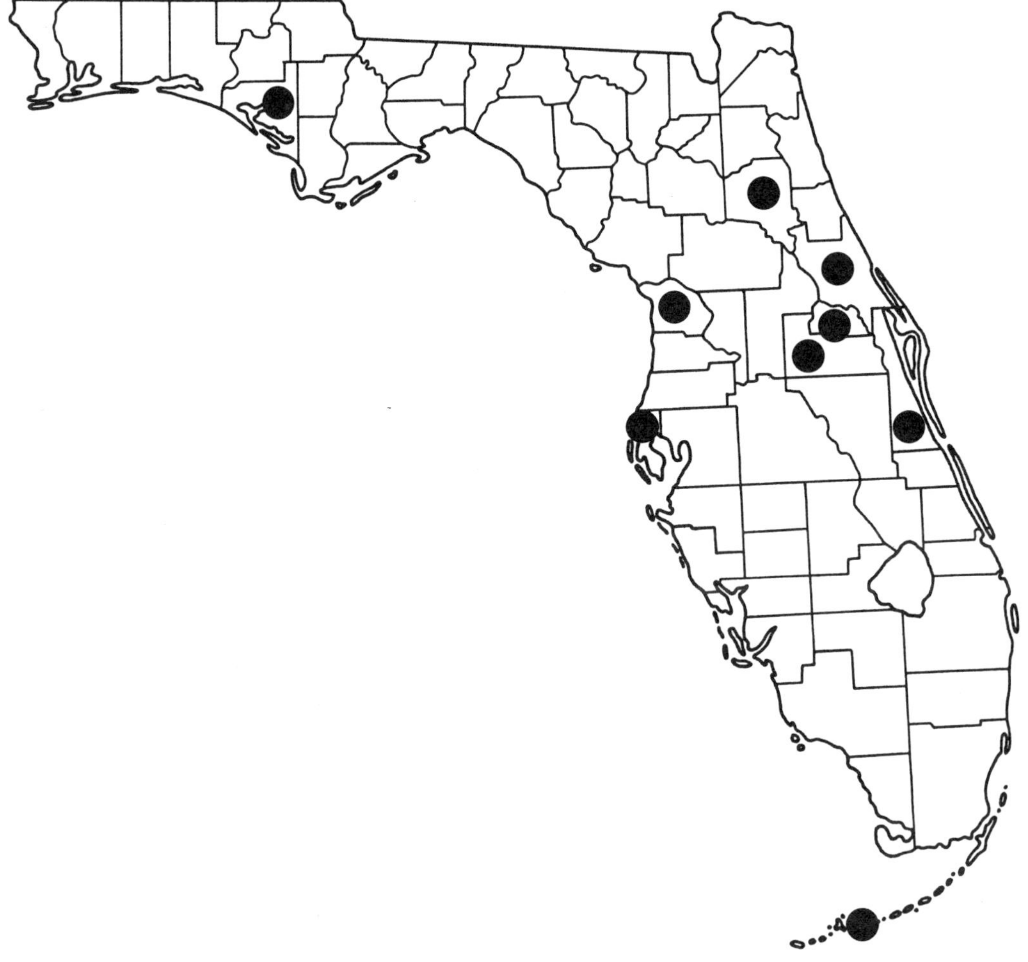

FIGURE 12.1. Distribution of morbidity and mortality of Ospreys in Florida due to electrocution, 1970–94. *Solid circles* = counties in which 1 or more electrocuted Ospreys were found. Data from Wiemeyer et al. 1980; Stroud 1985; Stroud 1986; Locke 1992; Collins and Gilliland 1995; Suto 1996.

nestling predation by crows and Bald Eagles occurred, but gave no supporting data, except for 1 incident in which the remains of a young Osprey were found in an active eagle nest. Delany (1986) found a USFWS aluminum band in the stomach of an American alligator from Newnan's Lake (Alachua County) in 1985; this band had been placed on an immature Osprey 4 months earlier in the same general area. Another incident of alligator predation on an Osprey was noted by Collins and Gilliland (1995) in Polk County during 1994.

IV. Inclement weather

High winds caused egg breakage or destruction of nests in Florida Bay (Ogden 1977), and certainly the results of hurricanes are most likely devastating, although we have little data on this. In September 1961 a number of birds was killed by Hurricane Donna in Florida Bay, including 1 Osprey (Robertson and Paulson 1961). According to the records of the Center for Birds of Prey in Maitland, severe storms were the cause of injury or death of 10 Ospreys

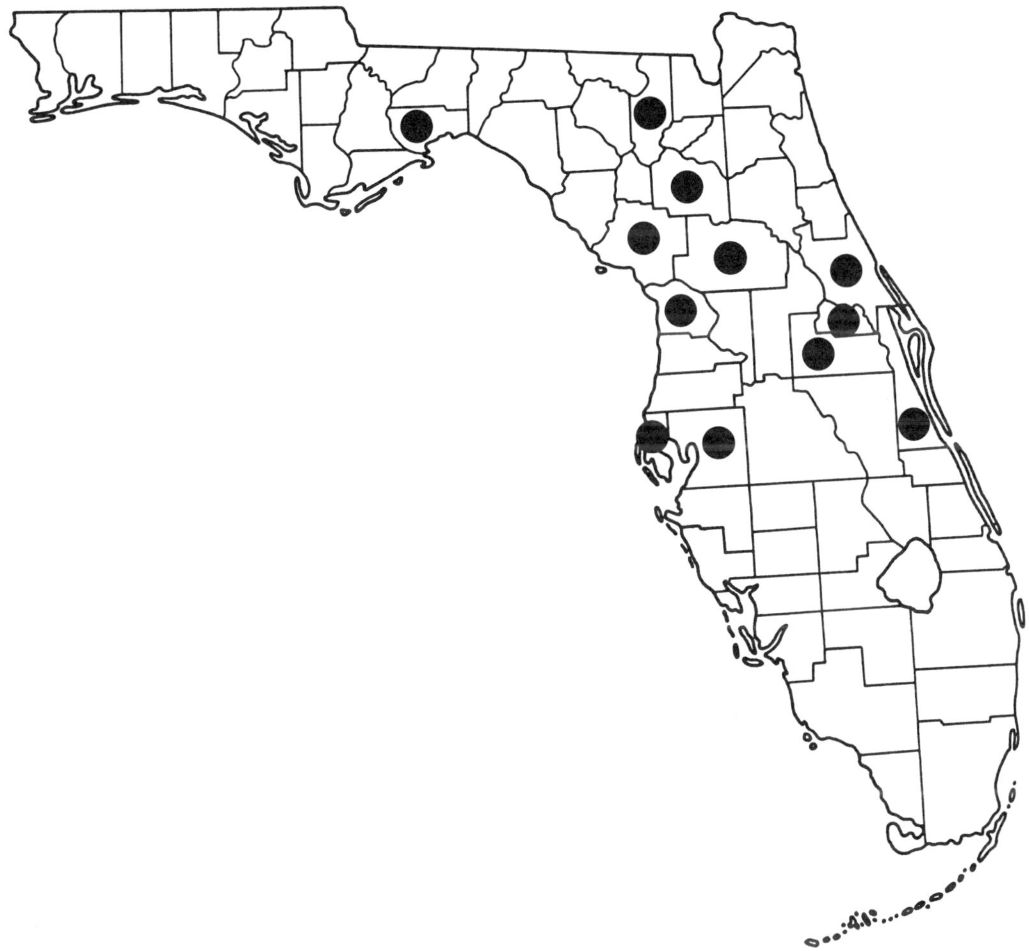

FIGURE 12.2. Distribution of emaciated Ospreys in Florida, 1970–94. *Solid circles* = counties in which 1 or more emaciated Ospreys were found. Data from Meteyer 1992; Collins and Gilliland 1995; Deem and Terrell 1996; Suto 1996.

during 1989–94 (table 12.1). Five of these were injured in the "storm of the century" during March of 1993 in Lake, Seminole, and Volusia counties (Collins and Gilliland 1995).

V. Electrocution

Electrocution due to contact with power lines and lightning strikes is an important cause of mortality in Ospreys in Florida. Leenhouts (1986) gave an account of an Osprey nest at Merritt Island National Wildlife Refuge (Bre-

vard County) being struck by lightning in June of 1986, resulting in the death of a nestling. This type of mortality may be more prevalent than is appreciated, since Ospreys commonly nest in manmade structures such as metal towers, which are prone to receiving lightning strikes. Lightning storms are very common in Florida at certain times of the year. Merritt Island National Wildlife Refuge, for instance, has between 10,000 and 30,000 lightning strikes each year (Leenhouts 1986).

Thirty-one cases of electrocution, most likely due to contact with power lines, have been

Table 12.3. Organochlorine residues reported from tissues of Ospreys from Florida

County	Date	Organ or tissue[a]	Residues (ppm wet weight)							Data source
			DDE	DDD	DDT	Total DDT	DDT equiv.[b]	Dieldrin	PCB	
Alachua										
	1975	UP	13	NG	NG	16	—	0.91	NG	Johnston (1978)
Brevard										
	1970	BR	2.2	0.29	ND	2.5	1.5	ND	1.3[f]	Wiemeyer et al. (1980)
		CS[c]	38	8.4	5.7	52	—	0.09	8.0[f]	Ibid.
Citrus										
	1975	BR	3.6	ND	3.8	7.4	4.0	NG	NG	Sundlof et al. (1986)
Collier										
	1968	BR	4.9	1.6	0.30	6.8	0.95	0.97	5.6[f]	Wiemeyer et al. (1980)
		CS	2.2	1.1	0.15	3.5	—	0.2	2.0[f]	Ibid.
Hillsborough										
	1974	BR	0.01	0.01	ND	0.02	ND	0.01	0.5	Sundlof et al. (1986)
		FA	0.81	1.3	ND	2.1[e]	—	0.07	1.2	Ibid.
		LV	0.02	0.01	ND	0.03	—	ND	0.06	Ibid.
		MU	0.11	0.60	ND	0.71[e]	—	0.01	0.32	Ibid.
	1975	BR	ND	ND	ND	ND	ND	NG	NG	Ibid.
Lake										
	1974	BR	0.02	ND	ND	0.02	ND	ND	0.05	Ibid.
		FA	0.91	0.43	0.12	1.5[e]	—	0.08	1.0	Ibid.
		LV	0.37	0.18	0.05	0.60	—	0.04	0.43	Ibid.
		MU	0.17	0.09	0.05	0.31	—	0.02	0.18	Ibid.
Marion										
	NG	FA	0.13	NG	NG	0.26	—	ND	NG	Johnston (1978)
		UP	1.6	NG	NG	1.8	—	ND	NG	Ibid.

County	Year	Tissue								Source
Monroe	1968	BR[d]	11	2.8	0.20	14	1.5	0.62	50[f]	Wiemeyer et al. (1980)
		CS[c]	3.9	0.87	0.10	4.9	—	0.19	20[f]	Ibid.
	1973	BR	0.23	ND	ND	0.23	0.15	0.15	0.84[f]	Ibid.
		CS	1.5	ND	ND	1.5	—	ND	2.0[f]	Ibid.
	1973	BR	ND	ND	ND	ND	ND	ND	ND	Ibid.
		CS	0.60	0.15	ND	0.75	—	ND	1.1[g]	Ibid.
	1973	FA	0.13	NG	NG	0.26	—	0.26	ND	Johnston (1978)
		UP	1.6	NG	NG	1.8	—	ND	NG	Ibid.
Pinellas	1974	FA	ND	NG	NG	ND	—	ND	NG	Ibid.
		UP	0.09	NG	NG	0.09	—	ND	NG	Ibid.
	1974	FA	0.33	NG	NG	0.33	—	ND	NG	Ibid.
		UP	0.32	NG	NG	0.32	—	ND	NG	Ibid.
	1976	FA	1.7	NG	NG	1.9	—	ND	NG	Ibid.
		UP	0.53	NG	NG	0.71	—	ND	NG	Ibid.
	1976	FA	1.5	NG	NG	2.5	—	ND	NG	Ibid.
		UP	1.4	NG	NG	1.4	—	ND	NG	Ibid.

Note: Lower limits for detection: Johnston (1978) = 0.01 ppm; Sundlof et al. (1986) = 0.01 ppm; Wiemeyer et al. (1980) = 0.5 ppm for PCBs and 0.1 ppm for organochlorine. ND = not detected, NG = not given by authors.

a. BR = brain, CS = carcass (entire bird except for head, wings, feet, skin, liver, and gastrointestinal tract), FA = fat, LV = liver, MU = muscle, UP = uropygial gland.
b. DDT equivalents = DDT + (0.067xDDE) + (0.20xDDD) according to the equation of Stickel et al. (1970).
c. Also contained 0.02 ppm heptachlor epoxide.
d. Also contained 0.07 ppm heptachlor epoxide.
e. These values have been corrected; numbers in the publication by Sundlof et al. (1986) are incorrect.
f. Aroclor 1260.
g. Aroclor 1254.

Table 12.4. Organochlorine residues from 3 fish taken from active Osprey nests on Seahorse Key, Levy County, Florida, 1972

Fish[a]	Organochlorine residues (ppm lipid basis)[b]				
	DDE	DDD	DDT	Dieldrin	PCB
Sea catfish	0.08	ND	0.05	ND	3.5
Ocellated flounder	1.8	ND	4.5	ND	227
Speckled trout	0.07	0.03	0.09	0.02	2.7

Source: Szaro (1978).
ND = none detected.
a. Scientific names were not given by the author; the sea catfish was probably *Arius felis,* the ocellated flounder was either *Bothus robinsi* or *Ancylopsetta quadrocellata,* the speckled trout was probably *Cynoscion nebulosus.*
b. Lowest limits of detection not given by author.

recorded for Ospreys in 9 counties. These included 27 cases that were submitted to 2 wildlife rehabilitation centers in Florida (table 12.1), as well as 3 cases examined at the National Wildlife Health Center (Stroud 1985–86; Locke 1992), and 1 examined at Patuxent Wildlife Research Center (Wiemeyer et al. 1980). The clustering of these cases in central Florida (figure 12.1), and especially Orange County, may be an artifact of collection since the Florida Audubon Society's Center for Birds of Prey is located in Orange County. These data, nevertheless, are an indication of the importance of electrocution as a mortality factor for Ospreys. As discussed in chapter 11 (Eagles), electrocution of raptors can be prevented by certain adaptations or modifications of the power line structures on which the birds tend to perch and also by changes in the grounding setup. Details on these practices have been published (Miller et al. 1975) and may help mitigate such mortality.

VI. Food stress and emaciation

The relationship of food stress to mortality in young Ospreys was studied in Florida Bay (Monroe County) in 1978–79 (Poole 1979, 1982). Nesting success was found to be related to food delivery to young in the nest, which was less in Florida Bay birds compared with Ospreys nesting in northern areas such as New York. This reduction in food delivery resulted in starvation of nestlings, particularly third chicks that hatched an average of 3.9 days later than the first 2 chicks in a nest. Sibling aggression, as mentioned above in the section on trauma, resulted in preferential feeding of the older chicks in the nest, but this occurred only when food was limited. This situation was studied further by Kushlan and Bass (1983), who found a 58% reduction in the number of Ospreys nesting in Florida Bay during the 8-year period 1973–80. This was attributed to "a reduction in the abundance or availability of food, especially during nesting" (Kushlan and Bass 1983). Ospreys in Florida Bay nest in the winter and spring months and lay eggs mainly between late November and early March (Ogden 1977). This means that the Ospreys in Florida Bay have fewer daylight hours during which to hunt for food in comparison with Ospreys nesting in more northern areas during summer months. Poole (1982) suggested that even a small reduction in the numbers of prey fishes in Florida Bay could result in a devastating effect on the growth and survival of nestling birds. It is of interest that during the same period of time the productivity of Brown Pelicans also decreased in Florida Bay, while Bald Eagles remained stable (Kushlan and Bass 1983).

An extremely emaciated Osprey was found dead in St. Mark's National Wildlife Refuge in

Table 12.5. Organochlorine residues reported from eggs of Ospreys from Florida

County	Year(s)	n[b]	Organochlorine residues (ppm lipid basis)[a]						Data source
			DDE	DDD	DDT	Total DDT	Dieldrin	PCB	
Brevard	1975	1	0.78	0.17	ND	0.95	0.05	2.3	Wiemeyer et al. (1988)
Levy	1972	3	8.3	3.6	0.30	12	0.26	30[c]	Szaro (1978)
Monroe	1969–72	15	1.2	NG	NG	—	NG	1.4	Ogden (1977)
	1973[d]	10	0.90	NG	NG	—	0.02	1.5	Blus et al. (1977)
	1973[d]	10	0.72	ND	ND	0.72	ND	1.3	Wiemeyer et al. (1988)

ND = none detected.
a. Values for Monroe County (1973) from Wiemeyer et al. (1988) are geometric means; others are mean values, but the authors did not state if they were arithmetic or geometric. The lowest limit of detection for the Wiemeyer et al. (1988) data = 0.5 ppm for PCBs and 0.1 ppm for organochlorine pesticides; limits not given by the other authors.
b. Number of clutches represented.
c. Aroclor 1254.
d. The values given by Blus et al. (1977) and Wiemeyer et al. (1988) for Ospreys in Monroe County may be from the same clutches; this is not clear from the papers.

March of 1992 (Meteyer 1992). No cause of the emaciation was determined. In addition there were 41 cases of emaciation as the primary diagnosis in 284 Ospreys submitted to 3 wildlife rehabilitation centers during 1988–94 (table 12.1). Again, the causes of this condition were not determined. The counties in which these emaciated Ospreys originated are indicated in figure 12.2. As was suggested for electrocution cases, the distribution of Ospreys with emaciation within the state may be an artifact, since most of the cases were near the 3 wildlife rehabilitation centers from which these data were obtained. Nevertheless, it is clear that food stress and emaciation are problems of considerable importance in Ospreys in Florida and should be further investigated.

VII. Organochlorines

Tissue residue data on organochlorines were collected from 16 Ospreys from 9 counties in Florida from 1968 to 1976 (table 12.3). PCBs were present in all birds that were analyzed specifically for them, with residues ranging from 0.05 to 50.00 ppm in brain samples. These values were considerably below lethal concentrations of 310 ppm and above (Stickel et al. 1984). Dieldrin was found in 8 of the 14 Ospreys for which it was tested. Residues were found in brains of 4 of these Ospreys and were all <1 ppm, which is below the hazardous range of 4 ppm or higher (Stickel et al. 1969).

Detectable concentrations of DDT and/or its metabolites were found in brain tissue of 7 of 9 Ospreys (table 12.3). DDT equivalents for these ranged from 0 to 4 ppm, well below the values that have been associated with mortality (Stickel et al. 1970). It is unlikely, therefore, that any of these Ospreys died as a direct result of DDT poisoning. The source of DDT in these Ospreys is of interest since the sale of this compound in the United States was stopped in 1972 (Klaassen 1980). Johnston (1978) gave 2 suggestions to explain this observation: (1) long-lived birds could accumulate and store pesticide residues in fat tissue for years, and (2) migratory raptors could obtain DDT in the West Indies or in Central America where DDT has not been banned and contaminated food is still available to them. These reasons may explain some of the contamination of Ospreys in Florida with DDT after 1972, but the bird from Citrus County in 1975 was a juvenile that had never undergone winter migration and had a fairly high concentration of DDE (3.6 ppm) and DDT (3.8 ppm) in brain tissue. The DDT was probably obtained in Florida in this case and most likely was related to illegal use of the insecticide (Sundlof et al. 1986). Another Osprey was obtained in 1975 from Alachua

Table 12.6. Measurements of eggshells of Ospreys from Florida, pre-1947–1973

County or site Year(s)	n[a]	Mean shell weight (gm)	Mean shell thickness (mm)	Thickness index[b]	% change in thickness[c]	Data source
"Florida"						
Pre-1947	83	7.16	0.498[d]	2.53	—	Anderson & Hickey (1972)
1949	3	6.33	NG	2.13	-16	Ibid.
1960	6	6.20	NG	2.20	-13	Ibid.
Monroe						
1970	11	NG	0.560	NG	+12	Ogden (1977)
1972	9	NG	0.555	NG	+11	Ibid.
1973	10	NG	0.501	NG	+1	Wiemeyer et al. (1988)
Levy						
1972	7	NG	0.460	NG	-8	Szaro (1978)

NG = not given by authors.
a. Number of clutches represented.
b. Thickness index = weight of eggshell (mg) divided by the product of the length x breadth (mm) (Ratcliffe 1967).
c. Compared with the pre-1947 norm of Anderson and Hickey (1972) of 0.498 mm.
d. For thickness, only 46 of the 83 eggs could be measured.

County and had almost 16 ppm total DDT, but the age of this bird was not known (Johnston 1978). Szaro (1978) analyzed 3 food fish taken from active nests at Seahorse Key (Levy County) (table 12.4). Low concentrations of DDE, DDD, DDT, and dieldrin were found; 1 fish (an ocellated flounder) had a residue of PCB of 227 ppm (lipid basis).

Eggs from Ospreys in 3 Florida counties were analyzed for organochlorine residues during 1969–75 (table 12.5). Residues of DDE, DDD, DDT, dieldrin, and PCB (Aroclor 1254) were found. DDE was found in all 39 eggs examined and varied from 0.35 to 8.34 ppm (Ogden 1977; Szaro 1978; Wiemeyer et al. 1988). The highest concentrations were in eggs from Seahorse Key in Levy County.

DDE residues are correlated closely with eggshell thinning in Ospreys (Wiemeyer et al. 1988; Poole 1989). Wiemeyer et al. (1988) examined Osprey eggs from 14 states between 1970 and 1979 and concluded that 10% shell thinning was associated with 2.0 ppm DDE, 15% with 4.2 ppm, and 20% with 8.7 ppm.

Table 12.7. Concentrations of mercury in tissues and eggs of Ospreys from Florida

County	Site	Year	Organ or tissue	Residue (ppm)	Data source
Marion	McIntosh	1978	Liver	6.4[a]	Gourlie (1984)
Monroe	Big Torch Key	1973	Liver	68[b]	Wiemeyer et al. (1988)
			Kidney	130[b]	Ibid.
	Plantation Key	1973	Liver	6.2[b]	Ibid.
			Kidney	6.5[b]	Ibid.
	Florida Bay	1972	Egg ($n = 1$)	0.07[a]	Ogden et al. (1974)
			Egg ($n = 1$)	0.21[a]	Ibid.
	Everglades Natl. Park	1973	Eggs ($n = 5$)	0.07[c,d]	Wiemeyer et al. (1988)

a. Lower limit of detection not given by authors.
b. Lower limit of detection = 0.001 ppm (wet weight basis).
c. Lower limit of detection = 0.02 ppm (wet weight basis).
d. Geometric mean.

Table 12.8. Concentrations of heavy metals (other than mercury) found in tissues of five Ospreys from Florida

County	Year	Organ	Concentrations (ppm wet weight)[a]					
			Arsenic	Cadmium	Copper	Chromium	Lead	Zinc
Franklin	1982	Liver	NA	NA	NA	NA	17	NA
Monroe	1973	Liver	7.2	3.6	5.7	0.008	<0.07	27
		Kidney	2.0	4.1	6.9	NA	NA	NA
	1973	Liver	1.9	0.11	2.2	0.004	<0.06	36
		Kidney	0.54	0.63	3.7	NA	NA	NA
	1986	Liver	NA	NA	NA	NA	0.1	NA
	1986	Liver	NA	NA	NA	NA	0.09	NA
Wakulla	1992	Liver	NA	NA	NA	NA	0.06	NA

Sources: 1973 Monroe County data from Wiemeyer et al. (1980); other data from National Wildlife Health Center records.
NA = not analyzed.
a. Lower limits of detection = 0.001 ppm for arsenic, chromium, and mercury; 0.10 ppm for copper and zinc; 0.005 ppm for cadmium; 0.05 ppm for lead.

Shell thicknesses of eggs from Ospreys in Florida were measured in 1949, 1960, and the early 1970s and compared with pre-1947 values (before the use of DDT was instituted). Values for the 1940s and 1960 were obtained from museum eggs (table 12.6). In 1949 and 1960 eggshell thicknesses were 16% and 13% below pre-1947 values, respectively. Osprey eggs that are 10–12% thinner seldom break, but clutches with more than about 16% are likely to lose at least 1 egg to breakage (Poole 1989). Judging from this information there may have been some breakage from eggshell thinning in Florida Ospreys in the late 1940s and 1950s, but there is no documentation that such a phenomenon occurred and the data that are available for 1949 and 1960 are based on a very small sample (table 12.6). Shells of eggs measured during the early 1970s were above normal in thickness, indicating that a recovery had occurred, except at Seahorse Key in Levy County where Szaro (1978) found that 7 eggs were 8% thinner than the pre-1947 values. Organochlorine residues (especially DDE and PCBs) in eggs from Seahorse Key at that time were fairly high (table 12.5) in comparison with eggs from Brevard and Monroe counties.

Ogden (1996), citing a personal communication from M. McMillian, stated that organochlorines were suspected to have caused a signif-icant decline in the numbers of nesting Ospreys on Lake Istokpoga in Highlands County during the early 1970s. No specific data were given, however, to back this up. Poole (1989) provides an excellent review of the effects of organochlorines on reproductive failures in Ospreys. Such failures were especially serious in populations in the northeastern United States where "DDT use and consequent shell thinning were greatest" (Poole 1989). However, Poole goes on to point out that the evidence for the implication of organochlorines as the cause of reproductive failures in Osprey populations is circumstantial. Experimental feeding studies such as were done for other raptors were not conducted on Ospreys. Poole goes on to say that "Nevertheless, as many uses of organochlorines were phased out during the 1960s and 1970s, these changes functioned as large scale experiments, albeit poorly controlled ones. . . . Ospreys showed correspondingly quick recoveries. By the late 1970s, nearly all previously contaminated North American Osprey populations were reproducing at close to pre-pesticide rates."

VIII. Organophosphates and carbamates

No information is available on this topic.

IX. Metals

Concentrations of mercury were determined for tissues of 3 Ospreys and 7 Osprey eggs (table 12.7). The Osprey from Big Torch Key had the highest concentration (68 ppm in liver), which may have been lethal to the bird (Wiemeyer et al. 1980). There are no specific experimental data on Ospreys to help with interpretation of these data, but 17 ppm in liver tissue has been associated with mortality in Red-tailed Hawks during experimental feeding trials (Eisler 1987). The concentrations found in eggs were below those linked with adverse effects on reproduction (Häkkinen and Häsänen 1980; Wiemeyer et al. 1988).

In addition to data on mercury, information has been reported on 6 other metals in tissues (table 12.8) and eggs (table 12.9) of Ospreys in Florida. These include arsenic, cadmium, chromium, copper, lead, and zinc. The concentrations of cadmium, chromium, copper, and zinc were probably normal and of no conse-

quence to Ospreys (Wiemeyer et al. 1980; Eisler 1985).

The concentrations of arsenic were somewhat elevated for the 2 Ospreys from Monroe County and may indicate that these birds had been exposed to arsenic contamination of the environment (Wiemeyer et al. 1980). One Os-

Table 12.9. Concentrations (ppm)[a] of heavy metals (other than mercury) found in 2 eggs of Ospreys from Florida Bay, Monroe County, Florida, 1972

Metal	Egg #1	Egg #2
Arsenic	ND	ND
Cadmium	ND	ND
Copper	0.78	0.54
Lead	ND	ND
Zinc	4.6	6.5

Source: Ogden et al. (1974).
ND = none detected.
a. Lowest limit of detection not given by authors.

Table 12.10. Bacteria identified from Ospreys in Florida

Species of bacteria	County	Year	Organs or tissue	Data source
Actinobacillus actinomycetemcomitans	Monroe	1973	Lungs[a]	Wiemeyer et al. (1980)
Edwardsiella tarda	Wakulla	1992	Intestines[f]	Meteyer (1992)
Enterobacter sp.	Lake	1974	Intestines	White & Forrester (1975)
Enterococcus sp.	Wakulla	1992	Intestines[f]	Meteyer (1992)
Escherichia coli	Brevard	1989	Spleen[c]	Franson (1990)
	Palm Beach	1972	Intestines	White & Forrester (1975)
	Pinellas	1975	Intestines	Ibid.
Klebsiella sp.	Brevard	1986	Lungs[b]	Duncan (1986)
Proteus sp.	Pinellas	1975	Liver	White & Forrester (1975)
Pseudomonas sp.	Palm Beach	1971	Intestines[c]	White & Forrester (1975)
Staphylococcus aureus	Monroe	1985	Feet[d]	Duncan (1986)
Staphylococcus sp.	Brevard	1986	Lungs[b]	Ibid.
	Collier	1968	Air sacs	Wiemeyer et al. (1980)
Streptococcus sp.	Monroe	1973	Lungs[a]	Ibid.
Yersinia kristensenii	Monroe	1986	Sm. intestines[e]	Duncan (1986)

a. This Osprey had acute hemorrhagic pneumonia.
b. This Osprey died of electrocution (lightning).
c. This Osprey died of trauma-induced injuries.
d. This Osprey had bumblefoot and septicemia and had been in captivity for an unspecified period of time.
e. This Osprey died of electrocution.
f. This Osprey was emaciated.

prey had a liver concentration of 7.2 ppm, but the significance of this to the health of the bird is not known. Wiemeyer et al. (1980) reported that high amounts of arsenic (17 ppm) contributed to the death of an Osprey from Maryland in 1973.

Lead was found in liver samples from 6 Ospreys. The concentration was highest (17 ppm) in 1 Osprey from St. Marks National Wildlife Refuge (Franklin County) in 1982. Although the cause of death was not determined for this bird (Stroud 1982), lead may have been a contributing factor. Experimental studies on lead exposure to Ospreys have not been conducted, but such studies with Bald Eagles have shown that concentrations of 10 ppm or more are indicative of lead poisoning (Pattee et al. 1981).

The significance of the concentrations of arsenic, cadmium, copper, lead, and zinc in Osprey eggs (table 12.9) is not known. The sublethal effects of these metals on Ospreys and Osprey reproduction should be studied.

X. Neoplasia

One case of suspected neoplasia has been reported from Ospreys in Florida. A possible osteogenic sarcoma of the right tibia-tarsus was found in a juvenile male Osprey from South Biscayne Bay (Dade County) in January, 1968 (Wiemeyer et al. 1980). No other information is available on this case and the significance of this finding is unknown.

XI. Biotoxins

No information is available on this topic.

XII. Viruses

Favorite (1960) examined an Osprey from the Orlando area in 1958 and found that it was seropositive (via the serum neutralization test) for both eastern equine encephalitis virus and

Highlands J virus. The pathological significance of these viruses to Ospreys is unknown.

Albers (1984) tested 8 Ospreys from a wildlife rehabilitation facility in Pinellas County in 1983 for evidence of rabies virus infection. All were seronegative.

XIII. Bacteria

Eleven species of bacteria have been isolated from Ospreys in Florida (table 12.10). Most of these are secondary findings, except for the isolations of *Actinobacillus actinomycetemcomitans* and *Streptococcus* sp. from an Osprey from Plantation Key (Monroe County) in 1973. This bird had an acute hemorrhagic pneumonia (Wiemeyer et al. 1980). Another Osprey from Key Deer National Wildlife Refuge (Monroe County), which had been captive for an unspecified period of time, had bumblefoot from which *Staphylococcus aureus* was isolated (Duncan 1986). These infections are probably not of significance to Osprey populations, although information on them is limited.

White and Forrester (1975) cultured 4 Ospreys for enteric pathogens such as *Salmonella* spp. and *Edwardsiella tarda*. The birds were from Lake County (*n* = 1, 1974), Palm Beach County (*n* = 1, 1972), and Pinellas County (*n* = 2, 1975). The Osprey from Palm Beach County and 1 from Pinellas County died from gunshot wounds, and the other 2 were found dead of unknown causes. All were negative for *Salmonella* spp. and *E. tarda*. Although these 4 birds were negative for *E. tarda,* this bacterium was identified from another Osprey, an emaciated female from Wakulla County in 1992 (Meteyer 1992).

XIV. Fungi

Aspergillosis caused by *Aspergillus fumigatus* was diagnosed at necropsy as a secondary finding in 1 of 3 Ospreys from Key Deer National Wildlife Refuge (Monroe County) in 1986

Table 12.11. Helminth infections in Ospreys from Florida

Helminth	Site in host[b]	County (or site)	Year(s)	No. Ospreys[a] Exam.	No. Ospreys[a] Inf.	Intensity Mean	Intensity Range
Trematoda							
Ascocotyle angrense	SI	Wakulla[e]	1992	1	1	20	—
Ascocotyle sp.	SI	Monroe[c]	1986	3	1	NG	NG
Clinostomum complanatum	PR	Alachua[f]	1998	1	1	1	—
Echinochasmus dietzevi	SI	Wakulla[e]	1992	1	1	—	—
Mesoophorodiplostomum pricei	SI	Wakulla[e]	1992	1	1	—	—
Mesostephanus appendiculatoides	SI	Alachua[f]	1998	1	1	431	—
Mesorchis denticulatus	SI	Lake	1974	1	1	232	—
		Marion	1978	1	1	9	—
		Pinellas	1975	1	1	2	—
		Wakulla[e]	1992	1	1	1	—
Microparyphium facetum	CL	Lake	1974	1	1	2	—
Neogogatea pandionis	SI	Lake	1974	1	1	169	—
Phagicola longa	SI	Wakulla[e]	1992	1	1	3	—
Posthodiplostomum minimum	SI	Alachua[f]	1998	1	1	76	—
Pygiodopsis pindoramensis	SI	Alachua[f]	1998	1	1	—	—
		Wakulla[e]	1992	1	1	4	—
Renicola ralli	KD	Citrus	1975	1	1	14	—
		Hillsborough	1976	1	1	167	—
		Marion	1978	1	1	107	—
Ribeiroia ondatrae	PR	Alachua[f]	1998	1	1	29	—
		Marion	1978	1	1	54	—
		Pinellas	1975	1	1	11	—
Scaphanocephalus expansus	SI	Bay[e]	1991	1	1	162	—
		Citrus	1975	1	1	17	—
		Hillsborough	1976	1	1	704	—
		Levy[g]	2000	1	1	74	—
Cestoda							
Cyclustera ibisae	SI	Wakulla[e]	1992	1	1	—	—
Paradilepis rugovaginosus	SI	Alachua[f]	1998	1	1	1	—
		Pinellas	1975	1	1	3	—
Nematoda							
Capillaria falconis	SI	Alachua[f]	1998	1	1	3	—
		Citrus	1975	1	1	1	—
		Wakulla[e]	1992	1	1	—	—
Cardiofilaria pavlovskyi	BC	Lake	1974	1	1	1	—
Contracaecum multipapillatum	PR	Alachua[d]	1986–87	2	2	ND	ND
		Citrus	1975	1	1	7	—
		Lake	1974	1	1	10	—
		Levy[g]	2000	1	1	1	—
		Wakulla[e]	1992	1	1	—	—
		No. Central Fla.[d]	1979	1	1	ND	ND
Contracaecum pandioni	PR	Alachua[f]	1998	1	1	1	—
		Lake	1974	1	1	1	—
Contracaecum sp.	PR	Alachua[d]	1980	1	1	ND	ND
		Alachua[f]	1998	1	1	3	—
		Brevard[c]	1986	1	1	NG	NG

(continued)

Table 12.11. *(continued)*

Helminth	Site in host[b]	County (or site)	Year(s)	No. Ospreys[a] Exam.	Inf.	Intensity Mean	Range
		Dade[e]	1993	1	1	—	—
		No. Central Fla.[d]	1985	1	1	ND	ND
Sexanoscara skrjabini	ES	Hillsborough	1976	1	1	2	—
Tetrameres microspinosa	PR	Alachua[f]	1998	1	1	1	—
		Hillsborough	1976	1	1	3	—
		Marion	1978	1	1	10	—
Acanthocephala							
Andracantha mergi	SI	Hillsborough	1976	1	1	4	—
Neoechinorhynchus chrysemydis	SI	Alachua[f]	1998	1	1	1	—
Polymorphis brevis	SI	Alachua[f]	1998	1	1	1	—

Sources: All data are from Kinsella and Forrester (1993) and Kinsella et al. (1996), except as otherwise noted.
a. The data from Kinsella et al. (1996) consist of 5 Ospreys, 1 each from Citrus (1975), Hillsborough (1976), Lake (1974), Marion (1978), and Pinellas (1975) counties. Unless indicated as positive under each of the helminths in the table, they were negative for that particular helminth.
b. Infection sites: BC = body cavity, CL = cloaca, ES = esophagus, KD = kidney, PR = proventriculus, SI = small intestine.
c. From Schorr (1986).
d. From Kinsella and Greiner (1993).
e. From Kinsella and Cole (1994).
f. From Kinsella and Foster (1998).
g. From Foster and Spalding (2000).

(Stroud 1986). The infection was classified as light and occurred in a bird that died of trauma. Lesions were in the lungs and air sacs. Aspergillosis was diagnosed in 8 Ospreys submitted to the Center for Birds of Prey in 1991 (*n* = 2), 1993 (*n* = 2) and 1994 (*n* = 4) (Collins and Gilliland 1995). Six of these had been in captivity for periods of time ranging from 1 to 9 days, while 1 was captive for 20 days and 1 for a month. Another Osprey from Putnam County was diagnosed with aspergillosis when admitted to the Veterinary Medical Teaching Hospital in 1993 (Deem and Terrell 1996). The geographic origins of these cases are shown in figure 12.3.

Fungal infections seem to be rare in free-living raptors (Keymer 1972) and there is some evidence that stress might increase the susceptibility of these birds to aspergillosis under certain conditions (Redig et al. 1980). In the midwestern United States aspergillosis was diagnosed in 3 of 21 Ospreys presented for treatment of injury or disease between 1974 and 1979 (Redig 1981).

XV. Protozoans

Only 1 protozoan has been found in Ospreys from Florida. Coccidia (species not identified) were found in the intestines of an Osprey from the Key Deer National Wildlife Refuge (Monroe County) in 1986 (Stroud 1986). The cause of death of this bird was determined to be electrocution by a power line accident; when examined at necropsy, the carcass was emaciated. The bird also had diarrhea and chronic enteritis, which may have been caused by the coccidial infection, but this diagnosis was complicated by the isolation of the bacterium *Yersinia kristensenii* from the intestines (Duncan 1986). Intestinal coccidiosis has been suggested as a contributory cause of mortality in other free-ranging raptors (Mathey 1966; Coon et al. 1970; Keymer et al. 1981), but the significance of this disease to Ospreys is unknown. It is possible that this coccidian infection represents a part of the life cycle of a species of *Sarcocystis* in which the Osprey is the definitive host and some other vertebrate is the intermediate host. More research is needed.

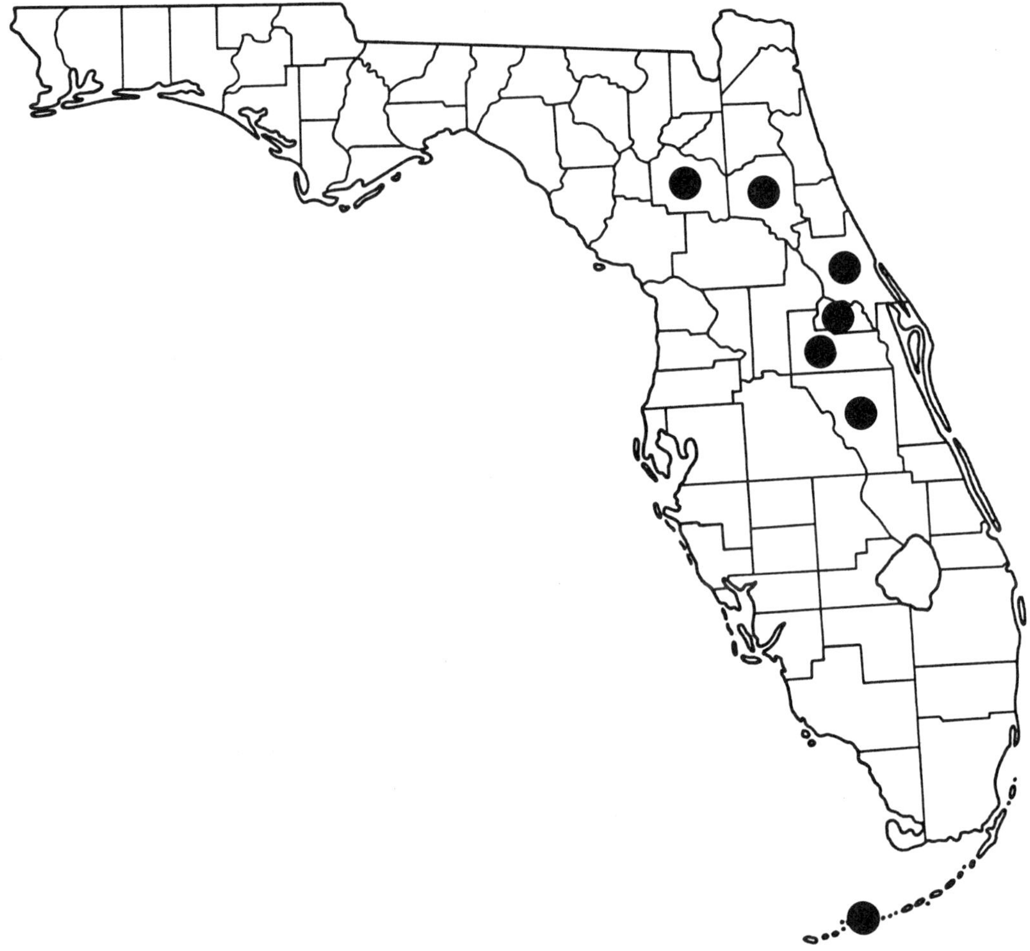

FIGURE 12.3. Distribution of morbidity and mortality of Ospreys in Florida due to aspergillosis, 1989–94. *Solid circles* = counties in which 1 or more infected Ospreys were found. Data from Stroud 1986; Collins and Gilliland 1995; Deem and Terrell 1996.

Blood films from 5 Ospreys were examined for blood protozoans. These included 2 birds from Alachua County (1990–91), 1 from Collier County (1990), 1 from Marion County (1990), and 1 from Nassau County (1991). All were negative (Forrester et al. 1994). Several species of blood protozoans of the genera *Haemoproteus*, *Leucocytozoon*, and *Hepatozoon* have been reported from Ospreys in other parts of their range (Bennett et al. 1982; Bishop and Bennett 1992); some of these might be found in Florida birds if a larger sample were examined.

XVI. Helminths

Information on the helminths of Ospreys in Florida is limited, but 14 trematodes, 2 cestodes, 6 nematodes, and 3 acanthocephalans have been found in a small sample of birds from 11 counties (table 12.11). Five of these species of helminths are considered specialists in Ospreys; i.e., they occur only in Ospreys, probably because of reproductive and ecological isolation of this host from other raptors (Kinsella et al. 1996). These include 2 trematodes (*Neogogatea pandionis* and *Scaphanocephalus expansus*), 1

Table 12.12. Parasitic arthropod infestations on Ospreys from Florida

Arthropod	County or site	Year	Data source
Feather mites			
Analloptes sp.	Lake	1974	Forrester & Mertins (1992)
Bonnetella fusca	Citrus	1975	Ibid.
	Hillsborough	1976	Ibid.
	Lake	1974	Ibid.
	Pinellas	1975	Ibid.
Chewing lice			
Kurodaia haliaeeti	Alachua	1934	Forrester et al. (1995)
	Dade	1991	Ibid.
	Highlands	1979	Ibid.
	Lake	1974	Ibid.
	Lee	1983	Ibid.
	Pinellas	1975	Ibid.
	"Florida"	NG	Peters (1936)
Louse flies			
Olfersia fumipennis	Glades	NG	Bequaert (1956)
	Indian River	1973	Wilson et al. (1990)
	St. Johns	1895	Johnson (1895, 1913, 1922)
	Wakulla	1973	Wilson et al. (1990)

NG = not given by authors.

cestode (*Paradilepis rugovaginosus*), and 2 nematodes (*Contracaecum pandioni* and *Sexanoscara skrjabini*). The other 20 species are generalists and are also found in other raptors, herons, pelicans, or turtles. The significance of these parasites to Ospreys has not been determined, but because intensities were fairly low, they are probably of limited pathologic consequence. However, additional research in this area should be undertaken with larger sample sizes.

XVII. Arthropods

Four species of parasitic arthropods have been found on Ospreys in Florida and include 2 feather mites, 1 chewing louse, and 1 louse fly (table 12.12). One of the feather mites (*Analloptes* sp.) is an undescribed species (Mertins 1992). The other feather mite (*Bonnetella fusca*) and the chewing louse (*Kurodaia haliaeeti*) are specific for Ospreys (Emerson 1972, Mertins 1992). The louse fly *Olfersia fumipennis* is primarily a parasite of Ospreys and has a worldwide distribution along with its host, but also occurs as a straggler on Bald Eagles (Wil-

son et al. 1990). Nothing is known about the significance of these ectoparasites to individual birds or to Osprey populations.

XVIII. Summary and conclusions

Excluding trauma, electrocution, and food stress, 58 morbidity/mortality factors and disease agents are known from Ospreys in Florida. These include 6 organochlorines, 7 metals, 1 tumor, 2 viruses, 11 bacteria, 1 fungus, 1 protozoan, 14 trematodes, 2 cestodes, 6 nematodes, 3 acanthocephalans, 2 feather mites, 1 chewing louse, and 1 louse fly.

Even though the list of morbidity and mortality factors and disease agents is fairly long, it is based on a comparatively small sample of Ospreys and, with the exception of information on food stress, our knowledge is limited. There do not appear to be any significant infectious or parasitic diseases affecting Osprey populations in Florida, even though individual cases of bacterial pneumonia, bumblefoot, and aspergillosis have been documented. In spite of the fact that the organochlorine threat is over, additional research should be conducted on other environmental contaminants such as mercury, lead, organophosphates, and carbamates in order to better understand their significance to Osprey populations, both as acute disease agents and as less obvious sublethal factors. Food stress appears to be an important problem, at least in Florida Bay, and should be monitored in other populations. Additional mortality due to human-related trauma (vehicle and power line collisions, gunshot) and electrocutions can be expected as the human population of the state grows and developers continue to invade more and more of the Osprey's choice habitat along lakes and streams.

XIX. Literature cited

Albers, H.F. 1984. Naturally occurring immunity to rabies in raptors. *Proc. Annu. Natl. Wildl. Rehabil. Symp.* 2:171–174.

Anderson, D.W., and J.J. Hickey. 1972. Eggshell changes in certain North American birds. *Proc. Int. Ornithol. Congr.* 15:514–540.

Audet, D.J., D.S. Scott, and S.N. Wiemeyer. 1992. Organochlorines and mercury in osprey eggs from the eastern United States. *J. Raptor Res.* 26:219–224.

Bennett, G.F., M. Whiteway, and C. Woodworth-Lynas. 1982. *A host-parasite catalogue of the avian Haematozoa.* Meml. Univ. Nfld. Occas. Pap. Biol. 5. 243 pp.

Bequaert, J.C. 1956. The Hippoboscidae or louse-flies (Diptera) of mammals and birds. Part II. Taxonomy, evolution and revision of American genera and species. *Entomol. Am.* 36:417–611.

Bishop, M.A., and G.F. Bennett. 1992. *Host-parasite catalogue of the avian Haematozoa,* suppl. 1. Meml. Univ. Nfld. Occas. Pap. Biol. 15. 210 pp.

Blus, L.J., S.N. Wiemeyer, J.A. Kerwin, R.C. Stendell, H.M. Ohlendorf, and L.F. Stickel. 1977. Impact of estuarine pollution on birds. In: *Estuarine pollution control and assessment, proceedings of a conference.* U.S. Environmental Protection Agency, Washington, D.C. pp. 57–71.

Collins, R., and C. Gilliland. 1995. Unpublished data. Florida Audubon Society, Center for Birds of Prey, Maitland.

Coon, N.C., L.N. Locke, E. Cromartie, and W.L. Reichel. 1970. Causes of bald eagle mortality, 1960–1965. *J. Wildl. Dis.* 6:72–76.

Deem, S.L., and S.P. Terrell. 1996. Unpublished data. University of Florida, Gainesville.

Deem, S.L., S.P. Terrell, and D.J. Forrester. 1998. A retrospective study of morbidity and mortality of raptors in Florida: 1988–1994. *J. Zoo Wildl. Med.* 29:160–164.

Delany, M.F. 1986. Bird bands recovered from American alligator stomachs in Florida. *N. Am. Bird Bander* 11:92–94.

Duncan, R.M. 1986. Unpublished data. National Wildlife Health Center, Madison, Wis.

Eisler, R. 1985. *Cadmium hazards to fish, wildlife, and invertebrates: a synoptic review.* U.S. Fish and Wildlife Service, Biological Report 85(1.2). 46 pp.

———. 1987. *Mercury hazards to fish, wildlife, and invertebrates: a synoptic review.* U.S. Fish and Wildlife Service, Biological Report 85(1.10). 90 pp.

Emerson, K.C. 1972. *Checklist of the Mallophaga of North America (north of Mexico).* Part II. *Suborder Amblycera.* Deseret Test Center, Dugway, Utah. 118 pp.

Favorite, F.G. 1960. Some evidence of local origin of EEE virus in Florida. *Mosq. News* 20:87–92.

Forrester, D.J. 1993. Unpublished data. University of Florida, Gainesville.

Forrester, D.J., H.W. Kale II, R.D. Price, K.C. Emerson, and G.W. Foster. 1995. Chewing lice (Mallophaga) from birds in Florida: A listing by host. *Bull. Fla. Mus. Nat. Hist.* 39:1–44.

Forrester, D.J., and J.W. Mertins. 1992. Unpublished data. University of Florida, Gainesville.

Forrester, D.J., S.R. Telford, Jr., G.W. Foster, and G.F. Bennett. 1994. Blood parasites of raptors in Florida. *J. Raptor Res.* 28:226–231.

Foster, G.W., and M.G. Spalding. 2000. Unpublished data. University of Florida, Gainesville.

Franson, J.C. 1990. Unpublished data. National Wildlife Health Center, Madison, Wis.

Gourlie, N. 1984. Unpublished data. Department of Environmental Regulation, Tallahassee, Fla.

Häkkinen, I., and E. Häsänen. 1980. Mercury in eggs and nestlings of the osprey (*Pandion hali-*

aetus) in Finland and its bioaccumulation from fish. *Ann. Zool. Fenn.* 17:131–139.

Henny, C.J. 1988. Osprey, *Pandion haliaetus*. In: *Handbook of North American birds*. Vol. 4. R.S. Palmer (ed.). Yale University Press, New Haven, Conn. pp. 73–101.

Henny, C.J., and R.G. Anthony. 1989. Bald eagle and osprey. *Natl. Wildl. Fed. Sci. Tech. Ser.* 12:66–82.

Johnson, C.W. 1895. Diptera of Florida. *Proc. Acad. Nat. Sci. Phila.*:303–340.

———. 1913. Insects of Florida. I. Diptera. *Bull. Am. Mus. Nat. Hist.* 32:37–90.

———. 1922. Notes on distribution and habits of some of the bird-flies, Hippoboscidae. *Psyche* 29:79–85.

Johnston, D.W. 1978. Organochlorine pesticide residues in Florida birds of prey, 1969–76. *Pestic. Monit. J.* 12:8–15.

Keymer, I.F. 1972. Diseases of birds of prey. *Vet. Rec.* 90:579–594.

Keymer, I.F., M.R. Fletcher, and P.I. Stanley. 1981. Causes of mortality in British kestrels (*Falco tinnunculus*). In: *Recent advances in the study of raptor diseases. Proceedings of the International Symposium on Diseases of Birds of Prey.* J.E. Cooper and A.G. Greenwood (eds.). Chiron, West Yorkshire, England. pp. 143–151.

Kinsella, J.M., and R.A. Cole. 1994. Unpublished data. University of Florida, Gainesville.

Kinsella, J.M., R.A. Cole, D.J. Forrester, and C.L. Roderick. 1996. Helminth parasites of the osprey, *Pandion haliaetus,* in North America. *J. Helminthol. Soc. Wash.* 63:262–265.

Kinsella, J.M., and D.J. Forrester. 1993. Unpublished data. University of Florida, Gainesville.

Kinsella, J.M., and G.W. Foster. 1998. Unpublished data. University of Florida, Gainesville.

Kinsella, J.M., and E.C. Greiner. 1993. Unpublished data. University of Florida, Gainesville.

Klaassen, C.D. 1980. Nonmetallic environmental toxicants: air pollutants, solvents and vapors, and pesticides. In: *Goodman and Gilman's The pharmacological basis of therapeutics.* 6th ed. A.G. Gilman, L.S. Goodman, and A. Gilman (eds.). Macmillan, New York. pp. 1649–1650.

Kushlan, J.A., and O.L. Bass. 1983. Decreases in the southern Florida osprey population, a possible result of food stress. In: *Biology and man-agement of bald eagles and Ospreys.* D.M. Bird (ed.). Harpell, Quebec. pp. 187–200.

Leenhouts, W.P. 1986. Osprey killed by lightning at Merrit Island National Wildlife Refuge, Florida. *Fla. Field Nat.* 15:22–23.

Locke, L.N. 1992. Unpublished data. National Wildlife Health Center, Madison, Wis.

Logan, T.H. 1997. Florida's endangered species, threatened species and species of special concern. Official lists. Florida Game and Fresh Water Fish Commission, Tallahassee. 14 pp.

Longstreet, R.J. 1910. A letter. *Oologist* 27:1.

Mathey, W.J. 1966. *Isospora buteonis* Henry 1932 in an American kestrel (*Falco sparverius*) and a golden eagle (*Aquila chrysaetos*). *Bull. Wildl. Dis. Assoc.* 2:20–22.

Mertins, J.W. 1992. Unpublished data. USDA, National Veterinary Services Laboratory, Ames, Iowa.

Meteyer, C.U. 1992. Unpublished data. National Wildlife Health Center, Madison, Wis.

Miller, D., E.L. Boeker, and R.S. Thorsell. 1975. Suggested practices for raptor protection on powerlines. Raptor Research Foundation, Provo, Utah. 19 pp.

Ogden, J.C. 1977. Preliminary report on a study of Florida Bay ospreys. *Trans. N. Am. Osprey Res. Conf., U.S. Natl. Park Serv. Trans. Proc. Ser.* 2:143–151.

———. 1996. Osprey. In: *Rare and endangered biota of Florida.* Vol. 5, *Birds.* J.A. Rodgers, Jr., H.W. Kale II, and H.T. Smith (eds.). University Press of Florida, Gainesville. pp. 170–178.

Ogden, J.C., W.B. Robertson, Jr., G.E. Davis, and T.W. Schmidt. 1974. Pesticides, polychlorinated biphenols and heavy metals in upper food chain levels, Everglades National Park and vicinity. National Technical Information Service, Dept. of the Interior, Atlanta, Ga.

Pattee, O.H., S.N. Wiemeyer, B.M. Mulhern, L. Sileo, and J.W. Carpenter. 1981. Experimental lead-shot poisoning in bald eagles. *J. Wildl. Manag.* 45:806–810.

Peters, H.S. 1936. A list of external parasites from birds of the eastern part of the United States. *Bird-Banding* 7:9–27.

Peterson, R.T. 1969. The osprey: endangered world citizen. *Natl. Geogr.* 136(1):52–67.

Poole, A.F. 1979. Sibling aggression among nestling ospreys in Florida Bay. *Auk* 96:415–417.

———. 1982. Brood reduction in temperate and sub-tropical ospreys. *Oecologia* (Berl.) 53:111–119.

———. 1989. *Ospreys: a natural and unnatural history.* Cambridge University Press, New York. 246 pp.

Ratcliffe, D.A. 1967. Decrease in eggshell weight in certain birds of prey. *Nature* 215:208–210.

Redig, P.T. 1981. Aspergillosis in raptors. In: *Recent advances in the study of raptor diseases. Proceedings of the international symposium on diseases of birds of prey.* J.E. Cooper and A.G. Greenwood (eds.). Chiron, West Yorkshire, England. pp. 117–122.

Redig, P.T., M.R. Fuller, and D.L. Evans. 1980. Prevalence of *Aspergillus fumigatus* in free-living goshawks (*Accipiter gentilis atricapillus*). *J. Wildl. Dis.* 16:169–174.

Robertson, W.B., Jr., and D.E. Paulson. 1961. Region reports: Florida region. *Audubon Field Notes* 15:26–35.

Robertson, W.B., Jr., and G.E. Woolfenden. 1992. *Florida bird species. An annotated list.* Spec. Publ. 6, Florida Ornithological Society, Gainesville. 260 pp.

Schorr, L.F. 1986. Unpublished data. National Wildlife Health Center, Madison, Wis.

Stickel, W.H., L.F. Stickel, and F.B. Coon. 1970. DDE and DDD residues correlated with mortality of experimental birds. In: *Inter-American conference on toxicology and occupational medicine, pesticide symposia.* W.P. Deichmann (ed.). Halos and Associates, Miami. pp. 287–294.

Stickel, W.H., L.F. Stickel, R.A. Dyrland, and D.L. Hughes. 1984. Aroclor 1254 residues in birds: lethal levels and loss rates. *Arch. Environ. Contam. Toxicol.* 13:7–13.

Stickel, W.H., L.F. Stickel, and J.W. Spann. 1969. Tissue residues of dieldrin in relation to mortality in birds and mammals. In: *Chemical fallout.* M.W. Miller and G.G. Berg (eds.). Charles C. Thomas, Springfield, Ill. pp. 174–204.

Stroud, R.K. 1982–86. Unpublished data. National Wildlife Health Center, Madison, Wis.

Sundlof, S.F., D.J. Forrester, N.P. Thompson, and M.W. Collopy. 1986. Residues of chlorinated hydrocarbons in tissues of raptors in Florida. *J. Wildl. Dis.* 22:71–82.

Suto, B.J. 1996. Unpublished data. Suncoast Seabird Sanctuary, Redington Beach, Fla.

Szaro, R.C. 1978. Reproductive success and foraging behavior of the osprey at Seahorse Key, Florida. *Wilson Bull.* 90:112–118.

Westall, M.A. 1990. Osprey. In: *Proc. Southeast Raptor Manag. Symp. Workshop.* Natl. Wildl. Fed. Sci. Tech. Ser. 14:22–28.

White, F.H., and D.J. Forrester. 1975. Unpublished data. University of Florida, Gainesville.

Wiemeyer, S.N., C.M. Bunck, and A.J. Krynitsky. 1988. Organochlorine pesticides, polychlorinated biphenyls, and mercury in Osprey eggs—1970–79—and their relationships to shell thinning and productivity. *Arch. Environ. Contam. Toxicol.* 17:767–787.

Wiemeyer, S.N., T.G. Lamont, and L.N. Locke. 1980. Residues of environmental pollutants and necropsy data for eastern United States ospreys, 1964–1973. *Estuaries* 3:155–167.

Wiemeyer, S.N., S.K. Schmeling, and A. Anderson. 1987. Environmental pollutant and necropsy data for ospreys from the eastern United States, 1975–1982. *J. Wildl. Dis.* 23:279–291.

Wilson, N.A., H.W. Kale II, and W.W. Baker. 1990. Unpublished data. University of Northern Iowa, Cedar Falls.

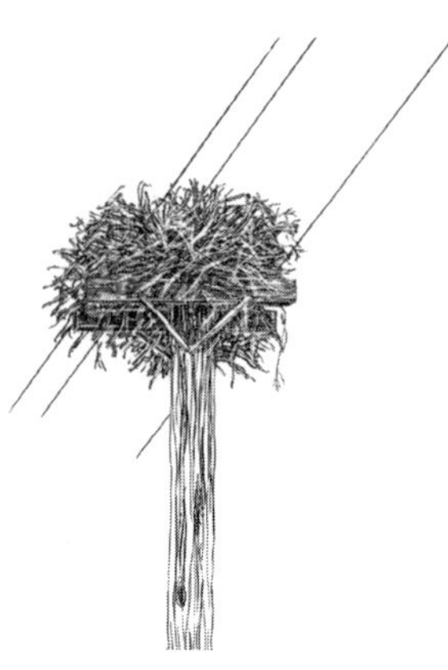

Kites

I. Introduction

Four species of kites breed in Florida; the Swallow-tailed Kite and the Mississippi Kite are long-range migrants (table 13.1). Most of the data available on morbidity and mortality of kites in Florida are concerned with the Snail Kite, with a moderate amount of information on the Swallow-tailed Kite and very few data on the Mississippi Kite. There is no information on parasites and diseases of White-tailed Kites in Florida.

The Snail Kite or Everglade Kite is a unique aquatic raptor with a very specialized diet consisting almost entirely of apple snails (*Pomacea paludosa*) with a few occasional exceptions (Sykes and Kale 1974; Sykes 1987a; Bennetts et al. 1994). The numbers of Snail Kites probably

were in the hundreds or even thousands prior to 1930, but by the mid-1900s population estimates had fallen to less than 100 birds (Bennetts et al. 1988): in 1950, for example, there were an estimated 60–100 Snail Kites in Florida (Sprunt 1950). These population declines were attributed to widespread drainage of the Everglades and subsequent losses of habitat and apple snails (Bent 1937; Stieglitz and Thompson 1967; Sykes 1979; Beissinger 1986; Takekawa and Beissinger 1989). In the 1970s and 1980s the numbers generally increased (Sykes 1979, 1983a; Bennetts et al. 1988). From 1986 to 1990, the numbers varied between 326 and 563 (Rodgers 1992). The mean for that 5-year period was 454, com-

Table 13.1. Distribution, occurrence, and abundance of kites in Florida[a]

Species of kite	Range	Seasonal occurrence	Relative abundance
Swallow-tailed Kite[b] *Elanoides forficatus*	Western panhandle	Summer resident & transient	Rare to uncommon
	Peninsula	Summer resident & transient	Uncommon to fairly common[c]
White-tailed Kite[d] *Elanus leucurus*	Everglades basin, parts of peninsula, & eastern panhandle	Resident	Rare
Snail Kite[e] *Rostrhamus sociabilis*	Southern Florida & St. Johns River	Resident	Rare
Mississippi Kite *Ictinia mississippiensis*	Panhandle & northern peninsula	Summer resident	Uncommon
	Panhandle	Spring & fall transient	Locally common
	Peninsula & Keys	Spring & fall transient	Rare

a. Modified from Robertson and Woolfenden (1992) and American Ornithologists' Union (1998).
b. Classified as threatened by the Florida Committee on Rare and Endangered Plants and Animals (Meyer and Collopy 1996).
c. More numerous southward; locally abundant in southern Florida prior to fall migration, where one or more roosts can have up to 50% of the U.S. population.
d. Classified as rare by the Florida Committee on Rare and Endangered Plants and Animals (Meyer 1996).
e. Classified as endangered by the Florida Game and Fresh Water Fish Commission and the U.S. Fish and Wildlife Service (Logan 1997) and by the Florida Committee on Rare and Endangered Plants and Animals (Rodgers 1996).

pared with a mean of 387 for the previous 5 years (Rodgers et al. 1988). In 1994 the statewide survey of wetlands in Florida known to be used by kites resulted in a total count of 996, which is the highest number recorded since such counts were initiated (Sykes et al. 1995). Population sizes are known to fluctuate widely from year to year and are related weakly to the amount of water present in freshwater marshes (Sykes et al. 1995). Droughts may result in lower survival (Beissinger 1986; Snyder et al. 1989; Takekawa and Beissinger 1989), but precise data on this are scant (Sykes et al. 1995). Currently the Snail Kite is classified as endangered by the Florida Game and Fresh Water Fish Commission, the U.S. Fish and Wildlife Service, and the Florida Committee on Rare and Endangered Plants and Animals (Logan 1997; Rodgers 1996).

Results of studies on various aspects of the life history of Snail Kites in Florida, dating back to 1844, have been published (Sykes 1984; Ben-netts et al. 1988, 1994; Sykes et al. 1995; Bennetts and Kitchens 1997) and include information on mortality factors, particularly predation and other aspects of nesting success. Bennetts et al. (1998) have provided a summary of the causes of mortality of post-fledging juvenile and adult Snail Kites in Florida over a 4-year period (1992–95). Deem et al. (1998) presented an overview of morbidity and mortality of raptors submitted to a rehabilitation center in Florida over a 7-year period (1988–94) and included data on 4 Mississippi Kites and 2 Swallow-tailed Kites. Sykes and Forrester (1983) published a review of the parasites of Snail Kites in Florida and other parts of its range. Information on genetic variation of populations of Snail Kites in Florida was published by Rodgers and Stangel (1996).

In contrast to the Snail Kite, the Swallow-tailed Kite is migratory and has a diverse diet of insects, frogs, anoles, snakes, and birds (Meyer and Collopy 1990). Less frequently it even eats

Table 13.2. Predation of Snail Kites in Florida

Predator County/site	Year(s)	No. and type of loss	Data source
Cottonmouths (*Agkistrodon piscivorous*)			
Lake Okeechobee	1956	1 nest[a]	Wachenfeld (1956)
Several[b]	1968–77	2 eggs (2 nests)	Sykes (1987b)
Several[c]	1966–83	1 nest	Snyder et al. (1989)
Cons. Area 2A	1993	1 nestling	Spalding & Bennetts (1994)
Everglades rat snakes (*Elaphe obsoleta*)			
Lake Okeechobee	1984	1 egg	Toner (1984)
Cons. Area 3A	1987	2 nestlings	Bennetts & Caton (1988)
Several[b]	1968–77	5 eggs, 4 young (9 nests)	Sykes (1987b)
Several[c]	1966–83	3 nests	Snyder et al. (1989)
American alligators (*Alligator mississippiensis*)			
Lake Kissimmee	1987–91	1 nest (nestlings)	Rodgers (1993a)
Boat-tailed Grackles (*Quiscalus major*)			
Lake Okeechobee	1956	"eggs"[a]	Wachenfeld (1956)
" "	1973	9 nests (eggs)	Chandler & Anderson (1974)
Several[c]	1966–83	1 nest (eggs)	Snyder et al. (1989)
Fish Crows (*Corvus ossifragus*)			
Several[b]	1968–77	5 eggs (2 nests)	Sykes (1987b)
Crows (*Corvus* sp.)			
"Florida"	1884	"nest"	Bailey (1884)
Several[b]	1968–77	1 egg	Sykes (1987b)
Barred Owls (*Strix varia*)			
E. Lake Tohopekaliga	1987–91	2 adults (on nest)	Rodgers (1993a)
Lake Kissimmee	1987–91	1 adult (on nest)	Ibid.
Raccoons (*Procyon lotor*)			
Several[c]	1966–83	1 nest (nestling)	Snyder et al. (1989)
Ants (*Crematogaster atkinsoni*)			
Lake Okeechobee	1972	1 nest (1 egg)[d]	Sykes & Chandler (1974)
Several[c]	1966–83	"Several nests" (eggs & nestlings)[d]	Snyder et al. (1989)
Dermestid beetles (*Dermestes nidum*)			
Cons. Area 3A	1978	2 nests[e] (3 nestlings)	Snyder et al. (1984)
Lake Okeechobee	1979	2 nests[e] (4 nestlings)	Ibid.
Cons. Area 3A	1979	7 nests[e] (11 nestlings)	Ibid.
" "	1980–83	6 nests[e] (1 nest failure)	Snyder et al. (1989)
" "	1986–87	2 nests (nestlings)	Bennetts et al. (1988)
Unknown			
Levy County	1981	1 juvenile	Schmeling (1981)
Osceola County	1981	2 nestlings	Thomas (1991)
Several[b]	1968–77	18 eggs, 3 young (21 nests)	Sykes (1987b)
Several[f]	1987–91	2 nests (nestlings)	Rodgers (1992)
"Florida"	1980	1 kite (age unknown)	Siegfried (1982)

a. Number of eggs in nest not determined.

b. Author presented combined data from several study areas (headwaters of St. Johns River, west side of Lake Okeechobee, Loxahatchee National Wildlife Refuge, Conservation Areas 2A and 3A).

c. Author presented combined data from several study areas (Lake Okeechobee, Conservation Area 3A, Lake Kissimmee, and Lake Tohopekaliga).

d. Ant predation caused nest desertion and subsequent loss of eggs and nestlings.

e. Nestlings had abdominal lesions; i.e., "crater-shaped holes measuring up to 7 mm in diameter at the surface and penetrating as deep as the body cavity in some cases" (Snyder et al. 1984), but most survived. There was only one case of nest failure.

f. One nest each from 5 areas (Lake Tohopekaliga, Lake Kissimmee, St. Johns marshes, Lake Okeechobee, and Conservation Area 3A).

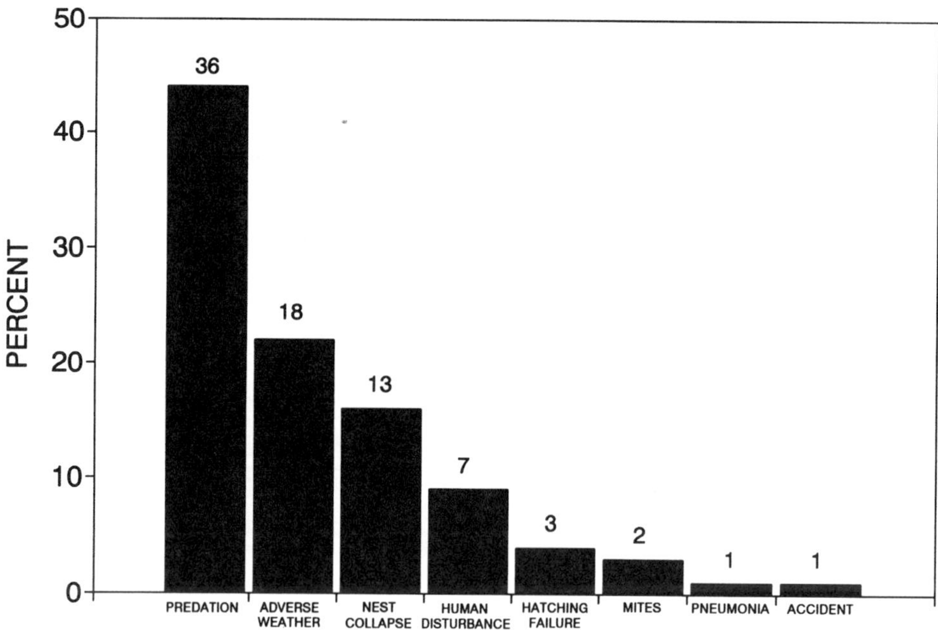

Figure 13.1. Bar graph illustrating the various factors responsible for the failure of 82 Snail Kite nests in southern Florida, 1968–77. Numbers on the tops of the bars are the actual numbers of nests lost for each category. Adverse weather includes strong winds, heavy rainfall, hailstorms, and low temperatures; human disturbance includes fires set by humans, airboats, and activity near nests. Data from Sykes 1987b.

fruit, bats, and small fish (Meyer 1995). The population in Florida consists of 450–750 breeding pairs, according to estimates reported by Meyer (1988, 1995) and Meyer and Collopy (1990). The Swallow-tailed Kite is currently classified as threatened by the Florida Committee on Rare and Endangered Plants and Animals (Meyer and Collopy 1996). Millsap (1987) described concentrations of fall migrants at Fisheating Creek (Glades County) consisting of large numbers of birds, sometimes totaling over 1,300.

Other than limited data on predation and nest destruction by high winds (Bent 1937; Snyder 1974; Meyer and Collopy 1990; Meyer 1995), little is known about morbidity and mortality factors of Swallow-tailed Kites. No comprehensive study of parasitic and infectious diseases has been conducted. Meyer (1995) cited some unpublished data of ours on helminths, chewing lice, and environmental contaminants.

II. Trauma

Structural collapse of Snail Kite nests has been documented as another significant mortality factor in certain habitats, especially when nests are built in cattail (*Typha* spp.) or bulrush (*Scirpus californicus*) (Wachenfeld 1956; Sykes and Chandler 1974; Beissinger 1986; Sykes 1987b; Snyder et al. 1989; Rodgers 1992; Bennetts et al. 1994). Eggs and nestlings fall out of nests that tilt or fall as a result of weak vegetation, wind, or changes in water level. Nestlings that fall into water in this manner often drown (Wachenfeld 1956). Beissinger (1986) and Snyder et al. (1989) rated such structural collapse of nests as the most important cause of nest failure. Snyder et al. (1989) estimated that as much as 64% of nest failures were due to this. They observed that only 3% of the nests in cattails were successful compared with 35% in southern willows (*Salix caroliniana*) that pro-

FIGURE 13.2. A nestling Snail Kite with abdominal lesions caused by larvae of *Dermestes nidum*. Most nestlings so afflicted survive and fledge. Courtesy of Noel F.R. Snyder.

vided stronger substrates. The use of artificial nest structures was initiated in 1973 with good success (Sykes and Chandler 1974). This technique involves the transfer of nests and their contents (usually eggs) into sturdy metal basketlike structures mounted on poles in appropriate nesting habitat. Snail Kites readily accept these structures and continue incubating their clutches. Snyder et al. (1989) found a 47% success rate for nests transferred from cattails to these artificial structures compared with the 44% success rate of nests built in strong willow substrates. In areas such as Water Conservation Area 3A, however, structural failure does not seem to be a serious problem since most nests are not constructed in cattails, but are built in stronger substrates such as willow, pond apple (*Annona glabra*), cypress (*Taxodium* sp.), punk tree (*Melaleuca quinquenerva*), and wax myrtle (*Myrica cerifera*) (Bennetts et al. 1988).

Shooting of Snail Kites during the waterfowl hunting season has been mentioned by several authors (Sprunt 1945; Stieglitz and Thompson 1967; Sykes et al. 1995) and may have contributed to the decline of this species (Sykes 1978), although this has been disputed (Beissinger 1988). Bennetts et al. (1998) found a dead juvenile Snail Kite in 1993 with evidence that it had been shot. Human persecution (i.e., shooting) has been suggested as the major cause of the decline of Swallow-tailed Kites (Robertson 1988). From 1987 to 1993 1 or more Swallow-tailed Kites were found each year in Florida with gunshot wounds (Meyer 1993). Two Swallow-tailed Kites (1 from Volusia County in 1990 and 1 from Lee County in 1991) were treated for gunshot trauma at the Florida Audubon Society's Center for Birds of Prey in Maitland, Florida (Collins and Gilliland 1995). Likewise, 2 kites, a Swallow-tailed Kite from Volusia County in 1994 and a Mississippi Kite from Alachua County in 1991, were submitted to the Veterinary Medical Teaching Hospital at the University of Florida for treatment of gunshot wounds (Deem and Terrell 1996).

Collision with vehicles was determined to be the cause of death of 3 Snail Kites (1 juvenile and 2 adults) in the study by Bennetts et al. (1998). They concluded that this type of mortality may occur when kites nest or forage near highways (Bennetts and Kitchens 1997). One

Table 13.3. Probable causes of mortality of post-fledging juvenile and adult Snail Kites in central and southern Florida, 1992–95

Probable cause	Juveniles		Adults		Totals	
	No.	%	No.	%	No.	%
Predation	10	32.2	7	43.8	17	36.2
Starvation	2	6.5	0	0.0	2	4.3
Disease	0	0.0	1	6.3	1	2.1
Vehicle collision	1	3.2	2	12.5	3	6.4
Gun shot	1	3.2	0	0.0	1	2.1
Unknown[a]	4	12.9	3	18.8	7	14.9
Unknown[b]	13	41.9	3	18.8	16	34.0
Totals	31	100.0	16	100.0	47	100.0

Source: Bennetts et al. (1998).
a. Carcasses were intact when found, indicating that predation was not a likely cause of death.
b. Carcasses were severely decomposed and had been disturbed such that it was not possible to determine the cause of death.

Swallow-tailed Kite was treated for injuries received when it flew into a building in Orange County in 1991 (Collins and Gilliland 1995).

Accidental deaths of kites due to falling from their nests undoubtedly occur, but documentation is rare. Sykes (1987b) found a nestling Snail Kite that had died after falling from its nest and becoming lodged by its neck in a fork of a small tree.

One suspected instance of siblicide of the runt of the brood was reported by Meyer (1993) in his study of Swallow-tailed Kites in southern Florida, but the significance of this phenomenon is poorly understood for this species.

Nine kites were treated for trauma of unknown causes in 2 wildlife rehabilitation facilities in Florida between 1988 and 1994 (Collins and Gilliland 1995; Deem and Terrell 1996). These included 5 Swallow-tailed Kites (1 each from Brevard County in 1989, Collier County in 1992, Highlands County in 1994, Levy County in 1993, and Volusia County in 1992) and 4 Mississippi Kites (2 from Alachua County in 1994 and 1 each from Columbia County in 1993 and Dixie County in 1991).

III. Predation

Predation appears to be an important cause of mortality in Snail Kites (Chandler and Anderson 1974; Beissinger 1986; Sykes 1987b, 1987c; Snyder et al. 1989; Bennetts et al. 1988, 1998; Rodgers 1992; Bennetts and Kitchens 1997). Sykes (1987b) documented the causes of the failure of 82 Snail Kite nests in southern Florida from 1968 through 1977 (figure 13.1). Predation accounted for 36 (44%) of the nesting failures in Sykes's study and involved Everglades rat snakes, cottonmouths, and Fish Crows. Other researchers in Florida have implicated Boat-tailed Grackles, Barred Owls, raccoons, ants, and dermestid beetle larvae (figure 13.2) as nest predators (Wachenfeld 1956; Chandler and Anderson 1974; Sykes and Chandler 1974; Snyder et al. 1984, 1989) (table 13.2). In addition to Barred Owls, several other raptors have been suspected as predators, including Great Horned Owls (Sykes 1987b; Bennetts et al. 1988), Northern Harriers (Beissinger 1988), and Turkey Vultures (Bennetts et al. 1988). Rat snakes (figure 13.3) are probably the most important predator (Sykes 1987b; Bennetts et al. 1988). Toner (1984) published a fascinating photograph taken by James Kern of a rat snake attempting to ingest a Snail Kite egg. Sykes (1987c) reported that the amount of nest predation was correlated inversely with the distance of the nest from upland habitats, including man-made levees and low dikes. He felt that these upland habitats and structures allowed predators easy access to kite nests and that nest sites within

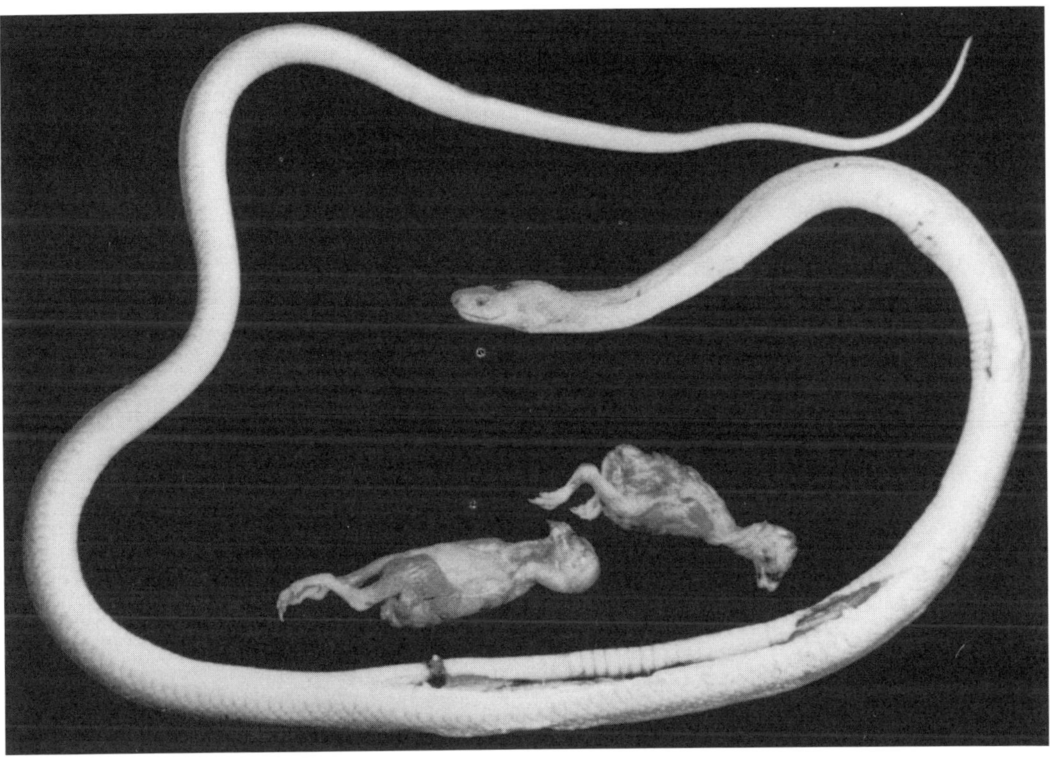

FIGURE 13.3. Photograph of a rat snake and the carcasses of 2 2–4-day-old nestlings collected at a Snail Kite nest in Conservation Area 3A (Dade County) in 1987. The snake was observed while in the process of ingesting one of the nestlings; the other nestling had already been eaten when the observers arrived at the nest site. Bennetts and Caton 1988; photo courtesy of Robert E. Bennetts.

200 meters of such upland sites acted as "reproductive sinks." Bennetts et al. (1988) studied this phenomenon further and found evidence in 1 year (1987) to support it, but not in another year (1988). After study of several hundred additional nests, Bennetts (1994) concluded that this relationship was not true. Predation appeared to be the leading cause of mortality of juvenile and adult Snail Kites in a 4-year radiotelemetry study conducted in southern Florida from 1992 to 1995 (Bennetts et al. 1998). Predation was judged to have been the cause of death of 17 of 47 (36%) juvenile and adult kites found dead (table 13.3). The predators involved were not determined, although feathers of a Great Horned Owl were found at the site of 1 carcass. A number of other mortalities were classified as unknown because of lack of information on the cause of death; some of these may have been predation cases as well.

Predators of eggs and young of Swallow-tailed Kites are probably of some significance. Raccoons, Bald Eagles, Great Horned Owls, and Red-shouldered Hawks have been suggested by Snyder (1974) and Meyer and Collopy (1990), but there is little quantitative information on this aspect (table 13.4). Only 2 of 30 Swallow-tailed Kite nesting failures and losses in 1988–89 in southern Florida were due to predation (figure 13.4).

IV. Inclement weather

Snail Kite populations are dependent on water levels in freshwater marsh habitats (Sykes

Table 13.4. Predation of Swallow-tailed Kites in Florida

Predator County or site	Year(s)	No. and type of loss	Data source
Raccoons (*Procyon lotor*)			
ENP[a]	1967–72	NG	Snyder (1974)
Bald Eagles (*Haliaeetus leucocephalus*)			
ENP[a]	1967–72	NG	Ibid.
Great Horned Owls (*Bubo virginianus*)			
ENP[a]	1967–72	NG	Ibid.
Red-shouldered Hawks (*Buteo lineatus*)			
Collier	1974	1 egg	Short (1974)
Lee	1989	1 egg[b]	Meyer & Collopy (1990), Meyer (1993)
Unknown			
Dade	1989	1 ad., 1 juv.	Ibid.
Collier	1992	1 fledgling	Ibid.
Collier	1993	2 fledgling	Ibid.

NG = not given.

a. Northwestern boundary of Everglades National Park (Collier County).

b. One additional egg in this nest was unharmed and eventually hatched and the young bird fledged.

1983b; Rodgers 1992). A number of researchers have focused on the significance of drought as a factor influencing the distribution and numbers of Snail Kites (Beissinger and Takekawa 1983; Sykes 1983b; Takekawa and Beissinger 1989; Snyder et al. 1989; Rodgers 1992). During these periodic droughts, populations of apple snails are reduced and therefore Snail Kites are put under conditions of food stress (Snyder et al. 1989). In addition, low water levels result in Snail Kites being forced to nest in nonwoody substrates at lake sites (e.g., cattails), which leads to increased mortality due to structural collapse of nests (Rodgers 1992). This can result in localized reductions in kite populations and movements of some birds to other areas in search of food and nesting habitats (Beissinger and Takekawa 1983; Sykes 1983a; Beissinger 1986; Takekawa and Beissinger 1989). This dispersal of Snail Kites is advantageous in that the chances of some catastrophic event having a serious impact on the population as a whole are reduced (Rodgers 1992).

Strong winds, rain and hail storms, and cold weather have been linked also to nesting failures of kites in Florida. Tilting and/or destruc-

tion of nests containing eggs and young during strong windstorms (with or without rain) are probably the most important of these and have been reported by numerous authors for Snail Kites (Chandler and Anderson 1974; Sykes and Chandler 1974; Toner 1984; Sykes 1987b; Rodgers 1992) and for Swallow-tailed Kites in Florida (Naggiar 1974; Snyder 1974; Meyer and Collopy 1990; Meyer 1993).

Sykes (1987b) recorded the failure of 18 Snail Kite nests due to adverse weather during his study of 1968–77 in southern Florida (figure 13.1). Of these, 10 were due to strong winds (without rain), 4 to a long period of heavy rain, 2 to low temperatures, 1 to heavy rain with hail, and 1 to heavy rain with strong winds. Pneumonia, probably as a sequel to exposure to cold weather and storms, was determined to be the cause of death of 3 5–8-week-old Snail Kite nestlings found dead in their nest at Lake Okeechobee in 1977 (Kerr 1977). A similar incident was reported by Sykes (1987b) for 2 other Snail Kite nestlings (location and date not given). Cold fronts that occur during the nesting season can cause nest desertion; entire colonies were observed to abandon their

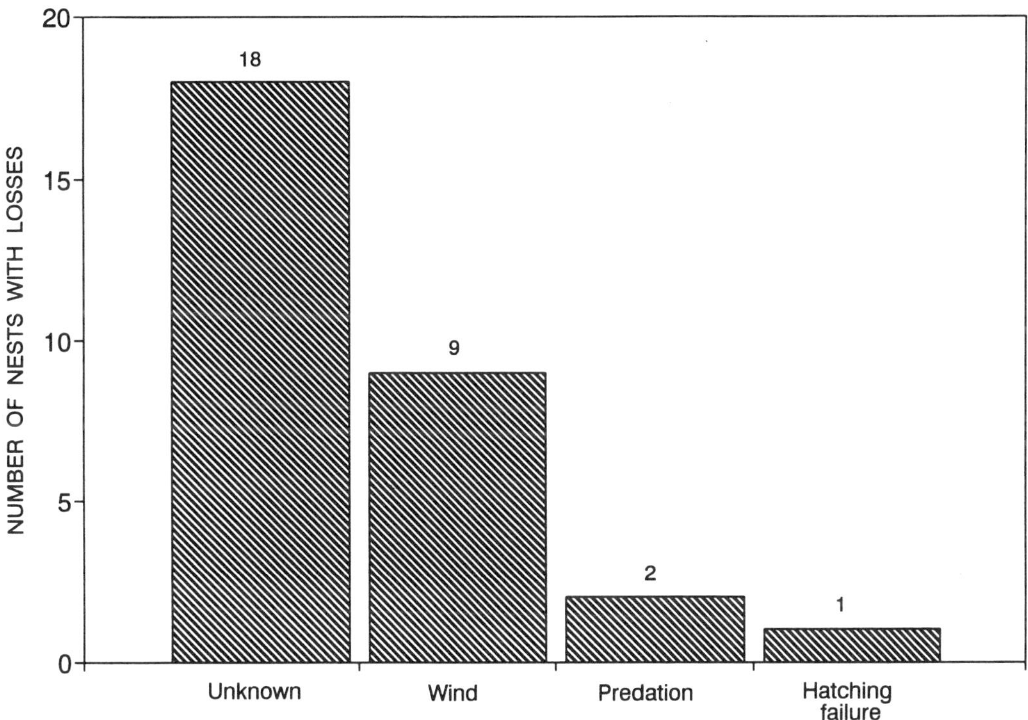

FIGURE 13.4. Bar graph illustrating the various factors responsible for losses of one or more eggs or young in 30 Swallow-tailed Kite nests in southern Florida during 1988–89. Numbers on top of the bars are the actual numbers of nests lost for each category. Data from Meyer and Collopy 1990.

nests after several days of cold weather (Beissinger 1986; Snyder et al. 1989). Apparently snail capture rates by foraging Snail Kites decrease significantly during cold weather (Cary 1985). Rodgers (1993b) reported the loss of 26 nests (20% of the total) on Lake Tohopekaliga during a 2-week period in 1991 when a series of cold fronts passed through the area. In March 1993 a major storm (called by some the "storm of the century") resulted in the loss of 2 nests on Lake Kissimmee and 37 nests on Lake Okeechobee (Rodgers 1993b). The losses at Lake Okeechobee involved 73% of the total nests under observation. However, in 1991 and 1993 Snail Kites recycled and successfully bred in both areas.

Bent (1937) claimed that hurricanes in southern Florida destroyed significant amounts of Spanish moss (*Tillandsia usneoides*), which Swallow-tailed Kites require for nesting mate-

rial, and that this caused the kites to move away from certain areas. Although others (Snyder 1974; Meyer and Collopy 1990) have found Spanish moss to be among the materials used for nest construction in Florida, the importance of Bent's observation has not been verified and its significance is not clear. Meyer and Collopy (1990) monitored 52 Swallow-tailed Kite nests during 1988–89 in southern Florida and found that 9 of 30 nests (in which there were 1 or more eggs or nestlings lost) had been damaged or destroyed by high winds (figure 13.4). In June of 1989 a dead 22–26-day-old nestling Swallow-tailed Kite was found on the ground below its nest in Big Cypress National Wildlife Refuge approximately 20 hours after a severe storm had passed through the area. The nestling was examined at necropsy and was in good nutritional state with moderate amounts of body fat, but had numerous ectoparasites (feather mites and chewing lice) and severe pneumonia (Spalding and Meyer 1989). The latter was the cause of death and was probably initiated by the stress of being blown out of the nest and the resulting exposure during the storm.

V. Emaciation

Emaciation was diagnosed in 2 of 31 juvenile Snail Kites found dead in southern Florida during a 4-year study, 1992–95, but was not seen in 16 adult birds found dead during the same time period (Bennetts et al. 1998). Their work was done during 4 consecutive nondrought years; the authors suggest that emaciation might be more common in periods of drought. The emaciated juveniles were found in marine environments, where there are no apple snails. The authors suggest that this condition might have been a result of dispersal of juveniles into areas with less predictable food supplies. We have no data on emaciation in other kites in Florida.

VI. Environmental contaminants

Low concentrations of DDE, DDD, DDT, dieldrin, and PCBs were present in various tissues of 4 of 10 Snail Kites obtained from several localities in southern Florida during 1966–1993 (table 13.5). Six of 19 Snail Kite eggs contained low concentrations of DDE, 4 contained small amounts of DDD, and 1 egg had small amounts of DDT and dieldrin; none was positive for PCBs (table 13.6). These organochlorine residues are all well below the concentrations associated with mortality in various avian species (Stickel et al. 1969, 1970, 1984; Rodgers 1997) and probably should be considered incidental background findings (Sykes 1985). There is a record of 1 die-off of Snail Kites in Surinam, South America, due to organochlorine poisoning (Vermeer et al. 1974). In fall 1971 50 kites were found dead in rice fields shortly after the fields were sprayed with sodium pentachlorophenate and endrin for control of snails (*Po-

macea glauca and *P. lineata*), the exclusive food of these kites in that area. Residues of sodium pentachlorophenate were 46, 20, and 11 ppm (wet weight) in liver, kidney, and brain samples from the dead birds, compared with 2–17 ppm in kites that survived exposure and <0.2 ppm in controls. No such incidents have been observed in Florida, but this case illustrates the susceptibility of kites to acute organochlorine contamination. Eggshell thinning has never been recognized as a problem by numerous researchers working on the reproductive biology of Snail Kites in Florida over the years; broken shells have not been observed in nests (Beissinger 1988; Sykes et al. 1995). Low concentrations of organochlorines (mean values of 0.06 ppm DDE, 0.03 ppm DDD, and 0.10 ppm DDT) were found in a sample of 30 apple snails collected from a canal in Conservation Area 2A near Loxahatchee National Wildlife Refuge in 1965 (Lamont and Reichel 1970).

The uropygial gland of 1 Swallow-tailed Kite from Marion County in 1975 had a low concentration of DDE (Johnston 1978) (table 13.5). Eggs of Swallow-tailed Kites in Florida have not been examined for environmental contaminants. We have no information on contaminants in White-tailed Kites or Mississippi Kites in Florida. However, Franson (1994) reported 2 sick and 14 dead Mississippi Kites, attributed to parathion poisoning at the edge of a golf course near a cotton field in Oklahoma. This was probably a case of secondary poisoning that occurred when the kites fed on contaminated insects.

Very small concentrations of mercury were detected in 9 of 10 Snail Kite eggs from Glades and Osceola counties, and mercury was present in kidney and liver samples from 1 Swallow-tailed Kite from Monroe County (table 13.7). Apple snails, the primary food source for Snail Kites, are probably the source of mercury contamination for Snail Kites. Sixty-two apple snails were tested for mercury in August 1992: 41 from the Everglades region, 10 from the Florida Panther National Wildlife Refuge, and 11 from areas north of Lake Okeechobee (Eisemann et al. 1997). Of those examined, low

Table 13.5. Organochlorine residues reported from tissues of kites in Florida

Site (county)	Year	Age	Sample	Residues (ppm wet weight)[a,b]					Data source
				DDE	DDD	DDT	Dieldrin	PCB	
Snail Kites									
Conservation Area 2A (Broward)									
	1967	Nestling	Carcass	0.20	0.05	T	T	NT	Lamont & Reichel (1970)
			Muscle	0.20	0.05	T	T	NT	Sykes (1985)
	1966	Adult	Carcass	0.28	0.06	0.09	T	NT	Lamont & Reichel (1970)
			Liver	0.43	0.12	0.26	0.09	NT	Ibid.
Lake Okeechobee (Glades)									
	1977	Nestling A[f]	Muscle	ND	ND	ND	ND	0.11	Sykes (1985)
		Nestling B[f]	Brain	ND	ND	ND	ND	ND	Reichel (1977)
			Carcass	ND	ND	ND	ND	0.11	Ibid.
		Nestling C	Brain	ND	ND	ND	ND	ND	Ibid.
			Carcass	ND	ND	ND	ND	ND	Ibid.
		Nestling D	Brain	ND	ND	ND	ND	ND	Ibid.
			Carcass	ND	ND	ND	ND	ND	Ibid.
Santa Fe River (Levy)									
	1981	Juvenile	Brain[c,d]	ND	ND	ND	ND	ND	Reichel (1981)
			Carcass[c,d]	0.33	ND	ND	ND	ND	Ibid.
Lake Tohopekaliga (Osceola)									
	1993	Nestling	Liver[e]	ND	ND	ND	ND	ND	Rodgers (1993a)
Loxahatchee NWR (Palm Beach)									
	1970	Nestling	Muscle	ND	ND	ND	ND	ND	Sykes (1985)
"South Florida"									
	1980	Unknown	Brain[c]	ND	ND	ND	ND	ND	Reichel (1982)
Swallow-tailed Kites									
Marion									
	1975	Unknown	Uropygial gland	0.48	ND	ND	ND	NT	Johnston (1978)

ND = not detected, NT = not tested
a. Lower limits for detection: Johnston (1978) = 0.01 ppm, all others = 0.05 ppm.
b. T = trace (<0.05 ppm).
c. Also tested for heptachlor epoxide, oxychlordane, cis-chlordane, trans-chlordane, cis-nonachlor, endrin, and toxaphene. None was detected.
d. Also tested for mirex and hexachlorobenzene. None was detected.
e. Tested for endrin, heptachlor epoxide, and toxaphene. None was detected.
f. Nestlings A and B might be the same bird. The records are not clear.

Table 13.6. Organochlorine residues reported from eggs of Snail Kites in Florida

Site (county) Egg no.[a]	Year	Residues (ppm wet weight)[b]					Data source
		DDE	DDD	DDT	Dieldrin	PCB	
Conservation Area 2A (Broward)							
1	1966	0.33	0.14	0.06	0.05	ND	Lamont & Reichel (1970); Sykes (1985)
Conservation Area 2B (Broward)							
2	1970	0.05	0.20	ND	ND	ND	Sykes (1985)
3	1970	0.34	ND	ND	ND	ND	Ibid.
Loxahatchee NWR (Palm Beach)							
4	1970	0.17	0.08	ND	ND	ND	Ibid.
5	1970	0.22	0.10	ND	ND	ND	Ibid.
6	1974	ND	ND	ND	ND	ND	Ibid.
7	1974	ND	ND	ND	ND	ND	Ibid.
8	1974	ND	ND	ND	ND	ND	Ibid.
9	1974	ND	ND	ND	ND	ND	Ibid.
Lake Okeechobee (Glades)							
10	1990	ND	ND	ND	ND	ND	Rodgers (1993a)
11	1993	ND	ND	ND	ND	ND	Ibid.
12	1993	ND	ND	ND	ND	ND	Ibid.
Lake Kissimmee (Osceola)							
13	1988	ND	ND	ND	ND	ND	Ibid.
14	1989	0.05	ND	ND	ND	ND	Ibid.
15	1990	ND	ND	ND	ND	ND	Ibid.
16	1993	ND	ND	ND	ND	ND	Ibid.
East Lake Tohopekaliga (Osceola)							
17	1989	ND	ND	ND	ND	ND	Ibid.
Lake Tohopekaliga (Osceola)							
18	1993	ND	ND	ND	ND	ND	Ibid.
19	1993	ND	ND	ND	ND	ND	Ibid.

ND = none detected.

a. Each sample consists of one egg. Eleven different clutches are represented: No. 1 is from one clutch, Nos. 2 and 3 from a second, Nos. 4 and 5 from a third, Nos. 6, 7, and 8 from a fourth, No. 9 from a fifth, and Nos. 10 through 19 are each from separate clutches.

b. Lowest limits of detection: Lamont & Reichel (1970) and Sykes (1985) = 0.05 ppm, Rodgers (1993a) = 0.5 ppm for PCBs and 0.05 ppm for DDE, DDD, DDT, and dieldrin. Samples number 10, 11, 13, 15, and 17 from Glades and Osceola counties were tested also for chlordane, endrin, heptachlor epoxide, toxaphene, and *trans*-nonachlor; samples number 12, 16, 18, and 19 were tested also for endrin, heptachlor epoxide, and toxaphene; and the samples from Broward and Palm Beach counties were tested also for heptachlorepoxide, oxychlordane, *cis*-chlordane, *trans*-nonachlor, *cis*-nonachlor, endrin, toxaphene, mirex and hexachlorobenzene. None was detected.

concentrations of mercury (mean = 0.063 ppm, wet weight, lowest limit of detection = 0.025 ppm) were detected in 53 snails. Ten snails had concentrations of 0.1 ppm or greater, with the highest concentration (0.14 ppm) being found in a snail from the Florida Panther National Wildlife Refuge. The highest concentrations were in snails from watersheds south of Lake Okeechobee. The authors concluded that apple snails could serve as "indicators of bioavailable mercury" (Eisemann et al. 1997). Two nestling Swallow-tailed Kites (1 from Collier County and 1 from Monroe County) were tested for mercury in 1989 (table 13.7). No mercury was detected in the kite from Collier County, but 0.06 ppm (wet weight) of mercury was meas-

Table 13.7. Concentrations of mercury and lead in eggs and tissues of kites from Florida

| County | Specific locality | Year | Organ/tissue | Concentrations (ppm wet weight)[a] | | Data source |
				Mercury	Lead	
Snail Kites[b]						
Glades	Lake Okeechobee	1990	Egg	0.12	0.17	Rodgers (1993a)
		1993	Egg	ND	ND	Ibid.
		1993	Egg	0.04	ND	Ibid.
Osceola	Lake Kissimmee	1988	Egg	0.35	0.06	Ibid.
		1989	Egg	0.08	ND	Ibid.
		1990	Egg	0.09	0.06	Ibid.
		1993	Egg	0.03	ND	Ibid.
Osceola	E. Lake Tohopekaliga	1989	Egg	0.11	0.07	Ibid.
		1993	Kidney	NT	ND	Thomas (1993)
Osceola	Lake Tohopekaliga	1993[c]	Liver	ND	ND	Rodgers (1993a)
		1993	Egg	0.18	ND	Ibid.
		1993	Egg	0.04	ND	Ibid.
Swallow-tailed Kites						
Collier	Big Cypress Nat. Preserve	1989[d]	Muscle	ND	NT	Spalding et al. (1993a)
Monroe	Long Pine Key	1988[e]	Kidney	0.06	NT	Ibid.
			Liver	0.06	NT	Ibid.

ND = none detected, NT = not tested.
a. Lowest limits of detection for egg data = 0.02 ppm and for other tissues = 0.05 ppm.
b. Each sample consists of 1 egg or liver sample each from a separate clutch or bird.
c. 1-week-old nestling.
d. Nestling (exact age unknown).
e. 27-day-old nestling.

ured in both kidney and liver tissues of the Monroe County bird.

Low concentrations of lead were detected in 4 of 10 Snail Kite eggs obtained from Glades and Osceola counties between 1988 and 1993 (table 13.7). No lead was detected in tissues from 2 other Snail Kites in Osceola County. Contrary to what was stated in Meyer (1995), tissues of Swallow-tailed Kites have not been tested for lead.

Two nestling Snail Kites from Lake Tohopekaliga were tested for arsenic in 1991; both contained very low concentrations in kidney samples: 51.0 and 73.9 ppb (Thomas 1991). Arsenic has been found in apple snails in other areas in Florida (Solid Waste Authority of Palm Beach County 1990) and most likely was the source of this contaminant in Snail Kite tissues.

Based on information obtained from other raptors, the heavy metals discussed above are probably of little or no significance (Wiemeyer et al. 1980; Pattee et al. 1981; Eisler 1987; Rodgers 1997), although the long-term effects of small amounts of metals on the health and reproduction of kites have not been studied.

VII. Neoplasia

No information is available on this topic.

VIII. Biotoxins

No information is available on this topic.

IX. Viruses

There is little information about viral diseases in kites in Florida. West Nile virus was identified by PCR and virus isolation from a Swal-

Table 13.8. Bacteria identified in young Snail Kites from Florida, 1991–93

Species of bacteria	County	No. of kites Examined	No. of kites Positive	Organ/tissue
Acinetobacter calcoaceticus	Osceola	1	1[a]	Lungs
Aeromonas hydrophilla	Osceola	2	1	Lungs
	Osceola	1	1	Liver & pericardial sac
	Broward	1	1	Oral mucosa
Alcaligenes sp.	Broward	1	1	Oral mucosa
Enterococcus faecium	Manatee	1	1	Liver & lungs
Enterococcus sp.	Osceola	1	1	Liver
	Osceola	2	2	Intestines
	Osceola	1	1	Abdominal wall
	Broward	1	1	Oral mucosa
Escherichia coli	Osceola	2	2	Intestines
	Osceola	1	1	Liver
	Osceola	1	1	Intestines
	Manatee	1	1	Intestines
Klebsiella pneumoniae	Manatee	1	1	Liver & lungs
Plesiomonas shigelloides	Osceola	2	1	Lungs
	Osceola	1	1	Abdominal wall
	Broward	1	1	Liver
Pseudomonas sp.	Broward	1	1	Oral mucosa
Staphylococcus sp.	Osceola	2	2	Lungs
	Osceola	1	1	Lungs & liver
	Broward	2	2	Oral mucosa
Vibrio cholerae Non-01	Broward	2	1	Oral mucosa

Source: Duncan (1991–93).
a. Also reported by Cole et al. (1995); presumably the same bird.

low-tailed Kite found dead in Alachua County in 2001 (Conti et al. 2002). It was assumed that the kite died because of the West Nile virus infection, but a necropsy was not performed to prove this. Three 5–8-week-old nestling Snail Kites were found dead in their nests on Lake Okeechobee in 1977 and were determined to have died of acute pneumonia of unknown etiology (Kerr 1977). A virus was suspected, but this was not confirmed by laboratory tests.

X. Bacteria

Ten species of bacteria have been isolated from young Snail Kites in Florida (table 13.8). These bacterial infections were not judged to have been the cause of death, and may have been secondary to other factors. The significance to kite populations is not known. There is no information on bacterial infections in Swallow-tailed Kites.

XI. Fungi

An unidentified species of *Aspergillus* was cultured from the lungs of an emaciated Snail Kite that died of parasitic bronchitis and airsacculitis (Cole et al. 1995). The kite was a fledgling found at East Lake Tohopekaliga (Osceola County) in 1993. Even though fungal infections are rare in free-ranging raptors (Keymer 1972), there is some evidence that stress might increase the susceptibility of these birds to as-

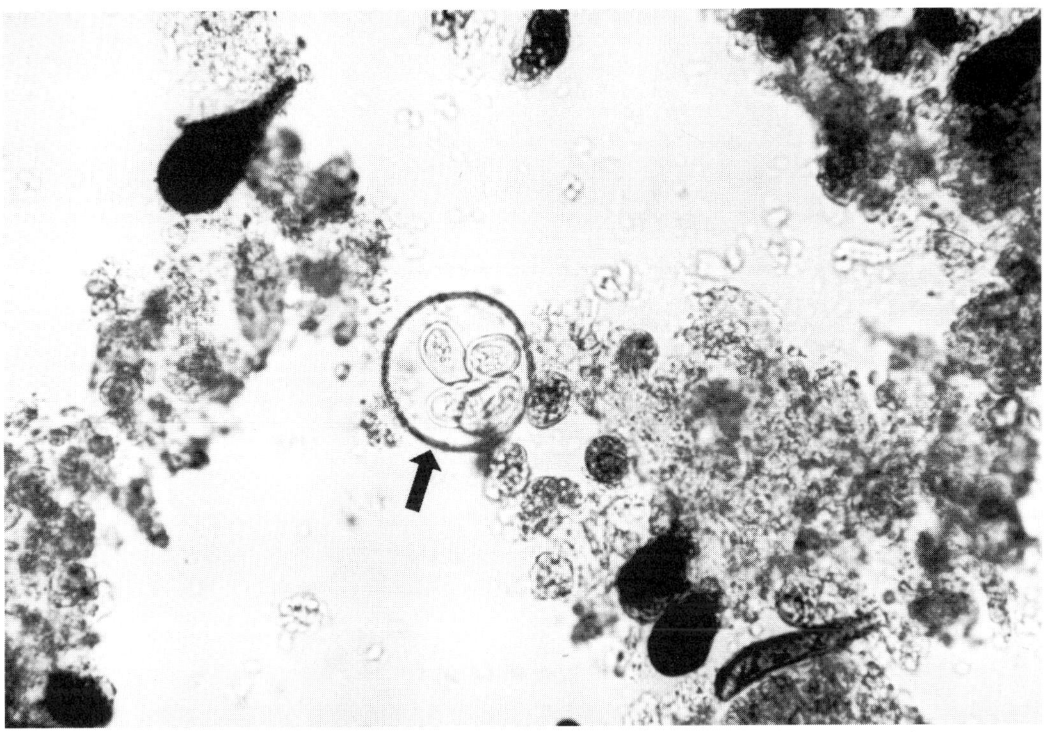

FIGURE 13.5. A partially sporulated oocyst *(arrow)* of an intestinal coccidian (probably of the genus *Eimeria*) from the feces of a nestling Snail Kite from Lake Okeechobee in 1974. Courtesy of Pamela P. Humphrey.

pergillosis (Redig et al. 1980). An illustration of this possibility was the case of a Swallow-tailed Kite found in Collier County in 1992, a trauma victim submitted to the Center for Birds of Prey in Maitland. It was treated and kept in captivity for almost 2 years before it developed aspergillosis and died (Collins and Gilliland 1995).

XII. Protozoans

Nothing is known about the parasitic protozoans infecting Swallow-tailed Kites and only 1 protozoan is known from Snail Kites. From 1973 to 1993 blood smears from 1 fledgling and 40 nestling Snail Kites were examined for parasitic protozoans (table 13.9). All were negative. Blood films from 2 Mississippi Kites were negative for blood protozoans (Telford et al.

1996). Both birds were adults and were examined in 1995; 1 was from Alachua County, but the locality of the other kite was not determined.

Two of 20 fecal samples from nestling Snail Kites were positive for oocysts of intestinal coccidia (table 13.9). Only a few oocysts were seen. These only partially sporulated and therefore could not be specifically identified (figure 13.5), but probably represent a species of *Eimeria* (Sykes and Forrester 1983). The significance of this coccidian to the health of Snail Kites is not known.

XIII. Helminths

Three species of helminths (2 trematodes and 1 nematode) are known from Snail Kites in Florida (table 13.10). The most important of

Table 13.9. Characteristics of protozoan infections in young Snail Kites from Florida

County Specific locality	Year	No. of kites[a]		Data source
		No. Examined	Positive	
Blood protozoans				
Broward	1994	9	0	Foster & Bennetts (1994)
Dade	1994	9	0	Ibid.
Glades				
Lake Okeechobee	1973	4	0	Sykes & Forrester (1983), Forrester (1993)
"	1974	7	0	Ibid.
"	1975	2	0	Ibid.
"	1991	6	0	Foster & Rodgers (1991)
"	1994	4	0	Foster & Bennetts (1994)
Indian River & St. Lucie				
St. Johns River headwaters	1973	6	0	Sykes & Forrester (1983), Forrester (1993)
"	1991	1	0	Foster & Rodgers (1991)
"	1994	2	0	Foster & Bennetts (1994)
Monroe				
Florida Bay	1993	1	0	Foster & Spalding (1993)
Osceola				
West Lake Tohopekaliga	1991	8	0	Foster & Rodgers (1991)
"	1991	1	0	Thomas (1991)
East Lake Tohopekaliga	1991	2	0	Foster & Rodgers (1991)
"	1993	1	0	Cole et al. (1995)
Lake Kissimmee	1991	2	0	Foster & Rodgers (1991)
"	1994	6	0	Foster & Bennetts (1994)
Intestinal coccidia				
Glades				
Lake Okeechobee	1973	4	0	Sykes & Forrester (1983), Forrester (1993)
"	1974	1	1[b]	Ibid.
"	1975	1	1[b]	Ibid.
"	1991	3	0	Foster & Rodgers (1991)
Indian River & St. Lucie				
St. Johns River headwaters	1974	1	0	Sykes & Forrester (1983), Forrester (1993)
"	1991	1	0	Foster & Rodgers (1991)
Osceola				
West Lake Tohopekaliga	1991	6	0	Ibid.
East Lake Tohopekaliga	1991	3	0	Ibid.

a. All were nestlings except 1 fledgling from East Lake Tohopekaliga in 1993.
b. Probably a species of the genus *Eimeria*. Complete identification was not possible because the oocysts only partially sporulated (see Figure 13.5).

these is probably *Bothrigaster variolaris*, observed in the bronchi and air sacs of 2 juvenile Snail Kites found dead in Osceola and Glades counties in 1993. In the kite from Osceola County the trematodes were probably a major contributing cause of debilitation and death (Cole et al. 1995). This parasite has been found several other times in Snail Kites from Cuba and Brazil (Travassos 1923; Dubois 1959).

Table 13.10. Helminth infections in kites from Florida

Host Species of helminth (Site)[a]	County site	Year	Number of helminths	Data source
Snail Kites				
Trematoda				
Bothrigaster variolaris (AS,LU)	Glades	1993	40	Cole et al. (1995)
	Osceola	1993	270[b]	Ibid.
Echinostoma sp. (immature) (ES,LI,CL)	Palm Beach	1993	26	Kinsella et al. (1994a)
Cestoda				
Unidentifiable fragments (PR)	Dade	1997	UN	Kinsella et al. (1999)
Nematoda				
Larval spirurids (PR)	Palm Beach	1993	3	Kinsella et al. (1994a)
Swallow-tailed Kites				
Trematoda				
Neodiplostomum attenuatum (SI)	Collier[c]	1989	13	Kinsella et al. (1994b)
	Collier[c]	1992	47	Ibid.
	Monroe[d]	1988	0	Ibid.
Stomylotrema vicarium (CL)	Collier[c]	1992	2	Ibid.
Cestoda				
Anomotaenia sp.[f] (SI)	"Florida"	NG	5	Stiles & Hassall (1894)
Nematoda				
Dispharynx nasuta (PR)	Collier[c]	1989	8	Kinsella et al. (1994b)
	Collier[c]	1992	5	Ibid.
	Monroe[d]	1988	31	Ibid.
Tetrameres robusta[g] (PR)	"Florida"	NG	1	Kinsella (1998)
Larval spirurids (PR)	Collier[c]	1992	17	Kinsella et al. (1994b)
Acanthocephala				
Centrorhynchus spinosus (SI)	Collier[c]	1989	11	Ibid.
	Collier[c]	1992	7	Ibid.
	Monroe[d]	1988	0	Ibid.
	"Florida"	NG	NG	USNPC[e]

NG = not given by author, UN = unknown.

a. Infection sites: AS = air sacs, LU = lungs, ES = esophagus, SI = small intestine, LI = large intestine, CL = cloaca, PR = proventriculus.

b. From the lungs and air sacs of an emaciated fledgling Snail Kite.

c. Nestling from Big Cypress National Preserve.

d. Nestling from Long Pine Key.

e. Specimen #82449 in the U.S. National Parasite Collection, Beltsville, Md.

f. Stiles and Hassall (1894) called this *Taenia vexata* and this record was repeated in Meyer (1995). The specimen had been deposited in the National Parasite Collection in Beltsville; upon examination it was identified as *Anomotaenia* sp. and may be an undescribed species (Kinsella 1998).

g. Found in the vial along with a tapeworm identified as *Taenia vexata* by Stiles and Hassall (1894), by J.M. Kinsella (1998) when he was studying the tapeworm. There is no mention of a nematode from the Swallow-tailed Kite in the reference by Stiles and Hassall (1894).

Six species of helminths are known from Swallow-tailed Kites in Florida (table 13.10). These include 2 trematodes, 1 cestode, 2 nematodes, and 1 acanthocephalan. The pathological significance of these parasites is unknown.

XIV. Arthropods

Several species of parasitic arthropods are known to infest kites in Florida and include mites, chewing lice, and mosquitoes (table 13.11). The lice are probably of little pathologic

Table 13.11. Parasitic arthropod infestations in kites from Florida

Host Species of arthropod	County or site	Year(s)	Data source
Snail Kites			
Feather mites			
Aetacarus sp.	Broward	1993	Spalding et al. (1994)
Dermonoton sp.	Broward	1993	Ibid.
Skin mites			
Ornithonyssus bursa	Palm Beach	1993	Ibid.
	St. Lucie	1974	Phillis et al. (1976),
			Sykes & Forrester (1983)
Mites (unidentified)	Broward	1969	Ibid.
	Broward	1986–87	Bennetts et al. (1988)
Chewing lice			
Colpocephalum turbinatum	Broward	1993	Forrester et al. (1995)
	Indian River	1973	Sykes & Forrester (1983)
	Osceola	1995	Forrester et al. (1995)
Craspedorrhynchus obscurus	Broward	1993	Ibid.
	Glades	1993	Cole et al. (1995)
	Osceola	1995	Forrester et al. (1995)
	Palm Beach	1993	Ibid.
	"No. America"[a]	NG	Malcomson (1960)
Falcolipeurus quadriguttatus	Glades	1993	Cole et al. (1995)
	Indian River	1973	Sykes & Forrester (1983)
	Indian River	1995	Forrester et al. (1995)
Diptera			
Anopheles crucians	Broward	1969	Sykes (1983c)
Anopheles walkeri	Broward	1969	Ibid.
Swallow-tailed Kites			
Feather mites			
Aetacarus sp.	Collier	1989	Spalding et al. (1993b)
Phoretic mites			
Macrocheles sp.	Collier	1989	Ibid.
Skin mites			
Ornithonyssus bursa	Collier	1989	Ibid.
	Collier	1992	Ibid.
Chewing lice			
Colpocephalum osborni	Collier	1989, 1992	Forrester et al. (1995)
	Monroe	1988	Ibid.
Cuculiphilus decoratus	Collier	1989	Ibid.
	Dade	1988	Ibid.
Degeeriella guimarcesi	Collier	1992	Ibid.
	"Florida"	NG	Clay (1958)
Diptera			
Mosquitoes (unidentified)	Monroe	1969	Snyder (1974)
Mississippi Kites			
Chewing lice			
Laemobothrion maximum	Leon	1926, 1960	Forrester et al. (1995)

NG = not given.

a. Exact locality not given by Malcomson (1960); since the Snail Kite occurs only in Florida in North America, it is assumed that this is a Florida record.

significance, but the mites and mosquitoes are known to harass and even kill nestling kites.

Sykes and Forrester (1983) related 2 incidents in which Snail Kite nestling mortality was related to severe mite infestation. The first was in Water Conservation Area 2A (Broward County) in 1969. The mites involved were not identified at that time, but were linked with the deaths of 3 nestlings. The second case was a nest in the northern part of Cloud Lake (St. Lucie County) "swarming with thousands of mites" and containing a weak 19–21-day-old nestling. One week later the nestling was gone and presumed dead; the nest was still infested with large numbers of mites. The nestling and a younger sibling were probably both killed by the infestation of mites, identified as tropical fowl mites (*Ornithonyssus bursa*), known to be capable of killing newly hatched domestic chicks (Baker et al. 1956).

Snyder (1974) reported that mosquitoes were problems for nesting Swallow-tailed Kites in the Everglades. He observed that "when the wind was low in early mornings, hordes of mosquitoes gathered around the treetop nests and caused the kites almost constant irritation." These mosquitoes were dispersed quickly when even a small wind came up. Snyder (1974) postulated that the placement of nests on the tops of tall mangroves might be an adaptation to decrease their exposure to mosquitoes and perhaps to other insect pests.

XV. Summary and conclusions

Predation, structural collapse of nests, and inclement weather (e.g., drought, storms, and cold weather) appear to be the most important mortality factors for Snail Kites in Florida. In addition 31 different contaminants, infectious disease agents, and parasites have been identified, although information on them is limited. These include 5 organochlorines, 3 heavy metals, 10 bacteria, 1 fungus, 1 protozoan, 2 trematodes, 1 nematode, and 8 arthropods. The pathologic significance of most of these is unknown. The respiratory tract trematode *Bothrigaster vario-*

laris and the tropical fowl mite *Ornithonyssus bursa* may be important pathogens when they occur in high numbers.

Information on morbidity and mortality factors affecting Swallow-tailed Kites is limited. Damage to nests during storms seems to be important; predation is probably of lesser significance. Fourteen contaminants and parasites have been identified, including 1 organochlorine, 1 metal, 1 virus, 2 trematodes, 1 cestode, 1 nematode, 1 acanthocephalan, and 6 arthropods. The importance of these is not known.

Information on Mississippi Kites consists of 5 records of trauma and 2 records of the occurrence of a species of chewing louse. There is no information on White-tailed Kites in Florida.

XVI. Literature cited

American Ornithologists' Union. 1998. *Check-list of North American birds*. 7th ed. American Ornithologists' Union, Washington, D.C. 829 pp.

Bailey, H.B. 1884. Breeding habits of the Everglade kite. *Auk* 1:95.

Baker, E.W., T.M. Evans, D.J. Gould, W.B. Hull, and H.L. Keegan. 1956. *A manual of parasitic mites of medical or economic importance*. Natl. Pest Control Assoc., New York. 170 pp.

Beissinger, S.R. 1986. Demography, environmental uncertainty, and the evolution of mate desertion in the snail kite. *Ecology* 67:1445–1459.

———. 1988. Snail kite. In: *Handbook of North American birds*. Vol. 4. R.S. Palmer (ed.). Yale University Press, New Haven, Conn. pp. 148–165.

Beissinger, S.R., and J.E. Takekawa. 1983. Habitat use by and dispersal of snail kites in Florida during drought conditions. *Fla. Field Nat.* 11:89–106.

Bennetts, R.E. 1994–95. Unpublished data. Coop. Fish Wildl. Res. Unit, University of Florida, Gainesville.

Bennetts, R.E., and E.L. Caton. 1988. An observed incident of rat snake predation of snail kite (*Rostrhamus sociabilis*) chicks in Florida. *Fla. Field. Nat.* 16:14–16.

Bennetts, R.E., M.W. Collopy, and S.R. Beissinger. 1988. *Nesting ecology of snail kites in Water Conservation Area 3A.* Department of Wildlife and Range Science, University of Florida, Gainesville. Florida Cooperative Fish and Wildlife Research Unit, Tech. Rep. 31. 174 pp.

Bennetts, R.E., M.W. Collopy, and J.A. Rodgers, Jr. 1994. The snail kite in the Florida Everglades: a food specialist in a changing environment. In: *Everglades: the ecosystem and its restoration.* S.M. Davis and J.C. Ogden (eds.). St. Lucie Press, Delray Beach, Fla. pp. 507–532.

Bennetts, R.E., and W.M. Kitchens. 1997. *The demography and movements of snail kites in Florida.* U.S. Geological Survey, Biological Research Division, Florida Cooperative Fish and Wilife Research Unit, Tech. Rep. 56. 169 pp.

Bennetts, R.E., M.R. Shannon, and W.M. Kitchens. 1998. Causes of mortality of post-fledging juvenile and adult snail kites in Florida. *Fla. Field Nat.* 26:84–87.

Bent, A.C. 1937. Life histories of North American birds of prey. Part 1. *U.S. Natl. Mus. Bull.* 167:1–409.

Cary, D.M. 1985. Climatological and environmental factors affecting the foraging behavior and ecology of Everglades kites. M.S. thesis, University of Miami, Coral Gables.

Chandler, R., and J.W. Anderson. 1974. Notes on Everglade kite reproduction. *Am. Birds* 28:856, 858.

Clay, T. 1958. Three new species of *Degeeriella* Neumann (Mallophaga) from the Falconiformes (Aves). *Proc. R. Entomol. Soc. Lond. Ser. B Taxon.* 27:1–7.

Cole, R.A., N.J. Thomas, and C.L. Roderick. 1995. *Bothrigaster variolaris* (Trematoda: Cyclocoelidae) infection in two Florida snail kites (*Rostrhamus sociabilis plumbeus*). *J. Wildl. Dis.* 31:576–578.

Collins, R., and C. Gilliland. 1995. Unpublished data. Florida Audubon Society, Center for Birds of Prey, Maitland.

Conti, L., R. Oliveri, and C. Blackmore. 2002. Unpublished data. Florida Department of Health, Tallahassee.

Deem, S.L., and S.P. Terrell. 1996. Unpublished data. University of Florida, Gainesville.

Deem, S.L., S.P. Terrell, and D.J. Forrester. 1998. A retrospective study of morbidity and mortality of raptors in Florida: 1988–1994. *J. Zoo Wildl. Med.* 29:160–164.

Dubois, G. 1959. Revision des Cyclocoelidae Kossack 1911 (Trematoda). *Rev. Suisse Zool.* 66:67–147.

Duncan, R.M. 1991–93. Unpublished data. National Wildlife Health Center, Madison, Wis.

Eisemann, J.D., W.N. Beyer, R.E. Bennetts, and A. Morton. 1997. Mercury residues in south Florida apple snails (*Pomacea paludosa*). *Bull. Environ. Contam. Toxicol.* 58:739–743.

Eisler, R. 1987. *Mercury hazards to fish, wildlife, and invertebrates: a synoptic review.* U.S. Fish and Wildlife Service, Biological Report 85(1.10). 90 pp.

Forrester, D.J. 1993. Unpublished data. University of Florida, Gainesville.

Forrester, D.J., H.W. Kale II, R.D. Price, K.C. Emerson, and G.W. Foster. 1995. Chewing lice (Mallophaga) from birds in Florida: a listing by host. *Bull. Fla. Mus. Nat. Hist.* 39:1–44.

Foster, G.W., and R.E. Bennetts. 1994. Unpublished data. University of Florida, Gainesville.

Foster, G.W., and J.A. Rodgers, Jr. 1991. Unpublished data. University of Florida, Gainesville.

Foster, G.W., and M.G. Spalding. 1993. Unpublished data. University of Florida, Gainesville.

Franson, J.C. 1994. Parathion poisoning of Mississippi kites in Oklahoma. *J. Raptor Res.* 28:108–109.

Johnston, D.W. 1978. Organochlorine pesticide residues in Florida birds of prey, 1969–76. *Pestic. Monit. J.* 12:8–15.

Kerr, S.M. 1977. Unpublished data. National Wildlife Health Center, Madison, Wis.

Keymer, I.F. 1972. Diseases of birds of prey. *Vet. Rec.* 90:579–594.

Kinsella, J.M. 1998. Unpublished data. University of Florida, Gainesville.

Kinsella, J.M., M.G. Spalding, and R.E. Bennetts. 1994a. Unpublished data. University of Florida, Gainesville.

Kinsella, J.M., M.G. Spalding, and V. Dreitz. 1999. Unpublished data. University of Florida, Gainesville.

Kinsella, J.M., M.G. Spalding, and K.D. Meyer.

1994b. Unpublished data. University of Florida, Gainesville.

Lamont, T., and W. Reichel. 1970. Organochlorine pesticide residues in whooping cranes and Everglade kites. *Auk* 87:158–159.

Logan, T.H. 1997. Florida's endangered species, threatened species and species of special concern. Official lists. Florida Game and Fresh Water Fish Commission, Tallahassee. 14 pp.

Malcomson, R.O. 1960. Mallophaga from birds of North America. *Wilson Bull.* 72:182–197.

Meyer, K.D. 1988. Kites. In: *Proc. Southeast Raptor Manag. Symp. Workshop.* National Wildlife Federation, Sci. Tech. Ser. 14:38–49.

———. 1993. Unpublished data. National Park Service, Big Cypress National Preserve, Ochopee, Fla.

———. 1995. Swallow-tailed kite (*Elanoides forficatus*). *Birds N. Am.* 138:1–23.

———. 1996. White-tailed kite. In: *Rare and endangered biota of Florida.* Vol. 5, *Birds.* J.A. Rodgers, Jr., H.W. Kale II, and H.T. Smith (eds.). University Press of Florida, Gainesville. pp. 309–314.

Meyer, K.D., and M.W. Collopy. 1990. *Status, distribution, and habitat requirements of the American swallow-tailed kite* (Elanoides forficatus) *in Florida.* Final Report. Florida Game and Fresh Water Fish Commission, Nongame Wildlife Section, Tallahassee. 146 pp.

———. 1996. American swallow-tailed kite. In: *Rare and endangered biota of Florida.* Vol. 5, *Birds.* J.A. Rodgers, Jr., H.W. Kale II, and H.T. Smith (eds.). University Press of Florida, Gainesville. pp. 188–196.

Millsap, B.A. 1987. Summer concentrations of American swallow-tailed kites at Lake Okeechobee, Florida, with comments on post-breeding movements. *Fla. Field Nat.* 15:85–92.

Naggiar, M. 1974. The swallow-tailed kite. *Fla. Wildl.* 28:19.

Pattee, O.H., S.N. Wiemeyer, B.M. Mulhern, L. Sileo, and J.W. Carpenter. 1981. Experimental lead-shot poisoning in bald eagles. *J. Wildl. Manag.* 45:806–810.

Phillis, W.A., H.L. Cromroy, and H.A. Denmark. 1976. New host and distribution records for the mite genera *Dermanyssus, Ornithonyssus,* and *Pellonyssus* (Acari: Mesostigmata: Laelapoidea) in Florida. *Fla. Entomol.* 59:89–92.

Redig, R.T., M.R. Fuller, and D.L. Evans. 1980. Prevalence of *Aspergillus fumigatus* in free-living goshawks (*Accipiter gentilis atricapillus*). *J. Wildl. Dis.* 16:169–174.

Reichel, W.L. 1977–82. Unpublished data. Patuxent Wildlife Research Center, Laurel, Md.

Robertson, W.B., Jr. 1988. American swallow-tailed kite. In: *Handbook of North American birds.* Vol. 4. R.S. Palmer (ed.). Yale University Press, New Haven, Conn. pp. 109–131.

Robertson, W.B., Jr., and G.E. Woolfenden. 1992. *Florida bird species. An annotated list.* Spec. Publ. 6, Florida Ornithological Society, Gainesville. 260 pp.

Rodgers, J.A., Jr. 1992. Annual snail kite survey and habitat assessment. Final Performance Report, Study no. 7520, Federal no. E-1-II-H-1. Florida Game and Fresh Water Fish Commission, Tallahassee. 46 pp.

———. 1993a. Unpublished data. Florida Game and Fresh Water Fish Commission, Gainesville.

———. 1993b. Effects of water fluctuations on snail kites nesting in the Kissimmee River basin. Annual Performance Report, Study no. 7524, Federal no. E-1-II-H-2. Florida Game and Fresh Water Fish Commission, Tallahassee. 11 pp.

———. 1996. Florida snail kite. In: *Rare and endangered biota of Florida.* Vol. 5, *Birds.* J.A. Rodgers, Jr., H.W. Kale II, and H.T. Smith (eds.). University Press of Florida, Gainesville. pp. 42–51.

———. 1997. Pesticide and heavy metal levels of waterbirds in the Everglades Agricultural Area of south Florida. *Fla. Field Nat.* 25:33–41.

Rodgers, J.A., Jr., S.T. Schwikert, and A.S. Wenner. 1988. The status of the snail kite in central and south Florida: 1981–1985. *Am. Birds* 42:30–35.

Rodgers, J.A., Jr., and F.W. Stangel. 1996. Genetic variation and population structure of the endangered snail kite in South Florida. *J. Raptor Res.* 30:111–117.

Schmeling, S.K. 1981. Unpublished data. National Wildlife Health Center, Madison, Wis.

Short, L.L. 1974. Egg-eating by red-shouldered hawk. *Oriole* 39:30–31.

Siegfried, L.M. 1982. Unpublished data. National Wildlife Health Center, Madison, Wis.

Snyder, N.F.R. 1974. Breeding biology of swallow-tailed kites in Florida. *Living Bird* 13:73–97.

Snyder, N.F.R., S.R. Beissinger, and R.E. Chandler. 1989. Reproduction and demography of the Florida Everglade (snail) kite. *Condor* 91:300–316.

Snyder, N.F.R., J.C. Ogden, J.D. Bittner, and G.A. Grau. 1984. Larval dermestid beetles feeding on nestling snail kites, wood storks, and great blue herons. *Condor* 86:170–174.

Solid Waste Authority of Palm Beach County. 1990. Ecotoxicological surveillance program of wildlife in the vicinity of a resource recovery facility: A baseline study. Unpublished report. West Palm Beach, Fla. 31 pp.

Spalding, M.G., and R.E. Bennetts. 1994. Unpublished data. University of Florida, Gainesville.

Spalding, M.G., G.W. Foster, and J.W. Mertins. 1994. Unpublished data. University of Florida, Gainesville.

Spalding, M.G., and K.D. Meyer. 1989. Unpublished data. University of Florida, Gainesville.

Spalding, M.G., K.D. Meyer, and S.F. Sundlof. 1993a. Unpublished data. University of Florida, Gainesville.

Spalding, M.G., K.D. Meyer, G.W. Foster, and J.W. Mertins. 1993b. Unpublished data. University of Florida, Gainesville.

Sprunt, A., Jr. 1945. The phantom of the marshes. *Audubon Mag.* 47:15–22.

———. 1950. Vanishing wings over the Everglades. *Audubon Mag.* 52:380–386.

Stickel, W.H., L.F. Stickel, and F.B. Coon. 1970. DDE and DDD residues correlated with mortality of experimental birds. In: *Inter-American conference on toxicology and occupational medicine, pesticide symposia.* W.P. Deichmann (ed.). Halos and Associates, Miami. pp. 287–294.

Stickel, W.H., L.F. Stickel, R.A. Dyrland, and D.L. Hughes. 1984. Aroclor 1254 residues in birds: lethal levels and loss rates. *Arch. Environ. Contam. Toxicol.* 13:7–13.

Stickel, W.H., L.F. Stickel, and J.W. Spann. 1969. Tissue residues of dieldrin in relation to mortality in birds and mammals. In: *Chemical fallout.* M.W. Miller and G.G. Berg (eds.). Charles C. Thomas, Springfield, Ill. pp. 174–204.

Stieglitz, W.O., and R. L. Thompson. 1967. *Status and life history of the Everglade kite in the United States.* U.S. Department of the Interior, Bureau of Sport Fisheries and Wildlife, Special Scientific Report—Wildlife 109. 21 pp.

Stiles, C.W., and A. Hassall. 1894. A preliminary catalogue of the parasites contained in the collections of the U.S. Bureau of Animal Industry, U.S. Army Medical Museum, Biological Department of the University of Pennsylvania (Coll. Leidy) and in Coll. Stiles and Coll. Hassall. *Vet. Mag.* 1:245–253, 331–354.

Sykes, P.W., Jr. 1978. Florida Everglade kite. In: *Rare and endangered biota of Florida.* Vol. 2, *Birds.* H.W. Kale II (ed.). University Presses of Florida, Gainesville. pp. 4–7.

———. 1979. Status of the Everglade kite in Florida, 1968–1978. *Wilson Bull.* 91:495–511.

———. 1983a. Recent population trends of the snail kite in Florida and its relationship to water levels. *J. Field Ornithol.* 54:237–246.

———. 1983b. Snail kite use of the freshwater marshes of south Florida. *Fla. Field Nat.* 11:73–88.

———. 1983c. Two species of mosquitoes feed on snail kites in Florida. *Fla. Field Nat.* 11:116.

———. 1984. The range of the snail kite and its history in Florida. *Bull. Fla. State Mus. Biol. Sci.* 29:211–264.

———. 1985. Pesticide concentrations in snail kite eggs and nestlings in Florida. *Condor* 87:438.

———. 1987a. The feeding habits of the snail kite in Florida, USA. *Colon. Waterbirds* 10:84–92.

———. 1987b. Some aspects of the breeding biology of the snail kite in Florida. *J. Field Ornithol.* 58:171–189.

———. 1987c. Snail kite nesting ecology in Florida. *Fla. Field Nat.* 15:57–84.

Sykes, P.W., Jr., and R. Chandler. 1974. Use of artificial nest structures by Everglade kites. *Wilson Bull.* 86:282–284.

Sykes, P.W., Jr., and D.J. Forrester. 1983. Parasites of the snail kite in Florida and summary of those reported for the species. *Fla. Field Nat.* 11:111–116.

Sykes, P.W., Jr., and H.W. Kale II. 1974. Everglade kites feed on nonsnail prey. *Auk* 91:818–820.

Sykes, P.W., Jr., J.A. Rodgers, Jr., and R.E. Bennetts. 1995. Snail kite (*Rostrhamus sociabilis*). *Birds N. Am.* 171:1–32.

Takekawa, J.E., and S.R. Beissinger. 1989. Cyclic drought, dispersal, and the conservation of the snail kite in Florida: lessons in critical habitat. *Conserv. Biol.* 3:302–311.

Telford, S.R., Jr., G.W. Foster, J.K. Nayar, S.P. Terrell, and D.J. Forrester. 1996. Unpublished data. University of Florida, Gainesville.

Thomas, N.J. 1991–93. Unpublished data. National Wildlife Health Center, Madison, Wis.

Toner, M. 1984. The kite hangs by a thread. *Natl. Wildl.* 22:38–43.

Travassos, L. 1923. Informacoes sobre a fauna helminthologica de Matto Grosso. *Folha Med.* 3:187–190.

Vermeer, K., R.W. Risebrough, A.L. Spaans, and L.M. Reynolds. 1974. Pesticide effects on fishes and birds in rice fields of Surinam, South America. *Environ. Pollut.* 7:217–236.

Wachenfeld, A.W. 1956. Present status of the Everglade kite. *Linn. Newsl.* 10:1.

Wiemeyer, S.N., T.G. Lamont, and L.N. Locke. 1980. Residues of environmental pollutants and necropsy data for eastern United States ospreys, 1964–1973. *Estuaries* 3:155–167.

CHAPTER FOURTEEN

Hawks

I. Introduction

Ten species of hawks occur in Florida (Robertson and Woolfenden 1992). Five of these are residents and 5 are transients that occur only in fall, winter, or spring months (table 14.1). None of the hawks is listed as threatened or endangered by the Florida Game and Fresh Water Fish Commission or the U.S. Fish and Wildlife Service (Logan 1997), but the Short-tailed Hawk is listed as rare, and Cooper's Hawk as a species of special concern, by the Florida Committee on Rare and Endangered Plants and Animals (Rodgers et al. 1996).

The numbers of Cooper's Hawks in eastern United States are believed to have declined from mid-1940 until the early 1970s because of the harmful effects of organochlorine pesticides (Henny and Wight 1972). This trend seems to have reversed since the use of these pesticides was restricted in the mid-1970s (Bednarz 1990). There are no accurate population data for Cooper's Hawks in Florida, but, based on anecdotal information from various sources, populations may have recovered to pre-DDT-era levels (Toland and Millsap 1996). Populations of Red-shouldered Hawks and Broad-winged Hawks in Florida are stable, and those of Red-tailed Hawks are increasing (Mitchell and Millsap 1990). Short-tailed Hawks are at their northern limits in Florida. Statewide surveys have not been conducted, but it is estimated that fewer than 500 individuals occur in the state, mainly in peninsular Florida (Millsap

Table 14.1. Distribution, occurrence, and abundance of hawks in Florida[a]

Species of hawk Range	Seasonal occurrence	Relative abundance
Northern Harrier (*Circus cyaneus*)		
Statewide	Winter transient	Rare to locally common
Sharp-shinned Hawk (*Accipiter striatus*)		
Statewide	Fall & winter transient	Common to very common
Cooper's Hawk[b] (*Accipiter cooperii*)		
Statewide	Resident (north Florida)	Rare to uncommon
	Fall & winter transient (south Florida)	
Northern Goshawk (*Accipiter gentilis*)		
Panhandle & northern peninsula	Fall and winter transient	Very rare (2 reports)
Red-shouldered Hawk (*Buteo lineatus*)		
Statewide	Resident	Fairly common
Broad-winged Hawk (*Buteo platypterus*)		
Statewide	Resident (south to Levy & Alachua counties)	Rare to fairly commom
	Fall and winter transient (south Florida, mostly near coasts)	Rare to abundant
	Transient (any season, south central Florida)	Very rare
Short-tailed Hawk[c] (*Buteo brachyurus*)		
Statewide	Resident	Rare to uncommon
Swainson's Hawk (*Buteo swainsoni*)		
Statewide	Winter & spring transient	Rare to locally common
Red-tailed Hawk (*Buteo jamaicensis*)		
Statewide	Resident (except Keys)	Uncommon to fairly common
	Fall & winter transient	Rare to locally common
Ferruginous Hawk (*Buteo regalis*)		
Panhandle & northern peninsula	Winter transient	Very rare (5 reports)

a. Modified from Robertson and Woolfenden (1992).
b. Listed as a species of special concern by the Florida Committee on Rare and Endangered Plants and Animals (Rodgers et al. 1996).
c. Listed as rare by the Florida Committee on Rare and Endangered Plants and Animals (Rodgers et al. 1996).

et al. 1989; Mitchell and Millsap 1990; Millsap et al. 1996).

There is little recent information on the status of the 5 transient species of hawks that occur in Florida (Adkisson 1990; Bildstein and Collopy 1990; Mitchell and Millsap 1990). Brown and Amadon (1968) reported that up to 300 Swainson's Hawks, mostly immatures, wintered in Florida, but there are no such current data. Mitchell and Millsap (1990) stated that Swainson's Hawk populations appeared to be stable in the southeastern United States, and a similar claim was made for Northern Harriers (Bildstein and Collopy 1990), although, again, we have no specific recent information for Florida. There is no information on the status of Sharp-shinned Hawks in Florida; Adkisson (1990) stated that its numbers may be increasing locally in various areas in the Southeast, but he pointed out that census data for

Table 14.2. Primary reasons given for submission of hawks to 3 wildlife rehabilitation centers in Florida, 1988–95

| Species of hawk | Number of cases per species of hawk | | | | | | |
	Trauma	Emaciation	Poisoning[a]	Electrocution	Infectious diseases	Inclement weather	Total number of hawks
Cooper's Hawk	70	12	7	0	0	0	89
Red-shouldered Hawk	178	24	21	6	3	1	233
Broad-winged Hawk	5	2	1	0	0	0	8
Short-tailed Hawk	1	0	0	0	0	0	1
Red-tailed Hawk	170	22	4	10	3	2	211
Sharp-shinned Hawk	84	9	20	0	0	2	115
Total cases	508	69	53	16	6	5	657
% of total cases	77	11	8	2	1	1	—

Sources: Collins and Gilliland (1995), 437 cases; Deem and Terrell (1996), 90 cases; Suto (1996), 130 cases.
a. Most of these were presumptive poisoning cases, since the poisons involved were not identified.

this hawk are very difficult to obtain and assess because of the hawk's small size and secretive behavior. Northern Goshawks and Ferruginous Hawks are very rare in Florida and nothing is known about their population status.

There is some published information on various morbidity and mortality factors of hawks, especially Red-shouldered and Red-tailed Hawks (e.g., the books edited by Fowler in 1986 and 1993 and the articles by Franson et al. 1996 and Deem et al. 1998), but less is available on other species. Books published by Cooper and Greenwood (1981) and Cooper (1985) emphasize clinical aspects of captive birds of prey, but also contain health and disease information relevant to free-ranging hawks.

Specific information on the diseases and parasites of hawks in Florida is limited. Most of the available data have come from studies on road-killed specimens and birds submitted to wildlife rehabilitation facilities. Important sources of specific information on hawks in Florida used in the preparation of this chapter were the files of Florida Audubon Society's Center for Birds of Prey in Maitland (provided by Resee Collins and Carol Gilliland), the files of Suncoast Seabird Sanctuary in St. Petersburg (provided by Barbara Suto), and the files of the Veterinary Medical Teaching Hospital at the University of

Florida (provided by Sharon L. Deem and Scott P. Terrell). These 3 data-sets comprise combined morbidity and mortality information on hawks submitted for rehabilitation purposes. Kinsella et al. (1995) published a paper on the helminths of 5 species of hawks in Florida. Data on blood parasites of the same 5 species of hawks were published by Forrester et al. (1994).

II. Trauma

Trauma is the leading cause of death for hawks in Florida (table 14.2). The types of trauma include roadkills, gunshot, collisions with buildings, tall towers, and power lines, and entrapment (tables 14.3–14.4).

Roadkills are probably the most important (tables 14.3–14.5). The distribution of morbidity and mortality caused by vehicles striking hawks is given in figure 14.1. There are 3 geographic clusters of data on the map, probably because of the presence of 3 wildlife rehabilitation facilities, from which much of these data were obtained, rather than amount of road use or speed limits in those counties. The reason for the large numbers of Red-shouldered Hawks and Red-tailed Hawks killed on highways is not clear. It is most likely a reflection of the high populations of these species in Florida, or their

Table 14.3. Types of trauma causing morbidity and mortality in hawks submitted to the Center for Birds of Prey, Maitland, Florida, 1989–94

| | Number of cases per species of hawk | | | | | | | | |
Species of hawk	Vehicle strike	Gunshot	Building strike	Entrapment[a]	Power line strike	Predation[b]	Misc.	Unknown	Total number of hawks
Cooper's Hawk	3	1	7	6	0	0	0	28	45
Red-shouldered Hawk	14	18	0	0	2	1	2[c]	82	119
Broad-winged Hawk	0	0	0	0	0	0	0	3	3
Short-tailed Hawk	0	1	0	0	0	0	0	0	1
Red-tailed Hawk	32	10	7	7	5	2	1[d]	65	129
Sharp-shinned Hawk	5	2	6	4	0	1	0	30	48
Total cases	54	32	20	17	7	4	3	208	345
% of total cases	16	9	6	5	2	1	1	60	—

Source: Collins and Gilliland (1995).
a. Includes birds caught in traps, fences, fishing lines, and buildings.
b. Domestic dogs.
c. One flew into a tree and one was hit by a golf ball.
d. Flew into a power pole.

Table 14.4. Data on hawks that died of trauma in Florida

Species of hawk County	Year	Age[a]	Sex[b]	Type of trauma	Data source
Cooper's Hawk					
Alachua	1979	IM	M	Road kill	Forrester (1994)
Monroe	1990	UN	UN	Struck window	Cox (1991)
Red-shouldered Hawk					
Alachua	1974	AD	M	Unknown	Forrester (1994)
Alachua	1979	AD	UN	Road kill	Ibid.
Alachua	1979	AD	M	Gunshot	Ibid.
Alachua	1993	AD	M	Road kill	Sileo (1993)
Alachua	1994	AD	M	Unknown	Franson (1994)
Bradford	1979	AD	F	Road kill	Forrester (1994)
Columbia	1985	AD	F	Road kill	Ibid.
Gadsden	1981	AD	F	Unknown	Ibid.
Gilchrist	1994	AD	F	Unknown	Sileo (1994)
Glades	1978	AD	F	Road kill	Forrester (1994)
Lake	1974	UN	M	Unknown	Ibid.
Marion	1977	AD	F	Road kill	Ibid.
Pasco	1977	AD	F	Road kill	Ibid.
Taylor	1977	AD	F	Road kill	Ibid.
Union	1977	AD	F	Road kill	Ibid.
Broad-winged Hawk					
St. Lucie	1978	IM	F	Road kill	Ibid.
Red-tailed Hawk					
Alachua	1974	IM	F	Road kill	Ibid.
Alachua	1975	IM	F	Gunshot	Ibid.
Alachua	1976	IM	M	Unknown	Ibid.
Alachua	1977	IM	F	Road kill	Ibid.
Alachua	1984	IM	UN	Gunshot	Ibid.
Bradford	1979	IM	F	Road kill	Ibid.
Leon	1959	UN	UN	Struck TV tower	Crawford (1981)
Leon	1976	UN	UN	Struck TV tower	Ibid.
Levy	1977	IM	M	Road kill	Forrester (1994)
Polk	1984	AD	F	Road kill	Ibid.
Santa Rosa	1985	IM	F	Gunshot	Howerth (1985)
Union	1975	IM	F	Unknown	Forrester (1994)
Sharp-shinned Hawk					
Gilchrist	1979	AD	F	Unknown	Ibid.
Orange	1992	AD	M	Struck building	Franson (1993)
Pinellas	1974	IM	M	Gunshot	Forrester (1994)
Polk	1973	IM	F	Gunshot	Ibid.

a. AD = adult, IM = immature, UN = unknown.
b. M = male, F = female, UN = unknown.

Table 14.5. Road-kill data: hawks in state parks and recreation areas of Florida

Species of hawk	No. hawks reported by year				
	1990	1991	1992	1993	Total
Cooper's Hawk	0	0	1	1	2
Red-shouldered Hawk	6	4	6	7	23
Broad-winged Hawk	1	0	0	0	1
Red-tailed Hawk	0	1	0	1	2
Northern Harrier	0	0	1	0	1
No. parks reporting	45	64	64	61	—

Sources: Stevenson (1994), Snyder (1994).

habit of hunting near the edges of highways and thereby being struck by vehicles as they fly over the road in pursuit of prey, or a combination of these 2 factors. Undoubtedly the numbers of all species of raptors killed by vehicles statewide are much higher than shown here, particularly on highways with high speed limits.

The distribution of 46 cases of morbidity and mortality of hawks due to gunshot is presented in figure 14.2. These cases of illegal shooting were recorded between 1973 and 1995 by Collins and Gilliland (1995), Deem and Terrell (1996), Forrester (1994), Howerth (1985), and Suto (1996). As mentioned above concerning roadkills, there is a geographic clustering of cases near 2 wildlife rehabilitation centers, 1 in Alachua County (University of Florida Veterinary Teaching Hospital), and 1 in Orange County (Center for Birds of Prey). Most of these hawks were Red-shouldered Hawks (*n* = 24) and Red-tailed Hawks (*n* = 16), probably a reflection of the higher populations of these species than of other hawks.

Collins and Gilliland (1995) have records of 4 Red-tailed Hawks (1 each from Orange, Osceola, Seminole, and Volusia counties) and 3 Red-shouldered Hawks (1 each from Orange, Polk, and Volusia counties) injured or killed by flying into power lines between 1989 and 1994.

Only 2 hawks, both Red-tailed, were recorded by Crawford (1981) in his 25-year study of >42,000 birds killed by striking a 1,010-foot TV tower in Leon County, 1955–80. Both hawks hit the tower during daylight hours. None was listed in several other studies

involving bird mortality and tall structures in various parts of Florida (Kale 1971; Taylor and Anderson 1973, 1974; Maehr et al. 1983; Taylor and Kershner 1986; Maehr and Smith 1988). Cox (1991) gave an interesting report of a Cooper's Hawk crashing into a window on Islamorada (Monroe County) while chasing a White-crowned Pigeon in October of 1990; both birds died as a result of the collision.

III. Predation

McEwan and Hirth (1980) found the remains of a Northern Harrier in 1 of 16 active Bald Eagle nests examined in a 4-county area (Alachua, Marion, Putnam, and Volusia) during 1975 and 1976. It is not clear, however, whether this was a case of predation or scavenging by the Bald Eagle. Predation on hawks by domestic dogs has been documented by Collins and Gilliland (1995). They recorded such predation on 2 Red-tailed Hawks from Seminole County in 1991-92, a Red-shouldered Hawk from Seminole County in 1994, and a Sharp-shinned Hawk from Orange County in 1994. Further details on these events are not available.

IV. Emaciation

Emaciation or nutritional deficiency was the primary diagnostic finding for 11% of 657 hawks representing 5 species that were submitted to 3 wildlife rehabilitation centers in Florida

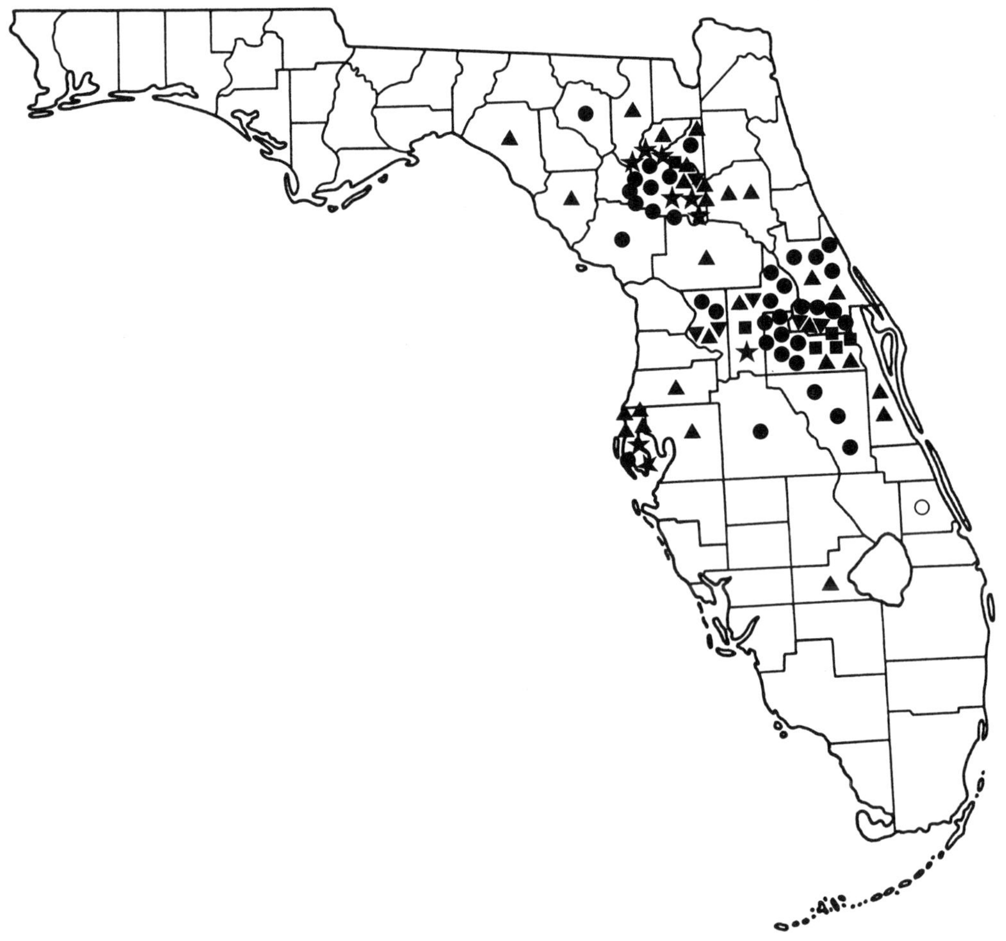

FIGURE 14.1. Distribution of morbidity and mortality of hawks in Florida due to vehicle strikes on highways, 1973–95. *Solid circle* = 1 Red-tailed Hawk, *open circle* = 1 Broad-winged Hawk, *solid star* = 1 Cooper's Hawk, *solid triangle* = 1 Red-shouldered Hawk, *solid square* = 1 Sharp-shinned Hawk. Data from Collins and Gilliland 1995; Deem and Terrell 1996; Forrester 1994; Sileo 1993; Suto 1996.

from 1988 to 1995 (table 14.2). In some of these cases there were other conditions such as trauma that were probably related to the emaciation and may have led to it, but the records are not clear on the relative significance of these or other secondary factors.

V. Electrocution

Eighteen cases of electrocution have been documented among hawks in Florida. Six were Red-shouldered Hawks from Alachua (1994), Bradford (1994), Citrus (1991), Orange (1991), Seminole (1993), and Volusia (1994) counties and 12 were Red-tailed Hawks from Alachua (1976, *n* = 2), Brevard (1993), Orange (1991-93, *n* = 5), Osceola (1991), Seminole (1994), and Volusia (1994) counties, and 1 from an unknown county (1991) (Collins and Gilliland 1995; Deem and Terrell 1996; Forrester 1994). Sixteen of these were hawks submitted to wildlife rehabilitation facilities (table 14.2).

As discussed previously (chapter 11, Eagles),

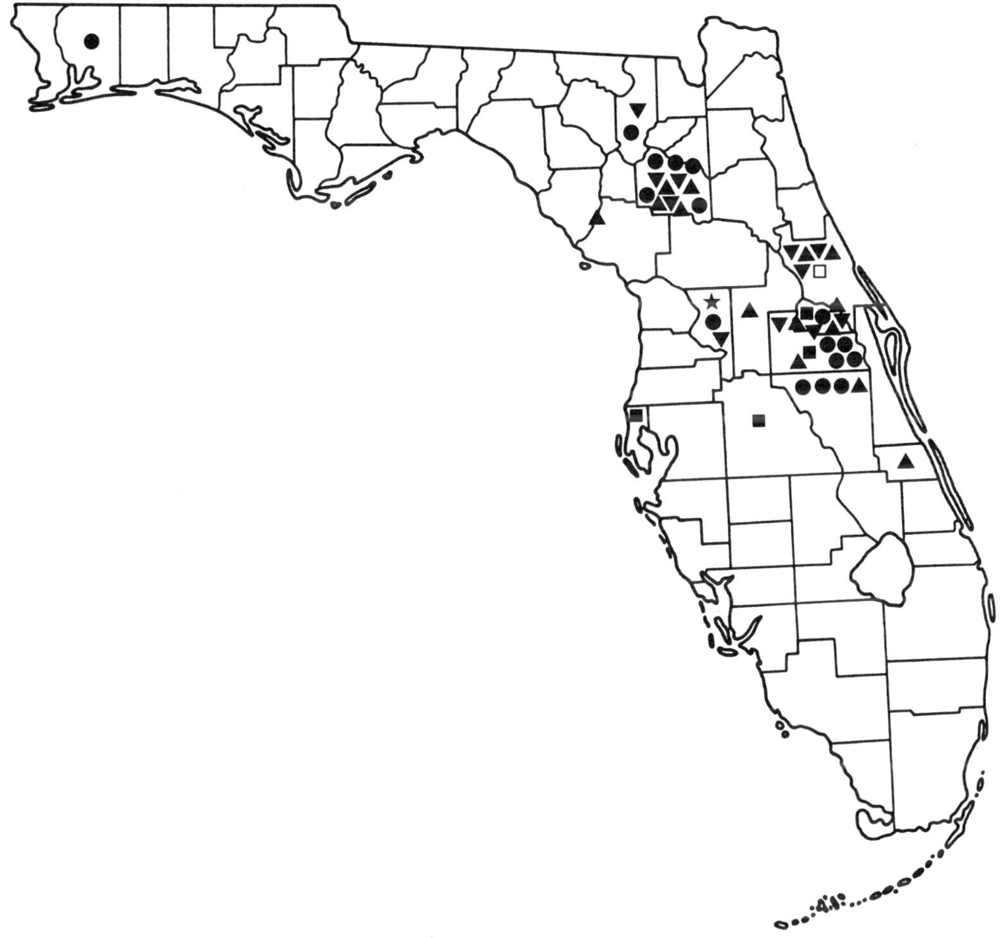

FIGURE 14.2. Distribution of morbidity and mortality of hawks in Florida due to gunshot wounds, 1973–95. *Solid circle* = 1 Red-tailed Hawk, *solid star* = 1 Cooper's Hawk, *solid triangle* = 1 Red-shouldered Hawk, *solid square* = 1 Sharp-shinned Hawk, *open square* = 1 Short-tailed Hawk. Data from Collins and Gilliland 1995; Deem and Terrell 1996; Forrester 1994; Howerth 1985; Suto 1996.

electrocution of raptors can be prevented by certain adaptations or modifications of the power line structures where the birds tend to perch and also by changes in the grounding setup. Details on these practices have been published (Miller et al. 1975) and may help mitigate such mortality.

VI. Inclement weather

Hurricane Donna swept through peninsular Florida in September 1961 and caused consid-

erable mortality among various birds, particularly in southern Florida (Robertson and Paulson 1961). Among the several hundred mostly wading birds found dead during systematic searches in Florida Bay and around Flamingo was 1 Red-shouldered Hawk.

On March 13, 1993, the "storm of the century" hit Florida with winds of more than 100 mph (Langridge 1993). A Red-shouldered Hawk's nest was destroyed near Tallahassee (Leon County). Beginning shortly after the storm and for the following several days, nu-

Table 14.6. Concentrations of chlorinated hydrocarbons in various tissues of hawks from Florida

Species of hawk County	Year	Age[b]	Sex	Sample[c]	Residues (ppm wet weight)[a]				Data source[e]
					DDE	ΣDDT	Dieldrin	PCB[d]	
Cooper's Hawks									
Alachua	1973	NG	F	UP	12	13	0.00	NA	A
Duval	1975	NG	NG	BR	8.9	8.9	2.4	0.32	B
				LV	23	23	5.6	2.0	
				MU	5.70	5.70	1.40	1.50	
Red-shouldered Hawks									
Alachua	1972	NG	F	AD	0.64	1.0	0.06	NA	A,C
				UP	0.18	0.18	0.00	NA	
	1973	IM	NG	AD	0.45	1.2	0.00	NA	A,C
				UP	0.24	0.60	0.00	NA	
	1974	AD	M	BR	0.09	0.09	0.00	0.37	B
				LV	0.23	0.23	0.01	0.31	
				MU	0.11	0.11	0.31	0.13	
	1974	AD	F	BR	0.57	0.57	0.61	2.7	B
				LV	1.5	1.9	1.1	3.5	
				MU	0.20	0.20	0.12	0.64	
	1976	AD	NG	AD	7.2	7.2	0.00	NA	A
				UP	0.39	0.39	0.00	NA	
Baker	1976	NG	F	UP	1.2	1.2	0.00	NA	A
Gadsden	1981	AD	F	BR	0.80	2.0	1.3	NA	B
Hillsborough	1975	AD	F	BR	0.13	0.13	2.6	1.5	B
				LV	0.23	0.23	3.4	3.5	
				MU	0.53	0.53	4.2	7.9	
Lake	1974	AD	M	BR	1.0	1.0	6.0	0.28	B
				MU	1.2	1.2	4.2	0.95	
	1974	AD	M	BR	0.21	0.21	0.27	0.46	B
				LV	0.71	0.71	0.74	0.58	
				MU	0.31	0.31	0.27	0.19	
Monroe	1972	NG	NG	BR,MU	0.09	0.10	0.01	0.78[n]	G
Pinellas	1974	IM	NG	BR	1.3	1.8	11	1.8	B
				LV	1.6	2.1	11	0.25	
				MU	2.5	3.0	14	0.31	
	1976	NG	F	AD	34	62	38	NA	A
				UP	0.80	1.80	0.80	NA	
Taylor	1977	AD	F	BR	1.3	1.3	NA	NA	B
Union	1977	AD	F	BR	0.50	0.50	NA	NA	B
Red-tailed Hawks									
Alachua	1973	IM	NG	AD	6.3	6.3	0.00	NA	A,C
				UP	0.10	0.10	0.00	NA	
	1973	IM	F	AD	3.5	3.7	0.89	NA	A,C
				UP	0.11	0.11	0.00	NA	
	1973	IM	NG	AD	0.37	0.48	0.04	NA	A,C
				UP	0.39	0.50	0.00	NA	
	1974	IM	NG	AD	0.59	1.3	0.00	NA	A
				UP	0.21	0.21	0.00	NA	
	1974	NG	F	AD	0.37	0.37	0.00	NA	A
				UP	0.00	0.00	0.00	NA	
	1974	IM	F	BR	0.02	0.02	0.01	0.14	B
				AD	1.9	2.3	0.32	1.6	
				LV	0.12	0.13	0.16	0.34	
				MU	0.09	0.12	0.02	0.11	

(continued)

Table 14.6. *(continued)*

Species of hawk County	Year	Age[b]	Sex	Sample[c]	Residues (ppm wet weight)[a]				Data source[e]
					DDE	ΣDDT	Dieldrin	PCB[d]	
	1975	AD	F	BR	0.04	0.04	0.03	0.08	B
				LV	0.04	0.05	0.02	0.10	
				MU	0.06	0.08	0.02	0.15	
	1976	NG	M	AD	4.3	5.9	0.85	NA	A
				UP	0.38	0.38	0.00	NA	
	1976	IM	NG	AD	6.1	7.2	2.4	NA	A
				UP	1.1	1.1	0.18	NA	
	1982	AD	M	BR[f]	0.96	NG	3.7	3.1[h]	D
				LV[g]	2.1	NG	4.4	4.6[h]	
Northern Harriers									
Seminole	1982	IM	F	CS[i]	6.3	6.3	0.83	1.2	E
				BR[j]	2.2	2.2	0.36	0.00	
Sharp-shinned Hawks									
Alachua	1973	NG	F	UP	12	13	0.00	NA	A
	1977	AD	F	BR	8.5	8.5	NA	NA	B
Orange	1982	IM	F	CS[k]	0.98	0.98	0.10	0.00	E
				BR[k]	0.67	0.67	0.12	0.00	
	1992	AD	F	BR[o]	0.00	0.00	0.00	0.00	F
	1992	AD	M	BR[o]	2.2	2.2	0.15	2.1	F
Pinellas	1974	AD	M	BR	0.25	0.25	0.04	0.05	B
				AD	22	23	1.5	2.5	
				LV	2.1	2.2	0.16	0.33	
				MU	0.70	0.70	0.08	0.26	
Polk	1971	AD	F	AD	19	19	0.45	6.3	B
				LV	0.24	0.24	0.01	0.23	
				MU	0.66	0.66	0.02	0.56	
Seminole	1982	AD	M	CS[l]	1.7	1.7	0.17	0.60	E
				BR[m]	5.5	5.6	0.41	1.90	

NA = not analyzed, NG = not given.

a. Lower limits of detection = 0.01 ppm for sources A, B, and C; unknown for source D; 0.01 ppm for pesticides and 0.05 ppm for PCBs for source E; 0.1 for pesticides and 0.5 for PCBs for source F; unknown for source G.

b. AD = adult, IM = immature.

c. AD = adipose tissue, BR = brain, LV = liver, MU = pectoral muscle, UP = uropygial gland, CS = carcass, based on the body of the bird minus the skin, feet, wings, liver, and gastrointestinal tract (Mulhern et al. 1970).

d. Unless otherwise noted, the PCBs most closely resembled Aroclor 1260 by chromatographic profile.

e. A = Johnston (1978), B = Sundlof et al. (1986), C = Johnston (1976), D = Couvillion (1982), E = Krynitsky (1984), F = Franson (1993), G = Ogden et al. (1974).

f. Additional residues included 0.19 ppm mirex and 1.2 ppm heptachlor epoxide.

g. Additional residue included 0.45 ppm mirex.

h. Aroclor 1254.

i. Additional residues included 0.12 ppm heptachlor epoxide, 0.48 ppm oxychlordane, 0.12 ppm *trans*-nonachlor, 0.73 ppm toxaphene. Other residues tested for, but not detected, included *cis*-chlordane, *cis*-nonachlor, and endrin.

j. Additional residues included 0.15 ppm oxychlordane and 0.33 ppm toxaphene. Other residues tested for, but not detected, included heptachlor epoxide, *cis*-chlordane, *trans*-nonachlor, *cis*-nonachlor, endrin, and PCBs.

k. Other residues tested for, but not detected, included heptachlor epoxide, oxychlordane, *cis*-chlordane, *trans*-nonachlor, *cis*-nonachlor, endrin, and toxaphene.

l. Additional residues included 0.16 ppm heptachlor epoxide and 0.17 ppm oxychlordane. Other residues tested for, but not detected, included *cis*-chlordane, *trans*-nonachlor, *cis*-nonachlor, endrin, and toxaphene.

m. Additional residues included 0.35 ppm heptachlor epoxide, 0.60 ppm oxychlordane, and 0.21 ppm *trans*-nonachlor. Other residues tested for, but not detected, included *cis*-chlordane, *cis*-nonachlor, endrin, and toxaphene.

n. Type of PCB not given by authors.

o. Other residues tested for, but not detected, included HCB, α-BHC, α-chlordane, β-BHC, endrin, Γ-BHC, Γ-chlordane, heptachlor epoxide, mirex, oxychlordane, toxaphene, and *trans*-nonachlor.

Table 14.7. Concentrations of mirex in stomach contents and various tissues of two Red-tailed Hawks examined before and after aerial application of 10-5 bait for fire ant control in Duval County, Florida, 1972

	Residues (ppm wet weight)[a]	
Tissue	Pre-treatment	1 month post-treatment
Brain	0.07	ND
Adipose tissue	0.01	0.82
Liver	ND	0.02
Muscle	0.02	0.32
Stomach contents	NT	ND

Source: Wheeler et al. (1977).
ND = not detected, NT = not tested.
a. Lowest limit of detection = 0.01 ppm.

merous raptors were submitted to the Center for Birds of Prey in Maitland. Among these was a Red-tailed Hawk from Orange County that had severe traumatic injuries (Collins and Gilliland 1995).

VII. Environmental contaminants

Tissue residue data on chlorinated hydrocarbons have been obtained for 5 species of hawks in Florida, including Cooper's Hawks, Red-shouldered Hawks, Red-tailed Hawks, Northern Harriers, and Sharp-shinned Hawks (tables 14.6 and 14.7). Most samples were collected in the 1970s, with a few in the 1980s and 1990s. Eight types of chlorinated hydrocarbons were identified, including DDT and its metabolites, dieldrin, PCBs, mirex, heptachlor epoxide, oxychlordane, *trans*-nonachlor, and toxaphene. Concentrations of DDT and its metabolites were detected in every hawk examined and dieldrin was present in 26 of 35 hawks. PCBs were detected in 14 of 15 birds that were tested. Neither Johnston (1978) nor Sundlof et al. (1986) found trends in their data to suggest that concentrations of DDE, dieldrin, or PCBs had declined in Florida over the years since the sale and use of these products were banned in the early 1970s. This lack of decline may be a sampling artifact due to the small numbers of birds examined in recent years. In addition, migra-

tory habits may have influenced the exposure of some, such as Red-shouldered Hawks and Red-tailed Hawks; they may have been exposed while migrating from the West Indies or Central America where some of these pesticides are still used (Johnston 1978).

The concentrations of dieldrin in brain tissues of 3 birds were considered hazardous or perhaps lethal. These included 2 Red-shouldered Hawks, 1 from Lake County with 6 ppm and 1 from Pinellas County with 11 ppm, and a Red-tailed Hawk from Bay County with almost 4 ppm. Concentrations greater than 9 ppm are considered lethal (Ohlendorf et al. 1981) and the Red-shouldered Hawk from Pinellas County may have died of dieldrin poisoning. None of the hawks contained concentrations of DDT and its metabolites or PCBs that were high enough to have been associated with mortality.

In a study of mirex residues in various species of wildlife at Dee Dot Ranch (Duval and St. Johns counties) during 1972–74, 2 Red-tailed Hawks were examined, 1 before and the other 1 month after application of 10-5 bait for fire ant control (table 14.7). Although the residues were higher in the hawk examined 1 month after treatment, they were all less than 1 ppm. The effects of low concentrations of mirex on hawks are not known.

Residues of chlorinated hydrocarbons in eggs of hawks in Florida have not been studied,

Table 14.8. Measurements of eggshells of hawks from Florida, pre- and post-1947

Species of hawk Time period	No. eggs sampled	Mean thickness index[a]	% change
Red-shouldered Hawks			
Pre-1947[b]	300	1.84	—
1947–49	17	1.78	0
1950–59	54	1.77	-4
1960–68	49	1.79	0
Red-tailed Hawk			
Pre-1947	76	2.24	—
1949–67	80	2.17	-3

Source: Anderson and Hickey (1972).
a. Thickness index = weight of eggshell (mg) divided by the product of the length x breadth (mm) (Ratcliffe 1967).
b. Authors stated that the pre-1947 values for Red-shouldered Hawks were from "Gulf Coast, Piedmont," which presumably included some data from Florida hawks, although this was not stated.

although there are some data on eggshell thinning. Anderson and Hickey (1972) reported 3 and 4% thinning of shells of Red-tailed and Red-shouldered Hawks, respectively, when eggs collected in Florida prior to 1947 (before the DDT era) were compared with those taken after that date (table 14.8). There are no recent data on eggshell thinning of hawks in Florida, but hopefully the situation has improved since the banning of the sale and use of DDT in the United States in the 1970s.

Although illness and death due to lead poisoning have been reported in Red-tailed Hawks in other parts of the United States (Reiser and Temple 1981), no lead was detected in 2 Cooper's Hawks, 2 Red-shouldered Hawks, or 1 Red-tailed Hawk examined from Alachua County in 1993–94 (table 14.9). Less than 1 ppm of mercury was found in the liver of 1 of the Cooper's Hawks and in a combined sample of brain and pectoral muscle of 1 of the Red-shouldered Hawks. The Red-shouldered Hawk also

Table 14.9. Concentrations of heavy metals in hawks from Florida

Species of hawk County	Year	Age[b]	Sex	Sample[c]	Concentrations (ppm wet weight)[a]			Data source
					Lead	Mercury	Arsenic	
Cooper's Hawks								
Alachua	1993	IM	M	LV	ND	0.2	NT	Locke (1993)
	1993	IM	F	LV	ND	NT	NT	Franson (1993)
Red-shouldered Hawks								
Alachua	1994	AD	M	LV	ND	NT	NT	Franson (1994)
	1994	AD	NG	LV	ND	NT	NT	Sileo (1994)
Monroe	1972	NG	NG	BR,PM	NT	0.78	0.15	Ogden et al. (1974)
Red-tailed Hawk								
Alachua	1994	AD	M	LV	ND	NT	NT	Sileo (1994)

ND = not detected, NG = not given by authors, NT = not tested.
a. Lower limit = 0.2 ppm for all data except those from Ogden et al. (1974), in which the lower limit was not given.
b. AD = adult, IM = immature.
c. BR = brain, LV = liver, PM = pectoral muscle.

had <1 ppm of arsenic. The effects of mercury and arsenic on populations of hawks in Florida are not known, but the low concentrations in these birds are probably of no significance.

VIII. Neoplasia

No information is available on this topic.

IX. Biotoxins

There are 2 records in the literature of hawks being killed by poisonous snakes in Florida. Brugger (1989) presented evidence of an adult male Red-tailed Hawk being killed by an eastern coral snake (*Micrurus fulvius*) in Alachua County in 1987. In this case the coral snake was partially melanistic; the author suggested that this abnormal coloration pattern might have reduced the hawk's avoidance response and led it to attempt to capture and ingest the snake.

The second record involved an adult female Cooper's Hawk from Tall Timbers Research Station (Leon County) killed by a rattlesnake in 1992 (Davidson 1992; Heckel et al. 1994). In this interesting incident the hawk was found dead near a stump hole inhabited by an eastern diamondback rattlesnake (*Crotalus adamanteus*) and a large cottonmouth (*Agkistrodon piscivorus*). Apparently the hawk had attempted to capture a Northern Bobwhite that tried to evade capture by running down the hole. Presumably the hawk was bitten when it approached or tried to reach into the hole for the bobwhite. These types of events probably are rare, but may be the cause of mortality from time to time. Heckel et al. (1994) cited 2 other cases of Red-tailed Hawks being killed by poisonous snakes in southern Georgia and linked this type of mortality to the inexperience of the hawks.

X. Viruses

There are 4 reports of viral infections in hawks from Florida. These include avian poxvirus, 2 arboviruses, and rabies virus. Avian pox has been reported from Red-tailed Hawks and Northern Goshawks in other areas (Graham and Halliwell 1986; Morishita et al. 1997), but we have only 1 report from Florida. Collins (1999) found an infected Red-tailed Hawk from Orange County in 1997. One dead Red-shouldered Hawk from Leon County was found positive via PCR and virus isolation techniques for West Nile virus in June of 2001. During the next 6 months infections were identified in 2 Cooper's Hawks (Hamilton and Leon counties), 2 Red-tailed Hawks (Dixie and Leon counties), and 2 Sharp-shinned Hawks (Dade and Hamilton counties) (Conti et al. 2002). It was assumed that these hawks died because of the West Nile virus infection, but this was not determined. Jennings et al. (1969) reported 1 Red-shouldered Hawk from the Tampa Bay area in 1962 to be seropositive (HI test) for St. Louis encephalitis (SLE) virus at a titer of 1:40. Spalding and McLean (1994) tested 1 Red-shouldered Hawk from Osceola County in 1993 for evidence of infection with eastern equine encephalitis virus; it was seronegative (SN test). SLE virus usually does not cause overt disease in wild birds (Karstad 1971) and the infection in a Red-shouldered Hawk was probably of little or no consequence, except in the possible spread of the virus to other animals and humans. West Nile virus, however, is pathogenic to hawks and a virus of significance, the extent of which is yet to be determined. It is also an important disease agent for horses and humans, where it can cause encephalitis and death. For more information on West Nile virus in Florida, see chapter 24, Perching Birds.

Albers (1984) tested 21 Red-tailed Hawks, 1 Sharp-shinned Hawk, and 1 Northern Harrier from Florida for serologic evidence of rabies virus infection. These raptors came from a wildlife rehabilitation facility in Pinellas County during 1982–83. Three of the Red-tailed Hawks were seropositive. None of the birds had signs of rabies and the pathological significance of these findings is not known. In general birds are probably not important epidemiologically in the spread of rabies virus. Infections in these cases were probably acquired from their prey (Albers 1984). Graham and Halliwell (1986: 412) reviewed the data on rabies virus infections in raptors and concluded that "perhaps more than the usual amount of care should be exercised by those who handle, treat, or perform necropsies on birds of prey."

XI. Bacteria

Salmonellosis was determined to be the cause of death in an immature female Red-shouldered Hawk from Alachua County submitted to the University of Florida Veterinary Medical Teaching Hospital in July 1985 (Spalding 1985). The bird was emaciated and died 4 days after being received. At necropsy it was observed to have pericarditis, epicarditis, ulcerative ventriculitis, and necrosis of spleen and kidneys. A pure culture of *Salmonella saint-paul* was obtained from the pericardial sac.

During 1973–81 White and Forrester (1981) cultured samples of livers and contents of large intestines from 26 hawks representing 5 species from 13 counties in Florida (table 14.10). Six of these hawks died of unknown causes, 2 died of aspergillosis, and the others died of anthropogenic causes, such as collisions with vehicles, electrocution, gunshot, and fence entanglement, and should be considered normal birds. Seven species of bacteria were identified. The significance of these infections to hawks is not known and in most cases the bacteria were either normal components of the intestinal flora, postmortem contaminants, or representative of secondary infections. The 2 *Salmonella* infections were serotyped as *S. typhimurium* and are of public health interest.

In 1992 an adult male Sharp-shinned Hawk from Orange County was submitted to the National Wildlife Health Center in Madison, Wisconsin, for examination. It had flown into a building and died of traumatic injuries 3 days after being obtained. At necropsy bacterial pneumonia was a secondary finding, but tissues were not cultured to determine which species of bacteria were involved (Franson 1993).

XII. Fungi

Aspergillosis is the only mycotic infection known from hawks in Florida. A Cooper's Hawk from Levy County died of aspergillosis in 1977 (Forrester 1994). This bird also had a *Pseudomonas* infection, as mentioned in section XI, Bacteria, above. The stress of the bacterial infection may have caused the expression of this mycotic infection. There are 9 cases of aspergillosis in Red-tailed Hawks ($n = 5$) and Red-shouldered Hawks ($n = 4$) in the records of the Center for Birds of Prey (Collins and Gilliland 1995), all in birds submitted because of other problems (mainly injuries due to trauma) and in captivity for 1–3 weeks before coming down with aspergillosis. Such events could have led to stress-related immunosuppression and subsequent increases in susceptibility to aspergillosis, as explained by Redig (1993). The importance of *Aspergillus* infections in various species of free-ranging hawks in Florida is not known.

XIII. Protozoans

Nine protozoan species have been identified from hawks in Florida (tables 14.11–14.16). Six of these are blood protozoans, 3 of which are illustrated in figure 14.3; some of these infections are discussed in reference to other species of raptors in Florida by Forrester et al. (1994). Mixed infections of 2 species of blood protozoans occurred in 14 birds (6 Red-tailed Hawks, 5 Red-shouldered Hawks, 1 Broadwinged Hawk, 1 Cooper's Hawk, and 1 Sharp-shinned Hawk). Four Red-tailed Hawks had concurrent infections of 3 species. In addition to the hawks listed in tables 14.11–14.15, 1 Northern Harrier from Gilchrist County (1995) was examined via blood smear and found negative for blood protozoans.

The mosquito vectors of *Plasmodium elongatum* and *P. forresteri* from Red-tailed Hawks were determined experimentally by Telford et al. (1997) and Nayar et al. (1998). They tested 10 species of Florida mosquitoes and found

Table 14.10. Bacteria identified from livers and large intestines of hawks[a] from Florida

Species of hawk Bacteria	County	Year	Organ cultured[b]	Cause of death[c]
Cooper's Hawk (*n* = 2)[d]				
Escherichia coli	Levy	1977	LI	AS
Pseudomonas sp.	Levy	1977	LV	AS
Red-shouldered Hawk (*n* = 11)[e]				
Escherichia coli	Bradford	1979	LI	RK
	Hillsborough	1975	LI	UK
	Hillsborough	1976	LI,LV	UK
	Lake	1974	LI	UK
Salmonella sp.	Pinellas	1974	LI	UK
Staphylococcus sp. (nonhemolytic)	Taylor	1976	LV	FE
Red-tailed Hawk (*n* = 10)[f]				
Enterococcus sp.	Alachua	1976	LI	UK
	Union	1975	LI	UK
Escherichia coli	Alachua	1974	LI	RK
	Alachua	1975	LI	GS
	Alachua	1976	LI	EL
	Bradford	1979	LI	RK
Proteus sp.	Alachua	1975	LI	GS
	Alachua	1976	LI	EL
	Levy	1976	LI	RK
Salmonella typhimurium	Alachua	1975	LI	GS
	Alachua	1976	Li	EL

Source: White and Forrester (1981).

a. In addition to the hawks listed in this table, 1 road-killed Broad-winged Hawk from St. Lucie County and 2 Sharp-shinned Hawks killed by gunshot in Hillsborough County (1974) and Polk County (1973) were cultured and were negative.

b. LI = large intestine, LV = liver.

c. AS = aspergillosis, RK = roadkill, UK = unknown, FE = entanglement in barbed wire fence, EL = electrocution, GS = gunshot.

d. One additional Cooper's Hawk from Duval County (1974), for which cause of death was unknown, was cultured and was negative.

e. Four additional Red-shouldered Hawks were cultured and were negative. All 4 were roadkills and were from Glades (1978), Marion (1977), Pasco (1977), and Taylor (1977) counties.

f. Two additional Red-tailed Hawks from Alachua County were cultured and were negative. One was a roadkill (1977) and the other a normal hawk that had been shot because it was preying on poultry (1980).

that 3 (*Culex nigripalpus*, *C. restuans*, and *C. salinarius*) transmitted *P. elongatum*, but that only 1 (*C. restuans*) was a vector of *P. forresteri*. Because of the breeding patterns and population peaks of these vectors, it can be concluded that *P. elongatum* is transmitted to hawks throughout the year but *P. forresteri* is transmitted mainly during early winter to the end of spring in southern Florida.

Very little is known about the pathogenic ef-fects of these infections on hawks in general and we have no information on Florida birds. Extremely high parasitemias of *Leucocytozoon toddi* have been observed in some hawks without being associated with mortality (Peirce 1981). Infections of *Plasmodium relictum* are considered pathogenic to some raptors (Garnham 1966; Kingston et al. 1976), but, again, we have no pertinent data from Florida. However, as suggested by Bennett et al. (1993: 998), these

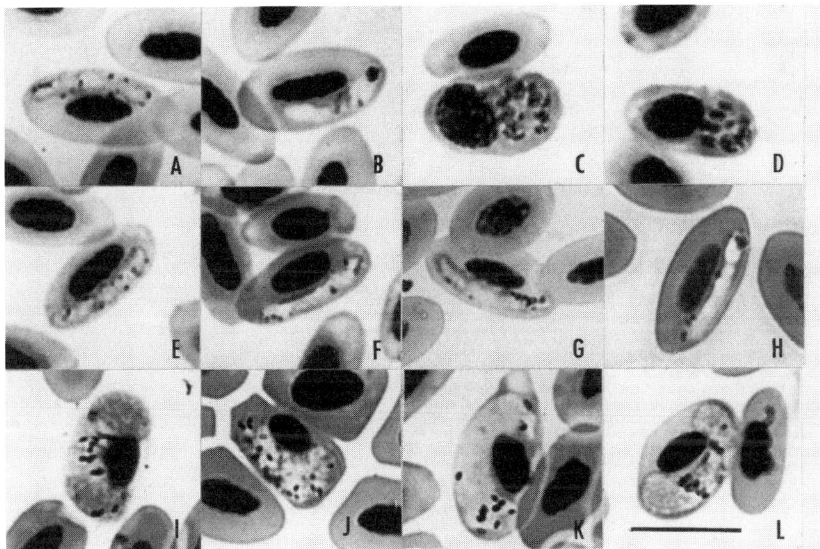

FIGURE 14.3. Blood parasites from hawks in Florida. *A*, macrogametocyte of *Plasmodium forresteri* from a Broad-winged Hawk; *B*, microgametocyte of *Plasmodium forresteri* from a Broad-winged Hawk; *C–D*, schizonts of *Plasmodium elongatum* from a Cooper's Hawk; *E*, macrogametocyte of *P. elongatum* from a Cooper's Hawk; *F*, microgametocyte of *P. elongatum* from a Cooper's Hawk; *G*, macrogametocyte of *P. elongatum* from a Red-tailed Hawk; *H*, microgametocyte of *P. elongatum* from a Red-tailed Hawk; *I*, macrogametocyte of *Haemoproteus elani* from a Broad-winged Hawk; *J*, microgametocyte of *H. elani* from a Broad-winged Hawk; *K*, macrogametocyte of *H. elani* from a Red-shouldered Hawk; *L*, microgametocyte of *H. elani* from a Red-tailed Hawk. Horizontal bar = 10 μm. Courtesy of Sam R. Telford, Jr.

protozoans "may play a significant role as concomitant infections with other diseases under certain conditions." Related to this, Olsen and Gaunt (1985) studied blood parasite infections in a series of 55 injured or orphaned raptors (including a number of hawks) that were treated at a rehabilitation center in Louisiana. They found that weakened or injured raptors infected with

Table 14.11. Protozoans identified in the peripheral blood of Cooper's Hawks in Florida

| | | | No. positive | |
| | | | --- | --- |
County	Year(s)	No. exam.	*Leucocytozoon toddi*	*Plasmodium elongatum*
Alachua	1979–95	7	0	0
Collier	1990	1	1	1
Martin	1996	1	0	0
Volusia	1996	1	0	0
All	1979–96	10	1	1

Source: Telford et al. (1996).

Table 14.12. Protozoans identified in the peripheral blood of Red-shouldered Hawks in Florida

County	Year(s)	No. exam.	Number positive[a]				
			H.e.	L.t.	P.e.	P.f.	P.r.
Alachua	1974–91	12	6[b]	1	0	0	1
	1993–95	10	3	0	2	0	0
Bradford	1990, 94	2	0	0	1	0	0
Citrus	1991	1	0	0	0	0	0
Clay	1977	1	0	0	0	0	0
Collier	1990	3	2	0	0	0	0
Lake	1974, 95	2	0	0	0	0	0
Levy	1995	2	1	0	0	0	0
Marion	1990, 95	3	1	0	0	0	0
Martin	1995	2	2	0	0	1[c]	0
Orange	1996	1	1	0	0	0	0
Putnam	1995	1	0	0	0	0	0
Seminole	1996	1	0	0	0	0	0
Sumter	1996	1	0	0	0	0	0
Union	1977	1	1	0	0	0	0
Volusia	1995–96	2	2	0	0	0	0
Unknown	1995–96	2	1	0	0	0	0
All	1974–96	47	20	1	3	1	1

Source: Telford et al. (1996).
a. H.e. = *Haemoproteus elani*, L.t. = *Leucocytozoon toddi*, P.e. = *Plasmodium elongatum*, P.f. = *Plasmodium forresteri*, P.r. = *Plasmodium relictum*. All infections were diagnosed by examination of smears of peripheral blood, unless otherwise indicated.
b. Two of these infections are also mentioned by Peirce et al. (1990).
c. Determined by isodiagnosis (Telford et al. 1997).

blood parasites had higher mortality or longer rehabilitation times than similarly injured, but uninfected birds. The taxonomy of the haemoproteids of Falconiformes has been reviewed recently by Peirce et al. (1990).

Infections of *Sarcocystis* sp. were recorded in 2 hawks from Alachua County in 1994 (table 14.16). Sarcocysts were observed in skeletal muscles of an adult female Red-shouldered Hawk that died of emaciation and trauma and in skeletal and heart muscles of an adult male Red-tailed Hawk that died of emaciation of un-

Table 14.13. Protozoans identified in the peripheral blood of Broad-winged Hawks in Florida

County	Year(s)	No. exam.	No. positive	
			Haemoproteus elani	*Plasmodium forresteri*
Collier	1990–91	2	1	1
Levy	1989	1	1	0
All	1989–91	3	2	1

Source: Telford et al. (1996).

Table 14.14. Protozoans identified in the peripheral blood of Red-tailed Hawks in Florida

County	Year(s)	No. exam.	No. positive[a]				
			H.e.	H.n.	L.t.	P.e.	P.f.
Alachua	1976–91	6	3	1	0	2	0
	1994–96	7	2	0	0	2[b]	1[c]
Bradford	1989	1	0	0	0	0	0
Brevard	1996	3	2	0	0	0	0
Columbia	1988	1	0	0	0	0	0
Hendry	1990	1	0	0	0	0	0
Hillsborough	1995	1	0	0	0	0	0
Levy	1988	2	1	0	0	1	0
Marion	1988,89	3	2	0	1	1	0
Martin	1995–96	3	3	0	0	2[d]	2
Orange	1995–96	3	1	0	0	0	0
Osceola	1996	2	0	0	0	0	0
Putnam	1996	1	1	0	0	0	0
Seminole	1995	3	1	0	0	0	0
Unknown	1994–96	7	3	0	1	2	0
All	1976–96	44	19	1	2	10	3

Source: Telford et al. (1996).
a. H.e. = *Haemoproteus elani*, H.n. = *Haemoproteus nisi*, L.t. = *Leucocytozoon toddi*, P.e. = *Plasmodium elongatum*, P.f. = *Plasmodium forresteri*. All infections were diagnosed by examination of smears of peripheral blood, unless otherwise indicated.
b. One of these birds was negative by blood smear, but positive by isodiagnosis (Telford et al. 1997).
c. Determined by isodiagnosis (Telford et al. 1997).
d. Both birds were negative by blood smear, but positive by isodiagnosis (Telford et al. 1997).

known cause (Sileo 1994). The species of *Sarcocystis* and its life cycle and pathogenicity are unknown.

Oocysts of an unknown species of coccidia were observed in the villi of the intestines of an immature male Cooper's Hawk from Alachua County in 1993 (Locke 1993). The cause of death of this hawk was not determined and the coccidial infection was probably incidental. A number of species of intestinal coccidia have been reported from raptors but little is known about their pathogenicity, even though there are suggestions that they may have been involved in causing morbidity and mortality (Keymer 1972; Pellerdy 1974; Ward 1986).

Trichomonosis was seen in a Red-tailed Hawk from Volusia County and a Red-shouldered Hawk from Sumter County (table 14.16). The Red-tailed Hawk (age unknown) was an electrocution case, euthanized at the Center for Birds of Prey shortly after being received. The Red-shouldered Hawk was a juvenile and extremely emaciated when received at the same center with a massive lesion in its throat, which was removed surgically. The hawk recovered and was released back into the wild 3 months later (Collins and Gilliland 1995).

According to Burridge et al. (1979) 1 Red-shouldered Hawk and 1 Northern Harrier (dates and localities not given) from Florida were seronegative for antibodies to *Toxoplasma gondii,* but, as mentioned in other chapters, these were tested via the IHA test, which may not be as sensitive in birds as in mammals and may result in false-negative results (Frenkel 1981). We have no other data on *T. gondii* infections in hawks in Florida, but Lindsay et al. (1993) found cysts in Red-shouldered Hawks, Red-tailed Hawks, and Cooper's Hawks in Alabama. *Toxoplasma gondii* infec-

Table 14.15. Protozoans identified in the peripheral blood of Sharp-shinned Hawks in Florida

| | | | Number positive | |
County	Year(s)	No. exam.	*Haemoproteus nisi*	*Leucocytozoon toddi*
Alachua	1978–89	9	1	1
Collier	1990	2	0	0
All	1978–90	11	1	1

Source: Telford et al. (1996).

tions probably occur in hawks in Florida as well and examination of additional samples would probably result in the demonstration of their presence. The significance of *T. gondii* infections in hawks is uncertain; Lindsay et al. (1991) observed no clinical signs or lesions in Red-shouldered Hawks experimentally infected with this protozoan. Additional research should be conducted on *T. gondii* in Florida to determine its distribution and effects on raptors and also to better understand the epidemiology of this organism of public health significance.

XIV. Helminths

The helminths of hawks in Florida have been studied to a limited extent. Information is available on 6 species, the most extensive on Red-shouldered Hawks (table 14.17) and Red-tailed Hawks (table 14.18). There are limited data on Sharp-shinned Hawks (table 14.19), Cooper's Hawks (table 14.20), and Broad-winged Hawks (table 14.21). In addition, 2 helminths (1 specimen of the nematode *Procyrnea* sp. and "numerous" specimens of the cestode *Cladotaenia* sp.) were reported from a Northern Harrier from Seminole County in 1982 (Tuggle 1983). Thirty-six species of helminths have been recorded and include 12 trematodes, 1 cestode, 22 nematodes, and 1 acanthocephalan. Nothing is known about the helminths of the other 4 species of hawks that occur in Florida. The strigeid trematodes were the most common and abundant helminths. Two of the species, *Neodiplostomum americanum* and *N. attenuatum*, occurred in all 5 of the buteos and are probably characteristic helminths of these hawks. The pathologic significance of the

Table 14.16. Protozoan infections (other than blood protozoans) reported from hawks in Florida

| Species of hawk County[a] | Year | No. exam. | Number positive | | | Data source |
			Intestinal coccidia[a]	Sarcocysts in muscle[a]	Trichomonosis[b]	
Cooper's Hawk						
Alachua	1993	1	1	NG	NG	Locke (1993)
Red-shouldered Hawk						
Alachua	1994	1	NG	1	NG	Sileo (1994)
Sumter	1994	1	NG	NG	1	Collins & Gilliland (1995)
Alachua	1995	1	NG	1	NG	Fischer (1995)
Red-tailed Hawk						
Alachua	1994	1	NG	1	NG	Sileo (1994)
Volusia	1994	1	NG	NG	1	Collins & Gilliland (1995)

NG = not given by author.
a. Determined by histopathological technique.
b. Determined by examination for characteristic gross lesions (= presumptive diagnosis).

Table 14.17. Helminth infections in Red-shouldered Hawks from Florida

| Helminth (Site)[a] | | No. hawks[b] | | Intensity | |
County or site	Year(s)	Exam.	Inf.	Mean	Range
Trematoda					
Baschkirovitrema incrassatum (SI)					
Alachua	1974–79	4	1	15	—
Brachylecithum rarum (LV)					
Bradford	1979	1	1	18	—
Columbia	1985	1	1	3	—
Hillsborough	1976	1	1	1	—
Pasco	1977	1	1	3	—
Taylor	1976	2	1	8	—
Microparyphium facetum (CL)					
Columbia	1985	1	1	1	—
Gadsden	1981	1	1	2	—
Neodiplostomum americanum (SI)					
Columbia	1985	1	1	35	—
Lake	1974	2	1	234	—
Pinellas	1974–76	2	1	210	—
"No. central Fla."[c]	1986	1	1	ND	—
Neodiplostomum attenuatum (SI)					
Alachua	1974–79	4	2	328	6–650
Hillsborough	1976	1	1	88	—
Lake	1974	2	2	248	45–450
Pasco	1977	1	1	1,022	—
Pinellas	1974–76	2	2	222	14–429
Taylor	1976–77	2	1	299	—
Union	1977	1	1	36	—
"No. central Fla."[c]	1986	1	1	ND	—
Ophiosoma microcephalum (SI)					
Alachua	1974–79	4	1	461	—
Hillsborough	1976	1	1	98	—
Lake	1974	2	2	8	3–14
Pinellas	1974–76	2	1	82	—
Union	1977	1	1	2	—
Parastrigea tulipoides (SI)					
Alachua	1974–79	4	2	3	2–3
Hillsborough	1976	1	1	15	—
Union	1977	1	1	8	—
Platynosomum illiciens (LV)					
Pinellas	1974–76	2	1	7	—
Nematoda					
Capillaria contorta (ES)					
Bradford	1979	1	1	1	—
Marion	1977	1	1	1	—
Taylor	1976	2	2	2	1–2
Capillaria falconis (SI)					
Alachua	1974–79	4	1	1	—
Columbia	1985	1	1	1	—

(continued)

Table 14.17. *(continued)*

| Helminth (Site)[a] | | No. hawks[b] | | Intensity | |
County or site	Year(s)	Exam.	Inf.	Mean	Range
Hillsborough	1976	1	1	1	—
Lake	1974	2	1	7	—
Pasco	1977	1	1	1	—
Taylor	1976–77	2	2	2	2
Union	1977	1	1	1	—
Gnathostoma sp. (larvae) (ES)					
Columbia	1985	1	1	1	—
Physaloptera sp. (PR)					
Taylor	1976	2	1	2	—
Union	1977	1	1	1	—
Porrocaecum sp. (SI)					
Alachua[d]	1994	1	1	NG	—
Procyrnea mansioni (ES,PR)					
Alachua	1974–79	4	1	2	—
Lake	1974	2	1	7	—
Marion	1977	1	1	1	—
Pinellas	1974–76	2	1	1	—
Taylor	1976–77	2	1	1	—
Union	1977	1	1	1	—
Spirurid larvae (ES,PR)					
Alachua	1974–79	4	2	5	1–8
Columbia	1985	1	1	2	—
Lake	1974	2	2	2	1–3
Marion	1977	1	1	4	—
Pasco	1977	1	1	1	—
Pinellas	1974–76	2	1	1	—
Union	1977	1	1	1	—
Synhimantus hamatus (PR)					
Alachua	1974–79	4	3	16	3–40
Gadsden	1981	1	1	1	—
Hillsborough	1976	1	1	35	—
Lake	1974	2	1	4	—
Pinellas	1974–76	2	1	1	—
Taylor	1976	2	2	2	1–3
Union	1977	1	1	2	—
Synhimantus sp. (PR)					
Hillsborough	1976	1	1	1	—
Tetrameres (*Microtetrameres*) sp.[e] (PR)					
Columbia	1985	1	1	3	—
Tetrameres (*Tetrameres*) sp. (PR)					
Alachua	1974–79	4	1	1	—
Columbia	1985	1	1	1	—
Hillsborough	1976	1	1	2	—
Lake	1974	2	2	4	1–7
Pinellas	1974–76	2	1	20	—

(continued)

Table 14.17. *(continued)*

| Helminth (Site)[a] | | No. hawks[b] | | Intensity | |
		Exam.	Inf.	Mean	Range
County or site	Year(s)				
Acanthocephala					
Centrorhynchus kuntzi (SI)					
Alachua	1974–79	4	2	9	3–14
Columbia	1985	1	1	5	—
Glades	1978	1	1	9	—
Hillsborough	1976	1	1	17	—
Lake	1974	2	1	4	—
Marion	1977	1	1	2	—
Pasco	1977	1	1	1	—
Pinellas	1974–76	2	2	4	2–7
Taylor	1976	2	2	20	19–21
Union	1977	1	1	15	—
"No. central Fla."[c]	1986	1	1	ND	—
Acanthocephalan larvae (SI)					
Taylor	1976	2	1	1	—

Sources: All data are from Kinsella et al. (1998), except as otherwise noted.
ND = not determined.
a. Infection sites: SI = small intestine, LV = liver, CL = cloaca, ES = esophagus, PR = proventriculus.
b. The data from Kinsella et al. (1998) consist of results from 18 Red-shouldered Hawks from 12 counties: Alachua (1974–79, *n* = 4), Bradford (1979, *n* = 1), Columbia (1985, *n* –1), Gadsden (1981, *n* = 1), Glades (1978, *n* = 1), Hillsborough (1976, *n* = 1), Lake (1974, *n* = 2), Marion (1977, *n* = 1), Pasco (1977, *n* = 1), Pinellas (1974–76, *n* = 2), Taylor (1976–77, *n* = 2), and Union (1977, *n* = 1). Unless indicated as positive under each helminth for a given county in the table, hawks from other counties were negative.
c. From Kinsella and Greiner (1993).
d. From Cole (1994).
e. An undescribed species; also found in a Cooper's Hawk.

helminths of hawks in Florida has not been studied, although the proventricular nematodes (*Tetrameres* spp.) and the intestinal acanthocephalan (*Centrorhynchus kuntzi*) may be harmful, especially to juvenile birds.

XV. Arthropods

Fifteen species of parasitic arthropods have been reported from hawks in Florida, including 3 feather mites, 1 flea, 10 chewing lice, and 1 louse fly. Data are available on 7 species of hawks. The most complete information is on Red-tailed Hawks (table 14.22) and on Red-shouldered Hawks (table 14.23), probably because both are widely distributed and fairly common throughout the state. The finding of an infestation of sticktight fleas (*Echidnophaga gallinacea*) on a Red-tailed Hawk from Seminole County in 1988 is of interest. The infestation by more than 50 fleas was probably contracted from another bird such as a domestic chicken that the hawk had eaten. Chickens are considered to be the normal hosts of this flea, although there are reports of infestations in other birds in Florida such as bobwhites, ducks, and American Kestrels (Layne 1971). Data on each of the other 5 species are limited to a single chewing louse record (table 14.24). Nothing is known about the significance of infestations by these arthropods on the health and welfare of hawks in Florida. The louse fly (*Icosta americana*) may be a vector of the blood protozoans *Haemoproteus elani* and *H. nisi* in Red-tailed Hawks, but this has not been

Table 14.18. Helminth infections in Red-tailed Hawks from Florida

| Helminth (Site)[a] | | No. hawks[b] | | Intensity | |
County or site	Year(s)	Exam.	Inf.	Mean	Range
Trematoda					
Ascocotyle sp. (SI)					
Citrus	1996	1	1	11	—
Echinoparyphium sp. (SI)					
Union	1975	1	1	1	—
Neodiplostomum americanum (SI)					
Alachua	1974–80	8	2	4	1–7
Citrus	1996	1	1	94	—
Union	1975	1	1	4	—
"No. central Fla."[c]	1981–87	8	4	ND	—
Neodiplostomum attenuatum (SI)					
Alachua	1974–80	8	4	90	22–228
Citrus	1996	1	1	1,299	—
Union	1975	1	1	76	—
"No. central Fla."[c]	1981–87	8	4	ND	—
Ophiosoma microcephalum (SI)					
Alachua	1974–80	8	1	6	—
"No. central Fla."[c]	1981–87	8	1	ND	—
Parastrigea tulipoides (SI)					
Citrus	1996	1	1	10	—
Phagicola longa (SI)					
Citrus	1996	1	1	103	—
Phagicola sp. (SI)					
Citrus	1996	1	1	6	—
Strigea falconis (SI)					
Alachua	1974–80	8	5	584	18–1,519
Citrus	1996	1	1	36	—
Union	1975	1	1	100	—
"No. central Fla."[c]	1981–87	8	5	ND	—
Nematoda					
Avioserpens galliardi (AS)					
Citrus	1996	1	1	1	—
Capillaria contorta (ES)					
Alachua	1974–80	8	2	3	1–4
Citrus	1996	1	1	2	—
Capillaria falconis (SI)					
Alachua	1974–80	8	4	5	1–12
Citrus	1996	1	1	2	—
"No. central Fla."[c]	1981–87	8	1	ND	—
Chandleronema longigutturata (PR)					
Citrus	1996	1	1	2	—
Cyathostoma americana (TR)					
Alachua	1974–80	8	1	1	—
Desportesius invaginatus (ES)					
Alachua	1974–80	8	3	3	1–5
Filarid (microfilariae) (LU)					
Alachua[d]	1994	1	1	ND	—

(continued)

Table 14.18. *(continued)*

| Helminth (Site)[a] | | No. hawks[b] | | Intensity | |
County or site	Year(s)	Exam.	Inf.	Mean	Range
Physaloptera buteonis (PR)					
"No. central Fla."[c]	1981–87	8	1	ND	—
Physaloptera sp. (PR)					
Alachua	1974–80	8	1	2	—
Citrus	1996	1	1	1	
Physocephalus sexalatus[e] (IN)					
"No. Florida"	NG	NG	NG	NG	—
Porrocaecum angusticolle (SI)					
Citrus	1996	1	1	4	—
"No. central Fla."[c]	1987	1	1	ND	—
Porrocaecum depressum (SI)					
Alachua	1974–80	8	3	11	1–30
Union	1975	1	1	2	—
"No. central Fla."[c]	1981–86	8	3	ND	—
Procyrnea mansioni (ES,PR)					
Alachua	1974–80	8	3	1	1–2
Citrus	1996	1	1	5	—
"No. central Fla."[c]	1981–87	8	2	ND	—
Spirurid larvae (ES,PR)					
Alachua	1974–80	8	2	26	1–50
Polk	1974	1	1	1	—
Synhimantus hamatus (PR)					
Alachua	1974–80	8	4	1	—
Citrus	1996	1	1	39	—
Union	1975	1	1	2	—
"No. central Fla."[c]	1981–86	8	1	ND	—
Synhimantus laticeps (PR)					
"No. central Fla."[c]	1981–86	8	1	ND	—
Tetrameres accipiter (PR)					
Alachua	1974–80	8	3	37	3–103
"No. central Fla."[c]	1981–87	8	2	ND	—
Acanthocephala					
Andracantha sp. (SI)					
Citrus	1996	1	1	1	—
Centrorhynchus kuntzi					
Alachua	1974–80	8	3	16	9–28
Citrus	1996	1	1	210	—
"No. central Fla."[c]	1981–87	8	5	ND	—

Sources: All data are from Kinsella et al. (1998), except as otherwise noted.
ND = not determined, NG = not given.
a. Infection sites: AS = air sacs, IN = intestine, SI = small intestine, ES = esophagus, TR = trachea, LU = lungs, PR = proventriculus.
b. The data from Kinsella et al. (1998) consisted of results from 13 Red-tailed Hawks from 6 counties: Alachua (1974–80, n = 8), Bradford (1979, n = 1), Citrus (1996, n = 1), Levy (1976, n = 1), Polk (1974, n = 1), and Union (1975, n = 1). Unless indicated as positive under each helminth for a given county in the table, hawks from other counties were negative.
c. From Kinsella and Greiner (1993).
d. From Sileo (1994); microfilariae were seen in histologic sections of the lung; adults were not located.
e. From Cram (1930).

Table 14.19. Helminth infections in Sharp-shinned Hawks from Florida

| Helminth (Site)[a] | | | | | |
| County or site | Year(s) | No. hawks[b] | | Intensity | |
		Exam.	Inf.	Mean	Range
Trematoda					
Neodiplostomum americanum (SI)					
Alachua	1977	2	2	172	52–291
Gilchrist	1979	1	1	27	—
Polk	1973	1	1	1	—
Neodiplostomum attenuatum (SI)					
Alachua	1977	2	1	15	—
Gilchrist	1979	1	1	2	—
Neodiplostomum pearsoni (SI)					
Gilchrist	1979	1	1	8	—
Parastrigea campanula (SI)					
Gilchrist	1979	1	1	15	—
Strigea falconis (SI)					
Gilchrist	1979	1	1	2	—
Nematoda					
Capillaria sp. (SI,LI)					
Alachua	1977	2	1	2	—
Gilchrist	1979	1	1	2	—
Physaloptera sp. (PR)					
Pinellas	1974	1	1	1	—
Porrocaecum angusticolle (SI)					
"No. central Fla."[c]	1987	1	1	ND	—
Procyrnea mansioni (ES,PR)					
Alachua	1977	1	1	1	—
Polk	1973	1	1	1	—
Spirurid (unidentified) (PR)					
Orange[d]	1992	1	1	ND	—
Tetrameres accipiter (PR)					
Alachua	1977	2	1	3	—
Gilchrist	1979	1	1	1	—

Sources: All data are from Kinsella et al. (1998), except as otherwise noted.
ND = not determined.
a. Infection sites: SI = small intestine, LI = large intestine, PR = proventriculus, ES = esophagus.
b. The data from Kinsella et al. (1998) consisted of results from 5 Sharp-shinned Hawks from 4 counties: Alachua (1977, *n* = 2), Gilchrist (1979, *n* = 1), Pinellas (1974, *n* = 1), and Polk (1973, *n* = 1). Unless indicated as positive under each helminth for a given county in the table, hawks from other counties were negative.
c. From Kinsella and Greiner (1993).
d. From Franson (1993).

Table 14.20. Helminth infections in 4 Cooper's Hawks from Florida

Helminth (Site)[a]	Number of helminths collected[d]			
	Alachua[b] (1979)	Alachua[c] (1993)	Citrus[b] (1997)	Duval[b] (1974)
Trematoda				
Neodiplostomum americanum (SI)	93	0	58	0
Neodiplostomum attenuatum (SI)	2	0	80	6
Mesoophorodiplostomum sp. (SI)	0	+	0	0
Ophiosoma microcephalum (SI)	0	0	0	9
Strigea falconis (SI)	30	0	4	0
Nematoda				
Capillaria sp. (SI,LI)	2	0	0	1
Cardiofilaria pavlovskyi (BC)	0	0	0	1
Cyathostoma americana (TR)	1	0	0	0
Porrocaecum depressum (SI)	0	0	0	2
Spirurid (unidentified) (ES,PR)	0	+	0	0
Spirurid larvae (ES,PR)	1	0	0	0
Synhimantus hamatus (PR)	3	0	0	3
Tetrameres (*Microtetrameres*) sp.[e] (PR)	0	0	0	1

a. Infection sites: SI = small intestine, LI = large intestine, BC = body cavity, TR = trachea, ES = esophagus, PR = proventriculus.
b. From Kinsella et al. (1998).
c. From Cole (1993).
d. + = helminth was present, but numbers were not determined.
e. An undescribed species; also found in a Red-shouldered Hawk.

Table 14.21. Helminth infections in Broad-winged Hawks from Florida

Helminth (Site)[a]	Number of helminths collected[b]
Trematoda	
Neodiplostomum americanum (SI)	1
Neodiplostomum attenuatum (SI)	165
Neodiplostomum pearsoni (SI)	5
Strigea falconis (SI)	16
Cestoda	
Cladotaenia globifera (SI)	2
Nematoda	
Desportesius invaginatus (ES)	4
Porrocaecum angusticolle (SI)	23
Procyrnea mansioni (ES,PR)	1
Synhimantus hamatus (PR)	12
Physaloptera acuticauda (PR)	1
Acanthocephala	
Centrorhynchus kuntzi (SI)	11

a. Infection sites: SI = small intestine, ES = esophagus, PR = proventriculus.
b. All helminths (except *Physaloptera acuticauda*) came from a Broad-winged Hawk from St. Lucie County, 1978 (Kinsella et al. 1998). The specimen of *P. acuticauda* came from a Broad-winged Hawk from Hawkinsville, Florida, 1885 (Walton 1927; Morgan 1943).

Table 14.22. Parasitic arthropod infestations on Red-tailed Hawks from Florida

Arthropod	County	Year	Data source
Feather mites			
Pseudalloptinus sp.	Alachua	1976	Forrester & Mertins (1992)
	Alachua	1980	Ibid.
Pterolichoidea	Alachua	1978	Ibid.
(species not determined)			
Chewing lice			
Colpocephalum napiforme	Dade	1939	Forrester et al. (1995)
Craspedorrhynchus americanus	Dade	1916	Ibid.
	St. Johns	1936	Ibid.
Degeeriella fulva	Alachua	1976	Ibid.
	Orange	1992	Ibid.
	Seminole	1988	Ibid.
Kurodaia fulvofasciata	Dade	1916	Ibid.
Louse flies			
Icosta americana	Alachua	1975	Wilson & Forrester (1990)
	Leon	1970	Wilson et al. (1990)
Fleas			
Echidnophaga gallinacea	Seminole	1988	Kinsella & Kale (1994)

Table 14.23. Parasitic arthropod infestations on Red-shouldered Hawks from Florida

Arthropod	County	Year(s)	Data source
Ticks[a]			
Feather mites			
Dermonoton sp.	Columbia	1985	Forrester & Mertins (1992)
Chewing lice			
Colpocephalum napiforme	Columbia	1985	Forrester et al. (1995)
Craspedorrhynchus buteonis	Collier	1956	Ibid.
	Lee	1983	Ibid.
	Pinellas	1929	Ibid.
	Taylor	1977	Ibid.
Degeeriella fulva	Bradford	1979	Ibid.
	Lee	1982–85	Ibid.
	Monroe	1979	Ibid.
	Monroe	NG	Ibid.
	Taylor	1976–77	Ibid.
Kurodaia fulvofasciata	Collier	1956	Ibid.

a. Travis (1941) examined 2 Red-shouldered Hawks from Florida (Orange, Osceola, or Collier County) in 1936–37; neither was infested with ticks.

Table 14.24. Infestations of chewing lice on Hawks (other than Red-shouldered and Red-tailed) from Florida

Species of hawk Chewing louse	County	Year	Data source
Northern Harrier			
Degeeriella fusca	Dade	1918	Forrester et al. (1995)
Cooper's Hawk			
Craspedorrhynchus subhaematopus	Leon	1925	Emerson (1960)
Northern Goshawk			
Colpocephalum nanum	Dade	1918	Forrester et al. (1995)
Broad-winged Hawk			
Laemobothrion maximum	Monroe	1987	Ibid.
Short-tailed Hawk			
Craspedorrhynchus sp.	Dade	1956	Ibid.

determined. Infestations of black flies have been reported to cause mortality in nestling Red-tailed Hawks in Wyoming (Smith et al. 1998), but this has not been seen in Florida, even though there are a large number of species of black flies found throughout the state (Pinkovsky and Butler 1978).

XVI. Summary and conclusions

Data are available on 8 of the 10 species of hawks that occur in Florida. Nothing is known about Swainson's Hawks or Ferruginous Hawks. The most complete information is on Red-shouldered Hawks, Red-tailed Hawks, and Cooper's Hawks. Ninety-one different conditions, disease agents, and parasites have been found as proven or potential morbidity/mortality factors of hawks in Florida. These include trauma (6), electrocution (1), toxicosis (2), chlorinated hydrocarbons (8), heavy metals (2), viruses (4), bacteria (8), fungi (1), protozoans (9), trematodes (12), cestodes (1), nematodes (22), acanthocephalans (1), feather mites (3), fleas (1), chewing lice (10), and louse flies (1). Trauma, especially that caused by collisions with motor vehicles, is probably the most important mortality factor for hawks in Florida.

XVII. Literature cited

Adkisson, C.S. 1990. Accipiters. *Proc. Southeast Raptor Manag. Symp. Workshop.* Natl. Wildl. Fed. Sci. Tech. Ser. 14:63–69.

Albers, H.F. 1984. Naturally occurring immunity to rabies in raptors. *Proc. Annu. Natl. Wildl. Rehab. Symp.* 2:171–174.

Anderson, D.W., and J.J. Hickey. 1972. Eggshell changes in certain North American birds. *Proc. Int. Ornithol. Congr.* 15:514–540.

Bednarz, J.C. 1990. Migration counts of raptors at Hawk Mountain, Pennsylvania, as indicators of population trends, 1934–1986. *Auk* 107:96–109.

Bennett, G.F., M.A. Peirce, and R.W. Ashford. 1993. Avian Haematozoa: mortality and pathogenicity. *J. Nat. Hist.* 27:993–1001.

Bildstein, K.L., and M.W. Collopy. 1990. Northern harrier. *Proc. Southeast Raptor Manag. Symp. Workshop.* Natl. Wildl. Fed. Sci. Tech. Ser. 14:70–77.

Brown, L., and D. Amadon. 1968. *Eagles, hawks, and falcons of the world.* McGraw-Hill, New York. 945 pp.

Brugger, K.E. 1989. Red-tailed hawk dies with coral snake in talons. *Copeia* 1989:508–510.

Burridge, M.J., W.J. Bigler, D.J. Forrester, and J.M. Hennemann. 1979. Serologic survey for

Toxoplasma gondii in wild animals in Florida. *J. Am. Vet. Med. Assoc.* 175:964–967.

Cole, R.A. 1993–94. Unpublished data. National Wildlife Health Center, Madison, Wis.

Collins, R. 1999. Unpublished data. Florida Audubon Society, Center for Birds of Prey, Maitland.

Collins, R., and C. Gilliland. 1995. Unpublished data. Florida Audubon Society, Center for Birds of Prey, Maitland.

Conti, L., R. Oliveri, and C. Blackmore. 2002. Unpublished data. Florida Department of Health, Tallahassee.

Cooper, J.E. 1985. *Veterinary aspects of captive birds of prey.* 2d ed. Standfast, Gloucestershire, England. 256+31 pp.

Cooper, J.E., and A.G. Greenwood (eds.). 1981. *Recent advances in the study of raptor diseases. Proceedings of the International Symposium on Diseases of Birds of Prey.* Chiron, West Yorkshire, England. 176 pp.

Couvillion, C.E. 1982. Unpublished data. Southeastern Cooperative Wildlife Disease Study, University of Georgia, Athens.

Cox, J. 1991. Field observations. Fall report: September–November 1990. *Fla. Field Nat.* 19:58–64.

Cram, E.B. 1930. Aberrant larvae of *Physocephalus sexalatus* in birds. *J. Parasitol.* 17:56.

Crawford, R.L. 1981. Bird casualties at a Leon County, Florida TV tower: a 25-year migration study. *Bull. Tall Timbers Res. Stn.* 22:1–30.

Davidson, W.R. 1992. Unpublished data. Southeastern Cooperative Wildlife Disease Study, University of Georgia, Athens.

Deem, S.L., and S.P. Terrell. 1996. Unpublished data. University of Florida, Gainesville.

Deem, S.L., S.P. Terrell, and D.J. Forrester. 1998. A retrospective study of morbidity and mortality of raptors in Florida: 1988–1994. *J. Zoo Wildl. Med.* 29:160–164.

Emerson, K.C. 1960. Two new species of *Craspedorrhynchus* (Mallophaga) from North America. *Proc. Biol. Soc. Wash.* 73:39–44.

Fischer, J.R. 1995. Unpublished data. Southeastern Cooperative Wildlife Disease Study, University of Georgia, Athens.

Forrester, D.J. 1994. Unpublished data. University of Florida, Gainesville.

Forrester, D.J., H.W. Kale II, R.D. Price, K.C. Emerson, and G.W. Foster. 1995. Chewing lice (Mallophaga) from birds in Florida: a listing by host. *Bull. Fla. Mus. Nat. Hist.* 39:1–44.

Forrester, D.J., and J.W. Mertins. 1992. Unpublished data. University of Florida, Gainesville.

Forrester, D.J., S.R. Telford, Jr., G.W. Foster, and G.F. Bennett. 1994. Blood parasites of raptors in Florida. *J. Raptor Res.* 28:226–231.

Fowler, M.E. (ed.). 1986. *Zoo and wild animal medicine.* 2d ed. W.B. Saunders, Philadelphia. 1,127 pp.

———(ed.). 1993. *Zoo and wild animal medicine: current therapy.* 3d ed. W.B. Saunders, Philadelphia. 617 pp.

Franson, J.C. 1993–94. Unpublished data. National Wildlife Health Center, Madison, Wis.

Franson, J.C., N.J. Thomas, M.R. Smith, A.H. Robbins, S. Newman, and P.C. McCartin. 1996. A retrospective study of postmortem findings in red-tailed hawks. *J. Raptor Res.* 30:7–14.

Frenkel, J.K. 1981. False-negative serologic tests for *Toxoplasma* in birds. *J. Parasitol.* 67:952–953.

Garnham, P.C.C. 1966. *Malaria parasites and other Haemosporidia.* Blackwell Scientific, Oxford. 1,114 pp.

Graham, D.L., and W.H. Halliwell. 1986. Viral diseases of birds of prey. In: *Zoo and wild animal medicine.* 2d ed. M.E. Fowler (ed.). W.B. Saunders, Philadelphia. pp. 408–413.

Heckel, J.-O., D.C. Sisson, and C.F. Quist. 1994. Apparent fatal snakebite in three hawks. *J. Wildl. Dis.* 30: 616–619.

Henny, C.J., and H.M. Wight. 1972. Population ecology and environmental pollution: red-tailed and Cooper's hawks. In: *Population ecology of migratory birds.* U.S. Fish and Wildlife Service, Wildlife Research Report 2:229–250.

Howerth, E.W. 1985. Unpublished data. Southeastern Cooperative Wildlife Disease Study, University of Georgia, Athens.

Jennings, W.L., W.G. Winkler, D.D. Stamm, P.H. Coleman, and A.L. Lewis. 1969. Serologic studies of possible avian or mammalian reservoirs of St. Louis encephalitis virus in Florida. *Fla. State Board Health Monogr. Ser.* 12:118–125.

Johnston, D.W. 1976. Organochlorine pesticide residues in uropygial glands and adipose tissue of wild birds. *Bull. Environ. Contam. Toxicol.* 16: 149–155.

———. 1978. Organochlorine pesticide residues in Florida birds of prey, 1969–76. *Pestic. Monit. J.* 12:8–15.

Kale, H.W. II. 1971. Florida region. *Am. Birds* 25:723–733.

Karstad, L. 1971. Arboviruses. In: *Infectious and parasitic diseases of wild birds.* J.W. Davis, R.C. Anderson, L. Karstad, and D.O. Trainer (eds.). Iowa State University Press, Ames. pp. 17–21.

Keymer, I.F. 1972. Diseases of birds of prey. *Vet. Rec.* 90:579–594.

Kingston, N., J.D. Remple, W. Burnham, R.W. Stable, and R.B. McGhee. 1976. Malaria in a captively-produced F1 gyrfalcon and in two F1 peregrine falcons. *J. Wildl. Dis.* 12:562–565.

Kinsella, J.M., G.W. Foster, and D.J. Forrester. 1995. Parasitic helminths of six species of hawks and falcons in Florida. *J. Raptor Res.* 29:117–122.

———. 1998. Unpublished data. University of Florida, Gainesville.

Kinsella, J.M., and E.C. Greiner. 1993. Unpublished data. University of Florida, Gainesville.

Kinsella, J.M., and H.W. Kale II. 1994. Unpublished data, University of Florida, Gainesville.

Krynitsky, A.J. 1984. Unpublished data. Patuxent Wildlife Research Center, Laurel, Md.

Langridge, H.P. 1993. Florida region. *Am. Birds.* 47:406–408.

Layne, J.N. 1971. Fleas (Siphonaptera) of Florida. *Fla. Entomol.* 54:35–51.

Lindsay, D.S., J.P. Dubey, and B.L. Blagburn. 1991. *Toxoplasma gondii* infections in red-tailed hawks inoculated orally with tissue cysts. *J. Parasitol.* 77:322–325.

Lindsay, D.S., P.C. Smith, F.J. Hoerr, and B.L. Blagburn. 1993. Prevalence of encysted *Toxoplasma gondii* in raptors from Alabama. *J. Parasitol.* 79:870–873.

Locke, L.N. 1993. Unpublished data. National Wildlife Health Research Center, Madison, Wis.

Logan, T.H. 1997. Florida's endangered species, threatened species and species of special concern. Official lists. Florida Game and Fresh Water Fish Commission, Tallahassee. 14 pp.

Maehr, D.S., and J.Q. Smith. 1988. Bird casualties at a central Florida power plant: 1982–1986. *Fla. Field Nat.* 16:57–80.

Maehr, D.S., A.G. Spratt, and D.K. Voigts. 1983. Bird casualties at a central Florida power plant. *Fla. Field Nat.* 11:45–49.

McEwan, L.C., and D.H. Hirth. 1980. Food habits of the bald eagle in north-central Florida. *Condor* 82:229–231.

Miller, D., E.L. Boeker, and R.S. Thorsell. 1975. Suggested practices for raptor protection on powerlines. Raptor Research Foundation, Provo, Utah. 19 pp.

Millsap, B.A., M. Robson, and D.E. Runde. 1989. Short-tailed hawk surveys. Annual Progress Report. Nongame Wildlife Section, Florida Game and Fresh Water Fish Commission, Tallahassee. 8 pp.

Millsap, B.A., M. Robson, and B.R. Toland. 1996. Short-tailed hawk. In: *Rare and endangered biota of Florida.* Vol. 5, *Birds.* J.A. Rodgers, Jr., H.W. Kale II, and H.T. Smith (eds.). University Press of Florida, Gainesville. pp. 315–322.

Mitchell, L.C., and B.A. Millsap. 1990. Buteos and golden eagle. *Proc. Southeast Raptor Manag. Symp. Workshop.* Natl. Wildl. Fed. Sci. Tech. Ser. 14:50–62.

Morgan, B.B. 1943. The Physalopterinae (Nematoda) of Aves. *Trans. Am. Microsc. Soc.* 62:72–80.

Morishita, T.Y., P.P. Aye, and D.L. Brooks. 1997. A survey of diseases of raptorial birds. *J. Avian Med. Surg.* 11:77–92.

Mulhern, B.M., W.L. Reichel, L.N. Locke, T.G. Lamont, A. Belisle, E. Cromartie, G.E. Bagley, and R.M. Prouty. 1970. Organochlorine residues and autopsy data from bald eagles, 1966–68. *Pestic. Monit. J.* 4:141–144.

Nayar, J.K., J.W. Knight, and S.R. Telford, Jr.

1998. Vector ability of mosquitoes for isolates of *Plasmodium elongatum* from raptors in Florida. *J. Parasitol.* 84:542–546.

Ogden, J.C., W.B. Robertson, Jr., G.E. Davis, and T.W. Schmidt. 1974. Pesticides, polychlorinated biphenols and heavy metals in upper food chain levels, Everglades National Park and vicinity. South Florida Environmental Project: Ecological Report no. DI-SFEP-74-16, Everglades National Park. 27 pp.

Ohlendorf, H.M., D.M. Swineford, and L.N. Locke. 1981. Organochlorine residues and mortality of herons. *Pestic. Monit. J.* 14:125–135.

Olsen, G.H., and S.D. Gaunt. 1985. Effect of hemoprotozoal infections on rehabilitation of wild raptors. *J. Am. Vet. Med. Assoc.* 187:1204–1205.

Peirce, M.A. 1981. Current knowledge of the Haematozoa of raptors. In: *Recent advances in the study of raptor diseases. Proceedings of the International Symposium on Diseases of Birds of Prey.* J.E. Cooper and A.G. Greenwood (eds.). Chiron, West Yorkshire, England. pp. 15–19.

Peirce, M.A., G.F. Bennett, and M. Bishop. 1990. The haemoproteids of the avian order Falconiformes. *J. Nat. Hist.* 24:1091–1100.

Pellerdy, L.P. 1974. *Coccidia and coccidiosis.* Verlag Paul Parey, Berlin. 959 pp.

Pinkovsky, D.D., and J.F. Butler. 1978. Black flies of Florida. I. Geographic and seasonal distribution. *Fla. Entomol.* 61:257–267.

Ratcliffe, D.A. 1967. Decrease in eggshell weight in certain birds of prey. *Nature* 215:208–210.

Redig, P.T. 1993. Avian aspergillosis. In: *Zoo and wild animal medicine: current therapy.* 3d ed. M.E. Fowler (ed.). W.B. Saunders, Philadephia. pp. 178–181.

Reiser, M.H., and S.A. Temple. 1981. Effects of chronic lead ingestion on birds of prey. In: *Recent advances in the study of raptor diseases. Proceedings of the International Symposium on Diseases of Birds of Prey.* J.E. Cooper and A.G. Greenwood (eds.). Chiron, West Yorkshire, England. pp. 21–25.

Robertson, W.B., Jr., and D.R. Paulson. 1961. Region reports: Florida region. *Audubon Field Notes* 15:26–35.

Robertson, W.B., and G.E. Woolfenden. 1992. *Florida bird species. An annotated list.* Spec. Publ. 6, Florida Ornithological Society, Gainesville. 260 pp.

Rodgers, J.A., Jr., H.W. Kale II, and H.T. Smith (eds.). 1996. *Rare and endangered biota of Florida.* Vol. 5, *Birds.* University Press of Florida, Gainesville. 688 pp.

Sileo, L. 1993–94. Unpublished data. National Wildlife Health Center, Madison, Wis.

Smith, R.N., S.L. Cain, S.H. Anderson, J.R. Dunk, and E.S. Williams. 1998. Blackfly-induced mortality of nestling red-tailed hawks. *Auk* 115:368–375.

Snyder, B. 1994. Unpublished data. Florida Department of Environmental Protection, Tallahassee.

Spalding, M.G. 1985. Unpublished data. University of Florida, Gainesville.

Spalding, M.G., and R.G. McLean. 1994. Unpublished data. University of Florida, Gainesville.

Stevenson, J.A. 1994. Unpublished data. Florida Department of Environmental Protection, Tallahassee.

Sundlof, S.F., D.J. Forrester, N.P. Thompson, and M.W. Collopy. 1986. Residues of chlorinated hydrocarbons in tissues of raptors in Florida. *J. Wildl. Dis.* 22:71–82.

Suto, B.J. 1996. Unpublished data. Suncoast Seabird Sanctuary, Redington Beach, Fla.

Taylor, W.K., and B.H. Anderson. 1973. Nocturnal migrants killed at a central Florida TV tower, autumns 1969–1971. *Wilson Bull.* 85:42–51.

———. 1974. Nocturnal migrants killed at a central Florida TV tower, autumn 1972. *Fla. Field Nat.* 2:40–43.

Taylor, W.K., and M.A. Kershner. 1986. Migrant birds killed at the Vehicle Assembly Building (VAB), John F. Kennedy Space Center. *J. Field Ornithol.* 57:142–154.

Telford, S.R., Jr., G.W. Foster, J.K. Nayar, S.P. Terrell, and D.J. Forrester. 1996. Unpublished data. University of Florida, Gainesville.

Telford, S.R., J.K. Nayar, G.W. Foster, and J.W.

Knight. 1997. *Plasmodium forresteri* n.sp., from raptors in Florida and southern Georgia: its distinction from *Plasmodium elongatum* morphologically within and among host species and by vector susceptibility. *J. Parasitol.* 83:932–937.

Toland, B.R., and B.A. Millsap. 1996. Cooper's hawk. In: *Rare and endangered biota of Florida.* Vol. 5, *Birds.* J.A. Rodgers, Jr., H.W. Kale II, and H.T. Smith (eds.). University Press of Florida, Gainesville. pp. 475–484.

Travis, B.V. 1941. Examinations of wild animals for the cattle tick *Boophilus annulatus microplus* (Can.) in Florida. *J. Parasitol.* 27:465–467.

Tuggle, B.N. 1983. Unpublished data. National Wildlife Health Center, Madison, Wis.

Walton, A.C. 1927. A revision of the nematodes of the Leidy Collections. *Proc. Acad. Nat. Sci. Phila.* 79:49–163.

Ward, F.P. 1986. Parasites and their treatment in birds of prey In: *Zoo and wild animal medicine.* 2d ed. M.E. Fowler (ed.). W.B. Saunders, Philadelphia. pp. 425–430.

Wheeler, W.B., D.P. Jouvenaz, D.P. Wojcik, W.A. Banks, C.H. VanMiddelem, C.S. Lofgren, S. Nesbitt, L. Williams, and R. Brown. 1977. Mirex residues in nontarget organisms after application of 10-5 bait for fire ant control, northeast Florida 1972–74. *Pestic. Monit. J.* 11:146–156.

White, F.H., and D.J. Forrester. 1981. Unpublished data. University of Florida, Gainesville.

Wilson, N.A., and D.J. Forrester. 1990. Unpublished data. University of Florida, Gainesville.

Wilson, N.A., H.W. Kale II, and W.W. Baker. 1990. Unpublished data. University of Northern Iowa, Cedar Falls.

Caracaras and Falcons

I. Introduction

Four species of falcons occur in Florida (Robertson and Woolfenden 1992). Two of these, the Crested Caracara and the American Kestrel, are residents, while the Merlin and the Peregrine Falcon are transients (table 15.1). All except the Merlin are listed as threatened or endangered by the Florida Game and Fresh Water Fish Commission, the U.S. Fish and Wildlife Service (Logan 1997), or the Florida Committee on Rare and Endangered Plants and Animals (Rodgers et al. 1996).

The population of Crested Caracaras in Florida was estimated at 400–500 individuals and considered to be stable over the 20-year period 1972–91 (Layne 1996). Morrison (1996) concluded that the numbers in Florida

might be even higher than indicated by published estimates. Populations of American Kestrels, however, are declining in Florida (Smallwood 1990). In north central Florida, for example, resident kestrels decreased by 82% between 1938–40 and 1981–83 (Hoffman and Collopy 1988; Collopy 1996). Populations of wintering Merlins appear to be stable or even increasing somewhat in Florida (Smallwood 1990; Smallwood and Meyer 1996). Populations of Peregrine Falcons in continental United States were almost extirpated by the 1960s because of organochlorine contamination, but were restored during the 1970s and 1980s by the implementation of a management program involving the release of captive-reared

Table 15.1. Distribution, occurrence, and abundance of falcons in Florida[a]

Species of falcon Range	Seasonal occurrence	Relative abundance
Crested Caracara[b] (*Caracara plancus*)		
South-central Florida	Resident	400–500 individuals
American Kestrel[c] (*Falco sparverius*)		
Panhandle and peninsula south to Lee, Glades, and Martin counties	Resident	Rare to locally common
Merlin (*Falco columbarius*)		
Statewide	Spring & fall transient	Rare to locally common
	Winter transient	Rare to uncommon
Peregrine Falcon[d] (*Falco peregrinus*)		
Statewide	Spring & fall transient	Rare to locally common
	Winter transient	Very rare to uncommon

a. Modified from Robertson and Woolfenden (1992) and American Ornithologists' Union (1998).
b. Classified as threatened by the Florida Game and Fresh Water Fish Commission, the U.S. Fish and Wildlife Service (Logan 1997), and the Florida Committee on Rare and Endangered Plants and Animals (Rodgers et al. 1996).
c. Classified as threatened by the Florida Game and Fresh Water Fish Commission (Logan 1997) and the Florida Committee on Rare and Endangered Plants and Animals (Rodgers et al. 1996).
d. Classified as endangered by the Florida Game and Fresh Water Fish Commission and the Florida Committee on Rare and Endangered Plants and Animals (Rodgers et al. 1996).

birds (Meyer and Smallwood 1996). In Florida there was a significant increase in the numbers of Peregrine Falcons seen during the annual Christmas bird counts conducted by the Audubon Society from 1970 to 1988 (Meyer and Smallwood 1996).

There is some published information on various morbidity and mortality factors of falcons, especially American Kestrels and Peregrine Falcons (e.g., the books edited by Fowler in 1986 and 1993). The American Kestrel has been labeled the raptorial "white mouse" since it has been used extensively in various field and laboratory toxicological studies, including bioaccumulation, lethal toxicity, and effects on reproduction, eggshell thickness, and related enzyme systems, as well as a number of other physiological and biochemical parameters (Wiemeyer and Lincer 1987). Balgooyen (1976) and Bird (1988a) gave brief overviews of the mortality factors affecting kestrels, as did Layne (1996) for caracaras. Trainer (1969) and Ratcliffe

Table 15.2. Primary reasons for submission of falcons to 3 wildlife rehabilitation centers in Florida, 1988–95

Species of falcons	Number of cases per species of falcons					Total number of cases
	Trauma	Poisoning	Emaciation	Infectious diseases	Electrocution	
Crested Caracara	2	0	0	0	0	2
American Kestrel	68	41	5	1	1	116
Merlin	19	2	1	0	0	22
Peregrine Falcon	18	0	4	2	1	25
Total falcons	107	43	10	3	2	165
% of total cases	65	26	6	2	1	—

Sources: Collins and Gilliland (1995), 90 cases; Deem and Terrell (1996), 11 cases; Suto (1996), 64 cases.

Table 15.3. Causes of mortality among 192 Crested Caracaras from south central Florida, 1994–98

| Age | Type of mortality | Number per county | | | | |
		DeSoto	Glades	Highlands	Okeechobee	Totals
Eggs	Failed to hatch	0	2	0	1	3
	Abandoned	2	5	2	0	9
	Predation	0	3	5	0	8
	Unknown	1	17	33	13	66[a]
Nestlings	Anomaly[b]	0	0	1	0	1
	Entangled (twine)[c]	0	0	1	0	1
	Fell from nest	4	1	0	0	5
	Predation	0	1	3	3	7
	Unknown	2	3	16	1	22
Fledglings	Road kill	1	7	7	7	22
	Hit by train	0	0	1	0	1
	Struck fence	0	0	0	1	1
	Predation	1	2	0	1	4
	Abandoned	0	0	2	0	2
	Unknown	0	2	4	8[d]	14
Juveniles	Road kill	0	3	2	3	8
	Gunshot	0	1	0	2	3
	Unknown	0	1	7	3	11
Adults	Road kill	0	0	2	1	3
	Gunshot	0	0	1	0	1
Totals	All	11	48	87	44	192

Source: Morrison (1999).
a. Two additional eggs (1 from Hendry County and 1 from Indian River County) disappeared from their nests, but are not listed in the table.
b. See section VIII, Anomalies, for a discussion of this bird.
c. A 7-week-old chick was entangled in bailing twine used for nest building material. It had a broken leg and was very small for its age. It was euthanized, since it probably would not have survived.
d. It was suspected that 2 of these fledglings had been poisoned, but this was not verified.

(1980) have provided reviews of the diseases and parasites of Peregrine Falcons, but, by far, the greatest bulk of information on this bird relates to organochlorine contamination and its effect on reproduction. For detailed summaries of this topic the reader is referred to Hickey (1969), Peakall (1976), Ratcliffe (1980), and Cade et al. (1988).

Specific information on the diseases and parasites of falcons in Florida is limited. Morrison (1996) and Layne (1996) provided overviews of the causes of mortality of Crested Caracaras in Florida. Kinsella et al. (1995) published a paper on the helminths of 22 American Kestrels in Florida. Data were published on blood para-sites of 3 American Kestrels and 1 Merlin (Forrester et al. 1994) and on a large sample of Crested Caracaras (Foster et al. 1998). Most of the available data have come from studies on road-killed specimens and birds submitted to wildlife rehabilitation facilities. Important sources of specific information on falcons in Florida used in the preparation of this chapter were the files of Florida Audubon Society's Center for Birds of Prey in Maitland (provided by Resee Collins and Carol Gilliland), the files of Suncoast Seabird Sanctuary in St. Petersburg (provided by Barbara Suto), and the files of the Veterinary Medical Teaching Hospital at the University of Florida (provided by Sharon L.

Table 15.4. Types of trauma causing morbidity and mortality in falcons submitted to the Center for Birds of Prey, Maitland, Florida, 1989–94

Species of falcon	Percent of cases per species of falcon							Total no. of falcons
	Vehicle strike	Power line strike	Building strike	Gunshot	Predation	Entrapment	Unk.	
Crested Caracara	1	0	0	0	0	0	1	2
American Kestrel	7	3	2	1	1	0	35	49
Merlin	1	2	0	0	0	0	8	11
Peregrine Falcon	0	2	0	0	0	1	14	17
Total falcons	9	7	2	1	1	1	58	79
% total cases	11	9	3	1	1	1	73	—

Source: Collins and Gilliland (1995).

Deem and Scott P. Terrell). These 3 datasets comprised combined morbidity and mortality information on falcons submitted for rehabilitation purposes. Some of the data from the UF dataset have been published in summary form (Deem et al. 1998).

II. Trauma

As with other raptors, trauma is an important morbidity and mortality factor for falcons (tables 15.2 and 15.3). The types of trauma include vehicle strikes, power line strikes, building strikes, gunshot, and entrapment (table 15.4). Vehicle strikes and power line strikes are the most significant types of trauma. The geographic distribution of known records of American Kestrels, Crested Caracaras, and Merlins killed on highways in Florida from 1971 to 1997 is shown in figure 15.1.

Layne (1996) stated that there has probably been an increase in the numbers of caracaras injured or killed by vehicles on the highways, especially since there has been such an increase in the number of roads and in high-speed traffic during recent years. In addition to the data presented in tables 15.3 and 15.4, we have other records of road-killed falcons. One of these was a Crested Caracara, an adult female from Glades County in January 1992 (Spalding

1992). From 1971 to 1979, deaths of 3 of 21 (14%) kestrels submitted for cause-of-death determinations were attributed to road-kill mortality (Forrester 1971–79). These birds originated from northern (n = 20) and southern (n = 1) Florida; 2 of the roadkills were from Alachua County and 1 from Citrus County. Six American Kestrels were found as roadkills on highways that pass through Florida state parks and recreation areas, during a 4–year survey conducted from 1990 to 1993 (Snyder 1994; Stevenson 1994). The numbers on a statewide basis are undoubtedly higher, especially on highways with high speed limits.

In addition to the data on trauma given in tables 15.2 and 15.4, 8 of 12 (67%) other Peregrine Falcons found dead in Florida between 1977 and 1993 were determined to be trauma cases, although the specific cause of the trauma was not determined (Meteyer 1992–93; Thomas 1985–93). Two of these were from Brevard County; 1 each was from Monroe, Osceola, Santa Rosa, St. Johns, St. Lucie, and Volusia counties.

No falcons were recorded by Crawford (1981) in his 25-year study of more than 42,000 birds killed by striking a 1,010-foot TV tower in Leon County, nor were any falcons listed in several other studies involving bird mortality and tall structures in various parts of Florida (Kale 1971; Taylor and Anderson

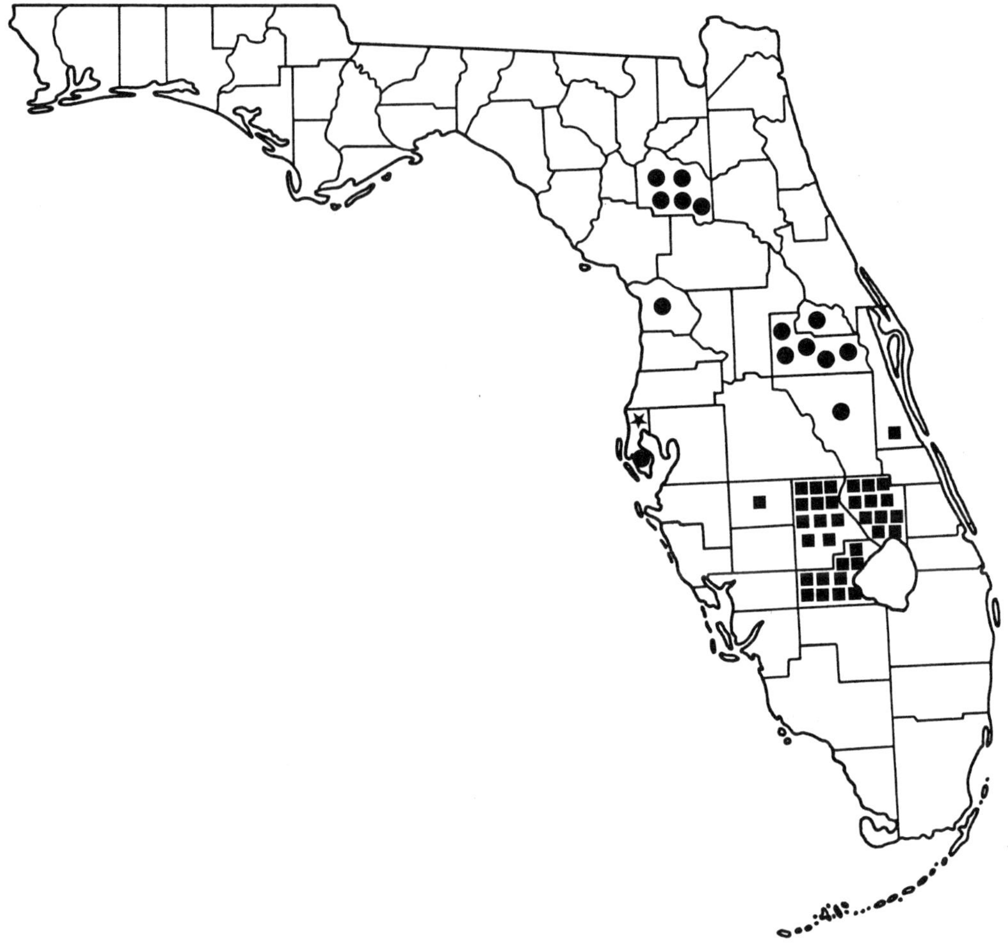

FIGURE 15.1. Distribution of morbidity and mortality of falcons in Florida due to vehicle strikes on highways, 1971–97. *Solid circle* = 1 American Kestrel, *solid star* = 1 Merlin, *solid square* = 1 Crested Caracara. Data from Collins and Gilliland 1995; Deem and Terrell 1996; Forrester 1971–79; Morrison 1999; Spalding 1992; Suto 1996.

1973, 1974; Maehr et al. 1983; Taylor and Kershner 1986; Maehr and Smith 1988).

Layne (1996) pointed out that illegal shooting and trapping were formerly of importance to caracaras, but that this was no longer a serious problem. Morrison (1996), however, stated that shooting was still a serious mortality factor for Crested Caracaras throughout their range. In a 2-year period (1994–95) 4 Crested Caracaras, 1 adult and 3 juveniles, were shot illegally in Glades, Highlands, and Okeechobee counties (Morrison 1999). Wojcik and Smallwood (1993) recorded 1 instance of kestrel eggs in a nest box being destroyed by shotgun in north central Florida.

III. Predation

Layne (1996) stated that Fish Crows and raccoons prey on eggs and young of Crested Caracaras. Morrison (1996) observed raccoons sleeping in caracara nests, but was not certain if the raccoons had destroyed the nest contents or if that had happened before they arrived. Bird (1988b) stated that several raptors (Goshawks, Red-tailed Hawks, Sharp-shinned Hawks, Cooper's Hawks, Peregrine Falcons, and Barn Owls) and crows (species not given) are known to prey on American Kestrels, but we have no such information for Florida. Fragments of an American Kestrel were identified in 1 of 1,098 pellets collected at a Northern Harrier roosting site in Leon County during the winter of 1925–26 (Stoddard 1931). Collins and Gilliland (1995) have a record of predation by a domestic dog on an American Kestrel in Seminole County during 1994. A snake (species not given) was reported as a predator of kestrel chicks in north central Florida in 1991 (Wojcik and Smallwood 1993). According to Palmer (1988), the Great Horned Owl is a serious predator on Peregrine Falcons, especially young birds, but since Peregrines don't breed in Florida, this may not be a significant problem in this part of its range.

Predation by the red imported fire ant (*Solenopsis invicta*) on kestrel chicks and pipping eggs has been reported by Wojcik and Smallwood (1993). Their study took place in north central Florida in 1991 during which 11 of 30 observed breeding attempts resulted in mortality of the young due to fire ant predation. In June 1993 2 nestling American Kestrels from Marion County were submitted to the Wildlife Clinic at the University of Florida College of Veterinary Medicine (Deem and Terrell 1996). These birds were from the same nest, in a rotten palm tree that fell over during a storm. Both birds had damaged toes and 1 had eye lesions, all attributed to fire ant attacks. Most of their toes became necrotic and fell off within 1 month. After treatment they were judged not suitable for release back into the wild and were maintained as permanent captives at 2 environmental educational facilities. The full significance of this type of morbidity and mortality at the population level is not known.

IV. Electrocution

There are only 3 records of electrocution as the cause of mortality in falcons in Florida. Wojcik and Smallwood (1993) reported that lightning struck a nest box containing eggs of an American Kestrel in north central Florida in 1991; this resulted in nesting failure. This type of mortality may be more prevalent than is appreciated. Lightning storms are very common at certain times of the year in Florida. Merritt Island National Wildlife Refuge, for instance, has between 10,000 and 30,000 lightning strikes each year (Leenhouts 1986).

An immature Peregrine Falcon was found dead at the base of a power pole in St. Lucie County in 1985, an apparent victim of electrocution (Thomas 1985). In 1992 another Peregrine Falcon was found electrocuted near some power lines in Brevard County (Collins and Gilliland 1995). As discussed in chapter 11 (Eagles), electrocution of raptors on power lines can be prevented by modifications of the power line structures on which the birds tend to perch and also by changes in the grounding setup. Details on these practices, which may help prevent this type of mortality, can be found in Miller et al. (1975).

V. Inclement weather

There are a number of reports of cold spells resulting in mortality of American Kestrels (Bird 1988a). However, only 1 such incident has been recorded for Florida. Layne (1980) stated that several hundred kestrels were found dead on the Marquesas Keys near Key West (Monroe County) after the passage of a cold front. The exact date of this finding is not known; the author stated that it occurred "in November or December 1975 or 1976." Sutton (1945) gave an anecdotal account of a Crested Caracara that was found dazed in Titusville (Brevard County) after a hurricane passed through the area in October of 1944. We have no information on the effects of inclement weather on other falcons in Florida.

Table 15.5. Concentrations of chlorinated hydrocarbons in various tissues of 3 Crested Caracaras and 1 Peregrine Falcon from Florida

Species of falcon County	Year	Age[b]	Sex	Sample[c]	Residue (ppm wet weight)[a]			Data source
					DDE	ΣDDT	PCB	
Crested Caracara								
Glades	1975	AD	F	AD	2.47	2.47	NA	Johnston (1978)
				UP	1.25	1.25	NA	
Highlands	1975	IM	F	AD	1.24	1.24	NA	Ibid.
				UP	0.48	0.48	NA	
	1976	IM	F	AD	3.25	3.25	NA	Ibid.
				UP	2.44	2.44	NA	
Peregrine Falcon								
Monroe	1977	IM	F	CS[d]	0.70	0.70	0.95[e]	Reichel (1978)
				BR[d]	0.00	0.00	0.00	

NA = not analyzed for this residue.
a. Lowest limit of detection: Crested Caracara = 0.01 ppm, Peregrine Falcon = 0.05 ppm.
b. AD = adult, IM = immature.
c. AD = adipose tissue, BR = brain, CS = carcass, based on the body of the bird minus the skin, feet, wings, liver, and gastrointestinal tract (Mulhern et al. 1970), UP = uropygial gland.
d. Also tested for the following residues and found negative: DDD, DDT, dieldrin, heptachlor epoxide, oxychlordane, *cis*-chlordane, *trans*-nonachlor, *cis*-nonachlor, endrin, toxaphene, hexachlorobenzene, and mirex.
e. Type of PCB not given by author.

VI. Environmental contaminants

Tissue residue data on chlorinated hydrocarbons have been obtained for 3 Crested Caracaras and 1 Peregrine Falcon (table 15.5) and 33 American Kestrels (table 15.6) from Florida. No such information is available for Merlins, however. All birds were collected between 1972 and 1977. Detectable concentrations of DDT and its metabolites were found in all 3 caracaras, in 26 of the kestrels, and in the Peregrine Falcon. Dieldrin was found only in kestrels and occurred in 19 of the 30 birds tested for this residue. One kestrel, an adult male from Pinellas County, had 680 ppm of dieldrin in its brain, a concentration approaching lethality (Sundlof et al. 1986). PCBs were present in tissues from all 15 kestrels tested for it and in the Peregrine Falcon; none was found in the Caracaras. None of the falcons contained concentrations of chlorinated hydrocarbons associated with acute mortality. The highest concentration of DDE in brain tissues, for example, was 1.70 ppm in a male kestrel of unknown age

from Pinellas County in December 1974. This amount is well below the lowest concentration associated with mortality in raptors; similar statements can be made for the concentrations of dieldrin and PCB (Sundlof et al. 1986).

Hoffman (1983) reported organochlorine residues in 6 of 8 American Kestrel eggs taken from 8 different clutches in north central Florida in 1982 (table 15.7). DDE residues were the most common and were found in eggs from Alachua and Marion counties; however, the concentrations were below those known to cause eggshell thinning, when kestrels were fed DDT experimentally in their diets (Porter and Wiemeyer 1969; Wiemeyer and Porter 1970; Lincer 1975). Anderson and Hickey (1972) found no eggshell thinning in eggs of American Kestrels collected in Florida during 1950–59, in comparison with eggs collected prior to 1947 (i.e., before the DDT era began) (table 15.8). However, a similar collection of eggs from Crested Caracaras in Florida in 1950–59 was 8% thinner (table 15.8). A few of these eggs exceeded 20% thin-

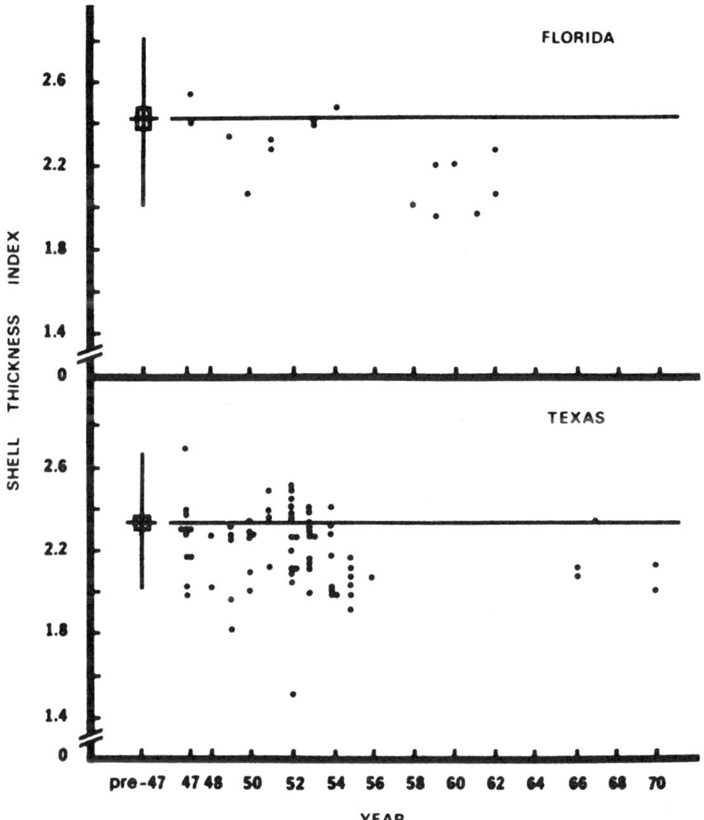

FIGURE 15.2. Eggshell thickness indexes of Crested Caracara eggs from Florida and Texas compared by years of collection. All pre-1947 values fell within the ranges indicated by the boxes at the left. Horizontal lines = mean thickness indexes for eggshells obtained before 1947. From Kiff et al. 1983; by permission of Lloyd F. Kiff.

ning (Kiff et al. 1983). These results are presented graphically in figure 15.2. There are no recent data on eggshell thicknesses of caracaras, but hopefully the problem has been resolved since the sale and use of DDT in the United States was banned in the 1970s (Environmental Protection Agency 1977). As opportunities are presented, eggshell thicknesses of this threatened species should be monitored, however.

Lead was detected in the livers of 4 of 10 Peregrine Falcons from Florida (table 15.9), collected from 8 different counties throughout the state during 1977–93. The concentrations were all <1 ppm and probably of little or no significance to the health of the birds. No data are available on other heavy metals from falcons in Florida.

VII. Neoplasia

An adult female Peregrine Falcon from Suwannee County was submitted to the Center for Birds of Prey in Maitland in October 1991 (Collins and Gilliland 1995). It had been banded as a nestling in the Northwest Territories, Canada, in July 1984. There was a tumor on 1 of the digits of its left foot. This tumor was removed and diagnosed histopathologi-

Table 15.6. Concentrations of chlorinated hydrocarbons in various tissues of American Kestrels from Florida

County	Date	Age[b]	Sex	Sample[c]	Residue (ppm wet weight)[a]				Data source[e]
					DDE	ΣDDT	Dieldrin	PCB[d]	
Alachua	1977	AD	M	BR	0.00	0.00	NA	NA	C
Broward	1973	UN	F	AD	0.79	0.79	0.00	NA	B
				UP	0.63	0.63	0.00	NA	
Citrus	1977	AD	M	BR	0.00	0.00	NA	NA	C
Indian River	1973	UN	M	AD	14.59	16.37	0.36	NA	A,B
				UP	3.15	3.15	0.00	NA	
	1973	UN	F	AD	1.77	2.09	0.00	NA	A,B
				UP	0.63	0.63	0.00	NA	
	1973	UN	F	AD	20.57	22.71	0.00	NA	A,B
				UP	4.63	4.63	0.00	NA	
		UN	F	AD	2.14	2.14	0.00	NA	B
				UP	1.66	1.66	0.00	NA	
	1973	UN	F	AD	9.07	9.07	0.00	NA	B
				UP	1.80	1.80	0.00	NA	
	1973	UN	M	UP	0.00	0.00	0.00	NA	B
	1974	UN	M	AD	0.00	0.00	0.00	NA	B
				UP	1.94	1.94	0.00	NA	
	1974	UN	M	AD	0.00	0.00	0.00	NA	B
				UP	0.00	0.00	0.00	NA	
Leon	1975	UN	M	AD	7.61	8.03	0.00	NA	B
				UP	1.44	2.20	0.00	NA	
Palm Beach	1971	AD	F	LV[f]	11.00	13.00	0.40	5.30	C
				MU	9.80	9.80	0.44	8.90	
Pinellas	1974	AD	M	BR	1.20	1.20	4.10	0.92	C
				LV	4.00	4.00	15.00	2.40	
				MU	2.20	2.20	7.20	2.10	
	1974	AD	M	BR	0.70	0.70	0.91	1.80	C
				LV	1.80	1.80	2.30	5.00	
				MU	2.80	2.80	2.70	8.10	
	1974	AD	F	BR	0.40	0.40	0.21	15.00	C
				LV	0.38	0.38	0.49	12.00	
				MU	0.70	0.70	0.53	19.00	
	1974	AD	M	BR[g]	0.71	1.00	1.40	0.51	C
				LV	3.30	3.30	5.80	1.60	
				MU	2.60	2.60	5.10	1.00	
	1974	AD	M	BR[h]	1.40	1.60	0.34	2.30	C
				LV	3.50	3.50	1.40	1.10	
				MU[i]	9.30	9.80	2.30	5.40	
	1974	AD	M	BR	0.18	0.18	0.32	1.00	C
				LV	1.30	1.30	1.50	12.00	
				MU	0.22	0.22	0.30	2.90	
	1974	UN	M	BR	1.70	1.70	1.40	5.20	C
				LV	1.90	1.90	1.70	7.30	
				MU[j]	3.60	4.30	2.60	13.00	
	1974	AD	F	BR	0.07	0.07	0.80	2.80	C
				LV	0.42	0.42	7.00	21.00	
				MU	0.44	0.44	9.50	28.00	

(continued)

Table 15.6. *(continued)*

| County | Date | Age[b] | Sex | Sample[c] | Residue (ppm wet weight)[a] | | | | Data source[e] |
					DDE	ΣDDT	Dieldrin	PCB[d]	
Pinellas	1975	AD	M	BR	0.90	0.90	1.10	1.40	C
				LV	1.50	1.50	1.90	2.40	
				MU	2.80	2.80	2.80	4.90	
	1975	AD	M	BR	0.93	0.93	6.80	0.49	C
				LV	2.10	2.10	1.00	2.00	
				MU	5.20	5.20	4.00	2.20	
	1975	AD	M	BR	0.84	0.84	4.00	4.10	C
				LV	1.50	1.50	7.00	6.30	
				MU	4.20	4.20	9.30	8.70	
	1975	UN	F	BR	1.30	1.30	1.70	2.30	C
				LV	6.40	6.40	6.00	11.00	
				MU	3.10	3.10	3.10	5.20	
	1975	UN	M	BR	1.00	1.00	0.96	3.60	C
				LV	1.90	1.90	1.50	6.70	
				MU	1.40	1.40	1.10	2.80	
	1975	UN	U	BR	0.86	0.86	0.75	2.80	C
				LV	2.40	2.40	2.50	5.60	
				MU	3.00	3.00	2.50	8.80	
	1975	UN	M	UP	4.12	4.12	3.09	NA	B
				BR	0.42	0.42	0.70	NA	
	1975	UN	M	UP	7.63	18.53	4.36	NA	B
				BR	0.30	0.74	0.82	NA	
	1975	UN	F	UP	2.06	2.06	1.03	NA	B
				BR	1.03	1.03	0.37	NA	
	1976	UN	M	AD	0.00	0.00	0.00	NA	B
				UP	0.00	0.00	0.00	NA	
	1976	AD	F	BR	0.00	0.00	NA	NA	C
Unknown	1972	UN	F	AD	0.79	0.79	0.00	NA	A
				UP	0.63	0.63	0.00	NA	

NA = not analyzed for this residue.
a. Lower limits of detection = 0.01 ppm.
b. AD = adult, UN = unknown.
c. AD = adipose tissue, BR = brain, LV = liver, MU = pectoral muscle, UP = uropygial gland.
d. Aroclor 1260.
e. Data sources: A = Johnston (1976), B = Johnston (1978), C = Sundlof et al. (1986).
f. Additional residue included 1.60 ppm of DDD.
g. Additional residues included 0.14 ppm of DDD and 0.19 ppm of DDT.
h. Additional residue included 0.22 ppm of DDD.
i. Additional residue included 0.54 ppm of DDD.
j. Additional residues included 0.28 ppm of DDD and 0.39 ppm of DDT.

cally as a chondrosarcoma (Kiehl 1991). The bird was euthanized because of poor prognosis of a chronic bumblefoot condition. This finding probably represents an isolated incident and is of little or no importance to populations of Peregrine Falcons in Florida.

VIII. Anomalies

In March 1995 a 7–week-old nestling Crested Caracara from Highlands County was euthanized because of observed malformations and examined at necropsy (Spalding and Morrison

Table 15.7. Organochlorine residues in eggs of American Kestrels from Florida, 1982

| | No. eggs | Residue (ppm wet weight)[a] | | | | |
		DDE	Heptachlor epoxide	Oxychlorine	cis-chlordane	trans-nonachlor
Alachua A[b]	1	0.14	ND	ND	ND	ND
	2	0.13	ND	ND	ND	ND
	3	0.14	ND	ND	ND	ND
	4	ND	ND	ND	ND	ND
Alachua B[c]	1	0.23	0.26	0.12	0.08	0.14
	2	ND	ND	ND	ND	ND
Marion	1	0.08	ND	ND	ND	ND
	2	0.10	ND	ND	ND	ND

Source: Hoffman (1983).
ND = none detected.
a. Lowest limits of detection were 0.1 ppm. Other residues tested for but not detected in any of the 8 eggs included DDD, DDT, dieldrin, *cis*-nonachlor, endrin, toxaphene, and PCB (Aroclor 1260).
b. Southwestern Alachua County.
c. 6 miles southwest of Gainesville.

1995). There was malformation of the distal spinal column, the tarsometatarsi, phalanges, and brachycephalia. In addition there was evidence of abnormalities of the caudal nervous system and histologic abnormalities in muscle. This was probably a rare instance due to a developmental accident or teratogenic agent.

IX. Biotoxins

Two fledgling Crested Caracaras from Highlands County were suspected of having died of a toxin of some type in 1997, but this was not proven (Morrison 1999).

X. Viruses

Falcons are susceptible to several viral diseases such as avian pox, Newcastle disease, inclusion body disease of falcons, and Marek's disease (Graham and Halliwell 1986). Some of these no doubt occur in falcons in Florida, but there is no information on such infections. West Nile virus was identified by PCR and virus isolation techniques in an American Kestrel from Hernando County and a Merlin from Palm Beach County in 2001 (Conti et al. 2002). It was assumed that these birds died because of the infection, but a complete necropsy was not performed to determine this.

Table 15.8. Measurements of eggshells of falcons from Florida, pre- and post-1947

| Species of falcon | Pre-1947 | | Post-1947 | | |
	Number of eggs sampled	Mean thickness index[a]	Number of eggs sampled	Mean thickness index[a]	% change
Crested Caracaras	20	2.43	14	2.23[b]	−8.2
American Kestrels	860	1.02	26	1.01[c]	0

Sources: For Crested Caracaras: Kiff et al. (1983); for American Kestrels: Anderson & Hickey (1972)
a. Thickness index = weight of eggshell (mg) divided by the product of the length x breadth (mm) (Ratcliffe 1967).
b. 1947–62.
c. 1950–59.

Table 15.9. Concentrations of lead in livers of Peregrine Falcons from Florida

County	Specific locality	Year	Age[a]	Sex	Concentrations (ppm wet weight)[b]
Brevard	Island NWR	1992	IM	F	ND
Collier	Marco Island	1984	IM	M	ND
Dade	Miami (sewage plant)	1985	AD	F	0.94
Monroe	Loggerhead Key	1977	IM	F	0.30
	Marathon	1991	IM	M	ND
	Key West	1991	IM	F	ND
Osceola	Kissimee	1993	IM	F	0.44
St. Johns	Anastasia	1987	IM	F	ND
St. Lucie	Jensen Beach	1985	IM	F	ND
Volusia	New Smyrna Beach	1993	IM	M	0.44

Source: Records of the National Wildlife Health Center, Madison, Wisconsin.
ND = none detected.
a. IM = immature, AD = adult.
b. Lower limit of detection = 0.2 ppm.

XI. Bacteria

During 1971–77 White and Forrester (1977) cultured samples of liver and large intestine from 18 American Kestrels from 5 counties in Florida (table 15.10). Five species of bacteria were identified. Two of these kestrels, 1 from Alachua County in 1977 and 1 from Pinellas County in 1974, were positive for *Salmonella* spp. The serovar from the kestrel in Pinellas County was not determined, but the isolation from Alachua County was identified as *S. typhimurium*. These organisms were isolated from the contents of the large intestines and probably represented the carrier state since they were not associated with lesions. The other bacteria were probably postmortem contaminants or normal components of the intestinal flora.

Bacteriological studies were conducted by Duncan (1993) on 5 Peregrine Falcons obtained from Florida in 1977 and 1991–93 (table 15.11). Fifteen species were identified in association with various lesions in these falcons. Many of these species were probably contaminant bacteria from wounds resulting from trauma. Three of the cases are of special interest

Table 15.10. Bacteria isolated from livers and large intestines of American Kestrels in Florida[a]

Bacteria	County	Year	Organ cultured[b]	Cause of death[c]
Enterobacter sp.	Hillsborough	1975	LI	UK
Escherichia coli	Citrus	1977	LI	RK
Salmonella typhimurium	Alachua	1977	LI	RK
Salmonella sp.	Pinellas	1974	LI	UK
Staphylococcus sp. (nonhemolytic)	Pinellas	1976	LV	UK

Source: White and Forrester (1977).
a. In addition to the American Kestrels listed in this table, 13 birds were cultured and were negative. One of these was a road-kill; the cause of death of the other 12 birds was not determined. The roadkill was from Alachua County in 1976, and the other birds were from Palm Beach County (*n* = 1 in 1976) and Pinellas County (*n* = 8 in 1974 and 3 in 1975).
b. LI = large intestine, LV = liver.
c. RK = roadkill, UK = unknown.

Table 15.11. Bacteria identified from 5 Peregrine Falcons in Florida

Species of bacteria

County	Year	Organ or tissue	Falcon identification[a]
Acinetobacter calcoaceticus			
Monroe	1977	Pericardial sac	A
Acinetobacter sp.			
Brevard	1993	Lungs	D
Bacillus sp.			
Brevard	1992	Neck tissue	C
Enterobacter aerogenes			
Brevard	1992	Neck tissue	C
Enterobacter sakazakii			
Brevard	1992	Pericardial sac, abdominal cavity, air sacs	C
Enterococcus sp.			
Brevard	1992	Neck tissue, pericardial sac, abdominal cavity, air sacs	C
Brevard	1993	Liver, spleen, lungs	D
Escherichia coli			
Brevard	1992	Neck tissues	C
Brevard	1993	Wing tissues, spleen, lungs	D
Monroe	1993	Throat, sinus	E
Mycoplasma sp.			
Monroe	1977	Sternum, pericardial sac	A
Monroe	1991	Sinus	E
Pasteurella hemolytica			
Brevard	1993	Liver	D
Pseudomonas aeruginosa			
Monroe	1977	Pericardial sac	A
Santa Rosa	1991	Air sacs	B
Pseudomonas fluorescens			
Monroe	1977	Lungs	A
Pseudomonas sp.			
Santa Rosa	1991	Liver	B
Salmonella brandenburg			
Santa Rosa	1991	Air sacs	B
Serratia liquefaciens			
Brevard	1992	Neck tissue	C
Staphylococcus aureus			
Monroe	1977	Small intestine	A
Staphylococcus epidermides			
Brevard	1993	Spleen	D
Vibrio alginolyticus			
Monroe	1991	Sinus	E

Source: Duncan (1977–93).

a. A = an immature female that died of mycoplasmal pneumonia (Sileo 1978); B = an adult male that died of airsacculitis, pneumonia, and peritonitis (Thomas 1992); C = an immature female that died of trauma and emaciation (Meteyer 1992); D = an immature male that died of trauma and secondary septicemia (Meteyer 1992); E = an immature female that died of necrotic stomatitis and sinusitis (Thomas 1991).

Table 15.12. Fungi identified from 5 Peregrine Falcons in Florida

Species of fungus	County	Year	Organ or tissue	Falcon identification[a]
Aspergillus fumigatus	Dade	1985	Air sacs	A
	Monroe	1991	Wing tissue[b]	B
	Brevard	1993	Lungs	C
Aspergillus sp.	Monroe	1991	Wing tissue	B
	Monroe	1991	Throat abscess	D
Candida albicans	Osceola	1993	Oral cavity	E
Mucor sp.	Monroe	1991	Wing tissue	B
	Monroe	1991	Sinus	D

Source: Duncan (1985–93).

a. A = an adult female that died of acute toxicosis (agent not identified) (Thomas 1985), B = an immature male that died of traumatic injuries (Thomas 1991), C = an immature male that died of traumatic injuries and subsequent bacteremia (Meteyer 1993), D = an immature female that died of necrotic stomatitis and sinusitis (Thomas 1991), E = an immature female that died of traumatic injuries (Thomas 1993).

b. Probably included air sac tissue.

because of the possible link between bacteria and cause of death. In 2, a species of *Mycoplasma* was isolated from Peregrine Falcons, 1 found dead of mycoplasmal pneumonia on Loggerhead Key in the Dry Tortugas in October 1977; the second a falcon that died of a necrotizing stomatitis and sinusitis near Key West in November 1991. The third case was the isolation of *Salmonella brandenburg* from the air sacs of a falcon with necrotizing airsacculitis and pneumonia. This bird had an injured wing and died after about a month in captivity.

The significance of these bacteria to the health of free-ranging populations of falcons in Florida is not known. The *Salmonella* species are of public health interest since they can infect humans.

XII. Fungi

Three species of fungi (*Aspergillus fumigatus, Candida albicans,* and *Mucor* sp.) have been identified from 5 free-ranging Peregrine Falcons in Florida (table 15.12). In all of these cases the fungal infections were incidental or only partially contributory to the cause of death. An additional Peregrine Falcon from Polk County was submitted to the Center for Birds of Prey in 1991 (Collins and Gilliland 1995), treated for a traumatic eye injury, and

transferred to another facility for evaluation of its suitability for release back into the wild. Four months later it was readmitted to the Center with respiratory problems. It died 10 days later and at necropsy was found to have severe aspergillosis involving the thoracic air sacs (Thomas 1991). *Aspergillus fumigatus* was cultured from air sac tissue (Duncan 1991). In this latter case the *Aspergillus* infection may have been present when the falcon was obtained from the wild and the stress of captivity may have led to the disease. There is reason to believe that stress increases the susceptibility of raptors to aspergillosis (Redig et al. 1980).

XIII. Protozoans

Information on protozoan infections in falcons from Florida is limited to several unpublished reports and 2 publications (Forrester et al. 1994; Foster et al. 1998). Two protozoan species have been reported (table 15.13), a blood protozoan (*Haemoproteus tinnunculi*) from Crested Caracaras and American Kestrels (figure 15.3) and an unidentified species of *Sarcocystis* (figure 15.4) from American Kestrels and Peregrine Falcons. Sarcocysts were observed in the heart and skeletal muscles of a Peregrine Falcon that had been

Table 15.13. Protozoan infections[a] reported from falcons in Florida

| Species of falcon | | | No. falcons | | |
Protozoan	County	Year(s)	Exam.	Infected	Data sources
Crested Caracara					
Haemoproteus					
tinnunculi	Alachua	1995	1	1	Telford et al. (1996)
	DeSoto	1994–96	10	6	Foster et al. (1998)
	Glades	1994–96	51	6	Ibid.
	Hendry	1994–96	4	2	Ibid.
	Highlands	1994–96	86	30	Ibid.
	Indian River	1994–96	2	0	Ibid.
	Martin	1995	1	1	Telford et al. (1996)
	Okeechobee	1994–96	67	19	Foster et al. (1998)
	Osceola	1994–96	2	0	Ibid.
	Polk	1994–96	1	0	Ibid.
American Kestrel					
Haemoproteus					
tinnunculi	Alachua	1987	1	1	Telford et al. (1996)
	Alachua	1995	1	1	Ibid.
	Collier	1990	2	1	Ibid.
Sarcocystis sp.	Alachua	1976–79	3	0	Forrester (1979)
	Alachua	1986	1	1	Wright (1986)
	Citrus	1977	1	0	Forrester (1977)
	Hillsborough	1976	1	0	Ibid.
	Palm Beach	1971	1	0	Ibid.
	Pinellas	1974–75	15	0	Ibid.
Merlin					
Blood protozoans	Collier	1990	1	0	Telford et al. (1996)
Peregrine Falcon					
Sarcocystis sp.	Brevard	1991	1	1	Thomas (1991)
	Brevard	1992	1	0	Meteyer (1992)
	Dade	1985	1	0	Thomas (1985)
	Monroe	1991	1	0	Thomas (1991)
	Santa Rosa	1991	1	0	Ibid.

a. Blood protozoan infections were determined by examination of thin blood films and *Sarcocystis* infections were determined by histopathological technique.

in captivity in Brevard County (1991) for almost a year prior to its death. It is not known if the infection was acquired in the wild or after it was taken into captivity.

The study reported by Foster et al. (1998) was based on blood films from 223 Crested Caracaras from 8 counties in south central Florida over a 3–year period (1994–96). *Haemoproteus tinnunculi* was the only blood protozoan seen on the smears. Its overall prevalence was 28%. Adults had a higher prevalence (50%) than did nestlings (20%). There were no differences in the prevalences of *H. tinnunculi* in adult birds during the sampling months each year (February through May) and from year to year. However, there were differences in prevalences in nestlings, which increased from 5% in February to 46% in May (figure 15.5). This increase in prevalence with age of the nestlings was postulated to be due a "func-

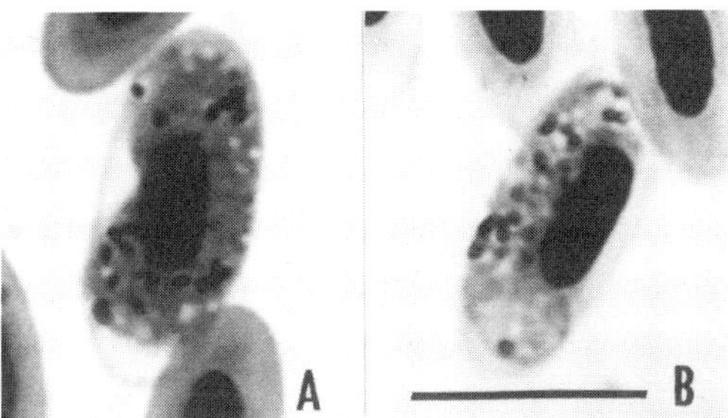

FIGURE 15.3. Macrogametocytes of *Haemoproteus tinnunculi* from an American Kestrel in Florida. Horizontal bar = 10 μm. Courtesy of Sam R. Telford, Jr.

tion of both an increase in the biting activity of arthropod vectors as the temperatures increase in the spring, as well as an increase in the population of arthropods which serve as vectors" (Foster et al. 1998).

The pathologic significance of these 2 protozoan infections to falcons in Florida is not known. A negative correlation between the body mass of female American Kestrels and parasitemia of *H. tinnunculi* was reported by Apanius and Kirkpatrick (1988) in Pennsylvania, but this phenomenon has not been verified, nor has it been studied in Florida.

Two other parasitic protozoans have been found in falcons in other parts of their range, the flagellate *Trichomonas gallinae* and the coccidian *Toxoplasma gondii*. It has been suggested that infections by *T. gallinae* were a possible contributing factor in the decline of Peregrine Falcons in parts of North America, although this idea has been challenged on epidemiologic grounds (Stabler 1969). Although infections of falcons with *T. gondii* have not been reported in Florida, they have been identified in Alabama (Lindsay et al. 1993) and may well be shown to be present in Florida populations if the appropriate techniques were applied. In the Alabama study 1 of 3 American Kestrels examined was positive.

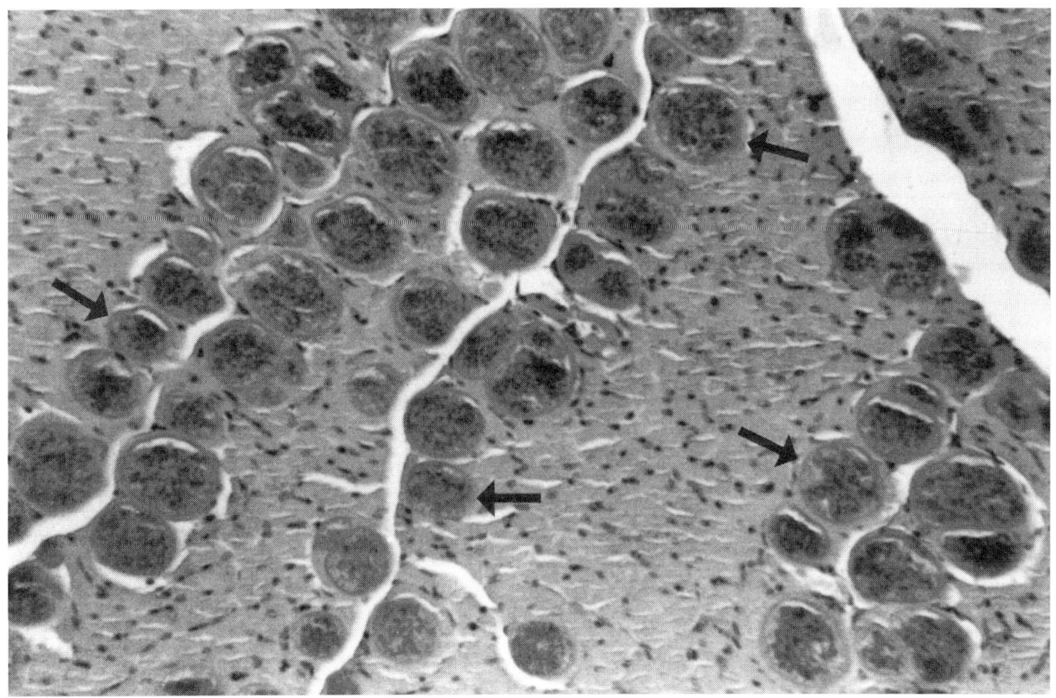

FIGURE 15.4. Sarcocysts *(arrows)* in pectoral muscle of an American Kestrel from Alachua County, 1986. Courtesy of Scott D. Wright.

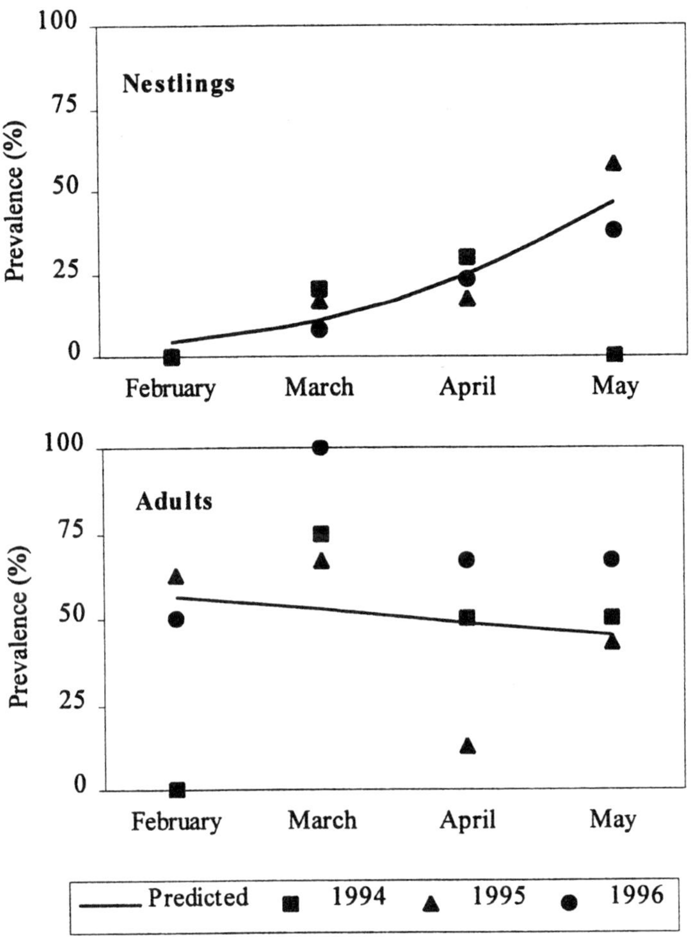

FIGURE 15.5. Predicted prevalence and observed annual prevalence of infections of *Haemoproteus tinnunculi* in nestling and adult Crested Caracaras from south central Florida, 1994–96. No nestlings were infected in February 1994, 1995, or 1996. From Foster et al. 1998; by permission of *Journal of Raptor Research*.

XIV. Helminths

The parasitic helminths of American Kestrels have been fairly well studied (Kinsella et al. 1995), but there is limited information on Crested Caracaras and Peregrine Falcons. Nothing is known about Merlins.

The helminths of kestrels in Florida include 5 species of trematodes, 10 nematodes, and 1 acanthocephalan (table 15.14). Kinsella et al. (1995) reported an average of only 1.7 species of helminths per infected bird in their sample of 22 examined from Florida. No kestrel had more than 3 species of helminths. In addition there was an almost complete absence of strigeid trematodes that are so characteristic of other raptors. They attributed these observations to the predominantly arthropod diet of kestrels.

The helminths of 11 Crested Caracaras included 5 species of trematodes, 1 nematode, and 1 acanthocephalan (table 15.15). Helminths were collected from 3 Peregrine Falcons. These

Table 15.14. Helminth infections in American Kestrels from Florida

Helminth (site in host)[a]	County	Year(s)	No. kestrels[b] Exam.	Infected	%	Intensity Mean	Range
Trematoda							
Brachylecithum nanum (LV)	Alachua	1977–80	4	1	—	2	—
	Pinellas	1975–77	15	2	13	3	2–4
Echinostoma trivolvis (LI)	Pinellas	1975–77	15	1	7	1	—
Platynosomum illiciens (LV)	Pinellas	1975–77	15	1	7	4	—
Prosthogonimus ovatus (CE)	Pinellas	1975–77	15	2	13	3	1–5
Strigea falconis (SI)	Pinellas	1975–77	15	1	7	1	—
Nematoda							
Capillaria falconis (SI)	Pinellas	1975–77	15	1	7	1	—
Cardiofilaria							
pavlovskyi (BC)	Alachua	1977–80	4	1	—	18	—
	Hillsborough	1976	1	1	—	2	—
	Palm Beach	1971	1	1	—	1	—
	Pinellas	1975–77	15	2	13	2	1–2
Cyrnea semilunaris (PR)	Alachua	1977–80	4	1	—	1	—
Cyrnea sp. (PR)	Alachua	1977–80	4	1	—	10	—
Dispharynx nasuta[c] (PR)	Alachua	1977–80	4	1	—	1	—
	Citrus	1977	1	1	—	4	—
	Pinellas	1975–77	15	7	47	36	1–125
Filaria sp.[d] (NG)	NG	NG	NG	NG	—	NG	—
Oxyspirura sp.[e] (EY)	Dade	1918	NG	NG	—	NG	—
Physaloptera sp. (PR)	Pinellas	1975–77	15	2	13	13	8–17
Procyrnea mansioni (ES,PR)	Alachua	1977–80	4	1	—	1	—
	Hillsborough	1976	1	1	—	1	—
	Pinellas	1975–77	15	2	13	1	—
Spirurid larvae (ES,PR)	Citrus	1977	1	1	—	3	—
	Palm Beach	1971	1	1	—	5	—
	Pinellas	1975–77	15	4	27	5	1–10
Acanthocephala							
Unidentified larvae (ES)	Pinellas	1975–77	15	1	7	1	—

Sources: All data are from Kinsella et al. (1993) and Kinsella et al. (1995), except as otherwise noted.
a. BC = body cavity, CE = ceacum, ES = esophagus, EY = eye, LI = large intestine, LV = liver, NG = not given, PR = proventriculus, SI = small intestine.
b. The data from Kinsella et al. (1993) consist of results from 22 American Kestrels from 5 counties: Alachua (1977–80, n = 4), Citrus (1977, n = 1), Hillsborough (1976, n = 1), Palm Beach (1974, n = 1), and Pinellas (1975–77, n = 15). Unless indicated as positive under each helminth for a given county in the table, kestrels from other counties were negative.
c. Kinsella and Greiner (1993) identified *D. nasuta* in one American Kestrel from north central Florida in 1987, but gave no data on prevalence and intensity.
d. From Walton (1927).
e. From the records of the U.S. National Parasite Collection, Beltsville, Md. (USNPC#-26147).

included 4 species of trematodes, 1 cestode, and 8 nematodes (table 15.16).

The significance of these helminth infections to falcon populations in Florida is not known, although some of them could be pathogenic if they occurred in high intensities.

XV. Arthropods

The ectoparasites of Crested Caracaras and American Kestrels in Florida have been fairly well studied. Three species of mites and 4 chewing lice are known from caracaras (table 15.17)

Table 15.15. Helminth infections in 11 Crested Caracaras[a] from Florida

Helminth (site in host)[b]	County	Year(s)	No. of caracaras Exam.	Inf.	Intensity Mean	Range
Trematoda						
Diplostomum sp. (SI)	Highlands	1992–96	5	1	1	—
Microparyphium facetum (CL)	Highlands	1992–96	5	1	1	—
Neodiplostomum americanum (SI)	Glades	1992, 1995	2	1	25	—
	Hendry	1996	1	1	5	—
	Highlands	1992–96	5	1	7	—
	Okeechobee	1994	1	1	1	—
Neodiplostomum attenuatum (SI)	Highlands	1992–96	5	2	2	1–3
	Okeechobee	1994	1	1	1	—
	Osceola	1997	1	1	3	—
Neodiplostomum sp. (SI)	Glades	1992, 1995	2	1[c]	4	—
Prosthogonimus ovatus (CL)	Glades	1992, 1995	2	1	1	—
	Highlands	1992–96	5	1	2	—
Nematoda						
Physaloptera sp. (PR)	Highlands	1995	1	1	2	—
Acanthocephala						
Centrorhynchus sp. (SI)	Highlands	1992–96	5	1	2	—
	Osceola	1997	1	1	1	—

Source: Kinsella et al. (1998).
a. All were nestlings or juveniles except as indicated. Eleven caracaras were examined: 2 from Glades County, 1 from Hendry County, 6 from Highlands County, 1 from Okeechobee County, and 1 from Osceola County. For each parasite, data are given for counties in which positive birds were found; negatives are not included in the table.
b. CL = cloaca, PR = proventriculus, SI = small intestine.
c. Adult female that died due to vehicle strike.

and 2 ticks, 1 flea, and 2 chewing lice have been reported from kestrels (table 15.18). Nothing is known of the significance of these infestations to the health and welfare of caracara and kestrel populations in Florida. There is no information on parasitic arthropods of Merlins or Peregrine Falcons in Florida.

XVI. Emaciation

Three of 12 Peregrine Falcons found dead in Florida between 1977 and 1993 were emaciated (Meteyer 1992; Thomas 1984–85). In 2 of these, an adult female from Dade County in 1985 and an immature female from Brevard County in 1992, emaciation was considered a secondary finding and in 1, an immature male from Collier County in 1984, the primary finding. Acute toxicosis was suspected but not proven in the case of the Dade County bird, and the bird from Brevard County also had evidence of trauma. All 3 falcons had marked pectoral muscle atrophy and a lack of subcutaneous, pericardial, and peritoneal fat. The emaciation in the first 2 birds may have been related to other conditions, but in the case of the bird from Collier County, there was no evidence of debilitating injury, disease, or an environmental contaminant problem. Between 1991 and 1995 3 Peregrine Falcons, 2 American Kestrels, and 1 Merlin from Pinellas County were submitted to the Suncoast Seabird Sanctuary in emaciated condition (Suto 1996). In addition, 1 emaciated Peregrine Falcon (Monroe County, 1991) and 3 emaciated American Kestrels (Orange County,

Table 15.16. Helminths collected from three Peregrine Falcons in Florida

Helminth (site in host)[a]	Number of helminths collected		
	St. Lucie Co. (1985)[b]	Santa Rosa Co. (1991)[c]	Monroe Co. (1991)[d]
Trematoda			
Echinochasmus sp. (SI)	0	0	6
Neodiplostomum accipitris (SI)	0	0	1
Phagicola sp. (SI)	0	0	1
Strigea falconis (SI)	0	0	12
Cestoda			
Cladotaenia foxi (SI)	2	0	3
Nematoda			
Contracaecum multipapillatum (PR)	0	0	5
Cyrnea semilunaris (PR)	2	0	0
Desportesius invaginatus (PR)	0	0	1
Physaloptera acuticauda (PR)	0	1	0
Procyrnea mansioni (PR)	0	20	0
Steptocara crassicauda (PR)	5	0	0
Synhimantus hamatus (PR)	0	0	2
Synhimantus laticeps (PR)	0	0	1

Source: Kinsella and Cole (1994).
a. PR = proventriculus, SI = small intestine.
b. An immature female that had been banded in Greenland as a nestling in July 1985; died of electrocution while perching on an electric pole near Jensen Beach in November of the same year (Thomas 1985).
c. A adult male that died of bacterial airsacculitis and pneumonia (Thomas 1991).
d. A immature female that died of necrotizing stomatitis and sinusitis near Key West (Thomas 1991).

1993; Seminole County, 1994; Sumter County, 1994) were submitted to the Center for Birds of Prey (Collins and Gilliland 1995). The kestrel from Sumter County had some signs of trauma, but the records are not clear for the other cases regarding other factors that may have contributed to the emaciation. In general, the significance and specific cause of emaciation in raptors such as these is usually not known. On a national scale emaciation was found to account for 8% of a sample of 1,429 Bald Eagles found dead from 1963 to 1984 (NWHRC 1985). For more details on this phenomenon, see chapter 11 (Eagles).

Primary screwworm (*Cochliomyia hominivorax*) was eliminated in Florida during the late 1950s and early 1960s (Diamant 1963). Prior to its elimination, this obligatory myiasis-producing parasite had a negative effect on warm-blooded wild animals, including many mammals, resulting in considerable mortality. Layne (1996) postulated that when it was eliminated, the reduction in available carrion might have been a factor in the decline in numbers of Crested Caracaras in Florida. If true, this would be a curious phenomenon: the successful control of a disease organism, harmful to some species of wildlife, resulting in harm to another species by its control.

XVII. Summary and conclusions

Some data are available on each of the 4 species of falcons that occur in Florida, although very little is known about Merlins. The most complete information is on American Kestrels, with lesser amounts of data on Crested Caracaras and Peregrine Falcons. Eighty-three different conditions, disease agents, and parasites have

Table 15.17. Parasitic arthropod infestations on Crested Caracaras from Florida

Species of arthropod	County or site	Year(s)	Data source
Feather mites			
Dubininia sp.[a]	Glades	1992	Mertins et al. (1996)
	Highlands	1992, 1994, 1995	Ibid.
	Okeechobee	1994	Mertins et al. (1999)
Hieracolichus sp.[a]	Glades	1992	Ibid.
	Highlands	1992, 1994–96[b]	Ibid.
	Okeechobee	1994	Mertins et al. (1999)
Tropical fowl mites			
Ornithonyssus bursa	Highlands	1976	FSCA[c]
	Indian River	1996	Mertins et al. (1996)
Chewing lice			
Acutifrons mexicanus	Highlands	1992	Forrester et al. (1995)
	"South Fla."	1974	Ibid.
Colpocephalum polybori	DeSoto	1996	Mertins et al. (1996)
	Glades	1974, 1976, 1978	Forrester et al. (1995)
		1981, 1992	Ibid.
		1996[d]	Mertins et al. (1996)
	Hendry	1996[e]	Ibid.
	Highlands	1974–76	Forrester et al. (1995)
		1978, 1981	Ibid.
		1992, 1994, 1995	Foster et al. (1995)
		1996[f]	Mertins et al. (1996)
	Indian River	1996	Ibid.
	Okeechobee	1992	Forrester et al. (1995)
		1996[g]	Mertins et al. (1996)
		1994	Mertins et al. (1999)
Cuculiphilus alternatus	Glades	1996[e]	Mertins et al. (1996)
Falcolipeurus josephi	"Florida"	NG	Tandan & Dhanda (1963)
Laemobothrion maximum	"Florida"	1945	Forrester et al. (1995)

NG = not given.
a. Undescribed species.
b. In 1996, 8 birds were positive.
c. FSCA = records from the Florida State Collection of Arthropods, Division of Plant Industry, Florida Department of Agriculture and Consumer Services, Gainesville.
d. Five birds were positive.
e. Two birds were positive.
f. Ten birds were positive.
g. In 1996, 3 birds were positive.

been identified as proven or potential morbidity/mortality factors of falcons in Florida. These include trauma (2), electrocution (2), inclement weather (1), chlorinated hydrocarbons (9), heavy metals (1), viruses (1), bacteria (18), fungi (3), protozoans (2), trematodes (14), cestodes (1), nematodes (15), acanthocephalans (2), ticks (2), mites (3), and chewing lice (7). Trauma due to predation and roadkills is prob-

ably the most important mortality factor, especially for Peregrine Falcons and Crested Caracaras, although the supporting data for this conclusion are limited. From the information that we have there do not appear to be any significant infectious or parasitic diseases affecting falcon populations, with the possible exception of mycoplasmosis in Peregrine Falcons, which should be further studied.

Table 15.18. Parasitic arthropod infestations on American Kestrels from Florida

Species of arthropod	County or site	Year(s)	Data source
Ticks			
Amblyomma americanum	Taylor[a]	1947–49	Taylor (1951)
	"Florida"	NG	Peters (1936)
	"Florida"[b]	NG	Clifford et al. (1961)
Amblyomma tuberculatum	Lake	1932[c]	NTC[d]
	"Florida"	1908[e]	NTC
Amblyomma sp.	Taylor[f]	1947–49	Taylor (1951)
Chewing lice			
Degeeriella carruthi	Alachua	1925	Forrester et al. (1995)
	Alachua	1987	Ibid.
	Brevard	1973	Ibid.
	Clay	1982	Ibid.
	Indian River	1970	Ibid.
Nosopon lucidum	Alachua	1987	Ibid.
Fleas			
Echidnophaga gallinacea	Marion	1957	FSCA[g]
	Marion	NG	Layne (1971)[h]

NG = not given.

a. One nymph was collected from one kestrel.

b. Larvae (numbers not given).

c. One larva from one kestrel.

d. NTC = records of the National Tick Collection, Georgia Southern University, Statesboro.

e. Three larvae from one kestrel. The records of the National Tick Collection state that these ticks were collected in Waco, Fla. (county unknown).

f. Two larvae were collected from one kestrel.

g. FSCA = records from the Florida State Collection of Arthropods, Division of Plant Industry, Florida Department of Agriculture and Consumer Services, Gainesville.

h. Collected from a captive kestrel.

XVIII. Literature cited

American Ornithologists' Union. 1998. *Checklist of North American birds.* 7th ed. American Ornithologists' Union, Washington, D.C. 829 pp.

Anderson, D.W., and J.J. Hickey. 1972. Eggshell changes in certain North American birds. *Proc. Int. Ornithol. Congr.* 15:514–540.

Apanius, V., and C.E. Kirkpatrick. 1988. Preliminary report of *Haemoproteus tinnunculi* infection in a breeding population of American kestrels (*Falco sparverius*). *J. Wildl. Dis.* 24:150–153.

Balgooyen, T.G. 1976. Behavior and ecology of the American kestrel (*Falco sparverius* L.) in the Sierra Nevada of California. *University Calif. Publ. Zool.* 103:1–83.

Bird, D.M. 1988a. Survival (American kestrel, *Falco sparverius*). In: *Handbook of North American birds.* Vol. 5. R.S. Palmer (ed.). Yale University Press, New Haven, Conn. p. 279.

———. 1988b. Some interspecific relations (American kestrel, *Falco sparverius*). In: *Handbook of North American birds.* Vol. 5. R.S. Palmer (ed.). Yale University Press, New Haven, Conn. pp. 287–288.

Cade, T.J., J.H. Enderson, C.G. Thelander, and C.M. White (eds.). 1988. *Peregrine falcon populations.* Peregrine Fund, Boise, Idaho. 949 pp.

Clifford, C.M., G. Anastos, and A. Elbl. 1961. The larval ixodid ticks of the eastern United States (Acarina-Ixodidae). *Misc. Publ. Entomol. Soc. Am.* 2:213–237.

Collins, R., and C. Gilliland. 1995. Unpublished data. Florida Audubon Society, Center for Birds of Prey, Maitland.

Collopy, M.W. 1996. Southeastern American kestrel. In: *Rare and endangered biota of Florida.* Vol. 5, *Birds.* J.A. Rodgers, H.W.

Kale II, and H.T. Smith (eds.). University Press of Florida, Gainesville. pp. 211–218.

Conti, L., R. Oliveri, and C. Blackmore. 2002. Unpublished data. Florida Department of Health, Tallahassee.

Crawford, R.L. 1981. Bird casualties at a Leon County, Florida TV tower: a 25–year migration study. *Bull. Tall Timbers Res. Stn.* 22:1–30.

Deem, S.L., and S.P. Terrell. 1996. Unpublished data. University of Florida, Gainesville.

Deem, S.L., S.P. Terrell, and D.J. Forrester. 1998. A retrospective study of morbidity and mortality of raptors in Florida: 1988–1994. *J. Zoo Wildl. Med.* 29:160–164.

Diamant, G. 1963. Screwworm eradication in southeastern United States. *Am. J. Public Health* 53:22–26.

Duncan, R.M. 1977–93. Unpublished data. National Wildlife Health Center, Madison, Wis.

Environmental Protection Agency. 1977. Suspended and cancelled pesticides. Washington, D.C. 16 pp.

Forrester, D.J. 1971–79. Unpublished data. University of Florida, Gainesville.

Forrester, D.J., H.W. Kale II, R.D. Price, K.C. Emerson, and G.W. Foster. 1995. Chewing lice (Mallophaga) from birds in Florida: a listing by host. *Bull. Fla. Mus. Nat. Hist.* 39:1–44.

Forrester, D.J., S.R. Telford, Jr., G.W. Foster, and G.F. Bennett. 1994. Blood parasites of raptors in Florida. *J. Raptor Res.* 28:226–231.

Foster, G.W., J.L. Morrison, C.S. Hartless, and D.J. Forrester. 1998. *Haemoproteus tinnunculi* in crested caracaras (*Caracara plancus audubonii*) from southcentral Florida. *J. Raptor Res.* 32: 159–162.

Foster, G.W., R.D. Price, and J.L. Morrison. 1995. Unpublished data. University of Florida, Gainesville.

Fowler, M.E. (ed.). 1986. *Zoo and wild animal medicine.* 2d ed. W.B. Saunders, Philadelphia. 1,127 pp.

———(ed.). 1993. *Zoo and wild animal medicine: current therapy.* 3d ed. W.B. Saunders, Philadelphia. 617 pp.

Graham, D.L., and W.H. Halliwell. 1986. Viral diseases of birds of prey. In: *Zoo and wild animal medicine.* 2d ed. M.E. Fowler (ed.). W.B. Saunders, Philadelphia. pp. 408–413.

Hickey, J.J. (ed.). 1969. *Peregrine falcon populations: their biology and decline.* University of Wisconsin Press, Madison. 596 pp.

Hoffman, M.L. 1983. Historical status and nest-site selection of the American kestrel (*Falco sparverius paulus*) in Florida. M.S. thesis, University of Florida, Gainesville. 100 pp.

Hoffman, M.L., and M.W. Collopy. 1988. Historical status of the American kestrel (*Falco sparverius paulus*) in Florida. *Wilson Bull.* 100: 91–107.

Johnston, D.W. 1976. Organochlorine pesticide residues in uropygial glands and adipose tissue of wild birds. *Bull. Environ. Contam. Toxicol.* 16: 149–155.

———. 1978. Organochlorine pesticide residues in Florida birds of prey, 1969–76. *Pestic. Monit. J.* 12:8–15.

Kale, H.W. II. 1971. Regional reports: Florida region. *Am. Birds* 25:723–725, 730–735.

Kiehl, A.R. 1991. Unpublished data. Doctor's and Physicians Laboratory, Leesburg, Fla.

Kiff, L.F., D.B. Peakall, M.L. Morrison, and S.R. Wilbur. 1983. Eggshell thickness and DDE residue levels in vulture eggs. In: *Vulture biology and management.* S.R. Wilbur and J.A. Jackson (eds.). University of California Press, Berkeley. pp. 440–458.

Kinsella, J.M., and R.A. Cole. 1994. Unpublished data. National Wildlife Health Center, Madison, Wis.

Kinsella, J.M., G.W. Foster, and D.J. Forrester. 1993. Unpublished data. University of Florida, Gainesville.

———. 1995. Parasitic helminths of six species of hawks and falcons in Florida. *J. Raptor Res.* 29:117–122.

Kinsella, J.M., and E.C. Greiner. 1993. Unpublished data. University of Florida, Gainesville.

Kinsella, J.M., M.G. Spalding, J.L. Morrison, and D.J. Forrester. 1998. Unpublished data. University of Florida, Gainesville.

Layne, J.N. 1971. Fleas (Siphonaptera) of Florida. *Fla. Entomol.* 54:35–51.

———. 1980. Trends in numbers of American kestrels on roadside counts in south central Florida from 1968 to 1976. *Fla. Field Nat.* 8:1–10.

———. 1996. Crested caracara. In: *Rare and endangered biota of Florida.* Vol. 5, *Birds.*

J.A. Rodgers, H.W. Kale II, and H.T. Smith. (eds.). University Press of Florida, Gainesville. pp. 197–210.

Leenhouts, W.P. 1986. Osprey killed by lightning at Merritt Island National Wildlife Refuge, Florida. *Fla. Field Nat.* 15:22–23.

Lincer, J.L. 1975. DDE-induced eggshell thinning in the American kestrel: a comparison of the field situation and laboratory results. *J. Appl. Ecol.* 12:781–793.

Lindsay, D.S., P.C. Smith, F.J. Hoerr, and B.L. Blagburn. 1993. Prevalence of encysted *Toxoplasma gondii* in raptors from Alabama. *J. Parasitol.* 79:870–873.

Logan, T.H. 1997. Florida's endangered species, threatened species and species of special concern. Official lists. Florida Game and Fresh Water Fish Commission, Tallahassee. 14 pp.

Maehr, D.S., and J.Q. Smith. 1988. Bird casualties at a central Florida power plant: 1982–1986. *Fla. Field Nat.* 16:57–80.

Maehr, D.S., A.G. Spratt, and D.K. Voigts. 1983. Bird casualties at a central Florida power plant. *Fla. Field Nat.* 11:45–49.

Mertins, J.W., G.W. Foster, M.G. Spalding, and J.L. Morrison. 1996–99. Unpublished data. USDA National Veterinary Services Laboratory, Ames, Iowa.

Meteyer, C.U. 1992–93. Unpublished data. National Wildlife Health Center, Madison, Wis.

Meyer, K.D., and J.A. Smallwood. 1996. Peregrine falcon. In: *Rare and endangered biota of Florida.* Vol. 5, *Birds.* J.A. Rodgers, H.W. Kale II, and H.T. Smith (eds.). University Press of Florida, Gainesville. pp. 52–60.

Miller, D., E.L. Boeker, and R.S. Thorsell. 1975. Suggested practices for raptor protection on powerlines. Raptor Research Foundation, Provo, Utah. 19 pp.

Morrison, J.L. 1996. Crested carcara (*Caracara plancus*). *Birds N. Am.* 249:1–28.

———. 1999. Unpublished data. Department of Wildlife Ecology and Conservation, University of Florida, Gainesville.

Mulhern, B.M., W.L. Reichel, L.N. Locke, T.G. Lamont, A. Belisle, E. Cromartie, G.E. Bagley, and R.M. Prouty. 1970. Organochlorine residues and autopsy data from bald eagles, 1966–68. *Pestic. Monit. J.* 4:141–144.

[NWHRC] National Wildlife Health Research Center. 1985. Bald eagle mortality from lead poisoning and other causes. U.S. Fish and Wildlife Service, Madison, Wis. 48 pp.

Palmer, R.S. 1988. Some raptor interactions (peregrine, *Falco peregrinus*). In: *Handbook of North American birds.* Vol. 5. R.S. Palmer (ed.). Yale University Press, New Haven, Conn. p. 372.

Peakall, D.B. 1976. The peregrine falcon (*Falco peregrinus*) and pesticides. *Can. Field Nat.* 90: 301–307.

Peters, H.S. 1936. A list of external parasites from birds of the eastern part of the United States. *Bird-Banding* 7:9–27.

Porter, R.D., and S.N. Wiemeyer. 1969. Dieldrin and DDT: effects on sparrow hawk eggshells and reproduction. *Science* 165:199–200.

Ratcliffe, D.A. 1967. Decrease in eggshell weight in certain birds of prey. *Nature* 215:208–210.

Ratcliffe, D. 1980. *The peregrine falcon.* Buteo Books, Vermillion, S. Dak. 416 pp.

Redig, R.T., M.R. Fuller, and D.L. Evans. 1980. Prevalence of *Aspergillus fumigatus* in free-living goshawks (*Accipiter gentilis atricapillus*). *J. Wildl. Dis.* 16:169–174.

Reichel, W.L. 1978. Unpublished data. Patuxent Wildlife Research Center, Laurel, Md.

Robertson, W.B., Jr., and G.E. Woolfenden. 1992. *Florida bird species. An annotated list.* Spec. Publ. 6, Florida Ornithological Society, Gainesville. 260 pp.

Rodgers, J.A., Jr., H.W. Kale II, and H.T. Smith (eds.). 1996. *Rare and endangered biota of Florida.* Vol. 5, *Birds.* University Press of Florida, Gainesville. 688 pp.

Sileo, L. 1978. Unpublished data. National Wildlife Health Center, Madison, Wis.

Smallwood, J.A. 1990. American kestrel and merlin. *Proc. Southeast Raptor Manag. Symp. Workshop.* Natl. Wildl. Fed. Sci. Tech. Ser. 14:29–37.

Smallwood, J.A., and K.D. Meyer. 1996. Merlin. In: *Rare and endangered biota of Florida.* Vol. 5, *Birds.* J.A. Rodgers, H.W. Kale II, and H.T. Smith (eds.). University Press of Florida, Gainesville. pp. 616–623.

Snyder, B. 1994. Unpublished data. Florida Department of Environmental Protection, Tallahassee.

Spalding, M.G. 1992. Unpublished data. University of Florida, Gainesville.

Spalding, M.G., and J.L. Morrison. 1995. Unpublished data. University of Florida, Gainesville.

Stabler, R.M. 1969. *Trichomonas gallinae* as a factor in the decline of the peregrine falcon. In: *Peregrine falcon populations, their biology and decline.* J.J. Hickey (ed.). University of Wisconsin Press, Madison. pp. 435–437.

Stevenson, J.A. 1994. Unpublished data. Florida Department of Environmental Protection, Tallahassee.

Stoddard, H.L. 1931. *The bobwhite quail, its habits, preservation and increase.* Charles Scribner's Sons, New York. 559 pp.

Sundlof, S.F., D.J. Forrester, N.P. Thompson, and M.W. Collopy. 1986. Residues of chlorinated hydrocarbons in tissues of raptors in Florida. *J. Wildl. Dis.* 22:71–82.

Suto, B.J. 1996. Unpublished data. Suncoast Seabird Sanctuary, Redington Beach, Fla.

Sutton, G.M. 1945. Behavior of birds during a Florida hurricane. *Auk* 62:603–606.

Tandan, B.K., and V. Dhanda. 1963. *Falcolipeurus josephi,* a new American mallophagan from caracaras of the genus *Polyborus,* and a key to allied species (Ischnocera: Philopteridae). *Ann. Entomol. Soc. Am.* 56:634–639.

Taylor, D.J. 1951. The distribution of ticks in Florida. M.S. thesis, University of Florida, Gainesville. 124 pp.

Taylor, W.K., and B.H. Anderson. 1973. Nocturnal migrants killed at a central Florida TV tower, autumns 1969–1971. *Wilson Bull.* 85:42–51.

———. 1974. Nocturnal migrants killed at a central Florida TV tower, autumn 1972. *Fla. Field Nat.* 2:40–43.

Taylor, W.K., and M.A. Kershner. 1986. Migrant birds killed at the Vehicle Assembly Building (VAB), John F. Kennedy Space Center. *J. Field Ornithol.* 57:142–154.

Telford, S.R., Jr., G.W. Foster, J.K. Nayar, S.P. Terrell, and D.J. Forrester. 1996. Unpublished data. University of Florida, Gainesville.

Thomas, N.J. 1985–93. Unpublished data. National Wildlife Health Center, Madison, Wis.

Trainer, D.O. 1969. Diseases in raptors: a review of the literature. In: *Peregrine falcon populations, their biology and decline.* J.J. Hickey (ed.). University of Wisconsin Press, Madison. pp. 425–433.

Walton, A.C. 1927. A revision of the nematodes of the Leidy Collections. *Proc. Acad. Nat. Sci. Phila.* 79:49–163.

White, F.H., and D.J. Forrester. 1977. Unpublished data. University of Florida, Gainesville.

Wiemeyer, S.N., and J.L. Lincer. 1987. The use of kestrels in toxicology. *Raptor Res. Rep.* 6:165–178.

Wiemeyer, S.N., and R.D. Porter. 1970. DDE thins eggshells of captive American kestrels. *Nature* 227:737–738.

Wojcik, D.P., and J.A. Smallwood. 1993. Red imported fire ant predation on kestrels. In: *Abstracts of the 1993 imported fire ant conference.* Department of Entomology, Clemson University, Clemson, S.C. pp. 120–121.

Wright, S.D. 1986. Unpublished data. University of Florida, Gainesville.

Northern Bobwhites

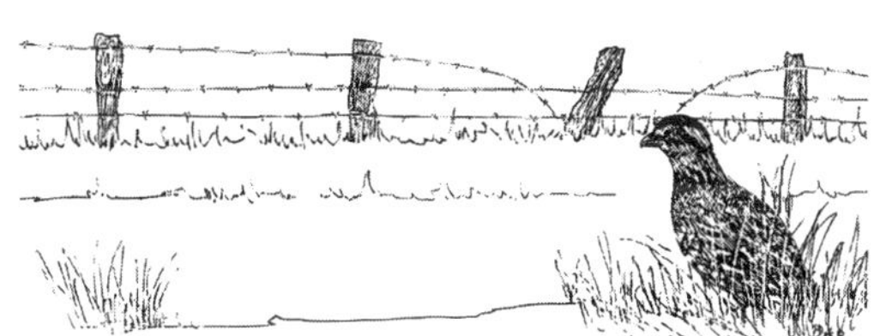

I. Introduction

The Northern Bobwhite, *Colinus virginianus*, is an important upland game bird in Florida. Two subspecies occur: the Florida Bobwhite (*C. v. floridanus*) in southern Florida and the Eastern Bobwhite (*C. v. virginianus*) in northern Florida. The Florida Bobwhite averages 1 ounce less in weight and is darker than the Eastern Bobwhite (Murray and Frye 1964).

The numbers of bobwhites throughout the United States have decreased drastically since the early 1950s, believed to be largely because of habitat changes associated with commercial agriculture and forestry. In the Southeast the expanding forest products industry has resulted in large, short-rotation pine plantations, beneficial to bobwhites for the first several years and then of lesser value as time passes (Crawford 1987). In Florida the numbers of bobwhites have decreased considerably. Annual harvest figures reflect population sizes to some degree and these have gone down steadily from a high of more than 3,000,000 in 1972 to an estimated 163,352 in 1997–98 (Linda 1998). The practice of releasing pen-reared bobwhites, which are the same species as Northern Bobwhites, for purposes of hunting, dog training, and field trials poses a disease risk that must be considered in an analysis of the effects of diseases on free-ranging populations of Northern Bobwhites in Florida (Davidson et al. 1982a; Martin 1984).

The voluminous literature on diseases and parasites of bobwhites has been summarized

(Kellogg and Calpin 1971; Kellogg and Doster 1972) and includes several references to data on bobwhites in Florida. The classic book by Stoddard (1931) contains several chapters on mortality factors, parasites, and diseases, based on material obtained from the Red Hills region in southern Georgia and northwestern Florida over a 4-year period (1924–27). In addition, Davidson et al. (1982a) have published a very useful overview of the significance of parasitism and disease in bobwhites in the southeastern United States. Much of the data included in the latter publication came from bobwhites studied at Tall Timbers Research Station (Leon County), which is also in the Red Hills region.

II. Trauma

Trauma (gunshot, impact, and predation) was the primary diagnostic finding in 38 of 127 (30%) bobwhites submitted to the Southeastern Cooperative Wildlife Disease Study (SCWDS) for determination of cause of morbidity or mortality during 1972–1981 (Davidson et al. 1982a). These birds originated from various southeastern states including Florida, although the specific numbers from Florida were not given.

Stoddard (1931) listed a number of cases of trauma due to bobwhites flying into fences, telegraph and telephone wires, and white buildings, but felt that none of these was a significant cause of mortality. Crawford (1981) listed 2 bobwhites as victims of mortality due to flying into a TV tower. His findings were based on >42,000 birds found at the base of such a tower near Tall Timbers Research Station (Leon County) over a 25-year period (1955–80).

As mentioned in the Introduction, above, an estimated 163,352 bobwhites were killed in Florida by hunters in 1997–98 (Linda 1998). Crippling losses due to hunting were studied over a 10-year period at Tall Timbers Research Station (Leon County) by Doster et al. (1982). They found that such mortality ranged from 15 to 29% (mean = 22%) of the total annual kill

and did not correlate with fluctuations in the density of bobwhite populations.

III. Predation

Stoddard (1931) gave detailed accounts of mortality of bobwhites due to predation by mammals, birds, and reptiles. From 1924 to 1927 he and his assistants observed 602 nests in the southern Georgia and northwestern Florida region and found that 223 (37%) of these nests were destroyed by predators. At least 16 different predators were identified, the most important being skunks, native fire ants, cotton rats, and domestic dogs (table 16.1).

Stoddard described ant predation as "one of the most pitiful forms of destruction encountered in the study, for the helpless baby bobwhites are literally eaten alive by the swarms of ants. Chicks have been removed from eggs alive and still squirming, with most of the flesh eaten off on one side of the face and along the vertebrae" (Stoddard 1931: 193). He stated that the ants involved were "small red, or thief, ants (*Solenopsis molesta*)," but E.V. Komarek, who worked with Stoddard, stated later (Komarek 1978) that the ants were not *S. molesta*, but were actually *S. geminata*, the native fire ant. Red imported fire ants (*S. invicta*) (RIFAs) were first identified in Florida in Escambia County in 1948, although there is evidence that they had been in the area for at least 10 years (Brown 1978). There is no published information on the effects of RIFAs on bobwhites in Florida, but harmful effects have been recorded from Texas (Allen et al. 1995; Giuliano et al. 1996; Mueller et al. 1999). These harmful effects include the killing and eating of bobwhite chicks at the time of hatching and reductions in chick survival and body mass. However, Brennan (1991, 1993), while acknowledging that fire ants do some damage, expressed doubt that RIFAs have significant negative effects on populations of bobwhites. This phenomenon should be investigated in Florida. Allen et al. (1998) have provided a brief overview of the ecological effects of RIFAs on native verte-

Table 16.1. Destruction of Northern Bobwhite nests by predation on eggs or young during the incubation or hatching period[a]

Predator	No. nests destroyed	Percent of total
Skunks[b]	65	29
Native fire ants[c]	24	11
Cotton rats	21	9
Domestic dogs	19	9
Snakes[d]	14	6
Domestic cats	13	6
Humans[e]	10	4
Virginia opossums	7	3
Blue Jays	6	3
Crows[f]	5	2
Domestic turkeys	4	2
Wild Turkeys	1	<1
Unknown	34	15
Total	223	—

Source: Stoddard (1931).
a. These data came from H.L. Stoddard's (1931) classic study of the life history of the Northern Bobwhite in the southern Georgia/northern Florida region.
b. Data combined for striped skunks and spotted skunks.
c. Stoddard (1931) stated that these were red or thief ants (*Solenopsis molesta*), but according to Komarek (1978), they were native fire ants (*S. geminata*).
d. Includes coachwhips, racers, and common kingsnakes.
e. Nests were robbed and eggs consumed.
f. Data combined for American Crows and Fish Crows.

brates, including the Northern Bobwhite, and the reader is referred to this article for more information on this topic.

With a few exceptions Stoddard's data on predation on young and adult bobwhites are anecdotal rather than quantitative. He lists the following as important predators in the Red Hills area: house cats, domestic dogs, Cooper's Hawks, Sharp-shinned Hawks, coachwhips (*Masticophis flagellum*), common kingsnakes (*Lampropeltis getulus*) and racers (*Coluber constrictor*). Fragments of Northern Bobwhites were found in 4 of 1,098 pellets collected from a Northern Harrier roost in Leon County during the winter of 1925–26 (Stoddard 1931). Remains of Northern Bobwhites were identified in stomach contents of 6 of 413 bobcats collected from 1977 to 1983 in Florida; specific localities were not given (Maehr and Brady 1986). The significance of predation as a mortality factor for bobwhite populations is not clear. Dimmick (1990) examined the literature on the effects of raptor predation on bobwhites and concluded that "there is little evidence that raptors remove sufficient numbers of bobwhites to depress the population below the habitat-imposed limitation." Nest predation, which is not related to raptors, seems to be a more important factor, but its significance to bobwhite populations in Florida is not known.

IV. Inclement weather

Stoddard (1931) stated that some bobwhite chicks may drown during rainstorms if they get into ditches and are separated from their hen and her brood, but gave no data to substantiate this claim. In September 1926, a hurricane came through one of Stoddard's study areas west of Pensacola (Escambia County) and provided him an opportunity to document the ef-

Table 16.2. Concentrations of mirex in stomach contents and various tissues of Northern Bobwhites examined before and after application of 10-5 bait for fire ant control in Duval County, Florida, 1972

| | Residues: ppm wet weight[a] (sample size) | | |
Tissue	Pretreatment	1 month posttreatment	6 months posttreatment
Brain	ND (3)	0.15 (1)	ND (2)
Fat	ND (3)	NT	NT
Liver	ND (3)	0.10 (1)	0.04 (2)
Muscle	ND (3)	ND	ND (2)
Stomach contents	ND (3)	NT	ND (2)

Source: Wheeler et al. (1977).
ND = none detected, NT = not tested.
a. Lowest limit of detection = 0.01 ppm.

fects of inclement weather on coveys of young birds. It was concluded that the losses were only slight.

The effects of inclement weather on nesting success are of much more concern. Stoddard (1931) recorded 36 of 120 (30%) nest desertions due to rain, floods, or drought. Many nests are built in low areas and flooded out during heavy rains, others are covered over with mud. Stoddard (1931: 185) stated that "Late in the season, however, bobwhites are so loath to desert their treasures that several instances came to light where buried eggs were dug out by their owners and hatched in due time. In one instance a hen remained on her eggs while mud was deposited over and around her." He also claimed that during times of drought many bobwhite eggs spoiled because of premature incubation caused by the heat.

V. Environmental contaminants

Experimental studies have shown that bobwhites are susceptible to chlorinated hydrocarbons (Dahlen and Haugen 1954; DeWitt 1955), organophosphates (Stromborg 1986a, 1986b), and other contaminants (Hoffman 1988). Effects include mortality, decreased hatchability of eggs, and decreased viability of chicks, depending on the compound and dosage. In addi-

tion, it has been shown experimentally, using pen-reared bobwhites, that DDT causes increased susceptibility to the harmful effects of infectious enterohepatitis (= histomoniasis or blackhead) (Thompson and Emerman 1974).

There is little information on the occurrence and effects of such contaminants on native bobwhites in the Southeast or in Florida. Percival et al. (1972) reported residues of DDT and mirex in tissues of native bobwhites in South Carolina, and Davidson et al. (1982a) reported azodrin as the primary cause of death in 4 bobwhites from Baker County in southern Georgia. Whisenhunt (1959) stated that the death of an unspecified number of bobwhites occurred in Florida (date and locality not given) because of the application of dieldrin at 2 pounds per acre. No further details were given and residue studies apparently were not conducted. Wheeler et al. (1977) reported only small concentrations of mirex residues in several bobwhites examined 1 and 6 months after application of mirex for fire ant control in Duval County (table 16.2). The effects of mirex on bobwhites in Florida are unknown. Baker (1965), however, conducted field tests and experimental studies with mirex and found no direct effects on adult or nearly grown bobwhites. He concluded that less obvious effects on reproduction and on newly hatched chicks should be investigated.

Murray and Frye (1964) reported that

there were numerous claims of mortality to bobwhites related to usage of insecticides in cotton and peanut fields in western Florida. They gave no further information, however, and apparently no residue studies were conducted.

The significance of contamination by organophosphorus (OP) and carbamate (CA) pesticides to bobwhite populations in Florida has not been investigated. As pointed out by Smith (1993) and White et al. (1990) the use of these types of pesticides has increased in recent years because they act quickly, are relatively short lived, and don't accumulate in nature, in contrast to organochlorines, which possess the opposite characteristics. A radiotelemetry study of the survival of bobwhites on a plantation in Georgia during 2 successive years was reported by White et al. (1990). During 1 year, only 6 applications of OP and CA pesticides were used on the crops being grown (peanuts, pecans, corn, wheat, and sorghum), whereas during the other year considerably more (19 applications) was applied. Mortality during the year of increased pesticide application was significantly higher in bobwhite populations than during the year of lower application. Brennan (1991) pointed out that the indirect effects of various pesticides such as organophosphates on bobwhite populations are poorly known; thus pesticides may adversely affect bobwhites by suppressing populations of arthropods that are important food items for them. Research on these aspects has not been conducted in Florida.

Stoddard (1931) reported mortality due to lead poisoning in pen-reared bobwhites maintained on ground over which shooting had been heavy in past years. He also recorded lead poisoning in a bobwhite from Texas that had been banded and released in Leon County. The bird was emaciated and partially paralyzed; at necropsy 2 eroded lead shot were found in the gizzard. It is obvious, therefore, that lead shot picked up as grit can result in lead poisoning in bobwhites, but Stoddard (1931) did not feel that such cases were very common. An adult female bobwhite from a plantation in Jefferson County was diagnosed to have died of chronic lead poisoning in 1992 (SCWDS records). The bird weighed 84 grams, had no subcutaneous body fat, and 1 eroded lead shot was found in the gizzard; lead residues in the liver were 125 ppm. Further details on lead poisoning can be found in chapters 9 (Whistling-Ducks, Swans, and Geese), 10 (Ducks), and 11 (Eagles) and in 2 publications on experimental studies using pen-reared bobwhites (McConnell 1968; Beyer et al. 1988).

VI. Neoplasia

No information is available on this topic.

VII. Biotoxins

Aflatoxicosis is a disease caused by mycotoxins produced by molds (*Aspergillus flavus* and *A. parasiticus*) that grow on corn, peanuts, cottonseed, and other substrates. Young ducklings and turkey poults have been found to be especially susceptible. Acute effects include hepatitis, necrosis of liver cells, prolonged blood clotting time, hemorrhage, and death. Prolonged ingestion of aflatoxins causes liver damage, depressed protein synthesis, and hepatic tumors (Pier 1980). Davidson et al. (1982a) reported on a 3-year study of aflatoxicosis at Tall Timbers Research Station in Leon County. They found that aflatoxins were present cyclically in the grain crops (mainly corn) of that area, were ingested by bobwhites, and sometimes produced slight to moderate pathologic effects. They also conducted some controlled feeding experiments with pen-reared bobwhites using concentrations of aflatoxin that occur normally in corn (0.5, 1.0, 2.0, and 4.0 ppm). Significant effects were seen only in birds receiving the highest concentration of aflatoxin. Among the effects was decreased egg production. The authors concluded, however, that aflatoxicosis alone was "not a regular or major mortality factor in wild bobwhites" (Davidson et al. 1982a).

Table 16.3. Avian pox infections in Northern Bobwhites from Florida

| County | Year(s) | Number of bobwhites | | | Data source |
		Exam.	Positive	%	
Alachua	1971–72	17	0	0	Forrester (1987)
Charlotte	1979–80	44	5	11	Davidson & Kellogg (1987)
	1980–81	104	5	5	Ibid.
	1981–82	82	18	22	Davidson & Kellogg (1982)
	1982–83	73	5	7	Davidson (1983)
	1983–84	307	10	3	Davidson (1984)
	1984–85	365	5	1	Doster & Davidson (1993)
	1985–86	541	21	4	Ibid.
	1986–87	692	38	5	Ibid.
	1987–88	456	53	12	Ibid.
	1988–89	555	72	13	Ibid.
	1989–90	468	68	15	Ibid.
	1990–91	204	31	15	Ibid.
	1991–92	0	—	—	Ibid.
	1992–93	247	51	21	Ibid.
Citrus	1978–79	18	0	0	Davidson et al. (1980a)
	1979–80	64	0	0	Davidson et al. (1982b)
	1980–81	26	0	0	Ibid.
Gadsden	1978–79	2	2	—	SCWDS (1979)
Glades	1971–72	28	0	0	Forrester (1987)
	1982–83	37	0	0	Ibid.
Jefferson	1978–79	114	4	4	Davidson et al. (1980a)
Leon[a]	1924–29	"several thousand"	NG	<2	Stoddard (1931)
Leon	1969–78	>8,300	NG	<1	Davidson et al. (1980a)
	1978–79	902	80	9	Ibid.
	1979–80	829	2	<1	Davidson et al. (1982b)
	1980–81	925	21	2	Ibid.
	1981–82	557	29	5	Davidson & Kellogg (1987)
	1982–83	736	3	<1	Ibid.
	1984	638	2	<1	Doster & Davidson (1993)
	1985	93	2	2	Ibid.
	1986	141	7	5	Ibid.
	1987	197	0	0	Ibid.
	1988	177	3	2	Ibid.
	1989	211	2	1	Ibid.
	1990	400	2	<1	Ibid.
	1991	622	2	<1	Ibid.
	1992	293	3	1	Ibid.
	1993	369	0	0	Ibid.
	1997	1	1	—	Quist (1997)
Marion	1971	19	0	0	Forrester (1987)
Osceola	1978–79	3	0	—	Davidson et al. (1980a)
	1979–80	58	1	2	Davidson et al. (1982b)
Santa Rosa	1980–81	23	2	9	Ibid.
Unknown	1980–81	59	0	0	Ibid.

NG = not given.
a. Stoddard's data came mostly from Leon County, but he also included in his figures data from several other southeastern states.

Table 16.4. Viral infections other than avian poxvirus in Northern Bobwhites from Florida

Virus	County or site	Year(s)	Exam.	No. bobwhites Positive	%	Method of diagnosis	Data source
Adenovirus (TR-59)							
	Leon	1976	37	2	5	Virus isolation	King et al. (1981)
Quail bronchitis virus (QBV)							
	Leon	1976	40	9	23	Serology	Ibid.
	NG	NG	+[a]	—		Virus isolation	Wiseman (1979)
	NG	NG	NG	+[b]	—	Serology	Ibid.
HE/MSD adenoviruses[c]							
	NG	NG	19	0	0	Serology	Domermuth et al. (1977)
Newcastle disease virus							
	Leon	1977–79	381	0	0	Serology	Davidson et al. (1982a), Davidson (1987)
Influenza A virus							
	Leon	1977–79	381	0	0	Virus isolation	Ibid.
SLE virus							
	Cent. Fla.	1960	14	1	7	Serology	Henderson et al. (1962)
	NG	1965–74	6	0	—	Serology	Bigler et al. (1975)
Everglades virus							
	NG	1965–74	4	0	—	Serology	Ibid.
EEE virus							
	Cent. Fla.	1960	3	1	—	Serology	Favorite (1960)
	NG	1965–74	6	0	—	Serology	Bigler et al. (1975)
	Osceola	1992–93	4	0	—	Serology	Spalding & McLean (1994)
Highlands J virus							
	Cent. Fla.	1960	14	1	7	Serology	Henderson et al. (1962)
	NG	1960	3	0	—	Serology	Bigler et al. (1975)

NG = not given.

a. Wiseman (1979) isolated QBV from bobwhites from Tall Timbers Research Station, but did not state how many bobwhites were used or how many were positive for the virus.

b. Wiseman (1979) found titers to QBV in 126 birds, but did not state how many were examined and found to be negative.

c. HE/MSD = Hemorrhagic enteritis of turkeys and marble spleen disease of pheasants.

VIII. Viruses

Six viruses have been identified from bobwhites in Florida (tables 16.3 and 16.4). These include avian poxvirus, 2 adenoviruses, and 3 arboviruses. Attempts to identify other viruses such as Newcastle disease virus, influenza A virus, and hemorrhagic disease/marble spleen disease virus by serologic studies or by isolation attempts have been unsuccessful (table 16.4).

The most important viral disease of bobwhites is avian pox. This disease is caused by poxviruses that are transmitted from bird to bird by physical contact with lesions, by inhalation of viral particles in dust, or by blood-sucking arthropods such as mosquitoes (Cunningham 1978; Davidson et al. 1982b). Infected birds have been documented in 6 counties in Florida (figure 16.1). Prior to 1978 pox was present in bobwhites in Florida, but occurred

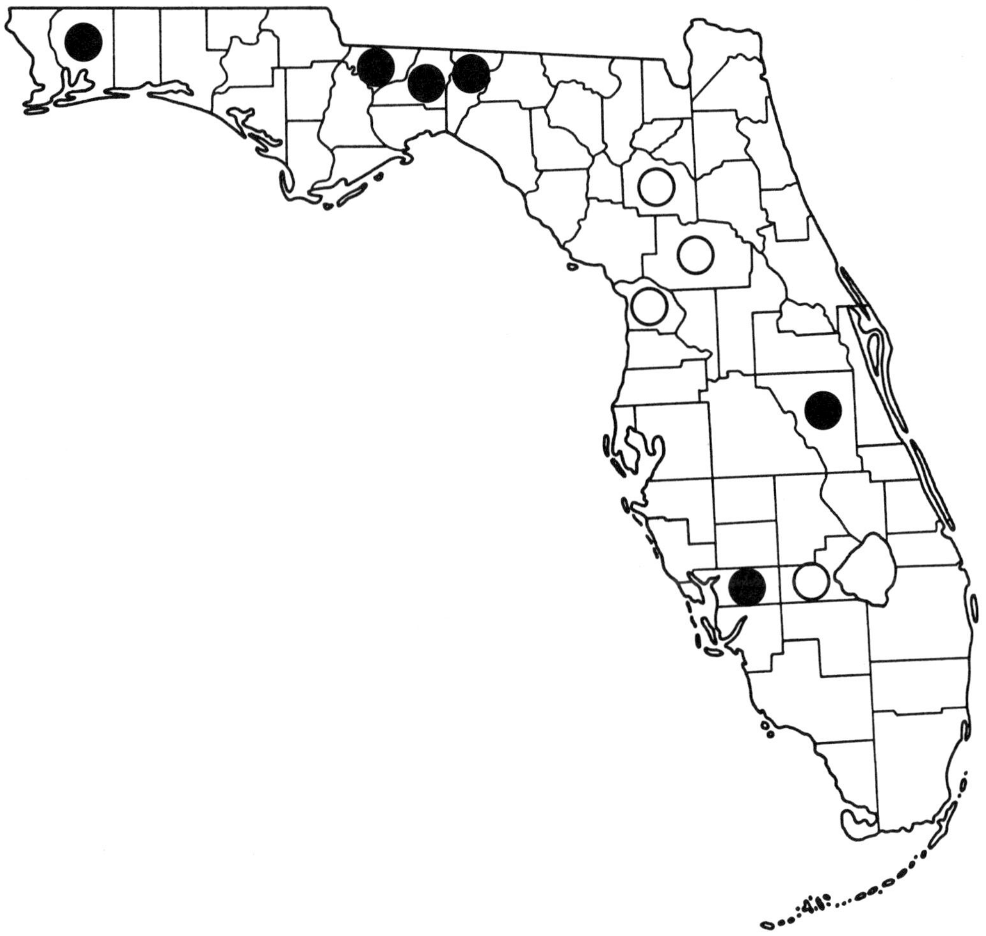

FIGURE 16.1. Distribution of avian pox in Northern Bobwhites in Florida. *Solid circles* = counties where pox-infected birds have been identified; *open circles* = counties where bobwhites have been examined and were negative for poxvirus infections. Data from Table 16.1.

sporadically at a low prevalence (around 1% to 2%) (table 16.3). During 1978–79 an outbreak occurred in the southwestern Georgia/north central Florida region (Davidson et al. 1980a), with an estimated 12-fold increase in the prevalence of pox. The epizootic was first detected in July 1978 and continued until at least March 1979. In Florida infected birds came from Gadsden, Leon, and Jefferson counties. During February 1979, 9% of 902 bobwhites in Leon County were infected. Although the prevalence varied greatly among specific localities, Davidson et al. (1982a) calculated that the overall

morbidity and mortality rates were approximately 2% and 1% respectively for the entire southwestern Georgia/north central Florida region, an area comprising 13,000 square kilometers. All segments of the population were affected equally since no differences in prevalence were noted for age or gender. During the 2 years following the outbreak, the prevalence of pox declined to about 1/3 of that during the epizootic and during the next 12 years prevalences varied from 0 to 5.2% (table 16.3).

Subsequent to the 1978–79 outbreak, bobwhites were examined from several other areas

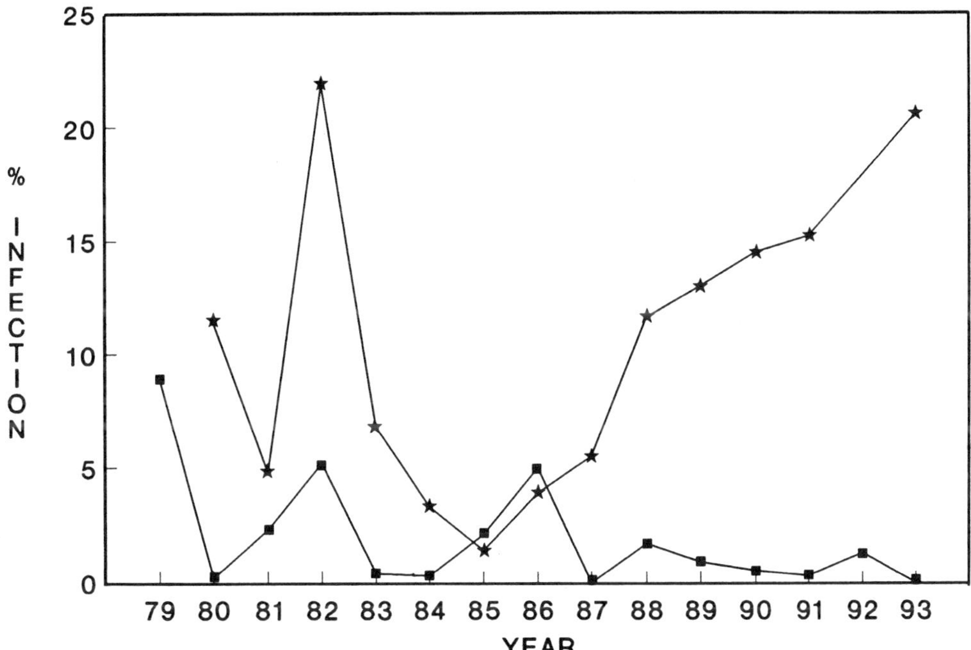

FIGURE 16.2. Prevalences of avian pox in Northern Bobwhites in Charlotte *(stars)* and Leon *(squares)* counties, 1979–93. Sample sizes are given in Table 16.3. No bobwhites were sampled in Charlotte County in 1992. Bobwhites were examined in Leon County during February of each year; bobwhites from Charlotte County were examined over a 4-month period (November–February) each year. Data courtesy of Gary L. Doster and William R. Davidson.

in Florida. Infected birds were found in Santa Rosa, Osceola, and Charlotte counties (table 16.3). A long-term study was conducted on the Cecil Webb Wildlife Management Area (Charlotte County) from 1979 through 1993 (Doster and Davidson 1993). The prevalence of infection varied from a low of 1.4% in 1984–85 to a high of 21.9% in 1981–82. After the high in 1981–82, prevalences declined for several years and then climbed again during 1988–93. Annual prevalences of pox in bobwhites from Charlotte County and Leon County are compared in figure 16.2. These data should be interpreted with caution, however, since the bobwhites in Leon County were sampled in February of each year, several months after the primary poxvirus transmission season. This means that many of the pox lesions would have had time to regress by the time the birds were examined each year and so the values should be

considered minimal for Leon County. On the other hand, the bobwhites from Charlotte County were sampled over the 4-month period of November–February each year; some of the birds would have been examined closer to the transmission season and therefore prevalences of pox lesions would have been higher. Other factors causing the higher prevalences of pox in Charlotte County compared with Leon County might include weather (particularly rainfall), mosquito populations, and the introduction of infected pen-reared bobwhites into the area. It is also possible that pox is enzootic at a higher level of prevalence in southern Florida (Charlotte County) than in the northern part of the state (Leon County).

Lesions caused by poxvirus in bobwhites are limited usually to the unfeathered areas of the legs, feet (figure 16.3), and skin around the eyes and beak (figure 16.4), and the oral (figure

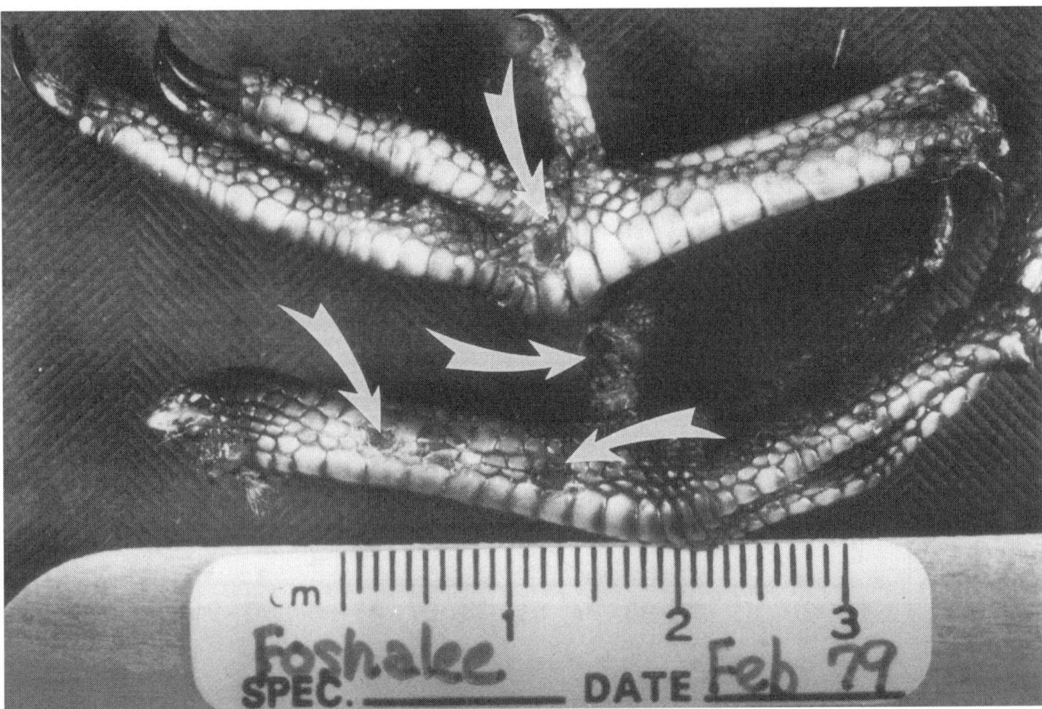

FIGURE 16.3. Lesions of avian pox *(arrows)* (= dry pox) on the feet of a Northern Bobwhite from Leon County, 1979. Courtesy of William R. Davidson.

16.5) and nasal cavities (Davidson et al. 1980a). The lesions on the feet and legs are generally not serious, whereas those involving the eyelids and the mucous membranes of the mouth, nasal passages, and upper respiratory tract are often of serious consequence, with blindness and impairment of feeding and breathing as possible results (Davidson et al. 1982b). The following description of the lesions observed during the 1978–79 epizootic in Georgia and Florida was provided by Davidson et al. (1980a):

Lesions on the legs consisted of circumscribed, raised hyperplastic nodules 1 to 5 mm. in diameter. Removal of loosely attached scabs from leg lesions left smooth light colored scars. Lesions involving the eyes included proliferative nodules externally on the eyelids and hyperplasia of the palpebral conjunctiva. Secondary infections and inflammation were common with eye le-

sions. Occasionally, the eyelids were completely sealed by dried exudate. The nasal passages and infraorbital sinuses often were completely occluded by hyperplastic nodules. Oral lesions ranged from discrete hyperplastic nodules to extensive diphtheritic ulcers and were most common on the base of the tongue, pharynx and palate. Lesions on the pharynx and palate occasionally occluded respiratory orifices. Birds with severe eye and mouth lesions often were emaciated.

About 77% of the infected bobwhites examined had lesions classified as mild, whereas 23% had lesions judged to be extensive. The latter resulted in impaired vision, breathing, and/or feeding (Davidson et al. 1980a). Infected birds also would be more vulnerable to predation. Mueller et al. (1993) investigated the effects of pox on free-ranging bobwhites and found that birds with wet pox infections

FIGURE 16.4. Lesions of avian pox *(arrows)* (= dry pox) on the eyelid and commisure of the beak of a Northern Bobwhite from Leon County, 1979. Courtesy of William R. Davidson.

had lower 6-week survival rates than bobwhites with dry pox or no pox infections (see chapter 17, Wild Turkeys, for descriptions of wet and dry pox). They also found that bobwhites with dry pox infections were not more susceptible to predation than uninfected birds. Their study took place in Grady County in southern Georgia, not far from Tall Timbers Research Station in Leon County.

Mosquitoes are important mechanical vectors of avian poxvirus in Wild Turkeys in southern Florida (Akey et al. 1981); a similar situation with pox infections in bobwhites may have existed in Leon County in 1978–79 and in Charlotte County in 1981–82. The circumstances causing or influencing the outbreak of pox in bobwhites from southwestern Georgia and north central Florida in 1978–79 are not completely understood. Avian pox was widespread and represented a severe problem in a number of

pen-reared bobwhite facilities in Florida and Georgia during the summer and fall of 1978. It is possible that the release of infected pen-reared bobwhites was related to the outbreak in wild birds, but Davidson et al. (1982b) concluded that there was no such relationship since pox occurred in increased prevalences in areas such as Tall Timbers Research Station (Leon County) where pen-reared birds were not present. However, the poxvirus may have spread to bobwhites at Tall Timbers Research Station from adjacent lands where pen-reared bobwhites were released. Davidson et al. (1982b) stated that the release of pen-reared birds may have resulted in an intensification of the disease problems in several localized areas. A number of environmental factors (temperature, rainfall, etc.) may have interacted to cause increased transmission of the poxvirus via mosquitoes. In addition other factors may have been important such as increased susceptibility of bobwhites to endemic strains of the virus or the introduction of new virulent strains into the populations.

Two related adenoviruses have been isolated from bobwhites at Tall Timbers Research Station. Quail bronchitis virus (QBV) was isolated from bobwhites by Wiseman (1979), although he gave no data on prevalence. King et al. (1981) found that 9 of 40 (23%) bobwhites were positive for QBV-neutralizing antibody. Wiseman also conducted serologic studies and found that young birds (7–9 months of age) had higher titers than older birds (17–21 months). QBV is known to cause acute respiratory disease accompanied by high mortality in pen-reared bobwhites, but such effects have not been documented in wild bobwhites (Davidson et al. 1982a). A second adenovirus (TR-59) was isolated from 2 of 37 bobwhites by King et al. (1981). This virus was associated with intranuclear inclusion bodies in the livers of many bobwhites, although Davidson et al. (1982a) stated that such lesions might be due to either QBV, TR-59 virus, or both. Small lesions consisting of necrotic cells, lymphocytic infiltration, and microgranulomas were associated also with the inclusion bodies in the liver. These inclusions were more prevalent in juveniles (46%) than in adults (27%), but there were no differences in the prevalences of infection between males and females. The overall prevalence of birds with inclusions varied from around 10% or less to 60% between 1975 and 1980. Clinical disease was not associated with infections by TR-59 virus or the presence of inclusion bodies; therefore the pathologic effects on bobwhite populations are not clear (Davidson et al. 1982a).

Only 1 bobwhite has been found seropositive for eastern equine encephalitis (EEE) virus, 1 for St. Louis encephalitis (SLE) virus, and 1 for Highlands J virus. These 3 birds were collected in central and south central Florida (table 16.4). Additional bobwhites examined from other parts of the state have been negative, but sample sizes were small. These viruses are probably of no pathological significance to bobwhite populations.

IX. Bacteria

A number of bacterial diseases have been reported from bobwhites, including avian tuberculosis, mycoplasmosis, erysipelas, fowl cholera, fowl typhoid, paratyphoid, pullorum disease, tularemia, and ulcerative enteritis or "quail disease" (Kellogg and Doster 1972). These diseases are more typical of pen-reared birds and have not been reported in free-ranging bobwhites in Florida (Stoddard 1931; Davidson et al. 1982a; Davidson 1986–87).

White and Forrester (1987) cultured livers and large intestines of 50 bobwhites from Alachua and Marion counties (*n* = 21), Glades County (*n* = 20), Charlotte County (*n* = 6), and Hillsborough County (*n* = 3) during 1971–74. These were either killed by hunters or live-trapped and therefore could be considered to be normal. None of the cultures was positive for enteric bacterial pathogens. *Escherichia coli* was cultured and identified from the contents of the large intestines of 2 birds, 1 from Alachua County in 1971 and 1 from Glades County in 1973. This bacterium is probably a normal component of the intestinal flora.

Davidson (1979) reported 1 case of bacterial infection of a bobwhite from Tall Timbers Research Station (Leon County) in 1979. The infection was not cultured so the species of bacteria was not identified. This was a secondary infection associated with trauma to the abdominal cavity. Fischer (1999) examined a Northern Bobwhite from Tall Timbers Research Station that had been found in the field in a weakened condition in 1999. At necropsy it was found to be emaciated and with necrotizing hepatitis. *Pseudomonas aeruginosa* was cultured from the liver (as well as the intestines), but its relationship to the hepatitis was not clear.

X. Fungi

Aspergillosis of the respiratory tract and candidiasis of the esophagus and crop have been reported from bobwhites in North America (Kellogg and Doster 1972). Four cases of mycotic infection were diagnosed from 127 wild bobwhites from southeastern United States during 1972–81 (Davidson et al. 1982a). Three of these were considered primary disease factors and 1 was secondary. These infections included aspergillosis and airsaculitis (*Mucor* sp.). None of the 4 birds was from Florida, however.

Fungal dermatitis was seen in an adult female bobwhite that was live-trapped at Tall Timbers Research Station in 1989 (Davidson 1989). The skin on the undersides of both wings was dry, yellow, and scaly. *Aspergillus* sp. and *Mucor* sp. were isolated from scrapings of the skin. Such fungal dermatoses are rare in bobwhites and this is the only such case seen in more than 10,000 bobwhites examined at Tall Timbers Research Station over the years (Davidson 1989).

XI. Protozoans

Four species of protozoans have been reported from bobwhites in Florida. These include 3 intestinal coccidia and 1 species of malaria.

Cram et al. (1931) reported coccidia in 19 of 78 (24%) bobwhites from the Tall Timbers Research Station area. While the species of coccidium involved was not given, a footnote was added stating that according to Tyzzer (1929), *Eimeria dispersa* was the species occurring in bobwhites in the southeastern United States. Ruff (1985) later found 3 distinct types of oocysts in feces from bobwhites from Tall Timbers Research Station. Two of these were unidentified species and 1 was described as a new species, *E. lettyae*. Prevalences of infection were not given. It was learned from experimental studies that *E. lettyae* was specific for bobwhites and would not infect domestic turkeys and other gallinaceous birds (Ruff 1985). Young infected pen-reared bobwhites were susceptible to reduced weight-gains and increased mortality, whereas mature infected birds did not die, but showed reduced egg production

and reduced fertility (Ruff and Wilkins 1987). The effects of coccidiosis on wild bobwhites are not known. Cram et al. (1931) examined large numbers of wild bobwhites in Leon County and found that none had coccidial lesions. They did state, however, that coccidiosis was a problem in pen-reared birds in the Tallahassee region, especially in chicks that suffered weight loss and mortality.

Bobwhites in Florida are known to be infected by only 1 species of blood protozoan, *Plasmodium hermani*. Forrester et al. (1987) examined 119 bobwhites from Alachua/Marion counties (*n* = 33), Leon County (*n* = 29), and Glades County (*n* = 57). One bobwhite from Fisheating Creek Wildlife Management Area (Glades County) was infected with *P. hermani*. This infection was detected by inoculation of whole blood into domestic turkey poults. *Plasmodium hermani*, as will be discussed in more detail in the following chapter, is transmitted by culicine mosquitoes and considered primarily a parasite of Wild Turkeys. This bobwhite strain of malaria was maintained in domestic turkeys and later transmitted via mosquitoes (*Culex nigripalpus* and *C. salinarius*) to other turkeys (Forrester et al. 1987). Strains of malaria from Wild Turkeys have been transmitted to pen-reared bobwhites by inoculation of infected whole blood (Telford and Forrester 1975) and by mosquito transmission (Nayar et al. 1982). These studies indicated that the Northern Bobwhite is a suitable and occasional host for *P. hermani* and may serve as a reservoir host in some areas for this Wild Turkey parasite. The significance of this parasite to natural populations of Northern Bobwhites is probably minor because of its low prevalence.

Nineteen bobwhites from various localities in Florida were examined serologically by the indirect hemagglutination (IHA) test for antibodies to *Toxoplasma gondii* (Burridge et al. 1979). All were negative. However, as pointed out elsewhere, the IHA test may not be as sensitive in birds as in mammals and may result in false-negative results (Frenkel 1981). There are no records of toxoplasmosis in wild bobwhites in

Florida. Dubey et al. (1993) found that experimentally inoculated pen-reared bobwhites were relatively resistant to clinical toxoplasmosis.

Infectious enterohepatitis (also known as histomoniasis or blackhead) is caused by the protozoan *Histomonas meleagridis* and is a serious problem in certain gallinaceous birds (Reid 1967; Kellogg and Reid 1970). This disease is transmitted by the cecal nematode *Heterakis gallinarum* and occurs in Wild Turkeys in a number of areas in the Southeast, including Florida; it will be discussed in more detail in chapter 17, Wild Turkeys. Cram et al. (1931) found lesions in Northern Bobwhites that were similar to those of infectious enterohepatitis, but did not identify *Histomonas meleagridis,* the etiologic agent of this disease, in these lesions. She did not give the origin of these birds, but they may have been from the Tall Timbers Research Station area in Leon County. However, other studies conducted at Tall Timbers Research Station resulted in examination of large numbers (>1,700) of bobwhites at necropsy (Kellogg and Prestwood 1968; Davidson et al. 1980b, 1982a, 1991; Forrester et al. 1984; Moore et al. 1986; Davidson 1987) and no cases of infectious enterohepatitis were seen. This is probably because bobwhites in Florida typically harbor the cecal worm *Heterakis isolonche* rather than *H. gallinarum. Heterakis isolonche* does not transmit *Histomonas meleagridis* (Davidson et al. 1978) and therefore infectious enterohepatitis is usually absent unless *H. gallinarum* (infected with *H. meleagridis*) is introduced into the bobwhite population. This introduction could be accomplished by the release of infected pen-reared bobwhites (Kellogg and Prestwood 1968) or by contamination from domestic chickens or other pen-reared game birds that commonly harbor *H. gallinarum.* Davidson et al. (1991) found very small numbers of *H. gallinarum* (1 or 2 per bird) in 5 of 700 bobwhites examined at Tall Timbers Research Station from 1971 to 1984. These were always mixed infections along with *H. isolonche* and might serve as a means for the spread of infectious enterohepatitis in the future. At the present time, however, this disease does not seem to be a problem in the bobwhite populations of Florida.

XII. Helminths

The helminth fauna of bobwhites in Florida has been well studied. Three investigations have resulted in a wealth of information. Two of these were at Tall Timbers Research Station in Leon County (Moore et al. 1986; Davidson et al. 1991) and 1 was a statewide study with special emphasis on the Cecil Webb Wildlife Management Area in Charlotte County (Forrester et al. 1984). The Charlotte County investigation and 1 of the Leon County studies were long term and involved hundreds of bobwhites.

Thirty-one species of helminths have been reported from bobwhites in Florida. These include 20 nematodes, 3 trematodes, 7 cestodes, and 1 acanthocephalan (tables 16.5 and 16.6). The gizzard nematode *Cyrnea colini* was the most prevalent helminth on a statewide basis (82.2%). However, while *C. colini* was the most abundant helminth in Charlotte County (abundance = 3), the cecal nematodes *Trichostrongylus cramae* (= *T. tenuis*) and *Heterakis isolonche* were by far the most abundant helminths in Leon County (abundances = 66 and 37, respectively). Six helminths (5 nematodes and 1 cestode) occurred in prevalences >20% and were distributed widely throughout the state; these therefore should be considered characteristic species of Northern Bobwhites in Florida (table 16.7). However, there are differences in the helminth faunas in various parts of Florida. Three of the characteristic species (*T. cramae, H. isolonche,* and *C. colini*) were more prevalent in Leon County than in other counties in the state and a fourth species (*Tetrameres pattersoni*) was more prevalent in Leon and Alachua counties (Forrester et al. 1984). *Ascaridia galli,* on the other hand, was prevalent in Charlotte County (36%) and virtually absent from other parts of the state, while *Cheilospirura spinosa* was fairly common in Leon County (23% prevalence) but, except for 1 infected bird in Mar-

Table 16.5. Nematode infections in Northern Bobwhites[a] from Florida

Nematode (Site)[b] County or site	Year(s)	Exam.	Infected	%	Mean	Range	Data source
Aproctella stoddardi (BC)							
Alachua & Wakulla	1924–27	NG	NG	—	NG	NG	Cram et al. (1931)[c]
Alachua	1971–72	33	15	45	2.8	1–6	Forrester et al. (1984), Forrester (1987)
Charlotte	1972–77	381	8	2	1.3[e]	1–3[e]	Ibid.
Glades	1971–73	31	8	26	2.1	1–5	Ibid.
Leon	1928	64	7	11	13.0	NG–49	Cram et al. (1931)[c], Cram (1939)
	1968–69	185	6	3	3.0	1–5	Davidson et al. (1980b)
	1971–75	21	4	19	5.0	1–17	Forrester et al. (1984), Forrester (1987)
	1971–84	700	31	4	1.8	NG–7	Davidson et al. (1991)
	1989	1	1	—	1.0	—	Hayes (1989)
Ascaridia compar (SI)							
"Florida"	1800s	NG	NG	—	NG	NG	Walton (1927)
Ascaridia galli (SI)							
Alachua	1971–72	33	1	3	1.0	1	Forrester et al. (1984), Forrester (1987)
Charlotte	1972–77	381	137	36	2.7	1–17	Ibid.
Capillaria obsignata (SI)							
Charlotte	1972–77	381	11	3	1.3	1–4	Ibid.
Leon	1971–75	21	1	5	1.0	1	Ibid.
Capillaria sp. (SI)							
Leon	1971–84	700	3	<1	2.0	NG–4	Davidson et al. (1991)
Cheilospirura spinosa (GZ)							
Marion	1971	14	1	7	1.0	1	Forrester et al. (1984), Forrester (1987)
Leon	1964	12	3	25	8.3	NG–17	Kellogg & Prestwood (1968)
	1968–69	185	109	59	4.0	1–35	Davidson et al. (1980b)
	1971–75	21	8	38	2.1	1–4	Forrester et al. (1984), Forrester (1987)
	1983–84	153	1	21	1.0	1	Moore et al. (1986)
	1971–84	700	129	18	3.3	NG–20	Davidson et al. (1991)
Cyrnea colini (PR,GZ)							
Alachua	1971–72	33	20	61	3.9	1–18	Forrester et al. (1984), Forrester (1987)
Charlotte	1972–77	381	290	76	4.5	1–26	Ibid.
Glades	1971–73	31	12	39	1.4	1–3	Ibid.
Leon	1924–27	228	160	70	3.3	NG–14	Cram et al. (1931)[c]
	1964	12	9	75	5.3	NG–11	Kellogg & Prestwood (1968)
	1968–69	185	146	79	4.5	1–21	Davidson et al. (1980b)
	1971–75	21	20	95	2.8	1–8	Forrester et al. (1984), Forrester (1987)
	1983–84	153	151	99	5.0	1–19	Moore et al. (1986)
	1971–84	700	632	90	4.5	NG–28	Davidson et al. (1991)
	1989	1	1	—	48.0	—	Hayes (1989)
Marion	1971	14	6	43	1.7	1–2	Forrester et al. (1984), Forrester (1987)

(continued)

Table 16.5. *(continued)*

Nematode (Site)[b] County or site	Year(s)	No. bobwhites		Intensity			Data source
		Exam.	Infected	%	Mean	Range	
Diplotriaenoides minutus (BC)							
"Florida"	1800s	NG	NG	—	NG	NG	Walton (1927)
Dispharynx nasuta (PR)							
Alachua	1971–72	33	6	18	2.8	1–5	Forrester et al. (1984), Forrester (1987)
Charlotte	1972–77	381	15	4	1.5	1–5	Ibid.
Glades	1971–75	36	9	25	4.3	1–23	Ibid.
	1983	29	12	41	2.5	1–6	Rickard (1983)
Hillsborough	1974	3	1	33	2.0	2	Forrester et al. (1984), Forrester (1987)
Leon	1964	12	1	8	4.0	4	Kellogg & Prestwood (1968)
	1968–69	185	10	6	1.0	1	Davidson et al. (1980b)
	1971–75	21	2	10	1.0	1	Forrester et al. (1984), Forrester (1987)
	1983–84	153	23	15	2.1	1–8	Moore et al. (1986)
	1971–84	700	72	10	1.5	NG–9	Davidson et al. (1991)
Gongylonema ingluvicola (ES,CR)							
Leon	1968–69	185	2	1	1.0	1	Davidson et al. (1980b)
	1983–84	153	1	1	1.0	1	Moore et al. (1986)
	1971–84	700	9	1	2.0	NG–7	Davidson et al. (1991)
Habronema bialatum (PR)							
"Florida"	1800s	NG	NG	—	NG	NG	Walton (1927)
Heterakis isolonche (= *H. bonasae*) (CE)							
Alachua	1971–72	33	3	9	3.0	1–7	Forrester et al. (1984), Forrester (1987)
Charlotte	1972–77	381	46	12	3.3	1–18	Ibid.
Hillsborough	1974	3	3	—	12.3	6–17	Ibid.
Marion	1971	14	1	7	2.0	2	Ibid.
Leon	1924–27	218	158	73	15.0	NG–313	Cram et al. (1931)[c]
	1964	12	11	92	17.6	NG–38	Kellogg & Prestwood (1968)
	1968–69	185	183	99	33.5	1–198	Davidson et al. (1980b)
	1971–75	21	19	90	21.4	1–56	Forrester et al. (1984), Forrester (1987)
	1983–84	153	152	99	37.7	1–186	Moore et al. (1986)
	1971–84	700	684	98	37.5	NG–288	Davidson et al. (1991)
	1989	1	1	—	218.0	—	Hayes (1989)
Heterakis gallinarum (CE)							
Leon	1971–84	700	5	<1	1.6	1–2	Davidson et al. (1991)
Heterakis vesicularis (CE)							
"Florida"	NG	NG	NG	—	NG	NG	Walton (1927)
Oxyspirura matogrosensis (EY)							
Leon	1971–84	700	3	<1	1.0	1	Davidson et al. (1991)
Strongyloides sp. (SI)							
Alachua	1971–72	33	1	3	5.0	5	Forrester et al. (1984), Forrester (1987)
Charlotte	1972–77	381	27	7	3.7	1–18	Ibid.
Leon	1968–69	185	25	14	5.0	1–17	Davidson et al. (1980b)
	1971–75	21	2	10	5.5	1–10	Forrester et al. (1984), Forrester (1987)

(continued)

Table 16.5. *(continued)*

Nematode (Site)[b] County or site	Year(s)	Exam.	Infected	%	Mean	Range	Data source
Spirurid larvae (PR)							
Alachua	1971–72	33	5	15	1.8	1–3	Ibid.
Charlotte	1972–77	381	11	3	1.4	1–4	Ibid.
Subulura sp. (CE)							
Leon	1968–69	185	3	2	1.0	1	Davidson et al. (1980b)
Tetrameres americana (PR)							
Leon	1924–27	228	5	2	4.6	NG–11	Cram et al. (1931)[c,d]
Tetrameres pattersoni (PR)							
Alachua	1971–72	33	12	36	3.8	1–8	Forrester et al. (1984), Forrester (1987)
Charlotte	1972–77	381	38	10	1.7	1–5	Ibid.
Glades	1971–75	31	1	3	10.0	10	Ibid.
Leon	1964	12	6[f]	50	4.7	NG–9	Kellogg & Prestwood (1968)
	1968–69	185	138	75	8.5	1–65	Davidson et al. (1980b)
	1971–75	21	7	33	2.7	1–6	Forrester et al. (1984), Forrester (1987)
	1983–84	153	2	1	1.5	1–2	Moore et al. (1986)
	1971–84	700	190	27	4.9	NG–38	Davidson et al. (1991)
Trichostrongylus cramae (= *T. tenuis*)[g] (CE)							
Alachua	1971–72	33	19	58	6.4	2–17	Forrester et al. (1984), Forrester (1987)
Charlotte	1972–77	381	46	12	3.6	1–17	Ibid.
Glades	1971–75	31	3	10	2.0	1–3	Ibid.
Hillsborough	1974	3	2	67	28.5	6–17	Ibid.
Leon	1927–28	90	60	67	32.0	NG–210	Cram et al. (1931)[c]
	1964	12	11	92	71.5	NG–464	Kellogg & Prestwood (1968)
	1968–69	185	130	70	32.0	1–361	Davidson et al. (1980b)
	1971–75	21	16	76	79.6	5–297	Forrester et al. (1984), Forrester (1987)
	1971–84	700	665	95	69.0	NG–1,455	Davidson et al. (1991)
	1983–84	153	131	86	61.9	1–470	Moore et al. (1986)
	1989	1	1	—	27.0	—	Hayes (1989)
Marion	1971	14	4	29	7.5	1–19	Forrester et al. (1984), Forrester (1987)

NG = not given.

a. A total of 1,860 bobwhites was examined, from the following counties: Alachua (*n* = 33), Charlotte (381), Glades (65), Hillsborough (3), Leon (1,364), Marion (14). For each parasite, data are given for counties in which positive birds were found; negatives are not included in the table.

b. Infection sites: BC = body cavity, CE = cecum, CR = crop, ES = esophagus, EY = eyes, GZ = gizzard, PR = proventriculus, SI = small intestine.

c. The bobwhites examined by Cram et al. (1931) were from "southwestern Georgia and northern Florida." Many (if not most) were probably from Leon County, Florida, but this is not known with certainty.

d. These may actually be *T. pattersoni* since numerous Northern Bobwhites from Leon County examined since Cram's work have had only *T. pattersoni* and not *T. americana* or a mixture of the 2 species.

e. Actual values may have been higher because only viscera were received for a large sample of bobwhites from Charlotte County.

f. Identified by Kellogg and Prestwood (1968) as *Tetrameres americana*. Davidson et al. (1980b) re-examined and found that they were actually *T. pattersoni*.

g. The species of *Trichostrongylus* in Northern Bobwhites from northern Florida has been studied by Durette-Desset et al. (1993) and Freehling and Moore (1993) and determined to be *T. cramae*, a species distinct from *T. tenuis* in the older literature.

Table 16.6. Infections by trematodes, cestodes, and acanthocephalans in Northern Bobwhites[a] from Florida

| Helminth (Site)[b] | | No. bobwhites | | | Intensity | | |
County or site	Year(s)	Exam.	Inf.	%	Mean	Range	Data source
Trematoda							
Brachylecithum nanum (LV)							
Leon	1971–84	700	1	<1	3.0	3	Davidson et al. (1991)
Brachylaima sp. (SI)							
Leon	1983–84	153	4	3	1.5	1–2	Moore et al. (1986)
Zonorchis petiolatus (SI)							
Charlotte	1972–77	381	1	<1	2.0	2	Forrester et al. (1984)
Cestoda							
Hymenolepis carioca (SI)							
Leon	1924–27	228	2	1	NG	NG	Cram et al. (1931)[c]
Hymenolepis sp. (SI)							
Leon	1968–69	185	1	<1	1.0	1	Davidson et al. (1980b)
Raillietina cesticillus (SI)							
Leon	1924–27	228	38	17	NG	NG	Cram et al. (1931)[c]
	1964	12	1	8	1.0	1	Kellogg & Prestwood (1968)
	1968–69	18	59	32	17.0	1–137	Davidson et al. (1980b)
	1983–84	153	38	25	13.9	1–86	Moore et al. (1986)
	1971–84	700	104	15	9.4	NG–107	Davidson et al. (1991)
	1989	1	1	—	18.1	—	Hayes (1989)
Raillietina colinia (SI)							
Leon	1964	12[d]	7	58	7.0	NG–8	Kellogg & Prestwood (1968)
	1968–69	185	88	48	7.0	1–98	Davidson et al. (1980b)
	1983–84	153	105	69	10.6	1–59	Moore et al. (1986)
	1971–84	700	264	38	3.6	NG–62	Davidson et al. (1991)
	1989	1	1	—	19.0	—	Hayes (1989)
Raillietina tetragona (SI)							
Leon	1924–27	228	16	7	NG	NG	Cram et al. (1931)[c]

(continued)

ion County, was not present in bobwhites examined elsewhere (Forrester et al. 1984). When indexes of similarity (Holmes and Podesta 1968) were calculated (figure 16.6) it was found that the helminth faunas of bobwhites in Marion and Glades counties were the least similar, but in general, the faunas throughout the state were quite similar, with a possible trend toward a north-to-south decrease in similarity (= increase in diversity). There was no difference in the helminth faunas of the 2 subspecies of bobwhites that occur in Florida (Forrester et al. 1984).

There was no relationship between host gender and prevalence or intensity of helminth infections in Charlotte County (Forrester et al. 1984), but in 1 of the Leon County studies, abundances of *H. isolonche* and *C. colini* were found to be higher in male bobwhites than in females (Davidson et al. 1991). In an earlier report on bobwhites from Leon County, Moore et al. (1987) reported no such relationship, but their sample was smaller ($n = 128$) than that of Davidson et al. (1991) ($n = 700$) and was obtained during a 1-year period, compared with the 14-year time frame of the Davidson study. These sampling variations may account for the observed differences.

In general, Davidson et al. (1980b), Forrester et al. (1984), Moore et al. (1986, 1987),

Table 16.6. *(continued)*

| Helminth (Site)[b] | | No. bobwhites | | | Intensity | | |
County or site	Year(s)	Exam.	Inf.	%	Mean	Range	Data source
Raillietina sp. (SI)							
Alachua	1971–72	33	11	33	NG	NG	Forrester et al. (1984), Forrester (1987)
Charlotte	1972–77	381	69	18	NG	NG	Ibid.
Glades	1971–75	31	3	10	NG	NG	Ibid.
Hillsborough	1974	3	1	33	NG	NG	Ibid.
Leon	1971–75	21	12	57	NG	NG	Ibid.
Marion	1971	14	8	57	NG	NG	Ibid.
Rhabdometra odiosa (SI)							
Leon	1925–28	228	7	3	NG	NG	Jones (1929), Cram et al. (1931)[c]
	1968–69	185	10	6	3.5	1–8	Davidson et al. (1980b)
	1983–84	153	1	<1	1.0	1	Moore et al. (1986)
	1971–84	700	3	<1	1.3	1–3	Davidson et al. (1991)
"Florida"	1800s	NG	4	—	NG	NG	Leidy (1887), Stiles & Hassall (1894)
Acanthocephala							
Mediorhynchus papillosum (SI)							
Leon	1968–69	185[e]	1	1	2.0	2	Byrd & Kellogg (1971), Davidson et al. (1980b)
	1983–84	153	1	1	1.0	1	Moore et al. (1986)

NG = not given by author.

a. See Table 16.5, note a, for details on sample sizes and distribution.

b. Infection sites: SI = small intestine, LV = liver.

c. The bobwhites examined by Cram et al. (1931) were from "southwestern Georgia and Northern Florida." Many (if not most) were probably from Leon County, Florida, but this is not known with certainty.

d. Identified by Kellogg and Prestwood (1968) as *Raillietina* sp. Davidson (1989) examined these specimens and found that they were actually *R. colinia*.

e. Described as *Mediorhynchus bakeri* by Byrd and Kellogg (1971). Schmidt and Kuntz (1977) examined these specimens and synonymized *M. bakeri* with *M. papillosus*.

Davidson (1987), and Davidson et al. (1991) found that host age might be an important factor influencing the prevalences and/or intensities of some helminths. In Leon County, 4 species (*T. cramae, C. spinosa, T. pattersoni,* and *Aproctella stoddardi*) occurred at higher prevalences and/or intensities in adult bobwhites than in juveniles (Davidson et al. 1991). The authors attributed this age effect to the combination of 2 factors: (1) longer exposure time to the infective stages of the helminths in the case of adult birds and (2) the length of time needed for the growth and maturation of the helminths once the birds are infected. The interaction of age and gender was important in the determination of the abundance of 2 species, *C. colini* and *H. isolonche* (Davidson et al. 1991). Adult females had much higher numbers of both species than did adult males, but in contrast, juvenile males had higher numbers than juvenile females. The reasons for these differences are not known.

Davidson et al. (1980b) and Moore et al. (1986) reported that most of the common helminths were acquired by juveniles by July and that by January intensities were almost as high as those of adult birds. As age increased there was a shift from a predominance of immature to adult stages of the helminths (figure 16.7). Davidson et al. (1980b) found 3 patterns

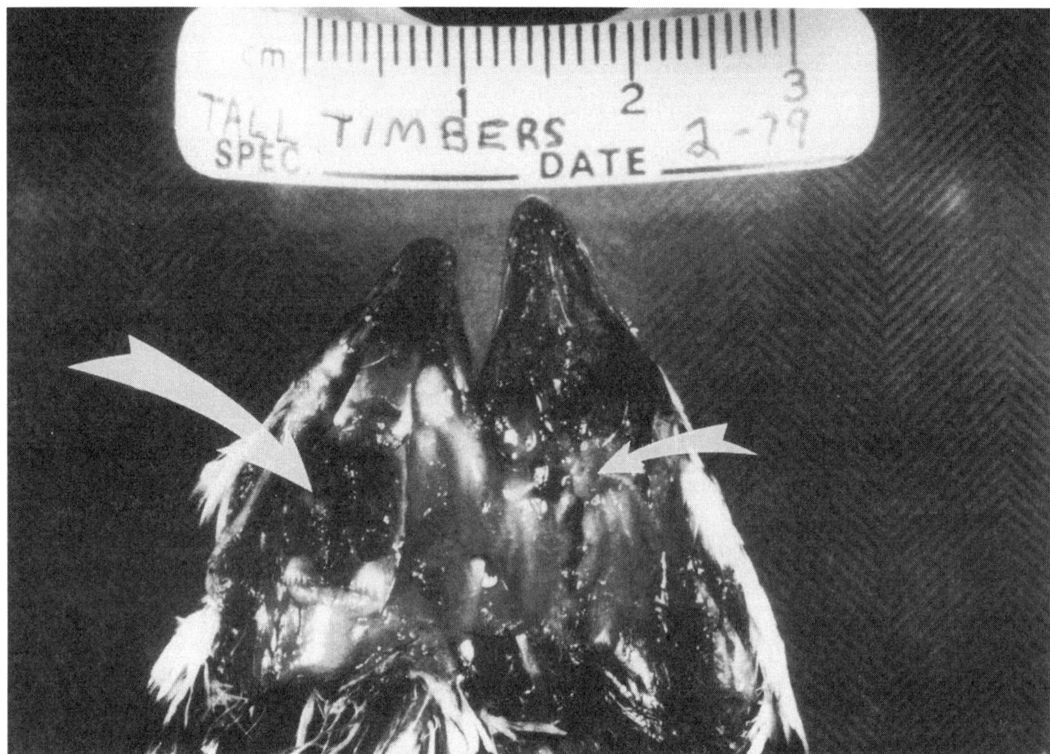

FIGURE 16.5. Lesions of avian pox (= wet pox) in the oral cavity of a Northern Bobwhite from Leon County, 1979. One lesion is occluding the epiglottis *(large arrow)*; one lesion is on the dorsal oral epithelium *(small arrow)*. Courtesy of William R. Davidson.

of seasonal transmission in adult bobwhites in Leon County: (1) distinct summer peaks such as seen for the nematodes *C. spinosa, C. colini,* and *T. pattersoni* (figure 16.8), probably due to the availability of arthropod intermediate hosts, (2) erratic fluctuations with no peaks as seen for the cestodes *Raillietina cesticillus* and *R. colinia,* probably due to the year round availability of coleopteran intermediate hosts, and (3) consistently high rates of transmission as evidenced by high prevalences accompanied by broad seasonal peaks as seen for the cecal nematodes *H. isolonche* and *T. cramae* (figure 16.9), perhaps due to the effects of weather on the survival and transmission of free-living infective stages of these direct cycle nematodes.

Year-to-year variations in prevalence were recorded for bobwhites in Charlotte County over a 5-year period from 1972–73 through 1976–77

(Forrester et al. 1984). Four of the 5 most common nematodes had significantly different prevalences from year to year and all 5 showed significant decreases in prevalence during 1974–75 (figure 16.10A–B). Only 1 of the nematodes (*C. colini*) had significant changes in intensity, especially during 1974–75, at which time there was a major decrease (figure 16.10C–D). Decreases in prevalence and intensity during 1974–75 were not due to weather, but more likely to decreased bobwhite density since during the 1973–74 and 1974–75 hunting seasons the harvest of bobwhites was low (Forrester et al. 1984).

Davidson et al. (1991) evaluated the relationships of helminths of bobwhites in Leon County over a 14-year period to host densities and land use. They found that the prevalences and/or abundances of 7 of the 9 most common species of helminth were correlated positively

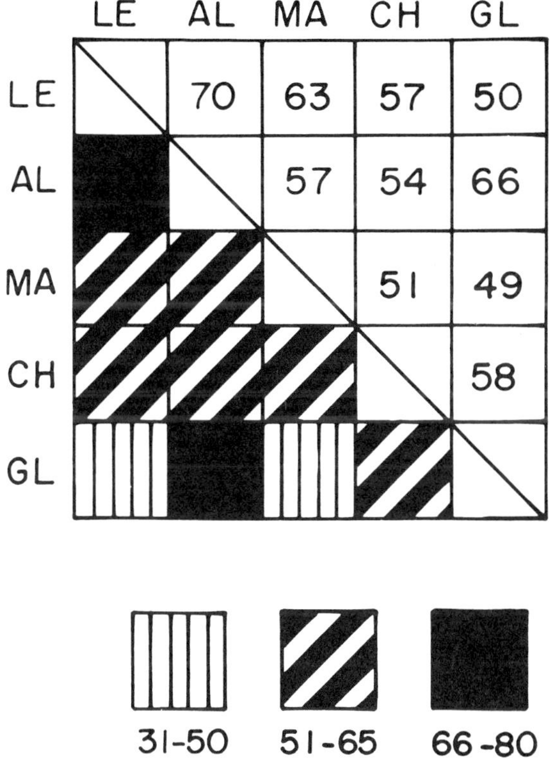

FIGURE 16.6. Trellis diagram of indexes of similarity for the helminth faunas of Northern Bobwhites from five counties in Florida. AL = Alachua County, CH = Charlotte County, GL = Glades County, LE = Leon County, MA = Marion County. From Forrester et al. 1984; by permission of *Proceedings of the Helminthological Society of Washington.*

with bobwhite density. These included 2 cestodes (*R. cesticillus* and *R. colinia*) and 5 nematodes (*C. spinosa, C. colini, H. isolonche, T. pattersoni,* and *T. cramae*). As an example, in figure 16.11 the prevalences and abundances of 1 of these helminths (*T. cramae*) are compared with bobwhite population densities and land use from 1971 to 1984. Two other nematodes (*Dispharynx nasuta* and *A. stoddardi*) did not vary in relation to host density, presumably because of the buffering effect of multiple definitive hosts being present in the area. The prevalences and abundances of *D. nasuta* are compared with bobwhite densities and land use from 1971 to 1984 in figure 16.12. There was an effect of land use (cultivated vs. fallow agricultural fields, which comprised about 20% of

each of the 2 study areas) on the prevalences and/or intensities of 1 tapeworm (*R. colinia*) and 2 nematodes (*C. spinosa* and *T. cramae*). *Trichostrongylus cramae,* for instance, occurred in higher abundances during years when agricultural fields on the study areas were fallow than when they had corn growing on them (figure 16.11). Davidson et al. (1991) suggested that this phenomenon was due to a number of factors such as (1) preferential fall/winter usage by bobwhites of fallow fields rather than harvested corn fields, due to the availability of better cover, (2) physical availability of nematode larvae to bobwhites feeding on fallow fields with light coverings of litter, (3) vegetation on fallow fields providing better microhabitat conditions for the development and survival of eggs

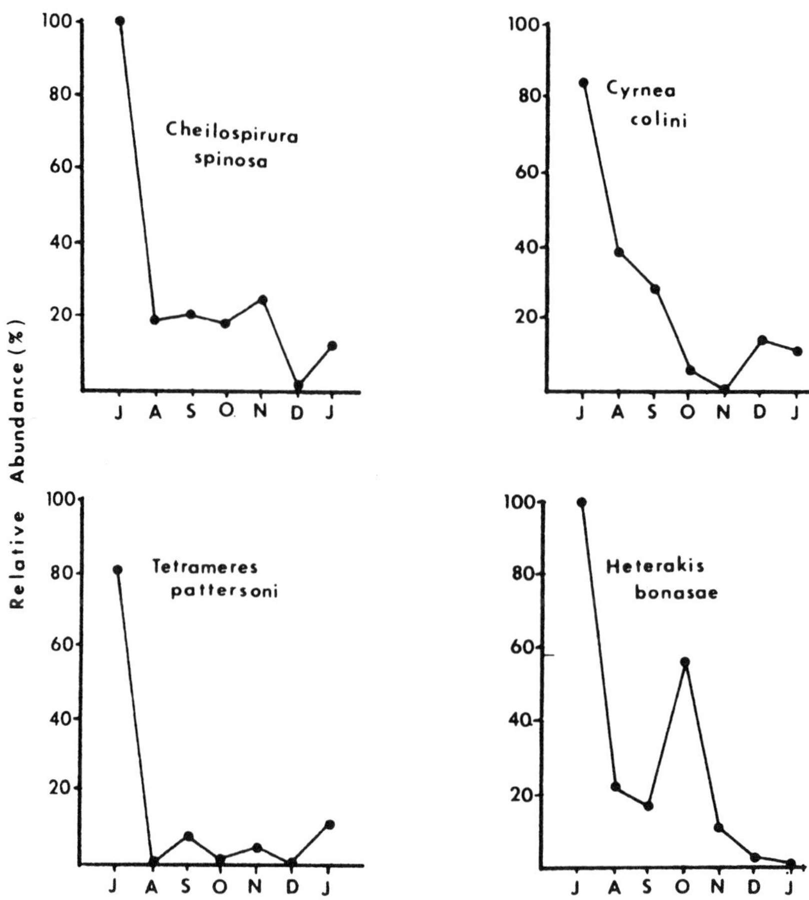

FIGURE 16.7. The relative abundance (percent of the population) of immature stages of *Cheilospirura spinosa, Cyrnea colini, Tetrameres pattersoni,* and *Heterakis bonasae* (now considered a synonym of *H. isolonche*) in juvenile Northern Bobwhites from Tall Timbers Research Station (Leon County), July 1968–January 1969. From Davidson et al. 1980b; by permission of *Journal of Wildlife Diseases.*

and larvae, and (4) lack of cultivation, allowing survival of more eggs and larvae on fallow fields. *Cheilospirura spinosa* occurred in higher prevalences and abundances during fallow years, whereas *R. colinia* occurred in higher prevalences and abundances during years when the fields were cultivated. The authors suggested that these 2 observations might be explained in terms of the ecological requirements of the arthropod intermediate hosts of these 2 nematodes; for the former, fallow fields were ideal, and for the latter, cultivation was best.

The relationship of the prevalence and in-

tensity of helminths of immature Northern Bobwhites at Tall Timbers Research Station to covey sizes was investigated in 1983–84 by Moore et al. (1988). They found that the only consistent association was with *T. cramae,* which occurred in higher intensities in large coveys than in small coveys. They pointed out that *T. cramae* is a monoxenous parasite (i.e., requires no intermediate host) and has the shortest life cycle (14 days) of the helminths occurring in bobwhites, and these factors may contribute to increased transmission rates in larger coveys. Moore and Simberloff (1990)

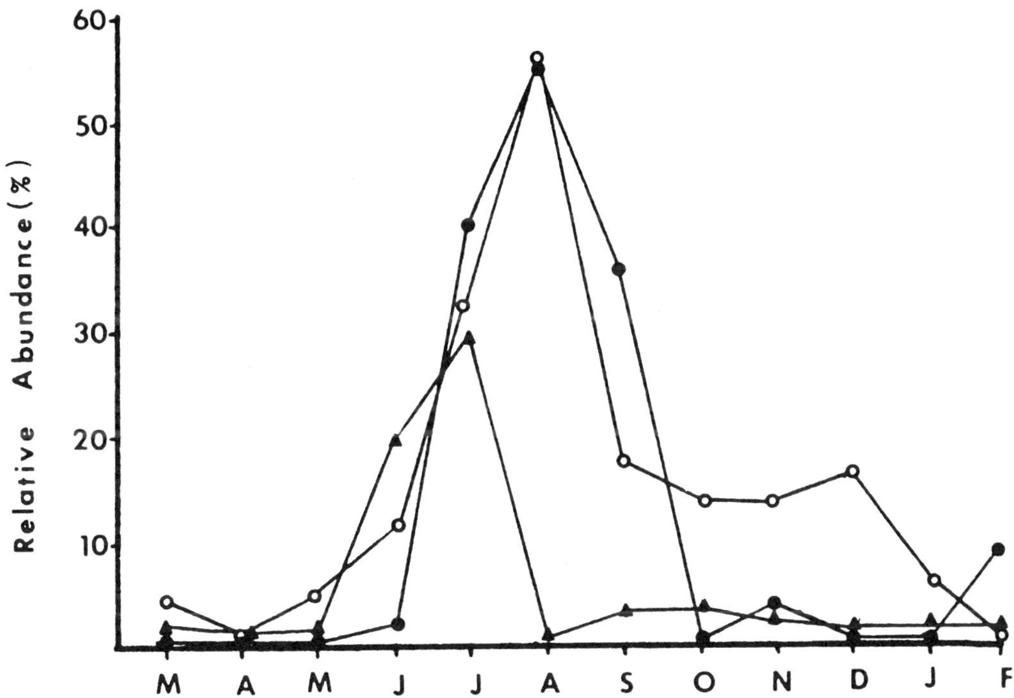

FIGURE 16.8. The relative abundance (percent of the population) of immature stages of *Cheilospirura spinosa* (*solid circles*), *Cyrnea colini* (*open circles*), and *Tetrameres pattersoni* (*solid triangles*) in adult Northern Bobwhites from Tall Timbers Research Station (Leon County), March 1968–February 1969. From Davidson et al. 1980b; by permission of *Journal of Wildlife Diseases*.

conducted additional analyses of their data collected from bobwhites at Tall Timbers Research Station and provided insights into the helminths as a community. These latter studies resulted in detailed information on associations of various species of helminths and interactions among helminths located in similar sites in the gastrointestinal tract. They concluded that the community interactions of the bobwhites at Tall Timbers Research Station did not fit any of the current models of parasite community structure. The reader is referred to their publication for more details on this topic.

Personnel of the Southeastern Cooperative Wildlife Disease Study studied the parasites of bobwhites at Tall Timbers Research Station for more than 25 years. In a publication that provides an overview of the significance of parasitism to bobwhites, Davidson et al. (1982a) concluded that parasitism was almost always "subclinical and that parasites which occur frequently in wild bobwhites have limited pathogenicity." They further concluded that parasitism was not important as a regulatory factor in bobwhite populations. Additional details of the subclinical effects of helminths on bobwhites at Tall Timbers Research Station were given by Davidson et al. (1991). They pointed out that during their long-term study they did not see any significant lesions associated with helminths. Localized inflammation and tissue damage (of minimal concern) were found in association with infections of several helminths, including *A. stoddardi, C. spinosa, C. colini, D. nasuta,* and *T. pattersoni.* In addition, Davidson et al. (1991) investigated further the suggestion by Dabney and Dimmick (1977) that parasitism might be associated with low body weights and

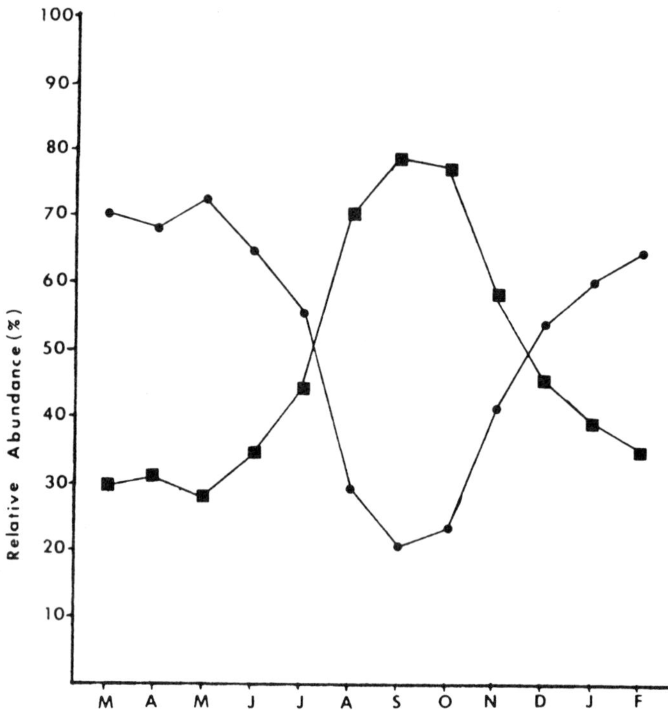

FIGURE 16.9. The relative abundance (percent of the population) of the cecal nematodes *Heterakis isolonche (squares)* and *Trichostrongylus cramae (circles)* in adult Northern Bobwhites from Tall Timbers Research Station (Leon County), March 1968–February 1969. From Davidson et al. 1980b; by permission of *Journal of Wildlife Diseases*.

fat reserves in bobwhites at Tall Timbers Research Station. A relationship was recognized between lower body weights of bobwhites (especially juveniles) and high intensities of *H. isolonche* or *T. cramae* (table 16.8). Davidson et al. (1991) concluded that intensities above 40 for *H. isolonche* or 200 for *T. cramae* might be related to lower body weights of juveniles, but cautioned that other factors such as bobwhite density, land use, or annual food supply might also influence intensities of parasites or body weights of bobwhites and should be taken into consideration when interpreting these data.

XIII. Arthropods

Twenty-seven species of parasitic arthropods have been found on bobwhites in Florida and include 9 ticks, 13 mites, 4 chewing lice, and 1 flea (tables 16.9–16.11). None of these is known to serve as a vector for other disease agents in bobwhites and according to Doster et al. (1980) and Davidson et al. (1982a), significant lesions were not found associated with any of these arthropods. Doster et al. (1980) reported superficial lesions at the point of attachment of ticks and "small crater-like lesions containing chiggers." Shaft mites were found to cause some feather damage, but this did not seem to be harmful to the health of the bobwhites. Stoddard (1931) stated that tick bites sometimes resulted in severe irritation and some swelling, but pointed out that for the most part, such reactions did not occur.

Some degree of ectoparasite control may be achieved by prescribed burning. Stoddard (1931) observed that ectoparasitism by ticks

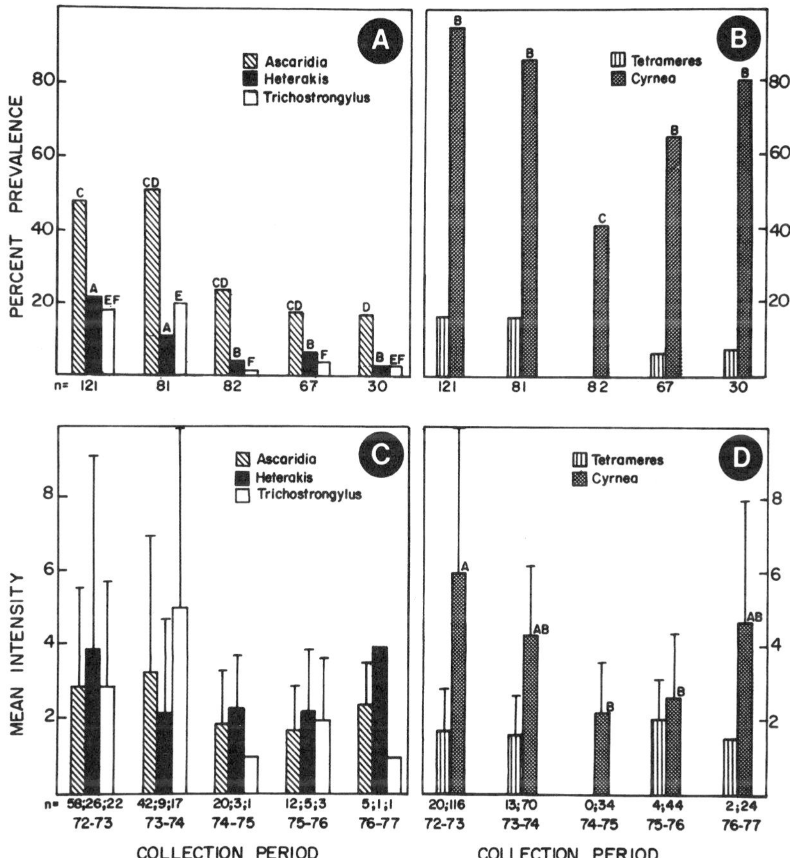

FIGURE 16.10. Year-to-year variations in prevalence and intensity of five species of nematodes in Northern Bobwhites from the Cecil Webb Wildlife Management Area (Charlotte County), 1972–77. Vertical lines = standard deviations for data on intensity. Different letters above bars represent values that are significantly different between certain years. Bars with letters in common or bars with no letters are not significantly different. Sample sizes for prevalence data (*A–B*) and mean intensities (*C–D*) are indicated by *n*. Sample sizes for data on intensity are different for each nematode during a given year because they are based on numbers of nematodes per infected bobwhite, which varies for each nematode. From Forrester et al. 1984; by permission of *Proceedings of the Helminthological Society of Washington.*

and chiggers was less severe on bobwhites in areas in Leon County where such burning of the woodlands was most intensive.

XIV. Disease risks from pen-reared bobwhites

In many areas of the Southeast, including Florida, pen-reared bobwhites (same species as the free-ranging bobwhites) are released for sporting purposes such as hunting, dog training, and field trials. This practice is often controversial because of disease problems, economics of the operation, and variable returns (Davidson et al. 1982a; Martin 1984). From a disease standpoint there is cause for concern. In table 16.12 significant infectious and parasitic diseases of pen-reared bobwhites are listed along with risk assessments as provided

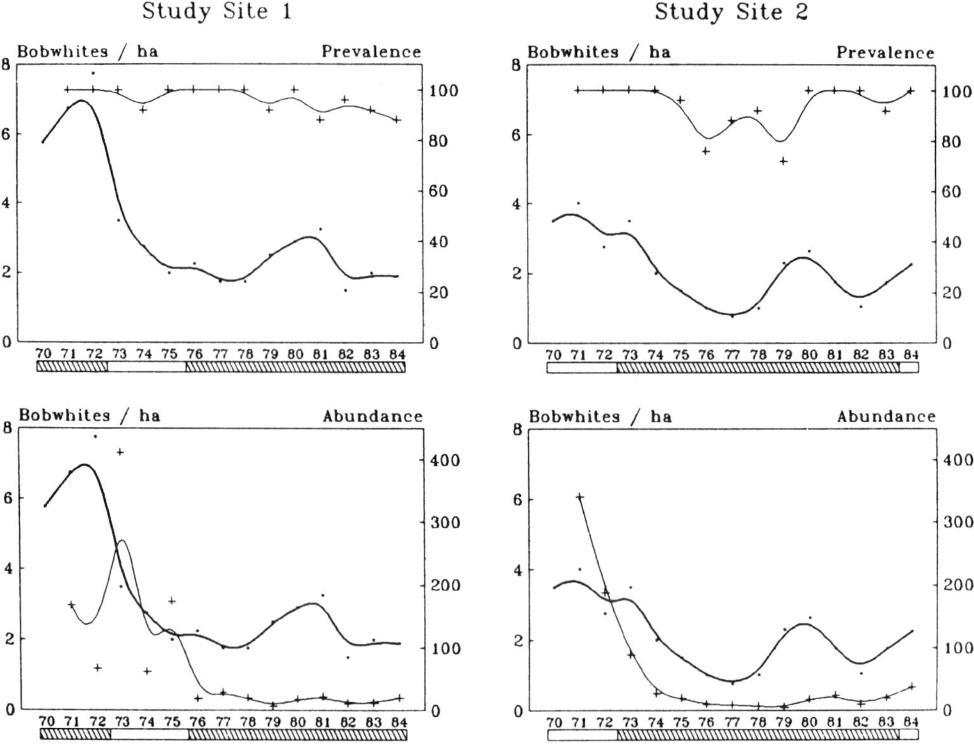

FIGURE 16.11. Comparison of the prevalences and abundances of the cecal nematode *Trichostrongylus cramae* with densities of Northern Bobwhite populations and land use at Tall Timbers Research Station (Leon County), 1971–84. Bobwhite densities are indicated by solid circles and prevalence or abundance values are shown by plus signs. The bars at the bottom of each graph indicate cultivation of agricultural fields on the study areas *(hatching)* or fields left fallow *(clear)*. From Davidson et al. 1991; courtesy of *Journal of Wildlife Diseases.*

by Davidson et al. (1982a). Avian pox and infectious enterohepatitis are of the most significance since they cause high morbidity and mortality, persist under field conditions, and can occur in association with releases of pen-reared bobwhites (Davidson et al. 1982a). The other diseases are problems mainly or exclusively of pen-reared birds and are essentially diseases of confinement. A survey of 27 groups of pen-raised bobwhites from Florida, Georgia, North Carolina, South Carolina, Virginia, and West Virginia was conducted by the Southeastern Cooperative Wildlife Disease Study (Davidson 1986–87). The following diseases/parasites were found, as expressed in the percentage of the groups infected: cecal worms

(*Heterakis gallinarum*) = 52%, avian pox = 22%, infectious enterohepatitis = 19%, crop mycosis = 11%, crop worms (*Capillaria contorta*) = 7%, and ulcerative enteritis = 4%. It was emphasized that 66% of the groups had at least 1 of the diseases or parasites that is a threat to native bobwhite populations.

In response to this situation it has been recommended that only healthy pen-reared birds be released (Landers et al. 1991). This goal can be accomplished by obtaining a random sample of 20 or 25 birds from a group intended for release and having them examined at necropsy by personnel at a diagnostic laboratory. Birds with infectious enterohepatitis and/or avian pox should not be released. In 1981 1 of 250 pen-reared

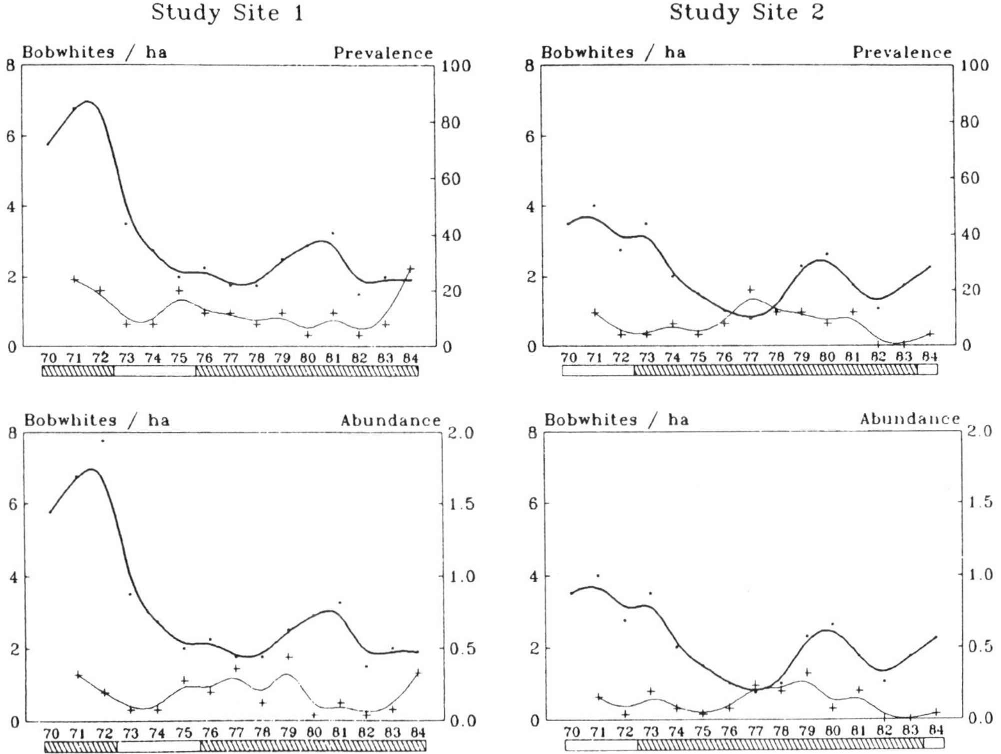

FIGURE 16.12. Comparison of the prevalences and abundances of the proventricular nematode *Dispharynx nasuta* with densities of Northern Bobwhite populations and land use in Tall Timbers Research Station (Leon County), 1971–84. Bobwhite densities are indicated by solid circles and prevalence or abundance values are shown by plus signs. The bars at the bottom of each graph indicate cultivation of agricultural fields on the study areas *(hatching)* or fields left fallow *(clear)*. From Davidson et al. 1991; courtesy of *Journal of Wildlife Diseases*.

bobwhites destined for release in Leon County was found to have airsacculitis (associated with *Escherichia coli* infection) and avian pox (SCWDS records). It was recommended that the remainder of the bobwhites not be released, even though 70 had already been turned loose. In 1982–84 a release of pen-reared bobwhites was planned on the Cecil Webb Wildlife Management Area in Charlotte County (Davidson 1982; Martin 1984). Bobwhites from several commercial game farms were examined until a source of disease-free birds was located. Such practices as illustrated here should be encouraged and will help prevent the introduction of diseases into populations of native bobwhites. For additional information on this topic the reader is referred to Davidson (1986–87) and Landers et al. (1991).

XV. Summary and conclusions

Seventy-six different disease agents and parasites have been found in wild bobwhites from Florida. These include 2 contaminants, 6 viruses, 2 bacteria, 1 mycotoxin, 2 fungi, 5 protozoans, 20 nematodes, 3 trematodes, 7 cestodes, 1 acanthocephalan, 9 ticks, 13 mites, 4 chewing lice, and 1 flea. The most significant of these diseases is avian pox. An epizootic of pox occurred in 1978–79 in the southwestern Geor-

Table 16.7. Prevalences, intensities, and abundances of the 6 characteristic helminths of Northern Bobwhites in Florida

| | Number of bobwhites | | | | |
Helminth	Exam.	Inf.	%	Mean intensity	Abundance
Cheilospirura spinosa	1,085	251	23	3.3	<1
Cyrnea colini	1,758	1,446	82	3.7	3
Heterakis isolonche	1,720	1,260	73	18.3	13
Raillietina colinia	1,050	464	44	7.0	3
Tetrameres pattersoni	1,516	394	26	4.7	1
Trichostrongylus cramae	1,623	1,087	67	35.8	24

Sources: Cram et al. (1931), Davidson et al. (1980b, 1991), Forrester (1987), Forrester et al. (1984), Kellogg and Prestwood (1968), Moore et al. (1986).

gia/north central Florida area. During that outbreak there was a 12-fold increase in the prevalence of pox, and morbidity and mortality rates were estimated to be 2% and 1% per month, respectively. In addition to north central Florida, pox has been diagnosed from bobwhites in central (Osceola County) and southern Florida (Charlotte County).

The release of pen-reared bobwhites into native habitat represents a disease threat to wild bobwhites. Two diseases, avian pox and infectious enterohepatitis, are common in pen-reared birds and are of particular concern. Measures should be taken to prevent the introduction of these 2 diseases into populations of wild bobwhites.

Table 16.8. Mean weights of adult and juvenile Northern Bobwhites from Tall Timbers Research Station (Leon County), compared with intensities of infection with *Heterakis isolonche* and *Trichostrongylus cramae*

| *Heterakis isolonche* | | | *Trichostrongylus cramae* | | |
| | Weight (grams)[a] | | | Weight (grams)[a] | |
Intensity	Adults	Juveniles	Intensity	Adults	Juveniles
<20	169.9 A, B (64)	165.8 A (241)	<50	169.5 A (122)	165.1 A (403)
21–40	166.8 A (51)	165.3 A (118)	51–100	166.8 A (28)	167.7 A (44)
41–80	168.6 A (43)	163.3 A, B (102)	101–200	163.7 B (18)	167.2 A (23)
81–160	166.8 A (29)	162.8 B (36)	201–400	163.9 B (11)	159.5 B (20)
>161	173.3 B (3)	160.1 B (9)	>401	170.2 A, C (11)	153.5 C (16)

Source: Davidson et al. (1991).
a. Number in parentheses = sample size. Column values with different capital letters are significantly different (ANOVA, $P < 0.05$).

Table 16.9. Tick infestations on Northern Bobwhites from Florida[a]

| Tick species | | | No. bobwhites | | | |
County or area	Year(s)	Stage of tick[b]	Exam	Pos.	%	Data source
Amblyomma americanum						
Leon	1968–75	NG	441	11	2	Doster et al. (1980)
Levy	1949–53	N	56	NG	—	Rogers (1953)
Amblyomma maculatum						
Charlotte	1974	NG	10	1	10	Doster et al. (1980)
Hillsborough	1947–49	N	NG	NG	—	Taylor (1951)
Leon	1968–75	NG	441	13	3	Doster et al. (1980)
Orange	NG	NG	NG	NG	—	Boardman (1929)
"Florida"	NG	L	NG	NG	—	Clifford et al. (1961)
Amblyomma tuberculatum						
Highlands	1967	NG	NG	NG	—	FSCA[c]
"North Florida"	1949–53	L	56	NG	—	Rogers (1953)
Dermacentor variabilis						
"North Florida"	1949–53	L	56	NG	—	Rogers (1953)
Haemaphysalis chordeilis						
Alachua	NG	NG	NG	NG	—	Boardman (1929)
Charlotte	1974	NG	10	4	40	Doster et al. (1980)
Gadsden	NG	NG	NG	NG	—	Boardman (1929)
Hillsborough	1947–49	L	NG	NG	—	Taylor (1951)
Haemaphysalis leporispalustris						
Hillsborough	1947–49	A,L	NG	NG	—	Taylor (1951)
Indian River	1967	L,N	31	NG	—	Wilson & Kale (1972)
Leon	1968–75	NG	441	29	7	Doster et al. (1980)
	1924–27	I	NG	NG	—	Stoddard (1931)
"North Florida"	1949–53	L,N	56	NG	—	Rogers (1953)
"Florida"	NG	NG	NG	NG	—	Boardman (1929)
"Florida"	NG	NG	NG	NG	—	Peters (1936)
Ixodes minor						
Leon	1968–75	NG	441	2	<1	Doster et al. (1980)
Ixodes sp.						
Leon	1927	N	NG	1	—	Stoddard (1931)
Rhipicephalus sp.[d]						
Jefferson & Wakulla	1924–27	I	NG	NG	—	Stoddard (1931)

NG = not given.

a. Travis (1941) examined two Northern Bobwhites from Orange, Osceola, or Collier County in 1936–37 and found no ticks.

b. A = adults, L = larvae, N = nymphs, I = immature (specific stage not given).

c. FSCA = records from the Florida State Collection of Arthropods, Division of Plant Industry, Florida Department of Agriculture and Consumer Services, Gainesville.

d. This finding is questionable. *Rhipicephalus sanguineus* is the only North American species of *Rhipicephalus* and found primarily on dogs and a few other species of mammals.

Table 16.10. Mite infestations on Northern Bobwhites from Florida

Mite species (location) County	Year(s)	No. bobwhites Exam.	Pos.	%	Data source
Eutrombicula alfreddugesi (skin)					
Leon	1924–27	NG	NG	—	Stoddard (1931)
Neoschoengastia americana (skin)					
Leon	1968–75	441	1	<1	Doster et al. (1980)
Neotrombicula whartoni (skin)					
Leon	1968–75	441	1	<1	Ibid.
Boydaia colini (nasal passages)					
Charlotte	1974	10	2	20	Ibid.
Leon	1968–75	441	117	27	Ibid.
Palm Beach	1969	5	2	—	Ibid.
Colinoptes cubanensis (nasal passages)					
Charlotte	1974	10	1	10	Ibid.
Leon	1968–75	441	203	46	Ibid.
Palm Beach	1969	5	1	—	Ibid.
Colinolichus virginianus[a] sp. (feathers)					
Charlotte	1974	10	10	100	Ibid.
Leon	1968–75	441	250	57	Ibid.
Palm Beach	1969	5	4	—	Ibid.
Megninia sp. (feathers)					
Charlotte	1974	10	6	60	Ibid.
Jefferson	1926	NG	NG	—	Stoddard (1931)
Leon	1968–75	441	284	64	Doster et al. (1980)
Palm Beach	1969	5	4	—	Ibid.
Colinophilus wilsoni[b] (feather shaft)					
Charlotte	1974	10	6	60	Ibid.
Leon	1968–75	441	269	61	Ibid.
Palm Beach	1969	5	3	—	Ibid.
Dermoglyphus sp. (feather shaft)					
Leon	1968–75	441	1	<1	Ibid.
Apionacarus wilsoni[c] (feather shaft)					
Charlotte	1974	10	6	60	Ibid.
Leon	1968–75	441	27	6	Ibid.
Microlichus sp. (skin)					
Leon	1968–75	441	1	<1	Ibid.
Rivoltasia sp. (skin)					
Leon	1968–75	441	2	<1	Ibid.
Ornithonyssus sylviarum (skin)					
Leon	1927	NG	NG	—	Stoddard (1931)

NG = not given.

a. Identified originally as *Pterolichus* sp. by Doster et al. (1980) and later described as a new genus and species (*Colinolichus virginianus*) by Gaud and Atyeo (1996).

b. Described from material collected from bobwhites at Tall Timbers Research Station, Leon County, February 8, 1971 (see Kethley 1973).

c. Described from material collected from bobwhites at Tall Timbers Research Station, Leon County, February 1971 (see Gaud and Atyeo 1975, 1977).

Table 16.11. Chewing lice and flea infestations on Northern Bobwhites from Florida

Species of arthropod County	Year(s)	No. bobwhites Exam.	Pos.	%	Data Source
Chewing lice					
Colinicola numidiana					
Charlotte	1974	10	1	10	Doster et al. (1980)
Jefferson	1925	NG	1	—	Stoddard (1931)
Leon	1968–75	441	114	26	Doster et al. (1980)
NG	NG	NG	NG	—	Emerson (1951)
Goniodes ortygis					
Charlotte	1974	10	3	30	Doster et al. (1980)
Jefferson	1926	NG	NG	—	Forrester et al. (1995)
Leon	1927	NG	NG	—	Ibid.
	1968–75	441	211	48	Doster et al. (1980)
Palm Beach	1969	5	3	—	Ibid.
NG	NG	NG	NG	—	Emerson (1951)
Menacanthus pricei					
Jefferson	1926	NG	NG	—	Stoddard (1931), Forrester et al. (1995)
Leon	1926	NG	1	—	Stoddard (1931)[a]
	1968–75	441	61	14	Doster et al. (1980)
	1968	NG	NG	—	Forrester et al. (1995)
Oxylipeurus clavatus					
Charlotte	1974	10	7	70	Doster et al. (1980)
Leon	1968–75	441	286	65	Ibid.
	NG	NG	NG[b]	—	Stoddard (1931)
Palm Beach	1969	5	3	—	Doster et al. (1980)
Fleas					
Echidnophaga gallinacea[c]					
Leon	1926	NG	NG	—	Fox (1940); Layne (1971)

NG = not given by authors.

a. Stoddard (1931) identified this as *Menacanthus* sp. Since other specimens taken from Northern Bobwhites in Jefferson County in 1926 were identified by K.C. Emerson as *M. pricei*, it is assumed that these are the same species.

b. This louse was of "frequent occurrence" according to Stoddard (1931).

c. Stoddard (1931) listed this species as being present on 4 Northern Bobwhites captured near houses or dwellings "where fleas of this species occur commonly on cats, dogs, or fowls," but said that the prevalence of this flea on bobwhites was rare. He did not give the specific locality of the four infested bobwhites; however, his study took place in the Red Hills region of southern Georgia and northwestern Florida (Jefferson and Leon counties) during 1924–27.

Table 16.12. Important infectious and parasitic diseases encountered commonly in pen-reared bobwhites and the significance of these diseases to free-ranging native Northern Bobwhites and other game birds

Disease	Etiologic agent	Risk to wild bobwhites and other game birds
Quail bronchitis	Quail bronchitis virus (QBV)	Unknown (occurs naturally in wild bobwhites in some areas)
Avian pox	Avian poxvirus	High risk (can initiate or exacerbate pox in wild bobwhites and possibly in other game birds)
Ulcerative enteritis (Quail disease)	*Clostridium colini*	Low risk (never reported from wild bobwhites)
Aspergillosis	*Aspergillus fumigatus*	Low risk (organism is ubiquitous)
Crop mycosis	*Candida albicans*	Low risk (organism is ubiquitous)
Infectious entero-hepatitis (Blackhead)	*Histomonas meleagridis*	High risk (pathogenic to wild bobwhites and Wild Turkeys)
Cryptosporidiosis	*Cryptosporidium* sp.	Unknown (occurs in pen-reared bobwhites)
Crop capillariasis	*Capillaria contorta*	Low risk (extremely rare in wild birds)
Dispharynxosis	*Dispharynx nasuta*	Low risk (parasite is ubiquitous)
None (Blackhead)	*Heterakis gallinarum*	High risk (important as vector for blackhead)

Sources: Adapted from Davidson et al. (1982a), Landers et al. (1991).

XVI. Literature cited

Akey, B.L., J.K. Nayar, and D.J. Forrester. 1981. Avian pox in Florida wild turkeys: *Culex nigripalpus* and *Wyeomyia vanduzeei* as experimental vectors. *J. Wildl. Dis.* 17:597–599.

Allen, C.R., R.S. Lutz, and S. Demarias. 1995. Red imported fire ant impacts on northern bobwhite populations. *Ecol. Appl.* 5:632–638.

———. 1998. Ecological effects of the invasive nonindigenous ant, *Solenopsis invicta*, on native vertebrates: the wheels on the bus. *Trans. N. Am. Wildl. Nat. Resour. Conf.* 63:56–65.

Baker, M.F. 1965. Studies on possible effects of mirex bait on the bobwhite quail and other birds. *Proc. Annu. Conf. Southeast. Assoc. Game Fish Comm.* 25:153–159.

Beyer, W.N., J.W. Spann, L. Sileo, and J.C. Franson. 1988. Lead poisoning in six captive species. *Arch. Environ. Contam. Toxicol.* 17:121–130.

Bigler, W.J., E. Lassing, E. Buff, A.L. Lewis, and G.L. Hoff. 1975. Arbovirus surveillance in Florida: wild vertebrate studies 1965–1974. *J. Wildl. Dis.* 11:348–356.

Boardman, E.T. 1929. Ticks of the Gainesville area. M.S. thesis, University of Florida, Gainesville. 57 pp.

Brennan, L.A. 1991. How can we reverse the northern bobwhite population decline? *Wildl. Soc. Bull.* 19:544–555.

———. 1993. Fire ants and northern bobwhites: a real problem or a red herring? *Wildl. Soc. Bull.* 21:351–355.

Brown, R.E. 1978. The imported fire ant in Florida. *Proc. Tall Timbers Res. Stn. Conf. Ecol. Anim. Control Habitat Manag.* 7:15–21.

Burridge, M.J., W.J. Bigler, D.J. Forrester, and J.M. Hennemann. 1979. Serologic survey for *Toxoplasma gondii* in wild animals in Florida. *J. Am. Vet. Med. Assoc.* 175:964–967.

Byrd, E.E., and F.E. Kellogg. 1971. *Mediorhynchus bakeri*, a new acanthocephalan (Gigantorhynchidae) from the bobwhite, *Colinus virginianus* (L.). *J. Parasitol.* 57:137–142.

Clifford, C.M., G. Anastos, and A. Elbl. 1961. The larval ixodid ticks of the eastern United States (Acarina-Ixodidae). *Misc. Publ. Entomol. Soc. Am.* 2:213–237.

Cram, E.B. 1939. Redescription and emendation of the genus *Aproctella* (Filariidae), nematodes from gallinaceous birds. *Proc. Helminthol. Soc. Wash.* 6:94–95.

Cram, E.B., M.F. Jones, and E.A. Allen. 1931. Internal parasites and parasitic diseases of the bobwhite. In: *The bobwhite quail, its habits, preservation and increase.* H.L. Stoddard. Charles Scribner's Sons, New York. pp. 229–313.

Crawford, B.T. 1987. Bobwhite quail. In: *Restoring America's wildlife, 1937–1987.* H. Kallman (ed.). U.S. Department of the Interior, Fish and Wildlife Service, Washington, D.C. pp. 299–303.

Crawford, R.L. 1981. Bird casualties at a Leon County, Florida TV tower: a 25-year migration study. *Bull. Tall Timbers Res. Stn.* 22:1–30.

Cunningham, C.H. 1978. Avian pox. In: *Diseases of poultry.* 7th ed. M.S. Hofstad, B.W. Calnek, C.F. Helmboldt, W.M. Reid, and H.W. Yoder, Jr. (eds.). Iowa State University Press, Ames. pp. 597–609.

Dabney, J.M., and R.W. Dimmick. 1977. Evaluating physiological condition of bobwhite quail. *Proc. Annu. Conf. Southeast. Assoc. Fish Wildl. Agencies* 31:116–122.

Dahlen, J.H., and A.O. Haugen. 1954. Acute toxicity of certain insecticides to the bobwhite quail and mourning dove. *J. Wildl. Manag.* 18:477–481.

Davidson, W.R. 1979–89. Unpublished data. Southeastern Cooperative Wildlife Disease Study, University of Georgia, Athens.

———. 1986–87. Pen-raised quail—are they typhoid Marys? *Tall Timbers Res. Stn. Rep.* 8:1, 6–7.

Davidson, W.R., G.L. Doster, and M.B. McGhee. 1978. Failure of *Heterakis bonasae* to transmit *Histomonas meleagridis*. *Avian Dis.* 22:627–632.

Davidson, W.R., and F.E. Kellogg. 1982–87. Unpublished data. Southeastern Cooperative Wildlife Disease Study, University of Georgia, Athens.

Davidson, W.R., F.E. Kellogg, and G.L. Doster. 1980a. An epornitic of avian pox in wild bobwhite quail. *J. Wildl. Dis.* 16:293–298.

———. 1980b. Seasonal trends of helminth parasites of bobwhite quail. *J. Wildl. Dis.* 16:367–375.

———. 1982a. An overview of disease and parasitism in southeastern bobwhite quail. *Proc. Natl. Bobwhite Quail Symp.* 2:57–63.

———. 1982b. Avian pox infections in southeastern bobwhites: historical and recent information. *Proc. Natl. Bobwhite Quail Symp.* 2:64–68.

Davidson, W.R., F.E. Kellogg, G.L. Doster, and C.T. Moore. 1991. Ecology of helminth parasitism in bobwhites from northern Florida. *J. Wildl. Dis.* 27:185–205.

DeWitt, J.B. 1955. Effects of chlorinated hydrocarbon insecticides upon quail and pheasants. *Agric. Food Chem.* 3:672–676.

Dimmick, R.W. 1990. Bobwhites, raptors and thresholds of security. *Proc. Southeast. Raptor Manag. Symp. Workshop.* Natl. Wildl. Fed. Sci. Tech. Ser. 14:158–161

Domermuth, C.H., D.J. Forrester, D.O. Trainer, and W.J. Bigler. 1977. Serologic examination of wild birds for hemorrhagic enteritis of turkey and marble spleen disease of pheasants. *J. Wildl. Dis.* 13:405–408.

Doster, G.L., and W.R. Davidson 1993. Unpublished data. Southeastern Cooperative Wildlife Disease Study, University of Georgia, Athens.

Doster, G.L., F.E. Kellogg, W.R. Davidson, and W.M. Martin. 1982. Hunter success and crippling losses for bobwhite quail. *Proc. Natl. Bobwhite Quail Symp.* 2:45–47.

Doster, G.L., N. Wilson, and F.E. Kellogg. 1980. Ectoparasites collected from bobwhite quail in the southeastern United States. *J. Wildl. Dis.* 16:515–520.

Dubey, J.P., M.D. Ruff, O.C.H. Kwook, S.K. Shea, G.C. Wilson, and P. Thulliez. 1993. Experimental toxoplasmosis in bobwhite quail (*Colinus virginianus*). *J. Parasitol.* 79:935–939.

Durette-Desset, M.-C., A.C. Chabaud, and J. Moore. 1993. *Trichostrongylus cramae* n.sp. (Nematoda), a parasite of bobwhite quail (*Colinus virginianus*). *Ann. Parasitol.* 68:43–48.

Emerson, K.C. 1951. A list of Mallophaga from gallinaceous birds of North America. *J. Wildl. Manag.* 15:193–195.

Favorite, F.G. 1960. Some evidence of local origin of EEE virus in Florida. *Mosq. News* 20:87–92.

Fischer, J.R. 1999. Unpublished data. Southeastern Cooperative Wildlife Disease Study, University of Georgia, Athens.

Forrester, D.J. 1987. Unpublished data. University of Florida, Gainesville.

Forrester, D.J., J.A. Conti, A.O. Bush, L.D. Campbell, and R.K. Frohlich. 1984. Ecology of helminth parasitism of bobwhites in Florida. *Proc. Helminthol. Soc. Wash.* 51:255–260.

Forrester, D.J., J.K. Nayar, and M.D. Young. 1987. Natural infection of *Plasmodium hermani* in the northern bobwhite, *Colinus virginianus,* in Florida. *J. Parasitol.* 73:865–866.

Forrester, D.J., H.W. Kale II, R.D. Price, K.C. Emerson, and G.W. Foster. 1995. Chewing lice (Mallophaga) from birds in Florida: a listing by host. *Bull. Fla. Mus. Nat. Hist.* 39:1–44.

Fox, I. 1940. *Fleas of eastern United States.* Iowa State College Press, Ames. 191 pp.

Freehling, M., and J. Moore. 1993. Host specificity of *Trichostrongylus tenuis* from red grouse and northern bobwhites in experimental infections of northern bobwhites. *J. Parasitol.* 79:538–541.

Frenkel, J.K. 1981. False-negative serologic tests for *Toxoplasma* in birds. *J. Parasitol.* 67:952–953.

Gaud, J., and W.T. Atyeo. 1975. Ovacaridae, une famille nouvelle de Sarcoptiformes plumicoles. *Acarologia* 17:169–176.

———. 1977. A new name for *Ovacarus* and Ovacaridae (Acarina: Analgoidea). *Acarologia* 18:568–569.

Giuliano, W.M., C.R. Allen, R.S. Lutz, and S. Demarais. 1996. Effects of red imported fire ants on northern bobwhite chicks. *J. Wildl. Manag.* 60:309–313.

Hayes, L.E. 1989. Unpublished data. Southeastern Cooperative Wildlife Disease Study, University of Georgia, Athens.

Henderson, J.R., N. Karabatsos, T.T.C. Bourke, R.C. Wallis, and R.M. Taylor. 1962. A survey for arthropod-borne viruses in south-central Florida. *Am. J. Trop. Med. Hyg.* 11:800–810.

Hoffman, D.J. 1988. Effects of Krenite® brush control agent (fosamine ammonium) on embryonic development in mallards and bobwhite. *Environ. Toxicol. Chem.* 7:69–75.

Holmes, J.C., and R. Podesta. 1968. The helminths of wolves and coyotes from the forested regions of Alberta. *Can. J. Zool.* 46:1193–1204.

Jones, M.F. 1929. Tapeworms of the genera *Rhabdometra* and *Paruterina* found in the quail and yellow-billed cuckoo. *Proc. U.S. Natl. Mus.* 75:1–8.

Kellogg, F.E., and J.P. Calpin. 1971. A checklist of parasites and diseases reported from the bobwhite quail. *Avian Dis.* 15:704–715.

Kellogg, F.E., and G.L. Doster. 1972. Diseases and parasites of the bobwhite. *Proc. Natl. Bobwhite Quail Symp.* 1:233–267.

Kellogg, F.E., and A.K. Prestwood. 1968. Gastrointestinal helminths from wild and pen-raised bobwhites. *J. Wildl. Manag.* 32:468–475.

Kellogg, F.E., and W.M. Reid. 1970. Bobwhites as possible reservoir hosts for blackhead in wild turkeys. *J. Wildl. Manag.* 34:155–159.

Kethley, J.B. 1973. A new genus and species of quill mites (Acarina: Syringophilidae) from *Colinus virginianus* (Galliformes: Phasianidae) with notes on developmental chaetotaxy. *Fieldiana Zool.* 65:1–8.

King, D.J., S.R. Pursglove, Jr., and W.R. Davidson. 1981. Adenovirus isolation and serology from wild bobwhite quail (*Colinus virginianus*). *Avian Dis.* 25:678–682.

Komarek, E.V. 1978. Comments on the "fire ant problem." *Proc. Tall Timbers Res. Stn. Conf. Ecol. Anim. Control Habitat Manag.* 7:1–8.

Landers, J.L., L.P. Simoneaux, and D.C. Sisson (eds.). 1991. *The effects of released, pen-raised bobwhites on wild bird populations.* Workshop Proceedings, August 16–17, 1990, Athens, Ga. 36 pp.

Layne, J.N. 1971. Fleas (Siphonaptera) of Florida. *Fla. Entomol.* 54:35–51.

Leidy, J. 1887. Tapeworms in birds. *J. Comp. Med. Surg.* 8:1–11.

Linda, S.B. 1998. Analysis of 1997–1998 statewide and wildlife management area hunter surveys: methods and results. Unpublished report. Florida Game and Fresh Water Fish Commission, Gainesville. 88 pp.

Maehr, D.S., and J.R. Brady. 1986. Food habits of bobcats in Florida. *J. Mamm.* 67:133–138.

Martin, S.A. 1984. Experimental release of pen-raised quail on the Field Trial Grounds, C.M. Webb Wildlife Management Area. Unpublished report. Florida Game and Fresh Water Fish Commission, Tallahassee. 15 pp.

McConnell, C.A. 1968. Experimental lead poisoning of bobwhite quail and mourning doves. *Proc. Annu. Conf. Southeast. Assoc. Game Fish Comm.* 21:208–219.

Moore, J., M. Freehling, and D. Simberloff. 1986. Gastrointestinal helminths of the northern bobwhite in Florida: 1968 and 1983. *J. Wildl. Dis.* 22:497–501.

Moore, J., M. Freehling, D. Horton, and O. Simberloff. 1987. Host age and sex in relation to intestinal helminths of bobwhite quail. *J. Parasitol.* 73:230–233.

Moore, J., and D. Simberloff. 1990. Gastrointestinal helminth communities of bobwhite quail. *Ecology* 71:344–359.

Moore, J., D. Simberloff, and M. Freehling. 1988. Relationships between bobwhite quail social-group size and intestinal helminth parasitism. *Am. Nat.* 131:22–32.

Mueller, B.S., W.R. Davidson, and J.B. Atkinson, Jr. 1993. Survival of northern bobwhite infected with avian pox. *Proc. Natl. Quail Symp.* 3:79–82.

Mueller, J.M., C.B. Dabbert, S. Demarais, and A.R. Forbes. 1999. Northern bobwhite chick mortality caused by red imported fire ants. *J. Wildl. Manag.* 63:1291–1298.

Murray, R.W., and O.E. Frye, Jr. 1964. *The bobwhite quail and its management in Florida.* Game Publication 2, Florida Game and Fresh Water Fish Commission, Tallahassee. 56 pp.

Nayar, J.K., M.D. Young, and D.J. Forrester. 1982. Experimental transmission by mosquitoes of *Plasmodium hermani* between domestic turkeys and pen-reared bobwhites. *J. Parasitol.* 68:874–876.

Percival, H.F., L.G. Webb, and J.K. Reed. 1972. Concentrations of selected chlorinated hydrocarbon insecticides in bobwhite quail in South Carolina. *Proc. Southeast. Assoc. Game Fish Comm.* 26:108–117.

Peters, H.S. 1936. A list of external parasites from birds of the eastern part of the United States. *Bird-Banding* 7:9–27.

Pier, A.C. 1980. Mycotoxicoses. In: *CRC handbook series in zoonoses.* Sec. A, *Bacterial, rickettsial, and mycotic diseases.* Vol. 2. J.H. Steele (ed.). CRC, Boca Raton, Fla. pp. 499–507.

Quist, C.F. 1997. Unpublished data. Southeastern Cooperative Wildlife Disease Study, University of Georgia, Athens.

Reid, W.M. 1967. Etiology and dissemination of the blackhead disease syndrome in turkeys and chickens. *Exp. Parasitol.* 21:249–275.

Rickard, L.G. 1983. Unpublished data. University of Florida, Gainesville.

Rogers, A.J. 1953. A study of the ixodid ticks of northern Florida, including the biology and life history of *Ixodes scapularis* Say (Ixodidae: Acarina). Ph.D. diss., University of Maryland, College Park. 191 pp.

Ruff, M.D. 1985. Life cycle and biology of *Eimeria lettyae* sp.n. (Protozoa: Eimeriidae) from the northern bobwhite, *Colinus virginianus. J. Wildl. Dis.* 21:361–370.

Ruff, M.D., and G.C. Wilkins. 1987. Pathogenicity of *Eimeria lettyae* Ruff, 1985 in the northern bobwhite (*Colinus virginianus*). *J. Wildl. Dis.* 23:121–126.

Schmidt, G.D., and R.E. Kuntz. 1977. Revision of *Mediorhynchus* Van Cleave 1916 (Acanthocephala) with a key to species. *J. Parasitol.* 63:500–507.

SCWDS. 1979. Unpublished data. Southeastern Cooperative Wildlife Disease Study, University of Georgia, Athens.

Smith, G.J. 1993. *Toxicology and pesticide use in relation to wildlife: organophosphorus and carbamate compounds.* C.K. Smoley, Boca Raton, Fla. 171 pp.

Spalding, M.G., and R.G. McLean. 1994. Unpublished data. University of Florida, Gainesville.

Stiles, C.W., and A. Hassall. 1894. A preliminary catalogue of the parasites contained in the collections of the U.S. Bureau of Animal Industry, U.S. Army Medical Museum, Biological Department of the University of Pennsylvania (Coll. Leidy) and in Coll. Stiles and Coll. Hassall. *Vet. Mag.* 1:245–253, 331–354.

Stoddard, H.L. 1931. *The bobwhite quail, its habits, preservation and increase.* Charles Scribner's Sons, New York. 559 pp.

Stromborg, K.L. 1986a. Reproductive toxicity of monocrotophos to bobwhite quail. *Poult. Sci.* 65:51–57.

———. 1986b. Reproduction of bobwhites fed different dietary concentrations of an organophosphate insecticide, methamidophos. *Arch. Environ. Contam. Toxicol.* 15:143–147.

Taylor, D.J. 1951. The distribution of ticks in Florida. M.S. thesis, University of Florida, Gainesville. 124 pp.

Telford, S.R., Jr., and D.J. Forrester. 1975. *Plasmodium (Huffia) hermani* sp.n. from wild turkeys (*Meleagris gallopavo*) in Florida. *J. Protozool.* 22:324–328.

Thompson, N.P., and R.L. Emerman. 1974. Interaction of p,p'-DDT with histomoniasis in bobwhites. *Bull. Environ. Contam. Toxicol.* 11:474–482.

Travis, B.V. 1941. Examination of wild animals for the cattle tick *Boophilus annulatus microplus* (Can.) in Florida. *J. Parasitol.* 27:465–467.

Tyzzer, E.E. 1929. Coccidiosis in gallinaceous birds. *Am. J. Hyg.* 10:269–383.

Walton, A.C. 1927. A revision of the nematodes of the Leidy Collections. *Proc. Acad. Nat. Sci. Phila.* 79:49–163.

Wheeler, W.B., D.P. Jouenaz, D.P. Wojcik, W.A. Banks, C.H. Van Middelem, C.S. Lofgren, S. Nesbitt, L. Williams, and R. Brown. 1977. Mirex residues in nontarget organisms after application of 10-5 bait for fire ant control, northeast Florida 1972–74. *Pestic. Monit. J.* 11:146–156.

Whisenhunt, M.H. 1959. Effect on wildlife of high-powered insecticides. *Fla. Nat.* 32:73–74, 88.

White, D.H., J.T. Seginak, and R.C. Simpson. 1990. Survival of northern bobwhites in Georgia: cropland use and pesticides. *Bull. Environ. Contam. Toxicol.* 44:73–80.

White, F.H., and D.J. Forrester 1987. Unpublished data. University of Florida, Gainesville.

Wilson, N., and H.W. Kale. 1972. Ticks collected from Indian River County, Florida (Acari: Metastigmata: Ixodidae). *Fla. Entomol.* 55:53–57.

Wiseman, S.P. 1979. Isolation, characterization, and distribution of an adenovirus of wild bobwhite quail. M.S. thesis, University of Georgia, Athens. 59 pp.

Wild Turkeys

I. Introduction

There are 2 subspecies of Wild Turkeys (*Meleagris gallopavo*) in Florida, the Florida Wild Turkey (*M. g. osceola*) in peninsular Florida and a population in northern Florida that intergrades with the Eastern Wild Turkey (*M. g. silvestris*) (Williams and Austin 1988; Williams 1992). Historically Florida had large populations of turkeys (Wright 1915), but these were drastically reduced in parts of the state and reached extremely low numbers by 1948. An intensive restocking program was conducted from 1949 to 1970 and resulted in the transplanting of more than 6,000 birds, mostly into areas not inhabited by turkeys (Powell 1965; Williams and Austin 1988). In 1989 there were an estimated 75,000 Florida turkeys and 25,000 intergrades in northern Florida (Kennamer and Kennamer 1990).

There are several useful reviews on the diseases and parasites of Wild Turkeys (Prestwood et al. 1973; Prestwood et al. 1975; Davidson and Wentworth 1992). The latter publication is extensive and provides an excellent general overview. Davidson et al. (1985) published a synopsis of the diseases diagnosed in Wild Turkeys in the southeastern United States over a 13-year period (1972–84), and a similar paper was published for turkeys in Florida for a 21-year period (1969–90) (Forrester 1992a). The book titled *Studies of the Wild Turkey in Florida* by Williams and Austin (1988) and the review titled "Florida Turkey" by Williams

(1992) are useful references on many aspects of the biology of turkeys in Florida.

II. Trauma

Morbidity and mortality of turkeys due to collisions with vehicles on roads are known to occur, but the extent is unknown. From 1969 through 1990 6 of 76 turkeys submitted for necropsy were judged to be roadkills (Forrester 1992a,b). These turkeys came from Alachua (*n* = 1), Glades (*n* = 3), Highlands (*n* = 1), and Levy (*n* = 1) counties. An additional turkey killed by a vehicle was documented in Hendry County in 1991 (Spalding and Forrester 1991). The full extent and significance of this type of trauma have not been assessed.

Large numbers of turkeys are killed each year during the hunting season and an unknown number is taken illegally by poachers, although the latter is probably of lesser consequence now in Florida than it was in the past because of better law enforcement and changes in land use (Powell 1965, 1967). At the present time legal hunting of turkeys occurs in each of Florida's 67 counties; in 1997 more than 12,000 were estimated to have been harvested by hunters (Linda 1998). Crippling losses due to hunting certainly occur, but are very difficult to document. Williams and Austin (1988) calculated that such losses equaled 25% of the legal harvest in the Lockloosa Wildlife Management Area in Alachua County over the 7-year period 1969–75. During a 21-year period, 1969–90, 4 turkeys in a sample of 76 that were submitted for determination of the cause of death had broken bones or old healing wounds that may have been caused by a hunter or perhaps by a predator (Forrester 1992a). These birds were from Alachua (1971), Duval (1974), Hendry (1970), and Lake (1989) counties (Forrester 1992b).

In 1977 a juvenile male turkey was captured at Fisheating Creek (Glades County) and held overnight in a holding pen. The next day it was dead and when examined at necropsy was found to have a ruptured auricle, resulting in extensive bleeding into the thoracic cavity (Forrester 1992b). This event may represent an unusual and isolated instance and has little or no relevance to wild populations. A single instance of aortic rupture in a live-captured Wild Turkey was reported by Davidson et al. (1985). In discussing this latter case further in relation to similar phenomena in domestic turkeys, Davidson and Wentworth (1992) mentioned that nutrition had been suggested as a contributory factor, but its exact role was not clear.

III. Predation

Predation on Wild Turkeys has been fairly well documented. The recent review by Miller and Leopold (1992) should be consulted for an evaluation of this type of trauma. Predation occurs on turkeys at all phases of their life cycle and can be divided into nest (or egg) predation, poult predation, and predation on older birds. The major egg predators of turkeys in North America are skunks, raccoons, opossums, feral dogs, coyotes, foxes, rodents, crows, and snakes (Williams 1981). In Florida the most important nest predators are raccoons, striped skunks, and spotted skunks, while others of secondary importance include opossums, gray foxes, bobcats, and American Crows (Williams and Austin 1988). Wild hogs have been suspected of destroying turkey nests, but were not found to be involved in such activities in southern Florida, according to Williams and Austin (1988), who conducted a long-term study at Fisheating Creek (Glades County). In the latter study it was found also that common long-nosed armadillos did not eat turkey eggs, but in 2 instances were observed rooting through nests and thereby rolling some of the eggs out of the nest and causing abandonment by the hens. In the Fisheating Creek study nest predation was higher (55%) in the cypress woods than in saw palmetto habitats (32%) or in habitats of heterogeneous plant composition and structure (21%) (Williams and Austin 1988).

McEwan and Hirth (1980) found the remains of a "young turkey" in 1 of 16 active

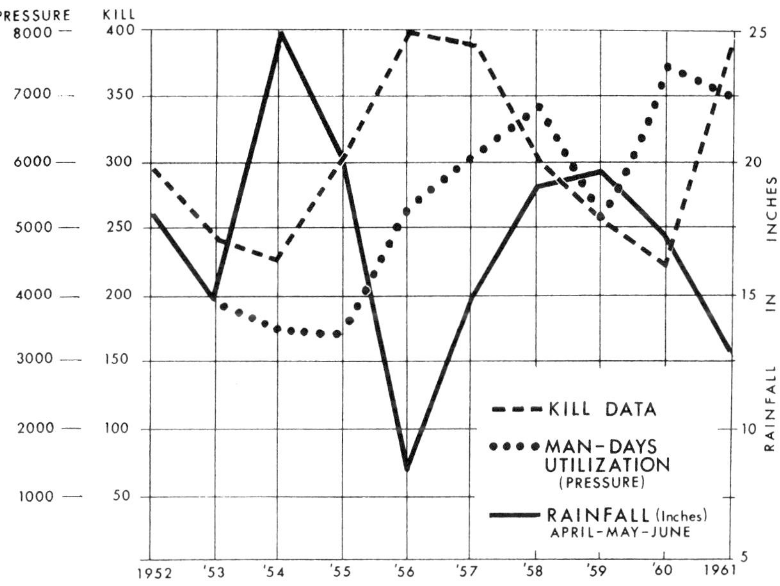

FIGURE 17.1. Numbers of Wild Turkeys killed by hunters in the fall, correlated with rainfall during the preceding spring months, at Fisheating Creek Wildlife Management Area (Glades County), 1952–61. From Powell 1965; courtesy of the Florida Game and Fresh Water Fish Commission.

Bald Eagle nests examined in Alachua, Marion, Putnam, and Volusia counties during 1975 and 1976. It is not clear, however, whether this was a Wild Turkey or a domestic turkey poult; also it is not known if this was a case of predation or scavenging by the Bald Eagle. Mortality of poults during the 14-night ground roosting period after hatching was about 74% at Fisheating Creek over a 17-year period, 1966–82 (Williams and Austin 1988). In 1980–81 Akey (1981) conducted a radiotelemetry study of poult mortality at Fisheating Creek and found that 63% of 52 poults were lost during the first 2 weeks of life, most due to predation. In 6 instances radio transmitters that had been attached to poults were located and retrieved from nests containing fledgling Red-shouldered Hawks. Other avian predators may have killed some of the poults; in several cases predation by raccoons was suspected because of their having been seen in the vicinity within a few minutes of the deaths of the poults. Gray foxes also have been implicated as predators of

poults in Florida (Williams 1992). Dusek (2000) observed a pair of adult Florida Sandhill Cranes kill, but not consume, a 3-day-old Wild Turkey poult in Osceola County in April of 2000.

Few specific data are available on predation on older turkeys. Fairly extensive studies of the food habits of black bears (Maehr and Brady 1984) and bobcats (Maehr and Brady 1986) have been conducted in Florida and there was no evidence of predation on turkeys. Maehr et al. (1990), however, found that 1 of 38 kills by Florida panthers in southwestern Florida from 1986 to 1989 was a turkey.

IV. Inclement weather

In his review of the influences of environmental factors on Wild Turkey populations, Healy (1992) stated that weather (i.e., "local and short-term atmospheric conditions at a specific time and place") and climate ("average

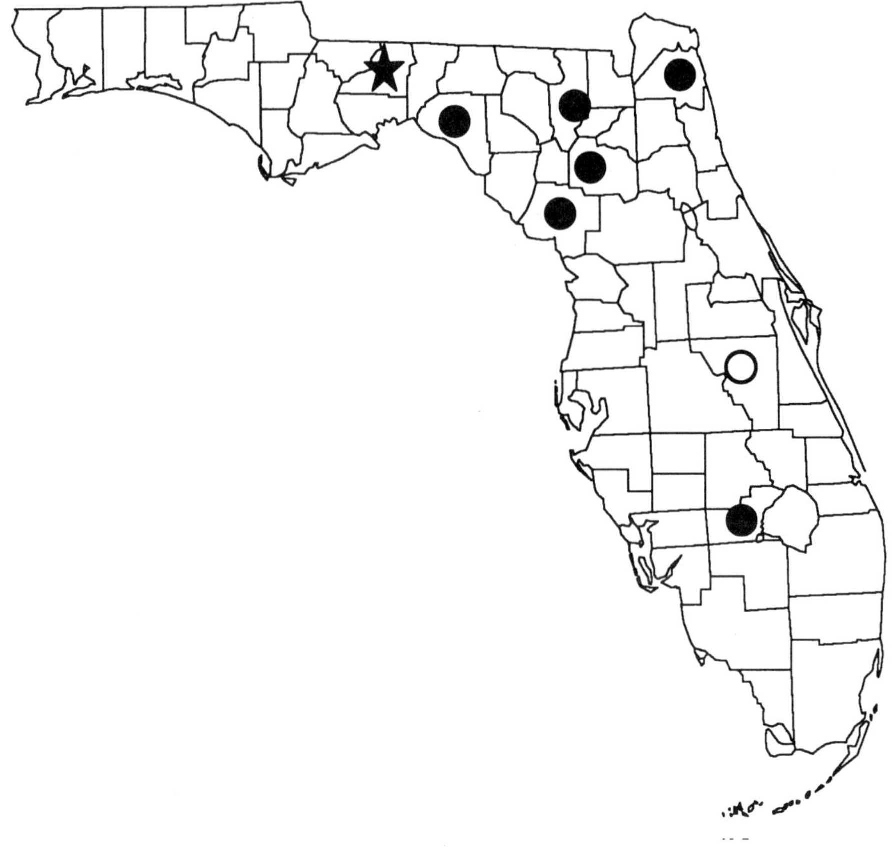

FIGURE 17.2. Distribution of neoplastic diseases in Wild Turkeys in Florida. *Star* = osteopetrosis, *solid circles* = lymphoproliferative disease, *open circle* = osteosarcoma. Data from Busch and Williams 1970; Forrester 1992; Frye 1958; Spalding and Forrester 1991.

weather conditions of a region over a period of years") exert "profound effects on wild turkey populations." The most important facet of weather as it affects Wild Turkeys in Florida appears to be abnormally high amounts of rainfall. Powell (1965) noted an inverse relationship between the numbers of turkeys harvested by hunters each fall and the amount of rainfall in the preceding spring months (April, May, and June) in Collier County and Glades County (figure 17.1) during the 1950s. He attributed the low fall harvests of turkeys following "wet" springs to losses of nests by flooding and losses of poults

by drowning. However, Williams and Austin (1988) observed many nests and broods during their long-term study at Fisheating Creek (Glades County) and found no evidence that rainfall adversely affected young poults directly by wetting and chilling them. They did record losses in 3 broods of poults that had crossed rain-swollen streams or ditches shortly after leaving their nests. They pointed out that by the time heavy summer rains set in, much of the turkey nesting has already been completed. Williams and Austin (1988) offered 3 possible explanations for the observations made by Powell (1965): (1) spring rains could cause high water levels that would prevent nesting or drown out nests, especially during the middle or latter part of the nesting season, (2) flooding waters might drive young poults into habitats where they would be more susceptible to predation, or (3) there might be a relationship between high water levels and increased transmission of turkey diseases. The latter idea will be addressed in subsequent sections of this chapter.

V. Environmental contaminants

Davidson and Wentworth (1992) reviewed the literature and found only a few published reports on environmental contaminants in Wild Turkeys; these were grouped into problems caused by agricultural pesticides and those caused by heavy metals. Pesticide problems involved dieldrin and heptachlor (Clawson 1959), dasnit (Nettles 1976), and diazinon and famfur (Davidson and Wentworth 1992) and were linked with accidental and intentional poisonings in several southeastern states. These incidents involved only small numbers of turkeys. One instance of lead poisoning in a turkey due to ingestion of spent shotgun pellets was reported in New York (Stone and Butkas 1978). Surveys of Wild Turkeys for various agricultural pesticides (in Illinois, Bridges and Andrews 1977) and heavy metals (in Virginia, Scanlon et al. 1979) resulted in finding only sublethal concentrations. It was concluded that Wild Turkeys can be adversely affected by environmental contaminants, but that such occurrences are rare and of little consequence (Davidson and Wentworth 1992). Nothing is known about the presence or significance of environmental contaminants in Wild Turkeys in Florida.

VI. Neoplasia

One bone tumor and 2 types of transmissible neoplastic diseases have been found in Wild Turkeys in Florida, namely osteosarcoma, lymphoproliferative disease, and osteopetrosis (table 17.1). The latter 2 are caused by RNA retroviruses (Calnek 1997).

An osteosarcoma was diagnosed in an adult female Wild Turkey found in a weakened condition on the Escape Ranch in Osceola County in 1995 (Spalding and Forrester 1996). Tumors involved the ribs and had metastasized to the lungs and liver. The severely emaciated turkey also had aspergillosis and leucocytozoonosis. This is the only report of an osteosarcoma in a Wild Turkey and its significance to turkey populations is not known.

Only 1 case of osteopetrosis has been reported (Frye 1958). This was in a turkey of unknown age and sex found in Leon County (figure 17.2) in June of 1957 and was an atypical form of the disease in that the metatarsi were not involved. Only the skeleton was available for examination; there was diffuse periosteal proliferation and ossification involving most of the long bones as well as the pelvis (figure 17.3). Frye pointed out that if the metatarsi had been affected the turkey would have become so debilitated that it probably would not have survived very long and the disease would not have reached the advanced stage.

Lymphoproliferative disease (LPD) is more common and has been reported from 5 counties in Florida (table 17.1). The first report of this disease in Wild Turkeys was published by Busch and Williams (1970), who at that time called it a "Marek's disease-like condition." They described 2 turkeys, 1 from Levy County and 1 from Glades County, that had lesions involving cell types of all forms of the lymphocytic series in the liver, spleen, kidneys, proventriculus, and nervous system. However, they did not make viral isolations or conduct serologic tests to conclusively identify the etiologic agent or agents involved. After that original report 9 other cases of LPD were diagnosed in Wild Turkeys in Florida (table 17.1). The disease was more common in adults than in juveniles, but was distributed equally between males and females. It was characterized by splenomegaly and sometimes an enlarged liver;

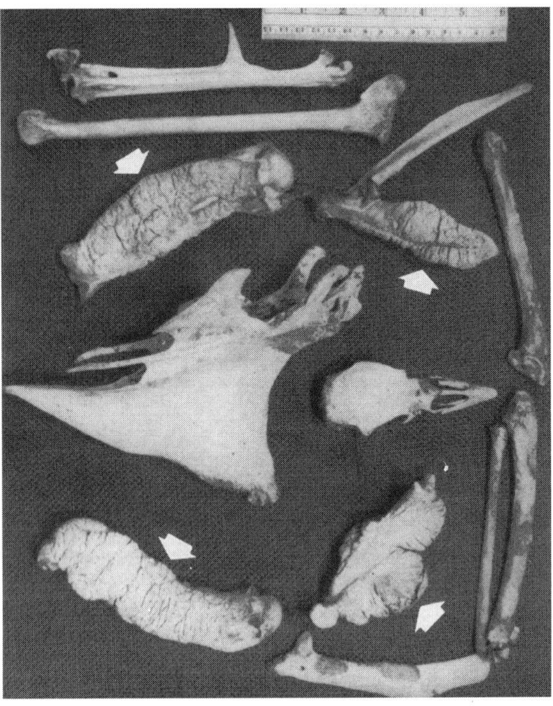

FIGURE 17.3. Skeleton of a Wild Turkey from Leon County with lesions *(arrows)* of osteopetrosis, 1957. From Frye 1958; courtesy of *Journal of Wildlife Management.*

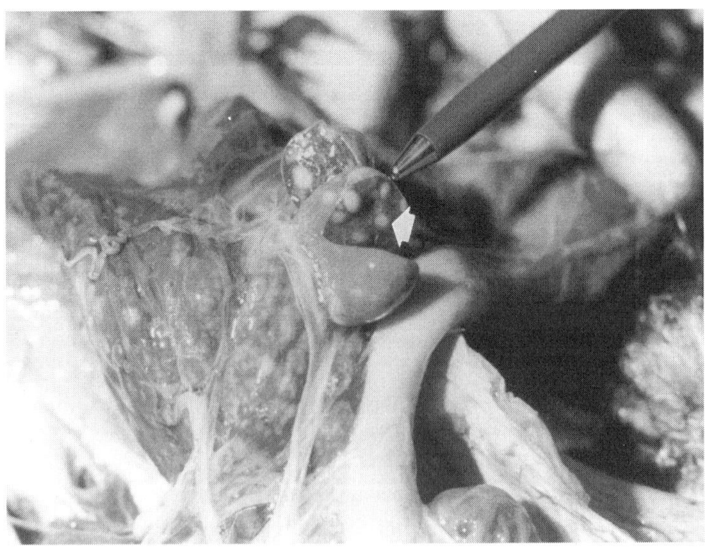

FIGURE 17.4. Spleen from a Wild Turkey from Taylor County with discrete gray-white lesions *(arrow)* of lymphoproliferative disease, 1978. Courtesy of Garry W. Foster.

Table 17.1. Neoplastic diseases diagnosed in Wild Turkeys from Florida

County	Year(s)	Number of cases	Age[a]	Sex	Data source
Lymphoproliferative disease					
Alachua	1978	1	AD	M	Forrester (1992b)
Columbia	1990	1	AD	M	Spalding & Forrester (1991)
Duval	1971	1	AD	F	Forrester (1992b)
Glades	1969	1	NG	F[b]	Busch & Williams (1970)
	1970	1	JU	M	Forrester (1992b)
	1970–71	2	AD	F	Ibid.
	1971	1	AD	M	Ibid.
	1977	1	JU	F	Ibid.
Levy	1969	1	NG	F[b]	Busch & Williams (1970)
	1977	1	AD	F	Forrester (1992b)
Taylor	1978	1	AD	M	Ibid.
Osteopetrosis					
Leon	1957	1[c]	NG	NG	Frye (1958)
Osteosarcoma					
Osceola	1995	1	AD	F	Spalding & Forrester (1996)

NG = not given by author.
a. AD = adult (>12 months), JU = juvenile (31 days–12 months).
b. Sex not given by Busch and Williams (1970); reported in a letter from Williams, July 28, 1969.
c. Diagnosis of the osteopetrotic form of avian leukosis was made by examination of the skeleton only (Frye 1958).

typically there were diffuse or discrete gray-white lesions in various internal organs such as liver, spleen, pancreas, thymus, kidneys, gonads, intestinal walls, lungs, and heart (figures 17.4–17.6).

Current information indicates that LPD is widespread throughout Florida from Taylor and Columbia counties in the north to Glades County in the southern part of the state (figure 17.2) and appears to be endemic, with occasional outbreaks such as described by Witter (1997) for domestic turkeys in the United States. One case has been reported in a Wild Turkey from South Carolina (Davidson et al. 1985; Davidson 1991). This disease may be reticuloendotheliosis, a type of lymphoproliferative disease caused by a retrovirus and described recently from Wild Turkeys in North Carolina (Ley et al. 1989) and Georgia (Hayes et al. 1992). Isolation and identification of the virus are necessary for a definitive diagnosis; this has not been done for any of the Florida cases. The significance of LPD in Wild Turkey populations in Florida is not known.

VII. Biotoxins

We have no data on biotoxins and their effects on Wild Turkeys in Florida. However, since aflatoxins (toxic metabolites of molds such as *Aspergillus flavus* or *A. parasiticus*) have been documented from moldy corn in Florida (Brownie et al. 1982), harmful effects of such toxins on Wild Turkeys could be expected. An experimental study was conducted in Georgia using 4-month-old pen-reared Wild Turkey poults given 100, 200, or 400 ppb aflatoxin in their feed (Quist et al. 2000). Aflatoxin-fed poults in all treatment groups had decreased weight gains and food consumption, compared with controls fed no aflatoxin. The authors concluded that exposure of Wild Turkeys to food contaminated with concentrations of aflatoxin

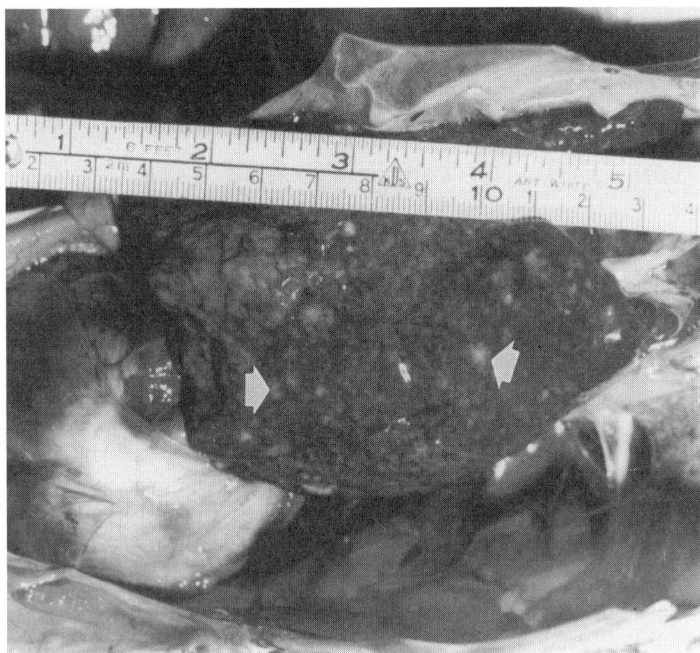

FIGURE 17.5. Liver from same turkey as in Figure 17.4 with small diffuse gray-white lesions *(arrows)* of lymphoproliferative disease. Courtesy of Garry W. Foster.

FIGURE 17.6. Liver of a Wild Turkey from Levy County with large discrete lesions *(arrow)* of lymphoproliferative disease, 1977. Courtesy of Garry W. Foster.

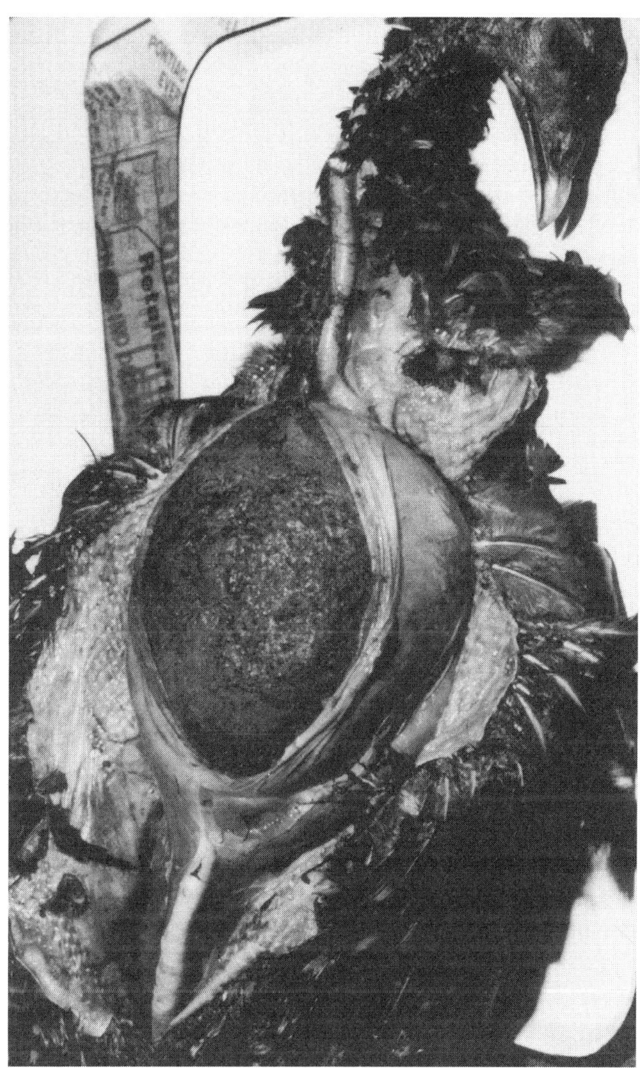

FIGURE 17.7. Crop impaction in an adult male Wild Turkey from Osceola County, 1993. Courtesy of Garry W. Foster.

greater than 100 ppb was harmful and should be avoided, if possible.

VIII. Anomalies

A number of anomalies have been encountered in Wild Turkeys in Florida, including crop impaction and abnormalities in bills, spurs, beards, and plumage. In November 1993 an adult male Wild Turkey that was unable to fly and unafraid of people was found in Osceola County (Spalding and Forrester 1993). It was hand-captured and died 2 hours later. At necropsy the turkey was emaciated and had hydropericardium and an impacted crop. The impaction was 15 cm in diameter and contained remains of grasshoppers, berries, and vegetation (figure 17.7). No food had passed into the stomach for some time, resulting in malnutrition and subsequent emaciation and hydropericardium. Crop impaction also occurs in domestic turkeys (Peckham 1978) and is believed to be caused by putting young birds out on range without readily available feed, leading them to feed on tough grasses; the fibrous material forms a ball in the crop that eventually becomes impacted and results in emaciation and death. This phenomenon is rare in Wild Turkeys and most likely has no population significance.

In July of 1975 a 3-month-old hen collected at Fisheating Creek (Glades County) as part of the disease study had a deformed upper bill (Forrester 1976). The deformity was judged to be not life threatening; the cause was unknown, perhaps being a result of trauma or a nutritional deficiency. It has been suggested that pox infections may cause bill deformities in birds (McClure 1989), but this has not been verified.

Supernumerary leg spurs were described in 2 turkeys, 1 from Leon County in 1962 and 1 from Wakulla County in 1965 (Williams 1967a). Williams and Austin (1969) reported spurs on 2 female turkeys from Glades County in 1968.

Turkeys with abnormally colored beards have been noted in Florida. In 1998 2 gobblers killed by hunters in Hendry County had beards with approximately the first inch at the base appearing scaly and colored whitish, gradually becoming light tan with increasing time post-mortem, while the remainder of the beard was the normal black coloration (Main 1998). There were reports of a number of turkeys observed with this condition in Hendry and Glades counties in 1998 (Main 1998). Williams (1996) discussed this anomaly and attributed such a condition to some factor such as disease that prevents the deposition of melanin in the beard feathers. Such feathers are weak and often break off in the area of the tan coloration.

There are several reports of abnormal plumage coloration. These include 3 shizochroistic pale turkeys from Baker County in 1962, 1 from Volusia County in 1962, and 1 from Bradford County in 1966 (Williams 1964; Williams 1969). In addition, an adult gobbler from Levy County had white specks throughout its body plumage and abnormal

white areas on the wings and tail; 2 jakes taken by hunters in Columbia County in 1992 had some blackish feathers (Williams 1996). In domestic turkeys, whitish areas on feathers are known to be caused by lysine deficiency in the diet, which prevents the normal formation of feather pigments (Scott and Krook 1972). A similar nutritional deficiency may be the cause of pale plumage in Wild Turkeys. Two brownish red turkeys, 1 from Nassau County in 1963 and 1 from Sarasota County in 1965, were reported by Williams (1967b). Changes in plumage coloration in populations of House Finches in southern California have been correlated spatially and temporally with pox infections (Zahn and Rothstein 1999). The mechanisms of this change have not been determined, but it is possible that the abnormal coloration in Wild Turkey feathers mentioned above could be connected with avian pox infections as well, since these infections are common in Florida (see section IX, Avian pox).

An adult hen with a fully feathered head was observed in Osceola County in 1971 by Williams (1972). The author stated that he had examined thousands of Wild Turkeys, primarily in Florida, and had never before seen this condition.

The significance of these anomalies to turkey populations is unknown. They are probably incidental findings.

IX. Avian pox

Avian pox is the most important infectious disease of Wild Turkeys in Florida (Davidson et al. 1985; Forrester 1991, 1992a). It is caused by a virus of the genus *Avipoxvirus* in the family Poxviridae. In addition to being significant in Wild Turkeys, it is an economically important disease in the domestic turkey industry (Tripathy and Reed 1997). The reviews by Karstad (1971b), Forrester (1991), Tripathy and Reed (1997), and Davidson and Wentworth (1992) contain details of many aspects of this disease.

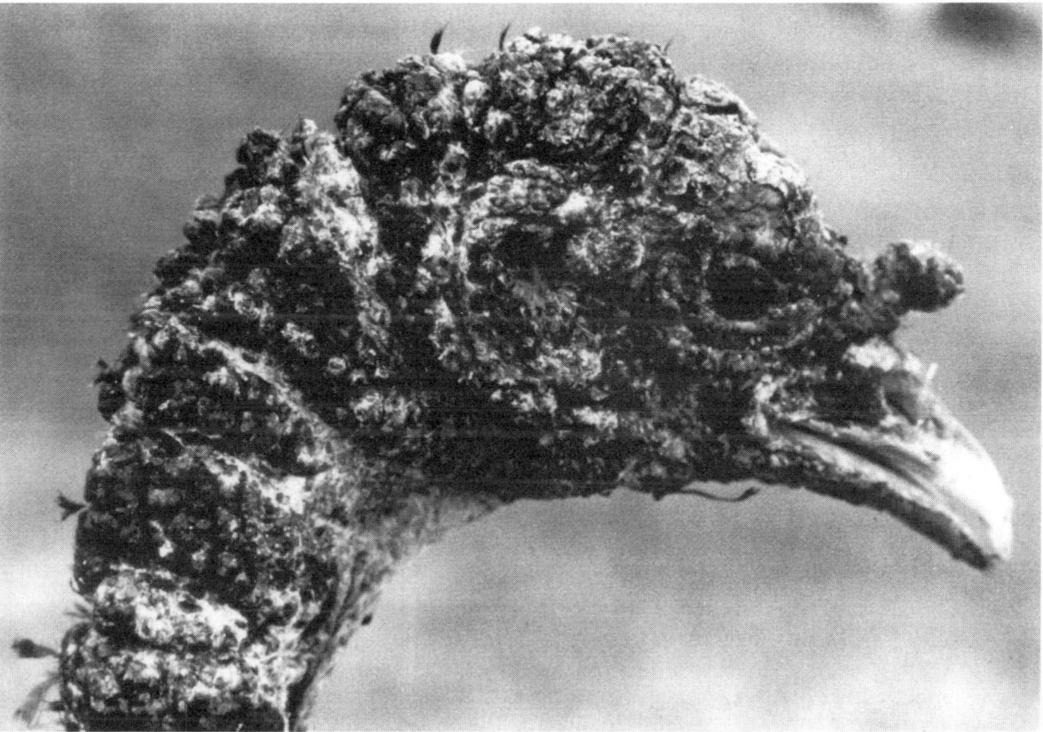

FIGURE 17.8. Cutaneous pox lesions on the head of a Wild Turkey. Note the lesion on the snood as well as others elsewhere on the unfeathered parts of the head. From Forrester 1991; photograph courtesy of William R. Davidson; by permission of *Bulletin of the Society of Vector Ecology.*

Pox infections vary in severity from mild cases, where there are only a few small lesions, to severe cases in which there are extensive lesions associated with visual problems, respiratory distress, emaciation, or weakness (Forrester 1991; Davidson and Nettles 1997). There are 2 forms of pox, cutaneous (= dry pox) and diphtheritic (= wet pox). The cutaneous form is the most common and is characterized by lesions located mainly on the unfeathered skin of the head and legs (figures 17.8 and 17.9). In the diphtheritic form lesions occur in the oral cavity, upper digestive tract, and upper respiratory tract (figure 17.10). Occasionally both cutaneous and diphtheritic lesions occur in the same turkey. Of the 58 turkeys found in Florida with pox infections during 1958–97 (table 17.2), 6 had both forms

of pox and 1 had only the diphtheritic form (Forrester 1992a,b). Turkeys with cutaneous pox are more likely to survive than birds with diphtheritic lesions. Cutaneous lesions, however, can interfere with vision and in turn result in increased susceptibility to predation. This problem is illustrated in figure 17.11, where the progression of cutaneous lesions on the head is followed over time in a young domestic turkey that had been infected naturally as a sentinel at Fisheating Creek during September 1978. By the 29th day after exposure (figure 17.11D) the bird was blind. By day 53 (figure 17.12) the pox scabs had fallen off and the infection had largely been resolved as the bird developed immunity. Although it was still blind, it did not die, since it was being maintained in captivity and could readily find its food and water. If this

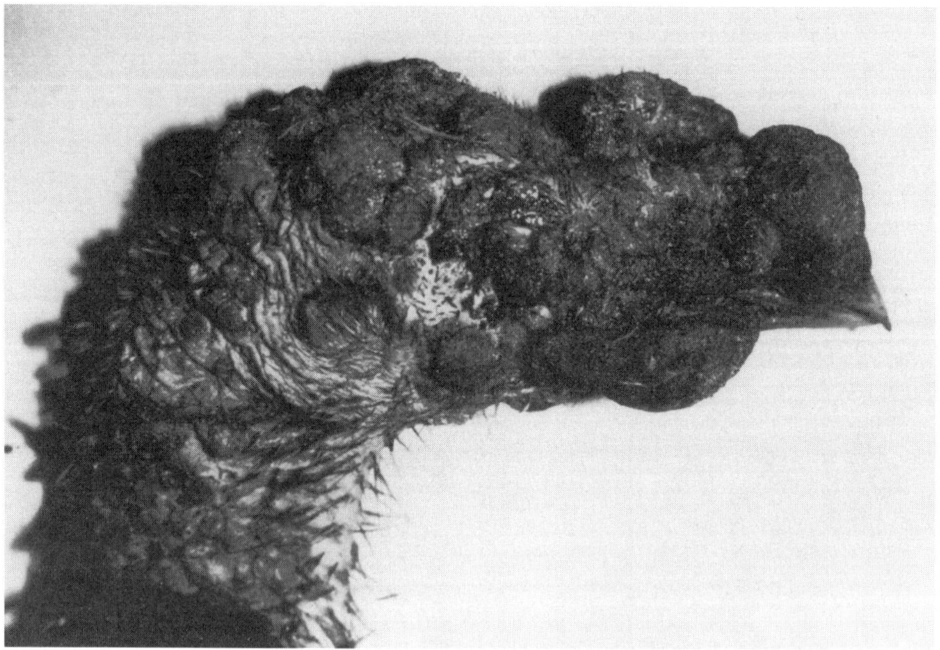

FIGURE 17.9. Severe case of cutaneous pox on head of a Wild Turkey from Duval County, 1977. Lesions of this type would most certainly result in a compromised host and could lead to emaciation and increased risk of predation.

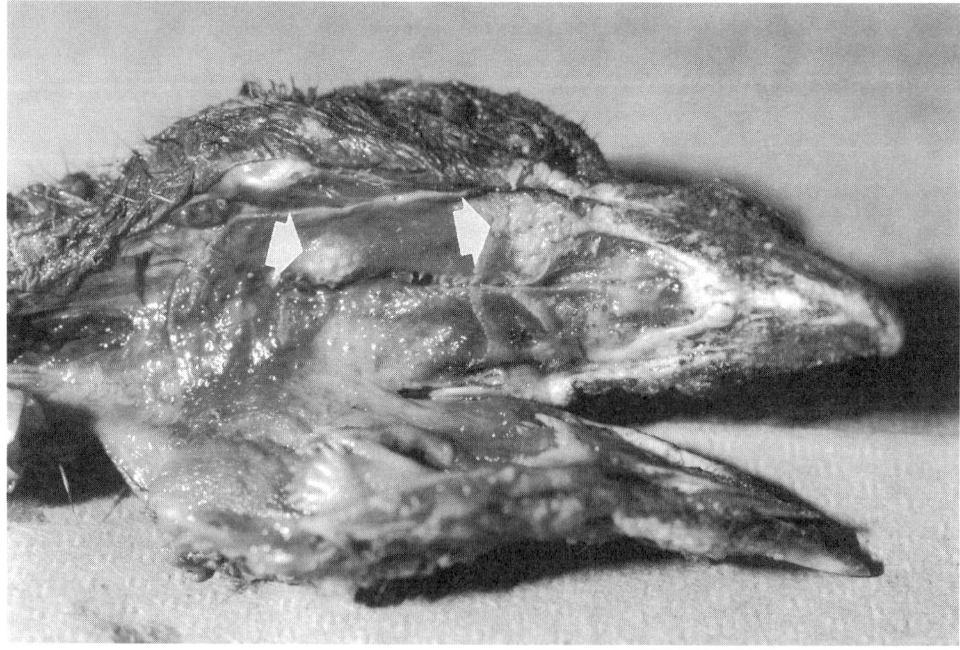

FIGURE 17.10. Diphtheric pox lesions *(arrows)* in the oral cavity of a Wild Turkey from Hendry County, 1991. Courtesy of Garry W. Foster.

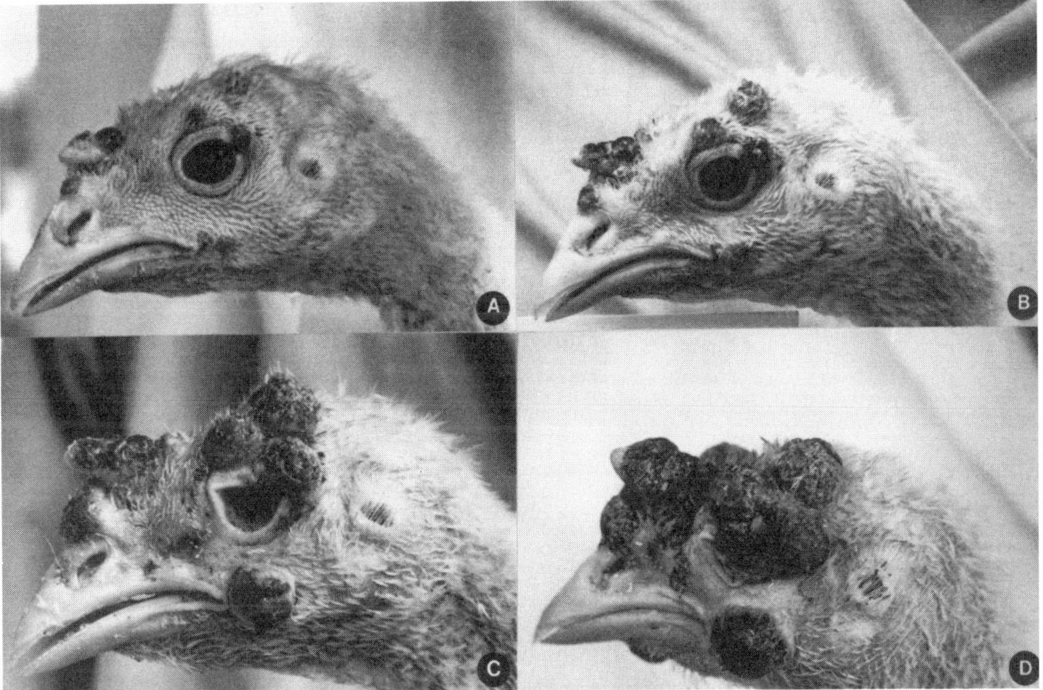

FIGURE 17.11. Development of cutaneous pox lesions on the head of a domestic turkey that had been infected naturally as a sentinel (probably by mosquitoes) at Fisheating Creek (Glades County) during September 1978. After being exposed as a sentinel for 2 weeks, the turkey was kept in an isolation room at the University of Florida and observed for the development of lesions. The photographs were taken at 8 days (*A*), 11 days (*B*), 15 days (*C*), and 29 days (*D*) after the bird was removed from the sentinel cage at Fisheating Creek and put in isolation. From Forrester 1991; photographs courtesy of Garry W. Foster; by permission of *Bulletin of the Society of Vector Ecology*.

had been a free-ranging Wild Turkey, it most certainly would have died of predation or some other complication.

Cutaneous and diphtheritic lesions have similar histopathologic characteristics that include epithelial hyperplasia, enlargement of cells, and inflammatory changes (Tripathy and Reed 1997). Cytoplasmic inclusions (= Bollinger bodies) are seen typically in tissues infected with poxviruses (figure 17.13) and are useful for diagnostic purposes.

Pox infections are widespread in Wild Turkeys throughout Florida and have been found in 24 counties from Bay County in the panhandle to Collier County in southern Florida (figure 17.14). It seems reasonable to assume that avian pox occurs throughout the state wherever Wild Turkeys occur, and that data from counties that are negative are most likely a reflection of sampling artifact.

From 1958 through 1997 58 cases of pox were documented in Wild Turkeys in Florida (table 17.2). These came from a sample of 1,385 turkeys from 31 counties and reflect a prevalence of 4.2%. This prevalence, however, should be interpreted with caution, since most of this sample was not taken randomly. The best data are from Fisheating Creek (Glades County), where 15 of 1,052 (1.4%) turkeys examined from 1968 to 1984 were positive for pox. Most of this sample consisted of birds collected for research purposes while a few

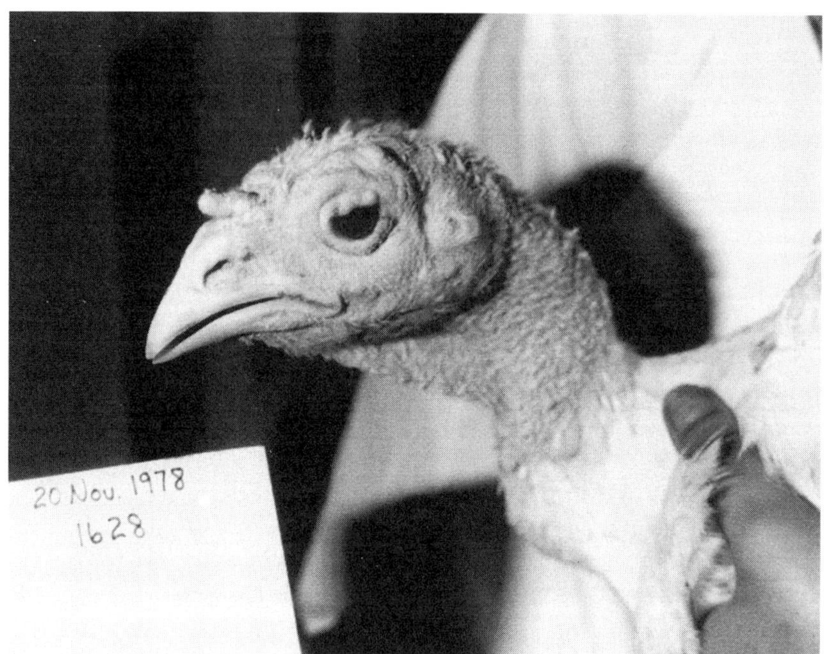

FIGURE 17.12. Same domestic turkey as in Figure 17.11 on the 53rd day after exposure. The scabby pox lesions had fallen off and the infection largely resolved. This bird was still blind, however. Courtesy of Garry W. Foster.

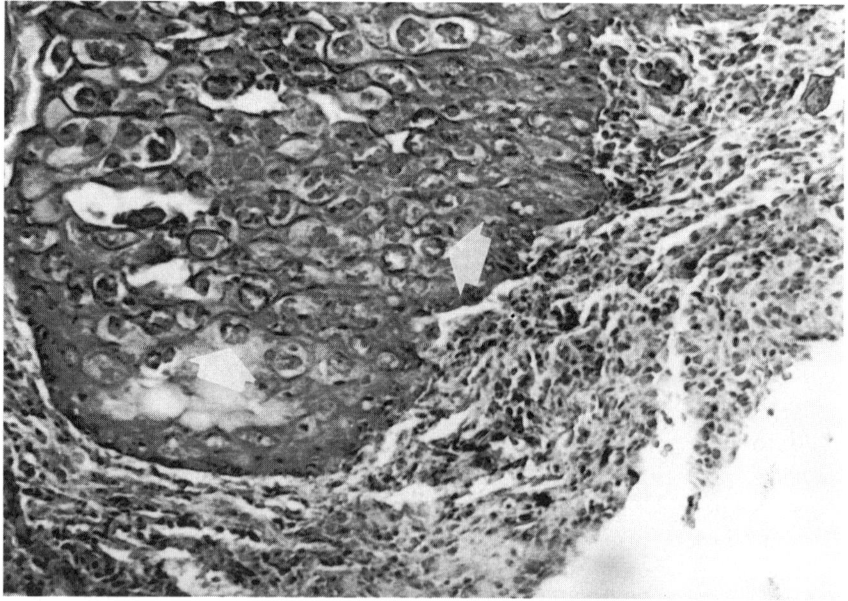

FIGURE 17.13. Photomicrograph of skin from a Wild Turkey in Florida showing cytoplasmic inclusions *(arrows)* of avian poxvirus in epidermal cells. Bar = 40 μm. From Forrester 1991; by permission of *Bulletin of the Society of Vector Ecology*.

Table 17.2. Distribution and prevalence of avian pox in Wild Turkeys from Florida

County	Year(s)	Number of turkeys			Data source
		Examined	Positive	Percent	
Alachua	1969–80	110	2	2	Forrester (1992b)
Bay	1984	1	1	—	Howerth (1984)
Collier	1980, 82	2	1	—	Forrester (1991)
Columbia	1990	1	1	—	Spalding & Forrester (1991)
Duval	1972–78	12	8	67	Forrester (1992b)
	1990	1	1	—	Eichholz (1991)
Flagler	1969, 83	6	2	—	Forrester (1991)
Gadsden	1971, 82	2	0	—	Ibid.
	1991	1	1	—	Ibid.
Glades	1968	1	1	—	Hanley & Austin (1968)
	1969–84	1051	14	1	Forrester (1992b)
Hamilton	1989	1	1	—	Ibid.
Hendry	1975, 85, 91	3	2	—	Ibid.
Lake	1986–89	5	1	—	Spalding & Forrester (1991)
Leon	1971–83	5	0	—	Forrester (1992b)
	1994	1	1	—	Smith (1994)
	1997	1	1	—	Cornish (1997)
Levy	1976–78	3	1	—	Forrester (1992b)
	1989–90	2	2	—	Eichholz (1991)
	1991	1	1	—	Spalding & Forrester (1991)
Liberty	1990	1	1	—	Eichholz (1991)
Madison	1971	2	1	—	Forrester (1992b)
Okeechobee	1988	1	1	—	Ibid.
Osceola	1970–76	87	1	1	Ibid.
	1997	1	1	—	Spalding & Forrester (1997)
Pasco	1979, 84	2	1	—	Forrester (1992b)
Putnam	1965	2	2	—	Eichholz (1991)
	1973	1	1	—	Kellogg (1973)
	1990	2	1	—	Spalding & Forrester (1991)
St. Johns	1972	1	1	—	Forrester (1992b)
Sumter	1974	1	1	—	Ibid.
Taylor	1974	1	1	—	Ibid.
Volusia	1958	1	1	—	Powell (1965)
Wakulla	1967	1	1	—	Payne (1967)
	1971, 85	10	1	10	Forrester (1992b)
Others[a]	1970–87	61	0	—	Forrester (1991)
Totals	1958–97	1385	58	4.2	

a. Includes negative data from 7 other counties (sample sizes in parentheses): Charlotte (27), Citrus (3), Escambia (2), Highlands (14), Jefferson (1), Manatee (12), and Orange (2).

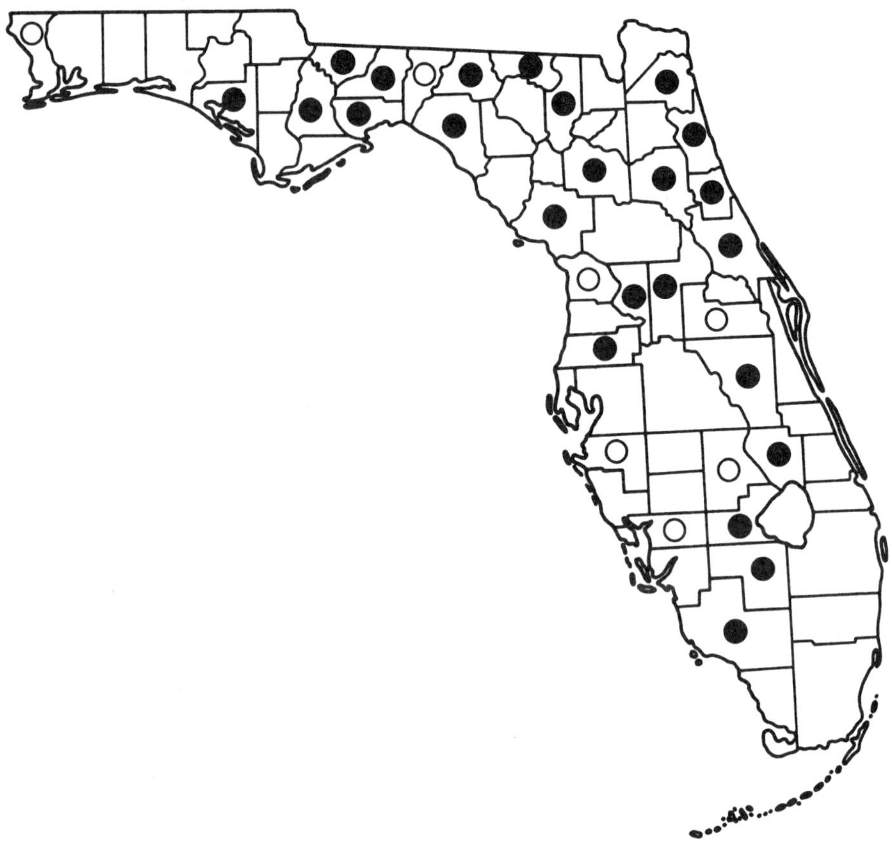

FIGURE 17.14. Distribution of avian pox in Wild Turkeys from Florida, 1958–97. Solid circles indicate counties where pox infections have been observed in Wild Turkeys; open circles are counties in which Wild Turkeys have been examined and found negative for pox. Data from Forrester 1991, 1992b; Powell 1965; Eichholz 1991; records of the Southeastern Cooperative Wildlife Disease Study.

were hunter-killed turkeys or turkeys found in diseased condition.

Transmission of the virus occurs mechanically by direct contact between birds or indirectly by contact with contaminated objects such as perches, food items, etc. (Karstad 1971b). However, blood-sucking arthropods, especially mosquitoes, are important mechanical vectors (DaMassa 1966; Forrester 1991) and may be the most significant means of transmission. In Florida Akey (1981) and Akey et al. (1981) demonstrated experimentally that 4 species of mosquitoes (*Culex nigripalpus*, *C. salinarius*, *Wyeomyia vanduzeei*,

and *Anopheles quadrimaculatus*) could transmit poxvirus (obtained originally from Wild Turkeys in southern Florida) to domestic turkey poults. All of these mosquitoes occur in southern Florida in Wild Turkey habitats and have population peaks during the late summer and early fall, when most pox infections occur (Nayar 1982; Akey et al. 1981; Forrester 1991). Apparently the viruses do not replicate within the mosquitoes, but once a mosquito is infected, it remains infected for extended periods, maybe for life. In the case of *C. nigripalpus*, viruses were shown to persist for at least 4 weeks, after which time they could still be

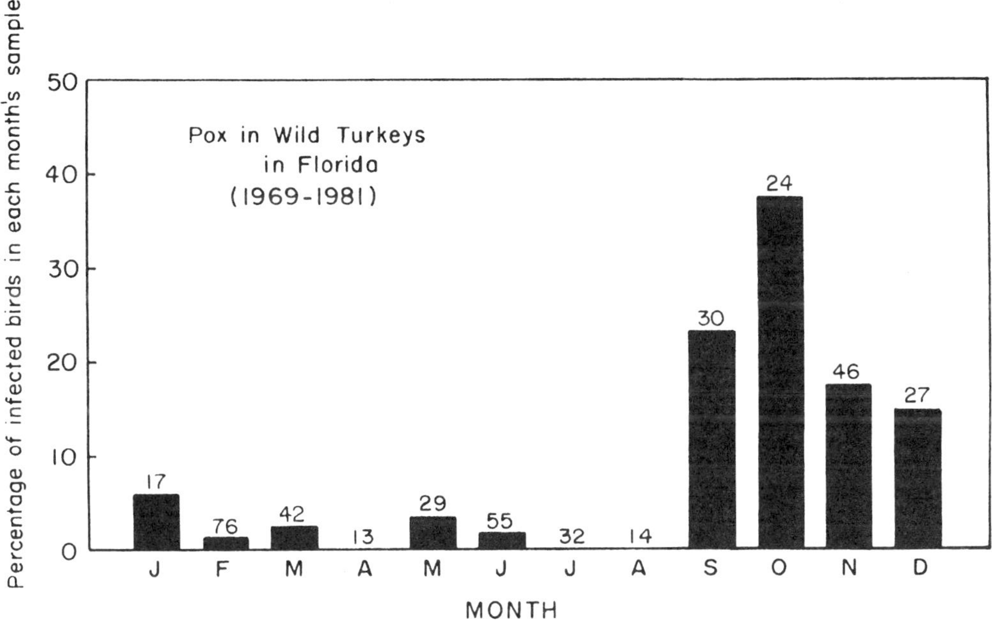

FIGURE 17.15. Seasonal occurrence of avian pox in Wild Turkeys from 12 counties in Florida, 1969–81. The numbers on top of each bar indicate sample sizes. From Forrester 1991; by permission of *Bulletin of the Society of Vector Ecology.*

transmitted to turkeys (Akey et al. 1981). Forrester and Nayar (1981) showed that wild-trapped specimens of *C. nigripalpus* and *W. vanduzeei* from turkey habitats at Fisheating Creek (Glades County) in September were carrying poxvirus. This was demonstrated by inoculating domestic poults with slurries prepared from batches of each species of mosquito; poults were then kept in isolation rooms and observed for the subsequent development of pox lesions (Forrester 1991).

The transmission of poxvirus to Wild Turkeys in Florida is seasonal; most infections occur during the fall, September to December, as shown in figure 17.15. This observation was confirmed by sentinel studies conducted with caged domestic turkeys at Fisheating Creek (Glades County) in 1976–77 (Forrester 1977, 1991). Domestic turkey poults, 2–4 weeks old, were reared at the University of Florida in an isolation room that protected them from blood-sucking arthropods, then transported in

insect-proof transport cages to Fisheating Creek where they were placed in holding cages (figures 17.16 and 17.17). The cages provided protection of the poults from predators, but allowed insects such as mosquitoes and other

FIGURE 17.16. View of a sentinel cage in which young domestic turkeys were placed for several weeks at a time in order to study the transmission of vector-borne diseases of Wild Turkeys in Florida.

FIGURE 17.17. Sentinel cage with domestic turkeys, food, and water containers. The wire screening on two sides was predator-proof, but allowed blood-seeking arthropods to enter the cage and feed on the turkeys.

FIGURE 17.18. Sentinel cage in typical ground site in the cypress swamp at Fisheating Creek, Glades County.

FIGURE 17.19. Sentinel cage in a cypress tree at Fisheating Creek (Glades County) in a location simulating the roosting sites of Wild Turkeys.

blood-seeking arthropods to feed on them. Two groups of 5 poults each were so exposed, 1 group in a cage on the ground in the cypress woods near the creek (figure 17.18) and another group in a cage in a cypress tree about 20 feet above the ground (figure 17.19), simulating a roosting site. The poults were provided food and water *ad libitum* and checked every 2 or 3 days. At the end of a 2-week period the poults were replaced by another group of poults, returned to isolation rooms at the University of Florida, and observed for the next 3–4 weeks to determine if pox lesions developed. By this method of exposure, the transmission of poxvirus was monitored on a 2-week basis continually for a 20-month period. During that time poxvirus was transmitted to up to 100% of the sentinel poults in both the

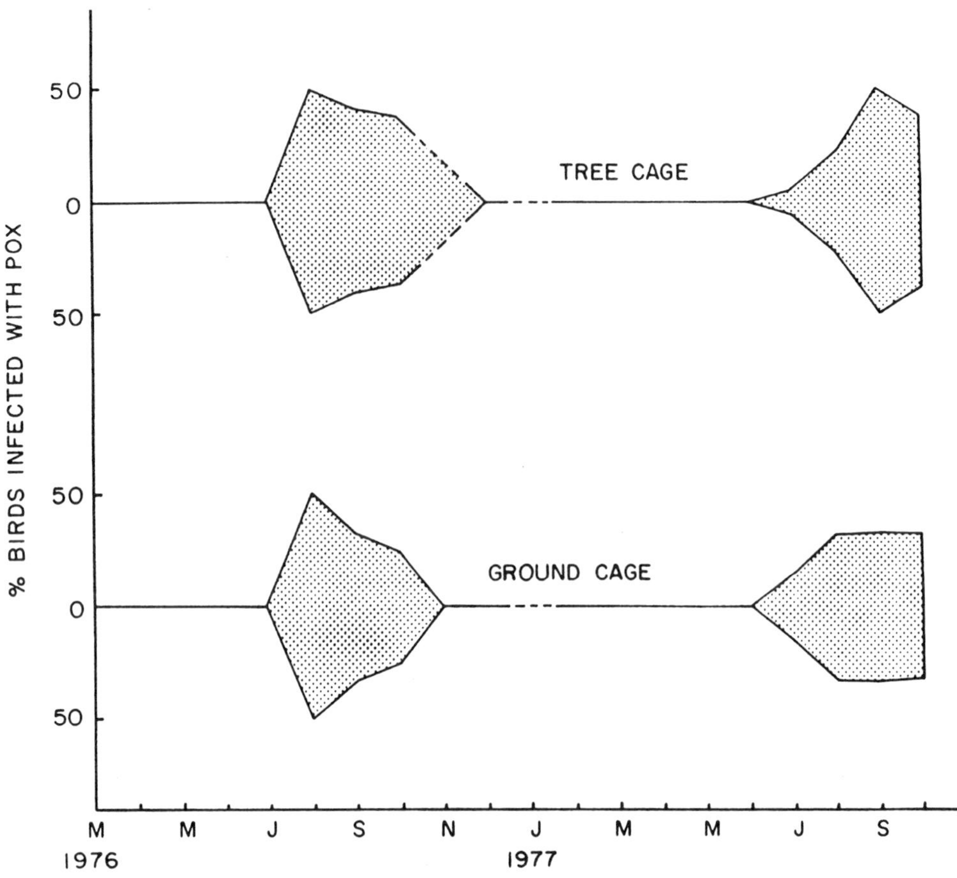

FIGURE 17.20. Seasonal transmission of avian pox to 2–4-week-old sentinel domestic turkeys at Fisheating Creek (Glades County), 1976–77. Broken lines indicate missing data due to the deaths of sentinel birds unrelated to disease. From Forrester 1991; by permission of *Bulletin of the Society of Vector Ecology.*

tree and ground cages during August, September, and October (figure 17.20), which is in close agreement with observations on Wild Turkeys (figure 17.15). In 1980 a second sentinel study was conducted at Fisheating Creek with similar results. Pox infections were not seen in these sentinels until August, but then continued until the end of November (Forrester 1981).

Seasonal transmission of poxvirus correlates with the rainy season in southern Florida and with peaks in populations of many species of mosquitoes, including *C. nigripalpus* and *W. vanduzeei* (Provost 1969; Akey et al. 1981;

Nayar 1982; Forrester 1991). Several other species of mosquitoes occur in smaller numbers along with *C. nigripalpus* and *W. vanduzeei* in Wild Turkey habitat in southern Florida. Nayar (1982) reported that representatives of 13 other species were captured in bait-can traps baited with live domestic turkey poults and 13 species in CDC light traps at Fisheating Creek from April to October of 1977 (table 17.3), although many were in very low numbers. Some of these other species may function as vectors of avian pox, although less significantly than *C. nigripalpus* and *W. vanduzeei* because of their lower abundance.

Table 17.3. Average number of adults of various species of mosquitoes collected at 2-week intervals at Fisheating Creek Wildlife Management Area, Glades County, April–October 1977

| | Average number of mosquitoes per trap | |
Species of mosquitoes	Bait cans[a]	CDC light traps
Aedes infirmatus	25	2
Aedes taeniorhynchus	6	0
Anopheles crucians	4	7
Anopheles quadrimaculatus	3	20
Coquillettidia perturbans	12	4
Culex erraticus	167	56
Culex nigripalpus	5909	186
Culex salinarius	2	1
Culiseta melanura	1	1
Mansonia dyari	90	148
Mansonia titillans	12	0
Psorophora ciliata	1	3
Psorophora columbiae	52	41
Psorophora ferox	8	0
Uranotaenia lowii	0	52
Uranotaenia sapphirina	0	54
Wyeomyia mitchellii	0	4
Wyeomyia vanduzeei	557	12

Source: Nayar (1982).
a. Live domestic turkey poults were used as bait to attract blood-seeking female mosquitoes (Nayar et al. 1980a, Forrester et al. 1980a).

Table 17.4. Prevalence of herpesviruses in Wild Turkeys from Florida

| | | Number of turkeys | | | Basis of | |
County	Year(s)	Exam.	Pos.	%	diagnosis	Data source
Alachua	1969–70	3[a]	1	—	Serology[b]	Colwell et al. (1973)
Glades	1969–70	27[c]	13	48	Serology and virus isolation	Ibid.
	1971–73	10	4	40	Virus isolation	Grant et al. (1975)
Osceola	1969–70	16[a]	9	56	Serology[b]	Colwell et al. (1973)
Totals	1969–73	56	27	48	—	

a. A herpesvirus was isolated from 1 of these birds.
b. Indirect hemagglutination test.
c. Herpesviruses were isolated from 2 of these birds. Colwell et al. (1973) reported erroneously that a herpesvirus was isolated from a turkey from Charlotte County (Babcock Ranch). The bird was actually from Glades County (Lykes Fisheating Creek Wildlife Management Area) (Forrester 1992b).

X. Other viruses

In addition to retroviruses and poxviruses, Wild Turkeys are susceptible to infection by a number of viruses that cause disease problems in domestic turkeys. Most of these, however, either do not occur naturally in Wild Turkey populations or occur only sporadically at low prevalences and are not known to be associated with morbidity or mortality (Davidson and Wentworth 1992). There is evidence for the occurrence of 4 other viruses in Wild Turkeys in Florida, a herpesvirus and 3 arboviruses, Eastern equine encephalitis (EEE) virus, St. Louis encephalitis (SLE) virus, and West Nile virus.

In a study of herpesviruses of Wild Turkeys in Florida conducted from 1969 to 1973 (Colwell et al. 1973, Grant 1972, Grant et al. 1975), herpesviruses were isolated from turkeys in Alachua and Glades counties and serological evidence was obtained from turkeys in Osceola County (table 17.4). Based on isolation and serological techniques, the prevalence was 48%. The virus could not be isolated from Wild Turkey embryos ($n = 6$), or from 1–5-day-old poults ($n = 15$), indicating that it is probably not transmitted through the eggs (Colwell et al. 1973). However, 3 5-day-old poults from the same brood at Fisheating Creek had high antibody titers (1:512) to the herpesvirus, indicating that they had received maternal antibodies. The pathogenicity of 1 isolate from a Wild Turkey from Fisheating Creek was tested experimentally in 3-day-old pen-reared Wild Turkey poults (Grant 1972). No gross or histopathologic signs of disease were seen in any of the exposed birds and attempts to reisolate the virus were unsuccessful. The experiment was complicated, however, by nutritional problems that resulted in a perosis-like malady in many of the poults and made interpretation of the results somewhat difficult. None of the free-ranging Wild Turkeys from which the herpesvirus was isolated (table 17.4) showed signs of disease. Furthermore, there is no evidence that this herpesvirus is related to the lymphoproliferative diseases seen in Wild Turkeys in Florida (Colwell et al. 1973; Forrester 1992a). Additional

research is needed on the possible significance of this herpesvirus as a disease agent.

Wild turkeys from Osceola County have been tested for 2 arboviruses that are associated with encephalitis in man and domestic animals. Four of 55 serum samples from Wild Turkeys in 1992–94 were seropositive for EEE virus and 8 of 54 of the same samples were seropositive for SLE virus (tested by plaque-reduction neutralization test in Vero cell cultures; Spalding and McLean 1994). Forrester and Wellings (1981), however, tested sera from 162 juvenile domestic turkeys that had been used in a Wild Turkey disease sentinel study at Fisheating Creek (Glades County). These samples came from turkeys exposed to mosquitoes and other blood-sucking arthropods in Wild Turkey habitat from April to November of 1980. Samples were tested for antibodies against EEE virus, Highlands J virus, Everglades virus, Keystone virus, and SLE virus. All samples were negative except for 3 that were seropositive for SLE virus. These positive samples came from birds that had been exposed for 2-week periods during late summer and fall (end of July through November), which corresponds with the time of population peaks of mosquito vectors in that area (Nayar 1982). These findings lead us to conclude that Wild Turkeys in Florida may serve some role as reservoirs of EEE and SLE viruses or perhaps other arboviruses of public health importance, such as has been recorded in Texas (Trainer 1973). Although there is no evidence that these arboviruses are harmful to Wild Turkeys (Karstad 1971a), additional research should be conducted to determine the significance of these preliminary observations.

In January 2002 a dead Wild Turkey from Calhoun County was found positive for West Nile virus using PCR and virus isolation techniques (Conti et al. 2002). The significance of this finding for Wild Turkey populations in Florida is not known.

Domermuth et al. (1977) conducted a serosurvey in Florida for the etiologic agent of hemorrhagic enteritis of turkeys. In that study 43 Wild Turkeys from 4 counties, including Alachua ($n = 5$), Glades ($n = 25$), Leon ($n = 1$),

Table 17.5. Bacterial diseases of Wild Turkeys in Florida

Disease condition

County	Year	Bacterium	Number of cases	Age[a]	Data source
Abscesses					
Glades	1970	Unknown	1	AD	Forrester (1970)
Glades	1980	*Corynebacterium* sp.	1	JU[b]	Akey (1981)
Duval	1974	Unknown	1	AD	Forrester (1974)
Bumblefoot					
Glades	1980	*Staphylococcus* sp.	1	AD	Forrester (1980)
Madison	1967	*Staphylococcus* sp.	1	JU	Doster (1967)
Dermatitis					
Citrus	1982	*Staphylococcus epidermidis*	1	AD	White & Forrester (1982)
Myocarditis/Septicemia					
Leon	1997	*Enterococcus* sp.[c] *Serratia liquefaciens*[c]	1	AD	Fischer (1997)
Pneumonia					
Glades	1971	Unknown	1	JU	Forrester (1971)
Sinusitis/Septicemia					
Levy	1991	*Staphylococcus aureus Corynebacterium* sp. *Escherichia coli*	1	AD	Spalding & Forrester (1991)
Salmonellosis					
Osceola[d]	1970	*Salmonella miami*	1	AD	White & Forrester (1970)
Osceola	1997	*S. tallahassee*	1	JU	Forrester & Spalding (2001)
Manatee[d]	1977	*S. braenderup*	1	AD	White & Forrester (1977)
Tuberculosis					
Jefferson	1993	*Mycobacterium* sp.[e]	1	AD	Smith (1993)

a. JU = juvenile (31 days–12 months), AD = adult (>12 months).
b. 42 days of age.
c. Probably postmortem invaders; Fischer (1997) felt that the cause of the bacterial septicemia was some other species of bacteria that could not be cultured.
d. Presumptive cases. See text for more details and Table 17.6 for information on various serovars of *Salmonella* isolated from Wild Turkeys in Florida.
e. Species not identified; probably *M. avium*.

and Osceola (*n* = 12), were tested (Forrester 1974). All were negative for this adenovirus that causes significant mortality in the domestic turkey industry.

XI. Bacteria

Eight types of bacterial diseases have been identified in Wild Turkeys in Florida (table 17.5). These include abscesses, bumblefoot, dermatitis, myocarditis/septicemia, pneumonia, sinusitis/ septicemia, salmonellosis, and tuberculosis. Since only 1–3 cases of each disease have been seen, they are probably not significant at the population level. These bacterial infections, except abscesses, pneumonia, and tuberculosis, have been reported previously from Wild Turkeys in other parts of the southeastern United States in low prevalences, as found in Florida (Davidson et al. 1982, 1985; Howerth 1985; Davidson and Wentworth 1992).

The case of sinusitis mentioned in table 17.5 is of interest because of the occurrence of *My-*

coplasma gallisepticum (MG) in low prevalences in Wild Turkeys (Davidson 1987; Davidson et al. 1988). Infections by MG cause sinusitis and airsacculitis, which result in considerable economic losses in the domestic turkey industry due to downgrading of carcasses, reduced feed and egg-production efficiency, and elevated medical costs (Ley and Yoder 1997). Mycoplasmosis has been reported from Wild Turkeys or semi-Wild Turkeys in California (Jessup et al. 1983), Colorado (Adrian 1984), and Georgia (Davidson et al. 1982). Experimental studies conducted on captive and pen-reared Wild Turkeys resulted in morbidity and decreased reproduction and fertility (Rocke and Yuill 1988; Rocke et al. 1988). These observations, along with results of several serologic surveys of Wild Turkey populations elsewhere in which evidence of MG infections was obtained (Rocke and Yuill 1987; Fritz et al. 1992), have stimulated interest among wildlife agencies involved in Wild Turkey relocation and restoration programs. Monitoring procedures have been established to prevent the dissemination of pathogenic mycoplasmas through infected Wild Turkeys to uninfected Wild Turkey populations or domestic turkey operations (Nettles and Thorne 1982; Nettles 1984; Amundson 1985; Davidson 1987).

MG does not seem to be a problem in Wild Turkeys in Florida. Only 1 case of sinusitis was seen in more than 1,400 turkeys examined from 1969 through 1991 (Forrester 1991) and attempts to isolate a mycoplasma from this turkey (an adult male from Levy County, 1991) were unsuccessful (Brown and Forrester 1992). In January and February 1984, 16 turkeys (5 juveniles and 11 adults) from Fisheating Creek Wildlife Management Area in Glades County and 12 turkeys (5 juveniles and 7 adults) from Perdida Wildlife Management Area in Escambia County were cultured for mycoplasmas using specialized media (Rocke and Forrester 1984); all cultures were negative. Davidson et al. (1988) conducted a serological survey of 291 Wild Turkeys from Georgia, Kentucky, Louisiana, North Car

olina, and Tennessee; all were seronegative for MG. From these results we conclude that Wild Turkeys in Florida and other parts of the southeast have limited exposure to MG. According to Davidson et al. (1988), when the disease does occur, it originates from domestic poultry, the most important source being "backyard" or "non-commercial" poultry and pen-reared gamebirds (Davidson 1987). Further evidence that the latter is true was obtained by Luttrell et al. (1991) in a study on Cumberland Island, Georgia. In addition Luttrell et al. (1992) examined 457 Wild Turkeys from South Carolina over a 5-year period and found no MG infections using serologic and culture techniques. They did, however, find evidence that other species of *Mycoplasma* were present in those turkeys and were able to culture *M. gallopavonis,* a nonpathogenic species, from 98% of a subsample of their birds. They pointed out that this species needs to be considered when turkeys are cultured in order to detect pathogenic mycoplasmas, since it is so widespread and grows more rapidly and vigorously in culture than does MG.

Salmonellosis is a term that encompasses 4 distinct diseases: pullorum disease, fowl typhoid, paratyphoid infection, and arizonosis, all of which are caused by bacteria of the genus *Salmonella* (Gast 1997). Pullorum disease and fowl typhoid are caused by nonmotile species of *Salmonella* and are relatively host specific, occurring primarily in domestic turkeys and chickens. Neither of these 2 diseases is known to occur in Wild Turkeys in the United States (Davidson and Wentworth 1992).

Paratyphoid infections and arizonosis, however, can develop as enteric and systemic salmonellosis, although this has not been reported frequently. These are caused by motile species of *Salmonella,* of which there are more than 2300 serovars (Gast 1997). Howerth (1985) reported the first clinical case of salmonellosis in a Wild Turkey from Alabama. *Salmonella typhimurium* was isolated from that bird and subsequently from 3 additional Wild Turkeys with salmonellosis: 1 from Alabama, 1 from Georgia, and 1 from Virginia (Davidson and

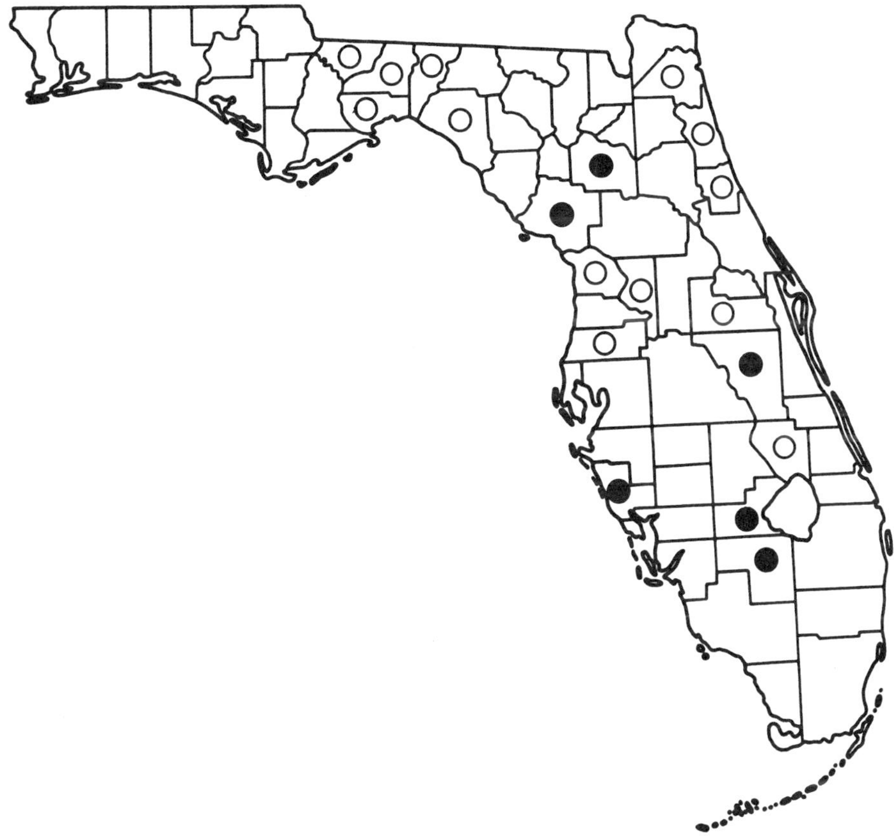

FIGURE 17.21. Distribution of *Salmonella* spp. in Wild Turkeys in Florida. Solid circles indicate counties where positive turkeys were found; open circles are counties in which turkeys were cultured, but were negative for *Salmonella* spp. Data from Akey 1981; White and Forrester 1992; White et al. 1981.

Wentworth 1992). In Florida, White and Forrester (1970) isolated *S. miami* from the liver of an adult male Wild Turkey from Lake X (Osceola County) in February 1970 (table 17.5). There were numerous white focal necrotic areas on the surface of the liver within which bacteria could be seen histologically; otherwise the turkey was normal. It had been captured during routine relocation activities and apparently died of an overdose of alphachloralose (Forrester 1970). This may have been a mild case of systemic salmonellosis in the early stages of development. A presumptive case of salmonellosis (enteric and systemic) was seen in an adult male turkey from Myakka River State Park (Manatee County) in March of 1977 (table

17.5). This bird was found in a weakened condition and died a few hours after being captured. At necropsy there were numerous inflammatory lesions throughout the body, including enteritis, myocarditis, epicarditis, and glossitis (Popp 1977). *Salmonella braenderup* was isolated from the intestines (White and Forrester 1977). Because the carcass of this animal was somewhat autolytic, a definitive diagnosis of salmonellosis could not be made, but the evidence pointed to this as the cause of morbidity in this turkey. A third Wild Turkey had a definitive case of salmonellosis (paratyphoid). This was an emaciated juvenile female found near Three Lakes Wildlife Management Area in Osceola County during January 1997

Table 17.6. Isolations of *Salmonella* spp. from Wild Turkeys in Florida

County	Year(s)	Number of turkeys	
		Examined	Positive
Alachua	1969–78	40	3[a]
	1980	1	0
Glades	1969–79	296	7[b]
	1975	1	1[c]
	1980	31	1[d]
	1981	29	2[e]
Hendry	1974–76	2	1[f]
Levy	1977–79	2	1[g]
Osceola	1969–79	36	6[h]
Sarasota	1971	2	1[i]
Others[j]	1969–82	37	0
Totals	1969–82	477	23 (4.8%)

Sources: Akey (1981), White and Forrester (1992), White et al. (1981).
a. *S. inverness, S. java,* and *S. muenchen* from an adult, a poult, and a juvenile, respectively.
b. *S. hartford* (from an adult), *S. manhattan* (from a 10-day-old poult), *S. miami* (from a 94-day-old juvenile), *S. newport* (from 2 juveniles), and *S. rubislaw* (from 2 juveniles).
c. *S. saint-paul* isolated from feces of poults (several days of age) in a nest at Fisheating Creek Wildlife Management Area.
d. *S. arizonae* from a 70-day-old juvenile.
e. *S.* sp. from a 32-day-old juvenile and *S. enteritidis* from an adult.
f. *S.* sp. from a juvenile.
g. *S. typhimurium* var. *copenhagen* from an adult.
h. *S. hartford, S. java, S. miami,* and *S. typhimurium* ($n = 3$), all from adults.
i. *S. braenderup* from an adult.
j. Includes negative data from 13 other counties (sample sizes in parentheses): Citrus (1), Duval (11), Flagler (4), Gadsden (3), Jefferson (1), Leon (2), Okeechobee (1), Orange (2), Pasco (1), St. Johns (1), Sumter (2), Taylor (5), and Wakulla (3).

(Forrester and Spalding 2001). The infection resulted in septicemia, typhlitis, enterocolitis, hepatitis, and nephritis. *Salmonella tallahassee* was identified in this turkey. The bird also had a concurrent pox infection on the head and within the oral cavity.

Although there are only the above 2 presumptive cases and 1 definitive case of salmonellosis known from Wild Turkeys in Florida, there is evidence that various serovars of *Salmonella* are present in turkey populations in the carrier state. From 1969 to 1982 intestinal contents of 478 Wild Turkeys from 19 counties in Florida were cultured for pathogenic bacteria (figure 17.21). Thirteen serovars were identified in 23 (4.8%) of the turkeys (table 17.6). A number of these isolations were from juveniles, 1 was from poult feces obtained from a Wild Turkey nest at Fisheating Creek, while another was from a 10-day-old poult

(White et al. 1981). The relationships of *Salmonella* infections in range cattle and other birds on the Fisheating Creek area in Glades County with those in Wild Turkeys were investigated also (White et al. 1981). Various serovars of *Salmonella* were isolated from intestinal contents of Eastern Towhees, Common Grackles, American Crows, and Northern Cardinals, as well as from range cattle. Four of the serovars (*braenderup, hartford, miami,* and *muenchen*) from these other birds and 1 (*saint-paul*) from range cattle were also isolated from Wild Turkeys (White and Forrester 1975; White et al. 1981), indicating that some of the serovars may be cycling between these hosts. These relationships need to be further investigated.

In addition to the pathogenic bacteria discussed above, 12 other species have been isolated from Wild Turkeys in Florida (table

17.7). The roles of these organisms as disease agents are not clear. Some may be postmortem contaminants or normal components of the intestinal tract flora, while others may be opportunistic pathogens.

Several other bacterial diseases, including coligranuloma-like diseases, avian chlamydiosis (= ornithosis), listeriosis, and various types of infections caused by species of *Fusobacterium* and *Clostridium,* have been reported occasionally from Wild Turkeys (Davidson et al. 1985; Davidson and Wentworth 1992). None of these has been identified from Wild Turkeys in Florida.

A rickettsial agent identified as *Aegyptianella pullorum* was detected in 24 of 300 Wild Turkeys examined by isodiagnostic techniques in southern Texas during 1983 and 1984 (Castle and Christensen 1985). This organism had not been reported previously from North America and may have been introduced into Texas by the importation of exotic birds. Its effects on Wild Turkeys are not known, although Old World strains are pathogenic to young domestic chickens (Gothe 1969). Fortunately there is no evidence that this rickettsia occurs in Wild Turkeys in Florida.

XII. Fungi

Davidson et al. (1985, 1989) and Davidson and Wentworth (1992) described 3 types of fungal infections in Wild Turkeys. These included aspergillosis, candidiasis, and tail-feather damage due to infections by several species of fungi. The tail-feather problem has not been seen in Wild Turkeys in Florida. Aspergillosis has been identified in 2 Wild Turkeys, an adult male from Leon County in 1994 and an adult female from Osceola County in 1995 (Smith 1994; Spalding and Forrester 1996). The turkey from Osceola County was emaciated and also had an osteosarcoma and a severe infection with *Leucocytozoon smithi.* Esophageal candidiasis was identified also in the turkey from Leon County mentioned above with aspergillosis; in addition this bird was in-

fected with avian poxvirus. Several cases of *Candida*-like infections have been observed in the upper digestive tracts of Florida turkeys infected with avian pox (Forrester 1992b). These *Candida* infections are probably secondary or incidental findings (Davidson et al. 1985).

Five Florida turkeys, from Citrus, Gadsden, Glades, Osceola, and Taylor counties, were diagnosed with mycotic dermatitis (Forrester 1992a). The species of fungi were not determined for any of these cases; culture attempts were unsuccessful. Several of these cases were severe and involved mainly the head and legs. These may have been secondary fungal invasions that represented sequelae to avian pox infections, although this could not be confirmed. Personnel of the Southeastern Cooperative Wildlife Disease Study have observed a few such infections in which pox inclusions were eventually demonstrated, but only after considerable searching on histologic sections (Davidson 1991).

Fungal elements were seen in the liver and spleen of 1 turkey (an adult male from Taylor County in 1978) that also had lymphoproliferative disease (Forrester 1992b). Attempts to culture and identify the species of fungi were unsuccessful. The significance of this finding is unknown.

XIII. *Haemoproteus* infections

The blood protozoan *Haemoproteus meleagridis* (figure 17.22) is one of the most common parasites of Wild Turkeys throughout their range (Forrester et al. 1974b). It was named by Levine (1961) based on a brief description published by Morehouse (1945) of a haemoproteid seen on blood smears from a domestic poult in Texas. Greiner and Forrester (1980) redescribed the species and gave details of the developmental morphology of its gametocytes. Atkinson (1986) provided additional data on the morphology of the gametocytes through experimental infections of Chukar Partridges (*Alectoris chukar*) and Ring-necked Pheasants (*Phasianus colchicus*), the only other birds known to be

Table 17.7. Miscellaneous bacteria isolated from Wild Turkeys in Florida

Bacteria	County	Year	Age[a]	Organ cultured[b]	Number Exam.	Number Pos.	Cause of death[c]
Arizona hinshawii	Glades	1980	PJ[e]	LI	24	1[d]	Collected
Bacillus sp.	Glades	1980	PJ[e]	LV	24	1[f]	Collected
Citrobacter freundii	Alachua	1980	AD	LI	1	1	Unknown
Enterobacter cloacae	Glades	1980	PJ[e]	LI	24	3[g]	Collected
Enterobacter sp.	Alachua	1978	AD	LI	1	1	Unknown
	Glades	1974	JU	LI	1	1	Collected
	Glades	1977	JU	LI	22	3	HK
	Glades	1977	AD	LI	15	4	HK
	Glades	1977	JU	LI	1	1	LPD
	Glades	1979	JU,AD	LI	36	1	HK
	Glades	1980	PJ[e]	LI	24	3[h]	Collected
	Glades	1981	PJ[i]	LI	26	2[j]	Collected
	Hendry	1976	AD	LI	1	1	Pox
Escherichia coli	Alachua	1978	AD	LI,LV	1	1	Unknown
	Alachua	1980	AD	LI,LV	1	1	Unknown
	Duval	1976–77	JU	LI	2	2	Pox
	Gadsden	1976	AD	LV,SI	1	1	Unknown
	Glades	1973	JU	LI	1	1	CM
	Glades	1974	JU	LI	1	1	Collected
	Glades	1976	AD	LI	1	1	CM
	Glades	1977	JU	LI	22	8	HK
	Glades	1977	AD	LI	15	6	HK
	Glades	1977	JU	LI	1	1	LPD
	Glades	1977	JU	LI	1	1	Pox
	Glades	1977	AD	LI	2	2	CM
	Glades	1978	JU	LI	1	1	CM
	Glades	1979	JU,AD	LI	36	12	HK
	Glades	1980	PJ[e,l]	LI	24	17	Collected
	Glades	1980	AD	LI	8	7	CM
	Glades	1981	PJ[i,l]	LI	26	9	Collected
	Glades	1981	AD	LI	1	1	CM
	Hendry	1976	AD	LI	1	1	Pox
	Jefferson	1980	AD	LI	1	1	Unknown
	Leon	1992	PO	LV,BR	2[k]	2	Unknown
	Levy	1978	AD	LI	1	1	Trauma
	Manatee	1977	AD	LI	1	1	Unknown
	Pasco	1979	JU	LI	1	1	Pox
Klebsiella oxytoca	Glades	1980	PJ[e]	LI	24	1[f]	Collected
Klebsiella sp.	Glades	1977	AD	LI	1	1	HK
	Osceola	1976	AD	LI	1	1	Pox
Proteus sp.	Alachua	1980	AD	LV	1	1	Unknown
	Glades	1980	PJ[e]	LV	24	1[f]	Collected
Pseudomonas sp.	Glades	1977	JU	LI	22	1	Collected
	Glades	1977	AD	LI	15	1	Collected
	Glades	1980	PJ[e]	LI	24	2[m]	Collected
Staphylococcus aureus	Osceola	1997	JU	LV	1	1	Salmonellosis

(continued)

Table 17.7. *(continued)*

Bacteria	County	Year	Age[a]	Organ cultured[b]	Number Exam.	Number Pos.	Cause of death[c]
Staphylococcus sp.	Leon	1992	PO	LV,BR	2[k]	2	Unknown
Staphylococcus sp. (nonhemolytic)	Glades	1980	AD	FT	1	1	CM
α-*Streptococcus* sp.	Leon	1992	PO	LV,BR	2[k]	1	Unknown
Streptococcus sp. (nonhemolytic)	Osceola	1997	JU	LV	1	1	Salmonellosis

Sources: Quist (1992), Spalding and Forrester (1997), White and Akey (1981), White and Forrester (1981).
a. PO = poults (<30 days), JU = juveniles (31 days–12 months), PJ = poults and juveniles, AD = adults (>12 months).
b. BR = brain, FT = foot, LI = large intestine, LV = liver, SI = small intestine.
c. CM = capture mortality, HK = hunter-kill, LPD = lymphoproliferative disease, Collected = birds were obtained via trapping or shooting for various research purposes.
d. 70-day-old juvenile.
e. PJ = poults and juveniles (27–71 days).
f. 71-day-old juvenile.
g. 28-day-old poult and 52- and 66-day-old juveniles.
h. 28-day-old poult and 65- and 75-day-old juveniles.
i. PJ = poults and juveniles 21 and 68 days of age, respectively.
j. 23-day-old poult and 67-day-old juvenile.
k. 5- and 7-day-old poults.
l. *E. coli* was isolated from the liver of 1 poult.
m. 38- and 47-day-old juveniles.

susceptible to this hemoparasite. Pre-erythrocytic development in skeletal and cardiac muscles was studied experimentally by Atkinson et al. (1986). The ultrastructure of the ookinetes and oocysts has been studied by Atkinson (1989, 1991b). In a recent comparative taxonomic study of the haemoproteids of the avian family Phasianidae, Bennett and Peirce (1989) stated that *H. meleagridis* is a valid species. Two excellent overviews of the pathogenicity and epizootiology of various species in the genus *Haemoproteus* have been published (Atkinson 1991a; Atkinson and van Riper 1991) and include data on *H. meleagridis.*

Haemoproteus meleagridis is distributed throughout Florida (figure 17.23) and infections have been found in turkeys from each county where an adequate sample has been taken. The prevalence was 85% for 1,027 Wild Turkeys (>1 month of age) examined by blood smear techniques, 1969–89 (table 17.8). Prevalences varied from 8 to 100% in counties where >10 samples were taken. At Fisheating Creek the prevalence was 87% (n = 813). The prevalence in turkeys from areas south of Hernando and Pasco counties was higher (87.6%, n = 936) than in turkeys to the north of those counties (59.3%, n = 91) (chi square = 50.06, P < .001). Both sexes of turkeys had similar prevalences of infection (table 17.9). Infections were present in turkeys older than 30 days in Glades County (table 17.10), but were not detected on blood films from younger poults. However, since the prepatent period for *H. meleagridis* is 17 days (Atkinson et al. 1988b), infections cannot be detected by the use of blood smear techniques until poults are at least 17 days of age, even though theoretically they could become infected at 1 day of age. Over a 16–year period (1969–84) prevalences in turkeys >1 month of age varied from year to year at Fisheating Creek, from a low of 67% in 1982 to 100% in 1969 and 1973 (table 17.11). During other years, prevalences varied between 76 and 96%. A statistically significant correlation

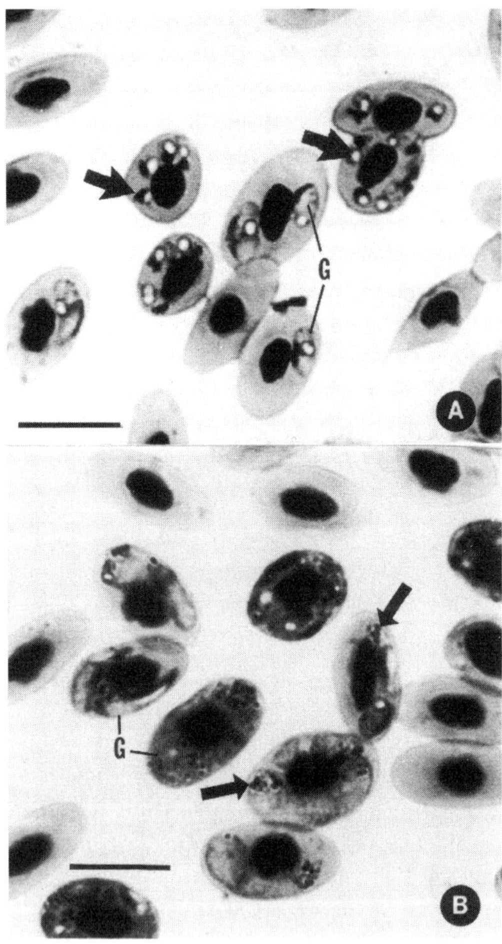

FIGURE 17.22. Gametocytes of *Haemoproteus meleagridis* in red blood cells of a domestic turkey. This bird had been infected experimentally with a strain of *H. meleagridis* obtained from Wild Turkeys in Florida. Bars = 10 μm. *A*, merozoites with large vacuoles and small nuclei *(arrows)* are shown within red blood cells. As merozoites develop into young gametocytes (*G*), they become elongated and sausage-shaped. Eventually they encircle the nucleus of the red blood cell. *B*, mature gametocytes (*G*) are shown within red blood cells and contain pigment granules *(arrows)*. From Atkinson 1991a; by permission of *Bulletin of the Society of Vector Ecology*.

C. arboricola, C. haematopotus, C. hinmani, and *C. knowltoni* (Atkinson et al. 1983; Atkinson 1988). *Culicoides edeni* was judged to be the primary vector by Atkinson (1988) for several reasons: (1) this species was found to be highly susceptible to sporogonic development of *H. meleagridis* (figure 17.25), (2) it was present in high prevalences in collections of biting midges feeding on turkeys (tables 17.12 and 17.13), and (3) *H. meleagridis* was isolated from naturally infected specimens (tables 17.14 and 17.15).

Transmission of *H. meleagridis* occurs in southern Florida on a year-round basis, as demonstrated by sentinel studies conducted in 1976–77 (figure 17.26) and in 1982–84 (figure 17.27). Infections were transmitted equally well in trees (roosting habitats) and on the ground in both feeding and resting areas and nesting habitats (Akey 1981; Atkinson 1985). In southern Florida the epizootiology of *H. meleagridis* is stable and holoendemic, meaning that the parasite occurs in high prevalences in the turkey populations (table 17.8) and the vectors (especially *Culicoides edeni*) occur in large and relatively stable populations, resulting in transmission throughout the year. In contrast, the situation in northern Florida, where the climate is more temperate (figure 17.28), is unstable and hyperendemic, with lower prevalences of infection in the turkey

(correlation coefficient = 0.723, P = 0.02) was noted at Fisheating Creek between the prevalence of *H. meleagridis* in turkeys examined during July, August, September, and October each year and the amount of rainfall during the preceding June (figure 17.24).

Haemoproteus meleagridis is transmitted by ceratopogonid midges of the genus *Culicoides* (Atkinson 1985, 1988, 1991a; Atkinson et al. 1983; Atkinson et al. 1988a), which require water or moist soil in which to breed (Harwood and James 1979). This latter requirement probably accounts for the correlation between rainfall and prevalences. Five species of *Culicoides* are known to serve as vectors of *H. meleagridis* in Florida: *C. edeni,*

Table 17.8. Distribution and prevalence[a] of *Haemoproteus meleagridis* in Wild Turkeys[b] from Florida, 1969–89

County	Number of turkeys		
	Examined	Positive	Percent
Alachua	57	33	58
Charlotte	24	24	100
Citrus	1	1	—
Duval	4	4	—
Escambia	13	1	8
Flagler	2	2	—
Gadsden	2	2	—
Glades	813	708	87
Lake	1	1	—
Leon	4	4	—
Levy	1	0	—
Okeechobee	20	18	90
Osceola	69	61	88
Sarasota	10	9	90
St. Johns	1	1	—
Taylor	2	2	—
Wakulla	3	3	—
Totals	1,027	874	85

Sources: Forrester (1992b), Forrester et al. (1974b).
a. As determined by blood smear techniques.
b. All turkeys were >1 month old.

populations and smaller and more variable vector populations, all of which are dependent on environmental conditions, host density, and vector bionomics (Atkinson 1985; Atkinson et al. 1988a).

The pathogenicity of *H. meleagridis* has been studied experimentally using domestic poults infected with sporozoites of a strain of the parasite obtained from Wild Turkeys at Fisheating Creek Wildlife Management Area (Atkinson 1985; Atkinson and Forrester 1987; Atkinson et al. 1988b). These infections re-

Table 17.9. Prevalence[a] of *Haemoproteus meleagridis* in male and female Wild Turkeys[b] in Florida, 1969–84

Sex	Number of turkeys		
	Examined	Positive	Percent
Males	416	361	87[c]
Females	597	501	84[c]
Both	1,013	862	85

Sources: Forrester (1992b), Forrester et al. (1974b).
a. As determined by blood smear techniques.
b. All turkeys were >1 month old.
c. Not significantly different (chi square = 1.363, $P = 0.243$).

Table 17.10. Prevalence[a] of *Haemoproteus meleagridis* in Wild Turkeys of various ages from Fisheating Creek Wildlife Management Area, Glades County, 1968–84

	No. of turkeys		
Age	Examined	Positive	Percent
2 days	6	0	—
3–4 days	42	0	0
5–6 days	31	0	0
7–13 days	33	0	0
14–30 days	25	0	0
1–7 months	212	170	80
8–12 months	288	271	94
>12 months	296	251	85
Total (all ages)	933	692	74

Source: Forrester (1992b).
a. As determined by blood smear.

Table 17.11. Annual prevalence[a] of *Haemoproteus meleagridis* in Wild Turkeys[b] at Fisheating Creek Wildlife Management Area, Glades County

	No. of turkeys		
Year	Examined	Positive	Percent
1969	12	12	100
1970	174	151	87
1971	73	62	85
1972	39	35	90
1973	24	24	100
1974	41	36	88
1975	50	48	96
1976	57	45	79
1977	66	60	91
1978	13	12	92
1979	42	34	81
1980	75	64	85
1981	101	88	87
1982	12	8	67
1983	17	16	94
1984	17	13	76
Totals	813	708	87

Source: Forrester (1992b).
a. As determined by blood smear techniques.
b. All turkeys were >1 month old.

Table 17.12. Numbers of engorged *Culicoides* captured in turkey-baited Bennett traps at Paynes Prairie State Preserve, Alachua County, May 1982–July 1984

Species of *Culicoides*	Site A		Site B		Both Sites	
	No.	%	No.	%	No.	%
C. edeni	292	42.7	393	52.8	685	48.0
C. hinmani	205	30.0	169	22.7	374	26.2
C. scanloni	21	3.1	92	12.4	113	7.9
C. arboricola	74	10.8	34	4.6	108	7.6
C. nanus	58	8.5	15	2.0	73	5.1
C. baueri	12	1.8	19	2.6	31	2.2
C. paraensis	6	0.9	10	1.3	16	1.1
C. haematopotus	7	1.0	7	0.9	14	1.0
C. crepuscularis	6	0.9	3	0.4	9	0.6
C. guttipennis	2	0.3	1	0.1	3	0.2
C. insignis	0	0.0	1	0.1	1	0.1
C. ousairani	1	0.1	0	0.0	1	0.1
Totals	684		744		1428	

Source: Atkinson (1988).
a. Located in a mixed deciduous forest.
b. Located in an ecotone at the edge of a mixed deciduous forest and an open, ungrazed field.

sulted in moderate to severe myositis (figures 17.29 and 17.30), lameness, and inhibited growth rates (figure 17.31). The myositis was associated with host reactions that contained mixed populations of inflammatory cells and red blood cells. Muscle fibers that surrounded developing megaloschizonts were necrotic and calcified. Seven days after infection, turkeys stood with drooped wings and had ruffled feathers, partially or completely closed eyes, and were lethargic. During the second week some birds had diarrhea and by day 15 most of the birds exhibited lameness, depression, emaciation, dehydration, and anorexia. Between days 19 and 22, 33% of the birds that had received high doses of sporozoites died. This mortality occurred 2–5 days after gametocytes first appeared in the peripheral blood. Such gametocytes appeared in the peripheral blood 17 days after infection and peak parasitemias

Table 17.13. Numbers of engorged *Culicoides* captured in turkey-baited Bennett traps[a] at Fisheating Creek Wildlife Management Area, Glades County, December 1982–November 1984

Species of *Culicoides*	No.	%
C. edeni	2,038	79.6
C. hinmani	475	18.6
C. knowltoni	35	1.4
C. arboricola	12	0.5
C. baueri	1	<0.1
Total	2,561	

Source: Atkinson (1988).
a. Located in a small live oak hammock surrounded by a dense cypress swamp.

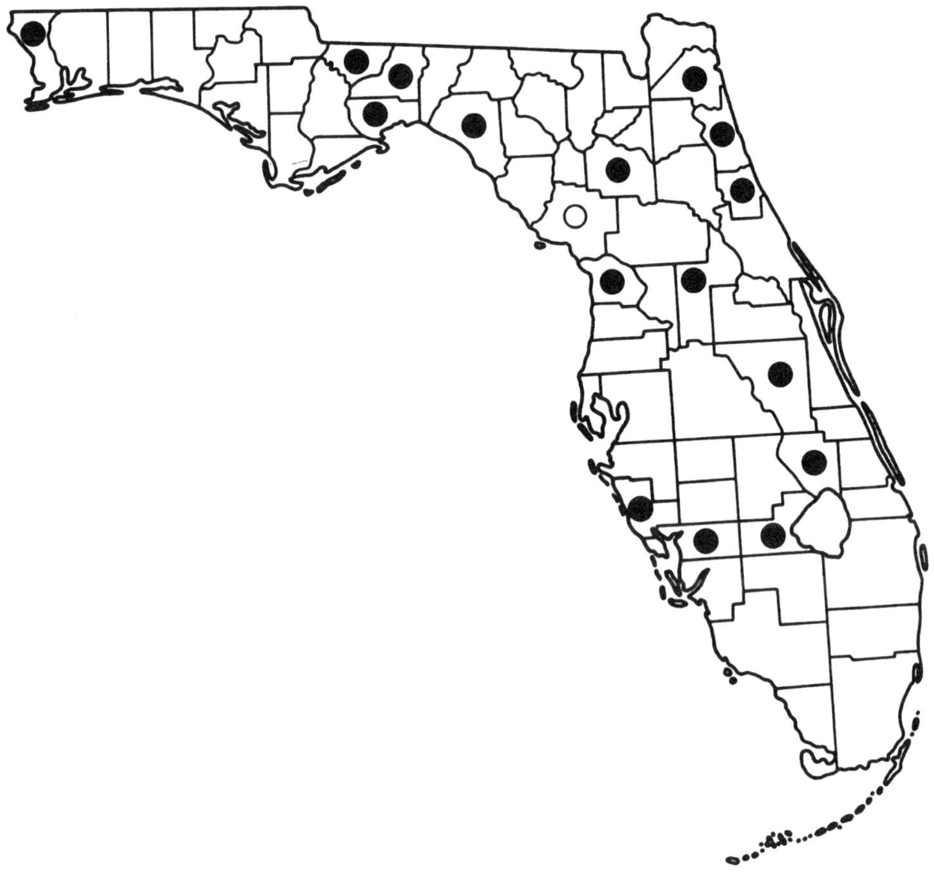

FIGURE 17.23. Distribution of *Haemoproteus meleagridis* in Wild Turkeys in Florida. Solid circles indicate counties where positive turkeys were found; open circle is a county where turkeys were negative. Data from Forrester 1992b; Forrester et al. 1974b.

occurred on day 21 (figure 17.32). The most severe pathologic effects, therefore, occurred before infections could be detected by blood smear techniques. Surprisingly, infected birds were not anemic, even though some exhibited >50% parasitemias of the circulating red blood cells.

Atkinson and Forrester (1987) presented evidence of comparable pathogenic effects (i.e., myopathy) in a Wild Turkey from Orange County (figure 17.33). The exact mechanism of this pathogenicity is not known, but Atkinson (1991a) suggested that it might be mediated by the release of cytokines. Never-theless, it is clear from the above observations that *H. meleagridis* should be considered a potentially important mortality factor, especially in turkeys in southern Florida where this parasite occurs in high prevalences on a year-round basis.

XIV. *Leucocytozoon* infections

Leucocytozoon smithi (figure 17.34) is a common parasite of Wild Turkeys throughout their range, although it is less prevalent than *Haemoproteus meleagridis* (see Forrester et al. 1974b).

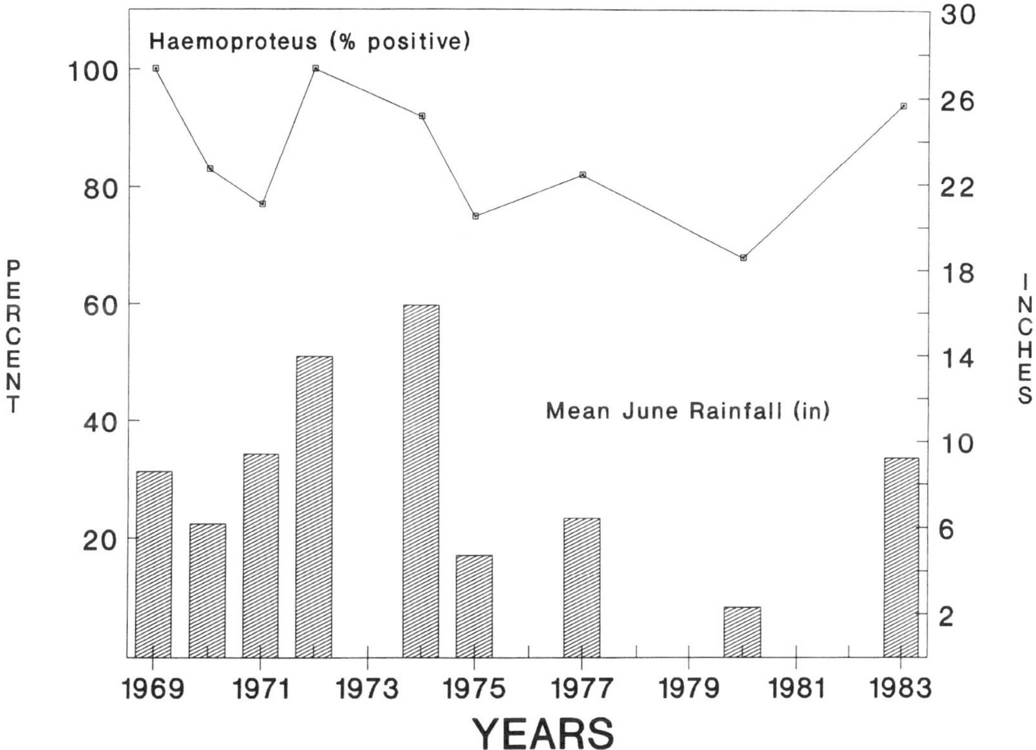

FIGURE 17.24. Comparison of the prevalences of *Haemoproteus meleagridis* in Wild Turkeys (>1 month old) at Fisheating Creek Wildlife Management Area (Glades County) during the months of July, August, September, and October, with rainfall during the preceding June, over a 15–year period, 1969–1983. Data from Forrester 1992b.

It was first discovered by Theobald Smith in domestic turkeys in Massachusetts and Rhode Island during the late 1800s (Smith 1895) and later named after him by Laveran and Lucet (1905). The initial report of *L. smithi* in Wild Turkeys was by Travis et al. (1939), in which 5 of 5 turkeys from Florida (specific locality not given) and 5 of 5 turkeys from Georgia were found infected. Two recent reviews on the genus *Leucocytozoon* have been published by Greiner (1991) and Atkinson and van Riper (1991) and include data on *L. smithi.*

Leucocytozoon smithi is host specific and has been found only in domestic and Wild Turkeys. This was confirmed experimentally by Byrd (1959) and Solis (1973) who exposed domestic chickens, Chukars, Ring-necked Pheas-

ants, Northern Bobwhites, White Pekin Ducks, and domestic turkeys to *L. smithi.* Only turkeys became infected, indicating that other birds are probably not involved in the epizootiology of this blood protozoan.

Leucocytozoon smithi is distributed throughout Florida (figure 17.35) and infections have been found in turkeys in most counties where an adequate sample has been examined (Travis et al. 1939; Simpson et al. 1956; Forrester et al. 1974b). The prevalence was 72% for 1,027 Wild Turkeys (>1 month of age) examined by blood smear techniques from 1969 to 1989 (table 17.16). Prevalences varied from 8% to 100% in counties where samples >10 were taken. At Fisheating Creek the prevalence was 74% (*n* = 813). The prevalence in turkeys from

Table 17.14. Isolations of *Haemoproteus meleagridis* from naturally infected specimens of *Culicoides edeni*, Paynes Prairie State Preserve, Alachua County

Year	Month	No. of pools	No. of midges	No. of positive pools[a]
1983				
	April	1	35	0
	June	1	83	0
	July	1	19	0
	August	1	10	0
	November	1	20	1
1984				
	March	1	12	0
	April	1	12	0
	May	2	152	1
Total		9	343	2

Sources: Atkinson (1985), Atkinson (1988).

a. Diagnosis accomplished by grinding up pools of midges, inoculating the slurry intraperitoneally into 1–2-week-old turkey poults, and subsequently examining these poults for development of infections by *H. meleagridis.*

areas south of Hernando and Pasco counties was higher (72.9%, *n* = 936) than in turkeys to the north of those counties (60.4%, *n* = 91) (chi square = 5.719, *P* = .017). Both sexes of turkeys had similar prevalences of infection (table 17.17). Infections were present in turkeys older than 30 days of age, but were absent from blood films from younger poults

Table 17.15. Isolations of *Haemoproteus meleagridis* from naturally infected specimens of *Culicoides edeni* at Fisheating Creek Wildlife Management Area, Glades County

Year	Month	No. of pools	No. of midges	No. of positive pools[a]
1983				
	April	2	98	2
	May	2	23	0
	June	1	17	0
	July	4	99	3
	August	2	42	0
	September	2	52	6
	November	7	101	1
	December	2	81	0
1984				
	January	1	9	0
	February	4	42	1
	March	9	117	5
	April	7	75	1
	May	6	56	2
Total		49	816	17

Sources: Atkinson (1985, 1988).

a. Diagnosis was accomplished by grinding up pools of midges, inoculating the slurry intraperitoneally into 1–2-week-old turkey poults, and subsequently examining these poults for development of infections by *H. meleagridis.*

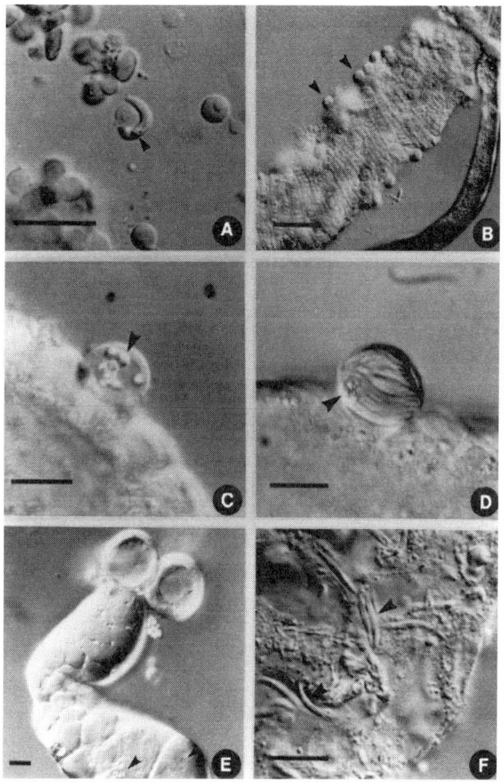

FIGURE 17.25. Sporogonic development of *Haemoproteus meleagridis* in its primary vector, *Culicoides edeni*. A, ookinete from the midgut of *C. edeni* 24 hours after the midge engorged on an infected turkey. Note pigment mass *(arrow)* near the posterior end of the organism. B, developing oocysts *(arrows)* on the midgut of *C. edeni* 4 days post–blood meal; C, 6-day-old degenerating oocyst containing large refractile granules *(arrow)*; D, a mature, 6-day-old oocyst packed with slender sporozoites parallel to one another. Note the small refractile residual body *(arrow)*. E, salivary gland of a specimen of *C. edeni* that had taken a blood meal 7 days earlier. Note the barely visible shadows *(arrows)* of sporozoites in secretory cells of the primary lobe. F, crushed salivary gland of a specimen of *C. edeni* that had taken a blood meal 7 days earlier. Note the numerous elongate sporozoites *(arrows)* located with the secretory cells. Bars in *A* and *C–F* = 10 µm; bar in *B* = 50 µm. Courtesy of Carter T. Atkinson.

(table 17.18). The prepatent period of *L. smithi* is 10–15 days (Pinkovsky et al. 1981) and infections cannot be detected in the peripheral blood before that time, although turkey poults could theoretically become infected at 1 day of age. Over the 16-year period 1969–1984 prevalences of *L. smithi* in turkeys >1 month of age varied from year to year at Fisheating Creek from a low of 49% in 1972 to 100% in 1983 (table 17.19). During other years prevalences varied between 53% and 92%. At Fisheating Creek a statistically significant correlation (correlation coefficient = 0.7884, *P* = 0.0166) was noted between the prevalence of *L. smithi* infections in turkeys during the months of July, August, September, and October and the depth of the creek during the preceding months of March and April (figure 17.36).

Leucocytozoon smithi is transmitted by black flies of the genus *Simulium* (see review by Greiner 1991), which require running water to breed (Harwood and James 1979). This latter requirement may be the reason for the correlation between creek levels and prevalences of *L. smithi* mentioned previously. Various species of *Simulium* have been shown to serve as vectors of *L. smithi* in a number of areas including Nebraska (Skidmore 1932), Wisconsin (Anderson and DeFoliart 1961), New York (Kiszewski and Cupp 1986), Virginia (Johnson et al. 1938), and South Carolina (Jones and Richey 1956; Noblet et al. 1972).

In Florida 3 species (*S. slossonae*, *S. congareenarum*, and *S. meridionale*) transmit *L. smithi* (Greiner and Forrester 1979; Pinkovsky et al. 1981). Of these 3, *S. slossonae* is considered the primary vector because (1) it occurs throughout most of Florida (figure 17.37) and is present in areas where infected Wild Turkeys are found (Forrester et al. 1974b; Pinkovsky and Butler 1978), (2) it is on the wing through-

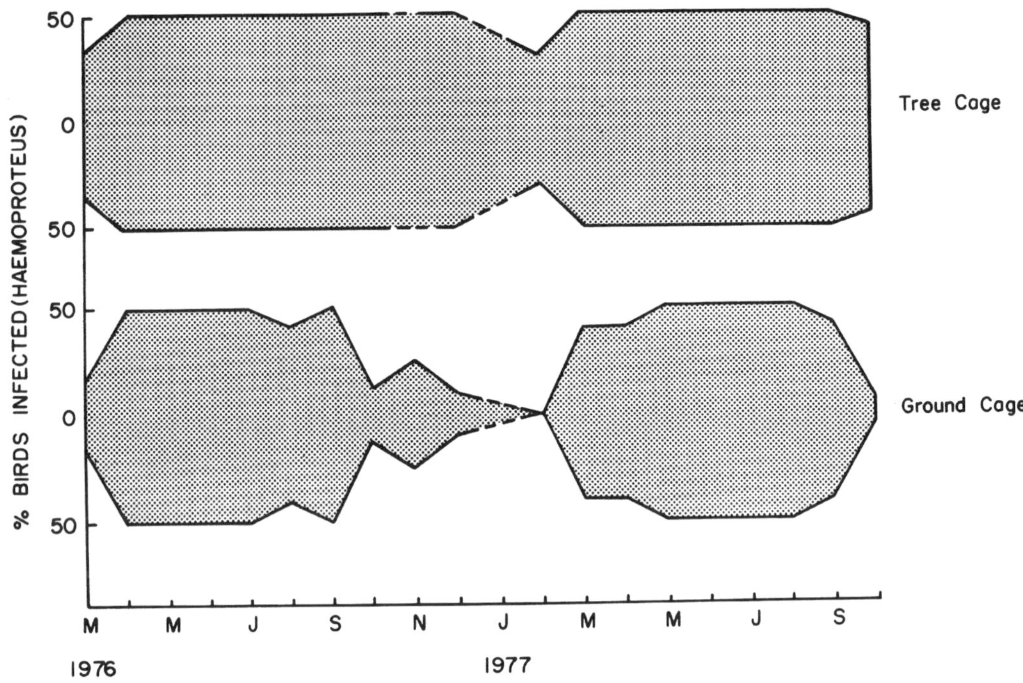

FIGURE 17.26. Kite diagram showing seasonal transmission of *Haemoproteus meleagridis* to sentinel domestic turkeys (2–4 weeks of age) exposed in Wild Turkey habitat at Fisheating Creek Wildlife Management Area (Glades County), 1976–77. *Broken lines* = missing data due to the deaths of sentinel birds unrelated to disease.

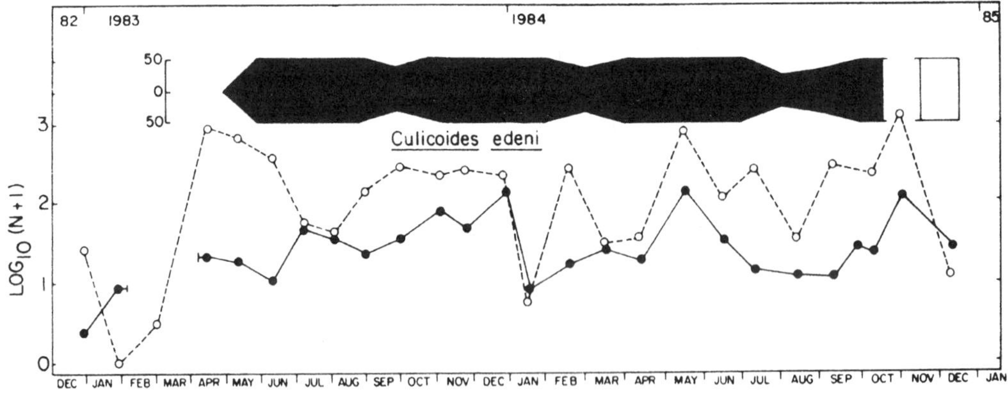

FIGURE 17.27. Seasonal transmission of *Haemoproteus meleagridis* compared with the abundance of *Culicoides edeni* at Fisheating Creek Wildlife Management Area (Glades County), 1982–84. Kite diagram at top indicates the prevalence of *H. meleagridis* in sentinel domestic turkeys (2–4 weeks of age) exposed for 1–3 days each month. *Solid circles* = geometric means of samples obtained each month utilizing turkey-baited Bennett traps; *open circles* = geometric means of samples collected in New Jersey light-traps. Sentinels were not exposed during October 1984; sentinels exposed during November 1984 did not become infected. Modified from Atkinson et al. 1988a; by permission of *Journal of Medical Entomology.*

Table 17.16. Distribution and prevalence[a] of *Leucocytozoon smithi* in Wild Turkeys[b] from Florida, 1969–89

County	Number of turkeys		
	Examined	Positive	Percent
Alachua	57	40	70
Charlotte	24	24	100
Citrus	1	0	—
Duval	4	4	—
Escambia	13	1	8
Flagler	2	2	—
Gadsden	2	1	—
Glades	813	600	74
Lake	1	0	—
Leon	4	2	—
Levy	1	1	—
Okeechobee	20	14	70
Osceola	69	44	64
Sarasota	10	1	10
St. Johns	1	0	—
Taylor	2	2	—
Wakulla	3	2	—
Totals	1,027	738	72

Sources: Forrester (1992b), Forrester et al. (1974b).
a. As determined by blood smear techniques.
b. All turkeys were >1 month old.

out the year (Pinkovsky and Butler 1978); (3) it has been observed to feed readily on turkeys (Pinkovsky et al. 1981); (4) it transmits *L. smithi* to turkeys under experimental conditions (table 17.20; Pinkovsky et al. 1981), and (5) sporozoites of *L. smithi* have been found in wild-caught specimens trapped in Wild Turkey habitat (table 17.21; Greiner and Forrester 1979). *Simulium congareenarum* is found in northern Florida and is probably important as a secondary vector from January to April since adults of this species are most numerous at that

Table 17.17. Prevalence[a] of *Leucocytozoon smithi* in male and female Wild Turkeys[b] in Florida, 1969–84

Sex	Number of turkeys		
	Examined	Positive	Percent
Males	416	305	73[c]
Females	597	423	71[c]
Both	1013	728	72

Sources: Forrester (1992b), Forrester et al. (1974b).
a. As determined by blood smear techniques.
b. All turkeys were >1 month old.
c. Not significantly different (chi square = 2.29, $P = 0.13$).

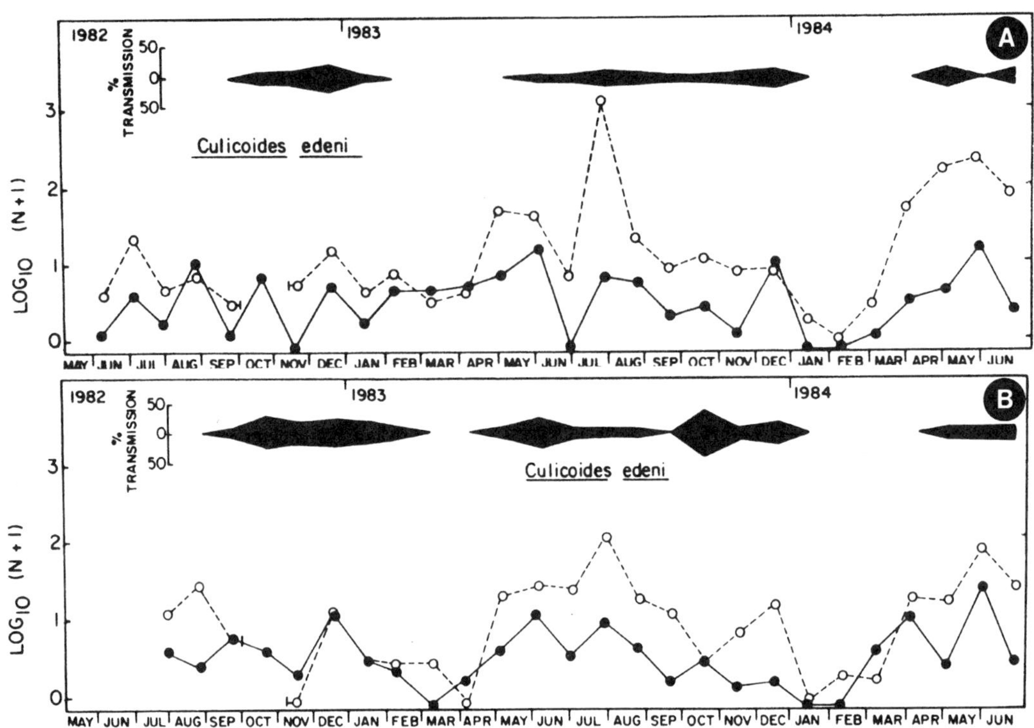

FIGURE 17.28. Seasonal transmission of *Haemoproteus meleagridis* compared with the abundance of *Culicoides edeni* at 2 sites in Paynes Prairie State Preserve (Alachua County), 1982–84. Site *A* was located in a mixed deciduous forest and Site *B* was in an ecotone at the edge of a mixed deciduous forest and open ungrazed field. Kite diagrams at the top of each graph indicate the prevalence of *H. meleagridis* in sentinel domestic turkeys (2–4 weeks of age) exposed for 2 weeks each month. Blank areas represent periods when there was no transmission. Solid circles = geometric means of samples obtained each month utilizing turkey-baited Bennett traps; open circles = geometric means of samples collected in New Jersey light-traps. Modified from Atkinson et al. 1988a; by permission of *Journal of Medical Entomology*.

time of year (Pinkovsky and Butler 1978). In addition to the above 2 species, *S. meridionale* may transmit *L. smithi* in the panhandle area, particularly along the Apalachicola River (Pinkovsky at al. 1981). Transmission of *L. smithi* in northern Florida occurs only during March, April, and May (figure 17.38), whereas in southern Florida it is extended over a longer period of time (i.e., May to February) (figure 17.39). The more restricted transmission period for northern Florida may be related to weather or to the distribution and numbers of vectors. Atkinson and van Riper (1991) suggested that this might be due to the absence of *S. slossonae* (which breeds throughout the year and produces several generations) from the Paynes Prairie study area, and the presence of only *S. congareenarum* (which produces only 1 generation per year in the spring months). This is a possibility, although *S. slossonae* is known to be distributed widely throughout Alachua County (Pinkovsky and Butler 1978).

Transmission of *L. smithi* to Wild Turkeys occurs mainly in the tree canopy while the birds are on the roost, although some transmission also occurs while birds are on the ground (figures 17.39 and 17.40). Spring relapse (an increase in the numbers of gametocytes in the peripheral blood of birds during the spring when vectors are abundant and active) has been demonstrated to occur in domestic turkeys infected with *L. smithi* in South Carolina (Alverson and Noblet 1977). Such a phenomenon has not been shown for Wild Turkeys, but if it does occur, it would be important in the epizootiology of *L. smithi,* particularly in northern Florida where transmission occurs only during the spring months.

Leucocytozoon smithi has been associated with significant mortality in domestic turkeys. In some flocks, losses as high as 75% have been reported (Skidmore 1932; Savage and Isa 1945; Stoddard et al. 1952). Such mortality has been noted from numerous areas, but especially in Virginia, South Carolina, Alabama, and Nebraska (Wehr 1962). Simpson et al. (1956) documented "heavy losses" on a turkey farm near Palatka where mortality occurred mainly in older birds; approximately 300 of 1,500 finished birds died with as many as 30 dying within a single 24-hour period. The authors also reported that 60 of 920 adult breeding turkeys died in another flock in Florida (specific area not given) over a period of 3 months. They concluded that *L. smithi* was not the only cause of these mortalities, but that it acted as a stress factor that exacerbated existing diseases, namely fowl cholera in the case of the first flock and leukosis in the second flock. Most of what is known about the pathogenic effect of *L. smithi* is based on observations in domestic turkeys. Signs of leucocytozoonosis in domestic turkeys include anorexia, excessive thirst, depression, and at times muscular incoordination and sudden death (Springer 1997). In young turkeys the progression of the disease can be rapid and fatal. Death is caused by obstruction of the circulatory system by multitudes of parasites and at necropsy this is seen as congestion of the liver, spleen, small intestine, and lungs as well as enlargement of the spleen and liver (Johnson et al. 1938). Birds that recover are less vigorous and may die when stressed (Springer 1997). Gobblers exhibit a re-

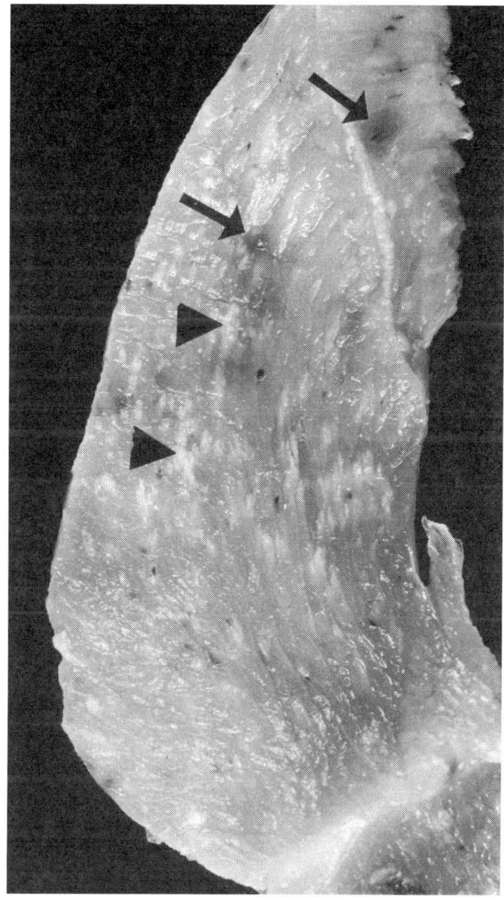

FIGURE 17.29. Pectoral muscle from a domestic turkey that died 19 days after experimental infection with 57,500 sporozoites of *Haemoproteus meleagridis*. The white streaks *(arrowheads),* dark flecks, and darkened hemorrhagic areas *(arrows)* correspond to megaloschizonts. From Atkinson et al. 1988b; by permission of *Journal of Parasitology.*

duction in mating activity and hens suffer a decrease in the production, weight, and hatchability of their eggs, although fertility is not affected (Johnson et al. 1938; Jones et al. 1972). Additional details on pathology and pathogenesis of infections by *L. smithi* in domestic turkeys can be found in the publications by Newberne (1955), Wehr (1962), and Siccardi et al. (1974).

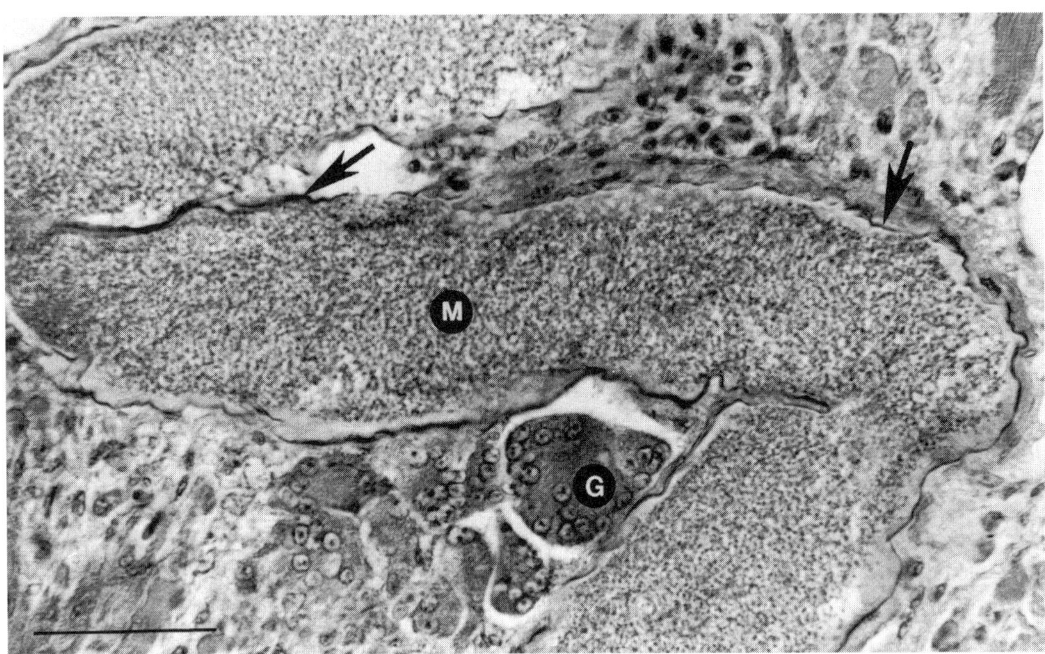

FIGURE 17.30. Megaloschizont of *Haemoproteus meleagridis* in an experimentally infected domestic turkey. The megaloschizont (M) is filled with merozoites and surrounded by a thick, hyaline wall (*arrows*). The surrounding muscle tissue contains infiltrates of inflammatory cells such as mononuclear cells and giant cells (G). Bar = 0.5 μm. From Atkinson 1991a; by permission of *Bulletin of the Society of Vector Ecology*.

Table 17.18. Prevalence[a] of *Leucocytozoon smithi* in Wild Turkeys of various ages from Fisheating Creek Wildlife Management Area, Glades County, Florida, 1968–84

Age	Number of turkeys		
	Examined	Positive	Percent
2 days	6	0	—
3–4 days	42	0	0
5–6 days	31	0	0
7–13 days	33	0	0
14–30 days	25	0	0
1–7 months	212	142	67
8–12 months	288	243	84
>12 months	296	205	69
All ages	933	590	63

Source: Forrester (1992b).
a. As determined by blood smear techniques.

Table 17.19. Annual prevalence[a] of *Leucocytozoon smithi* in Wild Turkeys[b] at Fisheating Creek Wildlife Management Area, Glades County, Florida

	Number of turkeys		
Year	Examined	Positive	Percent
1969	12	11	92
1970	174	148	85
1971	73	44	60
1972	39	19	49
1973	24	19	80
1974	41	28	68
1975	50	37	74
1976	57	40	70
1977	66	42	64
1978	13	9	69
1979	42	32	76
1980	75	55	73
1981	101	82	81
1982	12	8	67
1983	17	17	100
1984	17	9	53
Totals	813	600	74

Source: Forrester (1992b).
a. As determined by blood smear techniques.
b. All turkeys were >1 month old.

Less is known about the disease in Wild Turkeys. However, the observations reported by Byrd (1959) on naturally and experimentally infected pen-reared Wild Turkeys provide some insights. In that study, 2 groups of birds were used; 1 was 4 weeks of age and the other was 10 months. Domestic turkeys were used as controls. Pen-reared Wild Turkeys showed few effects due to the infections by *L. smithi;* the only sign was droopiness that occurred inconsistently. On the other hand, infections in the domestic controls were more severe and resulted in emaciation, lameness, depression, and death. At necropsy these latter birds had high

Table 17.20. Transmission of *Leucocytozoon smithi* in the laboratory by the bites of black flies from Florida

Species of *Simulium*	Source of black flies	Number of turkeys infected	Prepatent period (days)
S. slossonae	Wild-caught[a]	10	10–15
S. slossonae	Laboratory-reared	5	11–15
S. congareenarum	Wild-caught[a]	3	11–15
S. meridionale	Wild-caught[a]	1	12
Total	—	19	10–15

Source: Pinkovsky et al. (1981).
a. Collected from 7 counties throughout Florida.

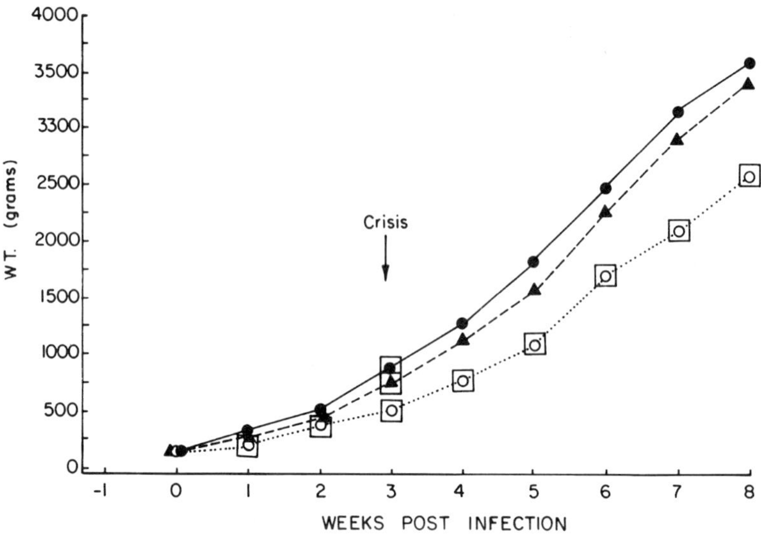

FIGURE 17.31. Mean weights for domestic turkeys infected experimentally with *Haemoproteus melea-gridis* at 1 week of age. Poults received 57,500 sporozoites (*open circles*), 4,400 sporozoites (*closed triangles*), or 0 sporozoites (*closed circles*) on day 0. Data points surrounded by an open square were significantly different when mean values of the 3 experimental groups were compared. From Atkinson et al. 1988b; by permission of *Journal of Parasitology*.

parasitemias, hemorrhage in the mucosa of the small intestine, and hypertrophy of the liver and spleen. Byrd (1959) concluded from his studies that *L. smithi* was not a serious primary pathogen for Wild Turkeys. This very well may be the case, and the significance of *L. smithi* to Wild Turkeys in Florida may be in its role as a stress factor or in synergistic effects in cases where other parasites and disease agents are present, as postulated by Borg (1953), Simpson et al. (1956), and others. Additional investigations are needed.

XV. Malaria

Four species of malarial parasites belonging to the genus *Plasmodium* have been reported from Wild Turkeys in North America (Forrester 1991), while only 1, *P. hermani* (figure 17.41), has been found in Wild Turkeys in Florida. It was first discovered in 1972 in turkeys from Fisheating Creek Wildlife Management Area (Glades County) and Lockloosa

Wildlife Management Area (Alachua County) through subinoculation of blood from Wild Turkeys into domestic poults (Forrester et al. 1974b) and was described as a new species in 1975 (Telford and Forrester 1975). Although *P. hermani* has been found once in a Northern Bobwhite from Glades County (Forrester et al. 1987) and pen-reared Northern Bobwhites are excellent experimental hosts (Nayar et al. 1982), this is essentially a parasite of Wild Turkeys. It has been transmitted experimentally to Red Knots (Forrester and Humphrey 1981) and to Common Canaries and domestic goslings (Telford and Forrester 1975), although the latter 2 were transient infections. The following were tested and found to be unsuitable experimental hosts: domestic chicks, zebra finches, and society finches (Telford and Forrester 1975).

Plasmodium hermani has been found in Wild Turkeys from 5 localities in Florida (figure 17.42) and probably is distributed throughout the state. Information on distribution is limited, however, because surveys based on blood smears

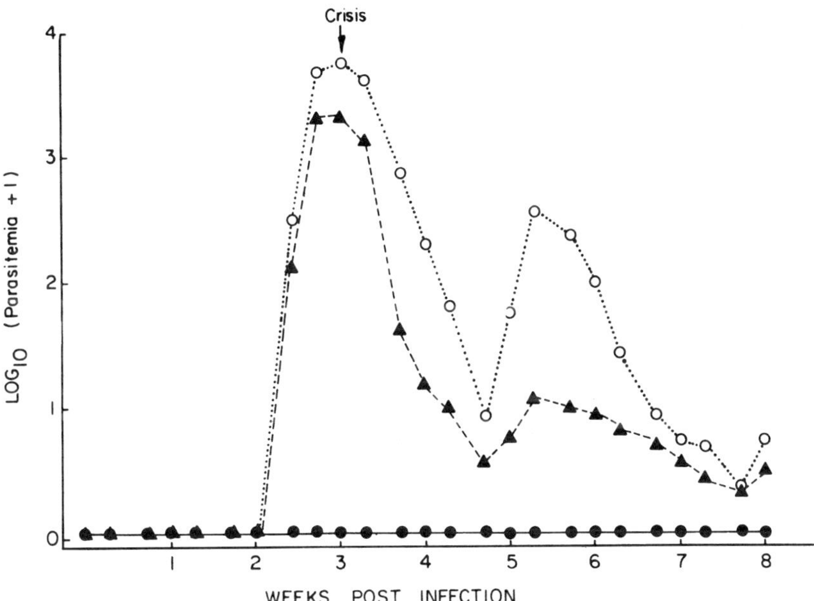

FIGURE 17.32. Average parasitemias for same groups of poults as in Figure 17.31. The time of crisis *(arrow)* occurred on the day when peak parasitemia was reached. From Atkinson et al. 1988b; by permission of *Journal of Parasitology.*

Table 17.21. Isolations of *Leucocytozoon smithi* from naturally infected specimens of *Simulium slossonae* at Fisheating Creek Wildlife Management Area (FEC), Glades County, and Lochloosa Wildlife Management Area, Alachua County, March 1978

		Number of black flies				
Area	Date[a]	Caught[b]	Dissected	Pos. for sporozoites	Homogenated and inoculated[c]	Inoculation results
FEC	March 3	5	3	1	ND	—
	March 13	11	11	0	ND	—
	March 14	17	17	1	ND	—
	March 15	29	2	0	27	Pos.[d]
Lochloosa	March 23	32	2	0	25	Neg.
	March 29	2	2	1	ND	—
	March 30	34	12	6	14	Pos.[e]

Source: Greiner and Forrester (1979).
a. Sentinel turkeys maintained at this time became positive for *L. smithi,* indicating that transmission was occurring.
b. In Bennett traps (Bennett 1960) baited with turkey poults.
c. Homogenates consisted of dissected salivary glands in 0.85% saline and were inoculated into uninfected turkey poults from which blood smears were made and examined for 30 days post-inoculation to detect gametocytes; ND = not done.
d. Patent infection occurred 14 days post-inoculation.
e. Patent infection occurred 19 days post-inoculation.

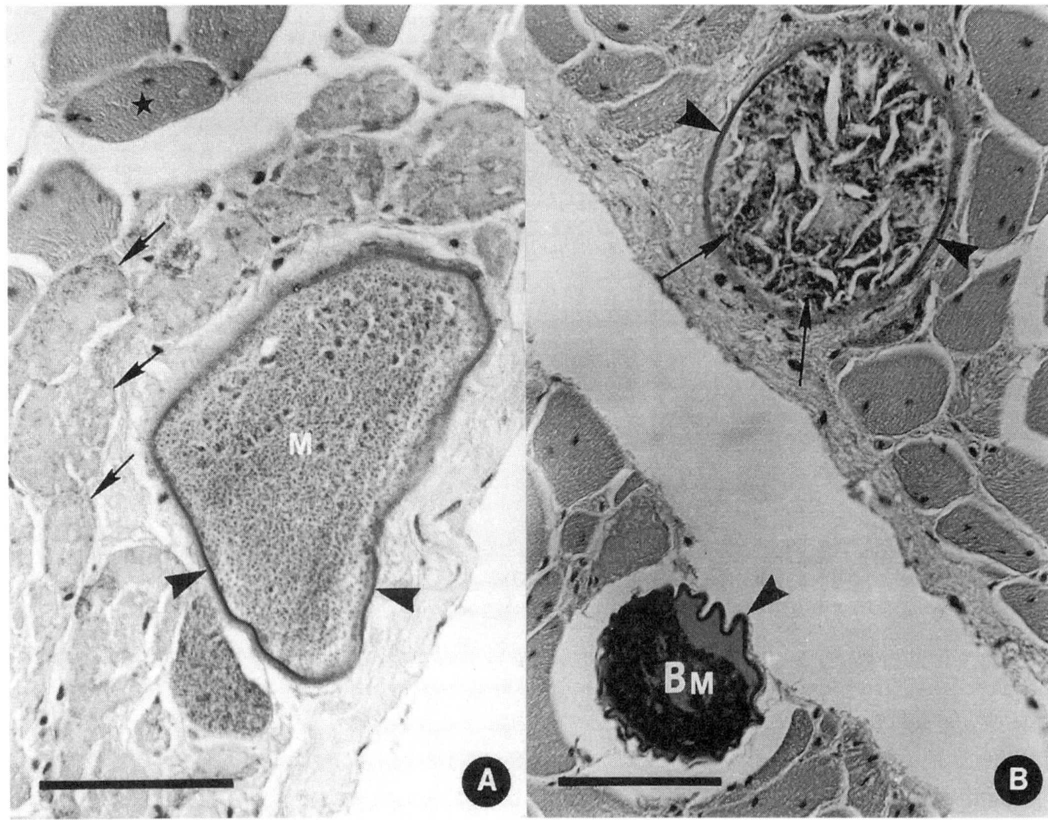

FIGURE 17.33. Natural infection of megaloschizonts of *Haemoproteus meleagridis* in the pectoral muscle of an adult male Wild Turkey from Lake Apopka (Orange County), November 1970. *A*, megaloschizont (*M*) surrounded by a thick, hyaline wall *(arrowheads)* and filled with merozoites. Muscle fibers around the megaloschizont are swollen, pale, and hyaline and contain basophilic granules *(arrows)*. Normal adjacent tissue is indicated by a solid star. *B*, 2 degenerating megaloschizonts surrounded by thick, hyaline walls *(arrowheads)* and containing either amorphous basophilic material (*BM*) or pale, eosinophilic material with scattered basophilic masses *(arrows)*. The partially collapsed wall of the megaloschizont in photograph *B* and the surrounding extracellular space as well as the large clefts within the interior of the other megaloschizont are all freezing artifacts that occurred before the tissue was fixed. From Atkinson and Forrester 1987; by permission of *Journal of Wildlife Diseases*.

cannot be used to accurately diagnose malaria in Wild Turkeys (Forrester 1991). Special isodiagnostic techniques are necessary and these require the inoculation of fresh whole blood from Wild Turkeys into domestic turkey poults or young pen-reared bobwhites and the examination of these recipients over a period of time to detect the subsequent development of malaria (Forrester et al. 1974b). These techniques are expensive and time consuming and require insect-proof facilities in which to rear and maintain recipient birds. For these reasons, minimal work has been done in determining more about the distribution of *P. hermani* in Wild Turkeys in Florida.

The overall prevalence of *P. hermani* in turkeys (>8 months of age) examined from 5 counties was 81% (*n* = 43) (table 17.22). Samples from Citrus, Jefferson, and Leon counties were too small to be used in detailed compar-

Table 17.22. Prevalence[a] of *Plasmodium hermani* in juvenile (8–12 months) and adult (>12 months) Wild Turkeys in Florida, 1972–83

County	Age class	Number of turkeys		
		Examined	Positive	Percent
Alachua	Juveniles	11	6	55
	Adults	3	2	—
Citrus	Juveniles	0	—	—
	Adults	1	1	—
Glades	Juveniles	14	13	93
	Adults	12	11	92
Jefferson	Juveniles	0	—	—
	Adults	1	1	—
Leon	Juveniles	0	—	—
	Adults	1	1	—
Totals	Juveniles	25	19	76
	Adults	18	16	89
	All	43	35	81

Sources: Forrester (1992b), Forrester et al. (1974b).
a. Prevalences determined by subinoculation of whole blood from Wild Turkeys into domestic turkey poults and subsequent examination of these poults for malarial infections.

isons, but turkeys in Glades County in southern Florida had higher prevalences (92–93%) than those in Alachua County in northern Florida (55–67%) (chi square = 5.007, p = 0.025). Both sexes of turkeys had similar prevalences of infection (table 17.23). The youngest turkey found infected was a 3.5-month-old gobbler from Fisheating Creek examined in July of 1975 (table 17.24; Forrester 1992b). Data on year-to-year changes in prevalence are limited, but samples from turkeys >8 months of age during 3 different years at Fisheating Creek varied from 89% to 100% (table 17.25).

Plasmodium hermani is transmitted by mosquitoes of the genus *Culex* (Young et al. 1977; Forrester et al. 1980a; Nayar and Forrester 1985; Forrester 1992a). One species, *C. nigripalpus,* serves as the primary natural vector, at least in southern Florida (Forrester et al. 1980a); 2 other species, *C. restuans* and *C. salinarius,* have been shown to be good experimental vectors (Nayar et al. 1981a; Nayar et al. 1981b). They may serve some role as natural vectors,

but this has not been proven. *Plasmodium hermani* undergoes sporogony in a fourth species, *Wyeomyia vanduzeei,* but this mosquito is unable to transmit infections to turkeys (Nayar et

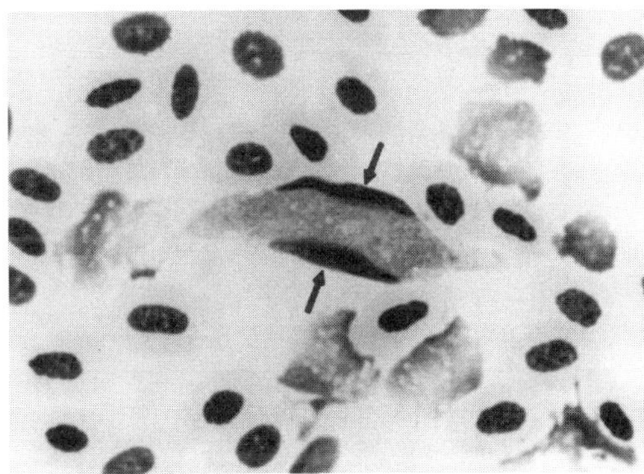

FIGURE 17.34. Macrogametocyte of *Leucocytozoon smithi* from a Wild Turkey from Florida. The host cell nucleus *(arrows)* has been split and is located on the periphery of the host cell/parasite complex. Splitting of the host cell nucleus is characteristic of *L. smithi.* Courtesy of Garry W. Foster.

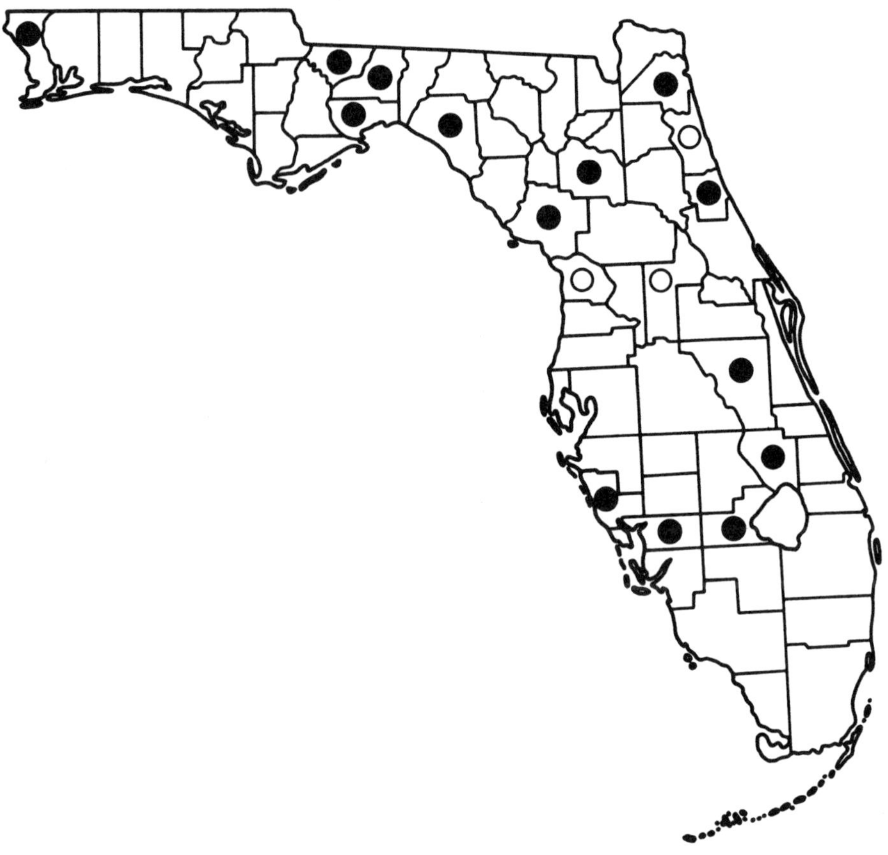

FIGURE 17.35. Distribution of *Leucocytozoon smithi* in Wild Turkeys in Florida. *Solid circles* = counties where positive turkeys were found; *open circles* = counties where turkeys were negative. Data from Forrester 1992b; Forrester et al. 1974b.

Table 17.23. Prevalence[a] of *Plasmodium hermani* in male and female Wild Turkeys[b] in Florida, 1972–83

	Number of turkeys		
Sex	Examined	Positive	Percent
Males	31	23	74[c]
Females	12	12	100[c]
Both	43	35	81

Source: Forrester (1992b).
a. Prevalences determined by subinoculation of whole blood from Wild Turkeys into domestic turkey poults and subsequent examination of these poults for malarial infections.
b. Only data from juveniles (8–12 months) and adults (>12 months) were used.
c. Not significantly different (chi square = 2.29, P = 0.13).

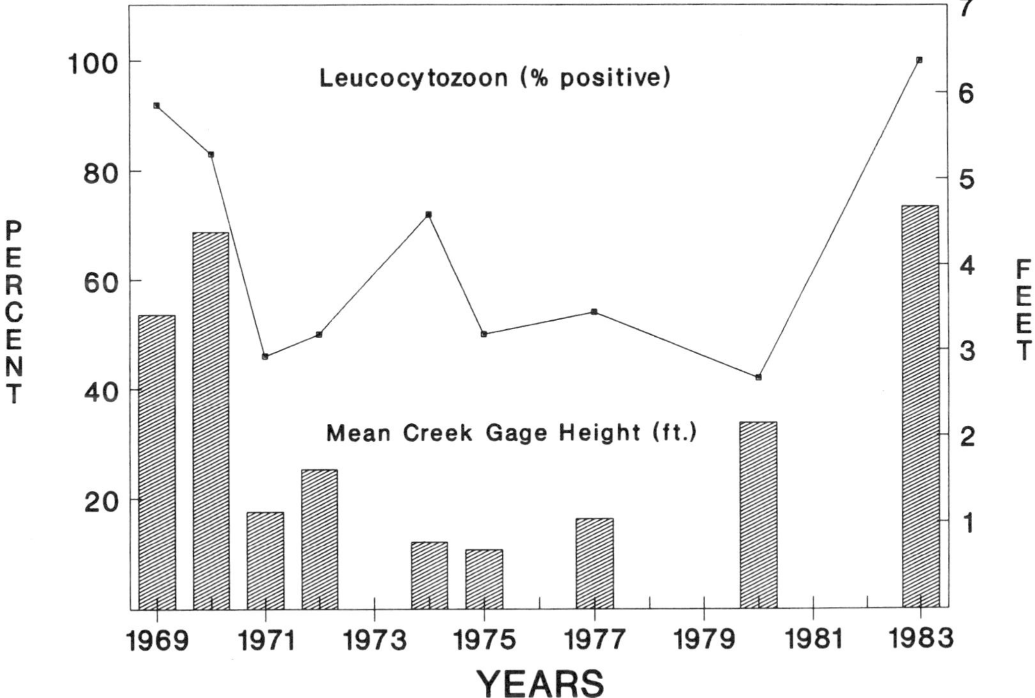

FIGURE 17.36. Comparison of prevalences of *Leucocytozoon smithi* in Wild Turkeys (>1 month old) at Fisheating Creek Wildlife Management Area (Glades County) during the months of July, August, September, and October, with the depth of water in Fisheating Creek during the preceding March and April, over a 15–year period, 1969–83. Data from Forrester 1992b.

al. 1980b). *Culex nigripalpus* is believed to be the primary vector for several reasons: (1) it was present in very high prevalences in collections of mosquitoes attempting to feed on bait turkeys in Wild Turkey habitat during the time of year when transmission of malaria was at its peak (table 17.3), (2) it is highly susceptible to sporogonic development of *Plasmodium hermani* (figure 17.43, table 17.26), (3) it serves as an excellent experimental vector of *P. hermani* (Young et al. 1977; Nayar et al. 1982; Nayar and Forrester 1985), and (4) *P. hermani* has been isolated from naturally infected specimens (table 17.27, figures 17.44–17.51; Forrester et al. 1980a).

Transmission of *P. hermani* is seasonal in Florida. A sentinel study was conducted in northern Florida during 1982–84 (Atkinson and van Riper 1991) and it was found that transmission occurred between August and December (figure 17.52). In southern Florida transmission occurred between June and October (figure 17.53) during a sentinel study in 1976–77 (Forrester 1991). In the latter study it was determined that transmission could occur on the ground or in the cypress canopy (= roosting habitat). The seasonality of transmission correlated closely to rainfall and peaks in populations of mosquito vectors. In southern Florida where this aspect has been well studied, breeding populations of *C. nigripalpus* are usually at their lowest during January through March, slowly build up from April through June, and then peak from July through October during the rainy season, after which they gradually decline (Provost 1969; Nayar 1982). Thus the peaks in rainfall, mosquito populations, and

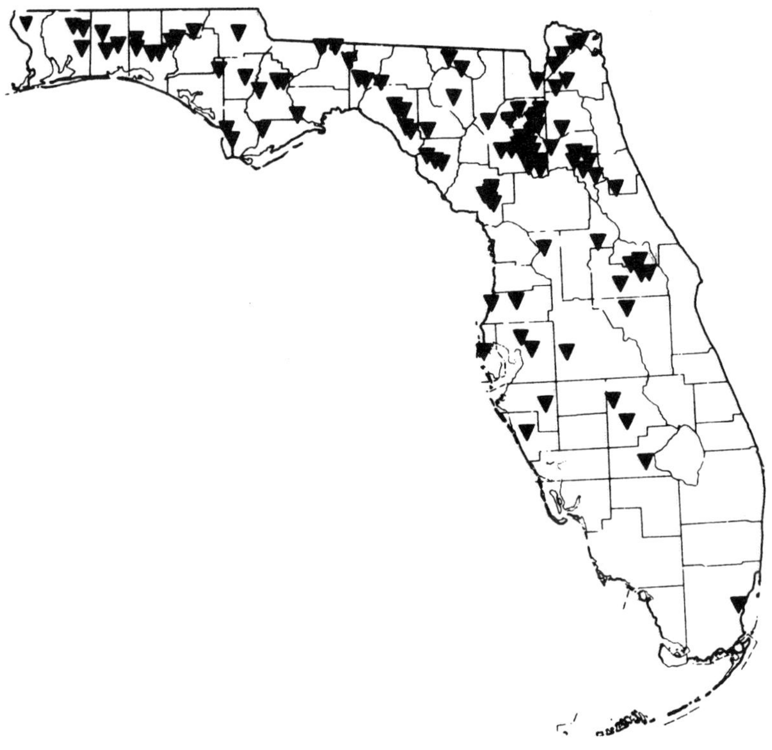

FIGURE 17.37. Distribution of *Simulium slossonae*, the primary vector of *Leucocytozoon smithi* in Florida. *Solid triangles* = sites where black flies of this species were collected. From Pinkovsky and Butler 1978; by permission of *Florida Entomologist*.

transmission of malaria are correlated in southern Florida.

The pathogenic effects of *P. hermani* have been investigated experimentally using both domestic and Wild Turkeys. A series of tests was conducted in which infections were established by the inoculation of infected blood (i.e., blood containing asexual stages of the parasite) into domestic poults and laboratory-reared Wild Turkey poults (Forrester et al. 1980b). Those infections resulted in anemia (figures 17.54 and 17.55), splenomegaly (figures 17.56 and 17.57), decreased growth rates (table 17.28, figure 17.58), but no mortality. Several hundred other blood-induced infections in domestic turkeys of various ages (1 week to several months) utilizing 44 different strains of *P. hermani* (33 from Glades County,

8 from Alachua County, and 1 each from Citrus, Jefferson, and Leon counties) never resulted in mortality that could be attributed to malaria (Forrester 1992b). A few birds died, however, when infections were established by sporozoites through the bites of mosquitoes (table 17.29). These included 1 5-day-old domestic turkey (Akey 1981) and 2 6-week-old pen-reared wild turkeys (Wright 1986). However, in all 3 cases there were other complications contributing to the mortality. The domestic poult had bacterial bronchopneumonia, 1 of the pen-reared Wild Turkeys had a dislocated hock, and the other pen-reared bird had neurologic problems probably induced by trauma. It is possible, therefore, that these 3 birds would not have died had they not been compromised by these additional conditions.

Table 17.24. Prevalence[a] of *Plasmodium hermani* in Wild Turkeys of various ages at Fisheating Creek Wildlife Management Area, Glades County, 1972–81

Age	Number of turkeys		
	Examined	Positive	Percent
10 days	7	0	—
15–20 days	5	0	—
21–25 days	7	0	—
27–28 days	4	0	—
30–35 days	3	0	—
38–50 days	14	0	0
51–58 days	11	0	0
2–2.5 months	13	0	0
3 months	4	0	—
3.5 months	1	1	—
8–12 months	14	13	93
>12 months	12	11	92

Sources: Akey (1981), Forrester (1992b), Forrester et al. (1974b).
a. Prevalences determined by subinoculation of whole blood from Wild Turkeys into domestic turkey poults and subsequent examination of these poults for malarial infections.

In the case of the domestic poult with bronchopneumonia, it is also possible that the malarial infection weakened the poult and made it more susceptible to the bacterial infection and its effects.

From the above it can be concluded that malaria by itself is not an important mortality factor, but that infections by *P. hermani* may predispose turkeys to predation and other problems, or function in an additive synergistic fashion. The infected poult in figure 17.58, for instance, not only was smaller and unthrifty in comparison with the uninfected control, but also exhibited several behavioral characteristics that were often noted in infected birds during the second and third weeks post-infection (Forrester et al. 1980b). These included apathy, lethargy, and postural changes (i.e., drooped

Table 17.25. Prevalence[a] of *Plasmodium hermani* in Wild Turkeys[b] at Fisheating Creek Wildlife Management Area, Glades County

Year	Number of turkeys		
	Examined	Positive	Percent
1972	18	16	89
1975	5	5	—
1979	3	3	—
Totals	26	24	92

Source: Forrester (1992b).
a. Prevalences determined by subinoculation of whole blood from Wild Turkeys into domestic turkey poults and subsequent examination of these poults for malarial infections.
b. Only data from juveniles (8–12 months) and adults (>12 months) were used.

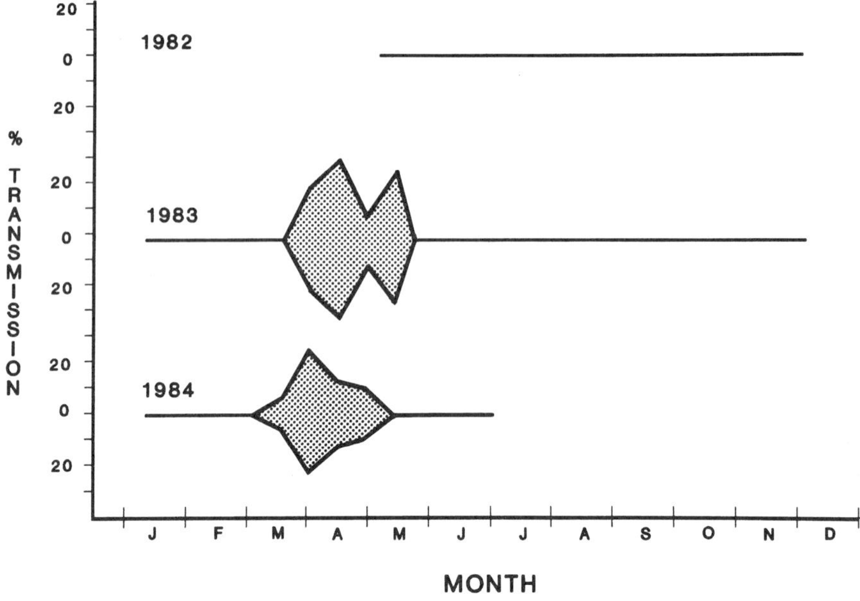

FIGURE 17.38. Seasonal transmission of *Leucocytozoon smithi* to sentinel domestic turkeys (2–4 weeks of age) exposed in Wild Turkey habitat at Paynes Prairie State Preserve (Alachua County), 1982–84. Modified from Atkinson and van Riper 1991; by permission of Carter T. Atkinson.

wings and retracted necks). In addition they emitted shrill, recurrent distress calls. In nature these factors would certainly act in a negative way and lead indirectly to mortality due to predation, inclement weather, or other diseases and parasites.

Wright (1986) studied the interactive effects of concurrent malaria and pox infections and found that experimental infections with both *Plasmodium hermani* and avian poxvirus were slightly more harmful overall to 1-week-old domestic poults than were single infections with either agent. This was determined by measuring a number of parameters, including mortality, weight gains, hematocrits, ratios of immature to mature erythrocytes, levels of parasitemia, and the severity of pox lesions. No ill effects were seen when 10-week-old domestic turkeys or 4-week-old Wild Turkey/domestic turkey hybrids were infected in a similar fashion. Mortalities observed in both age groups of domestic turkeys are presented in table 17.30. None of the older

birds died, whereas 3 (27%) of the pox-infected young birds and 2 (22%) of the pox and malaria-infected young birds died. Young birds infected with both pox and malaria had slower growth rates than did birds with single infections of either pox or malaria (figure 17.59) and these were statistically significant by the fifth week of infection. Young poults infected with both pox and malaria had higher parasitemias, which occurred in 3 peaks, compared with young poults infected only with malaria, which had lower parasitemias and only 1 peak (figure 17.60). The observed higher parasitemias occurring in 3 peaks in poults infected with both pox and malaria could have significant effects from pathological and epizootiological standpoints. This might have been due to higher numbers of malarial organisms in the blood, which in turn resulted in harmful effects to the young poults. In addition this could promote infections of blood-feeding mosquitoes and subsequently increase the spread of malaria in the Wild Turkey population.

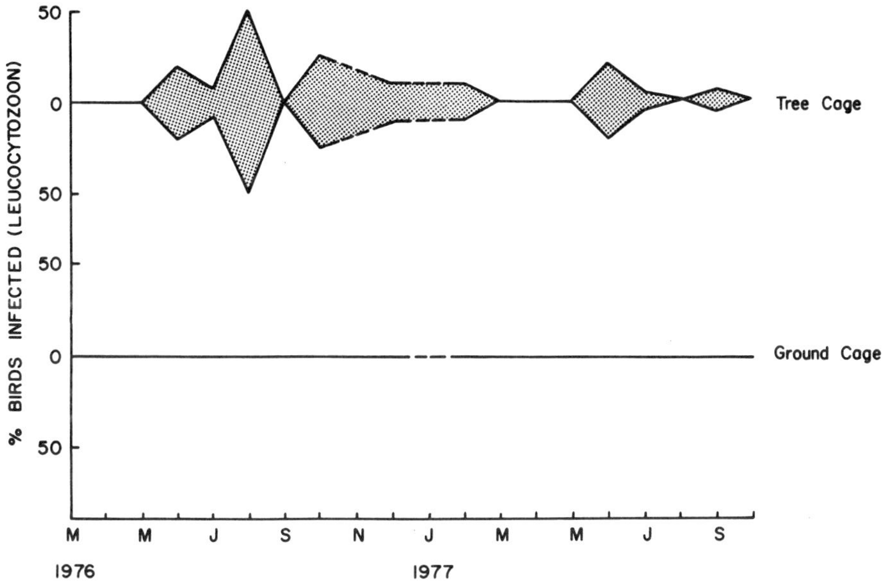

FIGURE 17.39. Seasonal transmission of *Leucocytozoon smithi* to sentinel domestic turkeys (2–4 weeks of age) exposed in Wild Turkey habitat at Fisheating Creek Wildlife Management Area (Glades County), 1976–77. *Broken lines* = missing data due to the deaths of sentinel birds unrelated to disease.

Table 17.26. Susceptibility of *Culex nigripalpus* to five isolates of *Plasmodium hermani* from Wild Turkeys in Florida

Isolates of P. hermani	No. passages in domestic turkeys prior to testing	Number of mosquitoes dissected	% midgut infected	No. oocysts per midgut	
				Mean	Range
P-27[a]	9	52	67	9.0	1–35
P-27	12	184	63	10.2	1–69
P-41[b]	2	66	79	9.1	1–62
P-41	6[c]	111	78	9.3	3–60
L-5[d]	11	73	79	11.2	2–55
L-5	36	121	88	12.8	1–54
W-1[e]	11	29	98	41.3	1–77
Le-1[f]	4–6	61	90	23.0	2–89

Sources: Nayar et al. (1980b, 1981a,b), Nayar and Forrester (1985), Young et al. (1977).
a. From Fisheating Creek Wildlife Management Area (Glades County), July 29, 1975.
b. From Fisheating Creek Wildlife Management Area (Glades County), August 2, 1979.
c. Includes one mosquito passage.
d. From Lochloosa Wildlife Management Area (Alachua County), February 23, 1972.
e. From vicinity of Waukeenah (Jefferson County), May 22, 1980.
f. From Talquin Wildlife Management Area (Leon County), June 14, 1983.

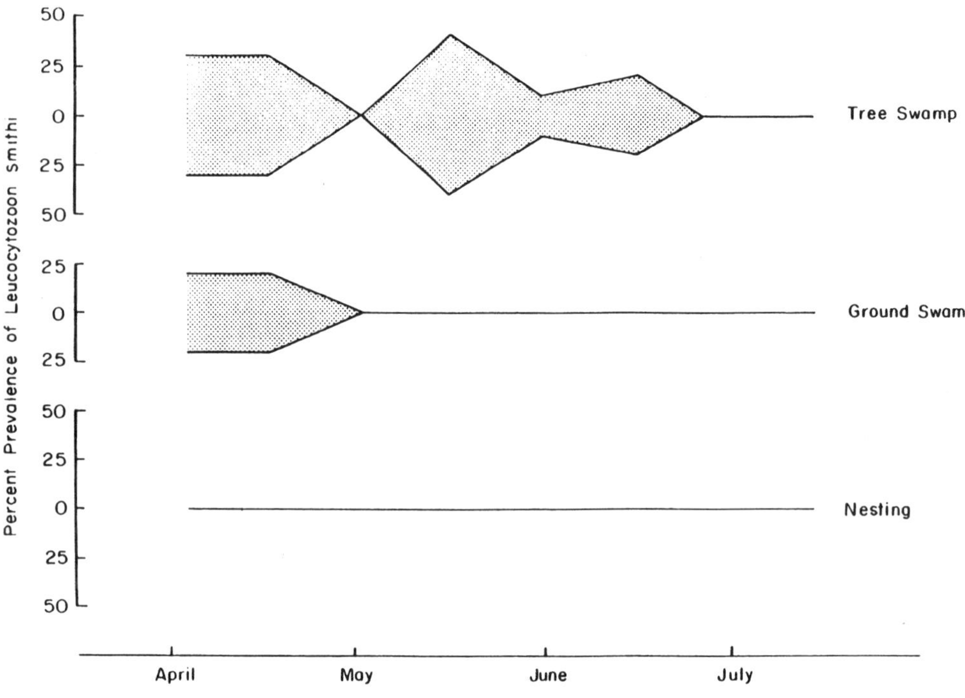

FIGURE 17.40. Transmission of *Leucocytozoon smithi* to sentinel domestic turkeys (2–4 weeks of age) exposed in cages located in a tree in roosting habitat, on the ground in the creek swamp, and on the ground in nesting habitat at Fisheating Creek Wildlife Management Area (Glades County) during the spring of 1980. From Akey 1981; by permission of Bruce L. Akey.

Table 17.27. Wild-caught mosquitoes from Fisheating Creek Wildlife Management Area (Glades County) tested for sporozoites of *Plasmodium hermani*, 1977 and 1978

Mosquito	Examination method[a]	Number of mosquitoes examined	Number of poults	
			Exposed	Positive
Culex nigripalpus	BF	5,124	98	0
	IN	9,747	193	3
Wyeomyia vanduzeei	BF	525	9	0
	IN	1,013	10	0
Culex erraticus	BF	361	3	0
	IN	85	2	0
Mansonia dyari	BF	0	—	—
	IN	251	7	0
Totals		17,106	322	3

Source: Forrester et al. (1980a).
a. Wild-caught mosquitoes were allowed to take a blood meal (BF) on domestic poults that were then monitored for malarial infections. In addition, batches of 50–100 mosquitoes were processed and inoculated (IN) into domestic poults that were then monitored for malarial infections.

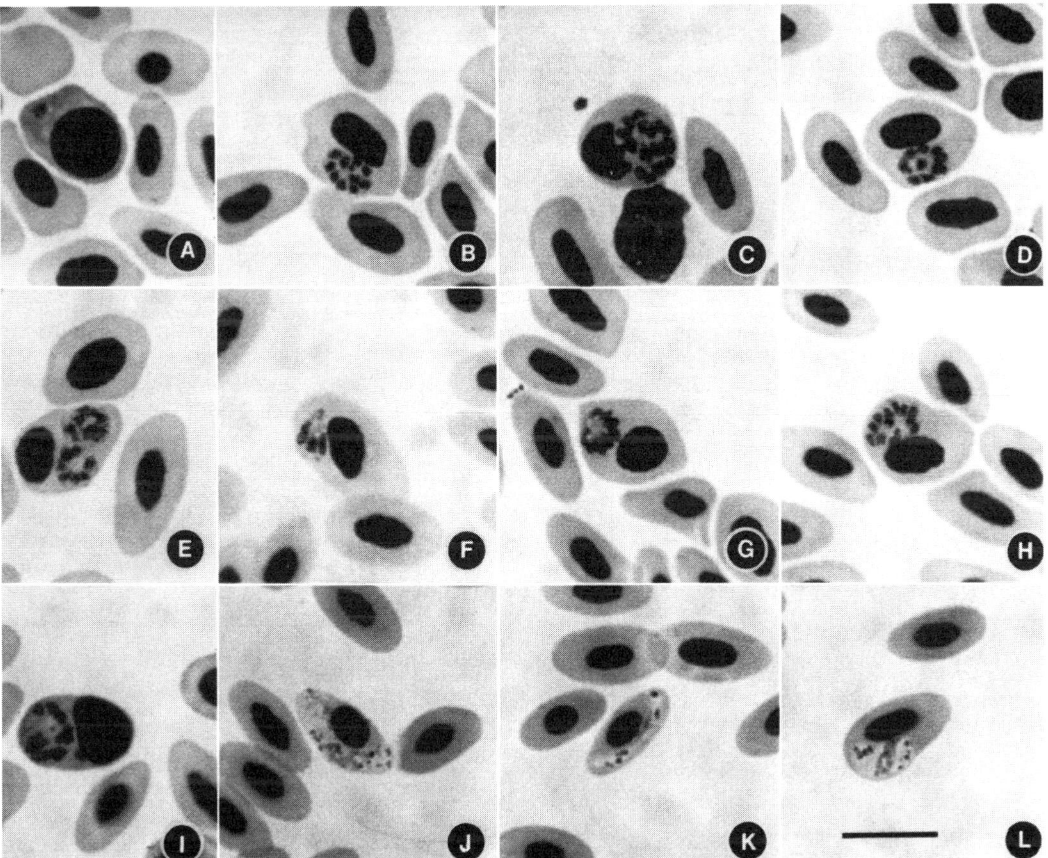

FIGURE 17.41. *Plasmodium hermani* in blood cells of a domestic turkey. This bird had been infected experimentally with a strain of *P. hermani* derived from Wild Turkeys in Florida. Bar = 10 μm. *A,* trophozoite in an immature erythrocyte; *B–D,* immature schizonts in immature erythrocytes; *E,* multiple infection (schizonts) in immature erythrocyte; *F,* mature schizont (fan-shaped) in immature erythrocyte; *G–H,* mature schizonts (rosettes) in immature erythrocytes; *I,* schizont in white blood cell; *J,* macrogametocyte in erythrocyte; *K,* microgametocyte (elongate form) in erythrocyte; *L,* microgametocyte (round form) in erythrocyte. Modified from Telford and Forrester 1975; by permission of *Journal of Protozoology.*

In a related study, Akey (1981) reported even higher levels of mortality than did Wright (1986). Akey used 5-day-old domestic poults and observed 78% mortality in birds infected with both pox and malaria, 21% in pox-only birds, and 6% in malaria-only birds; however, these higher mortality rates could have been caused by a nutritional deficiency or infections with other disease agents. A high percentage of Akey's birds had perosis, a leg deformity, and indication of choline deficiency; in addition he reported evidence of low-grade bacterial infections in most of his birds.

The above observations on experimental concurrent infections of pox and malaria, resulting increased negative effects, and the possible interaction of other disease agents and conditions with these infections provide insight into an understanding of the impact of malaria on Wild Turkeys. The indirect effect of malaria and its interaction with other disease agents might result in increased susceptibility

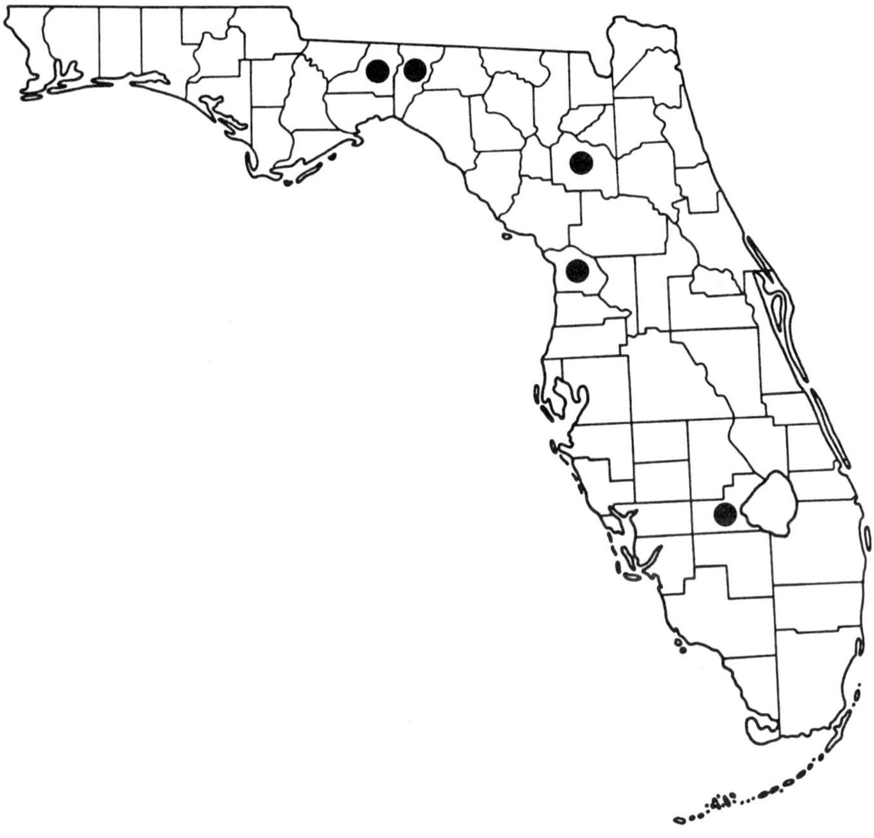

FIGURE 17.42. Distribution of malaria (*Plasmodium hermani*) in Wild Turkeys in Florida as determined by isodiagnostic techniques. *Circles* = counties where infected turkeys have been found.

to predation due to depressed growth rates, debilitation, and changes in behavior. These indirect effects might be as important as or even more important than if they caused direct mortality (Forrester 1991) and need to be studied further.

XVI. Other protozoans

In addition to the 3 blood protozoans discussed in previous sections (*Haemoproteus meleagridis, Leucocytozoon smithi,* and *Plasmodium hermani*) there is information on the occurrence of several other protozoans (table 17.31). These include *Eimeria* spp., *Histomonas meleagridis,* and *Tetratrichomonas gallinarum.*

Seven species of intestinal coccidia in the genus *Eimeria* have been reported from Wild Turkeys in the United States (Prestwood et al. 1971; Davidson and Wentworth 1992). Immunity-challenge studies are necessary to identify the species of turkey coccidia (Prestwood et al. 1971) and these specialized techniques have not been used in Florida. However, there is information on the prevalence of *Eimeria* spp. (as a group) for Wild Turkeys from 16 counties in Florida (table 17.32, figure 17.61). Sixty percent of 147 juvenile and adult turkeys examined from 1969 to 1999 were infected. The largest samples came from Alachua, Glades, and Osceola counties where 75%, 50%, and 80% were infected, respectively. At Fisheating Creek, infections were not found until poults

Table 17.28. Reduction in weight gains in domestic and wild turkey poults infected experimentally with *Plasmodium hermani* via transfer of infected blood

Type of poult	Number of poults infected	Age when infected	Infection dose[a]	Number of uninfected controls	% reduction in wt. gains of infected poults compared to uninfected controls
Domestic[b]					
	6	2 days	4.0×10^5	6	22
	18	15 days	3.1×10^6	18	13
	18	15 days	31.0×10^6	18	13
Wild[c]					
	4	12–18 hrs.	6.1×10^4	4	26
	6	3–5 days	6.1×10^4	5	15

Source: Forrester et al. (1980b).
a. Numbers of asexual stages (trophozoites, schizonts, and segmenters). This strain of *P. hermani* (P-1) was isolated from a Wild Turkey from Fisheating Creek Wildlife Management Area (Glades County) on February 3, 1972, and maintained in domestic turkeys until used in these experiments.
b. Amerine broad-breasted white poults.
c. Hatched in an incubator from eggs obtained from Wild Turkey nests at Fisheating Creek.

were >3 weeks of age (table 17.33). Coccidiosis has not been reported in free-ranging Wild Turkeys and mild infections (i.e., coccidiasis) may be helpful in stimulating immunity that results in protection from the disease (Davidson and Wentworth 1992).

Buchholz (1995) studied Wild Turkeys from Paynes Prairie State Preserve (Alachua County) and found a negative correlation between intensity of coccidial infections and long snoods and wide scullcaps. He concluded that these results were consistent with good genes models of female choice; i.e., there was a "parasite-driven explanation for the maintenance of female choice for male ornamentation in wild turkeys" (Buchholz 1995) as originally proposed by Hamilton and Zuk (1982). In other words, the mating choice of males by females may be related to the snood lengths of the males and may be a behavioral adaptation directed against parasitism. In support of his findings, Buchholz (1995) also cited a number of papers on domestic turkeys where it has been shown that coccidial infections resulted in delayed maturation, decreased egg and sperm production, and decreased egg fertility. His findings, although of interest, need to be further studied utilizing larger samples to verify these observations.

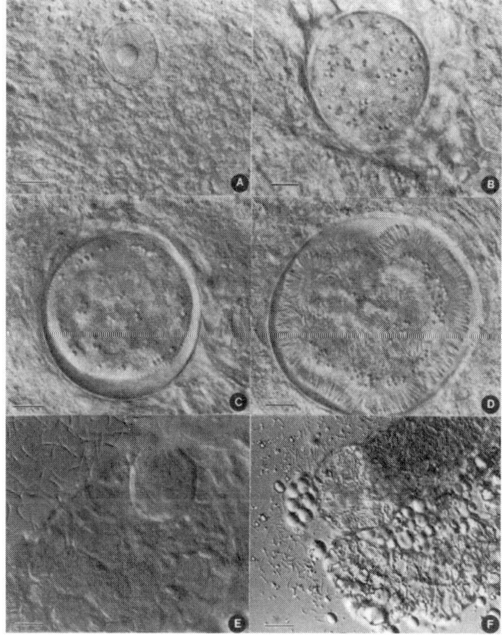

FIGURE 17.43. Sporogonic development of *Plasmodium hermani* in *Culex nigripalpus*. *A,* oocyst at 6–7 days; *B,* oocyst at 8 days; *C,* oocyst at 10 days; *D,* oocyst with sporozoites at 12 days; *E–F,* sporozoites in the salivary gland at 13 days. Bars in *A–E* = 14 µm; bar in *F* = 34 µm. From Forrester 1991; courtesy of J.K. Nayar; by permission of *Bulletin of the Society of Vector Ecology.*

FIGURE 17.44. Trap used to attract and capture mosquitoes alive for studies on the vectors of malaria in Wild Turkeys at Fisheating Creek Wildlife Management Area, Glades County. A 2–3-week-old domestic poult, reared in a mosquito-proof room after hatching, was placed in a wire cage to serve as an attractant for mosquitoes seeking a blood meal. Courtesy of Garry W. Foster.

Histomonas meleagridis is a protozoan parasite that can cause histomonosis (also known as infectious enterohepatitis or blackhead), a serious disease of domestic and Wild Turkeys. Davidson and Wentworth (1992) stated that this disease is referred to commonly in the technical and popular literature on Wild Turkeys, but that documented accounts of the disease are rare. Nevertheless, this disease is considered to be an important mortality factor in Wild Turkeys in the southeast (Davidson and Wentworth 1992; Hurst 1980).

The transmission of histomonosis is most unusual for a parasitic disease. The vector of the etiologic agent (i.e., the protozoan *H. meleagridis*) is the parasitic nematode *Heterakis gallinarum*, which infects the cecum of turkeys and other species of galliform birds. The histomonads are incorporated into the eggs of the nematode and pass out of the bird in its feces. Infected eggs can be ingested directly by an avian host or can be ingested by earthworms, which act as transport hosts. Once these earthworms are ingested by another turkey, the histomonads of *H. meleagridis* are released from the nematode egg and infect the new host and produce the disease. The disease is characterized by necrosis and ulceration of the mucosa of the ceca (which also has yellow cheese-like cores) and grayish-yellow to yellow-green circular depressed areas of necrosis in the liver. For more details see Reid (1967) and the excellent review by Davidson and Wentworth (1992).

Davidson et al. (1985) reported 11 cases of

FIGURE 17.45. Same trap as illustrated in Figure 17.44. The wire cage containing a live domestic poult is being inserted into an opening on the side of a lard-can bait trap that had been modified to prevent mosquitoes from taking a blood meal from the poult serving as an attractant (Nayar et al. 1980a). Photograph courtesy of Garry W. Foster.

Table 17.29. Mortality of domestic, pen-reared, and Wild Turkey poults in experimental infections of *Plasmodium hermani* established by sporozoites via mosquito bites

Type of turkey	Age	No. of turkeys exposed	No. of turkeys dead
Domestic[a]	5 days	18	1
	7 days	11	0
	14 days	22	0
	15 days	35	0
	2–6 weeks	22	0
	10 weeks	10	0
Pen-reared wild[b]	6 weeks	10	2
Wild[c]	1 day	1	0
	9 weeks	4	0

Sources: Akey (1981), Forrester and Nayar (1981), Nayar et al. (1980b, 1981a,b, 1982), Wright (1986).
a. Broad-breasted white, Nicholas strain.
b. Wild and domestic bronze hybrids (% of each unknown).
c. Florida Wild Turkeys obtained at 1 day of age from Fisheating Creek Wildlife Management Area, Glades County.

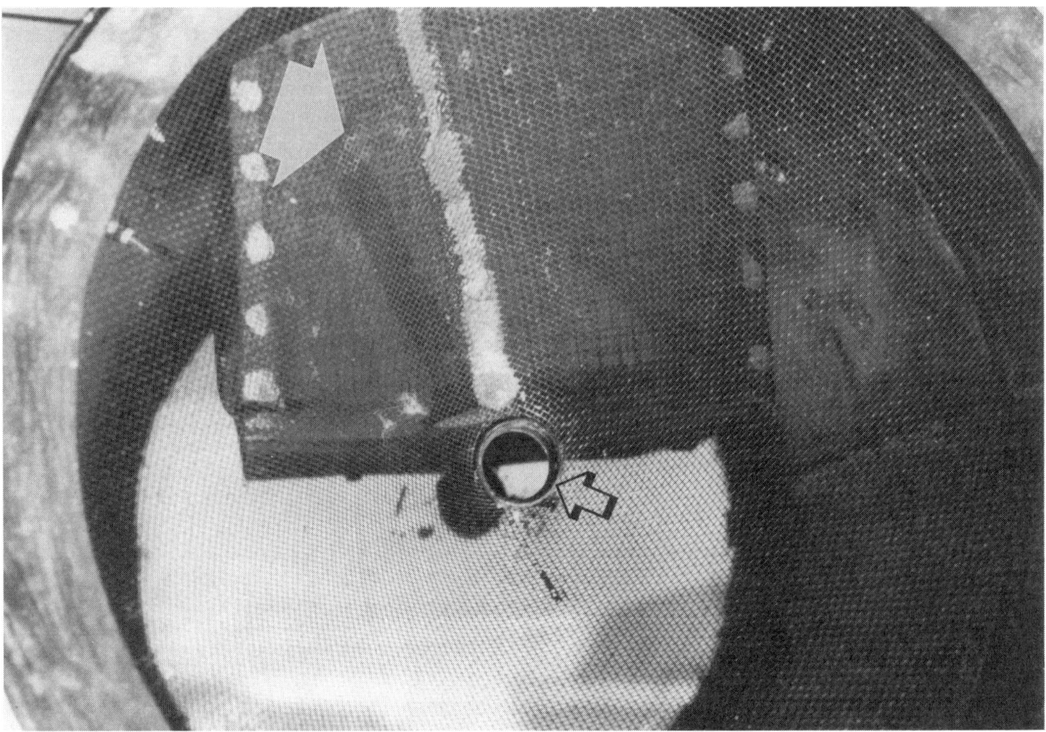

FIGURE 17.46. Same trap as illustrated in Figure 17.44. End view showing round opening *(small arrow)* through which mosquitoes entered when seeking a blood meal from the poult, protected within a mosquito-proof cage *(large arrow)*. Courtesy of Garry W. Foster.

histomonosis among 108 Wild Turkeys from Georgia, Tennessee, and Virginia. Hurst (1980) reported 3 cases in Wild Turkeys found during a 17-month period in Mississippi. The prevalence of histomonosis in Wild Turkeys in Florida is very low; from 1960 to 1982 only 3 cases were seen in 360 turkeys examined at necropsy (Forrester 1992b). These included a juvenile male in 1971 and an adult male in 1973 from Glades County and a juvenile female from Duval County in 1981 (figure 17.62) (Forrester 1992a,b). The 2 turkeys from Glades County were from isolated areas at Fisheating Creek with virtually no contact with domestic poultry; this might be evidence that the disease is enzootic in the Wild Turkey population there. In Duval County, however, the situation appears to be different. Histomonosis has been reported from domestic chickens

in the southern part of the county (Homer and Butcher 1991; Homer 1993) in an area about 10 miles from where the infected Wild Turkey was found (Forrester 1992b). In addition, domestic turkeys from Duval County are known to be infected with *Heterakis gallinarum*, the vector of histomonosis (90% prevalence; Maxfield et al. 1963); therefore, domestic poultry could have been the source of infection for the Wild Turkey in Duval County. The prevalence of *H. gallinarum* in Wild Turkeys throughout the state, however, is low (i.e., 9%, table 17.48) and may be the reason for the overall low prevalence of histomonosis (Forrester 1992a).

The practice of using commercial poultry litter as fertilizer for wildlife food plantings can have serious consequences for Wild Turkeys since this litter is a potential source of *His-*

Table 17.30. Mortality of 1- and 10-week-old domestic poults[a] infected[b] experimentally with avian pox, *Plasmodium hermani,* or both agents

Age	Experimental group	No. of turkeys in group	No. of turkeys dead at end of experiment[c]
1 week	Control	11	0
	Pox	11	3 (27%)
	Malaria	11	0
	Pox and malaria	9	2 (22%)
10 weeks	Control	11	0
	Pox	8	0
	Malaria	10	0
	Pox and malaria	8	0

Source: Wright (1986).
a. Broad-breasted white turkeys.
b. Via bites of infected mosquitoes (*Culex nigripalpus*).
c. The experiment lasted 7 weeks.

tomonas meleagridis. In many areas of southeastern United States (including Florida) poultry litter is spread in habitats where Wild Turkeys occur; turkeys as well as other wild birds are attracted to areas fertilized with this litter. In a recent evaluation of this problem (Waters 1992; Waters et al. 1994) it was concluded that litter from commercial broilers (approximately 7 weeks old) can be used without risk of transmission of histomonosis, but the authors recommended against the use of litter from breeder or layer operations since these are often contaminated with *H. meleagridis,* whereas broiler litter is not. The rationale for these recommendations is that broiler life spans are not sufficient for completion of the life cycle of the cecal worm. Waters (1992), for example, found that 2 breeder flocks in Florida (1 in Walton County and 1 in Leon County) had prevalences of infection by *Heterakis gallinarum* of 100% and 97% and intensities of 100 and 41 worms, respectively.

Infections by *Tetratrichomonas gallinarum* were detected in the ceca of 5 Wild Turkeys from 5 counties in northern and central Florida (table 17.31). This flagellate is common in domestic turkeys and other galliform birds such as chickens, quail, guineafowl, pheasants, and Chukars (Levine 1973). There is no evidence that this or-

Figure 17.47. Lard-can bait traps *(arrows)* hanging in cypress trees at Fisheating Creek Wildlife Management Area (Glades County) during a study of the mosquito vectors of malaria in Wild Turkeys, 1977–78. Courtesy of Garry W. Foster.

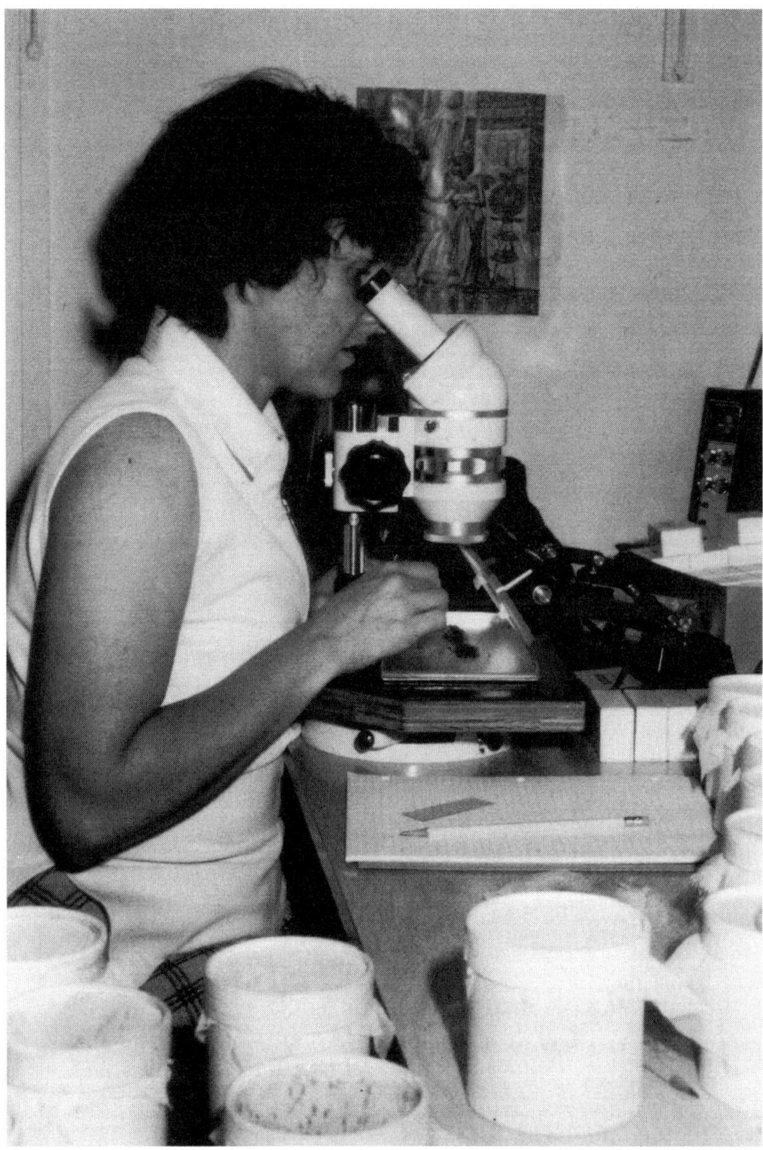

Figure 17.48. Technician Judy W. Knight identifying and sorting anesthetized mosquitoes caught alive in lard-can bait traps at Fisheating Creek Wildlife Management Area (Glades County) during a study of the mosquito vectors of malaria in Wild Turkeys, 1977–78. Courtesy of Garry W. Foster.

ganism is pathogenic and it probably should be considered a commensal rather than a parasite (Kemp 1978).

There is only a single report of a trypanosome from a Wild Turkey. Eve et al. (1972) found 1 organism on a blood film from 1 turkey in West Virginia. Trypanosomes have been reported from many species of birds (Bennett et al. 1982; Bishop and Bennett 1992), but are believed to be nonpathogenic (Levine 1973). Ex-

Table 17.31. Prevalence of miscellaneous protozoans reported from Wild Turkeys in Florida

Protozoan	County or site	Year(s)	Basis of diagnosis[a]	No. turkeys Exam.	Pos.	Data source
Eimeria spp.	Various[b]	1969–99	(1)	147	88	See Table 17.32
Histomonas meleagridis[c]	Duval	1971–77	(2)	4	1	Forrester (1992a,b)
	Glades	1960–81	(2)	250	2	Maxfield et al. (1963), Hon & Forrester (1972), Forrester (1992a,b), Akey & Forrester (1981)
Tetratrichomonas gallinarum	Citrus	1982	(3)	1	1	Forrester (1992b)
	Leon	1980	(3)	1	1	Ibid.
	Levy	1979	(3)	1	1	Ibid.
	Osceola	1976	(3)	1	1	Ibid.
	St. Johns	1972	(3)	1	1	Ibid.
Toxoplasma gondii	NG	NG	(4)	20	0	Burridge et al. (1979)
Trypanosoma sp.	Glades	1969–72	(6)	11	0	Forrester et al. (1974b)
	Various[d]	1969–89	(5)	1,027	0	Forrester et al. (1974b), Forrester (1992b)

NG = not given by authors.

a. Techniques used: (1) fecal flotation, (2) necropsy, (3) microscopic examination of fecal sample, (4) serology (indirect hemagglutination), (5) blood smear, (6) saline-neopeptone blood cultures of bone marrow.

b. Includes 15 counties; see Table 17.32 for details on sample sizes from each county.

c. An additional 106 turkeys from 19 counties were negative, 1969–84. Sample sizes: Alachua ($n = 24$), Charlotte (1), Citrus (1), Clay (11), Collier (1), DeSoto (1), Escambia (1), Flagler (4), Gadsden (3), Hendry (3), Highlands (17), Jefferson (1), Leon (2), Levy (10), Madison (1), Orange (1), Osceola (20), Sarasota (1), and Wakulla (3) (Forrester 1992b).

d. Includes 17 counties; see Table 17.8 for details on counties and sample sizes.

aminations of more than 1,000 Wild Turkeys in Florida were negative for *Trypanosoma* sp. (table 17.31).

No serologic evidence of the occurrence of *Toxoplasma gondii* was found in 20 Wild Turkeys from Florida examined by the use of the indirect hemagglutination (IHA) test (Burridge et al. 1979). However, as pointed out elsewhere, the IHA test may not be as sensitive in birds as in mammals and may result in false-negative results (Frenkel 1981). Dubey et al. (1993) found that the enzyme-linked immunosorbent assay (ELISA) and modified agglutination test (MAT) were more sensitive than the IHA test to detect *T. gondii* infections in domestic turkeys. There are no records of toxoplasmosis in Wild Turkeys in Florida, al-

though 2 fatal cases have been reported elsewhere, 1 from Georgia (Howerth and Rodenroth 1985) and 1 from West Virginia (Quist et al. 1995). Lindsay et al. (1994) found infections of *T. gondii* in heart tissues from 8 of 16 Wild Turkeys in Alabama. Serologic tests (MAT) were performed on these 16 turkeys and 1 additional bird; 12 were positive. Quist et al. (1995) also used the MAT test and found antibodies to *T. gondii* in 13 of 130 sera collected from Wild Turkeys in Georgia, Kentucky, Louisiana, Missouri, and North Carolina during 1984–89. These latter authors concluded that although turkeys are often infected with *T. gondii,* the disease toxoplasmosis is rare, most likely because of innate resistance.

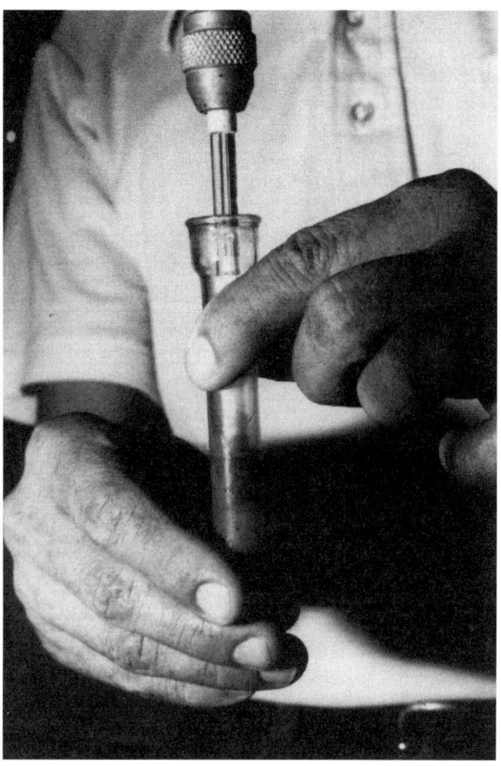

FIGURE 17.49. Mosquitoes being macerated in a Ten Broeck tissue grinder in saline in preparation for injection into domestic turkey poults during a study of the mosquito vectors of malaria in Wild Turkeys, 1977–78. Courtesy of Garry W. Foster.

XVII. Trematodes

At least 19 species of trematodes have been reported from Wild Turkeys in the United States (Davidson and Wentworth 1992). In Florida 10 species have been found (table 17.34), none of them common. The most abundant was the intestinal fluke *Echinoparyphium recurvatum*, found in turkeys from 6 counties, mainly in peninsular Florida (table 17.35, figure 17.63). The seasonal prevalence of *E. recurvatum* in 1970–71 was irregular in poults and juveniles at Fisheating Creek with no apparent trend (figure 17.64). Details on distribution, prevalence, and intensity of each of the other 9 species are given in tables 17.36–17.40. The Wild Turkey proba-

bly serves as a secondary or accidental host for most of these flukes, the primary hosts being other birds such as waterfowl or wading birds (Hon et al. 1975). The 2 most prevalent species (*E. recurvatum* and *Zygocotyle lunata*), in fact, are considered characteristic helminths of waterfowl (McDonald 1969) and have been commonly found in Mottled Ducks in Florida (Kinsella and Forrester 1972).

All of these trematodes require a snail as an intermediate host and some require a second intermediate host, usually an invertebrate (Davidson and Wentworth 1992). However, no work has been done on the life cycles of these parasites in Florida. Infective stages of *E. recurvatum* in other parts of the United States are known to encyst as metacercaria in various mollusks and also in anuran tadpoles (Kingston 1978), indicating that turkeys in Florida may become infected by ingestion of these food items along edges of streams or ponds.

The pathogenicity of these flukes in Wild Turkeys has not been studied, although Davidson and Wentworth (1992) stated that they "are not considered highly pathogenic and have not been associated with clinical disease." However, *E. recurvatum* is of potential concern since it has been found to cause severe enteritis, emaciation, anemia, anorexia, weight loss, reduced egg production, and mortality in

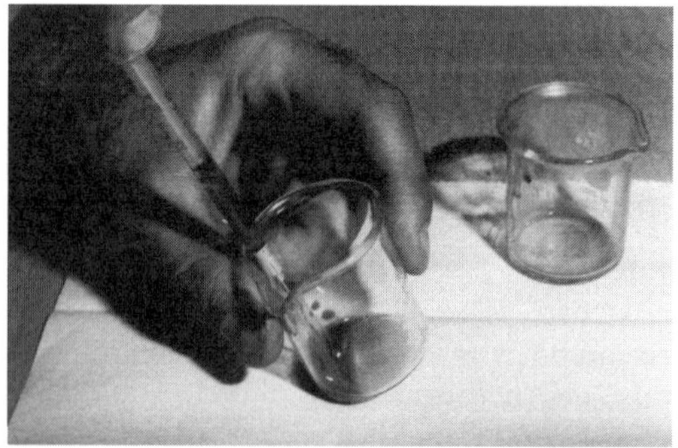

FIGURE 17.50. Mosquito extract being loaded into a syringe in preparation for injection into domestic turkey poults during a study of the mosquito vectors of malaria in Wild Turkeys, 1977–78. Courtesy of Garry W. Foster.

Table 17.32. Prevalence of coccidia (*Eimeria* spp.) in Wild Turkeys >1 month of age from Florida, 1969–99

County	Number of turkeys		
	Examined	Positive[a]	%
Alachua	12	9	75
Flagler	1	1	—
Glades	86	43	50
Hendry	1	1	—
Highlands	1	1	—
Lake	2	2	—
Leon	4	3	—
Levy	1	1	—
Osceola	25	20	80
Sarasota	6	5	—
Suwannee	1	1	—
Taylor	1	1	—
Others[b]	6	0	—
Totals	147	88	60

Sources: Akey (1981), Forrester (1992b), Forrester and Todd (1971), Dusenbury and Spalding (1999).
a. Presence of oocysts determined by fecal flotation test.
b. Includes negative data from 4 other counties (sample sizes in parentheses): Charlotte (1), Citrus (2), Duval (2), and Orange (1).

chickens and domestic turkeys (Kingston 1978). Annereaux (1940) reported that a 10-week-old domestic turkey poult infected with 267 specimens of *E. recurvatum* had a severe enteritis. Wild Turkeys in Florida have intensities that usually vary from 1 to 24 (table 17.35), although 1 juvenile gobbler from Fisheating Creek (Glades County) had 109 worms and another juvenile gobbler from Richloam Wildlife Management Area (Pasco County) had 924. In the latter case, the turkey was extremely emaciated and had a concurrent pox infection along with a large number of tapeworms (291), making it difficult to determine the significance of the fluke infections. Undoubtedly they contributed to the overall debilitation of this 1 bird, but their importance to Wild Turkey populations is not known.

Table 17.33. Prevalence of coccidia (*Eimeria* spp.) in Wild Turkeys from Fisheating Creek, Glades County, according to age, 1968–81

Age	Number of turkeys		
	Examined	Positive[a]	%
Poults (<30 days)	81	2[b]	2
Juveniles (1–12 months)	76	38	50
Adults (>12 months)	10	5	50

Sources: Akey (1981), Forrester (1992b), Forrester and Todd (1971).
a. Presence of oocysts determined by fecal flotation test.
b. 23 and 27 days of age.

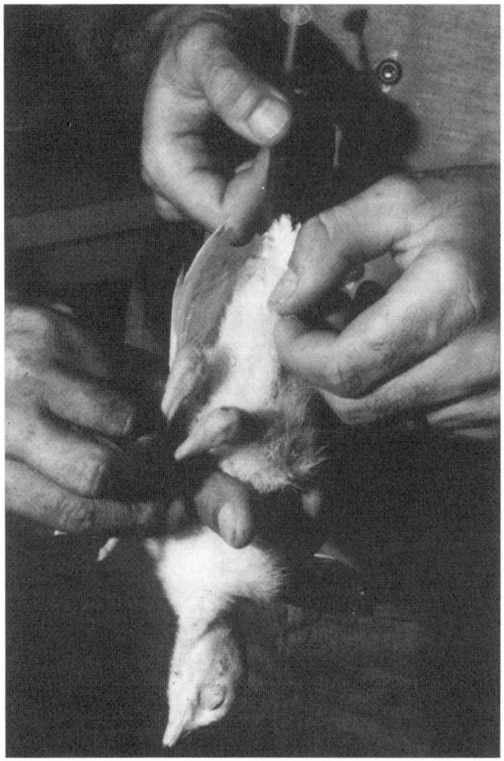

FIGURE 17.51. Mosquito extract being injected intraperitoneally into a domestic turkey poult during a study of the mosquito vectors of malaria in Wild Turkeys, 1977–78. Poults so injected were then monitored for 5 weeks to detect malaria infections. Courtesy of Garry W. Foster.

XVIII. Cestodes

Davidson and Wentworth (1992) reported that at least 13 species of tapeworms have been reported from Wild Turkeys in the United States. In Florida 7 species have been found (table 17.41). The 2 most common are *Metroliasthes lucida* (table 17.42) and *Raillietina georgiensis* (table 17.43). Data on distribution, prevalence, and intensity of other species are presented in tables 17.44–17.47. *Metroliasthes lucida* is especially prevalent (63%), having been found in turkeys from 16 counties (figure 17.65), and probably occurs throughout the state. When compared in a helminth profile, *M. lucida* stands

out as one of the most important components of the helminth fauna of turkeys at Fisheating Creek, second only to *Strongyloides* sp. (figure 17.66). Prevalences have varied at Fisheating Creek between 13% and 95% from year to year with some birds harboring several hundred specimens; 1 poult had more than 450. Infections are acquired at an early age; poults as young as 8 days were infected at Fisheating Creek (Hon et al. 1978). Prevalences of *M. lucida* in poults and juveniles rose sharply during the summer of 1970 and remained at high prevalences through February 1971 (figure 17.67).

All of the cestodes require intermediate hosts (usually terrestrial invertebrates) to complete their life cycles (Davidson and Wentworth 1992). There is no specific information for Florida, but in studies conducted elsewhere on domestic turkeys (Reid 1962), it has been determined that the following serve as intermediate hosts: grasshoppers (for *M. lucida*), ants (for *R. georgiensis*), and beetles (for *R. cesticillus* and *H. carioca*). Intermediate hosts for the other species of tapeworms in Wild Turkeys in Florida are unknown. The high summer prevalence of *M. lucida* illustrated in figure 17.67 was probably due to the heavy use of grasshoppers by poults and juveniles at that time. Grasshoppers have been found to be common in the diet of turkeys (Hon et al. 1978).

The pathogenicity of tapeworms has not been studied in Wild Turkeys. Information on domestic fowl has been extrapolated to Wild Turkeys (Davidson and Wentworth 1992) and indicates that these tapeworms probably have very little effect. *Raillietina georgiensis,* for example, has been reported to cause only mild enteritis in domestic turkeys and *R. cesticillus* has been linked with slightly decreased weight gains in domestic chicks (Reid 1962). Nothing is known about the other species of tapeworms found in Florida Wild Turkeys.

XIX. Gastrointestinal nematodes

Gastrointestinal nematodes are prevalent and common in Wild Turkeys; at least 22 species

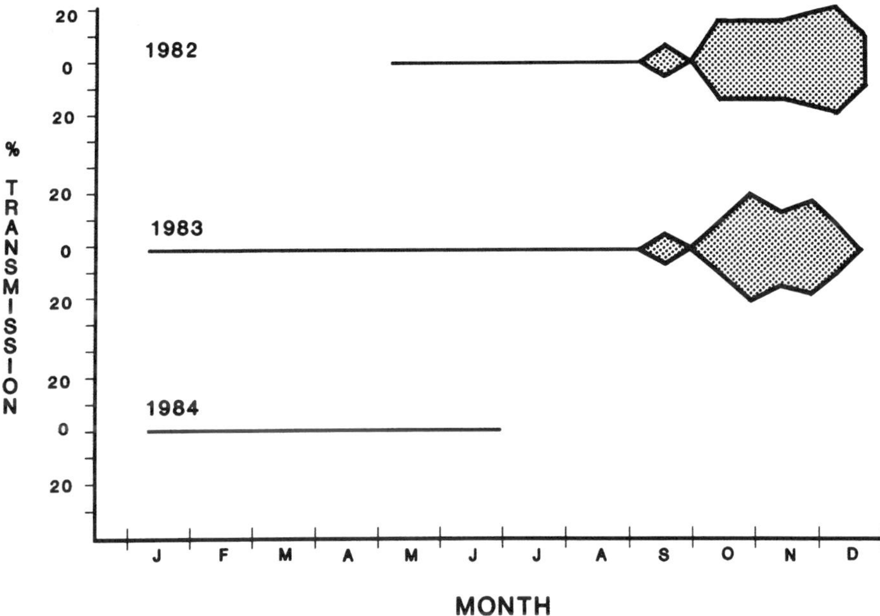

FIGURE 17.52. Seasonal transmission of *Plasmodium hermani* to sentinel domestic turkeys (2–4 weeks of age) exposed in Wild Turkey habitat at Paynes Prairie State Preserve (Alachua County), 1982–84. Modified from Atkinson and van Riper 1991; by permission of Carter T. Atkinson.

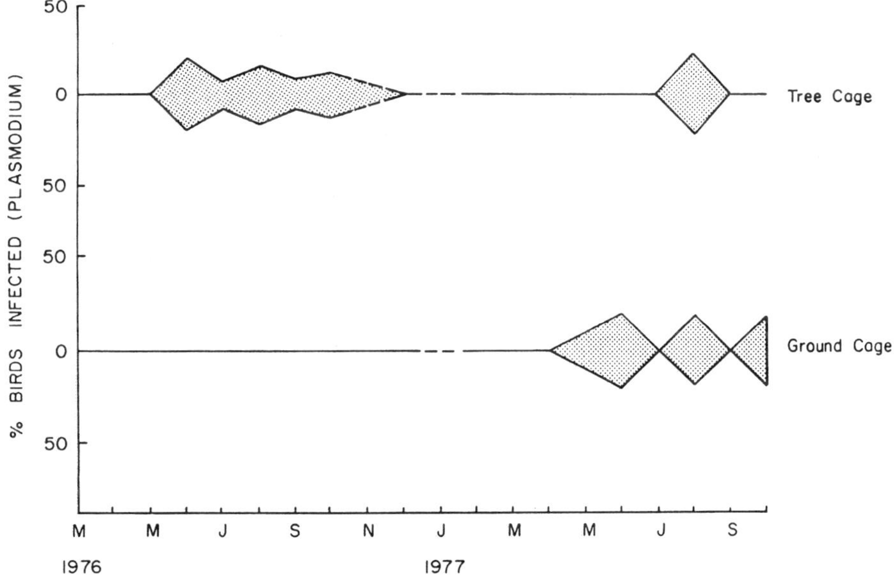

FIGURE 17.53. Seasonal transmission of *Plasmodium hermani* to sentinel domestic turkeys (2–4 weeks of age) exposed in Wild Turkey habitat at Fisheating Creek Wildlife Management Area (Glades County), 1976–77. *Broken lines* = missing data due to the deaths of sentinels unrelated to disease. From Forrester 1991; by permission of *Bulletin of the Society of Vector Ecology*.

Table 17.34. Prevalence, intensity, and abundance of digenetic trematodes reported from Wild Turkeys in Florida, 1960–84

| Infection site | Number of turkeys | | Intensity | | |
Trematode species	Exam.	Prevalence (%)	Mean	Range	Abundance
Small intestine					
Echinoparyphium recurvatum	500	7	70[a]	1–924	5[a]
Strigea elegans meleagris	500	3	11[b]	1–80	<1
Echinostoma trivolvis	500	<1	1	1	<1
Brachylaima virginianum	500	<1	3	3	<1
Ascocotyle sp.	500	<1	1	1	<1
Ceca					
Zygocotyle lunata	500	7	5	1–38	<1
Cloaca					
Stomylotrema vicarium	496	3	2	1–9	<1
Prosthogonimus ovatus	496	1	2	1–2	<1
Kidneys					
Tanaisia sp.	496	<1	3	3	<1
Liver					
Zonorchis sp.	496	<1	1	1	<1

Sources: Akey and Forrester (1981), Forrester (1992b), Hon and Forrester (1972), Maxfield et al. (1963).
a. The mean intensity is 4 and the abundance is <1 if the 1 turkey from Pasco County is omitted from the calculations; its count of 924 flukes is abnormally high.
b. The mean intensity is 3 if the 1 turkey from Pasco County is omitted from the calculations; its count of 80 flukes is abnormally high.

have been reported (Davidson and Wentworth 1992). In Florida 15 species have been found (table 17.48), several of which are pathogenic. Data on the distribution, prevalence, and intensity of these nematodes are presented in tables 17.49–17.57. The most prevalent gastrointestinal nematode is *Strongyloides* sp. (38%), followed by *Trichostrongylus cramae* (= *T. tenuis*) (32%) and *Dispharynx nasuta* (22%). *Strongyloides* sp. and *T. cramae* are the most abundant.

Maxfield et al. (1963) reported *Strongyloides avium* from Wild Turkeys in Glades County, but did not culture and examine the free-living adult stages, a process necessary for identification of the species. Hon et al. (1975) cultured feces from 1 young Wild Turkey from Glades County and found the free-living adult males to be morphometrically close to *S. avium* as described by Cram (1929). However, since there may be more than 1 species in Wild Turkeys and fecal samples from each bird were not cultured, the specific identity of the

Strongyloides in Florida populations remains unknown.

Infections by *Strongyloides* sp. have been found in Wild Turkeys from 10 counties (figure 17.68). In Glades County, where large numbers of turkeys were examined from 1970 to 1981, yearly prevalences varied from 33% to 88% when >20 turkeys were sampled each year (table 17.53). The helminth profile given for turkeys at Fisheating Creek in 1970–71 (figure 17.66) shows how dominant *Strongyloides* sp. is in comparison with other components of the fauna. Infections were acquired at an early age; poults as young as 21 days were found infected in 1970–71 (Hon et al. 1978) and 1980–81 (Akey 1981). Prevalences remained high throughout the year after infections were acquired by poults (figure 17.69). Intensities averaged 12 worms per bird and varied from 1 to 783. The turkey with 783 worms was a 5-month-old female collected in October 1970 (Hon and Forrester 1972).

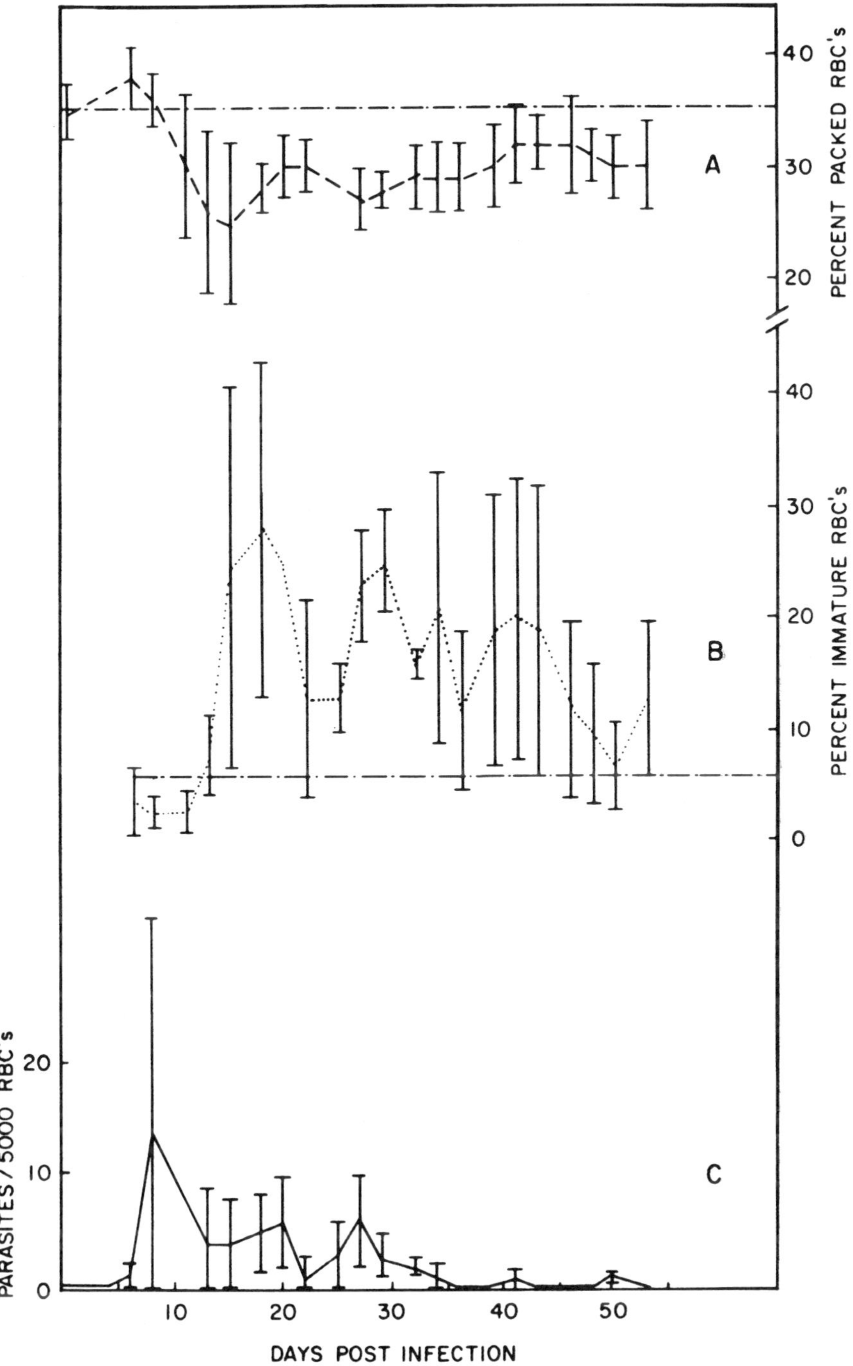

FIGURE 17.54. Hematocrits (*A*), percent immature erythrocytes (*B*), and parasitemias (*C*) of domestic turkey poults each infected with 31 million asexual stages of *Plasmodium hermani* at 15 days of age. Vertical bars = standard deviations, ·—· = mean control values. From Forrester et al. 1980b; by permission of *Journal of Wildlife Diseases*.

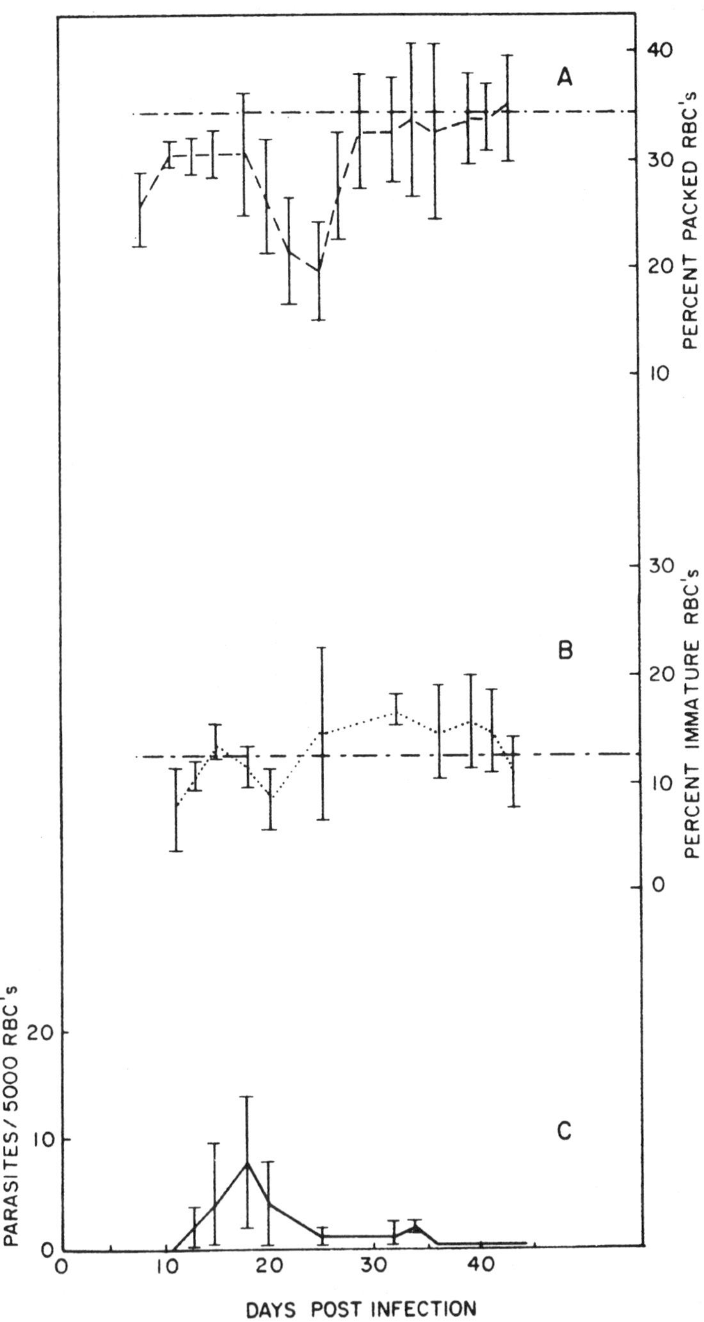

FIGURE 17.55. Hematocrits (*A*), percent immature erythrocytes (*B*), and parasitemias (*C*) of Wild Turkey poults infected with 61,000 asexual stages of *Plasmodium hermani* 12–18 hours after hatching. Vertical bars = standard deviations, ·—· = mean control values. From Forrester et al. 1980b; by permission of *Journal of Wildlife Diseases*.

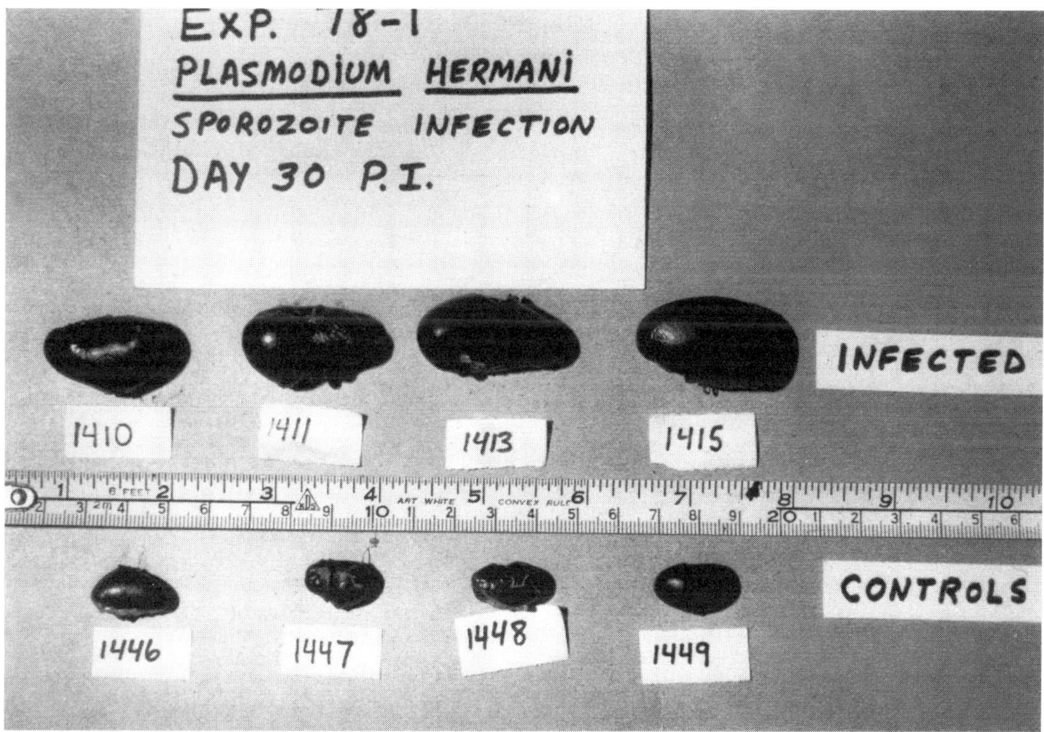

FIGURE 17.56. Enlarged spleens (splenomegaly) of domestic turkey poults 30 days after infection with *Plasmodium hermani*, compared with normal spleens from uninfected controls. From Forrester 1991; by permission of *Bulletin of the Society of Vector Ecology*.

The life cycle and transmission of species of *Strongyloides* differ from all other nematodes that are parasites of vertebrates (Levine 1980). The cycle has both parasitic and free-living adult stages. The parasitic form is a parthenogenetic female that occurs in the small intestine, large intestine, or cecum, where it produces embryonated eggs that pass out with the feces, hatch, and produce larvae. These larvae eventually become free-living adult males and females that mate and produce eggs from which infective larvae develop. These infective larvae are ingested with food or water and develop into parthenogenetic females (Ruff 1978). Intermediate hosts are not required to complete the cycle.

The pathogenicity of *Strongyloides* sp. in Wild Turkeys has not been studied; however, experimental work has been done with domestic chickens. Cecal lesions in chickens with low intensity infections were minimal, whereas infections of higher intensity resulted in thickened cecal walls and lumens filled with thick pasty contents (Cram 1929). These lesions regressed in birds that survived. The importance of infections of this common and widespread nematode in Wild Turkeys should be studied further.

The second most prevalent gastrointestinal nematode is *Trichostrongylus cramae* (= *T. tenuis*), which has been found in Wild Turkeys from 14 counties throughout Florida (figure 17.70). Prevalences and intensities were especially high in Osceola County (table 17.56). A rise in prevalence during the winter months (January and February) was recorded at Fisheating Creek in 1970–71 (figure 17.71). This nematode is also found in Northern Bobwhites in Florida (see chapter 16, Northern Bobwhites), where it is more than twice as prevalent (67%) as in Wild Turkeys (32%) and almost 5 times more abundant (24 vs. 5). It has

Table 17.35. Distribution, prevalence, and intensity of infections of the
intestinal fluke *Echinoparyphium recurvatum* in Wild Turkeys from Florida

| County or site | Number of turkeys | | | Intensity | |
Year(s)	Exam.	Pos.	%	Mean	Range
Glades					
1960–61	12	1	8	NG	NG
1968	1	0	—	—	—
1969	16	1	6	1	1
1970	81	10	12	3	1–14
1971	65	5	8	5	1–19
1972	16	6	38	24	2–109
1973	6	1	—	4	4
1974	11	0	0	—	—
1975	20	1	5	1	1
1976	1	0	—	—	—
1977	45	2	4	1	1
1978	5	0	—	—	—
1979	5	0	—	—	—
1980	31	0	0	—	—
1981	27	0	0	—	—
1982	5	0	—	—	—
1983	1	1	—	1	1
1984	4	0	—	—	—
Hendry					
1970	1	1	—	3	3
1974, 76	2	0	—	—	—
Madison					
1971	1	1	—	1	1
Manatee					
1977	1	1	—	11	11
Osceola					
1970	8	1	—	1	1
1971	9	0	—	—	—
1972	3	0	—	—	—
1975–76	3	0	—	—	—
Pasco					
1979	1	1	—	924	924
Others[a]					
1960–84	119	0	0	—	—

Sources: 1960–61 data from Maxfield et al. (1963), 1968–72 data from Hon and Forrester
(1972), 1980–81 Glades County data from Akey and Forrester (1981), remainder of data from
Forrester (1992b).
NG = not given by authors.
a. Includes negative data from 19 other counties (sample sizes in parentheses): Alachua (25),
Charlotte (1), Citrus (1), Clay (11), Collier (1), DeSoto (1), Duval (9), Escambia (2), Flagler (5),
Gadsden (3), Highlands (17), Jefferson (1), Leon (3), Levy (11), Sarasota (1), St. Johns (1),
Sumter (1), Taylor (2), Wakulla (3), and "Florida" (specific sites not given) (20).

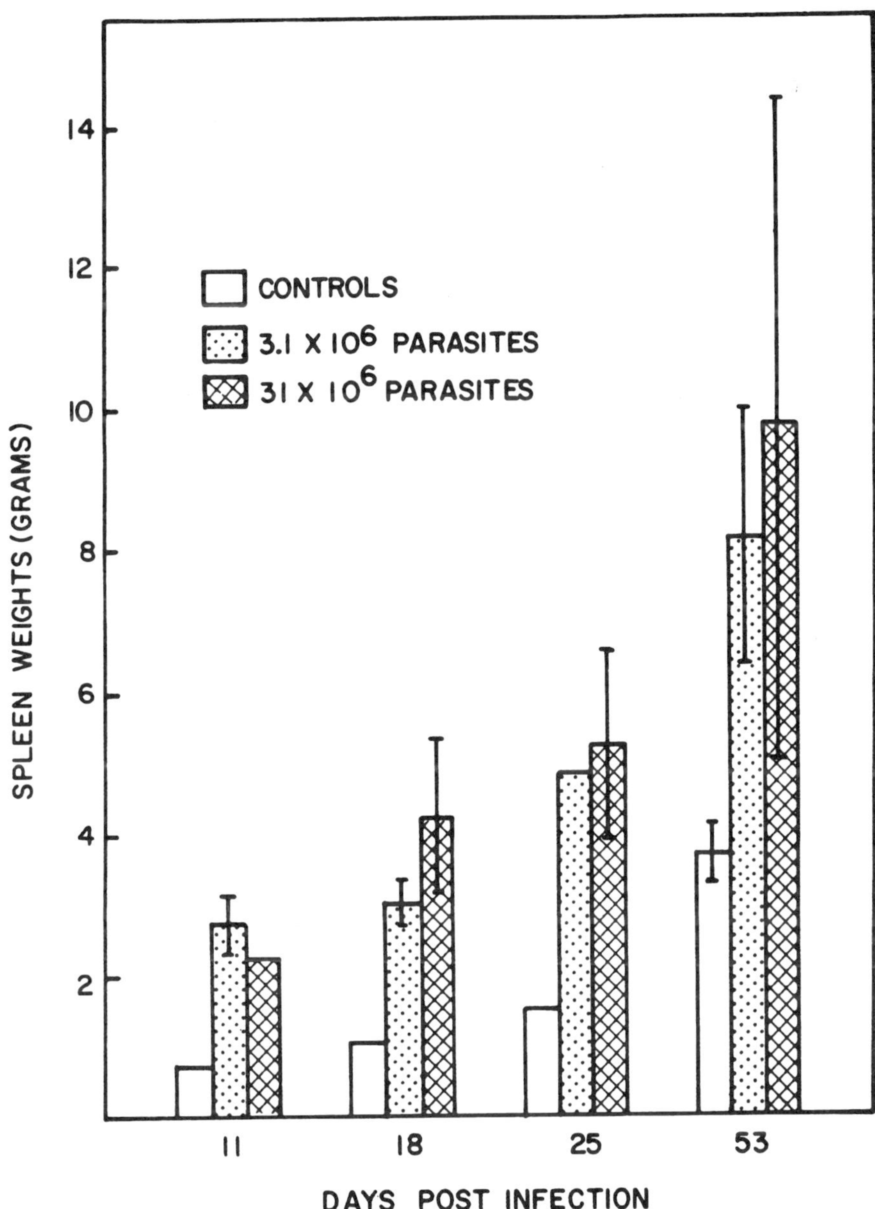

FIGURE 17.57. Weights of spleens from domestic turkey poults infected with 3,100,000 or 31,000,000 asexual stages of *Plasmodium hermani* at 15 days of age. Vertical bars = standard deviations. From Forrester et al. 1980b; by permission of *Journal of Wildlife Diseases*.

Table 17.36. Distribution, prevalence, and intensity of infections of the intestinal fluke *Strigea elegans meleagris* in Wild Turkeys from Florida

County	Year(s)	No. of turkeys			Intensity	
		Exam.	Pos.	%	Mean	Range
Alachua[a]	1969	6	0	—	—	—
	1970	7	1	—	1	1
	1971–80	12	0	0	—	—
Collier	1981	1	1	—	5	5
Duval	1971–76	5	0	—	—	—
	1977	2	1	—	1	1
	1978, 80	2	0	—	—	—
Flagler[a]	1969	4	3	—	7	4–14
	1983	1	0	—	—	—
Glades[a]	1960–61[b]	12	0	0	—	—
	1968–69	17	0	0	—	—
	1970	81	3	4	7	1–11
	1971	65	0	0	—	—
	1972	16	1	6	3	3
	1973–74	17	0	0	—	—
	1975	20	1	5	1	1
	1976–84	124	0	0	—	—
Levy	1960–61	8	0	—	—	—
	1976, 77	2	0	—	—	—
	1978	1	1	—	—	—
Osceola[a]	1970	8	2	—	1	1
	1971–76	15	0	0	—	—
Pasco	1979	1	1	—	80	80
Others[c]	1960–84	73	0	0	—	—

Sources: 1960–61 data from Maxfield et al. (1963), 1968–72 data from Hon and Forrester (1972), 1980–81 Glades County data from Akey and Forrester (1981), remainder of data from Forrester (1992b).
a. Specimens from turkeys from these counties were used by Dubois and Hon (1973) to redescribe *Strigea elegans meleagris*.
b. Maxfield et al. (1963) reported *Cotylurus flabelliformis*, a strigeid trematode resembling *Strigea elegans meleagris,* from 1 of 12 turkeys collected in Glades County. Since this species was not encountered during more extensive studies in Glades County, 1968–84, in which 337 turkeys were examined, it is assumed that Maxfield's report was in error and he actually recovered *S. e. meleagris*. This cannot be determined, however, since his specimens are not available for examination, and so his data are excluded from further consideration.
c. Includes data from 17 other counties; sample sizes are presented in Table 17.35.

a direct life cycle and transmission occurs when infective larvae are ingested. Lesions have not been found associated with infections by *T. cramae* (Hon et al. 1975) and the parasite is not considered pathogenic for Wild Turkeys (Davidson and Wentworth 1992). A closely related species (*T. tenuis*), however, has been reported to have direct and indirect effects on population cycles of Red Grouse (*Lagopus lagopus scoticus*) in England and Scotland (Hudson and Dobson 1991).

The proventricular spiruroid nematode *Dispharynx nasuta* (figure 17.72) is probably the most important nematode infecting Wild Turkeys in Florida because of its pathogenic effects on poults. It is has been found in Wild Turkeys from 8 counties (table 17.49, figure 17.73), but most likely it is more widespread. Since it is primarily a parasite of poults and juveniles less than 6 months of age (table 17.50)— age classes that have not been well sampled in Florida (except for Glades County)—our data

Table 17.37. Distribution, prevalence, and intensity of infections of the intestinal flukes *Echinostoma trivolvis*, *Brachylaima virginianum*, and *Ascocotyle* sp. in Wild Turkeys from Florida

| County | Year(s) | No. of turkeys | | | Intensity | |
		Exam.	Pos.	%	Mean	Range
Echinostoma trivolvis						
Glades	1960–61	12	0	0	—	—
	1968–69	17	0	0	—	—
	1970	81	1	1	1	1
	1971	65	1	2	1	1
	1972	16	1	6	1	1
	1973–84	161	0	0	—	—
Others[a]	1960–84	148	0	0	—	—
Brachylaima virginianum						
Hendry	1970	1	1	—	3	3
	1974, 76	2	0	—	—	—
Others[a]	1960–84	497	0	0	—	—
Ascocotyle sp.						
Wakulla	1971	3	1	—	1	1
Others[a]	1960–84	497	0	0	—	—

Sources: 1960–61 data from Maxfield et al. (1963), 1968–72 data from Hon and Forrester (1972), 1980–81 Glades County data from Akey and Forrester (1981), remainder of data from Forrester (1992b).
a. Includes negative data from 24 other counties; sample sizes are presented in Table 17.35.

Table 17.38. Distribution, prevalence, and intensity of infections of the intestinal fluke *Zygocotyle lunata* in Wild Turkeys from Florida

| County | Year(s) | No. of turkeys | | | Intensity | |
		Exam.	Pos.	%	Mean	Range
Glades	1960–61	12	7	58	NG	NG
	1968	1	1	—	1	—
	1968	16	1	6	3	—
	1970	81	7	12	4	1–20
	1971	65	11	17	2	1–6
	1972	16	2	13	1	1
	1973–74	17	0	0	—	—
	1975	20	1	5	38	—
	1976	1	0	—	—	—
	1977	45	2	4	1	1
	1978	5	1	—	1	—
	1979	5	1	—	1	—
	1980	31	1	3	1	—
	1981–84	37	0	0	—	—
Others[a]	1960–84	148	0	0	—	—

Sources: 1960–61 data from Maxfield et al. (1963), 1968–72 data from Hon and Forrester (1972), 1980–81 Glades County data from Akey and Forrester (1981), remainder of data from Forrester (1992b).
NG = not given by authors.
a. Includes data from 24 other counties; sample sizes are presented in Table 17.35.

FIGURE 17.58. Comparison of sizes of 2 domestic poults of the same age, one infected with *Plasmodium hermani (smaller bird on the left)*, the other an uninfected control. Note the droopy wings of the infected bird. From Forrester 1991; by permission of *Bulletin of the Society of Vector Ecology*.

on distribution probably reflect an underestimation. Poults as young as 3 days were infected with *D. nasuta* at Fisheating Creek in 1971 (Hon et al. 1978). Infections were common in poults during the early months of life; as the birds matured, the prevalence dropped sharply until, in November, it was nearly zero (figure 17.74). This was perhaps due to age-related resistance, immunity, or to changes in food habits, as suggested by Hon et al. (1978). Evidence that there may be an age-related resistance was obtained by Hon (1973), who tried to experimentally infect 2 domestic turkeys, 1 10 days old and the other 10 weeks old. An infection was established in the 10-day-old poult, but not in the older turkey.

The food-change hypothesis is supported by the observation that beginning in November there is a dramatic shift in the diet of juvenile turkeys in southern Florida from insects, arachnids, and other invertebrates (including crustaceans) and an assortment of plant material to a diet almost exclusively of plant material, with only a small trace of animal matter (Barwick et al. 1974). The significance of this diet-shift is related to the fact that *D. nasuta* requires terrestrial isopods (crustaceans, commonly called "sowbugs," "pillbugs," or "rolliups") as an intermediate host (Cram 1931). Sowbugs (figure 17.75) ingest embryonated eggs found in the feces of turkeys. These eggs hatch and develop to infective larvae in about 26 days. If the sowbug containing infective larvae of *D. nasuta* is eaten by a turkey, the larvae develop into adult nematodes in the proventriculus. Such intermediate hosts have been identified as *Oscelloscia floridana* and *Venezillo evergladensis* at Fisheating Creek (Hon and Forrester 1972; Hon 1973;

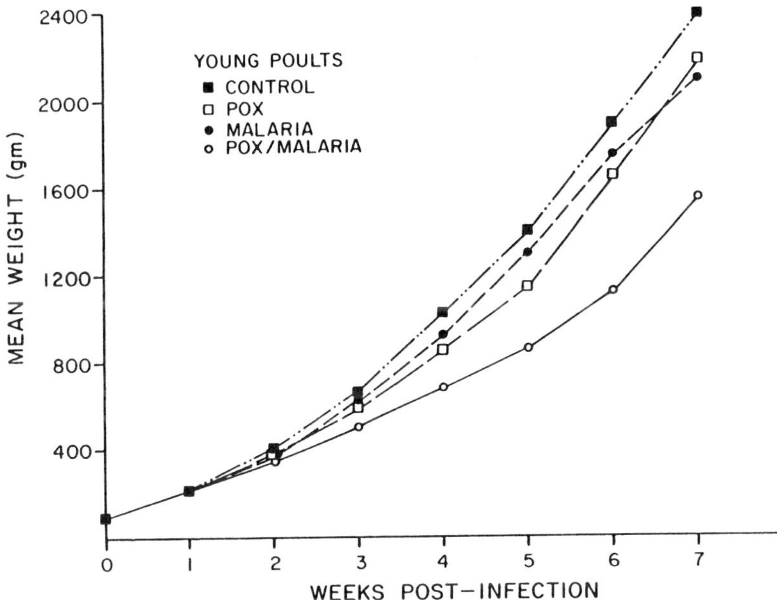

FIGURE 17.59. Comparison of mean weight gains of young domestic poults infected experimentally with avian pox, *Plasmodium hermani,* or pox and *P. hermani* concurrently over a 7-week period. From Wright 1986.

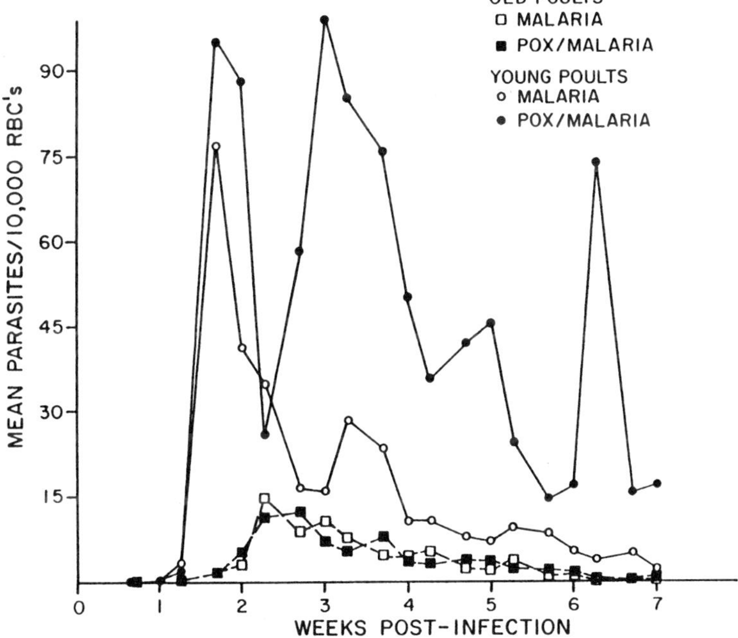

FIGURE 17.60. Comparison of the mean numbers of *Plasmodium hermani* (gametocytes and asexual stages) per 10,000 erythrocytes in young and old domestic poults infected experimentally with *P. hermani* and with concurrent infections of *P. hermani* and avian pox over a 7-week period. From Wright 1986.

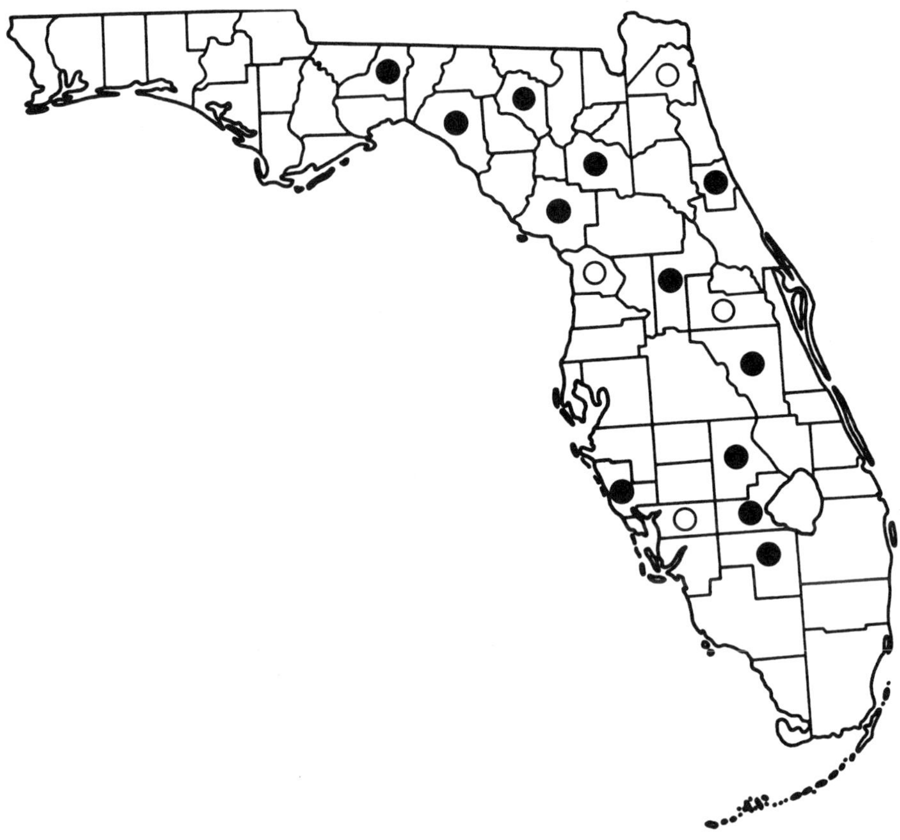

FIGURE 17.61. Distribution of coccidial infections (*Eimeria* spp.) in Wild Turkeys in Florida, 1969–99. *Closed circles* = counties where infected turkeys have been found; *open circles* = counties in which turkeys have been examined and found free of coccidia. Adapted using data from Akey 1981; Forrester 1992b; Forrester and Todd 1971; Dusenbury and Spalding 1999.

Hon et al. 1978; Rickard and Forrester 1984; Rickard 1985). Although both species of sowbug have been found to function as intermediate hosts under experimental laboratory conditions, only *O. floridana* has been found infected naturally in Wild Turkey habitat (Hon and Forrester 1972). In April 1972, a collection of 2,257 sowbugs (*O. floridana*) from Fisheating Creek was examined for infective larvae of *D. nasuta;* the prevalence of larvae was 1 per 133 sowbugs (Hon 1973). Identification of the larvae was confirmed by infection of a domestic poult and the establishment of a patent infection of *D. nasuta.* Judging from the results

of experimental infections in the laboratory (Hon 1973, Rickard and Forrester 1984), *O. floridana* may be the most suitable and important intermediate host of the 2, even though *V. evergladensis* is more common, at least at Fisheating Creek (Rickard and Forrester 1984).

Hon (1973) showed that larvae of *D. nasuta* could survive up to 6 months in the sowbug intermediate host under laboratory conditions. If this is also true for larvae in sowbugs in nature, it would explain how the parasite could survive through the winter until new susceptible turkey poults enter the population. Infection of sowbugs and turkeys probably occurs during times

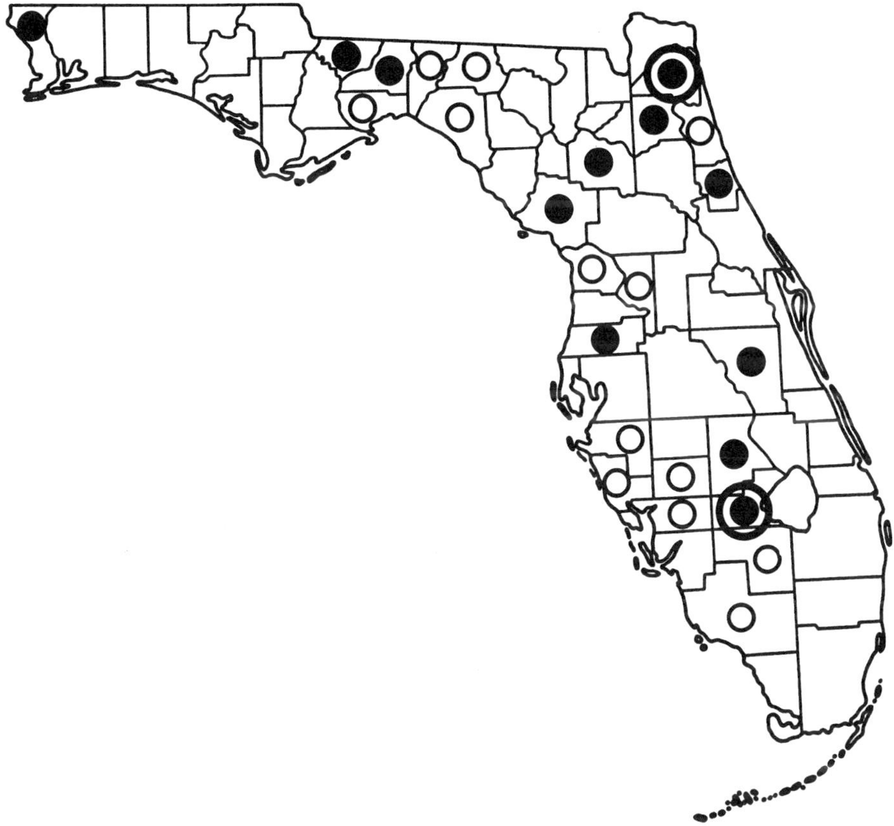

FIGURE 17.62. Distribution of histomonosis cases and the cecal nematode vector *Heterakis gallinarum* in Wild Turkeys in Florida, 1960–89. *Double circles* = counties where turkeys have been found with histomonosis and *H. gallinarum; closed circles* = counties with turkeys positive for *H. gallinarum* only; *open circles* = counties where neither *H. gallinarum* nor histomonosis have been found. Adapted using data from Maxfield et al. 1963; Akey and Forrester 1981; Forrester 1992a, 1992b; Hon and Forrester 1972.

when the sowbugs are most active, and thereby more exposed to turkey feces and to feeding birds. At Fisheating Creek this has been observed to occur from 1 hour before to 1 hour after sunset and during rain showers (Rickard and Forrester 1984). Even though weather with high humidity promotes activity of sowbugs, too much rainfall can decrease transmission. Hon et al. (1978) reported that in years of heavy rainfall and subsequent high water levels in the cypress swamps at Fisheating Creek, the prevalence and intensity of infections of poults by *D. nasuta* was decreased (figure 17.76). A

similar trend was observed also in poults in 1980 and 1981, 2 sequential wet and dry years at Fisheating Creek (Akey 1981). It was postulated that this was because sowbugs migrate vertically on cypress tree trunks and survive under loose bark above the water line when the swamp floor is flooded, making them inaccessible to foraging poults. In drier times the sowbugs are found on and around the bases of cypress trees, under debris and cattle feces, and on the open ground, putting them in easy reach of turkey poults (Hon et al. 1978).

Another important aspect of the epizootiol-

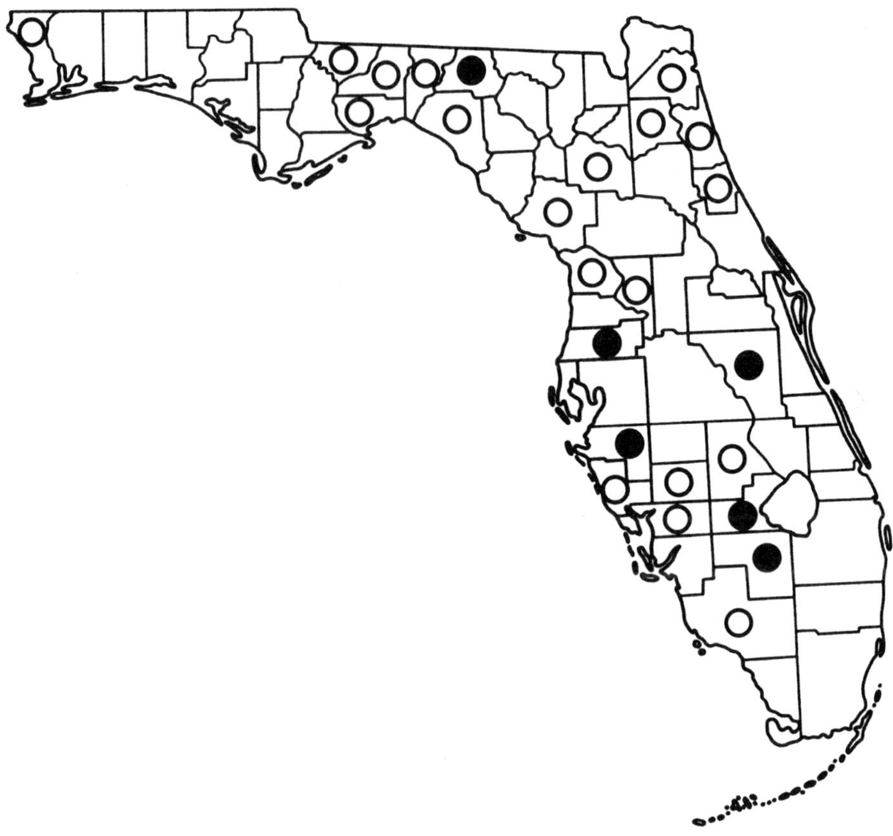

FIGURE 17.63. Distribution of the intestinal fluke *Echinoparyphium recurvatum* in Wild Turkeys in Florida, 1960–84. *Closed circles* = counties where infected turkeys have been found; *open circles* = counties in which turkeys have been examined and found free of *E. recurvatum*. Adapted using data from Akey and Forrester 1981; Forrester 1992b; Hon and Forrester 1972; Maxfield et al. 1963.

ogy of *D. nasuta* in Wild Turkeys concerns reservoir hosts. In addition to Wild Turkeys, at least 22 other avian species are known to harbor infections in Florida. These include another galliform, the Northern Bobwhite, and various members of 7 additional orders (table 17.51). Several passeriforms appear to be important, including Blue Jays, Carolina Wrens, Northern Mockingbirds, American Robins, and Boat-tailed Grackles. American Kestrels and Mourning Doves may also be important. Patent infections in these birds undoubtedly contribute to the maintenance of the parasite in sowbug populations in Wild Turkey habitats in Florida and

to the distribution of the nematode throughout the state.

The lesions caused by *D. nasuta* in birds have been described (Cram 1928; Goble and Kutz 1945; Bendell 1955; Hwang et al. 1961; Ruff 1978; Rickard 1985). The nematodes are usually found with their anterior ends buried in the mucosa and their posterior ends coiled in the lumen of the proventriculus (Ruff 1978; Rickard 1985). Infected proventriculi are often swollen, some as much as 3 times normal. "Raised lesions with multifocal petechial hemorrhages, excess mucus, mononuclear infiltrate, and epithelial desquamation" was a consistent finding of Rickard (1985), who studied *Dispharynx* infections in a number of Florida birds including domestic and Wild Turkeys. The reader is referred to Rickard's excellent paper (1985) for further details on the lesions caused by *D. nasuta* in natural and experimental infections. Although the possible effects of *D. nasuta* on turkey populations are discussed further in section XXIII, The Fisheating Creek study, there is little information available on this topic. Dead or dying poults are difficult to locate in Florida because of the climate and habitat and because of the activities of predators and scavengers. During the 1969–72 Fisheating Creek study reported by Hon et al. (1975), 1 sick 8–12-day-old poult was found weakened and unable to fly. It weighed 105 gm at necropsy. Eighteen *D. nasuta* were found in the proventriculus and were believed to have contributed to the poult's debilitation. Experimental studies are needed to determine how many larvae of *D. nasuta* are needed to cause morbidity and mortality in Wild Turkeys.

The other species of gastrointestinal nematodes are probably of little or no significance as pathogens (Davidson and Wentworth 1992). Certainly infections with large numbers of any of them could be harmful to individual turkeys, but there is no evidence that they are important from a population standpoint at this time. The cecal nematode *Heterakis gallinarum* might be an exception because of its role as the vector of *Histomonas meleagridis,* the etiologic agent of histomonosis. However, even though *H. gallinarum* is fairly widespread in Florida (figure 17.62), only 3 cases of histomonosis have been recognized in Wild Turkeys in the state during the past 25 years, which indicates that this disease is not important in Florida Wild Turkeys at the population level.

XX. Filarial worms

Four species of filarial nematodes have been reported from Wild Turkeys in the United States

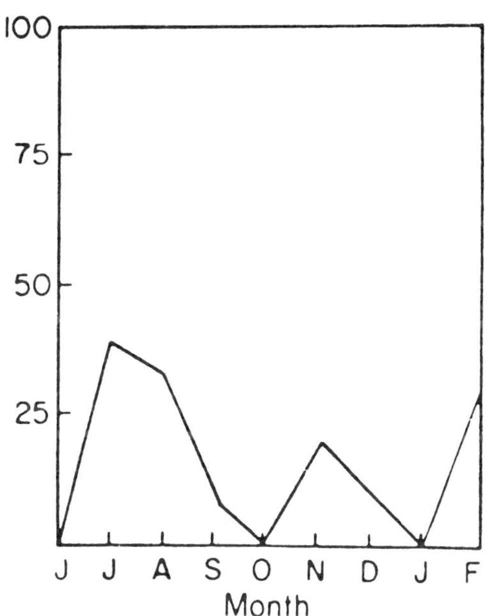

FIGURE 17.64. Seasonal prevalence of the intestinal fluke *Echinoparyphium recurvatum* in Wild Turkey poults and juveniles at Fisheating Creek (Glades County), June 1970–February 1971. Y axis represents prevalence (% infected). From Hon et al. 1978; by permission of *Proceedings of the Helminthological Society of Washington.*

(Hon et al. 1975; Davidson and Wentworth 1992). All 4 occur in turkeys in Florida in low prevalences and intensities (table 17.58). The characteristics of these infections and their distributions are given in tables 17.59–17.61. The values reported for 2 of these filarids (*Singhfilaria hayesi* and *Chandlerella* sp.) should be considered minimal since the adults are found in subcutaneous and connective tissues and are difficult to locate and easily overlooked at necropsy. The life cycles of all 4 species are unknown, but transmission is most likely via a blood-sucking arthropod, as is the case with other species in this group of nematodes. Nothing is known about the pathologic significance of these parasites in Wild Turkeys.

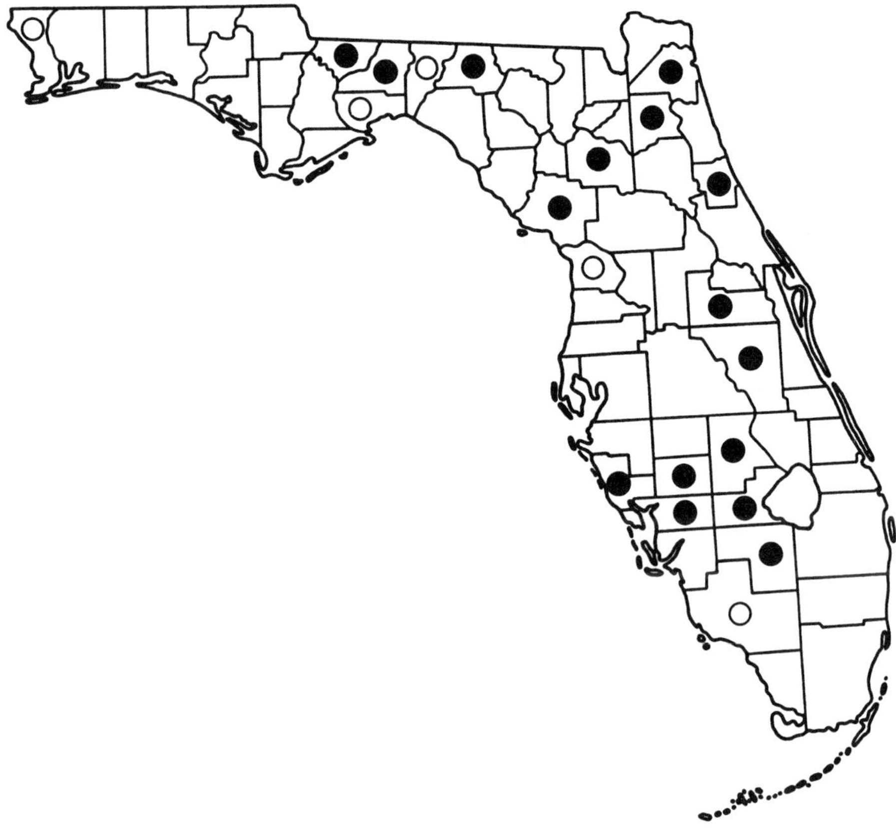

FIGURE 17.65. Distribution of the tapeworm *Metroliasthes lucida* in Wild Turkeys in Florida, 1960–83. *Closed circles* = counties where infected turkeys have been found; *open circles* = counties in which turkeys have been examined and found free of *M. lucida*. Adapted using data from Akey and Forrester 1981; Forrester 1992b; Hon and Forrester 1972; Maxfield et al. 1963.

XXI. Acanthocephalans

Acanthocephalans (also known as spiny-headed or thorny-headed worms) have been found in Wild Turkeys in Florida on 2 occasions (Hon et al. 1975; Forrester 1992b). Both cases were single infections in turkeys from Fisheating Creek. One specimen could not be identified because the proboscis was retracted; the other was an immature male of *Mediorhynchus papillosum,* which has been reported from several species of passeriforms and other avian hosts (Hon et al. 1975). These infections probably occurred accidentally when an arthropod intermediate host was ingested and are of little or no consequence to turkeys.

XXII. Arthropods

A number of species of arthropods has been reported as ectoparasites of Wild Turkeys in the United States. Davidson and Wentworth (1992) listed ticks (4 species), mites (4), chewing lice (8), and louse flies (3) and mentioned that several other flies (i.e., mosquitoes, midges, and blackflies) are known to be intermittent parasites. These are important as vectors of viruses

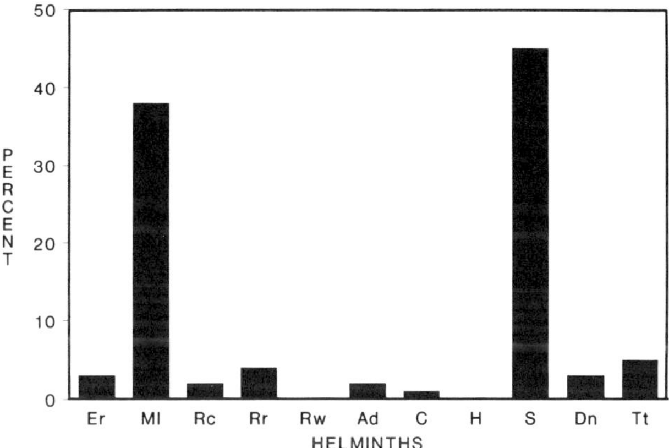

FIGURE 17.66. Helminth profile for Wild Turkeys from Fisheating Creek (Glades County), 1969–72. The Y axis is the percentage of the total number of helminths in all turkeys sampled represented by each species. Er = *Echinoparyphium recurvatum*; Ml = *Metroliasthes lucida*; Rc = *Raillietina cesticillus*; Rr = *Raillietina ransomi*; Rw = *Raillietina williamsi*; Ad = *Ascaridia dissimilis*; C = *Capillaria* spp.; H = *Heterakis gallinarum*; S = *Strongyloides* sp.; Dn = *Dispharynx nasuta*; Tt = *Trichostrongylus cramae* (= *T. tenuis*). From Hon et al. 1975; by permission of *Proceedings of the Helminthological Society of Washington*.

and blood protozoans, as discussed in previous sections.

In Florida 9 species of arthropods have been found on Wild Turkeys and include 1 tick, 4 feather mites, and 4 chewing lice. The lone star tick (*Amblyomma americanum*) has been found on Wild Turkeys from several localities in northern Florida (table 17.62), but is not a common parasite of Wild Turkeys in the state. Nymphs were the most numerous stage, although larvae and adults were seen also on a few occasions. The lone star tick is the most frequently reported tick on Wild Turkeys in other parts of the southeastern United States (Kellogg et al. 1969, Davidson and Wentworth 1992). It has a wide host range, with adults feeding primarily on large mammals and larvae and nymphs preferring birds and small mammals (Strickland et al. 1976). Although massive numbers of larvae of *A. americanum* can kill chickens (Strickland et al. 1976), this may not be a problem in Wild Turkeys because infestations are of low intensity. This tick transmits the etiologic agents of Q fever and Rocky Mountain spotted fever and can cause tick paralysis (Strickland et al. 1976); it is therefore of zoonotic importance. Jacobson

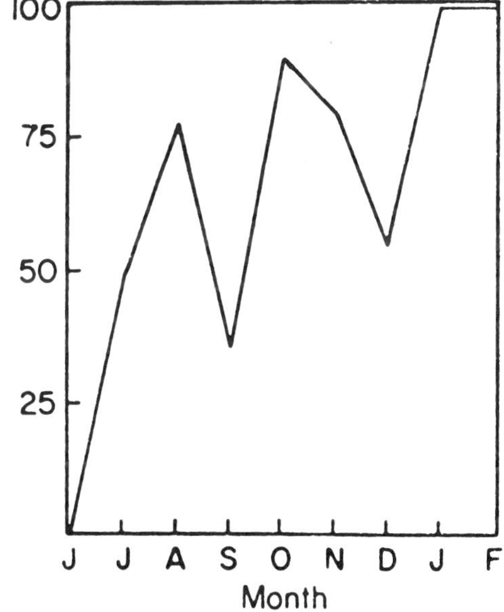

FIGURE 17.67. Seasonal prevalence of the tapeworm *Metroliasthes lucida* in Wild Turkey poults and juveniles at Fisheating Creek (Glades County), June 1970–February 1971. Y axis represents prevalence (% infected). From Hon et al. 1978; by permission of *Proceedings of the Helminthological Society of Washington*.

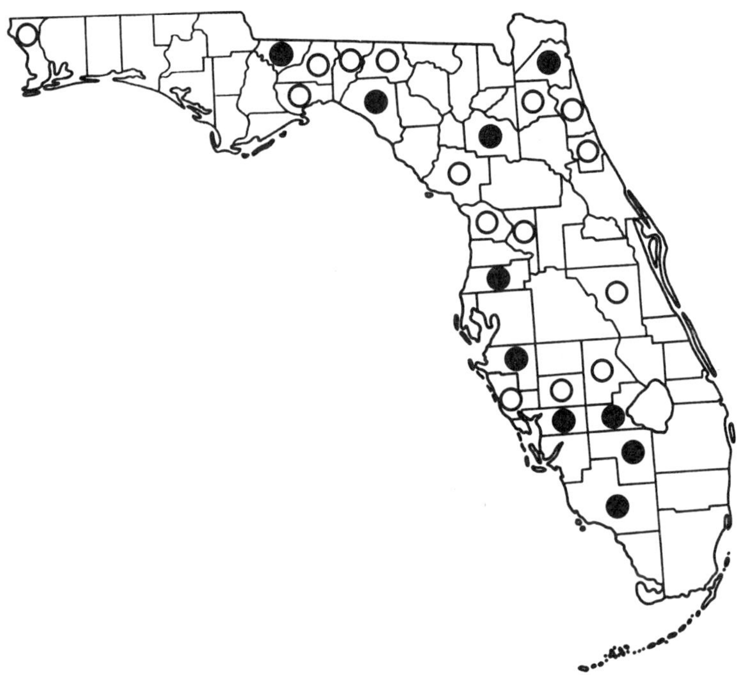

FIGURE 17.68. Distribution of the intestinal threadworm *Strongyloides* sp. in Wild Turkeys in Florida, 1960–84. *Closed circles* = counties where infected turkeys have been found; *open circles* = counties in which turkeys have been examined and found free of *Strongyloides* sp. Adapted from Akey and Forrester 1981; Forrester 1992b; Hon and Forrester 1972; Maxfield et al. 1963.

and Hurst (1979) reported that prescribed burning in Mississippi resulted in a reduction in infestation of Wild Turkey poults by this tick. For further information on the effects of burning on populations of *A. americanum*, the reader is referred to Davidson et al. (1994).

Four species of feather mites have been found on Wild Turkeys in Florida (table 17.63). The most prevalent species was *Megninia ginglymura*, occurring on 94% of 18 infested Wild Turkeys from 7 counties. These feather mites are not believed to be harmful to Wild Turkeys (Davidson and Wentworth 1992).

The most numerous and common ectoparasites of Wild Turkeys in Florida are chewing lice. Four species are known from turkeys in 15 counties throughout the state (table 17.64). *Chelopistes meleagridis* was the most numerous louse on infested birds (table 17.65). Wild Turkey poults as young as 5 days were found to

have lice, in a survey conducted at Fisheating Creek (table 17.66). Most likely the poults acquire chewing lice from the hens during the brooding process. Lice are not usually pathogenic to Wild Turkeys, but their presence in large numbers might indicate that another more serious problem is present because high intensities are often characteristic of a medically compromised host (Davidson and Wentworth 1992). We have seen several instances in Florida in which turkeys with avian pox or lymphoproliferative disease (or both) had louse infestations of high intensity.

XXIII. The Fisheating Creek study

A 15-year ecological study of the parasites and diseases of Wild Turkeys was conducted at Fisheating Creek Wildlife Management Area and

Refuge in Glades County between 1969 and 1984. This investigation was initiated by the Florida Game and Fresh Water Fish Commission (now called the Florida Fish and Wildlife Conservation Commission) in response to the observation that there had been a significant statewide decline in turkey populations during the mid to late 1960s (Williams and Austin 1988). The disease research was conducted concurrently and in cooperation with a study by Lovett E. Williams, Jr., and David H. Austin and their associates on various aspects of the nesting behavior and reproductive performance of Wild Turkeys at Fisheating Creek. The latter research has been reported in numerous published reports and was summarized in *Studies of the Wild Turkey in Florida* by Williams and Austin (1988). It is fortunate that this intensive and broad-based ecological study of turkeys was conducted at Fisheating Creek during that time, because that permitted ecological aspects and the significance of parasites and diseases to be investigated. The results of a number of disease studies at Fisheating Creek have been published in a variety of technical journal articles; many of these have been cited previously in this chapter along with other unpublished material. An attempt will be made in the present section to synthesize certain results of the 15-year Fisheating Creek study and to give a general hypothesis concerning the role and significance of parasites and diseases in the health and population status of Wild Turkeys at this location specifically and throughout Florida in general.

During the first several years of the study, and to a lesser degree throughout the entire 15-year period, an attempt was made to determine which parasites and disease agents were present in the turkey population at Fisheating Creek. This effort resulted in the discovery of a number of different infectious and parasitic disease agents. As time passed the list grew to at least 89 agents, including viruses, bacteria, fungi, protozoans, helminths, and arthropods (table 17.67). It was recognized that many of these agents have the potential to cause morbidity and mortality under appropriate ecological conditions and undoubtedly there are im-

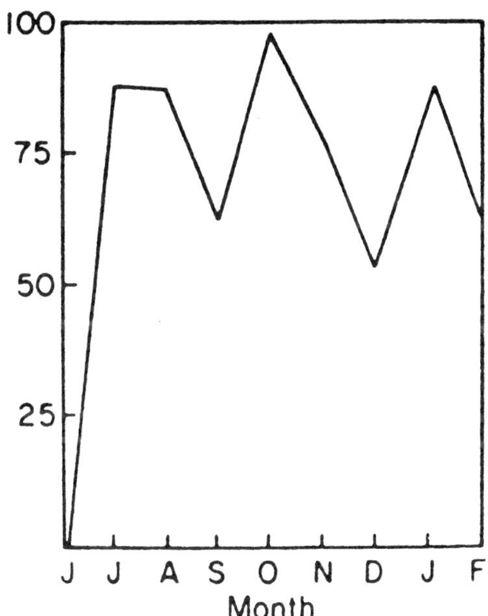

FIGURE 17.69. Seasonal prevalence of the intestinal threadworm *Strongyloides* sp. in Wild Turkey poults and juveniles at Fisheating Creek (Glades County), June 1970–February 1971. Y axis represents prevalence (% infected). From Hon et al. 1978; by permission of *Proceedings of the Helminthological Society of Washington.*

portant synergistic effects of some of these agents in concurrent infections, as previously discussed for poxvirus and malaria.

Since Williams and his associates noted that 30–50% of the poults disappeared from the population during the first several weeks after hatching, a decision was made to put special emphasis on the epizootiology and significance of the parasites and diseases of poults during their early life. Eventually it became clear that 5 agents (a virus, a nematode, and 3 species of blood protozoans) might be important pathogens, either individually or in concurrent infections. These included avian poxvirus, the spiruroid nematode *Dispharynx nasuta,* and 3 blood protozoans (*Haemoproteus meleagridis, Leucocytozoon smithi,* and *Plasmodium hermani*). All 5 have characteris-

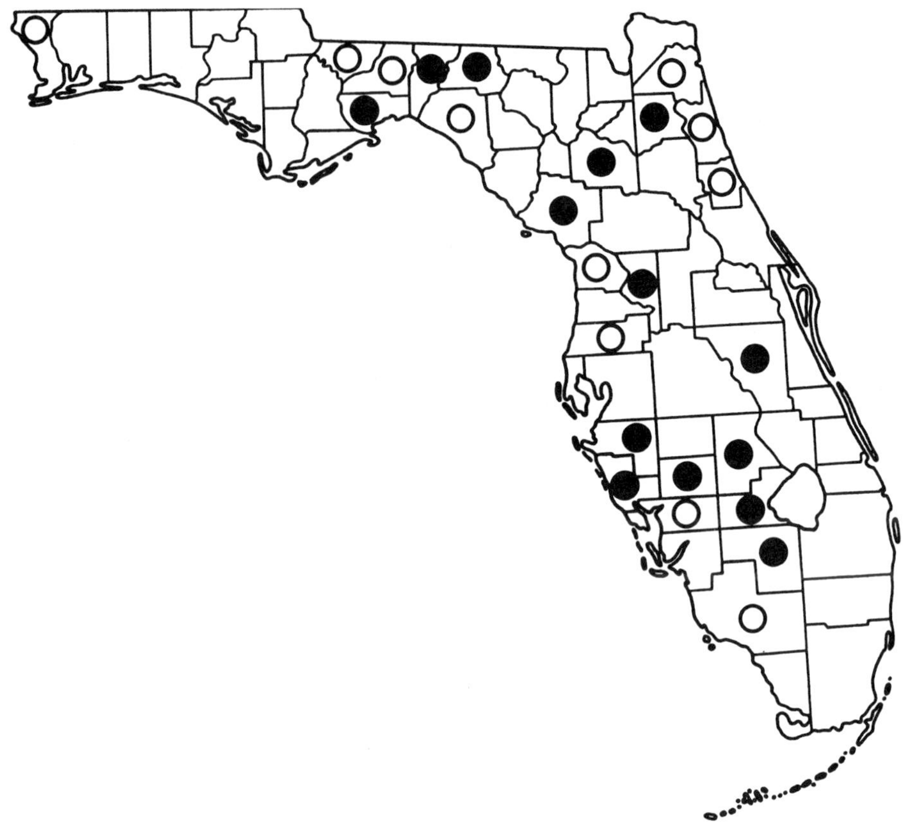

FIGURE 17.70. Distribution of the cecal nematode *Trichostrongylus cramae* (= *T. tenuis*) in Wild Turkeys in Florida, 1960–84. *Closed circles* = counties where infected turkeys have been found; *open circles* = counties in which turkeys have been examined and found free of *T. cramae*. Adapted from Akey and Forrester 1981; Forrester 1992b; Hon and Forrester 1972; Maxfield et al. 1963.

tics in common that contribute to their significance as important pathogens from a population standpoint. These include: (1) occur at high prevalences in some cohort of the turkey population, (2) infect young poults during the first several weeks or months of life, (3) are pathogenic in individual infections (and in the case of concurrent infections of poxvirus and malaria, the effects are additive), and (4) utilize life cycles that involve intermediate hosts or vectors that occur commonly at Fisheating Creek and whose development and activities are linked closely with rainfall and water levels in the creek swamp.

The nesting season of Wild Turkeys at Fisheating Creek begins in March when egg laying commences (figure 17.77). The majority (89%) of the nests are located in the saw palmetto prairie/grazed glade ecotone (58%) or the cypress woods (31%); the remaining nests (11%) are located in a number of other habitat types (Williams and Austin 1988). These 2 main nesting habitats are illustrated in figure 17.78 and make up 5% and 51% of the habitats in the study area, respectively. Hatching begins in April and ends in July, with the peak of hatching occurring sometime in May (figure 17.77). Shortly after hatching, hens take their

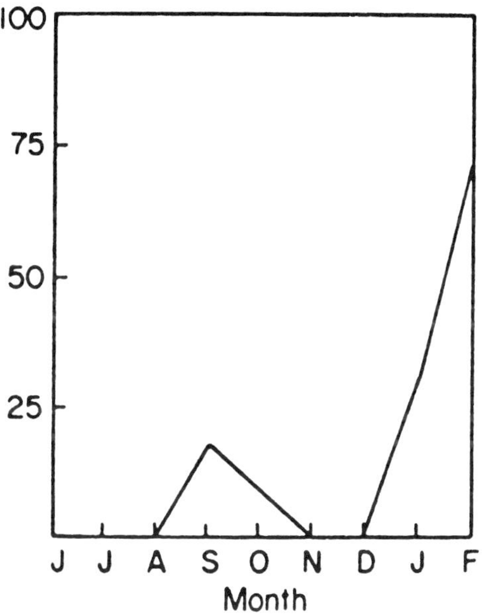

FIGURE 17.71. Seasonal prevalence of the cecal nematode *Trichostrongylus cramae* (= *T. tenuis*) in Wild Turkey poults and juveniles at Fisheating Creek (Glades County), June 1970–February 1971. Y axis represents prevalence (% infected). From Hon et al. 1978; by permission of *Proceedings of the Helminthological Society of Washington*.

broods into the cypress woods where they spend most of their time foraging and resting. For the first 12–19 days the hens roost with their poults on the ground at night, after which they roost in cypress trees. It is usually dry during the months that nesting and these early brood behavioral events occur. Normally the rainy season at Fisheating Creek is not until mid-summer to early fall, when the poults are about 4 months old. At that age they would be more likely to survive infections by blood parasites, avian poxviruses, or other disease agents because of the development of age resistance.

The following hypothetical scenario might explain how significant mortality of poults could occur at Fisheating Creek if ecological conditions varied from the norm. If the early part of the hatching and brood-rearing season were dry (i.e., normal), poults would have ready

access to sowbugs in the cypress woods and would become infected at a young age with the proventricular nematode *Dispharynx nasuta*. Then, if the rainy season came earlier than normal, such as in June or even May, causing accelerated reproduction and development of mosquitoes and other arthropod vectors, young poults would be exposed to the harmful effects of single or concurrent infections of avian pox, malaria, and infections by *Leucocytozoon smithi*, *Haemoproteus meleagridis* or other agents. Many of the poults could die as a result of the effects of these diseases (figure 17.79) or become weakened by them (figure 17.80) and die subsequently because of secondary effects of

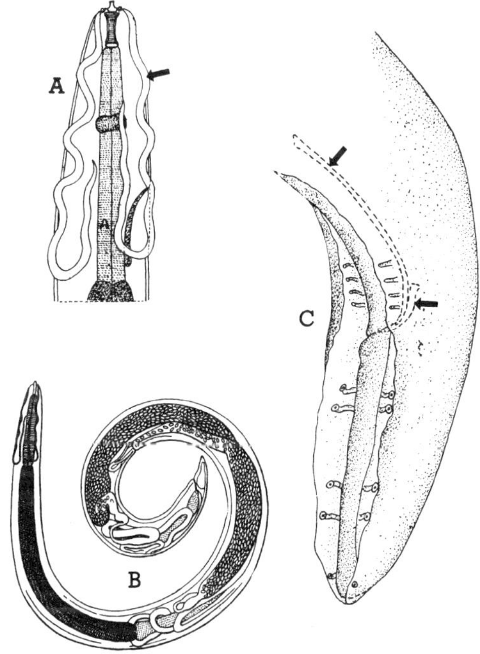

FIGURE 17.72. Line drawings of the proventricular nematode *Dispharynx nasuta* (from Ruff 1978; by permission of Iowa State University Press). *A*, anterior end of an adult showing characteristic cordons *(arrow)*, which are recurrent but do not anastomose (after Seurat 1916); *B*, adult female containing masses of embryonated eggs (after Piana 1897); *C*, posterior end of adult male showing unequal, dissimilar spicules *(arrow)* and pre- and postanal papillae (after Cram 1928).

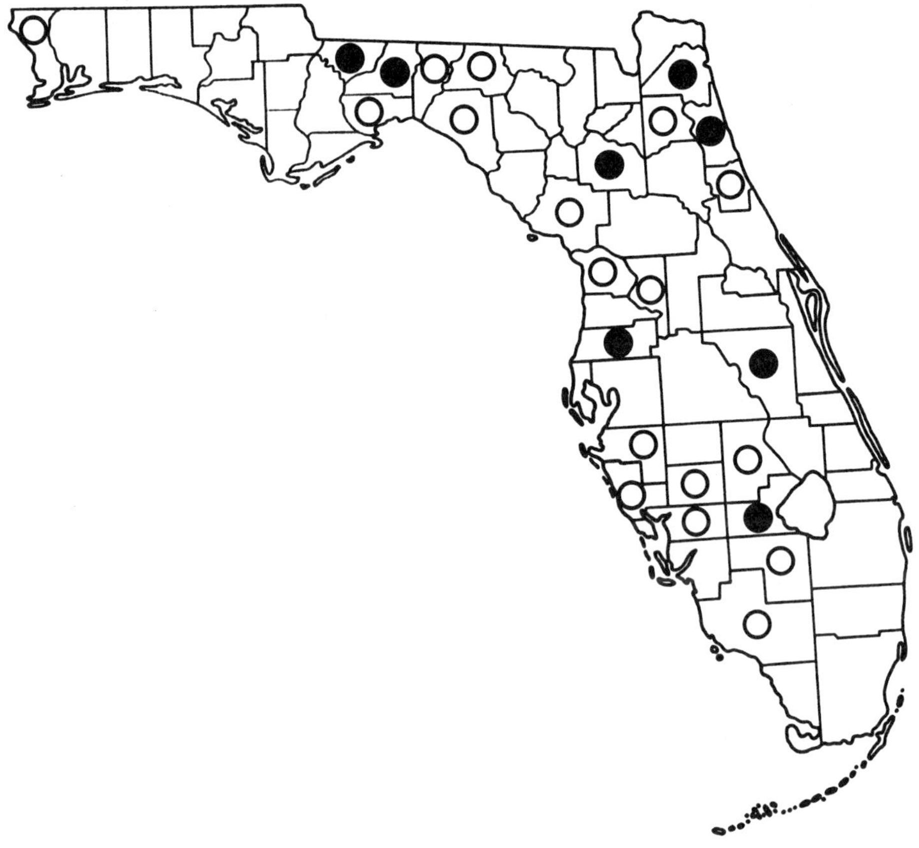

FIGURE 17.73. Distribution of the proventricular nematode *Dispharynx nasuta* in Wild Turkeys in Florida, 1960–84. *Closed circles* = counties where infected turkeys have been found; *open circles* = counties in which turkeys have been examined and found free of *D. nasuta*. Adapted from Akey and Forrester 1981; Forrester 1992b; Hon and Forrester 1972; Maxfield et al. 1963.

inclement weather or increased susceptibility to predation.

Based on the information available, it seems likely that this is what might have happened in the mid-1960s when the numbers of Wild Turkeys declined statewide. Rainfall during the April–July periods in 1965 and 1966 was abnormally high compared with the 50-year norm (figure 17.81), which may have caused the sequence of events outlined above, resulting in 2 sequential years of abnormally high poult mortality. This then would be reflected in poor harvests by hunters during the fall hunts when there were high amounts of rainfall in the preceding spring months, similar to that recorded by Powell (1965) in the 1950s and illustrated in figure 17.1. This hypothesis needs to be tested with field data obtained by radiotelemetry or other techniques, in order to determine the complete story about the effects of diseases, parasites, and weather on Wild Turkey populations in Florida.

XXIV. Disease risks from pen-reared Wild Turkeys

The problems associated with the release of pen-reared or game-farm Wild Turkeys for restocking purposes were reviewed by Davidson and Wentworth (1992). The practice incurs significant risks for 2 main reasons: first, pen-reared turkeys are genetically inferior; interbreeding with native Wild Turkeys results in the pollution of the gene pools and the loss of wildness (Leopold 1944; Lewis 1987). Second, pen-reared turkeys are potential sources of diseases that could be introduced to indigenous Wild Turkeys.

Schorr et al. (1988) conducted a study of 119 pen-reared Wild Turkeys obtained from 12 different game-farm facilities in 9 eastern states. These birds were examined and tested for a variety of pathogens, resulting in the identification of 33 species of parasites and 7 infectious disease agents. Of these, the agents of most concern were avian poxvirus, *Mycoplasma gallisepticum*, *Histomonas meleagridis*, *Syngamus trachea* (a nematode that occurs in the trachea), and *Salmonella* spp., since these would be potential threats to the health of native Wild Turkeys as well as domestic poultry.

Powell (1965) related an incident that occurred in 1958 in Volusia County where a group of sportsmen had released pen-reared Wild Turkeys. A photograph of a Wild Turkey

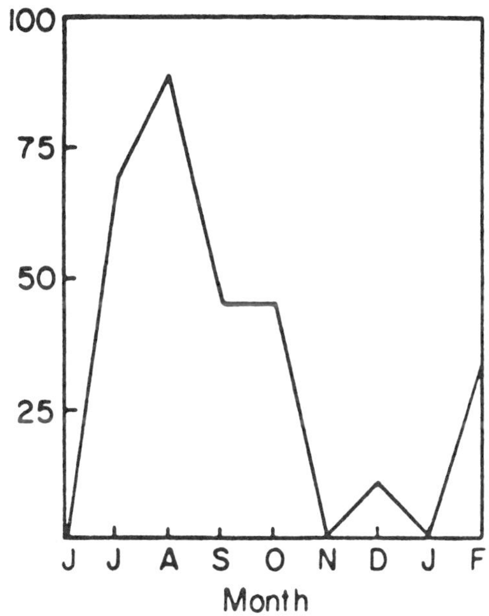

FIGURE 17.74. Seasonal prevalence of the proventricular nematode *Dispharynx nasuta* in Wild Turkey poults and juveniles at Fisheating Creek (Glades County), June 1970–February 1971. Y axis represents prevalence (% infected). From Hon et al. 1978; by permission of *Proceedings of the Helminthological Society of Washington*.

FIGURE 17.75. Sowbugs (*Venezillo evergladensis*), which serve as intermediate hosts of the proventricular nematode *Dispharynx nasuta* in Wild Turkeys in Florida. Courtesy of Lora G. Rickard.

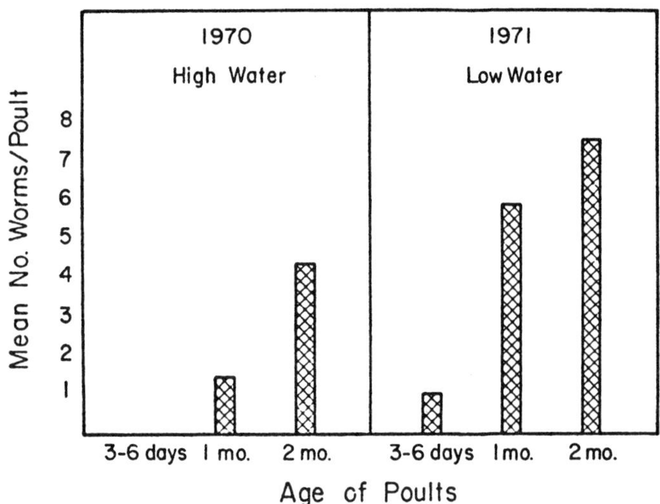

FIGURE 17.76. Intensity of infections by the proventricular nematode *Dispharynx nasuta* in Wild Turkey poults and juveniles at Fisheating Creek (Glades County) in 1970 (a year of high water in the creek swamp) and 1971 (a year of low water). From Hon et al. 1978; by permission of *Proceedings of the Helminthological Society of Washington*.

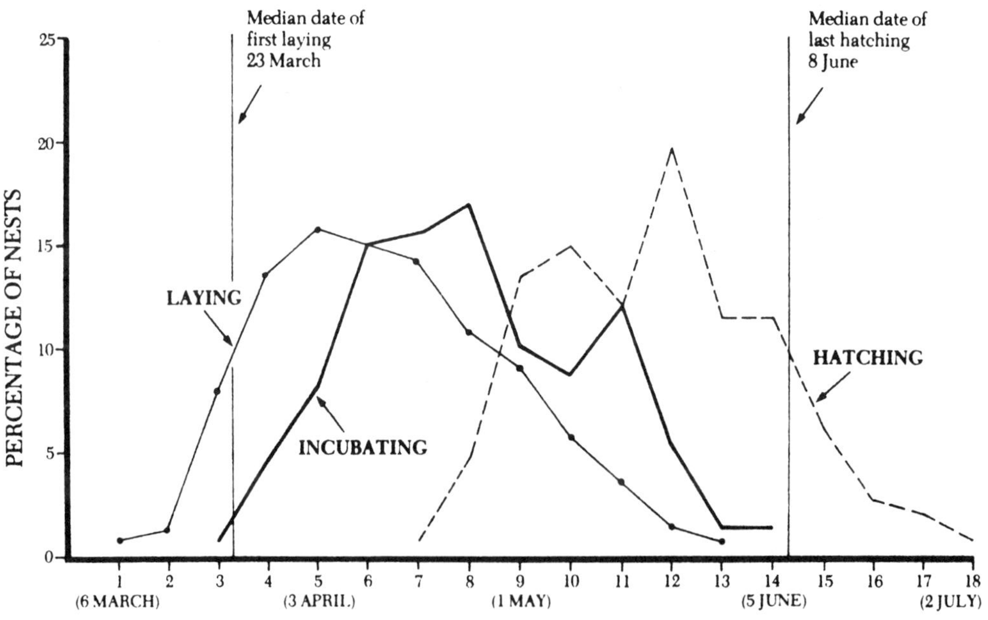

WEEK OF NESTING SEASON (AND CALENDAR DATE)

FIGURE 17.77. Graphic representation of the nesting season of Wild Turkey hens at Fisheating Creek (Glades County) and Lochloosa Wildlife Management Areas (Alachua County). From Williams and Austin 1988; by permission of University Press of Florida.

FIGURE 17.78. Aerial view of the Fisheating Creek study area in Glades County, showing principal plant associations. The area labeled "Prairie" is referred to as saw palmetto prairie in the text. From Williams and Austin 1988; by permission of University Press of Florida.

Figure 17.79. Dead Wild Turkey poult, Fisheating Creek (Glades County). Courtesy of Lovett E. Williams, Jr.

from the release area depicted a severe case of avian pox and was included in Powell's bulletin along with the statement that harvests of turkeys in that area had drastically decreased from between 60 and 70 each year before the release to only 2 after the release of the pen-reared turkeys. He implied that the poxvirus was introduced by the pen-reared birds; although this was not confirmed, it is certainly a real possibility. Schorr et al. (1988) did not include any samples of pen-reared turkeys from Florida in their study, but several such birds were examined by Forrester (1992b) in the early 1970s. The first 2 cases were from a game-farm in Bradford County in 1970 and were diagnosed as having avian pox. It was reported that 15 or so other turkeys along with several domestic chickens had lesions, but no mortality was noted. The second case was from a hunting preserve in Gadsden County in 1971 and involved a pen-reared poult that died of histomonosis.

The genetic and health risks associated with the practice of releasing pen-reared turkeys are obvious and such practices should be discouraged or prohibited. The use of live-trapped native Wild Turkeys to restock areas where Wild Turkeys occurred historically, but no longer occur, has been proven to be a sound and successful management practice and should be encouraged.

XXV. Summary and conclusions

Excluding trauma, Wild Turkeys in Florida are hosts to at least 134 different parasites, infectious disease agents, conditions, and morbidity/mortality factors. Of these, 84 are parasites (including 6 protozoans, 10 trematodes, 7 cestodes, 19 nematodes, 1 acanthocephalan, and 41 arthropods), 5 are viruses, 35 are bacteria, 2 are fungi, 3 are neoplasms, and 5 are anomalies. Our knowledge of some of these is incomplete, whereas for others a considerable amount is known. Some, such as avian poxvirus and the proventricular nematode *Dispharynx nasuta,* are primary pathogens, whereas others are either secondary pathogens or relatively nonpathogenic organisms or conditions. Others usually do not cause harmful effects directly, but when they occur concurrently with other disease agents the result can be heightened morbidity or mortality. An example would be the synergistic effects of avian pox and malaria (*Plasmodium hermani*) on poults. Other infections such as those caused by blood parasites or gastrointestinal nematodes can result in retarded growth and weakness that heightens the susceptibility of young turkeys to inclement weather, other diseases, or predation.

Under normal ecological conditions, Wild Turkeys in Florida appear to be well adapted to survival despite the many parasites and disease agents that are present within their populations. Some of these agents cause a limited amount of morbidity and mortality, but normally these effects are minimal at the population level. Examples include avian pox, lymphoproliferative disease, haemoproteid myositis, histomonosis, and dispharynxosis. Probably the most important of these is avian pox. There is some evidence, however, that in some years when rainfall and water levels in turkey habitats during the nesting season (April, May, and June) are abnormally high, arthropod-borne or -transmitted diseases (such as dispharynxosis, pox, malaria, and other blood parasites) can have significant population effects by decreasing poult survival. Such an event seems to have occurred in Florida during the mid to late 1960s, although definitive evidence for such a phenomenon is unconfirmed.

FIGURE 17.80. Sick Wild Turkey poult found in the cypress swamp at Fisheating Creek (Glades County). Courtesy of Lovett E. Williams, Jr.

With the exception of several of the bacteria (such as *Salmonella* spp. and *Mycobacterium* sp.), EEE virus, SLE virus, and West Nile virus, most of the parasites and diseases of Wild Turkeys are very host specific and of little public health concern. Virtually every one of them, however, will infect domestic turkeys and therefore there is the possibility of disease agents passing from Wild Turkeys to domestic turkeys, or the reverse, when the 2 populations come in contact. Fortunately there is little evidence that this is occurring to any significant degree in Florida, with the exception of several cases of avian pox and histomonosis in Wild Turkeys that seem to have been acquired through contact with domestic poultry.

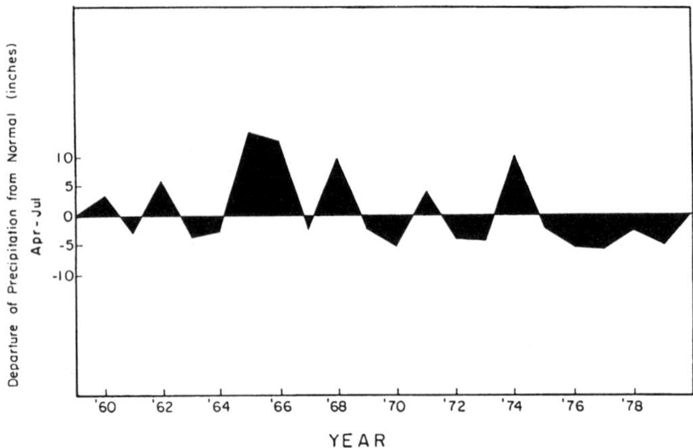

FIGURE 17.81. Departures of rainfall from the 50-year norm for Palmdale, Florida, near Fisheating Creek (Glades County), during April, May, June, and July each year from 1959 to 1979. From Forrester 1991; by permission of *Bulletin of the Society of Vector Ecology*.

Table 17.39. Distribution, prevalence, and intensity of infections of the cloacal flukes *Prosthogonimus ovatus* and *Stomylotrema vicarium* in Wild Turkeys from Florida

		No. of turkeys			Intensity	
County	Year(s)	Exam.	Pos.	%	Mean	Range
Prosthogonimus ovatus						
Duval	1971	1	0	—	—	—
	1972	2	1	—	2	2
	1974–81	6	0	—	—	—
Glades	1960–61	12	0	0	—	—
	1968–69	17	0	0	—	—
	1970	81	2	2	2	1–2
	1971	65	3	5	1	1
	1972–84	177	0	0	—	—
Others[a]	1960–84	135	0	0	—	—
Stomylotrema vicarium						
Glades	1960–61	12	0	0	—	—
	1968–69	17	0	0	—	—
	1970	81	1	1	1	—
	1971	65	11	17	3	1–9
	1972	16	1	6	1	1
	1973	6	0	—	—	—
	1974	11	1	9	1	1
	1975–84	144	0	0	—	—
Others[b]	1960–84	148	0	0	—	—

Sources: 1960–61 data from Maxfield et al. (1963), 1968–72 data from Hon and Forrester (1972), 1980–81 Glades County data from Akey and Forrester (1981), remainder of data from Forrester (1992b).
a. Includes negative data from 23 other counties; sample sizes are presented in Table 17.35, except that only 4 turkeys were examined from Osceola County in 1970.
b. Includes negative data from 24 other counties; sample sizes are presented in Table 17.35, except that only 4 turkeys were examined from Osceola County in 1970.

Table 17.40. Distribution, prevalence, and intensity of infections of the kidney fluke *Tanaisia* sp. and the liver fluke *Zonorchis* sp. in Wild Turkeys from Florida

| County | Year(s) | No. of turkeys | | | Intensity | |
		Exam.	Pos.	%	Mean	Range
Tanaisia sp.						
Glades	1968–70	98	0	0	—	—
	1971	65	1	2	3	3
	1972–84	189	0	0	—	—
Others[a]	1960–84	144	0	0	—	—
Zonorchis sp.						
Osceola	1970–71	17	0	0	—	—
	1972	3	1	—	1	1
	1975–76	3	0	—	—	—
Others[a]	1960–84	473	0	0	—	—

Sources: 1960–61 data from Maxfield et al. (1963), 1968–72 data from Hon and Forrester (1972), 1980–81 Glades County data from Akey and Forrester (1981), remainder of data from Forrester (1992b).
a. Includes negative data from 23 other counties; sample sizes are presented in Table 17.35, except that only 4 turkeys were examined from Osceola County in 1970.

Table 17.41. Prevalence, intensity, and abundance of tapeworms reported from Wild Turkeys in Florida, 1960–84

| Species of tapeworm | Number of turkeys | | Intensity | | Abundance |
	Exam.	Prevalence %	Mean	Range	
Metroliasthes lucida	426	63	20	1–453	13
Raillietina georgiensis	388	22	45	1–340	10
Raillietina ransomi	388	8	12	1–73	1
Davainea meleagridis	388	6	10	3–28	1
Raillietina cesticillus	388	3	11	1–39	<1
Hymenolepis carioca	388	1	1	—	<1
Raillietina williamsi	388	<1	NG	NG	—

Sources: Akey and Forrester (1981), Forrester (1992b), Hon and Forrester (1972), Maxfield et al. (1963).
NG = not given.

Table 17.42. Distribution, prevalence, and intensity of infections of the tapeworm *Metroliasthes lucida* in Wild Turkeys from Florida

County or site	Year(s)	No. turkeys			Intensity	
		Exam.	Pos.	%	Mean	Range
Alachua	1969–72	18	8	44	4	1–9
	1975–80	2	0	—	—	—
Charlotte	1970	1	1	—	13	—
Clay	1960–61	11	7	64	NG	NG
DeSoto	1978	1	1	—	31	—
Duval	1971–72	2	2	—	7	6–7
	1977	2	2	—	34	13–54
Flagler	1969	4	4	—	7	3–12
Gadsden	1971, 73	2	2	—	6	1–10
Glades	1960–61	12	7	58	NG	NG
	1968	1	0	—	—	—
	1969	13	8	62	8	1–56
	1970	81	43	53	49	1–453
	1971	64	28	44	28	2–190
	1972	16	14	88	23	1–66
	1973–74	8	1	—	30	—
	1975	12	2	17	22	2–42
	1977–78	43	41	95	26	1–171
	1980	26	22	85	36	5–91
	1981	26	22	85	57	11–130
Hendry	1970	1	1	—	3	—
	1974	1	1	—	22	—
	1976	1	0	—	—	—
Highlands	1960–61	16	13	81	NG	NG
	1983	1	0	—	—	—
Leon	1971	2	1	—	2	—
Levy	1960–61	8	1	—	NG	NG
	1977–78	2	0	—	—	—
Madison	1971	1	1	—	12	—
Orange	1970	1	1	—	1	—
Osceola	1970	7	6	—	43	3–105
	1971	9	8	—	36	1–100
	1972	3	1	—	4	—
Sarasota	1970	1	1	—	1	-
"Florida"	1960–61	20	17	85	NG	NG
Others[a]	1971–84	7	0	—	—	—

Sources: 1960–61 data from Maxfield et al. (1963), 1968–72 data from Hon and Forrester (1972), 1980–81 Glades County data from Akey and Forrester (1981), remainder of data from Forrester (1992b).

NG = not given by authors.

a. Includes negative data from 5 other counties (sample sizes in parentheses): Citrus (1), Collier (1), Escambia (1), Jefferson (1), and Wakulla (3).

Table 17.43. Distribution, prevalence, and intensity of infections of the tapeworm *Raillietina georgiensis* in Wild Turkeys from Florida

County	Year(s)	No. of turkeys			Intensity	
		Exam.	Pos.	%	Mean	Range
Alachua	1969–72	18	13	72	97	1–311
	1975–80	2	0	—	—	—
Clay	1960–61	11	8	73	NG	NG
Duval	1971, 78	2	1	—	8	—
Glades	1960–61	12	0	0	—	—
	1969	13	1	8	1	—
	1970	81	7	9	6	1–22
	1971	65	3	5	1	1–2
	1972	16	1	6	5	—
	1974	7	0	—	—	—
	1975	10	1	10	1	—
	1977–78	15	0	0	—	—
	1980	26	8	31	6	1–16
	1981	26	15	58	11	1–28
Highlands	1960–61	16	3	19	NG	NG
	1983	1	0	—	—	—
Leon	1971	2	2	—	31	—
Levy	1960–61	8	1	—	NG	NG
	1977	1	0	—	—	—
Osceola	1970	7	6	—	37	3–56
	1971	8	8	—	34	1–106
	1972	3	3	—	276	171–340
Wakulla	1971	3	3	—	112	71–135
"Florida"	1960–61	20	3	15	NG	NG
Others[a]	1969–84	15	0	0	—	—

Sources: 1960–61 data from Maxfield et al. (1963), 1969–72 data from Hon and Forrester (1972), 1980–81 Glades County data from Akey and Forrester (1981), remainder of data from Forrester (1992b).
NG = not given by authors.
a. Includes negative data from 10 other counties (sample sizes in parentheses): Charlotte (1), Citrus (1), Collier (1), Escambia (1), Flagler (4), Gadsden (2), Hendry (2), Madison (1), Orange (1), and Sarasota (1).

Table 17.44. Distribution, prevalence, and intensity of infections of the tapeworm *Raillietina ransomi* in Wild Turkeys from Florida

County or site	Year(s)	No. of turkeys			Intensity	
		Exam.	Pos.	%	Mean	Range
Gadsden	1971	1	1	—	10	—
	1982	1	0	—	—	—
Glades	1960–61	12	0	0	—	—
	1969	13	1	8	3	—
	1970	81	4	5	9	2–25
	1971	65	5	8	19	1–73
	1972	16	6	38	18	1–52
	1974	7	0	—	—	—
	1975	10	0	0	—	—
	1977–78	15	0	0	—	—
	1980	26	2	8	4	1–6
	1981	26	2	8	7	2–12
Hendry	1970	1	1	—	12	—
	1976	1	0	—	—	—
Highlands	1960–61	16	1	6	NG	NG
	1983	1	0	—	—	—
Leon	1971	2	2	—	25	19–30
Levy	1960–61	8	5	63	NG	NG
	1977	1	0	—	—	—
"Florida"	1960–61	20	1	5	NG	NG
Others[a]	1960–84	65	0	0	—	—

Sources: 1960–61 data from Maxfield et al. (1963), 1969–72 data from Hon and Forrester (1972), 1980–81 Glades County data from Akey and Forrester (1981), remainder of data from Forrester (1992b).
NG = not given by authors.
a. Includes negative data from 13 other counties; sample sizes are presented in Table 17.35.

Table 17.45. Distribution, prevalence, and intensity of infections of the tapeworm *Davainea meleagridis* in Wild Turkeys from Florida

County (or site)	Year(s)	No. of turkeys			Intensity	
		Exam.	Pos.	%	Mean	Range
Alachua	1969–72	18	1	6	3	—
	1975–80	2	0	—	—	—
Clay	1960–61	11	2	18	NG	NG
Glades	1960–61	12	2	17	NG	NG
	1969–70	94	0	0	—	—
	1971	65	2	3	4	3–4
	1972–75	33	0	0	—	—
	1977–78	15	0	0	—	—
	1980–81	52	0	0	—	—
Highlands	1960–61	16	10	63	NG	NG
	1983	1	0	—	—	—
Osceola	1970	7	0	—	—	—
	1971	8	1	—	28	—
	1972	3	0	—	—	—
"Florida"	1960–61	20	4	20	NG	NG
Others[a]	1960–84	31	0	0	—	—

Sources: 1960–61 data from Maxfield et al. (1963), 1969–72 data from Hon and Forrester (1972), 1980–81 Glades County data from Akey and Forrester (1981), remainder of data from Forrester (1992b).
NG = not given by authors.
a. Includes negative data from 14 other counties; sample sizes are presented in Table 17.35.

Table 17.46. Distribution, prevalence, and intensity of infections of the tapeworm *Raillietina cesticillus* in Wild Turkeys from Florida

County	Year(s)	No. of turkeys			Intensity	
		Exam.	Pos.	%	Mean	Range
Glades	1960–61	12	0	0	—	—
	1969	13	3	23	17	4–39
	1970	81	3	4	3	1–7
	1971	65	3	5	14	1–23
	1972	16	0	0	—	—
	1974	7	0	—	—	—
	1975	10	1	10	1	—
	1977–78	15	0	0	—	—
	1980	26	0	0	—	—
	1981	26	0	0	—	—
Others[a]	1960–84	117	0	0	—	—

Sources: 1960–61 data from Maxfield et al. (1963), 1969–72 data from Hon and Forrester (1972), 1980–81 Glades County data from Akey and Forrester (1981), remainder of data from Forrester (1992b).
a. Includes negative data from 18 other counties; sample sizes are presented in Table 17.35.

Table 17.47. Distribution, prevalence, and intensity of infections of the tapeworms *Hymenolepis carioca* and *Raillietina williamsi* in Wild Turkeys from Florida

County	Year(s)	No. of turkeys Exam.	Pos.	%	Intensity Mean	Range
Hymenolepis carioca						
Glades	1971	1	1	—	1	—
	1982	1	0	—	—	—
Leon	1971	2	1	—	1	—
Levy	1960–61	8	2	—	NG	NG
	1977	1	0	—	—	—
Others[a]	1960–84	375	0	0	—	—
Raillietina williamsi						
Levy	1960–61	8	1	—	NG	NG
Others[b]	1960–84	380	0	0	—	—

Sources: 1960–61 data from Maxfield et al. (1963), 1969–72 data from Hon and Forrester (1972), 1980–81 Glades County data from Akey and Forrester (1981), remainder of data from Forrester (1992b).
NG = not given by authors.
a. Includes negative data from 16 other counties; sample sizes are presented in Table 17.35.
b. Includes negative data from 18 other counties; sample sizes are presented in Table 17.35.

Table 17.48. Prevalences, intensity, and abundance of gastrointestinal nematodes reported from Wild Turkeys in Florida, 1960–84

Infection site	Nematode species	Number of turkeys Exam.	Prevalence %	Intensity Mean	Range	Abundance
Proventriculus	*Dispharynx nasuta*	545	22	4	1–27	<1
Gizzard	*Cyrnea* spp.[a]	500	19	3	1–16	<1
	Synhimantus sp.[b]	500	1	2	1–2	<1
	Cheilospirura spinosa	500	<1	1	—	<1
Small intestine	*Strongyloides* sp.[c]	500	38	12	1–783	5
	Ascaridia dissimilis	498	18	12	1–113	2
	Capillaria spp.[d]	500	13	3	1–22	<1
	Capillaria obsignata	500	<1	NG	NG	—
	Strongyloides avium	500	<1	NG	NG	—
	Ascaridia galli	500	<1	NG	NG	—
Cecum	*Trichostrongylus cramae*	499	32	17	1–211	5
	Heterakis gallinarum	499	9	8	1–52	1
	Aulonocephalus pennula	499	<1	4	—	<1

Sources: Akey and Forrester (1981), Forrester (1992b), Hon and Forrester (1972), Maxfield et al. (1963), Rickard (1985).
NG = not given by author.
a. Includes *Cyrnea colini* and *C. neeli,* considered here as one entity.
b. An undescribed species.
c. Also occurs in the large intestine and cecum.
d. Includes 2 undescribed species, considered here as one entity.

Table 17.49. Distribution, prevalence, and intensity of infections of the proventricular nematode *Dispharynx nasuta* in Wild Turkeys from Florida

County	Year(s)	No. of turkeys			Intensity	
		Exam.	Pos.	%	Mean	Range
Alachua	1969–72	18	2	11	6	1–10
	1974–75	5	1	—	3	—
	1978–80	2	0	—	—	—
Duval	1971–74	4	0	—	—	—
	1976–81	5	1	—	2	—
Gadsden	1971, 73	2	2	—	3	1–5
	1982	1	0	—	—	—
Glades	1960–61	12	0	0	—	—
	1968	1	0	—	—	—
	1969	16	1	6	2	—
	1970	81	25	31	3	1–8
	1971	65	27	42	5	1–18
	1972	16	2	13	1	—
	1973	6	1	—	4	—
	1974	11	2	18	1	—
	1975	19	4	21	8	1–15
	1976	1	0	—	—	—
	1977	45	3	7	4	1–10
	1978–79	10	0	0	—	—
	1980	31	17	55	4	1–11
	1981	27	24	89	5	1–15
	1982	7	0	—	—	—
	1983	1	1	—	2	—
	1984	49	3	6	11	2–27
Leon	1971	2	1	—	1	—
	1983	1	0	—	—	—
Osceola	1970	8	1	—	3	—
	1971	9	0	—	—	—
	1972	3	1	—	2	—
	1975–76	3	0	—	—	—
Pasco	1979	1	1	—	8	—
St. Johns	1972	1	1	—	1	—
Others[a]	1960–84	82	0	0	—	—

Sources: 1960–61 data from Maxfield et al. (1963), 1969–72 data from Hon and Forrester (1972), 1980–81 Glades County data from Akey and Forrester (1981), 1984 Glades County data from Rickard (1985), remainder of data from Forrester (1992b).

a. Includes negative data from 17 other counties; sample sizes are presented in Table 17.35.

Table 17.50. Prevalence, intensity, and abundance of infections of the proventricular nematode *Dispharynx nasuta* in various age classes of Wild Turkeys at Fisheating Creek (Glades County), Florida, 1970–83

Class	Age	No. of turkeys			Intensity		Abundance
		Exam.	Inf.	%	Mean	Range	
Poults	1–7 days	19	5	26	1.5	1–3	0.4
	8–14 days	11	7	64	6.0	1–18	3.8
	15–21 days	8	7	—	3.0	1–7	2.6
	22–30 days	13	13	100	6.8	1–16	6.8
Juveniles	1–7 months	128	67	52	3.8	1–15	2.0
	8–12 months	43	4	9	1.0	1	0.1
Adults	>12 months	69	1	1	10.0	—	0.1

Sources: Akey and Forrester (1981), Forrester (1992b), Hon and Forrester (1972).

Table 17.51. Prevalence, intensity, and abundance of infections of the proventricular nematode *Dispharynx nasuta* in various species of birds in Florida

Avian host	No. of birds			Mean intensity	Abundance
	Exam.	Pos.	%		
Falconiformes					
American Kestrels	22	9	41	28	11
Swallow-tailed Kites	2	2	—	20	20
Galliformes					
Wild Turkeys	545	120	22	4	<1
Northern Bobwhites	1553	151	10	2	<1
Gruiformes					
Greater Sandhill Cranes	34	1	3	2	<1
Florida Sandhill Cranes	15	1	7	1	<1
Whooping Cranes	27	3	11	1	<1
Charadriiformes					
Killdeer	15	3	20	3	<1
Columbiformes					
Mourning Doves	502	74	15	7	1
White-winged Doves	119	5	4	1	<1
Strigiformes					
Eastern Screech-Owls	32	3	9	19	2
Piciformes					
Red-bellied Woodpeckers	75	2	3	9	<1
Passeriformes					
Blue Jays	50	38	76	10	8
Florida Scrub-Jays	46	11	24	8	2
American Crows	78	7	9	6	<1
Fish Crows	3	3	—	2	2
Carolina Wrens	8	6	—	5	4
American Robins	2	1	—	12	6
Northern Mockingbirds	5	3	—	13	8
Northern Cardinals	32	6	19	3	<1
Eastern Towhees	20	3	15	2	<1
Red-winged Blackbirds	34	1	3	1	<1
Boat-tailed Grackles	3	1	—	100	33

Sources: Akey and Forrester (1981), Barrows and Hayes (1977), Conti and Forrester (1981, 1989), Davidson et al. (1980, 1991), Forrester (1987, 1989, 1992b), Forrester et al. (1974a, 1975, 1983, 1984), Foster et al. (2002), Hon and Forrester (1972), Kellogg and Prestwood (1968), Kinsella (1974, 1997), Kinsella and Forrester (1993), Maxfield et al. (1963), Moore et al. (1986), Rickard (1983, 1985), Spalding et al. (1996).

Table 17.52. Distribution, prevalence, and intensity of infections of the gizzard worms *Cyrnea* spp., *Synhimantus* sp., and *Cheilospirura spinosa* in Wild Turkeys from Florida

County	Year(s)	No. of turkeys			Intensity	
		Exam.	Pos.	%	Mean	Range
Cyrnea spp.[a]						
Alachua[b]	1969–72	18	10	56	4	1–6
	1974–75	5	3	—	1	1–2
	1978–80	2	0	—	—	—
Clay[b,c]	1960–61	11	3	27	NG	NG
Collier[c]	1981	1	1	—	1	—
Duval[b,c]	1971–74	4	2	—	2	1–3
	1976–81	5	3	—	4	1–9
Flagler[b]	1969	4	0	—	—	—
	1983	1	1	—	1	—
Gadsden[b]	1971, 73	2	1	—	16	—
	1982	1	0	—	—	—
Glades[b,c]	1960–61	12	0	0	—	—
	1968	1	0	—	—	—
	1969	16	1	6	1	—
	1970	81	32	40	3	1–8
	1971	65	8	12	2	1–9
	1972	16	2	12	1	—
	1973	6	0	—	—	—
	1974	11	1	9	2	—
	1975	19	2	11	1	—
	1976–80	87	0	0	—	—
	1981	27	1	4	1	—
	1982–84	12	0	0	—	—
Hendry	1970–76	3	2	—	7	1–12
Highlands	1960–61	16	1	6	NG	NG
	1983	1	0	—	—	—
Leon	1971	2	1	—	2	—
	1983	1	0	—	—	—
Osceola[b,c]	1970	8	4	—	3	1–7
	1971	9	8	—	4	1–11
	1972	3	3	—	4	4–5
	1975–76	3	2	—	2	1–2
Pasco[b]	1979	1	1	—	3	—
Wakulla	1971	3	1	—	1	—
"Florida"	1960–61	20	2	5	NG	NG
Others[d]	1960–84	23	0	0	—	—

(continued)

Table 17.52. *(continued)*

		No. of turkeys			Intensity	
County	Year(s)	Exam.	Pos.	%	Mean	Range
Synhimantus sp.						
Glades	1960–61	12	0	0	—	—
	1968–69	17	0	0	—	—
	1970	81	2	2	2	1–2
	1971	65	2	3	2	1–2
	1972–76	53	0	0	—	—
	1977	45	1	2	1	—
	1978–84	80	0	0	—	—
Others[e]	1960–84	147	0	0	—	—
Cheilospirura spinosa						
Glades	1960–61	12	0	0	—	—
	1968–69	17	0	0	—	—
	1970	81	1	1	1	—
	1971–84	243	0	0	—	—
Others[e]	1960–84	147	0	0	—	—

Sources: 1960–61 data from Maxfield et al. (1963), 1969–72 data from Hon and Forrester (1972), 1980–81 Glades County data from Akey and Forrester (1981), remainder of data from Forrester (1992b).

NG = not given by author.

a. A complex of 2 species (*C. colini* and *C. neeli*), considered here as 1 entity.

b. Counties from which *C. colini* have been identified.

c. Counties from which *C. neeli* have been identified (Davidson and Forrester 1977, Coyner and Forrester 1993, Davidson et al. 1977).

d. Includes negative data from 12 other counties; sample sizes are presented in Table 17.35.

e. Includes negative data from 24 other counties; sample sizes are presented in Table 17.35.

Table 17.53. Distribution, prevalence, and intensity of infections of the intestinal threadworm *Strongyloides* sp. in Wild Turkeys from Florida

County/site	Year(s)	No. of turkeys			Intensity	
		Exam.	Pos.	%	Mean	Range
Alachua	1969–72	18	1	6	19	—
	1974–75	5	0	—	—	—
	1978	1	0	—	—	—
	1980	1	1	—	22	—
Charlotte	1970	1	1	—	1	—
Collier	1981	1	1	—	6	—
Duval	1971–74	4	0	—	—	—
	1976–81	5	2	—	19	10–27
	1983	1	0	—	—	—
Gadsden	1971, 73	2	1	—	1	—
	1982	1	0	—	—	—
Glades	1960–61	12	1[a]	8	NG	NG
	1968	1	1	—	2	—
	1969	16	13	81	24	1–100
	1970	81	53	65	52	1–783
	1971	65	27	42	32	1–297
	1972	16	14	88	6	1–15
	1973	6	4	—	5	2–9
	1974	11	0	0	—	—
	1975	19	5	26	23	1–73
	1976	1	1	—	1	—
	1977	45	18	40	8	1–47
	1978	5	1	—	9	—
	1979	5	3	—	18	1–48
	1980	31	24	77	13	2–92
	1981	27	9	33	3	1–7
	1982	7	0	—	—	—
	1983	1	0	—	—	—
	1984	4	0	—	—	—
Hendry	1970–76	3	2	—	4	2–6
Manatee	1977	1	1	—	2	—
Pasco	1979	1	1	—	2	—
Taylor	1974	1	0	—	—	—
	1978	1	1	—	1	—
"Florida"	1960–61	20	2[b]	10	NG	NG
Others[c]	1960–84	81	0	0	—	—

Sources: 1960–61 data from Maxfield et al. (1963), 1969–72 data from Hon and Forrester (1972), 1980–81 Glades County data from Akey and Forrester (1981), remainder of data from Forrester (1992b).

NG = not given by author.

a. Maxfield et al. (1963: 269) state that *Strongyloides* sp. was found in 2 turkeys from Glades County, but show only 1 infection in Table 1, p. 262.

b. Maxfield et al. (1963) identified one of these infections as *Strongyloides avium,* but gave no details on his reasons for this identification. Hon et al. (1975) cultured feces from one young turkey and recovered free-living male specimens that conformed closely to *S. avium,* but since fecal samples from each bird were not cultured, it is not possible to attribute this species to all specimens.

c. Includes negative data from 15 other counties (sample sizes in parentheses): Citrus (1), Clay (11), DeSoto (1), Escambia (2), Flagler (5), Highlands (17), Jefferson (1), Leon (3), Levy (10), Madison (1), Osceola (23), Sarasota (1), St. Johns (1), Sumter (1), and Wakulla (3).

Table 17.54. Distribution, prevalence, and intensity of infections of the intestinal roundworm *Ascaridia dissimilis* in Wild Turkeys from Florida[a]

County or site[a]	Year(s)	No. of turkeys			Intensity	
		Exam.	Pos.	%	Mean	Range
Charlotte	1970	1	1	—	1	—
Clay	1960–61	11	3	27	NG	NG
Collier	1981	1	1	—	113	—
Gadsden	1971, 73	2	1	—	17	—
	1982	1	1	—	13	—
Glades	1960–61	12	7	58	NG	NG
	1968	1	1	—	3	—
	1969	15	4	27	2	—
	1970	81	25	29	3	1–10
	1971	65	19	29	3	1–8
	1972	16	0	0	—	—
	1973	6	2	—	2	1–2
	1974	11	0	0	—	—
	1975	18	2	11	5	1–8
	1976	1	0	—	—	—
	1977	45	1	2	1	—
	1978	5	0	—	—	—
	1979	5	1	—	1	—
	1980	31	0	0	—	—
	1981	27	2	7	4	3–5
	1982	7	0	—	—	—
	1983	1	0	—	—	—
	1984	4	0	—	—	—
Hendry	1970–76	3	1	—	2	—
Highlands	1960–61	16	2	13	NG	NG
	1983	1	0	—	—	—
Jefferson	1980	1	1	—	36	—
Leon	1971	2	1	—	2	—
	1983	1	0	—	—	—
Levy	1960–61	8	7	—	NG	NG
	1977–78	2	0	—	—	—
Taylor	1974	1	0	—	—	—
	1978	1	1	—	1	—
Wakulla	1971	3	1	—	1	—
"Florida"	1960–61	20	6	30	NG	NG
Others[b]	1969–83	72	0	0	—	—

Sources: 1960–61 data from Maxfield et al. (1963), 1969–72 data from Hon and Forrester (1972), 1980–81 Glades County data from Akey and Forrester (1981), remainder of data from Forrester (1992b).

NG = not given by author.

a. Maxfield et al. (1963) also reported 1 of 11 turkeys from Clay County was infected with *Ascaridia galli*; they gave no intensity data.

b. Includes negative data from 13 other counties; sample sizes are presented in Table 17.35.

Table 17.55. Distribution, prevalence, and intensity of infections of the intestinal nematodes *Capillaria* spp.[a] in Wild Turkeys from Florida

| County | Year(s) | No. of turkeys | | | Intensity | |
		Examined	Positive	%	Mean	Range
Alachua	1969–72	18	1	6	1	—
	1974–75	5	0	—	—	—
	1978–80	2	0	—	—	—
DeSoto	1978	1	1	—	3	—
Glades	1960–61	12	0	0	—	—
	1968	1	0	—	—	—
	1969	16	3	19	1	—
	1970	81	30	37	2	1–10
	1971	65	11	17	2	1–7
	1972	16	0	0	—	—
	1973	6	2	—	1	—
	1974	11	0	0	—	—
	1975	19	3	16	5	1–13
	1976	1	0	—	—	—
	1977	45	4	9	4	1–7
	1978	5	1	—	1	—
	1979–84	75	0	0	—	—
Hendry	1970–76	3	1	—	6	—
Highlands	1960–61	16	1	6	NG	NG
	1983	1	0	—	—	—
Levy	1960–61	8	1	—	NG	NG
	1977–78	2	2	—	13	3–22
Pasco	1979	1	1	—	1	—
Taylor	1974	1	0	—	—	—
	1978	1	1	—	1	—
Wakulla	1971	3	1	—	2	—
Others[b]	1960–84	85	0	0	—	—

Sources: 1960–61 data from Maxfield et al. (1963), 1969–72 data from Hon and Forrester (1972), 1980–81 Glades County data from Akey and Forrester (1981), remainder of data from Forrester (1992b).
NG = not given by author.
a. A complex of 2 undescribed species, considered here as one entity. Maxfield et al. (1963) also listed 1 of 16 turkeys from Highlands County (1960–61) and 1 of 8 from Levy County (1960–61) infected with *Capillaria obsignata*; no intensities were given.
b. Includes negative data from 16 other counties; sample sizes are presented in Table 17.35.

Table 17.56. Distribution, prevalence, and intensity of infections of the cecal nematode
Trichostrongylus cramae (= *T. tenuis*) in Wild Turkeys from Florida

County or site	Year(s)	No. of turkeys			Intensity	
		Exam.	Pos.	%	Mean	Range
Alachua	1969–72	18	10	56	8	1–19
	1974–75	5	2	—	4	1–7
	1978–80	2	0	—	—	—
Clay	1960–61	11	1	9	NG	NG
DeSoto	1978	1	1	—	7	—
Glades	1960–61	12	5	42	NG	NG
	1968	1	1	—	4	—
	1969	15	6	40	9	6–13
	1970	81	18	22	10	1–37
	1971	65	16	25	8	1–31
	1972	16	8	50	6	1–17
	1973	6	2	—	10	8–12
	1974	11	2	18	54	4–104
	1975	19	11	58	12	1–62
	1976	1	0	—	—	—
	1977	45	17	38	4	1–13
	1978	5	2	—	2	1–2
	1979	5	4	—	5	1–10
	1980	31	6	19	12	2–25
	1981	27	1	4	4	—
	1982	7	4	—	78	10–211
	1983	1	1	—	1	—
	1984	4	2	—	14	7–12
Hendry	1970–76	3	1	—	3	—
Highlands	1960–61	16	3	19	NG	NG
	1983	1	1	—	6	—
Jefferson	1980	1	1	—	2	—
Levy	1960–61	8	1	—	NG	NG
	1977–78	2	2	—	60	14–105
Madison	1971	1	1	—	2	—
Manatee	1977	1	1	—	19	—
Osceola	1970	8	7	—	89	20–201
	1971	9	8	—	32	2–76
	1972	3	3	—	46	25–77
	1975–76	3	2	—	50	36–64
Sarasota	1970	1	1	—	1	—
Sumter	1974	1	1	—	1	—
Wakulla	1971	3	1	—	1	—
"Florida"	1960–61	20	5	25	NG	NG
Others[a]	1969–83	29	0	0	—	—

Sources: 1960–61 data from Maxfield et al. (1963), 1969–72 data from Hon and Forrester (1972), 1980–81 Glades County data from Akey and Forrester (1981), remainder of data from Forrester (1992b).
NG = not given by authors.
a. Includes negative data from 11 other counties; sample sizes are presented in Table 17.35.

Table 17.57. Distribution, prevalence, and intensity of infections of the cecal nematodes *Heterakis gallinarum* and *Aulonocephalus pennula* in Wild Turkeys from Florida

County or site	Year(s)	No. of turkeys Examined	Positive	%	Intensity Mean	Range
Heterakis gallinarum						
Alachua	1969–72	18	2	11	1	—
	1974–75	5	0	—	—	—
	1978–80	2	0	—	—	—
Clay	1960–61	11	7	64	NG	NG
Duval	1971–74	4	0	—	—	—
	1976–81	5	1	—	3	—
Escambia	1984	2	1	—	1	—
Flagler	1969	4	0	—	—	—
	1983	1	1	—	2	—
Gadsden	1971, 73	2	2	—	17	10–24
	1982	1	1	—	6	—
Glades	1960–61	12	4	33	NG	NG
	1968–69	16	0	0	—	—
	1970	81	6	7	1	1–2
	1971	65	0	0	—	—
	1972	16	1	6	1	—
	1973–79	91	0	0	—	—
	1980	31	1	3	5	—
	1981–84	39	0	0	—	—
Highlands	1960–61	16	7	44	NG	NG
	1983	1	0	—	—	—
Leon	1937	NG	NG	—	NG	NG
Levy	1960–61	8	6	—	NG	NG
	1977–78	2	1	—	52	—
Osceola	1970	8	0	—	—	—
	1971	9	1	—	3	—
	1972	3	0	—	—	—
	1975–76	3	0	—	—	—
Pasco	1979	1	1	—	2	—
"Florida"	1960–61	20	2	10	NG	NG
Others[a]	1970–83	21	0	0	—	—
Aulonocephalus pennula						
Glades	1960–61	12	0	0	—	—
	1968–69	16	0	0	—	—
	1970	81	1	1	4	—
	1971–84	243	0	0	—	—
Others[b]	1960–84	147	0	0	—	—

Sources: 1960–61 data from Maxfield et al. (1963), 1969–72 data from Hon and Forrester (1972), 1980–81 Glades County data from Akey and Forrester (1981), Leon County data from U.S. National Parasite Collection, Accession #42919-3859-D; remainder of data from Forrester (1992b).

NG = not given by authors.

a. Includes negative data from 14 other counties; sample sizes are presented in Table 17.35.

b. Includes negative data from 24 other counties; sample sizes are presented in Table 17.35.

Table 17.58. Prevalence, intensity, and abundance of filarial nematodes reported from Wild Turkeys in Florida, 1968–84

		No. of turkeys		Intensity		
Infection site	Filarid species	Examined	Prevalence %	Mean	Range	Abundance
Subcutaneous tissue	*Singhfilaria hayesi*	350	6	2	1–9	<1
Body cavity	*Aproctella stoddardi*	428	2	2	1–2	<1
Connective tissue	*Chandlerella* sp.	320	<1	3	—	<1
Heart	*Splendidofilaria* sp.	375	<1	1	1–2	<1

Sources: Hon and Forrester (1972), Akey and Forrester (1981), Forrester (1992b).

Table 17.59. Distribution, prevalence, and intensity of infections of the filarial worm *Singhfilaria hayesi* in Wild Turkeys from Florida

		No. of turkeys			Intensity	
County	Year(s)	Examined	Positive	%	Mean	Range
Glades	1968	1	1	—	4	—
	1969	5	1	—	1	—
	1970	78	2	3	1	—
	1971	64	5	8	3	1–9
	1972	16	5	31	2	1–3
	1973–74	17	0	0	—	—
	1975	18	2	11	1	—
	1976	1	0	—	—	—
	1977	41	2	5	1	—
	1978–84	28	0	0	—	—
Leon	1971	2	2	—	4	2–6
	1983	1	0	—	—	—
Osceola	1970	8	1	—	1	—
	1971–76	15	0	0	—	—
Others[a]	1969–83	55	0	0	—	—

Sources: 1969–72 data from Hon and Forrester (1972), 1980–81 Glades County data from Akey and Forrester (1981), remainder of data from Forrester (1992b).
a. Includes negative data from 20 other counties (sample sizes in parentheses): Alachua (20), Charlotte (1), Citrus (1), DeSoto (1), Duval (9), Escambia (2), Flagler (1), Gadsden (2), Hendry (3), Highlands (1), Jefferson (1), Levy (2), Madison (1), Manatee (1), Pasco (1), Sarasota (1), St. Johns (1), Sumter (1), Taylor (2), Wakulla (3).

Table 17.60. Distribution, prevalence, and intensity of infections of the filarial worm *Aproctella stoddardi* in Wild Turkeys from Florida

County	Year(s)	No. of turkeys			Intensity	
		Examined	Positive	%	Mean	Range
Alachua	1969	5	0	—	—	—
	1970	7	2	—	2	1–2
	1971–80	13	0	0	—	—
Glades	1968–69	17	0	0	—	—
	1970	80	2	3	1	—
	1971	64	1	2	1	—
	1972–77	97	0	0	—	—
	1978	5	1	—	1	—
	1979–80	36	0	0	—	—
	1981	27	1	4	1	—
	1982–84	12	0	0	—	—
Osceola	1970	8	1	—	2	—
	1971	9	1	—	3	—
	1972–76	6	0	—	—	—
Sumter	1974	1	1	—	1	—
Others[a]	1969–84	41	0	0	—	—

Sources: 1969–72 data from Hon and Forrester (1972), 1980–81 Glades County data from Akey and Forrester (1981), remainder of data from Forrester (1992b).
a. Includes negative data from 19 counties (sample sizes in parentheses): Charlotte (1), Citrus (1), Collier (1), DeSoto (1), Duval (9), Escambia (2), Flagler (5), Gadsden (2), Hendry (3), Highlands (1), Jefferson (1), Leon (3), Levy (2), Madison (1), Manatee (1), Pasco (1), Sarasota (1), Taylor (2), Wakulla (3).

Table 17.61. Distribution, prevalence, and intensity of infections of the filarial worms *Splendidofilaria* sp. and *Chandlerella* sp. in Wild Turkeys from Florida

| County | Year(s) | No. of turkeys | | | Intensity | |
		Examined	Positive	%	Mean	Range
Splendidofilaria sp.						
Glades	1968–70	79	0	0	—	—
	1971	64	1	2	1	—
	1972–84	160	0	0	—	—
Hendry	1970	1	0	—	—	—
	1974	1	1	—	2	—
	1976	1	0	—	—	—
Osceola	1970	4	0	—	—	—
	1971	9	1	—	1	—
	1972–76	5	0	—	—	—
Others[a]	1969–84	51	0	0	—	—
Chandlerella sp.						
Glades	1968	1	0	—	—	—
	1970	55	1	2	3	—
	1971–84	205	0	0	—	—
Others[b]	1969–84	60	0	0	—	—

Sources: 1969–72 data from Hon and Forrester (1972), 1980–81 Glades County data from Akey and Forrester (1981), remainder of data from Forrester (1992b).

a. Includes negative data from 19 other counties (sample sizes in parentheses): Alachua (19), Citrus (1), Collier (1), DeSoto (1), Duval (8), Escambia (2), Flagler (1), Gadsden (2), Highlands (1), Jefferson (1), Leon (3), Levy (1), Madison (1), Manatee (1), Pasco (1), St. Johns (1), Sumter (1), Taylor (2), Wakulla (3).

b. Includes negative data from 19 other counties (sample sizes in parentheses): Alachua (18), Citrus (1), DeSoto (1), Duval (8), Escambia (2), Flagler (1), Gadsden (2), Hendry (3), Highlands (1), Jefferson (1), Leon (2), Levy (1), Madison (1), Manatee (1), Osceola (10), Pasco (1), Sumter (1), Taylor (2), Wakulla (3).

Table 17.62. Data on lone star ticks (*Amblyomma americanum*) collected from 6 Wild Turkeys in Florida

| County (or site)[a] | Year(s) | No. of each stage | | | Data source |
		Nymphs	Larvae	Adults	
Alachua	1971	1	0	0	Forrester & Keirans (1991)
	1976	1	0	0	Ibid.
Duval	1967	4	0	1	Gerrish (1968)
	1977	0	0	1	Forrester & Keirans (1991)
Levy	1978	3	2	0	Ibid.
"No. Fla."	1949–53	2	0	0	Rogers (1953)

a. Travis (1941) examined 2 Wild Turkeys from Orange/Osceola and Collier counties in 1936–37 but found no ticks.

Table 17.63. Number of infested Wild Turkeys examined and species of feather mites encountered in 7 counties in Florida, 1973–87

County	Number of turkeys infested	*Megninia ginglymura*	*Contolichus latus*	*Pterygocrusolichus chanayi*	*Megninia cubitalis*
Alachua	1	+	+	-	-
Citrus	1	+	-	+	+
Duval	3	+	+	-	-
Gadsden	1	+	+	-	-
Glades	9	+	+	+	+
Hendry	1	+	-	-	-
Levy	2	+	+	+	+
Total	18	94%[a]	61%	44%	11%

Source: Forrester and Atyeo (1992).
a. Percent of Wild Turkeys positive for each species of mite.

Table 17.64. Numbers of infested Wild Turkeys examined and species of chewing lice encountered in 15 counties in Florida, 1919–87

County	Number of turkeys infested	*Menacanthus stramineus*	*Chelopistes meleagridis*	*Oxylipeurus polytrapezius*[a]	*Oxylipeurus corpulentus*
Alachua	3	+	+	+	-
Citrus	1	+	-	+	+
Collier	1	-	-	+	-
Duval	5	+[b]	+[b]	+[b]	+
Escambia	1	+	+	+	-
Flagler	1	+	+	-	-
Gadsden	1	+	+	+	-
Glades	78	+	+	+	+
Hendry	2	+	+	-	+
Leon	1	+	+	+	-
Levy	2	+	+	+	+
Manatee	1	-	+	-	+
Orange	1	-	-	+	-
Osceola	3	+	+	-	+
Sarasota	1	-	-	+	+

Sources: Forrester and Emerson (1971), Forrester and Price (1992), Forrester et al. (1995).
a. Emerson (1951) reported this species from the Florida Wild Turkey, but did not give specific dates or localities.
b. Gerrish (1968) also identified these species collected from a Wild Turkey from Duval County in 1967.

Table 17.65. Numbers of adult chewing lice collected from 32 Wild Turkeys[a] from 11 counties[b] in Florida, 1970–83

Chewing louse	Number collected	% of total
Chelopistes meleagridis	240	42
Menacanthus stramineus	118	21
Oxylipeurus corpulentus	85	15
Oxylipeurus polytrapezius	124	22
Total	567	100

Source: Forrester and Price (1992).
a. Includes 3 poults, 13 juveniles, 14 adults, and 2 turkeys of unknown age.
b. Alachua (1), Citrus (1), Duval (5), Escambia (1), Flagler (1), Gadsden (1), Glades (16), Hendry (1), Levy (2), Manatee (1), Osceola (1), and unknown county (1).

Table 17.66. Occurrence of chewing lice in various age classes of Wild Turkeys from Fisheating Creek Wildlife Management Area, Glades County, 1970–71

Age	No. of turkeys infested	No. of turkeys infested with			
		Menacanthus stramineus	*Chelopistes meleagridis*	*Oxylipeurus polytrapezius*	*Oxylipeurus corpulentus*
5–6 days	2	2	0	0	0
5–6 wks	8	6	2	1	0
8–9 wks	2	2	0	0	0
10–16 wks	11	11	9	4	0
4 mos	7	6	4	0	0
5 mos	8	6	6	0	0
6 mos	8	4	8	1	0
7 mos	3	1	1	3	0
8 mos	6	1	5	2	0
9 mos	2	0	1	1	0
>12 mos	3	2	1	0	2
Totals	60	41 (68%)	37 (62%)	12 (20%)	2 (3%)

Sources: Forrester and Emerson (1971), Forrester and Price (1992).

Table 17.67. Numbers of species of infections and parasitic disease agents found in Wild Turkeys at Fisheating Creek, Glades County, 1969–84

Type of agent	No. of species present at Fisheating Creek
Viruses	4
Bacteria	18
Fungi	2[a]
Protozoans	5[b]
Trematodes	7
Cestodes	6
Nematodes	16
Acanthocephalans	1
Ticks	1
Mites	4
Chewing lice	4
Mosquitoes	15
Biting midges	5
Black flies	1
Total	89

Sources: Akey (1981), Akey and Forrester (1981), Atkinson (1988), Busch and Williams (1970), Colwell et al. (1973), Forrester (1992a,b), Forrester and Atyeo (1992), Forrester and Emerson (1971), Forrester and Price (1992), Forrester and Todd (1971), Forrester and Wellings (1981), Forrester et al. (1974b), Grant et al. (1975), Greiner and Forrester (1979), Hon and Forrester (1972), Maxfield et al. (1963), Nayar (1982), Rickard (1985), Telford and Forrester (1975), White and Forrester (1975).

a. The exact number of species involved is not known because the fungi involved were not cultured and identified; there are at least 2 species.

b. Since immunity-challenge studies were not conducted on the coccidia, this group is counted here as representing one species. However, since 7 species of coccidia have been reported from Wild Turkeys in the United States, there are probably more species in turkeys at Fisheating Creek.

XXVI. Literature cited

Adrian, W.J. 1984. Investigation of disease as a limiting factor in wild turkey populations. Ph.D. diss., Colorado State University, Fort Collins. 63 pp.

Akey, B.L. 1981. Mortality in Florida wild turkey poults (*Meleagris gallopavo osceola*). M.S. thesis, University of Florida, Gainesville. 78 pp.

Akey, B.L., and D.J. Forrester. 1981. Unpublished data. University of Florida, Gainesville.

Akey, B.L., J.K. Nayar, and D.J. Forrester. 1981. Avian pox in Florida wild turkeys: *Culex nigripalpus* and *Wyeomyia vanduzeei* as experimental vectors. *J. Wildl. Dis.* 17:597–599.

Alverson, D.R., and R. Noblet. 1977. Spring relapse of *Leucocytozoon smithi* (Sporozoa: Leucocytozoidae) in turkeys. *J. Med. Entomol.* 14: 132–133.

Amundson, T.E. 1985. Health management in wild turkey restoration programs. In: *Proc. Fifth Natl. Wild Turkey Symp.* National Wild Turkey Federation, Edgefield, S.C. pp. 285 294.

Anderson, J.R., and G.R. DeFoliart. 1961. Feeding behavior and host preferences of some black flies (Diptera: Simuliidae) in Wisconsin. *Ann. Entomol. Soc. Am.* 54: 716–729.

Annereaux, R.F. 1940. A note on *Echinoparyphium recurvatum* (von Linstow) parasitic in California turkeys. *J. Am. Vet. Med. Assoc.* 96:62–64.

Atkinson, C.T. 1985. The epizootiology and pathogenicity of *Haemoproteus meleagridis* Levine, 1961, from Florida turkeys. Ph.D. diss., University of Florida, Gainesville. 294 pp.

———. 1986. Host specificity and morphologic variation of *Haemoproteus meleagridis* Levine, 1961 (Protozoa: Haemosporina) in gallinaceous birds. *Can. J. Zool.* 64:2634–2638.

———. 1988. Epizootiology of *Haemoproteus meleagridis* (Protozoa: Haemosporina) in Florida: potential vectors and prevalence in naturally infected *Culicoides* (Diptera: Ceratopogonidae). *J. Med. Entomol.* 25:39–44.

———. 1989. Ultrastructure of the ookinetes of *Haemoproteus meleagridis* (Haemosporina: Haemoproteidae). *J. Parasitol.* 75:135–141.

———. 1991a. Vectors, epizootiology, and pathogenicity of avian species of *Haemoproteus* (Haemosporina: Haemoproteidae). *Bull. Soc. Vector Ecol.* 16:109–126.

———. 1991b. Sporogonic development of *Haemoproteus meleagridis* (Haemosporina: Haemoproteidae) in *Culicoides edeni* (Diptera: Ceratopogonidae). *Can. J. Zool.* 69:1880–1888.

Atkinson, C.T., and D.J. Forrester. 1987. Myopathy associated with megaloschizonts of *Haemoproteus meleagridis* in a wild turkey from Florida. *J. Wildl. Dis.* 23:495–498.

Atkinson, C.T., D.J. Forrester, and E.C. Greiner. 1988a. Epizootiology of *Haemoproteus meleagridis* (Protozoa: Haemosporina) in Florida: seasonal transmission and vector abundance. *J. Med. Entomol.* 25:45–51.

———. 1988b. Pathogenicity of *Haemoproteus meleagridis* (Haemosporina: Haemoproteidae) in experimentally infected domestic turkeys. *J. Parasitol.* 74:228–239.

Atkinson, C.T., E.C. Greiner, and D.J. Forrester. 1983. Experimental vectors of *Haemoproteus meleagridis* Levine from wild turkeys in Florida. *J. Wildl. Dis.* 19:366–368.

———. 1986. Pre-erythrocytic development and associated host responses to *Haemoproteus meleagridis* (Haemosporina: Haemoproteidae) in experimentally infected domestic turkeys. *J. Protozool.* 33:375–381.

Atkinson, C.T., and C. van Riper III. 1991. Pathogenicity and epizootiology of avian haematozoa: *Plasmodium*, *Haemoproteus*, and *Leucocytozoon*. In: *Bird-parasite interactions: ecology, evolution, and behaviour.* J.E. Loye and M. Zuk (eds.). Oxford University Press, Oxford. pp. 19–48.

Barrows, P.L., and F.A. Hayes. 1977. Studies on endoparasites of the mourning dove (*Zenaida macroura*) in the southeast United States. *J. Wildl. Dis.* 13:24–28.

Barwick, L.H., W.M. Hetrick, and L.E. Williams, Jr. 1974. Foods of young Florida wild turkeys. *Proc. Annu. Conf. Southeast. Assoc. Game Fish Comm.* 27:92–102.

Bendell, J.F. 1955. Disease as a control of a population of blue grouse, *Dendragapus obscurus fuliginosis* (Ridgeway). *Can. J. Zool.* 33:195–223.

Bennett, G.F. 1960. On some ornithophilic blood-sucking Diptera in Algonquin Park, Ontario, Canada. *Can. J. Zool.* 38:377–389.

Bennett, G.F., and M.A. Peirce. 1989. The haemoproteids of the avian family Phasianidae. *Can. J. Zool.* 67:1557–1565.

Bennett, G.F., M. Whiteway, and C. Woodworth-Lynas. 1982. *A host-parasite catalogue of the avian Haematozoa.* Meml. University Nfld. Occas. Pap. Biol. 5. 243 pp.

Bishop, M.A., and G.F. Bennett. 1992. *Host-parasite catalogue of the avian Haematozoa.* Suppl. 1. Meml. University Nfld. Occas. Pap. Biol. 15. 210 pp.

Borg, K. 1953. *On Leucocytozoon in Swedish capercaillie, black grouse and hazel grouse.* Berlingska Boktryckeriet, Lund. 109 pp.

Bridges, J.M., and R.D. Andrews. 1977. Agricultural pesticides in wild turkeys in southern Illinois. *Ill. State Acad. Sci.* 69:473–484.

Brown, M.B., and D.J. Forrester. 1992. Unpublished data. University of Florida, Gainesville.

Brownie, C.F., G.T. Edds, and R. Bortell. 1982. Mycotoxins in animal feeds and grain: a survey in 6 north central Florida counties—1981. *Fla. Vet. J.* 11:25–26.

Buchholz, R. 1995. Female choice, parasite load and male ornamentation in wild turkeys. *Anim. Behav.* 50:929–943.

Burridge, M.J., W.J. Bigler, D.J. Forrester, and J.M. Hennemann. 1979. Serologic survey for *Toxoplasma gondii* in wild animals in Florida. *J. Am. Vet. Med. Assoc.* 175:964–967.

Busch, R.H., and L.E. Williams, Jr. 1970. A Marek's disease-like condition in Florida turkeys. *Avian Dis.* 14:550–554.

Byrd, M.A. 1959. Observations on *Leucocytozoon* in pen-raised and free-ranging wild turkeys. *J. Wildl. Manag.* 23:145–156.

Calnek, B.W. 1997. Neoplastic diseases. In: *Diseases of poultry.* 10th ed. B.W. Calnek, H.J.

Barnes, C.W. Beard, L.R. McDougald, and Y.M. Saif (eds.). Iowa State University Press, Ames. pp. 367–368.

Castle, M.D., and B.M. Christensen. 1985. Isolation and identification of *Aegyptianella pullorum* (Rickettsiales, Anaplasmataceae) in wild turkeys from North America. *Avian Dis.* 29:437–445.

Clawson, S.G. 1959. A wild turkey population in an area treated with heptachlor and dieldrin. *Ala. Birdlife* 6:4–8.

Colwell, W.M., C.F. Simpson, L.E. Williams, Jr., and D.J. Forrester. 1973. Isolation of a herpesvirus from wild turkeys in Florida. *Avian Dis.* 17:1–11.

Conti, J.A., and D.J. Forrester. 1981. Interrelationships of parasites of white-winged doves and mourning doves in Florida. *J. Wildl. Dis.* 17:529–536.

———. 1989. Unpublished data. University of Florida, Gainesville.

Conti, L., R. Oliveri, and C. Blackmore. 2002. Unpublished data. Florida Department of Health, Tallahassee.

Cornish, T.E. 1997. Unpublished data. Southeastern Cooperative Wildlife Disease Study, University of Georgia, Athens.

Coyner, D.F., and D.J. Forrester. 1993. Unpublished data. University of Florida, Gainesville.

Cram, E.B. 1928. Nematodes of pathological significance found in some economically important birds in North America. U.S. Department of Agriculture Technical Bulletin 49. 10 pp.

———. 1929. A new roundworm parasite, *Strongyloides avium* of the chicken, with observations on its life history and pathogenicity. *N. Am. Vet.* 10:27–30.

———. 1931. Developmental stages of some nematodes of the Spiruroidea parasitic in poultry and game birds. U.S. Department of Agriculture Technical Bulletin 227. 27 pp.

DaMassa, A.J. 1966. The role of *Culex tarsalis* in the transmission of fowl-pox virus. *Avian Dis.* 10:57–66.

Davidson, W.R. 1987. Disease monitoring in wild turkey restoration programs. *Proc. Annu. Conf. West. Assoc. Fish Wildl. Agencies.* 67:113–118.

———. 1991. Unpublished data. Southeastern Cooperative Wildlife Disease Study, University of Georgia, Athens.

Davidson, W.R., and D.J. Forrester. 1977. Unpublished data. Southeastern Cooperative Wildlife Disease Study, University of Georgia, Athens.

Davidson, W.R., L.T. Hon, and D.J. Forrester. 1977. Status of the genus *Cyrnea* (Nematoda: Spiruroidea) in wild turkeys from the southeastern United States. *J. Parasitol.* 63:332–336.

Davidson, W.R., F.E. Kellogg, and G.L. Doster. 1980. Seasonal trends of helminth parasites of bobwhite quail. *J. Wildl. Dis.* 16:367–375.

Davidson, W.R., and V.F. Nettles. 1997. *Field manual of wildlife diseases in southeastern United States.* 2d ed. Southeastern Cooperative Wildlife Disease Study, University of Georgia, Athens. 417 pp.

Davidson, W.R., V.F. Nettles, C.E. Couvillion, and E.W. Howerth. 1985. Diseases diagnosed in wild turkeys (*Meleagris gallopavo*) of the southeastern United States. *J. Wildl. Dis.* 21:386–390.

Davidson, W.R., V.F. Nettles, C.E. Couvillion, and H.W. Yoder. 1982. Infectious sinusitis in wild turkeys. *Avian Dis.* 26:402–405.

Davidson, W.R., E.B. Shotts, J. Teska, and D.W. Moreland. 1989. Feather damage due to mycotic infections in wild turkeys. *J. Wildl. Dis.* 25:534–539.

Davidson, W.R., D.A. Siefken, and L.H. Creekmore. 1994. Influence of annual and biennial prescribed burning during March on the abundance of *Amblyomma americanum* (Acari: Ixodidae) in central Georgia. *J. Med. Entomol.* 31:72–81.

Davidson, W.R., D.A. Siefken, L.H. Creekmore, and C.T. Moore. 1991. Ecology of helminth parasitism in bobwhites from northern Florida. *J. Wildl. Dis.* 27:185–205.

Davidson, W.R., and E.J. Wentworth. 1992. Population influences: diseases and parasites. In: *The wild turkey—biology and management.* J.G. Dickson (ed.). Stackpole Books, Harrisburg, Pa. pp. 101–118.

Davidson, W.R., H.W. Yoder, M. Brugh, and V.F. Nettles. 1988. Serological monitoring of eastern wild turkeys for antibodies to *Mycoplasma* spp. and avian influenza viruses. *J. Wildl. Dis.* 24:348–351.

Domermuth, C.H., D.J. Forrester, D.O. Trainer, and W.J. Bigler. 1977. Serologic examination of wild birds for hemorrhagic enteritis of turkey and marble spleen disease of pheasants. *J. Wildl. Dis.* 13:405–408.

Doster, G.L. 1967. Unpublished data. Southeastern Cooperative Wildlife Disease Study, University of Georgia, Athens.

Dubey, J.P., M.E. Camargo, M.D. Ruff, G.C. Wilkins, S.K. Shen, O.C.N. Kwok, and P. Thulliez. 1993. Experimental toxoplasmosis in turkeys. *J. Parasitol.* 79:949–952.

Dubois, G., and L.T. Hon. 1973. Le strigeide du dindon sauvage (*Meleagris gallopavo* L.) au Texas et en Floride. *Bull. Soc. Neuchatel. Sci. Nat.* 96:89–95.

Dusek, R.J. 2000. Unpublished data. University of Florida, Gainesville.

Dusenbury, B.A., and M.G. Spalding. 1999. Unpublished data. University of Florida, Gainesville.

Eichholz, N.F. 1991. Unpublished data. Florida Game and Fresh Water Fish Commission, Quincy.

Emerson, K.C. 1951. A list of Mallophaga from gallinaceous birds of North America. *J. Wildl. Manag.* 15:193–195.

Eve, J.H., F.E. Kellogg, and R.W. Bailey. 1972. Blood parasites in wild turkeys of eastern West Virginia. *J. Wildl. Manag.* 36:624–627.

Fischer, J.R. 1997. Unpublished data. Southeastern Cooperative Wildlife Disease Study, University of Georgia, Athens.

Forrester, D.J. 1970–89. Unpublished data. University of Florida, Gainesville.

———. 1991. The ecology and epizootiology of avian pox and malaria in Wild Turkeys. *Bull. Soc. Vector Ecol.* 16:127–148.

———. 1992a. A synopsis of disease conditions in wild turkeys (*Meleagris gallopavo* L.) from Florida, 1969–1990. *Fla. Field Nat.* 20:29–56.

———. 1992b. Unpublished data. University of Florida, Gainesville.

Forrester, D.J., and W.T. Atyeo. 1992. Unpublished data. University of Florida, Gainesville.

Forrester, D.J., A.O. Bush, and L.E. Williams, Jr. 1975. Parasites of Florida sandhill cranes, *Grus canadensis pratensis. J. Parasitol.* 61:547–548.

Forrester, D.J., A.O. Bush, L.E. Williams, Jr., and D.J. Weiner. 1974a. Parasites of greater sandhill cranes (*Grus canadensis tabida*) on their wintering grounds in Florida. *Proc. Helminthol. Soc. Wash.* 41:55–59.

Forrester, D.J., J.A. Conti, A.O. Bush, L.D. Campbell, and R.K. Frohlich. 1984. Ecology of helminth parasitism of bobwhites in Florida. *Proc. Helminthol. Soc. Wash.* 51:255–260.

Forrester, D.J., J.A. Conti, J.D. Shamis, W.J. Bigler, and G.L. Hoff. 1983. Ecology of helminth parasitism of mourning doves in Florida. *Proc. Helminthol. Soc. Wash.* 50:143–152.

Forrester, D.J., and K.C. Emerson. 1971. Unpublished data. University of Florida, Gainesville.

Forrester, D.J., L.T. Hon, L.E. Williams, Jr., and D.H. Austin. 1974b. Blood protozoa of wild turkeys in Florida. *J. Protozool.* 21:494–497.

Forrester, D.J., and P.P. Humphrey. 1981. Susceptibility of the knot (*Calidris canutus*) to *Plasmodium hermani. J. Parasitol.* 67:747–748.

Forrester, D.J., P.P. Humphrey, S.R. Telford, Jr., and L.E. Williams, Jr. 1980b. Effects of blood-induced infections of *Plasmodium hermani* on domestic and wild turkey poults. *J. Wildl. Dis.* 16:237–244.

Forrester, D.J., H.W. Kale II, R.D. Price, K.C. Emerson, and G.W. Foster. 1995. Chewing lice (Mallophaga) from birds in Florida: a listing by host. *Bull. Fla. Mus. Nat. Hist.* 39:1–44.

Forrester, D.J., and J.E. Keirans. 1991. Unpublished data. University of Florida, Gainesville.

Forrester, D.J., and J.K. Nayar. 1981. Unpublished data. University of Florida, Gainesville.

Forrester, D.J., J.K. Nayar, and G.W. Foster. 1980a. *Culex nigripalpus:* a natural vector of wild turkey malaria (*Plasmodium hermani*) in Florida. *J. Wildl. Dis.* 16:391–394.

Forrester, D.J., J.K. Nayar, and M.D. Young. 1987. Natural infection of *Plasmodium hermani* in the northern bobwhite, *Colinus virginianus*, in Florida. *J. Parasitol.* 73:865–866.

Forrester, D.J., and R.D. Price. 1992. Unpublished data. University of Florida, Gainesville.

Forrester, D.J., and M.G. Spalding. 2001. Salmonellosis in a wild turkey from Florida. *Fla. Field Nat.* 29:51–53.

Forrester, D.J., and K.S. Todd, Jr. 1971. Unpublished data. University of Florida, Gainesville.

Forrester, D.J., and F.M. Wellings. 1981. Unpublished data. University of Florida, Gainesville.

Foster, G.W., J.M. Kinsella, E.L. Walters, M.S. Schrader, and D.J. Forrester. 2002. Parasitic helminths of red-bellied woodpeckers (*Melanerpes carolinus*) from the Apalachicola National Forest in Florida. *J. Parasitol.* (in press).

Frenkel, J.K. 1981. False-negative serologic tests for *Toxoplasma* in birds. *J. Parasitol.* 67:952–953.

Fritz, B.A., C.B. Thomas, and T.M. Yuill. 1992. Serological and microbial survey of *Mycoplasma gallisepticum* in wild turkeys (*Meleagris gallopavo*) from six western states. *J. Wildl. Dis.* 28:10–20.

Frye, O.E. 1958. A possible occurrence of avian leukosis in the wild turkey. *J. Wildl. Manag.* 22:94.

Gast, R.K. 1997. *Salmonella* infections. Introduction. In: *Diseases of poultry.* 10th ed. B.W. Calnek, H.J. Barnes, C.W. Beard, L.R. McDougald, and Y.M. Saif (eds.). Iowa State University Press, Ames. pp. 81–82.

Gaud, J., and W.T. Atyeo. 1996. Feather mites of the world (Acarina, Astigmata): the supraspecific taxa. Part I: text. *Ann. Zool. Sci.* 277:1–193.

Gerrish, R.R. 1968. Unpublished data. USDA, Agricultural Research Service, Beltsville, Md.

Goble, F.C., and H.L. Kutz. 1945. The genus *Dispharynx* (Nematoda: Acuariidae) in galliform and passeriform birds. *J. Parasitol.* 31:323–331.

Gothe, R. 1969. Zur Pathogenese der *Aegyptianella pullorum*—Infektion beim Huhn. *Z. Parasitenkd.* 31:3.

Grant, H.G. 1972. Isolation and characterization of a herpesvirus from wild turkeys in Florida. M.S. thesis, University of Florida, Gainesville. 49 pp.

Grant, H.G., K.D. Ley, and C.F. Simpson. 1975. Isolation and characterization of a herpesvirus from wild turkeys (*Meleagris gallopavo osceola*) in Florida. *J. Wildl. Dis.* 11:562–565.

Greiner, E.C. 1991. Leucocytozoonosis in waterfowl and wild galliform birds. *Bull. Soc. Vector Ecol.* 16:84–93.

Greiner, E.C., and D.J. Forrester. 1979. Prevalence of sporozoites of *Leucocytozoon smithi* in Florida blackflies. *J. Parasitol.* 65:324–326.

———. 1980. *Haemoproteus meleagridis* Levine 1961: Redescription and developmental morphology of the gametocytes in turkeys. *J. Parasitol.* 66:652–658.

Hamilton, W.D., and M. Zuk. 1982. Heritable true fitness and bright birds: a role for parasites? *Science* 218:384–387.

Hanley, J., and D.H. Austin. 1968. Report on accession no. 15580, December 11, 1968. Florida Department of Agriculture, Division of Plant Industry, Poultry Disease Diagnostic Laboratory, Dade City.

Harwood, R.F., and M.T. James. 1979. *Entomology in human and animal health.* Macmillan, New York. 548 pp.

Hayes, L.E., K.A. Langheinrich, and R.L. Witter. 1992. Reticuloendotheliosis in a wild turkey (*Meleagris gallopavo*) from coastal Georgia. *J. Wildl. Dis.* 28:154–158.

Healy, W.M. 1992. Population influences: environment. In: *The wild turkey: biology and management.* J.G. Dickson (ed.). Stackpole Books, Harrisburg, Pa. pp. 129–143.

Homer, B.L. 1993. Unpublished data. University of Florida, Gainesville.

Homer, B.L., and G.D. Butcher. 1991. Histomoniasis in leghorn pullets on a Florida farm. *Avian Dis.* 35:621–624.

Hon, L.T. 1973. An ecological study of the helminth parasites of wild turkeys in Florida. M.S. thesis, University of Florida, Gainesville. 115 pp.

Hon, L.T., and D.J. Forrester. 1972. Unpublished data. University of Florida, Gainesville.

Hon, L.T., D.J. Forrester, and L.E. Williams, Jr. 1975. Helminths of wild turkeys in Florida. *Proc. Helminthol. Soc. Wash.* 42:119–127.

———. 1978. Helminth acquisition by wild turkeys (*Meleagris gallopavo osceola*) in Florida. *Proc. Helminthol. Soc. Wash.* 45:211–218.

Howerth, E.W. 1984. Unpublished data. Southeastern Cooperative Wildlife Disease Study, University of Georgia, Athens.

———. 1985. Salmonellosis in a wild turkey. *J. Wildl. Dis.* 21:433–434.

Howerth, E.W., and N. Rodenroth. 1985. Fatal systemic toxoplasmosis in a wild turkey. *J. Wildl. Dis.* 21:446–449.

Hudson, P.J., and A.P. Dobson. 1991. The direct

and indirect effects of the caecal nematode *Trichostrongylus tenuis* on red grouse. In: *Bird-parasite interactions: ecology, evolution, and behaviour.* J.E. Loye and M. Zuk (eds.). Oxford University Press, Oxford. pp. 49–68.

Hurst, G.A. 1980. Histomoniasis in wild turkeys in Mississippi. *J. Wildl. Dis.* 16:357–358.

Hwang, J.C., N. Tolgay, W.T. Shalkop, and D.S. Jacquette. 1961. *Dispharynx nasuta* causing severe proventriculitis in pigeons. *Avian Dis.* 5:60–65.

Jacobson, H.A., and G.A. Hurst. 1979. Prevalence of parasitism by *Ambylomma americanum* on wild turkey poults as influenced by prescribed burning. *J. Wildl. Dis.* 15:43–47.

Jessup, D.A., A.J. DaMassa, R. Lewis, and F.R. Jones. 1983. *Mycoplasma gallisepticum* infection in wild-type turkeys living in close contact with domestic fowl. *J. Am. Vet. Med. Assoc.* 183:1245–1247.

Johnson, E.P., G.W. Underhill, J.A. Cox, and W.L. Threlkeld. 1938. A blood protozoan of turkeys transmitted by *Simulium nigroparvum* (Twinn). *Am. J. Hyg.* 27:649–665.

Jones, C.M., and D.J. Richey. 1956. Biology of the black flies in Jasper County, South Carolina, and some relationships to *Leucocytozoon* disease of turkeys. *J. Econ. Entomol.* 49:121–123.

Jones, J.E., B.D. Barnett, and J. Solis. 1972. The effect of *Leucocytozoon smithi* infection on production, fertility, and hatchability of broad breasted white turkey hens. *Poult. Sci.* 51:1543–1545.

Karstad, L. 1971a. Arboviruses. In: *Infectious and parasitic diseases of wild birds.* J.W. Davis, R.C. Anderson, L. Karstad, and D.O. Trainer (eds.). Iowa State University Press, Ames. pp. 17–21.

———. 1971b. Pox. In: *Infectious and parasitic diseases of wild birds.* J.W. Davis, R.C. Anderson, L. Karstad, and D.O. Trainer (eds.). Iowa State University Press, Ames. pp. 34–41.

Kellogg, F.E. 1973. Unpublished data. Southeastern Cooperative Wildlife Disease Study, University of Georgia, Athens.

Kellogg, F.E., and A.K. Prestwood. 1968. Gastrointestinal helminths from wild and pen-reared bobwhites. *J. Wildl. Manag.* 32: 468–475.

Kellogg, F.E., A.K. Prestwood, R.R. Gerrish, and G.L. Doster. 1969. Wild turkey ectoparasites collected in the southeastern United States. *J. Med. Entomol.* 6:329–330.

Kemp, R.L. 1978. Trichomonads, other flagellates, and protozoa. In: *Diseases of poultry.* 7th ed. M.S. Hofstad, B.W. Calnek, C.F. Helmboldt, W.M. Reid, and H.W. Yoder, Jr. (eds.). Iowa State University Press, Ames. pp. 841–846.

Kennamer, J.E., and M.C. Kennamer. 1990. Current status and distribution of the wild turkey, 1989. *Proc. Sixth Natl. Wild Turkey Symp.* National Wild Turkey Federation, Edgefield, S.C. pp. 1–12.

Kingston, N. 1978. Trematodes. In: *Diseases of poultry.* 7th ed. M.S. Hofstad, B.W. Calnek, C.F. Helmboldt, W.M. Reid, and H.W. Yoder, Jr. (eds.). Iowa State University Press, Ames. pp. 759–782.

Kinsella, J.M. 1974. Helminth fauna of the Florida scrub jay: host and ecological relationships. *Proc. Helminthol. Soc. Wash.* 41:127–130.

———. 1997. Unpublished data. University of Florida, Gainesville.

Kinsella, J.M., and D.J. Forrester. 1972. Helminths of the Florida duck, *Anas platyrhynchos fulvigula. Proc. Helminthol. Soc. Wash.* 39:173–176.

———. 1993. Unpublished data. University of Florida, Gainesville.

Kiszewski, A.E., and E.W. Cupp. 1986. Transmission of *Leucocytozoon smithi* by black flies in New York, U.S.A. *J. Med. Entomol.* 23:256–262.

Laveran, A., and A. Lucet. 1905. Deux hematozoaires de la perdix et du dindon. *C. R. Acad. Sci. Paris* 141:673–676.

Leopold, A.S. 1944. The nature of heritable wildness in turkeys. *Condor* 46:133–197.

Levine, N.D. 1961. *Protozoan parasites of domestic animals and of man.* Burgess, Minneapolis. 412 pp.

———. 1973. *Protozoan parasites of domestic animals and of man.* 2d ed. Burgess, Minneapolis. 406 pp.

———. 1980. *Nematode parasites of domestic animals and of man.* 2d ed. Burgess, Minneapolis. 477 pp.

Lewis, J.B. 1987. Success story: wild turkey. In: *Restoring America's wildlife.* H. Kallman (ed.). U.S. Department of the Interior, Fish and Wildlife Service, Washington, D.C. pp. 31–43.

Ley, D.H., M.D. Ficken, D.T. Cobb, and R.L. Witter. 1989. Histomoniasis and reticuloendotheliosis in a wild turkey (*Meleagris gallopavo*) in North Carolina. *J. Wildl. Dis.* 25:262–265.

Ley, D.H., and H.W. Yoder. 1997. *Mycoplasma gallisepticum* infection. In: *Diseases of poultry.* 10th ed. B.W. Calnek, H.J. Barnes, C.W. Beard, L.R. McDougald, and Y.M. Saif (eds.). Iowa State University Press, Ames. pp. 194–207.

Linda, S.B. 1998. Analysis of 1997–1998 statewide and wildlife management area hunter surveys: methods and results. Unpublished report. Florida Game and Fresh Water Fish Commission, Gainesville. 88 pp.

Lindsay, D.S., P.C. Smith, and B.L. Blagburn. 1994. Prevalence and isolation of *Toxoplasma gondii* from wild turkeys in Alabama. *J. Helminthol. Soc. Wash.* 61:115–117.

Luttrell, M.P., S.H. Kleven, and W.R. Davidson. 1991. An investigation of the persistence of *Mycoplasma gallisepticum* in an eastern population of Wild Turkeys. *J. Wildl. Dis.* 27:74–80.

Luttrell, M.P., T.H. Eleazer, and S.H. Kleven. 1992. *Mycoplasma gallopavonis* in eastern wild turkeys. *J. Wildl. Dis.* 28:288–291.

Maehr, D.S., and J.R. Brady. 1984. Food habits of Florida black bears. *J. Wildl. Manag.* 48:230–235.

———. 1986. Food habits of bobcats in Florida. *J. Mammal.* 67:133–138.

Maehr, D.S., R.C. Belden, E.D. Land, and L. Wilkins. 1990. Food habits of panthers in southwest Florida. *J. Wildl. Manag.* 54:420–423.

Main, M.B. 1998. Unpublished data. Southwest Florida Research and Education Center, University of Florida, Immokalee.

Maxfield, B.G., W.M. Reid, and F.A. Hayes. 1963. Gastrointestinal helminths from turkeys in southeastern United States. *J. Wildl. Manag.* 27:261–271.

McClure, H.E. 1989. Epizootic lesions of house finches in Ventura County, California. *J. Field Ornithol.* 60:421–430.

McDonald, M.E. 1969. *Catalogue of helminths of waterfowl (Anatidae).* U.S. Department of the Interior, Bureau of Sport Fisheries and Wildlife, Special Scientific Report—Wildlife 126. Washington, D.C. 692 pp.

McEwan, L.C., and D.H. Hirth. 1980. Food habits of the bald eagle in north-central Florida. *Condor* 82:229–231.

Miller, J.E., and B.D. Leopold. 1992. Population influences: predators. In: *The wild turkey: biology and management.* J.G. Dickson (ed.). Stackpole Books, Harrisburg, Pa. pp. 119–128.

Moore, J., M. Freehling, and D. Simberloff. 1986. Gastrointestinal helminths of the northern bobwhite in Florida: 1968 and 1983. *J. Wildl. Dis.* 22:497–501.

Morehouse, N.F. 1945. The occurrence of *Haemoproteus* sp. in the domesticated turkey. *Trans. Am. Microsc. Soc.* 64:109–111.

Nayar, J.K. 1982. *Bionomics and physiology of Culex nigripalpus (Diptera: Culicidae) of Florida: an important vector of diseases.* Florida Agricultural Experiment Station Bulletin 827. 73 pp.

Nayar, J.K., and D.J. Forrester. 1985. Susceptibility of *Culex nigripalpus* to several isolates of *Plasmodium hermani* from wild turkeys in Florida. *J. Am. Mosq. Control Assoc.* 1:253–255.

Nayar, J.K., M.W. Provost, and C.W. Hansen. 1980a. Quantitative bionomics of *Culex nigripalpus* (Diptera: Culicidae) populations in Florida. 2. Distribution, dispersal and survival patterns. *J. Med. Entomol.* 17:40–50.

Nayar, J.K., M.D. Young, and D.J. Forrester. 1980b. *Wyeomyia vanduzeei*, an experimental host for wild turkey malaria, *Plasmodium hermani.* *J. Parasitol.* 66:166–167.

———. 1981a. *Culex restuans:* an experimental vector for wild turkey malaria, *Plasmodium hermani. Mosq. News.* 41:748–750.

———. 1981b. *Plasmodium hermani:* experimental transmission by *Culex salinarius* and comparison with other susceptible Florida mosquitoes. *Exp. Parasitol.* 51:431–437.

———. 1982. Experimental transmission by mosquitoes of *Plasmodium hermani* between domestic turkeys and pen-reared bobwhites. *J. Parasitol.* 68:874–876.

Nettles, V.F. 1976. Organophosphate toxicity in wild turkeys. *J. Wildl. Dis.* 12:560–561.

———. 1984. Report of the fish and wildlife health committee. *Proc. Conv. Int. Assoc. Fish Wildl. Agencies.* Washington, D.C. pp. 89–101.

Nettles, V.F., and E.T. Thorne. 1982. Annual report of the wildlife disease committee. *Proc. Annu. Meet. U.S. Anim. Health Assoc.* 86:64–65.

Newberne, J.W. 1955. The pathology of *Leucocytozoon* infection in turkeys with a note on its tissue stages. *Am. J. Vet. Res.* 16:593–597.

Noblet, R., T.R. Adkins, and J.B. Kissam. 1972. *Simulium congareenarum* (Diptera: Simuliidae), a new vector of *Leucocytozoon smithi* (Sporozoa: Leucocytozoidae) in domestic turkeys. *J. Med. Entomol.* 9:580.

Payne, R.L. 1967. Unpublished data. Southeastern Cooperative Wildlife Disease Study, University of Georgia, Athens.

Peckham, M.C. 1978. Vices and miscellaneous diseases. In: *Diseases of poultry.* 7th ed. M.S. Hofstad, B.W. Calnek, C.F. Helmboldt, W.M. Reid, and H.W. Yoder, Jr. (eds.). Iowa State University Press, Ames. pp. 847–893.

Pinkovsky, D.D., and J.F. Butler. 1978. Black flies of Florida. I. Geographic and seasonal distribution. *Fla. Entomol.* 61:257–267.

Pinkovsky, D.D., D.J. Forrester, and J.F. Butler. 1981. Investigations on black fly vectors (Diptera: Simuliidae) of *Leucocytozoon smithi* (Sporozoa: Leucocytozoidae) in Florida. *J. Med. Entomol.* 18:153–157.

Popp, J.A. 1977. Unpublished data. Pathology Report no. C77-45, College of Veterinary Medicine, University of Florida, Gainesville.

Powell, J.A. 1965. *The Florida wild turkey.* Florida Game and Fresh Water Fish Commission Technical Bulletin 8. 28 pp.

———. 1967. Management of the Florida turkey and the eastern turkey in Georgia and Alabama. In: *The wild turkey and its management.* O.H. Hewitt (ed.). Wildlife Society, Washington, D.C. pp. 409–451.

Prestwood, A.K., F.E. Kellogg, and G.L. Doster. 1973. Parasitism and disease among southeastern wild turkeys. In: *Wild turkey management.* G.C. Sanderson and H.C. Schultz (eds.). University of Missouri Press, Columbia. pp. 159–167.

———. 1975. Parasitism among wild turkeys in the southeast. In: *Proc. Third Natl. Wild Turkey Symp.* L.K. Halls (ed.). Texas Chapter of the Wildlife Society, Austin. pp. 27–32.

Prestwood, A.K., F.E. Kellogg, G.L. Doster, and S.A. Edgar. 1971. Coccidia in eastern wild turkeys of the southeastern United States. *J. Parasitol.* 57:189–190.

Provost, M.W. 1969. The natural history of *Culex nigripalpus.* In: *St. Louis encephalitis in Florida. Fla. State Board Health Monogr. Ser.* 12:46–62.

Quist, C.F. 1992. Unpublished data. Southeastern Cooperative Wildlife Disease Study, University of Georgia, Athens.

Quist, C.F., D.I. Bounous, J.V. Kilburn, V.F. Nettles, and R.D. Wyatt. 2000. The effect of dietary aflatoxin on wild turkey poults. *J. Wildl. Dis.* 36:436–444.

Quist, C.F., J.P. Dubey, M.P. Luttell, and W.R. Davidson. 1995. Toxoplasmosis in wild turkeys: a case report and serologic survey. *J. Wildl. Dis.* 31:255–258.

Reid, W.M. 1962. *Chicken and turkey tapeworms: handbook to aid in identification and control of tapeworms found in the United States of America.* Georgia Agricultural Experiment Station, University of Georgia, Athens. 71 pp.

———. 1967. Etiology and dissemination of the blackhead disease syndrome in turkeys and chickens. *Exp. Parasitol.* 21:249–275.

Rickard, L.G. 1983. Unpublished data. University of Florida, Gainesville.

———. 1985. Proventricular lesions associated with natural and experimental infections of *Dispharynx nasuta* (Nematoda: Acuariidae). *Can. J. Zool.* 63: 2663–2668.

Rickard, L.G., and D.J. Forrester. 1984. Unpublished data. University of Florida, Gainesville.

Rocke, T.E., and D.J. Forrester. 1984. Unpublished data. University of Wisconsin, Madison.

Rocke, T.E., and T.M. Yuill. 1987. Microbial infections in a declining wild turkey population in Texas. *J. Wildl. Manag.* 51:778–782.

———. 1988. Serologic response of Rio Grande wild turkeys to experimental infections of *Mycoplasma gallisepticum. J. Wildl. Dis.* 24: 668–671.

Rocke, T.E., T.M. Yuill, and T.E. Amundson.

1988. Experimental *Mycoplasma gallisepticum* infections in captive-reared wild turkeys. *J. Wildl. Dis.* 24:528–532.

Rogers, A.J. 1953. A study of the ixodid ticks of northern Florida, including the biology and life history of *Ixodes scapularis* Say (Ixodidae: Acarina). Ph.D. diss., University of Maryland, College Park. 191 pp.

Ruff, M.D. 1978. Nematodes and acanthocephalans. In: *Diseases of poultry.* 7th ed. M.S. Hofstad, B.W. Calnek, C.F. Helmboldt, W.M. Reid, and H.W. Yoder, Jr. (eds.). Iowa State University Press, Ames. pp. 705–736.

Savage, A. and J.M. Isa. 1945. An outbreak of *Leucocytozoon* disease in turkeys. *Cornell Vet.* 35:270.

Scanlon, P.F., T.G. O'Brien, N.L. Schauer, and J.L. Coggin. 1979. Heavy metal levels in feathers of wild turkeys from Virginia. *Bull. Environ. Contam. Toxicol.* 21:591–595.

Schorr, L.F., W.R. Davidson, V.F. Nettles, J.E. Kennamer, P. Villegas, and H.W. Yoder. 1988. A survey of parasites and diseases of pen-raised wild turkeys. *Proc. Annu. Conf. Southeast. Assoc. Fish Wildl. Agencies* 42:315–328.

Scott, M.L., and L. Krook. 1972. Nutritional deficiency diseases. In: *Diseases of poultry.* 6th ed. M.S. Hofstad, B.W. Calnek, C.F. Helmboldt, W.M. Reid, and H.W. Yoder, Jr. (eds.). Iowa State University Press, Ames. pp. 50–80.

Seurat, L.G. 1916. Sur un nouveau dispharage des palmipedes. Compt. Rend. Soc. Biol., Paris. 79: 785–788.

Siccardi, F.J., H.O. Rutherford, and W.T. Derieux. 1974. Pathology and prevention of *Leucocytozoon smithi* infection of turkeys. *Avian Dis.* 18:21–32.

Simpson, C.F., D.W. Anthony, and F. Young. 1956. Parasitism of adult turkeys in Florida by *Leucocytozoon smithi* (Lavern and Lucet). *J. Am. Vet. Med. Assoc.* 129:573–576.

Skidmore, L.V. 1932. *Leucocytozoon smithi* infection in turkeys and its transmission by *Simulium occidentale* Yownsend. *Zentralbl. Bakteriol. Parasitenkd. Infektionskr. Hyg. Abt. I Orig.* 125:329–335.

Smith, K.E. 1993–94. Unpublished data. Southeastern Cooperative Wildlife Disease Study, University of Georgia, Athens.

Smith, T. 1895. An infectious disease among turkeys caused by a protozoa (infectious enterohepatitis). *U.S. Dep. Agric. Bull.* 8:7–38.

Solis, J. 1973. Nonsusceptibility of some avian species to turkey *Leucocytozoon* infection. *Poult. Sci.* 52:498–500.

Spalding, M.G., and D.J. Forrester. 1991–97. Unpublished data. University of Florida, Gainesville.

Spalding, M.G., J.M. Kinsella, S.A. Nesbitt, M.J. Folk, and G.W. Foster. 1996. Helminth and arthropod parasites of experimentally introduced whooping cranes in Florida. *J. Wildl. Dis.* 32:44–50.

Spalding, M.G., and R.G. McLean. 1994. Unpublished data. University of Florida, Gainesville.

Springer, W.T. 1997. Other blood and tissue protozoa. In: *Diseases of poultry.* 10th ed. B.W. Calnek, H.J. Barnes, C.W. Beard, L.R. McDougald, and Y.M. Saif (eds.). Iowa State University Press, Ames. pp. 900–911.

Stoddard, E.D., J.T. Tumlin, and D.E. Cooperrider. 1952. Recent outbreak of *Leukocytozoon* [sic] infection in adult turkeys in Georgia. *J. Am. Vet. Med. Assoc.* 12:190–191.

Stone, W.B., and S.A. Butkas. 1978. Lead poisoning in a wild turkey. *N.Y. Fish Game J.* 25:169.

Strickland, R.K., R.R. Gerrish, J.L. Hourrigan, and G.O. Schubert. 1976. *Ticks of veterinary importance.* Agriculture Handbook 485, APHIS, USDA, Washington, D.C. 122 pp.

Telford, S.R., Jr., and D.J. Forrester. 1975. *Plasmodium (Huffia) hermani* sp.n. from wild turkeys (*Meleagris gallopavo*) in Florida. *J. Protozool.* 22:324–328.

Trainer, D.O. 1973. Some diseases of wild turkeys from Texas and Wisconsin. In: *Wild turkey management, current problems and programs.* G.C. Sanderson and H.C. Schultz. (eds.). University of Missouri Press, Columbia. pp. 169–173.

Travis, B.V. 1941. Examinations of wild animals for the cattle tick *Boophilus annulatus microplus* (Can.) in Florida. *J. Parasitol.* 27: 465–467.

Travis, B.V., M.H. Goodwin, Jr., and E. Gambrell. 1939. Preliminary note on the occurrence of *Leucocytozoon smithi* Laveran and Lucet (1905) in turkeys in the southeastern United States. *J. Parasitol.* 25:278.

Tripathy, D.N., and W.M. Reed 1997. Pox. In: *Diseases of poultry.* 10th ed. B. W. Calnek, H.J. Barnes, C.W. Beard, L.R. McDougald, and Y.M. Saif (eds.). Iowa State University Press, Ames. pp. 643–659.

Waters, C.V. 1992. An evaluation of the risk of commercial poultry litter as a source of histomoniasis for wild turkeys and other susceptible galliform species. M.S. thesis, University of Georgia, Athens. 57 pp.

Waters, C.V., L.D. Hall, W.R. Davidson, E.A. Rollor III, and K.A. Lee. 1994. Status of commercial and noncommercial chickens as potential sources of histomoniasis among wild turkeys. *Wildl. Soc. Bull.* 22:43–49.

Wehr, E.E. 1962. Studies on leucocytozoonosis of turkeys, with notes on schizogony, transmission, and control of *Leucocytozoon smithi. Avian Dis.* 6:195–210.

White, F.H., and B.L. Akey. 1981. Unpublished data. University of Florida, Gainesville.

White, F.H., and D.J. Forrester. 1970–92. Unpublished data. University of Florida, Gainesville.

White, F.H., D.J. Forrester, and L.E. Williams, Jr. 1981. Isolations of *Salmonella* from wild turkeys in Florida. *J. Wildl. Dis.* 17:327–330.

Williams, L.E., Jr. 1964. A recurrent color aberrancy in the wild turkey. *J. Wildl. Manag.* 28:148–152.

———. 1967a. Wild turkeys (*Meleagris gallopavo*) with supernumerary leg spurs. *Auk* 84:113–114.

———. 1967b. Erythrism in the wild turkey. *Wilson Bull.* 79:239–240.

———. 1969. A pale mutant wild turkey in juvenal plumage. *Q. J. Fla. Acad. Sci.* 32:237–238.

———. 1972. A living wild turkey with a feathered head. *Auk.* 89:193–194.

———. 1981. *The book of the wild turkey.* Winchester Press, Tulsa, Okla. 181 pp.

———. 1992. Florida turkey. In: *The wild turkey: biology and management.* J.G. Dickson (ed.). Stackpole Books, Harrisburg, Pa. pp. 214–231.

———. 1996. *After the hunt with Lovett Williams.* Krause, Iola, Wis. 255 pp.

Williams, L.E., Jr., and D.H. Austin. 1969. Leg spurs on female wild turkeys. *Auk* 86:561–562.

———. 1988. *Studies of the wild turkey in Florida.* University of Florida Press, Gainesville. 255 pp.

Witter, R.L. 1997. Reticuloendotheliosis. In: *Diseases of poultry.* 10th ed. B.W. Calnek, H.J. Barnes, C.W. Beard, L.R. McDougald, and Y.M. Saif (eds.). Iowa State University Press, Ames. pp. 467–484.

Wright, A.H. 1915. Early records of the wild turkey. IV. *Auk* 32:207–224.

Wright, E.J. 1986. Interactive effects of turkeypox and malaria (*Plasmodium hermani*) on turkey poults. M.S. thesis, University of Florida, Gainesville. 78 pp.

Young, M.D., J.K. Nayar, and D.J. Forrester. 1977. Mosquito transmission of wild turkey malaria, *Plasmodium hermani. J. Wildl. Dis.* 13:168–169.

Zahn, S.N., and S.I. Rothstein. 1999. Recent increase in male house finch plumage variation and its possible relationship to avian pox disease. *Auk* 116:35–44.

CHAPTER EIGHTEEN

Rails, Coots, Gallinules, Moorhens, and Limpkins

I. Introduction

Six species of rails, the Purple Gallinule, the Common Moorhen, the American Coot, and the Limpkin occur in Florida (table 18.1). All but the Yellow Rail and Sora breed in Florida. The Mangrove Clapper Rail (*Rallus longirostris insularum*) and the Black Rail are being considered for listing by the U.S. Fish and Wildlife Service but not enough information exists to justify listing. The Black Rail is listed as rare by the Florida Committee on Rare and Endangered Plants and Animals (Rodgers et al. 1996). The Limpkin is listed as a species of special concern by the Florida Game and Fresh Water Fish Commission (Logan 1997) and by the Florida Committee on Rare and Endangered Plants and Animals (Rodgers et al. 1996).

The secretive nature of the rails makes observations and collection of health information difficult. Life history and population information is listed for the Black Rail (Eddleman et al. 1994; Runde 1996), Yellow Rail (Bookout 1995), King Rail (Meanley 1992), Clapper Rail (Eddleman and Conway 1998), Virginia Rail (Conway 1995), Sora (Melvin and Gibbs 1996), and Limpkin (Nesbitt et al. 1976; Bryan 1996). Information about the diet of Limpkins (Snyder and Snyder 1969), the Common Moorhen (Mulholland and Percival 1982; O'Meara et al. 1986), and Purple Gallinule (Mulholland and Percival 1982) has been reported in Florida.

Most of the information in this chapter comes from specimens collected for specific studies of

Table 18.1. Species of rails, gallinules, coots, and limpkins in Florida

Species		Distribution	Seasonal occurrence	Relative abundance
Yellow Rail	*Coturnicops noveboracensis*	Mainland	Winter	Rare
Black Rail	*Laterallus jamaicensis*	Throughout and winter transient	Resident	Rare[a]
Clapper Rail	*Rallus longirostris*	Coasts	Resident	Common[a]
King Rail	*Rallus elegans*	Throughout and winter transient	Resident	Rare[a]
Virginia Rail	*Rallus limicola*	Throughout	Winter	Rare[a]
Sora	*Porzana carolina*	Throughout	Winter	Common
Purple Gallinule	*Porphyrula martinica*	Throughout	Resident	Common[a]
Common Moorhen	*Gallinula chloropus*	Throughout	Resident	Common[a]
American Coot	*Fulica americana*	Throughout and winter transient	Resident	Common[a]
Limpkin	*Aramus guarauna*	Throughout	Resident	Common[a]

Source: Robertson & Woolfenden (1992).
a. Breeds in Florida.

contaminants and parasites. In only a few cases were sick or dead birds presented for etiologic diagnosis. For this reason, very little information about causes of death or illness (other than collisions) is available. General health information for this group of birds can be found in Friend and Franson (1999).

II. Trauma

Collision with tall manmade obstructions is common in this group of birds, probably because they fly at low altitudes during migration (Stevenson and Anderson 1994) (table 18.2). In the case of the rails, often these records provide the only information about migration and distribution that is available. Common Moorhens and American Coots were a relatively high proportion of the birds found dead (38%) under power lines in southern Florida (FPL and FGFWFC 1991).

Roadkills are relatively common, and likely underreported. Snyder (1994) conducted a survey of roadkills from 1990 to 1993 in and around state parks and found 3 Purple Gallinules, 3 Soras, 3 King Rails, 2 Virginia Rails, 2 Clapper Rails, 1 Common Moorhen, 1 American Coot, and 1 Limpkin, of a total 1562 birds found dead.

Hunters harvested 11,840 American Coots in 1994 and 13,056 in 1995 (Martin and Padding 1996). MacDonald and Martin (1971) estimated that 2,909–9,202 Common Moorhens, 8,300–141,900 American Coots, and 5,200 rails are harvested annually. Generally Clapper Rails are not distinguished from King Rails in hunting records. Limpkins are easily hunted, and this probably accounts for their declines in the early 1900s (Pearson in Bent 1926). Reduced hunting resulted in recovery by 1951 (Sprunt 1954).

Trauma, probably gunshot, was the cause of death of approximately 50–60 American Coots found dead on a lake in Leon County, 1997. The lake had approximately 400 coots on it at the time the dead birds were found. No other species were found dead (Spalding and Fischer 1997).

III. Habitat alteration

Habitat manipulation, such as ditching for mosquito control and limerock mining, can be

Table 18.2. Mortality of rails, coots, and gallinules from collision with manmade objects in Florida

Species County	Year(s)	Obstruction[a]	No. found dead	Total no. of all species	Data source
King Rail					
Brevard	1958	Tall building	1	NG	H
	1970–81	LVAB	7	5,046	J
	1971	LVAB	6	2,195	E
Escambia	1938–49	Bridge	1	740	K
Leon	1955–80	TV tower	5	42,384	A
Orange	1971	TV tower	1	NG	H
Palm Beach	1990–91	Power lines	1	165	B
St. Lucie	1971	Tower	1	289	E
Clapper Rail					
Brevard	1970–81	LVAB	9	5,046	J
	1971	LVAB	2	2,195	E
Citrus	1982–86	Power generating stacks	3	2,301	G
Leon	1955–80	TV tower	1	2,384	A
Orange	1969–72	TV tower	10	9,129	I
Virginia Rail					
Brevard	1970–81	LVAB	4	5,046	J
	1971	LVAB	2	2,195	E
Escambia	1938–49	Bridge	33	740	K
Leon	1955–80	TV tower	55	42,384	A
Orange	1969–72	TV tower	11	9,129	I
St. Lucie	1971	Tower	1	289	E
Sora					
Brevard	1970–81	LVAB	3	5,046	J
	1971	LVAB	1	2,195	E
Citrus	1982–86	Power generating stacks	1	2,301	G
Escambia	1938–49	Bridge	54	740	K
Leon	1955–80	TV tower	171	42,384	A
Orange	1969–72	TV tower	28	9,129	I
Yellow Rail					
Escambia	1938	Bridge	2	740	K
Leon	1955–80	TV tower	3	42,384	A
Orange	1972	TV tower	1	9,129	I
Black Rail					
Brevard	1970–81	LVAB	4	5,046	J
	1971	LVAB	3	2,195	E
	1901–12	Lighthouse	3	NG	D
Citrus	1982–86	Power generating stacks	1	2,301	G
Escambia	1885	Lighthouse	1	NG	D
	1938–49	Bridge	2	740	K
Leon	1955–80	TV tower	8	42,384	A
Volusia	1923	Light	1	1	F
Purple Gallinule					
Brevard	1970–81	LVAB	1	5,046	J
	1971	LVAB	1	2,195	E
Leon	1955–80	TV tower	11	42,384	A
Orange	1969–72	TV tower	3	9,129	I

(continued)

Table 18.2. *(continued)*

Species County	Year(s)	Obstruction[a]	No. found dead	Total no. of all species	Data source
Common Moorhen					
Brevard	1970–81	LVAB	2	5,046	J
Escambia	1938–49	Bridge	3	740	K
Gadsden	1971	Tower	1	198	E
Leon	1955–80	TV tower	20	42,384	A
Orange	1969–72	TV tower	12	9,129	I
Palm Beach	1990–91	Power lines	38	165	B
American Coot					
Alachua	1972	Vehicle	1	1	C
Escambia	1938–49	Bridge	12	740	K
Leon	1955–80	TV tower	94	42,384	A
Orange	1969–72	TV tower	3	9,129	I
Palm Beach	1990–91	Power lines	25	165	B

Sources: A = Crawford (1981), B = FPL and FGFWFC (1991), C = Kinsella & Forrester (1972), D = Sprunt (1954), E = Kale (1971), F = Longstreet (1955), G = Maehr & Smith (1988), H = Stevenson & Anderson (1994), I = Taylor & Anderson (1973, 1974), J = Taylor & Kershner (1986), K = Weston (1966).
NG = Not given.
a. LVAB = Lunar Vehicle Assembly Building

detrimental to rails that use a specific habitat, like the Black Rail (Runde 1996). Loss of the mangrove habitat in the Florida Keys is critical for the Keys subspecies of the Clapper Rail, the Mangrove Clapper Rail (Owre 1978). The riparian habitat of the Limpkin is rapidly being developed in Florida (Nesbitt 1978).

IV. Predation

Predation by alligators is a cause of death for coots and moorhens. Delany and Abercrombie (1986) reported American Coots in 1.4% of 350 American alligator stomachs from Alachua County, 1981–83. Delany (1986) found the band of an American Coot, banded in Wisconsin, in the stomach of an American alligator from Alachua County, 1982. Delany et al. (1988) found American Coot and Common Moorhen remains in 2.7 and 1.8%, respectively, of stomachs of 113 nuisance alligators collected from north central Florida in 1977.

In 1937 a largemouth bass was caught that had consumed a full-grown American Coot in Orange County (Cottam 1938). Bald Eagles were considered "worst enemies" for American Coots by C. J. Maynard in Bent (1926), and Nicholson (in Bent 1937) reported that coots were among the food items found in Bald Eagle nests. In a survey of Bald Eagle nests in north central Florida, American Coots comprised 11% of the prey items and were estimated at 19% of the biomass of food brought to chicks (McEwan and Hirth 1980). The authors also found remains of a single Common Moorhen in a Bald Eagle nest. Great Horned Owls captured at least 26 American Coots, the most common food item, and 3 Purple Gallinules and fed them to a tethered offspring in Welaka, Putnam County, in 1949 (Burns 1952). Upon examination of 1,098 pellets collected at a roost of Northern Harriers in Leon County in 1925–26, evidence of 2 American Coots was found (Stoddard 1931). Bobcats infrequently (<1% of prey items) took Common Moorhens and American Coots (Maehr and Brady 1986). A domestic cat killed a Common Moorhen chick in Dade County in 1995 (Sepulveda 1995), and a Sora in Wakulla County in 1939

Table 18.3. Contaminants in eggs and tissues from rails, gallinules, coots and limpkins from Florida[c]

Age Tissue Species	County/area	Year(s)	No. examined	DDE			Total DDTs			Dieldrin		
				No.[a]	Mean[b]	(Range)	No.[a]	Mean[b]	(Range)	No.[a]	Mean[b]	(Range)
Eggs												
Common Moorhen	Brevard, Glades, Lee, Marion	1972–73	15	11	0.21	(ND–0.80)	NA	—	—	0	ND	—
Limpkin	Glades	1973–74	14	4	0.18	(ND–5.6)	NA	—	—	0	ND	—
Purple Gallinule	Brevard, Glades, Lee, Marion	1972–73	10	6	0.18	(ND–0.95)	NA	—	—	0	ND	—
Adult												
Adipose												
Clapper Rail	"Florida"	1974	1	1	0.14	—	1	0.49	—	0	ND	—
Common Moorhen	"Florida"	1974	1	1	0.52	—	1	0.98	—	0	ND	—
Virginia Rail	"Florida"	1973	1	1	1.1	—	1	1.3	—	0	ND	—
Uropygial gland												
Clapper Rail	"Florida"	1974	1	1	0.11	—	1	0.27	—	0	ND	—
Common Moorhen	"Florida"	1974	1	1	0.13	—	1	0.13	—	0	ND	—
Virginia Rail	"Florida"	1973	1	0	ND	—	0	ND	—	0	ND	—

Sources: Klass et al. (1980) for egg values, Johnston (1976) for tissue values.

NA = not analyzed, ND = not detected.

a. Number with residue.

b. Geometric mean, ppm wet weight.

c. In addition to the compounds listed, Klass et al. (1980) did not detect DDT, DDD, mirex, heptachlor epoxide, oxychlordane, *cis*-chlordane (and/or *trans*-nonachlor), *cis*-nonachlor, hexachlorobenzene, toxaphene, endrin, or PCBs in any of the egg samples; Johnston (1976) did not detect PCBs in any of the adult samples. Johnston also found overall ratio of concentrations in adipose to uropygial gland was 2.2:1, and that adipose tissue generally had greater percent lipid. Wheeler et al. (1977) did not detect mirex in tissues of Common Moorhens (see Table 18.5).

(Stevenson and Anderson 1994). Limpkin nest predation has been attributed to "water-rats," raccoons, Fish Crows, and Boat-tailed Grackles (Walkinshaw 1982; Stevenson and Anderson 1994).

V. Inclement weather

A severe storm during the night in April 1909 caused hundreds if not thousands of birds, primarily perching birds, to fly into lighthouses and other lit areas on Key West and the Dry Tortugas, Monroe County. A single Purple Gallinule was included in those found dead (see chapter 24, Perching Birds, for more details) (Bennett 1909). Case et al. (1965) reported that unusual weather conditions associated with Hurricane Hilda and a cold front resulted in large numbers of migrating birds flying to lights and into buildings and vehicles in Brevard County, October 6–7, 1964. A total of 4,707 dead birds was counted over the 2-day period, mostly passerine birds (see chapter 24, Perching Birds). The list included 4 Clapper Rails, 4 Virginia Rails, and 3 Soras. A wounded Virginia Rail was found after a hurricane in October 1910 in Volusia County (Longstreet 1955). Walkinshaw (1982) attributed loss of eggs in 1 Limpkin nest to a severe windstorm in Polk County, and Bryan (1996) listed flooding as a cause of nest failure for this species. A Sora was found along with 65 egrets and herons on a beach in Collier County in April 1998 following a severe storm (Sileo 1997). It was suspected that the storm was responsible for the deaths; however, carcasses were too autolyzed to determine cause of death.

VI. Organochlorines

We could find only 3 studies of chemical contaminants in tissues of birds of this group. Johnston (1976) presented the concentrations of contaminants in adipose tissue and uropygial glands of Virginia and Clapper Rails and Common Moorhens in Florida (table 18.3).

For all of the species that he tested, DDTs and dieldrin concentrations in the adipose tissue were higher than in the uropygial gland (2.2:1 and 2.6:1, respectively). The uropygial gland can serve to excrete DDT. He found no PCBs in fat or uropygial gland. Concentrations were relatively low in these species compared with individuals in other groups (see chapter 5, Cormorants and Anhingas, chapter 15, Caracaras and Falcons, chapter 21, Gulls, Terns, and Skimmers, and chapter 23, Owls).

Klass et al. (1980) tested eggs of Limpkins, Purple Gallinules, Common Moorhens, and Clapper Rails collected from the east coast of the United States, including Florida, and found only DDE (table 18.3). No significant differences in shell thickness were found in museum eggs obtained before and after 1947 for any of the species (table 18.4). For the Common Moorhen and Clapper Rail, significantly thinner eggs were found in southern states, including Florida, when compared with northeastern and some midwestern states, but this was attributed to regional differences in the species rather than an effect of contamination. Significant correlations between shell thickness and DDE were not found (Klass et al. 1980).

Wheeler et al. (1977) measured residue concentrations of mirex in tissues of a wide variety of species on a ranch in Duval and St. Johns counties over a 2-year period beginning in 1972. There was no history of mirex use in the area. Mirex was applied as a 0.1% bait from the air at a rate of 1 lb/acre. Mirex was not detected in Common Moorhens in the pretreatment samples and the highest concentrations were in a pooled fat sample 1 month after treatment (table 18.5). No Moorhens were tested after 6 months. They found the highest concentrations in insectivorous mammals and birds (see chapter 5, Cormorants and Anhingas, chapter 24, Perching Birds, and chapter 25, Miscellaneous Birds).

VII. Organophosphates

Quist and Fischer (1996) reported the mortality of 28 American Coots due to poisoning

Table 18.4. Mean eggshell thickness of Clapper Rail, Purple Gallinule, Common Moorhen, and Limpkin eggs collected in Florida, pre- and post-1947

Species County/area	Year(s)	No. exam.[a]	Shell thickness (mm)	% Change
Clapper Rail				
"East coast"	1863–1946	55	0.244	—
	1947–1973	6	0.257	+5(NS)
"West coast"	1863–1946	38	0.241	—
	1947–1973	3	0.242	NS
Common Moorhen				
"Florida"	1863–1946	39	0.276	—
	1947–1973	23	0.281	+2(NS)
Limpkin				
"Florida"	1863–1946	124	0.360	—
	after 1947	151	0.357	NS
Dade and Indian River	1950	100	0.357	NS
Purple Gallinule				
"Florida"	1863–1946	89	0.218	—
	1947–1973	15	0.213	NS

Source: Klass et al. (1980).
NS = no significant difference.
a. Number of complete clutches measured.

with fenamiphos, an organophosphate systemic nematicide and insecticide, on a golf course in Bay County, 1996. Fenamiphos was identified in the gizzard contents, but brain acetylcholinesterase concentrations (19–20 μmol/min/g) were within the range of normal reported by Hill (1988).

VIII. Metals

Most of the information about metals in this group of birds comes from a study of trace metals in Common Moorhens at control and phosphate mine settling ponds (O'Meara et al. 1986) (see also chapter 5, Cormorants and An-

Table 18.5. Mirex residues (ppm wet weight) in tissues of Common Moorhens[a] after a single treatment of 10-5 bait, on a ranch in Duval and St. Johns counties, 1972–74

Tissue	Pretreatment	Months after treatment	
		1	6
Brain	ND	ND	0.06[b]
Fat	ND[b]	0.29	0.02[b]
Liver	ND	ND	ND[b]
Muscle	ND	0.03	ND[b]
Stomach contents	ND[b]	ND	ND[b]
Nestling: whole	ND	0.11[b]	NA[b]
Stomach contents	ND	ND[b]	NA[b]

Source: Wheeler et al. (1977)
NA = not analyzed, ND = not detected.
a. Sample size = 2 except where otherwise noted.
b. Sample size = 1.

Table 18.6. Metal concentrations in tissues of Common Moorhens from Florida, 1981–82

| Tissue | | Aluminum | | Bromine | | Copper | |
County	No. exam.[a]	Mean[b]	(SE[c])	Mean[b]	(SE[c])	Mean[b]	(SE[c])
Bone							
Alachua	20 (18)	140	(3)	8.1	(0.90)	5.7	(0.50)
Hamilton[d]	17	160	(27)	12	(1.1)	14	(1.5)
Osceola	17	190	(41)	14	(1.2)	8.8	(1.0)
Polk[d]	18 (17)	160	(14)	19	(2.5)	10	(1.9)
Kidney							
Alachua	14 (8)	3.0	(0.90)	3.5	(0.26)	3.1	(0.18)
Hamilton[d]	11	5.3	(1.0)	4.8	(0.29)	4.7	(0.34)
Osceola	11 (7)	4.1	(1.0)	5.8	(0.65)	4.3	(0.41)
Polk[d]	11 (10)	4.2	(0.86)	13	(5.9)	7.0	(2.8)
Liver							
Alachua	19 (16)	3.7	(0.48)	3.5	(0.42)	3.9	(0.62)
Hamilton[d]	18	6.3	(1.6)	3.4	(0.20)	5.8	(1.1)
Osceola	17 (15)	4.9	(1.0)	5.0	(0.41)	4.5	(0.68)
Polk[d]	17 (13)	12	(3.3)	9.5	(0.60)	8.2	(0.86)

| Tissue | | Manganese | | Lead | | Selenium | |
County		Mean[b]	(SE[c])	Mean[b]	(SE[c])	Mean[b]	(SE[c])
Bone							
Alachua	20	5.5	(0.40)	18	(4.3)	NA	—
Hamilton[d]	17	5.3	(0.60)	35	(22)	NA	—
Osceola	17	7.6	(0.90)	29	(24)	NA	—
Polk[d]	18	5.9	(0.40)	13	(6.8)	NA	—
Kidney							
Alachua	14	1.7	(0.11)	NA	—	1.1	(0.09)
Hamilton[d]	11	1.5	(0.16)	NA	—	1.8	(0.10)
Osceola	11	3.0	(0.51)	NA	—	1.5	(0.19)
Polk[d]	11	3.2	(0.97)	NA	—	3.3	(0.65)
Liver							
Alachua	19	1.4	(0.20)	NA	—	0.7	(0.08)
Hamilton[d]	18	1.1	(0.10)	NA	—	1.0	(0.10)
Osceola	17	2.2	(0.27)	NA	—	1.0	(0.18)
Polk[d]	17	1.7	(0.29)	NA	—	2.3	(0.22)

| Tissue | | Rubidium | | Vanadium | | Zinc | |
County		Mean[b]	(SE[c])	Mean[b]	(SE[c])	Mean[b]	(SE[c])
Bone							
Alachua	20 (18)	31.1	(3.6)	0.24	(0.020)	290	(10)
Hamilton[d]	17	29.1	(3.2)	0.47	(0.080)	340	(15)
Osceola	17	12.5	(1.5)	0.28	(0.030)	290	(12)
Polk[d]	18 (17)	36.0	(6.7)	0.74	(0.27)	300	(15)
Kidney							
Alachua	14 (8)	15	(1.5)	0.044	(0.013)	18	(0.90)
Hamilton[d]	11	9.5	(0.73)	0.078	(0.031)	24	(1.0)
Osceola	11 (7)	6.0	(0.78)	0.059	(0.016)	24	(2.2)
Polk[d]	11 (10)	23	(7.0)	0.089	(0.030)	37	(12)

(continued)

Table 18.6. *(continued)*

| Tissue | | Rubidium | | Vanadium | | Zinc | |
County	No. exam.[a]	Mean[b]	(SE[c])	Mean[b]	(SE[c])	Mean[b]	(SE[c])
Liver							
Alachua	19 (16)	21	(1.7)	0.028	(0.003)	36	(2.0)
Hamilton[d]	18	12	(1.1)	0.15	(0.033)	26	(1.3)
Osceola	17 (15)	6.5	(0.70)	0.035	(0.0075)	36	(3.0)
Polk[d]	17 (13)	25	(2.5)	0.15	(0.047)	53	(4.3)

Source: O'Meara et al. (1986).
NA = not analyzed.
a. Sample sizes for aluminum and vanadium are in parentheses.
b. Geometric mean. Bone values are ppm ash weight; liver and kidney values were reported as dry weights but were converted to wet weight values using conversion data presented in the report.
c. Standard error.
d. Collection sites in Polk and Hamilton counties were phosphate settling ponds.

hingas, and chapter 10, Ducks) (table 18.6). It was found that aluminum, bromine, copper, manganese, rubidium, selenium, and vanadium concentrations were significantly higher in livers collected at mined sites and all of the above except aluminum and vanadium were significantly higher in kidneys. Only aluminum and selenium were in higher concentrations than are usually found in other species. Aluminum was found in 2 common food items (duckweed, *Lemna minor,* and green algae, *Oedogonium* sp.) of Common Moorhens in concentrations higher than those demonstrated to cause reduced growth in young chickens (Subcommittee on Nutrient and Toxic Elements in Water 1974). No health effects that could be associated with toxicosis were noted in these birds which were collected by gunshot.

Gourlie (1984) measured liver concentrations of mercury in 4 Limpkins collected at the Rodman Reservoir and Oklawaha River, Putnam and Marion Counties in 1975–76. Values ranged from 0.20–0.75 ppm (geometric mean = 0.31 ppm) wet weight. These concentrations are low compared with fish-eating birds in Florida. Although we found no information about mercury contamination for other rails in Florida, Odom (1975) found mercury in breast muscles of Clapper Rails near Brunswick, Georgia, that exceeded the tolerance level at that time of 0.5 ppm wet weight. Concentrations of mercury were as high as 9.45 ppm.

IX. Radionuclides

Radium-226 concentrations were 12 times higher in bones of Common Moorhens collected at phosphate settling ponds when compared with natural lakes in central Florida, but below concentrations considered hazardous in humans (table 18.7) (Myers et al. 1989). Concentrations were generally lower than in ducks (see chapter 10, Ducks). Radium-226 concentrations were significantly higher in water and substrates at mined sites. They did not find a correlation between dietary and tissue contamination, probably because of foraging by Common Moorhens at other locations. See chapter 10, Ducks, for additional discussion.

X. Oiling

In an anonymous report (Marsh birds die 1948) a game warden in Nassau County estimated that 2/3 of marsh hens, gulls, and other saltwater birds died in a marsh there in 1948; a

Table 18.7. Concentrations of radium-226 in tissues of Common Moorhens in Florida

Tissue County	Year(s)	No. examined (No. pooled samples)	Mean[a]	Range	SE
Bone					
Alachua	1981–82	20(2)	3.5	3.3–3.7	NG
Hamilton[c]	1981–82	11(1)	5.9	5.9	NG
Osceola	1981–82	20(2)	0.7	0.7	NG
Polk[c]	1981–82	23(2)	8.9	8.1–9.6	NG
Kidney					
Hamilton[c]	1981–82	9(1)	0.049[b]	NG	NG
Polk[c]	1981–82	12(1)	0.049[b]	NG	NG
Liver					
Alachua	1981–82	20(2)	0.031[b]	NG	NG
Hamilton[c]	1981–82	20(2)	0.010[b]	NG	0.028[b]
Osceola	1981–82	20(2)	0.049[b]	NG	NG
Polk[c]	1981–82	23(2)	0.21[b]	NG	0.079[b]

Sources: Myers et al. (1989) for bone, O'Meara et al. (1986) for kidney and liver.
NG = not given.
a. Geometric mean, Becquerel/kg wet weight unless otherwise indicated.
b. Dry weight values presented in O'Meara et al. (1986) have been converted to wet weight values using conversion ratios presented in the report.
c. Polk and Hamilton County sites were phosphate-mine settling ponds.

"gummy substance" or grease was found on the marsh surface and on eggs.

XI. Neoplasia

No reports of tumors were found for this group of birds.

XII. Anomalies

A malformed 3-legged moorhen chick was collected by Mihalik (1993) from a wading bird colony on an artificial shell mound near a landfill in Palm Beach County, June 1993 (figure 18.1).

XIII. Biotoxins

Forrester et al. (1980) reported the first outbreaks of type C botulism in Florida in a shallow phosphate-mine settling pond in Hamilton County in May-June and again in December of 1979 (See chapter 10, Ducks). During the summer die-off, American Coots (48), Wood Ducks (26), and Common Moorhens (13) were the most common species found, of the 110 dead or moribund birds counted. Fewer numbers of 4 species of ducks, 2 Black-necked Stilts, 1 Snowy Egret, and 1 Semipalmated Sandpiper were also observed. Several hundred birds were estimated to have died over a 3-week period. The December die-off involved a small number of coots and Northern Shovelers during unseasonably warm weather. A mouse protection test, using serum from a coot, a moorhen, and 4 species of ducks, was utilized to confirm that type C botulism was the cause of the die-off. *Clostridium botulinum* type C was identified in the substrate of these ponds between April and October (Marion et al. 1983).

Clostridium botulinum type C toxin was identified from 2 of 6 Mottled Ducks in a large multispecies die-off following tropical storm Gordon on a dredge island in Tampa Bay, Hillsborough County, November 1994. A total of

FIGURE 18.1. Common Moorhen chick with an extra foot. Courtesy of M.B. Mihalik.

434 carcasses was found; the majority were ducks (see chapter 10, Ducks, for more complete details). One or more American Coots were among the birds found dead, but were not tested for botulinum toxins (Smith and Nettles 1994).

A large scale die-off during June–July 1971 in a shallow inundated area on the west shore of Lake Okeechobee, Glades County, involved 398 birds of 19 species, including ducks (see chapter 10, Ducks), ibises, rails, spoonbills, skimmers, terns, grebes, shorebirds, and wading birds (Jasmin et al. 1972). American Coots (35) and Purple Gallinules (2) were identified among those dead. Birds were weak and had severe hemorrhagic enteritis and acute hepatitis. It was concluded that *Clostridium perfringens* type C, causing necrotizing enteritis, was the most likely cause of mortality based on culture and mouse protection tests in other species.

XIV. Avian vacuolar myelinopathy

This disease is not known to occur in Florida, but has been diagnosed in coots in nearby states (see chapter 11, Eagles).

XV. Viruses

No information on viral infections is available for this group of birds.

Table 18.8. Bacteria cultured from rails and coots in Florida[a]

| Host species | | | | No. of birds | | |
Bacteria species	County	Year	Organ or tissue	Examined	Infected	Data source
American Coot						
Aeromonas sp.	Citrus	1985	Lung, liver	3	1	A
Enterococcus sp.	Bay	1997	NS	3	NS	B
Escherichia coli	Bay	1997	NS	3	NS	B
	Citrus	1985	Lung, liver	3	1	A
Klebsiella sp.	Bay	1997	NS	3	NS	B
Pasteurella multocida	Dade	1968	Intestine	7	7	C
Plesiomonas shigelloides	Citrus	1985	Liver	3	1	A
Proteus sp.	Citrus	1985	Lung, liver	3	1	A
Pseudomonas aeruginosa	Bay	1997	NS	3	NS	B
Vibrio cholera (non-01)	Citrus	1985	Lung, liver	3	2	A
Clapper Rail						
Citrobacter sp.	Duval	1975	Large intestine	1	1	D
Escherichia coli	Duval	1975	Large intestine	1	1	D
Salmonella rubislaw (group F)	Duval	1975	Feces	1	1	D

Sources: A = Thomas (1985), B = Quist and Fischer (1998), C = Klukas & Locke (1970), D = Forrester & White (1975). NS = not specified.
a. In addition to the cases listed, Forrester & White (1975) tested the liver and intestine of a Sora Rail from Duval County in 1975, liver and large intestines of a coot from Alachua County in 1973, and liver and large intestine of 2 Limpkins from Marion County in 1976 and found no pathogenic bacteria.

XVI. Bacteria

Avian cholera was the cause of a large die-off of coots from December 1967 to February 1968 in West Lake, Everglades National Park, Dade County (Klukas and Locke 1970). At the time, salinity in the lake was low, about 7 ppm. Between 5,000 and 6,000 American Coots were estimated to have died along with 50–100 birds of other species. Approximately 50,000 coots were estimated to be present in the area. Other species involved in the die-off included ducks, skimmers, and grebes. *Pasteurella multocida* was cultured from all of 7 coots examined (table 18.8) and also from a Blue-winged Teal, 2 Lesser Scaup, and a Pied-billed Grebe. Although a few individual birds of many other species have been infected with *P. multocida,* the 1967–68 event has been the only epizootic to occur in Florida. Avian cholera is a common annual cause of large-scale mortality of waterfowl and coots, especially in the central and western states of North America (Friend 1999).

Aeromonas and *Vibrio cholerae* (non-01) were isolated from 1 and 2 respectively of 6 American Coots collected from a die-off on Kings Bay, Crystal River, Citrus County in February of 1985 (table 18.8). A consistent pathogen could not be identified and the cause of the die-off remains undetermined. The die-off involved at least 40 American Coots and 2 Brown Pelicans. One coot was observed swimming and unable to hold its head up (Thomas 1985). The estimated coot population on Kings Bay at the time was 2,000. Tests for botulism and virus isolation were negative. No analyses for contaminants were performed.

Salmonella rubislaw was cultured from the feces of a Clapper Rail from Duval County,

1974, that was severely emaciated (Forrester and White 1975). It was not determined if the *Salmonella* caused disease. Other bacteria were isolated from tissues of the Clapper Rail, but were not associated with disease (table 18.8).

XVII. Fungi

No information about fungal infections was available.

XVIII. Protozoans

Protozoal parasites were rare in this group of birds. Quist and Fischer (1998) noted protozoa in the intestinal mucosa of an American Coot from Bay County in 1997 and thought it might be a coccidian parasite. *Toxoplasma gondii* antibodies were not detected in serum from a single American Coot from Florida (date and location not given) by Burridge et al. (1979). These were tested by inhibition hemagglutination, however, which may not be as sensitive in birds as in mammals, resulting in false negatives (Frenkel 1981).

Blood parasites appear to be regionally rare in this group of birds. *Atoxoplasma* sp. was found in blood from 1 of 44 Purple Gallinules examined in Alachua County in 1981 (Forrester and Bennett 1985). Unidentified blood parasites were found in 2 of 126 Common Moorhens examined from Alachua County, 1981–82 (Forrester and Bennett 1985). No blood parasites were found in: 1 Clapper Rail, Duval County, 1974; 1 juvenile Sora, Duval County, 1974; 1 Sora, Alachua County, 1981; 10 adult Common Moorhens, Columbia County, 1981; 7 adult and 4 juvenile Common Moorhens, Polk County, 1981; 11 adult and 1 juvenile Common Moorhens, Osceola County; American Coots from Glades (1, 1971), Alachua (2, 1975), and Hamilton (6, 1979) counties (Forrester and Bennett 1985); and 4 Limpkins, Marion County, 1975 (Conti et al. 1985). In contrast, *Plasmod-*

ium spp. and *Haemoproteus* spp. were present in >40% prevalence in Virginia and Sora Rails in Wisconsin (Lewandowski et al. 1998).

XIX. Helminths

The most extensive information for this group of birds is on helminth parasites. Parasite surveys have been made based on 56 Common Moorhens, 51 Purple Gallinules, 60 American Coots and 33 Clapper Rails (tables 18.9–18.10, 18.12–18.15). Small numbers of Soras and Limpkins have been examined also (tables 18.11, 18.16).

The greatest diversity of trematodes was found in Clapper Rails (24 species), followed by American Coots (8), Purple Gallinules (6), and Common Moorhens (6). *Athesmia heterolecithodes* was found in all of these except American Coots. Three other trematodes were shared by more than 1 of these species, suggesting overlap of prey items.

Cestodes had low diversity and prevalence in this group of birds and no species were shared by more than 1 host.

Acanthocephala were represented by 1 species in each of Clapper Rails, Purple Gallinules, Common Moorhens, and American Coots. The only evidence for morbidity or mortality related to helminth parasites in this group of birds is from an American Coot found dead in Wakulla County in February 1977. Numerous acanthocephalans, *Polymorphus trochus*, had perforated the small intestine and caecal walls, causing a bacterial peritonitis and enteritis (figure 18.2) (Nettles et al. 1996).

Nematodes were commonly shared by more than 1 member of this group, and were most diverse in Purple Gallinules and Common Moorhens. The most commonly shared nematodes were *Hystrichis tricolor* and *Tetrameres globosa*.

Kinsella et al. (1973) identified species of 3 genera of helminths, *Hadjelia*, *Diplotriaenia*, and *Zonorchis*, that have insect intermediate hosts in the Purple Gallinule, but not in Com-

Table 18.9. Trematodes of Clapper Rails from Florida, 1965–66

| Species (Location)[a] | No. of birds | | | Intensity | | |
County	Examined	Infected	%	Mean	Range	Data source
Athesmia heterolecithodes (LV)						
Indian River	10	4	40	NG	NG	A, B
Pinellas	12	2	17	NG	NG	A, B
Carneophallus spp. (SI)						
Indian River	10	3	30	NG	NG	A
Pinellas	12	2	17	NG	NG	A
Cloacitrema michiganensis (CL)						
Indian River	10	1	10	NG	NG	A
Monroe	11	2	18	NG	NG	A
Pinellas	12	2	17	NG	NG	A
Diacetabulum riggini (IN)						
Monroe	11	4	36	NG	NG	A
Diacetabulum sp. (IN)						
Monroe	11	3	27	NG	NG	A
Echinochasmus schwartzi (SI)						
Indian River	10	3	30	NG	NG	A
Pinellas	12	2	17	NG	NG	A
Echinostomatidae (IN)						
Monroe	11	5	45	NG	NG	A
Pinellas	12	8	67	NG	NG	A
Gynaecotyla adunca (IN)						
Indian River	10	6	60	NG	NG	A
Monroe	12	4	33	NG	NG	A
Levinseniella byrdi (CE)						
Indian River	10	1	10	NG	NG	A, C
Pinellas	12	3	25	NG	NG	A, C
Levinseniella deblocki (CE)						
Collier, Pinellas	NG	—	—	NG	NG	H
Levinseniella sp. (CE)						
Monroe	11	6	54	NG	NG	A
Longiductotrema floridensis (SI)[b]						
Monroe	NG	—	—	NG	NG	D
Lyperosomum sinuosum (PA)						
Indian River	10	3	30	NG	NG	A
Monroe	11	1	9	NG	NG	A
Pinellas	12	4	33	NG	NG	A
Maritrema patulus (SI)						
Monroe	11	6	54	NG	NG	A
Maritrema sp. (SI)						
Indian River	10	10	100	NG	NG	A
Monroe	11	2	18	NG	NG	A
Pinellas	12	12	100	NG	NG	A
Megalophallus pentadactylus (SI)						
Monroe	11	8	73	NG	NG	A
Megalophallus reamesi (SI,CE)[c]						
Monroe	13	9	64	NG	2–15	A, G
Microphallus sp. (IN)[b]						
Monroe	11	6	54	NG	NG	A
Notocotylus sp. (CE)						
Indian River	10	3	30	NG	NG	A

(continued)

Table 18.9. *(continued)*

| Species (Location)[a] | No. of birds | | | Intensity | | |
County	Examined	Infected	%	Mean	Range	Data source
Monroe	11	1	9	NG	NG	A
Pinellas	12	1	8	NG	NG	A
Odhneria raminellae (CE)						
Monroe	11	1	9	NG	NG	A
Ophthalmophagus sp. (NA)						
Pinellas	12	2	17	NG	NG	A
Pachytrema sanguineum (GB)						
Monroe	11	3	27	NG	NG	A
Parorchis acanthus (CL)						
Indian River	10	1	10	NG	NG	A
Monroe	11	1	9	NG	NG	A
Pinellas	12	8	67	NG	NG	A
Parvatrema sp. (IN)						
Pinellas	12	3	25	NG	NG	A
Philophthalmus hegeneri (EY)						
Monroe	11	3	27	NG	NG	A
Probolocoryphe glandulosa (SI)						
Indian River	10	2	20	NG	NG	A
Monroe	11	6	54	NG	NG	A, E
Pinellas	12	8	67	NG	NG	A, E
Prosthogonimus sp. (BU)						
Indian River	10	3	30	NG	NG	A
Renicola glandoloboides (KD)						
Pinellas	12[d]	—	—	NG	NG	F
Renicola hydranassae (KD)						
Monroe	11	1	9	NG	NG	A
Pinellas	12	2	17	NG	NG	A
Renicola ralli (KD)						
Indian River	10	3	30	NG	NG	A, F
Monroe	11	6	54	NG	NG	A, F
Pinellas	12[d]	—	—	NG	NG	A, F
Tanaisia fedtschenkoi (KD)						
Indian River	10	3	30	NG	NG	A
Pinellas	12	2	17	NG	NG	A
Tanaisia sp. (KD)						
Pinellas	12	1	8	NG	NG	A
unidentified larva (IN,BV)						
Monroe	11	2	18	NG	NG	A
unidentified larva I (SI)						
Monroe	11	2	18	NG	NG	A
unidentified larva II (SI)						
Pinellas	12	2	17	NG	NG	A

Sources: A = Heard (1967), B = Byrd et al. (1967), C = Heard (1968), D = Deblock & Heard (1969), E = Heard & Sikora (1969), F = Byrd & Heard (1970), G = Overstreet & Heard (1995), H = Heard & Kinsella (1994).
NG = data not given.
a. Location in host: BU = bursa, BV = blood vessels, CE = ceca, CL = cloaca, EY = eye, GB = gall bladder, IN = intestine, KD = kidney, LV = liver, NA = nares, PA = pancreas, SI = small intestine.
b. A microphalid from Monroe County in 1966 is described in Deblock & Heard (1969). This is probably the same parasite listed as *Microphallus* sp. in Heard (1967).
c. Reported as *Megalophallus* sp. by Heard (1967) and redescribed by Overstreet & Heard (1995). The number of birds examined includes 2 collected in 1968. The intermediate host is an isopod.
d. Seven of 12 birds from Pinellas County were infected with *Renicola* sp. The prevalences of *R. ralli* and *R. glandoloboides* were not specifically stated.

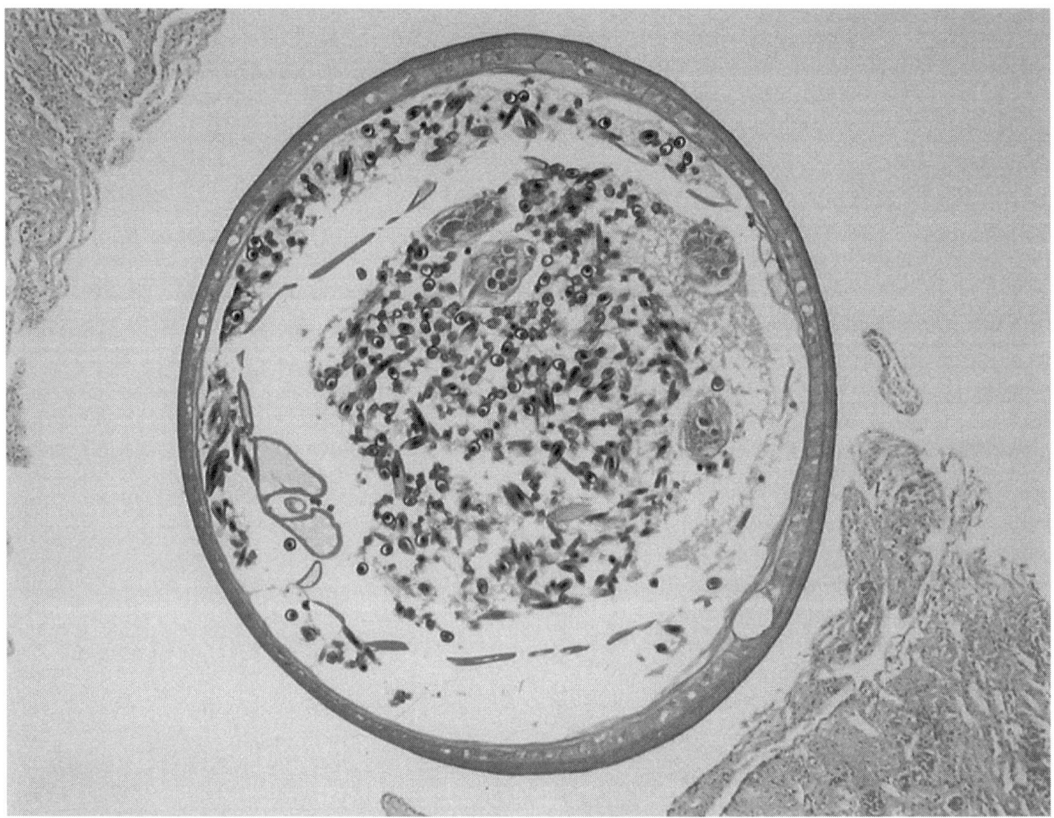

FIGURE 18.2. Cross section of an acanthocephalan, *Polymorphus trochus*, that caused perforation of the intestine of an American Coot.

mon Moorhens. Common Moorhens, which more commonly consume snails, were infected with *Cyclocoelum, Leucochloridium, Tanaisia,* and *Echinostoma*. Limpkins had lower helminth diversity than other gruiformes in Florida, probably because of their almost entirely restricted diet of snails and clams (Conti et al. 1985). For the same reason Limpkins did not share any helminths with other hosts in the group.

XX. Arthropods

Arthropods collected from this group of birds are summarized in table 18.17. We could find no indication of disease or mortality due to the presence of these ectoparasites.

XXI. Summary and Conclusions

Except for fairly extensive surveys of helminths of Clapper Rails, American Coots, Common Moorhens, and Purple Gallinules, and metal and radionuclide studies of Common Moorhens, little information about the health of this group of birds exists for Florida. Trauma from collision with tall structures is relatively common. Disease and poisonings due to clostridial biotoxins caused the largest numbers of nontrauma deaths. Avian cholera and biotoxins, produced by *Clostridium botulinum* and *Clostridium perfringens,* produced infrequent but large-scale mortality in coots, moorhens, and gallinules. American Coots died from a pesticide poisoning on a golf course. In a single case, acanthocephalans caused the death of a coot.

Table 18.10. Cestodes, nematodes, and acanthocephalans of Clapper Rails from Florida, 1966[d]

| Class | | No. of birds | | |
Species (Location)[a]	County	Examined	Infected	%
Cestoda				
Cyclustera sp. (SI)	Indian River	10	5	50
	Monroe	11	3	27
	Pinellas	12	5	42
Hymenolepis sp. (SI)				
immature	Monroe	11	2	18
unidentified	Monroe	11	1	9
Nematoda				
Capillaria sp. (CE,LI)	Indian River	10	1	10
	Monroe	11	2	18
	Pinellas	12	5	42
Skrjabinobronema sp. (GL)	Indian River	10	10	100
	Monroe	11	10	91
	Pinellas	12	12	100
Skrjabinoclava sp. (PR)	Monroe	11	2	18
Spirurid larva (GL)	Indian River	10	10	100
	Monroe	11	11	100
	Pinellas	12	11	92
Syncuaria sp. (ES)	Indian River	10	2	20
	Monroe	11	2	18
	Pinellas	12	1	8
Uncinaria-like hookworm[b] (SI)	Indian River	10	1	10
	Monroe	11	1	9
Acanthocephala				
Arhythmorhynchus frassoni[c] (CE,LI)	Indian River	10	7	70
	Pinellas	12	6	50
	Monroe	11	4	36

Source: Heard (1967).
a. Location in host: CE = ceca, ES = esophagus, GL = gizzard lining, LI = large intestine, PR = proventriculus, SI = small intestine.
b. First record of hookworms from birds. See also Yellow-crowned Night-Herons, Table 6.34.
c. Nickol & Heard (1970) described this species from material collected by Heard (1967).
d. Intensity data not given.

Table 18.11. Helminths of Soras in Florida

Class Species (location)[a]	County	Year	No. of birds		Intensity		
			Exam.	Inf.	Mean	Range	Data source
Trematoda							
Athesmia heterolecithoides (LV)	Alachua	1971–72	3	1	1	—	A
		1991	1	1	1	—	B
Notocotylus sp. (CE)	Alachua	1971–72	3	2	2	1–3	A
Prosthogonimus ovatus (LI)	Broward	1971	2	1	1	—	A
Cestoda							
Diorchis longihamulus (SI)	Alachua	1971–72	3	1	5	—	A
	Broward	1971	2	1	1	—	A
Nematoda							
Amidostomum quasifulicae (GL)	Alachua	1971–72	3	1	8	—	A
Hystrichis tricolor (PR)	Alachua	1971–72	3	1	1	—	A
		1991	1	1	1	—	B
Spirurid larvae (PR)	Alachua	1971–72	3	1	5	—	A
Strongyloides sp. (CE,DU,SI)	Alachua	1971–72	3	3	8	3–16	A
		1991	1	1	14	—	B
	Broward	1971	2	2	2	1–2	A
Tetrameres globosa (PR)	Alachua	1971–72	3	3	2	1–2	A
		1991	1	1	3	—	B
	Broward	1971	2	2	2	2	A
Acanthocephala							
Polymorphus trochus (SI)	Alachua	1971–72	3	1	9	—	A
	Broward	1971	2	1	7	—	A

Sources: A = Kinsella and Forrester (1972), B = McLaughlin and Kinsella (1996).
a. Location in host: CE = cecum, DU = duodenum, GL = gizzard lining, LI = large intestine, LV = liver, PR = proventriculus, SI = small intestine.

Table 18.12. Helminths of Purple Gallinules from Florida, 1970–1971

Class Species (Location)[a]	County	No. of birds			Intensity	
		Exam.[b]	Pos.	%	Mean	Range
Trematoda						
Athesmia heterolecithodes (CO)	Alachua	2	1	—	2	—
	Broward	15	2	13	3	2–3
	Putnam	34	6	17	6	1–14
Cyclocoelum mutabile (AS)	Broward	15	3	6	2	2–3
Immature echinostomes (SI)	Broward	15	5	33	36	1–129
	Putnam	34	1	3	1	—
Leucochloridium problematicum (CL)	Broward	15	2	13	11	7–15
	Putnam	34	1	3	5	—
Notocotylus pacifer (CE)	Alachua	2	1	—	2	—
	Broward	15	1	7	1	—
Tanaisia atra (KD)	Putnam	34	2	6	1	1
Zonorchis petiolatus (LV)	Putnam	34	1	3	2	—
Cestoda						
Diorchis longihamulus (SI)	Broward	15	1	7	—	—
Nematoda						
Amidostomum fulicae (GL)	Broward	15	3	20	1	1–2
Capillaria fulicae (CE)	Putnam	34	1	3	1	—
Diplotriaena sp. (CO)	Broward	15	2	13	6	1–10
Hadjelia neglecta (PR)	Broward	25	1	4	1	—
	Putnam	34	4	12	1	—
Hystrichis tricolor (PR)	Alachua	2	1	—	1	—
Strongyloides sp. (SI)	Alachua	2	2	—	3	2–3
	Broward	15	10	67	5	1–15
	Putnam	34	21	62	4	1–16
Sciadiocara chabaudi (GL)	Putnam	34	2	6	1	1
Tetrameres globosa (PR)	Alachua	2	1	—	1	—
	Broward	15	3	20	5	1–14
	Putnam	34	10	29	2	1–5
Unidentified filarids (CO)	Putnam	34	2	6	3	1–5
Acanthocephala						
Polymorphus matevossianae (SI)	Broward	15	1	7	1	—

Sources: Kinsella (1973, 1997), Kinsella et al. (1973).

a. Location in host: AS = air sacs, CE = ceca, CL = cloaca, CO = coelomic cavity, GL = gizzard lining, KD = kidney, LV = liver, PR = proventriculus, SI = small intestine.

b. Fifty-one Purple Gallinules were examined, from Alachua (2), Broward (15), and Putnam (34) counties; only those counties where each parasite was found are listed. Most birds were shot.

Table 18.13. Trematodes of Common Moorhens from Florida, 1970–71[c]

Species (Location)[a] County	No. of birds			Intensity	
	Examined[b]	Infected	%	Mean	Range
Athesmia heterolecithodes (LV)					
Alachua	22	7	32	10	1–25
Broward	6	1	—	1	—
Osceola	7	1	—	1	—
Putnam	18	3	17	11	1–30
Cyclocoelum mutabile (AS)					
Alachua	22	6	27	1	1–3
Broward	6	1	—	1	—
Marion	3	1	—	1	—
Osceola	7	1	—	4	—
Putnam	18	2	11	2	2
Echinostoma chloropodis (SI)					
Alachua	22	2	9	2	1–2
Immature echinostomes (SI)					
Alachua	22	4	18	4	1–11
Broward	6	1	—	1	
	18	1	6	1	—
Leucochloridium problematicum (CL)					
Alachua	22	6	27	4	1–10
Marion	3	1	—	1	—
Osceola	7	1	—	—	—
Notocotylus pacifer (CE)					
Alachua	22	17	77	10	1–35
Broward	6	3	—	9	1–16
Marion	3	1	—	2	—
Osceola	7	3	—	1	1–2
Putnam	18	4	22	2	1–2
Tanaisia atra (KD)					
Alachua	22	2	9	3	3
Broward	6	2	—	4	3–4
Putnam	18	3	17	11	2–25

Sources: Kinsella (1973, 1997), Kinsella et al. (1973).
a. Location in host: AS = air sacs, CE = ceca, CL = cloaca, KD = kidney, LV = liver, SI = small intestine.
b. Fifty-six Common Moorhens were examined, from Alachua (22), Broward (6), Marion (3), Osceola (7), and Putnam (18) counties; only those counties where each parasite was found are listed. Most were shot and all were >30 days old.
c. In addition to the cases listed, 1 Common Moorhen from Alachua County in 1998 had 1 *Athesmia heterolecithodes* and 1 *Notocotylus pacifer* (Kinsella 1998).

Table 18.14. Cestodes, nematodes and acanthocephalans of Common Moorhens from Florida, 1970–71[c]

Class Species (Location)[a]	County	No. of birds			Intensity	
		Examined[b]	Infected	%	Mean	Range
Cestoda						
Diorchis longihamulus (SI)	Alachua	22	8	36	3	1–8
	Broward	6	1	—	1	—
	Putnam	18	1	6	1	—
Nematoda						
Amidostomum quasifulicae (GL)	Alachua	22	19	86	8	1–32
	Broward	6	4	—	4	2–7
	Marion	3	1	—	14	—
	Osceola	7	6	—	8	1–16
	Putnam	18	8	44	7	1–12
Capillaria fulicae (CE)	Alachua	22	5	23	1	1
	Osceola	7	1	—	1	—
	Putnam	18	2	11	3	1–5
Capillaria sp. (GL)	Alachua	22	4	18	4	1–10
	Marion	3	1	—	2	—
	Osceola	7	1	—	2	—
	Putnam	18	3	17	1	1
Hystrichis tricolor (PR)	Alachua	22	6	27	2	1–3
	Marion	3	1	—	1	—
Strongyloides sp. (SI)	Alachua	22	18	82	11	2–30
	Broward	6	5	—	4	1–12
	Marion	3	2	—	2	1–3
	Osceola	7	4	—	2	1–5
	Putnam	18	9	50	3	1–9
Sciadiocara chabaudi (GL)	Alachua	22	1	4	1	—
Synhimantus sp. (PR)	Putnam	18	1	6	1	1
Tetrameres globosa (PR)	Alachua	22	2	9	11	6–16
Unidentified filarids (CO)	Putnam	18	3	17	2	1–4
Acanthocephala						
Polymorphus trochus (SI)	Alachua	22	13	59	3	1–11

Sources: Kinsella (1973, 1997), Kinsella et al. (1973).
a. Location in host: CE = ceca, CO = coelomic cavity, GL = gizzard lining, PR = proventriculus, SI = small intestine.
b. Fifty-six Common Moorhens were examined, from Alachua (22), Broward (6), Marion (3), Osceola (7), and Putnam (18) counties; only those counties where each parasite was found are listed. Most were shot and all were >30 days old.
c. In addition to the cases listed, 1 Common Moorhen from Alachua County in 1998 had 2 *Diorchis longihamulus* and 2 *Polymorphus trochus* (Kinsella 1998).

Table 18.15. Helminths of American Coots from Florida

Class Species (Location)[a]	County	Year(s)	No. of birds			Intensity	
			Examined	Infected	%	Mean	Range
Trematoda							
Conspicuum icteridorum (LV)	Alachua	1970–72	50	1	2	1	—
Cyclocoelum mutabile (AS,TR)	Alachua	1970–72	50	14	28	5	1–26
	Marion	1970–72	4	1	—	2	—
Cyclocoelum oceleum (AS)	Alachua	1970–72	50	5	10	2	1–3
Echinostoma attenuatum (SI)	Alachua	1970–72	50	10	20	4	1–9
	Marion	1970–72	4	3	—	2	2–3
Leucochloridium problematicum (CL)	Alachua	1970–72	50	3	6	22	1–55
	Marion	1970–72	4	1	—	7	—
Notocotylus pacifer (CE)	Alachua	1970–72	50	47	94	20	1–250
	Citrus	1970–72	6	5	—	9	2–26
	Marion	1970–72	4	3	—	6	2–12
Prosthogonimus ovatus (CL)	Alachua	1970–72	50	2	4	1	1
Tanaisia atra (KD)	Alachua	1970–72	50	2	4	1	1
Cestoda							
Diorchis americana (SI)	Alachua	1970–72	50	11	22	10	1–50
Diorchis ransomi (SI)	Alachua	1970–72	50	12	24	7	1–50
	Citrus	1970–72	6	1	—	1	—
	Marion	1970–72	4	1	—	3	—

Nematoda							
Amidostomum fulicae (GL)	Alachua	1970–72	50	42	84	11	1–30
	Citrus	1970–72	6	4	—	4	2–7
	Marion	1970–72	4	4	—	6	1–22
Capillaria fulicae (CE)	Alachua	1970–72	50	12	24	3	1–5
	Citrus	1970–72	6	1	—	19	—
	Marion	1970–72	4	1	—	1	—
Capillaria sp. (GL)	Alachua	1970–72	50	2	4	1	1
Hystrichis tricolor (PR)	Alachua	1970–72	50	2	4	1	1
Strongyloides sp. (SI)	Alachua	1970–72	50	30	60	14	1–137
	Citrus	1970–72	6	3	—	2	1–3
	Marion	1970–72	4	3	—	6	3–12
Tetrameres globosa (PR)	Alachua	1970–72	50	33	66	4	1–21
	Citrus	1970–72	6	5	—	3	1–6
	Marion	1970–72	4	2	—	2	1–4
Acanthocephala							
Polymorphus trochus (SI)	Alachua	1970–72	50	13	26	7	1–30
	Citrus	1970–72	6	4	—	6	1–15
	Wakulla	1977	1	1	—	NG	many[b]

Sources: Kinsella (1973, 1997) for Alachua, Citrus, and Marion counties; Kinsella & Nettles (1996) for Wakulla County.

NG = data not given by the author(s).

a. Location in host: AS = air sacs, CE = ceca, CL = cloaca, GL = gizzard lining, KD = kidney, LV = liver, PR = proventriculus, SI = small intestine, TR = trachea.

b. Resulted in perforation of intestine and death.

Table 18.16. Helminths of Limpkins from Florida

Class Species (Location)[a]	County	Year(s)	No. of birds		Intensity		Data source
			Examined	Infected	Mean	Range	
Trematoda							
Cyclocoelidae (ES,TR,LU,AS)	Marion	1975–76	9	9	429	2–3,554	A
	Monroe	1992	1	1	7	—	B
Echinostomatidae (SI,LI)	Marion	1975–76	9	3	2	1–3	A
	Monroe	1992	1	1	105	—	B
Lyperorchis lyperorchis (CL)	Marion	1975–76	9	5	4	1–12	A
	Monroe	1992	1	1	18	—	B
Prionosoma serratum (SI)	Marion	1975–76	9	1	2	2	A
	Monroe	1992	1	1	3	—	B
Nematoda							
Amidostomum acutum (GI,SI)	Marion	1975–76	9	5	6	2–10	A
	Monroe	1992	1	1	4	—	B
Strongyloides sp. (SI,CE,LI)	Marion	1975–76	9	3	59	25–78	A
	Monroe	1992	1	1	43	—	B

Sources: A = Conti et al. (1985), B = Spalding & Kinsella (1994).
a. Location in host: AS = air sacs, CE = ceca, CL = cloaca, ES = esophagus, GI = gizzard, LU = lung, LI = large intestine, SI = small intestine, TR = trachea.

Table 18.17. Ectoparasites collected from rails, coots, gallinules, moorhens, and Limpkins in Florida

Host	Species	County/area	Year(s)
Mites			
Clapper Rail	*Blankaartia pauli*	Indian River, Monroe, Pinellas	1966[a]
Sora	*Montchadskiana* sp.	Alachua	1991[b]
Chewing lice			
Yellow Rail	*Fulicoffula* sp.	Dade	1929
Clapper Rail	*Pseudomenopon scopulacorne*	Lee	1982, 1984
		Volusia	1939
	Rallicola californicus[c]	Lee	1982
		Volusia	1939
		"Florida"	NG[d]
		Volusia	1939
Sora	*Fulicoffula distincta*	Dade	1919
			1930[e]
	Pseudomenopon scopulacorne	Dade	1919
			1925
Purple Gallinule	*Pseudomenopon pilosum*	Lee	1982
	Rallicola elliotti	Lee	1979, 1982
Common Moorhen	*Laemobothrion chloropodis*	Alachua	1971
		Lee	1983
	Pseudomenopon pilosum	Alachua	1971
		Dade	1929
		Lee	1983
	Rallicola minutus	Dade	1929
		Lee	1982
American Coot	*Fulicoffula longipila*	Lee	1979
	Incidifrons transpositus	Alachua	1972
		Lee	1979, 1981
	Laemobothrion atrum	Alachua	1971
	Pseudomenopon pilosum	Alachua	1971–72
		Indian River	1973
		Lee	1979, 1981
	Rallicola advenus	Alachua	1972
		Indian River	1973
		Lee	1979, 1981–82
Limpkin	*Ibidoecus scolopaceus*	Hernando	1929[f]
		Indian River	1971
	Laemobothrion cubense	Hernando	1929
		Indian River	1971
		Lake	1980[g]
		Marion	1975–76[g]
		Monroe	1929
		"Florida"	1921
	Rallicola funebris	Indian River	1971
		Lake	1980[g]
		Marion	1975–76[g]
		Monroe	1929

Source: Forrester et al. (1995) unless otherwise noted.
a. Crossley & Atyeo (1972), Rohani & Cromroy (1979)
b. Forrester & Atyeo (1992)
c. Emerson (1955) recorded this as *Rallicola ortygometrae californicus.*
d. Emerson (1955)
e. Emerson (1960)
f. U.S. National Museum of Natural History, Entomological Collection, Smithsonian Institution, Washington, D.C.
g. Conti et al. (1985)

XXII. Literature cited

Bennett, F.M. 1909. A tragedy of migration. *Bird Lore* 11:110–113.

Bent, A.C. 1926. *Life histories of North American marsh birds.* U.S. Natl. Mus. Bull. 135.

———. 1937. *Life histories of North American birds of prey.* Part 1. U.S. Natl. Mus. Bull. 167. 409 pp.

Bookout, T.A. 1995. Yellow rail. *Birds N. Am.* 139:1–16.

Bryan, D.C. 1996. Limpkin. In: *Rare and endangered biota of Florida.* Vol. 5, *Birds.* J.A. Rodgers, Jr., H.W. Kale II, and H.T. Smith (eds.). University Press of Florida, Gainesville. pp. 485–496.

Burns, B.J. 1952. Food of a family of great horned owls, *Bubo virginianus,* in Florida. *Auk* 69:86–87.

Burridge, M.J., W.J. Bigler, D.J. Forrester, and J.M. Hennemann. 1979. Serologic survey for *Toxoplasma gondii* in wild animals in Florida. *J. Am. Vet. Med. Assoc.* 175:964–967.

Byrd, E.E., and R.W. Heard III. 1970. Two new kidney flukes of the genus *Renicola* Cohn, 1904, from the clapper rail, *Rallus longirostris* subspp. *J. Parasitol.* 56:493–497.

Byrd, E.E., A.K. Prestwood, F.E. Kellogg, and R.W.I. Heard. 1967. New hosts and locality records for the large liver fluke, *Athesmia heterolecithodes* (Braun, 1899) Looss, 1899 (Dicrocoeliidae) of birds and mammals. *J. Parasitol.* 53:1116–1117.

Case, L.D., H. Cruickshank, A.E. Ellis, and W.F. White. 1965. Weather causes heavy bird mortality. *Fla. Nat.* 38:29–30.

Conti, J.A., D.J. Forrester, and S.A. Nesbitt. 1985. Parasites of limpkins, *Aramus guarauna,* in Florida. *Proc. Helminthol. Soc. Wash.* 52:140–142.

Conway, C.J. 1995. Virginia rail. *Birds N. Am.* 173:1–20.

Cottam, C. 1938. Coot swallowed by fish. *Wilson Bull.* 50:60.

Crawford, R.L. 1981. Bird casualties at a Leon County, Florida TV tower: a 25-year migration study. *Bull. Tall Timbers Res. Stn.* 22:1–30.

Crossley, D.A., and W.T. Atyeo. 1972. A new species of chigger, *Blankaartia pauli* (Acarina: Trombiculidae), from the southeastern United States. *J. Med. Entomol.* 9:253–252.

Deblock, S., and R.W. Heard III. 1969. Contribution a l'etude des Microphallidae Travassos, 1920 (Trematoda). XIX—Description de *Maritrema prosthometra* n. sp. et de *Longiductotrema* nov. gen. parasites d'Oiseaux Ralliformes d'Amerique du Nord (Descriptions of North American rail parasites *Maritrema prosthometra* and *Longiductotrema* sp.). *Ann. Parasitol. Hum. Comp.* 44:415–424.

Delany, M.F. 1986. Bird bands recovered from American alligator stomachs in Florida. *N. Am. Bird Bander* 11:92–94.

Delany, M.F., and C.L. Abercrombie. 1986. American alligator food habits in north-central Florida. *J. Wildl. Manag.* 50:348–353.

Delany, M.F., A.R. Woodward, and I.H. Kochel. 1988. Nuisance alligator food habits in Florida. *Fla. Field Nat.* 16:90–96.

Eddleman, W.R., and C.J. Conway. 1998. Clapper rail. *Birds N. Am.* 340:1–32.

Eddleman, W.R., R.E. Flores, and M.L. Legare. 1994. Black rail. *Birds N. Am.* 123:1–20.

Emerson, K.C. 1955. A review of the genus *Rallicola* (Philopteridae, Mallophaga) found on Aramidae, Psophiidae and Rallidae. *Ann. Entomol. Soc. Am.* 48:284–299.

———. 1960. Two new species of *Fulicoffula* (Mallophaga) from the sora. *J. Kans. Entomol. Soc.* 33:162–165.

Forrester, D.J., and W.T. Atyeo. 1992. Unpublished data. University of Florida, Gainesville.

Forrester, D.J., and G.F. Bennett. 1985. Unpublished data. University of Florida, Gainesville.

Forrester, D.J., H.W. Kale II, R.D. Price, K.C. Emerson, and G.W. Foster. 1995. Chewing lice (Mallophaga) from birds in Florida: a listing by host. *Bull. Fla. Mus. Nat. Hist.* 39:1–44.

Forrester, D.J., K.C. Wenner, F.H. White, E.C. Greiner, W.R. Marion, J.E. Thul, and G.A. Berkhoff. 1980. An epizootic of avian botulism in a phosphate mine settling pond in northern Florida. *J. Wildl. Dis.* 16:323–327.

Forrester, D.J., and F.H. White. 1975. Unpublished data. University of Florida, Gainesville.

[FPL and FGFWFC] Florida Power and Light

Company and Florida Game and Fresh Water Fish Commission. 1991. Unpublished correspondence. Lake Harbor bird mortality project: daily inspections of test framing designs.

Frenkel, J.K. 1981. False-negative serologic tests for *Toxoplasma* in birds. *J. Parasitol.* 67:952–953.

Friend, M. 1999. Avian cholera. In: *Field manual of wildlife diseases. General field procedures and diseases of birds.* M. Friend and J.C. Franson (eds.). U.S. Department of the Interior, U.S. Geological Survey, Biological Research Division, Information and Technology Report 1999-001. Washington, D.C. pp. 75–92.

Friend, M., and J.C. Franson (eds.). 1999. *Field manual of wildlife diseases. General field procedures and diseases of birds.* U.S. Department of the Interior, U.S. Geological Survey, Biological Research Division, Information and Technology Report 1999-001. Washington, D.C. 426 pp.

Gourlie, N. 1984. Unpublished data. Florida Department of Environmental Regulation, Tallahassee.

Heard, R.W. III. 1967. Some helminth parasites of the clapper rail, *Rallus longiorostris* Boddaert, from the Atlantic and gulf coasts of the United States. M.S. thesis, University of Georgia, Athens. 32 pp.

———. 1968. Parasites of the clapper rail, *Rallus longiorostris* Boddaert. I. The current status of the genus *Levinseniella* with the description of *Levinseniella byrdi* n. sp. (Trematoda: Microphallidae). *Proc. Helminthol. Soc. Wash.* 35:62–67.

Heard, R.W. III, and J.M. Kinsella. 1994. *Levinseniella deblocki,* new species (Trematoda: Digenea: Microphallidae) from salt marshes along the eastern Gulf of Mexico with notes on its functional morphology and life history. *Gulf Res. Rep.* 9:97–103.

Heard, R.W. III, and W.B. Sikora. 1969. *Probolocoryphe otagaki,* 1958 (Trematoda: Microphallidae), a senior synonym of *Mecynophallus cable,* Connor, Balling, 1960, with notes on the genus. *J. Parasitol.* 55:674–675.

Hill, E.F. 1988. Brain cholinesterase activity of apparently normal wild birds. *J. Wildl. Dis.* 24:51–61.

Jasmin, A.M., D.E. Cooperrider, C.P. Powell, and J.N. Baucom. 1972. Enterotoxemia of wildfowl due to *Cl. perfringens* type C. *J. Wildl. Dis.* 8:79–84.

Johnston, D.W. 1976. Organochlorine pesticide residues in uropygial glands and adipose tissue of wild birds. *Bull. Environ. Contam. Toxicol.* 16:149–155.

Kale, H.W. II. 1971. Florida region. *Am. Birds* 25:723–733.

Kinsella, J.M. 1973. Helminth parasites of the American coot, *Fulica americana americana,* on its winter range in Florida. *Proc. Helminthol. Soc. Wash.* 40:240–242.

———. 1997–98. Unpublished data. University of Florida, Gainesville.

Kinsella, J.M., and D.J. Forrester. 1972. Unpublished data. University of Florida, Gainesville.

Kinsella, J.M., L.T. Hon, and P.B. Reed, Jr. 1973. A comparison of the helminth parasites of the common gallinule (*Gallinula chloropus cachinnans*) and the purple gallinule (*Porphyrula martinica*) in Florida. *Am. Midl. Nat.* 89:467–473.

Kinsella, J.M., and V.F. Nettles. 1996. Unpublished data. University of Florida, Gainesville.

Klass, E.E., H.M. Ohlendorf, and E. Cromartie. 1980. Organochlorine residues and shell thickness in eggs of the clapper rail, common gallinule, purple gallinule, and limpkin (Class Aves), eastern and southern United States, 1972–1974. *Pestic. Monit. J.* 14:90–94.

Klukas, R.W., and L.N. Locke. 1970. An outbreak of fowl cholera in Everglades National Park. *J. Wildl. Dis.* 6:77–79.

Lewandowski, K., C. Ribic, J. Skolada, and L. Sullivan. 1998. Prevalence of hematozoa in soras (*Porzana carolina*) and Virginia rails (*Rallus limicola*) in southeastern Wisconsin. Wildlife Disease Association Annual Conference, Madison, Wis.: 18 (abstract).

Logan, T.H. 1997. Florida's endangered species, threatened species, and species of special concern. Official lists. Florida Game and Fresh Water Fish Commission, Tallahassee. 14 pp.

Longstreet, R.J. 1955. Ornithology of the mosquitoes. *Fla. Nat.* 28:9–20.

MacDonald, D., and E. Martin. 1971. Trends in harvest of migratory game birds other than waterfowl, 1964–65 to 1968–69. U.S. Fish and Wildlife Service, Special Scientific Report—Wildlife 142.

Maehr, D.S., and J.R. Brady. 1986. Food habits of bobcats in Florida. *J. Mammal.* 67:133–138.

Maehr, D.S., and J.Q. Smith. 1988. Bird casualties at a central Florida power plant: 1982–86. *Fla. Field Nat.* 16:57–80.

Marion, W.R., T.E. O'Meara, G.D. Riddle, and H.A. Berkhoff. 1983. Prevalence of *Clostridium botulinum* type C in substrates of the phosphate-mine settling ponds and implications for epizootics of avian botulism. *J. Wildl. Dis.* 19:302–307.

Marsh birds die. 1948. *Fla. Wildl.* 2:20.

Martin, E.M., and P.I. Padding. 1996. Preliminary estimates of waterfowl harvest and hunter activity in the United States during the 1995 hunting season. Unpublished report. Office of Migratory Bird Management, U.S. Fish and Wildlife Service, Laurel, Md. 34 pp.

McEwan, L.C., and D.H. Hirth. 1980. Food habits of the bald eagle in north-central Florida. *Condor* 82:229–231.

McLaughlin, G.S., and J.M. Kinsella. 1996. Unpublished data. University of Florida, Gainesville.

Meanley, B. 1992. King rail. *Birds N. Am.* 3:1–12.

Melvin, S.M., and J.P. Gibbs. 1996. Sora. *Birds N. Am.* 250:1–20.

Mihalik, M.B. 1993. Unpublished data. Solid Waste Authority, Palm Beach County, West Palm Beach, Fla.

Mulholland, R., and H.F. Percival. 1982. Food habits of the common moorhen and purple gallinule in north-central Florida. *Proc. Annu. Conf. Southeast. Assoc. Game Fish Comm.* 36:527–536.

Myers, O.B., W.R. Marion, T.E. O'Meara, and C.E. Roessler. 1989. Radium-226 in wetland birds from Florida phosphate mines. *J. Wildl. Manag.* 53:1110–1116.

Nesbitt, S.A. 1978. Limpkin. In: *Rare and endangered biota of Florida*. Vol. 2, *Birds*. H.W. Kale II (ed.). University Press of Florida, Gainesville. pp. 86–88.

Nesbitt, S.A., D.T. Gilbert, and D.B. Barbour. 1976. Capturing and banding limpkins in Florida. *Bird-Banding* 47:164–165.

Nettles, V.F., F.E. Kellogg, J.M. Kinsella, and M.G. Spalding. 1996. Unpublished data. Southeastern Cooperative Wildlife Disease Study, Athens, Ga.

Nickol, B.B., and R.W. Heard III. 1970. *Arhythmorhynchus frassoni* from the clapper rail, *Rallus longirostris,* in North America. *J. Parasitol.* 56:204–206.

Odom, R.R. 1975. Mercury contamination in Georgia rails. *Proc. Annu. Conf. Southeast. Assoc. Game Fish Comm.* 28:649–658.

O'Meara, T.E., W.R. Marion, C.E. Roessler, G.S. Roessler, H.A. Van Rinsvelt, and O.B. Myers. 1986. *Environmental contaminants in birds: phosphate-mine and natural wetlands.* Florida Institute of Phosphate Research Publication 05-003-045, Bartow. 77 pp.

Overstreet, R.M., and R.W. Heard. 1995. A new species of *Megalophallus* (Digenea: Microphallidae) from the clapper rail, other birds, and the littoral isopod *Ligia baudiniana. Can. J. Fish. Aquat. Sci.* 52, Suppl. 1:98–104.

Owre, O.T. 1978. Mangrove clapper rail. In: *Rare and endangered biota of Florida*. Vol. 2, *Birds.* H.W. Kale II (ed.). University Press of Florida, Gainesville. pp. 113–114.

Quist, C.F., and J.R. Fischer. 1996–98. Unpublished data. Southeastern Cooperative Wildlife Disease Study. Athens, Ga.

Robertson, W.B., Jr., and G.E. Woolfenden. 1992. *Florida bird species. An annotated list.* Spec. Publ. 6, Florida Ornithological Society, Gainesville. 260 pp.

Rodgers, J.A., Jr., H.W. Kale II, and H.T. Smith (eds.). 1996. *Rare and endangered biota of Florida.* Vol. 5, *Birds.* University Press of Florida, Gainesville. 688 pp.

Rohani, I.B., and H.L. Cromroy. 1979. Taxonomy and distribution of chiggers (Acarina: Trombiculidae) in northcentral Florida. *Fla. Entomol.* 62:362–376.

Runde, D.E. 1996. Black rail. In: *Rare and endangered biota of Florida*. Vol. 5, *Birds.* J.A. Rodgers, Jr., H.W. Kale II, and H.T. Smith (eds.). University Press of Florida, Gainesville. pp 323–328.

Sepulveda, M.S. 1995. Unpublished data. University of Florida, Gainesville.

Sileo, L. 1997. Unpublished data. National Wildlife Health Research Center, Madison, Wis.

Smith, K., and V.F. Nettles. 1994. Unpublished data. Southeastern Cooperative Wildlife Disease Study, Athens, Ga.

Snyder, B. 1994. Unpublished data. Florida Department of Environmental Protection, Tallahassee.

Snyder, N.F., and H.A. Snyder. 1969. A comparative study of mollusc predation by limpkins, Everglade kites, and boat-tailed grackles. *Living Bird* 8:177–223.

Spalding, M.G., and J.R. Fischer. 1997. Unpublished data. University of Florida, Gainesville, and Southeastern Cooperative Wildlife Disease Study, Athens, Ga.

Spalding, M.G., and J.M. Kinsella. 1994. Unpublished data. University of Florida, Gainesville.

Sprunt, A., Jr. 1954. *Florida bird life.* Coward-McCann, New York. 527 pp.

Stevenson, H.M., and B.H. Anderson. 1994. *The birdlife of Florida.* University Press of Florida, Gainesville. 892 pp.

Stoddard, H.L. 1931. *The bobwhite quail, its habits, preservation and increase.* Charles Scribner's Sons, New York, 559 pp.

Subcommittee on Nutrient and Toxic Elements in Water. 1974. *Nutrients and toxic substances in water for livestock and poultry.* National Academy of Sciences, National Research Council. 93 pp.

Taylor, W.K., and B.H. Anderson. 1973. Nocturnal migrants killed at a central Florida TV tower, autumns 1969–71. *Wilson Bull.* 85:42–51.

———. 1974. Nocturnal migrants killed at a central Florida TV tower, autumn 1972. *Fla. Field Nat.* 2:40–43.

Taylor, W.K., and M.A. Kershner. 1986. Migrant birds killed at the Vehicle Assembly Building (VAB), John F. Kennedy Space Center. *J. Field Ornithol.* 57:142–154.

Thomas, N. 1985. Unpublished data. National Wildlife Health Center, Madison, Wis.

Walkinshaw, L.H. 1982. Observations on limpkin nesting. *Fla. Field Nat.* 10:45–64.

Weston, F.M. 1966. Bird casualities on the Pensacola Bay bridge (1938–1949). *Fla. Nat.* 39:53–55.

Wheeler, W.B., D.P. Jouvenaz, D.P. Wojcik, W.A. Banks, C.H. VanMiddelem, C.S. Lofgren, S. Nesbitt, L. Williams, and R. Brown. 1977. Mirex residues in nontarget organisms after application of 10-5 bait for fire ant control, northeast Florida, 1972–1974. *Pestic. Monit. J.* 11:146–156.

Cranes

I. Introduction

Two cranes, the Sandhill Crane (*Grus canadensis*) and the Whooping Crane (*Grus americana*), are found in Florida. Two subspecies of Sandhill Crane are common. The Florida Sandhill Crane (*G. c. pratensis*) is resident throughout the state, with most of the birds residing in northern and central Florida. The Greater Sandhill Crane (*G. c. tabida*) migrates from its breeding range in Michigan, eastern Minnesota, Wisconsin, and southern Ontario to northern and central Florida (Nesbitt and Williams 1979). Some captive-raised Greater Sandhill Cranes were released experimentally into Florida (Nesbitt and Carpenter 1993). The Lesser Sandhill Crane (*G. c. canadensis*) is rarely seen in Florida (Nesbitt 1992a). The history of the Whooping Cranes within Florida is sketchy, with no reliable reports after the 1920s (Nesbitt 1982). This crane was reintroduced recently into central Florida by soft-releasing captive-raised juvenile birds beginning in 1993 (U.S. Fish and Wildlife Service 1986, 1994; Nesbitt 1996a; Nesbitt et al. 1997, 2001); chicks were first produced in 2000.

The Florida Sandhill Crane is listed as threatened by the Florida Game and Fresh Water Fish Commission (Logan 1997) and the Florida Committee on Rare and Endangered Plants and Animals (Rodgers et al. 1996). The Whooping Crane is listed as a species of special concern by the Florida Game and Fresh Water Fish Commission, as a threatened/experimental population by the U.S. Fish and Wildlife Service (Logan 1997), and as recently extirpated by the Florida Committee on Rare and Endangered Plants and Animals (Rodgers et al. 1996).

Life history and some general disease information has been reported for the Florida Sandhill Crane (Williams and Phillips 1972; Carpenter and Derrickson 1987; Tacha et al. 1992; Carpenter 1993; Nesbitt 1985, 1996b; Nesbitt et al. 1991; Nesbitt and Tacha 1997), Greater Sandhill Crane (Windingstad 1988; Tacha et al. 1992) and Whooping Crane (Carpenter and Derrickson 1981; Nesbitt 1982, 1996a; Lewis 1995; Nesbitt et al. 2001), and all of these species (Gee et al. 1981; Carpenter 1993; Ellis et al. 1996; Friend and Franson 1999). Most of the information presented in this chapter stems from health evaluations associated with biological studies conducted by the Florida Game and Fresh Water Fish Commission (currently the Florida Fish and Wildlife Conservation Commission, FFWCC) in Alachua, Osceola, and Lake counties, and includes unpublished data collected by Spalding, Forrester, Nesbitt, Folk, and Williams, and clinical faculty at the University of Florida, College of Veterinary Medicine. Examinations include 865 Sandhill Crane observations and 1000 Whooping Crane observations from 1970 to 2000. These include examinations at death of 136 Sandhill Cranes and 107 Whooping Cranes.

II. Trauma

Trauma, associated with exposure to manmade objects and habitat change, was the greatest cause of mortality of Sandhill Cranes (table 19.1). However, this information is undoubtedly biased because birds found dead along roadsides are more visible than those that die of more natural causes in remote areas. Of the types of trauma that could be identified, vehicle strikes were most common, followed by entanglement in fences, collision with power lines, gunshot, kicks by cows or horses, and entanglement in wire or monofilament. Cranes may be attracted to roadsides to collect gravel, which aids digestion, since gravel is rare in Florida soils. Power line collisions may be a greater hazard for migratory Sandhills less familiar with local hazards than resident birds (Nesbitt and Gilbert 1976; Nesbitt 1996b). At least 1 if not more of the Sandhill Cranes that hit power lines were electrocuted. Their large wingspan and long legs make them more likely to span wires and thus susceptible to electrocution. One Sandhill Crane residing at an airport was hit and killed by an airplane (Folk et al. 2001), but this is the only such case of which we are aware. Hunting of Sandhill Cranes in Florida was suspended with the passage of the Migratory Bird Act in 1918, but some cranes undoubtedly continue to be shot illegally. The extent of this mortality is poorly documented, but unlikely to be extensive. One Sandhill Crane became entangled in monofilament line and lost a toe (Folk et al. 2001). Another died after becoming entangled in baling wire. A Sandhill Crane survived impalement by an arrow; the arrow went through the skin on the side of the body and later fell out on its own. Although mortality was never documented, several Sandhill and Whooping Cranes were captured to remove bits of black plastic mulch, rubber O-rings, or gunshot casings from their bills so that they could forage (Folk et al. 2001).

Trauma associated with manmade structures is probably assessed more completely using information collected from the Whooping Crane reintroduction, since every bird was monitored by radiotelemetry for several years. Of the 115 that died or disappeared between 1993–99, 8% died from interactions with manmade objects. Five were electrocuted when they hit power lines, 2 hit either a power line or a vehicle or both, 1 was shot, and 1 died secondarily to breaking its leg when trapped between 2 cattle feeders. Three Whooping Cranes died at one time from both electrocution and trauma when they hit a power line at night in Brevard County in 1997. Power line strikes have been a significant cause of mortality for Whooping Cranes elsewhere (Brown et al. 1987).

Table 19.1. Primary causes of mortality for 122 Sandhill and 115 Whooping Cranes in Florida[a]

Cause Subcategory	Sandhill Cranes		Whooping Cranes	
	Total	%	Total	%
Unidentified trauma	8	7	0	0
Road trauma	29[b]	24	7[c]	6
Trauma with other manmade objects				
Cattle trough	0	0	1	1
Fence	8[d]	7	0	0
Wire/Monofilament	2	2	0	0
Gunshot	4	3	1	1
Weather				
Lightning	2	2	0	0
Natural trauma				
Intraspecific aggression	3	2	0	0
Flying collision	1	1	0	0
Horse/Cow	2	2	0	0
Predation	14[e]	11	82[f]	71
Emaciation	2	2	0	0
Malformed beak	2	2	0	0
Neoplasia	9[g]	7	0	0
Multiple diseases	7	6	1	1
Biotoxins	14[h]	11	0	0
Protozoal diseases	4[i]	3	1	1
Aspergillosis	2	2	2	2
Unknown	9	7	20	17

Sources: Spalding et al. (2001), Folk et al. (2001), Nesbitt et al. (2001).

a. Best estimate of primary problem; i.e. a crane with peanut toxicosis that is killed by a predator is listed under peanut toxicosis. Table does not include birds that died in captivity of causes unrelated to the initial problem, but does include some birds that were part of reintroduction projects in Florida.

b. Includes 7 power line collisions, 1 resulting in electrocution; 15 vehicular collisions; and 6 cases for which the source of the trauma could not be identified.

c. Includes 5 electrocutions and 2 cases for which the source of trauma could not be identified.

d. One was electrocuted.

e. Four were killed by an avian predator, probably a Bald Eagle.

f. Five were killed by alligators, the remaining by bobcats or occasionally coyotes.

g. Eight were cholangiocarcinomas and 1 a chondrosarcoma.

h. Includes 12 birds that died from peanut toxicosis and 2 from venomous bites.

i. Includes 2 cases of disseminated visceral coccidiosis (DVC) and 2 cases of enterocolitis caused by *Hexamita* sp.

There have been other sublethal injuries: 2 Whooping Cranes became entangled in a fence, and another swallowed a fishing lure (figure 19.1). All of these would have died without intervention (Folk et al. 2001). Monofilament line or string became entangled on the legs and feet of 3 Whooping Cranes in Osceola County in 1996–97, causing swelling distal to the line (figure 19.2). These birds were captured and the monofilament line removed. One subsequently lost a hind toe (Folk et al. 2001). Whooping Cranes have been observed with shotgun shells and flattened aluminum cans temporarily caught on their bills. One Whooping Crane interacted with a power line in such a way that its radio transmitter was left hanging on the line and the bird escaped with no apparent injury (Folk et al. 2001).

More natural causes of traumatic injury include intraspecific aggression, accidents, ven-

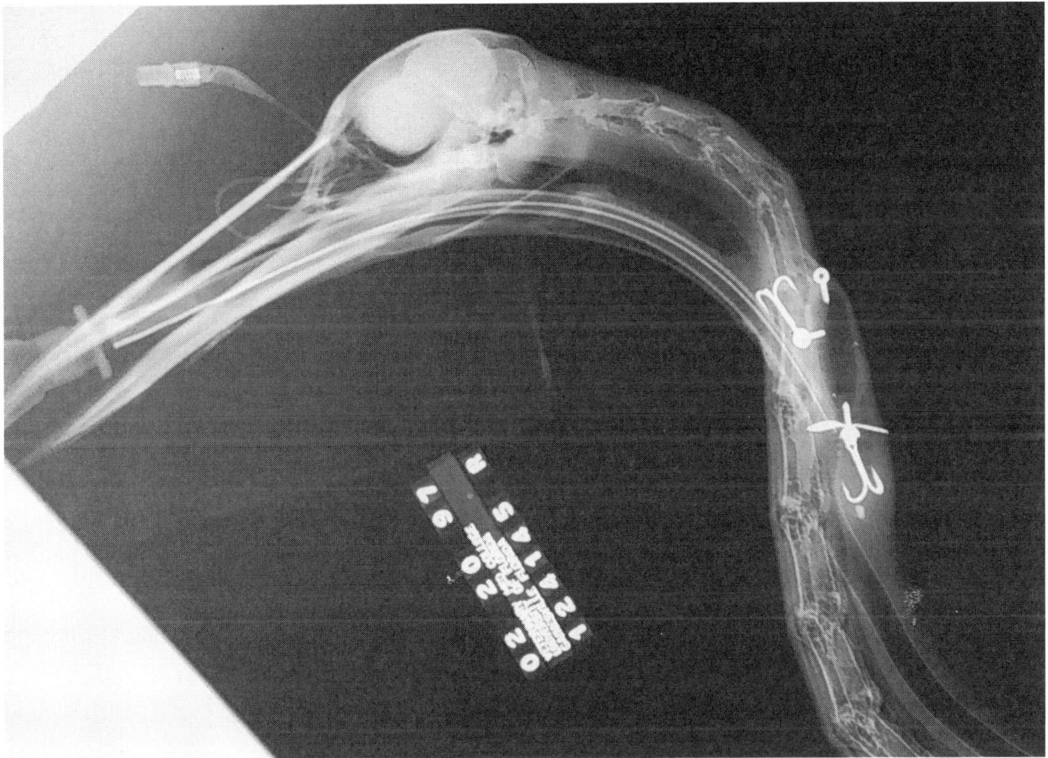

FIGURE 19.1. Radiograph of a fishing lure in the esophagus of a Whooping Crane from Lake County, 1997.

omous bites, and predation. A juvenile Sandhill Crane fell from the sky in Putnam County in 1990. A severe wound on the head, probably inflicted by another crane, may have caused it to fall and then die from additional injuries when it hit the ground (Spalding and Nesbitt 1998). Additionally a Sandhill Crane was observed to fall from a soaring flock into a tree in Levy County in 1998. It had a broken wing, suspected to have been caused by a collision with another soaring bird (Spalding and Nesbitt 1998). An experimentally released Sandhill Crane with a swollen leg, probably from a snake bite, was attacked and killed by other cranes when it became cornered within a fenced enclosure in Osceola County in 1992 (Spalding et al. 2001). In 1999 a fledgling Sandhill Crane, monitored with a radio transmitter, was found recumbent in Alachua County and died from being bitten by a snake

or other venomous animal (Spalding et al. 2001). In 2 cases Whooping Cranes in Osceola County were injured on the head, 1 losing its vision. It is likely that these injuries were due to intraspecific aggression.

Partial leg loss has not been uncommon. Three Sandhill Cranes in Alachua County, 1972–90, had amputations just below the hock and there was evidence of pododermatitis and degenerative changes in the opposite foot (figure 19.3). The cause was never apparent (Spalding and Nesbitt 1998). A Sandhill Crane captured in Lake County in 1994 had the last phalanx missing from all the toes of the right foot. There were also degenerative changes on the left foot. Frostbite is a possible, but unproven, cause of injury to legs and toes. Calle et al. (1982) discuss cases of distal extremity necrosis secondary to frostbite in captive birds. Mortality of Sandhill Cranes associated with

FIGURE 19.2. Whooping crane leg entangled in monofilament line. Courtesy of Martin J. Folk.

capture attempts was common before newer techniques utilizing drugged corn and other trap types (Folk et al. 1999) were developed. Most of this mortality was due to overdoses of alpha-chloralose (Williams and Phillips 1973) and rocket-netting, which caused physical injuries and secondary capture myopathy. Capture myopathy has also been reported associated with rocket-netting of Sandhill Cranes in Wisconsin (Windingstad et al. 1983).

III. Predation

Predation was the second most common cause of mortality (table 19.1); however, because of the poor detectability of predated carcasses predation was likely underestimated (Spalding et al. 2001). For Sandhill Cranes, 11% of those examined were killed by predators, mostly by bobcats. Wood et al. (1993) found remains of 3 Sandhill Cranes in 2 Bald Eagle nests in Alachua and Marion counties. They suspected that the eagles were taking cranes that were debilitated from peanut toxicosis or injured. Nesbitt and Badger (1995) suggested that predation, primarily by bobcats, is probably the most common cause of mortality for prefledged Florida Sandhill Cranes. They reported a case of coyote (a recent invader of Florida) predation of a prefledged Florida Sandhill Crane in Alachua County in 1993.

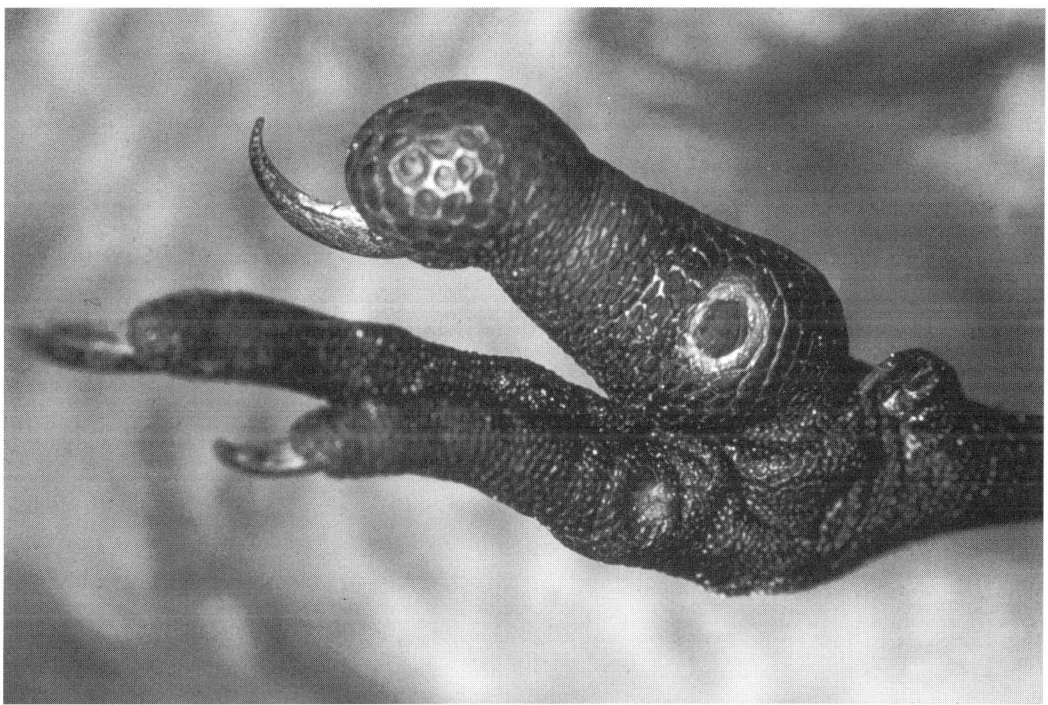

FIGURE 19.3. Plantar surface of the foot of a Sandhill Crane with loss of the lower portion of the opposite leg. Note the pododermatitis and degenerative changes. Courtesy of Stephen A. Nesbitt.

Prefledged young are susceptible also to predation by domestic dogs and cats (Nesbitt 1996b). Bennett and Bennett (1987) observed an alligator take a juvenile Sandhill Crane in the Okeefenokee Swamp in Georgia. Nesbitt (1992b) estimated the probability of survival of Florida Sandhill Cranes in north Florida to 80 days at 0.649.

Seventy-one percent of introduced Whooping Cranes that died were killed by predators, almost all of them bobcats, but in a few cases coyotes and alligators were suspected. The higher rate of predation mortality in Whooping Cranes may be due to a combination of factors, including more accurate monitoring but, more importantly, a lack of predator avoidance conditioning by parent birds, not available for experimentally released Whooping Cranes. Unlike many species, juvenile cranes remain in the company of their parents well beyond the age of fledging (67–75 days, Tacha et al. 1992), up to 321 days of age

(Nesbitt 1992b). Most (62%) of the first-year mortality of introduced Whooping Cranes occurred within the first 3 months of release (Nesbitt et al. 1997). An increase in alligator predation on recently released Whooping Cranes during 1999 may have been due to a reduction in wetland habitat caused by an ongoing drought (Nesbitt et al. 2001). Beginning in 1998, synchronous molt of flight feathers resulting in a period of flightlessness in 3-year-old birds was associated with increased bobcat predation during that time (Nesbitt et al. 2001).

IV. Inclement weather

An adult and juvenile Sandhill Crane were killed by lightning when standing close to a fence during a lightning storm in Osceola County in 1995 (Spalding et al. 2001). Reproduction in cranes appears to be closely related to water conditions in marshes. In 1999, a Whooping Crane nest flooded in Brevard County (Folk and Nesbitt 2000). Severe flooding caused submersion of nesting material and only a few Sandhill Crane chicks could be found in Alachua County during the 1998 nesting season. Flooding is the major cause of nest loss in Florida Sandhill Cranes (Nesbitt 1988). During 1998–2001, below-average yearly rainfall caused a severe and prolonged drought resulting in declines in lake levels and a reduction in roosting marshes in central Florida. During this period predation by alligators on Whooping Cranes increased, exposure of cranes to eastern equine encephalitis virus decreased, prevalence of oral disseminated visceral coccidiosis lesions increased, dispersal of birds from introduction sites increased, and breeding decreased (Nesbitt et al. 2001). Frostbite may occasionally cause distal extremity necrosis in cranes in Florida (see section II, Trauma).

V. Environmental contaminants

No information on organochlorines, organophosphates, and carbamates was available for cranes in Florida. See Lamont and Reichel (1970) and Lewis et al. (1992) for contaminants in Whooping Cranes outside of Florida.

Whooping Cranes frequently ingest unusual objects, probably because they are searching for large gastroliths, which are rare in Florida soils. Twenty-eight percent of 78 Whooping Cranes with intact stomachs at death had metal objects in their stomachs (Spalding et al. 2001). An additional 8% had glass or plastic objects in their stomachs. In contrast, only 3 of 212 (1%) of Sandhill Crane stomachs had metal objects in them. Sandhill Crane stomachs usually contained smaller quartz or limestone pieces than the larger stones found in Whooping Crane stomachs. After prolonged residence in Florida most Whooping Cranes lost most of their large stones.

Spalding et al. (1997) reported illness associated with metal objects in stomachs and elevated serum zinc concentrations in penned Whooping Cranes prior to their release in 1993. Illness appeared to be associated with serum zinc concentrations that were greater than 7 ppm; however, concentrations fluctuated widely. At least 7 of the 14 birds from the first release group had ingested metal objects. The pieces of metal were all of a type used in the construction of the rearing and release pens. The affected birds became lethargic and had abnormal posture and preening behavior. These birds improved following surgical removal of metal items; however, bobcats killed them after release. Metal ingestion probably contributed to the poor survival of this group of birds and was partly resolved by radiographing cranes prior to releasing and removing metal items laproscopically, removing metal from rearing and release pens, closely monitoring serum zinc concentrations, and using nonmetal release pens. However, wild Whooping Cranes continue to consume items that are both physically dangerous, such as a fishing lure (see section II, Trauma, above), and toxic, such as lead.

Three juvenile Florida Sandhill Cranes from Osceola County were tested to establish normal serum zinc and copper concentrations (table 19.2). A few clinical cases were tested also for blood lead concentrations and found to be mildly elevated, but no background survey of metals in Sandhill Cranes has been conducted in Florida.

A female Whooping Crane shot in Orange County in 1998 had a piece of metal in her stomach and 5.1 ppm lead in liver. This bird had nested but failed to lay eggs, then lost her mate prior to being shot. It is possible that the elevated lead concentration, which approached the range considered toxic for waterfowl (6–8 ppm, Locke and Thomas 1996) was responsible for her failure to lay eggs and maintain a pair bond. Windingstad et al. (1984) discussed lead poisoning in Sandhill Cranes and Snyder et al. (1992) reported lead poisoning of a Whooping Crane in New Mexico. A captive Sandhill Crane experimentally dosed with lead died with 26 ppm lead in liver (Windingstad et al. 1984).

Table 19.2. Concentrations of metals in tissues of Sandhill and Whooping Cranes from Florida

| Species Tissue | County | Year(s) | Zinc | | | Lead | | | Copper | | | Data source |
			No. Examined	Mean[a]	(Range)	No. Examined	Mean[a]	(Range)	No. Examined	Mean[a]	(Range)	
Sandhill Crane												
Blood	Alachua	1991	NA	—	—	1	1.3	—	NA	—	—	B
	Levy	1992	NA	—	—	1	0.51	—	NA	—	—	A
	Osceola	1992–93	3	2.8	(1.6–4.3)	NA	—	—	3	0.46	(0.41–0.59)	C
Whooping Crane												
Liver	Lafayette	1997	1	48	—	1	0.21	—	NA	—	—	A
	Lake	1997–99	3	33	(25–40)	1	0.66	—	2	6.5	(4.4–9.7)	A
	Orange	1998	1	31	—	1	5.1	—	NA	—	—	A
	Osceola	1993–98	20	49	(24–180)	9	0.30	(ND–1.2)	9	7.0	(3–19)	A
Blood	Osceola	1993–95	33	3.9	(0.54–17)	2	0.06	(0.05–0.07)	NA	—	—	A

| Species Tissue | County | Year(s) | Iron | | | Data source |
			No. Examined	Mean[a]	(Range)	
Liver	Osceola	1993–95	5	500	(210–1,100)	A

Sources: A = Spalding et al. (1997) and Spalding et al. (2001), B = Abou-Madi (1991), C = Lung (1992).
NA = not analyzed, ND = not detected.
a. Geometric mean, ppm wet weight.

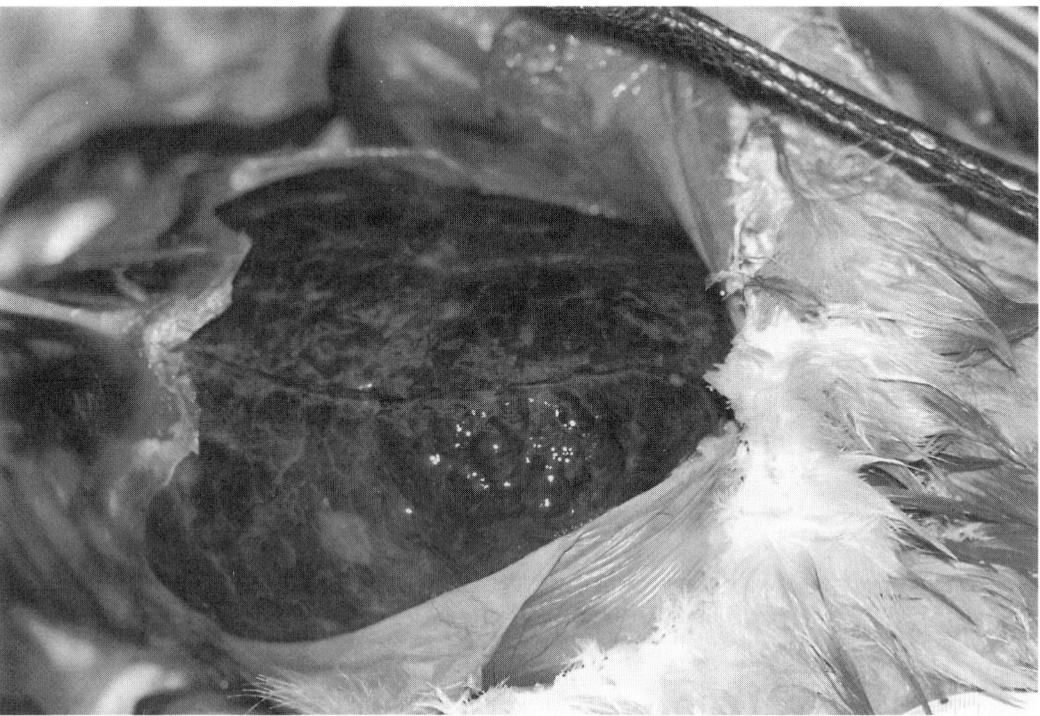

FIGURE 19.4. Exposed liver of a Sandhill Crane almost entirely replaced by a tumor of the bile ducts.

VI. Neoplasia

Bile duct tumors or cholangiocarcinomas were diagnosed in 9 Sandhill Cranes from Alachua and Marion counties, 1986–97 (Allen et al. 1985; Spalding et al. 2001) (table 19.1). This type of neoplasia appears to be more prevalent in Florida Sandhill Cranes than in Greater Sandhill Cranes. Six of the 9 were identified as Florida Sandhill Cranes, 2 were unidentified to subspecies, and only 1 appeared to be a Greater Sandhill Crane. All of the birds with tumors were emaciated, some with multiple diseases, and 2 were found with broken bones. Two of the birds were young—1 was 1 year old, the other less than 3 years—and 1 was at least 12 years old. The 1-year-old had been in captivity for a while, and thus may not be representative of the age distribution in wild birds; that bird also exhibited signs of acute mycotoxicosis (see section VIII, Biotoxins). The entire liver was enlarged and almost entirely replaced by large cystic spaces in the Greater Sandhill Crane found weak in Alachua County in 1997 (figure 19.4). In 2 other cases, the tumor had metastasized to other organs, including lungs, air sacs, and the serosal surface of the proventriculus.

Because of the high prevalence of bile duct carcinomas and their occurrence in young birds, a toxin or viral etiology has been suspected. Couvillion et al. (1991) speculated that chronic exposure of Mississippi Sandhill Cranes to aflatoxin could be the cause of the unusually high prevalence (31%) of neoplasms in that subspecies. Mycotoxins are known to be carcinogenic and immunosuppressive and to suppress reproduction in other birds (Carnaghan 1965; Giambrone et al. 1984). The very high prevalence in Florida Sandhill Cranes and not in Greater Sandhill Cranes indicates that the cause is likely to be found within Florida, and probably in north central Florida. Interestingly, Greater Sandhill Cranes appear to be more susceptible to acute mycotoxicosis (see section VIII, Biotoxins).

We found evidence of bile duct hyperplasia in an additional 12 Sandhill Cranes (Spalding et al. 2001). These pre-neoplastic changes were found in both resident and migratory Sandhill Cranes, and were restricted to birds collected from Alachua and Marion counties. Not enough known Greater Sandhill Cranes were examined to determine if there was a difference in prevalence between the 2 populations.

Cartilage tumors, possibly caused by a virus, are discussed in section IX, Viruses.

VII. Anomalies

Various degrees of beak malformation were present on 7 Sandhill Cranes. Most were in Florida Sandhill Cranes from Alachua and Citrus counties, 1986–98. Three of the birds (figure 19.5) had severely crossed bills that resulted in other health problems. One Sandhill Crane had duplicate toenails (figure 19.6). Two Sandhill Cranes from Alachua County in

FIGURE 19.5. A Sandhill Crane chick with a crossed bill.

FIGURE 19.6. A Sandhill Crane with duplicate toenails. Courtesy of Stephen A. Nesbitt.

FIGURE 19.7. A Sandhill Crane with crooked neck. Courtesy of Stephen A. Nesbitt.

1990–91, 1 each of the Florida and Greater subspecies, had malformed necks (figure 19.7). Both were 1 year old and 1 also had flaky skin; the other had malformed feathers. A cause was not determined.

VIII. Biotoxins

Acute mycotoxicosis (fusariomycotoxicosis) associated with ingestion of peanuts contaminated with tricothene resulted in illness and mortality in 28 Sandhill Cranes from Alachua, Levy, and Marion Counties during 1979–99 (Spalding et al. 2001). The clinical signs were characterized by a drooping head and neck, even during flight, as described by Roffe et al. (1989) and Windingstad et al. (1989). These signs were never observed in any Sandhill

Cranes known to be of the Florida subspecies. At necropsy, subcutaneous edema of the head and neck, petechial hemorrhage on the heart (figure 19.8), and a stomach full of peanuts were often found. If left in the field, these birds usually died or were taken by predators. Birds given supportive care frequently recovered. By closely monitoring color-marked bird in fields and at roost sites Nesbitt (1999) was able to estimate about 10% mortality due to direct toxicosis and indirect (predation) consequences of foraging in peanut fields in Alachua County during the 1980s. Several color-marked birds were known to recover spontaneously, without intervention. The number of birds affected by peanut toxicosis has declined significantly since then because farmers now plow the fields after harvesting the peanuts, reducing the availabil-

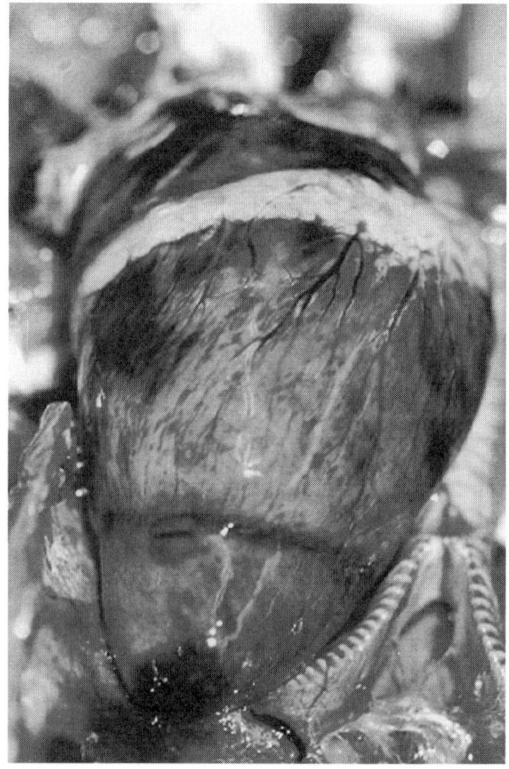

FIGURE 19.8. Petechial hemorrhages on the heart of a Sandhill Crane that died from peanut toxicosis.

FIGURE 19.9. A pox lesion on the bill of a young Florida Sandhill Crane.

ity of waste grains and peanuts. Mycotoxicosis has not been observed in either Whooping Cranes or Sandhill Cranes in central and southern Florida, where peanuts are infrequently grown. Large numbers of Sandhill Cranes have died elsewhere from peanut toxicosis, especially in Texas and New Mexico (Windingstad et al. 1989).

If bile duct tumors and biliary hyperplasia are caused by fungal toxins (see section VI, Neoplasia), then there appears to be a very interesting difference in the responses between migratory and resident populations of Sandhill Cranes. Tumors are much more common in Florida Sandhill Cranes, but acute mycotoxicosis, common in Greater Sandhill Cranes, has never been reported in this subspecies. These differences could be due to exposure to different toxins, exposure to different quantities of toxin, chronicity of exposure, or subspecific differences in response to a common toxin. Other etiologies have not been ruled out.

IX. Viruses

Poxviral infections appear to be rare and inconsequential in Sandhill Cranes. Cutaneous pox lesions were observed on 1 free-ranging Florida Sandhill Crane trapped in Alachua County in May 1973 and on 3 Florida Sandhill Cranes in July 1974 that had been pen-reared in Maryland and maintained in a pen for 9 months in Alachua County (Simpson et al. 1975). Two to 30 lesions were located on legs, feet, and heads of the cranes. Lesions were found also on the bills near the nares of 2 sibling fledgling Sandhill Cranes in Osceola County in 1993 (figure 19.9) (Spalding et al. 2001). An additional 22 birds in Alachua, Lake, Osceola, and Marion counties in 1987–96 had suspicious lesions; however, they were not confirmed by histopathology and in many cases may have been chrondromas, as described below. There have been no confirmed cases in Greater Sandhill Cranes or Whooping Cranes in Florida.

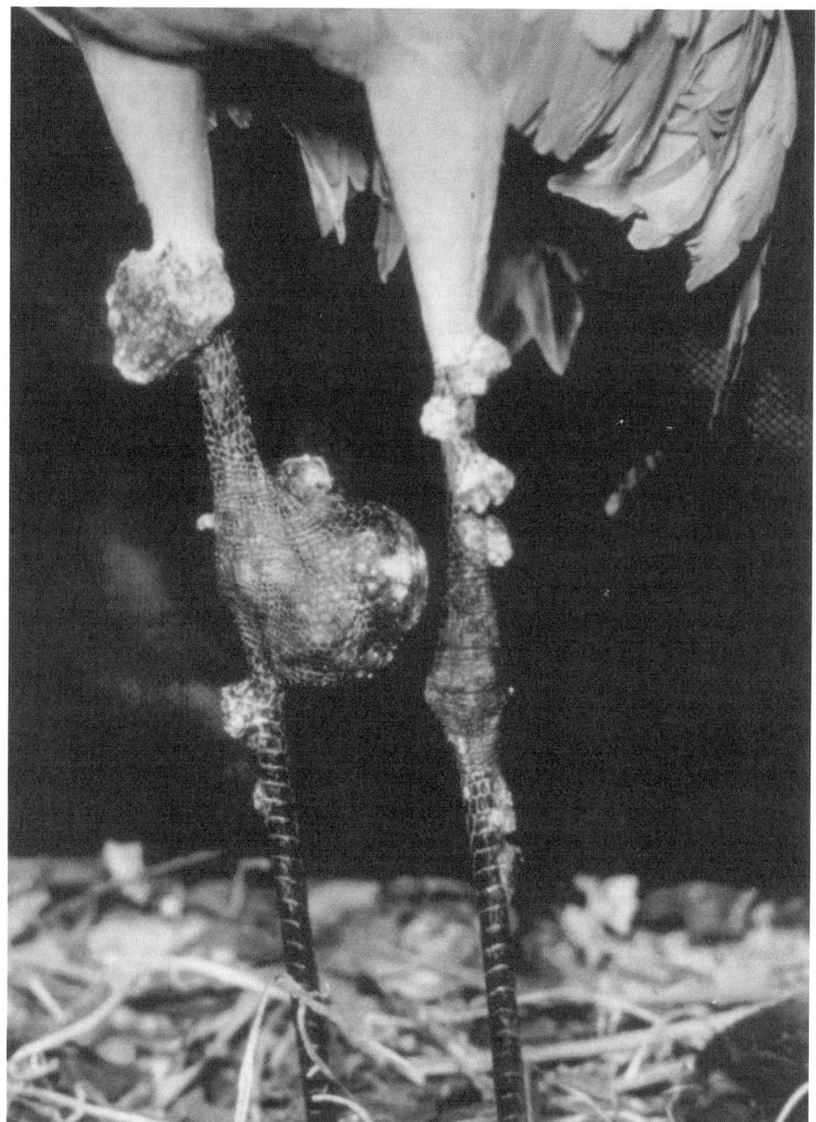

FIGURE 19.10. A Sandhill Crane with cartilage tumors on its legs. Courtesy of Stephen A. Nesbitt.

Large cutaneous cartilagenous nodules that grossly resembled poxvirus lesions were noted on the legs of a few free-flying Sandhill and Whooping Cranes in Alachua, Lake, and Osceola counties in 1989–95. Although the masses were usually below the epithelial layer the surface was frequently ulcerated; thus they grossly appeared to be of epidermal origin, like poxvirus lesions. The most severe case had nodules up to 3 cm in diameter (figure 19.10). This bird died from complications following attempted removal of some of the larger masses, which were confirmed histologically as multiple chondromas and chondrosarcomas. No poxviruses could be found in the tumors by electron microscopy (Spalding and Woodard

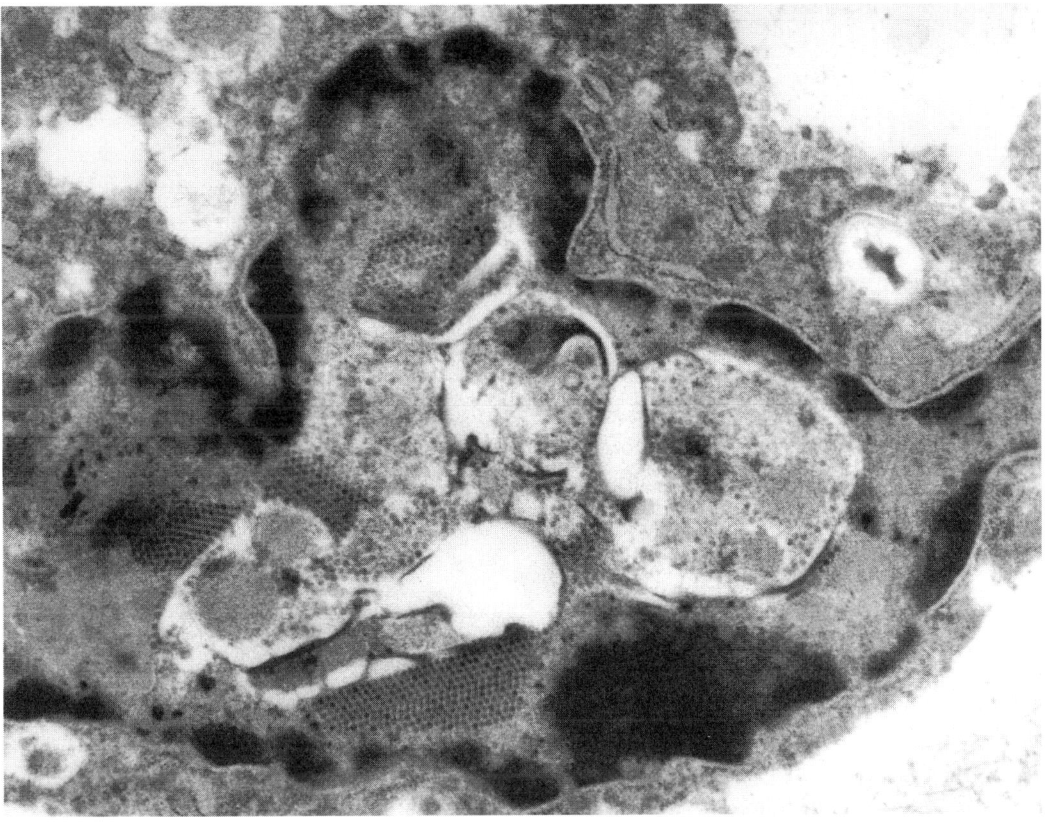

FIGURE 19.11. Electronmicrograph of intranuclear viral particles in a chondrocyte of a cartilage tumor on the leg of a Whooping Crane.

1992), however, an unidentified virus was found in several cranes with such lesions (figure 19.11) (Spalding and Woodard 1995).

Eastern equine encephalitis (EEE) virus caused mortality in a single captive Mississippi Sandhill Crane in Duval County in 1994 (Citino 1994). Other than this case, morbidity or mortality has not been associated with exposure to EEE virus in Sandhill Cranes, but mortality of captive Whooping Cranes in Maryland has occurred (Dein et al. 1986). A study of the epizootiology of EEE virus in north central Florida was begun in 1992 (Spalding et al. 2001) (table 19.3). Overall, 29% of Sandhill Cranes were seropositive, and the highest prevalences (>50%) were in Osceola County in 1992 and Lake County in 1995. Sentinel chickens and Northern Bobwhites tested positive in Osceola County in 1992 (4%) and in 1993 (90%), but not in 1994 (see also chapter 16, Northern Bobwhites). Wild birds, mostly passerines, tested positive in 1992 (3%), 1993 (12%), and 1994 (4%). Wild mammals were seropositive in 1992 (1%) and 1993 (31%). Gruiform blood was identified in <0.1% of blood meals collected from *Culiseta melanura*, the purported primary vector of EEE virus, in Vero Beach, Indian River County (Edman et al. 1972). Day and Stark (1996), using sentinel chickens in 40 counties in Florida, found that EEE virus was generally, but not exclusively, found in the panhandle and northern portions of the state whereas SLE virus was generally in the central and southern regions. Transmission of EEE virus to sentinel chickens declined significantly during the drought years

Table 19.3. Prevalence of Sandhill Cranes in Florida seropositive to arboviruses

Arbovirus Age	County	1992 No. Exam.	1992 % Positive	1993 No. Exam.	1993 % Positive	1994 No. Exam.	1994 % Positive	1995 No. Exam.	1995 % Positive	1992–95 No. Exam.	1992–95 % Positive
Eastern Equine Encephalitis (EEE)											
Juvenile (<1 yr)	Alachua	3	—	2	—	2	—	0	—	7	—
	Lake	0	—	4	—	4	—	7	—	15	53
	Osceola	20	0	21	14	1	—	0	—	42	7
Subadult & Adult (>1 yr)	Alachua	33	27	15	7	13	8	5	—	66	20
	Lake	0	—	17	29	24	33	26	54	67	40
	Osceola	28	54	11	27	1	—	0	—	40	45
TOTAL		84	29	70	20	45	22	38	55	237	29
St. Louis Encephalitis (SLE)											
Juvenile (<1 yr)	Alachua	3	—	2	—	2	—	0	—	7	—
	Lake	0	—	4	—	3	—	7	—	14	0
	Osceola	11	8	18	0	1	—	0	—	30	3
Subadult & Adult (>1 yr)	Alachua	21	33	6	—	12	8	9	—	48	21
	Lake	0	—	17	18	25	28	26	8	68	18
	Osceola	37	54	10	30	1	—	0	—	48	48
TOTAL		72	39	57	12	44	18	42	7	215	21

Source: Spalding et al. (2001).

of 1998–2000. Day and Stark (1998) also documented the transportation of western equine encephalitis and possibly EEE virus to Florida by infected Emus (*Dromaius novaehollandiae*).

Most of the Whooping Cranes introduced into Florida were vaccinated for EEE virus prior to arriving in Florida and again prior to release. Thus seroprevalence results had to be examined with knowledge of the vaccination history. Most Whooping Cranes had a decline in titer by 6 months following vaccination; since a killed vaccine was used, titers above 1:20 probably indicated natural exposure. Fifty-two Whooping Cranes were released without vaccination during 1994–1999 and the survival rate was similar to vaccinated birds released during the same time period. Only 1 of these birds seroconverted (>1:320) and that bird died from severe aspergillosis after being in Florida for 11 months. Forty percent of the 62 birds tested after >6 months post-vaccination had titers of 1:20 or greater. This was very similar to the seroprevalence of EEE virus in Sandhill Cranes in the same area. Two Whooping Cranes that became weak and emaciated in 1995 recovered with supportive care. These birds had elevated and declining titers to EEE virus during their recovery, and may have been ill from this virus.

Colwell et al. (1973) tested 1 Sandhill Crane by cell culture and that and another crane by indirect hemagglutination inhibition for turkey herpesvirus in 1970 and found them to be negative. A herpesvirus causing inclusion body disease of cranes has caused mortality of captive cranes, including a Greater Sandhill Crane, in Wisconsin (Docherty and Henning 1980; Docherty and Romaine 1983; Schuh et al. 1986). Sixty-five Florida Sandhill Cranes from Alachua (41), Levy (1), and Osceola (23) counties and 200 Whooping Cranes introduced into Florida tested negative for antibodies to this virus (Spalding et al. 2001).

X. Bacteria

Bacteria cultured from Sandhill Crane tissues are listed in table 19.4. A number of potentially pathogenic bacteria have been cultured from feces or large intestines of cranes with no apparent illness. White et al. (1973) examined 32 apparently normal Sandhill Cranes from north and central Florida and found no significant pathogens, except for 1 with *Edwardsiella tarda*, 1 with *Salmonella hartford*, and 1 with *S. java*. An additional bird with *S. hartford* was later identified by Forrester and White (1978). No *Salmonella* sp. isolates were detected in fecal samples from 11 Sandhill Cranes from Alachua County and 26 from Osceola County in 1992–94. Other *Salmonella* spp. have been reported in Sandhill Cranes outside of Florida, but have not been linked directly with adverse health effects (Francis and Huey 1964; Windingstad et al. 1977). Langenberg et al. (1996) tested various species of captive cranes in Wisconsin and wild Sandhill Cranes from Florida for *Campylobacter jejuni;* they found 2 or 3 of 7 from pooled samples of juveniles to be positive, but none of the wild adults. *Campylobacter jejuni* was commonly isolated from captive cranes, including some of the Whooping Cranes that were later released in Florida.

Mycobacterium flavescens was cultured from granulomatous lesions associated with the liver, spleen, and intestinal tract in a single severely debilitated Sandhill Crane in Alachua County in June 1981 (Buergelt 1981). A survey of feces from 15 Sandhill and 35 Whooping Cranes in 1992–97 for *M. avium* resulted in the finding of several nonchromogenic *Mycobacterium* listed in tables 19.4 and 19.5. *Mycobacterium avium* has been recovered outside of Florida from both Greater Sandhill Cranes and Whooping Cranes with granulomatous lesions (Thoen et al. 1977; Snyder et al. 1997) and was frequently found in Whooping Cranes that died from the experimentally introduced Rocky Mountain flock (Snyder et al. 1991).

Avian cholera outbreaks are extremely rare in Florida (see chapter 18, Rails, for additional information about this disease) and although cranes have been involved in these die-offs elsewhere (Zinkl et al. 1977), only 1 case of pasteurellosis has occurred in Florida cranes. A Whooping Crane injured by a bobcat in Lake

Table 19.4. Bacteria and fungi cultured from Sandhill Cranes collected in Florida

Class	Species	County	Year(s)	No. of birds	Tissue	Disease	Data source
Bacteria							
	Acinetobacter calcoaceticus	Alachua	1991	1	Trachea	Mycotoxicosis	A
	Campylobacter jejuni	Lake	1995	2	Feces	None	M
	Corynebacterium 2 spp.	Alachua	1990	1	Colon	Enterocolitis[a]	K
			1991	1	Trachea	Mycotoxicosis	A
	Diphtheroid	Alachua	1987	1	Liver	Multiple diseases	I
	Edwardsiella tarda	Highlands	1971	1	Large intestine	None	L
	Enterococcus 2 spp.	Alachua	1995	1	Humerus	Osteomyelitis	J
	Escherichia coli	Alachua	1987	1	Liver	Multiple diseases	I
					Feces	None	E
			1989	1	Ceca	Mycotoxicosis	F
		Clay	1984	1	Liver, spleen	None	B
	Flavobacterium sp.	Alachua	1991	1	Trachea	None	A
	Klebsiella oxytoca	Alachua	1991	1	Trachea	None	A
	Klebsiella-Enterobacter	Alachua	1995	1	Humerus	Osteomyelitis	J
	Micrococcus sp.	Alachua	1986	1	Feather	Malformed feathers	H
	Mycobacterium flavescens	Alachua	1981	1	Liver, spleen	Tuberculosis[a]	D
		Osceola	1993	2	Feces	None	N
	Mycobacterium terrae	Osceola	1993	1	Feces	None	N
	Proteus sp.	Alachua	1987	1	Leg abscess	Abscess	I
				1	Cloaca	None	I
	Pseudomonas aeruginosa	Alachua	1987	1	Liver	Multiple diseases	I

	Organism	County	Year	No.	Location	Disease	Source
	Pseudomonas sp.	Alachua	1970	1	Large intestine	None	G
			1991	1	Trachea	Mycotoxicosis	A
		Clay	1984	1	Liver	Multiple diseases	B
	Salmonella hartford	Alachua	1970, 1977	2	Large intestine	None	G
	Salmonella java	Glades	1970	1	Large intestine	None	G
	Serratia odorifera	Alachua	1987	1	Leg abscess	Abscess[a]	I
	Staphylococcus sp.	Alachua	1987	1	Liver	Multiple diseases	I
			1990	1	Feces	Enterocolitis	K
	Streptococcus sp.	Alachua	1987	1	Liver	Multiple diseases	I
	Streptococcus sp. α-hemolytic	Alachua	1991	1	Trachea	Mycotoxicosis	A
			1995	1	Humerus	Osteomyelitis[a]	J
	Streptomyces spp.	Alachua	1984	1	Air sac	Airsacculitis	C
Fungi							
	Aspergillus sp.	Alachua	1984	1	Lung	Pneumonia[a]	B
			1990	1	Air sac, lung	Pneumonia and airsacculitis[a]	K
			1997	1	Air sac	Airsacculitis[a]	N
	Aspergillus sp. (presumptive)	Alachua	1995	1	Air sac	Airsacculitis and pericarditis[a]	J

Sources: A = Roth et al. (1991), B = Allen et al. (1984), C = Calderwood-Mays et al. (1985), D = Buergelt (1981), E = Buergelt et al. (1987), F = Woodard & Layton (1989), G = Forrester & White (1978), H = Heard (1986), I = Russell & Bolon (1987), J = Homer et al. (1995), K = Buergelt et al. (1990), L = White et al. (1973), M = Langenberg et al. (1996), N = Spalding et al. (2001).

a. The organism listed was believed to have caused the disease.

Table 19.5. Bacteria and fungi cultured from Whooping Cranes in Florida

Class Species	County	Year(s)	No. of birds	Tissue	Disease	Data source
Bacteria						
Acinetobacter baumannii	Suwannee	1997	1	Trachea	None	A
Bacteroides ureolyticus	Osceola	1993	1	Lower intestine	None	B
Corynebacterium sp.	Suwannee	1997	1	Trachea	None	A
Enterobacter sp.	Suwannee	1997	1	Trachea	None	A
Enterococcus sp.	Osceola	1993	1	Upper intestine	None	B
Escherichia coli	Osceola	1993	1	Intestine	None	B
Klebsiella sp.	Suwannee	1997	1	Trachea	None	A
Mycobacterium (nonchromogenic)	Osceola	1994	1	Feces	None	A
Mycobacterium gordonae	Osceola	1994	1	Feces	None	A
Pasteurella multocida (somatic serotype 1)	Lake	2000	1	Blood, spleen	Septicemia[a,b]	A
Pasteurella multocida (A:2,5)	Osceola	1993	1	Liver, lung	Contaminant from predator?	B
Proteus sp.	Suwannee	1997	1	Trachea	None	A
Pseudomonas sp.	Suwannee	1997	1	Trachea	None	A
Salmonella java	Osceola	1993	1	Feces	None	A
Salmonella miami	Osceola	1994	1	Feces	None	A
Salmonella muenchen	Osceola	1993–97	8	Feces	None	A
Salmonella oranienburg	Osceola	1993	1	Feces	None	A
Salmonella sp. group C	Osceola	1994	1	Feces	None	A
Salmonella sp. group CI	Lake	1999	1	Feces	None	A
Salmonella thompson	Lake	1999	2	Feces	None	A
	Osceola	1995	3	Feces	None	A
Salmonella typhimurium	Osceola	1998	2	Feces	None	A
Salmonella virginiana	Osceola	1994	1	Feces	None	A
Staphylococcus sp.	Suwannee	1997	1	Trachea	None	A
Streptococcus sp.	Suwannee	1997	1	Trachea	None	A
Fungi						
Aspergillus fumigatus	Osceola	1995	1	Lung, air sac	Pneumonia and aspergillosis[a,c]	A
	Suwannee	1997	1	Lung, air sac	Pneumonia, airsacculitis, peritonitis, pericarditis[a]	A
Mucor sp.	Osceola	1998	1	Nasopharynx	Laryngitis	A

Sources: A = Spalding et al. (2001) and B = Meteyer (1993).
a. The organism listed was believed to have caused the disease.
b. Survived an attack by a bobcat and later died from injuries sustained in the attack.
c. Died while still in the release pen.

County in 2000 was septicemic with *Pasteurella multocida* (somatic serotype 1, the same serotype associated with large waterbird die-offs in North America [Friend 1999]). Areas of necrosis were present in the spleen. The crane died the next day from traumatic injuries sustained in the attack (Spalding et al. 2001). An isolation of a different serotype of *P. multocida* (capsular serogroup A: somatic serotype 2,5) was made from a Whooping Crane killed by a predator in Osceola County. The isolate was believed to be a contaminant from the mouth of a bobcat (Meteyer 1993).

XI. Fungi

Aspergillosis is relatively rare in wild cranes in Florida (tables 19.4 and 19.5). Infection with the fungus *Aspergillus* sp. caused pneumonia, airsacculitis and death in 1 young Sandhill Crane from Alachua County in 1990 (table 19.1). In captivity, aspergillosis is more common in cranes. An adult from Alachua County in 1995 developed aspergillosis while in captivity to repair a broken wing. Two Whooping Cranes have died with very severe aspergillosis in Florida, 1 prior to release in Osceola County in 1995; the second had been wild for 1 year in Suwanee County in 1997.

XII. Protozoans

Blood smears were examined from 383 Sandhill Cranes and 186 Whooping Cranes from Florida (Bennett et al. 1974, 1975; Forrester et al. 1974, 1975, 1976; Forrester and Bennett 1985; Telford et al. 1994; Spalding et al. 2001) (table 19.6). *Leucocytozoon grusi* and *Haemoproteus antigonis* were the most common blood parasites in Sandhill Cranes, but insufficient information exists to demonstrate any regional differences in Florida. *Leucocytozoon grusi* was originally described from Sandhill Cranes in Alachua County (Bennett et al. 1974). *Atoxoplasma* sp. was found only in cranes from Alachua County between the years 1974–76 (Forrester and Bennett 1985). *Plasmodium* sp. (*polare*-like) (figure 19.12) was described from 3 cranes in Alachua and Marion counties (Telford et al. 1994) and later observed in 2 additional birds from Alachua County in 1991–92 (Spalding 2000). All of these cranes were emaciated and debilitated severely, with the presence of other diseases or parasites (tuberculosis, coccidiosis, *H. antigonis*, *L. grusi,* and enteritis with yeast infection), indicating that these birds may have had a compromised immune system. Blood parasites similar in appearance to *Haemoproteus balearicae* were associated with severe anemia in sibling Sandhill Crane chicks in Osceola County in 1999 and in another chick from the same area (Dusek 2001). This parasite may have been introduced to wild Sandhill Cranes from captive exotic cranes in the area since it has been recorded previously in North America only from captive cranes (Halpern and Bennett 1983). Other than these, significant disease associated with hemoparasite infection has not been described for any Sandhill Cranes in Florida.

Haemoproteus antigonis was detected on 15 occasions from 9 Whooping Cranes (Spalding et al. 2001). An infection was present in 1 bird that was caught on 2 occasions, 2 years apart. In 1 case, the parasitemia was especially high, with >10 parasites per 100X field. This was the youngest Whooping Crane to be infected (16 months); it exhibited no sign of illness, and survived at least 6 years beyond that date. Although relatively high prevalences of *Leucocytozoon grusi* exist in Sandhill Cranes living in the same area, this parasite has not yet been observed in Whooping Cranes.

Sandhill Cranes frequently shed *Eimeria* sp. oocysts in feces. Both *Eimeria gruis* and *E. reichenowi* were recovered commonly from the feces of both Greater and Florida Sandhill Cranes (figure 19.13) (Courtney et al. 1975; Forrester et al. 1974, 1975, 1976; Spalding et al. 2001). Courtney et al. (1975) found 11 of 14 (79%) Florida Sandhills and 62 of 72 (86%) Greater Sandhills infected with *E. gruis*, and 12 of 14 (86%) Florida Sandhills and 66 of 72

Table 19.6. Blood parasites detected in cranes in Florida

Host Parasite species	County[a]	Year(s)	No. of birds examined	Number positive	% positive	Data source
Sandhill Crane						
Atoxoplasma sp.	Alachua	1970–99	274	13	5	A,B,D
Haemoproteus antigonis	Alachua	1970–99	274	31	11	A,B,D
	Duval	1996	1	1	—	B
	Glades	1970–71	40	1	2	A, D
	Highlands	1971	7	1	—	A
	Marion	1992–98	3	1	—	B
	Osceola	1991–98	122	13	5	B
Leucocytozoon grusi	Alachua	1970–99	274	34	12	A,B,D
	Lake	1974	1	1	—	A
	Levy	1991–92	2	1	—	B
	Marion	1992–98	3	1	—	B
	Osceola	1991–98	122	12	10	B
Plasmodium polare-like	Alachua	1970–99	274	4	1	B,C,D
	Marion	1992–98	3	1	—	C,B
Whooping Crane						
Haemoproteus antigonis	Osceola	1993–99	133	15	11	B
	Sumter	1997	1	1	—	B

Sources: A = Forrester et al. (1976), B = Spalding et al. (2001), C = Telford et al. (1994), D = Forrester & Bennett (1985).
a. In addition, blood smears from Sandhill Cranes from Citrus (1), Manatee (5), Putnam (1), and Sarasota (1) counties and Whooping Cranes from Lake (30), Orange (1), Brevard (2), Palm Beach (1), and Suwannee (1) counties were examined and found to be negative for hemoparasites.

(92%) Greater Sandhill Cranes infected with *E. reichenowi*. Spalding et al. (2001) found eimerian oocysts in 55% of the Sandhill Cranes examined, with 36% of these cranes shedding *E. gruis*, and 42% shedding *E. reichenowi* ova. These organisms could potentially be significant to young cranes; however, very few young birds have been examined. Disseminated visceral coccidiosis (DVC), a systemic infection with *Eimeria* sp., has caused mortality in captive Florida Sandhill and Whooping Cranes in Maryland (Carpenter and Novilla 1980; Novilla et al. 1981, 1989; Carpenter et al. 1984) and oral lesions have been observed in free-ranging Greater Sandhill Cranes outside of Florida (Carpenter et al. 1984). Ten cases of DVC were detected in Sandhill Cranes from Alachua, Marion, and Osceola counties in birds of all ages, 1974–99, and DVC was suspected to have caused death in 2 of these

(Spalding et al. 2001). Forty-eight percent of 295 Sandhill Crane examinations (288 individuals) and 41% of 482 Whooping Crane examinations (214 individuals) for oral granulomas (described by Carpenter et al. 1979) were positive but other causes for these oral lesions were not ruled out. In some cases infection with *Capillaria* sp. was known to have caused similar lesions (see section XIII, Helminths).

The picture is somewhat more complex with Whooping Cranes than with Sandhill Cranes (Spalding et al. 2001). Whooping and Sandhill Cranes apparently share the same species of *Eimeria* (Forrester et al. 1978). In Florida, recently released Whooping Cranes and occasionally older birds have access to feed that contains a coccidiostat. When just those birds that had been in Florida for >6 months were examined, 13% of 54 were shedding *Eimeria* oocysts, 9% *E. gruis*, and 7% *E. reichenowi* (Spalding et al.

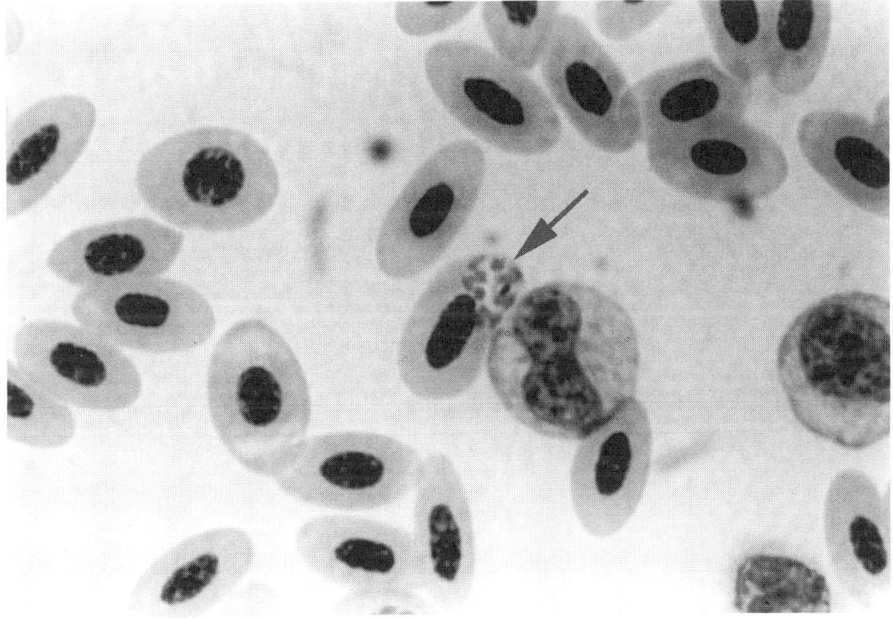

FIGURE 19.12. *Plasmodium* sp. (*polare*-like) in an erythrocyte of a Sandhill Crane.

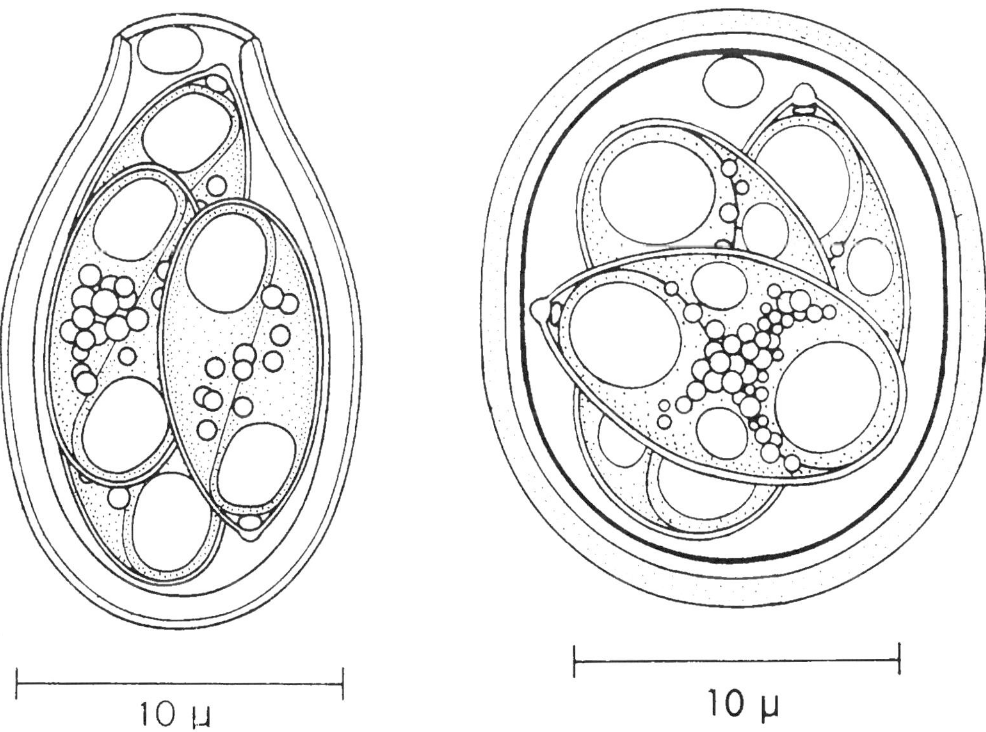

FIGURE 19.13. Oocysts of *Eimeria gruis (left)* and *E. reichenowi (right)* from the feces of Sandhill and Whooping Cranes. From Courtney et al. 1975; by permission of *Journal of Parasitology.*

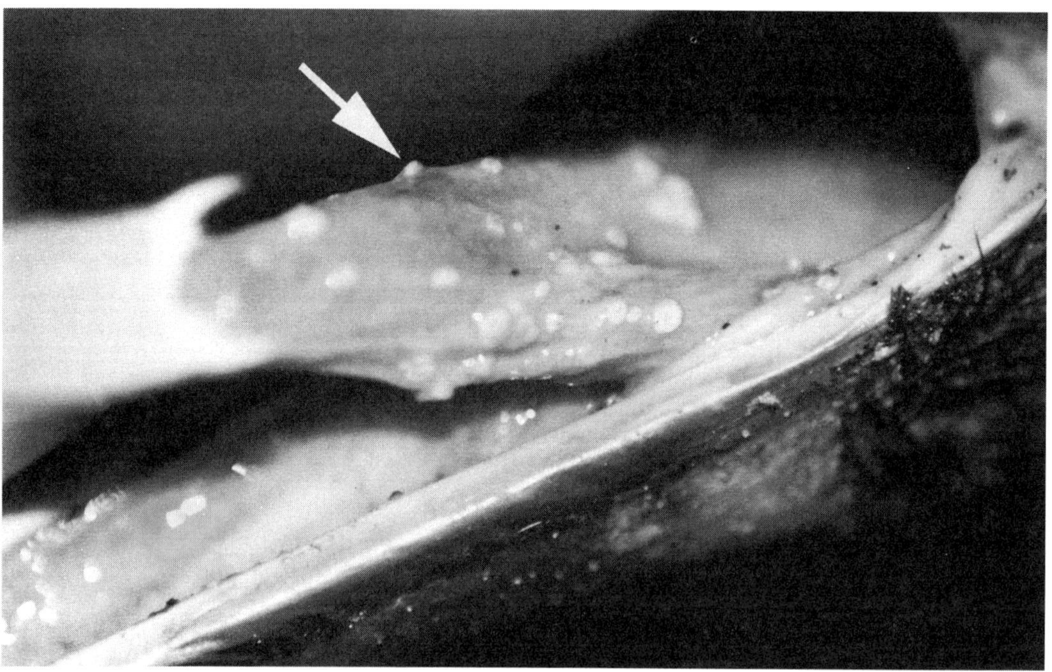

FIGURE 19.14. Oral granulomas *(arrow)* in a Whooping Crane caused by infection with *Eimeria* sp.

2001). Oral granulomas were commonly seen (figure 19.14). Apparently, infections with *E. gruis* and/or *E. reichenowi* become disseminated outside of the intestinal tract more commonly in Whooping Cranes, and also the infection more commonly results in mortality. Ninety-six percent of 71 Whooping Cranes examined after death had evidence of disseminated lesions (figure 19.15), but only in 2 of these were the lesions considered significant to the health of the bird. The protozoal organisms were rarely observed in the lesions (figure 19.16), possibly because of the coccidiostat in the feed.

Adelina sp. was present in the feces of 3 of 14 Florida Sandhill Cranes (Forrester et al. 1975) and 2 of 72 Greater Sandhill Cranes (Forrester et al. 1976) in 1970–75, in 3 Sandhill Cranes from Alachua County in 1986–93, and in 2 Whooping Cranes from Alachua and Sumter counties in 1997 (Spalding et al. 2001); however, the presence of this organism was considered a spurious finding from ingested invertebrates.

Sarcocysts, unidentified to species, were noted in striated muscle of 2 Greater Sandhill Cranes (1 died from disseminated visceral coccidiosis, the other from mycotoxicosis) from Alachua County in 1990 and 1995 (Spalding et al. 2001), and in the cardiac muscle of a Sandhill Crane also from Alachua County (Allen et al. 1985). Sarcocysts in a Sandhill Crane from Lake County in 2000 appeared to be similar to *Sarcocystis falcatula* (Spalding and Cheadle 2000). No adverse health effects associated with sarcocyst presence were observed.

Toxoplasma gondii antibodies were not detected in sera from 3 Sandhill Cranes from Florida (dates and locations not given) by Burridge et al. (1979). These were tested by inhibition hemagglutination, which may not be as sensitive in birds as in mammals, resulting in false negatives (Frenkel 1981).

Severe fibrinonecrotic enterocolitis and catarrhal typhlitis associated with the presence of *Hexamita* sp. (figure 19.17) was the cause of death of 2 captive-raised Florida Sandhill Cranes

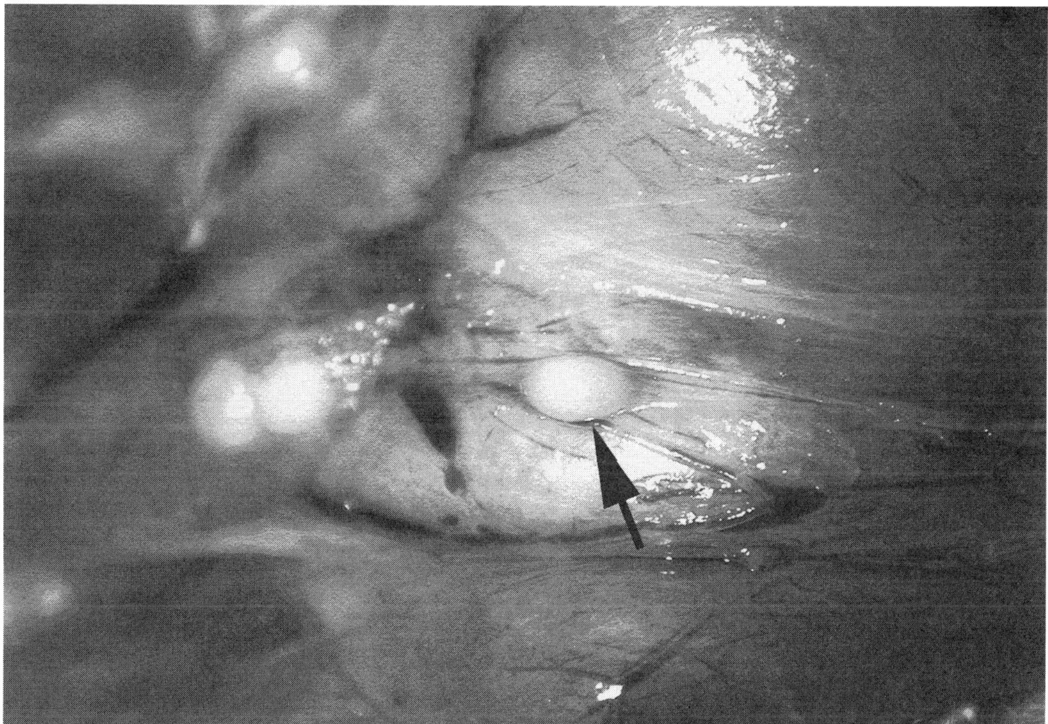

FIGURE 19.15. Disseminated visceral coccidiosis granuloma *(arrow)* on the surface of the cloaca of a Sandhill Crane.

held in a pen in Alachua County in 1991 prior to release (Spalding et al. 1994). They died 1 and 2 months after being transported to Florida. The source of the infection was not determined and subsequent fecal examinations of 105 Whooping Cranes and 52 Sandhill Cranes did not turn up any additional infections.

XIII. Helminths

Parasitic helminths recovered from Sandhill Cranes are listed in table 19.7. Although not listed in the table, analyses by subspecies were made by Forrester et al. (1974, 1975) and Spalding et al. (2001). *Orchipedium jolliei, Tetrameres grusi,* and spiruroid larvae were more prevalent in Greater Sandhills than in Florida Sandhill Cranes. A number of helminths occurred only in Florida Sandhill Cranes, i.e. unidentified echinostomes, *Stomylotrema vic-*

arium, Tanaisia fedtschenkoi, and physalopterid larvae. The 3 most prevalent parasites were *Strongyloides* sp., *O. jolliei,* and *T. grusi. Strongyloides* sp. and *Strigea gruis* were found in the largest numbers. *Cyathostoma variegatum* infection in the trachea of a captive-raised Sandhill Crane caused suffocation and death when the bird was transported to Florida in 1991 (Spalding et al. 1996). This was the only case for which parasitism was clearly associated with illness or death.

The parasites collected from introduced Whooping Cranes are listed in table 19.8. Spalding et al. (1996) recorded 14 helminth species during the first 3 years of the experimental reintroduction, and in the 5 years since then, an additional 15 species have been added (Spalding and Kinsella 2001). Three of the parasites, *Ascaridia pterophora, Eucoleus obtusiuscula,* and *C. variegatum,* were apparently introduced, since they had not been recorded

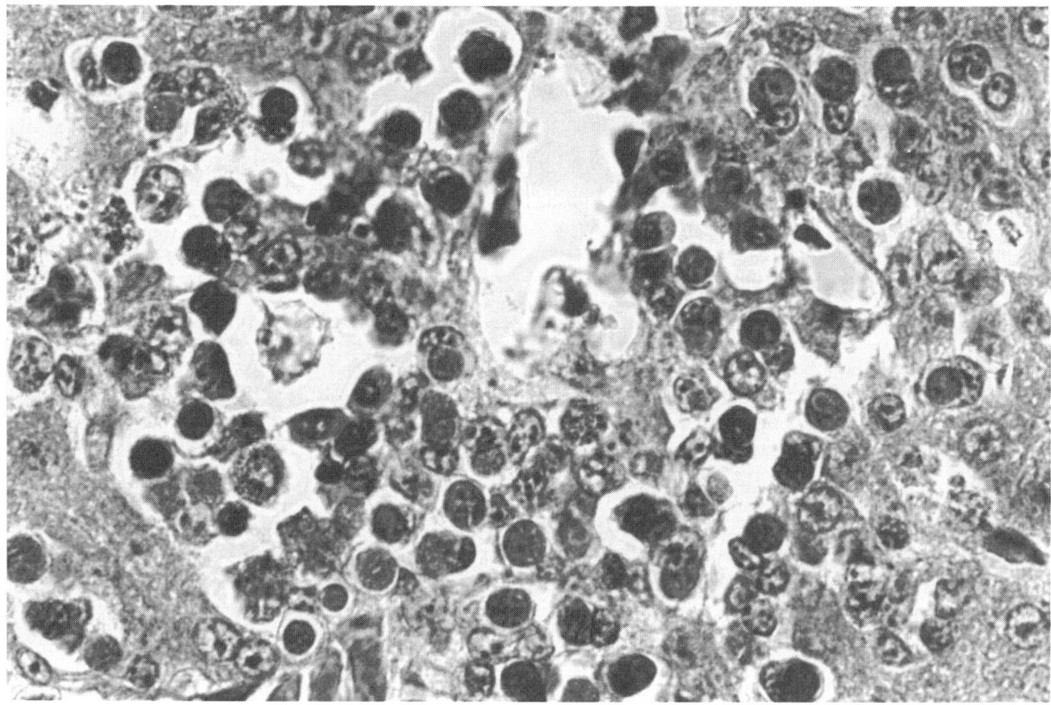

FIGURE 19.16. Photomicrograph of meronts of *Eimeria* sp. within a granulomatous lesion in the liver of a Whooping Crane.

from wild birds in North America before. They have been found in cranes from the source facilities in Wisconsin and Maryland. Two species, *A. pterophora* and *Strongyloides* sp., were found in higher numbers in emaciated Whooping Cranes, but no cause and effect relationship has been established.

Large numbers of encysted nematodes, probably *Physocephalus sexalatus,* were scattered throughout the wall of the small intestine and mesentery (figure 19.18) of a 16-month-old Whooping Crane killed by a bobcat (Varela et al. 2001). It had been released 9 months earlier in Orange County. The parasite undoubtedly had an impact on the health of this crane, but whether it predisposed the bird to predation is not known. This is a parasite of feral swine in Florida (Forrester 1992), transmitted to various aberrant hosts such as birds, reptiles, and amphibians by dung beetles.

Capillarid nematodes and their bipolar eggs (figure 19.19) were associated with inflammation of the epithelium and submucosa in the oral cavity, especially on the tongue, in 2 Whooping Cranes in Lake and Osceola counties (Spalding et al. 2001). The tongue was severely malformed in 1 case, probably as a result of the infection.

XIV. Arthropods

Arthropods collected from Sandhill Cranes and Whooping Cranes in Florida are listed in table 19.9. Four species of chewing lice were found commonly on both Florida and Greater Sandhill Cranes (Forrester et al. 1976), but only 2 of these have been found on introduced Whooping Cranes thus far. Previously, *Heleonomus assimilis* was known with assurance only from the Whooping Crane and the Siberian Crane (Emerson 1964).

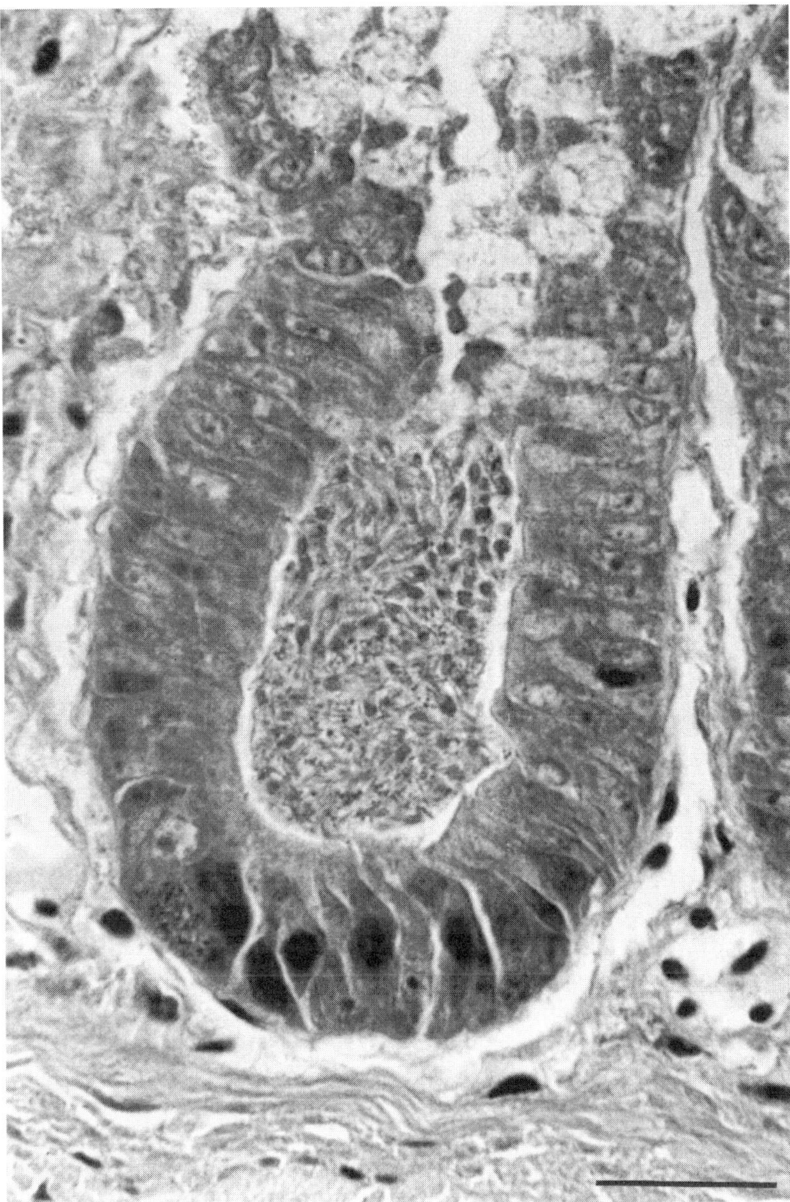

FIGURE 19.17. *Hexamita* sp. trophozoites in a crypt of Lieberkuhn of the large intestine. H&E. Bar = 20 μm. From Spalding et al. 1994; by permission of *Journal of Zoo and Wildlife Medicine*.

Atyeo and Windingstad (1979) described the kinds and distribution of feather mites on feathers of Greater Sandhill Cranes collected in Wisconsin and Indiana. To date, only 1 of these, *Geranolichus canadensis*, has been collected from Sandhill Cranes in Florida; however, most of the collections in Florida were made from dead birds. Epidermoptid mites (*Myialges nudus*) have been found on both Sandhill and Whooping Cranes in Florida, but

Table 19.7. Helminths of Sandhill Cranes from Florida

Class Species (location)[a]	County	Year(s)	No. birds			Intensity		Data source
			Examined	Positive	%	Mean	Range	
Trematoda								
Brachylaima fuscatum (LS)	Alachua	1970–98	48	6	13	3	1–5	A, B, C, D
	Desoto	1973	1	1	—	1	—	B
	Highlands	1971–74	5	1	—	37	—	A, B
	Osceola	1971–98	18	1	6	3	—	A, C, D
	Volusia	1992	1	1	—	11	—	D
Echinostomatidae (CE,LI)	Glades	1970–71	11	2	18	3	1–4	B
Orchipedium jolliei (LU,TR)	Alachua	1970–98	48	26	54	7	1–28	A, B, C, D
	Glades	1970–71	11	2	18	13	11–14	B
	Lake	1971–95	3	1	—	5	—	A, D
	Marion	1998	1	1	—	1	—	D
Prosthogonimus macrorchis (CL)	Osceola	1971–98	18	1	6	1	—	A, C, D
Stomylotrema vicarium (CE,CL,LS)	Desoto	1973	1	1	—	6	—	B
	Glades	1970–71	11	1	9	7	—	B
	Highlands	1971–74	5	1	—	43	—	A, B
	Osceola	1971–98	18	1	6	2	—	A, C, D
Strigea gruis (DU,LS)	Alachua	1970–98	48	4	8	112	11–258	A, B, C, D
	Glades	1970–71	11	1	9	19	—	B
	Highlands	1971–74	5	2	—	83	51–115	A, B
	Lake	1971–95	3	1	—	3	—	A, D
Tanaisia fedtschenkoi (KD)	Highlands	1971–74	5	1	—	9	—	A, B
Cestoda								
Unidentified scolex (DU)	Alachua	1970–98	48	1	2	1	—	A, B, C, D
Nematoda								
Amidostomum sp. (GI)	Lake	1971–95	3	1	—	1	—	A, D
Ascaridia sp. (LS,SI)	Alachua	1970–98	48	2	4	1	1	A, B, C, D
	Osceola	1971–98	18	1	6	3	—	A, C, D
Capillaria sp. (LS)	Alachua	1970–98	48	2	4	29	5–53	A, B, C, D
	Lake	1971–95	3	1	—	1	—	A, D
	Osceola	1971–98	18	1	6	1	—	A, C, D

Parasite (location)	County	Years						Sources
	Sarasota	1972	2	1	—	1	—	B
Cardiofilaria pavlovskyi (AS)	Alachua	1970–98	48	1	2	NG	—	A, B, C, D
Cyathostoma variegatum (TR)	Alachua	1970–98	48	2[b]	4	1	1	G
Dispharynx nasuta (PR)	Alachua	1970–98	48	3	6	14	1–38	A, B, C, D
	Monroe	1984	1	1	—	36	—	D
	Sarasota	1972	2	1	—	1	—	B
Paronchocerca sp.[c] (HT)	Glades	1970–71	11	1	9	1	—	B, E
Physalopterid larva (LS)	Sarasota	1972	2	1	—	1	—	B
Spiruroid larvae (GI,PR)	Alachua	1970–98	48	10	21	11	1–31	A, B, C, D
Strongyloides sp. (CE,DU,LS)	Alachua	1970–98	48	21	44	42	1–381	A, B, C, D
	Desoto	1973	1	1	—	6	—	B
	Glades	1970–71	11	7	64	14	6–34	B
	Highlands	1971–73	5	3	—	3	1–6	A, B
	Lake	1971–95	3	2	—	71	2–140	A, D
	Manatee	1973	2	1	—	16	—	C
	Monroe	1984	1	1	—	8	—	D
	Osceola	1971–98	18	1	6	1	—	A, C, D
	Sarasota	1972	2	2	—	12	3–20	B
Syngamus trachea (TR)	Alachua	1970–98	48	11	23	3	1–4	A, B, C, D
Synhimantus sp. (GI)	Alachua	1970–98	48	1	2	7	—	A, B, C, D
	Desoto	1973	1	1	—	1	—	B
	Glades	1970–71	11	2	18	2	1–2	B
Tetrameres grusi (PR)	Alachua	1970–98	48	25	52	29	1–153	A, B, C, D, F
	Glades	1970–73	11	2	18	16	5–26	B
	Lake	1971–95	3	1	—	2	—	A, D
	Sarasota	1972	2	1	—	2	—	B
Trichostrongylus cramae (DU)	Alachua	1970–98	48	2	4	10	6–13	A, B, C, D
	Lake	1971–95	3	1	—	6	—	A, D

Sources: A = Forrester et al. (1974), B = Forrester et al. (1975), C = Forrester and Kinsella (1980), D = Spalding and Kinsella (2001), E = Bartlett and Anderson (1987), F = Bush et al. (1973), G = Spalding et al. (1996)

NG = not given.

a. Location in host: AS = air sac, CE = cecum, CL = cloaca, DU = duodenum, GI = gizzard, HT = heart, KD = kidney, LI = large intestine, LS = lower small intestine, LU = lung, PR = proventriculus, SI = small intestine, TR = trachea.

b. Both cranes had been introduced from a captive flock in Laurel, Md. Mucus associated with this infection caused death of one of the cranes.

c. This filarid nematode was reported as *Chandlerella* sp. by Forrester et al. (1975); the specimen was reexamined by Bartlett and Anderson (1987) and determined to be a species of *Paronchocerca*.

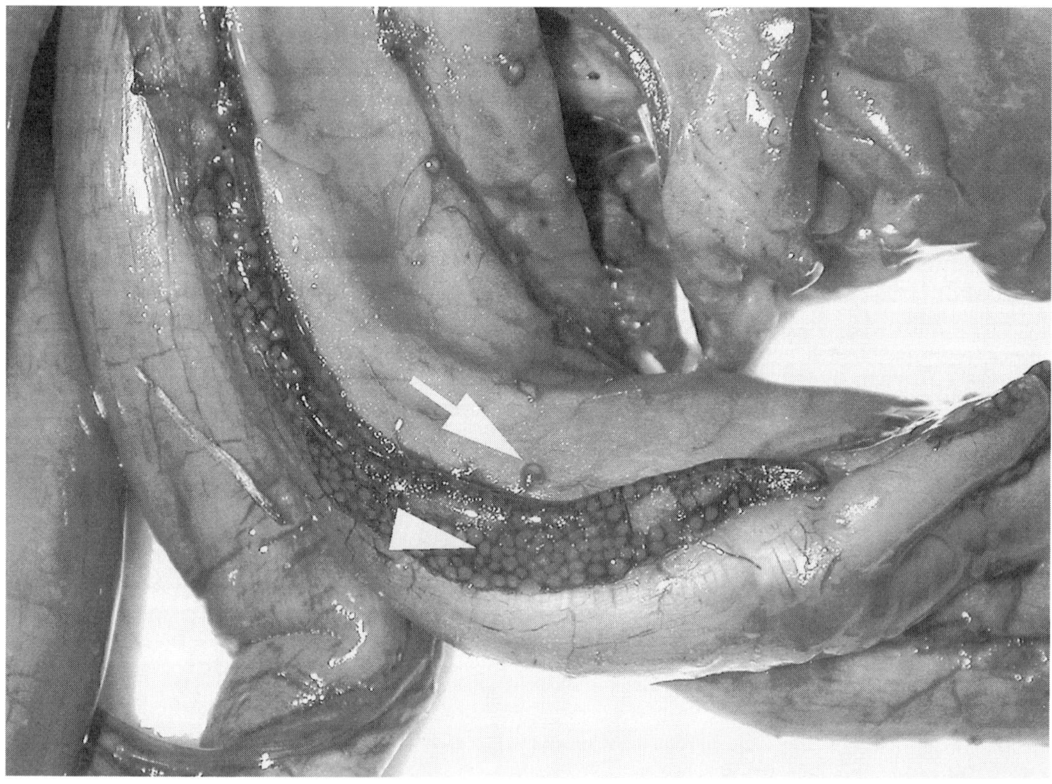

FIGURE 19.18. Duodenal loop of intestine of a Whooping Crane illustrating parasite granulomas on the surface of the pancreas (*arrow*) and in the mesentery *(arrowhead)* caused by *Physocephalus sexalatus.*

no associated lesions were noted. Gilardi et al. (2001) described mange in Laysan Albatross fledglings in Hawaii due to this mite.

Chiggers (*Blankaartia sinnamaryi*) occasionally cause firm raised papules on the skin of young Sandhill Cranes (Spalding, Wrenn, et al. 1997), but have not been reported from adults. The papules can be felt by palpation through the feathers, and visualized by wetting the feathers. Red/orange mites can be seen within a small depression on the top of the papule. A stylostome produced by the chigger extends into the dermis where it is surrounded by marked pyogranulomatous inflammation (figure 19.20). Numerous lesions were observed on the skin of a chick less than 1 week old from Alachua County that died from neck trauma (Spalding 2001).

Black flies (*Simulium slossonae*) have been observed swarming around the heads of penned Whooping Cranes and were occasionally seen on Sandhill Cranes. This fly may be the vector for *Leucocytozoon grusi,* a blood parasite common in Sandhill Cranes but never found in Whooping Cranes in Florida (Spalding et al. 2001).

The purpose of feather dusting by Sandhill Cranes is understood poorly (Nesbitt 1975). Cranes show a preference for clay soils rather than sand, so grain size may be important and function in repelling parasites. Kilham (1980) noted that Sandhill Cranes in Highlands County occasionally used ant mounds (red imported fire ants, *Solenopsis invicta*) as a source of dust, and he wondered whether the birds were "anting" or preferred the loose soil created by the ant mounds for dusting.

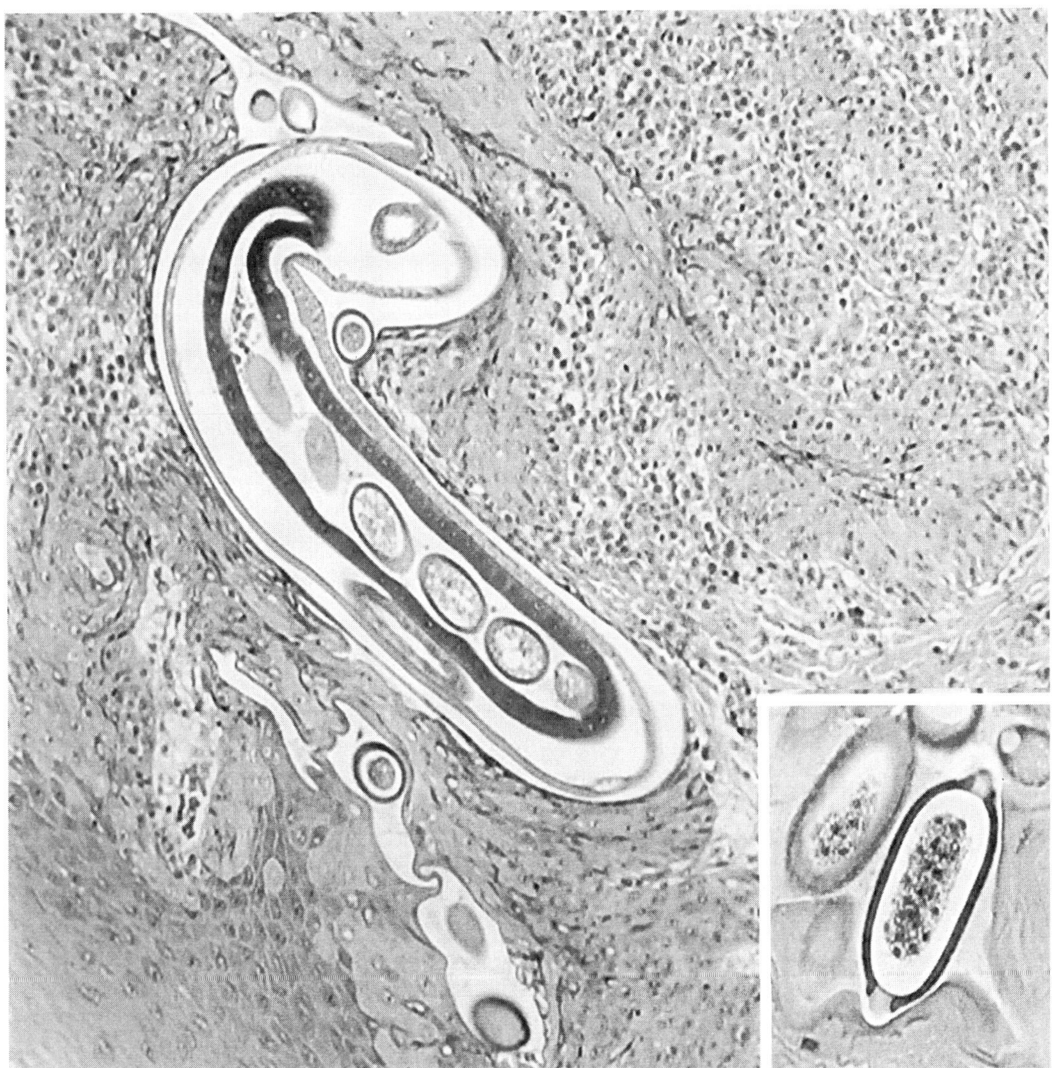

FIGURE 19.19. *Capillaria* sp. within the submucosa of the tongue of a Whooping Crane from Osceola County in 1995. Note the cross section of the nematode, eggs free within the tissue, and inflammatory infiltrate. The inset illustrates the typical bipolar egg.

XV. Emaciation

Emaciation with no other evident cause of death or illness has been observed in only 2 Sandhill Cranes (Spalding et al. 2001). This indicates that food is probably not a limiting factor for Sandhill Cranes in Florida.

XVI. Summary and Conclusions

Cranes are among the most extensively and systematically studied avian species in Florida. Certainly the Whooping Crane, which has been experimentally reintroduced, has been the species most closely monitored for health in the state.

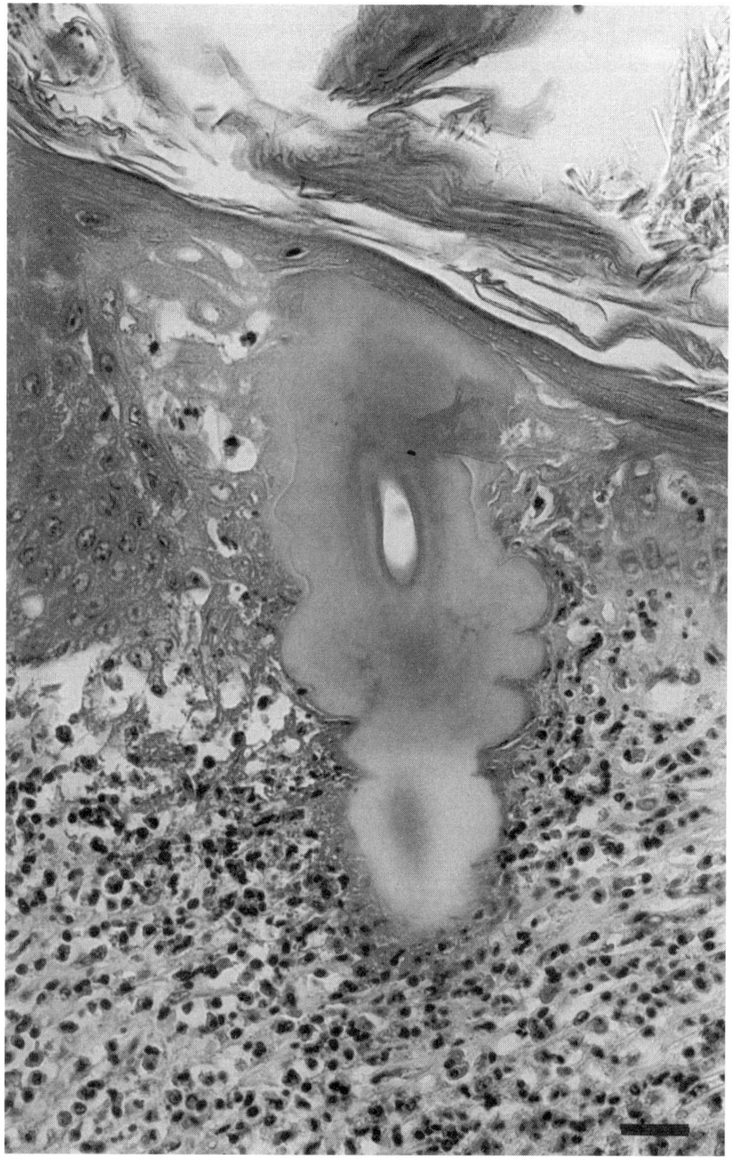

FIGURE 19.20. Photomicrograph of a stylostome of a chigger extending into the skin of a Sandhill Crane. Note the surrounding inflammatory infiltrate. Bar = 25 μm. From Spalding, Wrenn, et al. 1997; by permission of *Journal of Parasitology*.

As with many other birds, trauma associated with manmade objects causes the most documented mortality, with the exception of bobcat predation of recently reintroduced Whooping Cranes. Whooping Cranes are particularly sus-

ceptible to predation, believed to be a result of the lack of parental association in the natural environment during their development. For Sandhill Cranes peanut toxicosis has been an important mortality factor that has been re-

Table 19.8. Helminths of introduced Whooping Cranes in north central Florida[a]

| Class | Species (Location)[b] | No. Birds | | | Intensity | |
		Examined[c]	Positive	%	Mean	Range
Trematoda	*Brachylaema fuscatum* (SI)	71	5	7	11.6	1–33
	Cyclocoelum mutabile (AS)	84	3	4	6.7	1–17
	Lyperorchis lyperorchis (LS)	68	2	3	1	1
	Philophthalmus gralli[d] (CJ)	6	3	—	2	1–3
	Polycyclorchis eudocimi (AS)	84	2	2	1	1
	Prosthogonimus macrorchis (CL)	60	1	2	8	8
	Ribeiroia ondatrae (ES,ST)	81	1	1	3	3
	Strigea gruis (SI)	71	6	8	15.8	1–44
	Stomylotrema vicarium (LI,CL)	66	1	2	13	13
	Tanaisia atra (KD)	66	7	11	—	1–25
	larval cyclocoelids (OC,TR,LU)	74	1	1	11	11
Cestoda	*Cyclustera ibisae* (DU)	69	1	1	1	1
	Spirometra-like sparganum (TR)	71	1	1	1	1
Nematoda	*Ascaridia pterophora* (ST,IN)	81	23	28	13.4	1–61
	Capillaria fulicae (SI,CE)	72	6	8	2	1–4
	Capillaria sp. I[e] (OR)	24	2	8	ND	—
	Capillaria sp. II (LI)	64	1	2	1	1
	Chandleronema longigutturata (DU)	69	1	1	1	1
	Contracaecum multipapillatum (SI)	71	1	1	1	1
	Cyathostoma variegatum (TR)	83	1	1	1	1
	Dispharynx nasuta (PR)	75	5	7	7	1–28
	Eucoleus obtusiuscula (KO)	80	7	9	2.6	1–7
	Hadjelia neglecta (ST)	80	2	3	2	1–3
	Hystrichis tricolor (PR)	75	2	3	1.5	1–2
	Physaloptera sp. (ST)	80	1	1	1	1
	Physocephalus sp.[f]	84	1	1	ND	—
	Strongyloides sp. (SI)	71	15	21	16.8	1–102
	Tetrameres williamsi (PR)	75	1	1	1	1
	larval nematodes (ES,ST,IN)	82	6	7	1	1–2
Acanthocephala	*Centrorhynchus kuntzi* (SI)	71	1	1	1	1
	Polymorphus trochus (LS)	68	1	1	1	1
	Southwellina dimorpha (SI)	71	3	4	4	1–10
	larval acanthocephalans (SI,CE)	72	1	1	6	6

Sources: Spalding et al. (1996), Spalding & Kinsella (2001).
ND = not determined.
a. Eighty-four birds were screened for helminths from Brevard (4), Lafayette (1), Lake (12), Orange (1), Osceola (62), St. Lucie (1), Sumter (2), and Suwannee (1) counties, 1993–2000. All were treated with anthelmintics prior to their release in Florida; birds dying soon after release thus had abnormally low parasite faunas.
b. Location in host: AS = air sacs, CE = ceca, CJ = conjunctiva, CL = cloaca, DU = duodenum, ES = esophagus, IN = intestine, KD = kidney, KO = koilin, LI = large intestine, LU = lung, LS = lower small intestine, OR = oral cavity, PR = proventriculus, SI = small intestine, ST = stomach, TR = trachea.
c. Many of the carcasses were incomplete because of predation; therefore the number examined reflects the number of birds in which all or part of the normally inhabited host tissues were examined.
d. *Philophthalmus gralli* normally inhabits the conjunctiva of the eye. Eyes were examined specifically for this parasite only occasionally.
e. This species of *Capillaria* was seen only once histologically in an oral lesion. The tongue was malformed. The number examined reflects the number of birds in which oral lesions were examined histologically.
f. Probably *Physocephalus sexalatus* (Varela et al. 2002).

Table 19.9. Arthropods collected from cranes in Florida

Host Class	Species	County or area	Year(s)	Data source
Sandhill Cranes				
Chewing lice[a]	*Esthiopterum brevicephalum*	Alachua	1973–77	A
		Glades	1970	A
		Osceola	1986, 1993–94	A
		Sarasota	1972	A
		Taylor	1986	A
	Gruimenopon canadense	Alachua	1970, 73, 77, 79	A
		Glades	1970	A
		Osceola	1992–94	A
		Taylor	1986	A
		Volusia	1993	A
	Heleonomus assimilis	Alachua	1974, 77	A
		Glades	1970	A
		Highlands	1979	A
		Osceola	1986, 1992–94	A
	Saemundssonia sagulata	Alachua	1973–77	A
		Osceola	1994	A
		Sarasota	1972	A
		"Florida"	1882	B
Mites	*Geranolichus canadensis*	Taylor	1986	C
	Myialges nudus[b]	Osceola	1994	D
		Taylor	1986	C
	Petitota aluconis[c]	Taylor	1986	D
	Pseudogabucinia sp.[d]	Osceola	1994	D
Chiggers	*Blankaartia sinnamaryi*	Alachua	1996, 2001	E,H
		Osceola	1993	E
Hippoboscids	*Icosta albipennis*[e]	Osceola	1993	F
Whooping Cranes				
Chewing lice	*Gruimenopon canadense*	Brevard	1997	G
		Osceola	1996–98	G
	Heleonomus assimilis	Brevard	1997	G
		Osceola	1994–98	A,G
Ticks	*Ambylomma maculatum*	Osceola	1999	H
Mites	*Aetacarus* sp.[f]	Osceola	1996	G
	Apatelacarus sp.[g]	Osceola	1995–96	I
	Myialges nudus[b]	Osceola	1995	I
	Paralges sp.[h]	Osceola	1995	I
Hippoboscids	*Icosta* sp.	Osceola	1999	H

Sources: A = Forrester et al. (1995), B = Timmermann (1971), C = Forrester & Mertins (1995), D = Spalding & Mertins (1995), E = Spalding, Wrenn, et al. (1997), F = Spalding & Wilson (1993), G = Spalding & Mertins (1999), H = Spalding et al. (2001), I = Spalding & Mertins (1996).

a. All 4 species of chewing lice were found commonly on both Florida and Greater Sandhill Cranes (Forrester et al. 1976).

b. Often associated with a mange-like dermatitis (Fain 1965), but not in these cranes.

c. Classified previously in the genus *Neopetitota*.

d. An undescribed species of feather mite on the Florida Sandhill Crane, not *P. reticulata* previously described from Greater Sandhill Cranes.

e. Usually found on ardeids; this may be an accidental host.

f. An undescribed species of feather mite usually found on raptors. This could be a contaminant.

g. An undescribed feather mite of cranes.

h. An undescribed quill mite.

duced somewhat by changes in agricultural practices. Liver tumors are exceptionally high in Florida Sandhill Cranes when compared with typical frequencies for neoplasia in wild animals and may be caused by either naturally or unnaturally acquired toxins in their food. Pesticides and other toxins have been very poorly studied in Florida cranes. Ingestion of metal with subsequent toxicosis appears to be a unique problem for Whooping Cranes. Infectious agents have been well characterized, with 28 bacteria, 1 fungus, 3 viruses, 8 protozoa, 12 nematodes, 7 trematodes, 1 cestode, and 10 arthropods identified in Sandhill Cranes and 24 bacteria, 1 fungus, 2 viruses, 3 protozoa, 15 nematodes, 10 trematodes, 2 cestodes and 3 acanthocephala, and 3 arthropods identified in Whooping Cranes.

XVII. Literature cited

Abou-Madi, N. 1991. Unpublished data. University of Florida, Gainesville.

Allen, J.L., T.L. Gross, and S.L. Lin. 1984. Unpublished data. University of Florida, Gainesville.

Allen, J.L., H.D. Martin, and A.M. Crowley. 1985. Metastatic cholangiocarcinoma in a Florida sandhill crane. *J. Am. Vet. Med. Assoc.* 187:1215.

Atyeo, W.T., and R.M. Windingstad. 1979. Feather mites of the greater sandhill crane. *J. Parasitol.* 65:650–658.

Bartlett, C.M., and R.C. Anderson. 1987. *Chandlerella bushi* n. sp. and *Splendidofilaria caperata* Hibler, 1964 (Nematoda: Filarioidea) from *Fulica americana* (Gruiformes: Rallidae) in Manitoba, Canada. *Can. J. Zool.* 65:2799–2802.

Bennett, A.J., and L.A. Bennett. 1987. Evaluation of the Okefenokee Swamp as a site for developing a nonmigratory flock of whooping cranes. Unpublished report. Georgia Cooperative Fish and Wildlife Research Unit, Athens. 113 Pp.

Bennett, G.F., D.J. Forrester, E.C. Greiner, and A.G. Campbell. 1975. Avian haemoproteidae. 4. Description of *Haemoproteus telfordi* sp. nov., and a review of the haemoproteids of the families Gruidae and Otidae. *Can. J. Zool.* 53:72–81.

Bennett, G.F., R.A. Khan, and A.G. Campbell. 1974. *Leucocytozoon grusi* sp. n. (Sporozoa: Leucocytozoidae) from a sandhill crane, *Grus canadensis* (L.). *J. Parasitol.* 60:359–363.

Brown, W.M., R.C. Drewien, and E.G. Bizeau. 1987. Mortality of cranes and waterfowl from powerline collisions in the San Luis Valley, Colorado. In: *Proc. 1985 Crane Workshop.* J.C. Lewis (ed). Platte River Whooping Crane Maintenance Trust, Grand Island, Nebr. pp. 128–136.

Buergelt, C.D. 1981. Unpublished data. University of Florida, Gainesville.

Buergelt, C.D., P.E. Ginn, L.H. Kooistra, and R. Isaza. 1990. Unpublished data. University of Florida, Gainesville.

Buergelt, C.D., M.R. Wells, and R.L. Caligiuri. 1987. Unpublished data. University of Florida, Gainesville.

Burridge, M.J., W.J. Bigler, D.J. Forrester, and J.M. Hennemann. 1979. Serologic survey for *Toxoplasma gondii* in wild animals in Florida. *J. Am. Vet. Med. Assoc.* 175:964–967.

Bush, A.O., D.B. Pence, and D.J. Forrester. 1973. *Tetrameres (Gynaecophila) williamsi* sp. n. (Nematoda: Tetrameridae) from the white ibis, *Eudocimus albus*, with notes on *Tetrameres (Tetrameres) grusi* Shumakovich from the sandhill crane, *Grus canadensis. J. Parasitol.* 59:788–792.

Calderwood-Mays, M.B., A.M. Crowley, and J.L. Allen. 1985. Unpublished data. University of Florida, Gainesville.

Calle, P.P., R.J. Montalik, D.L. Janssen, M.K. Stoskopf, and J.D. Strandberg. 1982. Distal extremity necrosis in captive birds. *J. Wildl. Dis.* 18:473–479.

Carnaghan, R.B.A. 1965. Hepatic tumors in ducks fed a low level of toxic ground nut meal. *Nature* 208:308–312.

Carpenter, J.W. 1993. Infectious and parasitic diseases of cranes. In: *Zoo and wild animal medicine: current therapy.* 3d ed. M.E. Fowler (ed). W.B. Saunders, Philadelphia. pp. 229–237.

Carpenter, J.W., and S.R. Derrickson. 1981.

Whooping crane mortality at the Patuxent Wildlife Research Center 1966–1981. In: *Proc. 1981 Int. Crane Workshop*. J.C. Lewis (ed.). National Audubon Society, Tavernier, Fla. pp. 175–179.

———. 1987. Infectious and parasitic diseases of cranes: principles of treatment and prevention. In: *Proc. 1983 Int. Crane Workshop*. G.W. Archibald and R.F. Pasquier (eds.). International Crane Foundation, Baraboo, Wis. pp. 539–553.

Carpenter, J.W., and M.N. Novilla. 1980. The occurrence and control of disseminated visceral coccidiosis in captive cranes. *Annu. Proc. Am. Assoc. Zoo Vet.* pp. 99–101.

Carpenter, J.W., M.N. Novilla, R. Fayer, and G.C. Iverson. 1984. Disseminated visceral coccidiosis in Sandhill Cranes. *J. Am. Vet. Med. Assoc.* 185:1342–1346.

Carpenter, J.W., T.R. Spraker, C.H. Gardiner, and M.N. Novilla. 1979. Disseminated granulomas caused by an unidentified protozoan in sandhill cranes. *J. Am. Vet. Med. Assoc.* 175:948–951.

Citino, S. 1994. Unpublished data. White Oak Plantation, Yulee, Fla.

Colwell, W.M., C.F. Simpson, L.E. Williams, Jr., and D.J. Forrester. 1973. Isolation of a herpesvirus from wild turkeys in Florida. *Avian Dis.* 17:1–11.

Courtney, C.H., D.J. Forrester, J.V. Ernst, and S.A. Nesbitt. 1975. Coccidia of sandhill cranes, *Grus canadensis. J. Parasitol.* 61:695–699.

Couvillion, C.E., J.R. Jackson, R.P. Ingram, and C.P. McCoy. 1991. Potential natural exposure of Mississippi sandhill cranes to aflatoxin B1. *J. Wildl. Dis.* 27:650–656.

Day, J.F., and L.M. Stark. 1996. Transmission patterns of St. Louis encephalitis and eastern equine encephalitis viruses in Florida: 1978–1993. *J. Med. Entomol.* 33:132–139.

———. 1998. St. Louis encephalitis virus transmission to emus (*Dromaius novaehollandiae*) in Palm Beach County, Florida, with evidence of western equine encephalitis virus antibody transport to Florida by emus infected in other states. *Fla. Entomol.* 81:104–111.

Dein, F.J., J.W. Carpenter, G.G. Clark, R.J. Montali, C.L. Crabbs, T.F. Tsai, and D.E. Docherty. 1986. Mortality of captive whooping cranes caused by eastern equine encephalitis virus. *J. Am. Vet. Med. Assoc.* 189:1006–1010.

Docherty, D.E., and D.J. Henning. 1980. The isolation of a herpesvirus from captive cranes with inclusion body disease of cranes. *Avian Dis.* 24:278–283.

Doherty, D.E., and R.I. Romaine. 1983. Inclusion body disease of cranes: a serologic follow-up to the 1978 die-off. *Avian Dis.* 27:830–835.

Dusek, R. 2001. Unpublished data. University of Florida, Gainesville.

Edman, J.D., L.A. Webber, and H.W. Kale II. 1972. Host-feeding patterns of Florida mosquitoes II. Culiseta. *J. Med. Entomol.* 9:429–434.

Ellis, D.H., G.F. Gee, and C.M. Mirande. 1996. *Cranes: their biology, husbandry, and conservation.* National Biological Service, Washington D.C., and International Crane Foundation, Baraboo, Wis. 307 pp.

Emerson, K.C. 1964. *Checklist of the Mallophaga of North America (north of Mexico).* Part II, *Suborder Amblycera.* Dugway Proving Ground, Dugway, Utah. 104 Pp.

Fain, A. 1965. A review of the family Epidermoptidae Trouessart parasitic on the skin of birds (Acarina: Sarcoptiformes). Verhandelingen Van De Koninklijke Vlaamse Academie Voor Wetenschappen. *Letteren En Schone Kunsten Van Belgie, Klasse Wetens.* 27 Pt. I:1–176, Pt. II:1–144.

Folk, M.J., and S.A. Nesbitt. 2000. Unpublished data. Florida Fish and Wildlife Conservation Commission, Gainesville.

Folk, M.J., S.A. Nesbitt, and M.G. Spalding. 2001. Interactions of sandhill crane and whooping cranes with foreign objects in Florida. *Proc. N. Am. Crane Workshop* 8:195–197.

Folk, M.J., J.A. Schmidt, and S.A. Nesbitt. 1999. A trough-blind for capturing cranes. *J. Field Ornithol.* 70:251–256.

Forrester, D.J. 1992. *Parasites and diseases of wild mammals in Florida.* University Press of Florida, Gainesville. 459 pp.

Forrester, D.J., and G.F. Bennett. 1985. Unpublished data. University of Florida, Gainesville.

Forrester, D.J., A.O. Bush, and L.E. Williams, Jr. 1975. Parasites of Florida sandhill cranes, *Grus canadensis pratensis. J. Parasitol* 61:547–548.

Forrester, D.J., A.O. Bush, L.E. Williams, Jr., and

D.J. Weiner. 1974. Parasites of greater sandhill cranes (*Grus canadensis tabida*) on their wintering grounds in Florida. *Proc. Helminthol. Soc. Wash.* 41:55–59.

Forrester, D.J., J.W. Carpenter, and D.R. Blankinship. 1978. Coccidia of whooping cranes. *J. Wildl. Dis.* 14:24–27.

Forrester, D.J., and J.M. Kinsella. 1980. Unpublished data. University of Florida, Gainesville.

Forrester, D.J., H.W. Kale II, R.D. Price, K.C. Emerson, and G.W. Foster. 1995. Chewing lice (Mallophaga) from birds in Florida: a listing by host. *Bull. Fla. Mus. Nat. Hist.* 39:1–44.

Forrester, D.J., and J.W. Mertins. 1995. Unpublished data. University of Florida, Gainesville.

Forrester, D.J., and F.H. White. 1978. Unpublished data. University of Florida, Gainesville.

Forrester, D.J., F.H. White, and C.F. Simpson. 1976. Parasites and diseases of sandhill cranes in Florida. *Proc. Int. Crane Workshop* 1:284–290.

Francis, D.W., and W.S. Huey. 1964. Case report: salmonellae isolated from greater sandhill cranes (*Grus canadensis tabida*) in New Mexico. *Avian Dis.* 8:312–314.

Frenkel, J.K. 1981. False-negative serologic tests for *Toxoplasma* in birds. *J. Parasitol.* 67:952–953.

Friend, M. 1999. Avian cholera. In: *Field manual of wildlife diseases. General field procedures and diseases of birds.* M. Friend and J.C. Franson (eds.). U.S. Department of the Interior, U.S. Geological Survey, Biological Research Division, Information and Technology Report 1999-001. Washington, D.C. pp. 75–92.

Friend, M., and J.C. Franson (eds.). 1999. *Field manual of wildlife diseases. General field procedures and diseases of birds.* U.S. Department of the Interior, U.S. Geological Survey, Biological Research Division, Information and Technology Report 1999-001. Washington, D.C. 426 pp.

Gee, G.F., J.W. Carpenter, and G.L. Hensler. 1981. Species differences in hematological values of captive cranes, geese, raptors, and quail. *J. Wildl. Manage.* 45:463–483.

Giambrone, J.J., U.L. Diener, N.D. Davis, V.S. Panangala, and F.J. Hoerr. 1984. Effect of purified aflatoxin on turkeys. *Poult. Sci.* 64:859–865.

Gilardi, K.V.K., J.D. Gilardi, A. Frank, M.L. Goff, and W.M. Boyce. 2001. Epidermoptid mange in Laysan albatross fledglings in Hawaii. *J. Wildl. Dis.* 37:185–188.

Halpern, N., and G.F. Bennett. 1983. *Haemoproteus* and *Leucocytozoon* infections in birds of the Oklahoma City Zoo. *J. Wildl. Dis.* 19:330–332.

Heard, D. 1986. Unpublished data. University of Florida, Gainesville.

Homer, B.L., B. Sheppard, and A. Bennett. 1995. Unpublished data. University of Florida, Gainesville.

Kilham, L. 1980. Dusting by sandhill cranes in Florida. *Fla. Field Nat.* 8:19–20.

Lamont, T., and W. Reichel. 1970. Organochlorine pesticide residues in whooping cranes and Everglade kites. *Auk* 87:158–159.

Langenberg, J., M. Spalding, and J. Ramer. 1996. *Campylobacter jejuni* in captive and free-living cranes. *Proc. Am. Assoc. Zoo Vet.* 1996:487–488.

Lewis, J.C. 1995. Whooping crane. *Birds N. Am.* 153:1–28.

Lewis, J.C., R.C. Drewien, E. Kuyt, and C. Sanchez, Jr. 1992. Contaminants in habitat, tissues, and eggs of whooping cranes. *Proc. N. Am. Crane Workshop* 6:159–165.

Locke, L.N., and N.J. Thomas. 1996. Lead poisoning of waterfowl and raptors. In: *Noninfectious diseases of wildlife.* Iowa State University Press, Ames, Iowa. pp. 108–117.

Logan, T.H. 1997. Florida's endangered species, threatened species, and species of special concern. Official lists. Florida Game and Fresh Water Fish Comm., Tallahassee. 14 pp.

Lung, N.P. 1992. Unpublished data. University of Florida, Gainesville.

Meteyer, C. 1993. Unpublished data. National Wildlife Health Center, Madison, Wis.

Nesbitt, S.A. 1975. Feather staining in Florida sandhill cranes. *Fla. Field Nat.* 3:28–30.

———. 1982. The past, present, and future of the whooping crane in Florida. In: *Proc. 1981 Crane Workshop.* J.C. Lewis (ed.). National Audubon Society, Tavernier, Fla. pp. 151–154.

———. 1985. A technique for aging sandhill

cranes using wing molt: preliminary finding. In: *Proc. 1985 Crane Workshop*. Platte River Whooping Crane Maintenance Trust, Grand Island, Nebr. pp. 224–229.

———. 1988. Nesting, renesting and manipulating nesting of Florida sandhill cranes. *J. Wildl. Manag.* 52:758–763.

———. 1992a. A lesser sandhill crane in Florida. *Fla. Field Nat.* 20:15–17.

———. 1992b. First reproductive success and individual productivity in sandhill cranes. *J. Wildl. Manag.* 56:573–577.

———. 1996a. Whooping crane. In: *Rare and endangered biota of Florida*. Vol. 5, *Birds*. J.A. Rodgers, H. W. Kale II, and H. T. Smith (eds.). University Press of Florida, Gainesville. pp. 13–16.

———. 1996b. Florida sandhill crane. In: *Rare and endangered biota of Florida*. Vol. 5, *Birds*. J.A. Rodgers, H.W. Kale II, and H.T. Smith (eds.). University Press of Florida, Gainesville. pp. 219–229.

———. 1999. Unpublished data. Florida Fish and Wildlife Conservation Commission, Gainesville.

Nesbitt, S.A., and L.C. Badger. 1995. Coyote preys on young Florida sandhill crane. *Fla. Field Nat.* 23:15–16.

Nesbitt, S.A., and J.W. Carpenter. 1993. Survival and movements of greater sandhill cranes experimentally released in Florida. *J. Wildl. Manag.* 57:673–679.

Nesbitt, S.A., M.J. Folk, M.G. Spalding, J.A. Schmidt, S.T. Schwikert, J.M. Nicolich, M. Wellington, J.C. Lewis, and T.H. Logan. 1997. An experimental release of Whooping Cranes in Florida—the first three years. *Proc. N. Am. Crane Workshop* 7:79–85.

Nesbitt, S.A., M.J. Folk, K.A. Sullivan, S.T. Schwikert, and M.G. Spalding. 2001. An update of the whooping crane release project; through June 2000. *Proc. N. Am. Crane Working Group* 8, Albuquerque, N.M.: 62–73.

Nesbitt, S.A., and D.T. Gilbert. 1976. Powerlines and fences: hazards to birds. *Fla. Nat.* 49:23.

Nesbitt, S.A., C.T. Moore, and K.S. Williams. 1991. Gender prediction from body measurements of two subspecies of sandhill cranes. *Proc. N. Am. Crane Workshop* 6:38–42.

Nesbitt, S.A., and T.C. Tacha. 1997. Monogamy and productivity in sandhill cranes. *Proc. N. Am. Crane Workshop* 7:10–13.

Nesbitt, S.A., and L.E. Williams, Jr. 1979. Summer range and migration routes of Florida wintering greater sandhill cranes. *Wilson Bull.* 91:137–141.

Novilla, M.N., J.W. Carpenter, T.K. Jeffers, and S.L. White. 1989. Pulmonary lesions in disseminated visceral coccidiosis of sandhill and whooping cranes. *J. Wildl. Dis.* 25:527–533.

Novilla, M.N., J.W. Carpenter, T.R. Spraker, and T.K. Jeffers. 1981. Parental development of eimerian coccodia in sandhill and whooping cranes. *J. Protozool.* 28:248–255.

Rodgers, J.A., Jr., H.W. Kale II, and H.T. Smith (eds.). 1996. *Rare and endangered biota of Florida*. Vol. 5, *Birds*. University Press of Florida, Gainesville. 688 pp.

Roffe, T.J., R.K. Stroud, and R.M. Windingstad. 1989. Suspected fusariomycotoxicosis in sandhill cranes (*Grus canadensis*): clinical and pathological findings. *Avian Dis.* 33:451–457.

Roth, L., P.E. Ginn, and N. Abou-Madi. 1991. Unpublished data. University of Florida, Gainesville.

Russell, S., and B.H.D. Bolon. 1987. Unpublished data. University of Florida, Gainesville.

Schuh, J.C.L., L. Sileo, L.M. Siegfried, and T.M. Yuill. 1986. Inclusion body disease of cranes: comparison of pathologic findings in cranes with acquired versus experimentally induced disease. *J. Am. Vet. Med. Assoc.* 189:993–996.

Simpson, C.F., D.J. Forrester, and S.A. Nesbitt. 1975. Avian pox in Florida sandhill cranes. *J. Wildl. Dis.* 11:112–115.

Snyder, S.B., M.J. Richard, R.C. Drewien, N. Thomas, and J.P. Thilsted. 1991. Diseases of whooping cranes seen during annual migration of the Rocky Mountain flock. *Proc. Am. Assoc. Zoo Vet.* pp. 74–80.

Snyder, S.B., M.J. Richard, and C.U. Meteyer. 1997. Avian tuberculosis in a whooping crane: treatment and outcome. *Proc. N. Am. Crane Workshop* 7:253–255.

Snyder, S.B., M.J. Richard, J.P. Thilsted, R.C. Drewien, and J.C. Lewis. 1992. Lead poisoning in a whooping crane. In: *Proc. 1988 N. Am. Crane Workshop*. Florida Game and Fresh Water Fish Commission, Tallahassee, Nongame Wildlife Program Technical Report 12:207–210.

Spalding, M.G. 2000–01. Unpublished data. University of Florida, Gainesville.

Spalding, M.G., and M. Cheadle. 2000. Unpublished data, University of Florida, Gainesville.

Spalding, M.G., S.L. Erlandsen, and S.A. Nesbitt. 1994. *Hexamita*-like sp. associated with enteritis and death in captive Florida sandhill cranes (*Grus canadensis pratensis*). *J. Zoo Wildl. Med.* 25:281–285.

Spalding, M.G., and J.M. Kinsella. 2001. Unpublished data. University of Florida, Gainesville.

Spalding, M.G., J.M. Kinsella, S.A. Nesbitt, M.J. Folk, and G.W. Foster. 1996. Helminth and arthropod parasites of experimentally introduced whooping cranes in Florida. *J. Wildl. Dis.* 32:44–50.

Spalding, M.G., and J.W. Mertins. 1995–99. Unpublished data. University of Florida, Gainesville.

Spalding, M.G., and S.A. Nesbitt. 1998. Unpublished data. University of Florida, Gainesville.

Spalding, M.G., S.A. Nesbitt, M.J. Folk, and D.J. Forrester. 2001. Unpublished data. University of Florida and Florida Game and Fresh Water Fish Commission, Gainesville.

Spalding, M.G., S.A. Nesbitt, M.J. Folk, L.R. McDowell, and M.S. Sepulveda. 1997. Metal consumption by whooping cranes and possible zinc toxicosis. *Proc. N. Am. Crane Workshop* 7:237–242.

Spalding, M.G., and N.A. Wilson. 1993. Unpublished data. University of Florida, Gainesville.

Spalding, M.G., and J.C. Woodard. 1992. Chondrosarcoma in a wild great white heron from southern Florida. *J. Wildl. Dis.* 28:151–153.

———. 1995. Unpublished data. University of Florida, Gainesville.

Spalding, M.G., W.J. Wrenn, S. Schweikert, and J.A. Schmidt. 1997. Dermatitis in young Florida sandhill cranes (*Grus canadensis pratensis*) due to infestation by the chigger *Blankaartia sinnamaryi*. *J. Parasitol.* 83: 768–771.

Tacha, T.C., S.A. Nesbitt, and P.A. Vohs. 1992. Sandhill crane. *Birds N. Am.* 31:1–24.

Telford, S.R., S.A. Nesbitt, M.G. Spalding, and D.J. Forrester. 1994. A species of *Plasmodium* from sandhill cranes in Florida. *J. Parasitol.* 80:497–499.

Thoen, C.O., E.M. Himes, and R.E. Barrett. 1977. *Mycobacterium avium* serotype 1 infection in a sandhill crane (*Grus canadensis*). *J. Wildl. Dis.* 13:40–42.

Timmermann, G. 1971. Mallophagologische Killektaneen, 2. *Senckenb. Biol.* 52:41–47.

U.S. Fish and Wildlife Service. 1986. Whooping crane recovery plan. U.S. Fish and Wildlife Service, Albuquerque, N.M. 97 pp.

———. 1994. Whooping crane recovery plan. U.S. Fish and Wildlife Service, Albuquerque, N.M. 92 pp.

Varela, A., J.M. Kinsella, and M.G. Spalding. 2002. Presence of encysted immature nematodes in a released whooping crane (*Grus americana*). *J. Zoo Wildl. Med.* 32: 523–525.

White, F.H., C.F. Simpson, and L.E. Williams, Jr. 1973. Isolation of *Edwardsiella tarda* from aquatic animal species and surface waters in Florida. *J. Wildl. Dis.* 9:204–207.

Williams, L.E., Jr., and R.W. Phillips. 1972. North Florida sandhill crane populations. *Auk* 89: 541–548.

———. 1973. Capturing sandhill cranes with alpha-chloralose. *J. Wildl. Manag.* 37:94–97.

Windingstad, R.M. 1988. Nonhunting mortality in sandhill cranes. *J. Wildl. Manag.* 52:260–263.

Windingstad, R.M., R.J. Cole, P.E. Nelson, T.J. Roffe, R.R. George, and J.W. Dorner. 1989. *Fusarium* mycotoxins from peanuts suspected as a cause of sandhill crane mortality. *J. Wildl. Dis.* 25:38–46.

Windingstad, R.M., S.S. Hurley, and L. Sileo. 1983. Capture myopathy in a free-flying greater sandhill crane (*Grus canadensis tabida*) from Wisconsin. *J. Wildl. Dis.* 19:289–290.

Windingstad, R.M., S.M. Kerr, L.N. Locke, and J.J. Hurt. 1984. Lead poisoning of sandhill cranes (*Grus canadensis*). *Prairie Nat.* 16:21–24.

Windingstad. R.M., D.O. Trainer, and R. Duncan. 1977. *Salmonella enteritidis* and *Arizona hinshawii* isolated from wild sandhill cranes. *Avian Dis.* 21:704–707.

Wood, P.B., S.A. Nesbitt, and A. Steffer. 1993. Bald eagles prey on sandhill cranes in Florida. *J. Raptor Res.* 27:164–165.

Woodard, J.C., and A.W.C.V. Layton. 1989. Un-published data. University of Florida, Gainesville.

Zinkl, J.G., N. Dey, J.M. Hyland, J.J. Hurt, and K.L. Heddlestom. 1977. An epornitic of avian cholera in waterfowl and common crows in Phelps County, Nebraska in the spring, 1975. *J. Wildl. Dis.* 13:194–198.

Shorebirds

I. Introduction

Eight plovers, 1 oystercatcher, 1 jacana, 2 stilts, 33 sandpipers, and 3 phalaropes occur in Florida (table 20.1). Of these, only 4 plovers, the American Oystercatcher, the Black-necked Stilt, the American Avocet, the Willet, and the American Woodcock breed in Florida. The Piping Plover and the Cuban (Southeastern) Snowy Plover (*Charadrius alexandrinus tenuirostris*) are listed as threatened by the Florida Game and Fresh Water Fish Commission and the Piping Plover is additionally listed as threatened by the U.S. Fish and Wildlife Service. Both are listed as endangered by the Florida Committee on Rare and Endangered Plants and Animals (Rodgers et al. 1996). The American Oystercatcher is listed as a species of special concern by the Florida Game

and Fresh Water Fish Commission (Logan 1997) and threatened by the Florida Committee on Rare and Endangered Plants and Animals (Rodgers et al. 1996). The Wilson's Plover and American Avocet are listed as species of special concern by the Florida Committee on Rare and Endangered Plants and Animals (Rodgers et al. 1996).

Life history and population information are reported for Black-bellied Plovers (Paulson 1995), Cuban Snowy Plovers (Gore 1996), Snowy Plovers (Page et al. 1995), Piping Plovers (Haig 1992; Nicholls 1996), Wilson's Plovers (Sprandel 1996), American Avocets (Engstrom 1996; Robinson et al. 1997), American Oystercatchers (Nol and Humphrey 1994;

Table 20.1. Species of shorebirds in Florida

Class	Species		Distribution	Seasonal occurrence	Relative abundance
Plovers					
	Black-bellied Plover	*Pluvialis squatarola*	Throughout	Year-round[a]	Common
	Lesser Golden-Plover	*Pluvialis dominica*	Throughout	Spring/fall	Rare
	Snowy Plover	*Charadrius alexandrinus*	West coast	Year-round[a]	Uncommon
	Wilson's Plover	*Charadrius wilsonia*	Coastal	Year-round[a]	Common
	Semipalmated Plover	*Charadrius semipalmatus*	Throughout	Fall/winter/spring	Common
	Piping Plover	*Charadrius melodus*	Coastal	Fall/winter/spring	Locally common
	Killdeer	*Charadrius vociferus*	Throughout	Year-round[a]	Common
	Mountain Plover	*Charadrius montanus*	Coastal	Winter	Rare
Oystercatchers					
	American Oystercatcher	*Haematopus palliatus*	Coastal	Year-round[a]	Locally common
Stilts and Avocets					
	Black-necked Stilt	*Himantopus mexicanus*	Throughout	Year-round[a]	Locally common
	American Avocet	*Recurvirostra americana*	Throughout	Year-round[a]	Locally common
Jacanas					
	Northern Jacana	*Jacana spinosa*	South	Fall/spring	Few records
Sandpipers and allies					
	Greater Yellowlegs	*Tringa melanoleuca*	Throughout	Fall/winter/spring	Common
	Lesser Yellowlegs	*Tringa flavipes*	Throughout	Fall/winter/spring	Common
	Solitary Sandpiper	*Tringa solitaria*	Throughout	Spring/fall	Common
	Willet	*Catoptrophorus semipalmatus*	Throughout	Year-round[a]	Locally common
	Spotted Sandpiper	*Actitis macularia*	Coastal	Fall/winter/spring	Common
	Upland Sandpiper	*Bartramia longicauda*	Throughout	Fall/spring	Locally common
	Whimbrel	*Numenius phaeopus*	Coastal	Fall/winter/spring	Locally common
	Long-billed Curlew	*Numenius americanus*	Coastal	Winter	Uncommon
	Black-tailed Godwit	*Limosa limosa*	Coastal	Spring	Few records
	Hudsonian Godwit	*Limosa haemastica*	Throughout	Spring/fall	Rare

Bar-tailed Godwit	*Limosa lapponica*	Coastal	Winter/spring	Few records
Marbled Godwit	*Limosa fedoa*	Coastal	Year-round	Locally common
Ruddy Turnstone	*Arenaria interpres*	Throughout	Year-round	Locally common
Surfbird	*Aphriza virgata*	West coast	Spring	Few records
Red Knot	*Calidris canutus*	East coast	Year-round	Common
Sanderling	*Calidris alba*	Coastal	Year-round	Common
Semipalmated Sandpiper	*Calidris pusilla*	Throughout	Fall/winter/spring	Locally common
Western Sandpiper	*Calidris mauri*	Throughout	Fall/winter/spring	Common
Least Sandpiper	*Calidris minutilla*	Throughout	Fall/winter/spring	Common
White-rumped Sandpiper	*Calidris fuscicollis*	Coastal	Fall/spring	Locally common
Baird's Sandpiper	*Calidris bairdii*	Coastal	Fall/spring	Rare
Pectoral Sandpiper	*Calidris melanotos*	Throughout	Fall/spring	Locally common
Sharp-tailed Sandpiper	*Calidris acuminata*	Throughout	Irregular	Few records
Purple Sandpiper	*Calidris maritima*	Throughout	Winter	Rare
Dunlin	*Calidris alpina*	Throughout	Fall/winter/spring	Common
Curlew Sandpiper	*Calidris ferruginea*	Peninsula	Fall/winter/spring	Rare
Stilt Sandpiper	*Calidris himantopus*	Throughout	Fall/winter/spring	Locally common
Buff-breasted Sandpiper	*Tryngites subruficollis*	Throughout	Fall/spring	Uncommon
Ruff	*Philomachus pugnax*	Throughout	Fall/winter/spring	Rare
Short-billed Dowitcher	*Limnodromus griseus*	Throughout	Year-round	Common
Long-billed Dowitcher	*Limnodromus scolopaceus*	Throughout	Year-round	Locally common
Common Snipe	*Gallinago gallinago*	Throughout	Fall/winter/spring	Common
American Woodcock	*Scolopax minor*	Throughout	Year-round[a]	Uncommon
Phalaropes				
Wilson's Phalarope	*Phalaropus tricolor*	Throughout	Fall/spring	Locally common
Red-necked Phalarope	*Phalaropus lobatus*	Offshore	Fall/winter/spring	Locally common
Red Phalarope	*Phalaropus fulicaria*	Offshore	Fall/winter/spring	Common

Source: Robertson & Woolfenden (1992).
a. Breeds in Florida.

Below 1996), Greater Yellowlegs (Elphick and Tibbits 1998), Whimberel (Skeel and Mallory 1996), Solitary Sandpipers (Moskoff 1995), Spotted Sandpipers (Oring et al. 1997), Red Knots (Harrington 1982; Harrington et al. 1982), Western Sandpipers (Wilson 1994), Least Sandpipers (Cooper 1994), Dunlin (Warnock and Gill 1996), Stilt Sandpipers (Klima and Jehl 1998), and American Woodcock (Keppie and Whiting 1994).

General disease information for shorebirds is included in publications by Friend and Franson (1999) and Fairbrother et al. (1996). Bush (1990) provides an extensive analysis of helminth communities of Willets from 3 locations in Florida. Helminth parasites of the Common Snipe (Threlfall 1970) and Woodcock (Prestwood et al. 1969) are known outside of Florida.

II. Trauma

Crawford (1981) recorded birds found dead below a TV tower on Tall Timbers Research Station in Leon County between 1955 and 1980. It is presumed that most of these birds were killed during migration. During this period 42,384 birds representing 189 species were found, including 6 Killdeer, 3 Solitary Sandpipers, 1 Willet, 4 Spotted Sandpipers, 1 Least Sandpiper, 2 Upland Sandpipers, 41 Common Snipes, 4 American Woodcocks, and 2 Red Phalaropes.

Roadkills, whether by vehicular collision or by power line collision, are a relatively uncommon cause of death for this group of birds. A total of 165 birds was found dead under power lines across the Everglades, Palm Beach County, 1990–91. Of these a single Black-necked Stilt was found (FPL and FGFWFC 1991). Weston (1966) reported a single Wilson's (Common) Snipe among 740 birds found dead along the Pensacola Bay Bridge during 1938–49. Most of these mortalities occurred at night during migration. Mortality ceased when a new bridge, without an overhead cable, was built. Snyder (1994) conducted a survey of roadkills from 1990–93 in and around state parks; of a total of 1562 birds found dead, 4

were Sanderlings, 2 Short-billed Dowitchers, 1 Western Sandpiper, and 1 Snowy Plover. Two American Woodcock found dead in Alachua County in 1978 examined by Forrester (1978) had died of traumatic lesions. One was dead along a roadside, the other dead in a sheep pasture. Longstreet (1954) reported finding a dead Red Phalarope on the beach in Volusia County in January 1930. He also recounted a flock of Red-necked Phalaropes (Northern Phalaropes) flying into and out of lights at a baseball game in Volusia County in 1948. Twenty of these were later found dead (Longstreet 1949). A Dunlin that was collected during a botulism die-off (see below) had evidence of head trauma (Smith and Nettles 1994).

Hunters took an estimated 6,099 Woodcock in the 1968–69 season in Florida (Pursglove and Doster 1970). They harvested 33,652 and 42,940 Common Snipe in 1996 and 1997, respectively, and 0 and 707 Woodcock in 1996 and 1997, respectively (Linda 1998).

III. Habitat alteration

Snowy Plover numbers have decreased, probably because of human and dog activity on beaches (Stevenson and Anderson 1994; Gore 1996). They disappeared from nesting beaches in Pinellas County, 1972, probably for this reason (Kale 1972). The wintering, foraging, and roosting areas for the Piping Plover have been reduced by manmade structures such as seawalls and jetties (USDI 1985) and recreational activities (Nicholls 1996). Recreational use of beaches and adjacent waters results in frequent flushing and thus increased energy expenditure for many shorebirds (Below 1996). Rodgers and Smith (1997) discuss suggested approach distances to avoid human disturbance for some shorebirds.

IV. Predation

Stevenson and Anderson (1994) mention that Snowy Plovers and other shorebirds are captured by migrating flights of falcons and accipters along the coast of Florida. Upon examination of 1,098 pellets collected at a roost of Northern Harriers in Leon County in 1925–26, evidence of an American Woodcock was found (Stoddard 1931). A Common Snipe was found in the nest of a Bald Eagle in north central Florida (McEwan and Hirth 1980), and Nicholson (in Bent 1937) found a Killdeer carcass in a Bald Eagle nest.

American Oystercatchers are occasionally "captured" by oysters and held until drowned by rising tide (Stevenson and Anderson 1994). Largemouth bass (*Micropterus salmoides*) took young Black-necked Stilts that were frightened into Lake Okeechobee (Rand 1943).

American Oystercatcher nesting failure is caused by dogs, cats, raccoons, and foxes (Below 1996). A ghost crab (*Ocypode albicans*) was observed taking the eggs from a Snowy Plovers nest in Franklin County, 1990 (Stevenson and Anderson 1994).

V. Inclement weather

Killdeer appear to be especially susceptible to cold weather, perhaps because of their insectivorous prey (Stevenson and Anderson 1994).

VI. Environmental contaminants

Chlorinated hydrocarbons were measured in Woodcock tissues from 23 eastern and midwestern states and included samples from 5 birds collected in Alachua and Marion counties, 1970–71 (Clark and McLane 1974) (table 20.2). Concentrations were generally below those considered dangerous for human consumption, and higher in the south than in the north. Mirex concentrations were highest in birds from Mississippi and Louisiana.

In July of 1997 organophosphate poisoning (depressed cholinesterase activity) was diagnosed in 16 dead and dying Western Sandpipers, 4 Black Skimmers, and 2 Least Terns on Marco Island and Gordon Pass, Collier County (Meteyer 1997). Again in October of 1997 severely depressed brain cholinesterase activity, good body condition, and a lack of evidence of trauma or infectious disease indicated organophosphate poisoning of 78 Western Sandpipers found dead on Marco Island, Collier County (Meteyer 1997, Spalding 1998). Other species, including single or small numbers of Laughing Gulls, Black Skimmers, Sanderlings, Red Knots, Sandwich Terns, and a Piping Plover were also observed dead, but were not tested. Fisk (1976) suspected organophosphate insecticide contamination from Monocrotophos (trade name Azodrin) in her report of 175 Killdeer that died in a tomato field in Dade County in 1976, but no testing was done.

Mercury concentrations in liver of 4 Woodcock collected from Alachua and Marion Counties, 1970–71, was relatively high compared with those from more northern states, but low compared with fish-eating birds in Florida (table 20.2) (Clark and McLane 1974).

Liver iron concentration in a Dunlin from a botulism die-off in Hillsborough County in November 1994 was 3,800 ppm dry weight. This was higher than expected in this and some ducks that were also tested. Concentrations of other metals (copper, zinc, arsenic, lead) were not significant (Smith and Nettles 1994).

VII. Oiling

An oil spill on the coast of Wakulla County in December 1968 resulted in "many" ducks, snipe, and other birds so covered with oil that they were unable to fly. Smaller birds were unable to walk in the heavy oil (CSLP 1969). An oil spill near Tampa in 1970 killed "several thousand" birds, primarily Lesser Scaup, and smaller numbers of sandpipers and plovers (Stevenson 1970; Sims 1970). A spill also occurred in Jacksonville in the same year; however, a list of birds killed was not reported (Stevenson 1970).

Table 20.2. Chlorinated hydrocarbon and metal concentrations in tissues of shorebirds collected from Florida

Species Tissue	County	Year(s)	No. Exam.	DDE No.[a]	DDE Mean[b]	DDE (Range)	DDT No.[a]	DDT Mean[b]	DDT (Range)	DDD No.[a]	DDD Mean[b]	DDD (Range)	Data source
Red Knots													
Brain	Pinellas	1973–76	7	6	0.025	(ND–0.15)	ND[c]	—	—	1[c]	<0.001	(ND–0.020)	C
Intestine	Pinellas	1973	1	1	0.065	—	NA	—	—	NA	—	—	C
Liver	Pinellas	1973–76	7	5	0.020	(ND–0.38)	4[c]	0.009	(0.007–0.014)	2[c]	<0.001	—	C
Muscle	Pinellas	1973–76	7	7	0.14	(0.068–0.50)	3[c]	0.003	(ND–0.013)	2[c]	<0.001	—	C
American Woodcock													
Muscle[d]	Alachua/Marion	1970–71	5	5	11[e,f]	(3.0–22)[e,f]	NA	—	—	NA	—	—	A

Species Tissue	County	Year(s)	No. Exam.	PCBs No.[a]	PCBs Mean[b]	PCBs (Range)	Dieldrin No.[a]	Dieldrin Mean[b]	Dieldrin (Range)	Heptachlor epoxide No.[a]	Heptachlor epoxide Mean[b]	Heptachlor epoxide (Range)	Data source
Red Knots													
Brain	Pinellas	1974	4	4	0.44	(0.17–0.81)	ND	—	—	—	—	—	C
Liver	Pinellas	1974	4	4	0.076	(0.055–0.12)	4	0.036	(0.017–0.081)	—	—	—	C
Muscle	Pinellas	1974	4	4	0.11	(0.044–0.29)	4	0.11	(0.006–0.018)	—	—	—	C
American Woodcock													
Muscle[d]	Alachua/Marion	1970–71	5	5	7.4[f]	(0.82–16)[f]	5	0.98[f]	(0.34–1.6)[f]	5	2.3[f]	(0.0–8.7)[f]	A

Species Tissue	County	Year(s)	No. Exam.	Mirex No.[a]	Mirex Mean[b]	Mirex (Range)	Hg No.[a]	Hg Mean[b]	Hg (Range)	Fe No.[a]	Fe Mean[b]	Fe (Range)	Data source
Dunlin													
Liver	Hillsborough	1994	1	—	—	—	—	—	—	1	3,800[g]	—	B
American Woodcock													
Liver	Alachua/Marion	1970–71	4	NA	—	—	4	0.28	(0.18–0.36)	NA	—	—	A
Muscle[d]	Alachua/Marion	1970–71	5	1	NG	(NG–0.19)[f]	NA	—	—	NA	—	—	A

Sources: A = Clark & McLane (1974), B = Smith & Nettles (1994), C = Forrester & Thompson (1976).
NA = not analyzed, ND = not detected, NG = not given.
a. Number of individuals in which the specified parameter was detected.
b. Geometric mean. Mean and range include negative individuals (in which the parameter was not detected) and are ppm wet weight values unless otherwise noted.
c. *n* = 4
d. Breast muscle.
e. The values listed for DDE are the sum of DDE + DDD + DDT.
f. Lipid weight.
g. Dry weight.

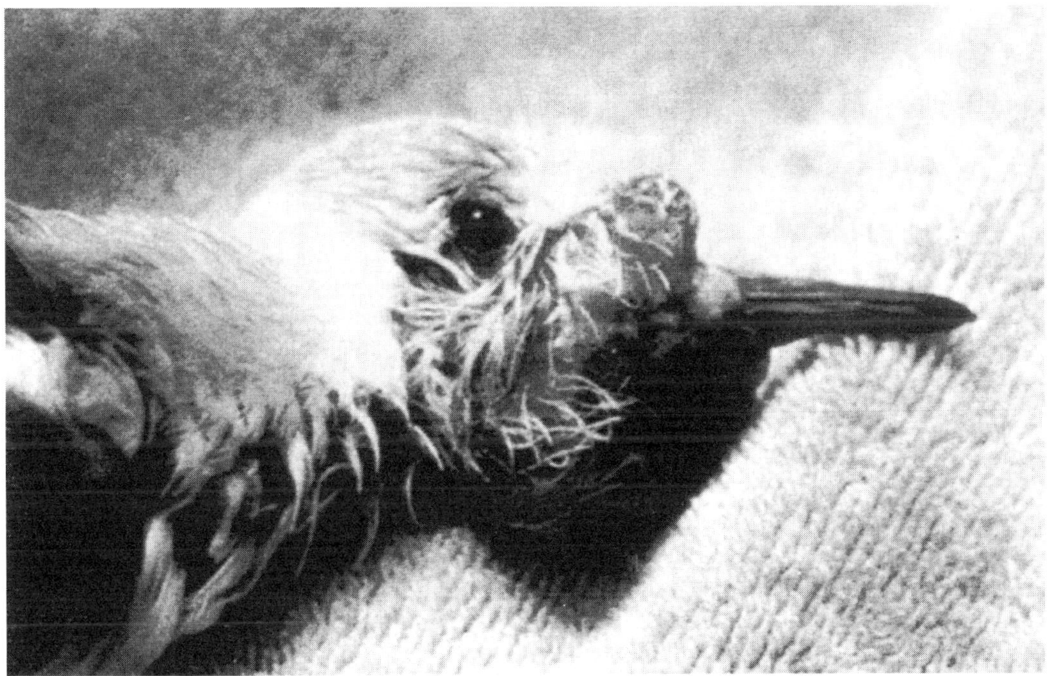

FIGURE 20.1. Sanderling with poxviral lesions. From Kreuder et al. 1999; by permission of *Journal of Wildlife Diseases*.

VIII. Neoplasia

We found no reports of neoplasia in this group of birds in Florida.

IX. Biotoxins

Two Ruddy Turnstones (1 sick and 1 dead) collected from a beach on Egmont Key National Wildlife Refuge, Hillsborough County, August 1999, tested positive for botulism type C. Sixteen turnstones were seen in the area and 1 of these was moribund (Converse 1999).

Shorebirds have been found dead along with other species in 2 other botulism die-offs in Florida; however, they have never been included in the testing and thus it is uncertain whether they died from botulism. Two Black-necked Stilts and a Semipalmated Sandpiper were found during the May–June botulism die-off in Hamilton County, 1979 (Forrester et al. 1980; see chapter 10, Ducks). A botulism epizootic following tropical storm Gordon in Hillsborough County, November 1994, included a Dunlin with evidence of head trauma and an American Avocet (Smith and Nettles 1994; see chapter 10, Ducks). Three of 5 Western Sandpipers from an organophosphate poisoning (see above, section VI, Environmental Contaminants) had weakly positive ELISA tests for botulism type C (see above) (Meteyer 1997).

Twenty-five Black-necked Stilts, a Long-billed Dowitcher, and a Lesser Yellowlegs were identified among the birds found dead on Lake Okeechobee, Glades County, in the summer of 1971. It was concluded that *Clostridium perfringens* type C, causing necrotizing enteritis, was the most likely cause of mortality based on culture and mouse protection tests (Jasmin et al. 1972; see chapter 10, Ducks, for more details). Shorebirds were not included in the species tested.

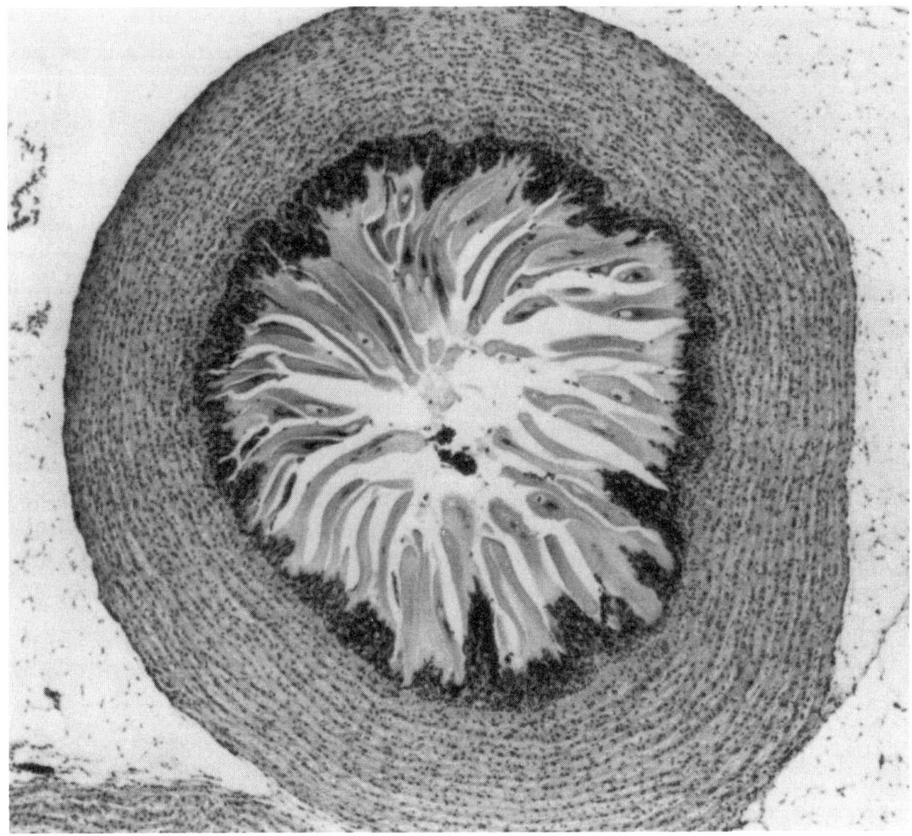

FIGURE 20.2. *Besnoitia*-like organisms within hypertrophied endothelial cells on the luminal surface of the descending aorta of a Red Knot. From Woodard et al. 1977; by permission of *Veterinary Pathology*.

X. Avian vacuolar myelinopathy

This disease is not known to occur in Florida, but has been diagnosed in Killdeer in nearby states (see chapter 11, Eagles).

XI. Viruses

Avian poxviral lesions were described grossly, histologically, and by electronmicroscopy in 3 Sanderlings brought to a rehabilitation facility in Lee County in February 1996 (Kreuder et al. 1999). The large (1–2.5 cm) masses were located on the wings, base of the beak, and/or base of the tongue of the 3 birds (figure 20.1),

causing them to be emaciated and weak. One of the masses bled profusely. This was the first report of this disease affecting shorebirds in North America; it has been reported for Dunlin elsewhere (Green 1969).

We found little evidence that shore birds have tested positive for less obvious viral infections. However, they rarely have been included in serologic studies, and viral infections rarely are considered or tested for when die-offs occur. As part of a survey, 2 of 5 Killdeer from central Florida during the spring and early summer of 1958 were positive (neutralized at least 1.5 logs) by the serum neutralization test for EEE virus antibody (Favorite 1960). No virus was isolated. One Spotted Sandpiper was

Table 20.3. Bacteria isolated from shorebirds in Florida

Bacteria Host species	County	Year(s)	Organ or tissue	Data source
Escherichia coli				
Killdeer[a]	Glades	1977–78	Intestine	Forrester & White (1978)
Red Knot	Pinellas	1973–74	Intestine	Woodard et al. (1977)
American Woodcock	Alachua	1978	Intestine	Forrester & White (1978)
Western Sandpiper	Collier	1997	Liver	Meteyer (1997)
Coliforms				
Red Knot	Pinellas	1973–74	Intestine	Woodard et al. (1977)
Proteus sp.				
Red Knot	Pinellas	1973–74	Intestine	Ibid.

a. Forrester & White (1978) cultured liver and intestine from 7 Killdeer in Glades County, 1977–78.

also tested and was negative. Charadriiform blood was identified in 5.5% of blood meals collected from *Culiseta melanura* in Indian River County (Edman et al. 1972). Edman et al. (1974) found that Willets were susceptible to bites of *Culex nigripalpus* in an experimental setting.

XII. Bacteria

No significant bacterial pathogens have been isolated from shorebirds in Florida (table 20.3). Avian cholera, caused by *Pasteurella multocida*, was diagnosed in a single American Oystercatcher in South Carolina (Blus et al. 1978). Red Knots were listed with the birds found dead during an avian cholera die-off in the Everglades; however, they were not included in the testing (Klukas and Locke 1970).

XIII. Fungi

We found no information for this group of birds.

XIV. Protozoans

A *Besnoitia*-like organism was associated with die-offs of 150 Red Knots found sick or dead during a 3-day period in December of 1973, and in 30 knots in October of 1974, and a few knots in August 1999 in Pinellas County (Simpson et al. 1976, 1977; Woodard et al. 1977, Spalding 1999). The infection caused hemorrhages on the serosal surface of the lower small intestine and colonic infarction. Protozoa within endothelial cells (figures 20.2–20.3) resulted in endaortitis and endarteritis. Schizonts were found near muscular arteries of the small intestine and within renal medullary tubular cells. Testing for botulism, organochlorine pesticides (table 20.2), and bacteria was found to be negative or insignificant in the earlier die-offs (Simpson et al. 1976, 1977; Woodard et al. 1977).

Blood parasites have not been observed in shorebirds in Florida, and are generally rare in shorebirds elsewhere (Greiner et al. 1975). Blood smears examined from the following birds were negative: 46 Red Knots, Pinellas County, 1973–76 (Forrester and Humphrey 1981); 8 Killdeer in 1977 and 22 in 1983, Glades County (Forrester and Bennett 1985); and 2 Lesser Yellowlegs, Glades County, 1983 (Forrester and Bennett 1985). Forrester and Humphrey (1981) were able to experimentally infect Red Knots captured during the above-mentioned die-off with *Plasmodium hermani* from Wild Turkeys. They remained parasitemic for up to 43 days. Blood from 22 Killdeer and 2 Lesser Yellowlegs inoculated into domestic turkey poults did not result in

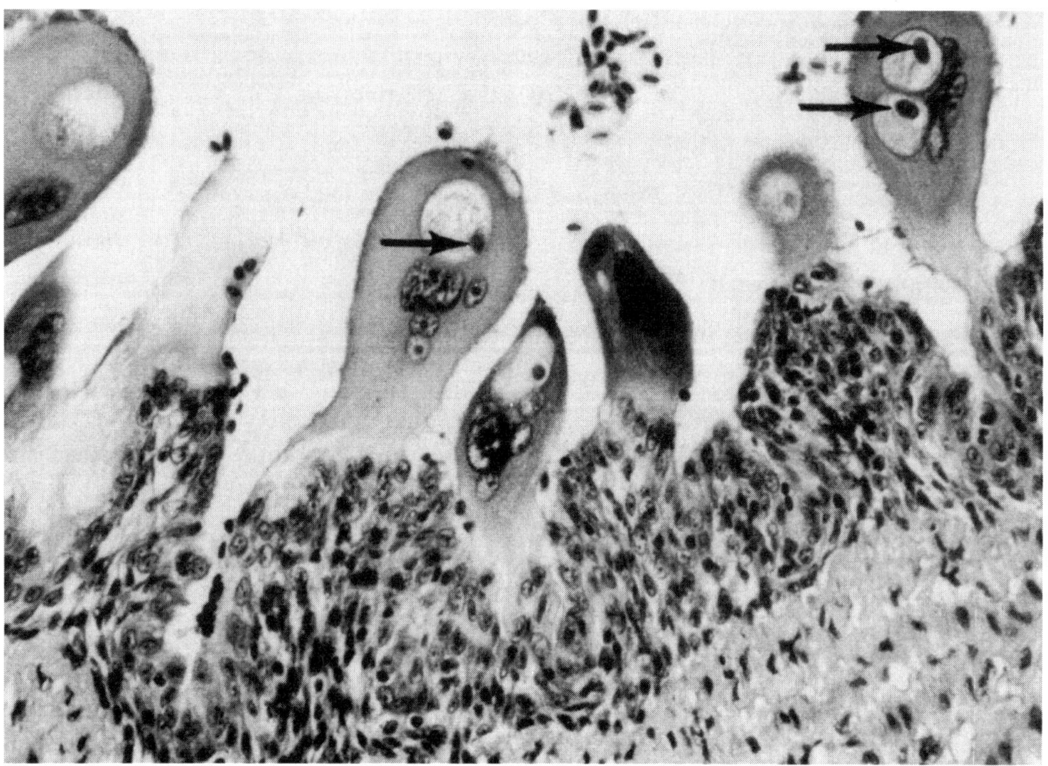

FIGURE 20.3. Higher-power photomicrograph of *Besnoitia*-like organisms within endothelial cells. The arrows point to parasites within large vacuoles. The intimal cells contain large multiple nuclei. From Woodard et al. 1977; by permission of *Veterinary Pathology*.

parasitemias of malarial organisms (Forrester 1994).

Toxoplasma gondii antibodies were not detected in serum from a single Lesser Yellowlegs from Florida (date and location not given) by Burridge et al. (1979). This was tested by IHA, which may not be as sensitive in birds as in mammals thus resulting in false negatives (Frenkel 1981).

XV. Helminths

Helminth parasites have been well documented for Killdeer (table 20.4), Willets (tables 20.5–20.6) and Red Knots (table 20.7) in Florida. Smaller numbers of miscellaneous shorebirds have also been examined for helminth parasites (table 20.8).

Bush (1990) reported similar numbers of helminth species in Willets collected from fresh and saltwater sites in the United States and Canada; however, he found significantly higher numbers of individual helminths in birds collected in saltwater areas.

Skriabinoclava pusillae (Acuarioidae) found in Dunlins, Western Sandpipers, and Semipalmated Sandpipers wintering in Marco Island, Collier County, is an indication that transmission may occur in this region (Wong and Anderson 1990).

Penner and Fried (1963) described the ocular trematode *Philophthalmus hegeneri* from cercariae found in the marine snail *Batillaria minima* along the Gulf Coast from Key West to Dunedin. The adult stage occurs in Royal Terns, Laughing Gulls, Yellow-crowned Night Herons, and Willets in Florida.

Table 20.4. Helminths from Killdeer collected in Glades County, Florida[a]

| Class | | No. of birds | | | Intensity | |
Species (Location)[b]	Year(s)	Examined	Positive	%	Mean	Range
Trematoda						
Stomylotrema vicarium (LI)	1983	10	2	20	1	1
Tanaisia fedtschenkoi (KD)	1977–78	8	1	—	2	2
	1983	10	1	10	1	1
Cestoda						
Megalacanthus sp. (SI)	1977–78	8	4	—	4.8	1–11
	1983	10	3	30	1.3	1–2
Progynotaenia americana (SI)	1977–78	8	7	—	8.3	1–15
	1983	10	1	10	4	1
Nematoda						
Capillaria vanelli (GI)	1983	10	3	30	4	3–6
Dispharynx nasuta (PR)	1977–78	8	2	—	3.5	1–6
	1983–84	12	1	8	4	—
Strongyloides sp. (SI)	1977–78	8	3	—	3.7	1–7
Larval spirurids (GL)	1983	10	6	60	11	2–19
Larval spirurids (LI)	1983	10	2	20	40	10–70
Acanthacephala						
Larval acanthacephalan (SI)	1983	10	1	10	1	1

Source: Kinsella & Ballwebber (1993).
a. All birds were shot.
b. Location in host: GI = gizzard, GL = gizzard lining, KD = kidney, LI = large intestine, PR = proventriculus, SI = small intestine.

XVI. Arthropods

Ectoparasites collected from shorebirds are summarized in tables 20.9 and 20.10. Most of these were identified from dead individuals. No information about the effects of the parasites on shorebirds was found.

XVII. Summary and Conclusions

Causes of mortality for shorebirds in Florida are very poorly documented. Trauma, especially for hunted species, appears to be the most common cause of mortality; however, reproductive and foraging site losses from disturbance of beach and other areas undoubtedly has even greater significance to many of these species, especially for the few that nest in Florida. Pesticide contamination was documented in 3 cases, but was rarely tested for.

The most significant disease-related mortality that has been documented for shorebirds in Florida is the *Besnoitia* like organism associated with 2 Red Knot die-offs. Shorebirds have been involved in biotoxin die-offs and oil spills, but usually other species are killed in greater numbers. A few cases of avian poxvirus resulted in mortality; however, the prevalence of this disease has not been investigated.

The majority of the detailed information available for this group of birds involves helminth parasites of Willets, Red Knots, and Killdeer. Thirty-seven trematodes, 10 cestodes, 10 nematodes, and 3 acanthocephalans were identified. Only 1 biotoxin, 2 bacteria, 2 viruses, no fungi, and 1 protozoan were detected from this group, reflecting a lack of effort in this area. No ticks, 8 mites, and 36 chewing lice were identified.

Table 20.5. Trematodes from Willets from Florida[h]

Species (location)[a]	County or area	Year(s)	No. of birds			Intensity[i]		Data source
			Examined	Positive	%	Mean	Range	
Ascocotyle ampullacea (SI)	St. Johns	1981–88	10	1	10	1	0–13	A
Ascocotyle sp.(SI)	St. Johns	1981–88	10	1	10	0.1	0–1	A
Cyclocoelum obscurum (CO)	Franklin	1958	6	3	—	84	2–150	E
Diacetabulum riggini (SI)	"Tampa Bay"	1981–88	10	NG	—	1	NG	A
	Taylor	1981–88	5	NG	—	76	NG	A
Echinostoma sp.	Franklin	1958	6	1	—	1	—	E
Endocotyle bushi	Levy	1972	1	1	—	2	—	D
Gynaecotyla adunca (SI)	St. Johns	1981–88	10	NG	—	2	NG	A
	Taylor	1981–88	5	NG	—	2	NG	A
Gynaecotyla riggini (IN)	Franklin	1958	6	1	—	1	—	E
Himasthla quissentensis (SI)	St. Johns	1981–88	10	NG	—	1	NG	A
	Taylor	1981–88	5	NG	—	0.2	NG	A
Levinseniella hunteri (CE)	Taylor	1981–88	5	NG	—	10	NG	A
Maritrema patulus (SI)	"Tampa Bay"	1981–88	10	NG	—	65	NG	A
	St. Johns	1981–88	10	NG	—	20	NG	A
	Taylor	1981–88	5	NG	—	173	NG	A
Mesorchis denticulatus[b] (SI)	St. Johns	1981–88	10	1	10	0.4	0–4	A
Microphallus nicolli (SI)	"Tampa Bay"	1981–88	10	NG	—	0.1	NG	A
	St. Johns	1981–88	10	NG	—	158	NG	A
	Taylor	1981–88	5	NG	—	197	NG	A
Microphallus turgidus (CE)	St. Johns	1981–88	10	NG	—	6	NG	A
	Taylor	1981–88	5	NG	—	11	NG	A
Notocotylus sp. (CE)	St. Johns	1981–88	10	1	10	0.1	0–1	A
Numeniotrema kinsellai (SI)	Taylor	1981–88	5	5	—	30	NG	A
	Taylor	1981	2	2	—	240	210–271	B
Odhneria odhneri (CE,SI)	Levy	1972	1	1	—	3	—	D

	"Tampa Bay"	1981–88	10	NG	—	151	NG	A
	St. Johns	1981–88	10	NG	—	208	NG	A
	Taylor	1981–88	5	NG	—	51	NG	A
Paragymnophallus kinsellai[c] (CE,LI)	"Tampa Bay"	1981–88	NG	NG	—	NG	NG	C
Parorchis acanthus[d] (LI)	"Tampa Bay"	1981–88	10	NG	—	0.3	NG	A
	St. Johns	1981–88	10	NG	—	0.5	NG	A
	Taylor	1981–88	5	NG	—	0.8	NG	A
Parvatrema sp. (SI)	Taylor	1981–88	5	NG	—	2	NG	A
Parvatrema borinquenae[e,f]	"Tampa Bay"	1981–88	10	NG	—	19	NG	A
Parvatrema bushi[f]	St. Johns	1981–88	10	NG	—	5	NG	A
Philophthalmus hegeneri[g] (OC)	Manatee to Charlotte	NG	NG	—	—	—	—	F
Probolocoryphe lanceolata (SI)	St. Johns	1981–88	10	3	30	6	NG	A
Stictodora cursitans	Taylor	1981–88	5	1	—	0.4	1	A
Stictodora hancocki (SI)	"Tampa Bay"	1981–88	10	NG	—	0.2	NG	A
	St. Johns	1981–88	10	NG	—	16	NG	A
	Taylor	1981–88	5	NG	—	912	NG	A
Stictodora sp.(SI)	St. Johns	1981–88	10	NG	—	3	NG	A
	Taylor	1981–88	5	NG	—	214	NG	A
Testiculoporus cedarkeyensis (SI)	Levy	1972	1	1	—	2	—	D
Trematode (BU,SI)	Taylor	1981–88	5	1	—	12	0–61	A

Sources: A = Bush (1990), B = Bush & Threlfall (1984), C = Ching (1995), D = Kinsella & Deblock (1997), E = MacInnis (1966), F = Penner & Fried (1963). NG = not given.

a. Location in host: BU = bursa, CE = ceca, CO = coelom and air sacs, LI = large intestine, OC = under nictitating membrane of the eye, SI = small intestine.

b. *Stephanoprora denticulata* is a synonym of *M. denticulatus*.

c. Described from material collected by Bush (1990).

d. Cercariae from snail *Cerithidia scalariformis* from Wakulla Co. developed into adult *P. acanthus* in chicks in 14 days (Lewis, in Loftin 1961).

e. Metacercariae found in the ladder horn snail (*Cerithidea sclariformis*) on the Gulf coast but not on the Atlantic coast (Ching 1995).

f. Bush (1990) reported these trematodes as *Parvatrema* sp.; they were later described as *P. bushi* by Ching (1995).

g. Penner & Fried (1963) described from adults reared from cercariae in the marine snail *Batillaria minima* from the southwest coast of Florida.

h. Not found in Florida, but in Willets elsewhere by Bush (1990): *Ascorhytis charadriformis, Maritrema laricola, Levinseniella gynmopocha, Stephonoprora* n.sp., *Maritrema prosthometra, Microphallus* n.sp., *Parvatrema borealis, Cloacitrema michiganense.*

i. Intensity values are given only for data sources B through F. Values from source A are abundance, not intensity.

Table 20.6. Cestodes, nematodes, and acanthocephalans from Willets from Florida, 1981–88[c]

Species (location)[a]	County or area	No. of birds			Abundance	
		Examined	Positive	%	Mean	Range
Cestoda						
Anomotaenia sp. (SI)	St. Johns	10	NG	—	52	NG
	"Tampa Bay"	10	NG	—	41	NG
	Taylor	5	NG	—	42	NG
Aploparaksis sp. (SI)	"Tampa Bay"	10	1	10	0.2	0–2
Dipelididae sp. 1 (SI)	St. Johns	10	1	10	0.3	0–3
Dipelididae sp. 2 (SI)	Taylor	5	1	—	2	0–8
Kowalewskiella cinguilifera (SI)	St. Johns	10	NG	—	0.5	NG
	"Tampa Bay"	10	NG	—	0.5	NG
Ophryocotyle insignus (SI)	St. Johns	10	NG	—	6	NG
	"Tampa Bay"	10	1	10	0.1	0–1
	Taylor	5	NG	—	24	NG
Nematoda						
Capillaria sp. A (SI)	St. Johns	10	NG	—	1	NG
	Taylor	5	NG	—	3	NG
Capillaria sp. B (SI)	St. Johns	10	NG	—	7	NG
	"Tampa Bay"	10	NG	—	2	NG
	Taylor	5	NG	—	16	NG
Skrjabinoclava inornatae[b] (PR)	Collier/Lee[b]	8[b]	2	—	15.5[d]	3–28[d]
Acanthocephala						
Parafilicollis sp. (SI)	St. Johns	10	NG	—	8	NG
	"Tampa Bay"	10	NG	—	1	NG
Prosthorhynchus sp. (SI)	St. Johns	10	4	40	2	NG

Source: Bush (1990) (except as noted for *Skrjabinoclava inornatae*).

NG = not given.

a. Location in host: PR = proventriculus, SI = small intestine.

b. Wong & Anderson (1990) examined 5 adult Western Willets, 1 Willet from Collier County, and 1 Willet from Lee County for *Skrjabinoclava* sp.

c. Not found in Willets collected in Florida, but in Willets elsewhere by Bush (1990): *Hymenolepididae* sp. 3, *Hymenolepis amphitricha, Spiruridae* sp. 1.

d. Intensity values.

Table 20.7. Helminth parasites of 32 Red Knots collected in Tampa Bay, 1973–76

| Class | | | Intensity | |
Species (location)[a]	No. positive	% positive	Mean	Range
Trematoda				
Cyclocoelum obscurum (BC)	1	3	1	—
Himasthla alincia (SI)	15	47	6	1–12
Parorchis acanthus (LI)	9	28	1	1–2
Parvatrema sp. (SI)	1	3	2	—
Odhneria odhneri (LI,CE)	10	31	10	1–46
Nematoda				
Capillaria cecumitis (LI)	14	44	9	1–26
Tetrameres calidris (PR)	2	6	1	1
Cestoda				
Aploparaksis sp. (SI)	5	16	2	1–2
Geraldolepis deblocki (SI)	1	3	2	—
Nadejdolepis paranitidulans (SI)	7	22	6	2–15

Source: Kinsella & Forrester (1994).

a. Location in host: BC = body cavity, CE = cecum, LI = large intestine, PR = proventriculus, SI = small intestine.

Table 20.8. Helminths of miscellaneous shorebirds from Florida

Class Species (location)[a]	Host	County	Year	No. of birds		Intensity		Data source
				Examined	Positive	Mean	Range	
Trematoda								
Androcotyla arenariae (SI)	Ruddy Turnstone	Monroe	1969	NG	—	NG	—	A
Cloacitrema michiganensis (CL)	Black-necked Stilt	Dade	1930	1	1	1	—	B
Cyclocoelum obscurum (CO,CL)	Red Knot	Franklin	NG	NG	—	—	—	C
	Short-billed Dowitcher	Franklin	NG	NG	—	—	—	C
Echinochasmus sp. (IN)	Semipalmated Plover	Franklin	NG	NG	—	—	—	C
Echinoparyphium sp. (IN)	Black-bellied Plover	Franklin	NG	NG	—	—	—	C
Gynaecotyla riggini (IN)	Ruddy Turnstone	Franklin	1955	1	1	5	—	D
Himasthla sp. (IN)	Dunlin	Wakulla	NG	NG	—	—	—	C
Longicollia sp. (IN)	Dunlin	Wakulla	NG	NG	—	—	—	C
Maritrema gratiosum (IN)	Short-billed Dowitcher	Franklin	NG	NG	—	—	—	C
Maritrema sp. (IN)	Short-billed Dowitcher	Wakulla	NG	NG	—	—	—	C
Megalophallus reamesi (IN)	Black-bellied Plover	Monroe	1968	2	NG	>60	—	E
	Ruddy Turnstone	Monroe	1968	2	NG	5	—	E
Parorchis acanthus[b] (CL)	Ruddy Turnstone	Franklin	NG	NG	—	—	—	C
	Short-billed Dowitcher	Franklin	NG	NG	—	—	—	C
		Pinellas	NG	NG	—	NG	—	F
Probolocoryphe glandulosa (SI)	Ruddy Turnstone	Monroe	1968	2	2	—	—	G
Psilochasmus sp. (IN)	Black-bellied Plover	Wakulla	NG	NG	—	—	—	C
Stephanoprora sp. (IN)	Black-bellied Plover	Wakulla	NG	NG	—	—	—	C
Tanaisia sp. (KD)	Short-billed Dowitcher	Wakulla	NG	NG	—	—	—	C

Parasite	Host	County	Year					Source
Cestoda								
Hymenolepis sp. (IN)	Short-billed Dowitcher	Franklin	NG	NG	—	—	—	C
Liga sp. (IN)	Black-bellied Plover	Wakulla	NG	NG	—	—	—	C
Nematoda								
Hystrichis sp. (PR)	Common Snipe	Alachua	1971	1	1	1	—	H
Skrjabinoclava bakeri[c] (PR)	Western Sandpiper	Collier	1987	7	6	15.0	3–41	I
			1992	1	1	21	—	J
			1997	2	2	25	18–32	K
Skrjabinoclava pusillae[c] (PR)	Dunlin	Collier	1987	5	4	3.0	1–6	I
	Western Sandpiper	Collier	1987	7	6	2.5	1–6	I
			1997	2	1	1	—	K
	Semipalmated Plover	Collier	1987	6	1	1	—	I
Skrjabinoclava tupacincai	Sanderling	Collier	1987	4	2	5.5	5–6	I
Skrjabinoclava sp.[c] (PR)	Sanderling	Collier	1987	4	1	2	—	I
	Western Sandpiper	Collier	1987	7	3	7	1–17	I
Tetrameres sp.[d] (PR)	American Woodcock	Alachua	1970	NG	—	—	—	L
	Common Snipe	Alachua	1971	1	1	6	—	H
Acanthocephala								
Arhythmorhynchus longicollis (SI)	Ruddy Turnstone	Monroe	1968–69	5	NG	3	NG	M

Sources: A = Deblock & Heard (1969), B = McIntosh (1938), C = Loftin (1961), D = Dery (1958), E = Overstreet & Heard (1995), F = Hutton (1964), G = Heard & Sikora (1969), H = Kinsella (1971), I = Wong & Anderson (1990), J = Foster et al. (2001), K = Spalding & Kinsella (1997), L = Mollhagen (1976), M = Nickol & Heard (1970). NG = not given.

a. Location from host: CL = cloaca, CO = coelom and air sacs, IN = intestine, KD = kidney, PR = proventriculus, SI = small intestine.

b. Cercariae from snail *Cerithidia scalariformes* from Wakulla Co. developed into adult *P. acanthus* in chicks in 14 days (Loftin 1961).

c. Only the upper gastrointestinal tract was examined by Wong & Anderson (1990).

d. Called *T. dubia* by Mollhagen (1976) but that name has not been published. Deposited in the U.S. National Parasite Collection, number 74406.

Table 20.9. Mites and ticks collected from shorebirds in Florida

Host	Species	County	Year	Data source
Black-necked Stilt	*Rhinonyssus himantopus*	Brevard	1954	C
Killdeer[a]	*Brephosceles longistriatus*	Glades	1983	B
	Brephosceles sp.[b,c]	Glades	1977	A
	Bychovskiata sp.[b,c]	Glades	1977	A
	Hemifreyana sp.	Glades	1983	B
	Ingrassia sp.	Glades	1977	A
		Glades	1983	B
	Limosilolichus sp.	Glades	1977	A
	Ptiloxenus sp.[c]	Glades	1983	B
Snowy plover	*Rhinonyssus coniventris*	Okaloosa	1951	C

Sources: A = Forrester & Atyeo (1992), B = Forrester & Mertins (1984), C = Strandtmann (1956).
a. Travis (1941) examined a single Killdeer from Orange, Osceola, or Collier County, 1933–37, and found no ticks.
b. This identification was uncertain; only females were present.
c. Undescribed species.

Table 20.10. Chewing lice collected from shorebirds in Florida

Host	Species	County	Year(s)
American Avocet	*Cirrophthirius testudinarius*	Dade	1956
	Quadraceps zephyra	Dade	1956
American Oystercatcher	*Actornithophilus grandiceps*	Dade	1919
	Quadraceps auratus	Dade	1919
	Saemundssonia haematopi	Dade	1919, 1929
American Woodcock	*Rhynonirmus parsonsae*	Dade	1930
Black-necked Stilt	*Quadraceps hemichrous*	Dade	1924
	Quadraceps semifissus	Dade	1924
Common Snipe	*Rhynonirmus scolopacis*	Dade	1929
Dunlin	*Actornithophilus umbrinus*	Indian River	1968
		Lee	1978
	Lunaceps nereis	Unknown	Unknown
	Carduiceps meinertzhageni	Lee	1978
Killdeer	*Actornithophilus hoplopteri*	Brevard	1957
		Glades	1977, 1983
	Austromenopon aegialitidis	Dade	1930
		Lee	1937
	Quadraceps boephilus	Dade	1930
		Glades	1977, 1983
		Lee	1937
	Saemundssonia conica conica	Dade	1930
Least Sandpiper	*Actornithophilus umbrinus*	Dade	1917
	Carduiceps zonarius	Volusia	1938
	Saemundssonia tringae	Dade	1917
Northern Jacana	*Rallicola exiguifrons*	Dade	1930
Piping Plover	*Quadraceps macrocephalus*	Indian River	1967
Red Knot	*Actornithophilus canuti*	Dade	1924
	Actornithophilus umbrinus	Lee	1985–86
		Pinellas	1973
		Volusia	1938
	Austromenopon lutescens	Dade	1924
		Pinellas	1973
	Carduiceps zonarius	Brevard	1957
		Lee	1983
		Pinellas	1975–76
		Volusia	1930, 1938
	Lunaceps drosti	Brevard	1957
		Dade	1924
		Lee	1983
		Pinellas	1973, 1975–76
	Saemundssonia platygaster islandica	Pinellas	1930
	Saemundssonia tringae	Pinellas	1931
Ruddy Turnstone	*Actornithophilus* sp.	Unknown	Unknown
	Quadraceps strepsilaris	Brevard	1957
Sanderling	*Actornithophilus umbrinus*	Indian River	1957, 1975
		Lee	1982, 1985
	Carduiceps zonarius	Indian River	1975
		Lee	1982

(continued)

Table 20.10. *(continued)*

Host	Species	County	Year(s)
	Lunaceps actophilus	Indian River	1975
		Lee	1982
Semipalmated Sandpiper	*Actornithophilus umbrinus*	Volusia	1957
	Lunaceps cabanisi	Volusia	1957
Short-billed Dowitcher	*Actornithophilus limarius*	Unknown	Unknown
	Quadraceps nigrolimbatus	Levy	1957
	Saemundssonia tringae	Levy	1957
Snowy Plover	*Quadraceps macrocephalus*	Hillsborough	1932
	Saemundssonia platygaster pastoris	Hillsborough	1932
Spotted Sandpiper	*Quadraceps ravus*	Dade	1917
Stilt Sandpiper	*Carduiceps zonarius*	Dade	1919
	Lunaceps nereis	Dade	1919
Western Sandpiper	*Actornithophilus umbrinus*	Indian River	1967, 1975
		Lee	1982
	Carduiceps zonarius	Lee	1982
	Lunaceps cabanisi	Lee	1982
Willet	*Actornithophilus lacustris*	Lee	1981–82, 1985
		Levy	1957
	Austromenopon sachtlebeni	Lee	1982, 1985
		Levy	1957
	Quadraceps carrikeri	Lee	1981–82, 1985
		Pinellas	1930
Wilson's Plover	*Quadraceps macrocephalus*	Pinellas	1934
	Saemundssonia platygaster platygaster	Monroe	1939

Source: Forrester et al. (1995).

XVIII. Literature Cited

Below, T.H. 1996. American oystercatcher. In: *Rare and endangered biota of Florida.* Vol. 5, *Birds.* J.A. Rodgers, Jr., H.W. Kale II, and H.T. Smith (eds.). University Press of Florida, Gainesville. pp. 230–235.

Bent, A.C. 1937. *Life histories of North American birds of prey.* Part 1. U.S. Natl. Mus. Bull. 167. 409 pp.

Blus, L.J., L.N. Locke, and E. Cromartie. 1978. Avian cholera and organochlorine residues in an American oystercatcher. *Estuaries* 1:128–129.

Burridge, M.J., W.J. Bigler, D.J. Forrester, and J.M. Hennemann. 1979. Serologic survey for *Toxoplasma gondii* in wild animals in Florida. *J. Am. Vet. Med. Assoc.* 175:964–967.

Bush, A.O. 1990. Helminth communities in avian hosts: determinants of pattern. In: *Parasite communities: patterns and processes.* G.A. Esch, A.O. Bush, and J. Aho (eds.). Chapman and Hall, New York. pp. 197–232.

Bush, A.O., and W. Threlfall. 1984. *Numeniotrema kinsellai* sp. n. (Digenea: Microphallidae) from the western willet *(Catoptrophorus semipalmatus inornatus)* on its wintering grounds in Florida. *J. Parasitol.* 70:355–357.

Ching, H.L. 1995. Four new gymnophallid digeneans from rice rats, willets, and molluscs in Florida. *J. Parasitol.* 81:924–928.

Clark, D.R., and M.A.R. McLane. 1974. Chlorinated hydrocarbon and mercury residues in woodcock in the United States, 1970–71. *Pestic. Monit. J.* 8:15–22.

Converse, K. 1999. Unpublished data. National Wildlife Health Center, Madison, Wis.

Cooper, J.M. 1994. Least sandpiper. *Birds N. Am.* 115:1–28.

Crawford, R.L. 1981. Bird casualties at a Leon County, Florida TV tower: a 25-year migration study. *Bull. Tall Timbers Res. Stn.* 22:1–30.

[CSLP] Center for Short-lived Phenomena. 1968. Annual progress report 1968. Center for Short-lived Phenomena, Smithsonian Institution, Cambridge, Mass. 152 pp.

Deblock, S., and R.W. Heard III. 1969. Contribution a l'etude des Microphallidae Travassos, 1920 (Trematoda). XIX—Description de *Maritrema prosthometra* n. sp. et de *Longiductotrema* nov. gen. parasites d'Oiseaux Ralliformes d'Amerique du Nord (Descriptions of North American rail parasites *Maritrema prosthometra* and *Longiductotrema* sp.). *Ann. Parasitol. Hum. Comp.* 44:415–424.

Dery, D.W. 1958. A revision of the genus *Gynaecotyla* (Microphallidae: Trematoda) with a description of *Gynaecotyla riggini* n. sp. *J. Parasitol.* 44:110–112.

Edman, J.D., L.A. Webber, and H.W. Kale II. 1972. Host-feeding patterns of Florida mosquitoes II. *Culiseta. J. Med. Entomol.* 9:429–434.

Edman, J.D., L.A. Webber, and A.A. Schmid. 1974. Effect of host defenses on the feeding pattern of *Culex nigripalpus* when offered a choice of blood sources. *J. Parasitol.* 60:874–883.

Elphick, C.S., and T.L. Tibbits. 1998. Greater yellowlegs. *Birds N. Am.* 355:1–24.

Engstrom, R.T. 1996. American avocet. In: *Rare and endangered biota of Florida.* Vol. 5, *Birds.* J.A. Rodgers, Jr., H.W. Kale II, and H.T. Smith (eds.). University Press of Florida, Gainesville. pp. 507–513.

Fairbrother, A., L.N. Locke, and G.L.E. Hoff (eds.). 1996. *Noninfectious diseases of wildlife.* 2d ed. Iowa State University Press, Ames. 219 pp.

Favorite, F.G. 1960. Some evidence of local origin of EEE virus in Florida. *Mosq. News* 20:87–92.

Fisk, E.J. 1976. A deadly rain of robins. *Fla. Nat.* 49:13–14.

Forrester, D.J. 1978–94. Unpublished data. University of Florida, Gainesville.

Forrester, D.J., and W.T. Atyeo. 1992. Unpublished data. University of Florida, Gainesville.

Forrester, D.J., and G.F. Bennett. 1985. Unpublished data. University of Florida, Gainesville.

Forrester, D.J., and P.P. Humphrey. 1981. Susceptibility of the knot (*Calidris canutus*) to *Plasmodium hermani. J. Parasitol.* 67:747–748.

Forrester, D.J., H.W. Kale II, R.D. Price, K.C. Emerson, and G.W. Foster. 1995. Chewing lice (Mallophaga) from birds in Florida: a listing by host. *Bull. Fla. Mus. Nat. Hist.* 39:1–44.

Forrester, D.J., and J.W. Mertins. 1984. Unpublished data. University of Florida, Gainesville.

Forrester, D.J., and N.P. Thompson. 1976. Unpublished data. University of Florida, Gainesville.

Forrester, D.J., K.C. Wenner, F.H. White, E.C. Greiner, W.R. Marion, J.E. Thul, and G.A. Berkhoff. 1980. An epizootic of avian botulism in a phosphate mine settling pond in northern Florida. *J. Wildl. Dis.* 16:323–327.

Forrester, D.J., and F.H. White. 1978. Unpublished data. University of Florida, Gainesville.

Foster, G.W., T. Below, and F. James. 2001. Unpublished data. University of Florida, Gainesville.

[FPL and FGFWFC] Florida Power and Light Company and Florida Game and Fresh Water Fish Commission. 1991. Unpublished correspondence. Lake Harbor bird mortality project: daily inspections of test framing designs.

Frenkel, J.K. 1981. False-negative serologic tests for *Toxoplasma* in birds. *J. Parasitol.* 67:952–953.

Friend, M., and J.C. Franson. 1999. *Field manual of wildlife diseases. General field procedures and diseases of birds.* U.S. Department of the Interior, U.S. Geological Survey, Biological Research Division, Information and Technical Report 1999-001. Washington, D.C. 426 pp.

Gore, J.A. 1996. Cuban snowy plover. In: *Rare and endangered biota of Florida.* Vol. 5, *Birds.* J.A. Rodgers, Jr., H.W. Kale II, and H.T. Smith (eds.). University Press of Florida, Gainesville. pp. 73–80.

Green, G. 1969. Suspected poxvirus of a Dunlin. *Br. Birds* 62:26–27.

Greiner, E.C., G.F. Bennett, E.M. White, and R.F. Coombs. 1975. Distribution of the avian hematozoa of North America. *Can. J. Zool.* 53:1762–1787.

Haig, S.M. 1992. Piping plover. *Birds of N. Am.* 2:1–18.

Harrington, B.A. 1982. Untying the enigma of the red knot. *Living Bird Q.* 1:4–7.

Harrington, B.A., D.C. Twichell, and L.E. Leddy. 1982. Knots in Florida: a migration mystery. *Fla. Nat.* 55:4–5.

Heard, R.W. III, and W.B. Sikora. 1969. *Probolocoryphe otagaki,* 1958 (Trematoda: Microphallidae), a senior synonym of *Mecynophallus cable,* Connor, Balling, 1960, with notes on the genus. *J. Parasitol.* 55:674–675.

Hutton, R.F. 1964. A second list of parasites from marine and coastal animals of Florida. *Trans. Am. Microsc. Soc.* 83:439–447.

Jasmin, A.M., D.E. Cooperrider, C.P. Powell, and J.N. Baucom. 1972. Enterotoxemia of wildfowl due to *Cl. perfringens* type C. *J. Wildl. Dis.* 8:79–84.

Kale, H.W. II. 1972. The spring migration: Florida region. *Am. Birds* 26:751–774.

Keppie, D.M., and R.M. Whiting, Jr. 1994. American woodcock. *Birds N. Am.* 100:1–20.

Kinsella, J.M. 1971. Unpublished data. University of Florida, Gainesville.

Kinsella, J.M., and L.G. Ballwebber. 1993. Unpublished data. University of Florida, Gainesville.

Kinsella, J.M., and S. Deblock. 1997. Microphallidae Travassos, 1920 (Trematoda). L. *Testiculoporus cedarkeyensis* n. g., n. sp. et *Endocotyle bushi* n. sp., parasites d'un *Catoptrophorus* (Aves) des Etats-Unis. *Syst. Parasitol.* 37:67–72.

Kinsella, J.M., and D.J. Forrester. 1994. Unpublished data. University of Florida, Gainesville.

Klima, J., and J.R. Jehl, Jr. 1998. Stilt sandpiper. *Birds N. Am.* 341:1–20.

Klukas, R.W., and L.N. Locke. 1970. An outbreak of fowl cholera in Everglades National Park. *J. Wildl. Dis.* 6:77–79.

Kreuder, C., A.R. Irizarry-Rovira, E.B. Janovitz, P.J. Deitschel, and D.B. DeNicola. 1999. Avian pox in sanderlings from Florida. *J. Wildl. Dis.* 35:582–585.

Linda, S.B. 1998. Analysis of 1997–1998 statewide and wildlife management area hunter surveys: methods and results. Unpublished report. Florida Game and Fresh Water Fish Commission, Gainesville. 88 pp.

Loftin, H. 1961. An annotated check-list of trematodes and cestodes and their vertebrate hosts from northwest Florida. *Q. J. Fla. Acad. Sci.* 23:302–314.

Logan, T.H. 1997. Florida's endangered species, threatened species, and species of special concern. Official lists. Florida Game and Fresh Water Fish Commission, Tallahassee. 14 pp.

Longstreet, R.J. 1949. Flock of northern phalaropes at Daytona Beach, Florida. *Auk* 66:204.

———. 1954. Ornithology of the mosquitoes. *Fla. Nat.* 27:9–20.

MacInnis, A.J. 1966. Trematodes from marine shorebirds from the northwest gulf coast of Florida. *Sonderdr. Zool. Anz.* 176:3–68.

McEwan, L.C., and D.H. Hirth. 1980. Food habits of the bald eagle in north-central Florida. *Condor* 82:229–231.

McIntosh, A. 1938. A new philophthalmid trematode of the spotted sandpiper from Michigan and of the black-necked stilt from Florida. *Proc. Helminthol. Soc. Wash.* 5:46–47.

Meteyer, C. 1997. Unpublished data. National Wildlife Health Center, Madison, Wis.

Mollhagen, T.R. 1976. A study of the systematics and hosts of the parasitic nematode genus *Tetrameres* (Habronematoidea: Tetrameridae). Ph.D. diss., Texas Tech. University, Lubbock. 546 pp.

Moskoff, W. 1995. Solitary sandpiper. *Birds N. Am.* 156:1–16.

Nicholls, J.L. 1996. Piping plover. In: *Rare and endangered biota of Florida.* Vol. 5, *Birds.* J.A. Rodgers, Jr., H.W. Kale II, and H.T. Smith (eds.). University Press of Florida, Gainesville. pp. 61–72.

Nickol, B.B., and R.W. Heard III. 1970. *Arhythmorhynchus frassoni* from the clapper rail, *Rallus longirostris,* in North America. *J. Parasitol.* 56:204–206.

Nol, E., and R.C. Humphrey. 1994. American oystercatcher. *Birds N. Am.* 82:1–24.

Oring, L.W., E.M. Gray, and J.M. Reed. 1997. Spotted sandpiper. *Birds N. Am.* 289:1–32.

Overstreet, R.M., and R.W. Heard. 1995. A new species of *Megalophallus* (Digenea: Microphallidae) from the clapper rail, other birds, and the littoral isopod *Ligia baudiniana. Can. J. Fish. Aquat. Sci.* 52 (Suppl. 1):98–104.

Page, G.W., J.X. Warriner, J.C. Warriner, and P.W.C. Paton. 1995. Snowy plover. *Birds N. Am.* 154:1–24.

Paulson, D.R. 1995. Black-bellied plover. *Birds N. Am.* 186:1–28.

Penner, L.R., and B. Fried. 1963. *Philophthalmus hegeneri* sp. n., an ocular trematode from birds. *J. Parasitol.* 49:974–977.

Prestwood, A.K., G.L. Doster, and F.A. Hayes. 1969. Studies of American woodcock parasite fauna. In: *Woodcock research and management programs 1967 and 1968*. W.H. Goudy (ed.). U.S. Department of the Interior, Bureau of Sport Fisheries and Wildlife, Special Scientific Report—Wildlife 123:24–27.

Pursglove, S.R., Jr., and G.L. Doster. 1970. Potentialities of the woodcock as a game resource bird in the southeastern U.S. *Proc. Annu. Conf. Southeast. Assoc. Game Fish Comm.* 24:223–231.

Rand, A.L. 1943. Bass eats yellowthroat, young stilts, and young ducks. *Auk* 60:95.

Robertson, W.B., Jr., and G.E. Woolfenden. 1992. *Florida bird species: an annotated list.* Spec. Publ. 6, Florida Ornithological Society, Gainesville. 260 pp.

Robinson, J.A., L.W. Oring, J.P. Skorupa, and R. Boettcher. 1997. American avocet. *Birds N. Am.* 275:1–32.

Rodgers, J.A., Jr., H.W. Kale II, and H.T. Smith (eds.). 1996. *Rare and endangered biota of Florida.* Vol. 5, *Birds.* University Press of Florida, Gainesville. 688 pp.

Rodgers, J.A., Jr., and H.T. Smith. 1997. Buffer zone distances to protect foraging and loafing waterbirds from human disturbance in Florida. *Wildl. Soc. Bull.* 25:139–145.

Simpson, C.F., J.C. Woodard, and D.J. Forrester. 1976. An aortitis of knots caused by a *Besnoitia*-like organism. *Proc. 34th Annu. Meet. Electron Microsc. Soc. Am.*, Miami Beach, 1976:144–145.

———. 1977. An epizootic among knots (*Calidris canutus*) in Florida. II. Ultrastructure of the causative agent, a *Besnoitia*-like organism. *Vet. Pathol.* 14:351–360.

Sims, H.W., Jr. 1970. Operation bird wash. *Fla. Nat.* 43:43–45.

Skeel, M.A., and E.P. Mallory. 1996. Whimbrel. *Birds N. Am.* 219:1–28.

Smith, K., and V.F. Nettles. 1994. Unpublished data. Southeastern Cooperative Wildlife Disease Study, University of Georgia, Athens.

Snyder, B. 1994. Unpublished data. Florida Department of Environmental Protection, Tallahassee.

Spalding, M.G. 1998–99. Unpublished data. University of Florida, Gainesville.

Spalding, M.G., and J.M. Kinsella. 1997. Unpublished data. University of Florida, Gainesville.

Sprandel, G.L. 1996. Wilson's plover. In: *Rare and endangered biota of Florida.* Vol. 5, *Birds.* J.A. Rodgers, Jr., H.W. Kale II, and H.T. Smith (eds.). University Press of Florida, Gainesville. pp. 497–509.

Stevenson, H.M. 1970. The winter season: Florida region. *Audubon Field Notes* 24:493–497.

Stevenson, H.M., and B.H. Anderson. 1994. *The birdlife of Florida.* University Press of Florida, Gainesville. 892 pp.

Stoddard, H.L. 1931. *The bobwhite quail, its habits, preservation and increase.* Charles Scribner's Sons, New York. 559 pp.

Strandtmann, R.W. 1956. The mesostigmatic nasal mites of birds. IV. The species and hosts of the genus *Rhinonyssus. Proc. Entomol. Soc. Wash.* 58:129–142.

Threlfall, W. 1970. A preliminary check list of the helminth parasites of the common snipe, *Capella gallinago* (Linnaeus). *Am. Midl. Nat.* 84:13–19.

Travis, B.V. 1941. Examinations of wild animals for the cattle tick *Boophilus annulatus microplus* (Can.) in Florida. *J. Parasitol.* 27:465–467.

[USDI] U.S. Department of the Interior. 1985. Coastal barrier resources system. Draft report to Congress. Washington, D.C. 466 pp.

Warnock, N.D., and R.E. Gill. 1996. Dunlin. *Birds N. Am.* 203:1–24.

Weston, F.M. 1966. Bird casualties on the Pensacola Bay bridge (1938–1949). *Fla. Nat.* 39:53–55.

Wilson, W.H. 1994. Western sandpiper. *Birds N. Am.* 90:1–20

Wong, P.L., and R.C. Anderson. 1990. Host and geographic distribution of *Skrjabinoclava* spp. (Nematoda: Acuarioidea) of nearctic wading birds (Aves: Charadriiformes) and evidence for transmission in marine habitats in staging and wintering areas. *Can. J. Zool.* 68:2539–2552.

Woodard, J.C., D.J. Forrester, F.H. White, J.M. Gaskin, and N.P. Thompson. 1977. An epizootic among knots (*Calidris canutus*) in Florida. I. Disease syndrome, histology and transmission studies. *Vet. Pathol.* 14:338–350.

Gulls, Terns, and Skimmers

I. Introduction

Although a total of 15 species of gulls has been recorded in Florida, only the Laughing Gull breeds here (table 21.1). Ten of the 14 terns and the Black Skimmer breed in Florida. The Roseate Tern is listed as threatened by both the Florida Game and Fresh Water Fish Commission and the U.S. Fish and Wildlife Service. The Least Tern is listed as threatened and the Black Skimmer as a species of special concern by the Florida Game and Fresh Water Fish Commission (Logan 1997). The Least Tern and Roseate Tern are both listed as threatened, the Sooty Tern, Royal Tern, Sandwich Tern, Caspian Tern, Brown Noddy, and Black Skimmer are all listed as species of special concern, and the Gull-billed Tern is listed as status undetermined by the

Florida Committee on Rare and Endangered Plants and Animals (Rodgers et al. 1996a).

A number of gull populations are increasing in Florida. Laughing Gull populations, both breeding and wintering, appear to be increasing (Schreiber and Schreiber 1977). Schreiber and Schreiber (1979) give information on mortality, reproductive condition, and molt for west coast Laughing Gulls, and Schreiber et al. (1979) and Schreiber and Schreiber (1980) report on their clutch size, hatching, and fledgling success. General information about Laughing Gulls and Herring Gulls in North America can be found in Burger (1996) and Pierotti and Good (1993). Ring-billed Gulls have expanded their range in Florida; the wintering population

Table 21.1. Species of gulls, terns, and skimmers that occur in Florida

Species	Range[b]	Seasonal occurrence	Relative abundance
Gulls			
Laughing Gull[a]	*Larus atricilla*	Coasts	Resident	Common
Franklin's Gull	*Larus pipixcan*	Coasts	Winter	Rare
Little Gull	*Larus minutus*	Coasts	Winter	Rare
Common Black-headed Gull	*Larus ridibundus*	Coasts	Winter	Very rare
Bonaparte's Gull	*Larus philadelphia*	Coasts & inland	Winter	Common
Band-tailed Gull	*Larus belcheri*	Gulf coast	Winter	Very rare
Ring-billed Gull	*Larus delawarensis*	Coasts	Winter	Common
Herring Gull	*Larus argentatus*	Coasts	Winter	Common
Thayer's Gull	*Larus thayeri*	Inland	Winter	Rare
Iceland Gull	*Larus glaucoides*	Coasts	Winter	Rare
Lesser Black-backed Gull	*Larus fuscus*	Coasts	Winter	Rare to common
Glaucous Gull	*Larus hyperboreus*	Coasts, especially NE	Winter	Rare to regular
Great Black-backed Gull	*Larus marinus*	Atlantic coast	Winter	Common
Sabine's Gull	*Xema sabini*	Atlantic coast	Fall & spring	Rare
Black-legged Kittiwake	*Rissa tridactyla*	Atlantic offshore	Winter	Common
Terns			
Gull-billed Tern[a]	*Sterna nilotica*	Mainland	Year-round	Common
Caspian Tern[a]	*Sterna caspia*	Coasts	Resident	Common
Royal Tern[a]	*Sterna maxima*	Coasts	Resident	Common
Sandwich Tern[a]	*Sterna sandvicensis*	Coasts	Resident	Locally common
Roseate Tern[a]	*Sterna dougallii*	Keys	Resident	Locally common
Common Tern[a]	*Sterna hirundo*	Coasts	Spring & fall	Regular transient
Arctic Tern	*Sterna paradisaea*	Atlantic & south Gulf coast	Spring	Rare
Forster's Tern	*Sterna forsteri*	Coasts	Year-round	Common
Least Tern[a]	*Sterna antillarum*	Coasts	Resident	Common
Bridled Tern[a]	*Sterna anaethetus*	Atlantic coast	Summer (rare breeder)	Common
Sooty Tern[a]	*Sterna fuscata*	Dry Tortugas (breeding) & coasts	Spring, winter, summer	Common
Black Tern	*Chlidonias niger*	Coasts	Fall & spring	Common
Brown Noddy[a]	*Anous stolidus*	Dry Tortugas (breeding) & coasts	Winter & summer	Common
Black Noddy	*Anous minutus*	Dry Tortugas	Summer	Rare
Skimmers			
Black Skimmer[a]	*Rynchops niger*	Coasts	Resident	Common

Source: Roberson & Woolfenden (1992).

a. Have nested in Florida.

b. Although most are found along the coasts, occurrence inland may be associated with large lakes, landfills, and storms.

has increased parallel to the recovery of the Great Lakes population (Ryder 1993). Ring-billed Gulls are the majority of the gulls using landfills and dumpsters throughout Florida (Clapp et al. 1983). Great Black-backed Gulls are seen more commonly and are more abundant on the northeast coast of Florida (Clapp et al. 1983; Good 1998).

Life history and population information in Florida is reported for Least Terns (Fisk 1978b; Gore 1996; Thompson et al. 1997), Roseate Terns (Robertson 1964, 1978b; Smith 1996), Sooty Terns (Robertson and Robertson 1996), Black Terns (Dunn and Agro 1995), Royal Terns (Egensteiner et al. 1996), Sandwich Terns (Rodgers et al. 1996b), Caspian Terns (Paul 1996b), Brown Noddies (Robertson 1996; Chardine and Morris 1996), Black Skimmers (Gochfeld and Burger 1994; Loftin and Smith 1996), and Gull-billed Terns (Parnell et al. 1995; Smith and Gore 1996). Robertson (1964) has provided a very interesting and thorough history of the terns nesting on the Dry Tortugas, Monroe County. Clapp et al. (1983) also provide many life history and observation records for seabirds in Florida. General health information for birds in this group can be found in Friend and Franson (1999).

II. Trauma

Compared with other types of birds, few seabirds collide with tall structures. Crawford (1981) recorded birds found dead below a TV tower near Tall Timbers Research Station in Leon County between 1955 and 1980. It was presumed that most of these birds were killed during migration. During this period a total of 42,384 birds of 189 species was found. Only 4 Herring Gulls were among those listed. Gulls and terns have not been found in other studies of mortality involving tall structures in Florida (Kale 1971; Maehr et al. 1983; Maehr and Smith 1988; Taylor and Anderson 1973, 1974; Taylor and Kershner 1986).

Roadkills are a common cause of death for this group of birds. Stevenson (1994) and Sny-der (1994) conducted a survey of roadkills, 1990–93, in and around state parks; a total 1,562 birds found dead included 82 Royal Terns, 23 Laughing Gulls, 12 Herring Gulls, 11 Ring-billed Gulls, 3 Common Terns, 2 Sandwich Terns, a Black Skimmer, and a Least Tern. Many of the Royal Terns, 2 Sandwich Terns, and 1 Black Skimmer found in this study were from a bridge crossing Sebastian Inlet, Indian River County, and had been banded north of Florida; they are further discussed by Smith et al. (1994). Both adults and chicks of Black Skimmers were killed on a highway when they nested on a causeway in Lee County, 1984 (Field notes 1988). Nine Sandwich Terns, 4 Laughing Gulls, an adult Sooty Tern, and a juvenile Black Skimmer were killed by vehicles when crossing a causeway during Hurricane Erin in Franklin County in August 1995 (McNair 1998). They had broken bones and low body weights. Even though larger numbers of terns were present during 2 other hurricanes in October, McNair concluded that no deaths were observed then because young birds were stronger flyers and adults were less attentive to juveniles at the time the hurricanes hit Florida.

Schreiber and Schreiber (1979) examined 220 Laughing Gull carcasses obtained from a rehabilitation center in Pinellas County. Trauma, in the form of broken wings, was reported in 57%, collision with a vehicle in 3%, and monofilament entanglement in 2%. Collision with power lines was suggested as the most important cause of broken wings.

Ingestion of foreign objects is reported less commonly in seabirds than in some of the larger fish-eating birds (see chapter 4, Pelicans, and chapter 6, Herons, Egrets, and Bitterns). A plastic bottle cap (5 × 6 cm) was found to cause the obstruction of the stomach of an emaciated unidentified gull in Duval County in 1990 (Franson 1991).

Intraspecific aggression was thought to be the cause of death in 12% of Laughing Gull chicks in Pinellas County (Schreiber and Schreiber 1980). These chicks, generally <8 days old, were found with bloody heads.

III. Predation

Schreiber and Schreiber (1980) found no evidence of predation of Laughing Gull chicks at a colony in Pinellas County. Van Velzen (1966) reported the recovery of a banded Royal Tern from the stomach of a tiger shark (*Galeocerdo cuvieri*), Lee County, 1965. Delany (1986) found the band of a Ring-billed Gull, banded in Ontario, in the stomach of an American alligator (*Alligator mississippiensis*) from Alachua County in 1981.

Roseate Tern chicks and eggs are taken by frigatebirds, dogs, cats, rats, and raccoons (Stevenson and Anderson 1994). Black rats (*Rattus rattus*), Laughing Gulls, and Cattle Egrets are predators of Roseate Terns on the Dry Tortugas in Monroe County (Robertson n.d., cited in Smith 1996; Robertson 1964).

Predators of eggs and chicks of Sooty Terns on the Dry Tortugas include Magnificent Frigatebirds, Cattle Egrets, Yellow-crowned Night-Herons, Peregrine Falcons, Purple Gallinules, Ruddy Turnstones, Herring Gulls, Laughing Gulls, and Short-eared Owls (Dinsmore 1972; Hoffman et al. 1979; Robertson 1964; Harrington 1974; Robertson and Robertson 1996). The effects of most of the predators have been minor, except for gull predation, which has increased recently (Robertson and Robertson 1996).

Rats greatly reduced the number of nesting Brown Noddies on the Dry Tortugas in 1938 (Robertson 1964) and the same author (Robertson 1978a) suggested that some of the large fluctuations in numbers of breeding Brown Noddies in the Dry Tortugas were due to rat predation.

Ruddy Turnstones poked holes in eggs of Royal Terns (2,000 nests) and fewer numbers of eggs of Gull-billed Terns and Black Skimmers, causing desertion of the colony in Duval County in 1977 (Ogden 1977; Loftin and Sutton 1979).

Fish Crows, American Kestrels, Red-shouldered Hawks, red foxes, cats, and raccoons are common predators of Least Terns in Florida (Ogden 1977; Fisk 1978b; Gore and Kinnison 1991). Burrowing Owls took Least Tern nestlings from a rooftop colony in Duval County (Ogden 1977). Red-shouldered Hawks took all of 25 Least Tern chicks from 1 colony in Monroe County (Paige 1968). Great White Herons and Great Horned Owls took Least Tern chicks in Collier County, 1990 (Paul 1991). A Ring-billed Gull was found in the nest of a Bald Eagle in north central Florida (McEwan and Hirth 1980), and terns were listed among the carcasses found in their nests by Nicholson (in Bent 1937).

IV. Human disturbance and habitat alteration

Historically, egg collection for food and museum specimens may have been an important source of mortality for seabirds. Sooty Tern eggs were collected in very large quantities by fisherman on the Dry Tortugas for sale in Cuba in 1832. Egg collection is believed to have resulted in the extirpation of nesting Royal and Sandwich Terns and the significant decrease of numbers of Least Terns and Sooty Terns on the Dry Tortugas (Robertson 1964). Eggs of Gull-billed Terns and Roseate Terns were taken frequently by egg collectors and for food, and collection of Sandwich Tern eggs may have contributed to their failure to breed in Florida for 45–70 years (Stevenson and Anderson 1994).

Least Terns were hunted for their plumage through the early 1900s and were almost extirpated from the east coast (Bent 1921). Roseate Terns were also shot for their plumage (Stevenson and Anderson 1994).

Disturbance of beaches, nesting and foraging habitat for many seabirds, can result in a wide range of negative effects, including loss of foraging habitat, crushing of eggs, predation on eggs, and abandonment of nests by frequently disturbed adults. There has been a significant loss of Least Tern nesting beaches due to vehicles being driven on beaches and causeways (Combs and LeBuff 1965). Boat traffic has resulted in the loss of Roseate Tern nests (Robertson 1964). Disturbance of Black Skimmers from their nests can result in thermal stress to chicks that are left unprotected and also attacks by other adults if chicks wander from the nest site (Loftin and Smith 1996). Schreiber et al. (1979) observed Laughing Gulls puncturing eggs of other Laughing Gulls when the colony was disturbed by the researchers' visitation. They believe that the high rate of disappearance of eggs between their visits was also caused by the Laughing Gulls themselves, as they saw no other evidence of predators.

Attraction to construction sites and gravel pits for nesting sites makes Least Terns, and to a lesser extent Black Skimmers, susceptible to trauma and nest loss from construction equipment, off-road vehicles, predators, and pedestrians (Skoog 1982, Field notes 1988). Rodgers and Smith (1997) discuss buffer distances to prevent disturbance for some of the seabirds.

Airplane overflights at low altitude accompanied by sonic booms may be the cause of hatching failure. For example, a complete hatching failure of Sooty Terns on the Dry Tortugas in 1969 was believed to be due to low-flying planes (Mead 1971; Austin et al. 1972; Robertson 1978a).

Many nesting seabirds, especially those with low population numbers, are affected severely by altered nesting or foraging habitat, for example, Roseate Terns (Smith 1996), and Gull-billed Terns (Smith and Gore 1996). Habitat alteration of beach and foraging areas is a significant threat to reproducing seabird populations in Florida, especially to those with few breeding sites in the state. On the other hand, the creation of spoil islands with appropriate configuration and vegetation might attract breeding and foraging seabirds, if managed properly: overgrowth of vegetation quickly makes them unsuitable. Sea level rise should have a negative impact on nesting seabirds with limited nesting habitat, such as Sooty Terns in Florida (Robertson and Robertson 1996).

Presently, the majority of Florida's Least Terns use flat gravel-covered roofs to nest (Hovis and Robson 1989; Gore 1991). A change in roof design to one that is unsuitable for nesting Least Terns could significantly reduce nesting habitat (Gore 1996). Rooftop nesting by Least Terns can result in a new set of hazards, such as eggs rolling off roofs, eggs being washed off

roofs, flooding, overheating, being covered by fill, and predation from Fish Crows, Laughing Gulls, grackles, and cats (Fisk 1978a; Fisk unpublished, cited in Clapp et al. 1983).

Greene and Kale (1976) observed Black Skimmer nests on a roof in Brevard County. The adults did not defend their nests as they normally would on a beach; some eggs had been punctured (Fish Crows seen in the area) and 1 broken eggshell had tar on it. They suspected that heat absorbed from the tar might have been the cause of the mortality of the eggs and nest abandonment.

V. Inclement weather

Longstreet (1953) mentions finding dead seabirds on the Volusia County coast following hurricanes in September 1926 (1 Sooty Tern and 5 Brown Noddies) and September 1948 (a live Sooty Tern). A list of birds found dead following Hurricane Donna in 1960 included a single Laughing Gull and at least 25 Brown Noddies (Robertson and Paulson 1961). Clapp et al. (1983) reported adult Sooty Terns being seen more commonly on the coast following hurricanes.

Hurricanes during the months of June–August are the largest single cause of mortality for Sooty Terns, and can eliminate entire year classes of chicks in the Dry Tortugas. For example, in 1966 Hurricane Alma reduced the productivity by about 50% (Robertson and Robertson 1996). Perhaps the best-studied effects of a hurricane were reported by White et al. (1976). At least 25% of Sooty Tern nestlings, especially those <10 days old, were dead before Hurricane Agnes, June 1972, reached its peak at the Dry Tortugas, probably because of exposure to rain and cold. Many failed to gain weight and wing growth was retarded. Disturbed seas may also have interfered with foraging by the adults, resulting in reduced food delivered to the young. Noddy Terns did not appear to be affected by the storm. Significant mortality among adults and young of the year also occurred during the early fall hurricanes: Donna in 1960, and Carla in 1961 (Robertson 1978a). Hurricanes can erode nesting beaches (Stevenson and Anderson 1994). Robertson (1978a) suggested that some of the large fluctuations in numbers of breeding Brown Noddies in the Dry Tortugas were due to destruction of nesting habitat by hurricanes. Hurricane Alma crossed over the Dry Tortugas in June 1966 with 125 mile-per-hour winds and waves that covered Bush Key. Robertson (pers. comm.) found hundreds of Sooty Terns buried in the sand or imprisoned within ground vegetation that had been rolled up by the waves. Rolls of vegetation were up to 5 feet tall and contained chicks in every possible orientation. For several days after the storm hundreds of chicks were released by people and many of them later returned to breed on the island, based on banding records. He also noticed increased frigatebird predation on chicks following several hurricanes.

Case et al. (1965) reported that unusual weather conditions associated with Hurricane Hilda and a cold front, Brevard County, October 6–7, 1964, resulted in large numbers of migrating birds attracted to lights and flying into buildings and vehicles. A total of 4,707 dead birds, mostly passerines, but including a single Sooty Tern, were counted over the 2-day period.

Storms, heavy rains, and high tides frequently destroyed eggs and nestlings of Laughing Gulls in Florida Bay, Monroe County (Frohring and Kushlan 1986) and caused mortality of chicks in Pinellas County (Schreiber and Schreiber 1980). Storms repeatedly destroyed eggs of Gull-billed and Royal Terns in Duval County (Ogden 1977, 1979), and Roseate Terns in the Tortugas (Robertson 1964). Storms have caused chick and nestling mortality in Black Skimmers nesting on causeways in Lee County in 1984 (Field notes 1988).

Ring-billed Gulls and Royal Terns were found as part of a large die-off of primarily Northern Gannets and Loons during the winter of 1993, thought to be due to inclement weather (see chapter 3, Oceanic Birds, and chapter 2, Loons and Grebes).

VI. Organochlorines

Information about chemical contaminants in the tissues of seabirds in Florida is sketchy and mostly from birds picked up during die-offs. We found contaminant measurements for only 10 birds and 7 eggs (table 21.2). There was no research conducted to investigate eggshell thinning in Florida. Eggshell thinning was associated with DDE contamination in Herring Gulls in other states (Hickey and Anderson 1968). A concentration of 3.9 ppm wet weight DDE in the brain of a Royal Tern tested as part of a die-off on Grassy Key, Monroe County, in 1977 was not considered to be high enough to cause death (Locke 1977). Much higher concentrations were detected in brains of Ring-billed Gulls believed to be poisoned by organochlorines in Ontario (Sileo et al. 1977). In that study an index was created by dividing each of the organochlorines by its lethal brain concentration and then adding them. An index of >1 indicated possible poisoning from the combined toxins.

VII. Organophosphates and carbamates

Organophosphate poisoning was suspected in a die-off of 30–40 Laughing and Ring-billed Gulls near the West Palm Beach landfill, January 1986, based on cholinesterase depression in 2 of each species. The specific compound was not identified (Stroud 1986).

In July of 1997 organophosphate poisoning (depressed cholinesterase activity) was diagnosed from 16 dead and dying Western Sandpipers, 4 Black Skimmers, and 2 Least Terns on Marco Island and Gordon Pass in Collier County (Meteyer 1997). Again in October of 1997 severely depressed brain cholinesterase activity was found in Western Sandpipers, but not in a Laughing Gull, from a die-off that was mostly Western Sandpipers (78) and smaller numbers of other seabirds, including Black Skimmers and Sandwich Terns (Spalding 1998, Converse 1998).

Carbamate poisoning was suspected in a

Table 21.2. Organochlorine residues in eggs and tissues of gulls, terns, and skimmers collected in Florida

Tissue Species	County or area	Year(s)	No. exam.	DDE			DDT			Dieldrin			Data source
				No.[a]	Mean[b]	(Range)	No.[a]	Mean[b]	(Range)	No.[a]	Mean[b]	(Range)	
Eggs													
Black Skimmer	"Charlotte Harbor"	1972	2	NG	4.5[c]	—	NA	—	—	NG	0.15[c]	—	B
Laughing Gull	"Charlotte Harbor"	1972	3	NG	12[c]	—	NA	—	—	NG	0.61[c]	—	B
Least Tern	"Charlotte Harbor"	1972	2	NG	3.2[c]	—	NA	—	—	NG	0.07[c]	—	B
Adipose													
Bridled Tern	"Florida"	1973	1	1	0.51	—	1	0.51	—	0	ND	—	A
Bonaparte's Gull	"Florida"	1974	1	1	10	—	1	20	—	0	ND	—	A
Caspian Tern	"Florida"	1973	1	1	53	—	1	53	—	0	ND	—	A
Common Tern	"Florida"	1973	1	1	2.0	—	1	3.4	—	0	ND	—	A
Laughing Gull	"Florida"	1974	1	1	16	—	1	26	—	1	1.8	—	A
Brain													
Royal Tern	Monroe	1977	1[f]	1	3.9	—	NA	—	—	1	0.15	—	D
Ring-billed Gull[d]	Manatee	1999	1	NA	—	—	NA	—	—	1	0.5	—	E
Sooty Tern	"Florida Bay"	1971	3[e]	1	0.015	(ND–0.16)	NA	—	—	3	0.10	(0.08–0.13)	C
Liver													
Ring-billed Gull[d]	Manatee	1999	1	1	1.4	—	NA	—	—	1	6.4	—	E
Muscle													
Sooty Tern	"Florida Bay"	1971	3[e]	3	0.042	(0.031–0.062)	NA	—	—	2	0.01	(ND–0.02)	C
Uropygial gland													
Bridled Tern	"Florida"	1973	1	1	0.69	—	1	0.69	—	0	ND	—	A
Bonaparte's Gull	"Florida"	1974	1	1	4.5	—	1	7.4	—	0	ND	—	A
Caspian Tern	"Florida"	1973	1	1	30	—	1	30	—	0	ND	—	A
Common Tern	"Florida"	1973	1	1	0.55	—	1	1.1	—	0	ND	—	A
Laughing Gull	"Florida"	1974	1	1	2.2	—	1	3.9	—	1	0.47	—	A

(continued)

Table 21.2. (continued)

Tissue / Species	County or area	Year(s)	No. exam.	PCB No.[a]	PCB Mean[b]	PCB (Range)	Oxychlordane No.[a]	Oxychlordane Mean[b]	Oxychlordane (Range)	Toxaphene No.[a]	Toxaphene Mean[b]	Toxaphene (Range)	Data source
Eggs													
Black Skimmer	"Charlotte Harbor"	1972	2	NG	2.1[c]	—	NA	—	—	NA	—	—	B
Laughing Gull	"Charlotte Harbor"	1972	3	NG	17[c]	—	NA	—	—	NA	—	—	B
Least Tern	"Charlotte Harbor"	1972	2	NG	12[c]	—	NA	—	—	NA	—	—	B
Adipose													
Bridled Tern	"Florida"	1973	1	0	ND	—	NA	—	—	NA	—	—	A
Bonaparte's Gull	"Florida"	1974	1	0	ND	—	NA	—	—	NA	—	—	A
Caspian Tern	"Florida"	1973	1	0	ND	—	NA	—	—	NA	—	—	A
Common Tern	"Florida"	1973	1	0	ND	—	NA	—	—	NA	—	—	A
Laughing Gull	"Florida"	1974	1	0	ND	—	NA	—	—	NA	—	—	A
Brain													
Royal Tern	Monroe	1977	1[f]	1	6.1	—	1	0.11	—	1	0.13	—	D
Sooty Tern	"Florida Bay"	1971	3[e]	1	0.17	(ND–1.9)	NA	—	—	NA	—	—	C
Muscle													
Sooty Tern	"Florida Bay"	1971	3[e]	0	ND	—	NA	—	—	NA	—	—	C
Uropygial gland													
Bridled Tern	"Florida"	1972	1	0	ND	—	NA	—	—	NA	—	—	A
Bonaparte's Gull	"Florida"	1974	1	0	ND	—	NA	—	—	NA	—	—	A
Caspian Tern	"Florida"	1973	1	0	ND	—	NA	—	—	NA	—	—	A
Common Tern	"Florida"	1973	1	0	ND	—	NA	—	—	NA	—	—	A
Laughing Gull	"Florida"	1974	1	0	ND	—	NA	—	—	NA	—	—	A

Source: A = Johnston (1976), B = Lincer & Salkind (1973), C = Ogden et al. (1974), D = Locke (1977), E = Gaydos (1999).
NA = not analyzed, ND = not detected, NG = not given.
a. Number with residue.
b. Geometric mean, ppm wet weight unless otherwise indicated.
c. Dry weight values. Lincer & Salkind (1973) also give extractable fat weights.
d. Collected during a multispecies die-off of undetermined cause.
e. In addition to the compounds listed, Ogden et al. (1974) did not detect DDD or DDT in muscle or brain of these 3 Sooty Terns from Florida Bay.
f. Locke (1977) measured 0.10 ppm *trans*-nonachlor and 0.13 ppm mirex and did not detect p,p'-DDD, p,p'-DDT, heptachlor epoxide, *cis*-chlordane, *cis*-nonachlor, endrin, or HCB in this Royal Tern from Monroe County.

die-off of about 200 gulls in a dump, St. Johns County, February 1994. Farmers in the area had been using baits for mole crickets. Birds were dying with bloody froth in the oral cavity. Three Ring-billed Gulls examined had depression of cholinesterase activity with reversal after incubation and intestinal hemorrhage or enteritis (Sileo 1994). The specific compound was not identified.

Cholinesterase activity has been measured frequently when gulls and terns are found dead, especially if they have been foraging in landfills. Negative results were obtained from Royal Terns in Lee County in 1987 (Stroud 1987), Ring-billed and unidentified gulls in Duval County in 1990 (Franson 1991), Laughing and Ring-billed Gulls in Polk County in 1993 (Fischer 1994), Laughing Gulls and Royal Terns in Lee County in 1992 (Franson 1993), and Laughing Gulls in Polk County in 1997 (Quist 1997).

VIII. Metals

Tests for metals have been conducted only occasionally in seabirds (n = 11) from Florida (table 21.3). The most extensive work was done with Sooty Terns in the Tortugas, Monroe County, and then only 4 birds and the eggs they were incubating were examined (Stoneburner et al. 1980). A significant correlation was found between mercury concentrations in the blood of the adult and the egg it was incubating. Except for the high concentrations of mercury in eggs, the tissue concentrations of mercury were relatively low when compared with those found in other fish-eating birds in Florida (see chapter 6, Herons, Egrets, and Bitterns). Mercury in eggs (mean = 7.9 ppm) was within the range of eggs of experimentally exposed ducks that had embryonic death, hatchling death, and demyelination and necrosis in brains (Heinz and Locke 1976). Mercury in tissues of Tortugas Sooty Terns was compared with findings at Lisianski Island in Hawaii. Higher concentrations of mercury were found in eggs

and feathers of the Florida birds; however, no significant differences were found in liver, kidney, and brain tissue between the 2 areas and the authors felt that feathers and eggs provided a mechanism to shed excess mercury (Stoneburner and Harrison 1981).

The cadmium concentrations in kidneys of Sooty Terns were within the range of those found in other seabirds with kidney lesions; however, the picture is complicated by concurrent mercury contamination (see discussion in Furness 1996). Pelagic seabirds apparently have both high concentrations of cadmium in kidneys and renal lesions, even in birds that are apparently healthy and reproducing.

The liver concentrations of selenium in Sooty Terns reported by Stoneburner et al. (1980) are above those considered toxic for birds (Heinz 1996); however, such high concentrations are frequently observed concurrently with elevated mercury concentrations in fish-eating birds (see chapter 6, Herons, Egrets, and Bitterns). Egg selenium concentrations were also above the level noted to cause deformities and hatching failure (3 pmm, Heinz 1996), but such observations were not made in this colony (see section XI, Anomalies).

IX. Oiling

In Nassau County in 1948 2/3 of "gulls and other salt water birds" died and eggs were covered with "grease" (Marsh birds die 1948). Oiled Ring-billed and Herring Gulls are commonly seen at landfills in Florida, and they probably acquired the oil in the dumps (Hoffman in Clapp et al. 1983). Two of 220 (1%) Laughing Gulls examined at Suncoast Seabird Sanctuary, Pinellas County, between 1974 and 1976 were oiled (Schreiber and Schreiber 1979). Gulls were included in a long list of species that were affected by an oil spill during the winter of 1970 in Tampa Bay, but no numbers or species identifications were given (Stevenson 1970).

Robertson and Robertson (1978) reported that only 3% of more than 45,000 Sooty Terns

Table 21.3. Metal concentrations in eggs and tissues of terns from Florida

Species Tissue	County or area	Year	No. exam.	Cadmium Mean[a]	(SE[b] or Range)	Mercury Mean[a]	(SE[b] or Range)	Selenium Mean[a]	(SE[b] or Range)	Arsenic Mean[a]	(SE[b] or Range)	Data source
Royal Tern												
Liver	Collier	1993	1	NA	—	4.1	—	NA	—	NA	—	C
Sandwich Tern												
Liver	Collier	1993	3	NA	—	1.6	(0.97–2.4)	NA	—	NA	—	C
Sooty Tern												
Egg	Monroe	1977	4	2.3	(±0.1)	7.9	(±0.6)	3. 8	(±0.2)	ND	—	A
Adult												
Blood	Monroe	1977	4	0.85	(±0.09)	4.5	(±2.0)	8.8	(±0.5)	ND	—	A
Bone	Monroe	1977	4	12	(±2.2)	0.84	(±0.07)	5.2	(±0.7)	ND	—	A
Brain	Monroe	1977	4	3.1	(±0.4)	0.85	(±0.05)	4.7	(±0.3)	ND	—	A
Fat	Monroe	1977	4	4.0	(±1.0)	1.7	(±0.1)	4. 4	(±0.9)	ND	—	A
Feathers	Monroe	1977	4	4.2	(±0.4)	5.4	(±0.9)	19	(±0.9)	ND	—	A
Feces	Monroe	1977	4	3.6	(±0.2)	0.59	(±0.09)	9.6	(±0.6)	ND	—	A
Kidney	Monroe	1977	4	24	(±1.1)	1.6	(±0.2)	41	(±3.5)	ND	—	A
Liver	Monroe	1977	4	3.8	(±0.4)	1.5	(±0.1)	23	(±1.2)	ND	—	A
	"Florida Bay"	1972	3	NA	—	0.053	(0.05–0.06)	NA	—	NA	—	B
Muscle	Monroe	1977	4	1.1	(±0.1)	0.64	(±0.07)	6.7	(±0.6)	ND	—	A
Stomach contents	Monroe	1977	4	2.8	(±0.5)	0.28	(±0.01)	4.2	(±0.3)	ND	—	A

Sources: A = Stoneburner et al. (1980), B = Ogden et al. (1974), C = Fischer (1994).
NA = not analyzed, ND = not detected.
a. Geometric mean, ppm wet weight.
b. Standard error.

captured for banding were oiled, most only slightly. They found no significant difference in weights or return rates of oiled birds, but did notice an increase in numbers of oiled birds over time. Since that report they have observed severely oiled birds, including 1 that could not fly (Robertson in Clapp et al. 1983). Oiled Brown Noddies are occasionally seen at the Dry Tortugas, but their brown plumage may make oil difficult to see (Robertson in Clapp et al. 1983). In the last 15 years oiling has decreased to <1% and oil has decreased on beaches and the surface of the ocean (Robertson pers. comm.). Three oiled Royal Terns and an unspecified number of Common Terns from Florida have been reported to the Bird Banding Laboratory (Clapp et al. 1983). Although there are no records of such, Clapp et al. (1983) considered the Least Tern population in the southeastern United States to be at the highest risk from oil industry development in Florida.

There are 2 records of banded Black Skimmers that died from oiling in Florida (Bird Banding Laboratory in Clapp et al. 1983).

Exposure to oil can cause direct mortality and reduced reproductive success. Experimental oiling of the breasts of incubating Laughing Gulls resulted in embryonic mortality (King and LeFever 1979).

X. Neoplasia

We found no reports of neoplasia among gulls, terns, and skimmers in Florida.

XI. Anomalies

Anomalies were extremely rare in Sooty Terns, which have been studied intensively, and have not been reported in other species. A single Sooty Tern chick about 18 days old found on Bush Key, the Dry Tortugas, Monroe County, had extra toes in the hock region. This was the first deformed chick observed among more than 125,000 chicks handled over an 8-year period (Austin 1969). Young Sooty Terns on

the Dry Tortugas, Monroe County, in 1971 molted partly developed primaries and had other minor developmental anomalies. Chlorinated hydrocarbons were relatively low in samples from these birds, and a cause for the anomalies was never discovered (Ogden et al. 1974; Robertson 1978a). Egg selenium and mercury concentrations were high (see section VIII, Metals).

XII. Biotoxins

Botulism caused by ingestion of a toxin produced by the bacterium *Clostridium botulinum* is primarily a disease of ducks and is relatively rare in the southeast (see chapter 10, Ducks, for a more detailed account). Mortality from both type C and type E is frequently noted for gulls in the central and western United States (Rocke and Friend 1999). Type C botulism toxin has been documented in Florida in seabirds 3 times, type E has been documented in pelicans only once (see chapter 4, Pelicans).

Clostridium botulinum type C was found in 6% of substrate samples collected in 4 of 6 phosphate settling ponds and was not found in a natural pond in Hamilton County. Bacteria were present during April through October in a wide variety of ecological conditions (Marion et al. 1983). In Florida, gulls have been involved in a number of multispecies mortality events and were the majority of birds in a die-off associated with a landfill.

A single Ring-billed Gull was examined from a die-off of primarily cormorants and pelicans on a golf course at Patrick Air Force Base, Brevard County, April 1982 (see chapter 4, Pelicans, and chapter 5, Cormorants and Anhingas, for more details). The Ring-billed Gull tested positive for botulism type C; the 1 pelican tested was negative (Stroud 1982).

Botulism type C was diagnosed in 2 unidentified and 1 Ring-billed Gull, of 4 gulls tested from a die-off of Herring and Ring-billed Gulls, an Osprey, and a Brown Pelican near Mayport Naval Air Station in Duval County, December 1990. One of the birds examined was totally paralyzed, a second had only neck movement (Franson 1991).

Repeated die-offs of gulls have been associated with a landfill in Polk County. One hundred thirty-two dead gulls were found on Lake Lena and 25 on Lake Ariana, more than 1,000 dead gulls were observed from the air, and 15–20 dead Common Grackles were found on a parking lot between December 1992 and February 1993. Live gulls in the area were flying from the landfill. Type C botulism toxin was identified in a pooled sample of 4 subadult Laughing Gulls and 2 adult female Ring-billed Gulls (Fischer 1994) collected in February. A second presumptive botulism die-off occurred at this same location in February 1997 (see section XX, Unresolved die-offs).

Clostridium botulinum type C toxin was identified from 2 of 6 Mottled Ducks in a large multispecies die-off on a dredge island in Tampa Bay, Hillsborough County, in November 1994 following tropical storm Gordon. A total of 434 carcasses was found; the majority were ducks (see chapter 10, Ducks, for more complete details). One or more Herring Gulls and Laughing Gulls were among the birds found dead, but were not tested for botulism (Smith and Nettles 1994).

A second biotoxin affecting seabirds in Florida is *Clostridium perfringens* type C, which causes a hemorrhagic enteritis. Only 1 epizootic has been reported. A large-scale die-off during June–July 1971 in a shallowly inundated area on the west shore of Lake Okeechobee, Glades County, involved 398 birds of 19 species including ducks (see chapter 10, Ducks), ibises, rails, spoonbills, skimmers, terns, grebes, shorebirds, and wading birds (Jasmin et al. 1972). Sixteen Black Terns, 16 Black Skimmers, and 2 Gull-billed Terns were identified among the dead and moribund. Birds were weak and had severe hemorrhagic enteritis and acute hepatitis. It was concluded that *Clostridium perfringens* type C was the most likely cause of mortality based on culture and mouse protection tests. Cultures were negative for *C. botulinum*. Two Black Skimmers included in the testing were protected only by *C. perfringens* type C anti-

toxin. Two Black Terns were protected only by *C. perfringens* type B antitoxin. *Clostridium perfringens* was isolated from the liver of a Laughing Gull during an unresolved die-off in Polk County in 1997, but this finding was inconsistent among the birds tested, and the type was not determined (see section XX, Unresolved die-offs).

We found no reports of gull or tern mortality associated with red tide events in Florida. Nisbet (1983) reported paralytic shellfish poisoning of 70 Common Terns and lesser numbers of other terns and gulls in Massachusetts in 1978 associated with the dinoflagellate *Gonyaulax excavata* toxin in sand-launce (*Ammodytes americanus*).

XIII. Viruses

A previously undescribed virus was isolated from soft ticks of the *Ornithodoros capensis* group (now = *Carios denmarki*) that were collected from Bush Key, Monroe County, in 1962. The virus could not be identified using immune sera of known viruses available at the time (Hughes et al. 1964). These ticks were collected from Brown Noddy nests and vegetation on the island. *Carios denmarki* were found in high numbers in a portion of Sooty Tern colonies that were abandoned in the Seychelles (Converse et al. 1975). Soldado virus (Nairovirus) was isolated from sick and healthy chicks and from domestic chickens that died from experimental exposure to ticks from those colonies. Soldado virus (Nairovirus) has also been isolated from ticks collected in a nesting colony in Trinidad (Jonkers et al. 1973), and from *O. maritimus* ticks collected from a Yellow-legged Gull (*Larus cachinnans*) in Morocco (Chastel et al. 1995). Viruses have also been collected from these same ticks associated with other seabird colonies (see chapter 4, Pelicans, and chapter 6, Herons, Egrets, and Bitterns).

Jennings et al. (1969) reported that 2 of 6 Laughing Gulls collected by shooting in the Tampa Bay area in 1962, during a St. Louis encephalitis (SLE) epidemic, had hemagglutination-inhibition antibody for Group B arbovirus. It was assumed that positive results indicated exposure to SLE virus; however, this was not confirmed by serum neutralization or virus isolation.

West Nile virus was identified in 2 Glaucous Gulls (1 from Collier County and 1 from St. Lucie County) by PCR and virus isolation techniques in 2001 (Conti et al. 2002). It was assumed that these gulls had died of the infection, but a complete necropsy was not performed.

Jacobson et al. (1980) described an avian poxvirus infection that caused proliferative lesions on the legs and toe webs of a Royal Tern collected northeast of Jacksonville, Duval County.

Adenovirus was isolated from some Royal and Sandwich Terns during an unresolved die-off in Collier County in 1993 (Fischer 1994). This was considered to be an incidental finding since there were no associated lesions.

Avian influenza viruses (type A) have been isolated from Ring-billed Gulls in the Atlantic flyway (Maryland) (Graves 1992), but we found no records of testing for this virus in Florida.

XIV. Bacteria

Table 21.4 lists bacteria isolated from seabird tissues in Florida. Significant mortality from bacterial diseases has not been identified for seabirds in Florida, but a few individual cases have been reported.

Avian cholera epizootics, caused by systemic infection with *Pasteurella multocida,* have been reported only 1 time from Florida, in Dade County in 1968 (see chapter 18, Rails). A single Black Skimmer tested during that die-off in January 1968 was not positive for *P. multocida* (Klukas and Locke 1970). Although *Pasteurella multocida* was cultured from the intestine of 1 of the Royal Terns that was part of a die-off in Lee County in 1987, there was no histologic evidence of systemic disease (see section XX, Unresolved die-offs).

White et al. (1973) isolated *Edwardsiella tarda* from the large intestine of a Ring-billed

Gull that was unable to fly from Lake Apopka, Orange County, in 1971. No intestinal lesions were noted. They also isolated *E. tarda* from pelicans with necrotic enteritis, surface waters, and sick fish in north central Florida.

A colony of 1,250–1,300 Roseate Terns on Tank Island off Key West, Monroe County, failed during the summer of 1988. The colony was less than 1 mile from the Key West sewage outfall. *Escherichia coli* was cultured from a few of the eggs but no direct cause and effect relationship was established (Paul 1988).

A heavy growth of *Clostridium perfringens* was cultured from the liver of 1 Laughing Gull collected during an unresolved die-off of more than 145 birds near a landfill in Polk County in 1977 (see section XX, Unresolved die-offs) (Fischer 1997). Although the most likely cause of this die-off was botulism toxin, tests were not positive. *Clostridium perfringens* has caused mortality in other species (see section XII, Biotoxins) and has been associated with the consumption of garbage.

Salmonella sp. group B was isolated from the intestine of 1 of 3 gulls examined when a number of Laughing Gulls were found dead or dying at a landfill in Polk County, November 1997 (Quist 1997). Hudson et al. (2000) further characterized this isolate as *S. typhimerium* with virulence associated genes, but not with antibiotic resistant genes. *Escherichia coli* was isolated from the liver of all 3 gulls tested and from the intestine of 2. Tests for botulism, strychnine, and brain cholinesterase depression were negative. Salmonellosis has been reported as the cause of death of Herring and Black-backed Gulls and Black Skimmers in New York (Brand et al. 1988).

Chlamydiosis is listed as an occasional disease of gulls and terns (Franson 1999) in North America, but has not been reported in Florida.

XV. Fungi

Aspergillus sp., frequently *A. fumigatus*, is usually considered a secondary invader in birds stressed for some other reason and is found

Table 21.4. Bacteria and fungi isolated from gulls, terns, and skimmers in Florida

Organism	Host species	County	Year	Tissue	Data source
Bacteria					
Acinetobacter calcoaceticus	Royal Tern	Martin	1988	Liver, kidney, pharynx[a]	A
Clostridium perfringens	Laughing Gull	Polk	1997	Liver	B
"Diphtheroid"	Royal Tern	Lee	1987	Intestine	C
Edwardsiella tarda	Ring-billed Gull	Orange	1971	Large intestine	D
	Royal Tern	Brevard	1972	Liver, intestine[a]	E
Enterobacter agglormerans	Captive Royal Tern	Duval	NG	Lung	F
Enterobacter sp.	Laughing Gull	Polk	1997	Liver, intestine	B
Enterococcus sp.	Laughing Gull	Polk	1997	Liver, intestine	B
Escherichia coli	Black Skimmer	Collier	1997	Liver	G
	Laughing Gull	Polk	1997	Liver, intestine	H
			1997	Liver, intestine	B
	Ring-billed Gull	St. Johns	1994	Liver, intestine	I
	Roseate Tern	Monroe	1988	Eggs	J
	Royal Tern	Lee	1987	Intestine	C
Klebsiella sp.	Laughing Gull	Polk	1997	Liver, intestine	B
Moraxella phenylpyruvicii	Royal Tern	Martin	1988	Pharynx[a]	A
Pasteurella multocida	Royal Tern	Lee	1987	Intestine	C
Pasteurella pneumotropica	Black Skimmer	Collier	1997	Liver	G
Proteus mirabilis	Captive Royal Tern	Duval	NG	Lung	F
Proteus sp.	Laughing Gull	Polk	1997	Liver, intestine	B
Salmonella typhimurium	Laughing Gull	Polk	1997	Intestine	H, K
Yersinia enterocolitica	Ring-billed Gull	Palm Beach	1986	Intestine	L
Fungi					
Aspergillus fumigatus	Herring Gull	Palm Beach	1972	NG	M
	Gull (unspecified)	Brevard	1987	Lung/air sac, spleen[a]	N
	Laughing Gull[b]	Lee	1992	Lung, liver[a]	O
		Pinellas[b]	1974–76	NG	P
	Ring-billed Gull	Polk	2000	Lung/air sac[a]	Q
	Royal Tern	Lee	1992	Lung[a]	O

Sources: A = Roffe & McAllister (1988), B = Fischer (1997), C = Stroud (1987), D = White et al. (1973), E = Forrester & White (1972), F = Jacobson et al. (1980), G = Meteyer (1997), H = Quist (1997), I = Sileo (1994), J = Paul (1988), K = Hudson et al. (2000), L = Stroud (1986), M = Locke (1972), N = Roffe (1987), O = Franson (1993), P = Schreiber & Schreiber (1979), Q = Spalding (2000).

NG = not given.

a. Believed to be a significant pathogen.

b. Thirty percent of 220 Laughing Gull carcasses from a rehabilitation center in Pinellas County had *Aspergillus* lesions (Schreiber & Schreiber 1979). There was no information about time spent in captivity.

commonly in wild birds held in captivity (figures 21.1 and 21.2). Aspergillosis has been a common cause of mortality for gulls and was the most common finding in gulls found at a botulism die-off in New York (Brand et al. 1988). Aspergillosis has occasionally been found in gulls and terns in Florida (table 21.4).

Aspergillus fumigatus was isolated from the lung and liver of 2 Laughing Gulls and from the lung of a Royal Tern collected in Lee County, December 1992. About a dozen birds were reported to be lethargic, weak, and having hyperextended legs. Brain cholinesterase levels were normal, and botulism tests were negative. All were moderately emaciated. The gulls had typical gross lesions of aspergillosis, the tern did not (Franson 1993).

Lesions consistent with *Aspergillus* sp. were present in 30% of 220 Laughing Gulls examined after death at Suncoast Seabird Sanctuary

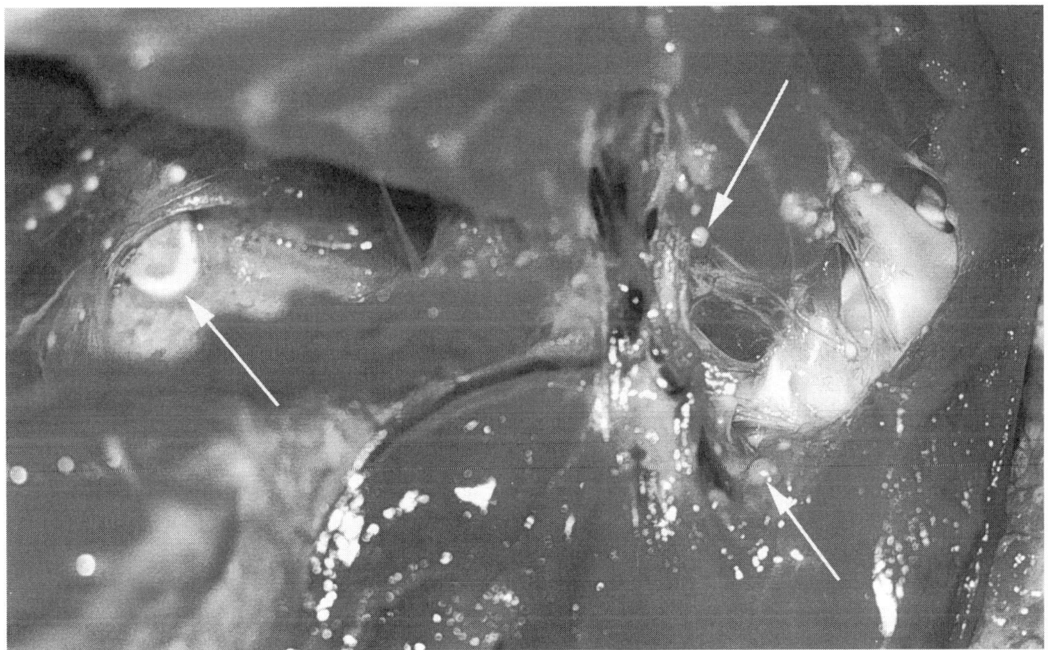

FIGURE 21.1. Fungal colonies *(arrows)* of *Aspergillus fumigatus* in the lungs and air sacs of a Ring-billed Gull.

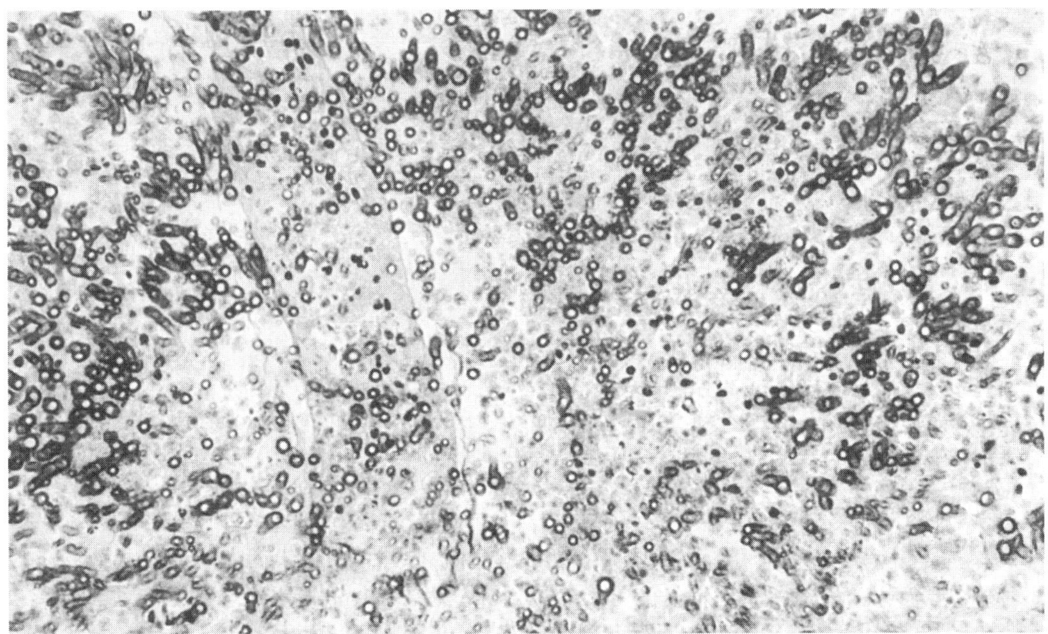

FIGURE 21.2. Photomicrograph of fungal hyphae, *Aspergillus fumigatus,* in the lung of a Ring-billed Gull. GMS stain.

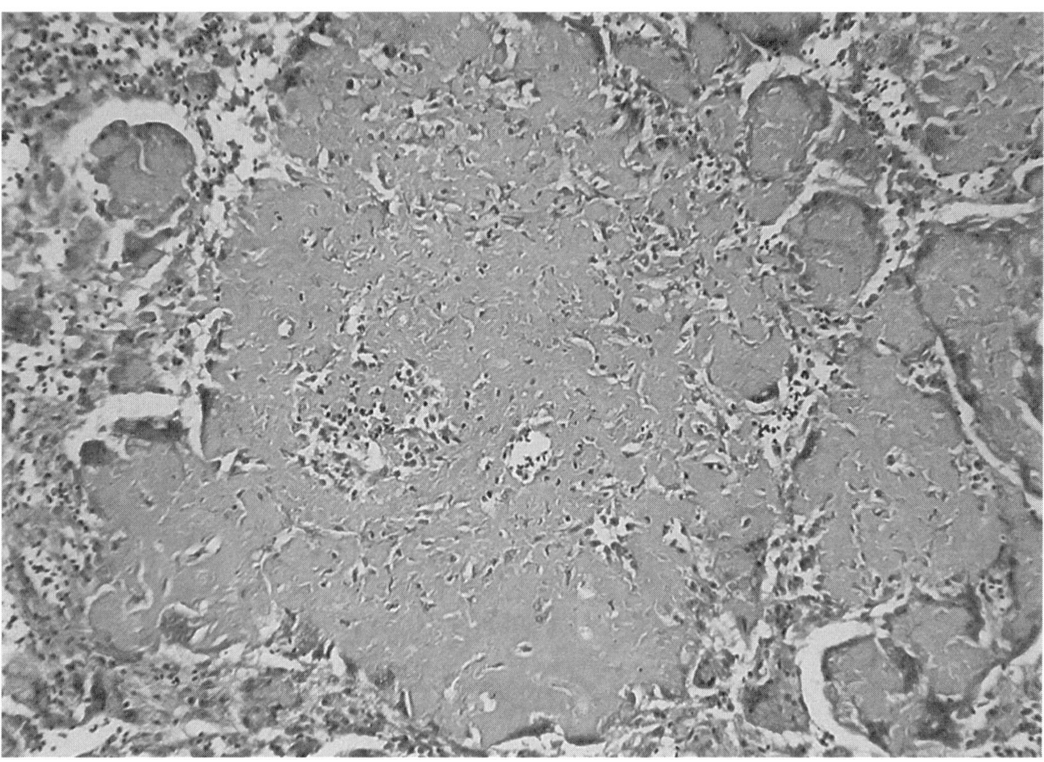

FIGURE 21.3. Photomicrograph of large amyloid deposits in the spleen of a Ring-billed Gull from Polk County in 2000 with aspergillosis and splenomegaly. Note the multinucleated giant cells associated with the perimeter of the deposits, which displace much of the splenic parenchyma.

in Pinellas County (Schreiber and Schreiber 1979). Information about how long the birds had been in captivity was not available, so it could not be determined whether aspergillosis had been acquired while in captivity.

Very severe airsacculitis and pneumonia, and splenic amyloidosis (figure 21.3) were found in all 3 Ring-billed Gulls examined from a multi-species die-off on an urban lake in Polk County in the winter and spring of 2000 (Spalding 2000). Because of the severe nature of the lesions, it was not possible to determine if another toxin or pathogen predisposed these birds to the fungal infection. Unfortunately other species from this die-off were not examined. Spleenic amyloidosis has previously been reported in Herring Gulls kept in captivity (Hoffman and Leighton 1985), and thus may be a response to stress in these species. Amyloid

deposits were also noted inconsistently in Ring-billed Gulls suspected to be poisoned by organochlorines (Sileo et al. 1977).

XVI. Protozoans

Toxoplasma gondii antibodies were detected in serum from 2 of 13 Ring-billed Gulls (1:128, 1:256), in 2 of 33 Laughing Gulls (1:64, 1:128), but not in 2 Herring Gulls, 5 Royal Terns, and 4 Black Skimmers from Florida (dates and locations not given) tested by Burridge et al. (1979). These were tested by indirect hemagglutination, which may not be as sensitive in birds as in mammals, resulting in false negatives (Frenkel 1981).

No blood parasites were observed in blood smears from Black Skimmers (Lake Okeechobee,

1971, *n* = 6; Levy County, 1980, *n* = 9) and Ring-billed Gulls (Orange County, 1971, *n* = 3; Pinellas County, 1974, *n* = 1; Brevard County, 1977, *n* = 1) (Forrester and Bennett 1985).

Sarcocystis falcatula-like sarcocysts were found in the pectoral muscle of a Ring-billed Gull that died in Polk County from aspergillosis in 2000 (Spalding and Cheadle 2000). An unidentified protozoan was found in the collecting tubules of the kidney of a Least Tern that was part of an organophosphate die-off in Collier County (see above) (Meteyer 1997).

XVII. Helminths

Helminths collected from seabirds in Florida are listed in tables 21.5–21.7. Most of those recorded were incidental findings at necropsy during investigation of the cause of die-off events. Only 3 parasitological surveys, with low sample sizes, were conducted. Kinsella (1972) looked at 5 Black Skimmers, Hutton and Sogandares-Bernal (1960) looked at 8 Laughing Gulls, and miscellaneous seabirds were examined by MacInnis (1966). Nematodes are conspicuously rare in gulls and terns, with only an unidentified capillarid nematode from terns. Stroud (1987) reported capillarid eggs in Royal Tern feces from Lee County in 1987. Three species of nematodes were found in Black Skimmers. No records of acanthocephalans were found in the entire group.

Adult trematodes were identified histologically in the kidneys of a Royal tern in Martin County in 1998 (Roffe and McAllister 1988) and a Royal Tern and a Sandwich Tern examined in Collier County in 1993 (Fischer 1993). Trematode eggs were found in the lumens of collecting tubules of a Least Tern, also in Collier County in 1997 (Meteyer 1997). There was no associated inflammation in any of these cases. These are probably *Renicola glandoloba,* which has been reported from terns in Florida (see table 21.6).

Metacercariae in a half beak fish, *Hyporhamphus unifasciatus,* were very similar to *Galactosomum spinetum* collected from Black Skim-

mers and are believed to be the same species (Sogandares-Bernal and Hutton 1960). Hutton and Sogandares-Bernal (1959) experimentally infected a Ring-billed Gull with *Mesostephanus appendiculatoides.* Penner and Fried (1963) described the ocular trematode *Philophthalmus hegeneri* from cercariae found in the marine snail *Batillaria minima* along the Gulf coast. The adult occurs in Royal Terns, Laughing Gulls, Yellow-crowned Night Herons, and Willets in Florida.

A schistosome parasite (*Ornithobilharzia canaliculata*), was found in the venous portal system of Royal Terns in Florida. The cercaria, found in the marine gastropod *Batillaria minima,* produced a dermatitis in humans similar to that called "gulf itch" (Penner 1953). Stroud (1987) suspected the presence of schistosome trematodes from histologic examination of the intestine of a Royal Tern from Lee County in 1987, and thought they might have been the cause of encephalitis in the bird although the trematodes were not seen in the brain. He also noted an unidentified trematode from the right ventricle of the heart of another Royal Tern.

XVIII. Arthropods

Ectoparasites collected from seabirds in Florida are listed in tables 21.8–21.9. We found no reports of significant effects of arthropods other than ants on gulls and terns, although harmful tick infestations have certainly been reported elsewhere (Duffy 1991). Schreiber et al. (1979) attributed deaths of 2–3 pipping Laughing Gull eggs each year to ants entering the pipped eggshell on a landfill island in Pinellas County. Imported fire ants (*Solenopsis invicta*) have been reported to cause significant (92%) reduction in waterbird reproduction (including Laughing Gulls, Gull-billed Terns, and Forster's Terns) on Texas spoil islands (Drees 1994). *Carios denmarki* ticks have been found in low numbers in a tern colony in Monroe County (Denmark and Clifford 1962). They occurred in high numbers in portions of Sooty Tern colonies that were abandoned in the Seychelles (Converse et al.

Table 21.5. Helminths of gulls from Florida

Host Species (location)[a]	County/area	Year(s)	No. birds Examined	No. birds Positive	Intensity Mean	Intensity Range	Data source
Herring Gull							
Trematoda							
Cryptocotyle lingua (SI)	Pinellas	1958	1	1	3	—	A
Gymnophallus deliciosus (SI)	Hillsborough	1955–62	NG	—	—	—	B
Stictodora lariformicola (LI)	Pinellas	NG	NG	—	—	—	C
Mesorchis denticulatus[b] (SI)	Pinellas	1958	1	1	1	—	A,B
Laughing Gull							
Trematoda							
Cardiocephalus megaloconus[c] (SI)	Franklin	1958	NG	—	—	—	D,E
Cotylurus aquavis (IN)	Franklin	NG	—	—	—	—	F
Galactosomum fregatae[d] (CL)	Pinellas	1958	8	1	—	—	A
Galactosomum spinetum (SI)	Franklin	1957	NG	—	—	—	G
Gynaecotyla adunca (SI)	Pinellas	1955–62	NG	—	—	—	B
Pachytrema sanguineum (SI)	Pinellas	1958	6	2	2	1–2	A
Philophthalmus hegeneri[e] (OC)	Lee	1986	1	1	—	—	H
	Manatee to Charlotte	NG	NG	—	—	—	I
Philophthalmus larsoni[f] (OC)	Sarasota	NG	—	—	—	—	J
Renicola glandoloba (KD)	Franklin	1958	2	2	4	4	K
	"Tampa Bay"	1955–62	NG	—	—	—	B
Stictodora lariformicola (SI)	Pinellas	NG	NG	—	—	—	C
Cestoda							
Paricterotaenia sp. (SI)	Franklin	NG	NG	—	—	—	F
Ring-billed gull							
Trematoda							
Echinostomatidae (IN)	St. Johns	1994	3	1	NG	—	L
Microphallus pygmaeus (SI)	Pinellas	1955–62	NG	—	—	—	B
Stictodora lariformicola (SI)	Pinellas	NG	NG	—	—	—	C

Sources: A = Hutton & Sogandares-Bernal (1960), B = Hutton (1964), C = Sogandares-Bernal & Walton (1965), D = MacInnis (1960), E = Dubois (1966), F = Loftin (1961), G = Pearson (1973), H = Forrester et al. (1986), I = Penner & Fried (1963), J = Penner & Trimble (1970), K = MacInnis (1966), L = Sileo (1994).

NG = not given.

a. Location in host: CL = cloaca, IN = intestine, KD = kidney, LI = large intestine, OC = under nictitating membrane of eye, SI = small intestine.

b. *Stephanophora denticulata* is a synonym of *M. denticulatus*.

c. MacInnis (1960) previously incorrectly reported this as *C. medioconiger*; Dubois (1966) reexamined it. This species is not included for Laughing Gulls in MacInnis (1959, 1966).

d. *Galactosomum* sp. reported by Hutton and Sogandares-Bernal (1960) in *Larus atricilla* was named *G. fregatae* by Pearson (1973).

e. Penner & Fried (1963) described this species from adults reared from cercariae in the marine snail *Batillaria minima* from the southwest coast of Florida.

f. Described from adults reared from megalurous cercariae collected from the marine snail *Cerithium muscarum* from Sarasota County (Penner & Trimble 1970).

Table 21.6. Helminths of terns from Florida

Host Helminth	Species (location)[a]	County	Year(s)	No. of birds		Intensity		Data source
				Examed	Positive	Mean	Range	
Common Tern								
Trematoda	*Renicola glandoloba* (KD)	Franklin	1958	1	1	~90	—	A
Forster's Tern								
Trematoda	*Cardiocephalus megaloconus*[b] (SI)	Franklin	1958	NG	—	—	—	B
	Renicola sp. (KD)	Franklin	NG	NG	—	—	—	C
	Tetracladium sp. (LI)	Franklin	NG	—	—	—	—	C
Gull-billed Tern								
Trematoda	*Galactosomum spinetum* (CO)	Franklin	1958	1	1	1	—	A
	Renicola glandoloba (KD)	Franklin	1958	1	1	6	—	A
Royal tern								
Trematoda	*Cardiocephalus megaloconus* (SI)	Franklin	1958	2	1	7	0–7	A
	Ornithobilharzia canaliculata (BV)	Sarasota/Manatee	NG	NG	—	NG	—	D
	Pachytrema sanguineum (GB)	Franklin	1958	1	1	2	—	A
	Pachytrema sp.	Pinellas	1955–62	—	—	—	—	E
	Philophthalmus andersoni (OC)	Manatee	1958	1	1	NG	—	F
	Philophthalmus hegeneri[c] (OC)	Manatee to Charlotte Charlotte	NG	NG	—	—	—	G
	Renicola glandoloba (KD)	Franklin	1958	2	2	70	50–90	A
		"Tampa Bay"	1955–62	NG	—	—	—	E
	Stictodora lariformicola (LI)	Pinellas	NG	NG	NG	NG	NG	H
	Stephanoprora sp. (IN)	Franklin	NG	NG	NG	NG	NG	C
	Stictodora sp.	Pinellas	1955–62	NG	NG	NG	NG	E
Nematoda	*Capillaria* sp. (IN)	Lee	1992	1	1	NG	—	I

Table 21.7. Helminths of Black Skimmers from Florida

Host Helminth Species (location)[a]	County	Year(s)	No. birds		Intensity		Data source
			Exam.	Pos.	Mean	Range	
Trematoda							
Clinostomum attenuatum (OR)	Glades	1971	5[b]	2	1	1–2	A
Diplostomum spathaceum (SI)	Glades	1971	5[b]	3	2	1–5	A
Galactosomum spinetum[c] (IN)	Charlotte	1955–62	NG	—	—	—	B
Neostictodora huttoni	Charlotte	1959	NG	—	—	—	B
Mesoophorodiplostomum pricei (SI)	Glades	1971	5[b]	3	7	1–15	A
Microparyphium facetum (CL)	Glades	1971	5[b]	2	2	1–1	A
Renicola ralli (KD)	Glades	1971	5[b]	3	18	2–50	A
Stictodora martini (IN)	Charlotte	1959	NG	—	—	—	B
Mesorchis denticulatus (SI)	Charlotte	NG	NG	—	—	—	B
Cestoda							
Cyclustera ibisae[d] (SI)	Glades	1971	5[b]	5	33	1–100	A
	Glades	1971	5[b]	5	31	5–75	A
Nematoda							
Capillaria sp. (SI)	Glades	1971	5[b]	4	6	1–12	A
Cosmocephalus obvelata (ES)	Glades	1971	5[b]	1	1	1	A
Paracuaria macdonaldi (GL)	Glades	1971	5[b]	1	1	1	A

Sources: A = Kinsella (1972), B = Sogandares-Bernal (1959).
NG = not given.
a. Location in host: CL = cloaca, ES = esophagus, GL = gizzard lining, KD = kidney, IN = intestine, OR = oral cavity, SI = small intestine.
b. These 5 Black Skimmers examined by Kinsella (1972) were collected during a die-off on Lake Okeechobee that was possibly due to a bacterial toxin (Jasmin et al. 1972).
c. Sogandares-Bernal & Hutton (1960) found metacercariae in the half-beak *Hyporhampus unifasciatus* in John's Pass, Fla., and reported this as *G. spinectum*, a synonym. Pearson (1973) further described the species partly on the basis of these specimens.
d. Only immature forms were found and this is likely an accidental host. The host from which it is described is White Ibis (Schmidt & Bush 1972). The cestodes were identified as *Parvitaenia ibisae*, a synonym.

1975) and were not found in the healthy parts of the colonies. In addition to the direct effect of ticks on birds, ticks may transmit viruses (see section XIII, Viruses).

XIX. Emaciation

Emaciation, with no evident cause, is a common finding in seabirds that are involved in die-offs (see section V, Inclement weather, section XX, Unresolved die-offs, and also chapter 2, Loons and Grebes). Schreiber and Schreiber (1980) felt that 21% of Laughing Gull chicks found dead in a colony in Pinellas County died of starvation, yet this was a colony located near a landfill where food resources were probably unlimited.

XX. Unresolved die-offs

A severely emaciated Royal Tern with mild enteritis was found dead during a die-off of pelicans, gulls, and a kingfisher in 1977 at Grassy Key, Monroe County. DDE in the brain (table 21.2) was relatively high, but not considered high enough to cause death. *Salmonella* sp. and other pathogenic bacteria were not isolated from liver, blood, or intestines (Locke 1977).

Three Royal Terns, 2 Laughing Gulls, and a Double-crested Cormorant were examined from a die-off of about 12 birds on Boca Grande Beach, Lee County, in 1987 (Stroud 1987). Pesticide intoxication was suspected because sand fleas were also found dead and because Baygon (propoxur, carbamate) and Bay-

Table 21.8. Chewing lice collected from gulls in Florida

Host	Species	County/area	Year(s)
Bonaparte's Gull	*Saemundssonia lari*	St. Lucie	1973
Great black-backed Gull	*Saemundssonia lari*	St. Lucie	1979
Herring Gull	*Actornithophilus piceus lari*	Lee	1981
		Volusia	1940
	Austromenopon transversum	Franklin	1937
		Lee	1982
		Volusia	1940
	Quadraceps ornatus striolatus	Lee	1984
	Quadraceps punctatus regressus	Lee	1984
		Volusia	1940
	Saemundssonia lari	Brevard	1977
		Franklin	1937
		Lee	1981, 1984
		Orange	1978
		Volusia	1940
Laughing Gull	*Actornithophilus piceus lari*	Lee	1980
		Pinellas	1965
	Austromenopon transversum	Lee	1980–81
	Quadraceps punctatus punctatus	Dade	1929
		Lee	1980–82
	Saemundssonia lari	Lee	1980–81
		Pinellas	1965
		St. Lucie	1976–77
Ring-billed Gull	*Austromenopon transversum*	Lee	1982
	Quadraceps punctatus sublingulatus	Lee	1980, 1982
	Saemundssonia lari	Indian River	1968, 1976
		Lee	1980–83
		Levy	1957

Source: Forrester et al. (1995).

tex (fenthion, organophosphate) had been used for insect control in the area. This was not substantiated by cholinesterase inhibition tests. All birds were emaciated. All other findings were inconsistent among birds, including parasites and bacteria. The Laughing Gull had an abscess in the pectoral muscle.

In April 1987 3 Brown Pelicans, an unidentified gull, and a Laughing Gull were examined from a die-off at a dump site and a shellfish processing plant in Vero Beach, Brevard County. There was no consistent finding. Both gulls were emaciated and the unidentified gull had aspergillosis (Roffe 1987).

Schreiber et al. (1979) noticed an increase in the number of Laughing Gull eggs that failed to hatch (16%) in a landfill island in Pinellas County in 1975. A large proportion (80%) of those eggs had no embryonic development. They did not find an explanation for this failure to develop, but smaller clutch sizes and smaller egg sizes were also noted in the colony.

Eight Royal and 13 Sandwich Terns were found dead during October–November 1993 in Collier County during a die-off of about 60–70 birds. Trematodes were found in the kidneys of 1 of each of the species, and cestodes were found in the duodenum of a Royal Tern. An adenovirus was isolated from some of these terns, but this was considered to be an inciden-

Table 21.9. Ectoparasites collected from terns, noddies, and skimmers in Florida

Parasite Host	Species	County/area	Year(s)
Chewing lice			
Arctic Tern	*Saemundssonia lockeyi*	Volusia	1977
Black Skimmer	*Quadraceps elongatus*	Dade	1929
		Indian River	1968, 1978
		Lee	1981, 1983–84
		Nassau	1974
		Orange	1983
		Seminole	1983
		St. Johns	1964
		St. Lucie	1981
	Saemundssonia anisorhamphos	Lee	1983
Black Tern	*Quadraceps phaeonotus*	Dade	1919
	Saemundssonia lobaticeps	Dade	1919, 1955
Bridled Tern	*Saemundssonia meridiana*	St. Lucie	1976
Brown Noddy	*Actornithophilus incisus*	Lee	1985
	Quadraceps separatus	Lee	1985
		Monroe	1977
	Saemundssonia remota	Monroe	1976
Caspian Tern	*Quadraceps caspius*	Indian River	1975, 1978
	Saemundssonia sp.	Indian River	1975, 1978
Common Tern	*Austromenopon atrofulvum*	Martin	1974
	Quadraceps sellatus	Indian River	1967, 1981
		Martin	1974
	Saemundssonia sternae	Indian River	1967, 1981
		Martin	1974
Forster's Tern	*Actornithophilus piceus piceus*	Pinellas	1931
	Saemundssonia parvigenitalis	Pinellas	1931
		Pinellas[a]	NG
Least Tern	*Actornithophilus piceus piceus*	Pinellas	1976
	Quadraceps nychthemerus	Indian River	1966
		Pinellas	1976
	Saemundssonia melanocephala	Collier	1955
		Indian River	1966
Royal Tern	*Actornithophilus piceus piceus*	Pinellas	1929
	Austromenopon atrofulvum	Lee	1980–81
	Quadraceps praestans	Brevard	1972
		Lee	1978, 1980–82
		Nassau	1974
	Saemundssonia laticaudata	Brevard	1972
		Indian River	1968, 1979
		Lee	1980–82, 1984
		Nassau	1974
		Pinellas	1930
		Pinellas[a]	NG
		St. Lucie	1972

(continued)

Table 21.9. *(continued)*

Parasite Host	Species	County/area	Year(s)
Sandwich Tern	*Austromenopon atrofulvum*	Lee	1980–81
	Quadraceps longicollis	Lee	1981
	Saemundssonia laticaudata	Lee	1982
Sooty Tern	*Austromenopon atrofulvum*	Monroe	1977–78
	Quadraceps birostris	Indian River	1970, 1974
		Lee	1982
		Monroe	1963
			1968, 1976–78
	Saemundssonia albemarlensis	Indian River	1970
		Lee	1982
		Monroe	1963
			1968
			1976–77
Mites			
Brown Noddy nest	*Paracheyletia bakeri*	Monroe[b]	1962
Ticks			
Brown Noddy nests	*Carios denmarki*[c]	Monroe[d]	1961–62
Sooty Tern (foot)	*Carios denmarki*[c]	Monroe[d]	1961

Source: Forrester et al. (1995) unless otherwise noted.
NG = not given.
a. *Source:* Ward (1955).
b. *Source:* Records from the Florida State Collection of Arthropods, Division of Plant Industry, Department of Agriculture and Consumer Services, Gainesville. This species is probably not a parasite.
c. Originally described as *Ornithodoros capensis* and later described as a new species, *O. denmarki* by Kohls et al. (1965). Klompen & Oliver (1993) placed this tick in the genus *Carios*.
d. *Source:* Records from the Florida State Collection of Arthropods, Division of Plant Industry, Department of Agriculture and Consumer Services, Gainesville, and Denmark & Clifford (1962).

tal finding. Botulism toxin tests were negative. Brain cholinesterase and liver mercury concentrations (table 21.3) were not unusual. Intestine and liver were negative for *Salmonella, Pasteurella multocida,* and other pathogens (Fischer 1994).

A die-off of more than 145 gulls at 2 lakes in February 1997 was associated with foraging at a Polk County landfill (Fischer 1997). The cause of death was not determined in the 4 Laughing Gulls and 1 Ring-billed Gull examined. A heavy growth of *Clostridium perfringens* was cultured from the liver of 1 of the Laughing Gulls. This organism has caused mortality in other birds and has been associated with the consumption of garbage. Mouse inoculation tests were negative, but ELISA tests were equivocal, for botu-linum toxin. At a previous die-off at this site in 1994, botulism toxin type C was identified (see section XII, Biotoxins).

XXI. Gulls as a public health hazard

Histoplasma capsulatum has been cultured from the soil in a gull colony and researchers working in that colony in Michigan contracted histoplasmosis (Southern 1986). The risk appears to be associated with disturbance of the soil, handling objects that have been soiled by feces, and the use of blinds without floors that allow for contact with the organism.

In England, the prevalence of *Salmonella* carriers among gulls has increased over time.

It is thought that the gulls were exposed to *Salmonella* from foraging at sewage outfalls along the coast and in landfills. Thus they can carry the organism from the landfills to other locations, where humans, livestock, and wildlife may be exposed (Coulson et al. 1983; Butterfield et al. 1983). Quessy and Messier (1992) recovered 10 serotypes of *Salmonella* from feces of 23 of 264 (9%) Ring-billed Gulls in Canada and suggested that gulls played a minor part in distributing *Salmonella*. Sellin et al. (2000) found high frequencies of antibiotic-resistant enterococci in gulls in Sweden where this organism was an emerging human health concern. He suggested that the bacteria were acquired from feeding in landfills and were being spread by these migrating birds to other locations.

Soldado virus (Nairovirus), although not yet definitively isolated from birds in Florida, can cause a self-limiting febrile illness and puritus in humans exposed to infected ticks by working in gull and tern colonies (Chastel et al. 1995).

The extensive use of landfills by gulls, especially in winter, was documented in Florida by Patton (1988). Gulls attracted to landfills can also be a public nuisance, hazard to aircraft, agricultural pests, and a threat to other species of birds (Blokpoel and Tessier 1986). Donoghue (1996) reports that the group of birds most often hit by planes are gulls.

XXII. Summary and Conclusions

Because the focus of development in Florida has been largely on coastal habitats, human disturbance and habitat alteration have significant but poorly measured effects on this group of birds in Florida. This development also adds to the mortality of chicks from predation and of adults from vehicular collision. The use of landfills by gulls increases the exposure to some diseases, such as pesticide toxicosis, botulism, and salmonellosis, and also implicates gulls as vectors for diseases that are potentially harmful to humans, livestock, and other wildlife. Helminth parasites (25 trematodes, 2 cestodes, 3 nematodes, no acanthocephalans), 1 protozoal parasite, contaminants (8 organochlorines, 3 metals), bacteria (16), and viruses (6) of gulls, terns, and skimmers have been very poorly studied.

XXIII. Literature cited

Austin, O.L., Jr. 1969. Extra toes on a sooty tern chick. *Auk* 86:352.

Austin, O.L., Jr., W.B. Robertson, Jr., and G.E. Woolfenden. 1972. Mass hatching failure in Dry Tortugas sooty terns (*Sterna fuscata*). Abstract. *Proc. 15th Int. Ornithol. Congr.* 1970:627.

Bent, A.C. 1921. *Life histories of North American gulls and terns.* U.S. Natl. Mus. Bull. 113.

———. 1937. *Life histories of North American birds of prey.* Part 1. U.S. Natl. Mus. Bull. 167. 409 pp.

Blokpoel, H., and G.D. Tessier. 1986. The ring-billed gull in Ontario: a review of a new problem species. Canadian Wildlife Service, Occasional Paper 57.

Brand, C.J., R.M. Windingstad, L.M. Siegfried, R.M. Duncan, and R.M. Cook. 1988. Avian morbidity and mortality from botulism, aspergillosis, and salmonellosis at Jamaica Bay Wildlife Refuge, New York, U.S.A. *Colon. Waterbirds* 11:284–292.

Burger, J. 1996. Laughing gull. *Birds N. Am.* 225:1–28.

Burridge, M.J., W.J. Bigler, D.J. Forrester, and J.M. Hennemann. 1979. Serologic survey for *Toxoplasma gondii* in wild animals in Florida. *J. Am. Vet. Med. Assoc.* 175:964–967.

Butterfield, J., J.C. Coulson, S.V. Kearsey, J.H. McCoy, and G.E. Spain. 1983. The herring gull *Larus argentatus* as a carrier of salmonella. *J. Hyg. Lond.* 91:429–436.

Case, L.D., H. Cruickshank, A.E. Ellis, and W.F. White. 1965. Weather causes heavy bird mortality. *Fla. Nat.* 38:29–30.

Chardine, J.W., and R.D. Morris. 1996. Brown noddy. *Birds N. Am.* 220:1–24.

Chastel, C., C.H. Bailly, H.D. Bach, G.I. Lay, M.C. Legrand, F.I. Goff, and C. Vermeil. 1995.

Arbovirus transmis par des tiques au Maghreb. *Bull. Soc. Pathol. Exot.* 88:81–85.

Clapp, R.B., D. Morgan-Jacobs, and R.C. Banks. 1983. *Marine birds of the southeastern United States and Gulf of Mexico. Part III: Charadriiformes.* U.S. Fish and Wildlife Service, Biological Service Program FWS-OBS 83/30. 853 pp.

Combs, W., and C. LeBuff. 1965. Sanibel-Captiva conserves. *Fla. Nat.* 38:104.

Conti, L., R. Oliveri, and C. Blackmore. 2002. Unpublished data. Florida Department of Health, Tallahassee.

Converse, J.D., H. Hoogstraal, M.I. Moussa, C.J. Feare, and M.N. Kaiser. 1975. Soldado virus (Hughes group) from *Ornithodoros (Alectorobius) capensis* (Ixodoidea: Argasidae) infesting sooty tern colonies in the Seychelles, Indian Ocean. *Am. J. Trop. Med. Hyg.* 24:1010–1018.

Converse, K. 1998. Unpublished data. National Wildlife Health Center, Madison, Wis.

Coulson, J.C., J. Butterfield, and C. Thomas. 1983. The herring gull *Larus argentatus* as a likely transmitting agent of *Salmonella montevideo* to sheep and cattle. *J. Hyg. Lond.* 91:437–443.

Crawford, R.L. 1981. Bird casualties at a Leon County, Florida TV tower: a 25-year migration study. *Bull. Tall Timbers Res. Stn.* 22:1–30.

Delany, M.F. 1986. Bird bands recovered from American alligator stomachs in Florida. *N. Am. Bird Bander* 11:92–94.

Denmark, H.A., and C.M. Clifford, Jr. 1962. A tick of the *Ornithodoros capensis* group established on Bush Key, Dry Tortugas, Florida. *Fla. Entomol.* 45:139–142.

Dinsmore, J.J. 1972. Sooty tern behavior. *Bull. Fla. State Mus.* 16:129–179.

Donoghue, J.A. 1996. Sharing the skies. *Air Transport World* November: 55–60.

Drees, B.M. 1994. Red imported fire ant predation on nestlings of colonial waterbirds. *Southwest. Entomol.* 19:355–359.

Dronen, N.O., Jr., and L.R. Penner. 1975. Concerning *Philopthalamus andersoni* sp. n. (Trematoda: Philopthalmidae), another ocular helminth from birds which develops in a marine gastropod. *University Conn. Occas. Pap. Biol. Sci. Ser.* 2:217–224.

Dubois, G. 1966. Du statut de Quelques Strigeata la rue, 1926 (Trematoda). II. *Bull. Soc. Neuchatel. Sci. Nat.* 89:19–56.

Duffy, D.C. 1991. Ants, ticks, and nesting seabirds: dynamic interactions? In: *Bird-parasite interactions.* J.E. Loye and M. Zuk (eds.). Oxford University Press, Oxford. pp. 242–256.

Dunn, E.H., and D.J. Agro. 1995. Black tern. *Birds N. Am.* 147:1–24.

Egensteiner, E.D., H.T. Smith, and J.A. Rodgers. 1996. Royal tern. In: *Rare and endangered biota of Florida.* Vol. 5, *Birds.* J.A. Rodgers, Jr., H.W. Kale II, and H.T. Smith (eds.). University Press of Florida, Gainesville. pp. 532–540.

Field notes. 1988. *The Skimmer,* Newsletter of Nongame and Endangered Wildlife. Florida Game and Fresh Water Fish Commission. Tallahassee. Summer 4:6.

Fischer, J. 1993–97. Unpublished data. Southeastern Cooperative Wildlife Disease Study, Athens, Ga.

Fisk, E.J. 1978a. Roof-nesting terns, skimmers, and plovers in Florida. *Fla. Field Nat.* 6:1–8.

———. 1978b. Least tern. In: *Rare and endangered biota of Florida.* Vol. 2, *Birds.* H.W. Kale II (ed.). University Press of Florida, Gainesville. pp. 40–43.

Forrester, D.J., and F.H. White. 1972. Unpublished data. University of Florida, Gainesville.

Forrester, D.J., and G.F. Bennett. 1985. Unpublished data. University of Florida, Gainesville.

Forrester, D.J., H.W. Kale II, R.D. Price, K.C. Emerson, and G.W. Foster. 1995. Chewing lice (Mallophaga) from birds in Florida: a listing by host. *Bull. Fla. Mus. Nat. Hist.* 39:1–44.

Forrester, D.J., P.R. Douglass, and P.J. Douglass. 1986. Unpublished data. University of Florida, Gainesville.

Franson, J.C. 1990–93. Unpublished data. National Wildlife Health Center, Madison, Wis.

———. 1999. Chlamydiosis. In: *Field manual of wildlife diseases. General field procedures and diseases of birds.* M. Friend and J.C. Franson (eds.). U.S. Department of the Interior, U.S. Geological Survey, Biological Research Division, Information and Technical Report 1999-001. Washington, D.C. pp. 111–114.

Frenkel, J.K. 1981. False-negative serologic tests for *Toxoplasma* in birds. *J. Parasitol.* 67:952–953.

Friend, M., and J.C. Franson. 1999. *Field manual of wildlife diseases. General field procedures and diseases of birds.* U.S. Department of the Interior, U.S. Geological Survey, Biological Research Division, Information and Technology Report 1999-001. Washington, D.C. 426 pp.

Frohring, P.C., and J.A. Kushlan. 1986. Nesting status and colony site variability of laughing gulls in southern Florida. *Fla. Field Nat.* 14:1–17.

Furness, R.W. 1996. Cadmium in birds. In: *Interpreting environmental contaminants in animal tissues.* W.N. Beyer, G.H. Heinz, and A. Redmon (eds.). Lewis, Boca Raton, Fla. 494 pp.

Gaydos, J.R. 1999. Unpublished data. Southeastern Cooperative Wildlife Disease Study, Athens, Ga.

Gochfeld, M., and J. Burger. 1994. Black skimmer. *Birds N. Am.* 108:1–28.

Good, T.P. 1998. Great black-backed Gull. *Birds N. Am.* 330:1–32.

Gore, J.A. 1991. Distribution and abundance of nesting least terns and black skimmers in northwest Florida. *Fla. Field Nat.* 19:65–72.

———. 1996. Least tern. In: *Rare and endangered biota of Florida.* Vol. 5, *Birds.* J.A. Rodgers, Jr., H.W. Kale II, and H.T. Smith (eds.). University Press of Florida, Gainesville. pp. 236–246.

Gore, J.A., and M.J. Kinnison. 1991. Hatching success in roof and ground colonies of least terns. *Condor* 93:759–762.

Graves, I.L. 1992. Influenza viruses in birds of the Atlantic flyway. *Avian Dis.* 36:1–10.

Greene, L.L., and H.W. Kale II. 1976. Roof-nesting by black skimmers. *Fla. Field Nat.* 4:15–17.

Harrington, B.A. 1974. Colony visitation behaviour and breeding ages of sooty terns (*Sterna fuscata*). *Bird-Banding* 45:115–144.

Heinz, G.H. 1996. Mercury poisoning in wildlife. In: *Noninfectious diseases of wildlife.* A. Fairbrother, L. N. Locke, and G. L. Hoff (eds.). Iowa State University Press, Ames. pp 118–127.

Heinz, G.H., and L.N. Locke. 1976. Brain lesions in mallard ducklings from parents fed methylmercury. *Avian Dis.* 20:9–17.

Hickey, J.J., and D.W. Anderson. 1968. Chlorinated hydrocarbons and eggshell changes in raptorial and fish-eating birds. *Science* 162:271–273.

Hoffman, A.M., and F.A. Leighton. 1985. Hemograms and microscopic lesions of herring gulls during captivity. *J. Am. Vet. Med. Assoc.* 187:1125–1128.

Hoffman, W., W.B. Robertson, Jr., and P.C. Patty. 1979. Short-eared owl on Bush Key, Dry Tortugas, Florida. *Fla. Field Nat.* 7:29–30.

Hovis, J.A., and M.S. Robson. 1989. Breeding status and distribution of the least tern in the Florida Keys. *Fla. Field Nat.* 17:61–66.

Hudson, C.R., C. Quist, M.D. Lee, K. Keyes, S.V. Dodson, C. Morales, S. Sanchez, D.G. White, and J.J.M. Maurer. 2000. Genetic relatedness of *Salmonella* isolates from nondomestic birds in southeastern United States. *J. Clin. Microbiol.* 38:1860–1865.

Hughes, L.E., C.M. Clifford, L.A. Thomas, H.A. Denmark, and C.B. Philip. 1964. Isolation and characterization of a virus from soft ticks (*Ornithodorus capensis*) collected on Bush Key, Dry Tortugas, Florida. *Am. J. Trop. Med. Hyg.* 13:118–122.

Hutton, R.F. 1964. A second list of parasites from marine and coastal animals of Florida. *Trans. Am. Microsc. Soc.* 83:439–447.

Hutton, R.F., and F. Sogandares-Bernal. 1959. *Studies of the trematode parasites encysted in Florida mullets.* Special Scientific Report 1, Florida State Board of Conservation Marine Laboratory no. 59-12.

———. 1960. Studies on helminth parasites from the coast of Florida. II. Digenetic trematodes from shore birds of the west coast of Florida. *Bull. Mar. Sci. Gulf Caribb.* 10:40–54.

Jacobson, E.R., B.L. Raphael, H.T. Nguyen, E.C. Greiner, and T. Gross. 1980. Avian pox infection, aspergillosis and renal trematodiasis in a royal tern. *J. Wildl. Dis.* 16:627–631.

Jasmin, A.M., D.E. Cooperrider, C.P. Powell, and J.N. Baucom. 1972. Enterotoxemia of wildfowl due to *Cl. perfringens* type C. *J. Wildl. Dis.* 8:79–84.

Jennings, W.L., W.G. Winkler, D.D. Stamm, P.H. Coleman, and A.L. Lewis. 1969. Serologic studies of possible avian or mammalian reservoirs of St. Louis encephalitis virus in Florida. *Fla. State Board Health. Monogr. Ser.* 12:118–125.

Johnston, D.W. 1976. Organochlorine pesticide residues in uropygial glands and adipose tissue of wild birds. *Bull. Environ. Contam. Toxicol.* 16:149–155.

Jonkers, A.H., J. Casals, T.H.G. Aitkin, and L. Spence. 1973. Soldado virus, a new agent from Trinidadian *Ornithodoros* ticks. *J. Med. Entomol.* 10:517–519.

Kale, H.W. II. 1971. Florida region. *Am. Birds* 25:723–733.

King, K.A., and C.A. LeFever. 1979. Effects of oil transferred from incubating gulls to their eggs. *Mar. Pollut. Bull.* 10:319–321.

Kinsella, J.M. 1972. Helminth parasites of the black skimmer, *Rynchops nigra,* from Lake Okeechobee, Florida. *J. Parasitol.* 58:780.

Klompen, J.S.H., and J.H. Oliver, Jr. 1993. Systematic relationships in the soft ticks (Acari: Ixodida: Argasidae). *System. Entomol.* 18:313–331.

Klukas, R.W., and L.N. Locke. 1970. An outbreak of fowl cholera in Everglades National Park. *J. Wildl. Dis.* 6:77–79.

Kohls, G.M., D.E. Sonenshine, and C.M. Clifford. 1965. The systematics of the subfamily Ornithodorinae (Acarina: Argasidae). II. Identification of the larvae of the western hemisphere and descriptions of three new species. *Ann. Entomol. Soc. Am.* 58:331–364.

Lincer, J.L., and D. Salkind. 1973. A preliminary note on organochlorine residues in the eggs of fish-eating birds of the west coast of Florida. *Fla. Field Nat.* 1:19–22.

Locke, L.N. 1972. Unpublished data. Patuxent Wildlife Research Center, Laurel, Md.

———. 1977. Unpublished data. National Wildlife Health Research Center, Madison, Wis.

Loftin, H. 1961. An annotated check-list of trematodes and cestodes and their vertebrate hosts from northwest Florida. *Q. J. Fla. Acad. Sci.* 23:302–314.

Loftin, H., and H.T. Smith. 1996. Black skimmer. In: *Rare and endangered biota of Florida.* Vol. 5, *Birds.* J.A. Rodgers, Jr., H.W. Kale II, and H.T. Smith (eds.). University Press of Florida, Gainesville. pp. 571–578.

Loftin, H., and S. Sutton. 1979. Ruddy turnstones destroy royal tern colony. *Wilson Bull.* 91:133–135.

Logan, T.H. 1997. Florida's endangered species, threatened species, and species of special concern. Official lists. Florida Game and Fresh Water Fish Commission, Tallahassee. 14 pp.

Longstreet, R.J. 1953. Ornithology of the mosquitoes. *Fla. Nat.* 26:175–183.

MacInnis, A.J. 1959. Some helminth parasites, mostly from marine birds from the northwest gulf coast of Florida. M.S. thesis, Florida State University, Tallahassee. 37 pp.

———. 1960. Some digenetic trematodes from birds from the northwest gulf coast of Florida. Abstract. *Assoc. Southwest. Biol. Bull.* 7:34.

———. 1966. Trematodes from marine shorebirds from the northwest gulf coast of Florida. *Sonderbd. Zool. Anz.* 176:3–68.

Marsh birds die. 1948. *Fla. Wildl.* 2:20.

Maehr, D.S., and J.Q. Smith. 1988. Bird casualties at a central Florida power plant: 1982–86. *Fla. Field Nat.* 16:57–80.

Maehr, D.S., A.G. Spratt, and D.K. Voigts. 1983. Bird casualties at a central Florida power plant. *Fla. Field Nat.* 11:45–49.

Marion, W.R., T.E. O'Meara, G.D. Riddle, and H.A. Berkhoff. 1983. Prevalence of *Clostridium botulinum* type C in substrates of the phosphate-mine settling ponds and implications for epizootics of avian botulism. *J. Wildl. Dis.* 19:302–307.

McEwan, L.C., and D.H. Hirth. 1980. Food habits of the bald eagle in north-central Florida. *Condor* 82:229–231.

McNair, D.B. 1998. Sandwich tern mortality caused by vehicle collision associated with Hurricane Erin. *Fla. Field Nat.* 26:97–99.

Mead, C. 1971. Bang go the eggs. *Br. Trust Ornithol. News* 44:8.

Meteyer, C. 1997. Unpublished data. National Wildlife Health Center, Madison, Wis.

Nisbet, I.C.T. 1983. Paralytic shellfish poisoning: effects on breeding terns. *Condor* 85:338–345.

Ogden, J.C. 1977. The nesting season: Florida region. *Am. Birds* 31:1128–1130.

———. 1979. The nesting season: Florida region. *Am. Birds* 33:855–858.

Ogden, J.C., W.B. Robertson, G.E. Davis, and T.W. Schmidt. 1974. *Pesticides, polychlorinated biphenyls and heavy metals in upper food chain levels, Everglades National Park and vicinity.* U.S. Department of the Interior, National Technical Information Service, Atlanta. 24 pp.

Paige, B.B. 1968. The least tern in man's world. *Fla. Nat.* 41:14–16.

Parnell, J.F., R.M. Erwin, and K.C. Molina. 1995. Gull-billed tern. *Birds N. Am.* 140:1–20.

Patton, S.R. 1988. Abundance of gulls at Tampa Bay landfills. *Wilson Bull.* 100:431–442.

Paul, R.T. 1988. The nesting season: Florida region. *Am. Birds* 42:1278–1281.

———. 1991. The nesting season: Florida region. *Am. Birds* 45:91–94.

———. 1996b. Caspian tern. In: *Rare and endangered biota* of *Florida.* Vol. 5, *Birds.* J.A. Rodgers, Jr., H.W. Kale II, and H.T. Smith (eds.). University Press of Florida, Gainesville. pp. 551–558.

Pearson, J.C. 1973. A revision of the subfamily Haplorchinae Looss, 1899 (Trematoda: Heterophyidae) II. Genus *Galactosomum. Philos. Trans. R. Soc. Lond. B Biol. Sci.* 266:341–447.

Penner, L.R. 1953. The biology of a marine dermatitis-producing schistosome cercaria from *Batillaria minima* (Gmelin). *J. Parasitol.* 39:19–20.

Penner, L.R., and B. Fried. 1963. *Philophthalmus hegeneri* sp. n., an ocular trematode from birds. *J. Parasitol.* 49:974–977.

Penner, L.R., and J.J. Trimble III. 1970. *Philophthalmus larsoni* sp. n., an ocular trematode from birds. *Univ. Conn. Occas. Pap. Biol. Sci. Ser.* 1:265–273.

Pierotti, R.J., and T.P. Good. 1993. Herring gull. *Birds N. Am.* 124:1–28.

Quessy, S., and S. Messier. 1992. Prevalence of *Salmonella* spp., *Campylobacter* spp. and *Listeria* spp. in ring-billed gulls (*Larus delawarensis*). *J. Wildl. Dis.* 28:526–531.

Quist, C. 1997. Unpublished data. Southeastern Cooperative Wildlife Disease Study, University of Georgia, Athens.

Robertson, M.J., and W.B. Robertson, Jr. 1978. Occurrence and effects of chronic, low-level oil contamination in a population of sooty terns (*Sterna fuscata*). Unpublished manuscript. South Florida Research Center, Everglades National Park. 42 pp.

Robertson, W.B., Jr. 1964. The terns of the Dry Tortugas. *Bull. Fla. State Mus.* 8:1–94.

———. 1978a. Sooty tern. In: *Rare and endangered biota of Florida.* Vol. 2, *Birds.* H.W. Kale II (ed.). University Press of Florida, Gainesville. pp. 89–91.

———. 1978b. Roseate tern. In: *Rare and endangered biota of Florida.* Vol. 2, *Birds.* H.W. Kale II (ed.). University Press of Florida, Gainesville. pp. 39–40.

———. 1996. Brown noddy. In: *Rare and endangered biota of Florida.* Vol. 5, *Birds.* J.A. Rodgers, Jr., H.W. Kale II, and H.T. Smith (eds.). University Press of Florida, Gainesville. pp. 559–570.

———. n.d. Roseate tern. Unpublished report. U.S. National Park Service, Everglades National Park, Homestead, Fla.

Robertson, W.B., Jr., and D.R. Paulson. 1961. Region reports: Florida region. *Audubon Field Notes* 15:26–35.

Robertson, W.B., Jr., and M.J. Robertson. 1996. Sooty tern. In: *Rare and endangered biota of Florida.* Vol. 5, *Birds.* J.A. Rodgers, Jr., H.W. Kale II, and H.T. Smith (eds.). University Press of Florida, Gainesville. pp. 514–531.

Robertson, W.B., Jr., and G.E. Woolfenden. 1992. *Florida bird species: an annotated list.* Spec. Publ. 6, Florida Ornithological Society, Gainesville. 260 pp.

Rocke, T.E., and M. Friend. 1999. Avian botulism. In: *Field manual of wildlife diseases. General field procedures and diseases of birds.* M. Friend and J.C. Franson (eds.). U.S. Department of the Interior, U.S. Geological Survey, Biological Research Division, Information and Technical Report 1999-001. Washington, D.C. pp. 271–281.

Rodgers, J.A., Jr., and H.T. Smith. 1997. Buffer zone distances to protect foraging and loafing waterbirds from human disturbance in Florida. *Wildl. Soc. Bull.* 25:139–145.

Rodgers, J.A., Jr., H.W. Kale II, and H.T. Smith (eds.). 1996a. *Rare and endangered biota of Florida*. Vol. 5, *Birds*. University Press of Florida, Gainesville. 688 pp.

Rodgers, J.A., H.T. Smith, and R.T. Paul. 1996b. Sandwich tern. In: *Rare and endangered biota of Florida*. Vol. 5, *Birds*. J.A. Rodgers, Jr., H.W. Kale II, and H.T. Smith (eds.). University Press of Florida, Gainesville. pp. 541–550.

Roffe, T.J. 1987. Unpublished data. National Wildlife Health Center, Madison, Wis.

Roffe, T.J., and H.A. McAllister. 1988. Unpublished data. National Wildlife Health Center, Madison, Wis.

Ryder, J.P. 1993. Ring-billed gull. *Birds N. Am.* 33:1–28.

Schmidt, G.D., and A.O. Bush. 1972. *Parvitaenia ibisae* sp. n. (Cestoidea: Dilepididae), from birds in Florida. *J. Parasitol.* 58:1095–1097.

Schreiber, E.A., and R.W. Schreiber. 1977. Gulls wintering in Florida: Christmas bird count analysis. *Fla. Field Nat.* 5:35–40.

———. 1980. Breeding biology of laughing gulls in Tampa Bay, Florida. Part II. Nesting parameters. *J. Field Ornithol* 51:340–355.

Schreiber, E.A., R.W. Schreiber. and J.J. Dinsmore. 1979. Breeding biology of laughing gulls in Tampa Bay, Florida. Part I. Nesting, egg, and incubation parameters. *Bird-Banding* 50:304–321.

Schreiber, R.W., and E.A. Schreiber. 1979. Notes on measurements, mortality, moult, and gonad condition in Florida west coast laughing gulls. *Fla. Field Nat.* 7:19–23.

Sellin, M., H. Palmgren, T. Broman, S. Bergstrom, and B. Olsen. 2000. Letter: Involving ornithologists in the surveillance of vancomycin-resistant enterococci. *Emerg. Infect. Dis.* 6:87–88.

Sileo, L. 1994. Unpublished data. National Wildlife Health Center, Madison, Wis.

Sileo, L., L. Karstad, R. Frank, M.V.H. Holdrinet, E. Addison, and H.E. Braun. 1977. Organochlorine poisoning of ring-billed gulls in southern Ontario. *J. Wildl. Dis.* 13:313–322.

Skoog, P. 1982. Highways and endangered wildlife in Florida: impacts and recommendations. Florida Game and Fresh Water Fish Commission, Environmental Report FL-ER 27-84.

Smith, H.T. 1996. Roseate tern. In: *Rare and endangered biota of Florida*. Vol. 5, *Birds*. J.A. Rodgers, Jr., H.W. Kale II, and H.T. Smith (eds.). University Press of Florida, Gainesville. pp. 247–257.

Smith, H.T., and J.A. Gore. 1996. Gull-billed tern. In: *Rare and endangered biota of Florida*. Vol. 5, *Birds*. J.A. Rodgers, Jr., H.W. Kale II, and H.T. Smith (eds.). University Press of Florida, Gainesville. pp. 624–632.

Smith, H.T., W.J.B. Miller, R.E. Roberts, C.V. Tamborski, W.W. Timmerman, and J.S. Weske. 1994. Banded royal terns recovered at Sebastian Inlet, Florida. *Fla. Field Nat.* 22:81–83.

Smith, K., and V.F. Nettles. 1994. Unpublished data. Southeastern Cooperative Wildlife Disease Study, Athens, Ga.

Snyder, B. 1994. Unpublished data. Florida Department of Environmental Protection, Tallahassee.

Sogandares-Bernal, F. 1959. Four trematodes from the black skimmer, *Rynchops nigra* Linn., (Aves: Rynchopidae), in Gasparilla Sound, Florida, including the description of a new genus and two new species. *Q. J. Fla. Acad. Sci.* 22:125–132.

Soganderes-Bernal. F., and R.F. Hutton. 1960. Notes on the probable partial life-history of *Galactosomum spinetum* (Braun, 1901) (Trematoda) from the west coast of Florida. *Proc. Helminthol. Soc. Wash.* 27:75–77.

Soganderes-Bernal. F., and D.W. Walton. 1965. *Stictodora lariformicola* n. sp. (Trematoda: Heterophyidae) from Floridian piscivorous birds. *Proc. Helminthol. Soc. Wash.* 32:115–117.

Southern, W.E. 1986. Histoplasmosis associated with a gull colony: health concerns and precautions. *Colon. Waterbirds* 9:121–123.

Spalding, M.G. 1998–2000. Unpublished data. University of Florida, Gainesville.

Spalding, M.G., and M. Cheadle. 2000. Unpublished data. University of Florida, Gainesville.

Stevenson, H.M. 1970. The winter season: Florida region. *Audubon Field Notes* 24:493–497.

Stevenson, H.M., and B.H. Anderson. 1994. *The birdlife of Florida*. University Press of Florida, Gainesville. 892 pp.

Stevenson, J.A. 1994. Unpublished data. Florida

Department of Environmental Protection. Tallahassee.

Stoneburner, D.L., and C.S. Harrison. 1981. Heavy metal residues in sooty tern tissues from the Gulf of Mexico and north central Pacific Ocean. *Sci. Total Environ.* 17:51–58.

———, P.C. Patty, and W.B. Robertson, Jr. 1980. Evidence of heavy metal accumulations in Sooty Terns. *Sci. Total Environ.* 14:147–152.

Stroud, R.K. 1982–1987. Unpublished data. National Wildlife Health Center, Madison, Wis.

Taylor, W.K., and B.H. Anderson. 1973. Nocturnal migrants killed at a central Florida TV tower, autumns 1969–71. *Wilson Bull.* 85:42–51.

———. 1974. Nocturnal migrants killed at a central Florida TV tower, autumn 1972. *Fla. Field Nat.* 2:40–43.

Taylor, W.K., and M.A. Kershner. 1986. Migrant birds killed at the Vehicle Assembly Building (VAB), John F. Kennedy Space Center. *J. Field Ornithol.* 57:142–154.

Thompson, B.C., J.A. Jackson, J. Burger, L.A. Hill, E.M. Kirsch, and J.L. Atwood. 1997. Least tern. *Birds N. Am.* 290:1–32.

Van Velzen, W.T. 1966. Royal tern recovery from tiger shark. *East. Bird Banding Assoc. News* 29:210.

Ward, R.A. 1955. Biting lice of the genus *Saemundssonia* (Mallophaga: Philopteridae) occurring on terns. *Proc. U.S. Natl. Mus.* 105:155–165.

White, F.H., C.F. Simpson, and L.E. Williams, Jr. 1973. Isolation of *Edwardsiella tarda* from aquatic animal species and surface waters in Florida. *J. Wildl. Dis.* 9:204–208.

White, S.C., W.B. Robertson, Jr., and R.E. Ricklefs. 1976. The effect of Hurricane Agnes on growth and survival of tern chicks in Florida. *Bird-Banding* 47:54–71.

Woolfenden, G.E. 1978. Snowy plover. In *Rare and endangered biota of Florida*. Vol. 2, *Birds*. H.W. Kale II (ed.). University Press of Florida, Gainesville. pp. 8–10.

Pigeons and Doves

I. Introduction

Twelve species of columbids (pigeons and doves) occur in Florida (table 22.1). Six of these breed in the state, whereas the other 6 are periodic visitors. Three columbids, the Rock Dove, the Eurasian Collared-Dove, and the White-winged Dove, are exotic species that were introduced and now are residents.

For most of the columbids in Florida there is little or no information on diseases and parasites. Mourning Doves and White-winged Doves are exceptions and have been studied fairly well. Investigations on Mourning Doves have resulted in publications on population status, movement, and migration (Aldrich 1952; Winston 1954; Marion et al. 1981), reproduction (Schnoes 1980; Marion and

Schnoes 1982), and capture techniques using oral anesthetics (Martin 1967; Williams and Phillips 1972). Zwart (1986) published a brief overview of the diseases and parasites of doves and pigeons with emphasis on captive birds. Conti (1993) provided an excellent review of the parasites and diseases (including environmental contaminants) of Mourning Doves in North America. Information available on the effects of inclement weather, predation, and trauma was reviewed for Mourning Doves in the United States by Sadler (1993). Wiley and Wiley (1979) published a monograph on the biology of the White-crowned Pigeon in Puerto Rico and included information on predation and parasitism.

Table 22.1. Distribution, occurrence, and abundance of doves and pigeons in Florida

Species of columbid		Range	Seasonal occurrence	Relative abundance
Rock Dove (= Domestic Pigeon)[a]	*Columba livia*	Statewide	Resident	Occasional to abundant
Scaly-naped Pigeon	*Columba squamosa*	Dade Co. & Key West	Nonresident	Only 3 records
White-crowned Pigeon[b]	*Columba leucocephala*	Southern peninsula	Resident	Rare to abundant
Band-tailed Pigeon	*Columba fasciata*	Southern peninsula	Nonresident	Occasional
European Turtle-Dove	*Streptopelia turtur*	Upper Keys	Nonresident	Only 1 record
Eurasian Collared-Dove[a]	*Streptopelia decaocto*	Statewide	Resident	Rare to abundant
White-winged Dove[a,c]	*Zenaida asiatica*	Northern Florida	Nonresident	Occasional
		Southern peninsula	Resident	Occasional to abundant
Zenaida Dove	*Zenaida aurita*	Peninsular Florida	Nonresident	Rare
Mourning Dove	*Zenaida macroura*	Statewide	Resident	Rare to abundant
Common Ground-Dove	*Columbina passerina*	Statewide	Resident	Rare to abundant
Key West Quail-Dove	*Geotrygon chrysia*	Southern peninsula	Nonresident	Occasional
Ruddy Quail-Dove	*Geotrygon montana*	Keys & Dry Tortugas	Nonresident	Only 5 reports

Sources: Stevenson (1986) and Robertson and Woolfenden (1992), except for data on the Eurasian Collared-Dove, which are from Romagosa (1999).
a. Introduced (exotic).
b. Classified as threatened by the Florida Game and Fresh Water Fish Commission (Logan 1997) and the Florida Committee on Rare and Endangered Plants and Animals (Rodgers et al. 1996).
c. Although not classified as an exotic species, the White-winged Dove is a relative newcomer to Florida. A breeding population is well established in Dade County near Homestead, having been first recognized there in 1959 (Saunders 1980).

II. Trauma

Trauma other than crippling losses due to hunting is undoubtedly responsible for some morbidity and mortality among doves and pigeons in Florida; however, limited specific data exist on this topic. Crawford (1981) listed only 1 Rock Dove, 89 Mourning Doves, and 6 Common Ground-Doves as victims of mortality due to flying into a 1,010-foot TV tower. These findings were based on more than 42,000 birds found dead at the base of a tower near Tall Timbers Research Station (Leon County) over a 25-year period (1955–80). In another study conducted in Orange County, 2 Common Ground-Doves were among 1,347 birds killed as a result of colliding with a 1,484-foot TV tower while flying at night between August and November 1972 (Taylor and Anderson 1973, 1974).

Mourning Doves and White-winged Doves have been hunted legally in Florida for many years, but there has been a decrease in the numbers of doves killed by hunters over the past 9 years. In 1988 more than 2.1 million birds were killed; in 1997–98, an estimated 689,870 were harvested by hunters (Linda 1998). This decline may not be related to a decrease in the numbers of doves, but might rather reflect the number of hunters and the number of hunter days, both of which have also declined. Crippling losses due to hunting of Mourning Doves vary from 2% to 35% of the total kill in Florida (Winston 1954; Shaw 1956). These values are comparable to those obtained in other states such as Georgia (33%, Nelson 1957), Utah (16–24%, Dahlgren 1955), and Illinois (5–25%, Hanson and Kossack 1963). Such percentages take on greater significance when one realizes that they may be considerably lower than actual losses. This type of information is difficult to obtain and most likely the values are underestimated.

An unusual case of trauma was encountered

FIGURE 22.1. Rock Dove found in Alachua County, April 1993, with its body penetrated by a blowgun dart. Courtesy of Scott P. Terrell.

by personnel in the University of Florida Veterinary Medical Teaching Hospital in April of 1993 (Terrell 1993). A Rock Dove was found in Alachua County with a blowgun dart completely penetrating its sternum and protruding out its back (figure 22.1). The amazing outcome of this story is that the dart was removed and the bird recovered and was released, apparently no worse for the experience!

Bigler et al. (1977) examined the effects of trauma and stress associated with the removal of samples of blood from Mourning Doves during trapping and release operations. They found no influence on survival based on recaptures and band returns of 2,557 doves over an 8-year period in Orange County.

III. Predation

Information on predation of Mourning Doves in North America has been reviewed by Sadler (1993). He cited data from the U.S. Fish and Wildlife Service on the recovery of 51,632 bands from Mourning Doves prior to 1984. Of these bands, 577 (1.1%) were linked to predation. The predators included domestic cats, domestic dogs, and several raptors, rodents, and snakes.

For Florida there are a number of reports of direct observations relating to predation on columbids (table 22.2). Predators include Sharp-shinned Hawks, Northern Harriers, Red-winged Blackbirds, Florida Scrub-Jays, and green tree snakes. In addition there are statements in the literature concerning other predators of columbids in Florida, although specific data were not given to back up the claims. These reports include Short-tailed Hawks feeding on White-winged Doves and Eurasian Collared-Doves (Ogden 1992; Robertson and Woolfenden 1992), Sharp-shinned Hawks and Cooper's Hawks preying on Mourning Doves (Stevenson and Anderson 1994), Blue Jays and flying squirrels feeding on eggs and nestlings of Mourning Doves (Hutt 1964), domestic dogs and cats eating the eggs and nestlings of Common Ground-Doves (Stevenson and Anderson 1994), and raccoons preying on nest contents of White-crowned Pigeons (Strong et al. 1991).

Table 22.2. Direct observation of predation on pigeons and doves in Florida

Species of columbid	Type of prey	County	Year(s)	Predator	Data source
Rock Dove	Adults	Alachua	1982	Sharp-shinned Hawk	Conti (1992)
White-crowned Pigeon	Eggs	Monroe	1986–90	Red-winged Blackbird	Bancroft et al. (1990)
Mourning Dove	Adults[a]	Leon	1925–26	Northern Harrier	Stoddard (1931)
	Eggs	Alachua	1982	Green tree snake	Conti (1992)
Common Ground-Dove	Adults	Collier	1992	Florida Scrub-Jay	McGinity (1997)
	Eggs	Highlands	NG	Florida Scrub-Jay	Hailman (1989)

NG = not given.

a. Fragments of Mourning Doves were found in 8 of 1,098 pellets collected from a Northern Harrier roost in Leon County during the winter of 1925–26.

Raccoons, in fact, appear to be an important factor in limiting the use of certain Florida Keys for nesting by White-crowned Pigeons (Strong et al. 1991). Bancroft et al. (1990) conducted a 5-year study (1986–90) of White-crowned Pigeons in southern Florida and reported that "[p]redation was by far the greatest source of mortality of eggs and nestlings," causing 65–80% of the losses that occurred on West Butternut Key, Middle Butternut Key, and Bottle Key in Florida Bay. Predation was higher during years of poor fruit production. This may have been because of the parents' need to spend increased amounts of time getting food, away from the nests, thus allowing predation to occur at a higher than normal level.

Travis (1938) claimed that it was "a well established fact" that the numbers of certain ground nesting birds have decreased because of predation by native fire ants (*Solenopsis geminata*) on their eggs. He gave as an example the general decrease in numbers of Common Ground-Doves in Leon County in the early 1900s, but gave only circumstantial information to support that idea. This topic needs to be investigated further before conclusions can be drawn as to the importance of fire ant predation on columbids in Florida.

IV. Brood parasitism

Stevenson and Anderson (1994) stated that White-winged Doves were subject to brood parasitism by cowbirds, but gave no details on the species of cowbirds involved or other information on the extent of this phenomenon in Florida. For more on brood parasitism the reader is referred to chapter 24, Perching Birds.

V. Inclement weather

Sadler (1993) reviewed the available information on mortality of Mourning Doves due to adverse winter weather. He described extensive mortality due to a severe blizzard over a large area, including parts of Alabama, Arkansas, Kentucky, Louisiana, Tennessee, and Texas, in January of 1951. It was estimated that more than 63,000 doves died in Tennessee alone (Schultz 1954). Hurricanes and other severe storms in Florida most likely result in mortality of columbids, but we found only 1 record of such harmful effect, a Mourning Dove identified as a casualty of Hurricane Hilda in Brevard County during October 6–8, 1964; it was found among 4,707 other birds, mostly passeriforms (Case et al. 1965).

VI. Environmental contaminants

Dahlen and Haugen (1954) gave experimental acute dosages of aldrin, dieldrin, toxaphene, and lindane to pen-reared bobwhites and to Mourn-

Table 22.3. Residues of organochlorine pesticides and PCBs in pectoral muscles of 10 Mourning Doves collected in Highlands County, 1970–71

Residue	No. of birds positive	PPM (wet weight)		PPM (lipid weight)	
		Mean	Range	Mean	Range
DDE	10	0.09	0.01–0.25	7.76	0.99–25.69
DDD	10	0.01	<0.01–0.03	0.73	<0.44–3.30
DDT	10	0.01	<0.01–0.15	0.66	<0.38–1.19
Dieldrin	5	0.01	0.01–0.01[a]	NG	<0.22–<0.51
Heptachlor epoxide	9	NG	<0.01–0.01	NG	<0.38–0.65
Mirex	3	0.01	0.01–0.01[a]	0.69	0.52–0.76
PCB	10	0.10	0.06–0.16	8.98	3.92–15.77

Source: Kreitzer (1974).
NG = not given.
a. All doves reported at this value.

ing Doves that had been obtained from the wild. They found that Mourning Doves were approximately 3 times more resistant to the harmful effects of these insecticides than were bobwhites. LD-50 values for the doves were as follows: aldrin, 15–17 ppm; dieldrin, 44–46; toxaphene, 200–250; and lindane 350–400. However, experimentally administered PCBs (10 ppm in the feed) have been documented to change courtship behavior and to cause a general decrease in reproductive efficiency in Mourning Doves (Tori and Peterle 1983; Koval et al. 1987). The reader is referred to the review by Conti (1993) for more details about environmental contaminants and their effects on doves.

The effects on various species of wildlife of a DDT spraying program for mosquito eradication in Dade County during 1965 were evaluated by Lehner et al. (1967). Organochlorine residues in brains of 2 Rock Doves and 1 Common Ground-Dove obtained from the spray zone were determined. One of the Rock Doves was an adult found dead; residues of DDT and its metabolites were not detected in brain tissues of this bird, but a small amount (0.22 ppm wet weight) of DDE was found in brain tissue of a healthy nestling. The Common Ground-Dove was found sick 8 days after the spray application and contained 17, 35, and 13 ppm of DDT, DDD, and DDE, respectively, in its brain.

The bird was observed having convulsions shortly before it died. The authors concluded that this ground dove died of DDT poisoning, most likely from the spray used in the eradication program.

Samples of pectoral muscle from 10 Mourning Doves from Florida were analyzed in 1970–71 for residues of organochlorine pesticides, PCBs, and mercury (Kreitzer 1974). Concentrations of the organochlorine compounds were <1 ppm (wet weight) (table 22.3) and considered to be of little or no hazard to the doves or to humans or animals who might consume them. Concentrations of mercury were low also (<0.05 ppm, wet weight) and considered to be of little concern. The long-term effects of sublethal concentrations of these contaminants are not known, however.

VII. Neoplasia

No information is available on this topic.

VIII. Biotoxins

Two plants, the Brazilian pepper tree (= Florida holly) (*Schinus terebinthifolius*) and creeping (or trailing) indigo (*Indigofera spi-*

cata) are toxic to mammals and birds (Morton 1978, 1989). Both have been introduced into Florida. In 1983 several Mourning Doves and a large number of American Robins were found dead in Tampa near a Brazilian pepper tree (Forrester 1989). Berries of the plant were found in the gastrointestinal tract of several of these birds (see chapter 24, Perching Birds, for more details on this incident and experimental studies conducted in connection with it). Fatalities of Rock Doves that had ingested the seeds of creeping indigo in southern Florida have been reported (Morton 1989). These occurred in October of 1987. At necropsy, hemorrhages were present in the lungs, liver, and kidneys.

The full significance of these plants as mortality factors in populations of columbids in Florida is not known. It is possible, however, that some die-offs of birds attributed to pesticides may in fact have been due to ingestion of poisonous berries and seeds such as those of the Brazilian pepper and creeping indigo, respectively, and perhaps others.

IX. Viruses

Six viruses have been identified from columbids in Florida (table 22.4). These include avian poxvirus, pigeon paramyxovirus type 1, and 4 arboviruses: eastern equine encephalitis (EEE) virus, St. Louis encephalitis (SLE) virus, West Nile virus, and Highlands J virus. Attempts to identify Everglades virus and avian adenoviruses in columbids from Florida have been unsuccessful.

The prevalence of EEE virus in a large sample of Mourning Doves was <1%, leading Bigler et al. (1975) to conclude that these hosts are insignificant in the epizootiology of this arbovirus in Florida. Jennings et al. (1969) conducted a serosurvey in the Tampa Bay area in 1962, however, and found that SLE virus was fairly common in Mourning Doves and Rock Doves. Common Ground-Doves were also positive and may be an important reservoir host for this virus as well as Highlands J virus. Although columbids do not appear to be affected adversely by SLE virus (Kemp 1981), it has been suggested that these birds may be useful as indicators of virus activity and thereby of value in the surveillance of this important zoonotic virus (Bigler et al. 1975). Reisen et al. (1992), however, showed that the antibody responses to SLE virus were transient, making the utility of Rock Doves in epidemiological studies of limited value. Rock Doves may play some role in the maintenance of SLE virus in nature, as do other wild birds (Kemp 1981); however, Reisen et al. (1992) conducted experimental studies on Rock Doves and found that they rarely developed viremias, thus this role may be minimal. The pathologic significance of Highlands J virus to columbids in Florida is unknown. This virus is not thought to be pathogenic to birds, but there is some evidence that mortality can result occasionally, at least in passeriforms (Johnson 1960). We have no such data for columbids in Florida, however.

West Nile virus and pigeon paramyxovirus type 1 may be important pathogens of columbids in Florida. During the summer and fall of 2001, West Nile virus was identified by PCR and virus isolation techniques from 71 Common Ground-Doves, 37 Mourning Doves, 14 Rock Doves, and 2 White-crowned Pigeons that were found dead in various counties in Florida (table 22.4). It was assumed that these birds died of the West Nile virus infections, but necropsies were not performed to prove this. In the same time period there were 3 epizootics of Eurasian Collared-Doves caused by paramyxovirus type 1. These occurred in Pinellas County from May to July, when 2,000–3,000 died, in the panhandle (Bay, Okaloosa, and Santa Rosa counties) from August to October, when at least 400 died, and in Ramrod Key (Monroe County) in November, when 12–15 died (Spalding et al. 2002).

As discussed in previous chapters, pox can be a very important disease in birds (Karstad 1971). Lesions that block vision (figure 22.2) or interfere with feeding can result in fatalities. Lesions on other parts of the body such as legs or toes (figure 22.3) may not be as serious. In-

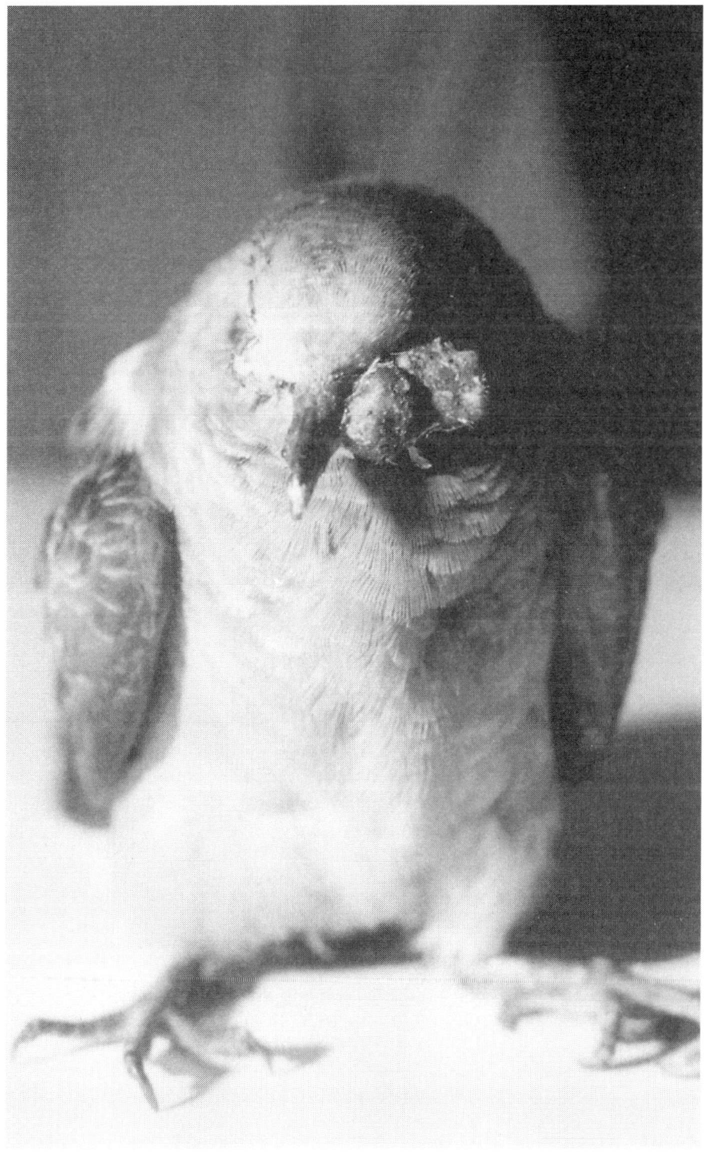

Figure 22.2. Mourning Dove from Alachua County, 1978, with pox lesions around its eyes. Courtesy of Garry W. Foster.

fections have been recorded in Mourning Doves and Rock Doves (table 22.4) in Alachua, Dade, Duval, and Palm Beach counties, but are probably much more widespread. These were cutaneous infections and involved the eyelids, bills, and toes. The diphtheritic form of the disease has not been seen in Florida. As in other hosts, pox infections in Mourning Doves in Florida are most common in the late summer and early fall (August–October) (Locke 1961, Forrester 1989), which is during the peak of the mosquito season. The distribution and effects of pox infections on populations of columbids in Florida are not known. Infections undoubtedly occur in other species of columbids besides Mourning Doves. Concurrent infec-

Table 22.4. Viral infections in doves from Florida

| Species of columbid Virus[a] | County/site | Year(s) | No. of birds | | | Data source |
			Exam.[b]	Inf.	%	
Rock Dove						
Pox	Duval	1978	NG	1	—	Hoff (1978)
SLE	Hillsborough	1962	108	61	56	Jennings et al. (1969)
	Orange	1962	11	3[d]	33	Gainer et al. (1964)
West Nile	Many[c]	2001	NG	14	—	Conti et al. (2002)
Mourning Dove						
Pox	Alachua	1976	ND	1	—	Forrester (1989)
	Alachua	1978	92	3	3	Conti & Forrester (1989)
	Alachua	1981	ND	1	—	Forrester (1989)
	Dade	1978	53	1	2	Conti & Forrester (1989)
	Orange	1999	1	1	—	Terrell (1999)
	Palm Beach	1951	NG	NG	20	Locke (1961), Winston (1954)
	Palm Beach[e]	1960	NG	6	—	Locke (1961)
SLE	Okaloosa	1980	31	1	3	McCaig et al. (1994)
	Orange	1962	111	31	28	Jennings et al. (1969)
	Orange	1969	NG	1	—	Bigler et al. (1975)
	"Florida"	1965–74	3637	10	<1	Ibid.
EEE	"Florida"	1965–74	3637	4	<1	Ibid.
Highlands J	"Florida"	1965–74	21	0	0	Ibid.
Everglades	"Florida"	1965–74	3616	0	0	Ibid.
Adenovirus	"Florida"	NG	30	0	0	Domermuth et al. (1977)
West Nile	Many[f]	2001	NG	37	—	Conti et al. (2002)
Common Ground-Dove						
SLE	Dade	1969	4	0	—	Bigler (1969)
	Highlands	1960	10	1	10	Henderson et al. (1962)
		1961	14	0	0	Ibid.
	Hillsborough	1962	2	1	—	Jennings et al. (1969)
	Osceola	1994	(16)	(0)	(0)	Spalding & McLean (1994)
EEE	Dade	1969	4	0	—	Bigler (1969)
	Highlands	1993–94	24	0	0	Henderson et al. (1962)
	Osceola	1993–94	(18)	(0)	(0)	Spalding & McLean (1994)
	Central Fla.	1958	(3)	(1)	(—)	Favorite (1960)

(continued)

tions of pox and other diseases such as trichomonosis may be significant.

X. Bacteria

Little is known about bacterial infections in columbids from Florida. During 1973–78 3 Mourning Doves from Alachua County, 8 Mourning Doves from Orange County, and 20 White-winged Doves from Dade County were cultured for pathogenic bacteria (White and Forrester 1989). Intestinal contents from all doves and livers from the White-winged Doves and the 3 Mourning Doves from Alachua County were examined. *Escherichia coli* was identified in the intestinal contents of 14 of the 20 White-winged Doves and *Enterococcus* sp. in 1. Other samples were negative. The pathologic significance of these findings is unknown.

Table 22.4. *(continued)*

Species of columbid / Virus[a]	County/site	Year(s)	No. of birds Exam.[b]	Inf.	%	Data source
Common Ground-Dove						
Highlands J	Dade	1969	4	0	0	Bigler (1969)
	Highlands	1960	10	2	20	Henderson et al. (1962)
		1961	14	1	7	Ibid.
	Central Fla.	1958	(3)	(0)	(—)	Favorite (1960)
Everglades	Dade	1969	4	0	0	Bigler (1969)
Adenovirus	"Florida"	NG	11	0	0	Domermuth et al. (1977)
West Nile	Many[g]	2001	NG	71	—	Conti et al. (2002)
White-crowned Pigeon						
West Nile	Monroe	2001	NG	1	—	Ibid.
Eurasian Collared-Dove						
Paramyxovirus-1	Bay	2001	1	1	—	Romero & Spalding (2002)
	Monroe	2001	1	1	—	Ibid.
	Okaloosa	2001	1	1	—	Ibid.
	Pinellas	2001	4	4	—	Ibid.
	Santa Rosa	2001	3	3	—	Schmitt (2001)

ND = not determined, NG = not given.

a. EEE = Eastern equine encephalitis, SLE = St. Louis encephalitis.

b. Diagnoses of poxvirus infections were presumptive (based on finding typical gross lesions), except for Locke (1961), who examined several lesions histologically and diagnosed poxvirus by finding inclusion bodies; serologic testing for SLE, EEE, Highlands J, and Everglades viruses was by hemagglutination-inhibition (HI); parentheses indicate serum neutralization (SN) tests. Bigler et al. (1975) did primary screens using HI tests and confirmed some of the positive samples via SN tests; diagnosis of adenovirus infections was by precipitin tests for antibodies to hemorrhagic enteritis of turkeys/marble spleen disease of pheasants group of avian adenoviruse; diagnosis of West Nile viruses was by PCR and virus isolation techniques.

c. Bay (1), Broward (1), Clay (2), Dade (4), Duval (3), Gadsden (1), Leon (1), and Orange (1).

d. In addition to the serologic results given here, SLE virus was isolated from 2 of these birds by Gainer et al. (1964).

e. Locality may have been Sebring (Highlands Co.), according to records of Patuxent Wildlife Research Center, Laurel, Md.

f. Bay (4), Bradford (1), Charlotte (2), Collier (1), Dade (1), Duval (1), Gilchrist (1), Gulf (2), Hernando (1), Marion (1), Monroe (1), Okaloosa (3), Pasco (7), Putnam (3), Santa Rosa (4), Seminole (3), and Suwannee (1).

g. Alachua (2), Bay (10), Brevard (1), Broward (2), Charlotte (2), Citrus (1), Clay (4), Collier (1), Dixie (2), Duval (4), Escambia (4), Glades (1), Hernando (1), Highlands (1), Lake (1), Leon (3), Levy (2), Manatee (3), Marion (2), Monroe (2), Okaloosa (7), Okeechobee (1), Palm Beach (1), Pasco (4), Polk (1), Sarasota (2), Suwannee (1), Taylor (2), and Wakulla (1).

XI. Fungi

There is little information available on the fungi or yeasts that inhabit the upper gastrointestinal tract of columbids in Florida, except for 1 study on nestling White-crowned Pigeons (Kocan and Hasenclever 1972; Hasenclever and Kocan 1973). In that study 23 of 41 (56%) squabs were positive for *Saccharomyces telluris,* although none of the infected birds showed signs of disease. This yeast is probably not pathogenic and may be a normal component of the flora of the upper digestive tract of pigeons and doves.

XII. Trichomonosis

Trichomonas gallinae is the most important parasite of columbids in Florida. This flagellate (figure 22.4) is the etiologic agent of trichomonosis, or canker, a disease that can cause epizootics of considerable proportion (Haugen and Keeler 1952). The Rock Dove is considered to be the primary host of *T. gallinae,* but the protozoan occurs also in a number of other avian hosts (Kocan and Herman 1971). In Florida it has been found in Rock Doves, White-crowned Pigeons, Eurasian Collared-

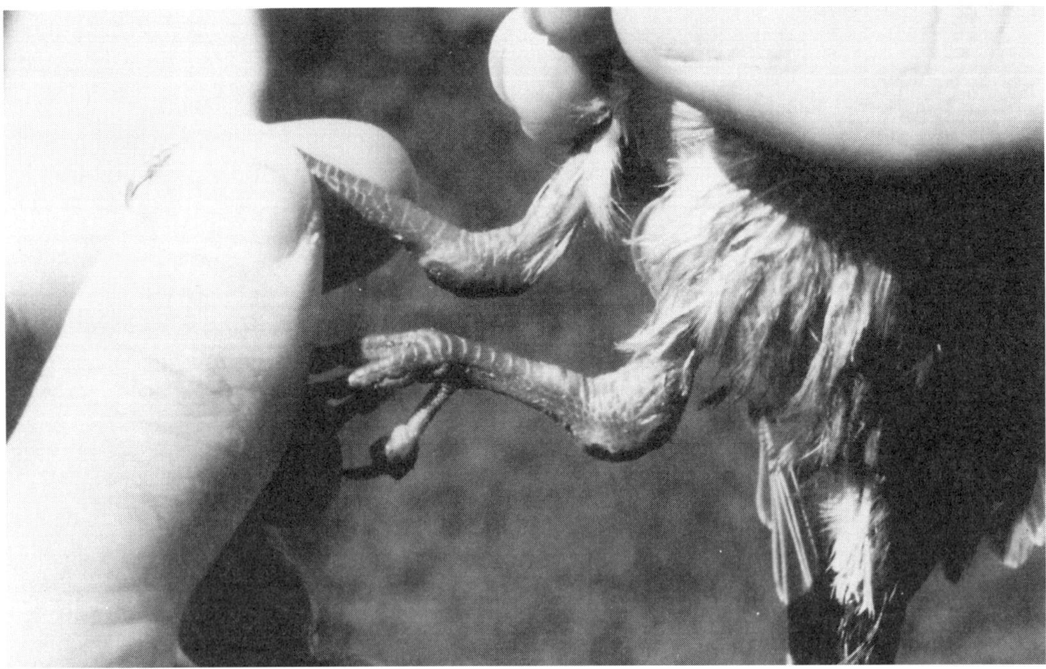

FIGURE 22.3. Mourning Dove from Alachua County, 1976, with pox lesions on its legs and toes. Courtesy of Garry W. Foster.

Doves, White-winged Doves, Mourning Doves, and Common Ground-Doves (table 22.5). Raptors are susceptible and become infected secondarily by feeding on infected columbids (see chapters 14, Hawks, and 23, Owls). Transmission from dove to dove occurs via ingestion of contaminated food or water. During courtship, columbids often exhibit cross-feeding and billing behavior and these practices can result in transmission of the organism. Regurgitative feeding of young by adult columbids, however, is the primary method of transmission (Kocan and Herman 1971).

Trichomonosis is a disease of the upper digestive tract and oral cavity. Lesions are yellowish, well circumscribed masses (figure 22.5), either on the floor or the roof of the mouth or sometimes in the pharyngeal region (Kocan and Herman 1971). As the infection progresses the lesions become large cheesy masses that can impair breathing and feeding and result in emaciation and death due to star-

vation. Lesions resembling trichomonosis can also be caused by aspergillosis or pox; it is important to eliminate the possibility of these other 2 diseases before a diagnosis of trichomonosis is made (Kocan and Herman 1971). Since *T. gallinae* can occur as nonvirulent or mildly virulent strains, the finding of trichomonads in the upper gastrointestinal tract of a columbid must be interpreted with caution. Stabler (1948, 1951) provided data showing that immunity develops after a pigeon is infected with a nonvirulent strain or a pathogenic strain (and survives). He found that subsequent exposure of such birds to a pathogenic strain of *T. gallinae* did not result in trichomonosis, but that the bird could then become a carrier. In the case of a bird that has a nonvirulent strain and is subsequently infected with a pathogenic strain, that bird becomes a carrier for both strains and can transmit 1 strain on 1 occasion and the other strain on another occasion.

Trichomonosis can result in epizootics of

Table 22.5. *Trichomonas gallinae* infections in doves and pigeons from Florida

Species of columbid County or site	Year(s)	No. of birds			Basis of diagnosis[a]	Data source
		Exam.	Inf.	%		
Rock Dove						
Alachua	1976	13	9	69	CU	Shamis (1977)
Alachua	1978	19	13	68	CU	Conti & Forrester (1989)
Alachua	1981	8	8	—	CU	Ibid.
White-crowned Pigeon						
Monroe	1969	12	12	100	OS	Kocan & Sprunt (1971)
Monroe	1970	41	36	88	CU	Ibid.
Eurasian Collared-Dove						
Dade	1993	1	1	—	MO	Spalding & Forrester (1993)
Pinellas	2001	5	5	—	OB,MO	Spalding & Forrester (2001)
White-winged Dove						
Broward & Dade	1982–83	25	25	100	OS,CU	Conti et al. (1985)
Dade	1977–78	67	65	97	OS,CU	Conti (1980), Conti & Forrester (1981)
Mourning Dove						
Alachua	1976	40	0	0	OS,CU	Shamis (1977)
Alachua	1978	89	1	1	OS,CU	Conti (1980), Conti & Forrester (1981)
Charlotte	1974	10	0	0	OS	Barrows (1975), Barrows & Hayes (1977)
Dade	1978	53	9	17	OS,CU	Conti (1980), Conti & Forrester (1981)
Palm Beach	1989	1	1	—	MO	Hayes (1989)
Pinellas	1951	NG	"several"	—	OB	Winston (1954)
Polk	1951	NG	1	—	OS	Herman (1951)
Common Ground-Dove						
"Florida"	NG	4	3	75	OS	Stabler & Holt (1962)

NG = not given by authors.

a. OS = oral swab, OB = direct observation for lesions, MO = microscopic observation of lesions, CU = culture on special media.

considerable magnitude. A review of the occurrence of this disease in Mourning Doves and other birds was published by Stabler and Herman (1951), who indicated that it was distributed throughout the United States. Probably the most severe die-off in recent times was one that occurred during 1950–51 in the southeastern United States. Mortality was reported in Kentucky, North Carolina, and Alabama (Stabler 1954), with the greatest numbers of dead birds reported in Alabama (Haugen 1952; Haugen and Keeler 1952), where mortality was estimated at 50,000–100,000 doves (Haugen

and Keeler 1952) and occurred in 43 of the state's 67 counties (Haugen 1952). Florida's dove populations apparently were not affected seriously by this epizootic, although several cases of trichomonosis were seen in Pinellas and Palm Beach counties during the summer of 1951 (Winston 1954). Since that time only a few scattered cases have been recorded in Mourning Doves in Florida; these were in Alachua County and included 1 dove in 1978, 1 in 1979, and 1 in 1980 (Conti 1980; Conti and Forrester 1981, 1989).

During May, June, and July of 2001, an epi-

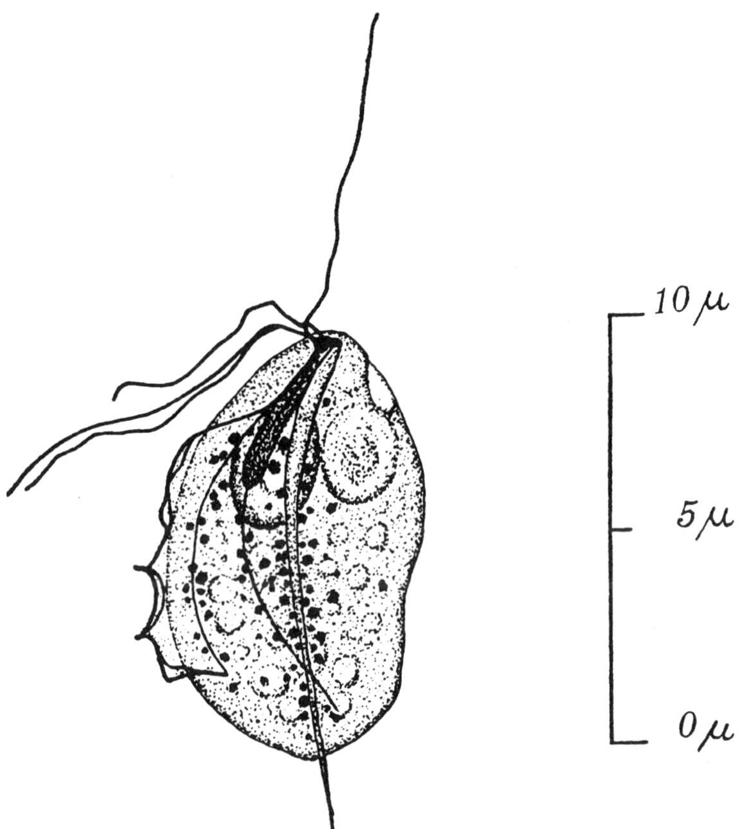

FIGURE 22.4. Drawing of *Trichomonas gallinae*, the etiologic agent of trichomonosis or canker in doves and pigeons. From Stabler 1947; by permission of *Journal of Parasitology.*

zootic caused by *Trichomonas gallinae* and pigeon paramyxovirus type 1 occurred among Eurasian Collared-Doves in urban areas of St. Petersburg (Pinellas County) (Spalding et al. 2002). It was estimated that at least 1,000 doves died. Trichomonad lesions were primarily in the liver, whereas typical cankers in the mouth, throat, and crop were rare. This may be the largest epizootic of trichomonosis on record for Florida, but this outbreak was complicated by concurrent infections of paramyxovirus. As noted above, Winston (1954) reported several cases of trichomonosis in Mourning Doves in Pinellas County in 1951. Since Rock Doves are the primary hosts of *T. gallinae*, it is probable that they were carriers of 1 or more virulent strains in Pinellas County and these were passed on to the Eurasian Collared-Doves via contaminated food or water.

Infections of *T. gallinae* in White-winged Doves in southern Florida have received special attention because of the possibility of the spread of virulent strains from this introduced columbid to the indigenous Mourning Dove (Conti 1980; Conti and Forrester 1981; Conti et al. 1985). In 1959 breeding populations of White-winged Doves were discovered in the citrus-growing areas of southern Florida (Saunders 1980). Conti and Forrester (1981) examined a series of White-winged Doves and Mourning Doves from Dade County (near Homestead) in 1977 and 1978. They deter-

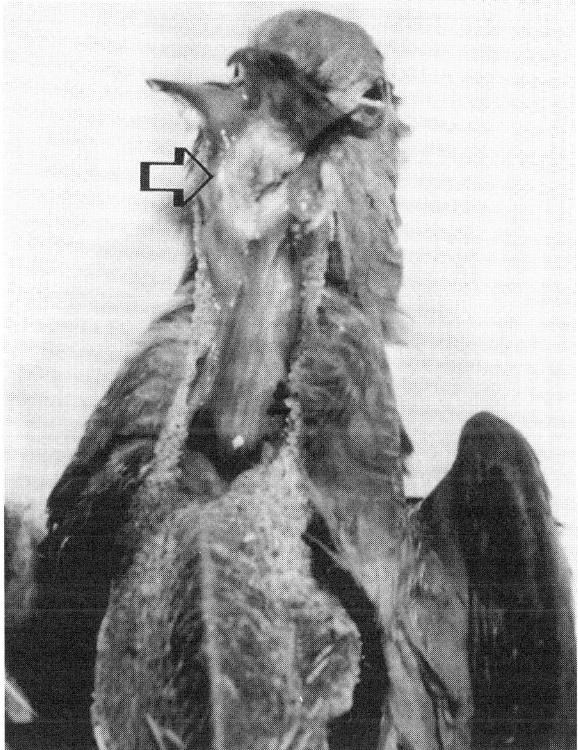

FIGURE 22.5. Oral cavity lesion or canker *(arrow)* in a Mourning Dove infected with *Trichomonas gallinae.* Courtesy of Ellis C. Greiner.

mined that there was a high prevalence (97%) of *T. gallinae* in White-winged Doves in contrast to a low prevalence (17%) in Mourning Doves. At the same time a study in Alachua County, where White-winged Doves do not occur, was conducted and the prevalence of *T. gallinae* in Mourning Doves there was only 1%. It was speculated that White-winged Doves might be carriers of strains of *T. gallinae* and through contact during feeding and drinking in mixed flocks might spread virulent trichomonads to susceptible Mourning Doves. Therefore, isolates of *T. gallinae* from 25 White-winged Doves trapped in Broward and Dade counties were transferred experimentally to young Mourning Doves and Rock Doves (Conti et al. 1985). Twenty-three of 25 Mourning Doves and 8 of 8 Rock Doves developed infections with *T. gallinae,* but none developed

trichomonosis. It was concluded that the strains studied were avirulent and that expanding populations of White-winged Doves in southern Florida were probably not a threat to Mourning Dove populations because of their particular strains of *T. gallinae.* Conti et al. (1985), however, cautioned that only 25 strains from 2 populations were studied, out of many that exist in Florida, and there still could be virulent strains present. They urged that future management strategies consider the possibility of outbreaks of trichomonosis involving White-winged Doves and Mourning Doves. Since avirulent strains have been changed into virulent isolates by genetic manipulation in the laboratory (Stabler et al. 1964; Honigberg et al. 1971), similar transformations could occur in nature (Conti et al. 1985).

Trichomonosis is a serious disease and should

be regarded as a potential threat to populations of pigeons and doves in Florida, as McLaurin (1955) pointed out a number of years ago. Several authors have postulated that there might have been a relationship between the demise of the Passenger Pigeon and trichomonosis (Stabler and Herman 1951; Haugen 1952; Stabler 1954). Of course, this can never be proven, but the epizootic of 1950–51 in Alabama occurred in an area where immense flocks of Passenger Pigeons were known to roost historically (Haugen 1952). In the fall of 1881 it was reported that millions of Passenger Pigeons blackened the skies in Alabama. The following year and subsequently thereafter no Passenger Pigeons were reported in that area (Haugen 1952). If the Passenger Pigeon did succumb to trichomonosis, the source of the trichomonads could have been Rock Doves, which were introduced into North America by the colonists in the early 1600s (Stabler 1954).

XIII. Other protozoans

In addition to *Trichomonas gallinae,* 7 other species of protozoans have been reported from columbids in Florida (tables 22.6 and 22.7). These include 1 ciliate, 3 coccidia, and 3 hematozoans.

The ciliate *Colpoda steinii* is probably not a parasite, but rather a commensal, and therefore of little importance to doves and pigeons. Kocan (1968) followed up on the earlier work of Toepfer (1964) and showed that the birds most likely obtained the cysts of these free-living ciliates from contaminated grain. The ciliates did not last longer than 1 day in the oral cavities of doves and pigeons.

Intestinal coccidia of the genus *Eimeria* have been reported from Mourning Doves, White-winged Doves, and Common Ground-Doves in Florida (table 22.6). Although the coccidium in Common Ground-Doves was identified as *E. labbeana,* the species in Mourning Doves and White-winged Doves was not determined and may be undescribed (Conti 1980). Nothing is

known about the pathologic significance of these coccidia to pigeon and dove populations in Florida. Barrows and Hayes (1977) reported a thickened duodenal mucosa in each of 2 Mourning Doves from the southeast (specific locality not given) that had numerous tissue stages of a species of *Eimeria.* They felt that intestinal coccidia were of only incidental importance to Mourning Doves in the southeast.

A species of *Sarcocystis* was found in low prevalence in pectoral muscles of Mourning Doves and White-winged Doves (table 22.6). The species involved was not determined. Doves are the intermediate hosts since sarcocysts were found in them; a carnivore is probably the definitive host, but this has not been determined. The pathologic significance of this parasite is unknown.

Three species of blood protozoans were found in columbids in Florida (table 22.7). Bishop and Bennett (1992) listed a fourth blood protozoan (*Haemoproteus orizivorae*) from a Mourning Dove from Florida; Bennett (1994), however, indicated that this was an error. One of the 3 species, *Leucocytozoon marchouxi,* was found in only 1 Mourning Dove from Orange County (Shamis and Forrester 1977). This infected dove was probably a migratory dove overwintering in Florida.

Two species of *Haemoproteus* were found, the most prevalent being *H. columbae. Haemoproteus sacharovi* was found in low prevalences and only in Mourning Doves, whereas *H. columbae* was common in Rock Doves, Common Ground-Doves, Mourning Doves, and White-winged Doves. In much of the literature on Mourning Doves the latter species is referred to as *H. maccallumi,* but recent studies by Bennett and Peirce (1990) have shown that *H. maccallumi* is a synonym of *H. columbae.* Shamis and Forrester (1977) reported that Mourning Doves from southern Florida had higher prevalences and parasitemias of *H. columbae* than Mourning Doves from northern Florida. The life cycles and transmission ecology of these 2 species of *Haemoproteus* have not been studied in Florida. The vectors may be hippoboscids or

Table 22.6. Ciliates and coccidia infecting doves from Florida

Species of columbid / Protozoan	County/ site	Year(s)	No. of birds — Exam.	Inf.	%	Basis of diagnosis[a]	Data source
Rock Dove							
Coccidia							
Toxoplasma gondii	"Florida"	NG	112	0	0	SE[b]	Burridge et al. (1979)
Mourning Dove							
Ciliates							
Colpoda steinii	Pinellas	1963	29	18	62	OS	Toepfer (1964)
Coccidia							
Eimeria sp.	Alachua	1978	45	15	33	FL	Conti (1980), Conti & Forrester (1981)
	Charlotte	1974	10	0	0	DS	Barrows (1975), Barrows & Hayes (1977)
	Dade	1978	53	52	98	FL	Conti (1980), Conti & Forrester (1981)
	Suwannee	1974	8	0	—	BS	Barrows (1975), Barrows & Hayes (1977)
Sarcocystis sp.	Alachua	1978	45	4	9	DI	Conti (1980), Conti & Forrester (1981)
	Charlotte	1974	10	0	0	HP	Barrows (1975), Barrows & Hayes (1977)
	Dade	1978	44	3	7	DI	Conti (1980), Conti & Forrester (1981)
	Suwannee	1974	8	0	—	HP	Barrows (1975), Barrows & Hayes (1977)
Toxoplasma gondii	"Florida"	NG	10	0	0	SE[b]	Burridge et al. (1979)
White-winged Dove							
Coccidia							
Eimeria sp.	Dade	1977–78	67	4	6	FL	Conti (1980), Conti & Forrester (1981)
Sarcocystis sp.	Dade	1977–78	67	7	10	DI	Conti (1980), Conti & Forrester (1981)
Common Ground-Dove							
Coccidia							
Eimeria labbeana	"Florida"	NG	4	2	—	NG	Stabler & Holt (1962)
Toxoplasma gondii	"Florida"	NG	13	0	0	SE[b]	Burridge et al. (1979)

NG = not given by authors.

a. OS = oral swab, DS = direct smear of intestinal contents, FL = flotation of intestinal contents, HP = histopathology, DI = artificial digestion of pectoral muscle, SE = serology (indirect hemagglutination).

b. Indirect hemagglutination test may not be as sensitive in birds as in mammals; may result in false-negative results (Frenkel 1981).

Table 22.7. Blood protozoan infections in doves from Florida

| Species of columbid | | | No. of birds | | | Basis of | |
Protozoan	County/site	Year(s)	Exam.	Inf.	%	diagnois[a]	Data source
Rock Dove							
Haemoproteus columbae							
	Alachua	1971–78	108	102	94	BS	Conti & Forrester (1989)
White-crowned Pigeon							
Haemoproteus columbae							
	Monroe	1989	4	1	—	BS	Spalding & Foster (1989)
White-winged Dove							
Haemoproteus columbae							
	Dade	1977–78	127	117[b]	92	BS	Conti (1980), Conti & Forrester (1981)
Haemoproteus sacharovi							
	Dade	1977–78	127	0	0	BS	Ibid.
Leucocytozoon marchouxi							
	Dade	1977–78	127	0	0	BS	Ibid.
Plasmodium sp.							
	Dade	1977–78	127	0	0	SI[d]	Ibid.
	Dade	1977–78	18[c]	0	0	SI[d]	Ibid.
Trypanosoma sp.							
	Dade	1977–78	182	0	0	BS,BM	Ibid.
Mourning Dove							
Haemoproteus columbae							
	Alachua	1973–76	159	134[b]	84	BS	Shamis & Forrester (1977)
	Alachua	1978	88	23[b]	26	BS	Conti (1980), Conti & Forrester (1981)
	Charlotte	1974	10	10[b]	100	BS	Barrows (1975), Barrows & Hayes (1977)
	Dade	1977–78	53	52[b]	98	BS	Conti (1980), Conti & Forrester (1981)
	Gadsden	1973–76	19	2[b]	11	BS	Shamis & Forrester (1977)
	Glades	1973–76	23	22[b]	96	BS	Ibid.
	Highlands	1973–76	15	14[b]	93	BS	Ibid.
	Leon	1973–76	19	5[b]	26	BS	Ibid.
	Orange	1973–76	683	669[b]	98	BS	Ibid.
	"Florida"	1964	10	10[e]	100	BS	Knisley & Herman (1967)
Haemoproteus sacharovi							
	Alachua	1973–76	159	10	6	BS	Shamis & Forrester (1977)
	Alachua	1978	88	1	1	BS	Conti (1980), Conti & Forrester (1981)
	Charlotte	1974	1	0	0	BS	Barrows (1975), Barrows & Hayes (1977)
	Dade	1978	53	0		BS	Conti (1980),
	Gadsden	1973–76	19	0	0	BS	Shamis & Forrester (1977)
	Glades	1973–76	23	7	35	BS	Ibid
	Highlands	1973–76	15	1	8	BS	Ibid.
	Leon	1973–76	19	0	0	BS	Ibid.
	Orange	1973–76	683	7	1	BS	Ibid.
	"Florida"	1964	10	0	0	BS	Knisley & Herman (1967)

(continued)

Table 22.7. *(continued)*

Species of columbid Protozoan	County/site	Year(s)	No. of birds Exam.	Inf.	%	Basis of diagnois[a]	Data source
Leucocytozoon marchouxi							
	Orange	1973–76	683	1	<1	BS	Shamis & Forrester (1977)
	Various[f]	1973–78	396[f]	0	0	BS	Bar rows (1975), Shamis &Forrester (1977), Conti (1980), Conti & Forrester (1981)
Plasmodium sp.							
	Alachua	1976	40[g]	0	0	SI[h]	Shamis (1977), Shamis & Forrester (1977)
	Alachua	1978	15[g]	0	0	SI[d]	Conti (1980), Conti & Forrester (1981)
	Dade	1977–78	13[g]	0	0	SI[d]	Ibid.
	Various[f,i]	1964–78	1079[f,i]	0	0	BS	Knisley & Herman (1967), Barrows (1975), Shamis & Forrester (1977), Conti (1980), Conti & Forrester (1981)
Trypanosoma sp.							
	Alachua	1978	92[g]	0	0	BM	Conti (1980), Conti & Forrester (1981)
	Dade	1978	53[g]	0	0	BM	Ibid.
	Various[f,i]	1964–78	1079[f,i]	0	0	BS	Knisley & Herman (1967), Barrows (1975), Shamis (1977), Shamis & Forrester (1977), Conti (1980), Conti & Forrester (1981)
Common Ground-Dove							
Haemoproteus columbae							
	Alachua	1974–75	4	2	—	BS	Forrester (1989)
	Dade	1978	1	1	—	BS	Ibid.
	Glades	1975	22	18	82	BS	Ibid.
	Osceola	1993	1	1	—	BS	Spalding et al. (1993)
	"Florida"	NG	4	4	—	NG	Stabler & Holt (1962)

NG = not given.

a. BS = blood smear, SI = subinoculation of blood, BM = bone marrow squash.

b. Referred to as *H. maccallumi* in the original literature, but studies by Bennett and Peirce (1990) have shown that *H. maccullumi* is a synonym of *H. columbae*.

c. Examined by blood smears also and included in the 127 listed above.

d. Whole blood was inoculated into Rock Doves for isodiagnostic purposes (Herman et al. 1966).

e. The authors stated that this was "*Haemoproteus* of the *columbae* type," the parasites probably were *H. columbae*.

f. Includes the following counties (number of birds examined in parentheses): Charlotte (10), Gadsden (19), Leon (19), Alachua (247), Highlands (15), Glades (23), Dade (53), unknown county (10).

g. Examined by blood smears also and included among the 1079 listed in this table.

h. Whole blood was inoculated into Pekin ducklings for isodiagnostic purposes (Herman et al. 1966).

i. Plus Orange County (*n* = 683).

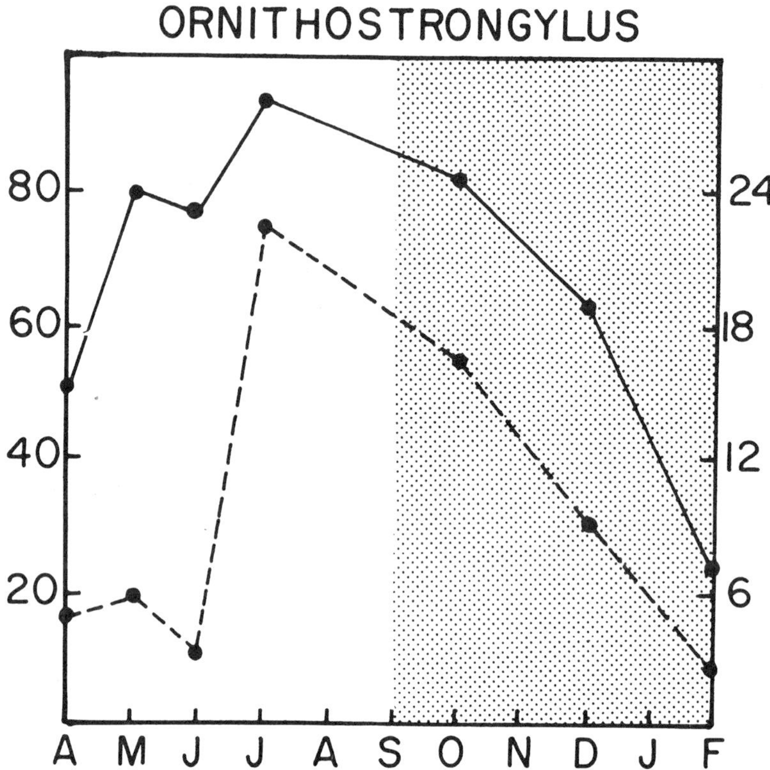

FIGURE 22.6. Seasonal variations in prevalence *(solid line, left Y axis)* and intensity *(broken line, right Y axis)* of *Ornithostrongylus* spp. in Mourning Doves from Orange County, April 1974–February 1975. Stippled section indicates time of year when migratory doves and doves hatched in Florida are present; clear section indicates time that only doves hatched in Florida are present. Adapted from Forrester et al. 1983; by permission of *Proceedings of the Helminthological Society of Washington.*

ceratopogonid midges such as *Culicoides*. Two of the 3 species of hippoboscids (*Pseudolynchia canariensis* and *P. brunnea*) reported from columbids in Florida (table 22.14) have been shown capable of transmitting *H. columbae* (see Garnham [1966] for a review of this topic). These may serve as vectors in Florida, but that has not been determined. Several authors have presented evidence that *H. columbae* is pathogenic (Kadner 1941; Coatney and Hickman 1952; Garnham 1966; Markus and Oosthuizen 1971; Earle et al. 1993), but critical experimental and field studies have not been conducted.

At the present time, the significance of these hemoproteids to doves and pigeons in Florida remains unknown.

XIV. Helminths

Sixteen species of helminths have been reported from columbids in Florida. These include 7 nematodes, 2 trematodes, and 7 cestodes (tables 22.8–22.10). Although there is information on helminths from 5 species of columbids (i.e. Mourning Doves, White-winged Doves, Common Ground-Doves, Rock Doves, and Eurasian Collared-Doves), the most complete data are from Mourning Doves (Conti and Forrester 1981; Forrester et al. 1983) and White-winged Doves (Conti 1980; Conti and Forrester 1981).

From October 1973 to August 1978, 455 Mourning Doves were collected from 8 counties in Florida and examined for helminths (Forrester et al. 1983). Fourteen species were identified and included 7 nematodes, 2 trematodes, and 5 cestodes. The number of species per host varied from 1 to 5 and the total number of helminths per infected bird ranged from 1 to 163 (mean = 14). Two of the nematodes (*Ornithostrongylus quadriradiatus* and *Ascaridia columbae*) are core species and were the most prevalent. Of the 5,518 specimens obtained from all doves, 99.6% were nematodes, with 72% being *O. quadriradiatus* and *O. iheringi*. There were no effects of geographic location, age, or gender of the doves on prevalences of helminths, but there were differences in intensity. The doves from southern Florida (Dade and Glades counties) had higher intensities (all species of helminths combined) than doves from areas further north. *Ascaridia columbae* was more abundant in adult doves than in juveniles. There were differences in prevalence and intensity of *Ornithostrongylus* spp. according to season in doves from Orange County that were sampled on a year-round basis (figure 22.6). The peaks in both prevalence and intensity occurred during the spring and summer months (the rainy season), when the Mourning Dove populations were made up predominantly of birds that were hatched in Florida (Marion et al. 1981; Forrester et al. 1983).

The helminth fauna was fairly diverse on a statewide basis (Simpson's Index = 0.34), but on a county basis doves from the southern part of the state had less diverse faunas than those from the northern part (Forrester et al. 1983). When indexes of similarity were calculated for doves from each county, it was shown that the faunas of most doves in the state were similar, with the least similarity oc-

curring between doves from Alachua and Dade counties (figure 22.7).

From October 1977 through October 1978 119 White-winged Doves and 47 Mourning Doves from Dade County were examined for helminths (Conti 1980; Conti and Forrester 1981). These were studied to determine the interrelationships of the helminths of the indigenous Mourning Dove with those of the introduced White-winged Dove. A sample of 49 Mourning Doves from Alachua County, where no White-winged Doves occur, was also studied for comparative purposes. Most doves were infected with 1 or 2 species of helminths and only a small number were free of helminths (table 22.11). White-winged Doves and Mourning Doves from Dade County had higher intensities than Mourning Doves from Alachua County. Simpson's Index of Diversity was higher for doves (both species) from Dade County than for those from Alachua County. This indicated that there was some concentration of dominance in Mourning Doves from Alachua County. Nematodes were by far the most abundant type of helminth and constituted more than 90% of the helminths in each population.

In figure 22.8 the relative abundance of each helminth is shown diagrammatically for each of the 3 populations of doves. In this figure each species of helminth is represented by its percentage of the total number of helminths collected. This presentation shows again that nematodes make up a major component of the fauna and also that *Ornithostrongylus* spp. are very significant quantitatively in Mourning Doves in Dade County. White-winged Doves in Dade County were found to have 2 nematodes of importance, *Ornithostrongylus* spp. and *Ascaridia columbae*. In Alachua County, however, the situation was different, with 4 genera of nematodes being of quantitative significance. In addition it can be seen that the *Ornithostrongylus* species are less numerous than in Dade County birds and *Dispharynx nasuta* is of more significance.

Little is known about the pathogenic effects of helminths on columbids in Florida. How-

Table 22.8. Nematode infections in columbids from Florida

Host species Nematoda	Site in host[a]	County/ site	Year(s)	No. of birds		Intensity		Range	Data source
				Exam.	Inf.	%	Mean		
Rock Dove									
Tetrameres columbicola									
	PR	Alachua	NG	6	NG	—	NG	NG	Simpson et al. (1984)
Eurasian Collared-Dove									
Ascaridia columbae									
	SI	Palm Beach	1998	5	3	—	61	27–106	Kinsella et al. (1998)
Ornithostrongylus sp.									
	SI	Palm Beach	1998	5	2	—	1	1	Ibid.
White-winged Dove									
Aproctella stoddardi									
	BC	Dade	1977–78	119	0	0	0	—	Conti & Forrester (1981)
Ascaridia columbae									
	SI	Dade	1977–78	119	52	44	9	1–47	Ibid.
Capillaria obsignata									
	SI	Dade	1977–78	119	10	1	2	2	Ibid.
Dispharynx nasuta									
	PR	Dade	1977–78	119	5	4	1	1–3	Ibid.
Ornithostrongylus spp.[b]									
	SI	Dade	1977–78	119	94	79	10	1–104	Conti & Forrester (1981, 1989)
Tetrameres columbicola									
	PR	Dade	1977–78	119	1	<1	2	—	Conti & Forrester (1981)
Mourning Dove									
Aproctella stoddardi									
	BC	Alachua	1973–78	52	10	19	3	1–9	Conti & Forrester (1981, 1989)
	BC	Charlotte	1974	10	0	0	0	—	Barrows (1975)
	BC	Dade	1973–78	47	0	0	0	—	Conti & Forrester (1981, 1989)
	BC	Gadsden	1973–78	15	5	33	2	1–5	Forrester et al. (1983), Forrester (1989)
	BC	Glades	1973–78	30	0	0	0	—	Ibid.
	BC	Highlands	1973–78	15	2	13	2	1–2	Ibid.
	BC	Leon	1973–78	15	2	13	2	1–3	Ibid.
	BC	Orange	1973–78	191	15	8	2	1–6	Ibid.
	BC	Polk	1973–78	90	13	14	5	1–34	Ibid.
	BC	Suwannee	1974	8	6	—	17	1–62	Barrows (1975)

Ascaridia columbae								
	SI	Alachua	1973–78	52	19	37	4	1–14	Conti & Forrester (1981, 1989)
	SI	Charlotte	1974	10	2	20	8	4–12	Barrows (1975)
	SI	Dade	1973–78	47	5	11	7	1–16	Conti & Forrester (1981, 1989)
	SI	Gadsden	1973–78	15	0	0	0	—	Forrester et al. (1983), Forrester (1989)
	SI	Glades	1973–78	30	9	30	2	1–5	Ibid.
	SI	Highlands	1973–78	15	6	40	2	1–3	Ibi
	SI	Leon	1973–78	15	1	7	1	—	Ibid.
	SI	Orange	1973–78	191	78	41	4	1–34	Ibid.
	SI	Polk	1973–78	90	21	23	1	1–3	Ibid
	SI	Suwannee	1974	8	0	—	0	—	Barrows (1975)
Capillaria obsignata								
	SI	Charlotte	1974	10	0	0	0	—	Barrows (1975)
	SI	Orange	1973–78	191	4	2	6	1–8	Forrester et al. (1983), Forrester (1989)
	SI	Suwannee	1974	8	0	—	0	—	Barrows (1975)
	SI	Various[c]	1973–78	264[c]	0	0	0	—	Forrester et al. (1983), Forrester (1989), Conti & Forrester (1981)
	SI	"Florida"	NG	NG	1	—	NG	—	Stiles & Hassall (1894)
Dispharynx nasuta								
	PR	Alachua	1973–78	52	18	35	6	1–25	Conti & Forrester (1981, 1989)
	PR	Alachua Glades	1984	7	0	—	0	—	Rickard (1985)
	PR	Charlotte	1974	10	0	—	0	—	Barrows (1975)
	PR	Dade	1973–78	47	2	4	8	1–15	Conti & Forrester (1981, 1989)
	PR	Gadsden	1973–78	15	1	7	4	—	Forrester et al. (1983), Forrester (1989)
		Glades	1973–78	30	0	0	0	—	Ibid.
		Glades	1975	22	1	<1	12	—	Forrester (1989)
	PR	Highlands	1973–78	15	0	0	0	—	Forrester et al. (1983), Forrester (1989)
	PR	Leon	1973–78	15	2	13	2	1–2	Ibid.
	PR	Orange	1973–78	191	44	23	14	1–144	Ibid.
Ornithostrongylus quadriradiatus								
	SI	Charlotte	1974	10	3	30	2	1–2	Barrows (1975)
	SI	Suwannee	1974	8	5	—	3	—	Ibid.

(continued)

Table 22.8. *(continued)*

Host species Nematoda	Site in host[a]	County/ site	Year(s)	No. of birds		Intensity		Range	Data source
				Exam.	Inf.	%	Mean		
Ornithostrongylus spp.[b]									
	SI	Alachua	1973–78	52	16	31	2	1–9	Conti & Forrester (1981, 1989)
	SI	Dade	1973–78	47	46	98	19	1–71	Ibid.
	SI	Gadsden	1973–78	15	6	40	7	1–23	Forreste r et al. (1983), Forrester (1989)
	SI	Glades	1973–78	30	27	90	30	2–160	Ibid.
	SI	Highlands	1973–78	15	11	73	11	1–69	Ibid .
	SI	Leon	1973–78	15	15	100	18	2–65	Ibid.
	SI	Orange	1973–78	191	122	64	11	1–133	Ibid .
	SI	Polk	1973–78	90	63	70	8	1–52	Ibid.
Tetrameres columbicola									
	PR	Alachua	1973–78	52	1	2	1	—	Conti Forrester (1981, 1989)
	PR	Charlotte	1974	10	0	0	0	—	Barrows (1975)
	PR	Dade	1973–78	47	1	2	1	—	Conti & Forrester (1981, 1989)
	PR	Orange	1973–78	191	17	9	2	1–9	Forrester et al. (1983), Forrester (1989)
	PR	Suwannee	1974	8	0	—	0	—	Barrows (1975)
	PR	Various[d]	1973–78	165[d]	0	0	0	—	Forrester et al. (1983), Forrester (1989)
Common Ground-Dove *Aproctella stoddardi*									
	BC	"Florida"	1885	NG	1[e]	—	N G	—	Walton (1927)
Dispharynx nasuta									
	PR	Glades	1975	21	0	0	0	—	Forrester (1989)
Ornithostrongylus quadriradiatus									
	FE	"Florida"	NG	4[f]	3	—	N G	NG	Stabler & Holt (1962)

NG = not given.

a. Infection sites: BC = body cavity, SI = small intestine, PR = proventriculus, FE = feces.

b. A complex of 2 species, *O. quadriradiatus* and *O. iheringi* in a ratio of 14:1, based on males only.

c. Includes doves from the following counties: Gadsden ($n = 15$), Leon ($n = 15$), Alachua ($n = 52$), Polk ($n = 90$), Highlands ($n = 15$), Glades ($n = 30$), Dade ($n = 47$).

d. Includes doves from the following counties: Gadsden ($n = 15$), Leon ($n = 15$), Polk ($n = 90$), Highlands ($n = 15$), Glades ($n = 30$).

e. Walton (1927) reported a "Filaria" from a cyst in the hepatic region of one Common Ground-Dove. This may have been *Aproctella stoddardi*, but not confirmed.

f. Diagnosis was by identification of eggs in feces, not by examination of adult worms from the small intestines.

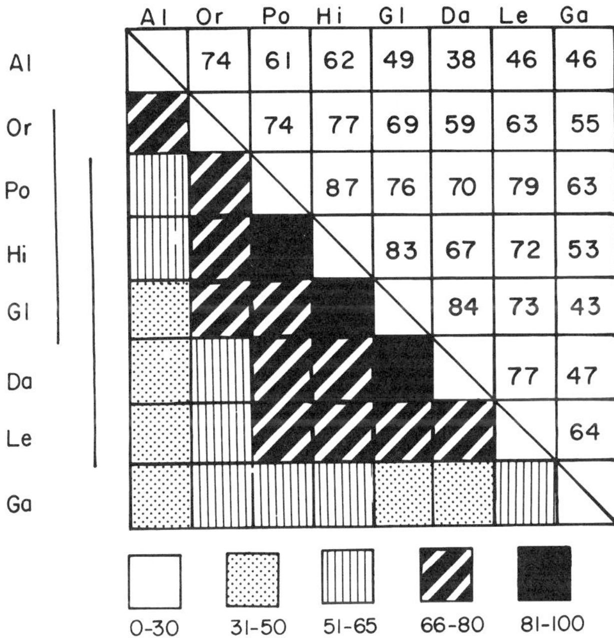

FIGURE 22.7. Trellis diagram of indexes of similarity for the helminth faunas of Mourning Doves from 8 counties in Florida. Al = Alachua County, Or = Orange County, Po = Polk County, Hi = Highlands County, Gl = Glades County, Da = Dade County, Le = Leon County, Ga = Gadsden County. Vertical lines indicate clusters of closely related indexes. From Forrester et al. 1983; by permission of *Proceedings of the Helminthological Society of Washington*.

ever, it is possible to gain some insight into probable effects from studies and observations made elsewhere on several of the nematodes. Large numbers of *O. quadriradiatus* have caused anemia and catarrhal enteritis in Rock Doves (Cuvillier 1937). In addition, *Ascaridia columbae* has been associated with perforation of the intestine, peritonitis, and death in Rock Doves (Mozgovoi 1953) and distention of the duodenum and localized hyperemia of the mucosa in Mourning Doves (Barrows and Hayes 1977). In some cases *A. columbae* can cause remarkable distention (figures 22.9 and 22.10) and obstruction of the intestine that can lead to emaciation and death (Wehr 1971). *Dispharynx nasuta* has been found associated with proventriculitis and death in Ruffed Grouse (*Bonasa umbellus*), Blue Grouse (*Dendragapus obscurus*), Wild Turkeys, and Rock Doves (Goble and Kutz 1945; Bendell 1955; Hwang et al. 1961; Hon et al. 1975). Conti (1980) reported excessive amounts of mucous secretions around clusters of *D. nasuta* in Mourning Doves in Florida (figure 22.11). Hwang et al. (1961) reported similar reactions in Rock Doves. *Aproctella stoddardi* was reported by Barrows and Hayes (1977) to be associated with granulomatous pericarditis and adhesions of the liver and small intestine to the body wall of Mourning Doves. *Tetrameres columbicola* has been linked with proventriculitis, emaciation, and death in Rock Doves (Ewing et al. 1967; Flatt and Nelson 1969; Simpson et al. 1984). Although some of the nematodes can cause pathologic effects as mentioned above, the overall significance of those effects on populations of columbids is not known.

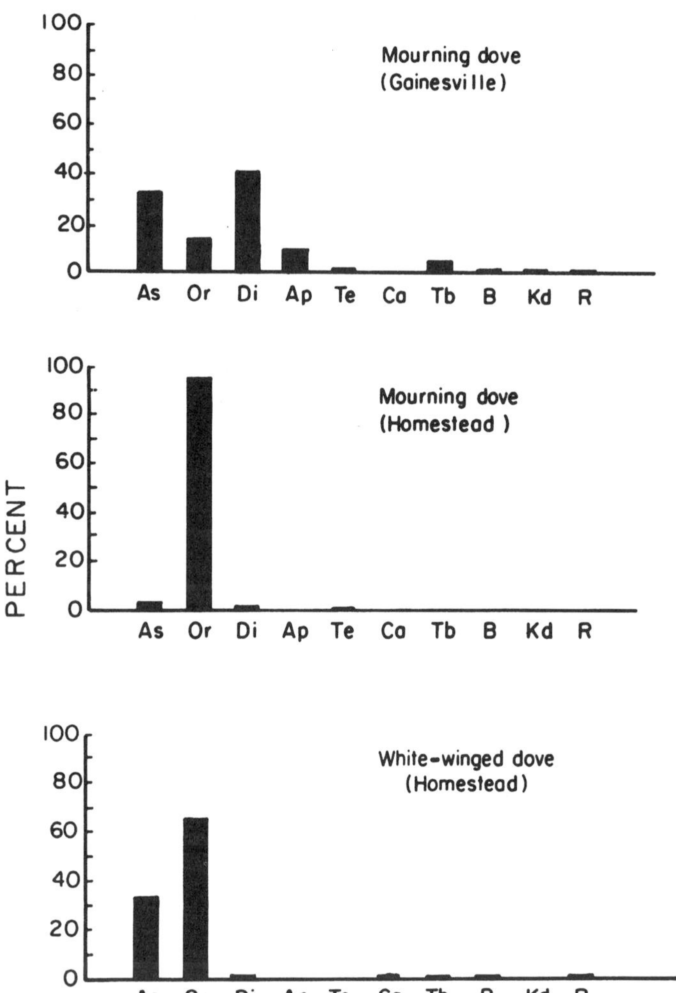

FIGURE 22.8. Helminth profiles for 3 populations of Mourning Doves in central Florida, 1977–78. As = *Ascaridia columbae*, Or = *Ornithostrongylus* spp. (a complex of 2 species: *O. quadriradiatus* and *O. iheringi* in a ratio of 14:1), Di = *Dispharynx nasuta*, Ap = *Aproctella stoddardi*, Te = *Tetrameres columbicola*, Ca = *Capillaria obsignata*, Tb = *Tanaisia bragai*, B = *Brachylaima* sp., Kd = *Killigrewia delafondi*, R = *Raillietina* spp. (a complex of at least 2 species). From Conti and Forrester 1981; by permission of *Journal of Wildlife Diseases*.

Seven White-crowned Pigeons (3 adults, 2 juveniles, and 2 nestlings) from Florida Bay (Monroe County) were examined for helminth parasites in 1988–89. All were negative (Spalding and Forrester 1990). Their very specialized diet of mainly fruits and berries (Owre 1978) may account for the lack of helminths.

XV. Arthropods

Twenty-seven species of parasitic arthropods have been found on columbids in Florida and include 1 tick, 9 mites, 13 chewing lice, and 4 louse flies (tables 22.12–22.14). The most complete data on intensity were from Mourning

FIGURE 22.9. The gizzard *(top)* and small intestine of a Mourning Dove from Florida. Note the distensions of the intestine *(arrows)*, which contain large numbers of the nematode *Ascaridia columbae*. Courtesy of Joseph A. Conti.

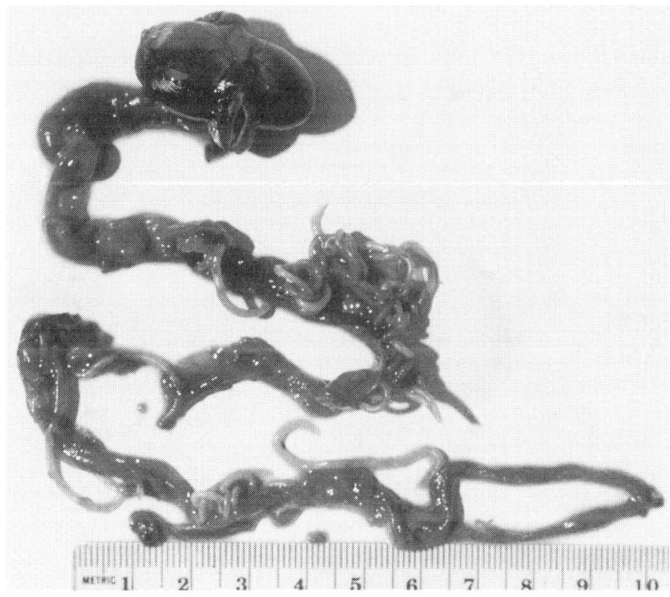

FIGURE 22.10. The same gizzard and small intestine of a Mourning Dove pictured in Figure 22.9, with the intestine opened up to show the mass of *Ascaridia columbae* inside. From Conti 1993; by permission of Wildlife Management Institute, Washington, D.C.

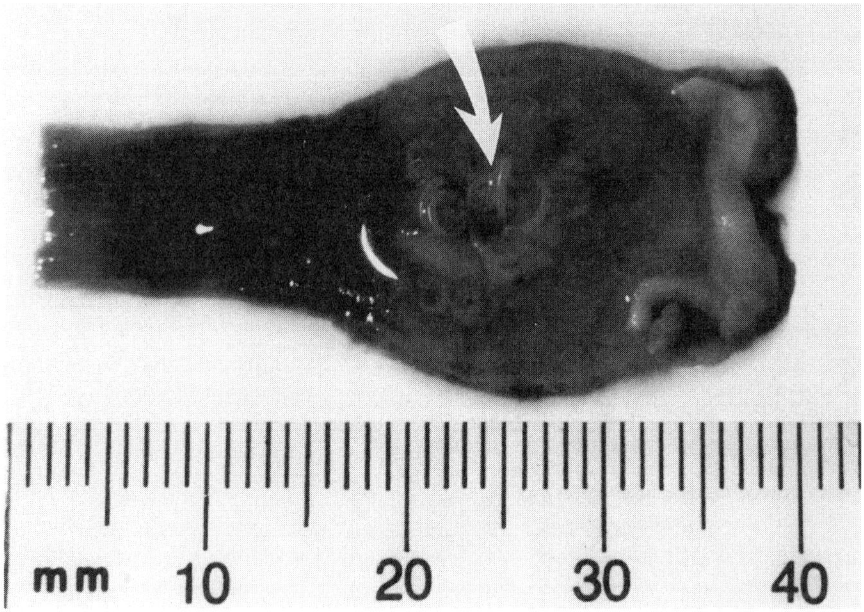

FIGURE 22.11. The opened proventriculus of a Mourning Dove from Florida, showing numerous coiled specimens of *Dispharynx nasuta* (*arrow*) with a mucous exudate. From Conti 1993; by permission of Wildlife Management Institute, Washington, D.C.

Doves and White-winged Doves. These were given by Conti (1980) and Conti and Forrester (1981), who found that the ectoparasites occurred in small numbers. They concluded that these parasitic arthropods are probably of little pathologic consequence. Two of the hippoboscids (*Pseudolynchia canariensis* and *P. brunnea*) may serve as vectors of haemoproteid blood parasites, as mentioned earlier.

An interesting notation about hippoboscids or "pigeon flies" appeared in an old issue of the *American Pigeon Journal,* concerning infestations that occurred in a racing pigeon loft in Jacksonville, Florida:

I have noticed some of these pests among my birds lately and believe they came from a pair of birds which I bought recently, at what I thought was a very low price, and which I now wish were back with their former owner. These are in reality blood suckers as I have caught several and they look more like a tick than anything else, with the exception that they have wings and can fly very fast. They are also very quick on their feet and will usually run in under the feathers of a bird rather than fly off of it. These parasites can be found on the squabs and I have noticed lately that quite a number of my squabs are small and runty and believe this is due to these flies. (Gilbert 1922)

XVI. Rock Doves as pests and public health hazards

Since its introduction into North America in the early 1600s (Stabler 1954) the Rock Dove or Domestic Pigeon has been recognized as a nuisance or pest in many areas, including Florida. Large populations can result in the defacement and destruction of property and the consumption of grain and other food intended for domestic animals or man. In addition they are known to play an important role in the spread of more than 40 diseases of veterinary

Table 22.9. Trematode infections in columbids from Florida

Host species Trematoda Site in host[a]	County/site	Year(s)	No. of birds			Intensity		Data source
			Examined	Infected	%	Mean	Range	
Rock Dove								
Austrobilharzia penneri[b]								
BV	Franklin	1959	NG	NG	—	NG	NG	Short & Holliman (1961)
White-winged Dove								
Tanaisia bragai								
KI	Dade	1977–78	119	1	<1	3	—	Conti & Forrester (1981)
Brachylaima sp.								
SI	Dade	1977–78	119	1	<1	2	—	Ibid.
Mourning Dove								
Tanaisia bragai								
KI	Alachua	1973–78	52	1	2	12	—	Conti & Forrester (1981, 1989)
KI	Charlotte	1974	10	0	0	0	—	Barrows (1975)
KI	Suwannee	1974	8	0	—	0	—	Ibid.
KI	Various[c]	1973–78	403[c]	0	0	0	—	Forrester et al. (1983), Forrester (1989), Conti & Forrester (1981)
Brachylaima sp.								
SI	Alachua	1973–78	52	1	2	1	—	Conti & Forrester (1981, 1989)
SI	Charlotte	1970	10	0	0	0	—	Barrows (1975)
SI	Suwannee	1970	8	0	—	0	—	Ibid.
SI	Various[c]	1973–78	403[c]	0	0	0	—	Conti & Forrester (1981), Forrester et al. (1983), Forrester (1989)

NG = not given.

a. Infection sites: BV = blood vessels, KI = kidney, SI = small intestine.

b. An experimental infection of some Rock Doves via cercariae obtained from marine snails (*Certhidea scalariformis*) collected at Shell Point, summer 1959. The adult trematodes obtained from this experimental feeding were described as a new species by Short and Holliman (1961). The natural definitive host or hosts are not known.

c. Includes doves from the following counties: Gadsden ($n = 15$), Leon ($n = 15$), Orange ($n = 191$), Polk ($n = 90$), Highlands ($n = 15$), Glades ($n = 30$), Dade ($n = 47$).

Table 22.10. Cestode infections in columbids from Florida

Host species Cestoda Site in host[a]	County/site	Year(s)	No. of birds			Intensity		Data source
			Examined	Infected	%	Mean	Range	
Eurasian Collared-Dove								
Fuhrmannetta crassula								
SI	Palm Beach	1998	5	2	—	7	1–14	Kinsella et al. (1998)
White-winged Dove								
Hymenolepididae[b]								
SI	Dade	1977–78	119	0	0	0	—	Conti & Forrester (1981)
Killigrewia delafondi								
SI	Dade	1977–78	119	0	0	0	—	Ibid.
Raillietina spp.[b]								
SI	Dade	1977–78	119	4	3	3	1–7	Ibid.
Mourning Dove								
Hymenolepis sp.								
SI	Charlotte	1974	10	1	10	1	—	Barrows (1975)
SI	Suwannee	1974	8	1	—	1	—	Ibid.
Hymenolepididae[b]								
SI	Gadsden	1973–78	15	2	13	1	—	Forrester et al. (1983), Forrester (1989)
SI	Leon	1973–78	15	1	7	1	—	Ibid.
SI	Polk	1973–78	90	3	3	1	—	Ibid.

Site[a]	Locality	Years	n					Reference
SI	Various[c]	1973–78	335[c]	0	0	0	—	Forrester et al. (1983), Forrester 1989), Conti & Forrester (1981)
Killigrewia delafondi								
SI	Alachua	1973–78	52	1	2	1	—	Conti & Forrester (1981, 1989)
SI	Charlotte	1974	10	0	0	0	—	Barrows (1975)
SI	Polk	1973–78	90	2	2	3	—	Forrester et al. (1983), Forrester (1989)
SI	Suwannee	1974	8	2	—	2	—	Barrows (1975)
SI	Various[d]	1973–78	313[d]	0	0	0	—	Forrester et al. (1983), Forrester (1989), Conti & Forrester (1981)
Raillietina spp.[b]								
SI	Alachua	1973–74	52	1	2	1	—	Conti & Forrester (1981, 1989)
SI	Charlotte	1974	10	0	0	0	—	Barrows (1975)
SI	Suwannee	1974	8	0	—	0	—	Ibid.
SI	Various[e]	1973–78	403[e]	0	0	0	—	Forrester et al. (1983) Forrester (1989), Conti & Forrester (1989)

a. Infection site: SI = small intestine.
b. A complex of at least 2 species; specific identification was not possible because of the poor condition of the specimens.
c. Includes doves from the following counties: Alachua ($n = 52$), Orange ($n = 191$), Highlands ($n = 15$), Glades ($n = 30$), Dade ($n = 47$).
d. Includes doves from the following counties: Gadsden ($n = 15$), Leon ($n = 15$), Orange ($n = 191$), Highlands ($n = 15$), Glades ($n = 30$), Dade ($n = 47$).
e. Includes doves from the following counties: Gadsden ($n = 15$), Leon ($n = 15$), Orange ($n = 191$), Polk ($n = 90$), Highlands ($n = 15$), Glades ($n = 30$), Dade ($n = 47$).

Table 22.11. Characteristics of the helminth faunas of White-winged Doves and Mourning Doves in Florida, 1977–78

	White-winged Doves (Dade County)	Mourning Doves (Dade County)	Mourning Doves (Alachua County)
Total number of doves examined	119	47	49
Number of doves free of helminths	8	0	9
Total number of helminths collected	1,450	936	262
Percent of total number of helminths composed of:			
Nematoda	99.0	100.0	94.3
Trematoda	0.3	0.0	5.0
Cestoda	0.7	0.0	0.8
Mean number of helminths per infected dove (range)	13.1 (1–105)	19.9 (1–87)	6.6 (1–27)
Mean number of species of helminths per infected dove (range)	1.4 (1–3)	1.1 (1–3)	1.6 (1–4)
Simpson's Index of Diversity[a]	0.46	0.74	0.23

Sources: Conti (1980), Conti and Forrester (1981).

a. Indicates whether a population has an equal distribution of helminth species (low values) or is dominated by a few species (high values closer to 1).

Table 22.12. Tick and mite infestations on doves and pigeons from Florida[a]

Species of columbid Arthropod	Year(s)	County	No. of birds Examined	No. of birds Infected	%
White-winged Dove					
Ticks					
Ixodidae (larvae)	1977–78	Dade	119	1	<1
Mites					
Diplaegidia sp.	1977–78	Dade	119	59	50
Falculifer sp.	1977–78	Dade	119	3	3
Dermoglyphus sp.	1977–78	Dade	119	4	3
Cheyletiellidae	1977–78	Dade	119	1	<1
Tinaminyssus triangulus	1977–78	Dade	119	1	<1
White-crowned Pigeon					
Mites					
Falculifer sp.	1988–89	Monroe[a]	7	1	—
Hypodectes propus	1988–89	Monroe[a]	7	1	—
Neottialges sp.	1988–89	Monroe[a]	7	1	—
Syringophilus columbae	1969	Monroe[b]	NG	1	—
Mourning Dove					
Ticks[c]					
Mites					
Diplaegidia sp.	1977–78	Alachua	49	27	55
	1977–78	Dade	47	15	32
Falculifer sp.	1977–78	Alachua	49	31	63
	1977–78	Dade	47	15	32
Dermoglyphus sp.	1977–78	Alachua	49	1	2
Tinaminyssus zenaidurae	1977–78	Alachua	49	5	10
	1977–78	Dade	47	1	2

Sources: White-winged Doves and Mourning Doves: Conti (1980), Conti and Forrester (1981); White-crowned Pigeons: Kocan (1969), Spalding et al. (1989), Pence et al. (1997).

NG = not given.

a. The positive White-crowned Pigeon was from Bottle Key.

b. The positive White-crowned Pigeon was from Key West.

c. Travis (1941) examined 2 Mourning Doves in 1936–37 from Orange, Osceola, and/or Collier counties. Both were negative for ticks.

Table 22.13. Chewing lice infestations on doves and pigeons from Florida

Species of columbid Species of louse	Year(s)	County/site	No. of birds			Data source[a]
			Examined	Infected	%	
Rock Dove						
Campanulotes compar	1972	Indian River	NG	NG	—	A
Columbicola columbae	1972	Indian River	NG	NG	—	A[c]
Scaly-naped Pigeon						
Columbicola waggermani	1930	Dade	NG	NG	—	A[c]
Hohorstiella sp.	1930	Dade	NG	NG	—	A
White-crowned Pigeon						
Columbicola waggermani	1988	Monroe	NG	NG	—	A[c]
Physconelloides sp.	1988	Monroe	NG	NG	—	A
White-winged Dove						
Bonomiella sp.	1977–78	Dade	119	51	43	B
Columbicola macrourae	1977–78	Dade	119	24	20	B
Hohorstiella spp.[b]	1977–78	Dade	119	12	10	B
Physconelloides wisemani	1977–78	Dade	119	6	5	B
Mourning Dove						
Bonomiella columbae	1977–78	Alachua	49	8	16	B
	1977–78	Dade	47	15	32	B
Columbicola macrourae	1977–78	Alachua	49	9	18	B[c]
	1977–78	Dade	47	25	53	B[c]
Hohorstiella spp.[b]	1977–78	Dade	47	1	2	
Physconelloides zenaidurae	1977–78	Alachua	49	3	6	B
	1977–78	Dade	47	15	32	B
	NG	"Florida"	NG	NG	—	C
Common Ground-Dove						
Columbicola passerinae	NG	"Florida"	NG	NG	—	D
Physconelloides passerinae	1939	Dade	NG	NG	—	A
	1936	Lee	NG	NG	—	A
	NG	"Florida"	NG	NG	—	C

NG = not given.
a. A = Forrester et al. (1995), B = Conti (1980) and Conti and Forrester (1981), C = Hill and Tuff (1978), D = Peters (1936).
b. A complex of 2 species, *H. paladinella* and an undescribed species.
c. Clayton and Price (1999) gave records of these lice from Florida on these hosts, but no dates or specific localities.

Table 22.14. Characteristics of louse fly (Hippoboscidae) infestations on columbids from Florida

Host species Louse fly	Year(s)	No. birds Examined	Infected	%	Data source
Rock Dove					
Microlynchia pusilla					
Indian River	1971	NG	NG	—	Wilson et al. (1990)
Pseudolynchia canariensis					
Alachua	NG	NG	NG	—	Bequaert (1955)
Alachua	1955	NG	NG	—	Wilson et al. (1990)
Alachua	1975–76	NG	3	—	Wilson & Forrester (1990)
Dade	NG	NG	NG	—	Bequaert (1955)
Duval	NG	NG	NG	—	Gilbert (1922), Bequaert (1955)
Escambia	NG	NG	NG	—	Bequaert (1955)
Hillsborough	1929, 1960	NG	NG	—	Wilson et al. (1990)
Hillsborough	1980	ND	1	—	Forrester (1989)
Indian River	1968, 1972	NG	NG	—	Wilson et al. (1990)
Jefferson	NG	NG	NG	—	Bequaert (1955)
Monroe	NG	NG	NG	—	Knab (1916), Bequaert (1955)
Orange	NG	NG	NG	—	Bishopp (1929), Bequaert (1955)
Palm Beach	1931	NG	NG	—	Wilson et al. (1990)
Palm Beach	NG	NG	NG	—	Bequaert (1955)
Polk	1948	NG	NG	—	Wilson et al. (1990)
Seminole	NG	NG	NG	—	Bequaert (1955)
White-winged Dove					
Stilbometopa podopostyla					
Dade	1977–78	119	1	<1	Conti (1980), Conti & Forrester (1981)
Pseudolynchia canariensis					
Dade	1977–78	119	1	<1	Ibid.
Mourning Dove					
Stilbometopa podopostyla					
Alachua	1977–78	49	0[a]	—	Ibid.
Dade	1977–78	47	0[a]	—	Ibid.
Microlynchia pusilla					
Dade	NG	NG	NG	—	Bequaert (1955)
Hillsborough	1974	ND	1	—	Wilson & Forrester (1990)
Indian River	1972, 1979	NG	NG	—	Wilson et al. (1990)
Orange	1982	NG	NG	—	Ibid.
Pinellas	1965	NG	NG	—	Ibid.
Pseudolynchia canariensis					
Alachua	1977–78	49	0[a]	—	Conti (1980), Conti & Forrester (1981)
Dade	1877–78	47	0[a]	—	Ibid.
Pseudolynchia brunnea					
Orange	1974	NG	NG	—	FSCA[b]

(continued)

Table 22.14. *(continued)*

Host species Louse fly	Year(s)	No. birds Examined	Infected	%	Data source
Species of Dove not given					
Microlynchia pusilla					
Dade	1958	NG	NG	—	Wilson et al. (1990)
Pseudolynchia canariensis					
Jefferson	NG	NG	NG	—	Maa (1966)
Pinellas	1929	NG	NG	—	Wilson et al. (1990)
Seminole	1959	NG	NG	—	Ibid.

ND = not determined, NG = not given.

a. Conti (1980) noticed hippoboscids crawling on the feathers of Mourning Doves from Alachua and Dade counties, but did not collect or identify them, nor give any prevalence data.

b. FSCA = Florida State Collection of Arthropods, Division of Plant Industry, Department of Agriculture and Consumer Services, Gainesville.

and public health importance (Weber 1979), including aspergillosis, candidiasis, chlamydiosis, histoplasmosis, encephalitis, salmonellosis, and toxoplasmosis. Although specific studies of the health hazards of Rock Doves in Florida have not been conducted, there is little doubt that this is an important issue. The monograph by Weber (1979) should be consulted for more details on the diseases transmitted by Rock Doves and for suggestions for control of such problems.

XVII. Summary and conclusions

Twelve species of columbids occur in Florida; 6 are resident breeding birds and 6 are transients or occasional visitors. Data on parasites and diseases exist for 7 of these columbids. The most complete information is on Mourning Doves and White-winged Doves, although some data exist for Rock Doves, Scaly-naped Pigeons, Common Ground-Doves, White-crowned Pigeons, and Eurasian Collared-Doves.

In addition to brood parasitism and trauma due to predation, collisions with TV towers, and other causes, 70 different disease agents and parasites have been found in columbids in Florida. These include poisonous plants (2), en-

vironmental contaminants (8), viruses (6), bacteria (2), fungi (1), protozoans (8), nematodes (7), trematodes (2), cestodes (7), ticks (1), mites (9), chewing lice (13), and louse flies (4). The most important diseases are trichomonosis, avian pox, pigeon paramyxovirus type 1, and West Nile virus. Other pathogenic parasites include *Haemoproteus columbae*, *Ascaridia columbae*, *Dispharynx nasuta*, and *Tetrameres columbicola*.

The Rock Dove is a serious pest and public health hazard. Although specific information is not available for Florida, it is known that this introduced pigeon can transmit more than 40 diseases of veterinary and public health importance.

XVIII. Literature cited

Aldrich, J.W. 1952. The source of migrant mourning doves in southern Florida. *J. Wildl. Manag.* 16:447–456.

Bancroft, G.T., R. Bowman, R.J. Sawicki, and A.M. Strong. 1990. Relationship between the reproductive ecology of the white-crowned pigeon and the fruiting phenology of tropical hardwood hammock trees. Final Project Report, Project no. GFC-86-031, submitted to the Florida Game and Fresh Water Fish Commission. National Audubon Society, Tavernier, Fla. 168 pp.

Barrows, P.L. 1975. Studies on endoparasites of the mourning dove (*Zenaida macroura* L.). M.S. thesis, University of Georgia, Athens. 107 pp.

Barrows, P.L., and F.A. Hayes. 1977. Studies on endoparasites of the mourning dove (*Zenaida macroura*) in the southeast United States. *J. Wildl. Dis.* 13:24–28.

Bendell, J.F. 1955. Disease as a control of a population of blue grouse, *Dendragapus obscurus fuliginosus* (Ridgway). *Can. J. Zool.* 33:195–223.

Bennett, G.F. 1994. Unpublished data. Memorial University of Newfoundland, St. Johns.

Bennett, G.F., and M.A. Peirce. 1990. The haemoproteid parasites of the pigeons and doves (family Columbidae). *J. Nat. Hist.* 24:311–325.

Bequaert, J.C. 1955. The Hippoboscidae or louseflies (Diptera) of mammals and birds. Part II. Taxonomy, evolution and revision of American genera and species. *Entomol. Am.* 35:233–416.

Bigler, W.J. 1969. Unpublished data. Florida Department of Health and Rehabilitative Services, Tallahassee.

Bigler, W.J., G.L. Hoff, and L.A. Scribner. 1977. Survival of mourning doves unaffected by withdrawing blood samples. *Bird-Banding* 48:168.

Bigler, W.J., E. Lassing, E. Buff, A.L. Lewis, and G.L. Hoff. 1975. Arbovirus surveillance in Florida: wild vertebrate studies, 1965–1974. *J. Wildl. Dis.* 11:348–356.

Bishop, M.A., and G.F. Bennett. 1992. *Host-parasite catalogue of the avian Haematozoa.* Suppl. 1. Meml. University Nfld. Occas. Pap. Biol. 15. 210 pp.

Bishopp, F.C. 1929. The pigeon fly: an important pest of pigeons in the United States. *J. Econ. Entomol.* 22:974–980.

Burridge, M.J., W.J. Bigler, D.J. Forrester, and J.M. Hennemann. 1979. Serologic survey for *Toxoplasma gondii* in wild animals in Florida. *J. Am. Vet. Med. Assoc.* 175:964–967.

Case, L.D., H. Cruickshank, A.E. Ellis, and W.F. White. 1965. Weather causes heavy bird mortality. *Fla. Nat.* 38:29–30.

Clayton, D.H., and R.D. Price. 1999. Taxonomy of New World *Columbicola* (Phthiraptera: Philopteridae) from the Columbiformes (Aves), with descriptions of five new species. *Ann. Entomol. Soc. Am.* 92:675–685.

Coatney, G.R., and B.B. Hickman. 1952. The course of sporozoite-induced *Haemoproteus columbae* infection in the pigeon. *J. Parasitol.* 38:12.

Conti, J.A. 1980. Interrelationships of parasites of white-winged doves and mourning doves in Florida. M.S. thesis, University of Florida, Gainesville. 69 pp.

———. 1992. Unpublished data. University of Florida, Gainesville.

———. 1993. Diseases, parasites and contaminants. In: *Ecology and management of the mourning dove.* T.S. Baskett, M.W. Sayre, R.E. Tomlinson, and R.E. Mirarchi (eds.). Stackpole Books, Harrisburg, Pa. pp. 205–224.

Conti, J.A., and D.J. Forrester. 1981. Interrelationships of parasites of white-winged doves and mourning doves in Florida. *J. Wildl. Dis.* 17:529–536.

———. 1989. Unpublished data. University of Florida, Gainesville.

Conti, J.A., R.K. Frohlich, and D.J. Forrester. 1985. Experimental transfer of *Trichomonas gallinae* (Rivolta, 1878) from white-winged doves to mourning doves. *J. Wildl. Dis.* 21:229–232.

Conti, L., R. Oliveri, and C. Blackmore. 2002. Unpublished data. Florida Department of Health, Tallahassee.

Crawford, R.L. 1981. Bird casualties at a Leon County, Florida TV tower: a 25-year migration study. *Bull. Tall Timbers Res. Stn.* 22:1–30.

Cuvillier, E. 1937. The nematode, *Ornithostrongylus quadriradiatus*, a parasite of the domesticated pigeon. USDA, Technical Bulletin 569. 36 pp.

Dahlen, J.H., and A.D. Haugen. 1954. Acute toxicity of certain insecticides to the bobwhite quail and mourning dove. *J. Wildl. Manag.* 18:477–481.

Dahlgren, R.B. 1955. Factors affecting mourning

dove populations in Utah. M.S. thesis, Utah State Agricultural College, Logan. 93 pp.

Domermuth, C.H., D.J. Forrester, D.O. Trainer, and W.J. Bigler. 1977. Serologic examination of wild birds for hemorrhagic enteritis of turkey and marble spleen disease of pheasants. *J. Wildl. Dis.* 13:405–408.

Earle, R.A., S.S. Bastianello, G.F. Bennett, and R.C. Krecek. 1993. Histopathology and morphology of the tissue stages of *Haemoproteus columbae* causing mortality in Columbiformes. *Avian Pathol.* 22:67–80.

Ewing, S.A., J.L. West, and A.L. Malle. 1967. *Tetrameres* sp. (Nematode: Spiruridae) found in pigeons (*Columba livia*) in Kansas and Oklahoma. *Avian Dis.* 11:407–412.

Favorite, F.G. 1960. Some evidence of local origin of EEE virus in Florida. *Mosq. News* 20:87–92.

Flatt, R.E., and L.R. Nelson. 1969. *Tetrameres americana* in laboratory pigeons (*Columba livia*). *Lab. Anim. Care* 19:853–856.

Forrester, D.J. 1989. Unpublished data. University of Florida, Gainesville.

Forrester, D.J., J.A. Conti, J.D. Shamis, W.J. Bigler, and G.L. Hoff. 1983. Ecology of helminth parasitism of mourning doves in Florida. *Proc. Helminthol. Soc. Wash.* 50:143–152.

Forrester, D.J., H.W. Kale II, R.D. Price, K.C. Emerson, and G.W. Foster. 1995. Chewing lice (Mallophaga) from birds in Florida: a listing by host. *Bull. Fla. Mus. Nat. Hist.* 39:1–44.

Frenkel, J.K. 1981. False-negative serologic tests for *Toxoplasma* in birds. *J. Parasitol.* 67:952–953.

Gainer, J.H., W.G. Winkler, A.L. Lewis, W.L. Jennings, and P.H. Coleman. 1964. Isolations of St. Louis encephalitis virus from domestic pigeons, *Columba livia. Am. J. Trop. Med. Hyg.* 13:472–474.

Garnham, P.C.C. 1966. *Malaria parasites and other haemosporidia.* Blackwell Scientific, Oxford. 1,114 pp.

Gilbert, T.J. 1922. Another experience with "blood suckers." *Am. Pigeon J.* 11:383.

Goble, F.C., and H.L. Kutz. 1945. The genus *Dispharynx* (Nematoda: Acuariidae) in galliform and passeriform birds. *J. Parasitol.* 31:323–331.

Hailman, J.P. 1989. Common ground-dove's injury-feigning distracts Florida scrub-jay. *Auk* 106:742.

Hanson, H.C., and C.W. Kossack. 1963. *The mourning dove in Illinois.* Illinois Department of Conservation, Technical Bulletin 2. Southern Illinois University Press, Carbondale. 133 pp.

Hasenclever, H.F., and R.M. Kocan. 1973. Serotypes in *Saccharomyces telluris:* Their relation to source of isolation. *Infect. Immun.* 7:610–612.

Haugen, A.O. 1952. Trichomoniasis in Alabama mourning doves. *J. Wildl. Manag.* 16:164–169.

Haugen, A.O., and J. Keeler. 1952. Mortality of mourning doves from trichomoniasis in Alabama during 1951. *Trans. N. Am. Wildl. Conf.* 17:141–151.

Hayes, L.E. 1989. Unpublished data. Southeastern Cooperative Wildlife Disease Study, University of Georgia, Athens.

Henderson, J.R., N. Karabatsos, T.T.C. Bourke, R.C. Wallis, and R.M. Taylor. 1962. A survey for arthropod-borne viruses in south-central Florida. *Am. J. Trop. Med. Hyg.* 11:800–810.

Herman, C.M. 1951. Unpublished data. Patuxent Wildlife Research Center, Laurel, Md.

Herman, C.M., J.O. Knisley, Jr., and E.L. Snyder. 1966. Subinoculation as a technique in the diagnosis of avian *Plasmodium. Avian Dis.* 10:541–547.

Hill, W.W., and D.W. Tuff. 1978. A review of the Mallophaga parasitizing the Columbiformes of North America north of Mexico. *J. Kansas Entomol. Soc.* 51:307–327.

Hoff, G.L. 1978. Unpublished data. Florida Department of Health and Rehabilitative Services, Tallahassee.

Hon, L.T., D.J. Forrester, and L.E. Williams, Jr. 1975. Helminths of wild turkeys in Florida. *Proc. Helminthol. Soc. Wash.* 42:119–127.

Honigberg, B.M., M.C. Livingston, and R.M. Stabler. 1971. Pathogenicity transformation of *Trichomonas gallinae.* 1. Effects of homogenates and of mixtures of DNA and RNA from a virulent strain on pathogenicity of an avirulent strain. *J. Parasitol.* 57:929–938.

Hutt, A. 1964. The mourning dove. *Fla. Wildl.* 18:12–16.

Hwang, J.C., N. Tolgay, W.T. Shalkop, and D.S. Jacquette. 1961. Case report: *Dispharynx nasuta* causing severe proventriculitis in pigeons. *Avian Dis.* 5:60–65.

Jennings, W.L., W.G. Winkler, D.D. Stamm, P.H. Coleman, and A.L. Lewis. 1969. Serologic studies of possible avian or mammalian reservoirs of St. Louis encephalitis virus in Florida. *Fla. State Board Health Monogr. Ser.* 12:118–125.

Johnson, H.N. 1960. Public health in relation to birds: arthropod-borne viruses. *Trans. N. Am. Wildl. Conf.* 25:121–133.

Kadner, C.G. 1941. Pigeon malaria in California. *Science.* 93:281.

Karstad, L. 1971. Pox. In: *Infectious and parasitic diseases of wild birds.* J.W. Davis, R.C. Anderson, L. Karstad, and D.O. Trainer (eds.). Iowa State University Press, Ames. pp. 34–41.

Kemp, G.E. 1981. Saint Louis encephalitis (SLE). In: *CRC handbook series in zoonoses.* Section B, *Viral zoonoses.* Vol. 1. G.W. Beran (ed.). CRC Press, Boca Raton, Fla. pp. 71–83.

Kinsella, J.M., G.W. Foster, M.G. Spalding, and C.M. Romagosa. 1998. Unpublished data. University of Florida, Gainesville.

Knab, F. 1916. Four European Diptera established in North America. *Ins. Inscit. Mens.* 4:1–4.

Knisley, J.O., Jr., and C.M. Herman. 1967. *Haemoproteus,* a blood parasite in domestic pigeons and mourning doves in Maryland. *Chesapeake Sci.* 8:200–202.

Kocan, R.M. 1968. Probable origin of ciliates seen in oral swabbings of doves and pigeons. *J. Parasitol.* 54:1033.

———. 1969. Unpublished data. Patuxent Wildlife Research Center, Laurel, Md.

Kocan, R.M., and H.F. Hasenclever. 1972. Normal yeast flora of the upper digestive tract of some wild columbids. *J. Wildl. Dis.* 8:365–368.

Kocan, R.M., and C.M. Herman. 1971. Trichomoniasis. In: *Infectious and parasitic disease of wild birds.* J.W. Davis, R.C. Anderson, L. Karstad, and D.O. Trainer (eds.). Iowa State University Press, Ames. pp. 282–290.

Kocan, R.M., and A. Sprunt IV. 1971. The white-crowned pigeon: a fruit eating pigeon as a host for *Trichomonas gallinae. J. Wildl. Dis.* 7:217–218.

Koval, P.J., T.J. Peterle, and J.D. Harder. 1987. Effects of polychlorinated biphenyls on mourning dove reproduction and circulating progesterone levels. *Bull. Environ. Contam. Toxicol.* 39:663–670.

Kreitzer, J.F. 1974. Residues of organochlorine pesticides, mercury, and PCB's in mourning doves from eastern United States. *Pestic. Monit. J.* 7:195–199.

Lehner, P.N., T.O. Boswell, and F. Copeland. 1967. An evaluation of the effects of the *Aedes aegypti* eradication program on wildlife in south Florida. *Pestic. Monit. J.* 1:29–34.

Linda, S.B. 1998. Analysis of 1997–1998 statewide and wildlife management area hunter surveys: methods and results. Unpublished report. Florida Game and Fresh Water Fish Commission, Gainesville. 88 pp.

Locke, L.N. 1961. Pox in mourning doves in the United States. *J. Wildl. Manag.* 25:211–212.

Logan, T.H. 1997. Florida's endangered species, threatened species and species of special concern. Official lists. Florida Game and Fresh Water Fish Commission, Tallahassee. 14 pp.

Maa, T.C. 1966. On the genus *Pseudolynchia* Bequaert (Diptera: Hippoboscidae). *Pac. Insects Monogr.* 10:125–138.

Marion, W.R., T.E. O'Meara, and L.D. Harris. 1981. Characteristics of the mourning dove harvest in Florida. *J. Wildl. Manag.* 45:1062–1066.

Marion, W.R., and M.S. Schnoes. 1982. Seasonality of mourning dove nesting in Florida. *Proc. Annu. Conf. Southeast. Assoc. Fish Wildl. Agencies* 36:543–551.

Markus, M.B., and J.H. Oosthuizen. 1971. Pathogenicity of *Haemoproteus columbae. Trans. R. Soc. Trop. Med. Hyg.* 66:186–187.

Martin, L.L. 1967. Comparison of methoxymol, alphachloralose and two barbiturates for capturing doves. *Proc. Annu. Conf. Southeast. Assoc. Game Fish Comm.* 21:193–200.

McCaig, L.F., H.T. Janowski, R.A. Gunn, and T.F. Tsai. 1994. Epidemiologic aspects of a St. Louis encephalitis outbreak in Fort Walton Beach,

Florida in 1980. *Am. J. Trop. Med. Hyg.* 50: 387–391.

McGinity, J. 1997. Florida scrub-jay kills common ground-dove. *Fla. Field Nat.* 25:101–102.

McLaurin, E. 1955. Last chance for doves. *Fla. Wildl.* 9:8–10, 43.

Morton, J.F. 1978. Brazilian pepper—its impact on people, animals and the environment. *Econ. Bot.* 32:353–359.

———. 1989. Creeping indigo (*lndigofera spicata* Forsk.) (Fabaceae)—a hazard to herbivores in Florida. *Econ. Bot.* 43:314–327.

Mozgovoi, A.A. 1953. Ascaridata of animals and man and the diseases caused by them. Part I. In: *Essentials of nematology.* Vol 2. K.I. Skrjabin (ed.). Israel Program for Scientific Translations, Jerusalem. pp. 238–243.

Nelson, D.J. 1957. Some aspects of dove hunting in Georgia. *J. Wildl. Manag.* 21:58–61.

Ogden, J.C. 1992. The winter season: Florida region. *Am. Birds* 46:225–257.

Owre, O.T. 1978. White-crowned pigeon (*Columba leucocephala* Linnaeus, family Columbidae, order Columbiformes). In: *Rare and endangered biota of Florida.* Vol. 2, *Birds.* H.W. Kale II (ed.). University Press of Florida, Gainesville. pp. 43–45.

Pence, D.B., M.G. Spalding, J.F. Bergan, R.A. Cole, S. Newman, and P.N. Gray. 1997. New records of subcutaneous mites (Acari: Hypoderatidae) in birds, with examples of potential host colonization events. *J. Med. Entomol.* 34:411–416.

Peters, H.S. 1936. A list of external parasites from birds of the eastern part of the United States. *Bird-Banding* 7:9–27.

Reisen, W.K., J.L. Hardy, and S.B. Presser. 1992. Evaluation of domestic pigeons as sentinels for detecting arbovirus activity in southern California. *Am. J. Trop. Med. Hyg.* 46:69–79.

Rickard, L.G. 1985. Proventricular lesions associated with natural and experimental infections of *Dispharynx nasuta* (Nematoda: Acuariidae). *Can. J. Zool.* 63:2663–2668.

Robertson, W.B., Jr., and G.E. Woolfenden. 1992. *Florida bird species: an annotated list.* Spec. Publ. 6, Florida Ornithological Society, Gainesville. 260 pp.

Rodgers, J.A., Jr., H.W. Kale II, and H.T. Smith (eds.). 1996. *Rare and endangered biota of Florida.* Volume 5, *Birds.* University Press of Florida, Gainesville. 688 pp.

Romero, C.H., and M.G. Spalding. 2002. Unpublished data. University of Florida, Gainesville.

Romagosa, C.M. 1999. Unpublished data. Department of Wildlife Ecology and Conservation, University of Florida, Gainesville.

Sadler, K.C. 1993. Other natural mortality. In: *Ecology and management of the mourning dove.* T.S. Baskett, M.W. Sayre, R.E. Tomlinson, and R.E. Mirarchi (eds.). Stackpole Books, Harrisburg, Pa. pp. 225–230.

Saunders, G.B. 1980. The origin of white-winged doves breeding in south Florida. *Fla. Field Nat.* 8:50–51.

Schmitt, B.J. 2001. Unpublished data. USDA, APHIS, National Veterinary Services Laboratory, Ames, Iowa.

Schnoes, M.S. 1980. The importance of fall nesting of mourning doves (*Zenaida macroura*) in Florida. M.S. thesis, University of Florida, Gainesville. 56 pp.

Schultz, V. 1954. The effects of a severe snow and ice storm on game populations in Tennessee. *J. Tenn. Acad. Sci.* 29:24–35.

Shamis, J.D. 1977. An ecological study of haematozoan and trichomonad parasites of mourning doves in Florida. M.S. thesis, University of Florida, Gainesville. 69 pp.

Shamis, J.D., and D.J. Forrester. 1977. Haematozoan parasites of mourning doves in Florida. *J. Wildl. Dis.* 13:349–355.

Shaw, M.H. 1956. We're under-harvesting our dove crop. *Fla. Wildl.* 9:18–21, 34.

Short, R.B., and R.B. Holliman. 1961. *Australbilharzia penneri,* a new schistosome from marine snails. *J. Parasitol.* 47:447–452.

Simpson, C.F., J.W. Carlisle, and J.A. Conti. 1984. *Tetrameres columbicola* (Nematoda: Spiruridae) infection of pigeons: ultrastructure of the gravid female in glands of the proventriculus. *Am. J. Vet. Res.* 45:1184–1192.

Spalding, M.G., and D.J. Forrester. 1990–2001. Unpublished data. University of Florida, Gainesville.

Spalding, M.G., and G.W. Foster. 1989. Unpublished data. University of Florida, Gainesville.

Spalding, M.G., and R.G. McLean. 1994. Unpublished data. University of Florida, Gainesville.

Spalding, M.G., R.G. McLean, M.C. Garvin, and G.F. Bennett. 1993. Unpublished data. University of Florida, Gainesville.

Spalding, M.G., J.W. Mertins, and G.W. Foster. 1989. Unpublished data. University of Florida, Gainesville.

Spalding, M.G., C.H. Romero, and D.J. Forrester. 2002. Unpublished data. University of Florida, Gainesville.

Stabler, R.M. 1947. *Trichomonas gallinae,* a pathogenic trichomonad of birds. *J. Parasitol.* 33:207–213.

———. 1948. Protection in pigeons against virulent *Trichomonas gallinae* acquired by infection with milder strains. *J. Parasitol.* 34:150–153.

———. 1951. Effect of *Trichomonas gallinae* from diseased mourning doves on clean domestic pigeons. *J. Parasitol.* 37:473–478.

———. 1954. *Trichomonas gallinae:* a review. *Exp. Parasitol.* 3:368–402.

Stabler, R.M., and C.M. Herman. 1951. Upper digestive tract trichomoniasis in mourning doves and other birds. *Trans. N. Am. Wildl. Conf.* 16:145–162.

Stabler, R.M., and P.A. Holt. 1962. The parasites of four eastern ground doves. *Proc. Helminthol. Soc. Wash.* 29:76.

Stabler, R.M., B.M. Honigberg, and V.M. King. 1964. Effect of certain laboratory procedures on virulence of the Jones' barn strain of *Trichomonas gallinae* for pigeons. *J. Parasitol.* 50:36–41.

Stevenson, H.M. 1986. *A checklist of Florida's birds.* Florida Game and Fresh Water Fish Commission, Tallahassee. 16 pp.

Stevenson, H.M., and B.H. Anderson. 1994. *The birdlife of Florida.* University Press of Florida, Gainesville. 892 pp.

Stiles, C.W., and A. Hassall. 1894. A preliminary catalogue of the parasites contained in the collections of the United States Bureau of Animal Industry, United States Army Medical Museum, Biological Department of the University of Pennsylvania (Coll. Leidy) and in Coll. Stiles and Coll. Hassall. *Vet. Mag.* 1:245–253, 331–354.

Stoddard, H.L. 1931. *The bobwhite quail, its habits, preservation and increase.* Charles Scribner's Sons, New York. 559 pp.

Strong, A.M., R.J. Sawicki, and G.T. Bancroft. 1991. Effects of predator presence on nesting distribution of white-crowned pigeons. *Wilson Bull.* 103:415–425.

Taylor, W.K., and B.H. Anderson. 1973. Nocturnal migrants killed at a central Florida TV tower, autumns 1969–1971. *Wilson Bull.* 85:42–51.

———. 1974. Nocturnal migrants killed at a central Florida TV tower, autumn 1972. *Fla. Field Nat.* 2:40–43.

Terrell, S.P. 1993–99. Unpublished data. University of Florida, Gainesville.

Travis, B.V. 1938. Fire ant problem in the southeast with special reference to quail. *Trans. N. Am. Wildl. Conf.* 3:705–708.

———. 1941. Examinations of wild animals for the cattle tick *Boophilus annulatus microplus* (Can.) in Florida. *J. Parasitol.* 27:465–467.

Toepfer, E.W., Jr. 1964. *Colpoda steinii* in oral swabbings from mourning doves (*Zenaida macroura* L.). *J. Parasitol.* 50:703.

Tori, G.M., and T.J. Peterle. 1983. Effects of PCBs on mourning dove courtship behavior. *Bull. Environ. Contam. Toxicol.* 30:44–49.

Walton, A.C. 1927. A revision of the nematodes of the Leidy Collections. *Proc. Acad. Nat. Sci. Phila.* 79:49–163.

Weber, W.J. 1979. *Health hazards from pigeons, starlings and English sparrows.* Thompson, Fresno, Cal. 138 pp.

Wehr, E.E. 1971. Nematodes. In: *Infectious and parasitic diseases of wild birds.* J.W. Davis, R.C. Anderson, L. Karstad, and D.O. Trainer (eds.). Iowa State University Press, Ames. pp. 185–233.

White, F.H., and D.J. Forrester. 1989. Unpublished data. University of Florida, Gainesville.

Wiley, J.W., and B.N. Wiley. 1979. The biology of

the white-crowned pigeon. *Wildl. Monogr.* 64:1–54.

Wilson, N.A., and D.J. Forrester. 1990. Unpublished data. University of Florida, Gainesville.

Wilson, N.A., H.W. Kale II, and W.W. Baker. 1990. Unpublished data. University of Northern Iowa, Cedar Falls.

Winston, F.A. 1954. *Status, movement and management of the mourning dove in Florida.* Florida Game and Fresh Water Fish Commission, Technical Bulletin 2. 86 pp.

Williams, L.E., Jr., and R.W. Phillips. 1972. Tests of oral anesthetics to capture mourning doves and bobwhites. *J. Wildl. Manag.* 36:968–971.

Zwart, P. 1986. Pigeons and doves. In: *Zoo and wild animal medicine.* 2d ed. M.E. Fowler (ed.). W.B. Saunders, Philadelphia. pp. 439–445.

Owls

I. Introduction

Nine species of owls occur in Florida, 1 barn owl (family Tytonidae) and 8 typical owls (family Strigidae) (Robertson and Woolfenden 1992). Of these, 5 are year-round residents and are distributed throughout Florida (table 23.1). Three other species migrate into the state during the fall and winter months and are more restricted in their distribution. One of the resident species, the Burrowing Owl, is listed as a species of special concern by the Florida Game and Fresh Water Fish Commission (Logan 1997) and the Florida Committee on Rare and Endangered Plants and Animals (Millsap 1996). Smith et al. (1990) reviewed the distribution and status of owl populations in the southeastern states including Florida

and provided information on various mortality factors.

For an overview of diseases and parasites of owls the reader is referred to the books edited by Fowler (1986, 1993) for discussions on physiology, hematology, blood chemistry, anesthesia, viral diseases, bacterial diseases, mycotic infections, parasites, toxic and metabolic conditions, and neoplasms. Cooper (1985) has published a book with emphasis on clinical aspects of captive birds of prey and Morishita et al. (1997) contributed a review article on the diseases of raptors; both of these contain information on owls. Mortality data have been reviewed for several species of owls in North America, including Northern Saw-whet Owls,

Table 23.1. Occurrence, distribution, and abundance of owls in Florida[a]

Species of owl	Range	Seasonal occurrence	Relative abundance
Barn Owl (*Tyto alba*)	Statewide	Resident	Rare to fairly common
Flammulated Owl (*Otus flammeolus*)	Pinellas Co.	—	1 record
Eastern Screech-Owl (*Otus asio*)	Statewide	Resident	Rare to locally common
Great Horned Owl (*Bubo virginianus*)	Statewide	Resident	Rare to fairly common
Burrowing Owl (*Athene cunicularia*)	Statewide	Resident	Locally common
Barred Owl (*Strix varia*)	Statewide	Resident	Occasional to fairly common
Long-eared Owl (*Asio otus*)	Southern Fla.	Winter	Occasional
Short-eared Owl (*Asio flammeus*)	Peninsular Fla.	Fall & winter	Rare
Northern Saw-whet Owl (*Aegolius acadicus*)	Northern Fla.	Fall & winter	Occasional

Modified from Robertson and Woolfenden (1992) and American Ornithologists' Union (1998).

Eastern Screech Owls, and Great Horned Owls (Franson and Little 1996; Loos and Kerlinger 1993; Stewart 1969). The literature on the effects of pesticides on owls in North America was reviewed by Blus (1996).

Information on the diseases and parasites of owls in Florida is limited. What is known has come in large part from studies on road-killed specimens and birds submitted to wildlife rehabilitation facilities for treatment of traumatic injuries and other problems. Some general information on morbidity and mortality of owls in Florida, particularly north central Florida, has been published (Deem et al. 1998).

II. Trauma

Trauma, for owls in Florida as for other raptors, is the leading cause of death (table 23.2). The types of trauma include vehicle strikes, entrapment, entanglement in fences, gunshot, and others (table 23.3).

Vehicle strikes are the most important. Over a 4-year period (1990–93), 78 owls representing 6 species were recorded as roadkills on highways passing through Florida state parks and recreation areas (table 23.4). The majority of these were Barred Owls (*n* = 37) and Eastern Screech-Owls (*n* = 30). The statewide distribution of morbidity and mortality caused by vehicles is given in table 23.5 for 5 species of owls submitted to 3 wildlife rehabilitation centers from 1988 to 1995. The most commonly killed owls were Eastern Screech-Owls (*n* = 207) and Barred Owls (*n* = 137). The same clustering of cases is evident from examination of these data, as was seen for other raptors. This is most likely an artifact of sampling due to the geographic locations of the 3 wildlife rehabilitation centers from which these data were obtained. Kale (1978) stated that Barn Owls are struck by vehicles on highways at night, but gave no specific data. Ogden (1992) reported 2 road-killed Barn Owls in Broward County in 1992. Apparently the owls had been nesting under a turnpike bridge.

The distribution of 14 cases of morbidity or mortality of owls due to gunshot is presented in table 23.6. These cases of illegal shooting were recorded between 1988 and 1995 by Collins and Gilliland (1995) and Suto (1996).

Table 23.2. Primary reasons given for submission of owls to 3 wildlife rehabilitation centers in Florida, 1988–95

Species of owl	No. of cases per species of owl								Total number of owls
	Trauma	Poison-ing[a]	Emaci-ation	Separated from parents	Inclement weather	Infec-tious diseases	Electro-cution	Misc.	
Barn Owl	42	10	8	0	0	0	3	0	63
Eastern Screech-Owl	978	117	85	112	10	9	3	5[b]	1,319
Great Horned Owl	127	14	11	0	10	6	3	3[c]	174
Burrowing Owl	15	1	0	0	0	1	0	0	17
Barred Owl	291	13	16	0	5	1	4	0	330
Long-eared Owl	0	0	1	0	0	0	0	0	1
Short-eared Owl	1	0	0	0	0	0	0	0	1
Northern Saw-whet Owl	0	1	0	0	0	0	0	0	1
Total cases	1,454	156	121	112	25	17	13	8	1,906
% of total cases	76	8	6	6	1	<1	<1	<1	—

Sources: Collins and Gilliland (1995), 1,131 cases; Deem and Terrell (1996), 184 cases; Suto (1996), 591 cases.
a. Most of these were presumptive poisoning cases since the poisons involved were not identified.
b. Three owls with deformed feet, 1 with a deformed beak, and 1 with cataracts. The deformed feet and beak were probably congenital, whereas the cause of the cataracts is unknown. Cataracts are common in owls (Greenwood and Barnett 1981); they can be a result of trauma, old age, congenital factors, toxic agents, or nutritional deficiencies (Kern 1997).
c. One owl with congenital cardiomyopathy and 2 with cataracts.

In addition to the above surveys there are other bits and pieces of information concerning trauma in owls in Florida. Regan (1996) found a dead Barn Owl near a power line in Palm Beach County in 1991 and assumed it had been killed by colliding with the power line or by electrocution, but no necropsy was done to determine the actual cause of death. Crawford (1981) reported 1 Great Horned Owl and 1 Barred Owl as trauma casualties at a TV tower in Leon County. His study spanned 25 years and involved more than 42,000 avian specimens picked up on a daily basis at the foot of the 1,010-foot TV tower. A number of other reports have been published on casualties of birds at construction cranes, radio and TV towers, tall buildings, and smokestacks in Florida, but none mentioned owls as being among the birds killed (Kale 1971; Taylor and Anderson 1973, 1974; Maehr et al. 1983; Taylor and Kershner 1986; Maehr and Smith 1988). Apparently such occurrences are rare for owls.

III. Predation

Kale (1978) listed Great Horned Owls, yellow rat snakes, raccoons, and bobcats as probable predators of Barn Owls, but gave no specific information. Fragments of an Eastern Screech-Owl were found in 1 of 1,098 pellets collected from a Northern Harrier roost in Leon County during the winter of 1925–26 (Stoddard 1931). Bent (1938) listed skunks and opossums as predators of the eggs of Burrowing Owls in Florida. Grimes (1936) gave an interesting account of his firsthand observation of a Great Horned Owl and a 46-inch common black snake in "mortal combat." He made the observation in 1934 near Jacksonville. Apparently the owl tried to capture the snake; in the process the snake coiled itself around the owl in such a way that the owl could not kill it. After watching the struggle for a few minutes the author stated that the owl "flew feebly away a hundred yards, barely

Table 23.3. Types of trauma causing morbidity and mortality in owls submitted to the Center for Birds of Prey in Maitland, Florida, 1989–94

| | Number of cases per species of owl | | | | | | | | | Total number |
Species of owl	Vehicle strike	Predation	Entrapment	Fence entanglement	Gunshot	Building strike	Fishing line entanglement	Misc.	Unknown	of cases
Barn Owl	5[a]	1[b]	1	1	0	0	1	1[d]	18	28
Eastern Screech-Owl	153	43[c]	36	1	2	5	1	4[e]	349	594
Great Horned Owl	22	0	0	6	4	1	1	0	58	92
Burrowing Owl	1	0	1	0	0	1	0	0	8	11
Barred Owl	113	1[b]	0	2	4	0	2	0	74	196
Short-eared Owl	0	0	0	1	0	0	0	0	0	1
Total owls	294	45	38	11	10	7	5	5	507	922
% of total cases	32	5	4	1	1	<1	<1	<1	55	—

Sources: Collins and Gilliland (1995).
a. Four were struck by motor vehicles on highways, 1 by an airplane.
b. Domestic dog.
c. Five were attacked by domestic dogs, 38 by domestic cats.
d. Injured by a horse.
e. One flew into a power line, 1 flew into a tree, 1 was injured by a lawnmower, and 1 was hit by a baseball.

Table 23.4. Road-kill data for owls in state parks and recreation areas of Florida

Species of owl	No. of owls reported by year				
	1990	1991	1992	1993	Totals
Barn Owl	0	4	1	0	5
Eastern Screech-Owl	7	3	12	8	30
Great Horned Owl	0	1	1	0	2
Burrowing Owl	0	2	0	0	2
Barred Owl	8	7	14	8	37
Short-eared Owl	0	2	0	0	2
No. parks reporting	45	64	64	61	—

Sources: Snyder (1994), Stevenson (1994).

clearing the vegetation, and with both ends of the black snake dangling down a few inches. She came to rest on a stump and eyed me wearily." The owl flew again with the snake still attached and landed on a log in a small pond. Some crows then harassed the owl until the author killed both the owl and the snake and ended their ordeal.

Collins and Gilliland (1995) recorded 39 cases of predation of owls by domestic cats and 7 cases of predation by domestic dogs in central Florida between 1989 and 1994 (table 23.7). Three species of owls were involved and included 1 Barn Owl, 1 Barred Owl, and 44 Eastern Screech-Owls. Details concerning the circumstances of these instances of predation are not known.

IV. Electrocution

Thirteen cases of electrocution have been documented among owls in Florida. These include 3 Barn Owls, 3 Eastern Screech-Owls, 3 Great Horned Owls, and 4 Barred Owls (table 23.8).

As discussed previously in chapter 11, Eagles, electrocution of raptors can be prevented by certain adaptations or modifications of the power line structures on which the birds perch, and also by changes in the grounding setup. Detailed information on these recommended changes have been published (Miller et al. 1975).

V. Inclement weather

Little is known about the harmful effects of inclement weather on owls in Florida. In March of 1992 4 young Burrowing Owls were killed during a hailstorm in Orange County (Collins and Gilliland 1995). Langridge (1993) mentioned that Great Horned Owl nestlings were blown out of their nests in central Florida during the "storm of the century" in March 1993. Over a several-day period after the storm, numerous injured and dead raptors were brought in from Orange and Seminole counties to the Center for Birds of Prey in Maitland (Collins and Gilliland 1995). These included nestlings of Eastern Screech-Owls (n = 6), Great Horned Owls (n = 9), and Barred Owls (n = 4), as well as 4 adult screech owls and 1 adult Great Horned Owl. Undoubtedly the total numbers of owls injured and killed by this storm and others like it in the past were considerable.

VI. Environmental contaminants

In table 23.2, 156 cases of morbidity or mortality due to poisoning are listed. These diagnoses were presumptive and the poisons involved not identified. The only published account of mortality of Florida owls due to toxins was that by Weston (1946), who reported a Burrowing Owl

Table 23.5. Distribution of morbidity and mortality of owls resulting from vehicle strikes on highways in Florida, 1988–95

Species of owl	County	No. of owls
Barn Owl (*n* = 5)	Brevard	2
	Orange	2
	Osceola	1
Eastern Screech-Owl (*n* = 207)	Alachua	13
	Hillsborough	5
	Indian River	2
	Lake	9
	Levy	2
	Orange	88
	Osceola	6
	Pasco	1
	Pinellas	32
	Seminole	30
	Sumter	1
	Volusia	14
	Unknown	4
Great Horned Owl (*n* = 31)	Alachua	6
	Bradford	1
	Gilchrist	1
	Lake	3
	Orange	9
	Osceola	3
	Pinellas	1
	Seminole	2
	Volusia	4
	Unknown	1
Burrowing Owl (*n* = 2)	Alachua	1
	Lake	1
Barred Owl (*n* = 137)	Alachua	12
	Brevard	3
	Charlotte	1
	Citrus	3
	Dixie	1
	Flagler	1
	Hillsborough	1
	Lake	7
	Marion	3
	Orange	23
	Osceola	10
	Pasco	2
	Pinellas	3
	Polk	1
	Putnam	1
	Sarasota	1
	Seminole	32
	St. Johns	2
	Sumter	3
	Volusia	22
	Unknown	5

Sources: Collins and Gilliland (1995), 290 cases; Deem and Terrell (1996), 47 cases; Suto (1996), 45 cases.

Table 23.6. Distribution of morbidity and mortality of owls resulting from gunshot wounds in Florida, 1988–95

Species of owl	County	Number of owls
Eastern Screech-Owl	Citrus	1
	Hillsborough	1
	Orange	1
	Pinellas	2
Great Horned Owl	Alachua	1
	Orange	1
	Osceola	1
	Seminole	1
Barred Owl	Hillsborough	1
	Orange	3
	Volusia	1

Sources: Collins and Gilliland (1995), 10 cases; Suto (1996), 4 cases.

Table 23.7. Distribution of morbidity and mortality of owls resulting from attacks by domestic cats and dogs in central Florida, 1989–94

Species of owl	County	No. of owls injured or killed by	
		Domestic cats	Domestic dogs
Barn Owl	Orange	0	1
Eastern Screech-Owl	Lake	3	0
	Orange	29	4
	Osceola	1	0
	Seminole	6	0
	Volusia	0	1
Barred Owl	Orange	0	1

Source: Collins and Gilliland (1995).

Table 23.8. Distribution of morbidity and mortality of owls resulting from electrocution in central Florida, 1990–94

Species of owl	County	No. of owls
Barn Owl	Orange	1
	Osceola	1
	Seminole	1
Eastern Screech-Owl	Orange	2
	Unknown	1
Great Horned Owl	Orange	1
	Osceola	1
	Seminole	1
Barred Owl	Orange	2
	Seminole	2

Sources: Collins and Gilliland (1995), 12 cases; Suto (1996), 1 case.

Table 23.9. Concentrations of chlorinated hydrocarbons in various tissues of omnivorous owls in Florida

Species of owl County	Year	Age	Sex	Organ or tissue[b]	Residues (ppm wet weight)[a]				Data source
					DDE	ΣDDT	Dieldrin	PCB[c]	
Barn Owl									
Hillsborough	1975	AD	M	LV	0. 11	0.11	0.18	0.45	Sundlof et al. (1986)
				MU	1.5	1.5	0.26	2.1	
				AD	2.3	2.3	0.72	3.6	
Indian River	1969	NG	M	AD	8.3	9.3	1.7	NA	Johnston (1976, 1978)
				UP	1.3	1.3	0.00	NA	
	1976	NG	M	UP	0.00	0.00	0.00	NA	Johnston (1978)
Great Horned Owl									
Alachua	1973	AD	M	LV	8.9	8.9	0.00	1.6	Sundlof et al. (1986)
	1974	AD	F	BR	26	26	1.1	12	Ibid.
				LV	53	53	2.0	26	
				MU	1.7	1.7	0.60	0.83	
	1975	NG	F	AD	5.4	9.2	3.0	NA	Johnston (1978)
				UP	3.3	4.3	0.47	NA	
Dixie	1975	NG	NG	UP	17	17	0.00	NA	Ibid .
Duval	1974	AD	F	BR	6.5	6.8	0.36	2.5	Sundlof et al. (1986)
				LV	12	12	5.4	15	
				MU	2.3	2.3	0.13	0.59	
	1976	NG	NG	BR	0.00	0.00	NA	NA	Ibid.
Leon	1973	NG	NG	AD	2.1	3.6	0.08	NA	Johnston (1978)
				UP	8.2	8.7	0.00	NA	
Marion	1976	AD	F	AD	9.7	12	6.2	NA	Ibid .
				UP	0.81	0.81	0.00	NA	
Barred Owl									
Alachua	1973	AD	F	LV	28	28	3.8	64	Sundlof et al. (1986)
				MU	7.9	8.7	0.87	15	
	1974	AD	F	UP	24	77	0.00	NA	Johnston (1978)
	1975	AD	F	AD	1.1	1.1	0.00	NA	Ibid.
				UP	0.37	0.37	0.40	NA	
	1976	AD	M	BR	0.00	0.00	NA	NA	Sundlof et al. (1986)
	1976	AD	F	BR	0.90	0.90	NA	NA	Ibid.
	1977	AD	F	BR	0.00	0.00	NA	NA	Ibid.
Dixie	1973	NG	NG	AD	5.8	6.9	0.21	NA	Johnston (1976, 1978)
				UP	5.1	5.7	0.14	NA	
Hernando	1975	AD	NG	BR	0.00	0.00	NA	NA	Sundlof et al. (1986)
Hillsborough	1975	AD	M	BR	0.55	0.56	0.00	0.36	Ibid.
				LV	1.3	1.3	0.00	0.55	
				MU	0.43	0.48	0.00	0.22	
Lake	1974	NG	NG	BR	0.08	0.08	0.10	0.06	Ibid.
				LV	0.23	0.24	0.28	0.17	
				MU	0.08	0.08	0.08	0.06	
Leon	1975	AD	M	AD	7.6	8.0	0.00	NA	Johnston (1978)
				UP	1.4	2.2	0.00	NA	
Levy	1976	AD	NG	BR	0.00	0.00	NA	NA	Sundlof et al. (1986)
Marion	1973	AD	F	BR	0.31	0.34	0.23	0.29	Ibid.
				AD	30	31	0.00	2.3	

(continued)

Table 23.9. *(continued)*

Species of owl County	Year	Age	Sex	Organ or tissue[b]	Residues (ppm wet weight)[a]				Data source
					DDE	ΣDDT	Dieldrin	PCB[c]	
				LV	1.7	1.8	0.00	0.23	
				MU	0.93	0.96	0.00	0.04	
	1991	AD	F	LV	0.41	NG	NG	NG	SCWDS[d]
Pasco	1975	AD	F	BR	0.00	0.00	NA	NA	Sundlof et al. (1986)
	1976	AD	NG	AD	1.1	1.1	0.00	NA	Johnston (1978)
				UP	0.73	0.73	0.00	NA	
Putnam	1975	AD	F	AD	2.1	2.1	0.00	1.8	Sundlof et al. (1986)
				LV	0.34	0.34	0.00	0.81	
				MU	0.12	0.12	0.00	0.15	
Sumter	1976	AD	NG	BR	0.00	0.00	NA	NA	Ibid.

NA = not analyzed, NG = not given.
a. Lower limits of detection = 0.01 ppm.
b. AD = adipose tissue, BR = brain, LV = liver, MU = pectoral muscle, UP = uropygial gland.
c. The chromatographic profile of the PCBs most closely resembled that of Aroclor 1260.
d. SCWDS = records of the Southeastern Cooperative Wildlife Disease Study, University of Georgia, Athens.

found dead at the Naval Air Station in Pensacola (Escambia County) in 1944. He concluded that the owl died of "some kind of poison." The stomach contained several cockroaches that Weston felt might have been the "source of this poison," but apparently no laboratory tests were conducted to confirm this idea.

Three studies have been published on residues of chlorinated hydrocarbons in tissues of owls in Florida (Johnston 1976, 1978; Sundlof et al. 1986). Concentrations of DDT and its metabolites, dieldrin, and PCBs are given in tables 23.9 and 23.10 for Barn Owls, Great Horned Owls, Barred Owls, and Eastern Screech-Owls, from samples taken between 1969 and 1981. There was considerable variation in the concentrations of various compounds between individual birds; none of the owls, however, contained concentrations believed to be high enough to cause death. The highest concentration of DDE was in an adult female Eastern Screech-Owl found in Pinellas County, May 1975. There were 69 ppm of DDE in the brain of this owl, less than half the lowest concentration associated with mortality (Sundlof et al. 1986). The highest concentration of PCBs, 64 ppm in the liver of an adult fe-

male Barred Owl from Alachua County (June 1973), was well below the amount determined to be lethal for raptors (Wiemeyer and Cromartie 1981). Concentrations of dieldrin were also below estimated lethal amounts (>9 ppm in brain, Ohlendorf et al. 1981). Neither Johnston (1978) nor Sundlof et al. (1986) found evidence that concentrations of DDE and dieldrin had diminished between 1971 and 1981. Because sales of DDT and dieldrin were halted in the United States in 1972 and 1974, respectively, it was suspected that a decline in residues of these organochlorines would be seen. No trends were recognized by Sundlof et al. (1986) when they grouped their data according to the dietary habits of the owls (i.e., omnivores vs. insectivores). These results may have been influenced by the relatively small numbers of birds examined, however.

The impact of chlorinated hydrocarbons on populations of owls in Florida is not understood. The above data would indicate that there may be no deleterious effects; with the banning of the use of DDT, dieldrin, and PCBs in the United States, the outlook is hopeful. However, the effects of sublethal concentrations of these environmentally persistent chemicals on owls

Table 23.10. Concentrations of chlorinated hydrocarbons in various tissues of Eastern Screech-Owls from Florida

| | | | | | Residues (ppm wet weight)[a] | | | | |
County	Year	Age	Sex	Organ or tissue[b]	DDE	ΣDDT	Dieldrin	PCB[c]	Data source
Alachua	1971	AD	F	AD	6.2	6.2	0.0	NA	Johnston (1978)
				UP	1.2	1.2	0.0	NA	
	1974	AD	F	BR	13	13	1.0	3.8	Sundlof et al. (1986)
				LV	45	45	2.4	5.8	
				MU	20	21	1.2	3.8	
	1974	AD	F	LV	1.7	1.8	0.02	0.58	Ibid.
				MU	1.1	1.2	0.01	0.38	
	1975	NG	NG	AD	0.30	0.30	0.0	NA	Johnston (1978)
				UP	3.5	3.5	0.0	NA	
	1978	AD	F	BR	0.0	0.0	NA	NA	Sundlof et al. (1986)
	1978	AD	F	BR	0.15	0.15	NA	NA	Ibid.
Citrus	1977	AD	F	BR	0.0	0.0	NA	NA	Ibid.
Hillsborough	1974	NG	NG	LV	0.15	0.15	0.02	0.09	Ibid.
				MU	0.13	0.13	0.01	0.11	Ibid.
	1975	AD	M	LV	2.2	2.2	0.09	0.92	Ibid.
				MU	2.0	2.0	0.04	1.7	
	1975	JU	M	BR	5.5	5.5	NA	NA	Ibid.
	1975	AD	M	BR	2.4	2.4	NA	NA	Ibid.
Indian River	1975	NG	NG	UP	11	11	0.0	NA	Johnston (1978)
Lafayette	1976	AD	M	BR	11	11	NA	NA	Sundlof et al. (1986)
Levy	1973	NG	NG	AD	0.26	0.26	0.0	NA	Johnston (1976, 1978)
				UP	0.0	0.0	0.0	NA	
Pasco	1975	JU	M	BR	1.6	2.4	NA	NA	Sundlof et al. (1986)
Pinellas	1975	AD	F	BR	69	69	NA	NA	Ibid.
	1975	AD	M	BR	0.56	0.56	NA	NA	Ibid.
	1975	AD	F	BR	0.0	0.0	NA	NA	Ibid.
	1975	AD	F	BR	0.10	0.10	NA	NA	Ibid.
	1976	AD	M	UP	50	50	0.0	NA	Johnston (1978)
	1976	AD	M	UP	1.3	1.3	0.0	NA	Ibid.
	1976	JU	NG	BR	2.0	2.6	NA	NA	Sundlof et al. (1986)
	1976	JU	M	BR	18	20	NA	NA	Ibid.
	1976	JU	NG	BR	3.2	3.2	NA	NA	Ibid.
	1976	AD	M	BR	13	13	NA	NA	Ibid.
Putnam	1972	AD	F	BR	0.0	0.0	NA	NA	Ibid.

NA = not analyzed, NG = not given.
a. Lowest limit of detection = 0.01 ppm.
b. AD = adipose tissue, BR = brain, LV = liver, MU = pectoral muscle, UP = uropygial gland.
c. The chromatographic profile of the PCBs most closely resembled that of Aroclor 1260.

Table 23.11. Measurements of eggshells of Great Horned Owls from Florida, pre-1947–1968[a]

Date	n[b]	Mean shell weight (gm)	Mean shell thickness (mm)	Thickness index[c]	% change in thickness[d]
Pre-1947	67	5.05	0.354[e]	1.89	—
1949–52	27	4.50	ND	1.66	–12
1957–68	8	4.11	ND	1.56	–17

Source: Anderson and Hickey (1972).
ND = no data available.
a. Specific locality not given.
b. Number of clutches represented.
c. Thickness index = weight of eggshell (mg) divided by the product of the length and breadth (mm) (Ratcliffe 1967).
d. Compared with the pre-1947 norm of 0.354 mm.
e. For thickness, only 20 of the 67 eggs could be measured.

and their relationships to other diseases have not been investigated. Henny (1972) studied the effects of pesticides on mortality and recruitment rates of a number of birds, with reference to changes during the 25-year period after 1945, when DDT and other organochlorines were used heavily. One of his datasets was for Great Horned Owls in Florida; his calculations showed no differences in postfledging mortality or in recruitment rates during the "prepesticide era" (prior to 1946) compared with the "pesticide era" (1960–68). This was in spite of the findings of Anderson and Hickey (1972) that eggshells of Great Horned Owls in Florida were 12–17% thinner after 1946 than before that date (table 23.11). The situation for other owls is unknown.

There is almost no information on the effects of heavy metals on owls in Florida. A Barred Owl from Marion County (1991) had 0.34 ppm of lead and 0.37 ppm of mercury in its liver (SCWDS records). The significance of these findings is unknown. Experimental data on chronic lead poisoning are available for Eastern Screech-Owls (Beyer et al. 1988) and will be useful in future studies and for diagnosing lead poisoning in that species.

We have no information on organophosphates and carbamates and their effects on owls in Florida. Blus (1996) summarized information on a number of mortality incidents in North America due to anticholinesterase pesticides, involving Great Horned Owls, Barn Owls, Short-eared Owls, and Eastern Screech-Owls, but none of these occurred in Florida.

VII. Avian vacuolar myelinopathy

This disease is not known to occur in Florida, but has been diagnosed in a Great Horned Owl from South Carolina (see chapter 11, Eagles).

VIII. Neoplasia

The Center for Birds of Prey in Maitland has a record of a malignant tumor beneath the eye of a Burrowing Owl (Collins and Gilliland 1995), an adult male from Orange County that had been in captivity for a considerable time; its exact age was not known. It was euthanized in 1991 and at necropsy an invasive squamous cell carcinoma was diagnosed in the lung (Kiehl 1991). The significance of this finding to populations of Burrowing Owls in Florida is not known.

IX. Anomalies

Information on anomalies in owls in Florida is extremely limited. The Center for Birds of Prey in Maitland has records of 5 owls with anom-

Table 23.12. Avian poxvirus infections in owls from Florida

Species of owl	County	Year	Data source
Eastern Screech-Owl	Orange	1993	Collins & Gilliland (1995)
		1994	Ibid.
		1995	Collins (1999)
	Seminole	1994	Deem et al. (1997)
	St. Johns	1991	Collins & Gilliland (1995)
		1995	Deem et al. (1997)
	Sumter	1989	Collins & Gilliland (1995)
Great Horned Owl	Hillsborough	1995	Suto (1996)
	Marion	1998	Collins (1999)
	Orange	1994	Collins & Gilliland (1995)
		1995	Collins (1999)
Barred Owl	Alachua	1995	Deem et al. (1997)
	Marion	1995	Ibid.
	Unknown	1989	Collins & Gilliland (1995)

alies (Collins and Gilliland 1995). Three of these were Eastern Screech-Owls with deformed feet; 2 from Orange County (1991 and 1994) and 1 from Osceola County (1991). An additional Eastern Screech Owl from Orange County (1991) had a beak deformity. A Great Horned Owl from Sarasota County (1992) was described as having congenital cardiomyopathy. No further details are available on these cases.

X. Biotoxins

No information is available on this topic.

XI. Viruses

Three viruses have been identified from owls in Florida. Bigler et al. (1975) obtained serologic evidence of eastern equine encephalitis (EEE) virus in 1 of 5 owls. The species of owls involved and localities were not given in the 1975 paper, but Bigler (1997) indicated that there had been 2 Barred Owls and 3 Eastern Screech-Owls and that the positive bird was one of the screech owls. They also examined 5 owls for antibodies to Saint Louis encephalitis (SLE) virus, 3 for Everglades virus, and 3 for Highlands J virus. None was positive. Spalding and McLean (1994) tested 2 Eastern Screech Owls, 1 Barred Owl, and 1 Great Horned Owl from Osceola County in 1993–94 for evidence of EEE virus infections. One of the screech owls and the Great Horned Owl were seropositive (SN test). EEE viral infections are probably inconsequential to owls, but such infections may play a role in the epidemiology of this infection of zoonotic and veterinary significance. Spalding and McLean (1994) also tested 1 Barred Owl from Osceola County for evidence of SLE infections. It was seronegative. West Nile virus was identified by PCR and virus isolation techniques in 1 Great Horned Owl (Hillsborough County) and 1 Short-eared Owl in Citrus County (Conti et al. 2002). These 2 owls were presumed to have died because of West Nile virus infection but necropsies were not done to prove this.

Fourteen cases of avian poxvirus infections have been reported in 3 species of owls from Florida (table 23.12). Several authors have stated that there are no published reports of avian pox in owls (Graham and Halliwell 1986; Ritchie and Carter 1995; Morishita et al. 1997), although Chiocco (1992) reported a case in a Long-eared Owl in Italy. Deem et al. (1997) have discussed 4 of the Florida cases, 2 in East-

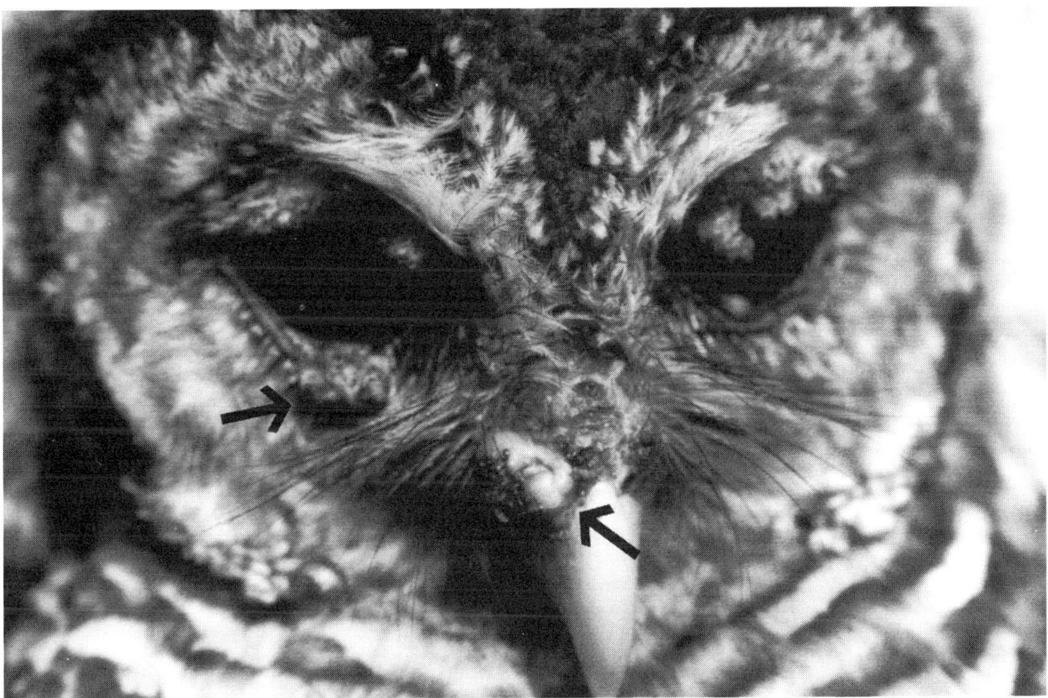

FIGURE 23.1. Pox lesions on the cere and eyelids *(arrows)* of a Barred Owl from Marion County, 1995. From Deem et al. 1997; by permission of *Journal of Wildlife Diseases*.

ern Screech-Owls and 2 in Barred Owls. They provided photographs of lesions on the eyelids and cere (figure 23.1) and on the foot of a Barred Owl (figure 23.2). They also provided an electron microphotograph of the poxvirus inclusion bodies (figure 23.3). Pox lesions on the feet and around the eyes can result in serious debilitation and mortality for raptors that depend on seeing and capturing live prey in order to survive (Morishita et al. 1997) and this may apply to owls as well. The overall effects of pox infections on owl populations are not known.

Two Eastern Screech-owls (localities and dates not given) were examined serologically for antibodies to hemorrhagic/marble spleen disease virus by Domermuth et al. (1977); both were negative.

Albers (1984) tested 5 Great Horned Owls, 9 Barred Owls, and 5 Barn Owls for serologic evidence of rabies infection. These birds were examined during 1982–83 and came from Pinellas County. All were seronegative. Al-

though birds are not thought to be of epidemiologic significance in the spread of rabies, there is some evidence to suggest that this needs further investigation, particularly with respect to owls. Jorgenson et al. (1976) fed the carcass of a rabid skunk to a captive Great Horned Owl and demonstrated that the owl produced antibodies to rabies virus and there was evidence of the presence of rabies virus in its pharynx. There were no clinical signs of rabies, but the findings indicated that the owl might have been able to initiate a rabies infection. Additional serologic surveys of owls and other raptors are needed.

XII. Bacteria

During 1971–78 White and Forrester (1989) cultured samples of liver and large intestine from 47 owls (representing 4 species) from 11 counties in Florida (table 23.13). The cause of

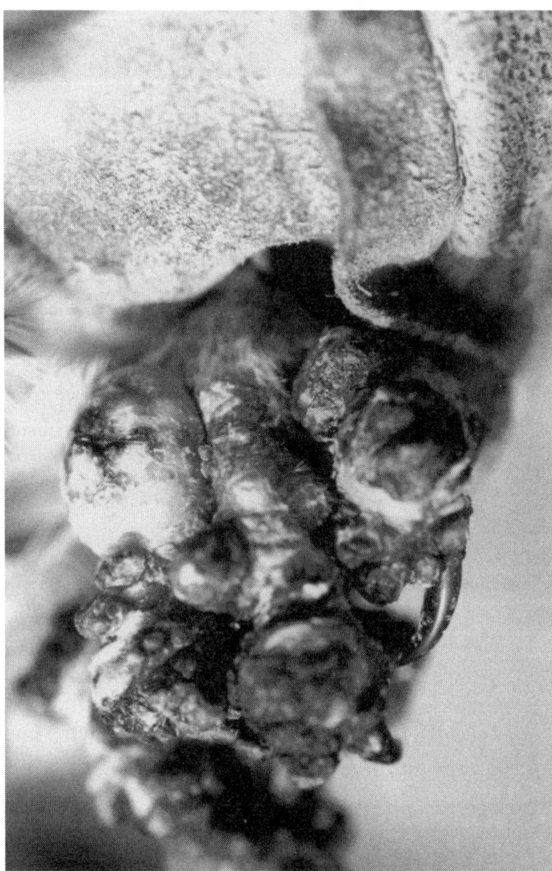

FIGURE 23.2. Pox lesions on the foot of a Barred Owl
from Marion County, 1995. From Deem et al. 1997;
by permission of *Journal of Wildlife Diseases.*

death of most of these owls was not known; some were roadkills and 2 had died of gunshot wounds. Six species of bacteria were identified. The significance of these infections to owls is unknown. In none of these cases was the cause of death attributed to the bacterial infections, with the possible exception of the 2 cases involving *Pseudomonas* infections. Most of these bacterial infections were probably secondary in nature or normal components of the intestinal flora; some were most likely postmortem contaminants.

A nestling Barn Owl from Orange County with an infected sinus above the left eye was submitted to the Center for Birds of Prey and subsequently cultured for bacteria. *Staphylo-coccus aureus* was identified along with some diphtheroids (Collins and Gilliland 1995). The significance of this finding to Barn Owl populations is not known. This probably represents a onetime event.

In September of 1995 an adult female Great Horned Owl was found moribund in Citrus County. It died after 1 day in captivity and at necropsy was found to have acute pneumonia and hepatitis (Smith 1996). *Klebsiella pneumoniae* was isolated from the lungs and *Enterobacter cloacae* from the liver. *Klebsiella pneumoniae* is known to be a pathogen and was considered a likely cause of death. This was considered to be a "single animal event" and of

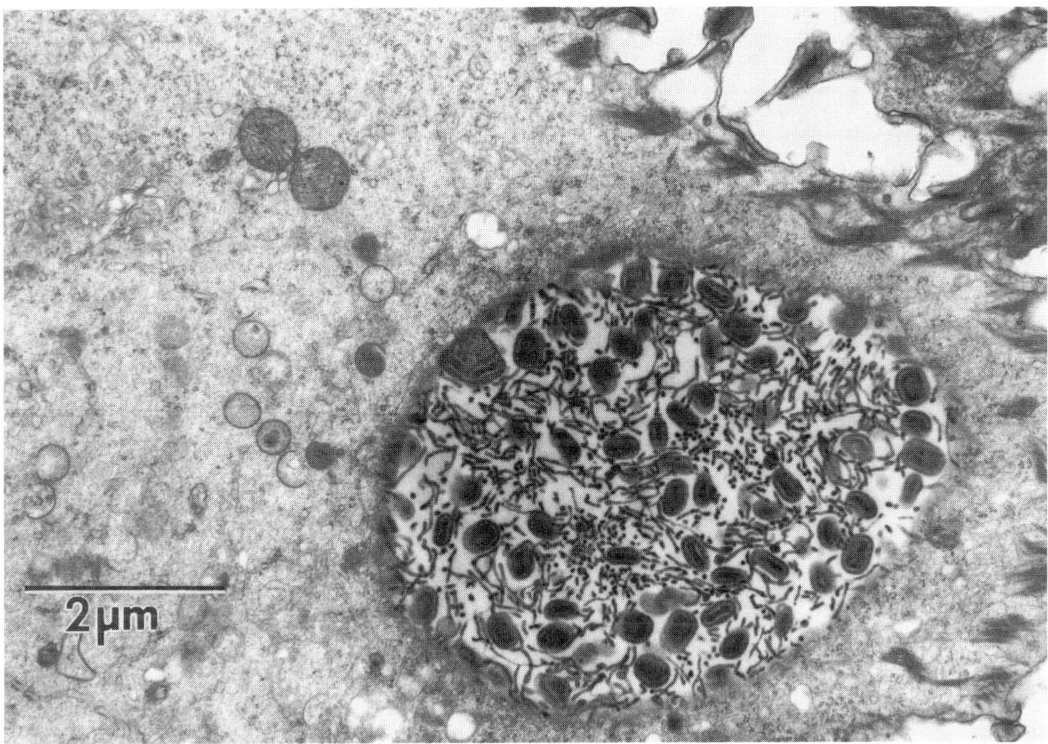

FIGURE 23.3. Electron micrograph of a lesion from the foot of an Eastern Screech-Owl from Seminole County, 1994. Note swollen keratinocyte containing many poxvirus particles. From Deem et al. 1997; by permission of *Journal of Wildlife Diseases*.

little importance to owl populations. The *Enterobacter cloacae* was probably a postmortem contaminant (Smith 1996).

XIII. Fungi

Aspergillosis, although found in a number of diurnal raptors, is not often seen in owls (Keymer 1972; Redig 1978). The exception may be the Great Horned Owl. Aspergillosis was diagnosed in 3 Great Horned Owls submitted to the Center for Birds of Prey in Maitland (Collins and Gilliland 1995), originating from Citrus County (1994), Osceola County (1989), and Seminole County (1993). All were at the Center for only a short time, 24 hours for 2, 12 days for the third; the fungal infections were therefore probably not related to the stress associated with captivity. The owl from Seminole County was a nestling that had fallen out of its nest and died only 1 day after being submitted. Redig (1981) found the disease in only 1 of 251 Great Horned Owls submitted to a rehabilitation center in Minnesota, while Franson and Little (1996) found no cases in 132 Great Horned Owls submitted from 24 states to the National Wildlife Health Center for determination of cause of death. The effect of aspergillosis on Great Horned Owls in Florida is most likely minimal.

XIV. Protozoans

Eleven protozoans have been identified from owls in Florida (tables 23.14–23.17). Nine of these are blood protozoans, most commonly *Haemoproteus syrnii* (56 infections, 3 host species), *H. noctuae* (34 infections, 3 host

Table 23.13. Bacteria identified from livers and large intestines of owls in Florida

Species of owl	Bacteria	County	Year	Organ cultured[a]	Cause of death[b]
Barn Owl	*Enterobacter* sp.	Columbia	1977	LI	RK
(*n* = 6)[c]		Okeechobee	1978	LI	RK
	Enterococcus sp.	Alachua	1976	LI,LV	NE
		Okeechobee	1978	LI	RK
	Escherichia coli	Alachua	1976	LI	RK
		Alachua	1976	LI,LV	NE
		Columbia	1977	LI	RK
		Okeechobee	1978	LI	RK
Barred Owl	*Enterobacter* sp.	Alachua	1976	LI	RK
(*n* = 13)[d]					
	Enterococcus sp.	Levy	1976	LI	RK
		Sumter	1976	LI	RK
	Escherichia coli	Alachua	1976	LI	RK
		Hillsborough	1977	LI	GS
	Proteus sp.	Marion	1974	LI	UK
Great Horned Owl	*Escherichia coli*	Duval	1974	LI	GS
(*n* = 5)[e]					
	Proteus sp.	Alachua	1974	LI	UK
	Pseudomonas sp.	Alachua	1973	LI	UK
		Hillsborough	1976	LI	UK
Eastern Screech-Owl	*Enterobacter* sp.	Citrus	1977	LI	RK
(*n* = 22)[f]		Pinellas	1977	LI	UK
		Pinellas	1977	LI	UK
	Enterococcus sp.	Pasco	1977	LI	TR
		Pasco	1977	LI	UK
		Pinellas	1977	LI	UK
		Pinellas	1977	LI	UK
		Pinellas	1977	LI	UK
		Pinellas	1977	LI	UK
	Escherichia coli	Alachua	1978	LI	RK
		Citrus	1977	LI	RK
		Marion	1977	LV	RK
		Pasco	1977	LI	UK
		Pinellas	1977	LI	UK
		Pinellas	1977	LI	UK
	Staphylococcus sp. (nonhemolytic)	Pinellas	1977	LV	UK

Source: White and Forrester (1989).

a. LI = large intestine, LV = liver.

b. RK = roadkill, NE = young from destroyed nest, GS = gunshot, UK = unknown, TR = trauma (type unknown).

c. Two additional Barn Owls were cultured and were negative. One was a roadkill from Palm Beach County (1977) and the other from Pinellas County (1975), cause of death unknown.

d. Eight additional Barred Owls were cultured and were negative. Two were roadkills from Alachua County (1976, 1977), 1 was from Alachua County (1971), cause of death unknown, 2 were roadkills from Hernando County (1975, 1977), 1 was a roadkill from Pasco County (1975), 1 was from Pinellas County (1975), cause of death unknown, and 1 was a roadkill from Putnam County (1975).

e. One additional Great Horned Owl from Pinellas County (1976), cause of death unknown, was cultured and was negative.

f. Ten additional Eastern Screech-Owls, cause of death unknown, were cultured and were negative. These included 2 from Alachua County (1975, 1977) and 8 from Pinellas County (1 in 1975, 7 in 1977).

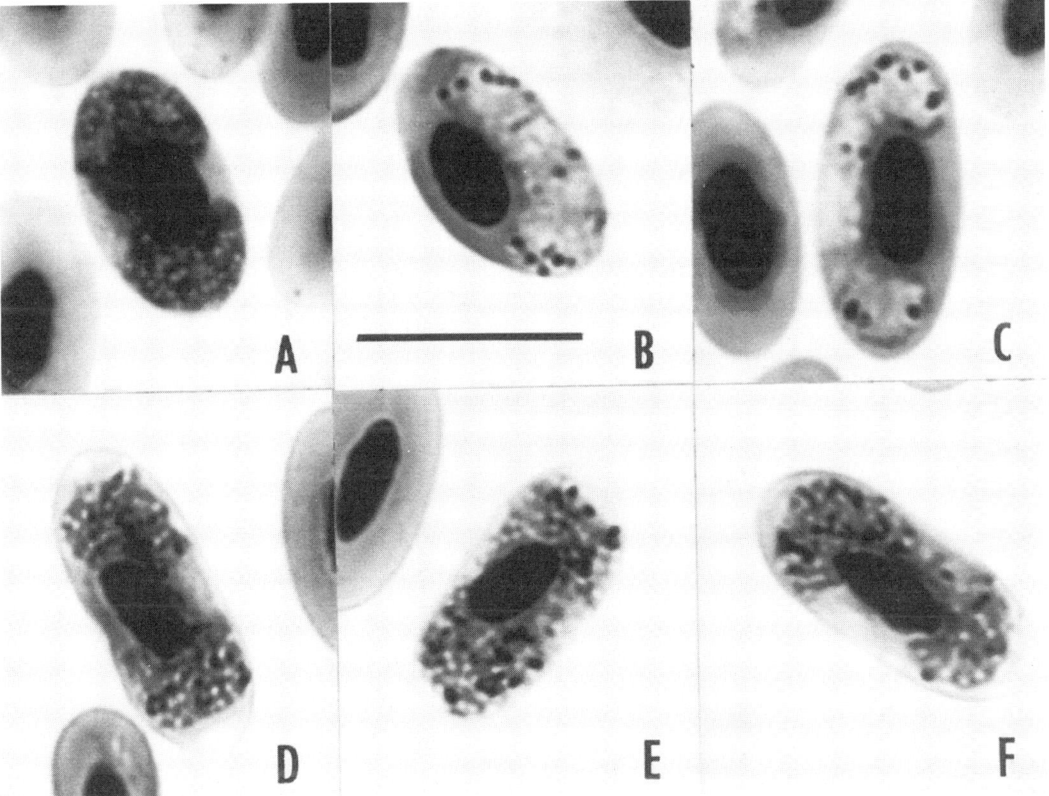

FIGURE 23.4. Haemoproteid blood parasites from owls in Florida. *A, Haemoproteus noctuae* from a Barred Owl; *B–C, H. syrnii* from an Eastern Screech-Owl; *D–E, H. noctuae* from a Great Horned Owl; *F, H. syrnii* from a Great Horned Owl. All parasites are macrogametocytes. Horizontal bar = 10 μm. Courtesy of Sam R. Telford, Jr.

species), and *Plasmodium forresteri* (7 infections, 3 host species). Mixed infections of 2, 3, and 4 species of blood protozoans occurred in a number of owls; the most common were infections of *Haemoproteus syrnii* and *H. noctuae*, in 22 Barred Owls and 7 Great Horned Owls, indicating that these 2 blood parasites may have the same arthropod vector or that the vectors have the same ecological requirements. The species of *Haemoproteus* and *Plasmodium* (except *P. relictum*) are illustrated in figures 23.4 and 23.5. The pathologic significance of these infections in owls is not known. There is some evidence that one of the species, *Leucocytozoon ziemanni*, may cause a reduction in clutch size of female owls during times of food

shortage (Korpimäki et al. 1993), but this has not been studied in Florida. Evans and Otter (1998) concluded that a mixed infection of *Haemoproteus noctuae* and *Leucocytozoon ziemanni* resulted in the death of a juvenile Snowy Owl (*Nyctea scandiaca*). The taxonomy of the haemoproteids of owls has been discussed by Bishop and Bennett (1989). Telford et al. (1997) gave detailed descriptions of strains of *Plasmodium forresteri* and *P. elongatum* isolated from Barred Owls in Florida. The mosquito vectors of a strain of *Plasmodium elongatum* from an Eastern Screech-Owl were determined experimentally by Nayar et al. (1998). They tested 6 species of Florida mosquitoes and found that 3 (*Culex nigripalpus, C.*

Table 23.14. Protozoan infections[a] in Barred Owls[b] from Florida

Protozoan	County/site	Year(s)	Examined	Infected	%
Haemoproteus syrnii[c]	Alachua	1979–91	13	9	69
	Alachua	1995–96	12	7	58
	Brevard	1995	1	1	—
	Collier	1990–91	5	4	—
	Gilchrist	1989	1	1	—
	Gilchrist	1995	1	1	—
	Glades	1980	1	1	—
	Marion	1973, 1990	2	1	—
	Putnam	1988, 1991	2	1	—
	St. Johns	1989	2	2	—
	Orange	1996	2	1	—
	Osceola	1993	1	1	—
	Seminole	1995–96	2	2	—
	Sumter	1996	1	1	—
	Volusia	1996	4	3	—
	Unknown Others[d]	1995	1	1	—
Haemoproteus noctuae[c]	Alachua	1979–91	13	5	38
		1995–96	12	3	25
	Brevard	1995	1	1	—
	Collier	1990–91	5	2	—
	Gilchrist	1995	1	1	—
	Glades	1980	1	1	—
	Marion	1973, 1990	2	1	—
	Orange	1996	2	2	—
	Seminole	1995–96	2	2	—
	St. Johns	1989	2	2	—
	Volusia	1996	4	2	—
	Unknown Others[d]	1995	1	1	—

(continued)

restuans, and *C. salinarius*) transmitted *P. elongatum*. Because of the breeding patterns and population peaks of these vectors it can be concluded that *P. elongatum* is transmitted to screech-owls throughout the year in Florida. Other details on infections of blood protozoans in owls in Florida have been presented by Forrester et al. (1994).

Trichomonas gallinae was identified in 1 of 3 nestling Eastern Screech-Owls found in 1976 when a tree was cut down near Newberry (Alachua County). The infected owl was depressed (figure 23.6) and unable to open its mouth very wide. It died on May 10, 1976, and at necropsy was diagnosed as having trichomonosis (= frounce). There was a large yellowish caseous lesion in the mouth cavity and throat (figure 23.7) and the lesion had progressed into the brain (Forrester 1989). Screech Owls feed extensively on insects, although they are known to prey also on Rock Doves and other birds (Bent 1938). In this case the nestling had most likely eaten an infected dove or pigeon brought to it by its parent. This disease is probably the earliest known parasitic disease of wildlife for which there are records (Kocan and Hermann 1971). Stabler (1954) referred to a 1619 reference on this disease as recognized by

Table 23.14. *(continued)*

Protozoan	County/site	Year(s)	Examined	Infected	%
Haemoproteus sp.	Alachua	1995–96	12	1	8
Plasmodium circumflexum[c]	Marion	1973, 90	2	1	—
	Others[d]				
Plasmodium elongatum[c]	Alachua	1979–91	13	1	8
	Brevard	1995	1	1	—
	Others[d]				
Plasmodium forresteri	Alachua	1995–96	12	3	25
	Gilchrist	1995	1	1	—
	Volusia	1996	4	1	—
	Others[d]				
Plasmodium relictum[c]	Alachua	1979–91	13	4	31
	Others[d]				
Trypanosoma confusum	Alachua	1979–91	13	1	8
	Others[d]				
Trichomonas gallinae	Alachua	1990–91	NG	2	—
	Osceola	1992	NG	1	—

Sources: Blood parasites: Telford et al. (1993, 1996); *Trichomonas gallinae*: Collins and Gilliland (1995). NG = not given by authors (Collins and Gilliland 1995).

a. Infections of blood protozoans were diagnosed by examination of thin blood smears; *Trichomonas* infections were diagnosed by examination for characteristic lesions (= presumptive diagnosis).

b. In addition to the data in the table, 4 Barred Owls were examined by blood smear and were negative: 1 from Bradford County (1991), 1 from Hillsborough County (1975), 2 from Martin County (1973, 1990).

c. Listed also in Bishop and Bennett (1992) as being found in Barred Owls in Florida, but without specific localities, dates, or prevalences.

d. Fifty-three owls were examined, from the following counties: Alachua (n = 25, 1979–96), Bradford (n = 1, 1991), Brevard (n = 1, 1995), Collier (n = 5, 1990–91), Gilchrist (n = 2, 1989, 1995), Glades (n = 1, 1980), Hillsborough (n = 1, 1975), Marion (n = 2, 1973, 1990), Putnam (n = 2, 1988, 1991), Orange (n = 2, 1996), Osceola (n = 1, 1993), Seminole (n = 2, 1995–96), St. Johns (n = 2, 1989), Sumter (n = 1, 1996), Volusia County (n = 4, 1996), unknown (n = 1, 1995). For each parasite, data are given for counties in which positive birds were found; negatives are not included in the table.

falconers. Apparently they realized back then that the infection was acquired by captive hawks eating infected pigeons. Five other cases of trichomonosis were found in the records of the Center for Birds of Prey in Maitland (Collins and Gilliland 1995). Three were in Barred Owls, 1 in a Great Horned Owl, and 1 in an Eastern Screech-Owl (tables 23.14–23.16). For more information on trichomonosis the reader is referred to chapter 22, Pigeons and Doves.

Oocysts of a species of *Eimeria* were seen in feces of a Great Horned Owl from Alachua County in 1978 (Forrester 1989). The species was not determined. This may have been a case of spurious parasitism, the oocysts being from a rodent that the owl had eaten. Henry (1932) described a species of *Isospora* (*I. buteonis*) from a Great Horned Owl in California, but there are no known species of *Eimeria* from Great Horned Owls (or any other owls, for that matter) in North America (Pellerdy 1974).

No serologic evidence of the occurrence of *Toxoplasma gondii* was found in 3 Eastern Screech-Owls from Florida examined by the use of the indirect hemagglutination (IHA) test (Burridge et al. 1979). However, as pointed out elsewhere, the IHA test may not be as sensitive in birds as in mammals and may result in false

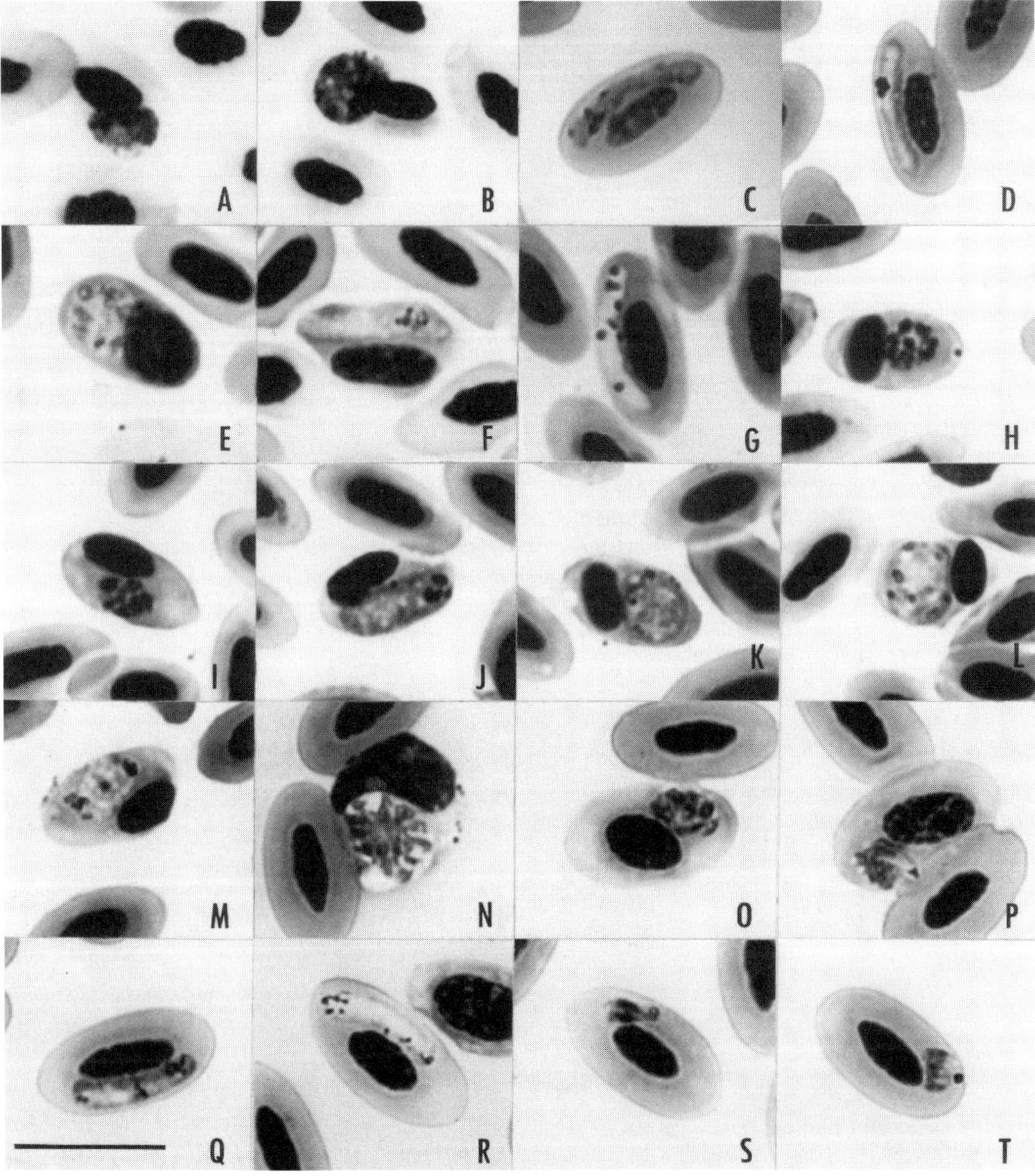

FIGURE 23.5. Species of *Plasmodium* from owls in Florida. *A–B,* binucleate schizonts of *Plasmodium circumflexum* from a Barred Owl; *C–D,* gametocytes of *P. elongatum* from a Barred Owl; *E,* schizont of *P. elongatum* from an Eastern Screech-Owl; *F–G,* gametocytes from an Eastern Screech-Owl; *H–I,* schizonts of *P. subpraecox* from an Eastern Screech-Owl; *J–L,* macrogametocytes of *P. subpraecox* from an Eastern Screech-Owl; *M,* microgametocyte of *P. subpraecox* from an Eastern Screech-Owl; *N–P,* schizonts of *P. elongatum* from a Great Horned Owl; *Q–R,* gametocytes of *P. elongatum* from a Great Horned Owl; *S–T,* schizonts of *Plasmodium forresteri* from a Great Horned Owl. Horizontal bar = 10 μm. Courtesy of Sam R. Telford, Jr.

FIGURE 23.6. Two nestling Eastern Screech-Owls obtained from a nesting cavity in a fallen tree in Alachua County, 1976. The nestling on the left (*A*) is normal; the one on the right (*B*) has trichomonosis or frounce, caused by *Trichomonas gallinae*. The sick nestling died 14 days after this photograph was taken. Courtesy of Garry W. Foster.

FIGURE 23.7. A nestling Eastern Screech-Owl from Alachua County that died of trichomonosis in 1976. This is the same bird shown on the right (*B*) in Figure 23.6, above. Note the large caseous lesion (*arrow*) in the oral cavity. Courtesy of Garry W. Foster.

Table 23.15. Protozoan infections in Great Horned Owls from Florida[a]

Protozoan	County	Year(s)	No. of owls		Basis of diagnosis[b]
			Exam.	Inf.	
Haemoproteus syrnii[c,d]	Alachua	1974–89	9	4	BS
	Citrus	1995	1	1	BS
	Collier	1990	1	1	BS
	Columbia	1996	1	1	BS
	Lake	1996	1	1	BS
	Martin	1995–96	6	2	BS
	Orange	1996	1	1	BS
	Volusia	1996	1	1	BS
	Unknown	1994–95	3	1	BS
Haemoproteus noctuae[c,d]	Alachua	1974–89	9	4	BS
	Lake	1996	1	1	BS
	Martin	1995–96	6	2	BS
	Orange	1996	1	1	BS
	Volusia	1996	1	1	BS
	Unknown	1994–95	3	1	BS
Plasmodium elongatum[c,d]	Alachua	1974–89	9	1	BS
Plasmodium relictum[c,d]	Alachua	1974–89	9	1	BS
Plasmodium forresteri[d]	Alachua	1974–89	9	1	BS
	Lake	1996	1	1	BS
Eimeria sp.	Alachua	1974, 78	2	1	FE
Trichomonas gallinae	Orange	1992	NG	1	LE

Sources: Blood protozoans: Telford et al. (1993, 1996); *Eimeria*: Forrester (1989); *Trichomonas gallinae*: Collins and Gilliland (1995).
NG = not given.
a. In addition to the data in the table, 2 Great Horned Owls were examined by blood smear and were negative: 1 from Alachua County (1995), 1 from Hernando County (1996).
b. BS = thin blood smear, FE = flotation of fecal sample to visualize oocysts, LE = examined for characteristic lesion (= presumptive diagnosis).
c. Listed also in Bishop and Bennett (1992) as found in Great Horned Owls in Florida, without specific localities, dates, or prevalences.
d. Twenty-four owls were examined, from the following counties: Alachua (*n* = 9, 1974–89), Citrus (*n* = 1, 1995), Collier (*n* = 1, 1990), Columbia (*n* = 1, 1996), Lake (*n* = 1, 1996), Martin (*n* = 6, 1995–96), Orange (*n* = 1, 1996), Volusia (*n* = 1, 1996), unknown (*n* = 3, 1994–95). For each parasite, data are given for counties in which positive birds were found; negatives are not included in the table.

negative results (Frenkel 1981). Lindsay et al. (1993) found cysts of *T. gondii* in 1 of 50 Great Horned Owls and 4 of 15 Barred Owls in Alabama; infections might be identified in Florida owls if larger samples were obtained and appropriate techniques applied.

XV. Helminths

The helminths of owls that breed in Florida have been studied to some extent. Barred Owls (table 23.18) and Eastern Screech-Owls (table 23.19) have been fairly well sampled, but only limited information is available for Barn Owls (table 23.20), Great Horned Owls (table 23.21), and Burrowing Owls (table 23.22). Thirty-four species of helminths have been recorded and include 12 trematodes, 2 cestodes, 18 nematodes, and 2 acanthocephalans. Nothing is known about other owls that occur but do not breed in Florida, except for one report of the nematode *Porrocaecum depressum* in a Short-eared Owl (date and specific locality in Florida not given)

Table 23.16. Protozoan infections in Eastern Screech-Owls from Florida[a]

Protozoan	County (or site)	Year(s)	No. of owls Exam.	Inf.	Basis of diagnosis[b]
Haemoproteus syrnii[c,d]	Alachua	1974–91	6	1	BS
		1996	1	1	BS
	Collier	1990–91	4	3	BS
	Seminole	1995	1	1	BS
Haemoproteus noctuae[d]	Seminole	1995	1	1	BS
Plasmodium elongatum[c,d]	Collier	1990–91	4	1	BS
Plasmodium forresteri[d]	Seminole	1995	1	1	BS
Plasmodium subpraecox[c,d]	Hillsborough	1975, 1991	2	1	BS
Trichomonas gallinae	Alachua	1976	3	1	CU
	Volusia	1993	NG	1	LE
	Others[e]	1974–78	32	0	LE

Sources: Blood protozoans: Telford et al. (1993, 1996); *Trichomonas:* Forrester (1989), Collins and Gilliland (1995). NG = not given by authors.

a. In addition to the data in the table, 2 Eastern Screech-Owls were examined by blood smear for protozoans and were negative: 1 from Levy County, 1995 (Telford et al. 1996), 1 from Osceola County, 1993 (Garvin and Spalding 1995).

b. BS = thin blood smear, CU = axenic culture (Diamond 1957), LE = examined for characteristic gross lesions (= presumptive diagnosis).

c. Listed also in Bishop and Bennett (1992) as found in Eastern Screech-Owls in Florida, without specific localities, dates, or prevalences.

d. Fourteen owls were examined, from the following counties: Alachua (*n* = 7, 1974–96), Collier (*n* = 4, 1990–91), Seminole (*n* = 1), Hillsborough (*n* = 2). For each parasite, data are given for counties in which positive birds were found; negatives are not included in the table.

e. Includes owls from 8 counties: Alachua (*n* = 6), Citrus (*n* = 2), Hillsborough (*n* = 3), Lafayette (*n* = 1), Marion (*n* = 1), Pasco (*n* = 3), Pinellas (*n* = 14), Putnam (*n* = 2).

by Walton (1927). There is also a curious report of the acanthocephalan *Centrorhynchus spinosus* from a Great Gray Owl (*Strix nebulosa*) in Florida by Leidy (1887) and later referred to by Van Cleave (1924). No date or specific locality was given. Since the Great Gray Owl does not occur in Florida, this might have been a captive bird brought in from elsewhere, or the identification of the owl might have been in error.

The life cycles and effects of these helminths

Table 23.17. Blood protozoan infections[a] in Barn Owls and Burrowing Owls from Florida

Species of owl Protozoan	County	Year(s)	No. of owls Exam.	Inf.
Barn Owl				
Leucocytozoon ziemanni	Alachua	1976	2	1
		1995	2	0
	Collier	1991	1	0
	Pinellas	1975	1	0
Burrowing Owl				
Blood protozoans	Citrus	1975–80	1	0
	Hillsborough	1975–80	29	0

Sources: Barn Owls: Telford et al. (1993, 1996); Burrowing Owls: Dooris et al. (1981).
a. Diagnosed by examination of thin blood smears.

Table 23.18. Helminth infections in Barred Owls from Florida

			No. of owls[a]		Intensity	
Helminth	County or site	Year(s)	Exam.	Inf.	Mean	Range
Trematoda						
Brachylaema mcintoshi						
	Alachua	1971–92	8	1	4	—
	Hillsborough	1975–76	2	2	3	1–5
	Levy	1972–76	2	1	1	—
Brachylecithum rarum						
	Alachua	1971–92	8	1	12	—
	Marion	1973–82	3	1	5	—
Clinostomum complanatum						
	Pinellas	1987	1	1	20	—
Neodiplostomum americanum						
	Alachua	1971–92	8	4	29	7–62
	Hillsborough	1975–76	2	2	16	5–26
	Levy	1972–76	2	1	25	—
	Marion	1973–82	3	2	2	1–3
	Pasco	1975	1	1	12	—
	Sumter	1976	1	1	1	—
	Taylor	1976	1	1	42	—
	"No. central Fla."	1986	2	2	ND	ND
Neodiplostomum reflexum						
	Alachua	1971–92	8	3	12	1–29
	Hillsborough	1975–76	2	2	82	80–83
	Pasco	1975	1	1	16	—
	Suwannee	1976	1	1	2	—
	Taylor	1976	1	1	83	—
Strigea elegans						
	Pasco	1975	1	1	4	—
	Taylor	1976	1	1	3	—
	"No. central Fla."	1986	1	1	ND	ND
Nematoda						
Capillaria dispar						
	Levy	1972–76	2	1	2	—
Capillaria falconis						
	Marion	1973–82	3	1	2	—
	Taylor	1976	1	1	2	—
Capillaria tenuissima						
	Alachua	1971–92	8	2	9	6–11
	Hernando	1975–77	2	1	3	—
	Hillsborough	1975–76	2	2	7	2–12
	Pasco	1975	1	1	9	—
	Putnam	1975	1	1	11	—
	Suwannee	1981	1	1	7	—
	Taylor	1976	1	1	5	—
Chandlerella longigutturata						
	Alachua	1971–92	8	1	2	—
	Hernando	1975–77	2	1	4	—
	Levy	1972–76	2	1	1	—

(continued)

Table 23.18. *(continued)*

Helminth	County or site	Year(s)	No. of owls[a]		Intensity	
			Exam.	Inf.	Mean	Range
	Pasco	1975	1	1	1	—
	Putnam	1975	1	1	5	—
	Taylor	1976	1	1	4	—
Excisa excisiformis						
	Suwannee	1981	1	1	2	—
Porrocaecum depressum						
	Marion	1973–82	3	1	4	—
	"No. central Fla."	1985	1	1	ND	ND
	"No. central Fla."	1988	1	1	ND	ND
Strongyloides sp.						
	Alachua	1971–92	8	1	3	—
	Marion	1973–82	3	1	1	—
Tetrameres microspinosa						
	Alachua	1971–92	8	1	1	—
	Levy	1972–76	2	1	8	—
	Pasco	1975	1	1	2	—
	Sumter	1976	1	1	1	—
Tetrameres strigiphila[b]						
	Marion	1973–82	3	1	40	—
Acanthocephala						
Centrorhynchus spinosus						
	Alachua	1971–92	8	7	16	1–41
	Hernando	1975–77	2	2	18	7–28
	Hillsborough	1975–76	2	2	1	—
	Lake	1974	1	1	13	—
	Levy	1972–76	2	1	22	—
	Marion	1973–82	3	2	18	12–24
	Pasco	1975	1	1	4	—
	Putnam	1975	1	1	1	—
	Taylor	1976	1	1	2	—
	"No. central Fla."	1986	4	4	ND	ND
	"No. central Fla."	1988	1	1	ND	ND

Sources: Acanthocephalans: Humphlett and Forrester (1989); No. central Fla.: Kinsella and Greiner (1993); other: Kinsella and Forrester (1993).

ND = not determined.

a. Thirty-four owls were examined, from the following counties: Alachua (n = 8, 1971–92), Hernando (n = 2, 1975–77), Hillsborough (n = 2, 1975–76), Lake (n = 1, 1974), Levy (n = 2, 1972–76), Marion (n = 3, 1973–82), Pasco (n = 1, 1975), Putnam (n = 1, 1975), Sumter (n = 1, 1976), Suwannee (n = 1, 1981), Taylor (n = 1, 1976), "No. central Fla." (n = 11, 1985–88). For each helminth, data are given for counties in which positive birds were found; negatives are not included in the table.

b. Described originally as *T. strigiphila* from specimens obtained from a Barred Owl in Ocala, 1973 (Pence et al. 1975). Mollhagen (1976) studied these specimens further and declared *T. strigiphila* a synonym of *T. paradoxa,* but in an unpublished thesis rather than in a recognized journal. This is not valid and the name *T. strigiphila* stands.

Table 23.19. Helminth infections in Eastern Screech-Owls from Florida

Helminth	County or site	Year(s)	No. of owls[a] Exam.	No. of owls[a] Inf.	Intensity Mean	Intensity Range
Trematoda						
Neodiplostomum americanum						
	Pinellas	1975–76	14	1	5	—
Neodiplostomum reflexum						
	Alachua	1974–78	6	1	17	—
Plagiorchis speotyonis						
	Pasco	1975–76	3	1	4	—
Cestoda						
Choanotaenia speotytonis						
	Alachua	1974–78	6	1	14	—
	Citrus	1977–78	2	1	16	—
Nematoda						
Capillaria sp.						
	Hillsborough	1975	3	1	1	—
	Pinellas	1975–76	14	2	1	—
	Putnam	1975–77	2	1	1	—
Excisa excisiformis						
	Alachua	1974–78	6	2	1	—
	Pinellas	1975–76	14	1	6	—
	Putnam	1975–77	2	1	3	—
Dispharynx nasuta						
	Pinellas	1975–76	14	3	19	7–41
	"No. central Fla."	1991	1	1	ND	ND
Larval spirurids						
	Alachua	1974–78	6	2	1	—
	Pasco	1975–76	3	1	1	—
	Pinellas	1975–76	14	1	1	—
Physocephalus sexalatus						
	No. Florida	NG	NG	NG	NG	—
Strongyloides sp.						
	Hillsborough	1975	3	1	1	—
	Pinellas	1975–76	14	1	4	—
Subulura reclinata						
	Alachua	1974–78	6	1	2	—
Acanthocephala						
Centrorhynchus spinosus						
	Alachua	1974–78	6	2	9	4–14
	Citrus	1977–78	2	1	2	—
	Hillsborough	1975	3	2	2	1–3
	Marion	1977	1	1	1	—
	Pinellas	1975–76	14	1	2	—
	"No. central Fla."	1991	1	1	ND	ND

Sources: Acanthocephalans: Humphlett and Forrester (1989); No. central Fla.: Kinsella and Greiner (1993); *Physocephalus sexalatus:* Cram (1930); other: Kinsella and Forrester (1993).

ND = not determined, NG = not given.

a. Thirty-four owls were examined from the following counties: Alachua (*n* = 6, 1974–78), Citrus (*n* = 2, 1977–78), Hillsborough (*n* = 3, 1975), Lafayette (*n* = 1, 1976), Marion (*n* = 1, 1977), Pasco (*n* = 3, 1975–76), Pinellas (*n* = 14, 1975–76), Putnam (*n* = 2, 1975–77), "No. central Fla." (*n* = 2, 1991). For each helminth, data are given for counties in which positive birds were found; negatives are not included in the table.

Table 23.20. Helminth infections in Barn Owls from Florida

Helminth	County or site	Year(s)	No. of owls[a]		Intensity	
			Exam.	Inf.	Mean	Range
Trematoda						
Neodiplostomum americanum						
	Alachua	1976–84	5	3	4	3–7
	Others[a]	1975–78	4	0	—	—
Nematoda						
Dispharynx affinis						
	Alachua	1976–84	5	1	5	—
	Gilchrist	1977	1	0	—	—
	Palm Beach	1977–78	2	1	5	—
	Pinellas	1975	1	1	1	—
Larval spirurids						
	Alachua	1976–84	5	1	1	—
	Others[a]	1975–78	4	0	—	—
Porrocaecum depressum						
	Alachua	1976–84	5	1	1	—
	Others[a]	1975–78	4	0	—	—
Microfilariae[b]						
	Alachua	1976	2	0	—	—
	Pinellas	1975	1	1	ND	ND

Sources: Kinsella and Forrester (1993) unless otherwise indicated.
ND = not determined.
a. Negative birds include 1 from Gilchrist County (1977), 2 from Palm Beach County (1977–78), 1 from Pinellas County (1975).
b. Determined by examination of thin blood smears (Telford et al. 1993); adults were not found.

(and others) on owls in Florida are unknown and should be investigated.

XVI. Arthropods

Twenty-eight species of parasitic arthropods have been reported from owls in Florida, including 1 tick, 12 mites, 10 chewing lice, 4 louse flies, and 1 bird fly. The 1 record of a tick was 1 larva of *Haemaphysalis leporispalustrus* collected from an "owl" (species not given) in northern Florida (county not given) between 1949 and 1953 by Rogers (1953). Data are available from all 5 species of owls that breed in Florida (tables 23.23–23.27). In addition there is a report of a chewing louse (*Strigiphilus cursor*) from a Short-eared Owl from Dade County

in 1916 in the records of the U.S. National Museum, Smithsonian Institution, Washington, D.C. (Forrester et al. 1995). Nothing is known of the significance of these arthropod infestations to the health and welfare of owl populations in Florida. Cannings (1986) suggested that the bird fly *Carnus hemapterus* (a close relative of *C. floridensis*) was harmful to nestling Northern Saw-whet Owls, but Kirkpatrick and Colvin (1989) found no evidence that this fly caused a reduction in the survival of Barn Owl nestlings. In addition Dawson and Bortolotti (1997) found that *C. hemapterus* was not pathogenic to nestling American Kestrels. The louse fly (*Icosta americana*) may be a vector of the blood protozoans *Haemoproteus noctuae* and *H. syrnii* in Barred Owls, Great Horned Owls, and Eastern Screech-Owls, but this has not been determined.

Table 23.21. Helminth infections in Great Horned Owls from Florida

| Helminth | County or site | Year(s) | No. of owls[a] | | Intensity | |
			Exam.	Inf.	Mean	Range
Trematoda						
Brachylaima mcintoshi						
	Hillsborough	1976	1	1	5	—
Neodiplostomum americanum						
	Alachua	1973–89	5	4	766	2–2,353
	Columbia	1981	1	1	11	—
	Duval	1974	1	1	46	—
	Hillsborough	1976	1	1	24	—
	Jefferson	1981	1	1	17	—
	"No. central Fla."	1985	1	1	ND	ND
Neodiplostomum reflexum						
	Alachua	1973–89	5	2	1,769	5–3,533
	Columbia	1981	1	1	8	—
	Jefferson	1981	1	1	8	—
Stomylotrema vicarium						
	Alachua	1973–89	5	2	12	2–21
	Columbia	1981	1	1	4	—
	Duval	1974	1	1	14	—
	Hillsborough	1976	1	1	13	—
	Jefferson	1981	1	1	2	—
	"No. central Fla."	1985	1	1	ND	ND
Strigea elegans						
	Alachua	1973–89	5	4	12	2–27
	Duval	1974	1	1	108	—
	Hillsborough	1976	1	1	3	—
	Jefferson	1981	1	1	2	—
	Suwannee	1979	1	1	2	—
	"No. central Fla."	1985	1	1	ND	ND
Cestoda						
Raillietina bumi						
	"No. central Fla."	1985	1	1	ND	ND
Nematoda						
Capillaria falconis						
	Alachua	1973–89	5	1	14	—
	Duval	1974	1	1	—	—
	Jefferson	1981	1	1	4	—
	Suwannee	1979	1	1	7	—
Capillaria tenuissima						
	Alachua	1973–89	5	3	10	1–23
	Columbia	1981	1	1	21	—
	Duval	1974	1	1	15	—
	Hillsborough	1976	1	1	2	—
	Jefferson	1981	1	1	8	—
	Suwannee	1979	1	1	6	—
Chandlerella longigutturata						
	Jefferson	1981	1	1	1	—
Excisa excisiformis						
	Alachua	1973–89	5	1	7	—
	Duval	1974	1	1	3	—

(continued)

Table 23.21. *(continued)*

Helminth	County or site	Year(s)	No. of owls[a]		Intensity	
			Exam.	Inf.	Mean	Range
Lemdana wernaarti						
	Alachua	1973–89	5	1	1	—
Porrocaecum depressum						
	Alachua	1973–89	5	2	8	1–14
	Columbia	1981	1	1	2	—
	Jefferson	1981	1	1	18	—
	Suwannee	1979	1	1	1	—
	"No. central Fla."	1985	1	1	ND	ND
	"No. central Fla."	NG	1	1	ND	ND
Synhimantus laticeps						
	Alachua	1973–89	5	1	1	—
Synhimantus affinis						
	Duval	1974	1	1	1	—
Larval spirurids						
	Suwannee	1979	1	1	4	—
Acanthocephala						
Centrorhynchus kuntzi						
	Alachua	1973–89	5	1[b]	2	—
	Duval	1974	1	1	1	—
	Jefferson	1981	1	1	1	—
Centrorhynchus spinosus						
	Alachua	1973–89	5	5[b]	13	1–340
	Columbia	1981	1	1	2	—
	Duval	1974	1	1	162	—
	Jefferson	1981	1	1	2	—
	Suwannee	1979	1	1	10	—
	"No. central Fla."	1985	1	1	ND	ND

Sources: Acanthocephalans: Humphlett and Forrester (1989); No. central Fla.: Kinsella and Greiner (1993); other: Kinsella and Forrester (1993).

ND = not determined.

a. Fifteen owls were examined, from the following counties: Alachua ($n = 5$, 1973–89), Columbia ($n = 1$, 1981), Duval ($n = 1$, 1974), Hillsborough ($n = 1$, 1976), Jefferson ($n = 1$, 1981), Pinellas ($n = 1$, 1976), Suwannee ($n = 1$, 1979), "No. central Fla." ($n = 4$, 1985–86). For each helminth, data are given for counties in which positive birds were found; negatives are not included in the table.

b. Nickol (1983) examined "ten of several dozen" specimens from 1 of these Alachua County owls and found a mixed infection of 2 *Centrorhynchus kuntzi* and 8 *C. spinosus*.

XVII. Pentastomes

A nymphal pentastome (probably *Porocephalus crotali*) was found in the intestines of a Barred Owl from north central Florida (exact location unknown) in 1985 (Kinsella and Greiner 1993). This was probably a case of accidental parasitism in which the owl ingested an infected rodent. Barred Owls feed primarily on small mammals such as rats and mice (Snyder and Wiley 1976), several species of which are known to be infected with nymphs of *P. crotali*. In Florida infections have been found in cotton rats, rice rats, Florida mice, and cotton mice (Forrester 1992).

Table 23.22. Helminth infections in Burrowing Owls from Florida

Helminth	County	Year(s)	No. of owls Exam.	Inf.	Mean intensity
Trematoda					
Neodiplostomum americanum					
	Hillsborough	1971	1	0	—
	Lee	1989	6	0	—
	Monroe	1988–89	2	1	1
Maritrema patulus					
	Hillsborough	1971	1	0	—
	Lee	1989	6	0	—
	Monroe	1988–89	2	1	2
Maritrema sp.					
	Hillsborough	1971	1	0	—
	Lee	1989	6	0	—
	Monroe	1988–89	2	1	1
Prosthogonimus ovatus					
	Hillsborough	1971	1	0	—
	Lee	1989	6	1	1
	Monroe	1988–89	2	0	—
Microphallus pygmaeum					
	Hillsborough	1971	1	0	—
	Lee	1989	6	0	—
	Monroe	1988–89	2	1	1
Nematoda					
Subulura forcipata					
	Hillsborough	1971	1	0	—
	Lee	1989	6	1	1
	Monroe	1988–89	2	0	—
Capillaria sp.					
	Hillsborough	1971	1	0	—
	Lee	1989	6	0	—
	Monroe	1988–89	2	1	1

Source: Kinsella and Foster (1994).

XVIII. Emaciation

Emaciation or nutritional deficiency was the primary diagnostic finding for 8% of 1,906 owls (representing 5 species) that were submitted to 3 wildlife rehabilitation centers in Florida from 1988 to 1995 (table 23.2). In some of these cases there were other conditions, such as trauma, probably related to the emaciation and that may have led to it, but the records are not clear on the relative significance of these or other secondary factors. The cause or causes of the emaciation in these cases was essentially not determined. Franson and Little (1996) classified 42 (32%) of 132 Great Horned Owls examined at necropsy as emaciation cases. They were able to identify factors that might have contributed to the emaciation in only 16 of these owls, including ocular lesions, oral lesions, a beak deformity, intestinal impaction, puncture wounds, joint dislocations and old fractures, and dieldrin residues. No contributing factors were identified for the remaining 26 emaciated owls.

Table 23.23. Parasitic arthropods reported from Barn Owls from Florida

Arthropod	County	Year(s)	Data source
Feather mites			
Dermonoton parallelus	Alachua	1984	Forrester & Mertins (1992)
Dermonoton sp.[a]	Alachua	1984	Ibid.
Glaucalges attenuatus	Alachua	1983–84	Ibid.
	Pinellas	1975	Ibid.
Kramerella quadrata	Alachua	1983–84	Ibid.
	Pinellas	1975	Ibid.
Pandalura strigisoti	Alachua	1983–84	Ibid.
	Pinellas	1975	Ibid.
Chewing lice			
Kurodaia subpachygaster	Alachua	1983–84	Forrester et al. (1995)
	Lee	1986	Ibid.
Louse flies			
Icosta americana	Brevard	1974	Wilson et al. (1990)

a. An undescribed species, also found on Barred Owls (see table 23.27).

Table 23.24. Parasitic arthropods reported from Burrowing Owls from Florida

Arthropod	County/area	Year(s)	Data source
Ticks[a]			
Feather mites			
Dermonoton parallelus	Lee[b]	1989	Foster & Mertins (1994)
Glaucalges sp.	Lee[c]	1989	Ibid.
Kramerella sp.	Lee[d]	1989	Ibid.
Chewing lice			
Colpocephalum pectinatum	Hendry	1933	Forrester et al. (1995)
	Sarasota	1918	Ibid.
	"Florida"	NG	Peters (1936)
Strigiphilus speotyti	Collier	1933	Forrester et al. (1995)
	Hendry	1933	Ibid.
	Lee[e]	1989	Ibid.
	Monroe	1988	Ibid.

NG = not given.
a. Travis (1941) examined 2 Burrowing Owls from Florida (Orange, Osceola, or Collier County) in 1936–37; Foster and Mertins (1994) examined 7 in 1989; none was infested with ticks.
b. 7 of 7 owls were infested; mean intensity = 17 (range = 1–95).
c. 3 of 7 owls were infested with this undescribed species; mean intensity = 24 (range = 2–61).
d. 2 of 7 owls were infested; intensities were 1 and 3.
e. 1 of 7 owls was infested; intensity = 1.

Table 23.25. Parasitic arthropods reported from Eastern Screech-Owls from Florida

Arthropod	County	Year(s)	Data source
Mites			
Ornithonyssus sylviarum	Hillsborough	1985	Forrester & Milstrey (1989)
Louse flies			
Icosta americana	Alachua	1976	Wilson et al. (1990)
	Brevard	NG	Bequaert (1955)
	Dade	NG	Johnson (1913)
	Escambia	1929	FSCA[a], Wilson et al. (1990)
	Hernando	NG	Bequaert (1955)
	Indian River	1967, 1972	Wilson et al. (1990)
	Manatee	NG	Bequaert (1955)
	St. Johns	NG	Johnson (1922), Bequaert (1955)
	St. Lucie	1966	Wilson et al. (1990)
Pseudolynchia canariensis	Escambia	1929	FSCA[a], Wilson et al. (1990)
Bird flies			
Carnus floridensis	NG	NG	Bequaert (1942), Grimaldi (1997)

NG = not given.
a. FSCA = records of the Florida State Collection of Arthropods, Division of Plant Industry, Department of Agriculture and Consumer Services, Gainesville.

Table 23.26. Parasitic arthropods reported from Great Horned Owls from Florida

Arthropod	County	Year(s)	Data source
Feather mites			
Dermonoton parallelus	Alachua	1973	Forrester & Mertins (1992)
	Duval	1974	Ibid.
Glaucalges attenuatus	Alachua	1973–74	Ibid.
	Columbia	1981	Ibid.
	Duval	1974	Ibid.
	Jefferson	1981	Ibid.
Kramerella sp. 1[a]	Alachua	1973	Ibid.
	Jefferson	1981	Ibid.
Pandalura strigisoti	Alachua	1973	Ibid.
	Columbia	1981	Ibid.
	Jefferson	1981	Ibid.
Chewing lice			
Colpocephalum brachysomum	Lee	1981, 1983	Forrester et al. (1995)
	Marion	1956	Emerson (1961)
Kurodaia magna	Dade	1917	Forrester et al. (1995)
Strigiphilus oculatus	Marion	1956	Ibid.
Strigiphilus syrnii	Alachua	1973	Ibid.
	Jefferson	1981	Ibid.
Louse flies			
Icosta americana	Alachua	1954	FSCA[b], Wilson et al. (1990)
	Brevard	NG	Bequaert (1955)
	Dade	NG	Ibid.
	Leon	1963–75	Wilson et al. (1990)
	Levy	1979	Ibid.
Ornithoica vicina	Leon	1963	Ibid.

NG = not given.
a. An undescribed species.
b. FSCA = records of the Florida State Collection of Arthropods, Division of Plant Industry, Department of Agriculture and Consumer Services, Gainesville.

Table 23.27. Parasitic arthropods reported from Barred Owls from Florida

Arthropod	County/site	Year(s)	Data source
Ticks[a]			
Feather mites			
Dermonoton sp.[b]	Alachua	1973–85	Forrester & Mertins (1992)
	Marion	1982	Ibid.
	Suwannee	1981	Ibid.
	Taylor	1976	Ibid.
Kramerella sp. 2	Alachua	1985	Ibid.
Petitota aluconis	Alachua	1985	Ibid.
	Hillsborough	1976	Ibid.
	Taylor	1976	Ibid.
Pandalura strigisoti	Alachua	1976	Ibid.
	Hillsborough	1976	Ibid.
	Marion	1982	Ibid.
	Suwannee	1981	Ibid.
	Taylor	1976	Ibid.
Quill mites			
Bubophilus ascalaphus	Alachua	1985	Ibid.
Macronyssid mites			
Ornithonyssus sp.	"Florida"	NG	Peters (1936)
Chewing lice			
Kurodaia magna	Alachua	1976, 1985	Forrester et al. (1995)
	Hernando	1975	Ibid.
	Hillsborough	1975	Ibid.
	Marion	1982	Ibid.
	Monroe	1980	Ibid.
	Orange	1936	Ibid.
	Taylor	1976	Ibid.
Strigiphilus syrnii	Alachua	1976–85	Ibid.
	Dade	1917	Ibid.
	Hendry	1984	Ibid.
	Hillsborough	1975–76	Ibid.
	Leon	1960	Ibid.
	Marion	1982	Ibid.
	Monroe	1980	Ibid.
	Orange	1936	Ibid.
	Taylor	1976	Ibid.
	"Florida"	NG	Clayton & Price (1984)
Louse flies			
Icosta americana	Alachua	NG	Bequaert (1955)
	Brevard	NG	Ibid.
	Dade	NG	Ibid.
	Highlands	1972	Wilson et al. (1990)
Olfersia sordida	Alachua	NG	Bequaert (1956)

NG = not given.

a. Travis (1941) examined 1 Barred Owl from Florida (either Orange, Osceola, or Collier County) in 1936–37; it was negative for ticks.

b. An undescribed species; also found on Barn Owls (see Table 23.23).

XIX. Summary and conclusions

Of the 9 species of owls that occur in Florida, data on parasites and diseases exist for 8. Most of the information is on Eastern Screech-Owls, Great Horned Owls, and Barred Owls, but even those data are limited. In addition to inclement weather, 103 different conditions, disease agents, and parasites have been found in owls in Florida. These include trauma (12), electrocution (1), chlorinated hydrocarbons (4), heavy metals (2), neoplasia (1), viruses (2), bacteria (8), fungi (1), protozoans (11), trematodes (12), cestodes (2), nematodes (17), acanthocephalans (2), pentastomes (1), ticks (1), mites (12), chewing lice (10), louse flies (4), and bird flies (1). Trauma, especially as a result of collisions with motor vehicles, is probably the most important mortality factor for owls in Florida. Other diseases and parasites such as pox, trichomonosis or hematozoan infections may be important, but have not been studied extensively in Florida.

XX. Literature cited

Albers, H.F. 1984. Naturally occurring immunity to rabies in raptors. *Proc. Annu. Natl. Wildl. Rehabil. Symp.* 2:171–174.

American Ornithologists' Union. 1998. *Check-list of North American birds.* 7th ed. American Ornithologists' Union, Washington, D.C. 829 pp.

Anderson, D.W., and J.J. Hickey. 1972. Eggshell changes in certain North American birds. *Proc. Int. Ornithol. Congr.* 15:514–540.

Bent, A.C. 1938. *Life histories of North American birds of prey.* Part 2, *Hawks, falcons, caracaras, and owls.* Bull. U.S. Natl. Mus. 170. Dover, New York. 482 pp.

Bequaert, J.C. 1942. *Carnus hemapterus* Nitzsch, an ectoparasitic fly of birds, new to America (Diptera). *Bull. Brooklyn Entomol. Soc.* 38: 140–149.

———. 1955. The hippoboscidae or louse-flies (Diptera) of mammals and birds. Part II. Taxonomy, evolution and revision of American genera and species. *Entomol. Am.* 35:233–416.

———. 1956. The Hippoboscidae or louse-flies (Diptera) of mammals and birds. Part II. Taxonomy, evolution and revision of American genera and species. *Entomol. Am.* 36:417–611.

Beyer, W.N., J.W. Spann, L. Sileo, and J.C. Franson. 1988. Lead poisoning in six captive avian species. *Arch. Environ. Contam. Toxicol.* 17: 121–130.

Bigler, W.J. 1997. Unpublished data. Florida Department of Health, Tallahassee.

Bigler, W.J., E. Lassing, E. Buff, A.L. Lewis, and G.L. Hoff. 1975. Arbovirus surveillance in Florida: wild vertebrate studies, 1965–1974. *J. Wildl. Dis.* 11:348–356.

Bishop, M.A., and G.F. Bennett. 1989. The haemoproteids of the avian order Strigiformes. *Can. J. Zool.* 67:2676–2684.

———. 1992. *Host-parasite catalogue of the avian Haematozoa.* Suppl. 1. Meml. University Nfld. Occas. Pap. Biol. 15. 210 pp.

Blus, L.J. 1996. Effects of pesticides on owls in North America. *J. Raptor Res.* 30:198–206.

Burridge, M.J., W.J. Bigler, D.J. Forrester, and J.M. Hennemann. 1979. Serologic survey for *Toxoplasma gondii* in wild animals in Florida. *J. Am. Vet. Med. Assoc.* 175:964–967.

Cannings, R.J. 1986. *Carnus hemapterus* (Diptera: Carnidae), an avian nest parasite new to British Columbia, Canada. *J. Entomol. Soc. B.C.* 83: 38.

Chiocco, D. 1992. Pox virus del gufo: isolamento e prove di immunità crociata nei polli. *Acta Med. Vet.* 38:261–266.

Clayton, D.H., and R.D. Price. 1984. Taxonomy of the *Strigiphilus cursitans* group (Ischnocerca: Philopteridae), parasites of owls (Strigiformes). *Ann. Entomol. Soc. Am.* 77:340–363.

Collins, R. 1999. Unpublished data. Florida Audubon Society, Center for Birds of Prey, Maitland.

Collins, R., and C. Gilliland. 1995. Unpublished data. Florida Audubon Society, Center for Birds of Prey, Maitland.

Conti, L., R. Oliveri, and C. Blackmore. 2002. Unpublished data. Florida Department of Health, Tallahassee.

Cooper, J.E. 1985. *Veterinary aspects of captive birds of prey.* 2d ed. Standfast Press, Gloucestershire, England. 256+31 suppl. pp.

Cram, E.B. 1930. Aberrant larvae of *Physocephalus sexalatus* in birds. *J. Parasitol.* 17:56.

Crawford, R.L. 1981. Bird casualties at a Leon County, Florida TV tower: a 25-year migration study. *Bull. Tall Timbers Res. Stn.* 22:1–30.

Dawson, R.D., and G. R. Bortolotti. 1997. Ecology of parasitism of nestling American Kestrels by *Carnus hemapterus* (Diptera: Carnidae). *Can. J. Zool.* 75:2021–2026.

Deem, S.L., D.J. Heard, and J.H. Fox. 1997. Avian pox in eastern screech owls from Florida. *J. Wildl. Dis.* 33:323–327.

Deem, S.L., and S.P. Terrell. 1996. Unpublished data. University of Florida, Gainesville.

Deem, S.L., S.P. Terrell, and D.J. Forrester. 1998. A retrospective study of morbidity and mortality of raptors in Florida: 1988–1994. *J. Zoo Wildl. Med.* 29:160–164.

Diamond, L.S. 1957. The establishment of various trichomonads of animals and man in axenic cultures. *J. Parasitol.* 43:488–490.

Domermuth, C.H., D.J. Forrester, D.O. Trainer, and W.J. Bigler. 1977. Serologic examination of wild birds for hemorrhagic enteritis of turkeys and marble spleen disease in pheasants. *J. Wildl. Dis.* 13:405–408.

Dooris, G.M., P.M. Dooris, and W.D. Courser. 1981. The absence of Haematozoa in burrowing owls of the Tampa Bay area, Florida. *Fla. Field Nat.* 9:9.

Emerson, K.C. 1961. Three new species of Mallophaga from the great horned owl. *Proc. Biol. Soc. Wash.* 74:187–192.

Evans, M., and A. Otter. 1998. Fatal combined infection with *Haemoproteus noctuae* and *Leucocytozoon ziemanni* in juvenile snowy owls (*Nyctea scandiaca*). *Vet. Rec.* 143:72–76.

Forrester, D.J. 1989. Unpublished data. University of Florida, Gainesville.

———. 1992. *Parasites and diseases of wild mammals in Florida.* University Press of Florida, Gainesville. 459 pp.

Forrester, D.J., H.W. Kale II, R.D. Price, K.C. Emerson, and G.W. Foster. 1995. Chewing lice (Mallophaga) from birds in Florida: a listing by host. *Bull. Fla. Mus. Nat. Hist.* 39:1–44.

Forrester, D.J., and J.W. Mertins. 1992. Unpublished data. University of Florida, Gainesville.

Forrester, D.J., and E.G. Milstrey. 1989. Unpublished data. University of Florida, Gainesville.

Forrester, D.J., S.R. Telford, Jr., G.W. Foster, and G.F. Bennett. 1994. Blood parasites of raptors in Florida. *J. Raptor Res.* 28:226–231.

Foster, G.W., and J.W. Mertins. 1994. Unpublished data. University of Florida, Gainesville.

Fowler, M.E. (ed.). 1986. *Zoo and wild animal medicine.* 2d ed. W.B. Saunders, Philadelphia. 1,127 pp.

———(ed.). 1993. *Zoo and wild animal medicine: current therapy.* 3d ed. W.B. Saunders, Philadelphia. 617 pp.

Franson, J.C., and S.E. Little. 1996. Diagnostic findings in 132 great horned owls. *J. Raptor Res.* 30:1–6.

Frenkel, J.K. 1981. False-negative serologic tests for *Toxoplasma* in birds. *J. Parasitol.* 67:952–953.

Garvin, M.C., and M.G. Spalding. 1995. Unpublished data. University of Florida, Gainesville.

Graham, D.L., and W.H. Halliwell. 1986. Viral diseases of birds of prey. In: *Zoo and wild animal medicine.* 2d ed. M.E. Fowler (ed.). W.B. Saunders, Philadelphia. pp. 408–413.

Greenwood, A.G., and K.C. Barnett. 1981. The investigation of visual defects in raptors. In: *Recent advances in the study of raptor diseases. Proceedings of the International Symposium on Diseases of Birds of Prey.* J.E. Cooper and A.G. Greenwood (eds.). Chiron, West Yorkshire, England. pp. 131–135.

Grimaldi, D. 1997. The bird flies, genus *Carnus*: species revision, generic relationships, and a fossil *Meoneura* in amber (Diptera: Carnidae). *Am. Mus. Novit.* 3190:1–30.

Grimes, S.A. 1936. Great horned owl and common black snake in mortal combat. *Fla. Nat.* 9:77–78.

Henny, C.J. 1972. An analysis of the population dynamics of selected avian species with special reference to changes during the modern pesticide era. U.S. Fish and Wildlife Service, Wildlife Research Report 1. 99 pp.

Henry, D.P. 1932. *Isospora buteonis* sp. nov. from the hawk and owl, and notes on *Isospora lacazii* (Labbé) in birds. *University Calif. Publ. Zool.* 37:291–300.

Humphlett, P.K., and D.J. Forrester. 1989. Unpublished data. University of Florida, Gainesville.

Johnson, C.W. 1913. Insects of Florida. I. Diptera. *Bull. Am. Mus. Nat. Hist.* 32:37–90.

———. 1922. Notes on distribution and habits of some of the bird-flies, Hippoboscidae. *Psyche* 29:79–85.

Johnston, D.W. 1976. Organochlorine pesticide residues in uropygial glands and adipose tissue of wild birds. *Bull. Environ. Contam. Toxicol.* 16:149–155.

———. 1978. Organochlorine pesticide residues in Florida birds of prey, 1969–76. *Pestic. Monit. J.* 12:8–15.

Jorgenson, R.D., P.M. Gough, and D.L. Graham. 1976. Experimental rabies in a great horned owl. *J. Wildl. Dis.* 12:444–447.

Kale, H.W. II. 1971. Regional reports: Florida region. *Am. Birds* 25:723–725, 730–735.

———. 1978. Barn owl. *Fla. Nat.* 51:2–6.

Kern, T.J. 1997. Disorders of the special senses. In: *Avian medicine and surgery.* R.B. Altman, S.L. Clubb, G.M. Dorrestein, and K. Quesenberry (eds.). W.B. Saunders, Philadelphia. pp. 563–589.

Keymer, I.F. 1972. Diseases of birds of prey. *Vet. Rec.* 90:579–594.

Kiehl, A.R. 1991. Unpublished data. Doctor's and Physicians Laboratory, Leesburg, Fla.

Kinsella, J.M., and D.J. Forrester. 1993. Unpublished data. University of Florida, Gainesville.

Kinsella, J.M., and G.W. Foster. 1994. Unpublished data. University of Florida, Gainesville.

Kinsella, J.M., and E.C. Greiner. 1993. Unpublished data. University of Florida, Gainesville.

Kirkpatrick, C.E., and B.A. Colvin. 1989. Ectoparasitic fly *Carnus hemapterus* (Diptera: Carnidae) in a nesting population of common barn-owls (Strigiformes: Tytonidae). *J. Med. Entomol.* 26:109–112.

Kocan, R.M., and C.M. Herman. 1971. Trichomoniasis. In: *Infectious and parasitic diseases of wild birds.* J.W. Davis, R.C. Anderson, L. Karstad, and D.O. Trainer (eds.). Iowa State University Press, Ames. pp. 282–290.

Korpimäki, E., H. Hakkarainen, and G.F. Bennett. 1993. Blood parasites and reproductive success of Tengmalm's owls: detrimental effects on females but not males? *Funct. Ecol.* 7:420–426.

Langridge, H.P. 1993. Florida region. *Am. Birds.* 47:406–408.

Leidy, J. 1887. Notice of some parasitic worms. *Proc. Acad. Nat. Sci. Phila.* 39:20–24.

Lindsay, D.S., P.C. Smith, F.J. Hoerr, and B.L. Blagburn. 1993. Prevalence of encysted *Toxoplasma gondii* in raptors from Alabama. *J. Parasitol.* 79:870–873.

Logan, T.H. 1997. Florida's endangered species, threatened species and species of special concern. Official lists. Florida Game and Fresh Water Fish Commission, Tallahassee. 14 pp.

Loos, G., and P. Kerlinger. 1993. Road mortality of saw-whet and screech-owls on the Cape May peninsula. *J. Raptor Res.* 27:210–213.

Maehr, D.S., and J.Q. Smith. 1988. Bird casualties at a central Florida power plant: 1982–1986. *Fla. Field Nat.* 16:57–80.

Maehr, D.S., A.G. Spratt, and D.K. Voigts. 1983. Bird casualties at a central Florida power plant. *Fla. Field Nat.* 11:45–49.

Miller, D., E.L. Boeker, and R.S. Thorsell. 1975. Suggested practices for raptor protection on powerlines. Raptor Research Foundation, Provo, Utah. 19 pp.

Millsap, B.A. 1996. Florida burrowing owl. In: *Rare and endangered biota of Florida.* Vol. 5, *Birds* J.A. Rodgers, Jr., H.W. Kale II, and H.T. Smith (eds.). University Press of Florida, Gainesville. pp. 579–587.

Mollhagen, T.R. 1976. A study of the systematics and hosts of the parasitic nematode genus *Tetrameres* (Habronematoidea: Tetrameridae). Ph.D. diss., Texas Tech University, Lubbock, Texas. 546 pp.

Morishita, T.Y., P.P. Aye, and D.L. Brooks. 1997. A survey of diseases of raptorial birds. *J. Avian Med. Surg.* 11:77–92.

Nayar, J.K., J.W. Knight, and S.R. Telford. 1998. Vector ability of mosquitoes for isolates of *Plasmodium elongatum* from raptors in Florida. *J. Parasitol.* 84:542–546.

Nickol, B.B. 1983. *Centrorhynchus kuntzi* from the USA with description of the male and redescription of *C. spinosus* (Acanthocephala:

Centrorhynchidae). *J. Parasitol.* 69:221–225.

Ogden, J.C. 1992. Florida region. *Am. Birds.* 46:255–257.

Ohlendorf, H.M., D.M. Swineford, and L.N. Locke. 1981. Organochlorine residues and mortality of herons. *Pestic. Monit. J.* 14:125–135.

Pellerdy, L.P. 1974. *Coccidia and coccidiosis.* Verlag Paul Parey, Berlin. 959 pp.

Pence, D.B., T. Mollhagen, and D.J. Forrester. 1975. *Tetrameres* (*Gynaecophila*) *strigiphila* sp.n. from the Florida barred owl, *Strix varia georgia,* with notes on the status of the subgenus *Gynaecophila* (Nematoda: Tetrameridae). *J. Parasitol.* 61:494–498.

Peters, H.S. 1936. A list of external parasites from birds of the eastern part of the United States. *Bird-Banding* 7:9–27.

Ratcliffe, D.A. 1967. Decrease in eggshell weight in certain birds of prey. *Nature* 215:208–210.

Redig, P.T. 1978. Mycotic infections of birds of prey. In: *Zoo and wild animal medicine.* M.E. Fowler (ed.). W.B. Saunders, Philadelphia. pp. 273–276.

———. 1981. Aspergillosis in raptors. In: *Recent advances in the study of raptor diseases. Proceedings of the International Symposium on Diseases of Birds of Prey.* J.E. Cooper and A.G. Greenwood (eds.). Chiron, West Yorkshire, England. pp. 117–122.

Regan, T.W. 1996. Unpublished data. Florida Game and Fresh Water Fish Commission, West Palm Beach.

Ritchie, B.W., and K. Carter. 1995. *Avian viruses: function and control.* Wingers, Lake Worth, Fla. pp. 285–311.

Robertson, W.B., Jr., and G.E. Woolfenden. 1992. *Florida bird species. An annotated list.* Spec. Publ. 6, Florida Ornithological Society, Gainesville. 260 pp.

Rogers, A.J. 1953. A study of the ixodid ticks of northern Florida, including the biology and life history of *Ixodes scapularis* Say (Ixodidae: Acarina). Ph.D diss., University of Maryland, College Park. 191 pp.

Smith, D.G., D.H. Ellis, and B.A. Millsap. 1990. Owls. In: *Proc. Southeast. Raptor Manag.*

Symp. Workshop. Natl. Wildl. Fed. Sci. Tech. Ser. 14, Washington, D.C. pp. 89–117.

Smith, K.E. 1996. Unpublished data. Southeastern Cooperative Wildlife Disease Study, University of Georgia, Athens.

Snyder, B. 1994. Unpublished data. Florida Department of Environmental Protection, Tallahassee.

Snyder, N.F., and J.W. Wiley. 1976. Sexual age dimorphism in hawks and owls of North America. *Ornithol. Monogr.* 20:1–96.

Spalding, M.G., and R.G. McLean. 1994. Unpublished data. University of Florida, Gainesville.

Stabler, R.M. 1954. *Trichomonas gallinae:* A review. *Exp. Parasitol.* 3:368–402.

Stevenson, J.A. 1994. Unpublished data. Florida Department of Environmental Protection, Tallahassee.

Stewart, P.A. 1969. Movements, population fluctuations, and mortality among great horned owls. *Wilson Bull.* 81:155–162.

Stoddard, H.L. 1931. *The bobwhite quail, its habits, preservation and increase.* Charles Scribner's Sons, New York. 559 pp.

Sundlof, S.F., D.J. Forrester, N.P. Thompson, and M.W. Collopy. 1986. Residues of chlorinated hydrocarbons in tissues of raptors in Florida. *J. Wildl. Dis.* 22:71–82.

Suto, B.J. 1996. Unpublished data. Suncoast Seabird Sanctuary, Redington Beach, Fla.

Taylor, W.K., and B.H. Anderson. 1973. Nocturnal migrants killed at a central Florida TV tower, autumns 1969–1971. *Wilson Bull.* 85:42–51.

———. 1974. Nocturnal migrants killed at a central Florida TV tower, autumn 1972. *Fla. Field Nat.* 2:40–43.

Taylor, W.K., and M.A. Kershner. 1986. Migrant birds killed at the Vehicle Assembly Building (VAB, John F. Kennedy Space Center). *J. Field Ornithol.* 57:142–154.

Telford, S.R., Jr., G.F. Bennett, G.W. Foster, and D.J. Forrester. 1993. Unpublished data. University of Florida, Gainesville.

Telford, S.R., G.W. Foster, J.K. Nayar, S.P. Terrell, and D.J. Forrester. 1996. Unpublished data. University of Florida, Gainesville.

Telford, S.R., J.K. Nayar, G.W. Foster, and J.W.

Knight. 1997. *Plasmodium forresteri* n.sp., from raptors in Florida and southern Georgia: its distinction from *Plasmodium elongatum* morphologically within and among host species and by vector susceptibility. *J. Parasitol.* 83:932–937.

Travis, B.V. 1941. Examinations of wild animals for the cattle tick *Boophilus annulatus microplus* (Can.) in Florida. *J. Parasitol.* 27: 465–467.

Van Cleave, H.J. 1924. A critical study of the Acanthocephala described and identified by Joseph Leidy. *Proc. Acad. Nat. Sci. Phila.* 76: 279–334.

Walton, A.C. 1927. A revision of the nematodes of the Leidy Collections. *Proc. Acad. Nat. Sci. Phila.* 79:49–163.

Weston, F.M. 1946. Additions to the Florida list. *Auk* 63:451–452.

White, F.H., and D.J. Forrester. 1989. Unpublished data. University of Florida, Gainesville.

Wiemeyer, S.N., and E. Cromartie. 1981. Relationships between brain and carcass organochlorine residues in ospreys. *Bull. Environ. Contam. Toxicol.* 27:499–505.

Wilson, N.A., H.W. Kale II, and W.W. Baker. 1990. Unpublished data. University of Northern Iowa, Cedar Falls.

Perching Birds

I. Introduction

On a worldwide basis, the order Passeriformes, commonly known as perching birds or songbirds, is a cosmopolitan group comprising 60 families and more than 5,000 species of birds (Stevenson and Anderson 1994). They are represented in Florida by 26 families and 194 species (table 24.1). Nineteen species or subspecies have been declared endangered (n = 4), threatened (n = 1), rare (n = 6), species of special concern (n = 6), or species with undetermined status (n = 2) (table 24.2). One subspecies (the Dusky Seaside Sparrow, *Ammodramus maritimus nigrescens*) recently became extinct (Kale 1996). Robertson and Woolfenden (1992) and Stevenson and Anderson (1994) provide a com-

plete list of passeriforms reported from the state and give details on the status and distribution of each species.

Asterino (1996) has published a useful overview of the diseases of wild passeriforms. Additional data can be found in Davis et al. (1971), Friend and Franson (1999), Loye and Zuk (1991), Fairbrother et al. (1996), and Beyer et al. (1996). There is a wealth of information, particularly clinical data, in several treatises on captive birds (Fowler 1978, 1986; Petrak 1982; Ritchie et al. 1994; Rosskopf and Woerpel 1996; Altman et al. 1997), some of it possibly applicable to free-ranging birds.

Table 24.1. Families and numbers of species of perching birds found in Florida

Family	Common name	No. of species
Tyrannidae	Tyrant Flycatchers	23
Laniidae	Shrikes	1
Vireonidae	Vireos	10
Corvidae	Jays and Crows	4
Alaudidae	Larks	1
Hirundinidae	Swallows	10
Paridae	Chickadees and Titmice	2
Sittidae	Nuthatches	3
Certhiidae	Creepers	1
Troglodytidae	Wrens	7
Pycnonotidae	Bulbuls	1
Regulidae	Kinglets	2
Sylviidae	Gnatcatchers	1
Turdidae	Thrushes	9
Mimidae	Mockingbirds and Thrashers	6
Sturnidae	Starlings	1
Motacillidae	Pipits	2
Bombycillidae	Waxwings	1
Parulidae	Wood-Warblers	42
Coerebidae	Bananaquits	1
Thraupidae	Tanagers	4
Emberizidae	Emberizids	31
Cardinalidae	Cardinals, Grosbeaks, Buntings	8
Icteridae	Blackbirds and Orioles	16
Fringillidae	Fringilline and Cardueline Finches	6
Passeridae	Old-world Sparrows	1
Totals		194

Sources: Robertson and Woolfenden (1992), American Ornithologists' Union (1988).

II. Trauma

Casualties resulting from nocturnally migrating birds striking manmade structures (radio, TV, and cell phone towers, lighthouses, smokestacks, buildings, bridges, etc.) in various areas of North America have been studied and analyzed (Weir 1976; Avery et al. 1980). In the early 1980s it was estimated that more than a million birds died each year in the United States because of collisions with towers (Avise and Crawford 1981). Evans (1998) estimated that annual losses of songbirds attributed to tower impacts in eastern North America in the late 1990s were between 2 and 4 million birds.

Most likely this figure will increase as more and more towers and tall buildings are constructed. In 1998 there were an estimated 75,000 manmade obstructions more than 200 feet high in the United States, with another 100,000 due to be built in the next 10 years (Evans 1998).

There are two ways that birds are killed at communications towers (Evans 1999). One is by "blind collision": flying during times of poor visibility and failing to see a structure in time to avoid a collision. The other involves a situation with water particles in fog or a low cloud ceiling reflecting light, resulting in an illuminated area around the tower. The migrating birds tend to continue flying in the lighted

Table 24.2. Perching birds of Florida classified as threatened, rare, species of special concern, or status undetermined

Status Species of bird		Range	Seasonal occurrence	Relative abundance
Endangered				
Kirtland's Warbler	*Dendroica kirtlandii*	Statewide except Keys	Fall & Spring	Very rare (15 reports)
Bachman's Warbler	*Vermivora bachmanii*	NA	NA	Possibly extinct
Florida Grasshopper Sparrow	*Ammodramus savannarum floridanus*	South central peninsula	Resident	<500
Cape Sable Seaside Sparrow	*Ammodramus maritimus mirabilis*	Dade and Monroe Co.	Resident	~2,800
Threatened				
Florida Scrub-Jay	*Aphelocoma coerulescens*	Mid-peninsula	Resident	9,500–11,000
Rare				
West Indian Cave Swallow	*Petrochelidon fulva fulva*	Dade County	Resident	~100
Black-whiskered Vireo	*Vireo altiloquus*	Coastal areas in south Florida	Summer	Rare to locally common
Worm-eating Warbler	*Helmitheros vermivorus*	Statewide	Summer & Fall	Rare
		South & central Fla.	Winter	Very rare
Louisiana Waterthrush	*Seiurus motacilla*	Statewide	Transient	Rare to very rare
		Panhandle	Summer	Rare
American Redstart	*Setophaga ruticilla*	Western panhandle	Summer	Uncommon
		Statewide	Transient	Uncommon to common
Cuban Yellow Warbler	*Dendroica petechia gundlachi*	Southern peninsula	Resident	Fairly common
Species of Special Concern				
Marian's Marsh Wren	*Cistothorus palustris marianae*	Northern Gulf Coast	Resident	2,000–4,000
Worthington's Marsh Wren	*Cistothorus palustris griseus*	Northern Atlantic Coast	Resident	4,000–6,000
MacGillivray's Seaside Sparrow	*Ammodramus maritimus macgillivraii*	Duval County	Resident	1,500–2,000
Scott's Seaside Sparrow	*Ammodramus maritimus peninsulae*	Northern Gulf Coast	Resident	10,000–20,000
Louisiana Seaside Sparrow	*Ammodramus maritimus fisheri*	Santa Rosa County	Resident	"low hundreds"
White-breasted Nuthatch	*Sitta carolinensis*	Leon and Jefferson Co.	Resident	Rare to fairly common
Status Undetermined				
Florida Prairie Warbler	*Dendroica discolor paludicola*	Statewide	Winter	Rare to common
		Northern Florida	Summer	Rare to fairly common
Painted Bunting	*Passerina ciris*	Statewide	Winter	Very rare to locally common
		NE coast & parts of panhandle	Summer	Locally abundant

Sources: Modified from Robertson and Woolfenden (1992), Rodgers et al. (1996), Logan (1997).
NA = not applicable.

FIGURE 24.1. Locations of 2,342 communication towers throughout the state of Florida as of November 2, 1998. Heights vary from 200 feet to >800 feet. Adapted from Evans 1999; by permission of William R. Evans.

area, since they can no longer navigate by the stars. They are killed when they collide with the tower or its guy wires or even with each other. Evans (1999) points out that "the lights apparently do not attract birds from afar, but rather tend to hold birds that pass within a certain illuminated vicinity." Most of the mortality that occurs at tall towers is due to this second phenomenon, called the "phototactic mechanism" by Bill Evans. In Florida there were 2,342 towers more than 200 feet in height as of November 2, 1998 (figure 24.1): 1,126 at 200–299 feet, 1,009 at 300–499 feet, 134 at 500–799 feet, and 73 at >800 feet (Evans 1999). The higher towers are the ones that cause the most mortality.

There are numerous publications that document bird-strike mortality in Florida, but these records probably represent only a fraction of what occurs (Bagg 1957, 1969; Cunningham 1964a,b, 1965a; Edscorn 1974, 1975; Robertson and Ogden 1969; Stevenson 1956, 1958a,c,

FIGURE 24.2. Photograph of Dr. Herbert L. Stoddard, Sr., demonstrating his "Volkswagen technique" for checking bird casualties at the base of the WCTV tower near Tall Timbers Research Station in Leon County. Part of the tower can be seen on the right, and many of the guy wires are visible. Courtesy of Tall Timbers Research Station, Tallahassee.

1959, 1960, 1962, 1966, 1973). Five long-term studies varying from 4 to 25 years in duration have been conducted on mortality at TV towers in Leon County (figure 24.2; Stoddard 1962; Stoddard and Norris 1967; Crawford 1974, 1978, 1980, 1981a; Crawford and Stevenson 1984; Crawford and Engstrom 2001) and Orange County (Taylor and Anderson 1973, 1974), a power plant with 2 tall chimneys in Citrus County (Maehr and Smith 1988), a massive building (the Vehicle Assembly Building at the Kennedy Space Center) in Brevard County (Taylor and Kershner 1986), and the Pensacola Bridge in Escambia and Santa Rosa counties (Weston 1966). From these 5 studies, casualties were recorded for more than 57,000 passeriforms representing 123 species from 15 families (tables 24.3 and 24.4). Although there were differences in the findings of these studies, probably related to the duration of the investigations, type and geographic location of the structures involved, migratory routes of various passeriforms, patterns of weather, and other factors, the highest numbers of casualties were among the vireos (Red-eyed Vireos and White-eyed Vireos) and wood warblers (Common Yellowthroats, Ovenbirds, Palm Warblers, Yellow-rumped Warblers, and Northern Parulas). By far the greatest losses were in Red-eyed Vireos and Common Yellowthroats, for which 7,710 and 5,668 individuals were found dead, respectively.

In addition, there are other reports based on more limited data sets (Stimson 1966; Robertson 1971; Kale 1971). The species of passeriforms reported in these studies were the same as those in the 5 larger studies, with the exception of 1 Black-whiskered Vireo (*Vireo altiloquus*) found by Kale (1971) in St. Lucie County.

The numbers of birds striking manmade objects and being killed in a single night can be considerable (Johnston and Haines 1957). Stoddard (1962) estimated that 4,000–7,000 birds of at least 62 species were killed at the WCTV tower near Tall Timbers during the night of October 8–9, 1955. Maehr et al. (1983) estimated that approximately 3,000 birds representing 31 species were killed during the night of September 23, 1982, at a power plant in Citrus County. The most common species were White-eyed Vireos (52%), Northern Parulas (13%), and Red-eyed Vireos (10%). In another short-term incident, Roberts and Tamborski (1993) identified 617 birds of 9 species that were killed by striking a U.S. Coast Guard LORAN tower in Martin County during the night of October 8, 1991. The majority of these birds (95%) were Blackpoll Warblers. Maehr and Smith (1988) reported a peak in total mortality in September and October at their Citrus County study site (figure 24.3) and 2 peaks in the numbers of species, a small one in April and a larger one in September and October (figure 24.4). They stated that their data with some exceptions were

similar to those of Stoddard and Norris (1967) and Crawford (1974) in Leon County and concluded that these mortality patterns were due to a concentration of migration activity during the fall months. Taylor and Kershner (1986) found the opposite in Brevard County; i.e., spring mortality outnumbered that which occurred in the fall. The largest kills have been found to be correlated with the arrival of cold fronts and associated inclement weather (Cunningham 1965a; Taylor and Anderson 1973; Crawford 1981b; Taylor and Kershner 1986), although some mortality has been measured during normal weather conditions. Verheijen (1981) stated that bird kills at lighted manmade structures were related to nights at or near a full moon, but Crawford (1981c) concluded that this was not the case in Florida.

Maehr et al. (1983) point out that although many authors have discussed various details of passeriform deaths due to striking manmade structures, few have addressed the problem of mitigating these massive mortalities. There seems to be agreement that lighting is involved during overcast conditions (Weir 1976), but there are conflicting opinions on exactly what can be considered "safe lighting" (Maehr et al. 1983). Taylor (1981) suggests that strobe lights are less of a problem to migrating birds during overcast conditions than are colored lights, but the findings of Maehr et al. (1983) contradict this. However, Maehr et al. (1983) point out that there were complications in their study, construction lights at the base of the smokestacks that were on during the night and possibly influencing migrating songbirds. Further studies are needed to determine more about this type of mortality and how it can be alleviated.

Much information has been derived from the collection and study of the carcasses of these casualties. Crawford (1981a) notes that over the years thousands of carcasses of birds killed at the TV tower in Leon County have been made into study skins and skeletons and deposited in various museums; numerous papers have resulted on such topics as migration, systematics, energetics, and pesticides. In addition Crawford (1971) studied the impact of

Table 24.3. Perching birds killed by collisions with manmade structures in Florida, 1955–86

| Family | No. of birds found dead/county | | | |
Species	Leon[a]	Citrus[b]	Orange[c]	Brevard[d]
Tyrannidae				
Eastern Kingbird (*Tyrannus tyrannus*)	27	1	1	0
Great Crested Flycatcher (*Myiarchus crinitus*)	36	0	0	0
Eastern Phoebe (*Sayornis phoebe*)	13	0	0	3
Yellow-bellied Flycatcher (*Empidonax flaviventris*)	3	0	0	0
Acadian Flycatcher (*Empidonax virescens*)	109	23	1	3
Willow Flycatcher (*Empidonax traillii*)	10	0	0	0
Alder Flycatcher (*Empidonax alnorum*)	11	0	0	0
Least Flycatcher (*Empidonax minimus*)	2	0	0	0
Eastern Wood-Pewee (*Contopus virens*)	48	1	1	1
Laniidae				
Loggerhead Shrike (*Lanius ludovicianus*)	1	0	0	0
Vireonidae				
White-eyed Vireo (*Vireo griseus*)	1,364	988	95	4
Yellow-throated Vireo (*Vireo flavifrons*)	288	23	8	10
Blue-headed Vireo (*Vireo solitarius*)	113	4	2	2
Red-eyed Vireo (*Vireo olivaceus*)	7,159	344	153	41
Philadelphia Vireo (*Vireo philadelphicus*)	16	0	3	0
Warbling Vireo (*Vireo gilvus*)	1	0	0	0
Corvidae				
Blue Jay (*Cyanocitta cristata*)	5	0	0	0
American Crow (*Corvus brachyrhynchos*)	1	0	0	0
Hirundinidae				
Tree Swallow (*Tachycineta bicolor*)	13	0	0	0
Bank Swallow (*Riparia riparia*)	1	0	0	0
Barn Swallow (*Hirundo rustica*)	4	0	0	0
Purple Martin (*Progne subis*)	5	0	0	0
Sittidae				
Red-breasted Nuthatch (*Sitta canadensis*)	1	0	0	0
Certhiidae				
Brown Creeper (*Certhia americana*)	7	0	0	0
Troglodytidae				
House Wren (*Troglodytes aedon*)	385	0	118	0
Winter Wren (*Troglodytes troglodytes*)	26	0	0	0
Bewick's Wren (*Thryomanes bewickii*)	3	0	0	0
Carolina Wren (*Thryothorus ludovicianus*)	1	0	0	0
Marsh Wren (*Cistothorus palustris*)	131	7	194	6
Sedge Wren (*Cistothorus platensis*)	251	2	51	1
Regulidae				
Ruby-crowned Kinglet (*Regulus calendula*)	1,015	5	49	1
Golden-crowned Kinglet (*Regulus satrapa*)	60	0	0	0
Sylviidae				
Blue-gray Gnatcatcher (*Polioptila caerulea*)	1	0	0	0
Turdidae				
Veery (*Catharus fuscescens*)	1,114	1	28	3
Eastern Bluebird (*Sialia sialis*)	3	0	0	0
American Robin (*Turdus migratorius*)	143	0	0	0
Wood Thrush (*Hylocichla mustelina*)	344	0	3	0
Hermit Thrush (*Catharus guttatus*)	184	0	2	0
Swainson's Thrush (*Catharus ustulatus*)	534	0	18	0
Grey-cheeked Thrush (*Catharus minimus*)	302	0	9	0

(*continued*)

Table 24.3. *(continued)*

| Family | No. of birds found dead/county | | | |
Species	Leon[a]	Citrus[b]	Orange[c]	Brevard[d]
Mimidae				
Northern Mockingbird (*Mimus polyglottos*)	35	0	1	0
Gray Catbird (*Dumetella carolinensis*)	1,144	21	174	64
Brown Thrasher (*Toxostoma rufum*)	290	0	1	1
Sturnidae				
European Starling (*Sturnus vulgaris*)	23	0	0	1
Motacillidae				
American Pipit (*Anthus rubescens*)	3	0	0	0
Bombycillidae				
Cedar Waxwing (*Bombycilla cedrorum*)	75	0	0	0
Parulidae				
Black-and-white Warbler (*Mniotilta varia*)	632	25	186	511
Prothonotary Warbler (*Protonotaria citrea*)	444	2	3	1
Swainson's Warbler (*Limnothlypis swainsonii*)	177	1	59	45
Worm-eating Warbler (*Helmitheros vermivorus*)	177	7	42	148
Golden-winged Warbler (*Vermivora chrysoptera*)	69	0	1	0
Blue-winged Warbler (*Vermivora pinus*)	53	0	0	0
Tennessee Warbler (*Vermivora peregrina*)	549	3	15	2
Orange-crowned Warbler (*Vermivora celata*)	318	0	3	0
Nashville Warbler (*Vermivora ruficapilla*)	4	0	0	0
Northern Parula (*Parula americana*)	1,469	228	414	151
Yellow Warbler (*Dendroica petechia*)	119	2	13	0
Magnolia Warbler (*Dendroica magnolia*)	520	18	13	11
Cape May Warbler (*Dendroica tigrina*)	60	0	139	363
Black-throated Blue Warbler (*Dendroica caerulescens*)	64	5	981	437
Yellow-rumped Warbler (*Dendroica coronata*)	2,797	2	68	2
Black-throated Green Warbler (*Dendroica virens*)	65	1	7	0
Cerulean Warbler (*Dendroica cerulea*)	141	0	4	0
Blackburnian Warbler (*Dendroica fusca*)	374	6	15	1
Yellow-throated Warbler (*Dendroica dominica*)	262	4	42	4
Chestnut-sided Warbler (*Dendroica pensylvanica*)	483	8	4	0
Bay-breasted Warbler (*Dendroica castanea*)	466	0	14	0
Blackpoll Warbler (*Dendroica striata*)	163	0	10	657
Pine Warbler (*Dendroica pinus*)	217	3	31	2
Prairie Warbler (*Dendroica discolor*)	886	44	110	31
Palm Warbler (*Dendroica palmarum*)	2,305	120	536	14
Ovenbird (*Seiurus aurocapillus*)	834	27	938	1,218
Northern Waterthrush (*Seiurus noveboracensis*)	467	5	206	91
Louisiana Waterthrush (*Seiurus motacilla*)	75	0	3	1
Kentucky Warbler (*Oporornis formosus*)	530	7	2	3
Connecticut Warbler (*Oporornis agilis*)	7	2	0	11
Mourning Warbler (*Oporornis philadelphia*)	4	0	0	0
Common Yellowthroat (*Geothlypis trichas*)	1,408	219	3,199	706
Yellow-breasted Chat (*Icteria virens*)	100	4	5	1
Hooded Warbler (*Wilsonia citrina*)	1,147	31	2	5
Wilson's Warbler (*Wilsonia pusilla*)	5	0	0	0
Canada Warbler (*Wilsonia canadensis*)	12	0	0	0
American Redstart (*Setophaga ruticilla*)	791	66	708	178
Thraupidae				
Western Tanager (*Piranga ludoviciana*)	1	0	0	0
Scarlet Tanager (*Piranga olivacea*)	153	1	1	0

(continued)

Table 24.3. *(continued)*

| Family | No. of birds found dead/county | | | |
Species	Leon[a]	Citrus[b]	Orange[c]	Brevard[d]
Summer Tanager (*Piranga rubra*)	292	1	1	2
Emberizidae				
Eastern Towhee (*Pipilo erythrophthalmus*)	183	0	2	0
Savannah Sparrow (*Passerculus sandwichensis*)	1,002	1	32	5
Grasshopper Sparrow (*Ammodramus savannarum*)	401	4	12	56
Henslow's Sparrow (*Ammodramus henslowii*)	43	0	0	0
Saltmarsh Sharp-tailed Sparrow (*Ammodramus caudacutus*)	11	0	1	1
Le Conte's Sparrow (*Ammodramus leconteii*)	5	0	0	0
Seaside Sparrow (*Ammodramus maritimus*)	0	0	0	1
Vesper Sparrow (*Pooecetes gramineus*)	167	0	0	0
Bachman's Sparrow (*Aimophila aestivalis*)	72	0	2	0
Chipping Sparrow (*Spizella passerina*)	670	0	2	0
Clay-colored Sparrow (*Spizella pallida*)	6	0	0	0
Field Sparrow (*Spizella pusilla*)	96	0	0	0
White-crowned Sparrow (*Zonotrichia leucophrys*)	13	0	1	0
White-throated Sparrow (*Zonotrichia albicollis*)	551	0	0	0
Fox Sparrow (*Passerella iliaca*)	27	0	0	0
Lincoln's Sparrow (*Melospiza lincolnii*)	9	0	0	0
Swamp Sparrow (*Melospiza georgiana*)	581	2	50	17
Song Sparrow (*Melospiza melodia*)	274	0	0	0
Dark-eyed Junco (*Junco hyemalis*)	37	0	0	0
Cardinalidae				
Northern Cardinal (*Cardinalis cardinalis*)	54	0	0	0
Rose-breasted Grosbeak (*Pheucticus ludovicianus*)	65	0	3	1
Blue Grosbeak (*Guiraca caerulea*)	84	0	0	0
Indigo Bunting (*Passerina cyanea*)	864	10	13	22
Painted Bunting (*Passerina ciris*)	1	0	4	32
Dickcissel (*Spiza americana*)	13	0	0	0
Icteridae				
Bobolink (*Dolichonyx oryzivorus*)	761	1	174	84
Eastern Meadowlark (*Sturnella magna*)	87	0	0	0
Red-winged Blackbird (*Agelaius phoeniceus*)	399	1	0	0
Orchard Oriole (*Icterus spurius*)	76	0	1	0
Baltimore Oriole (*Icterus galbula*)	31	0	7	1
Rusty Blackbird (*Euphagus carolinus*)	11	0	0	0
Common Grackle (*Quiscalus quiscula*)	4	0	0	0
Brown-headed Cowbird (*Molothrus ater*)	220	0	0	0
Fringillidae				
Purple Finch (*Carpodacus purpureus*)	8	0	0	0
Pine Siskin (*Carduelis pinus*)	3	0	0	0
American Goldfinch (*Carduelis tristis*)	15	0	1	0
Evening Grosbeak (*Coccothraustes vespertinus*)	2	0	0	0
Passeridae				
House Sparrow (*Passer domesticus*)	6	0	0	0
Totals	40,797	2,281	8,992	4,957

a. A 1,010-foot TV tower (WCTV); records span 25 years, 1955–80 (Crawford 1981a).
b. An electrical power plant (Crystal River Generating Facility) with 2 chimneys, one 500 feet and one 604 feet tall; records span 4.5 years, 1982–86 (Maehr and Smith 1988).
c. A 1,483-foot TV tower (WDBO); records span 4 years, 1969–72 (Taylor and Anderson 1973, 1974).
d. A massive building 210 feet in height (Vehicle Assembly Building and the John F. Kennedy Space Center on Merritt Island); records span 12 years, 1970–81 (Taylor and Kershner 1986).

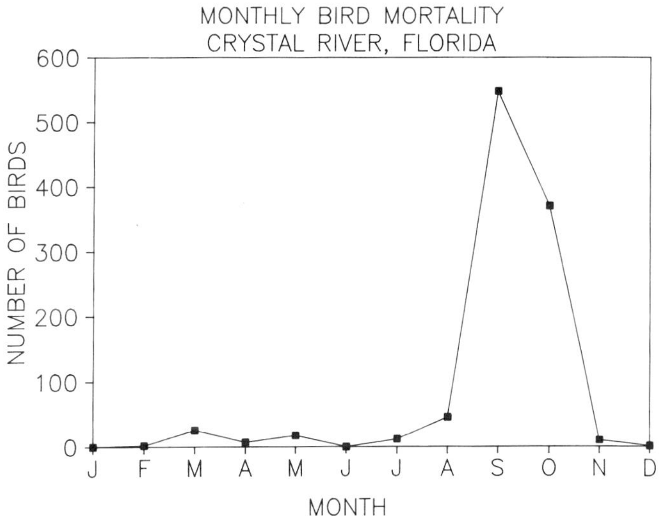

FIGURE 24.3. Seasonal variation in total mortality of birds at the Crystal River Generating Facility in Citrus County, 1983–86. From Maehr and Smith 1988; by permission of *Florida Field Naturalist*.

predation and scavenging of birds injured or killed at the Leon County tower. The bibliography by Avery et al. (1980) contains close to 80 references that resulted from material obtained from this tower.

Weston (1966) kept track of the numbers and species of birds that died in collisions with the 3-mile-long Pensacola Bay Bridge over a 12-year period, 1938–49. The mortality was due to birds striking power lines, support standards, or the cables supporting the standards. The power lines were located 20 feet above the roadway of the bridge, which was 34 feet above normal water level. The researchers' coverage was spotty in

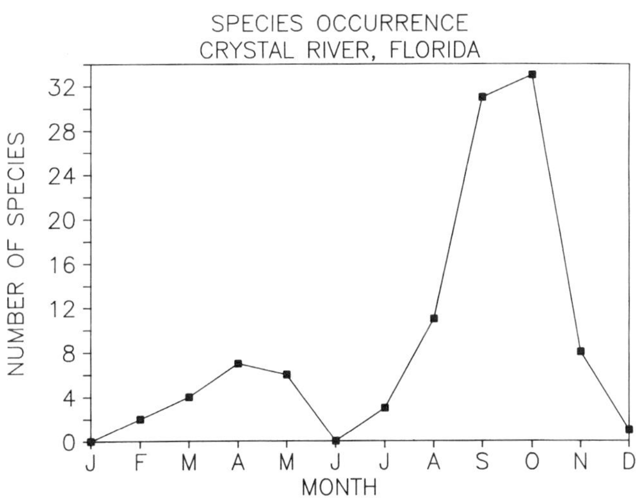

FIGURE 24.4. Seasonal variation in the numbers of species of birds killed at the Crystal River Generating Facility in Citrus County, 1983–86. From Maehr and Smith 1988; by permission of *Florida Field Naturalist*.

Table 24.4. Perching birds killed by collisions with the Pensacola Bay Bridge (Escambia and Santa Rosa counties) over a 12-year period

Family Species	No. of birds found dead		
	1938–43	1944–49	Totals
Tyrannidae			
Great Crested Flycatcher	3	6	9
Yellow-bellied Flycatcher	1	0	1
Acadian Flycatcher	3	0	3
Alder Flycatcher	1	0	1
Eastern Wood-Pewee	2	3	5
Flycatcher (unidentifiable)	1	1	2
Vireonidae			
White-eyed Vireo	2	1	3
Red-eyed Vireo	9	4	13
Troglodytidae			
House Wren	3	4	7
Marsh Wren	5	18	23
Sedge Wren	4	4	8
Wren (unidentifiable)	0	2	2
Regulidae			
Ruby-crowned Kinglet	1	0	1
Turdidae			
Veery	0	1	1
American Robin	0	2	2
Wood Thrush	0	1	1
Hermit Thrush	4	1	5
Swainson's Thrush	4	7	11
Grey-cheeked Thrush	2	2	4
Thrush (unidentifiable)	0	4	4
Mimidae			
Northern Mockingbird	0	1	1
Gray Catbird	5	3	8
Parulidae			
Black-and-white Warbler	1	2	3
Prothonotary Warbler	3	2	5
Swainson's Warbler	0	1	1
Golden-winged Warbler	1	0	1
Blue-winged Warbler	1	1	2
Tennessee Warbler	3	2	5
Orange-crowned Warbler	1	2	3
Nashville Warbler	0	1	1
Northern Parula	1	1	2
Yellow Warbler	4	3	7
Magnolia Warbler	2	8	10
Yellow-rumped Warbler	9	34	43
Black-throated Green Warbler	1	9	10
Blackburnian Warbler	2	0	2
Chestnut-sided Warbler	1	1	2
Pine Warbler	0	2	2
Prairie Warbler	1	1	2

(continued)

Table 24.4. *(continued)*

| Family | No. of birds found dead | | |
Species	1938–43	1944–49	Totals
Palm Warbler	2	6	8
Ovenbird	9	18	27
Northern Waterthrush	17	11	28
Louisiana Waterthrush	0	1	1
Kentucky Warbler	1	1	2
Common Yellowthroat	89	47	136
Yellow-breasted Chat	12	7	19
Hooded Warbler	1	2	3
Wilson's Warbler	1	0	1
American Redstart	4	10	14
Warbler (unidentifiable)	6	19	25
Emberizidae			
Eastern Towhee	0	1	1
Savannah Sparrow	17	20	37
Grasshopper Sparrow	1	2	3
Henslow's Sparrow	0	1	1
Saltmarsh Sharp-tailed Sparrow	3	2	5
Vesper Sparrow	1	0	1
Bachman's Sparrow	0	1	1
Field Sparrow	0	1	1
White-throated Sparrow	0	11	11
Swamp Sparrow	14	26	40
Song Sparrow	0	4	4
Sparrow (unidentifiable)	0	9	9
Cardinalidae			
Rose-breasted Grosbeak	0	1	1
Indigo Bunting	8	19	27
Dickcissel	0	1	1
Icteridae			
Bobolink	1	0	1
Passeridae			
House Sparrow	1	1	2
Totals	269	357	626

Source: Weston (1966).

that they picked up carcasses in an irregular fashion, but nevertheless they were able to gain some insight into the extent of this type of mortality over time. Over the 12-year period they identified 626 passeriforms representing 62 species in 7 families (table 24.4). No relationship between weather or lighting on the bridge was determined. October and November were the months with the most casualties. Most of the warblers were found in September and October, while most of the sparrows were found in November. In 1950 the overhead power cables were removed and no more mortality was noted.

Many passeriforms are injured or die after striking windows or overhead wires. We have heard of a number of such incidents in Florida, but the importance of this type of mortality has not been quantitatively documented as has been

                                    Parasites and Diseases of Wild Birds in Florida

Table 24.5. Road-kill data on perching birds in Florida state parks and recreation areas

| Family | No. of birds reported as road-kill | | | | |
Species	1990	1991	1992	1993	Totals
Tyrannidae					
Eastern Phoebe	1	0	0	1	2
Eastern Kingbird	0	0	1	1	2
Laniidae					
Loggerhead Shrike	0	0	1	0	1
Vireonidae					
White-eyed Vireo	0	1	2	0	3
Black-whiskered Vireo[b]	0	0	1	0	1
Corvidae					
Blue Jay	2	2	1	1	6
American Crow	3	3	1	0	7
Fish Crow[a]	1	0	0	0	1
Hirundinidae					
Purple Martin	0	2	0	0	2
Tree Swallow	0	0	0	2	2
Barn Swallow	0	0	3	0	3
Troglodytidae					
Carolina Wren	0	1	1	1	3
Turdidae					
Eastern Bluebird	0	0	1	0	1
Hermit Thrush	1	0	0	2	3
American Robin	8	12	4	30	54
Mimidae					
Gray Catbird	14	21	21	24	80
Northern Mockingbird	1	3	6	14	24
Brown Thrasher	4	1	0	1	6
Parulidae					
Bachman's Warbler[c]	0	1	0	0	1
Northern Parula	0	0	0	1	1
Yellow Warbler	0	1	0	0	1
Magnolia Warbler	0	0	0	1	1
Yellow-rumped Warbler	11	0	2	3	16
Yellow-throated Warbler	0	0	1	0	1
Pine Warbler	1	0	1	0	2
Palm Warbler	3	3	10	4	20
Black-and-white Warbler[d]	0	0	1	0	1
American Redstart	1	0	2	0	3
Ovenbird	0	1	1	2	4
Kentucky Warbler	1	0	0	0	1
Connecticut Warbler	0	0	1	0	1
Common Yellowthroat	0	3	1	3	7
Yellow-breasted Chat	0	0	0	1	1
Thraupidae					
Summer Tanager	0	1	0	0	1
Emberizidae					
Eastern Towhee	1	1	1	2	5
Chipping Sparrow	0	0	0	2	2
Swamp Sparrow	0	4	0	0	4

(continued)

Table 24.5. *(continued)*

Family Species	No. of birds reported as road-kill				
	1990	1991	1992	1993	Totals
Cardinalidae					
Northern Cardinal	14	16	19	28	77
Indigo Bunting	0	0	1	1	2
Icteridae					
Red-winged Blackbird	1	0	1	0	2
Boat-tailed Grackle[e]	3	11	7	2	23
Common Grackle	2	1	0	4	7
Fringillidae					
American Goldfinch	0	0	1	0	1
Number of parks reporting	45	64	64	61	—

Sources: Snyder (1994), Stevenson (1994).
a. *Corvus ossifragus* Wilson.
b. *Vireo altiloquus* (Vieillot).
c. *Vermivora bachmanii* (Audubon).
d. *Mniotilta varia* (L.).
e. *Quiscalus major* Vieillot.

done for tower strikes. Avise and Crawford (1981) stated that these types of incidents have not been reported as often, but may actually result in even greater mortality than is known from birds striking towers and tall buildings.

The number of passeriforms killed on highways is difficult to assess because of a number of factors, and as a result such mortality is probably underestimated. For example, Stewart (1971) compared the persistence of carcasses of House Sparrows (*Passer domesticus*) on an interstate highway (high speed) with those on a paved country road (lower speed) in North Carolina. He found that within 2 hours there was no trace of the carcasses placed on the interstate, whereas those on the country road were still recognizable for at least 6 or 7 hours. Over a 4-year period (1990–93), 386 passeriforms representing 43 species were recorded as roadkills on highways passing through Florida state parks and recreation areas (table 24.5). The largest numbers of these were Gray Catbirds (*n* = 80), Northern Cardinals (*n* = 77), American Robins (*n* = 54), Northern Mockingbirds (*n* = 24), and Boat-tailed Grackles (*n* = 23). On a statewide basis, the number of passeriforms killed is certainly

higher, especially on highways with high speed limits such as freeways and interstate highways. In some instances, road mortality can be quite high. Fisk (1976) reported finding "many" dead American Robins killed by vehicles along roads in Dade County over a 10-year period. In the fall of 1988, hundreds of Purple Martins were killed by cars crossing the Ochlockonee Bridge in Wakulla County where the birds had roosted overnight (Gore 1988). Stevenson and Anderson (1994) stated that Boat-tailed Grackles feed along the shoulders of roads south of Lake Okeechobee and are commonly killed by vehicles as they fly across the roads. For some populations of rare and endangered species, road mortality could be significant. Florida Scrub-Jays, for example, are attracted to roadsides where they can hunt for insects and cache acorns and are thereby put in harm's way. In addition individual territories can span both sides of a road and territorial disputes can occur across the road, putting the birds in danger of being struck by a passing vehicle. Dreschel et al. (1990) discussed such phenomena and cited examples in Brevard and Highlands counties. Pranty (2000) documented the mortality of a Florida Grasshopper Sparrow that was struck by a vehicle in Highlands County in July of 1996.

In the 1970s a Palm Warbler was found caught in the web of a golden orb weaver (*Nephila clavipes*) in a live oak–cabbage palm hammock in Glades County (Layne 1997). The warbler was dead when found and hanging by a 2-foot-long strand of web about 6 feet above ground. This spider has a very strong web; several strands were wrapped tightly around the primaries of 1 wing. Layne suspected that the bird flew into the net and totally destroyed it with its struggles, since he saw no sign of other remains of a web. This type of mortality most likely represents an incidental and unusual event.

III. Predation

Predation, particularly predation of nest contents, is a significant source of mortality for passeriforms (Bancroft 1986). A summary of direct evidence of predation on 47 species representing 11 families of passeriforms in Florida is presented in table 24.6. Stoddard (1931) identified remains of 27 species of passeriforms in 1,098 pellets collected at a Northern Harrier roost in Leon County during the winter of 1925–26. The most common species were Song Sparrows, Eastern Meadowlarks, and Savannah Sparrows, the remains of which were found in 64, 26, and 23 pellets, respectively.

The best studied species include Florida Scrub-Jays, Eastern Bluebirds, Boat-tailed Grackles, Seaside Sparrows, and Great Crested Flycatchers, with the most extensive investigations being those Glen Woolfenden and his associates conducted during a long-term study of the Florida Scrub-Jay at Archbold Biological Station in Highlands County. Their conclusions are based on actual observations as well as circumstantial evidence. Predators of adults include coachwhips, indigo snakes, Sharp-shinned Hawks, Cooper's Hawks, Northern Harriers, Merlins, bobcats, and feral house cats (Woolfenden and Fitzpatrick 1996). The list of predators of eggs

and nestlings is longer and includes the above species as well as Florida Scrub-Jays themselves, Blue Jays, American Crows, Fish Crows, Eastern Screech-Owls, Great Horned Owls, Red-tailed Hawks, cotton rats, raccoons, and other snakes (species not given) (Fitzpatrick et al. 1991; Mumme 1987; Lohrer 1980; Schaub et al. 1992; Schoech 1999; Webber 1980; Wescott 1970; Woolfenden 1974; Woolfenden and Fitzpatrick 1996). Woolfenden and Fitzpatrick (1984) concluded that predation was "the primary cause of nest failure in the Florida Scrub-Jay, accounting for 67% of egg loss and 85% of nestling loss." Further studies were reported by Schaub et al. (1992), who found that nest predation increased as the breeding season progressed (figure 24.5); it was more severe on eggs and young nestlings than on older nestlings and was attributed to the activities of diurnal snakes and birds. Nocturnal mammalian predators were of lesser importance. Schaub et al. (1992) reported predation rates of 67% for eggs and 85% for nestlings over a 10-year period from 1969 to 1979. These values are much higher than those obtained in studies of Eastern Bluebirds, Boat-tailed Grackles, and Seaside Sparrows in Florida (table 24.7).

Predation on songbirds by free-ranging domestic cats is a serious problem (Coleman and Temple 1993). Domestic cats that are well fed by their owners still have strong urges to hunt (Adamec 1976) and about 20% of their prey consists of birds (Fitzgerald 1988), most of them passeriforms. It has been estimated that hundreds of millions of birds are killed each year in the United States by domestic cats (Coleman et al. 1997). Figures for Florida are not available, but undoubtedly are significant, making domestic cat predation an important mortality factor in some songbird populations.

Predation on passeriforms by ants, especially the red imported fire ant (*Solenopsis invicta*), has been reported in Alabama (Mount 1981) and Texas (Sikes and Arnold 1986; Wilson and Silvy 1988), but not in Florida. Yosef and Lohrer (1995) were not able to demonstrate clearly that red fire ants had caused changes in

Table 24.6. Direct evidence[a] of predation on perching birds in Florida

Family Species	Type of prey	County/area	Year(s)	Predator	Data source
Tyrannidae					
Eastern Kingbird	Nestlings	Hillsborough, Pasco	1966–72	Rat snake[b]	White & Woolfenden (1973)
	Nestlings	Hillsborough, Pasco	1966–72	Corn snake[c]	Ibid.
	Nestlings	Highlands	1981	Swallow-tailed Kite	Lohrer & Lohrer (1984)
Great Crested Flycatcher	Eggs	Orange	1979–89	Rat snake	Taylor & Kershner (1991)
	Eggs	Orange	1979–89	Corn snake	Ibid.
	Eggs	Orange	1979–89	Indigo snake[d]	Ibid.
	Nestling	Orange	1979–89	Indigo snake	Ibid.
	Nestling	Orange	1979–89	Rat snake	Ibid.
	Adult	Orange	1979–89	Rat snake	Ibid.
Laniidae					
Loggerhead Shrike	Nestling	Highlands	1979	Swallow-tailed Kite	Lohrer & Winegarner (1980)
Vireonidae					
Red-eyed Vireo	Adult	Dry Tortugas	1970	Cattle Egret	Harrington & Dinsmore (1975)
Corvidae					
Florida Scrub-Jay	Nestlings	Highlands	1967–70	Coachwhip[e]	Westcott (1970)
	Fledglings	Highlands	1979	Coachwhip	Webber (1980)
	Nestlings	Highlands	1998	Coachwhip	Schoech (1999)
	NG	Highlands	1981	Bobcat[f]	Schaub (1990)
	Fledglings	Highlands	1989	Indigo snake	Mumme (1987)
	Eggs	Highlands	1969–72	Florida Scrub-Jay	Woolfenden (1974)
	NG	Highlands	1971	Northern Harrier	Schaub (1990)
Blue Jay	Nestlings	Highlands	1979	Coachwhip	Lohrer (1980)
	Fledglings	"Florida"	NG	Racer[g]	Nicholson (1929)
	NG	Highlands	1967–79	Bobcat	Wassmer et al. (1988)
Hirundinidae					
Tree Swallow	Adult	Dade	1940	American Kestrel	Christy (1940)
Barn Swallow	Adult	Dry Tortugas	1970	Cattle Egret	Harrington & Dinsmore (1975)
Troglodytidae					
Carolina Wren	NG	Leon	1925–26	Northern Harrier	Stoddard (1931)
	NG	"Florida"	1976–83	Bobcat	Maehr & Brady (1986)
House Wren	NG	"Florida"	1976–83	Bobcat	Ibid.
Bewick's Wren	NG	Leon	1925–26	Northern Harrier	Stoddard (1931)

(continued)

Table 24.6. (*continued*)

Family Species	Type of prey	County/ area	Year(s)	Predator	Data source
Marsh Wren	NG	Leon	1925–26	Northern Harrier	Ibid.
Sedge Wren	NG	Highlands	1997–98	Loggerhead Shrike	Pranty (1999)
Turdidae					
American Robin	NG	Putnam	1949	Great Horned Owl	Burns (1952)
	NG	"Florida"	1976–83	Bobcat	Maehr & Brady (1986)
	NG	Leon	1925–26	Northern Harrier	Stoddard (1931)
Hermit Thrush	NG	Leon	1925–26	Northern Harrier	Ibid.
Eastern Bluebird	NG	Leon	1925–26	Northern Harrier	Ibid.
Mimidae					
Northern Mockingbird	Nestling	Highlands	1978	Swallow-tailed Kite	Lohrer & Winegarner (1980)
	Nestling	Highlands	1987	Swallow-tailed Kite	Schaub (1990)
	Adult	Highlands	1989	Florida Scrub-Jay	Curry (1990)
Gray Catbird	NG	Leon	1925–26	Northern Harrier	Stoddard (1931)
	NG	"Florida"	1976–83	Bobcat	Maehr & Brady (1986)
Brown Thrasher	NG	"Florida"	1976–83	Bobcat	Ibid.
	NG	Leon	1925–26	Northern Harrier	Stoddard (1931)
Sturnidae					
European Starling	NG	Leon	1925–26	Northern Harrier	Stoddard (1931)
Parulidae					
Common Yellowthroat	Adult	Lake Okeechobee	1942	Largemouth bass[h]	Rand (1943)
	Adult	"Florida"	1976–83	Bobcat	Maehr & Brady (1986)
	Adult	Leon	1925–26	Northern Harrier	Stoddard (1931)
Prairie Warbler	Adult	Leon	1925–26	Northern Harrier	Ibid.
Blackpoll Warbler	Adult	Dry Tortugas	1962	Cattle Egret	Cunningham (1965b)
	Adult	Dry Tortugas	1970	Cattle Egret	Harrington & Dinsmore (1975)
Yellow-rumped Warbler	Adult	Hendry	1958	Cattle Egret	Palmer (1962)
	Adult	Dry Tortugas	1962	Cattle Egret	Cunningham (1965b)
	Adult	Everglades N.P.	1963	Cattle Egret	Ibid.
Black-throated Blue Warbler	Adult	Dry Tortugas	1970	Cattle Egret	Harrington & Dinsmore (1975)
Ovenbird	Adult	Dry Tortugas	NG	Cattle Egret	Fogarty & Hetrick (1973)
Emberizidae					
Seaside Sparrow	Eggs	Levy	1979–80	Rice rat	Post (1981)
	Eggs	Levy	1979–80	Fish Crow	Ibid.
	Adult	Monroe	1971	Short-tailed Hawk	Kale (1971)

Species	Age	County	Year	Predator	Reference
Florida Grasshopper Sparrow	Adult	Highlands	1998	Loggerhead Shrike	Pranty (2000)
	Adult	Highlands	1996	Loggerhead Shrike	Vickery (1996), Dean (1999)
Swamp Sparrow	NG	Highlands	1976–83	Bobcat	Maehr & Brady (1986)
	NG	Leon	1925–26	Northern Harrier	Stoddard (1931)
Savannah Sparrow	NG	Leon	1925–26	Northern Harrier	Ibid.
Le Conte's Sparrow	NG	Leon	1925–26	Northern Harrier	Ibid.
Henslow's Sparrow	NG	Leon	1925–26	Northern Harrier	Ibid.
Vesper Sparrow	NG	Leon	1925–26	Northern Harrier	Ibid.
Bachman's Sparrow	NG	Leon	1925–26	Northern Harrier	Ibid.
Chipping Sparrow	NG	Leon	1925–26	Northern Harrier	Ibid.
Field Sparrow	NG	Leon	1925–26	Northern Harrier	Ibid.
White-throated Sparrow	NG	Leon	1925–26	Northern Harrier	Ibid.
Song Sparrow	NG	Leon	1925–26	Northern Harrier	Ibid.
Cardinalidae					
Northern Cardinal	NG	"Florida"	1976–83	Bobcat	Maehr & Brady (1986)
	NG	Leon	1925–26	Northern Harrier	Stoddard (1931)
Indigo Bunting	Adult	Monroe	1984	Cattle Egret	Paul (1984)
Eastern Towhee	NG	"Florida"	1976–83	Bobcat	Maehr & Brady (1986)
	NG	Leon	1925–26	Northern Harrier	Stoddard (1931)
Icteridae					
Red-winged Blackbird	Eggs	Dade	1982	Rice rat[i]	NeSmith & Cox (1985)
	Nestlings	NG	NG	Common Grackle	Stevenson & Anderson (1994)
	NG	Leon	1925–26	Northern Harrier	Stoddard (1931)
Boat-tailed Grackle	Eggs	Hillsborough	1978–81	Rice rat	Bancroft (1986)
Common Grackle	NG	"Florida"	1976–83	Bobcat	Maehr & Brady (1986)
	NG	Leon	1925–26	Northern Harrier	Stoddard (1931)
Brown-headed Cowbird	Fledgling	Alachua	1986	Domestic dog	Paul (1986)
Eastern Meadowlark	NG	Highlands	1967–79	Bobcat	Wassmer et al. (1988)
	NG	Leon	1925–26	Northern Harrier	Stoddard (1931)
Passeridae					
House Sparrow	NG	Leon	1925–26	Northern Harrier	Ibid.

NG = not given.

a. Observations of predation as it occurred or the finding of prey in stomachs or fecal samples of predators.

b. *Elaphe obsoleta.*

c. *Elaphe guttata.*

d. *Drymarchon corais.*

e. *Masticophis flagellum.*

f. *Felis rufus.*

g. *Coluber constrictor.*

h. *Micropterus salmoides.*

i. *Oryzomys palustris.*

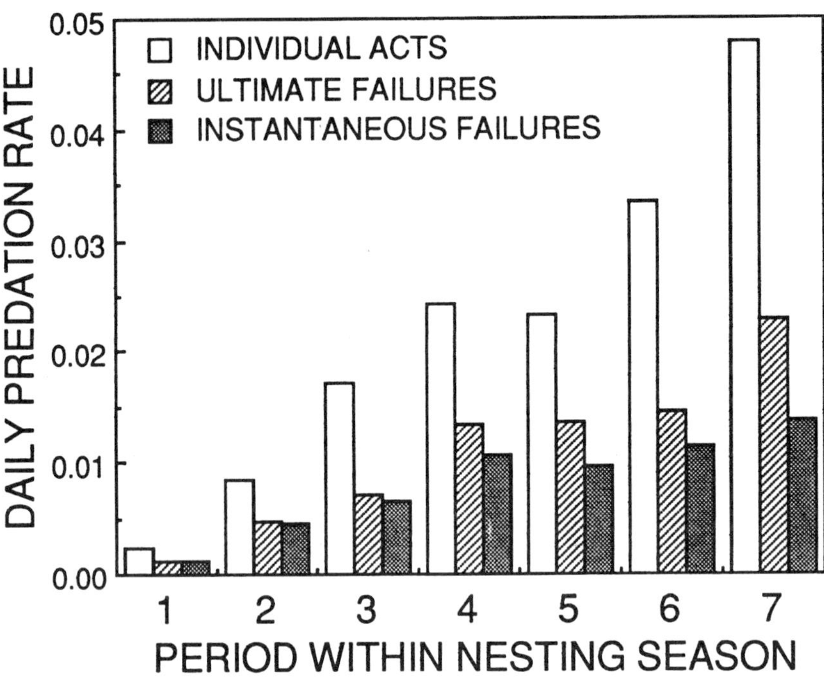

FIGURE 24.5. Daily predation rates on nests of Florida Scrub-Jays at Archbold Biological Station (Highlands County) at 2-week intervals during the nesting seasons of 1974–79 and 1981–87, combined. Intervals are: (1) 9–24 March, (2) 25 March–8 April, (3) 9–23 April, (4) 24 April–8 May, (5) 9–23 May, (6) 24 May–7 June, (7) 8–23 June. From Schaub et al. 1992; by permission of *Auk*.

the ecosystem detrimental to populations of Loggerhead Shrikes in Highlands County. However, since fire ants are widespread in Florida, this topic should be investigated further for shrikes and other passeriforms.

IV. Electrocution

Electrocutions of American Crows and Loggerhead Shrikes were reported in South Jacksonville (Duval County) in the summer of 1921 (Hallinan 1922). No data were given on numbers of birds killed, but the author did provide some details of how the electrocutions took place. Apparently the crows perched on insulators, metal cross-arms, or static ground bayonets, and the expanse of their wings caused "a short-circuit between the conductors or a ground connection between the conductors and the static ground system." The electrocution of the shrikes was a result of the impalement of prey items such as grasshoppers or lizards on the sharp end of the tie wire. This produced a flashover and resulted in the electrocution of the shrikes. The significance of electrocution on populations of crows, shrikes, and other passeriforms in Florida is not known.

V. Brood parasitism

Some passeriform birds are brood or nest parasites; that is, they lay their eggs in the nests of other species of passeriforms that in turn incubate, hatch, and rear their young for them. In the process, the eggs and/or young of the "parasitized" species are sometimes ejected

Table 24.7. Mortality of eggs and nestlings of perching birds in Florida resulting from predation

Species of bird	County	Year(s)	% Mortality		Data source
			Eggs	Nestlings	
Corvidae					
Florida Scrub-Jay	Highlands	1969–79	67	85	Schaub et al. (1992)
Turdidae					
Eastern Bluebird	Hillsborough, Pasco	1970	15	23	White & Woolfenden (1973)
Emberizidae					
Seaside Sparrow	Levy	1979–80	40[a]		Post (1981)
Icteridae					
Boat-tailed Grackle	Hillsborough	1978–81	15	13	Bancroft (1986)

a. Value is for eggs and nestlings combined.

from the nest or, if they remain in the nest and hatch, die of starvation since they receive less food than the larger and more aggressive "parasitic" nestlings. Close to 1% of all avian species on a global basis are brood parasites and include species of honey guides, cuckoos, finches, and cowbirds (Payne 1977). The reader is referred to Johnsgard (1997) and Rothstein and Robinson (1998) for more information on brood parasitism. Determination of the species of chewing lice found on fledgling cowbirds has been used to identify the hosts parasitized by their adult parents (Hahn et al. 2000).

In Florida, 2 species of cowbirds, the Shiny Cowbird and the Brown-headed Cowbird, are known to be brood parasites. Both are fairly recent invaders. The Shiny Cowbird is a more recent invader from South America and was first seen in the state in 1985. It now breeds in Florida and its numbers are increasing, but is less numerous than the Brown-headed Cowbird. For the most part it is restricted to coastal counties in the southern peninsula (Robertson and Woolfenden 1992; Stevenson and Anderson 1994). Brown-headed Cowbirds were irregular visitors from the north in the early 1900s (Howell 1932), but their numbers have gradually increased over the years. Breeding was first recorded in 1957 (Monroe 1957), increased moderately in the 1970s, then increased dramatically during the 1980s. At the present they are fairly common and sometimes locally abundant residents throughout the state, although they breed mainly in the coastal areas in the southern half of peninsular Florida (Robertson and Woolfenden 1992).

There are only 2 published records of brood parasitism by Shiny Cowbirds in Florida, both in broods of Red-winged Blackbirds; one was in Dade County in 1991 (Robertson and Woolfenden 1992; Stevenson and Anderson 1994; Cruz et al. 1998) and one in Seminole County in 1989 (Cruz et al. 1998). Brown-headed Cowbirds, however, have been observed as brood parasites in nests of 18 species representing 3 families (table 24.8). Stevenson and Anderson (1994) stated that gnatcatchers seemed to be one of the favorite hosts in Florida, but gave no quantitative data. Cruz et al. (1998) pointed out that there is a third species, the Bronzed Cowbird, that since 1995 has been a regular winter visitor throughout Florida and could become established as a breeding species in the future. Cruz et al. (1998) also evaluated the potential impact of cowbirds on Florida's avifauna. They listed, in addition to the specific records of cowbird parasitism, another 26 Florida passeriforms that are vulnerable to brood parasitism to some degree and that may be at risk in the future. The overall significance of brood parasitism on populations of perching birds in Florida has not been determined and certainly is worthy of

Table 24.8. Occurrences of brood parasitism by cowbirds in nests of perching birds of Florida[a]

Host	County/site	Year(s)	Data source
Vireonidae			
White-eyed Vireo	Escambia	1963	J.R. Stewart (1963)
	Leon	NG	Stevenson & Anderson (1994)
	Okaloosa	1963	Ibid.
	Pasco	NG	Cruz et al. (1998)
Red-eyed Vireo	Alachua	1986	Paul (1986)
Black-whiskered Vireo	Monroe	1989	Kale (1989)
Yellow-throated Vireo	Leon	1995–97	Cox (1997)
Sylviidae			
Blue-grey Gnatcatcher	Duval	1984	Paul (1984)
		1988	Cruz et al. (1998)
	Leon	1976	Kale (1976)
	Levy	1990	Cruz et al. (1998)
Turdidae			
Eastern Bluebird	Duval	1982	Paul (1982)
Wood Thrush	Leon	1992–97	Cox (1997)
Parulidae			
Northern Parula	Alachua	1988	Cruz et al. (1998)
	Escambia	1983	Ibid.
	Hamilton	1988	Ibid.
	Volusia	1987	Ibid.
	NG	NG	Stevenson & Anderson (1994)
Pine Warbler	Duval	1988	Cruz et al. (1998)
	NG	NG	Stevenson & Anderson (1994)
Prairie Warbler	Monroe	1989	Kale (1989), Paul (1989)
	Lee	1990	Cruz et al. (1998)
	Pinellas[b]	1987	Paul (1987), Atherton & Atherton (1988)
	Sarasota[b]	1988	Atherton & Atherton (1988)

(continued)

further attention, particularly since all 3 species of cowbirds (Bronzed Cowbirds, Brown-headed Cowbirds, and Shiny Cowbirds) are increasing in their numbers and in their geographic distribution.

VI. Inclement weather

Considerable mortality among passeriforms in Florida has occurred over the years due to various types of inclement weather, including storms, hurricanes, floods, droughts, and low temperatures. Some records of this type of mortality date back into the late 1800s and early 1900s and constitute fascinating if not sad reading.

In 1909 an account was given in an article titled "A tragedy of migration" of the results of a violent thunderstorm in the Florida Keys (Bennett 1909). Actual counts were not obtained, but hundreds of migratory birds were estimated to have died during the night of April 14, 1909. These included 14 species of emberizids (Indigo and Painted Buntings, Summer and Scarlet Tanagers, Ovenbirds, Orchard Orioles, Black-throated Blue Warblers, Hooded Warblers, Common Yellowthroats, Cape May Warblers, Worm-eating Warblers, Blackpoll Warblers, Prairie Warblers, and Blue Grosbeaks), 1 tyrannid (Eastern Kingbirds), and 1 vireonid (Warbling Vireos). Many of the dead birds were found near lighthouses and the author

Table 24.8. *(continued)*

Host	County/site	Year(s)	Data source
Yellow-throated Warbler	Alachua	1976	Cruz et al. (1998)
Common Yellowthroat	Pasco	1987	Ibid.
Thraupidae			
Summer Tanager	Alachua	1986	Paul (1986)
Emberizidae			
Eastern Towhee	Alachua	1980	Edscorn (1980)
	Hernando	1990	Cruz et al. (1998)
Cardinalidae			
Northern Cardinal	Collier[b]	1989	Paul (1989)
	Escambia	1959	Cruz et al. (1998)
	Washington	1987	Ibid.
	NG	NG	Stevenson & Anderson (1994)
Painted Bunting	NE Florida	NG	Kropp & Cox (1996)
Icteridae			
Orchard Oriole	Leon	NG	Stevenson & Anderson (1994)
Red-winged Blackbird	Dade[c]	1991	Robertson & Woolfenden (1992), Stevenson and Anderson (1994), Cruz et al. (1998)
	Lee[b]	1994	Cruz et al. (1998)
	Seminole[c]	1989	Ibid.
	NG	NG	Stevenson & Anderson (1994)

NG = not given.

a. Unless otherwise noted, these are records of parasitism by Brown-headed Cowbirds.

b. Species of cowbird not determined.

c. The parasitic species was identified as a Shiny Cowbird.

claimed that each light had "acted as a magnet and drawn to itself countless thousands of birds of many kinds, overpowered by the storm and forced to seek asylum." One lighthouse keeper on Loggerhead Key related that "By midnight, a few had struck the plate-glass panes of the lantern, and . . . from 2 o'clock until dawn they came in such masses that he actually could not see out through the glass panes! He said that they were all on the lee side of the lantern, away from the wind, and did not fly against the glass at full speed, but rather fluttered and beat against it, bruising and wounding themselves and each other, and thus causing death to many. Occasionally, one coming at higher speed would strike hard enough to kill itself on the spot."

A hurricane that passed through southern Florida in September of 1935 was reputed to have killed "all" of the Cape Sable Seaside Sparrows in the Cape Sable area (Monroe County) (Sprunt 1954). Sutton (1945) gave an anecdotal account of a Palm Warbler that was found dead in Titusville (Brevard County) after a hurricane passed through the area in October 1944. A remarkable event occurred 20 years later during a 3-day period in October of 1964 (Case et al. 1965; Cunningham 1965a): the tail end of Hurricane Hilda moved into Brevard County and brought with it torrential rains followed by continuous drizzle, haze, and winds between 20 and 30 mph. The low ceiling caused migrating birds to fly low and become disoriented; large numbers perished by collisions with vehicles, tall buildings and towers. More than 4,700 dead birds were counted; 3,550 were passeriforms representing 30 species from 4 families (table 24.9). The most numerous species were

Table 24.9. Numbers of perching birds found dead after a severe storm in Brevard County, October 6–8, 1964

Family	Species	No. dead
Vireonidae	White-eyed Vireo	1
	Yellow-throated Vireo	2
	Red-eyed Vireo	4
Regulidae	Ruby-crowned Kinglet	1
Turdidae	Swainson's Thrush	2
	Gray-cheeked Thrush	1
Mimidae	Gray Catbird	2
Parulidae	Black-and-white Warbler	90
	Swainson's Warbler	10
	Worm-eating Warbler	3
	Northern Parula	155
	Magnolia Warbler	14
	Cape May Warbler	3
	Black-throated Blue Warbler	445
	Black-throated Green Warbler	2
	Yellow-throated Warbler	31
	Blackpoll Warbler	322
	Prairie Warbler	4
	Palm Warbler	145
	Ovenbird	373
	Northern Waterthrush	56
	Kentucky Warbler	1
	Connecticut Warbler	4
	Mourning Warbler	1
	Common Yellowthroat	1,354
	Hooded Warbler	1
	American Redstart	497
Cardinalidae	Rose-breasted Grosbeak	3
	Painted Bunting	1
Icteridae	Bobolink	22
	Total	3,550

Source: Case et al. (1965).

Common Yellowthroats, American Redstarts, Black-throated Blue Warblers, and Ovenbirds. The authors stated that this list was only a small fraction of the total kill in that area, which certainly must have been massive.

Storms were found to damage nests and uproot vegetation in which nests of Boat-tailed Grackles were constructed in Hillsborough County during a 4-year study (1978–81) (Bancroft 1986). This type of mortality accounted for 8 of 460 (1.7%) eggs and 24 of 400 (6%)

nestlings that were lost during the study. The "storm of the century" on March 13, 1993, stopped nesting of Florida Scrub-Jays in central Florida, delaying it for 2–3 weeks (Langridge 1993).

Drought was blamed for the death of "large numbers" of migrating passeriforms in the lower Keys in May of 1962 (Paulson and Stevenson 1962). Many dead and dying birds were seen floating on the water and on the roads of the Keys. Observers on the Dry Tortu-

Table 24.10. Mortality events related to cold weather among passeriforms overwintering in Florida

Bird	County (or area)	Year(s)	Extent of mortality	Data source
Vireonidae				
Black-whiskered Vireo	Pinellas	1970–80	NG	Stevenson & Anderson (1994)
Hirundinidae				
Purple Martin	Osceola	1940	"thousands"	Baker (1940)
	Brevard	1958	"many"[a]	Stevenson (1958b)
	W. Panhandle	1960	"two large colonies"	Weston (1965)
	Pinellas	1973	"numerous"	Woolfenden (1973)
Tree Swallow	Dade	1895	>200	Smith (1895)
	Dade	1895	"a dozen"	Cory (1895)
	Dade	1899	"hundreds"	Slosson (1899)
	Dade	1940	>500	Weber (1940)
	Dade	1940	>100	Christy (1940)
	Broward	1958	"many"[a]	Stevenson (1958b)
Barn Swallow	Wakulla	1962	1	Stevenson & Anderson (1994)
Turdidae				
Eastern Bluebird	Leon	1895	"disastrous"	Stevenson & Anderson (1994), Williams (1904)
	Leon	1899	"disastrous"	Ibid.
	NG	1911	"damaging"	Stevenson & Anderson (1994)
	W. Panhandle	1940	"numbers"	Weston (1965)
	W. Panhandle	1957	"numbers"	Ibid.
American Robin	Broward	1958	"many"[a]	Stevenson (1958b)
Mimidae				
Gray Catbird	Dade	1940	1	Christy (1940)
	Broward	1958	"many"[a]	Stevenson (1958b)
Northern Mockingbird	Dade	1940	1	Christy (1940)
Parulidae				
Yellow-throated Warbler	NG	1957–58	"decimation"	Stevenson & Anderson (1994)
	NG	1963–64	"decimation"	Ibid.
	NG	1977–78	"decimation"	Ibid.
Yellow-rumped Warbler	Dade	1940	"large numbers"	Ruff (1940)
	Osceola	1940	"thousands"	Baker (1940)
	Broward	1958	"many"[a]	Stevenson (1958b)
	Pinellas	1973	"numerous"	Woolfenden (1973)

NG = not given.

a. Author stated that >450 birds were found but did not give numbers of each species separately.

gas identified 22 species of birds, including 23 Bobolinks and 21 Blackpoll Warblers. This mortality was attributed to drought conditions that had in turn led to a significant decrease in the numbers of insects needed to replenish the energy reserves of these migrating birds.

Destruction of Seaside Sparrow nests due to flooding associated with high spring tides and storms in a salt marsh in Gulf Hammock (Levy County) was documented by Post (1981). Eleven nests were flooded in 1979 and 3 in 1980. Water levels have significant effects on populations of Cape Sable Seaside Sparrows in Everglades National Park (Lockwood et al. 1997; Curnutt et al. 1998; Nott et al. 1998). Breeding seasons are controlled by rainfall and water levels. Predation increases and nest success decreases with the onset of summer rains. Observations were made of 2 nests in 1996 that were flooded; 2 eggs in 1 nest and 3 nestlings in another were lost (Lockwood et al. 1997). Seasonal flooding was reported as a likely cause of nesting failure for Florida Grasshopper Sparrows, although no details

Table 24.11. Organochlorine residues[a] in carcasses[b] of European Starlings from Bay County, Florida

| Year(s) | No. birds in pooled sample | Mean (ppm wet weight)[c] | | | | | | Data source |
		DDE	DDD	DDT	Dieldrin	Heptachlor epoxide	BHC	
1967–68	30	1.20	0.02	0.03	0.14	0.25[d]	ND	A
1970	10	0.73	0.01	0.02	0.09	0.10	0.01	B
1972	10	1.10	TR	0.02	0.31	0.09	TR	C
1974	10	0.73	0.01	0.03	0.26	0.07	0.02	D
1976	10	0.23	NG	0.06	0.09	0.04	NG	E
1979	9	0.15	NG	ND	0.1	0.02	NG	F
1982	10	0.18	NG	NG	0.02	0.06	NG	G

Year(s)	No. birds in pooled sample	PCBs	Oxychlordane	HCB	Chlordane isomers	Mirex	Data source
1967–68	30	NT	NT	NT	NG	NG	A
1970	10	0.26	NT	NT	NG	0.31	B, H
1972	10	0.29	0.10	0.04	NG	NG	C
1974	10	0.15	0.07	TR	NG	NG	D
1976	10	0.28	NG	ND	0.07	NG	E
1979	9	0.21	NG	ND	0.15	0.18	F
1982	10	0.28	0.07	0.01	NG	0.06	G

Sources: A = Martin (1969), B = Martin and Nickerson (1972), C = Nickerson and Barbehenn (1975), D = White (1976), E = White (1979), F = Cain and Bunck (1983), G = Bunck et al. (1987), H = Oberheu (1972).
ND = none detected, NG = not given, NT = not tested for this compound, TR = trace (<0.002 ppm).
a. In addition, 0.04 ppm of *trans*-nonachlor was detected in 1982.
b. Ground-up carcasses minus the skin, feathers, beak, feet, and wings.
c. Type of mean not stated. Limits of detection: Reference A = not given, Reference B = ranged 0.005–0.01 ppm, Reference C = ranged 0.005–0.1 ppm, Reference D = 0.005 ppm for all pesticides and 0.01 ppm for PCBs, Reference E = 0.01 ppm, Reference F = 0.01 ppm, Reference G = 0.01 ppm.
d. Based on 20 birds.

were given (Vickery 1996). Pranty (2000) documented 2 instances of flooding of Florida Grasshopper Sparrow nests in Highlands County. One resulted in the loss of 3 nestlings and an egg after a rainstorm in July 1997; the second, in the loss of 4 eggs after a severe thunderstorm in June 1999.

During some years cold weather is a significant mortality factor for passeriforms wintering in Florida, resulting in losses in the hundreds and even thousands. The full extent of this mortality is not known, but there are anecdotal records of such effects on 10 species representing 5 families (table 24.10). Purple Martins, Tree Swallows, Eastern Bluebirds, and Yellow-throated Warblers seem to be especially vulnerable. It is postulated that this mortality is related to the scarcity of insects during severe cold weather and consequent starvation (Stevenson and Anderson 1994). Passeriforms arriving in Florida after extensive migration experience energy depletion; the inability to readily replenish this energy and to thermoregulate during cold weather can spell disaster.

VII. Organochlorines

Residues of 13 organochlorines have been detected in tissues of 42 species representing 11 families of perching birds from Florida. The bulk of our knowledge comes from 4 studies: a surveillance study on European Starlings, one on migratory birds killed by striking tall TV

Table 24.12. Organochlorine residues[a] in carcasses[b] of European Starlings from Madison County, Florida

Year(s)	No. birds in pooled sample	Mean (ppm wet weight)[c]						Data source
		DDE	DDD	DDT	Dieldrin	Heptachlor epoxide	BHC	
1967–68	30	5.6	0.01	0.03	0.06	0.39[d]	ND	A
1970	10	3.8	0.01	0.04	0.04	0.03	0.06	B
1972	10	1.6	TR	0.04	0.31	0.09	TR	C
1974	10	0.42	TR	0.02	0.01	TR	TR	D
1976	10	0.9	NG	ND	0.11	0.07	NG	E
1979	10	0.35	NG	ND	0.01	0.02	NG	F
1982	10	0.26	NG	NG	ND	0.05	NG	G

Year(s)	No. birds in pooled sample	PCBs	Oxychlordane	HCB	Chlordane isomers	Mirex	Data source
1967–68	30	NT	NT	NT	NG	NG	A
1970	10	0.47	NT	NT	NG	0.01	B, H
1972	10	0.41	0.1	0.44	NG	NG	C
1974	10	0.10	0.02	TR	NG	NG	D
1976	10	ND	NG	ND	0.18	NG	E
1979	10	0.12	NG	ND	0.11	0.05	F
1982	10	0.06	0.07	ND	NG	0.01	G

Sources: A = Martin (1969), B = Martin and Nickerson (1972), C = Nickerson and Barbehenn (1975), D = White (1976), E = White (1979), F = Cain and Bunck (1983), G = Bunck et al. (1987), H = Oberheu (1972).
ND = none detected, NG = not given, NT = not tested for this compound, TR = trace (<0.002 ppm).
a. In addition, 0.015 ppm of Lindane was detected in 1967–68, and 0.04 ppm of *trans*-nonachlor was detected in 1982.
b. Ground-up carcasses minus the skin, feathers, beak, feet, and wings.
c. Type of mean not stated. Limits of detection: Reference A = not given, Reference B = ranged 0.005–0.01 ppm, Reference C = ranged 0.005–0.1 ppm, Reference D = 0.005 ppm for all pesticides and 0.01 ppm for PCBs, Reference E = 0.01 ppm, Reference F = 0.01 ppm, Reference G = 0.01 ppm.
d. Based on 10 birds.

towers, one dealing with the aerial application of mirex for control of fire ants, and one on the effects of a DDT spraying program for mosquito control in Dade County during 1965 on various species of wildlife.

The European Starling is the most thoroughly studied species and was part of the National Contaminant Biomonitoring Program initiated by the U.S. Fish and Wildlife Service in 1967 (Martin 1969; Martin and Nickerson 1972; Oberheu 1972; Nickerson and Barbehenn 1975; White 1976, 1979; Cain and Bunck 1983; Bunck et al. 1987). Data on European Starlings from 4 counties in Florida (Bay, Madison, Polk, and Highlands) were collected in 2–3-year intervals over a 16-year period from 1967 to 1982 (tables 24.11–24.14). Mean values for most contaminants were <1 ppm with the exception of DDE, which ranged as high as 5.6 ppm in Madison County. There was a general decline in concentrations of DDE from 1967–68 to 1982, which may have been a reflection of the 1972 ban on the use of DDT in the United States.

Another study was conducted by D.W. Johnston (1974, 1975, 1976) on residues in fat tissues of migratory songbirds killed by colliding with tall TV towers during nocturnal flights in northern Florida from 1964 to 1973 (see also section II, Trauma). Residues of DDE, ΣDDT, and dieldrin were reported for 17 species representing 4 families of passeriforms (table 24.15).

Table 24.13. Organochlorine residues[a] in carcasses[b] of European Starlings from Polk County, Florida

| Year(s) | No. birds in pooled sample | Mean (ppm wet weight)[c] | | | | | | Data source |
		DDE	DDD	DDT	Dieldrin	Heptachlor epoxide	BHC	
1967–68	29	1.9	0.05	0.15	0.09	0.04	ND	A
1970	10	0.39	0.01	0.03	0.03	0.04	0.01	B
1972	10	0.46	TR	0.02	0.33	ND	TR	C
1974	10	0.01	TR	0.02	0.13	0.01	0.01	D
1976	0	—	—	—	—	—	—	E
1979	10	0.17	NG	ND	ND	0.01	NG	F
1982	10	0.18	NG	NG	ND	0.02	NG	G

Year(s)	No. birds in pooled sample	PCBs	Oxychlordane	HCB	Chlordane isomers	Mirex	Data source
1967–68	29	NT	NT	NT	NG	NG	A
1970	10	0.38	NT	NT	NG	0.34	B, H
1972	10	0.41	ND	TR	NG	NG	C
1974	10	0.13	0.02	ND	NG	NG	D
1976	0	—	—	—	—	—	E
1979	10	0.1	NG	ND	0.08	0.06	F
1982	10	0.07	0.02	ND	NG	0.05	G

Sources: A = Martin (1969), B = Martin and Nickerson (1972), C = Nickerson and Barbehenn (1975), D = White (1976), E = White (1979), F = Cain and Bunck (1983), G = Bunck et al. (1987), H = Oberheu (1972).

a. In addition, a trace of lindane was detected in 1967–68.

b. Ground-up carcasses minus the skin, feathers, beak, feet, and wings.

c. Type of mean not stated. Limits of detection: Reference A = not given, Reference B = ranged 0.005–0.01 ppm, Reference C = ranged 0.005–0.1 ppm, Reference D = 0.005 ppm for all pesticides and 0.01 ppm for PCBs, Reference E = 0.01 ppm, Reference F = 0.01 ppm, Reference G = 0.01 ppm.

All samples were positive for DDE, with concentrations as high as 12 ppm in a Palm Warbler, 14 ppm in an Ovenbird, and 15 ppm in a Savannah Sparrow. Dieldrin was present in 14 of the 17 species, with the highest concentration (1.1 ppm) being found in a Red-eyed Vireo. As for Starlings, a decline in the concentrations of DDT and its metabolites was noted over time (figure 24.6) and was attributed to the decreased usage of DDT in the United States during that time period (Johnston 1974). Johnston (1975) compared the concentrations of DDE in fat with those in other organs (liver and brain) in some of these birds and found that concentrations in the same bird decreased progressively from fat to liver to brain.

The third major investigation was conducted on the 20,000-acre Dee Dot ranch in Duval and St. Johns counties in 1972–74 (Wheeler et al. 1977). In that study, residues of mirex were examined in a series of passeriforms before and after the aerial application of 10-5 bait for control of red and black imported fire ants (*Solenopsis invicta* and *S. richteri*). Varying numbers of 21 species of birds from 7 families were collected and tested for mirex residues before the application occurred and then at 1 month, 3 months, 6 months, 9 months, 1 year, 18 months, and 2 years after application (table 24.16). The insectivorous species (Eastern Bluebirds, Pine Warblers, Brown-headed Nuthatches, and Bachman's Sparrows) had the most dramatic accumulation of mirex in their tissues, whereas omnivores (Blue Jays and American Crows) accumulated mirex to a lesser degree. The concentrations of mirex generally decreased from 9 to 24 months after aerial application and reached pre-application levels.

Table 24.14. Organochlorine residues in carcasses of European Starlings from Highlands County, Florida[a]

Year(s)	No. birds in pooled sample	Mean (ppm wet weight)[b]						Data source
		DDE	DDD	DDT	Dieldrin	Heptachlor epoxide	BHC	
1967–68	24	0.9	0.01	0.07	0.58	0.04[c]	ND	A
1970	10	0.14	0.01	0.04	0.01	0.01	0.01	B
1972	10	0.42	0.01	0.03	0.02	TR	ND	C
1974	10	0.18	TR	0.02	0.02	TR	TR	D
1976	10	0.67	NG	ND	0.01	ND	NG	E
1979	10	0.14	NG	ND	ND	0.01	NG	F
1982	2	0.11	NG	NG	ND	ND	NG	G

Year(s)	No. birds in pooled sample	PCBs	Oxychlordane	HCB	Chlordane isomers	Mirex	Data source
1967–68	24	NT	NT	NT	NG	NG	A
1970	10	0.43	NT	NT	NG	NG	B, H
1972	10	0.32	0.01	ND	NG	NG	C
1974	10	0.08	0.01	ND	NG	NG	D
1976	10	ND	NG	ND	ND	NG	E
1979	10	ND	NG	ND	0.01	ND	F
1982	2	0.03	ND	ND	NG	0.03	G

Sources: A = Martin (1969), B = Martin and Nickerson (1972), C = Nickerson and Barbehenn (1975), D = White (1976), E = White (1979), F = Cain and Bunck (1983), G = Bunck et al. (1987), H = Oberheu (1972).
ND = none detected, NG = not given, NT = not tested for this compound, TR = trace (<0.002 ppm).
a. Ground-up carcasses minus the skin, feathers, beak, feet, and wings.
b. Type of mean not stated. Limits of detection: Reference A = not given, Reference B = ranged 0.005–0.01 ppm, Reference C = ranged 0.005–0.1 ppm, Reference D = 0.005 ppm for all pesticides and 0.01 ppm for PCBs, Reference E = 0.01 ppm, Reference F = 0.01 ppm, Reference G = 0.01 ppm.
c. Based on 10 birds.

FIGURE 24.6. Total amount of DDT and its metabolites in 10 species of migratory passeriforms in northern Florida, 1964–73. Regression line A was calculated for 7 species that glean insects from arboreal surfaces: American Redstarts, Black-throated Blue Warblers, Yellow-rumped Warblers, Palm Warblers, Black-and-white Warblers, Red-eyed Vireos, and White-eyed Vireos. Regression line B was calculated for these 7 species plus 3 others that are partly frugivorous or take insects near the ground: Common Yellowthroats, Ovenbirds, and Gray Catbirds. From Johnston 1974; by permission of *Science*.

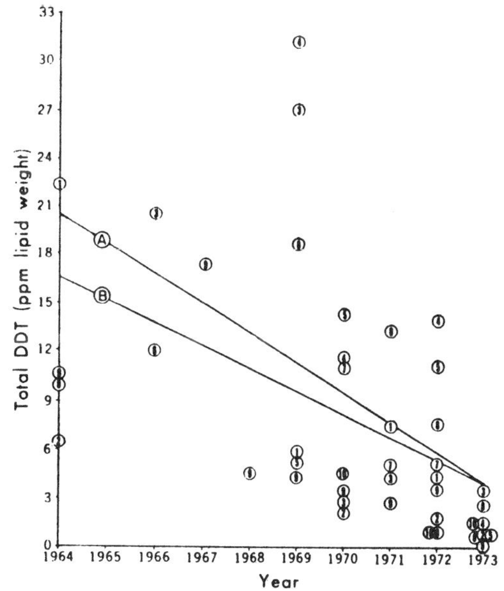

Table 24.15. Organochlorine residues in adipose tissue of perching birds from Florida

Species of bird	County	Year	Residues (ppm wet weight)[a]		
			DDE	ΣDDT	Dieldrin
Vireonidae					
White-eyed Vireo	Orange	1970	3.60	3.90	0.03
	Brevard	1971	1.90	3.00	0.01
	Duval	1972	2.10	4.30	0.05
Red-eyed Vireo	Leon	1966	4.80	8.60	1.10
	Leon	1970	0.50	0.81	ND
	Duval	1972	3.00	3.20	0.17
	Leon	1972	0.08	0.50	ND
	Leon	1972	0.02	0.18	ND
	Leon	1973	0.36	0.61	ND
Turdidae					
Swainson's Thrush	Leon	1969	1.30	1.70	ND
	Leon	1971	0.45	0.51	ND
	Leon	1972	1.30	1.60	0.10
Veery	Leon	1969	0.21	0.67	0.01
	Leon	1971	0.05	0.05	ND
Mimidae					
Gray Catbird	Leon	1972	0.21	0.33	ND
	Leon	1972	0.09	0.27	ND
	Leon	1973	0.88	1.70	ND
Parulidae					
Black-and-white Warbler	Leon	1969	3.40	3.60	ND
	Leon	1970	4.20	9.00	0.07
	Brevard	1971	6.20	8.90	ND
	Brevard	1971	1.60	2.60	ND
	Brevard	1972	0.41	1.50	ND
	Leon	1972	2.70	5.40	0.12
	NG	1972	1.70	2.30	ND
	Leon	1973	0.29	0.50	0.05
Yellow-rumped Warbler	Leon	1966	3.30	7.10	0.13
	Leon	1966	3.00	5.80	0.11
	Leon	1969	3.20	5.30	ND
	Leon	1969	5.20	7.60	0.05
	Leon	1970	1.40	2.60	0.12
	Leon	1971	1.40	3.30	ND
	Leon	1973	2.10	2.50	0.07
Blackpoll Warbler	Duval	1971	0.14	0.21	ND
	Duval	1971	0.06	0.06	ND
	Brevard	1972	0.28	0.37	ND
Black-throated Blue Warbler	Orange	1970	1.10	1.60	ND
	Duval	1972	0.55	1.40	ND
	Duval	1972	0.14	0.87	ND
	Orange	1972	0.28	1.20	ND
	Orange	1972	0.66	1.10	ND
	Orange	1972	0.22	0.66	ND
	Orange	1972	0.78	1.60	ND
	Duval	1972	2.40	3.80	0.06
	Duval	1972	1.60	2.60	ND
	Duval	1973	0.28	0.41	0.04
Palm Warbler	Leon	1969	12.00	25.00	0.85
	Leon	1970	3.10	4.40	0.13
	Duval	1971	2.60	5.10	0.29
	Duval	1972	2.40	8.80	0.23
	NG	1972	0.98	0.98	ND
	Duval	1971	3.50	6.60	0.38
	Leon	1973	0.61	0.70	0.21

(continued)

Table 24.15. *(continued)*

Species of bird	County	Year	Residues (ppm wet weight)[a]		
			DDE	ΣDDT	Dieldrin
Ovenbird	Duval	1964	1.70	3.30	ND
	Duval	1964	1.40	2.70	ND
	Duval	1964	4.00	5.40	0.06
	Duval	1964	2.10	3.20	0.03
	Duval	1967	5.20	8.60	0.03
	Duval	1968	1.80	3.10	0.02
	Duval	1968	0.40	1.20	ND
	Duval	1969	14.00	40.00	NG
	Leon	1969	1.90	2.40	0.02
	Leon	1969	2.70	3.10	0.02
	Orange	1970	2.10	2.70	0.03
	Orange	1970	2.10	3.00	ND
	Brevard	1971	1.70	2.30	0.03
	Duval	1971	1.30	1.90	ND
	Duval	1971	1.10	2.40	ND
	Duval	1972	1.80	2.80	ND
	NG	1972	0.48	0.69	ND
	Duval	1973	0.32	0.38	0.02
Northern Waterthrush	Brevard	1971	5.10	11.00	ND
	Duval	1972	4.10	9.70	0.79
	NG	1972	4.80	5.00	ND
Common Yellowthroat	Duval	1964	3.10	3.60	ND
	Duval	1964	3.50	6.50	0.03
	Duval	1969	7.10	8.00	ND
	Leon	1969	3.60	5.50	ND
	Duval	1971	4.60	6.20	ND
	Brevard	1972	4.10	4.40	ND
	Leon	1972	3.60	4.20	ND
	Leon	1972	2.20[b]	3.30[b]	0.19[b]
	Duval	1972	2.20	3.30	0.11
	Duval	1972	3.40	4.50	0.11
	NG	1972	1.00	1.00	ND
	Leon	1973	1.50	1.90	0.11
American Redstart	Duval	1964	6.70	10.00	0.27
	Duval	1964	6.10	10.00	0.66
	Brevard	1969	2.00	4.70	0.51
	Leon	1971	3.00	5.60	0.37
	Leon	1971	5.30	12.00	NG
	Brevard	1972	3.00	3.30	ND
	Brevard	1972	0.97	0.97	ND
	Leon	1972	1.50	2.69	0.08
	Duval	1972	2.30	6.00	ND
	NG	1972	2.90	2.90	ND
Emberizidae					
Savannah Sparrow	NG	1966	15.00	18.00	0.18
White-throated Sparrow	NG	1966	2.30	2.50	ND
	NG	1966	7.70	15.00	ND
Icteridae					
Bobolink	Leon	1971	0.50	0.90	0.08

Source: Johnston (1975), unless otherwise indicated.
ND = none detected, NG = not given.
a. Limit of detection = 0.01 ppm.
b. A pool of 10 birds (Johnston 1976). Uropygial glands from the same 10 birds had 1.9 ppm of DDE and 1.9 ppm of ΣDDT; no residue of dieldrin was detected.

Table 24.16. Concentrations (ppm wet weight)[a] of mirex in stomach contents and carcasses[b] or selected tissues of various perching birds before and after application of 10-5 bait for fire ant control in Duval County, Florida, 1972

Species of bird	Pre-treatment[c]	Posttreatment[c]						
		1 month	3 months	6 months	9 months	1 year	18 months	2 years
Tyrannidae								
Eastern Phoebe								
Carcass	NT	NT	NT	1.3(2)	NT	NT	NT	NT
Stomach contents	NT	NT	NT	0.07(2)	NT	NT	NT	NT
Corvidae								
Blue Jay								
Carcass	ND(3)	0.20(1)	0.17(3)	0.10(6)	0.05(2)	0.06(2)	ND(1)	0.04(3)
Stomach contents	ND(3)	ND(1)	NT	NT	NT	NT	ND(1)	NT
American Crow								
Brain	NT	0.18(2)	0.21(1)	0.06(1)	0.35(1)	ND(1)	0.03(1)	NT
Fat	0.05(1)	0.39(2)	1.(1)	NT	NT	0.39(1)	0.18(1)	NT
Liver	ND(1)	0.39(2)	0.65(1)	0.16(1)	0.58(1)	0.44(1)	0.06(1)	NT
Muscle	ND(1)	0.15(2)	0.32(1)	0.09(1)	0.20(1)	ND(1)	0.06(1)	NT
Stomach contents	ND(1)	0.15(2)	NT	0.09(1)	NT	NT	NT	NT
Paridae								
Carolina Chickadee								
Carcass	ND(3)	0.20(3)	0.18(2)	0.11(1)	0.09(3)	0.12(2)	0.03(2)	NT
Stomach contents	ND(3)	ND(2)	NT	ND(1)	NT	NT	NT	NT
Sittidae								
Brown-headed Nuthatch								
Carcass	ND(3)	0.96(3)	0.47(2)	0.46(3)	0.24(3)	0.67(3)	0.10(2)	0.03(4)
Stomach contents	0.20(3)	0.01(3)	NT	0.01(3)	NT	NT	ND(2)	ND(4)
Troglodytidae								
House Wren								
Carcass	NT	NT	NT	4.1(2)	NT	NT	NT	NT
Stomach contents	NT	NT	NT	0.55(2)	NT	NT	NT	NT
Turdidae								
American Robin								

Carcass	NT	NT	NT	0.08(1)	NT	NT	ND(2)	NT
Stomach contents	NT	NT	NT	0.05(1)	NT	NT	NT	NT
Hermit Thrush								
Carcass	NT	NT	NT	1.2(1)	NT	NT	NT	NT
Stomach contents	NT	NT	NT	0.14(1)	NT	NT	NT	NT
Eastern Bluebird								
Carcass	ND(3)	8.3(1)	2.4(1)	2.3(2)	3.7(3)	NT	0.11(3)	0.21(2)
Stomach contents	0.01(3)	1.3(3)	NT	1.4(2)	NT	NT	ND(3)	NT
Parulidae								
Pine Warbler								
Carcass	ND(4)	2.6(7)	0.45(3)	0.49(3)	0.19(2)	0.05(3)	0.01(4)	0.02(1)
Stomach contents	1.05(4)	0.07(7)	NT	0.29(3)	NT	NT	NT	NT
Yellow-rumped Warbler								
Carcass	NT	NT	NT	ND(1)	NT	NT	NT	NT
Stomach contents	NT	NT	NT	ND(1)	NT	NT	NT	NT
Palm Warbler								
Carcass	NT	NT	NT	0.33(1)	NT	NT	NT	NT
Stomach contents	NT	NT	NT	ND(1)	NT	NT	NT	NT
American Redstart								
Carcass	NT	NT	0.81(1)	NT	NT	NT	NT	NT
Stomach contents	NT	NT	NT	NT	NT	NT	NT	NT
Emberizidae								
Eastern Towhee								
Carcass	NT	NT	NT	NT	NT	NT	ND(1)	NT
Stomach contents	NT	NT	NT	NT	NT	NT	ND(1)	NT
Swamp Sparrow								
Carcass	NT	NT	NT	0.11(2)	NT	NT	NT	NT
Stomach contents	NT	NT	NT	ND(2)	NT	NT	NT	NT
Song Sparrow								
Carcass	NT	NT	NT	ND(1)	NT	NT	NT	NT
Stomach contents	NT	NT	NT	ND(1)	NT	NT	NT	NT
Chipping Sparrow								
Carcass	NT	NT	NT	0.08(2)	NT	NT	NT	NT
Stomach contents	NT	NT	NT	ND(2)	NT	NT	NT	NT

(continued)

Table 24.16. (*continued*)

Species of bird	Pre-treatment[c]	Posttreatment[c]						
		1 month	3 months	6 months	9 months	1 year	18 months	2 years
American Tree Sparrow								
Carcass	NT	NT	NT	0.15(1)	NT	NT	NT	NT
Stomach contents	NT	NT	NT	0.15(1)	NT	NT	NT	NT
Savannah Sparrow								
Carcass	NT	NT	NT	ND(1)	NT	NT	NT	NT
Stomach contents	NT	NT	NT	ND(1)	NT	NT	NT	NT
Bachman's Sparrow								
Carcass	ND(4)	3.7(5)	1.2(3)	NT	0.70(8)	1.1(3)	0.03(3)	0.30(3)
Stomach contents	0.03(3)	0.25(5)	NT	NT	NT	NT	ND(3)	NT
Icteridae								
Red-winged Blackbird								
Carcass	NT	0.48(2)	NT	ND(1)	0.05(3)	ND(1)	0.03(2)	0.02(5)
Stomach contents	NT	0.02(2)	NT	ND(1)	NT	NT	0.01(2)	NT
Common Grackle								
Carcass	ND(3)	NT	NT	NT	NT	0.68(2)	NT	NT
Stomach contents	ND(2)	NT	NT	0.16(1)	NT	NT	NT	NT
Brain	NT	NT	NT	0.12(1)	NT	NT	NT	NT
Fat	NT	NT	NT	10(1)	NT	NT	NT	NT
Liver	NT	NT	NT	0.07(1)	NT	NT	NT	NT
Muscle	NT	NT	NT	0.14(1)	NT	NT	NT	NT

Source: Wheeler et al. (1977).
ND = none detected, NT = not tested.
a. Lowest limit of detection = 0.01 ppm.
b. Ground-up carcasses minus the skin, feathers, beak, feet, and wings.
c. Number in parentheses = sample size.

The fourth investigation was conducted in 1965 and consisted of an evaluation of the effects on various wildlife of a DDT spraying program designed to eradicate the mosquito, *Aedes aegypti,* in Dade County (Lehner et al. 1967). Sick and dead birds (including House Sparrows) were collected from areas that had been sprayed (some more than once) and healthy House Sparrows and Northern Mockingbirds were collected from an unsprayed area. The residues of DDT and its metabolites in brain tissues of nestling House Sparrows in the sprayed and nonsprayed zones are presented in table 24.17. DDE was the most prevalent residue in both sets of birds. Concentrations were highest, however, in nestling House Sparrows from the sprayed zone, although the authors stated that this was not the case! Residues in brains of 7 other species of passeriforms from sprayed and nonsprayed areas are given in table 24.18. Most of the residues were probably not at lethal concentrations, with the exception of the Northern Cardinal; DDT equivalents for this bird were close to 15, which would probably indicate lethality (Blus 1996).

It is difficult to interpret the significance of the above findings in relation to the health and reproduction of songbirds in Florida. The aerial application of mirex was suspected by the lay public to have caused a decline in numbers of songbirds in Sarasota County in the 1960s, but this was not supported by facts. It was pointed out by Ashdown (1969) that declines in songbird populations were also occurring in other areas of the state where there was no treatment with mirex for the control of fire ants and that the declines actually began before the aerial treatments were instituted. In the case of the starling study, results were based on whole carcasses; separate values for brain, liver, and other tissues were not determined. In fact, the starling study was not designed for that purpose, but rather as a monitoring study to assess the general trends of pesticide residues in the environment (Dustman et al. 1971). However, the concentrations of DDE, dieldrin, and mirex were found to be well below those considered lethal for songbirds (Stickel et al. 1973, 1984; Peakall 1996). We have evidence of mortality as a result of acute organochlorine poisoning of an Eastern Meadowlark from Nassau County (Forrester and Thompson 1975). The bird was observed with tremors prior to death in February 1975. Residues of DDE, ΣDDT, dieldrin, and PCBs for that bird were determined and are given in table 24.19. Concentrations of dieldrin in brain (8.5 ppm) and liver (27 ppm) were within the lethal ranges reported for Eastern Meadowlarks by Stickel et al. (1969).

The sublethal effects of low concentrations like those found in birds in the 3 major studies above are understood only in part. Even though there are claims in the published literature that organochlorines have had harmful effects on various songbirds in Florida (Whisenhunt 1959; Stimson 1964; Stevenson and Anderson 1994), these are based on circumstantial information and assumptions; hard core data are lacking. Eggshell thinning has not been recognized in these passeriforms (Johnston 1974), nor were decreases in eggshell thickness found in a series of pre-1947 (i.e., pre-DDT era) eggs of Loggerhead Shrikes (Morrison 1979) and American Crows (Anderson and Hickey 1972) that were measured and compared with post-1947 samples (table 24.20).

VIII. Organophosphates

Five organophosphates have been linked with eight mortality incidents in Florida involving passeriforms (table 24.21). One occurred in a potato field near Homestead (Dade County) in January 1972 and resulted in the death of an estimated 10,000 American Robins (Stevenson 1972, Fisk 1976). The insecticide involved was identified as monocrotophos (= azodrin), used on many agricultural crops such as cotton, sugarcane, peanuts, corn, and tobacco. The incident was investigated by a pesticide company and determined to be the result of improper application of monocrotophos to

Table 24.17. Residues of DDT and its metabolites in the brain tissue of nestling House Sparrows collected from sprayed and nonsprayed areas in Dade County, 1965

Area	Sample number	DDT	DDD	p,p'-DDE
			Residues (ppm)[a]	
Sprayed[b]	1	ND	ND	0.40
	2	ND	ND	ND
	3	0.63	0.25	1.10
	4	ND	0.39	0.25
	5[c]	ND	ND	1.60
	6[d]	ND	ND	2.70
	7	ND	ND	0.33
	8	ND	ND	0.86
	9[e]	ND	ND	1.50
	10[e]	2.80	ND	5.70
	11[f]	0.55	ND	1.20
	12	ND	ND	0.56
	13	ND	ND	0.12
	14	0.31	ND	0.26
	15	ND	ND	0.65
	16	0.18	ND	0.57
	17	ND	ND	0.21
	18[g]	ND	ND	0.66
Nonsprayed	1	0.34	ND	0.39
	2	ND	ND	0.22
	3	ND	ND	0.66
	4[h]	ND	ND	0.25
	5	ND	ND	0.28
	6	ND	ND	0.16
	7	ND	ND	0.72
	8	ND	ND	0.81
	9	ND	ND	0.43
	10	ND	ND	0.70
	11	ND	ND	0.32
	12	ND	ND	0.18
	13	0.80	ND	0.33
	14	ND	ND	0.25
	15	ND	ND	0.18

Source: Lehner et al. (1967).
ND = none detected.
a. Authors did not state if the values were on a wet weight, dry weight, or lipid weight basis.
b. The area was sprayed twice with 1.25% DDT by weight.
c. Sample represents pooled sample of 3 nestlings.
d. Sample represents pooled sample of 5 nestlings.
e. Nestling found dead. All others were collected alive.
f. Sample represents pooled sample of 4 nestlings.
g. Sample also contained 0.19 ppm of o,p'-DDE.
h. Sample also contained 0.10 ppm of o,p'-DDE.

Table 24.18. Residues of DDT and its metabolites in brain tissue of various passeriforms collected from sprayed and nonsprayed areas in Dade County, 1964–65

Family Species	Condition when collected[a]	No. of times area sprayed	Residues (ppm)[b]			
			DDT	DDD	p,p'-DDE	o,p'-DDE
Laniidae						
Loggerhead Shrike	H	4	ND	ND	28	0.80
Corvidae						
American Crow	S	NG	ND	1.1	16	0.42
Blue Jay	D	2	ND	ND	0.09	ND
Mimidae						
Northern Mockingbird	H	0	ND	ND	0.49	ND
	H	0	ND	ND	0.62	ND
	H	0	ND	ND	2.0	ND
	H	0	ND	ND	1.4	ND
	D	1	ND	ND	1.3	0.21
	D	2	0.56	1.3	14	1.6
	S	2	ND	ND	1.3	ND
	S	4	8.7	2.7	4.0	0.48
	H	4	ND	ND	1.4	ND
	H	4	ND	ND	ND	ND
	H	4	ND	ND	2.6	0.12
Parulidae						
Yellow-rumped Warbler	D	1	ND	0.55	2.3	ND
Cardinalidae						
Northern Cardinal	S	1	11	16	11	0.42
Icteridae						
Red-winged Blackbird	H	0	ND	0.12	1.3	ND

Source: Lehner et al. (1967).
ND = none detected, NG = not given by authors.
a. H = healthy, S = sick, D = dead.
b. Authors did not state if the values were on a wet weight basis.

field margins and surrounding hedges, where food-stressed Robins were feeding on berries of the Brazilian pepper tree, *Schinus terebinthifolius* (Smith 1993). Monocrotophos is extremely toxic and there are numerous reports of its involvement in the mortality of various species of birds (Smith 1993). A second incident involving the same insecticide, but smaller numbers of American Robins, occurred in a tomato field in Dade County in the winter of 1975–76 (Fisk 1976).

Each year an estimated 25 wildlife mortality incidents of various sizes occur in the United States because of organophosphates (Fairbrother 1996). However, in addition to acute exposures, sublethal effects could be as important or more so. A number of studies have been conducted to investigate the effects of organophosphates on reproductive success of passeriforms and these have been reviewed by Fairbrother (1996). No such studies have been done in Florida, but Patnode and White (1991) found that songbirds nesting at the periphery of pecan orchards in southern Georgia were adversely affected by multiple exposures to organophosphates and carbamates. They studied Northern Cardinals, Northern Mockingbirds, and Brown Thrashers and found that the daily survival rates of eggs and nestlings varied inversely with exposure. They concluded that

Table 24.19. Residues of organochlorines in tissues of an Eastern
Meadowlark that died in spasms in Nassau County, 1975

| | ppm (wet weight) | | | |
Tissue	DDE	ΣDDT	Dieldrin	PCBs[a]
Brain	0.84	0.84	8.5	3.3
Liver	2.5	3.3	27	9.1
Muscle	0.55	0.74	4.9	2.2

Source: Forrester and Thompson (1975).
a. Aroclor 1254.

"increasing exposure to pesticides may reduce songbird productivity." This topic deserves more attention in relation to Florida songbirds.

IX. Heavy metals

Information on heavy metals in perching birds of Florida is limited to data obtained from European Starlings over a 3-year period in the early 1970s. Concentrations of 5 metals were identified and included mercury, lead, cadmium, arsenic, and selenium (table 24.22). All were below 1 ppm except for lead in Alachua County in 1970 (1.2 ppm). These concentrations were below those considered toxic to birds (Franson 1996; Beyer et al. 1988; Furness 1996; Eisler 1988; Heinz 1996), although sublethal effects might be significant and should be investigated in Florida.

X. Neoplasia

No information is available on this topic.

XI. Anomalies

Beak anomalies have been documented in Brown Thrashers and European Starlings in Florida. There are 11 published sightings or ex-

Table 24.20. Measurements of eggshells of Loggerhead Shrikes and American Crows from Florida,
pre- and post-1947

Species of bird	Time period	No. of eggs sampled	Mean thickness index[a]	% change
Loggerhead Shrike	pre-1947	81	0.52	—
	1950–68	66	0.52	0
American Crow	pre-1947	149	0.98	—
	1948	9	1.00	0
	1950–53	18	0.98	0
	1959	25	0.98	0

Sources: Shrike data from Morrison (1979), crow data from Anderson and Hickey (1972). Specific locations in Florida were not given.
a. Thickness index = weight of eggshell (mg) divided by the product of the length x breadth (mm) (Ratcliffe 1967).

Table 24.21. Mortality of perching birds in Florida resulting from organophosphate poisoning

Organophosphate Species	County	Year(s)	Extent of mortality	Data source
Chlorpyrifos				
Boat-tailed Grackle	Highlands	1988	1	Wentworth (1988)
American Robin	Santa Rosa	1993	100+	Quist (1993)
	Okaloosa	1997	4	Teglas (1997)
Diazinon				
Common Grackle	Escambia	1993	9	Quist (1993)
Famfur				
Common Grackle	Marion	1991	12	Quist (1992)
Fenamiphos				
American Robins	Martin	1990	NG[a]	Hayes (1990)
Cedar Waxwings	Martin	1990	NG[a]	Ibid.
Monocrotophos				
American Robin	Dade	1972	~10,000	Stevenson (1972), Fisk (1976)
	Dade	1975–76	Many[b]	Fisk (1976)

a. The author did not say how many of each species died, but stated that "58 robins and cedar waxwings were found dead."
b. "Collected two pails of birds."

aminations of Brown Thrashers with abnormally elongated bills from 9 counties between 1927 and 1975 and 1 European Starling from Orange County in 1975 (table 24.23). These occurrences were in 2 clusters, one in northeastern Florida (Duval, Clay, and Putnam counties) and another in central Florida (Orange, Seminole, and Brevard counties) (figure 24.7). In most of the cases the abnormality consisted of an elongated and decurved maxilla and a partial sickling and extension of the mandible (Brown 1976). In 1 Brown Thrasher from Seminole County, however, the tip of the tongue was divided into 4 parts and the inner surface of the gizzard was "greatly contorted with thick and hard walls" (Taylor and Anderson 1972). Two of the thrashers, 1 from Bay County (Stevenson and Anderson 1994) and 1 from Pinellas

Table 24.22. Concentrations of heavy metals from carcasses of European Starlings from Florida[a]

Year	County	ppm (wet weight)[b]				
		Mercury	Lead	Cadmium	Arsenic	Selenium
1970	Alachua	NT	1.20	NT	NT	NT
	Bay	ND	NT	NT	NT	NT
	Hardee	ND	NT	NT	NT	NT
	Madison	ND	NT	NT	NT	NT
	Polk	ND	NT	NT	NT	NT
1971	Alachua	0.21	0.52	ND	0.03	NT
1973	Alachua	ND	0.65	0.05	0.12	0.10

Sources: 1970 data: Martin (1972), 1971 data: Martin and Nickerson (1973), 1973 data: White et al. (1977).
ND = none detected, NT = not tested.
a. Pools from 10 samples, each sample = a ground up carcass minus the skin, feathers, beak, feet, and wings.
b. Limits of detection = 0.05 ppm for mercury in 1970, 0.01 ppm in 1971, 1973; 0.1 ppm for lead; 0.05 ppm for cadmium; 0.01 ppm for arsenic in 1971, 0.05 ppm in 1973; 0.05 ppm for selenium in 1973.

Table 24.23. Records of elongated bill deformities in passeriforms in Florida

Species of bird	County	Year(s)	No. of birds seen	Data source
Mimidae				
Brown Thrasher	Bay[a]	1964	1	Stevenson & Anderson (1994)
	Clay	1964	1	Stitt (1968)
	Dade	1967	1	Steffee (1968)
	Duval	1927–67	3	Ibid.
	Hillsborough	1975	1	Brown (1976)
	Orange	NG[b]	1	Stimson (1968)
	Pinellas[a]	1946	1	Cobb (1946)
	Putnam	1967	"numbers"	Steffee (1968)
	Seminole	1971	1	Taylor & Anderson (1972)
	NG	1974	1	Brown (1976)
	NG	1975	1	Ibid.
Sturnidae				
European Starling	Orange	1975	1	Stevenson & Anderson (1994)

NG = not given.

a. The birds involved were originally identified as Long-billed Thrashers (*Toxostoma longirostra*) and presumed to be vagrants since this species does not occur in Florida. However, it was later concluded that these were actually Brown Thrashers with elongated bill deformities (Cobb 1946; Stevenson 1976; Robertson and Woolfenden 1992; Stevenson and Anderson 1994).

County (Cobb 1946), were identified originally as vagrant Long-billed Thrashers, but later considered to be Brown Thrashers with elongated bills (Robertson and Woolfenden 1992; Stevenson and Anderson 1994). It is curious that this phenomenon is limited mainly to Brown Thrashers. Brown (1976) speculated that it may have a genetic basis or perhaps a connection with environmental contaminants. Beak deformities are known from a number of captive-reared avian species, including pet birds and poultry, and are thought to be related to certain nutritional deficiencies during the embryonic stage or to genetic mutations (Pence 1996). McClure (1989) suggested that deformed mandibles of House Finches in California were due to pox infections, but this has not been studied further. Pox infections have not been recorded in Brown Thrashers in Florida, but have been seen commonly in Northern Mockingbirds from Dade, Hillsborough, and Pinellas counties, where thrashers with elongated bills have been seen. The significance of this bill deformity to populations of Brown Thrashers in Florida is not known. In the cases mentioned here, the thrashers appeared to be in good health, although they were forced to assume abnormal head postures in order to feed, which might have increased their susceptibility to predation. Northern Flickers also have been found with bill anomalies in Florida and are discussed in chapter 25 (Miscellaneous Birds). P.A. Stewart (1963) reported bill deformities in 4 of 8,275 Brown-headed Cowbirds trapped in Alabama during the winter of 1960–61. All affected birds were females. He also noted an abnormal toenail in 1 cowbird and partial albinism in 4 birds.

XII. Biotoxins

Plants belonging to 12 different families have fruits that have been implicated as the cause of intoxication in birds (Dennis 1987). In Florida there are several reports of the intoxication and death of American Robins after consumption of the berries of the Brazilian pepper tree (table 24.24). This exotic plant was introduced into Florida from Brazil in the early 1900s and is

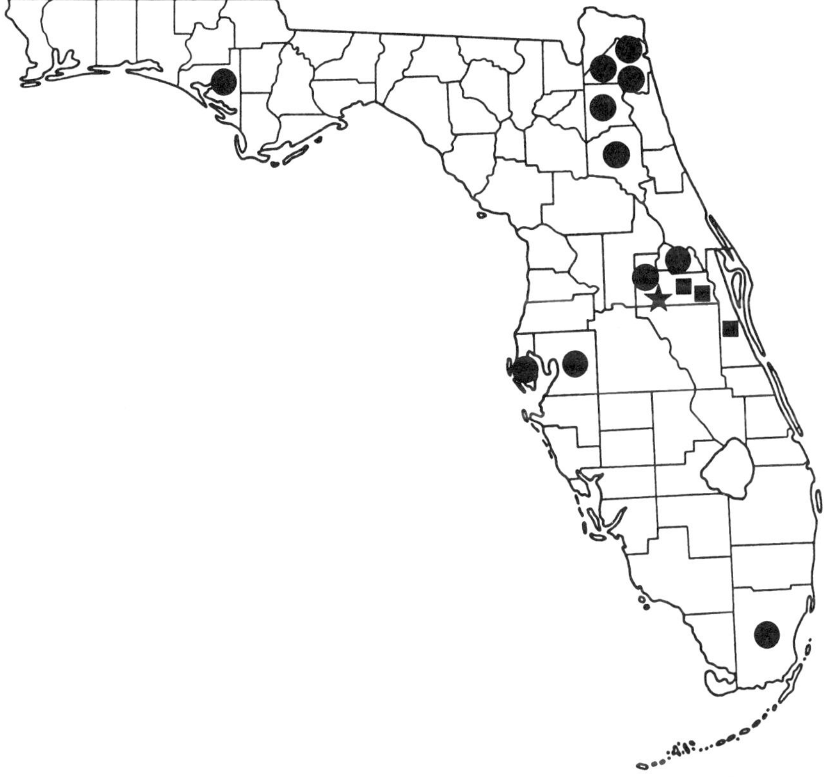

FIGURE 24.7. Distribution of birds found with bill anomalies in Florida, 1927–75. *Circles* = Brown Thrashers; *star* = European Starling; *squares* = Northern Flickers. Data from Brown 1976; Cobb 1946; Steffee 1968; Stevenson and Anderson 1994; Stimson 1968; Stitt 1968; Taylor 1973; Taylor and Anderson 1972.

now widespread, especially in southern Florida (Morton 1978). Its berries contain a number of toxic ingredients, including triterpenes, hydrocarbons, and more than 50 other components (Lloyd et al. 1977). Blassingame (1955) gave a colorful description of the intoxication of many American Robins on Anna Marie Key off the coast of Manatee County in 1954. The actual numbers afflicted were not given, only the observation that "Thousands of birds descended on the island . . . and soon launched a colossal binge." Many were killed after flying into windows and buildings or being hit by passing cars. The robins were feeding on fallen and fermenting Brazilian pepper tree berries as well as palm berries (species not given).

As noted in table 24.24 and in chapter 22 (Pigeons and Doves), 10 American Robins and 2 Mourning Doves were found dead near a Brazilian pepper tree in Tampa in 1983 (Forrester 1983). All of the dead birds had Brazilian pepper tree berries in their gastrointestinal tracts and were believed to have died from poisoning by the berries. Subsequently, berries were collected from the pepper tree near where the robins and doves were found and used in a feeding experiment to test the effects of this type of biotoxin on European Starlings and Northern Cardinals. Most of the starlings died, whereas the cardinals were unaffected, indicating differences in species susceptibility (Forrester 1983).

Table 24.24. Mortality of American Robins in Florida resulting from
Brazilian pepper berry poisoning

County	Year	Extent of mortality	Data source
Dade	1978	115	Navarrete (1978)
Hillsborough	1983	10	Forrester (1983)
Manatee	1954	"many"	Blassingame (1955)
Sarasota	1988	33	Wright & Forrester (1988)

Fourteen Cedar Waxwings were found dead beneath a Carolina laurelcherry (*Prunus caroliniana*) tree in a residential area of Orange Park (Clay County) in March of 1999. This acute mortality was attributed to ingestion of the berries of the laurelcherry, the seeds of which are known to contain the cyanide-producing compound amygdalin (Forrester and Foster 1999).

XIII. Viruses

Seven viruses have been identified in perching birds from Florida. These include poxvirus and 6 arboviruses, namely St. Louis encephalitis (SLE) virus, eastern equine encephalomyelitis (EEE) virus, Highlands J virus, Keystone virus, Everglades virus, and West Nile virus.

Poxvirus infections have been reported from a large number of passeriforms on a worldwide basis (Karstad 1971b; Kirmse 1967). In Florida, however, infections have been documented in only 6 species (table 24.25). These infections are probably more widespread since many of the species of passeriforms that occur in Florida have been found infected in other parts of North America (Kirmse 1967). The most commonly infected passeriform in Florida seems to be the Northern Mockingbird. Kale and Jennings (1966) reported on a sizable outbreak that occurred among Northern Mockingbirds in the Tampa Bay area (Hillsborough, Manatee, and Pinellas counties) during the summer of 1964. They found infections in more than 51% of a sample of 199 birds that they captured and examined in Pinellas County during a field study of movements of immature

mockingbirds. The population significance of these infections is not known; most infections in wild birds have been reported to be self-limiting and mild, but severe lesions on the eyelids can lead to mortality (Karstad 1971b). It has been postulated recently that pox infections are the cause of color variation in plumage of House Finches in California (Zahn and Rothstein 1999), but this has neither been verified elsewhere nor studied in Florida passeriforms. Few details on lesions were given for the cases seen in Florida passeriforms. Bigler (1997) reported that the lesions on the infected mockingbirds he examined were only on the legs and toes. The lesions on the fledgling mockingbird from Flagler County in 1997 were on the unfeathered skin of the head, neck, and hips (Spalding 1997). This bird came from a rehabilitation facility and several other mockingbirds in the facility eventually became infected with poxvirus, while birds of other species remained uninfected. In this case the transmission of the virus might have been by direct contact rather than by mosquitoes.

Of the 6 arboviruses that occur in songbirds in Florida, the most important are EEE virus, SLE virus, Highlands J virus, and West Nile virus. There is little information on Keystone virus and Everglades virus. Bigler et al. (1975) reported the isolation of Keystone virus from a Blue Jay in Alachua County in 1966, but Taylor et al. (1971) found no serologic evidence of Keystone virus in passeriforms (species and numbers sampled not given). They worked in and around a freshwater swamp in Hillsborough County in 1966–67, where virus activity was high in the mosquito population. There

Table 24.25. Records of pox infections in passeriforms in Florida

Species of bird	County	Year	No. of cases	Data source
Corvidae				
Blue Jay	Indian River	1970	1	Locke (1970)
Mimidae				
Northern Mockingbird	Dade	1969	2	Bigler (1997)
	Flagler	1997	1	Spalding (1997)
	Hillsborough	1964	"numerous"	Kale & Jennings (1966)
	Manatee	1964	"numerous"	Ibid.
	Palm Beach	1970	1	Locke (1970)
	Pinellas	1964	102[a]	Kale & Jennings (1966)
Emberizidae				
Chipping Sparrow	NG	NG	NG	Stevenson & Anderson (1994)
Icteridae				
Red-winged Blackbird	Dade	1971	"common"	Fisk (1972)
Common Grackle	Orange	1999	19[b]	Terrell (1999)
Passeridae				
House Finch	Dade	1971	"few"	Fisk (1972)

NG = not given.

a. During June, July, and August 199 Northern Mockingbirds were examined; 102 (51%) had pox lesions.

b. Eighteen birds had cutaneous lesions only, one had both cutaneous and "wet pox" lesions in the oral cavity.

are no known isolations of Everglades virus from passeriforms in Florida. Bigler et al. (1975) reported that 2 of 1,004 passeriforms (a Northern Mockingbird and a House Sparrow) were seropositive via the hemagglutination-inhibition test, but these results could not be confirmed by the serum neutralization test, indicating that this may have been a spurious observation.

EEE virus and West Nile virus are probably the most important arboviruses associated with encephalitis in horses and humans in Florida (Bigler et al. 1976). The majority of equine infections of EEE virus in the United States during the past 25 years have occurred in Florida (Gibbs and Tsai 1994) and in 1991 there were 159 confirmed cases (Rubin 1991). The case fatality rate in humans is high (>30%) and for that reason outbreaks in horses result in considerable concern among public health officials and the general public (Gibbs and Tsai 1994). Florida by far has more human cases of EEE than any other state; 38 of 122 confirmed cases in the United States between 1964 and 1989 were in Florida (Gibbs and Tsai 1994). In 1991

there were 10 cases reported nationally, including 5 in Duval, Bradford, Leon, and St. Johns counties during June and July (Centers for Disease Control 1991); 2 of these were fatal.

EEE virus is transmitted by a number of species of mosquitoes, the most important being *Culiseta melanura,* a freshwater swamp mosquito (Harwood and James 1979). In Florida *C. melanura* is the most significant mosquito vector, but eight other species have been found to be involved as well (Bigler et al. 1976). Transmission occurs throughout the year in Florida with peaks between May and August (Bigler et al. 1976). Songbirds are important in the epizootiology of this disease and serve as reservoir hosts for the virus (figure 24.8). In Florida EEE virus has been isolated from 13 species of passeriforms representing eight families (table 24.26). In addition, there is serological evidence for infections in a larger number of species (at least 26 species from 10 families) (table 24.27). Songbirds usually are not adversely affected by the virus, although other birds, particularly introduced species such as pheasants, chukars, emus, ostriches,

Table 24.26. Records of isolations of eastern equine encephalitis virus from passeriform birds in Florida

| Species of bird | County or site | Year(s) | No. of birds | | Data source |
			Examined	Positive	
Laniidae					
Loggerhead Shrike	Orange	1966	NG	1	Bigler et al. (1975)
Vireonidae					
Red-eyed Vireo	Pinellas	1960	NG	1	Chamberlain et al. (1963)
Corvidae					
Florida Scrub-Jay	Highlands	1995	86[a]	1	Garvin et al. (1997)
Paridae					
Tufted Titmouse	"central Fla."	1958	4	1	Favorite (1960)
Troglodytidae					
Carolina Wren	"central Fla."	1958	5	1	Ibid.
Turdidae					
Eastern Bluebird	"central Fla."	1958	16	1	Ibid.
Mimidae					
Gray Catbird	Collier	1965	NG	1	Lord & Calisher (1970)
Parulidae					
American Redstart	Tampa Bay area	1963–70	56	1	Wellings et al. (1972)
Emberizidae					
Eastern Towhee	Tampa Bay area	1960	22	1	Favorite (1960)
Cardinalidae					
Northern Cardinal	Pinellas	1960	NG	1	Chamberlain et al. (1963)
Icteridae					
Rusty Blackbird	"central Fla."	1960	6	1	Favorite (1960)
Common Grackle	Highlands	1960	231	1	Henderson et al. (1962)
Eastern Meadowlark	"central Fla."	1958	59	1	Favorite (1960)

NG = not given.

a. Of these birds, 55 were nestlings and 31 were adults. The isolation was from a nestling.

and House Sparrows, are extremely susceptible (Karstad 1971a; Gibbs and Tsai 1994). EEE virus is also pathogenic to the Whooping Crane (as was discussed in chapter 19, Cranes), whereas a closely related species, the Sandhill Crane, does not appear to be harmed by the virus. The reviews by Bigler et al. (1976) and Gibbs and Tsai (1994) should be consulted for more details on EEE virus in Florida and other parts of the United States.

Prior to the summer of 1999, West Nile virus (WNV), an Old World flavivirus related to SLE virus, was known to occur in Africa, Europe, the Middle East, and western Asia, but not in the western hemisphere (Petersen and Roehrig 2001). The virus is transmitted by mosquitoes and infects humans, horses, and a variety of birds (APHIS 2000). In the summer of 1999 it appeared in the New York City borough of Queens; between August and October, 62 human cases of WNV encephalitis were identified, 7 of whom died (Rappole et al. 2000). Twenty-five infected horses were recognized also and 9 of these died or were euthanized (APHIS 2000). In addition, throughout the summer large numbers of wild birds (>18,000) were found dead in New York, New Jersey, and Connecticut. A sample of these was tested and WNV infections were found in 20 species of birds, most of them (89%) American Crows (Eidson et al. 2001). Although the means of the introduction of WNV are unknown (probably by an infected mosquito, bird, human, or some other vertebrate), its ori-

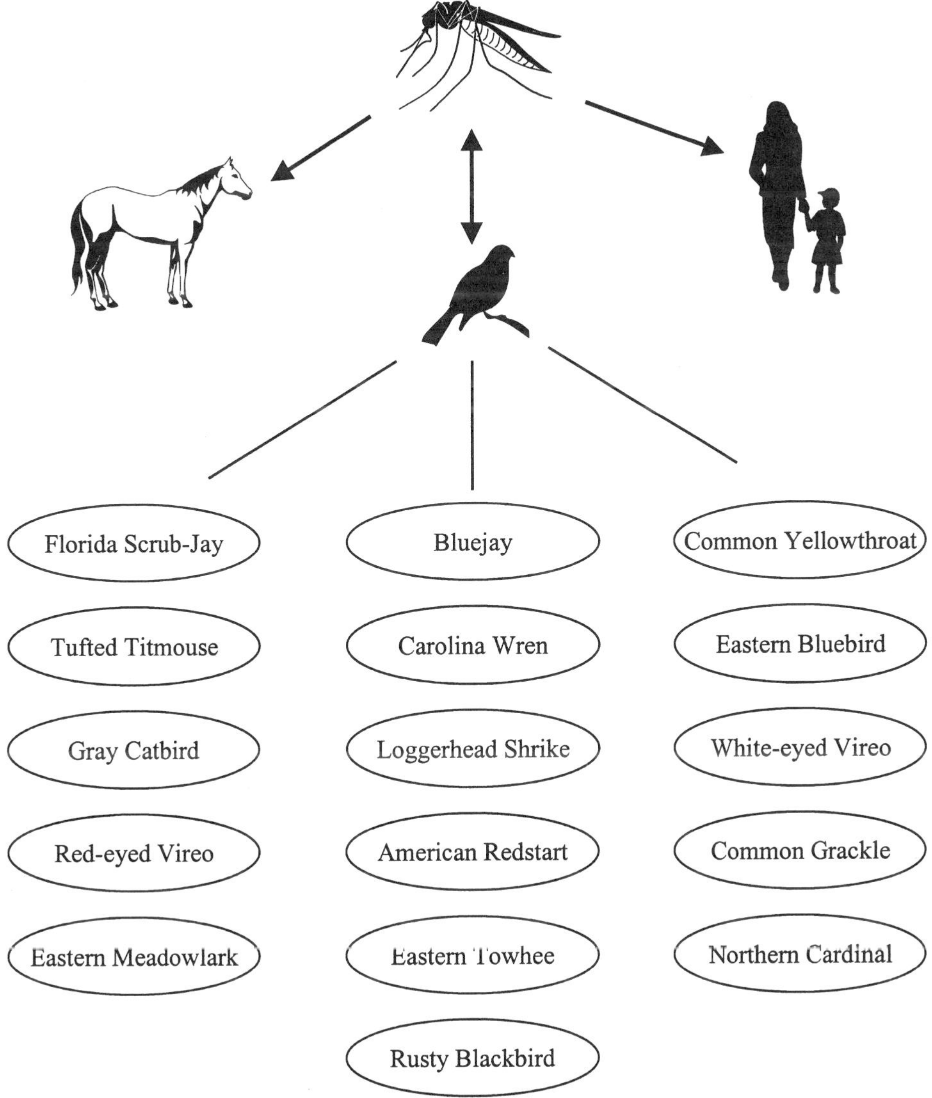

FIGURE 24.8. The role of songbirds in the life cycle of eastern equine encephalitis virus in Florida.

gin is thought to be Israel because of close genetic relationships between WN viruses isolated from that country and those from New York (Petersen and Roehrig 2001). The following year (2000) there was widespread activity of WNV in the eastern United States with infections again in humans, horses, and wild birds (Marfin et al. 2001). Infected birds representing 63 species were identified in 12 states and the District of Columbia. North Carolina was the southernmost state with infected birds during that year. Of the birds tested, 93% were corvids (American Crows, Fish Crows, Common Ravens, and Blue Jays). West Nile virus was identified in a dead Red-shouldered Hawk on June 2, 2001, in Leon County by using PCR and virus isolation techniques (Conti et al. 2002). This represents the first documented oc-

Table 24.27. Prevalence and distribution of eastern equine encephalitis virus infections in passeriforms from Florida as determined by serologic studies

Family / Species	County/site	Year(s)	No. of birds[a] Examined	Positive	%	Data source
Tyrannidae						
Great Crested Flycatcher	"central Fla."	1958	(16)	(2)	(13)	Favorite (1960)[b]
Laniidae						
Loggerhead Shrike	"central Fla."	1958	(14)	(7)	(50)	Ibid.
	"Florida"	1965–74	11	2	18	Bigler et al. (1975)
Vireonidae						
White-eyed Vireo	Osceola	1992–94	(77)	(7)	(9)	Spalding & McLean (1994)
Corvidae						
Blue Jay	"central Fla."	1958	(7)	(2)	(—)	Favorite (1960)
	Highlands	1960–61	133	42	32	Henderson et al. (1962)[c]
	Highlands	1994–95	355	53	15	Garvin et al. (1997)
	Highlands	1994–95	(243)	(83)	(34)	Ibid.
	"Florida"	1965–74	102	15	15	Bigler et al. (1975)[d]
	"Florida"	1965–74	(10)	(6)	(60)	Ibid.
Florida Scrub-Jay	Highlands	1960–61	13	3	23	Henderson et al. (1962)
	Highlands	1994–95	198	32	16	Garvin et al. (1997)
	Highlands	1994–95	(137)	(21)	(15)	Ibid.
American Crow	Osceola	1994	(1)	(1)	(—)	Spalding & McLean (1994)
Crow (species not given)	"central Fla."	1958	(7)	(2)	(—)	Favorite (1960)
Paridae						
Tufted Titmouse	"central Fla."	1958	(4)	(1)	(—)	Ibid.
	Osceola	1992–94	(11)	(1)	(—)	Spalding & McLean (1994)[e]
Troglodytidae						
Carolina Wren	Osceola	1992–94	(37)	(3)	(8)	Ibid.

Turdidae						
Eastern Bluebird	"central Fla."	1958	(16)	(6)	(38)	Favorite (1960)
Mimidae						
Brown Thrasher	"central Fla."	1958	(3)	(1)	(—)	Ibid.
	Osceola	1993–94	(6)	(2)	(—)	Spalding & McLean (1994)
Gray Catbird	"central Fla."	1958	(2)	(1)	(—)	Favorite (1960)
Northern Mockingbird	"central Fla."	1958	(4)	(2)	(—)	Ibid.
Species not given	"Florida"	1965–74	236	24	10	Bigler et al. (1975)
	"Florida"	1965–74	(11)	(1)	(9)	Ibid.
Parulidae						
Common Yellowthroat	Osceola	1992–94	(48)	(2)	(4)	Spalding & McLean (1994)
Thraupidae						
Summer Tanager	"central Fla."	1958	(1)	(1)	(—)	Favorite (1960)
Emberizidae						
Eastern Towhee	"central Fla."	1959	(22)	(4)	(18)	Ibid.
	Highlands	1960–61	37	7	19	Henderson et al. (1962)
	Osceola	1992–94	(41)	(6)	(15)	Spalding & McLean (1994)
Cardinalidae						
Northern Cardinal	"central Fla."	1958	(29)	(10)	(34)	Favorite (1960)
	"Florida"	1965–74	328	14	4	Bigler et al. (1975)
	"Florida"	1965–74	(7)	(4)	(—)	Ibid.
	Highlands	1960	24	1	4	Henderson et al. (1962)
	Osceola	1992–94	(50)	(6)	(12)	Spalding & McLean (1994)
Icteridae						
Boat-tailed Grackle	"central Fla."	1958	(6)	(3)	(—)	Favorite (1960)
Common Grackle	Highlands	1960–61	183	30	16	Henderson et al. (1962)
Eastern Meadowlark	"central Fla."	1958	(59)	(25)	(42)	Favorite (1960))

(continued)

Table 24.27. (continued)

Family Species	County/site	Year(s)	No. of birds[a]			Data source
			Examined	Positive	%	
Red-winged Blackbird	"central Fla."	1958	(16)	(3)	(19)	Ibid.
	Highlands	1960–61	42	2	5	Henderson et al. (1962)
Rusty Blackbird	"central Fla."	1958	(6)	(2)	(—)	Favorite (1960)
Brewer's Blackbird	"central Fla."	1958	(9)	(1)	(—)	Ibid.
Passeridae						
House Sparrow	"Florida"	1965–74	223	3	1	Bigler et al. (1975)
	"Florida"	1965–74	(1)	(0)	(—)	Ibid.

a. Results are from hemagglutination-inhibition tests; parentheses indicate serum neutralization tests.

b. Favorite (1960) also tested specimens of the following species (sample sizes in parentheses) and found them seronegative for EEE virus: Eastern Kingbird (6), Blue-gray Gnatcatcher (2), Yellow-throated Warbler (1), Northern Parula 3), Northern Waterthrush (1), Blackpoll Warbler (1), House Sparrow (1), Bachman's Sparrow (1).

c. Henderson et al. (1962) also tested specimens of the following species (sample sizes in parentheses) and found them seronegative for EEE virus: Carolina Wren (14), White-eyed Vireo (12).

d. Bigler et al. (1975) also tested specimens from the following families (results of individual species not given) (sample sizes in parentheses) and found them seronegative for EEE virus: Tyrannidae (20), Hirundinidae (8), Paridae (9), Troglodytidae (29), Muscicapidae (39), Vireonidae (29), Emberizidae (343).

e. Spalding and McLean (1994) also tested specimens of the following species (sample sizes in parentheses) and found them seronegative for EEE virus: Acadian Flycatcher (3), Great Crested Flycatcher (1), Blue Jay (8), Blue-gray Gnatcatcher (1), Veery (3), Swainson's Thrush (12), Northern Mockingbird (3), Gray Catbird (1), Red-eyed Vireo (3) Northern Parula (2), Black-throated Blue Warbler (6), Yellow-throated Warbler (1), Prairie Warbler (4), Black-and-white Warbler (7), American Redstart (1), Prothonotary Warbler (4), Worm-eating Warbler (4), Ovenbird (38), Northern Waterthrush (31), Kentucky Warbler (1), Hooded Warbler (1), Eastern Meadowlark (1).

currence of this virus in Florida. Infections were identified during the remainder of 2001 in 54 additional species of birds found dead, including 33 species of perching birds (table 24.28). Infections were found in birds from 63 of the 67 counties in the state, from Escambia County in the panhandle to Nassau County in northeastern Florida and Monroe and Dade counties in the south. It was assumed that these birds died because of West Nile virus infections, but necropsies were not performed in order to prove this. WNV has been isolated in several species of mosquitoes within the genus *Culex*. The principal species of *Culex* found in Florida, *Culex quinquefasciatus, C. nigripaplus* and *C. salinarius,* are presumed to be important vectors.

The impact of WNV on birds in Florida is unknown at this writing (April 2002), but it could be significant, especially among corvids. WNV rarely causes mortality in birds in other parts of the world, but in the United States is very pathogenic, especially to crows but also to other perching birds and representatives of a number of additional avian orders (Steele et al. 2000). Affected birds exhibit neurologic signs and at necropsy many have meningoencephalitis and myocarditis along with other changes that have led to death. The numbers of American Crows in parts of eastern New York before and after the epidemic in 1999 were analyzed by Eidson et al. (2001) using the National Audubon Society's Christmas bird counts taken at the epicenter. There were decreases in the numbers of crows sighted in 1999 compared with 1998, especially in the boroughs of Queens (69%) and the Bronx (65%). Only time will tell what the effects on Florida's avian populations might be. The threatened Florida Scrub-Jay is particularly at risk because of its low numbers and limited distribution in scrub habitat. Nothing is known about its susceptibility to WNV, although it is likely to be high, as has been shown to be the case among other corvids in the northeastern states.

Infections by SLE virus are of public health importance and a number of outbreaks in humans have occurred in Florida during 1958,

1962, 1964, 1977, and 1990 (Luby 1994). In Florida the principal vector is the mosquito *Culex nigripalpus* (see review in Nayar 1982). Unlike EEE infections, horses are not involved, but a number of birds including songbirds serve as reservoir hosts of SLE virus, along with wild mammals such as raccoons and cotton rats (McLean and Bowen 1980). There are no records of the isolation of SLE virus from songbirds in Florida, but there is serologic evidence for infections in at least 12 species representing 5 families of passeriforms (table 24.29). McLean and Bowen (1980) listed several songbirds as important reservoirs of SLE virus in various parts of the United States, including House Sparrows, Blue Jays, American Robins, Northern Mockingbirds, and House Finches. All 5 of these species are seasonally common and widespread in Florida and there are SLE seropositive data for 2, Blue Jays and House Sparrows, that are year-round residents. SLE infections are not known to be lethal to songbirds (McLean and Bowen 1980). More information on SLE virus can be found in the monograph published by the Florida State Board of Health (1969), in the book edited by Monath (1980), and in a briefer and more recent review by Luby (1994).

Highlands J virus infects horses and humans, but rarely causes disease (Hoff et al. 1978; Johnston and Peters 1996). In Florida the virus is endemic throughout the state, except for the southern and southeastern portions of the peninsula, and is transmitted by *Culiseta melanura* during every month except January and February (Hoff et al. 1978). The virus has been isolated from 3 species of passeriforms (table 24.30) and 8 species have been found to be seropositive (table 24.31). Passeriforms become infected, but the virus is usually not pathogenic to them. One report of a mortality incident attributed to Highlands J virus came from California in 1955, when the virus was isolated from 3 nestling House Sparrows found dead in their nests (Johnson 1960). Other birds such as domestic turkeys and chukars, however, are susceptible (Guy et al. 1993; Eleazer and Hill 1994). A natural outbreak attributed to west-

Table 24.28. Distribution of West Nile virus infections in passeriforms from Florida as determined by PCR and virus isolation studies in 2001.

Family	Species	County	No. infected
Laniidae	Loggerhead Shrike	Duval	1
		Wakulla	1
Vireonidae	Black-whiskered Vireo	Duval	1
Corvidae	Blue Jay	Many[a]	204
	American Crow	Columbia	20
		Gulf	1
		Jefferson	1
	Fish Crow	Gulf	1
		Jefferson	3
		Levy	1
		Marion	1
		Wakulla	1
	Crow (unidentified)	Many[b]	395
Hirundinidae	Purple Martin	Wakulla	1
Paridae	Carolina Chickadee	Leon	1
		Wakulla	1
Troglodytidae	Carolina Wren	Putnam	1
Turdidae	Eastern Bluebird	Lee	1
		Leon	1
		Suwannee	1
	Hermit Thrush	Levy	1
	American Robin	Bradford	1
Mimidae	Gray Catbird	Citrus	2
		Clay	1
		Escambia	1
		Gulf	1
		Lake	1
		Manatee	1
		Marion	2
		Palm Beach	1
	Northern Mockingbird	Alachua	1
		Bay	1
		Clay	1
		Dade	1
		Escambia	1
		Jefferson	1
		Lake	1
		Leon	1
		Monroe	1
		Sumter	1
	Brown Thrasher	Bay	2
		Dade	1
		Duval	3
		Gadsden	1
		Holmes	1
		Lake	2
		Leon	1
		Marion	1
		Monroe	1
Sturnidae	European Starling	Monroe	1
Motacillidae	Sprague's Pipit	Monroe	1
Parulidae	Canada Warbler	Santa Rosa	1
	Northern Parula	Baker	1
		Bradford	1
		Citrus	2
		Lake	1
		Marion	1

(continued)

Table 24.28. *(continued)*

Family	Species	County	No. infected
		Monroe	1
	Yellow Warbler	Bradford	1
		Clay	3
		Santa Rosa	1
	Hooded Warbler	Bradford	1
Emberizidae			
	Song Sparrow	Alachua	3
		Bay	1
		Dade	1
		Leon	3
		Marion	2
		Okaloosa	1
		Putnam	1
	Savannah Sparrow	Duval	1
Cardinalidae	Northern Cardinal	Many[c]	18
Icteridae	Red-winged Blackbird	Alachua	1
	Rusty Blackbird	Hernando	1
	Brown-headed Cowbird	Bradford	1
		Gadsden	1
		Monroe	1
		Okaloosa	1
	Baltimore Oriole	Levy	1
	Boat-tailed Grackle	Alachua	1
	Common Grackle	Charlotte	2
		Leon	2
		Okaloosa	1
		Okeechobee	1
		Palm Beach	1
		Polk	2
		Suwannee	1
Fringillidae	American Goldfinch	Charlotte	1
		Duval	1
		Lake	1
	House Finch	Bay	1
		Collier	1
		Duval	4
		Marion	1
		Pinellas	1
		Santa Rosa	1
	Purple Finch	Leon	1
Passeridae	House Sparrow	Leon	1
		Okaloosa	2
		Union	1

Source: Conti et al. (2002).

a. Alachua (2), Bay (23), Bradford (3), Brevard (1), Calhoun (4), Citrus (2), Clay (6), Columbia (3), Dade (2), Duval (44), Escambia (3), Franklin (3), Gadsden (2), Gilchrist (2), Gulf (5), Hernando (1), Highlands (1), Holmes (8), Jackson (4), Jefferson (2), Lake (3), Leon (26), Levy (2), Liberty (6), Madison (2), Marion (2), Martin (1), Nassau (3), Okaloosa (5), Orange (1), Santa Rosa (1), Suwannee (7), Taylor (4), Union (2), Wakulla (12), Walton (2), Washington (2).

b. Alachua (43), Baker (1) Bay (13), Bradford (10), Calhoun (4), Citrus (3), Clay (25), Columbia (4), Dade (1), Dixie (1), Duval (17), Flagler (1), Franklin (1), Gadsden (18), Gilchrist (8), Hamilton (11), Holmes (3), Jackson (5), Jefferson (4), Lafayette (7), Lake (2), Lee (1), Leon (42), Levy (12), Liberty (3), Madison (6), Marion (22), Martin (1), Nassau (18), Osceola (1), Pasco (1), Polk (2), Putnam (7), Suwannee (30), Taylor (6), Union (5), Volusia (2), Wakulla (39), Walton (3), Washington (12).

c. Duval (3), Gadsden (1), Hernando (1), Holmes (1), Jackson (1), Lafayette (1), Leon (2), Levy (1), Marion (1), Martin (1), Nassau (1), Palm Beach (1), Seminole (1), Suwannee (1).

Table 24.29. Prevalence and distribution of St. Louis encephalitis (SLE) virus infections in passeriforms from Florida as determined by serologic studies

| Family | | | No. of birds[a] | | | |
Species	County/site	Year(s)	Examined	Positive	%	Data source
Corvidae						
Blue Jay	Highlands	1960–61	133	16	12	Henderson et al. (1962)[b]
	"Tampa Bay"	1962	6	2	—	Jennings et al. (1969)
	Highlands	1994–95	355	0	—	Garvin et al. (1997)
	Highlands	1994–95	(77)	(10)	(13)	Ibid.
Florida Scrub-Jay	Highlands	1960–61	13	1	8	Henderson et al. (1962)
	Highlands	1994–95	198	1	<1	Garvin et al. (1997)
	Highlands	1994–95	(12)	(1)	(8)	Ibid.
Species not given	"Florida"	1965–74	102	2	2	Bigler et al. (1975)[c]
	"Florida"	1965–74	(2)	(0)	(—)	Ibid.
Troglodytidae						
Carolina Wren	"Tampa Bay"	1962	7	4	—	Jennings et al. (1969)
Mimidae						
Species not given	"Florida"	1965–74	236	4	2	Bigler et al. (1975)
	"Florida"	1965–74	(4)	(0)	(—)	Ibid.
Emberizidae						
Eastern Towhee	Highlands	1960–61	37	1	3	Henderson et al. (1962)
Cardinalidae						

Northern Cardinal	Highlands	1960	24	1	4	Ibid.
	"Tampa Bay"	1962	32	8	25	Jennings et al. (1969)
	"Florida"	1965–74	328	1	<1	Bigler et al. (1975)
	"Florida"	1965–74	(1)	(0)	(—)	Ibid.
Icteridae						
Common Grackle	Highlands	1960–61	183	10	5	Henderson et al. (1962)
Red-winged Blackbird	Highlands	1960–61	42	3	7	Henderson et al. (1962)
	"Tampa Bay"	1962	118	5	4	Jennings et al. (1969)
Eastern Meadowlark	Osceola	1993	(1)	(1)	(—)	Spalding & McLean (1994)[d]
Species not given	"Florida"	1965–74	343	2	1	Bigler et al. (1975)
	"Florida"	1965–74	(1)	(0)	(—)	Ibid.
Passeridae						
House Sparrow	"Tampa Bay"	1962	93	5	5	Jennings et al. (1969)
	"Florida"	1965–74	223	2	1	Bigler et al. (1975)
	"Florida"	1965–74	(2)	(0)	(—)	Ibid.

a. Results are from hemagglutination-inhibition tests; parentheses indicate serum neutralization tests.
b. Henderson et al. (1962) also tested specimens of the following species (sample size in parentheses) and found them seronegative for SLE virus: Carolina Wren (14), White-eyed Vireo (12).
c. Bigler et al. (1975) also tested specimens from the following families (results of individual species not given) (sample sizes in parentheses) and found them seronegative for SLE virus: Tyrannidae (20), Hirundinidae (8), Paridae (9), Troglodytidae (29), Muscicapidae (39), Laniidae (11), Vireonidae (29), Emberizidae (19).
d. Spalding and McLean (1994) tested the following birds from Osceola County in 1992–94 for SLE virus antibodies (via serum neutralization tests) (numbers examined in parentheses): Acadian Flycatcher (2), Great Crested Flycatcher (1), American Crow (1), Blue Jay (5), Tufted Titmouse (3), Carolina Wren (15), Blue-gray Gnatcatcher (1), Swainson's Thrush (9), Gray Catbird (1), Brown Thrasher (1), White-eyed Vireo (33), Red-eyed Vireo (2), Black-throated Blue Warbler (4), Prairie Warbler (1), Black-and-white Warbler (5), Prothonotary Warbler (1), Worm-eating Warbler (2), Ovenbird (11), Northern Waterthrush (11), Common Yellowthroat (33), Northern Cardinal (15), Eastern Towhee (15). All were seronegative.

Table 24.30. Isolations of Highlands J virus from passeriform birds in Florida

| Family | | | No. of birds | | |
Species	County	Year(s)	Examined	Positive	Data source
Corvidae					
Blue Jay	Highlands	1960	263	2	Henderson et al. (1962)
Mimidae					
Gray Catbird	Pinellas	1960	NG	1	Chamberlain et al. (1963)
Parulidae					
Black-and-white Warbler	Hendry	1961	NG	1	Chamberlain et al. (1969)

NG = not given.

ern equine encephalitis virus was reported in Florida among pen-reared chukars in the early 1960s (Ranck et al. 1965), but was most certainly Highlands J virus.

A die-off of Florida Scrub-Jays occurred at Archbold Biological Station in Highlands County in 1979–80 in a closely observed and well studied population (Woolfenden and Fitzpatrick 1984, 1991; Fitzpatrick et al. 1991). This mortality involved about half of the adults and 92 of 93 juveniles of that breeding year (figure 24.9). Overall, 128 of 184 jays in the study area died; of these, 41% died during September and October. The cause of this die-off was not determined, since the authors were not able to retrieve carcasses or locate sick birds to examine. However, they considered the circumstantial evidence—high amounts of rainfall that year and resultant increased mosquito populations—and speculated that arbovirus infections were a possibility. Although, as mentioned above, arboviruses do not usually cause mortality in indigenous songbirds, there are field and experimental observations that indicate some passeriforms may die of EEE under certain ecological conditions (see discus-

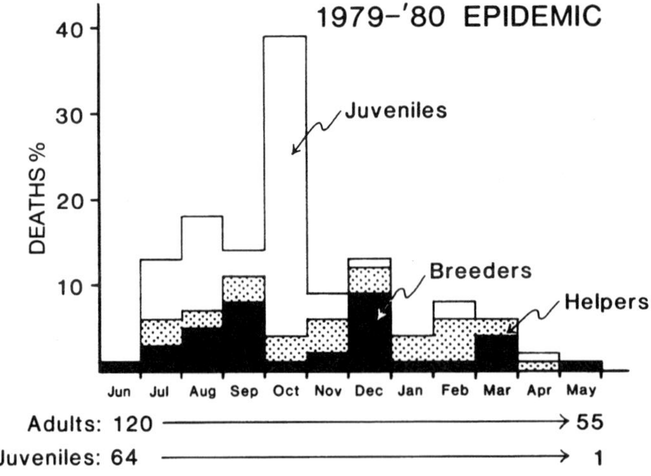

FIGURE 24.9. Epizootic curve for the Florida Scrub-Jay die-off in 1979–80 at Archbold Biological Station, Highlands County. Adapted from Woolfenden and Fitzpatrick 1991; by permission of Oxford University Press.

Table 24.31. Prevalence and distribution of Highlands J (HJ) virus infections in passeriforms from Florida as determined by serologic studies

Family Species	County/site	Year(s)	No. of birds[a]			Data source
			Examined	Positive	%	
Corvidae						
Blue Jay	Highlands	1960–61	133	46	35	Henderson et al. (1962)[b]
	Highlands	1994–95	(224)	(58)	(26)	Garvin et al. (1997)
Florida Scrub-Jay	Highlands	1960–61	13	5	38	Henderson et al. (1962)
	Highlands	1994–95	(116)	(28)	(24)	Garvin et al. (1997)
Species not given	"Florida"	1965–74	42	2	5	Bigler et al. (1975)[c]
	"Florida"	1965–74	(2)	(0)	(—)	Ibid.
Mimidae						
Species not given	"Florida"	1965–74	94	4	4	Ibid.
	"Florida"	1965–74	(2)	(0)	(—)	Ibid.
Emberizidae						
Eastern Towhee	Highlands	1960–61	37	5	14	Henderson et al. (1962)
Icteridae						
Common Grackle	Highlands	1960–61	183	30	16	Ibid.
Eastern Meadowlark	"central Fla."	1958	59	2	3	Favorite (1960)[d]
Red-winged Blackbird	Highlands	1960–61	42	2	5	Henderson et al. (1962)
Passeridae						
House Sparrow	"Florida"	1965–74	22	3	14	Bigler et al. (1975)
	"Florida"	1965–74	(3)	(0)	(—)	Ibid.

a. Results are from hemagglutination-inhibition tests; parentheses indicate serum neutralization tests.
b. Henderson et al. (1962) also tested specimens of the following species (sample sizes in parentheses) and found them seronegative for HJ virus: Northern Cardinal (24), Carolina Wren (14), White-eyed Vireo (12).
c. Bigler et al. (1975) also tested specimens from the following families (results of individual species not given) (sample sizes in parentheses) and found them seronegative for HJ virus: Tyrannidae (12), Paridae (4), Troglodytidae (10), Muscicapidae (31), Laniidae (8), Vireonidae (6), Emberizidae (213), Fringillidae (115).
d. Favorite (1960) also tested specimens of the following species (sample sizes in parentheses) and found them seronegative for HJ virus: Crested Flycatcher (16), Eastern Kingbird (6), Crow (species not given) (7), Blue Jay (7), Tufted Titmouse (4), Carolina Wren (5), Blue-gray Gnatcatcher (2), Eastern Bluebird (16), Gray Catbird (2), Northern Mockingbird (4), Brown Thrasher (3), Loggerhead Shrike (14), Yellow-throated Warbler (1), Northern Parula (3), Northern Waterthrush (1), Blackpoll Warbler (1), Red-winged Blackbird (16), Rusty Blackbird (6), Brewer's Blackbird (9), Boat-tailed Grackle (6), Summer Tanager (1), Northern Cardinal (29), Eastern Towhee (22), House Sparrow (1), Bachman's Sparrow (1).

Table 24.32. Bacteria identified in perching birds[a] from Florida

Species of bird Bacteria	County/site	Year	Tissue/organ cultured[c]	Data source
Corvidae				
Blue Jay				
Proteus sp.	Broward	1993	LV	Duncan (1993)
Mycoplasma sturni	Broward	1994	EY	Ley et al. (1998)
Hirundinidae				
Purple Martin				
Escherichia coli	Alachua[b]	1975	LI	White & Forrester (1975)
Mimidae				
Northern Mockingbird				
Proteus sp.	Broward[d]	1993	LI	Ley et al. (1998)
Serratia liquefaciens	Broward[e]	1993	LU	Ibid.
Staphylococcus sp.	Broward[e]	1993	LU	Ibid.
Mycoplasma sturni	Broward[e]	1994	EY	Ibid.
Bombycillidae				
Cedar Waxwing				
Enterobacter sp.	Palm Beach[f]	1990	LV	Mahnke & Hayes (1990)
Parulidae				
Yellow-rumped Warbler				
Escherichia coli	Duval[j]	1979	LI	White & Forrester (1979)
Emberizidae				
Field Sparrow				
Salmonella typhimurium	Volusia	1977	NG	Duncan (1977)
Savannah Sparrow				
Enterobacter sp.	"Florida"	1980	LI	White & Forrester (1980)
Pasteurella sp.	"Florida"	1980	LV	Ibid.
Cardinalidae				
Northern Cardinal				
Enterobacter sp.	"Florida"[g]	1980	LI	White & Forrester (1980)
Salmonella typhimurium	Marion[h]	1973	LV	White & Forrester (1973)
	Gilchrist	1998	SP	Terrell et al. (1998)

(continued)

sion by McLean et al. 1985). Because 1979–80 was a year of high activity of EEE virus in Florida, the virus may have spilled over from nearby enzootic areas into the Scrub-Jays via infected mosquitoes and caused the observed mortality in a highly susceptible population.

XIV. Bacteria

Nine species of bacteria have been isolated from songbirds in Florida between 1973 and 1998 (table 24.32). Except for *Salmonella ty-* *phimurium, Mycoplasma sturni,* and *Pasteurella* sp., the significance of these bacterial infections is not known. Many are probably part of the normal gut flora, or are secondary invaders or contaminants.

Numerous serotypes of *Salmonella* have been reported from free-ranging birds (Steele and Galton 1971). *Salmonella typhimurium* has been linked commonly with outbreaks of salmonellosis in songbirds (Wobeser and Finlayson 1969; Friend 1999b) and is of public health significance (Williams 1980). Mortality of passeriforms due to salmonellosis was doc-

Table 24.32. *(continued)*

Species of bird Bacteria	County/site	Year	Tissue/organ cultured[c]	Data source
Icteridae				
Brown-headed Cowbird				
Escherichia coli	Columbia	1979	LI	White & Forrester (1979)
Salmonella typhimurium	Marion	1973	LV	White & Forrester (1973)
	Gilchrist	1998	SP	Terrell et al. (1998)
Common Grackle				
Proteus sp.	Broward	1993	SP	Duncan (1993)
Red-winged Blackbird				
Enterobacter sp.	Indian River	1979	LI	White & Forrester (1979)
Klebsiella sp.	Columbia	1979	LI,LV	Ibid.
Salmonella typhimurium	Marion[i]	1973	LV	White & Forrester (1973)
Salmonella (Group B)	Osceola	2000	SI	Spalding & McCracken (2000)
Passeridae				
House Sparrow				
Salmonella typhimurium	Marion	1973	LV	White & Forrester (1973)

NG = not given.
a. Isolations made from the one bird examined, unless otherwise indicated.
b. Isolations made from 4 of 8 Purple Martins.
c. EY = eye, LI = large intestine, LV = liver, LU = lung, SI = small intestine, SP = spleen.
d. Isolations were made from 3 of 3 Northern Mockingbirds.
e. Isolation made from 1 of 2 Northern Mockingbirds.
f. Isolation made from 1 of 2 Cedar Waxwings.
g. Isolation made from 1 of 2 Northern Cardinals.
h. Isolations made from 3 Northern Cardinals.
i. Isolations made from 5 Red-winged Blackbirds.
j. Isolation made from 1 of 2 Yellow-rumped Warblers.

umented at 4 backyard feeding stations in Florida: 1 in Marion County (Nesbitt and White 1974), 1 in Gilchrist County (Terrell et al. 1998), 1 in Osceola County (Spalding and McCracken 2000), and the fourth in Citrus County (NWHC 2000). In the Marion County incident the problem extended over the 4-year period 1971–74 and involved Blue Jays, Tufted Titmice, Brown Thrashers, House Sparrows, Red-winged Blackbirds, Common Grackles, Northern Cardinals, Chipping Sparrows, and White-throated Sparrows. Most of the deaths were in late winter and early spring. Many birds had subcutaneous pustular lesions in their pectoral muscles similar to those described in House Sparrows by Wobeser and Finlayson (1969). *Salmonella typhimurium* was isolated from birds with lesions and from other affected birds that lacked lesions. The homeowner had a history of providing large quantities of bird seed on the ground; in some areas the accumulated seed was several inches thick. The Gilchrist County incident occurred over a 2-week period in January of 1998 and involved Brown-headed Cowbirds and Northern Cardinals. Twelve cowbirds and 6 cardinals were documented to have died, but mortality was thought to have been more extensive that that. An unknown number of cardinals and approximately 1,000 cowbirds were reported to be using the feeders at the time of the outbreak. Gross lesions seen at necropsy included enlarged spleens and granulomas in the walls of the esophagus. Necrotizing and inflammatory lesions with intralesional bacterial rods were seen in the spleen, liver, lung, kidney, esophagus, brain, and pectoral muscle. *Salmonella typhimurium* (serotype *Copenhagen*) was isolated from the spleens of 1 cowbird and 1 cardinal. The homeowner in this second case had been feeding birds in her backyard for several years and there was a large accumulation of waste seed and feces on the ground under the feeders. Mortality ceased after the feeders were disinfected with 10% bleach solution and the waste material on the ground cleaned up. The third incident occurred at the Bull Creek Wildlife Management Area (Osceola County) in January 2000 (Spalding and McCracken 2000). Fifteen birds were found dead over a 3-week period near bird feeders and included Red-winged Blackbirds, Northern Cardinals, Brown-headed Cowbirds, and Common Grackles. The fourth case took place in Floral City (Citrus County) over a period of several weeks in January and February of 2000 and involved a small number of Northern Cardinals. Such mortality problems as these can be prevented or controlled by refraining from scattering seeds on the ground and by the use of bird feeders that are kept clean and moved to new locations from time to time (Friend 1999b).

In February 1994 House Finches in suburban Washington, D.C. were found for the first time with a mycoplasmal disease characterized by conjunctivitis, sinusitis, and rhinitis (Fischer et al. 1997). Over the next 2–3 years the disease spread to the north, south, and west until it was found throughout virtually the entire eastern population of House Finches. Between 1994 and 1998 it was found in American Goldfinches and in 25 other species of birds in the eastern and midwest United States, including passeriforms, apodiforms, piciforms, and columbiforms (Hartup et al. 2001). The etiologic agent of this disease is a strain of *Mycoplasma gallisepticum,* distinct from other related strains that are important pathogens in domestic turkeys and chickens (Ley et al. 1997). The agent has been isolated also from finches with no signs of the disease (Luttrell et al. 1996b). The source of this strain and its introduction into the finch population is not known (Fischer et al. 1997), although domestic poultry or pen-reared Wild Turkeys may have been involved (Luttrell et al. 1996b). Transmission of the disease in domestic poultry is by direct contact, by airborne droplets or dust, or through eggs (Yoder 1991). Conjunctivitis in House Finches was first seen in Florida in November of 1995 in northern Leon County (Dhondt 1997). Since then, there have been reports of affected House Finches in Gadsden County in 1996 (Dhondt 1997) and in Alachua County in 1997 (Straub 1997), and in House Sparrows from Brevard County in 1995 (Hartup et al. 2001; Hartup 2001), although isolations of *M. gallisepticum* have not been reported. The impact of this newly recognized disease problem in populations of House Finches and American Goldfinches is yet to be determined.

Beginning in July 1994 several abandoned nestling and fledgling Northern Mockingbirds and Blue Jays that were being housed in a wildlife rehabilitation facility in Fort Lauderdale (Broward County) developed signs of conjunctivitis (Ley et al. 1998). These birds had been received as normal birds; clinical signs and gross lesions appeared sometime later, in some cases within 2 or 3 days. *Mycoplasma sturni* was identified from 1 of 2 jays and 6 of 9 mockingbirds. This species of *Mycoplasma* had been isolated and identified previously as a new species from a European Starling in Connecticut, where it was seen also in a Northern Mockingbird (Forsyth et al. 1996; Frasca et al. 1997). The occurrence and distribution of this pathogenic bacterium in free-ranging populations of passeriforms in Florida have not been studied and its effects at the population level are unknown.

The species of *Pasteurella* cultured from the liver of a Savannah Sparrow (table 24.32) was not determined, but may have been *P. multocida,* the causative agent of avian cholera or pasteurellosis. *Pasteurella multocida* has been found in a number of songbirds in the United States (Rosen 1971). Scavenger species such as crows and ravens sometimes acquire avian cholera from feeding on contaminated car-

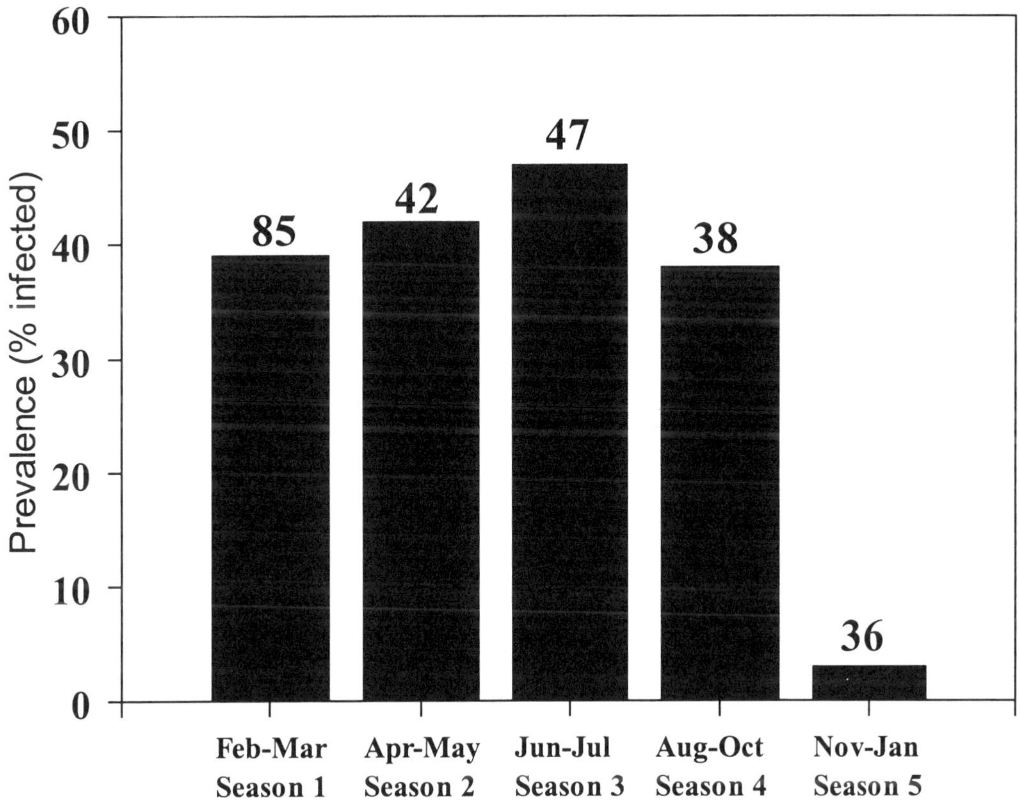

FIGURE 24.10. Seasonal prevalences of *Haemoproteus danilewskyi* in Blue Jays >1 year old, 1992–95, Archbold Biological Station, Highlands County. Numbers on top of the bars are sample sizes. From Garvin 1996.

casses of birds dying of the disease, but infections are less common in other songbirds (Friend 1999a). Sanders (1938) found experimentally that Fish Crows in Florida were susceptible to *P. multocida* and that some died of cholera, while others survived with chronic infections. He concluded that Fish Crows might be involved in the dissemination of the bacteria to other animals in nature. Avian cholera, however, is not common in Florida, only 1 outbreak having been recorded in the state (see chapter 18, Rails, Coots, Gallinules, Moorhens, and Limpkins). That outbreak occurred in Everglades National Park in 1967–68 and involved coots and ducks (Klukas and Locke 1970). Red-winged Blackbirds and American Crows were observed feeding on coot carcasses during the dieoff, but their role in disseminating the disease was not determined.

XV. Fungi

Funderburg (1968) reported a fungal infection in an emaciated Northern Parula found dead in Polk County in November of 1967. The lesion was described as a large smoothly rounded growth on the upper bill, extending from near the tip back onto the feathers at the base of the bill. He stated that the growth was examined microscopically and determined to be caused by a fungus, although no histopathologic details were given. An attempt was made to culture and identify the fungus without success. The author felt that the fungal infection was responsible for the death of the warbler. The lesion might have been the result of a traumatic wound that was secondarily infected by a saprophytic fungus as described by Asterino (1996).

A Prairie Warbler from Winter Haven (Polk County) died of aspergillosis in August of 1997 (Cornish 1997). The left lung of this bird was covered by a round, raised, yellow-white plaque measuring 1 cm in diameter. Similar plaques were also seen in the air sacs. The cause of death was judged to be respiratory insufficiency due to the severe infection with *Aspergillus fumigatus*. The significance of this finding is not known, although this was probably an isolated case and of no importance at the population level. Transmission of aspergillosis, which is rare in free-ranging passeriforms, is believed to occur via the inhalation or ingestion of spores during exposure to damp grains, seeds, or other foods that are contaminated with the fungus (Asterino 1996).

XVI. Protozoans

Eighteen species of parasitic protozoans have been identified from 16 species of passeriforms in Florida. These include 13 species of blood protozoans (eight species of *Haemoproteus*, 2 species of *Plasmodium*, 1 species of *Atoxoplasma* and 2 species of *Trypanosoma*), 3 species of tissue cyst-forming coccidia (*Toxoplasma gondii* and 2 species of *Sarcocystis*), and 2 species of intestinal coccidia (*Isospora* spp.).

Details on the distribution and prevalence of the blood protozoans are presented in table 24.33. With one exception, virtually nothing is known about the epizootiology and pathogenicity of these protozoans in passeriforms of Florida, largely because they have not been studied. The one exception is a study of *Haemoproteus danilewskyi* in a Blue Jay population at Archbold Biological Station (Highlands County) conducted during 1992–95 (Garvin and Tarvin 1996). A wealth of excellent information was obtained in that study, including data on the prevalence and intensity

of infections in the Blue Jay population, the vectors involved and their seasonal abundance and distribution, and the effects of the parasite on Blue Jays, including some data on reproductive success. The overall prevalence of infection in the population was 27%. There were no year-to-year variations or differences related to sex, but infections were more prevalent in older birds during June and July (47%) and lowest during November, December, and January (3%) in birds older than 1 year (figure 24.10). Most infected birds had low intensities, with high intensities occurring in only a few birds. Intensities did not vary between sexes and seasons, but were higher in young birds. Three species of biting midges (*Culicoides edeni*, *C. arboricola*, and *C. knowltoni*) are believed to be vectors; *Culicoides edeni* was judged to be the most important in this study because of its abundance and seasonal correlation with the highest prevalence of *H. danilewskyi* in the population (figure 24.11).

Experimentally infected Blue Jays had various histopathologic changes in liver, lung, and spleen; these were considered of minor significance, but such conditions may become more important during periods of stress or increased energy demands. Schizont stages of *H. danilewskyi* were observed in the pulmonary capillaries of 1 jay. Infected birds also had elevations in the numbers of lymphocytes, monocytes, eosinophils, heterophils, and basophils compared with controls. Five weeks after infection there was a significant decrease in the numbers of red blood cells, as reflected by PCV values. The data on the effects of the parasite on reproductive success were not conclusive. No differences were seen in the numbers of young fledged by infected vs. uninfected females. However, prevalences and intensities of infection were higher in males that fledged 1 or 2 young than those that fledged 3 or 4. This could be important since male Blue Jays have a significant role in providing food for incubating and brooding females and for young nestlings. Therefore, if males incur greater energy demands because of this feeding activity, the additional costs of infections by *Haemo-*

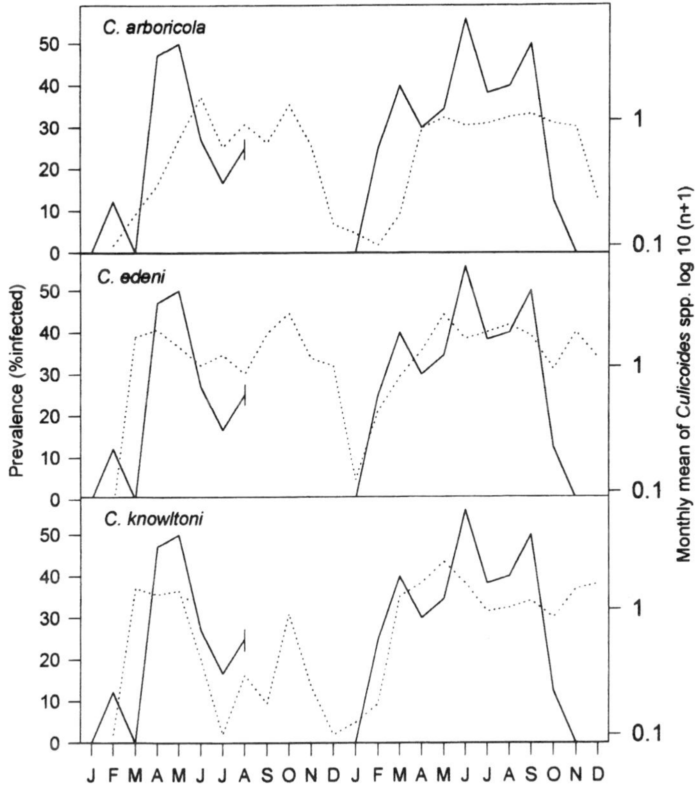

FIGURE 24.11. Mean monthly prevalences of *Haemoproteus danilewskyi* (solid line) in Blue Jays compared with abundances of the vectors *Culicoides arboricola, C. edeni,* and *C. knowltoni* (dotted line), 1994–95, Archbold Biological Station, Highlands County. From Garvin 1996.

proteus danilewskyi could be of importance. The reader is referred to Garvin (1996) for additional details of this study.

The significance of blood protozoans as morbidity and mortality factors in populations of passeriforms has been discussed in several reviews (Fallis and Desser 1977; Seed and Manwell 1977; Atkinson and van Riper 1991; Bennett et al. 1993). Bennett et al. (1993) pointed out that there are "remarkably few reports of mortality caused by blood parasites in wild birds." They felt that this was probably due to the scarcity of carcasses of small birds and the efficient activities of scavengers. Three species of *Plasmodium* (*P. cathemerium, P. relictum,* and *P. circumflexum*) are the blood parasites most commonly linked with mortality in passeriforms (Seed and Manwell 1977). Perhaps the best known example of such harmful effects is the significant reduction in numbers and the actual extinction of certain species of native Hawaiian land birds due to infections by *P. relictum* and avian pox (Warner 1968; van Riper et al. 1986). This example, however, is unusual in that neither of these disease agents is native to Hawaii; both have been introduced to the islands by the activities of humans or by migratory birds, where native avian populations have no history of exposure to such diseases. There are anecdotal reports of mortality in American Robins, House Sparrows, and Common Grackles due to malaria (*P. cathemerium*

Table 24.33. Blood protozoans identified[a] from songbirds in Florida

Family Species Blood protozoan	County/site	Year(s)	No. of birds[a]			Data source
			Examined	Positive	%	
Vireonidae						
White-eyed Vireo						
Haemoproteus vireonis	Osceola	1992–93	21	1	5	Spalding et al. (1993)
Corvidae						
American Crow						
Haemoproteus danilewskyi	Glades	1975	42	2	5	Dusek & Forrester (2002)
Haemoproteus picae	Glades	1975	42	8	19	Ibid.
	Glades	1983	1	1	—	Forrester & Bennett (1997)[b]
Haemoproteus sp.	Glades	1975	42	1	2	Dusek & Forrester (2002)
Trypanosoma avium	Glades	1975	42	3	7	Ibid.
Trypanosoma ontarioensis	Glades	1975	42	1	2	Ibid.
Trypanosoma sp.	Glades	1975	42	1	2	Ibid.
Fish Crow						
Haemoproteus danilewskyi	Alachua	1999	46	1	2	Dusek & Forrester (2002)
	Duval	1975	NG	NG	—	Bishop & Bennett (1990)
Haemoproteus picae	Alachua	1997	1	1	—	Spalding (1997)
		1999	46	8	17	Dusek & Forrester (2002)
	Duval	1975	NG	NG	—	Bishop & Bennett (1990)
Plasmodium relictum	Alachua	1999	46	6	13	Dusek & Forrester (2002)
Trypanosoma avium	Alachua	1999	46	10	22	Ibid.
Blue Jay						
Haemoproteus danilewskyi	Highlands	1992–95	539	146	27	Garvin (1996)
Haemoproteus picae	Glades	1975	15	1	7	Forrester & Bennett (1997)
Plasmodium circumflexum	Highlands	1992–95	539	1	<1	Garvin (1996)
Trypanosoma avium	Highlands	1992–95	539	2	<1	Ibid.
Florida Scrub-Jay						
Haemoproteus danilewskyi	Highlands	1994–96	277	3[c]	1	Ibid.
	Levy	1998	2	1	—	Foster & Webber (1998)
Sittidae[f]						
Turdidae						
Swainson's Thrush						
Haemoproteus fallisi	Osceola	1993	3	1	—	Spalding et al. (1993)[d]
Mimidae						
Brown Thrasher						
Haemoproteus beckeri	Osceola	1993	2	2	—	Ibid.

Northern Mockingbird						
Haemoproteus beckeri	Osceola	1993	2	1	—	Ibid.
Plasmodium circumflexum	Alachua	1979	6	2	—	Forrester & Bennett (1997)
Parulidae						
Common Yellowthroat						
Haemoproteus parulis	Osceola	1992–93	8	1	—	Spalding et al. (1993)
Trypanosoma sp.	Osceola	1992–93	8	1	—	Ibid.
Northern Waterthrush						
Haemoproteus sp.	Osceola	1992–93	9	1	—	Ibid.
Trypanosoma ontarioensis	Osceola	1992–93	9	1	—	Ibid.
Emberizidae[e]						
Eastern Towhee						
Haemoproteus coatneyi	Osceola	1992–93	14	1	7	Ibid.
Haemoproteus sp.	Glades	1975	21	1	5	Forrester & Bennett (1997)
Cardinalidae						
Northern Cardinal						
Haemoproteus sp.	Alachua	1979	2	1	—	Ibid.
Haemoproteus sp.	Glades	1975	16	3	19	Ibid.
Plasmodium vaughani	Osceola	1992–93	12	1	8	Spalding et al. (1993)
Trypanosoma avium	Glades	1975	16	1	6	Forrester & Bennett (1997)
Trypanosoma ontarioensis	Glades	1975	16	1	6	Ibid.
Icteridae						
Common Grackle						
Haemoproteus quiscalus	Alachua	1979	4	2	—	Forrester & Bennett (1997)
	Glades	1975	26	2	8	Ibid.
Brown-headed Cowbird						
Haemoproteus quiscalus	Alachua	1995	202	32	16	Greiner & Zeng (2000)
Plasmodium sp.	Alachua	1995	202	3	1	Ibid.
Atoxoplasma sp.	Alachua	1995	202	3	1	Ibid.
Boat-tailed Grackle						
Haemoproteus quiscalus	Lee	2000	5	1	—	Dusek et al. (2000)

NG = not given.

a. All data were obtained from examination of stained blood smears.

b. Forrester and Bennett (1997) also examined the following birds and found them negative (sample sizes in parentheses): Blue Jay (7), Brown Thrasher (6), Gray Catbird (1), Loggerhead Shrike (2), Blue Grosbeak (1), Common Grackle (4), Red-winged Blackbird (43), Savannah Sparrow (1), House Sparrow (17).

c. One was a hatch-year bird, the other 2 were 3 months old.

d. Spalding et al. (1993) also examined the following birds and found them negative (sample sizes in parentheses): Great Crested Flycatcher (1), Acadian Flycatcher (1), Tree Swallow (5), Blue Jay (1), Tufted Titmouse (6), Carolina Wren (8), Veery (1), Red-eyed Vireo (1), Black-and-white Warbler (1), Northern Parula (1), Ovenbird (10), Prothonotary Warbler (1), Yellow-throated Warbler (1).

e. Foster and Dean (1997) examined 29 Florida Grasshopper Sparrows from the following counties (sample sizes in parentheses) in 1997 and found them all negative for blood protozoans: Highlands (2), Osceola (1), Okeechobee (20), Polk (6).

f. Foster and Slater (1999) examined 12 Brown-headed Nuthatches from Collier County in 1997–98 and found them all negative.

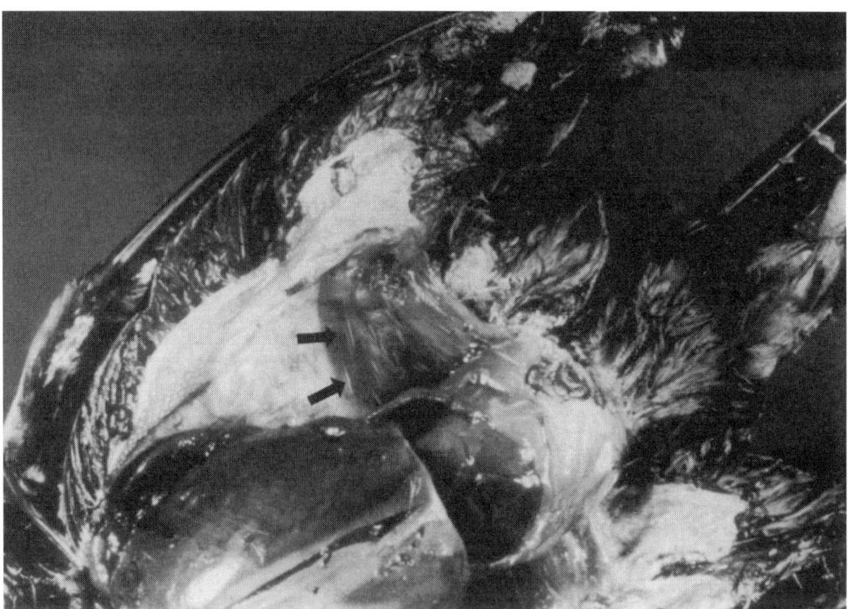

FIGURE 24.12. Sarcocysts *(arrows)* in musculature of a Common Grackle from Marion County, 1992. Courtesy of Charlotte F. Quist.

and *P. relictum*) (Stone et al. 1971; Beier et al. 1981), but other long-term studies of several species of passeriforms (Great Tits, Collared Flycatchers, Pied Flycatchers, Barn Swallows, Cliff Swallows, Brown-headed Cowbirds, Common Grackles, Red-winged Blackbirds, and Purple Martins) have not resulted in the identification of blood protozoans as significant mortality factors (Allander and Bennett 1992; Bennett 1993; Davidar and Morton 1993; Weatherhead and Bennett 1991, 1992).

Sublethal effects of blood protozoans on passeriforms have been investigated; the results and conclusions of such studies are varied. Bennett et al. (1988) found that infections by several species of *Haemoproteus* and *Leucocytozoon* in 15 species of passeriforms had no effects on body mass. Valkiunas (1993), however, observed reduced body masses in Chaffinches (*Fringilla coelebs*) infected with *H. fringillae* during the 3-day period of peak parasitemia compared with uninfected controls. During this period the birds were also less active and exhibited increased fear of humans. In addition, the author noted that blood proto-

zoans had a negative effect on the accumulation of migratory fat in a number of other species of passeriforms during spring migration. Davidar and Morton (1993) found that although female Purple Martins infected with *H. progneri* had clutch sizes similar to those of uninfected birds, they had higher breeding success; i.e., more young were fledged.

Details on the distribution and prevalence of intestinal and tissue cyst-forming coccidia identified in passeriforms from Florida are presented in table 24.34. Serum samples from 792 passeriforms in Florida representing 12 species were tested by the IHA test for antibodies to *Toxoplasma gondii* (Burridge et al. 1979). One of 133 Northern Mockingbirds and 1 of 27 Common Grackles were seropositive. However, as pointed out elsewhere, the IHA test may not be as sensitive in birds as in mammals and may result in false-negative results (Frenkel 1981). We have no information on toxoplasmosis in free-ranging passeriforms in Florida. However, clinical infections have been recognized in caged passeriforms and attributed to "reactivation of a latent infection because of

Table 24.34. Intestinal and tissue cyst-forming coccidia identified[a] from songbirds in Florida

Family Species Species of coccidia	County/site	Year(s)	No. of birds[a]			Data source
			Examined	Positive	%	
Corvidae						
Florida Scrub-Jay						
Isospora sp.	Highlands	1994–96	553	9[b]	2	Garvin and Tarvin (1996)
Mimidae						
Northern Mockingbird						
Toxoplasma gondii	"Florida"	NG	133	1[c]	<1	Burridge et al. (1979)[d]
Emberizidae						
Florida Grasshopper Sparrow[e]						
Isospora sp.	Okeechobee	1997	6	1	—	Foster et al. (1997)
	Osceola	1997	2	1	—	Ibid.
Icteridae						
Boat-tailed Grackle						
Sarcocystis sp.	Alachua	1972	NG	6	—	Simpson & Forrester (1973)
Brown-headed Cowbird						
Sarcocystis falcatula	Alachua	NG	NG	1	—	Dame et al. (1995)
	Alachua	1996	268	11	4	Luznar et al. (2001)
Common Grackle						
Sarcocystis sp.	Escambia	1993	5	3	—	Quist (1993)
Sarcocystis sp.	Marion	1991	12	NG	—	Quist (1992)
Toxoplasma gondii	"Florida"	NG	27	1[c]	4	Burridge et al. (1979)

NG = not given.

a. Identification was by histologic analysis, unless otherwise noted.

b. Three were hatch-year birds, 6 were adults (Garvin and Tarvin 1996). Identification was by microscopic examination of oocysts that had been allowed to sporulate in potassium dichromate.

c. Identification was by the indirect hemagglutination test.

d. Samples from the following species (sample sizes in parentheses) were seronegative: Blue Jay (50), Carolina Wren (19), Gray Catbird (17), White-eyed Vireo (13), Ovenbird (11), Red-winged Blackbird (84), Boat-tailed Grackle (15), Brown-headed Cowbird (10), Northern Cardinal (204), House Sparrow (209).

e. Identification was by microscopic examination of oocysts that had been allowed to sporulate in potassium dichromate. One Florida Grasshopper Sparrow from Polk County was also examined in 1997 and found negative for oocysts (Foster et al. 1997).

the stress of captivity" (Dubey and Beattie 1988). The species of *Sarcocystis* that occur in passeriforms (figure 24.12) were once thought to be of importance because of their possible relationship with equine protozoal myeloencephalitis (EPM) in horses (Dame et al. 1995; Fenger et al. 1997). However, Dubey and Lindsay (1998), Cutler et al. (1999), and Tanhauser et al. (1999) have shown subsequently that the species of *Sarcocystis* in passeriforms are unlikely to be the etiologic agents of EPM. The effects of these *Sarcocystis* spp. on passeriforms are unknown, as is the significance of the isosporan infections in the Florida Scrub-Jays and Florida Grasshopper Sparrows. These coccidia probably represent undescribed species, since most passeriforms have their own species of intestinal *Isospora* and none have been reported previously from these hosts.

XVII. Helminths

Forty-six species of parasitic helminths have been identified from passeriforms in Florida (tables 24.35–24.37). These include 11 species of trematodes, 5 cestodes, 26 nematodes, and 4 acanthocephalans. The majority of the information is on corvids and emberizids. Blue Jays and Florida Scrub-Jays are the best-studied species, especially Scrub-Jays (table 24.36).

The most comprehensive dataset on helminths of any species of passeriform in Florida is that of Kinsella (1974). This information was obtained from 45 Florida Scrub-Jays in south central Florida (Highlands County [*n* = 41] and Indian River County [*n* = 4]) during 1973. Fifteen species of helminths were identified and included 5 trematodes, 1 cestode, eight nematodes, and 1 acanthocephalan. However, only 6 of these species (*Brachylecithum nanum*, *Microtetrameres spiculata*, *Dispharynx nasuta*, *Oxyspirura pusillae*, *Cardiofilaria inornata*, and *Mediorhynchus robustus*) were considered to make up the "normal" helminth fauna of Florida Scrub-Jays, with the other species regarded as incidental parasites of this host. Ex-

cept for *M. spiculata* (which is known only from Blue Jays), these helminths have been identified in a variety of other birds and are not specific for Scrub-Jays. Kinsella concluded that the helminth fauna of the Florida Scrub-Jay represents a "complex interrelationship with the faunas of the birds with which it coexists and the intermediate hosts of the worms." He further stated that Scrub-Jays have a relatively rich fauna in terms of intensities and numbers of species, compared with other corvids. This is somewhat surprising since this jay is sedentary, has a restricted range in Florida, and is found in relatively xeric habitats.

With the exception of *Dispharynx nasuta* and *Tetrameres paucispina*, the pathogenic effects of these helminths are unknown. The life cycle, epizootiology, and pathogenicity of *D. nasuta* have been discussed in detail in chapter 17 (Wild Turkeys) in relation to its significance especially to Wild Turkey poults. This nematode has been found in 11 species of passeriforms in Florida as well as in turkeys, bobwhites, and doves. The lesions produced by *D. nasuta* in Brown-headed Cowbirds, Boat-tailed Grackles, American Crows, Blue Jays, Northern Mockingbirds, Northern Cardinals, and Carolina Wrens have been described by Rickard (1985): infected proventriculi were enlarged, in some cases 3 times normal size, and had elevated lesions that involved 10–40% of the mucosal surface (figure 24.13). There were hemorrhages in the tissue around the worms and often there was an excessive amount of clear to yellowish mucus present. Tissue reactions to the worms were less severe in Northern Mockingbirds, Carolina Wrens, and Blue Jays than in the other passeriforms studied, indicating that these species may be more tolerant. The overall pathological significance of infections of *D. nasuta* to populations of passeriforms needs clarification. Kinsella (1974) reported finding 2 emaciated 3-month-old Florida Scrub-Jays in Highlands County in 1973 with 28 and 30 *D. nasuta* in their proventriculi, but he was not able to determine if other factors were involved in causing the observed debilitation.

Tetrameres paucispina was found associated

with reddish nodules in the proventriculi of several Common Grackles (figure 24.14) from Marion and Escambia counties in 1991 and 1993 (Quist 1992–93). The overall pathological significance of these lesions in grackles is not known. Infections by large numbers of other species of *Tetrameres* are known to cause emaciation, anemia, and death in pigeons and ducks (Wehr 1971).

XVIII. Arthropods

One hundred and two species of parasitic arthropods have been identified from passeriforms in Florida. These include 10 species of ticks, 24 feather mites, 5 skin mites, 3 nasal mites, 5 chiggers, 44 chewing lice, 4 louse flies, 1 bird fly, 3 myiasis-producing flies, 1 flea, and 2 cimicid bugs (tables 24.38–24.44).

The vast majority of ticks infesting passeriforms are in nymphal or larval stages; the conspecific adults do not occur on birds, but instead feed on mammals or reptiles. The one exception is *Ixodes brunneus,* all stages of which utilize birds as hosts. The pathologic sig-

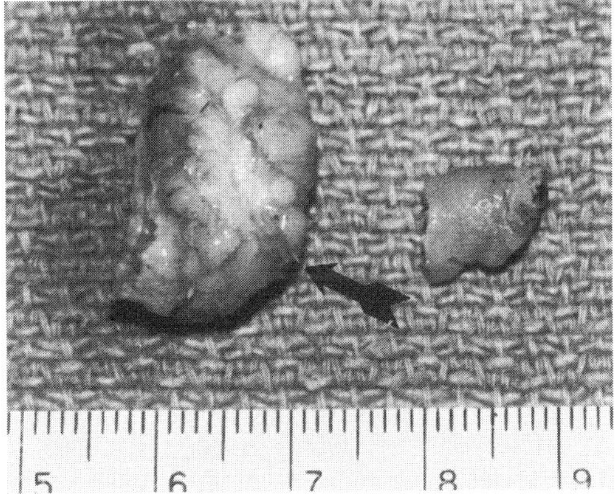

FIGURE 24.13. Comparison of proventriculi from Boat-tailed Grackles of comparable ages in Florida. The proventriculus on the left is from a bird infected with *Dispharynx nasuta;* the one on the right is uninfected. The arrow points to the posterior end of an embedded nematode. The scale is in centimeters. From Rickard 1985; by permission of *Canadian Journal of Zoology.*

nificance of infestations by nymphs and larval ticks is not known for passeriforms in Florida. Undoubtedly they cause irritation and swelling

FIGURE 24.14. Nodules *(arrowhead)* in the proventriculus of a Common Grackle from Marion County in 1992 caused by *Tetrameres paucispina.* Courtesy of Charlotte F. Quist.

Table 24.35. Helminths reported from passeriforms in Florida

Family
  Host species
    Helminth (location in host)[a]
	County/site	Year	No. of birds			Intensity		Data source[b]

			Examined	Positive	%	Mean	Range	
Laniidae								
Loggerhead Shrike								
Trematoda								
Brachylecithum nanum (LV)	Highlands	1973	1	1	—	4	—	C
Nematoda								
Acuaria sp. (GL)	Highlands	1973	1	1	—	4	—	C
Dicheilonema cylindrica (SQ)	"Florida"	NG	NG	NG	—	NG	—	A
Physocephalus sexalatus[e] (IN)	No. Fla.	NG	NG	NG	—	NG	—	F
Porrocaecum semiteres (NG)	"Florida"	1885	NG	1	—	1	—	A
Corvidae (See table 24.36)								
Sittidae								
Nuthatch (species not given)								
Nematoda								
Capillaria sp. (IN)	"Florida"	NG	NG	1	—	1	—	A
Troglodytidae								
Carolina Wren								
Nematoda								
Dispharynx nasuta (PR)[c]	Alachua	1984	8	6	—	5	1–9	B

Turdidae								
American Robin								
Nematoda								
Dispharynx nasuta (PR)	Dade	1972	2	1	—	12	—	C
Hermit Thrush								
Nematoda								
Capillaria sp. (SI)	Dade	1994	2	1	—	3	—	C
Acanthocephala								
Plagiorhynchus cylindraceus (SI)	Dade	1994	2	1	—	2	—	C
Swainson's Thrush								
Trematoda								
Brachylaima sp. (IN)	Franklin	NG	NG	NG	—	NG	—	D
Mimidae								
Brown Thrasher								
Nematoda								
Acuaria sp. (GL)	Polk	1973	1	1	—	1	—	E
Northern Mockingbird								
Nematoda								
Dispharynx nasuta (PR)[d]	Alachua	1984	5	3	—	13	1–32	B

Thraupidae, Emberizidae, Cardinalidae, and Icteridae: see Table 24.37.

NG = not given by authors.

a. IN = intestines, SI = small intestine, PR = proventriculus, GL = gizzard lining, LV = liver, SQ = subcutaneous tissue, NG = not given by author.

b. A = Walton (1927), B = Rickard (1984), C = Kinsella (1997), D = Loftin (1960), E = Kinsella (1973), F = Cram (1930).

c. The proventriculi of the following species from Alachua County were examined in 1984 for the presence of *Dispharynx nasuta* and were negative (sample sizes in parentheses): Veery (1), Gray Catbird (3), European Starling (1), White-eyed Vireo (1), House Sparrow (3) (Rickard 1984).

d. The proventriculus of 1 Northern Mockingbird from Glades County was examined in 1975 for the presence of *Dispharynx nasuta* and was negative (Forrester 1975).

e. Aberrant third stage larvae.

Table 24.36. Helminths reported from corvids (crows and jays) in Florida

Host species Helminth (location in host)[a]	County/site	Year	No. of birds			Intensity		Data source[b]
			Examined	Positive	%	Mean	Range	
American Crow								
Trematoda								
Echinostoma sp. (SI)	Franklin	NG	NG	NG	—	NG	—	A
Nematoda								
Dispharynx nasuta (PR)	Glades	1975	42	2	5	2	1–2	B
	Glades	1984	36	5	14	11	1–26	C
Fish Crow								
Trematoda								
Brachylaima fuscatum (CE)	Monroe	1929	NG	NG	—	NG	—	D
Cestoda								
Paroniella sp. (SI)	Alachua	1997	3	2	—	2	1–2	E
Sparganum (unidentified) (SQ)	Alachua	1997	3	1	—	1	—	E
Nematoda								
Cardiofilaria pavlovskyi (AS)	Alachua	1997	3	3	—	19	6–33	E
Dispharynx nasuta (PR)	Alachua	1997	3	3	—	2	1–3	E
Strongyloides sp. (SI)	Highlands	1973	1	1	—	25	—	F
Unidentified microfilariae	Alachua	1999	46	2	4	—	—	K
Blue Jay								
Trematoda								
Brachylecithum nanum (LV)	Alachua	1999	5	3	—	5	1–10	F
Lutztrema microstomum (LV)	Alachua	1999	5	2	—	6	1–10	F
Lyperosomum oswaldoi (LV)	Alachua	1999	5	2	—	1	—	F
Mosesia sp. (SI)	Alachua	1999	5	1	—	8	—	F
Cestoda								
Oligorchis sp. (SI)	Franklin	NG	NG	NG	—	NG	—	A
Nematoda								
Diplotriaena tricuspis (AS)	Highlands	1993	4	1	—	1	—	G
Dispharynx nasuta (PR)	Alachua	1984	6	5	—	6	2–13	H
	Glades	1975	21	15	71	14	1–77	B
	Glades	1984	23	18	78	5	1–16	C
	Highlands	1973	5	3	—	3	2–5	F
	Highlands	1993	4	2	—	1	—	G
Microtetrameres sp. (PR)	Glades	1975	21	1	5	2	—	E
Physaloptera sp. (larva) (PR)	Highlands	1973	5	1	—	1	—	F
Squamofilaria sp. (TC)	Highlands	1973	5	1	—	1	—	F
Strongyloides sp. (SI)	Highlands	1973	5	1	—	1	—	F
Viguiera pari (GL)	Highlands	1973	5	1	—	5	—	F

Parasite (location)[a]	County	Year	No. examined	No. infected	Total	Mean	Range	Reference[b]
Acanthocephala								
Mediorhynchus robustus (SI)	Highlands	1993	4	1	—	3	—	G
Florida Scrub-Jay								
Trematoda								
Brachylecithum americanum (LV)	Highlands	1973	41	1	2	9	—	I
Brachylecithum nanum (LV)	Highlands	1973	41	9	22	3	1–8	I
	Indian River	1973	4	2	—	3	1–5	I
Echinostoma trivolvis (SI)								
	Highlands	1973	41	1	2	3	—	I
	Indian River	1973	4	1	—	1	—	I
Mosesia sp. (SI)	Highlands	1973	41	1	2	1	—	I
Stomylotrema vicarium (CL)	Highlands	1973	41	1	2	2	—	I
Cestoda								
Oligorchis cyanocitti (SI)	Highlands	1973	41	1	2	1	—	I
Sparganum (unidentifed) (SI)	Highlands	1993	6	1	—	1	—	G
Nematoda								
Acuaria quiscula (GL)	Highlands	1973	41	22	54	2	1–6	I
	Indian River	1973	4	1	—	1	—	I
Aprocta sp. (AS)	Highlands	1973	41	1	2	1	—	I
Cardiofilaria pavlovskyi (AS)	Highlands	1973	41	8	20	1	—	I
	Highlands	1993	6	1	—	4	—	G
	Indian River	1973	4	1	—	1	—	I
Diplotriaena multituberculata (TC)	"Florida"	NG	NG	1	—	1	—	J
Diplotriaena tricuspis (AS)	Highlands	1973	41	1	2	1	—	I
Diplotriaenoides hepaticus (LV)	"Florida"	NG	NG	1	—	1	—	J
Dispharynx nasuta (PR)	Glades	1975	2	0	—	—	—	B
	Highlands	1973	41	13	32	7	1–30	I
	Highlands	1993	6	1	—	2	—	G
	Indian River	1973	4	3	—	19	3–49	I
Microtetrameres spiculata (PR)	Highlands	1973	41	24	59	5	1–23	I
	Highlands	1993	6	1	—	1	—	G
Oxyspirura pusillae (NM)	Highlands	1973	41	9	22	3	1–11	I
Strongyloides sp. (SI)	Highlands	1973	41	1	2	1	—	I
	Highlands	1993	6	1	—	1	—	G
Acanthocephala								
Mediorhynchus robustus (SI)	Highlands	1973	41	15	37	4	1–15	I
	Highlands	1993	6	1	—	4	—	G

NG = not given.

a. Location in host: AS = air sacs, CE = caecum, CL = cloaca, GL = under gizzard lining, LV = liver, NM = under nictitating membrane, PR = proventriculus, SI = small intestine, SQ = subcutaneous, TC = thoracic cavity.

b. A = Loftin (1960), B = Forrester (1975), C = Rickard (1984, 1985), D = U.S. National Parasite Collection #69754, E = Kinsella (1997), F = Kinsella (1973), G = Kinsella & Garvin (1993), H = Rickard (1984), I = Kinsella (1974, 1997), J = Walton (1927), K = Dusek & Forrester (2002).

Table 24.37. Helminths reported from thraupids, emberizids, cardinalids, and icterids in Florida

Host species Helminth (location in host)[a]	County/site	Year	No. of birds			Intensity		Data source[b]
			Examined	Positive	%	Mean	Range	
Thraupidae								
Summer Tanager								
Nematoda								
Porrocaecum angusticolle (PR)	"Florida"	NG	NG	1	—	1	—	C
Emberizidae								
Eastern Towhee								
Trematoda								
Brachylecithum nanum (LV)	Highlands	1973	1	1	—	25	—	B
Lyperosomum byrdi (LV)	Highlands	1973	4	1	—	8	—	L
Nematoda								
Diplotriaena tricuspis (NG)	"Florida"	1885	NG	1	—	1	—	C
Dispharynx nasuta (PR)	Glades	1975	20	3	15	2	1–3	F
Florida Grasshopper Sparrow								
Nematoda								
Unidentified microfilariae	Polk	1997	6	1	—	—	—	Q
Acanthocephala								
Mediorhynchus papillosum (IN)	Osceola	1997	1	1	—	7	—	A
Bachman's Sparrow								
Trematoda								
Stomylotrema vicarium (CL)	Osceola	1997	1	1	—	7	—	A
Cardinalidae								
Northern Cardinal								
Cestoda								
Orthoskrjabinia rostellata (SI)	Sumter	2000	1	1	—	2	—	R
Nematoda								
Dispharynx nasuta (PR)	Alachua	1984	14	5	36	5	1–12	M
	Glades	1975	18	1	6	1	—	F
Icteridae								
Boat-tailed Grackle								
Trematoda								
Brachylecithum americanum (LV)	Alachua	1972	1	1	—	25	—	B
Echinostoma trivolvis (SI)	Alachua	1972	1	1	—	1	—	B
Cestoda								
Dicranotaenia farcimosa (SI)	Alachua	1972	1	1	—	8	—	B

	County	Year						Ref.
Nematoda								
Capillaria sp. (SI)	Alachua	1972	1	1	—	25	—	B
Diplotriaena tricuspis (NG)	"Florida"	NG	NG	1	—	3	—	C
Dispharynx nasuta (PR)	Alachua	1984	3	1	—	100	—	M
Filaria obtunsa (NG)	"Florida"	NG	NG	1	—	1	—	D
Viktorocara garridoi (GL)	Levy	1972	1	1	—	6	—	B
Common Grackle								
Trematoda								
Conspicuum icteridorum (LV)	Alachua	1972	2	2	—	3	2–4	B
	Alachua	1983	1	1	—	5	—	B
Nematoda								
Acuaria sp. (GL)	Alachua	1983	2	1	—	3	—	B
Chandlerella quiscali (CV)	Marion	1991	12	NG	—	NG	—	E
Dispharynx nasuta (PR)	Glades	1975	25	0	—	—	—	F
Filaria obtunsa (NG)	"Florida"	NG	NG	NG	—	5	—	G
Tetrameres paucispina (PR)	Highlands	1973	1	1	—	9	—	H
Tetrameres sp. (PR)	Escambia	1993	5	3	—	NG	—	I
	Marion	1991	12	5	42	NG	—	E
	Seminole	NG	NG	NG	—	NG	—	J
Eastern Meadowlark								
Nematoda								
Diplotriaena agelaius (BC)	Nassua	1975	1	1	—	4	—	B
Porrocaecum sp. (larva) (ME)	"Florida"	NG	NG	1	—	1	—	C
Acanthocephala								
Mediorhynchus grandis (NG)	"Florida"	NG	NG	1	—	1	—	K
Red-winged Blackbird								
Trematoda								
Brachylaima americanum (LV)	Franklin	NG	NG	NG	—	NG	—	N
Nematoda								
Diplotriaena agelaius (AS)	Alachua	1987	6	1	—	2	—	O
Diplotriaena sp. (BC)	Seminole	1928	NG	NG	—	NG	—	P
Dispharynx nasuta (PR)	Alachua	1984	13	0	—	—	—	M
	Glades	1975	21	1	5	4	—	F
Brown-headed Cowbird								
Nematoda								
Cardiofilaria pavlovskyi (AS)	Osceola	2000	2	1	—	1	—	S

NG = not given.

a. Location in host: AS = air sacs, BC = body cavity, CL = cloaca, CV = cranial vault, GL = gizzard lining, IN = intestine, LV = liver, ME = mesentery, PR = proventriculus, SI = small intestine.

b. A = Kinsella et al. (1997), B = Kinsella (1997), C = Walton (1927), D = U.S. National Parasite Collection #M180-D, E = Quist (1992), F = Forrester (1975), G = Leidy (1886), H = Mollhagen (1976), I = Quist (1993), J = Wehr (1934), K = U.S. National Parasite Collection #5197.02, L = Denton & Krissinger (1975), M = Rickard (1984), N = Loftin (1960), O = Spalding & Forrester (1987), P = U.S. National Parasite Collection #34401, Q = Foster and Dean (1997), R = Kinsella et al. (2000), S = Kinsella and Foster (2000).

Table 24.38. Ticks reported from passeriforms in Florida

Host species Tick species	County/site	Year(s)	Stage of tick[b]	No. of birds			Data source[a]
				Examined	Positive	%	
Laniidae							
Loggerhead Shrike							
Haemaphysalis leporispalustris	"No. Florida"	1949–53	N,L	2	NG	—	G
Vireonidae							
White-eyed Vireo							
Amblyomma longirostre	Leon	1957	N	NG	NG	—	K
Corvidae							
Blue Jay							
Amblyomma americanum	Highlands	1993–95	L	539	2	<1	A
Amblyomma tuberculatum	Highlands	1993–95	N,L	539	17	3	A
Haemaphysalis leporispalustris	Highlands	1993–95	N,L	539	7	1	A
Florida Scrub-Jay							
Amblyomma americanum	Highlands	1993–95	N,L	279	3	1	A
Amblyomma maculatum	Highlands	1993–95	L	279	1	<1	A
Amblyomma tuberculatum	Highlands	1982	L	1	1	—	B
	Highlands	1993–95	L	279	12	4	A
Ixodes scapularis	Highlands	1993–95	N,L	279	2	<1	A
Haemaphysalis leporispalustris	Highlands	1993–95	N,L	279	37	13	A
Hirundinidae							
Purple Martin							
Ornithodoros sp.	Okeechobee	1988	U	NG	NG	—	U
Troglodytidae							
Carolina Wren							
Ixodes minor	Polk	1992	N	1	1	—	C
Haemaphysalis chordeilis	"Florida"	NG	L	1	1	—	D
Sedge Wren							
Haemaphysalis chordeilis	Highlands	1997	N	1	1	—	E
Turdidae							
American Robin							
Amblyomma tuberculatum	Indian River	1969	L	NG	1	—	F
	"No. Florida"	1949–53	L	5	1	—	G

Ixodes brunneus	Indian River	1970	A	NG	1	—	F
Haemaphysalis leporispalustris	"No. Florida"	1949–53	N,L	5	NG	—	G
Hermit Thrush							
Amblyomma tuberculatum	"Florida"	NG	L	NG	NG	—	H
Mimidae							
Brown Thrasher							
Haemaphysalis chordeilis	Indian River	1969	N	NG	1	—	F
	Levy	1965	U	NG	NG	—	I
Haemaphysalis leporispalustris	Alachua	1965	U	NG	NG	—	I
	Alachua	1974–75	U	NG	NG	—	I
	"No. Florida"	1949–53	N,L	2	NG	—	G
Ixodes scapularis	"No. Florida"	1949–53	L	2	1	—	G
Gray Catbird							
Haemaphysalis leporispalustris	Lee	1981	N	NG	1	—	J
Northern Mockingbird							
Haemaphysalis leporispalustris	Alachua	1965	U	NG	NG	—	I
Parulidae							
Pine Warbler							
Haemaphysalis leporispalustris	"No. Florida"	1949–53	L	8	NG	—	G
Common Yellowthroat							
Amblyomma americanum	"Florida"	NG	N	NG	1	—	D
Emberizidae							
Eastern Towhee							
Haemophysalis chordeilis	"Florida"	NG	L	NG	NG	—	H
	"Florida"[c]	1936–37	U	1	1	—	O
Haemaphysalis leporispalustris	"Florida"[c]	1936–37	U	1	1	—	O
	"Florida"	NG	L	NG	NG	—	H
	Highlands	1994	N	1	1	—	P
	"No. Florida"	1949–53	N,L	18	NG	—	G
Savannah Sparrow							
Haemaphysalis chordeilis	Highlands	1997	N	8	2	—	E
Eastern Grasshopper Sparrow							
Amblyomma maculatum	Highlands	1996–97	N,L	16	3	19	E
	Osceola	1996–97	N	15	10	67	E
Haemaphysalis chordeilis	Highlands	1996–97	N,L	16	7	44	E
	Osceola	1996–97	N,L	15	11	73	E

(continued)

Table 24.38. (continued)

Host species Tick species	County/site	Year(s)	Stage of tick[b]	No. of birds			Data source[a]
				Examined	Positive	%	
Florida Grasshopper Sparrow							
Amblyomma maculatum	Orange	1937	N	NG	NG	—	Q
	"Florida"[c]	1936–37	U	1	1	—	O
	"Florida"	NG	N	NG	2	—	D
	Highlands	1989–92	N	73	1	1	R
	Highlands	1997	N	16	3	19	E
	Osceola	1996–97	N	12	8	67	E
Haemaphysalis chordeilis	"Florida"	NG	N	NG	2	—	D
	Highlands	1989–92	N	73	1	1	R
	Highlands	1997	N	16	3	19	E
	Osceola	1996–97	N	12	6	50	E
Haemaphysalis leporispalustris	Osceola	1998	A	1	1	—	E
Grasshopper Sparrow (subspecies unknown)							
Haemaphysalis chordeilis	Monroe	1954	U	NG	NG	—	I
Haemaphysalis leporispalustris	Monroe	1954	U	NG	NG	—	I
Henslow's Sparrow							
Amblyomma maculatum	Highlands	1997	L	13	0	0	E
	Osceola	1997	L	1	1	—	E
Haemaphysalis chordeilis	Highlands	1997	L	13	1	8	E
	Osceola	1997	L	1	0	0	E
Bachman's Sparrow							
Amblyomma maculatum	Highlands	1997	L,N	74	2	3	E
	Osceola	1997	—	1	0	—	E
Haemaphysalis chordeilis	Highlands	1997	L	74	3	4	E
	Osceola	1997	—	1	0	—	E
Chipping Sparrow							
Amblyomma tuberculatum	Alachua	1972	L	1	1	—	S
Ixodes brunneus	Alachua	1993	A	1	1	—	S
Swamp Sparrow							
Amblyomma maculatum	Alachua	1976	U	1	1	—	T
	Highlands	1997	—	1	0	—	E
	Osceola	1997	N	2	1	—	E

Cardinalidae							
Northern Cardinal							
Amblyomma americanum	"No. Florida"	1949–53	N,L	4	3	—	G
Haemaphysalis leporispalustris	"No. Florida"	1949–53	N,L	4	NG	—	G
Icteridae							
Common Grackle							
Amblyomma tuberculatum	Highlands	1993	L	4	3	—	L
Eastern Meadowlark							
Amblyomma maculatum	Orange	NG	U	NG	NG	—	M
	Clay	1937	L	NG	1	—	D
Amblyomma tuberculatum	Pinellas	1937	L	NG	1	—	N
	"No. Florida"	1949–53	L	15	NG	—	G
Haemaphysalis chordeilis	Alachua	NG	U	NG	NG	—	M
		1951–52	N,L,A	NG	NG	—	G
	Clay	1937	L	NG	1	—	D
	Gadsden	NG	L	NG	NG	—	M
	Highlands	1997	A	5	1		E
	"Florida"[c]	1936–37	U	55	NG	—	O
	"Florida"	NG	L	NG	NG	—	H
Haemaphysalis leporispalustris	"Florida"	NG	U	NG	NG	—	M
	"Florida"[c]	1936–37	L	55	NG	—	O
	"No. Florida"	1949–53	N,L	15	NG	—	G
	Clay	1937	L	NG	NG	—	D
Red-winged Blackbird							
Amblyomma maculatum	Orange	NG	N,L	NG	NG	—	M
Haemaphysalis chordeilis	Indian River	1968	N,L	NG	NG	—	F
Haemaphysalis leporispalustris	Indian River	1968	N,L	NG	NG	—	I,F

NG = not given.

a. A = Garvin et al. (1996a), B = Keirans & Kethley (1982), C = Foster & Keirans (1994), D = Bishopp & Trembley (1945), E = Mertins & Dean (1997–01), F = Wilson & Kale (1972), G = Rogers (1953), H = Clifford et al. (1961), I = FSCA[d], J = RML[e] #115947, K = RML[e] #43241, L = Garvin et al. (1993–94), M = Boardman (1929), N = RML[e] #56906, O = Travis (1941), P = Garvin et al. (1993–94), Q = RML[e] #57680, R = Delany & Forrester (1997) (Note: Vickery (1996) referred to some of the same records published by Delany & Forrester (1997) and cited them as "unpublished data"), S = Foster & Keirans (1994), T = Forrester (1976), U = Loye & Regan (1991).

b. N = nymph, L = larva, A = adult, U = unknown.

c. Exact locality not given; in either Orange, Osceola, or Collier County. Travis (1941) also examined the following passeriforms (sample sizes in parentheses) and found no ticks: American Crow (2), Brown-headed Nuthatch (1), Northern Mockingbird (1), American Robin (2), Eastern Bluebird (3), Loggerhead Shrike (3), Common Grackle (sample size not given).

d. FSCA = Records from the Florida State Collection of Arthropods, Division of Plant Industry, Department of Agriculture and Consumer Services, Gainesville.

e. Record from the U.S. National Tick Collection, Georgia Southern University, Statesboro.

Table 24.39. Feather mites reported from passeriforms in Florida

Host species Species of feather mite	County/site	Year(s)	Data source[a]
Laniidae			
Loggerhead Shrike			
Proctophyllodes ludovicianus	Taylor	1960	Atyeo & Braasch (1966)
Vireonidae			
Red-eyed Vireo			
Proctophyllodes stoddardi	Leon	1958	Ibid.
Corvidae			
Blue Jay			
Proctophyllodes occidentalis	"Florida"	NG	Ibid.
Proctophyllodes sp.	Highlands	1994	Garvin et al. (1994–96)
Pterodectes sp.	Highlands	1992–93	Ibid.[b]
Florida Scrub-Jay			
Pterodectes sp.	Highlands	1994–95	Ibid.[c]
Hirundinidae			
Tree Swallow			
Trouessartia sp.	Broward	1957	FSCA
Sittidae			
Brown-headed Nuthatch			
Analges sp.	Collier	1998	Mertins & Slater (1999)
Mesalgoides sp.[g]	Collier	1998	Ibid.
Turdidae			
American Robin			
Proctophyllodes hipposideros	Alachua	1973	FSCA
Eastern Bluebird			
Trouessartia sp.	Osceola	1998	Mertins & Dusek (1999)
Hermit Thrush			
Proctophyllodes hyocichlae	Alachua	1994	Mertins and Foster (1995)
Mimidae			
Brown Thrasher			
Proctophyllodes sp.	Levy	1965	FSCA
Parulidae			
Ovenbird			
Syringophiloides seiurus	"Florida"	NG	Kethley (1970)
Yellow-rumped Warbler			
Aulobia dendroicus	"Florida"	NG	Kethley (1970)
Proctophyllodes quadrisetosus	Alachua	1968	FSCA
Proctophyllodes sp.	Alachua	1965	Ibid.
Pine Warbler			
Proctophyllodes dendroicae	"Florida"	NG	Atyeo & Braasch (1966)
Emberizidae			
Eastern Towhee			
Aulonastus pipili	Highlands	1960	Kethley (1970)
Trouessartia sp.	Highlands	1998	Mertins & Dean (2001)
Bachman's Sparrow			
Trouessartia capensis	Highlands	1997–98	Ibid.
Savannah Sparrow			
Analges sp.	Highlands	1997–98	Ibid.
Trouessartia sp.	Highlands	1997–98	Ibid.
Proctophyllodes polyxenes	Highlands	1997–98	Ibid.

(continued)

Table 24.39. *(continued)*

Host species Species of feather mite	County/site	Year(s)	Data source[a]
Eastern Grasshopper Sparrow			
Trouessartia capensis	Highlands	1997	Mertins & Dean (1997)
	Osceola	1997	Mertins & Dean (1997)[d]
Florida Grasshopper Sparrow			
Mesalgoides sp.[g]	Brevard	1997	Mertins & Dean (1999)
	Osceola	1998	Ibid.
Pterodectes sp.	Osceola	1998	Ibid.
Trouessartia capensis	Osceola	1997	Ibid.
	Highlands	1997	Mertins & Dean (1997)[e]
	Osceola	1997	Mertins & Dean (1997)[f]
	Osceola	1998	Mertins & Dean (1999)
Swamp Sparrow			
Proctophyllodes sp.	Osceola	1997	Mertins & Dean (1997)
Proctophyllodes polyxenes	Osceola	1998	Mertins & Dean (2001)
Cardinalidae			
Indigo Bunting			
Syringophilopsis passerina	"Florida"	NG	Kethley (1970)
Northern Cardinal			
Analges sp.	Marion	1981	FSCA
	Alachua	1998	Mertins & Foster (1999)
Proctophyllodes longiphyllus	Alachua	1998	Ibid.
	Marion	1981	FSCA
Proctophyllodes sp.	Alachua	1965	Ibid.
Pterodectes sp.	Alachua	1998	Mertins & Foster (1999)
	Marion	1981	FSCA
Trouessartia sp.	Alachua	1998	Mertins & Foster (1999)
Icteridae			
Eastern Meadowlark			
Analges sp.	Osceola	1998	Mertins & Dean (1998)
Proctophyllodes trisetosus	Highlands	1998	Mertins & Dean (2001)
Xolalgoides sp.[g]	Osceola	1998	Mertins & Dean (1998)

NG = not given.

a. FSCA = Records from the Florida State Collection of Arthropods, Division of Plant Industry, Department of Agriculture and Consumer Services, Gainesville.

b. Collected from 3 Blue Jays; probably represent an undescribed species.

c. Collected from 21 Florida Scrub-Jays; not the same species as found in Blue Jays and probably represent an undescribed species.

d. Collected from 3 Eastern Grasshopper Sparrows.

e. Collected from 2 Florida Grasshopper Sparrows.

f. Collected from 4 Florida Grasshopper Sparrows.

g. Probably an undescribed species.

in some cases and may be of importance when attached to eyelids; in other cases when present in large numbers they could cause anemia. None is known to serve as a vector of a disease agent of significance to perching birds. On the other hand, passeriforms as hosts of immature stages of ticks are important in the epizootiology of several infectious diseases of humans, domestic animals, and wildlife, including Rocky Mountain spotted fever, tularemia, Q

Table 24.40. Skin and nest mites reported from passeriforms in Florida

Host species Species of mite[h]	County/site	Year(s)	Data source[a]
Tyrannidae			
Great Crested Flycatcher			
Ornithonyssus bursa[g]	NG	NG	Stevenson & Anderson (1994)
Corvidae			
Blue Jay			
Ornithonyssus sylviarum[g]	Alachua	1966	FSCA, Phillis et al. (1976)
	Indian River	1970	Phillis et al. (1976)
Turdidae			
American Robin			
Knemidokoptes jamaicensis[b]	Alachua	1990–96	Brugger et al. (1997), Pence et al. (1999)
	Santa Rosa	1993	Quist (1993)
Mimidae			
Brown Thrasher			
Ornithonyssus sylviarum	Alachua	1973	Phillis et al. (1976)
Gray Catbird			
Microlichus americanus	"Florida"	1960	Fain (1965)
Northern Mockingbird			
Ornithonyssus bursa[c]	Alachua	1969–70	Phillis & Cromroy (1972)
Ornithonyssus sylvarium	Alachua	1960	FSCA, Phillis et al. (1976)
	Alachua	1966	Ibid.
	Alachua	1969–70	Phillis & Cromroy (1972)[d]
Pellonyssus passeri[g]	Alachua	1960	Strandtmann et al. (1963)
	Alachua	1969–70	Phillis & Cromroy (1972)[d]
	Alachua	1972	Phillis et al. (1976)
Emberizidae			
Eastern Grasshopper Sparrow			
Ornithonyssus sylviarum	Osceola	1997	Mertins & Dean (1997)

(continued)

fever, and Lyme disease (Strickland et al. 1976). More details on this topic in relation to Florida can be found in Forrester (1992).

There are records of 2 infestations of adult female *Ixodes brunneus* on passeriforms from Florida, 1 from an American Robin and 1 from a Chipping Sparrow (table 24.38). One of these was diagnosed as a presumptive case of tick paralysis (Foster and Keirans 1994). This involved an adult female Chipping Sparrow found huddled under a house in Alachua County in November of 1993. The bird was moribund and sitting on the ground with its eyes closed and head drooping to one side. It died within minutes of being captured by hand and subsequently an engorged female *I. brunneus* was found attached to its neck. A number of cases of tick paralysis have been linked with infestations of passeriforms by this species of tick in the southeastern United States (Gregson 1973; Pitts and Hayes 1990; Mullen and Hribar 1991; Luttrell et al. 1996a). These included Cedar Waxwings, Purple Finches, House Finches, American Goldfinches, Common Grackles, Eastern Bluebirds, Red-winged Blackbirds, Chipping Sparrows, White-throated Sparrows, Yellow-rumped Warblers, American Robins, and Dark-eyed Juncos. Tick

Table 24.40. (continued)

Host species 　Species of mite[h]	County/site	Year(s)	Data source[a]
Icteridae			
Boat-tailed Grackle			
Knemidokoptes jamaicensis	Alachua	1990–96	Brugger et al. (1997), Pence et al. (1999)
	Highlands	NG	Stedman (1981)[e]
Pellonyssus passeri	Alachua	1973	Phillis et al. (1976)
Common Grackle			
Knemidokoptes jamaicensis	Alachua	1990–96	Brugger et al. (1997), Pence et al. (1999)
	Highlands	NG	Stedman (1981)[e]
Red-winged Blackbird			
Knemidokoptes jamaicensis	Alachua	1990–96	Brugger et al. (1997), Pence et al. (1999)
	Highlands	NG	Stedman (1981)[e]
	Sarasota	1980	Ibid.
Brown-headed Cowbird			
Knemidokoptes jamaicensis	Alachua	1997	Greiner (1997)
Passeridae			
House Sparrow			
Pellonyssus passeri[f]	Alachua	1969–70	Phillis & Cromroy (1972)
	Alachua	1972–73	Phillis et al. (1976)
	Alachua	1977	Forrester & Atyeo (1992)

NG = not given.

a. FSCA = Records from the Florida State Collection of Arthropods, Division of Plant Industry, Florida Department of Agriculture and Consumer Services, Gainesville.

b. In 1990, none of 24 American Robins examined when captured were infested (as determined by visual inspection); after 2 weeks in captivity, 13 of 15 of these original 24 robins had developed lesions characteristic of infestation by *K. jamaicensis*. In 1991, 7 of 55 American Robins were infested (Brugger et al. 1997).

c. Found in 5 of 5 nests examined shortly after the young fledged.

d. Found in 1 of 5 nests examined shortly after the young fledged.

e. Presumptive diagnosis based on characteristic lesions; specimens of the mites were not examined.

f. Found in 10 of 10 nests examined shortly after the young fledged.

g. Nest mites.

h. Skin mites unless noted otherwise.

paralysis is an acute flaccid ascending paralysis thought to be caused by neurotoxins secreted by the salivary glands of adult female ticks during feeding (Gregson 1973; Luttrell et al. 1996a). The disease is widespread, occurring in humans and domestic animals as well as in wild mammals and birds (Gregson 1973), and has been associated with at least 60 species of hard (ixodid) and soft (argasid) ticks (Gothe and Neitz 1991). One case was documented of a young man and his dog from Lake City, both of whom died of tick paralysis in 1951, although the tick species involved was not determined (Alexander 1952). At the present time the diagnosis of this condition is difficult and is usually based on rapid recovery of the host after removal of the tick or ticks. In wild birds, diagnosis is usually presumptive because most birds found with the condition are already dead. The presence of engorged adult females of a species of tick known to cause tick paralysis and the absence of other detrimental circum-

Table 24.41. Nasal mites reported from passeriforms in Florida

Host species Species of mite	County	Year(s)	Data source
Hirundinidae			
Barn Swallow			
Ptilonyssus echinatus	Dade	1954	George (1961)[a]
Tree Swallow			
Ptilonyssus tachycinetae	Dade	1954	Ibid.
Parulidae			
Blackpoll Warbler			
Ptilonyssus sairae	Leon	1974–75	Pence & Casto (1976)
Black-throated Blue Warbler			
Ptilonyssus sairae	Leon	1974–75	Ibid.
Cape May Warbler			
Ptilonyssus sairae	Leon	1974–75	Ibid.
Cerulean Warbler			
Ptilonyssus sairae	Leon	1974–75	Ibid.
Common Yellowthroat			
Ptilonyssus sairae	Leon	1974–75	Ibid.
Northern Parula			
Ptilonyssus sairae	Leon	1974–75	Ibid.
Ovenbird			
Ptilonyssus sairae	Leon	1974–75	Ibid.
Swainson's Warbler			
Ptilonyssus sairae	Leon	1974–75	Ibid.
Icteridae			
Bobolink			
Ptilonyssus sairae	Leon	1974–75	Ibid.

a. Also cited in the host list published by Pence (1975).

stances or disease complications are the foundation of such a diagnosis (Luttrell et al. 1996a). Although mortality of some individual birds most certainly occurs, the overall impact of tick paralysis on populations of passeriforms in Florida is unknown and would be difficult to evaluate. For additional information on tick paralysis, the reader is referred to Gregson (1973) and Gothe and Neitz (1991).

No information is available on the effects of the various species of feather mites, chiggers, chewing lice, or louse flies on passeriforms in Florida. Most likely some of these ectoparasites when present in high intensities under appropriate conditions can cause skin irritation, feather damage, feather loss, and/or general decline of condition as has been observed in some pet and aviary birds (Greve 1996) and in poultry (Loomis 1978), but we have no specific data for free-ranging birds in Florida.

There is no information on the effects of nasal mites on perching birds in Florida. One nasal mite in particular, *Sternostoma trachaecolum*, occurs not only in the nasal cavities, but also in the trachea, air sacs, bronchi, and lungs and is a pathogen, resulting in a wasting disease that can be fatal (Kettle 1995). According to Pence (1975), *S. trachaecolum* has been identified in a number of migratory songbirds from eastern United States, some of which overwinter in Florida. This mite most likely occurs in the state, although it has not yet been detected here in free-ranging passeriforms.

Table 24.42. Chiggers reported from passeriforms in Florida

Host species Species of chigger	County	Year(s)	Data source
Corvidae			
Florida Scrub-Jay			
Eutrombicula alfreddugesi	Highlands	1994	Garvin et al. (1994–96)[a]
Eutrombicula lipovskyana	Highlands	1994	Ibid.
Emberizidae			
Eastern Towhee			
Neotrombicula whartoni	Orange	1950	Brennan & Wharton (1950)
Icteridae			
Eastern Meadowlark			
Eutrombicula batatas	Collier	1933	Ewing (1943)
Eutrombicula multisetosa	Lee	1936	Ibid.

a. Found on 2 jays.

Nothing is known about the effects of *Ornithonyssus bursa* and *O. sylviarum* on free-ranging passeriforms in Florida, but based on observations of the effects of infestations of these species of haematophagous nest mites on domestic poultry, pathogenic manifestations could be expected. In poultry, infestations result in listlessness, reduced egg production, and poor growth and development of young birds (Harwood and James 1979). *Knemidokoptes jamaicensis* causes mange (scaly-leg or podoknemidokoptosis) on the unfeathered parts of the bird's legs and feet (figure 24.15) and has been identified in American Robins, Boat-tailed Grackles, Common Grackles, Brown-headed Cowbirds, and Red-winged Blackbirds in Florida (table 24.40). Brugger et al. (1997) noticed that American Robins had obvious lesions when examined at capture, but that infestations in Boat-tailed Grackles, Common Grackles, and Red-winged Blackbirds were not obvious until they had been subjected to the stresses of captivity for several weeks. The spherical-shaped mites burrow into the cornified layer of the epithelium (figure 24.16), where they lay their eggs and stimulate a host response consisting of cellular proliferation, scales, and crusts (Pence 1970; Greve 1996). The entire life cycle of the mite occurs in the skin lesion and when this is scraped, adults,

eggs, larvae, and nymphs can be seen (Greve 1996). Transmission is by direct contact (Loomis 1978). Lesions can be severe and lead to distorted feet, toes, and claws, making walking and perching difficult; secondary bacterial infection can further complicate the disease (Greve 1996). This disease can result in mor-

FIGURE 24.15. Scaly-leg disease in an Eastern Towhee from Louisiana caused by an infestation of *Knemidokoptes jamaicensis*. Top photograph is a diseased leg; bottom photograph is a normal leg. From Pence 1970; by permission of *Journal of Medical Entomology*.

Table 24.43. Chewing lice reported from passeriforms in Florida

Host species Species of louse	County/site	Year(s)	Data source
Tyrannidae			
Great Crested Flycatcher			
Menacanthus sp.[a]	Clay	1997	Mertins & Miller (1997)
Philopterus sp.	NG	NG	Peters (1936)
Laniidae			
Loggerhead Shrike			
Menacanthus camelinus	"Florida"	NG	Peters (1936)
Philopterus coarctatus	"Florida"	NG	Ibid.
Vireonidae			
White-eyed Vireo			
Menacanthus curuccae	Leon	1962	Forrester et al. (1995)
Philopterus sp.	Leon	1959, 1962	Ibid.
Ricinus vireoensis	Indian River	1966	Ibid.
	Leon	1962	Ibid.
Black-whiskered Vireo			
Philopterus sp.	Collier	1955	Ibid.
Corvidae			
Blue Jay			
Machaerilaemus sp.	Highlands	1992–95	Garvin et al. (1996b)[b]
Menacanthus eurysternus	Highlands	1992	Forrester et al. (1995)
	Highlands	1992–95	Garvin et al. (1996b)[c]
Myrsidea sp.	Glades	1983	Forrester et al. (1995)
	Highlands	1994	Ibid.
	Highlands	1992–95	Garvin et al. (1996b)[b]
Philopterus cristatus	Highlands	1992–94	Forrester et al. (1995)
	Highlands	1992–95	Garvin et al. (1996b)[d]
Florida Scrub-Jay			
Menacanthus eurysternus	Highlands	1982	Layne & Emerson (1994)
Myrsidea sp.	Highlands	1981–82	Ibid.
	Highlands	1992–93	Forrester et al. (1995)
	Highlands	1994–96	Garvin et al. (1996b)[e]
	Martin	1963	Forrester et al. (1995)
Philopterus sp.	Highlands	1994	Ibid.
American Crow			
Myrsidea interrupta	Highlands	1982	Ibid.
	Osceola	1933	Ibid.
	Pinellas	1929	Ibid.
	"Florida"	NG	Peters (1936)
Philopterus ocellatus	Pinellas	1929	Forrester et al. (1995)
Fish Crow			
Colpocephalum fregili	Dade	1937	Ibid.
Myrsidea t. timmermanni	Dade	1937	Ibid.
	Indian River	1971	Ibid.
	Lee	1981	Ibid.
	Pinellas	1929–30	Ibid.
	"Florida"	NG	Peters (1936)
Philopterus sp.	"Florida"	NG	Ibid.
Hirundinidae			
Purple Martin			
Brueelia subis	Lee	1981	Forrester et al. (1995)
Myrsidea dissimilis	Duval	1982	Ibid.
	Franklin	1980	Ibid.
	Indian River	1973	Ibid.
	Lee	1981, 1986	Ibid.
	Wakulla	1981	Ibid.
Tree Swallow			
Philopterus major	Broward	1957	Ibid.
	Dade	1917, 1928	Ibid.
	Seminole	1929	Ibid.

(continued)

Table 24.43. *(continued)*

Host species Species of louse	County/site	Year(s)	Data source
	"Florida"	NG	Peters (1936)
Bank Swallow			
Myrsidea latifrons	Dade	1926	Forrester et al. (1995)
Philopterus sp.	Dade	1928	Ibid.
Barn Swallow			
Myrsidea rustica	Dade	1926	Ibid.
Philopterus microsomaticus		1926	Ibid.
Sittidae			
Brown-headed Nuthatch			
Myrsidea sp.	Collier	1998	Mertins & Slater (1999)
Troglodytidae			
Carolina Wren			
Philopterus sp.	"Florida"	NG	Peters (1936)
Turdidae			
Eastern Bluebird			
Philopterus sialii	"Florida"	NG	Ibid.
	Osceola	1998	Mertins & Dusek (1999)
Ricinus sp.	"Florida"	NG	Peters (1936)
Swainson's Thrush			
Myrsidea incerta	Indian River	1966	Forrester et al. (1995)
Hermit Thrush			
Brueelia antiqua	Alachua	1994	Ibid.
American Robin			
Brueelia iliaci	Pinellas	1935	Ibid.
Philopterus sp.	"Florida"	NG	Peters (1936)
Sturnidoecus simplex	Indian River	1972	Forrester et al. (1995)
	Pinellas	1935	Peters (1935), Forrester et al. (1995)
Mimidae			
Gray Catbird			
Brueelia brunneinucha	Lee	1986	Forrester et al. (1995)
Myrsidea sp.	Indian River	1967, 73–74	Ibid.
Northern Mockingbird			
Brueelia brunneinucha	Alachua	1954	Ibid.
	Alachua	NG	Williams (1983)
Menacanthus eurysternus	Indian River	1965, 68	Forrester et al. (1995)
Myrsidea sp.	Alachua	1954	Ibid.
	Indian River	1965, 71	Ibid.
	Lee	1984	Ibid.
Philopterus sp.	Alachua	1954	Ibid.
Brown Thrasher			
Brueelia dorsale	Lee	1984	Ibid.
Myrsidea sp.	Lee	1981	Ibid.
Sturnidae			
European Starling			
Menacanthus eurysternus	Lee	1979	Forrester et al. (1995)
Parulidae			
Blue-winged Warbler			
Ricinus picturatus	Dade	1918	Ibid.
Nashville Warbler			
Ricinus picturatus	Dade	1926	Ibid.
Yellow Warbler			
Ricinus dendroicae	Dade	1917	Ibid.
Yellow-rumped Warbler			
Machaerilaemus sp.	Indian River	1972	Ibid.
Prairie Warbler			
Ricinus dendroicae	Leon	1958, 61–62	Ibid.
Canada Warbler			
Ricinus emersoni	Dade	1928	Ibid.

(continued)

Table 24.43. *(continued)*

Host species Species of louse	County/site	Year(s)	Data source
Emberizidae			
Eastern Towhee			
Myrsidea melanorum	Dade	1928	Peters (1936)
Bachman's Sparrow			
Ricinus wolfi	Dade	1928	Ibid.
Chipping Sparrow			
Brueelia sp.	Alachua	1954	Forrester et al. (1995)
Vesper Sparrow			
Ricinus fringillae	Dade	1926	Ibid.
Savannah Sparrow			
Machaerilaemus maestrus	Indian River	1969	Ibid.
Ricinus diffusus	Dade	1924, 29	Ibid.
Saltmarsh Sharp-tailed Sparrow			
Philopterus sp.	"Florida"	NG	Peters (1936)
Seaside Sparrow			
Philopterus sp.	Dixie	1980	Forrester et al. (1995)
	Taylor	1980	Ibid.
	Walton	1980	Ibid.
	"Florida"	NG	Peters (1936)
Ricinus diffusus	Dade	1929	Forrester et al. (1995)
Song Sparrow			
Ricinus diffusus	Dade	1930	Ibid.
Cardinalidae			
Northern Cardinal			
Myrsidea sp.	Alachua	1954	Ibid.
	Indian River	1973	Ibid.
Philopterus sp.	"Florida"	NG	Peters (1936)
Icteridae			
Red-winged Blackbird			
Brueelia ornatissima	"Florida"	NG	Peters (1936)
Myrsidea fuscomarginata	Indian River	1972–73	Forrester et al. (1995)
Philopterus agelaii	"Florida"	NG	Peters (1936)
Eastern Meadowlark			
Brueelia picturata	Alachua	1954	Forrester et al. (1995)
	Indian River	1970	Ibid.
Menacanthus sturnellae	Alachua	1954	Price (1977), Forrester et al. (1995)
Philopterus sp.	"Florida"	NG	Peters (1936)
Boat-tailed Grackle			
Menacanthus quiscali	"Florida"	NG	Ibid.
Myrsidea balteri	Alachua	1976	Forrester et al. (1995)
	Duval	1975	Ibid.
	Indian River	1968, 71	Ibid.
	Pinellas	1929	Clay (1968), Forrester et al. (1995)
Philopterus sp.	Alachua	1976	Forrester et al. (1995)
Common Grackle			
Myrsidea fuscomarginata	Alachua	1954	Clay (1968), Forrester et al. (1995)
	Highlands	1993	Forrester et al. (1995)
	Indian River	1973	Ibid.
	Pinellas	1929	Clay (1968)
Philopterus quiscali	"Florida"	NG	Peters (1936)

NG = not given.

a. near *Menacanthus distinctus.*

b. 1 of 539 Blue Jays examined were infested.

c. 2 of 539 Blue Jays examined were infested.

d. 76 of 539 Blue Jays examined were infested.

e. 7 of 279 (3%) Florida Scrub-Jays examined were infested with adult lice; nits (presumed to be those of *Myrsidea* sp.) were present on 76 (27%) of the jays.

bidity and mortality, but the effects on populations have not been determined. More details on podoknemidokoptosis in passeriforms have been provided by Pence et al. (1999).

Several cases of apparent facultative myiasis have been documented in Florida. *Phaenicia coeruleiviridis* was identified in an American Crow from Alachua County and in an Eastern Bluebird nestling from Highlands County, and *Synthesiomyia nudiseta* was found in 2 Eastern Bluebird nestlings from Highlands County (table 24.44). These infestations might have been sequelae to invasion of wounds initiated by trauma, but this was not clear. In addition, *Philornis porteri* (= *Neomusca porteri*), an obligate bird parasite, was found in Great Crested Flycatchers in Clay, Highlands, and Orange counties and in Eastern Bluebirds in Highlands and Polk counties. In the spring of 1973, Kinsella and Winegarner (1974) identified larvae of *P. porteri* from 11 nests located in artificial nesting boxes at Archbold Biological Station (Highlands County). Numbers per nest ranged from 1 to 83, with the average number of larvae per nestling being 7.1. The larvae were most commonly seen in subcutaneous cysts located between the bases of the wing feathers and on the head, but some were also on the chin, throat, back, and tail. Kinsella and Winegarner (1974) stated that because "only 1 nestling died during the study, it seems doubtful that the flies had any serious effect on the host population." Spalding et al. (2002) reported *P. porteri* from the nest of an Eastern Bluebird and from the ear canal and jaw muscles of 1 Eastern Bluebird nestling and in the abdominal musculature of another. These nestlings were among 59 found dead in nesting boxes on the Avon Park Air Force Range in Highlands and Osceola counties in June 1999. The role of *P. porteri* in this mortality event was not determined, but the authors felt that they might have been involved to some extent. Other species of *Philornis* have been found to be pathogenic to nestling passeriforms (Uhazy and Arendt 1986; Nores 1995). Calliphorid flies of the genus *Protocalliphora*

cause nestling mortality under certain conditions (Sabrosky et al. 1989), but this genus does not occur in Florida. There is one report of the larvae of *Protocalliphora* sp. from nestling Great Crested Flycatchers in Florida (Taylor and Kershner 1991), but we feel that this was a misidentification and that the specimens were most likely *P. porteri*. In that report young birds in 4 nests in Orange County were observed to be "heavily parasitized" in 1989. The population significance of these observations is not clear and more information is needed.

The sticktight flea (*Echidnophaga gallinacea*) was found on 5 of 539 Blue Jays examined at Archbold Biological Station (Highlands County) over a 3-year period (1992–95) (Garvin and Tarvin 1996). This is a serious pest of poultry (Loomis 1978) and in the cases reported from Archbold, the infested jays were known to have access to domestic chickens or poultry waste used as fertilizer in orange groves, which were the most likely source of the infestations. These are probably incidental findings and of limited pathological significance.

The Mexican chicken bug (*Ornithocoris pallidus*) is a hemipteran and related to the well-known human bedbugs (*Cimex* spp.) (Usinger 1966). It has been found associated with chicken nests and Wood Duck nesting boxes in Florida (Blatchley 1928; Eggeman et al. 1996; Usinger 1959), and was also recorded from artificial nesting houses of Purple Martins in Pensacola (Santa Rosa County) in 1957 (Usinger 1959). Because of its blood-sucking habits, infestations of large numbers of *O. pallidus* are known to cause reduced egg production, weight loss, and death of domestic poultry (Turner 1971). The effects on Purple Martins in Florida are not known.

The Cliff Swallow bug (*Oeciacus vicarius*) was reported from several Purple Martins in a colony in Okeechobee County in 1988 (Loye and Regan 1991). Ten juvenile martins were observed to jump from their nesting boxes, presumably in response to being infested with numerous cliff swallow bugs. These birds sub-

Table 24.44. Diptera (flies) reported from passeriforms in Florida

Species of fly Host species	County	Year(s)	Data source
Hippoboscids (louse-flies)			
Mimidae			
Gray Catbird			
Ornithoica vicina	Leon	1968	Wilson et al. (1990)
Parulidae			
Ovenbird			
Ornithoctona fusciventris	Leon	1973	Ibid.
Emberizidae			
Savannah Sparrow			
Ornithomya fringillina	Leon	1974	Ibid.
Cardinalidae			
Northern Cardinal			
Microlynchia pusilla	Indian River	1972	Ibid.
Icteridae			
Red-winged Blackbird			
Microlynchia pusilla	Indian River	1972	Ibid.
Bird flies			
Tyrannidae			
Great Crested Flycatcher			
Carnus floridensis[a]	Clay	1997	Mertins & Miller (1997)
Myiasis-producing flies			
Tyrannidae			
Great Crested Flycatcher			
Philornis porteri	Clay	1997	Mertins & Miller (1997)
	Highlands	1973	Kinsella & Winegarner (1974)
		1999	Spalding & Mertins (1999)
	Orange[c]	1989	Taylor & Kershner (1991)
Corvidae			
American Crow			
Phaenicia coeruleiviridis	Alachua	1972	FSCA[b]
Turdidae			
Eastern Bluebird			
Phaenicia coeruleiviridis	Highlands, Polk	1999	Mertins & Spalding (2001)
Philornis porteri	Highlands, Polk	1999	Ibid.
Synthesiomyia nudiseta	Highlands, Polk	1999	Ibid.

a. Two birds, 1 with 1 fly, and 1 with 2.

b. FSCA = Records from the Florida State Collection of Arthropods, Division of Plant Industry, Department of Agriculture and Consumer Services, Gainesville. This specimen was identified originally as *P. cuprina*; J.W. Mertins (2001) examined it and identified as *C. coeruleiviridis*.

c. Taylor and Kershner (1991) stated that this was *Protocalliphora* sp.; probably a misidentification since this genus does not occur in Florida (Sabrosky et al. 1989). We are assuming that this was actually *Philornis porteri*.

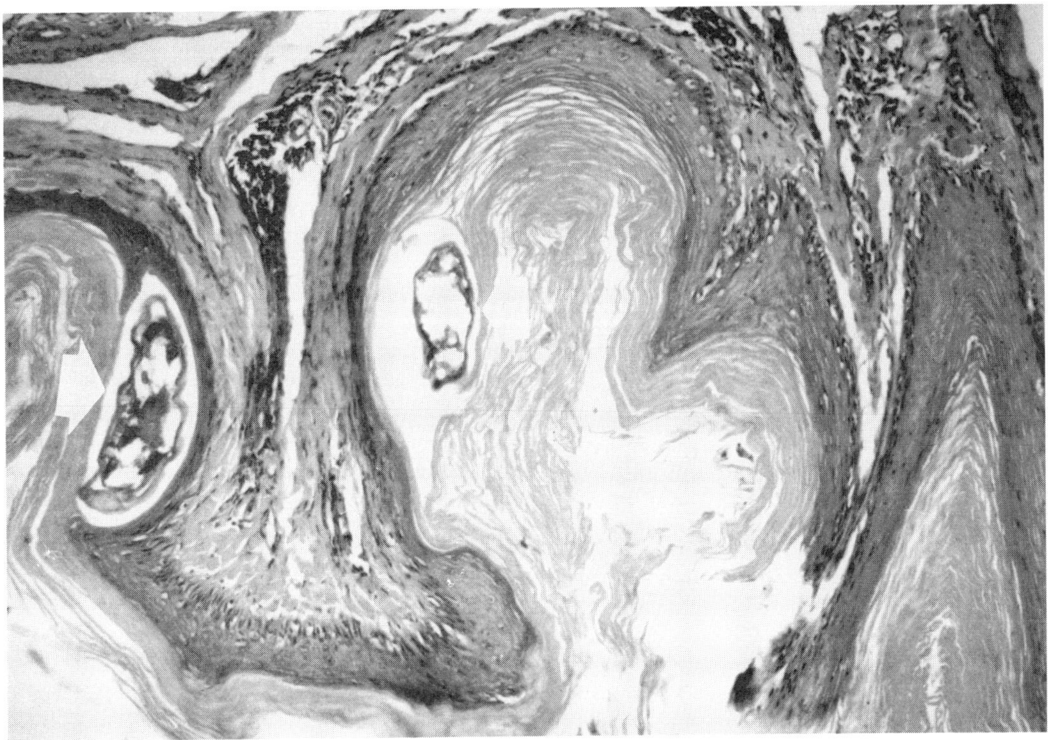

FIGURE 24.16. Histologic section of a scaly-leg lesion from the leg of an American Robin from Alachua County, 1992. The arrowhead indicates location of adult mites (*Knemidokoptes jamaicensis*). Courtesy of Bruce L. Homer.

sequently died of exposure or were killed by domestic cats. These events resulted in the eventual abandonment of 2 houses by the adult birds, which fluttered in front of the entrances to the infested nesting boxes but would not enter (Loye and Regan 1991). The authors suggested that this cimicid bug is a serious threat to all cavity-nesting birds.

XIX. Starvation

Starvation has already been discussed in earlier sections in relation to the sequential effects of brood parasitism and inclement weather. Another aspect of starvation will be addressed in this section.

Food availability has long been recognized to have an important influence on reproduction in passeriform populations (Lack 1954). Two studies on Florida passeriforms, one on Boat-tailed Grackles and one on Florida Scrub-Jays, have resulted in information on starvation. Bancroft (1986) conducted a 4-year study (1978–81) of Boat-tailed Grackles in Hillsborough County and found that starvation was the most common cause of mortality among nestlings (13% of 153 birds) among 605 nests followed. In a 10-year study (1970–80) of Florida Scrub-Jays in Highlands County, 13% of all nestling losses were due to starvation (Woolfenden and Fitzpatrick 1984). The authors concluded that starvation did "not result from delivery of insufficient food, but instead from competition for food among brood mates." This was attributed to the fact that hatching is not synchronized in Florida Scrub-Jays; some young hatch 1 or 2 days later than other brood mates and eventually starve because they are not able to compete successfully for the food brought to the nest by the parents.

XX. Perching birds as pests and public health hazards

A number of passeriforms, especially blackbirds, grackles, cowbirds, European Starlings, and House Sparrows, are responsible for crop losses as well as for defacement and destruction of property. They may also be of public health importance as hosts and disseminators of diseases and parasites. The environmental impact statement published in 1994 by the USDA (Animal Damage Control Program 1994) and the monograph by Weber (1979) contain a wealth of data on this topic and should be consulted for additional information.

Losses and damage to the following crops due to passeriforms have been recorded for Florida, although exact dollar values have not been determined: corn (blackbirds and crows), peanuts (crows), soybeans (blackbirds), blueberries (European Starlings, American Robins, Cedar Waxwings, Northern Mockingbirds, Brown Thrashers, Eastern Bluebirds, orioles, and crows), melons (blackbirds and crows), and strawberries (American Robins, Cedar Waxwings, and blackbirds) (Animal Damage Control Program 1994; Avery and Constanin 1997). All native migratory birds are protected by federal law under the Federal Migratory Bird Treaty Act, but under a special provision, certain species may be killed only when committing or about to commit depredations. These include blackbirds, grackles, cowbirds, and crows (Avery and Constanin 1997). Some producers might institute control measures and kill birds responsible for such losses, but precise data on the extent of this are not available. In Florida, a permit from the U.S. Fish and Wildlife Service is needed for a farmer to kill American Robins or Cedar Waxwings that cause crop damage. In 1997 4 such permits were issued to allow the killing of a total of 300

waxwings, although the exact number actually taken was not determined (Avery and Constantin 1997). Other birds such as House Sparrows and European Starlings are not protected by federal law and may be killed without obtaining a permit if they are causing depredation problems. There are no data available on the numbers of these birds killed each year, but they are estimated to be fairly low (Avery and Constantin 1997).

Information on public health hazards associated with 2 introduced species (House Sparrows and European Starlings) has been summarized by Weber (1979). He listed 26 such diseases for House Sparrows and 24 for European Starlings, including toxoplasmosis, chlamydiosis, encephalitis, and salmonellosis, which are of veterinary and human health importance. Histoplasmosis associated with blackbird roosts has been a problem in some states (Animal Damage Control Program 1994), but there is no specific information on this problem in relation to passeriforms in Florida. Several of these diseases have been discussed in this chapter under sections XIII, Viruses, XIV, Bacteria, and XVI, Protozoans. Concern over the contamination of foods and food processing plants and equipment by feces of passeriforms has been expressed and such contamination might be a problem in Florida, but the extent and significance of this is not clear. Even though *Salmonella typhimurium* has been isolated from 5 species of passeriforms in Florida (table 24.32), the epizootiological relationship of these findings to human infections in Florida has not been determined. In Florida *Escherichia coli* has been identified in Purple Martins, Brown-headed Cowbirds, and Yellow-rumped Warblers (table 24.32), but the identity of the strains involved and their pathogenicity were not determined. A pathogenic strain of *Escherichia coli* (O157:H7) that causes bloody diarrhea in humans has been identified from deer (Keene at al. 1997; Doster 1997). This organism also occurs in a number of domestic animals (cattle, sheep, pigs, and poultry) and may be present in other species of wildlife such as passeriforms, although this has not been determined.

XXI. Summary and conclusions

Excluding trauma, electrocution, brood parasitism, inclement weather, anomalies, and starvation, 209 different morbidity/mortality factors and disease agents are known from passeriforms in Florida. These include 13 organochlorines, 5 organophosphates, 5 metals, 2 biotoxins, 7 viruses, 9 bacteria, 2 fungi, 19 protozoans, 11 trematodes, 5 cestodes, 26 nematodes, 4 acanthocephalans, 10 ticks, 23 feather mites, 5 skin mites, 3 nasal mites, 5 chiggers, 44 chewing lice, 4 louse flies, 1 bird fly, 3 myiasis-producing flies, 1 flea, and 2 cimicid bugs. Probably the overall best-studied passeriform species in Florida are Blue Jays and Florida Scrub-Jays.

The most important mortality factor is trauma caused by striking manmade structures, especially radio and TV towers, by nocturnally migrating passeriforms. Long-term studies at 4 sites in Florida resulted in the documentation of mortality of more than 57,000 birds representing 123 species. Inclement weather, especially storms and periods of low temperature, has also resulted in large mortality events numbering in the hundreds and even thousands of birds. Organophosphates have been incriminated in 8 die-offs; most of these involved small numbers of birds, but in one incident more than 10,000 American Robins died. Four mortality incidents caused by ingestion of Brazilian pepper berries by American Robins have been observed; one of these involved 115 birds. Eastern equine encephalitis virus was suspected to have caused mortality in a population of Florida Scrub-Jays, but definitive evidence for this was not available. Mortality caused by West Nile virus infections may prove to be significant at the population level, especially in the Corvidae family. Limited information on parasitic infections of pathologic significance included data on 2 nematodes (*Dispharynx nasuta* and *Tetrameres paucispina*), 1 tick (*Ixodes brunneus*), 1 skin mite (*Knemidokoptes jamaicensis*), and 1 cimicid bug (*Oeciacus vicarius*). Starvation was determined to be significant in populations of Florida Scrub-Jays and Boat-tailed Grackles.

Several agents are of public health concern and include West Nile virus, Eastern equine encephalitis virus, St. Louis encephalitis (SLE) virus, Everglades virus, *Salmonella typhimurium, Toxoplasma gondii, Eutrombicula alfreddugesi, Ornithonyssus bursa, O. sylviarum, Ornithocoris pallidus,* and all of the ticks (except *Ixodes brunneus*). Except for SLE virus and Everglades virus, all of these are also of veterinary concern.

XXII. Literature cited

Adamec, R.E. 1976. The interaction of hunger and preying in the domestic cat (*Felis catus*): an adaptive hierarchy. *Behav. Biol.* 18:263–272.

Alexander, R.M. 1952. Tick paralysis. Report of a case in Florida. *J. Am. Med. Assoc.* 149:931–932.

Allander, K., and G.F. Bennett. 1992. Unpublished data. Memorial University of Newfoundland, St. Johns.

Altman, R.B., S.L. Clubb, G.M. Dorrestein, and K. Quesenberry (eds.). 1997. *Avian medicine and surgery.* W.B. Saunders, Philadelphia. 1,070 pp.

American Ornithologists' Union. 1998. *Check-list of North American birds.* 7th ed. American Ornithologists' Union, Washington, D.C. 829 pp.

Anderson, D.W., and J.J. Hickey. 1972. Eggshell changes in certain North American birds. *Proc. Int. Ornithol. Congr.* 15:514–540.

Animal Damage Control Program. 1994. Final environmental impact statement. Vol. 2. APHIS, USDA, Washington, D.C.

APHIS. 2000. APHIS, USDA, Veterinary Services Factsheet. 2 pp.

Ashdown, P. 1969. For the defense. *Agric. Chem.* 24:10–12.

Asterino, R. 1996. Diseases and care of wild passerines. In: *Diseases of cage and aviary birds.* 3d ed. W.J. Rosskopf, Jr. and R.W. Woerpel (eds.). Williams and Wilkins, Baltimore. pp. 965–980.

Atherton, L.S., and B.H. Atherton. 1988. Florida region. *Am. Birds* 42:60–63.

Atkinson, C.T., and C. van Riper 1991. Pathogenicity and epizootiology of avian haematozoa: *Plasmodium, Leucocytozoon,* and *Haemoproteus.* In: *Bird-parasite interactions: ecology, evolution, and behaviour.* J.E. Loye and M. Zuk (eds.). Oxford University Press, Oxford. pp. 19–48.

Atyeo, W.T., and N.L. Braasch. 1966. The feather mite genus *Proctophyllodes* (Sarcopiformes: Proctophyllodidae). *Bull. Univ. Nebr. State Mus.* 5:1–354.

Avery, M.L., and B.U. Constantin. 1997. Unpublished data. USDA, APHIS, Wildlife Services, Gainesville, Fla.

Avery, M.L., P.F. Springer, and N.S. Dailey. 1980. *Avian mortality at man-made structures: an annotated bibliography* (revised). U.S. Fish and Wildlife Service, Biological Services Program, National Power Plant Team, FWS/OBS-80/54.

Avise, J.C., and R.L. Crawford. 1981. A matter of lights and death. *Nat. Hist.* 90:6–14.

Bagg, A.M. 1957. The changing seasons. *Audubon Field Notes* 11:312–325.

———. 1969. The changing seasons. *Audubon Field Notes* 23:4–12.

Baker, J.H. 1940. The director reports to you. *Bird-Lore* 42:183–192.

Bancroft, G.T. 1986. Nesting success and mortality of the Boat-tailed Grackle. *Auk* 103:86–99.

Beier, J.C., J. Strandberg, M.K. Stoskopf, and C. Craft. 1981. Mortality in robins (*Turdus migratorius*) due to avian malaria. *J. Wildl. Dis.* 17:247–250.

Bennett, F.M. 1909. A tragedy of migration. *Bird-Lore* 11:110–113.

Bennett, G.F. 1993. Unpublished data. Memorial University of Newfoundland, St. Johns.

Bennett, G.F., J.R. Caines, and M.A. Bishop. 1988. Influence of blood parasites on the body mass of passeriform birds. *J. Wildl. Dis.* 24:339–343.

Bennett, G.F., M.A. Peirce, and R.W. Ashford. 1993. Avian Haematozoa: mortality and pathogenicity. *J. Nat. Hist.* 27:993–1001.

Beyer, W.N., G.H. Heinz, and A.W. Redmon-Norwood (eds.). 1996. *Environmental contaminants in wildlife. Interpreting tissue concentrations.* Lewis, Boca Raton, Fla. 494 pp.

Beyer, W.N., J.W. Spann, L. Sileo, and J.C. Franson. 1988. Lead poisoning in six captive avian species. *Arch. Environ. Contam. Toxicol.* 17: 121–130.

Bigler, W.J. 1997. Unpublished data. Florida Department of Health, Tallahassee.

Bigler, W.J., E.B. Lassing, E.E. Buff, A.L. Lewis, and G.L. Hoff. 1975. Arbovirus surveillance in Florida: wild vertebrate studies 1965–1974. *J. Wildl. Dis.* 11:348–356.

Bigler, W.J., E.B. Lassing, E.E. Buff, E.C. Prather, E.C. Beck, and G.L. Hoff. 1976. Endemic eastern equine encephalomyelitis in Florida: a twenty-year analysis, 1955–74. *Am. J. Trop. Med. Hyg.* 25:884–890.

Bishop, M.A., and G.F. Bennett. 1990. The haemoproteids of the avian families Corvidae (crows and jays) and Sturnidae (starlings and mynas) (Passeriformes). *Can. J. Zool.* 68:2251–2256.

Bishopp, F.C., and H.L. Trembley. 1945. Distribution and hosts of certain North American ticks. *J. Parasitol.* 31:1–54.

Blassingame, W. 1955. Tourist on a bender. *Fla. Wildl.* December: 31, 50.

Blatchley, W.S. 1928. The Mexican chicken bug in Florida. *Fla. Entomol.* 12:43–44.

Blus, L.J. 1996. DDT, DDD, and DDE in birds. In: *Environmental contaminants in wildlife. Interpreting tissue concentrations.* W.N. Beyer, G.H. Heinz, and A.W. Redmon-Norwood (eds.). Lewis, Boca Raton, Fla. pp. 49–71.

Boardman, E.T. 1929. Ticks of the Gainesville area. M.S. thesis, University of Florida, Gainesville. 57 pp.

Brennan, J.M., and G.W. Wharton. 1950. Studies on North American chiggers. 3. The subgenus *Neotrombicula. Am. Midl. Nat.* 44:153–197.

Brown, L.N. 1976. Prevalence of bill abnormalities in Florida brown thrashers. *Fla. Field Nat.* 4:11–13.

Brugger, K.E., A.P. Nol, and M.L. Avery. 1997. Unpublished data. USDA, APHIS, Wildlife Services, Gainesville, Fla.

Bunck, C.M., R.M. Prouty, and A.J. Krynitsky. 1987. Residues of organochlorine pesticides and polychloribiphenyls in starlings (*Sturnus vulgaris*), from the continental United States, 1982. *Environ. Monit. Assess.* 8:59–75.

Burns, B.J. 1952. Food of a family of great horned owls, *Bubo virginianus,* in Florida. *Auk* 69:86–87.

Burridge, M.J., W.J. Bigler, D.J. Forrester, and J.M. Hennemann. 1979. Serologic survey for *Toxoplasma gondii* in wild animals in Florida. *J. Am. Vet. Med. Assoc.* 175:964–967.

Cain, B.W., and C.M. Bunck. 1983. Residues of organochlorine compounds in starlings (*Sturnus vulgaris*), 1979. *Environ. Monit. Assess.* 3:161–172.

Case, L.D., H. Cruickshank, A.E. Ellis, and W.F. White. 1965. Weather causes heavy bird mortality. *Fla. Nat.* 38:29–30.

Centers for Disease Control. 1991. Eastern equine encephalomyelitis—Florida, eastern United States. *Mortality, Morbidity Weekly Report* 40:533.

Chamberlain, R.W., D.D. Stamm, and P.H. Coleman. 1963. Laboratory findings. *Fla. State Board Health Monogr. Ser.* 5:65–66.

Chamberlain, R.W., W.D. Sudia, T.H. Work, P.H. Coleman, V.F. Newhouse, and J.G. Johnston, Jr. 1969. Arbovirus studies in south Florida, with emphasis on Venezuelan equine encephalomyelitis virus. *Am. J. Epidemiol.* 89:197–210.

Christy, B.H. 1940. Mortality among tree swallows. *Auk* 57:404–405.

Clay, T. 1968. Contributions towards a revision of *Myrsidea* Waterston (Menoponoidae: Mallophaga). III. *Bull. Br. Mus. (Nat. Hist.) Entomol.* 21:205–243.

Clifford, C.M., G. Anastos, and A. Elbl. 1961. The larval ixodid ticks of the eastern United States (Acarina: Ixodidae). *Misc. Pub. Entomol. Soc. Am.* 2:213–237.

Cobb, M.P. 1946. Report of spring migration. *Fla. Nat.* 19:67–68.

Coleman, J.S., and S.A. Temple. 1993. Rural residents' free-ranging domestic cats: a survey. *Wildl. Soc. Bull.* 21:381–390.

Coleman, J.S., S.A. Temple, and S.R. Craven. 1997. Cats and wildlife. A conservation dilemma. University of Wisconsin–Extension website, Madison: http://wildlife.wisc.edu/extension/catfly3.htm.

Conti, L., R. Oliveri, and C. Blackmore. 2002. Unpublished data. Florida Department of Health, Tallahassee.

Cornish, T.E. 1997. Unpublished data. Southeastern Cooperative Wildlife Disease Study, Athens, Ga.

Cory, C.B. 1895. Florida notes. *Auk* 12:187–188.

Cox, J. 1997. Unpublished data. Florida Game and Fresh Water Fish Commission, Tallahassee.

Cram, E.B. 1930. Aberrant larvae of *Physocephalus sexalatus* in birds. *J. Parasitol.* 17:56.

Crawford, R.L. 1971. Predation on birds killed at a TV tower. *Oriole* 36:33–35.

————. 1974. Bird casualties at a northwest Florida TV tower: October 1966–September 1973. *Bull. Tall Timbers Res. Stn.* 18:1–27.

————. 1978. Autumn bird casualties at a northwest Florida TV tower: 1973–1975. *Wilson Bull.* 90:335–345.

————. 1980. Wind direction and the species composition of autumn TV tower kills in northwest Florida. *Auk* 97:892–895.

————. 1981a. Bird casualties at a Leon County, Florida TV tower: a 25-year migration study. *Bull. Tall Timbers Res. Stn.* 22:1–30.

————. 1981b. Weather, migration and autumn bird kills at a north Florida TV tower. *Wilson Bull.* 93:189–195.

————. 1981c. Bird kills at a lighted man-made structure: often on nights close to a full moon. *Am. Birds* 35:913–914.

Crawford, R.L., and R.T. Engstrom. 2001. Characteristics of avian mortality at a north Florida television tower: a 29-year study. *J. Field Ornithol.* 72:380–388.

Crawford, R.L., and H.M. Stevenson. 1984. Patterns of spring and fall migration in northwest Florida. *J. Field Ornithol.* 55:196–203.

Cruz, A., W. Post, J.W. Wiley, C.P. Ortega, T.K. Nakamura, and J.W. Prather. 1998. Potential impacts of cowbird range expansion in Florida. In: *Parasitic birds and their hosts. Studies in co-evolution.* S.I. Rothstein and S.K. Robinson (eds.). Oxford University Press, New York. pp. 313–336.

Cunningham, R.L. 1964a. Fall migration: Florida region. *Audubon Field Notes* 18:24–28.

————. 1964b. Fall migration: Florida region. *Audubon Field Notes* 18:442–446.

———. 1965a. Fall migration: Florida region. *Audubon Field Notes* 19:28–33.

———. 1965b. Predation on birds by the Cattle Egret. *Auk* 82:502–503.

Curnutt, J.L., A.L. Mayer, T.M. Brooks, L. Manne, O.L. Bass, Jr., D.M. Fleming, M.P. Nott, and S.L. Pimm. 1998. Population dynamics of the endangered Cape Sable seaside-sparrow. *Anim. Conserv.* 1:11–21.

Curry, R.L. 1990. Florida scrub jay kills a mockingbird. *Condor* 92:256–257.

Cutler, T.J., R.J. MacKay, P.E. Ginn, E.C. Greiner, R. Porter, C.A. Yowell, and J.B. Dame. 1999. Are *Sarcocystis neurona* and *Sarcocystis falcatula* synonymous? A horse infection challenge. *J. Parasitol.* 85:301–305.

Dame, J.B., R.J. MacKay, C.A. Yowell, T.J. Cutler, A. Marsh, and E.C. Greiner. 1995. *Sarcocystis falcatula* from passerine and psittacine birds: synonymy with *Sarcocystis neurona*, agent of equine protozoal myeloencephalitis. *J. Parasitol.* 81:930–935.

Davidar, P., and E.S. Morton. 1993. Living with parasites: prevalence of a blood parasite and its effect on survivorship in the purple martin. *Auk* 110:109–116.

Davis, J.W., R.C. Anderson, L. Karstad, and D.O. Trainer (eds.) 1971. *Infectious and parasitic diseases of wild birds.* Iowa State University Press, Ames. 344 pp.

Dean, T.F. 1999. Unpublished data. University of Massachusetts, Amherst.

Delany, M.F., and D.J. Forrester. 1997. Ticks from Florida grasshopper sparrows. *Fla. Field Nat.* 25:58–59.

Dennis, J.V. 1987. If you drink, don't fly. *Birder's World* 1:15–19.

Denton, J.F., and W.A. Krissinger. 1975. *Lyperosomum byrdi* sp. n. (Digenea: Dicrocoeliidae) from the rufous-sided towhee, *Pipilo erythrophthalmus* (L.), with a revised synopsis of the genus. *Proc. Helminthol. Soc. Wash.* 42:38–42.

Dhondt, A.A. 1997. Unpublished data. Cornell University, Ithaca, N.Y.

Doster, G.L. (ed.). 1997. *E. coli* O157:H7 . . . are deer involved? *SCWDS Briefs* 13:1–2.

Dreschel, T.W., R.B. Smith, and D.R. Breininger. 1990. Florida Scrub Jay mortality on roadsides. *Fla. Field Nat.* 18:82–83.

Dubey, J.P., and C.P. Beattie. 1988. *Toxoplasmosis of animals and man.* CRC, Boca Raton, Fla. 220 pp.

Dubey, J.P., and D.S. Lindsay. 1998. Isolation in immunodeficient mice of *Sarcocystis neurona* from opossum (*Didelphis virginiana*) faeces, and its differentiation from *Sarcocystis falcatula.* *Int. J. Parasitol.* 28:1823–1828.

Duncan, R.M. 1978–94. Unpublished data. National Wildlife Health Center, Madison, Wis.

Dusek, R.J., and D.J. Forrester. 2002. Blood parasites of American crows (*Corvus brachyrhynchos*) and fish crows (*Corvus ossifragus*) in Florida, U.S.A. *Comp. Parasitol.* 69:92–96.

Dusek, R.J., G.W. Foster, and D.J. Forrester. 2000. Unpublished data. University of Florida, Gainesville.

Dustman, E.H., W.E. Martin, R.G. Heath, and W.L. Reichel. 1971. Monitoring pesticides in wildlife. *Pestic. Monit. J.* 5:50–52.

Edscorn, J.B. 1974. The fall migration: Florida region. *Am. Birds* 28:40–44.

———. 1975. The fall migration: Florida region. *Am. Birds* 29:44–48.

———. 1980. The nesting season: Florida region. *Am. Birds* 34:887–889.

Eggeman, D.R., C.M. Ohme, and S. Halbert. 1996. Unpublished data. Florida Game and Fresh Water Fish Commission, Tallahassee.

Eidson, M., N. Komar, F. Sorhage, R. Nelson, T. Talbot, F. Mostachari, R. McLean, and the West Nile Virus Avian Mortality Surveillance Group. 2001. Crow deaths as a sentinel surveillance system for West Nile virus in the northeastern United States, 1999. *Emerg. Infect. Dis.* 7:615–620.

Eisler, R. 1988. Arsenic hazards to fish, wildlife, and invertebrates: a synoptic review. U.S. Fish and Wildlife Service, Biological Report 85(1.12). 92 pp.

Eleazer, T.H., and J.E. Hill. 1994. Highlands J virus—associated mortality in chukar partridges. *J. Vet. Diagn. Invest.* 6:98–99.

Evans, W. 1998. 150 pounds of Lapland longspurs. *Fla. Ornithol. Soc. Newsl.* 24:2–3.

———. 1999. Towerkill.com website: http://www.towerkill.com/FL.html.

Ewing, H.E. 1943. The American chiggers (larvae of the Trombiculinae) of the genus *Acariscus*, new genus. *Proc. Entomol. Soc. Wash.* 45:57–66.

Fain, A. 1965. A review of the family Epidermoptidae Trouessart parasitic on the skin of birds (Acarina: Sarcoptiformes). Part 1. *Verh. K. Vlaam. Acad. Wet. Lett. Schone Kunsten Belg. Kl. Wet.* 27:1–176.

Fairbrother, A. 1996. Cholinesterase-inhibiting pesticides. In: *Noninfectious diseases of wildlife.* 2d ed. A. Fairbrother, L.N. Locke, and G.L. Hoff (eds.). Iowa State University Press, Ames. pp. 52–60.

Fairbrother, A., L.N. Locke, and G.L. Hoff (eds.). 1996. *Noninfectious diseases of wildlife.* 2d ed. Iowa State University Press, Ames. 219 pp.

Fallis, A.M., and S.S. Desser. 1977. On species of *Leucocytozoon, Haemoproteus,* and *Hepatocystis.* In: *Parasitic protozoa.* Vol. 3. J.P. Kreier (ed.). Academic Press, New York. pp. 239–266.

Favorite, F.G. 1960. Some evidence of local origin of EEE virus in Florida. *Mosq. News* 20:87–92.

Fenger, C.K., D.E. Granstrom, A.A. Gajadhar, N.M. Williams, S.A. McCrillis, S. Stamper, J.L. Langemeier, and J.P. Dubey. 1997. Experimental infection of equine protozoal myeloencephalitis in horses using *Sarcocystis* sp. sporocysts from the opposum (*Didelphis virginiana*). *Vet. Parasitol.* 68:199–213.

Fischer, J.R., D.E. Stallknecht, M.P. Luttrell, A.A. Dhondt, and K.A. Converse. 1997. Mycoplasmal conjunctivitis in wild songbirds: the spread of a new contagious disease in a mobile host population. *Emerg. Inf. Dis.* 3:67–72.

Fisk, E.J. 1972. Injuries and disease observed at a south Florida banding station, 1971. *EBBA News* 35:62.

———. 1976. A deadly rain of robins. *Fla. Nat.* 49:13–14.

Fitzgerald, B.M. 1988. Diet of domestic cats and their impact on prey populations. In: *The domestication of the cat.* D.C. Turner and P. Bateson (eds.). Cambridge University Press, Cambridge. pp. 123–147.

Fitzpatrick, J.W., G.E. Woolfenden, and M.T. Kopeny. 1991. Ecology and development-related habitat requirements of the Florida scrub jay (*Aphelocoma coerulescens coerulescens*). Florida Game and Fresh Water Fish Commission, Nongame Wildlife Program, Technical Report 8. 49 pp.

Florida State Board of Health. 1969. *St. Louis encephalitis in Florida.* Ten years of research, surveillance and control programs. *Fla. State Board Health Monogr. Ser.* 12:1–125.

Fogarty, M.J., and W.M. Hetrick. 1973. Summer foods of cattle egrets in north central Florida. *Auk* 90:268–280.

Forrester, D.J. 1975–83. Unpublished data. University of Florida, Gainesville.

———. 1992. *Parasites and diseases of wild mammals in Florida.* University Press of Florida, Gainesville. 459 pp.

Forrester, D.J., and W.T. Atyeo. 1992. Unpublished data. University of Florida, Gainesville.

Forrester, D.J., and G.F. Bennett. 1997. Unpublished data. University of Florida, Gainesville.

Forrester, D.J., and G.W. Foster. 1999. Unpublished data. University of Florida, Gainesville.

Forrester, D.J., H.W. Kale II, R.D. Price, K.C. Emerson, and G.W. Foster. 1995. Chewing lice (Mallophaga) from birds in Florida: a listing by host. *Bull. Fla. Mus. Nat. Hist.* 39:1–44.

Forrester, D.J., and N.P. Thompson. 1975. Unpublished data. University of Florida, Gainesville.

Forsyth, M.H., J.G. Tully, T.S. Gorton, L. Hinckley, S. Frasca, Jr., H.J. Van Kruiningen, and S.J. Geary. 1996. *Mycoplasma sturni* sp. nov., from the conjunctiva of a European starling (*Sturnus vulgaris*). *Int. J. Syst. Bacteriol.* 46:716–719.

Foster, G.W., and T.F. Dean. 1997. Unpublished data. University of Florida, Gainesville.

Foster, G.W., T.F. Dean, and D.J. Forrester. 1997. Unpublished data. University of Florida, Gainesville.

Foster, G.W., and J.E. Keirans. 1994. Unpublished data. University of Florida, Gainesville.

Foster, G.W., and G.L. Slater. 1999. Unpublished data. University of Florida, Gainesville.

Foster, G.W., and T.A. Webber. 1998. Unpublished data. University of Florida, Gainesville.

Fowler, M.E. (ed.). 1978. *Zoo and wild animal medicine.* W.B. Saunders, Philadelphia. 951 pp.

———(ed.). 1986. *Zoo and wild animal medicine.* 2d ed. W.B. Saunders, Philadelphia. 1,127 pp.

Franson, J.C. 1996. Interpretation of tissue lead residues in birds other than waterfowl. In: *Environmental contaminants in wildlife. Interpreting tissue concentrations.* W.N. Beyer, G.H. Heinz, and A.W. Redmon-Norwood (eds.). Lewis, Boca Raton, Fla. pp. 265–279.

Frasca, S., Jr., L. Hinckley, M.H. Forsyth, T.S. Gorton, S.J. Geary, and H.J. Van Kruinigen. 1997. Mycoplasmal conjunctivitis in a European starling. *J. Wildl. Dis.* 33:336–339.

Frenkel, J.K. 1981. False-negative serologic tests for *Toxoplasma* in birds. *J. Parasitol.* 67:952–953.

Friend, M. 1999a. Avian cholera. In: *Field manual of wildlife diseases. General field procedures and diseases of birds.* M. Friend and J.C. Franson (eds.). U.S. Department of the Interior, U.S. Geological Survey, Biological Research Division, Information and Technology Report 1999-001. Washington, D.C. pp. 75–92.

———. 1999b. Salmonellosis. In: *Field manual of wildlife diseases. General field procedures and diseases of birds.* M. Friend and J.C. Franson (eds.). U.S. Department of the Interior, U.S. Geological Survey, Biological Research Division, Information and Technology Report 1999-001. Washington, D.C. pp. 99–109.

Friend, M., and J.C. Franson (eds.). 1999. *Field manual of wildlife diseases. General field procedures and diseases of birds.* U.S. Department of the Interior, U.S. Geological Survey, Biological Research Division, Information and Technology Report 1999-001. Washington, D.C. 426 pp.

Funderburg, J.B. 1968. Fungus infection kills parula warbler. *Fla. Nat.* 41:171.

Furness, R.W. 1996. Cadmium in birds. In: *Environmental contaminants in wildlife. Interpreting tissue concentrations.* W.N. Beyer, G.H. Heinz, and A.W. Redmon-Norwood (eds.). Lewis, Boca Raton, Fla. pp. 389–404.

Garvin, M.C. 1996. *Haemoproteus danilewskyi* Kruse, 1890, in blue jays in south-central Florida: epizootiology, pathogenicity and effect on host reproductive success. Ph.D. diss., University of Florida, Gainesville. 110 pp.

Garvin, M.C., J.F. Day, K.A. Tarvin, G.E. Woolfenden, and L.M. Stark. 1997. Unpublished data. Archbold Biological Station, Lake Placid, Fla.

Garvin, M.C., and K.A. Tarvin. 1996. Unpublished data. Archbold Biological Station, Lake Placid, Fla.

Garvin, M.C., K.A. Tarvin, G.E. Woolfenden, and L.A. Durden. 1996a. Unpublished data. Archbold Biological Station, Lake Placid, Fla.

Garvin, M.C., K.A. Tarvin, G.E. Woolfenden, and J.E. Keirans. 1993–94. Unpublished data. Archbold Biological Station, Lake Placid, Fla.

Garvin, M.C., K.A. Tarvin, G.E. Woolfenden, and J.W. Mertins. 1994–96. Unpublished data. Archbold Biological Station, Lake Placid, Fla.

Garvin, M.C., K.A. Tarvin, G.E. Woolfenden, and R.D. Price. 1996b. Unpublished data. Archbold Biological Station, Lake Placid, Fla.

George, J.E. 1961. The nasal mites of the genus *Ptilonyssus* (Acarina: Rhinonyssidae) occurring in some North American passeriform birds. *J. Kans. Entomol. Soc.* 34:105–132.

Gibbs, E.P.J., and T.F. Tsai. 1994. Eastern encephalitis. In: *Handbook of zoonoses.* Section B, Viral. 2d ed. G.W. Beran (ed.). CRC, Boca Raton, Fla. pp. 11–24.

Gore, J. 1988. Field notes. *Skimmer* 4:6.

Gothe, R., and A.W.H. Neitz. 1991. Tick paralyses: pathogenesis and etiology. In: *Advances in disease vector research.* K.F. Harris (ed.). Springer-Verlag, New York. pp. 177–204.

Gregson, J.D. 1973. *Tick paralysis. An appraisal of natural and experimental data.* Canadian Department of Agriculture Monograph 9, Ottawa, Ontario. 109 pp.

Greiner, E.C. 1997. Unpublished data. University of Florida, Gainesville.

Greiner, E.C., and Q.Y. Zeng. 2000. Unpublished data. University of Florida, Gainesville.

Greve, J.H. 1996. Parasites of the skin. In: *Diseases of cage and aviary birds.* 3d ed. W.J. Rosskopf, Jr., and R.W. Woerpel (eds.) Williams and Wilkins, Baltimore. pp. 623–626.

Guy, J.S., M.D. Ficken, H.J. Barnes, D.P. Wages, and L.G. Smith. 1993. Experimental infection of young turkeys with eastern equine encephalitis virus and Highlands J virus. *Avian Dis.* 37:389–395.

Hahn, D.C., R.D. Price, and P.C. Osenton. 2000. Use of lice to identify cowbird hosts. *Auk* 117:943–951.

Hallinan, T. 1922. Bird interference on high tension electric transmission lines. *Auk* 39:573.

Harrington, B.A., and J.J. Dinsmore. 1975. Mortality of transient cattle egrets at Dry Tortugas, Florida. *Bird-Banding* 46:7–14.

Hartup, B.K. 2001. Unpublished data. Cornell University, Ithaca, N.Y.

Hartup, B.K., A.A. Dhondt, K.V. Sydenstricker, W.M. Hochachka, and G.V. Kollias. 2001. Host range and dynamics of mycoplasmal conjunctivitis among birds in North America. *J. Wildl. Dis.* 37:72–81.

Harwood, R.F., and M.T. James. 1979. *Entomology in human and animal health.* Macmillan, New York. 548 pp.

Hayes, L.E. 1990. Unpublished data. Southeastern Cooperative Wildlife Disease Study, Athens, Ga.

Heinz, G.H. 1996. Selenium in birds. In: *Environmental contaminants in wildlife. Interpreting tissue concentrations.* W.N. Beyer, G.H. Heinz, and A.W. Redmon-Norwood (eds.). Lewis, Boca Raton, Fla. pp. 447–458.

Henderson, J.R., N. Karabatsos, T.T.C. Bourke, R.C. Wallis, and R.M. Taylor. 1962. A survey for arthropod-borne viruses in south-central Florida. *Am. J. Trop. Med. Hyg.* 11:800–810.

Hoff, G.L., W.J. Bigler, E.E. Buff, and E. Beck. 1978. Occurrence and distribution of western equine encephalomyelitis in Florida. *J. Am. Vet. Med. Assoc.* 172:351–352.

Howell, A.H. 1932. *Florida bird life.* Coward-McCann, New York. 579 pp.

Jennings, W.L., W.G. Winkler, D.D. Stamm, P.H. Coleman, and A.L. Lewis. 1969. Serologic studies of possible avian or mammalian reservoirs of St. Louis encephalitis virus in Florida. *Fla. Board Health Monogr. Ser.* 12:118–125.

Johnsgard, P.A. 1997. *The avian brood parasites: deception at the nest.* Oxford University Press, New York. 409 pp.

Johnson, H.N. 1960. Public health in relation to birds: arthropod-borne viruses. *Trans. N. Am. Wildl. Conf.* 25:121–133.

Johnston, D.W. 1974. Decline of DDT residues in migratory songbirds. *Science* 186:841–842.

———. 1975. Organochlorine pesticide residues in small migratory birds, 1964–73. *Pestic. Monit. J.* 9:79–88.

———. 1976. Organochlorine pesticide residues in uropygial glands and adipose tissue of wild birds. *Bull. Environ. Contam. Toxicol.* 16:149–155.

Johnston, D.W., and T.P. Haines. 1957. Analysis of mass bird mortality in October, 1954. *Auk* 74:447–458.

Johnston, R.E., and C.J. Peters. 1996. Alphaviruses. In: *Fields virology.* 3d ed. B.N. Fields, D.M. Knipe, and P.M. Howley (eds.). Lippincott-Raven, Philadelphia. pp. 843–898.

Kale, H.W. II. 1971. Regional reports: Florida region. *Am. Birds* 25:723–735.

———. 1976. The spring migration: Florida region. *Am. Birds* 30:828–832.

———. 1989. Florida birds. *Fla. Nat.* 62:14.

———. 1996. Dusky seaside sparrow. In: *Rare and endangered biota of Florida.* Vol. 5, *Birds.* J.A. Rodgers, Jr., H.W. Kale II, and H.T. Smith (eds.). University Press of Florida, Gainesville. pp. 7–12.

Kale, H.W. II, and W.L. Jennings. 1966. Movements of immature mockingbirds between swamp and residential areas of Pinellas County, Florida. *Bird-Banding* 37:113–120.

Karstad, L. 1971a. Arboviruses. In: *Infectious and parasitic diseases of wild birds.* J.W. Davis, R.C. Anderson, L. Karstad, and D.O. Trainer (eds.). Iowa State University Press, Ames. pp. 17–21.

———. 1971b. Pox. In: *Infectious and parasitic diseases of wild birds.* J.W. Davis, R.C. Anderson, L. Karstad, and D.O. Trainer (eds.). Iowa State University Press, Ames. pp. 34–41.

Keene, W.E., E. Sazie, J. Kok, D.H. Rice, D.D. Hancock, V.K. Balan, T. Zhao, and M.P. Doyle. 1997. An outbreak of *Escherichia coli* O157:H7 infections traced to jerky made from deer meat. *J. Am. Med. Assoc.* 277:1129–1231.

Keirans, J.E., and J.B. Kethley. 1982. Unpublished data. Georgia Southern University, Statesboro.

Kethley, J.B. 1970. A revision of the family Syringophilidae. *Contrib. Am. Entomol. Inst.* 5:1–76.

Kettle, D.S. 1995. *Medical and veterinary entomology.* 2d ed. CAB International, Wallingford, England. 725 pp.

Kinsella, J.M. 1974. Helminth fauna of the Florida scrub jay: host and ecological relationships. *Proc. Helminthol. Soc. Wash.* 41:127–130.

———. 1973–97. Unpublished data. University of Florida, Gainesville.

Kinsella, J.M., and G.W. Foster. 2000. Unpublished data. University of Florida, Gainesville.

Kinsella, J.M., G.W. Foster, and R.J. Dusek. 2000. Unpublished data. University of Florida, Gainesville.

Kinsella, J.M., G.W. Foster, M.G. Spalding, and T.F. Dean. 1997. Unpublished data. University of Florida, Gainesville.

Kinsella, J.M., and M.C. Garvin. 1993. Unpublished data. University of Florida, Gainesville.

Kinsella, J.M., and C.E. Winegarner. 1974. Notes on the life history of *Neomusca porteri* (Dodge), parasitic on nestlings of the great crested flycatcher in Florida. *J. Med. Entomol.* 11:633.

Kirmse, P. 1967. Pox in wild birds: an annotated bibliography. *Wildl. Dis.* 49:1–10.

Klukas, R.W., and L.N. Locke. 1970. An outbreak of fowl cholera in Everglades National Park. *J. Wildl. Dis.* 6:77–79.

Kropp, A., and J. Cox. 1996. Painted bunting conservation program—other surveys (7615). Annual performance report. Florida Game and Fresh Water Fish Commission, Tallahassee. 33 pp.

Lack, D. 1954. *The natural regulation of animal numbers.* Clarendon Press, Oxford. 343 pp.

Langridge, H.P. 1993. Florida region. *Am. Birds* 47:406–408.

Layne, J.N. 1997. Unpublished data. Archbold Biological Station, Lake Placid, Fla.

Layne, J.N., and K.C. Emerson. 1994. Unpublished data. Archbold Biological Station, Lake Placid, Fla.

Lehner, P.N., T.O. Boswell, and F. Copeland. 1967. An evaluation of the effects of the *Aedes aegypti* eradication program on wildlife in south Florida. *Pestic. Monit. J.* 1:29–34.

Leidy, J. 1886. Notices of nematoid worms. *Proc. Acad. Nat. Sci. Phila.* 38:308–313.

Ley, D.H., J.E. Berkhoff, and S. Levisohn. 1997. Molecular epidemiologic investigations of *Mycoplasma gallisepticum* conjunctivitis in songbirds by random amplified polymorphic DNA analyses. *Emerg. Inf. Dis.* 3:375–380.

Ley, D.H., S.J. Geary, J.E. Berkhoff, J.M. McLaren, and S. Levisohn. 1998. *Mycoplasma sturni* from blue jays and northern mockingbirds with conjunctivitis in Florida. *J. Wildl. Dis.* 34:403–406.

Lloyd, H.A., T.M. Jaouni, S.L. Evans, and J.F. Morton. 1977. Terpenes of *Schinus terebinthifolius. Phytochemistry* 16:1301–1302.

Locke, L.N. 1970. Unpublished data. National Wildlife Health Center, Madison, Wis.

Lockwood, J.L., K.H. Fenn, J.L. Curnutt, D. Rosenthal, K.L. Balent, and A.L. Mayer. 1997. Life history of the endangered Cape Sable seaside sparrow. *Wilson Bull.* 109:720–731.

Loftin, H. 1961. An annotated check-list of trematodes and cestodes and their vertebrate hosts from northwest Florida. *Q. J. Fla. Acad. Sci.* 23:302–314.

Logan, T.H. 1997. Florida's endangered species, threatened species and species of special concern. Official lists. Florida Game and Fresh Water Fish Commission, Tallahassee. 14 pp.

Lohrer, C.E., and F.E. Lohrer. 1984. Persistent predation by American swallow-tailed kites on eastern kingbirds. *Fla. Field Nat.* 12:42–43.

Lohrer, F.E. 1980. Eastern coachwhip predation on nestling blue jays. *Fla. Field Nat.* 8:28–29.

Lohrer, F.E., and C.E. Winegarner. 1980. Swallow-tailed kite predation on nestling mockingbird and loggerhead shrike. *Fla. Field Nat.* 8:47–48.

Loomis, E.C. 1978. External parasites. In: *Diseases of poultry.* 7th ed. M.S. Hofstad, B.W. Calnek, C.F. Helmboldt, W.M. Reid, and H.W. Yoder, Jr. (eds.). Iowa State University Press, Ames. pp. 667–704.

Lord, R.D., and C.H. Calisher. 1970. Further evidence of southward transport of arboviruses by migratory birds. *Am. J. Epidemiol.* 92:73–78.

Loye, J.E., and T.W. Regan. 1991. The cliff swallow bug *Oeciacus vicarius* (Hemiptera: Cimicidae) in Florida: ectoparasite implications for hole-nesting birds. *Med. Vet. Entomol.* 5:511–513.

Loye, J.E., and M. Zuk (eds.). 1991. *Bird-parasite interactions. Ecology, evolution, and behaviour.* Oxford University Press, New York. 406 pp.

Luby, J.P. 1994. St. Louis encephalitis. In: *Handbook of zoonoses.* Section B, Viral. 2d ed. G.W. Beran (ed.). CRC, Boca Raton, Fla. pp. 47–58.

Luttrell, M.P., L.H. Creekmore, and J.W. Mertins. 1996a. Avian tick paralysis caused by *Ixodes brunneus* in the southeastern United States. *J. Wildl. Dis.* 32:133–136.

Luttrell, M.P., J.R. Fischer, D.E. Stallknecht, and S.H. Kleven. 1996b. Field investigation of *Mycoplasma gallisepticum* infections in house finches (*Carpodacus mexicanus*) from Maryland and Georgia. *Avian Dis.* 40:335–341.

Luznar, S.L., M.L. Avery, J.B. Dame, R.J. MacKay, and E.C. Greiner. 2001. Development of *Sarcocystis falcatula* in its intermediate host, the brown-headed cowbird (*Molothrus ater*). *Vet. Parasitol.* 95:327–334.

Maehr, D.S., and J.R. Brady. 1986. Food habits of bobcats in Florida. *J. Mammal.* 67:133–138.

Maehr, D.S., and J.Q. Smith. 1988. Bird casualties at a central Florida power plant: 1982–1986. *Fla. Field Nat.* 16:57–64.

Maehr, D.S., A.G. Spratt, and D.K. Voigts. 1983. Bird casualties at a central Florida power plant. *Fla. Field Nat.* 11:45–49.

Mahnke, G., and L.E. Hayes. 1990. Unpublished data. Southeastern Cooperative Wildlife Disease Study, Athens, Ga.

Marfin, A.A., L.R. Petersen, M. Eidson, J. Miller, J. Hadler, C. Farello, B. Werner, G.L. Campbell, M. Layton, P. Smith, E. Bresnitz, M. Cartter, J. Scaletta, G. Obiri, M. Bunning, R.C. Craven, J.T. Roehrig, K.J. Julian, S.R. Hinten, D.J. Gubler, and the ArboNET Cooperative Surveillance Group. 2001. Widespread West Nile virus activity, eastern United States, 2000. *Emerg. Infect. Dis.* 7:730–735.

Martin, W.E. 1969. Organochlorine insecticide residues in starlings. *Pestic. Monit. J.* 3:102–114.

———. 1972. Mercury and lead residues in Starlings—1970. *Pestic. Monit. J.* 6:27–32.

Martin, W.E., and P.R. Nickerson. 1972. Organochlorine residues in Starlings—1970. *Pestic. Monit. J.* 6:33–40.

———. 1973. Mercury, lead, cadmium, and arsenic residues in Starlings—1971. Pestic. Monit. J. 7:67–72.

McClure, H.E. 1989. Epizootic lesions of house finches in Ventura County, California. *J. Field Ornithol.* 60:421–430.

McLean, R.G., and G.S. Bowen. 1980. Vertebrate hosts. In: *St. Louis encephalitis.* T.P. Monath (ed.). American Public Health Association, Washington, D.C. pp. 381–450.

McLean, R.G., G. Frier, G.L. Plarham, D.B. Francy, T.P. Monath, E.G. Campos, A. Therrien, J. Kerschner, and C.H. Calisher. 1985. Investigations of the vertebrate hosts of eastern equine encephalitis during an epizootic in Michigan, 1980. *Am. J. Trop. Med. Hyg.* 34:1190–1202.

Mertins, J.W. 2001. Unpublished data. USDA, APHIS, National Veterinary Services Laboratory, Ames, Iowa.

Mertins, J.W., and T.F. Dean. 1997–01. Unpublished data. USDA, APHIS, National Veterinary Services Laboratory, Ames, Iowa.

Mertins, J.W., and R.J. Dusek. 1999. Unpublished data. USDA, APHIS, National Veterinary Services Laboratory, Ames, Iowa.

Mertins, J.W., and G.W. Foster. 1995–99. Unpublished data. USDA, APHIS, National Veterinary Services Laboratory, Ames, Iowa.

Mertins, J.W., and K.E. Miller. 1997. Unpublished data. USDA, APHIS, National Veterinary Services Laboratory, Ames, Iowa.

Mertins, J.W., and G.L. Slater. 1999. Unpublished data. USDA, APHIS, National Veterinary Services Laboratory, Ames, Iowa.

Mertins, J.W., and M.G. Spalding. 2001. Unpublished data. USDA, APHIS, National Veterinary Services Laboratory, Ames, Iowa.

Mollhagen, T.R. 1976. A study of the systematics and hosts of the parasitic nematode genus *Tetrameres* (Habronematoidea: Tertameridae). Ph.D. diss., Texas Tech University, Lubbock. 546 pp.

Monath, T.P. (ed.). 1980. *St. Louis encephalitis.* American Public Health Association, Washington, D.C. 680 pp.

Monroe, B.L., Jr. 1957. A breeding record of the cowbird in Florida. *Fla. Nat.* 30:124.

Morrison, M.L. 1979. Loggerhead shrike eggshell thickness in California and Florida. *Wilson Bull.* 91:468–469.

Morton, J.F. 1978. Brazilian pepper—its impact on people, animals and the environment. *Econ. Bot.* 32:353–359.

Mount, R.H. 1981. The red imported fire ant, *Solenopsis invicta* (Hymenoptera: Formicidae), as a possible serious predator of some native southeastern vertebrates: direct observations and subjective impressions. *J. Ala. Acad. Sci.* 52:71–78.

Mullen, G.R., and L.J. Hribar. 1991. Tick paralysis observed in wild birds. Alabama Agricultural Experiment Station, *Highlights Agric. Res.* 38:12.

Mumme, R.L. 1987. Eastern indigo snake preys on juvenile Florida Scrub Jay. *Fla. Field Nat.* 15:53–54.

Navarrete, V. 1978. Berry banquet leaves 115 birds dead in south Dade. *Miami Herald,* March 2, 1978, sec. C.

Nayar, J.K. 1982. Bionomics and physiology of *Culex nigripalpus* (Diptera: Culicidae) of Florida: an important vector of diseases. Florida Agricultural Experiment Stations Bulletin 827, Gainesville. 73 pp.

Nesbitt, S.A., and F.H. White. 1974. A *Salmonella typhimurium* outbreak at a bird feeding station. *Fla. Field Nat.* 2:46–47.

NeSmith, C.C., and J. Cox. 1985. Red-winged blackbird nest usurpation by rice rats in Florida and Mexico. *Fla. Field Nat.* 13:35–36.

Nicholson, D.J. 1929. Black snakes as bird killers. *Wilson Bull.* 41:190.

Nickerson, P.R., and K.R. Barbehenn. 1975. Organochlorine residues in starlings, 1972. *Pestic. Monit. J.* 8:247–254.

Nores, A.I. 1995. Botfly ectoparasitism of the brown cacholote and the fire-gatherer. *Wilson Bull.* 107:734–738.

Nott, M.P., O.L. Bass, Jr., D.M. Fleming, S.E. Killeffer, N. Fraley, L. Manne, J.L. Curnutt, T.M. Brooks, R. Powell, and S.L. Pimm. 1998. Water levels, rapid vegetational changes, and the endangered Cape Sable seaside sparrow. *Anim. Conserv.* 1:23–32.

NWHC. 2000. Unpublished data. National Wildlife Health Center, Madison, Wis.

Oberheu, J.C. 1972. The occurrence of mirex in starlings collected in seven southeastern states—1970. *Pestic. Monit. J.* 6:41–42.

Palmer, R.S. 1962. *Handbook of North American birds.* Vol. 1. Yale University Press, New Haven, Conn. 567 pp.

Patnode, K.A. and D.H. White. 1991. Effects of pesticides on songbird productivity in conjunction with pecan cultivation in southern Georgia: a multiple-exposure experimental design. *Environ. Toxicol. Chem.* 10:1479–1486.

Paul, R.T. 1982. Florida region. *Am. Birds* 36:967–970.

———. 1984. Florida region. *Am. Birds* 38:1011–1013.

———. 1986. Florida region. *Am. Birds* 40:1193–1197.

———. 1987. Florida region. *Am. Birds* 41:1425–1428.

———. 1989. Florida region. *Am. Birds* 43:1307–1310.

Paulson, D.R., and H.M. Stevenson. 1962. Region reports: Florida region. *Audubon Field Notes* 16:398–404.

Payne, R.B. 1977. The ecology of brood parasitism in birds. *Annu. Rev. Ecol. Syst.* 8:1–28.

Peakall, D.B. 1996. Dieldrin and other cyclodiene pesticides in wildlife. In: *Environmental contaminants in wildlife. Interpreting tissue concentrations.* W.N. Beyer, G.H. Heinz, and A.W. Redmon-Norwood (eds.). Lewis, Boca Raton, Fla. pp. 73–97.

Pence, D.B. 1970. *Knemidokoptes jamaicensis* Turk from the rufous-sided towhee in Louisiana. *J. Med. Entomol.* 7:686.

———. 1975. Keys, species and host list, and bibliography for nasal mites of North American birds (Acarina: Rhinonyssinae, Turbinoptinae, Speleognathinae, and Cytoditidae). *Spec. Publ. Mus. Texas Tech. University* 8:3–148.

Pence, D.B., and S.D. Casto. 1976. Studies on the variation and morphology of the *Ptilonyssus "sairae"* complex (Acarina: Rhinonyssinae) from North American passeriform birds. *J. Med. Entomol.* 13:71–95.

Pence, D.B., R.A. Cole, K.E. Brugger, and J.R. Fischer. 1999. Epizootic podoknemidokoptiasis in American robins. *J. Wildl. Dis.* 35:1–7.

Pence, P.A. 1996. Developmental anomalies. In: *Diseases of cage and aviary birds.* 3d ed. W.J. Rosskopf, Jr. and R.W. Woerpel (eds.). Williams and Wilkins, Baltimore. pp. 464–469.

Peters, H.S. 1935. Two new biting lice (Mallophaga: Philopteridae) from birds in the United States. *Proc. Entomol. Soc. Wash.* 37:146–149.

———. 1936. A list of external parasites from birds of the eastern part of the United States. *Bird-Banding* 7:9–27.

Petersen, L.R., and J.T. Roehrig. 2001. West Nile virus: a reemerging global pathogen. *Emerg. Infect. Dis.* 7:611–614.

Petrak, M.L. (ed.). 1982. *Diseases of cage and aviary birds.* 2d ed. Lea and Febiger, Philadelphia. 680 pp.

Phillis, W.A., and H.L. Cromroy. 1972. A preliminary survey of ectoparasitic mites (Acari) of the house sparrow and mockingbird in Florida. *Fla. Entomol.* 55:256.

Phillis, W.A., H.L. Cromroy, and H.A. Denmark. 1976. New host and distribution records for the mite genera *Dermanyssus, Ornithonyssus* and *Pellonyssus* (Acari: Mesostigmata: Laelapoidea) in Florida. *Fla. Entomol.* 59:89–92.

Pitts, T.D., and L.E. Hayes. 1990. Eastern bluebird mortality due to tick paralysis. *Sialia* 12:3–4.

Post, W. 1981. The influence of rice rats *Oryzomys palustris* on the habitat use of the seaside sparrow *Ammospiza maritima*. *Behav. Ecol. Sociobiol.* 9:35–40.

Pranty, B. 1999. Unpublished data. Avon Park Air Force Range, Fla.

———. 2000. Three sources of Florida grass-

hopper sparrow mortality. *Fla. Field Nat.* 28:27–29.

Price, R.D. 1977. The *Menacanthus* (Mallophaga: Menoponidae) of the Passeriformes (Aves). *J. Med. Entomol.* 14:207–220.

Quist, C.F. 1992–93. Unpublished data. Southeastern Cooperative Wildlife Disease Study, Athens, Ga.

Ranck, F.M., Jr., J.H. Gainer, J.E. Hanley, and S.L. Nelson. 1965. Natural outbreak of eastern and western encephalitis in pen-raised chukars in Florida. *Avian Dis.* 9:8–20.

Rand, A.L. 1943. Bass eats yellowthroat, young stilts and young ducks. *Auk.* 60:95.

Rappole, J.H., S.R. Derrickson, and Z. Hubálek. 2000. Migratory birds and spread of West Nile virus in the western hemisphere. *Emerg. Infect. Dis.* 6:319–328.

Ratcliffe, D.A. 1967. Decrease in eggshell weight in certain birds of prey. *Nature* 215:208–210.

Rickard, L.G. 1984. Unpublished data. University of Florida, Gainesville.

———. 1985. Proventricular lesions associated with natural and experimental infections of *Dispharynx nasuta* (Nematoda: Acuariidae). *Can. J. Zool.* 63:2663–2668.

Ritchie, B.W., G.J. Harrison, and L.R. Harrison (eds.). 1994. *Avian medicine: principles and applications.* Wingers Publishing, Lake Worth, Fla. 1,384 pp.

Roberts, R.E., and C.V. Tamborski. 1993. Blackpoll warbler mortality during fall migration at a tower in southeastern Florida. *Fla. Field Nat.* 21:118–120.

Robertson, W.B., Jr. 1971. Region reports: Florida region. *Am. Birds* 25:44–49.

Robertson, W.B., Jr., and J.C. Ogden. 1969. Fall migration: Florida region. *Audubon Field Notes* 23:35–40.

Robertson, W.B., Jr., and G.E. Woolfenden. 1992. *Florida bird species. An annotated list.* Spec. Publ. 6, Florida Ornithological Society, Gainesville. 260 pp.

Rodgers, J.A., Jr., H.W. Kale II, and H.T. Smith (eds.). 1996. *Rare and endangered biota of Florida.* Vol. 5, *Birds.* University Press of Florida, Gainesville. 688 pp.

Rogers, A.J. 1953. A study of the ixodid ticks of

northern Florida, including the biology and life history of *Ixodes scapularis* Say (Ixodidae: Acarina). Ph.D. diss., University of Maryland, College Park. 191 pp.

Rosen, M.N. 1971. Avian cholera. In: *Infectious and parasitic diseases of wild birds.* J.W. Davis, R.C. Anderson, L. Karstad, and D.O. Trainer (eds.). Iowa State University Press, Ames. pp. 59–74.

Rosskopf, W.J., Jr., and R.W. Woerpel (eds.). 1996. *Diseases of cage and aviary birds.* 3d ed. Williams and Wilkins, Baltimore. 1,088 pp.

Rothstein, S.I., and S.K. Robinson. 1998. *Parasitic birds and their hosts. Studies in coevolution.* Oxford University Press, New York. 444pp.

Rubin, H.L. 1991. Report of the Animal Diseases Diagnostic Laboratory. Florida Department of Agriculture and Consumer Services, Division of Animal Industry, Kissimmee.

Ruff, F.J. 1940. Mortality among myrtle warblers near Ocala, Florida. *Auk* 57:405–406.

Sabrosky, C.W., G.F. Bennett, and T.L. Whitworth. 1989. *Bird blow flies (Protocalliphora) in North America (Diptera: Calliphoridae) with notes on the Palearctic species.* Smithsonian Institution Press, Washington, D.C. 312 pp.

Sanders, D.A. 1938. Hemorrhagic septicemia: the significance of *Pasteurella bovispetica* encountered in the blood of some Florida cattle. University of Florida Agricultural Experiment Station Bulletin 322:5–24.

Schaub, R. 1990. Predation on the eggs and nestlings of Florida scrub jays. M.S. thesis. University of South Florida, Tampa. 66 pp.

Schaub, R., R.L. Mumme, and G.E. Woolfenden. 1992. Predation on the eggs and nestlings of Florida scrub jays. *Auk* 109:585–593.

Schoech, S.J. 1999. Florida scrub-jay nestlings preyed upon by an eastern coachwhip. *Fla. Field Nat.* 27:57–58.

Seed, T.M., and R.D. Manwell. 1977. Plasmodia of birds. In: *Parasitic protozoa.* Vol. 3. J.P. Kreier (ed.). Academic Press, New York. pp. 311–357.

Sikes, P.J., and K.A. Arnold. 1986. Red imported fire ant (*Solenopsis invicta*) predation on cliff swallow (*Hirundo pyrrhonata*) nestlings in east-central Texas. *Southwest. Nat.* 31:105–106.

Simpson, C.F., and D.J. Forrester. 1973. Electron microscopy of *Sarcocystis* sp.: cyst wall, micropore, rhoptries, and an unidentified body. *Int. J. Parasitol.* 3:467–470.

Slosson, A.T. 1899. A tragic St. Valentine's day. *Bird-Lore* 1:45.

Smith, G.J. 1993. *Toxicology and pesticide use in relation to wildlife: organophosphorus and carbamate compounds.* C.K. Smoley, Boca Raton, Fla. 171 pp.

Smith, H.M. 1895. Mortality among white-bellied swallows in Florida. *Auk* 12:183–184.

Snyder, B. 1994. Unpublished data. Florida Department of Environmental Protection, Tallahassee.

Spalding, M.G. 1997. Unpublished data. University of Florida, Gainesville.

Spalding, M.G., and D.J. Forrester. 1987. Unpublished data. University of Florida, Gainesville.

Spalding, M.G., and R. McCracken. 2000. Unpublished data. University of Florida, Gainesville.

Spalding, M.G., and R.G. McLean. 1994. Unpublished data. University of Florida, Gainesville.

Spalding, M.G., R.G. McLean, M.C. Garvin, and G.F. Bennett. 1993. Unpublished data. University of Florida, Gainesville.

Spalding, M.G., J.W. Mertins, P.B. Walsh, K.C. Morin, D.E. Dunmore, and D.J. Forrester. 2002. Burrowing fly larvae associated with mortality of eastern bluebirds in Florida. *J. Wildl. Dis.* 38:776–783.

Spalding, M.G., and J.W. Mertins. 1999. Unpublished data. University of Florida, Gainesville.

Sprunt, A., Jr. 1954. *Florida bird life.* Coward-McCann, New York. 527 pp.

Stedman, A.F. 1981. Scaly-leg (knemidokoptiasis) infected red-winged blackbird banded at Casey Key, Florida. *Inland Bird Banding* 53:28–29.

Steele, J.H., and M.M. Galton. 1971. Salmonellosis. In: *Infectious and parasitic diseases of wild birds.* J.W. Davis, R.C. Anderson, L. Karstad, and D.O. Trainer (eds.). Iowa State University Press, Ames. pp. 51–58.

Steele, K.E., M.J. Linn, R.J. Schoepp, N. Komar,

T.W. Geisbert, R.M. Manduca, P.P. Calle, B.L. Raphael, T.L. Clippinger, T. Larsen, J. Smith, R.S. Lanciotti, N.A. Panella, and T.S. McNamara. 2000. Pathology of fatal West Nile virus infections in native and exotic birds during the 1999 outbreak in New York City, New York. *Vet. Pathol.* 37:208–224.

Steffee, N.D. 1968. Numbers of deformed brown thrashers (*Toxostoma rufum*) reported from scattered locations. *Fla. Nat.* 41:126–127.

Stevenson, H.M. 1956. Florida region. *Audubon Field Notes* 10:18–22.

———. 1958a. Florida region. *Audubon Field Notes* 12:21–26.

———. 1958b. Florida region. *Audubon Field Notes* 12:271–276.

———. 1958c. Florida region. *Audubon Field Notes* 12:344–348.

———. 1959. Florida region. *Audubon Field Notes* 13:21–25.

———. 1960. Florida region. *Audubon Field Notes* 14:379–383.

———. 1962. Florida region. *Audubon Field Notes* 16:21–25.

———. 1966. Florida region. *Audubon Field Notes* 20:20–35.

———. 1972. Florida region. *Am. Birds* 26:592–596.

———. 1973. Florida region. *Audubon Field Notes* 27:45–49.

———. 1976. *Vertebrates of Florida. Identification and distribution.* University Press of Florida, Gainesville. 607 pp.

Stevenson, H.M., and B.H. Anderson. 1994. *The birdlife of Florida.* University Press of Florida, Gainesville. 892 pp.

Stevenson, J.A. 1994. Unpublished data. Florida Department of Environmental Protection, Tallahassee.

Stewart, J.R., Jr. 1963. Central southern region. *Audubon Field Notes* 17:465–467.

Stewart, P.A. 1963. Abnormalities among brownheaded cowbirds trapped in Alabama. *Bird-Banding* 34:199–202.

———. 1971. Persistence of remains of birds killed on motor highways. *Wilson Bull.* 83:203–204.

Stickel, W.H., J.A. Galen, R.A. Dyrland, and D.L. Hughes. 1973. Toxicity and persistence of mirex in birds. In: *Pesticides and the environment: a continuing controversy.* W.B. Deichmann (ed.). Intercontinental Medical Book Corp., New York. pp. 437–467.

Stickel, W.H., L.F. Stickel, R.A. Dyrland, and D.L. Hughes. 1984. DDE in birds: lethal residues and loss rates. *Arch. Environ. Contam. Toxicol.* 13:1–6.

Stickel, W.H., L.F. Stickel, and J.W. Spann. 1969. Tissue residues of dieldrin in relation to mortality in birds and mammals. In: *Chemical fallout. Current research on persistent pesticides.* M.W. Miller and G.G. Berg (eds.). Charles C. Thomas, Springfield, Ill. pp. 174–204.

Stimson, L.A. 1964. Black-whiskered vireos in deciduous woods. *Fla. Nat.* 37:24–25.

———. 1966. A remarkable ten days at the Dry Tortugas. *Fla. Nat.* 39:149–150.

———. 1968. Letter to the editor. *Fla. Nat.* 41:127.

Stitt, W.T. 1968. More on deformed thrashers. *Fla. Nat.* 41:171.

Stoddard, H.L. 1931. *The bobwhite quail, its habits, preservation and increase.* Charles Scribner's Sons, New York. 559 pp.

———. 1962. Bird casualties at a Leon County, Florida TV tower, 1955–1961. *Bull. Tall Timbers Res. Stn.* 1:1–94.

Stoddard, H.L., and R.A. Norris. 1967. Bird casualties at a Leon County, Florida TV tower: an eleven-year study. *Bull. Tall Timbers Res. Stn.* 8:1–104.

Stone, W.B., B.L. Weber, and F.J. Parks. 1971. Morbidity and mortality of birds due to avian malaria. *N.Y. Fish Game J.* 18:62–63.

Strandtmann, R.W., B. McDaniel, and M.A. Price. 1963. New host and distribution records for *Pellonyssus passeri* Clark and Yunker (Acarina: Dermanyssidae). *J. Parasitol.* 49:38.

Straub, L. 1997. Unpublished data. Gainesville, Fla.

Strickland, R.K., R.R. Gerrish, J.L. Hourrigan, and G.O. Schubert. 1976. *Ticks of veterinary importance.* Agriculture Handbook 485, APHIS, USDA, Washington, D.C. 122 pp.

Sutton, G.M. 1945. Behavior of birds during a Florida hurricane. *Auk* 62:603–606.

Tanhauser, S.M., C.A. Yowell, T.J. Cutler, E.C. Greiner, R.J. Mackay, and J.B. Dame. 1999. Multiple DNA markers differentiate *Sarcocystis neurona* and *Sarcocystis falcatula*. *J. Parasitol.* 85: 221–228.

Taylor, D.J., A.L. Lewis, J.D. Edman, and W.L. Jennings. 1971. California group arboviruses in Florida. Host-vector relations. *Am. J. Trop. Med. Hyg.* 20:139–145.

———. 1981. No longer a big killer. *Fla. Nat.* 54:4–5, 10.

Taylor, W.K., and B.H. Anderson. 1972. Brown thrasher (*Toxostoma rufum*) with a bill abnormality in Seminole County. *Fla. Nat.* 45:129.

———. 1973. Nocturnal migrants killed at a central Florida TV tower, autumns 1969–1971. *Wilson Bull.* 85:42–51.

———. 1974. Nocturnal migrants killed at a central Florida TV tower, autumn 1972. *Fla. Field Nat.* 2:40–43.

Taylor, W.K., and M.A. Kershner. 1986. Migrant birds killed at the Vehicle Assembly Building (VAB, John F. Kennedy Space Center). *J. Field Ornithol.* 57:142–154.

——— 1991. Breeding biology of the great crested flycatcher in central Florida. *J. Field Ornithol.* 62:28–39.

Teglas, M.B. 1997. Unpublished data. Southeastern Cooperative Wildlife Disease Study, Athens, Ga.

Terrell, S.P. 1999. Unpublished data. University of Florida, Gainesville.

Terrell, S.P., M.G. Spalding, and D.J. Forrester. 1998. Diagnosis and management of a small epizootic of salmonellosis in passerines. *Fla. Field Nat.* 26:114–116.

Travis, B.V. 1941. Examinations of wild animals for the cattle tick *Boophilus annulatus microplus* (Can.) in Florida. *J. Parasitol.* 27:465–467.

Turner, E.C., Jr. 1971. Fleas and lice. In: *Infectious and parasitic diseases of wild birds*. J.W. Davis, R.C. Anderson, L. Karstad, and D.O. Trainer (eds.). Iowa State University Press, Ames. pp. 175–184.

Uhazy, L.S., and W.J. Arendt. 1986. Pathogenesis associated with philornid myiasis (Diptera: Muscidae) on nestling pearly-eyed thrashers (Aves: Mimidae) in the Luquillo rain forest, Puerto Rico. *J. Wildl. Dis.* 22:224–237.

Usinger, R.L. 1959. New species of Cimicidae (Hemiptera). *Entomologist* 92:218–222.

———. 1966. Monograph of Cimicidae (Hemiptera-Heteroptera). *Thomas Say Found.* 7:1–585.

Valkiunas, G. 1993. Pathogenic influence of haemosporidians and trypanosomes on wild birds in the field conditions: facts and hypotheses. *Ekologija* 1963:47–60.

van Riper III, C., S.G. van Riper, M.L. Goff, and M. Laird. 1986. The epizootiology and ecological significance of malaria in Hawaiian land birds. *Ecol. Monogr.* 56:327–344.

Verheijen, F.J. 1981. Bird kills at lighted man-made structures: not on nights close to a full moon. *Am. Birds* 35:251–254.

Vickery, P.D. 1996. Grasshopper sparrow (*Ammodramus savannarum*). *Birds N. Am.* 239: 1–23.

Walton, A.C. 1927. A revision of the nematodes of the Leidy Collections. *Proc. Acad. Nat. Sci. Phila.* 79:49–163.

Warner, R.E. 1968. The role of introduced diseases in the extinction of the endemic Hawaiian avifauna. *Condor* 70:101–120.

Wassmer, D.A., D.D. Guenther, and J.N. Layne. 1988. Ecology of the bobcat in south-central Florida. *Bull. Fla. State Mus. Biol. Sci.* 33:159–228.

Weatherhead, P.J., and G.F. Bennett. 1991. Ecology of red-winged blackbird parasitism by Haematozoa. *Can. J. Zool.* 69:2352–2359.

———. 1992. Ecology of parasitism of brown-headed cowbirds by Haematozoa. *Can. J. Zool.* 70:1–7.

Webber, T.A. 1980. Eastern coachwhip predation on juvenile scrub jays. *Fla. Field Nat.* 8:29–30.

Weber, J.A. 1940. Destruction of tree swallows. *Auk* 57:405.

Weber, W.J. 1979. *Health hazards from pigeons, starlings and English sparrows*. Thompson, Fresno, Cal. 138 pp.

Wehr, E.E. 1934. Coexistence of adult male and female *Tetrameres* (Nematoda: Spiruridae) in proventriculus of the Florida grackle. *Proc. Helminthol. Soc. Wash.* 1:50.

———. 1971. Nematodes. In: *Infectious and parasitic diseases of birds.* J.W. Davis, R.C. Anderson, L. Karstad, and D.O. Trainer (eds.). Iowa State University Press, Ames. pp. 185–233.

Weir, R.D. 1976. *Annotated bibliography of bird kills at man-made obstacles: a review of the state of the art and solutions.* Department of Fisheries and the Environment, Canadian Wildlife Service, Ontario Region.

Wellings, F.M., A.L. Lewis, and L.V. Pierce. 1972. Agents encountered during arboviral ecological studies: Tampa Bay area, Florida, 1963 to 1970. *Am. J. Trop. Med. Hyg.* 21:201–213.

Wentworth, E.J. 1988. Unpublished data. Southeastern Cooperative Wildlife Disease Study, Athens, Ga.

Wescott, P.W. 1970. Ecology and behavior of the Florida scrub jay. Ph.D. diss., University of Florida, Gainesville. 85 pp.

Weston, F.M. 1965. Survey of the birdlife of northwestern Florida. *Bull. Tall Timbers Res. Stn.* 5:1–147.

———. 1966. Bird casualties on the Pensacola Bay Bridge (1938–1949). *Fla. Nat.* 39:53–55.

Wheeler, W.B., D.P. Jowenaz, D.P. Wojcik, W.A. Banks, C.H. Van Middelem, C.S. Lofgren, S. Nesbitt, L. Williams, and R. Brown. 1977. Mirex residues in nontarget organisms after application of 10-5 bait for fire ant control, northeast Florida 1972–74. *Pestic. Monit. J.* 11:146–156.

Whisenhunt, M.H. 1959. Effect on wildlife of high powered insecticides. *Fla. Nat.* 32:73–74, 88.

White, D.H. 1976. Nationwide residues of organochlorines in starlings, 1974. *Pestic. Monit. J.* 10:10–17.

———. 1979. Nationwide residues of organochlorine compounds in starlings (*Sturnus vulgaris*), 1976. *Pestic. Monit. J.* 12:193–197.

White, D.H., J.R. Bean, and J.R. Longcore. 1977. Nationwide residues of mercury, lead, cadmium, arsenic, and selenium in starlings, 1973. *Pestic. Monit. J.* 11:35–39.

White, F.H., and D.J. Forrester. 1973–80. Unpublished data. University of Florida, Gainesville.

White, S.C., and G.E. Woolfenden. 1973. Breeding of the eastern bluebird in central Florida. *Bird-Banding* 44:110–123.

Williams, L.P., Jr. 1980. Salmonellosis. In: *Handbook of zoonoses.* Section A, *Bacterial, rickettsial, and mycotic diseases.* Vol. 2. J.H. Steele. (ed.). CRC, Boca Raton, Fla. pp. 11–34.

Williams, N.S. 1983. Three new species of *Brueelia* (Mallophaga: Philopteridae) from the Mimidae (Aves: Passeriformes). *Proc. Biol. Soc. Wash.* 96:599–604.

Williams, R.W. 1904. A preliminary list of the birds of Leon County, Florida. *Auk* 21:449–462.

Wilson, D.E., and N.J. Silvy. 1988. Impact of the imported fire ants on birds. In: *The red imported fire ant: assessment and recommendations.* Proceedings of the Governor's Conference. Sportsmen Conservationists of Texas, Austin. pp. 70–74.

Wilson, N.A., and H.W. Kale. 1972. Ticks collected from Indian River County, Florida (Acari: Metastigmata: Ixodidae). *Fla. Entomol.* 55:53–57.

Wilson, N.A., H.W. Kale, and W.W. Baker. 1990. Unpublished data. University of Northern Iowa, Cedar Falls.

Wobeser, G.A., and M.C. Finlayson. 1969. *Salmonella typhimurium* infection in house sparrows. *Arch. Environ. Health* 19:882–884.

Woolfenden, G.E. 1973. The winter season: Florida region. *Am. Birds* 27:603–607.

———. 1974. Nesting and survival in a population of Florida scrub jays. *Living Bird* 1973:25–49.

Woolfenden, G.E., and J.W. Fitzpatrick. 1984. *The Florida scrub jay. Demography of a cooperative-breeding bird.* Princeton University Press, Princeton, N.J. 406 pp.

———. 1991. Florida scrub jay ecology and conservation. In: *Bird population studies, relevance to conservation and management.* C.M. Perrine, J.D. Labreton, and G.J.M. Hirons (eds.). Oxford University Press, Oxford. pp. 542–565.

———. 1996. Florida scrub-jay. In: *Rare and endangered biota of Florida.* Vol. 5, *Birds.* J.A. Rodgers, Jr., H.W. Kale II, and H.T. Smith (eds.). University Press of Florida, Gainesville. pp. 267–280.

Wright, S.D., and D.J. Forrester. 1988. Unpublished data. University of Florida, Gainesville.

Yoder, H.W., Jr. 1991. *Mycoplasma gallisepticum* infection. In *Diseases of poultry.* 9th ed. B.W. Calnek, H.J. Barnes, C.W. Beard, W.M. Reid, and H.W. Yoder, Jr. (eds.). Iowa State University Press, Ames. pp. 198–212.

Yosef, R., and F.E. Lohrer. 1995. Loggerhead shrikes, red fire ants and red herrings? *Condor* 97:1053–1056.

Zahn, S.N., and S.I. Rothstein. 1999. Recent increase in male house finch plumage variation and its possible relationship to avian pox disease. *Auk* 116:35–44.

Miscellaneous Birds

I. Introduction

This chapter covers 6 additional orders of birds, represented in Florida (table 25.1) by 33 species, including 4 species of parakeets, 3 cuckoos, 2 anis, 3 nighthawks, 2 nightjars, 3 swifts, 7 hummingbirds, 1 kingfisher, and 9 woodpeckers. One of these, the Red-cockaded Woodpecker, has been declared threatened by the Florida Game and Fresh Water Fish Commission and endangered by the U.S. Fish and Wildlife Service (Logan 1997) and the Florida Committee on Rare and Endangered Plants and Animals (Rodgers et al. 1996). Three others have been classified as rare (the Mangrove Cuckoo and the Antillean Nighthawk) or species of special concern (the Hairy Woodpecker) by the Florida Committee on Rare and

Endangered Plants and Animals (Rodgers et al. 1996). The Carolina Parakeet, *Conuropsis carolinensis*, formerly occurred throughout Florida except for the Keys but is now considered extinct (Rodgers 1996), and an additional species, the Ivory-billed Woodpecker, *Campephilus principalis*, is probably in the same category (Jackson 1996).

Several treatises on captive birds contain information on some of the birds covered in the present chapter, particularly Budgerigars and other psittaciforms (Fowler 1978, 1986, 1993; Petrak 1982; Ritchie et al. 1994; Rosskopf and Woerpel 1996; Altman et al. 1997). Some of the data in these books may be applicable to free-ranging birds.

Table 25.1. Distribution, occurrence, and abundance of miscellaneous birds in Florida

Species		Range	Seasonal occurrence	Relative abundance
Psittaciformes				
Budgerigar	*Melopsittacus undulatus*	Central Gulf Coast[a]	Resident[b]	"Hundreds"
Monk Parakeet	*Myiopsitta monachus*	Statewide, esp. coasts	Resident[b]	>1,000
White-winged Parakeet	*Brotogeris versicolurus*	Dade Co., Tampa Bay	Resident[b]	"Hundreds"
Cuculiformes				
Black-billed Cuckoo	*Coccyzus erythropthalmus*	Statewide	Transient	Very rare to rare
Yellow-billed Cuckoo	*Coccyzus americanus*	Statewide	Transient	Common to abundant
		Statewide	Summer	Uncommon to fairly common
		Southern peninsula & Keys	Winter transient	Rare
Mangrove Cuckoo	*Coccyzus minor*	Dade Co., southern Gulf Coast	Resident	Rare to locally common
Smooth-billed Ani	*Crotophaga ani*	Southern 2/3 of peninsula	Resident	Rare to locally common
Groove-billed Ani	*Crotophaga sulcirostris*	Panhandle & northern peninsula	Winter transient	Rare
Caprimulgiformes				
Lesser Nighthawk	*Chordeiles acutipennis*	Panhandle coast, Dry Tortugas	Transient	Rare
Common Nighthawk	*Chordeiles minor*	Statewide	Transient	Common to abundant
Antillean Nighthawk	*Chordeiles gundlachii*	Southern peninsula	Summer	Uncommon
		Keys	Resident	Uncommon
Chuck-will's-widow	*Caprimulgus carolinensis*	Statewide	Transient	Fairly common
		Mainland, upper Keys	Summer	Uncommon to common
		Keys	Winter transient	Rare
Whip-poor-will	*Caprimulgus vociferus*	Statewide except Keys	Winter transient	Rare to fairly common
		Keys	Winter transient	Occasional
Apodiformes				
White-collared Swift	*Streptoprocne zonaris*	Panhandle	Winter transient	1 record
Chimney Swift	*Chaetura pelagica*	Statewide	Transient	Rare to fairly common

Antillean Palm-Swift	*Tachornis phoenicobia*	Key West	Summer	1 record
Buff-bellied Hummingbird	*Amazilia yucatanensis*	Panhandle	Winter transient	Very rare
Bahama Woodstar	*Calliphlox evelynae*	Southern Florida	Transient	5–6 reports
Ruby-throated Hummingbird	*Archilochus colubris*	Panhandle, northern peninsula	Resident	Uncommon to fairly common
		Southern peninsula	Resident	Rare
		Statewide	Transient	Uncommon to fairly common
Black-chinned Hummingbird	*Archilochus alexandri*	Panhandle, northern peninsula	Winter transient	Rare
Anna's Hummingbird	*Calypte anna*	Leon Co.	Winter transient	1 report
Calliope Hummingbird	*Stellula calliope*	Okaloosa Co.	Winter transient	1 report
Rufous Hummingbird	*Selasphorus rufus*	Statewide except Keys	Winter transient	Rare
Coraciiformes				
Belted Kingfisher	*Ceryle alcyon*	Statewide	Resident	Uncommon to common
Piciformes				
Red-headed Woodpecker	*Melanerpes erythrocephalus*	Panhandle, northern peninsula	Resident	Uncommon to fairly common
		Southern peninsula except Keys	Winter & Spring	Very rare
Golden-fronted Woodpecker	*Melanerpes aurifrons*	Pensacola	Transient	1 record
Red-bellied Woodpecker	*Melanerpes carolinus*	Statewide	Resident	Fairly common to common
Yellow-bellied Sapsucker	*Sphyrapicus varius*	Statewide	Transient & Winter	Uncommon to fairly common
Downy Woodpecker	*Picoides pubescens*	Statewide except Keys	Resident	Uncommon to locally common
Hairy Woodpecker	*Picoides villosus*	Statewide	Resident	Very rare to uncommon
Red-cockaded Woodpecker	*Picoides borealis*	Panhandle, northern peninsula	Resident	Rare[c]
Northern Flicker	*Colaptes auratus*	Statewide except lower Keys	Resident	Uncommon to locally common
		Lower Keys	Winter transient	Rare
Pileated Woodpecker	*Dryocopus pileatus*	Statewide except lower Keys	Resident	Uncommon to fairly common
		Lower Keys	Transient	Very rare

Sources: Robertson and Woolfenden (1992), Stevenson and Ancerson (1994), American Ornithologists' Union (1998).
a. Also reported in the Keys, the Panhandle, and other parts of mainland Florida.
b. An established exotic.
c. According to Robertson & Woolfenden (1992), "population much-reduced and increasingly fragmented."

Table 25.2. Miscellaneous birds killed by collisions with manmade structures in Florida, 1955–86

Order Species	No. of birds found dead/county			
	Leon[a]	Citrus[b]	Orange[c]	Brevard[d]
Psittaciformes	0	0	0	0
Cuculiformes				
Black-billed Cuckoo	47	0	2	0
Yellow-billed Cuckoo	475	15	38	20
Caprimulgiformes				
Common Nighthawk	22	0	1	0
Chuck-will's-widow	6	0	0	1
Apodiformes				
Chimney Swift	31	0	0	0
Coraciiformes				
Belted Kingfisher	2	0	0	0
Piciformes				
Red-headed Woodpecker	3	0	0	0
Red-bellied Woodpecker	2	0	0	0
Yellow-bellied Sapsucker	97	2	8	1
Red-cockaded Woodpecker	2	0	0	0
Northern Flicker	19	0	1	0

a. A 1,010-foot tower (WCTV); records span 25 years, 1955–80 (Crawford 1981).

b. An electrical power plant (Crystal River Generating Facility) with 2 chimneys, one 500 feet and one 604 feet tall; records span 4.5 years, 1982–86 (Maehr and Smith 1988).

c. A 1,483-foot TV tower (WDBO); records span 4 years, 1969–72 (Taylor and Anderson 1973, 1974).

d. A massive building 210 feet in height (Vehicle Assembly Building and the John F. Kennedy Space Center, Merritt Island); records span 12 years, 1970–81 (Taylor and Kershner 1986).

II. Trauma

As mentioned in chapter 24, Perching Birds, nocturnally migrating birds striking manmade structures such as radio and TV towers, lighthouses, smokestacks, and buildings account for significant numbers of casualties. Mortality data for 11 species representing 5 orders of miscellaneous birds are summarized in table 25.2. This information resulted from 4 long-term studies conducted in Leon, Citrus, Orange, and Brevard counties. Cuckoos, especially Yellow-billed Cuckoos, as well as Yellow-bellied Sapsuckers were the most common casualties, particularly in Leon County. The reader is referred to chapter 24 (Perching Birds) for more details on this phenomenon and a discussion of possibilities for mitigating this type of mortality. In addition, Weston (1966) reported a Yellow-bellied Sapsucker killed by striking cables associ-ated with the 3-mile Pensacola Bay Bridge (Santa Rosa and Escambia counties) in 1945. His data were collected over a 12-year period (1938–49), when he tabulated 740 specimens of dead birds representing 75 species, most of them passeriforms.

Mortality of birds due to being struck by vehicles on highways that pass through Florida state parks and recreation areas was studied over the 4-year period 1990–93. In table 25.3 data are presented on 12 species representing 5 avian orders. The most common casualties were the nighthawks and nightjars, especially Chuck-will's-widows, most likely because of their habit of foraging for insects over roads at night (Stevenson and Anderson 1994).

The eggs and young of Chimney Swifts occupying nests built in house chimneys are often destroyed because of accidental spilling of nest contents onto the hearths below. Baker

Table 25.3. Road-killed miscellaneous birds in Florida state parks and recreation areas, 1990–93

| Order | No. of birds reported killed | | | | |
Species	1990	1991	1992	1993	Totals
Psittaciformes					
"Yellow Parakeet"[a]	0	0	0	1	1
Cuculiformes					
Yellow-billed Cuckoo	1	3	4	1	9
Mangrove Cuckoo	1	0	3	0	4
Caprimulgiformes					
Common Nighthawk	0	5	0	4	9
Chuck-will's-widow	7	24	10	4	45
Whip-poor-will	1	0	0	2	3
Apodiformes	0	0	0	0	0
Coraciiformes					
Belted Kingfisher	0	1	0	1	2
Piciformes					
Red-headed Woodpecker	0	2	0	0	2
Red-bellied Woodpecker	0	1	1	2	4
Yellow-bellied Sapsucker	0	0	0	1	1
Northern Flicker	0	1	1	0	2
Pileated Woodpecker	0	1	0	1	2

Sources: Snyder (1994), Stevenson (1994)
a. Species unknown. Probably an escaped exotic, found in Talbot Islands GEOPark, Duval County, July 1993.

(1932) reported this type of mortality in Orange County in 1931.

III. Predation

Predation is no doubt a significant mortality factor for these birds in Florida as it is for passeriforms, although few such data are available. Shapiro (1979) reported Red-bellied Woodpeckers and House Sparrows as predators of clutches of Budgerigars nesting in Pasco County in 1978. She also observed a domestic cat and a Red-shouldered Hawk each prey on an adult Budgerigar. Pranty (2001) witnessed a Sharp-shinned Hawk capture a Budgerigar from a bird feeder in Pasco County in 1979. European Starlings have been observed removing the eggs of Northern Flickers from their nest cavities so as to utilize the cavities for their own nests (Stevenson and Anderson 1994). Woolfenden and Rohwer (1969) reported such an incident in Pasco County in 1966. Fragments of Northern Flickers were found in 3 of 1,098 pellets collected from a Northern Harrier roost in Leon County during the winter of 1925–26 (Stoddard 1931). Burns (1952) found remains of 1 Northern Flicker in a pellet from a Great Horned Owl nesting in Putnam County in 1949 (Burns 1952). Predation was suspected as the cause of death of an adult Red-cockaded Woodpecker from St. Marks National Wildlife Refuge (Wakulla County) in 1987 (Thomas 1987) and of 2 nestlings from Osceola National Forest (Baker County) in 1992 (Thomas 1992).

IV. Inclement weather

There are a number of records of mortality in Florida among cuculiforms, caprimulgiforms, apodiforms, and piciforms due to cold weather, storms, and drought conditions (table 25.4). As

Table 25.4. Mortality related to inclement weather among miscellaneous birds in Florida

Order Species	County/ area	Year(s)	Extent of mortality	Data source
Cold weather				
Cuculiformes				
Smooth-billed Ani	"Florida"	1977–80s	"numbers reduced"	Robertson & Woolfenden (1992)
	"Florida"	1986–91	"numbers decreased"	Stevenson & Anderson (1994)
Caprimulgiformes				
Chuck-will's-widow	"Everglades"	1940	"several"	Weber (1940)
	Brevard	1958	>1	Stevenson (1958)
	Leon	1958	1	Ibid.
Whip-poor-will	"Everglades"	1940	"several"	Weber (1940)
	Dade	1940	1	Christy (1940)
	Brevard	1958	1	Stevenson (1958)
Apodiformes				
Ruby-throated Hummingbird	St. Johns	1890	"a number"	Stevenson & Anderson (1994)
Severe storms / hurricanes				
Cuculiformes				
Yellow-billed Cuckoo	Brevard	1964	6	Case et al. (1965)
Apodiformes				
Ruby-throated Hummingbird	"Fla. Keys"	1909	"numbers"	Bennett (1909)
Piciformes				
Red-cockaded Woodpecker	BCNP[a]	1992	3 colonies	FGFWFC (1992)
Drought				
Cuculiformes				
Yellow-billed Cuckoo	"Fla. Keys"	1962	7	Paulson & Stevenson (1962)

a. Big Cypress National Preserve (Collier, Dade, and Monroe counties).

is the case with passeriforms (see chapter 24, Perching Birds), most of the records of such events are on the effects of cold weather. Populations of Smooth-billed Anis in particular appear to be limited seriously by severe freezes in southern Florida (Robertson and Woolfenden 1992; Stevenson and Anderson 1994). The oldest record dates back to 1890 when "a number" of Ruby-throated Hummingbirds succumbed during a late spring frost in St. Augustine (St. Johns County) (Stevenson and Anderson 1994).

The account by Bennett (1909) of mortality in the Florida Keys as a result of a violent thunderstorm in 1909 was discussed in chapter 24, Perching Birds. This event resulted in the death of many passeriforms, but also included a number of Ruby-throated Hummingbirds. Six Yellow-billed Cuckoos were identified as casualties of Hurricane Hilda in Brevard County during October 6–8, 1964 (Case et al. 1965), among 4,707 birds, mostly passeriforms, that died as a result of the storm. In August 1992 3 colonies of Red-cockaded Woodpeckers in the Lostman's Pines Unit of Big Cypress National Preserve were destroyed by Hurricane Andrew. These colonies represented the southernmost occurrence of this threatened species in Florida (FGFWFC 1992).

Table 25.5. Concentrations of mirex in stomach contents and carcasses of Red-bellied Woodpeckers before and after application of 10-5 bait for fire ant control in Duval County, Florida, 1972

| | | Concentrations (ppm wet weight)[a] | | | | | | |
| | | | | Post-treatment | | | | |
Sample	Pre-treatment	1 month	3 months	6 months	9 months	1 year	18 months	2 years
Carcass[b]	ND	0.21	NT	0.12	0.23	0.11	0.01	0.02
Stomach contents	0.02[c]	0.01	NT	0.53	NT	NT	ND	NT

Source: Wheeler et al. (1977).
ND = none detected, NT = not tested
a. Each value represents the mean concentration determined from 2 birds. Lowest limit of detection = 0.01 ppm.
b. Ground-up carcass minus the skin, feathers, beak, feet, and wings.
c. Mean concentration determined from 3 birds.

Paulson and Stevenson (1962) described a mortality event in the lower Florida Keys in May of 1962 in which large numbers of birds were found dead, including 7 Yellow-billed Cuckoos. It was postulated that a drought had caused a reduction in insect populations critical for these birds following their arrival after spring migration, when they need to replenish their energy stores.

V. Environmental contaminants

Data on environmental contaminants are limited to information from 4 species: Red-bellied Woodpeckers, Yellow-billed Cuckoos, Black-billed Cuckoos, and Common Nighthawks. The effects on various species of wildlife of a DDT spraying program for mosquito eradication in Dade County during 1965 were evaluated by Lehner et al. (1967). The brain tissue of 1 adult Red-bellied Woodpecker found sick in an area that had been sprayed 3 times was tested for residues of organochlorines. Residues of DDT, DDD, or o,p'-DDE were not detected in brain tissues of this bird, but a small amount (0.17 ppm wet weight) of p,p'-DDE was found.

Residues of mirex in carcasses and stomach contents were determined for 15 Red-bellied Woodpeckers before and after aerial application of 10-5 bait for control of red and black imported fire ants on the Dee Dot Ranch in Duval County during 1972–74 (Wheeler et al. 1977). Concentrations were all less than 1 ppm (table 25.5). Residues of DDT and dieldrin were detected in adipose tissues of cuckoos and nighthawks from several localities in Florida during 1970–73 and ranged from 0.01 to 2.4 ppm (table 25.6). Grocki and Johnston (1974) found higher concentrations in adult cuckoos compared with immatures. They also found that concentrations of DDT were higher in cuckoos collected in the autumn (mean = 1.1 ppm) than in those collected in the spring (mean = 0.42 ppm) and speculated that this difference might have been due to excretion of organochlorines during migration or during the winter months, or that it might have been due to the movement of pesticides from adipose tissue to the brain or muscles during times of fat utilization.

The significance of these residues to the health and reproduction of these birds is unknown. The reader is referred to the discussion of this topic in relation to songbird populations in chapter 24, Perching Birds, for further information and thoughts along this line.

VI. Neoplasia

No information is available on this topic.

Table 25.6. Organochlorine residues in adipose tissues of cuckoos and nighthawks from Florida, 1970–73

Species of bird Age/Sex[b]	Date	ppm wet weight[a]				
		DDE	DDD	DDT	ΣDDT	Dieldrin
Yellow-billed Cuckoo[c]						
A/M	Oct. 1971	1.1	0.26	1.0	2.4	0.06
A/M	Sept. 1970	0.18	0.04	0.15	0.37	0.21
A/U	Sept. 1970	0.25	0.07	0.27	0.59	0.01
I/M	Oct. 1972	0.06	ND	0.06	0.12	0.03
I/M	Oct. 1971	0.28	ND	0.20	0.48	ND
I/M	Oct. 1971	0.24	ND	0.68	0.92	ND
I/M	Oct. 1971	0.55	ND	0.50	1.06	0.01
I/M	Oct. 1971	0.10	ND	0.19	0.29	ND
U/M	May 1972	0.04	ND	ND	0.04	ND
U/M	Apr. 1973	0.13	ND	ND	0.13	0.02
U/F	May 1972	ND	ND	ND	ND	ND
U/F	Apr. 1973	0.41	ND	ND	0.41	ND
Black-billed Cuckoo[c,d]						
I/M	"Autumn"	0.36	0.03	0.03	0.42	0.01
Common Nighthawk						
A/U	Aug. 1970[e]	0.35	NG	NG	0.47	0.16
A/U	Sept. 1970[f]	1.13	NG	NG	2.25	ND

Source: Johnston (1975) except as otherwise noted.
NG = not given.
a. Limits of detection = 0.01 ppm.
b. A = adults, I = immatures, U = unknown; M = males, F = females.
c. Exact localities for the data from cuckoos were not given. Grocki and Johnston (1974) reported that the birds came from Leon and Orange counties, but data on cuckoos from these localities were not identified as such in the text. Data on cuckoos are for individual birds.
d. Data from Grocki and Johnson (1974).
e. Means of 6 birds from Indian River County.
f. Means of 2 birds from Orange County.

VII. Anomalies

Three instances of beak anomalies in Northern Flickers were reported by Taylor (1973). Two of these birds came from the Orlando area (Orange County) and 1 from Merritt Island (Brevard County) in 1972–73 (see figure 24.5 in chapter 24, Perching Birds). The abnormality in the bird from Merritt Island was not described in detail, but the 2 birds from Orlando had 2 different conditions. In one, a female, the bill was elongated and crossed; the maxilla was decurved and the mandible slightly recurved. The author determined by radiographic techniques that the abnormality was "due entirely to abnormal growth of the rhamphotheca." The other flicker, a male, had a normal maxilla, but its mandible was straight, compressed, and elongated. The total lengths of the mandibles from the female and the male were 49.0 and 51.0 mm, respectively, compared against a mean length of 35.5 mm for 4 normal Northern Flickers. The cause or causes of these anomalies are unknown as is the significance of this phenomenon to flicker populations. Bill anomalies have been reported also in a Brown Thrasher and a European Starling from Orange County (Stimson 1968; Stevenson and Anderson 1994; see chapter 24, Perching Birds).

VIII. Biotoxins

No information is available on this topic.

IX. Viruses

Five viruses have been identified in piciforms in Florida. These include poxvirus and 4 arboviruses: eastern equine encephalitis (EEE) virus, St. Louis encephalitis (SLE) virus, West Nile virus, and Highlands J virus. No information is available on viruses of free-ranging psittaciforms or coraciiforms in Florida.

Poxvirus infection was reported from a Northern Flicker from Indian River County in 1970 by Locke and Kale (1970), but details on the severity of the lesion(s) are not available. Kirmse (1967) studied poxvirus infections in Northern Flickers from southern Ontario in detail. He found that the pox lesions were persistent for up to 13 months and that the virus was extremely host specific. Attempts to infect other Northern Flickers were successful, but numerous wild birds, including 4 Yellow-bellied Sapsuckers and 1 Downy Woodpecker, and several species of domestic birds and mammals were refractory. Two of the naturally infected Northern Flickers observed by Kirmse (1967) developed unilateral keratitis and blindness, indicating that this disease can be serious. The impact on flicker populations in Florida, however, is unknown.

Serologic evidence of infections by EEE, SLE, and Highlands J viruses have been published (tables 25.7–25.9). Of these 3, EEE virus has been the most prevalent; seropositive birds included Red-headed Woodpeckers, Red-bellied Woodpeckers, a Pileated Woodpecker, and a Northern Flicker. There are no data to indicate that Everglades virus infects these birds (table 25.10). In addition, Taylor et al. (1971) found no evidence of Keystone viruses in Yellow-billed Cuckoos, Chuck-will's-widow, Northern Flickers, or Red-bellied Woodpeckers that were examined during their 2-year study of this virus in a freshwater swamp in Hillsborough County. West Nile virus has been identified using PCR and virus isolation techniques in Yellow-billed Cuckoos, Ruby-throated Hummingbirds, Red-headed Woodpeckers, and Yellow-bellied Sapsuckers found dead in 2001 (table 25.11). It was assumed they died because of the West Nile virus infections, but necropsies were not performed to prove this. For more details the reader is referred to the discussion in chapter 24, Perching Birds, on the role of birds in Florida as reservoirs of EEE virus, West Nile virus, and other arboviruses. The significance of these viral infections to populations of piciforms in Florida has not been studied.

X. Bacteria

Information on bacteria is limited to a small amount of data on 3 species of piciforms. The 6 species of bacteria listed in table 25.12 were found during necropsy examinations of a Pileated Woodpecker, 2 Red-cockaded Woodpeckers, and a Northern Flicker. The significance of these findings is unknown. The isolations of *Enterococcus faecalis* and *Klebsiella pneumoniae* from livers and *Escherichia coli* and *Serratia marcescens* from brains might be of concern, although they also could reflect postmortem changes. No data are available on the bacteria of free-ranging psittaciforms, cuculiforms, caprimulgiforms, apodiforms, or coraciiforms in Florida.

XI. Fungi

No information is available on this topic.

XII. Protozoans

Twenty-five percent of 196 Red-bellied Woodpeckers examined in the Apalachicola National Forest (Liberty and Wakulla counties) in 2000–01 were positive for *Haemoproteus velans*. No protozoans have been reported from psittaciforms, cuculiforms, caprimulgiforms, apodiforms, or coraciiforms in Florida. One

Table 25.7. Prevalence and distribution of eastern equine encephalitis virus infections in miscellaneous birds in Florida as determined by serologic tests

Family Species	County/site	Year(s)	No. of birds[a]			Data source
			Examined	Positive	%	
Cuculiformes						
Yellow-billed Cuckoo	"central Fla."	1958	(1)	(0)	(—)	Favorite (1960)
	Osceola	1993	(1)	(0)	(—)	Spalding & McLean (1994)
Species not given	"Florida"	1965–74	10	0	0	Bigler et al. (1975)
Caprimulgiformes						
Chuck-will's-widow	"central Fla."	1958	(2)	(0)	(—)	Favorite (1960)
	Highlands	1960–61	1	0	—	Henderson et al. (1962)
Whip-poor-will	Osceola	1994	3	0	—	Spalding & McLean (1994)
Species not given	"Florida"	1965–74	6	0	—	Bigler et al. (1975)
Piciformes						
Red-headed Woodpecker	"central Fla."	1958	(21)	(8)	(38)	Favorite (1960)
	Highlands	1960–61	18	0	—	Henderson et al. (1962)
Red-bellied Woodpecker	"central Fla."	1958	(12)	(2)	(17)	Favorite (1960)
	Highlands	1960–61	10	0	—	Henderson et al. (1962)
	Osceola	1993	(1)	(0)	(—)	Spalding & McLean (1994)
Red-cockaded Woodpecker	"central Fla."	1958	(1)	(0)	(—)	Favorite (1960)
Pileated Woodpecker	"central Fla."	1958	(3)	(1)	(—)	Favorite (1960)
Downy Woodpecker	Highlands	1960–61	1	0	—	Henderson et al. (1962)
	Osceola	1992–94	(14)	(0)	(—)	Spalding & McLean (1994)
Northern Flicker	"central Fla."	1958	(4)	(0)	(—)	Favorite (1960)
	Highlands	1960–61	5	0	—	Henderson et al. (1962)
	Osceola	1993	(1)	(1)	(—)	Spalding & McLean (1994)
Species not given	"Florida"	1965–74	51	2	4	Bigler et al. (1975)

a. Results from hemagglutination-inhibition tests; parentheses indicate serum neutralization tests.

Table 25.8. Prevalence and distribution of St. Louis encephalitis virus in miscellaneous birds from Florida as determined by serologic tests

Family Species	County/site	Year(s)	No. of birds[a]			Data source
			Examined	Positive	%	
Cuculiformes						
Yellow-billed Cuckoo	Osceola	1992–94	(1)	(0)	(—)	Spalding & McLean (1994)
Species not given	"Florida"	1965–74	10	0	0	Bigler et al. (1975)
Caprimulgiformes						
Chuck-will's-widow	Highlands	1960–61	1	0	—	Henderson et al. (1962)
Whip-poor-will	Osceola	1994	3	0	—	Spalding & McLean (1994)
Species not given	"Florida"	1965–74	6	1	—	Bigler et al. (1975)
	"Florida"	1965–74	(1)	(1)	(—)	Ibid.
Piciformes						
Red-bellied Woodpecker	Highlands	1960–61	10	0	0	Henderson et al. (1962)
	Osceola	1992–94	(1)	(0)	(—)	Spalding & McLean (1994)
Red-headed Woodpecker	Highlands	1960–61	18	0	0	Henderson et al. (1962)
Downy Woodpecker	Highlands	1960–61	1	0	—	Ibid.
	Osceola	1994	(5)	(0)	(—)	Spalding & McLean (1994)
Northern Flicker	Highlands	1960–61	5	0	—	Henderson et al. (1962)
	Osceola	1992–94	(1)	(0)	(—)	Spalding & McLean (1994)
Species not given	"Florida"	1965–74	51	1	2	Bigler et al. (1975)
	"Florida"	1965–74	(1)	(0)	(—)	Ibid.

a. Results from hemagglutination-inhibition tests; parentheses indicate serum neutralization tests.

Table 25.9. Prevalence and distribution of Highlands J virus in miscellaneous birds in Florida as determined by serologic tests, 1960–74

Family / Species	County/site	Year(s)	No. of birds[a] Examined	Positive	%	Data source
Cuculiformes						
Yellow-billed Cuckoo	"central Fla."	1958	(1)	(0)	(—)	Favorite (1960)
Caprimulgiformes						
Chuck-will's-widow	"central Fla."	1958	(2)	(0)	(—)	Ibid.
	Highlands	1960–61	1	0	—	Henderson et al. (1962)
Species not given	"Florida"	1965–74	1	0	—	Bigler et al. (1975)
Piciformes						
Red-bellied Woodpecker	"central Fla."	1958	(12)	(0)	(0)	Favorite (1960)
	Highlands	1960–61	10	0	0	Henderson et al. (1962)
Red-headed Woodpecker	"central Fla."	1958	(21)	(0)	(0)	Favorite (1960)
	Highlands	1960–61	18	0	0	Henderson et al. (1962)
Red-cockaded Woodpecker	"central Fla."	1958	(1)	(0)	(—)	Favorite (1960)
Pileated Woodpecker	"central Fla."	1958	(3)	(0)	(—)	Ibid.
Downy Woodpecker	Highlands	1960–61	1	0	—	Henderson et al. (1962)
Northern Flicker	"central Fla."	1958	(4)	(0)	(—)	Favorite (1960)
	Highlands	1960–61	5	0	—	Henderson et al. (1962)
Species not given	"Florida"	1965–74	27	1	4	Bigler et al. (1975)

a. Results from hemagglutination-inhibition tests; parentheses indicate serum neutralization tests.

Table 25.10. Prevalence and distribution of Everglades virus in miscellaneous birds in Florida as determined by serologic tests, 1960–74

Family / Species	County/site	Year(s)	No. of birds[a]			Data source
			Examined	Positive	%	
Cuculiformes						
Species not given	"Florida"	1965–74	10	0	0	Bigler et al. (1975)
Caprimulgiformes						
Chuck-will's-widow	Highlands	1960–61	1	0	—	Henderson et al. (1962)
Species not given	"Florida"	1965–74	5	0	—	Bigler et al. (1975)
Piciformes						
Red-bellied Woodpecker	Highlands	1960–61	10	0	0	Henderson et al. (1962)
Red-headed Woodpecker	Highlands	1960–61	18	0	0	Ibid.
Downy Woodpecker	Highlands	1960–61	1	0	—	Ibid.
Northern Flicker	Highlands	1960–61	5	0	—	Ibid.
Species not given	"Florida"	1965–74	24	0	0	Bigler et al. (1975)

a. Results from hemagglutination-inhibition tests.

Table 25.11. Distribution of West Nile virus infections in Apodiformes, Piciformes, and Cuculiformes from Florida as determined by PCR and virus isolation studies in 2001

Order	Species	County	No. infected
Cuculiformes	Yellow-billed Cuckoo	Clay	2
		Citrus	1
		St. Lucie	1
Apodiformes	Ruby-throated Hummingbird	Bay	1
		Suwannee	1
Piciformes	Red-headed Woodpecker	Duval	1
		Leon	1
		Okaloosa	1
		Wakulla	1
	Yellow-bellied Sapsucker	Clay	1

Source: Conti et al. (2002).

Yellow-billed Cuckoo, 1 Chuck-will's-widow, 7 Downy Woodpeckers, and 4 Northern Flickers were examined for blood protozoans (table 25.13). All smears were negative.

XIII. Helminths

Nothing is known about the helminths of apodiforms, but limited information is available on the helminths of psittaciforms, cuculiforms, caprimulgiforms, and coraciiforms in Florida. Kinsella (1997–99) identified the helminths of 2 Yellow-billed Cuckoos, 1 Chuck-will's-widow, 1 Common Highthawk, 7 Belted Kingfishers, 2 Red-bellied Woodpeckers, and 3 Pileated Woodpeckers (table 25.14). The Red-bellied Woodpecker, from which 1 species of trematode, 2 species of cestodes, 8 species of nematodes, and 1 species of acanthocephalan were identified, is the best-studied species.

Nothing is known about the pathogenicity of these helminths, with the possible exception of *Dispharynx nasuta* and the unidentified species of *Tetrameres*, both discussed in chap-

Table 25.12. Bacteria identified from piciforms[a] from Florida

Species of bird Bacteria	County	Year	Tissue or organ cultured	Data source
Pileated Woodpecker				
Enterobacter sp.	Columbia	1978	Large intestine	White & Forrester (1980)
Red-cockaded Woodpecker				
Enterococcus faecalis	Baker	1992	Liver	Duncan & Thomas (1992)
Enterococcus sp.	Wakulla	1986	Intestine	Duncan & Thomas (1986)
Escherichia coli	Wakulla	1986	Intestine, brain	Ibid.
Klebsiella pneumoniae	Baker	1992	Liver	Duncan & Thomas (1992)
Serratia marcescens	Wakulla	1986	Brain	Duncan & Thomas (1986)
Northern Flicker				
Pseudomonas sp.	Dade	1975	Large intestine	White & Bigler (1975)

a. Isolations made from the one bird examined.

Table 25.13. Cuculiforms and piciforms from Florida examined for blood protozoans, 1975–93

Family Host species	County	Year	No. of birds examined[a]	Data source
Cuculiformes				
Yellow-billed Cuckoo	Alachua	1977	1	Forrester & Bennett (1997)
Caprimulgiformes				
Chuck-will's-widow	Alachua	1983	1	Ibid.
Piciformes				
Red-bellied Woodpecker	Alachua	1983	1	Ibid.
	Osceola	1993	1	Spalding et al. (1993)
	Wakulla/ Liberty	2000–01	196[b]	Schrader et al. (2002)
Downy Woodpecker	Osceola	1992	2	Ibid.
	Osceola	1993	5	Ibid.
Northern Flicker	Alachua	1975	1	Forrester & Bennett (1997)
	Alachua	1979	1	Ibid.
	Alachua	1985	1	Ibid.
	Osceola	1993	1	Spalding et al. (1993)

a. All birds were examined by the blood smear technique and found negative for blood protozoans, unless otherwise noted.
b. 49 were infected with *Haemoproteus velans*.

ter 24, Perching Birds, to which the reader is referred for more information.

XIV. Arthropods

As can be seen from table 25.15, there is some information about ticks, mites, chewing lice, dipterans, and cimicid bugs from psittaciforms, cuculiforms, caprimulgiforms, apodiforms, coraciiforms, and piciforms.

The piciforms are the best studied, with records of 12 species of arthropods on them. These include 1 tick, 4 mites, 6 chewing lice, and 1 bird fly from woodpeckers, sapsuckers, and flickers. Nothing is known about the effects of these infestations on birds in Florida.

XV. Summary and conclusions

Little is known about the parasites and diseases of free-ranging psittaciforms in Florida. There is a small amount of information on predation of Budgerigars and on trauma resulting from collision with vehicles as well as data on 3 species of parasites from 10 Monk Parakeets. Both trauma and inclement weather appear to be of some importance as mortality factors in cuculiforms, caprimulgiforms, apodiforms, and piciforms. Trauma also might be of some significance to coraciiforms. Bill anomalies have been reported from piciforms, but their significance is not known.

In addition to these mortality factors, there are limited data on environmental contaminants, viruses, bacteria, protozoans, helminths, and parasitic arthropods from cuculiforms, caprimulgiforms, apodiforms, coraciiforms, with piciforms having the most information, namely 38 disease agents, including 2 organochlorines, 5 viruses, 6 bacteria, 1 protozoan, 12 helminths, and 12 arthropods. The following numbers of disease agents (in parentheses) were identified from various avian groups: cuculiforms (14), coraciiforms (11), caprimulgiforms (6), and apodiforms (3). The significance of these agents on populations of these birds in Florida is not known.

Table 25.14. Helminths reported from miscellaneous birds in Florida

Family / Host species	Helminth (location in host)[a]	County/area	Year	No. of birds Examined	Positive	%	Intensity Mean	Range
Psittaciformes								
Monk Parakeet								
Cestoda	*Raillietina* sp.[b] (SI)	Dade	1999	5	1	—	NG	NG
		Pinellas	1999	1	0	—	NG	NG
Nematoda	*Ascaridia* sp.[c] (SI)	Dade	1999	5	0	—	NG	NG
		Pinellas	1999	1	1	—	NG	NG
Cuculiformes								
Yellow-billed Cuckoo								
Trematoda	*Brachylecithum tuberculatum* (LV)	Alachua	1972	2	1	—	15	—
Cestoda	*Paruterina similis* (SI)	Alachua	1972	2	2	—	15	5–25
Nematoda	*Cyrnea piayae* (SI)	Alachua	1972	2	2	—	11	7–15
	Microtetrameres sp. (PR)	Alachua	1972	2	1	—	4	—
	Subulura reclinata (VE)	Alachua	1972	2	1	—	20	—
	Viguiera coccyzae (GL)	Alachua	1972	2	2	—	3	315–50
Caprimulgiformes								
Chuck-will's-widow								
Cestoda	*Raillietina* sp.[d] (SI)	Alachua	1983	1	1	—	50	—
Common Nighthawk								
Nematoda	*Subulura* sp. (CE)	Alachua	1970	1	1	—	2	—
Coraciiformes								
Belted Kingfisher								
Trematoda	*Amphimerus elongatus* (LV)	Collier	1994	1	1	—	1	—
	Cathaemasia reticulata (AS)	Levy	1971	5	1	—	13	—
	Crassiphiala bulboglossa (SI)	Levy	1971	5	3	—	26	1–75
		Collier	1994	1	1	—	9	—
	Uvulifer semicircumcisus (SI)	Levy	1971	5	2	—	18	10–25
Nematoda	*Aproctella stoddardi* (AS)	Levy	1971	5	4	—	8	1–14
	Aviculariella alcyona (GZ)	Levy	1971	5	4	—	9	2–15
		Collier	1994	1	1	—	10	—
	Capillaria sp. (ES)	Levy	1971	5	1	—	2	—
	Gnathostoma sp. (larva) (PR)	Collier	1994	1	1	—	1	—

	Locality	Year					
Monopetalonema alcedinis (BC)	Alachua	1982	1	1	—	3	—
	Collier	1994	1	1	—	6	—
	Levy	1971	5	2	—	2	1–2
Tetrameres sp. (PR)	Levy	1971	5	1	—	1	—

Piciformes[e]

Red-bellied Woodpecker

	Locality	Year					
Trematoda							
Brachylaima fuscatum (SI)	Liberty[f]	1997–2001	74	1	1	9	—
Cestoda							
Orthoskrjabinia rostellata (SI)	Leon, Wakulla[f]	1997–2001	74	3	41	—	—
Raillietina centuri (SI)	Liberty[f]	1997–2001	74	1	1	1	—
Nematoda							
Acuaria sp. (GZ)	Leon[f]	1997–2001	74	1	1	1	—
Aproctella stoddardi (BC)	Leon, Liberty, Wakulla[f]	1997–2001	74	8	11	1	1–2
Diplotriaena americana (BC)	Leon[f]	1997–2001	74	1	1	3	—
Dispharynx nasuta (PR)	Alachua	1974	1	1	—	16	—
	Wakulla[f]	1997–2001	74	1	1	3	—
Oxyspirura pusillae (EY)	Highlands	1973	1	1	—	2	—
	Wakulla[f]	1997–2001	41	1	2	6	—
Procyrnea pileata (GZ,GL)	Liberty, Wakulla[f]	1997–2001	74	6	8	4	1–10
Pseudoaprocta samueli (BC)	Leon, Liberty	1997–2001	74	7	9	6	1–24
Tridentocapillaria tridens (SI)	Leon, Wakulla[f]	1997–2001	74	2	3	1	1–2
Acanthocephala							
Mediorhynchus centurorum (SI)	Leon, Liberty, Wakulla[f]	1997–2001	74	8	11	5	1–22

Pileated Woodpecker

	Locality	Year					
Cestoda							
Raillietina sp. (SI)	"Florida"	NG	NG	1[g]	—	3	—
	Alachua	1994	1	1	—	168	—
Nematoda							
Capillaria longistriata (SI)	Columbia	1978	1	1	—	1	—

Source: Kinsella (1997–98), unless otherwise indicated.

NG = not given.

a. Location in host: AS = air sacs, BC = body cavity, CE = caeca, ES = esophagus, EY = eye, GL = under gizzard lining, GZ = gizzard, LV = liver, PR = proventriculus, SI = small intestine, VE = ventriculus.

b. Greiner and Kinsella (2001).

c. Greiner (2001).

d. Stiles and Hassall (1894) reported "*Taenia* sp." from a Chuck-will's-widow from Florida (date and exact locality not given) that may have been *Raillietina* sp.

e. Rickard (1984) examined the proventriculi of one Downy Woodpecker and one Red-headed Woodpecker from Alachua County in 1984. Both were negative for helminths.

f. Foster et al. (2002).

g. Identified as *Taenia vexata* by Joseph Leidy and found in the U.S. National Parasite Collection under USNPC#-05174 with no date of collection or more specific locality than Florida given. Examined by J. M. Kinsella in 1999 and determined to be an unidentifiable species of *Raillietina* (Kinsella 1999).

Table 25.15. Parasitic arthropods reported from miscellaneous birds in Florida

Host species Species of arthropod	County/site	Year(s)	Data source
Psittaciformes			
Monk Parakeet			
Chewing lice			
Paragoniocotes fulvofasciatus	Dade	1999	Greiner & Mertins (2000)
Cuculiformes			
Black-billed Cuckoo			
Chewing lice			
Cuculicola erthryopthalmus	Dade	1928	Emerson (1964), Forrester et al. (1995)
	St. Lucie	1967	Forrester et al. (1995)
Yellow-billed Cuckoo			
Chewing lice			
Cuculiphilus snodgrassi	Indian River	1970–73	Ibid.
	Lee	1954, 1982	Ibid.
Cuculoecus coccygii	Dade	1917	Ibid.
	Lee	1986	Ibid.
	St. Lucie	1973	Ibid.
Caprimulgiformes			
Chuck-will's-widow			
Hippoboscids			
Olfersia sp.	St. Johns	NG	Johnson (1895, 1913, 1922)
Pseudolynchia brunnea	Dade	NG	Bequaert (1955)
	Franklin	NG	Ibid.
	Monroe	NG	Ibid.
	St. Johns[a]	NG	Ibid.
Whip-poor-will			
Hippoboscids			
Pseudolynchia brunnea	Dade	NG	Ibid.
Common Nighthawk			
Hippoboscids			
Pseudolynchia brunnea	Brevard	NG	Ibid.
Apodiformes			
Chimney Swift			
Chewing lice			
Dennyus dubius	Dade	1926	Forrester et al. (1995)
	Lee	1984	Ibid.
	Leon	1936	Ibid.
Cimicid bugs			
Cimexopsis nyctalis	Leon	1936	Lee (1955), Usinger (1966)
Coraciiformes			
Belted Kingfisher			
Feather mites			
Proterothrix sp.[b]	Collier	1994	Foster & Mertins (1999)
Mites (hypopi)			
Alcedinectes alcyon	Glades	1994	Pence & Gray (1996)
Piciformes			
Red-bellied Woodpecker			

(continued)

Table 25.15. *(continued)*

Host species Species of arthropod	County/site	Year(s)	Data source
Ticks			
Amblyomma maculatum	"Florida"[c]	1936–37	Travis (1941)
Chewing lice			
Penenirmus auritus	Alachua	1953	Forrester et al. (1995)
	"Shell Point"	1936	Ibid.
Yellow-bellied Sapsucker			
Chewing lice			
Penenirmus auritus	Dade	1917	Ibid.
	Pinellas	1930	Ibid.
	NG	NG	Peters (1936)
Hairy Woodpecker			
Chewing lice			
Brueelia straminea	Dade	1918	Forrester et al. (1995)
Menacanthus pici	Dade	1918	Ibid.
Penenirmus auritus	Dade	1926	Ibid.
Red-cockaded Woodpecker			
Mites			
Analges sp.[e]	Polk	1994	Garvin & Mertins (1994)
Chewing lice			
Brueelia straminea	Collier	1937	Forrester et al. (1995)
	Polk	1994	Ibid.
Northern Flicker			
Chewing lice			
Picicola porisma	Dade	1916	Ibid.
	Indian River	1977	Ibid.
Penenirmus jungens	Dade	1916–17	Ibid.
Pileated Woodpecker			
Feather mites			
Capitolichus sp.[e]	Alachua	1994	Forrester & Mertins (1994)
Mesalgoides sp.[e]	Alachua	1994	Ibid.
Pterotrogus sp.[e]	Alachua	1994	Ibid.
Chewing lice			
Menacanthus pici	Indian River	1919	Forrester et al. (1995)
Picicola marginatulus	Alachua	1994	Ibid.
"Woodpecker"[d]			
Bird Flies			
Carnus floridensis	Wakulla	1992	Grimaldi (1997)

NG = not given.

a. This record is from St. Augustine (St. Johns County); according to Wilson et al. (1990), may be the same record referred to by Johnson (1895, 1913, 1922) as *Olfersia* sp.

b. An undescribed species (Foster and Mertins 1999).

c. Exact locality not given; in either Orange, Osceola, or Collier County. Travis (1941) also examined 1 Chuck-will's-widow and 2 Red-cockaded Woodpeckers and found no ticks on them.

d. Species not given.

e. Probably an undescribed species, according to Mertins (2001).

XVI. Literature cited

Altman, R.B., S.L. Clubb, G.M. Dorrestein, and K. Quesenberry (eds.). 1997. *Avian medicine and surgery.* W.B. Saunders, Philadelphia. 1,070 pp.

American Ornithologists' Union. 1998. *Check-list of North American birds.* 7th ed. American Ornithologists' Union, Washington, D.C. 829 pp.

Baker, M.F. 1932. Dooryard birds in Florida. *Fla. Nat.* 6:3–5.

Bennett, F.M. 1909. A tragedy of migration. *Bird-Lore* 11:110–113.

Bequaert, J.C. 1955. The hippoboscidae or louseflies (Diptera) of mammals and birds. Part 2. Taxonomy, evolution and revision of American genera and species. *Entomol. Am.* 36:233–416.

Bigler, W.J., E.B. Lassing, E.E. Buff, A.L. Lewis, and G.L. Hoff. 1975. Arbovirus surveillance in Florida: wild vertebrate studies 1965–1974. *J. Wildl. Dis.* 11:348–356.

Burns, B.J. 1952. Food of a family of great horned owls, *Bubo virginianus,* in Florida. *Auk* 69:86–87.

Case, L.D., H. Cruickshank, A.E. Ellis, and W.F. White. 1965. Weather causes heavy bird mortality. *Fla. Nat.* 38:29–30.

Christy, B.H. 1940. Mortality among tree swallows. *Auk* 57:404–405.

Conti, L., R. Oliveri, and C. Blackmore. 2002. Unpublished data. Florida Department of Health, Tallahassee.

Crawford, R.L. 1981. Bird casualties at a Leon County, Florida TV tower: a 25-year migration study. *Bull. Tall Timbers Res. Stn.* 22:1–30.

Duncan, R.M., and N.J. Thomas. 1986–92. Unpublished data. National Wildlife Health Center, Madison, Wis.

Emerson, K.C. 1964. A new species of Mallophaga from the black-billed cuckoo. *Entomol. News* 75:69–71.

Favorite, F.G. 1960. Some evidence of local origin of EEE virus in Florida. *Mosq. News* 20:87–92.

[FGFWFC] Florida Game and Fresh Water Fish Commission. 1992. Effects of Hurricane Andrew on fish and wildlife in south Florida. Unpublished. Tallahassee. 19 pp.

Forrester, D.J., and G.F. Bennett. 1997. Unpublished data. University of Florida, Gainesville.

Forrester, D.J., H.W. Kale II, R.D. Price, K.C. Emerson, and G.W. Foster. 1995. Chewing lice (Mallophaga) from birds in Florida: a listing by host. *Bull. Fla. Mus. Nat. Hist.* 39:1–44.

Forrester, D.J., and J.W. Mertins. 1994. Unpublished data. University of Florida, Gainesville.

Foster, G.W., J.M. Kinsella, E.L. Walters, M.S. Schrader, and D.J. Forrester. 2002. Parasitic helminths of red-bellied woodpeckers (*Melanerpes carolinus*) from the Apalachicola National Forest in Florida. *J. Parasitol.* (in press).

Foster, G.W., and J.W. Mertins. 1999. Unpublished data. University of Florida, Gainesville.

Fowler, M.E. (ed.). 1978. *Zoo and wild animal medicine.* W.B. Saunders, Philadelphia. 951 pp.

———(ed.). 1986. *Zoo and wild animal medicine.* 2d ed. W.B. Saunders, Philadelphia. 1,127 pp.

———(ed.). 1993. *Zoo and wild animal medicine: current therapy.* 3d ed. W.B. Saunders, Philadelphia. 617 pp.

Garvin, M.C., and J.W. Mertins. 1994. Unpublished data. University of Florida, Gainesville.

Greiner, E.C. 2001. Unpublished data. University of Florida, Gainesville.

Greiner, E.C., and J.M. Kinsella. 2001. Unpublished data. University of Florida, Gainesville.

Greiner, E.C., and J.W. Mertins. 2000. Unpublished data. University of Florida, Gainesville.

Grimaldi, D. 1997. The bird flies, genus *Carnus*: species revision, generic relationships, and a fossil *Meoneura* in amber (Diptera: Carnidae). *Am. Mus. Novit.* 3190:1–30.

Grocki, D.R.J., and D.W. Johnston. 1974. Chlorinated hydrocarbon pesticides in North American cuckoos. *Auk* 91:186–188.

Henderson, J.R., N. Karabatsos, T.T.C. Bourke, R.C. Wallis, and R.M. Taylor. 1962. A survey for arthropod-borne viruses in south-central Florida. *Am. J. Trop. Med. Hyg.* 11:800–810.

Jackson, J.A. 1996. Ivory-billed woodpecker. In: *Rare and endangered biota of Florida.* Vol. 5, *Birds.* J.A. Rodgers, H.W. Kale II, and H.T. Smith. (eds.). University Press of Florida. Gainesville. pp. 103–112.

Johnson, C.W. 1895. Diptera of Florida. *Proc. Acad. Nat. Sci. Phila.* pp. 303–340.

———. 1913. Insects of Florida. I. Diptera. *Bull. Am. Mus. Nat. Hist.* 32:37–90.

———. 1922. Notes on distribution and habits of some of the bird-flies, Hippoboscidae. *Psyche* 29:79–85.

Johnston, D.W. 1975. Organochlorine pesticide residues in small migratory birds, 1964–73. *Pestic. Monit. J.* 9:79–88.

Kinsella, J.M. 1997–99. Unpublished data. University of Florida, Gainesville.

Kirmse, P. 1967. Host specificity and long persistence of pox infection in the flicker (*Colaptes auratus*). *J. Wildl. Dis.* 3:14–20.

Lee, D.R. 1955. New locality records for *Cimexopsis nyctalis* list, the chimney swift bug (Hemiptera: Cimicidae). *Bull. Brooklyn Entomol. Soc.* 50:51–52.

Lehner, P.N., T.O. Boswell, and F. Copeland. 1967. An evaluation of the effects of the *Aedes aegypti* eradication program on wildlife in south Florida. *Pestic. Monit. J.* 1:29–34.

Locke, L.N., and H.W. Kale. 1970. Unpublished data. National Wildlife Health Center, Madison, Wis.

Logan, T.H. 1997. Florida's endangered species, threatened species and species of special concern. Official lists. Florida Game and Fresh Water Fish Commission, Tallahassee. 14 pp.

Maehr, D.S., and J.Q. Smith. 1988. Bird casualties at a central Florida power plant: 1982–1986. *Fla. Field Nat.* 16:57 61.

Mertins, J.W. 2001. Unpublished data. USDA, APHIS, National Veterinary Services Laboratory, Ames, Iowa.

Paulson, D.R., and H.M. Stevenson. 1962. Region reports: Florida region. *Audubon Field Notes* 16:398–404.

Pence, D.B., and P.N. Gray. 1996. Redescription of *Alcedinectes alcyon* (Acari: Hypoderatidae) from the belted kingfisher (Aves: Coraciiformes; Alcedinidae). *J. Med. Entomol.* 33:772–776.

Peters, H.S. 1936. A list of external parasites from birds of the eastern part of the United States. *Bird-Banding* 7:9–27.

Petrak, M.L. (ed.). 1982. *Diseases of cage and aviary birds.* 2d ed. Lea and Febiger, Philadelphia. 680 pp.

Pranty, B. 2001. *The Budgerigar in Florida: Rise and fall of an exotic psittacid.* *No. Amer. Birds* 55: 389–397.

Rickard, L.G. 1984. Unpublished data. University of Florida, Gainesville.

Ritchie, B.W., G.J. Harrison, and L.R. Harrison (eds.). 1994. *Avian medicine: principles and applications.* Wingers, Lake Worth, Fla. 1384 pp.

Robertson, W.B., Jr., and G.E. Woolfenden. 1992. *Florida bird species. An annotated list.* Spec. Publ. 6, Florida Ornithological Society, Gainesville. 260 pp.

Rodgers, J.A. 1996. Carolina parakeet. In: *Rare and endangered biota of Florida.* Vol. 5, *Birds.* J.A. Rodgers, H.W. Kale II, and H.T. Smith. (eds.). University Press of Florida. Gainesville. pp. 4–6.

Rodgers, J.A., H.W. Kale II, and H.T. Smith (eds.). 1996. *Rare and endangered biota of Florida.* Volume 5, *Birds.* University Press of Florida, Gainesville. 688 pp.

Rosskopf, W.J., Jr., and R.W. Woerpel (eds.). 1996. *Diseases of cage and aviary birds.* 3d ed. Williams and Wilkins, Baltimore. 1,088 pp.

Schrader, M.S., E.L. Walters, E.C. Greiner, and F.C. James. 2002. Seasonal prevalence of a haematozoan parasite of the red-bellied woodpecker and its association with host condition. *Auk* (in press).

Shapiro, A.E. 1979. Status, habitat utilization, and breeding biology of the feral Budgerigar *(Melopsittacus undulatus)* in Florida. M.S. thesis, University of Florida, Gainesville. 62 pp.

Snyder, B. 1994. Unpublished data. Florida Department of Environmental Protection, Tallahassee.

Spalding, M.G., and R.G. McLean. 1994. Unpublished data. University of Florida, Gainesville.

Spalding, M.G., R.G. McLean, M.C. Garvin, and G.F. Bennett. 1993. Unpublished data. University of Florida, Gainesville.

Stevenson, H.M. 1958. Region report: Florida region. *Audubon Field Notes* 12:271–276.

Stevenson, H.M., and B.H. Anderson. 1994. *The birdlife of Florida.* University Press of Florida, Gainesville. 892 pp.

Stevenson, J.A. 1994. Unpublished data. Florida Department of Environmental Protection, Tallahassee.

Stiles, C.W., and A. Hassall. 1894. A preliminary

catalogue of the parasites contained in the collections of the U.S. Bureau of Animal Industry, U.S. Army Medical Museum, Biological Department of the University of Pennsylvania (Coll. Leidy) and in Coll. Stiles and Coll. Hassall. *Vet. Mag.* 1:245–253, 331–354.

Stimson, L.A. 1968. Letter to the editor. *Fla. Nat.* 41:127.

Stoddard, H.L. 1931. *The bobwhite quail, its habits, preservation and increase.* Charles Scribner's Sons, New York. 559 pp.

Taylor, D.J., A.L. Lewis, J.D. Edman, and W.L. Jennings. 1971. California group arboviruses in Florida. Host-vector relations. *Am. J. Trop. Med. Hyg.* 20:139–145.

Taylor, W.K. 1973. Bill deformities in two Florida birds. *Fla. Field Nat.* 1:6–8.

Taylor, W.K., and B.H. Anderson. 1973. Nocturnal migrants killed at a central Florida TV tower, autumns 1969–1971. *Wilson Bull.* 85:42–51.

———. 1974. Nocturnal migrants killed at a Central Florida TV tower, autumn 1972. *Fla. Field Nat.* 2:40–43.

Taylor, W.K., and M.A. Kershner. 1986. Migrant birds killed at the Vehicle Assembly Building (VAB, John F. Kennedy Space Center). *J. Field Ornithol.* 57:142–154.

Thomas, N.J. 1987–92. Unpublished data. National Wildlife Health Center, Madison, Wis.

Travis, B.V. 1941. Examinations of wild animals for the cattle tick *Boophilus annulatus microplus* (Can.) in Florida. *J. Parasitol.* 27:465–467.

Usinger, R.L. 1966. Monograph of Cimicidae (Hemiptera-Heteroptera). *Thomas Say Found.* 7:1–585.

Weber, J.A. 1940. Destruction of tree swallows. *Auk* 57:405.

Weston, F.M. 1966. Bird casualties on the Pensacola Bay bridge (1938–1949). *Fla. Nat.* 39:53–55

Wheeler, W.B., D.P. Jowenaz, D.P. Wojcik, W.A. Banks, C.H. Van Middelem, C.S. Lofgren, S. Nesbitt, L. Williams, and R. Brown. 1977. Mirex residues in nontarget organisms after application of 10-5 bait for fire ant control, northeast Florida 1972–74. *Pestic. Monit. J.* 11:146–156.

White, F.H., and W.J. Bigler. 1975. Unpublished data. University of Florida, Gainesville.

White, F.H., and D.J. Forrester. 1980. Unpublished data. University of Florida, Gainesville.

Wilson, N.A., H.W. Kale, and W.W. Baker. 1990. Unpublished data. University of Northern Iowa, Cedar Falls.

Woolfenden, G.E., and S.A. Rohwer. 1969. Breeding birds in a Florida suburb. *Bull. Fla. State Mus.* 13:1–83.

CHAPTER TWENTY-SIX

Summary and Conclusions

The first rule of intelligent tinkering is to save all the parts.
—Aldo Leopold, *A Sand County Almanac, and Sketches Here and There*

A large number and variety of diseases, parasites, and other morbidity and mortality factors have been found in free-ranging birds of Florida; however, the record is incomplete. Even though some information is available on 311 (68%) of the 457 species of birds found in Florida, the coverage of each species is variable. Very little is known about some birds and nothing at all about 146 species (table 26.1), mostly perching birds, shorebirds, waterfowl, gulls, terns, or oceanic birds that are transients or rare or uncommon statewide. The low relative abundance and lack of study of these animals in Florida may account for the absence of data on them. Other species such as Wood Ducks, Mottled Ducks, Wild Turkeys, Northern Bobwhites, and Mourning Doves have been well studied, probably because of their abundance and economic importance. The federal and state survival status (endangered, threatened, or species of special concern) of some species, such as Common Loons, Brown Pelicans, Sandhill Cranes, Whooping Cranes, Bald Eagles, and other raptors, has resulted in increased research on them. On the other hand, the endangered status of some species has resulted in the opposite effect. For example, it was very difficult for some time to work on Wood Storks in Florida because of their highly protected status. Only recently have we been able to obtain permission to secure specimens to examine.

Table 26.1. Summary of major health problems of significance or potential significance and numbers of diseases or disease agents identified for each species of bird in Florida and estimations of investigative effort for each species

Group Species	Total number examined	Parasitic or infectious agents	Toxins	Number of diseases or agents identified	Major health problem/ factor of significance or potential significance[b]
		Numbers examined for[a]			
Loons					
Red-throated Loon	*	*	—	1	Unknown
Pacific Loon	—	—	—	—	Unknown
Common Loon	* * * *	* * * *	* * * *	121	Emaciation syndrome, oiling, aspergillosis
Grebes					
Least Grebe	—	—	—	—	Unknown
Pied-billed Grebe	* * *	* *	—	20	Unknown
Horned Grebe	* * * *	*	—	4	Unknown
Eared Grebe	—	—	—	—	Unknown
Western Grebe	—	—	—	—	Unknown
Albatrosses					
Yellow-nosed Albatross	—	—	—	—	Unknown
Shearwaters and Petrels					
Black-capped Petrel	*	*	—	4	Unknown
Cory's Shearwater	*	*	—	3	Unknown
Greater Shearwater	* * *	* *	—	20	Inclement weather
Sooty Shearwater	*	—	—	—	Unknown
Manx Shearwater	—	—	—	—	Unknown
Audubon's Shearwater	*	*	—	4	Unknown
Storm-Petrels					
Wilson's Storm-Petrel	*	—	—	—	Unknown
Leach's Storm-Petrel	* *	*	—	10	Unknown
Band-rumped Storm-Petrel	—	—	—	—	Unknown
Tropicbirds					
White-tailed Tropicbird	*	*	—	1	Unknown
Red-billed Tropicbird	*	*	—	1	Unknown
Boobies and Gannets					
Masked Booby	*	*	—	2	Unknown
Brown Booby	*	*	—	1	Unknown
Red-footed Booby	*	—	—	2	Unknown
Northern Gannet	* * * *	* *	* *	45	Inclement weather, aspergillosis
Pelicans					
American White Pelican	* *	* *	—	29	Organochlorines
Brown Pelican	* * * *	* * * *	* * * *	142	Fishing gear entanglement, inclement weather, organochlorines(P)
Cormorants					
Great Cormorant	—	—	—	—	Unknown
Double-crested Cormorant	* * * *	* * * *	* * * *	96	Fishing gear entanglement, trauma, emaciation

(continued)

Table 26.1. *(continued)*

Group Species	Total number examined	Numbers examined for[a] Parasitic or infectious agents	Toxins	Number of diseases or agents identified	Major health problem/ factor of significance or potential significance[b]
Anhingas					
Anhinga	****	***	****	50	Unknown
Frigatebirds					
Magnificent Frigatebird	***	*	—	10	Unknown
Bitterns					
American Bittern	**	*	*	21	Unknown
Least Bittern	**	*	**	22	Unknown
Herons					
Great Blue Heron (blue)	****	****	***	64	Eustrongylidosis, power line strikes, mercury poisoning (P)
Great Blue Heron (white)	****	****	***	61	Eustrongylidosis, mercury poisoning(P)
Great Egret	****	****	****	101	Eustrongylidosis, power line strikes, mercury poisoning(P)
Snowy Egret	****	****	***	48	Eustrongylidosis, power line strikes, mercury poisoning(P)
Little Blue Heron	****	****	***	68	Eustrongylidosis, power line strikes, mercury poisoning(P)
Tricolored Heron	****	****	**	56	Eustrongylidosis
Reddish Egret	**	**	*	42	Unknown
Cattle Egret	****	****	***	44	Roadkills
Green Heron	****	*	**	35	Unknown
Black-crowned Night-Heron	***	**	**	49	Unknown
Yellow-crowned Night-Heron	**	*	**	42	Unknown
Ibises					
White Ibis	****	****	****	94	Inclement weather
Scarlet Ibis	—	—	—	—	Unknown
Glossy Ibis	***	*	*	20	Unknown
White-faced Ibis[c]	**	—	—	1	Unknown
Spoonbills					
Roseate Spoonbill	****	****	***	84	Predation
Storks					
Wood Stork	****	****	****	94	Power line strikes, mercury poisoning(P)
American Vultures					
Black Vulture	****	****	*	42	Trauma
Turkey Vulture	****	****	**	31	Unknown
Flamingos					
Greater Flamingo	*	*	—	6	Unknown

(continued)

Table 26.1. *(continued)*

Group Species	Total number examined	Numbers examined for[a] Parasitic or infectious agents	Toxins	Number of diseases or agents identified	Major health problem/ factor of significance or potential significance[b]
Whistling-Ducks[e]					
Fulvous Whistling-Duck	**	**	**	45	Unknown
Black-bellied					
Whistling-Duck	—	—	—	—	Unknown
Swans					
Tundra Swan	—	—	—	—	Unknown
Geese[e]					
Greater White-fronted					
Goose	—	—	—	—	Unknown
Snow Goose	—	—	—	—	Unknown
Ross's Goose	—	—	—	—	Unknown
Brant	—	—	—	—	Unknown
Canada Goose	***	***	—	10	Lead poisoning
Perching Ducks[e]					
Muscovy Duck	*	*	—	1	Unknown
Wood Duck	****	****	**	66	Botulism, lead poisoning
Dabbling Ducks[e]					
Green-winged Teal	****	**	—	4	Botulism
American Black Duck	****	*	***	5	Unknown
Mottled Duck	****	****	**	70	Lead poisoning, botulism
Mallard	****	**	****	14	Botulism
White-cheeked Pintail	—	—	—	—	Unknown
Northern Pintail	****	*	—	2	Botulism, lead poisoning
Blue-winged Teal	****	**	—	10	Botulism, avian cholera
Cinnamon Teal	—	—	—	—	Unknown
Northern Shoveler	****	**	*	25	Botulism
Gadwall	****	*	—	3	Botulism
Eurasian Wigeon	—	—	—	—	Unknown
American Wigeon	****	*	—	2	Unknown
Diving Ducks[e]					
Canvasback	****	—	—	—	Lead poisoning
Redhead	****	*	—	2	Botulism, lead poisoning
Ring-necked Duck	****	**	—	10	Botulism, lead poisoning
Greater Scaup	*	*	—	2	Unknown
Lesser Scaup	***	**	*	26	Brevitoxicosis, oiling, botulism, avian cholera
Eiders[e]					
Common Eider	—	—	—	—	Unknown
King Eider	—	—	—	—	Unknown
Sea Ducks					
Harlequin Duck	—	—	—	—	Unknown
Oldsquaw[e]	—	—	—	—	Unknown
Black Scoter[e]	****	—	—	1	Unknown
Surf Scoter[e]	*	*	—	1	Unknown

(continued)

Table 26.1. *(continued)*

Group Species	Numbers examined for[a]			Number of diseases or agents identified	Major health problem/ factor of significance or potential significance[b]
	Total number examined	Parasitic or infectious agents	Toxins		
White-winged Scoter[e]	—	—	—	—	Unknown
Common Goldeneye[e]	*	—	—	—	Unknown
Bufflehead[e]	* *	*	—	1	Unknown
Mergansers[e]					
Hooded Merganser	*	*	—	2	Unknown
Common Merganser	—	—	—	—	Unknown
Red-breasted Merganser	* *	* *	* *	20	Oiling
Stiff-tailed Ducks[e]					
Ruddy Duck	* * * *	*	—	6	Lead poisoning, botulism
Masked Duck	—	—	—	—	Unknown
Ospreys					
Osprey	* * * *	* * * *	* * *	70	Food stress, trauma
Kites					
Swallow-tailed Kite	* *	* *	*	16	Inclement weather
White-tailed Kites	—	—	—	—	Unknown
Snail Kite	* * * *	* * * *	* *	34	Trauma, inclement weather
Mississippi Kite	* *	* *	—	2	Unknown
Eagles					
Bald Eagle	* * * *	* * * *	* * * *	112	Trauma, electrocution, toxins, pox
Golden Eagle	*	*	*	8	Unknown
Hawks					
Northern Harrier	*	*	*	8	Trauma
Sharp-shinned Hawk	* * *	* *	*	27	Trauma
Cooper's Hawk	* * *	* *	*	34	Trauma
Northern Goshawk	*	*	—	1	Unknown
Red-shouldered Hawk	* * * *	* * *	* *	52	Trauma
Broad-winged Hawk	* *	*	—	17	Trauma
Short-tailed Hawk	*	*	—	2	Unknown
Swainson's Hawk	—	—	—	—	Unknown
Red-tailed Hawk	* * * *	* * *	* *	62	Trauma
Ferruginous Hawk	—	—	—	—	Unknown
Caracaras and falcons					
Crested Caracara	* * * *	* * * *	*	26	Trauma
American Kestrel	* * * *	* * * *	* *	44	Trauma
Merlin	*	*	—	6	Unknown
Peregrine Falcon	* *	* *	* *	42	Trauma
Turkeys					
Wild Turkey[e]	* * * *	* * * *	—	136	Pox, dispharynxosis, blood protozoans
Quail					
Northern Bobwhite[e]	* * * *	* * * *	* *	76	Pox
Rails					
Yellow Rail	*	*	—	2	Unknown
Black Rail	* *	—	—	2	Unknown

(continued)

Table 26.1. *(continued)*

Group Species	Total number examined	Parasitic or infectious agents	Toxins	Number of diseases or agents identified	Major health problem/ factor of significance or potential significance[b]
		Numbers examined for[a]			
Clapper Rail[e]	* * *	* *	*	55	Striking tall structures
King Rail[e]	* *	—	—	3	Striking tall structures
Virginia Rail[e]	* * * *	—	*	7	Striking tall structures
Sora[e]	* * * *	* *	—	16	Striking tall structures
Gallinules					
Purple Gallinule	* * *	* * *	* *	39	Biotoxins
Common Moorhen[e]	* * * *	*	* *	53	Biotoxins
Coots					
American Coot[e]	* * * *	* *	*	34	Avian cholera, botulism
Limpkins					
Limpkin	* *	* *	* *	29	Habitat alteration
Cranes					
Sandhill Crane	* * * *	* * * *	*	74	Roadkills, power line strikes, entanglement in fences
Whooping Crane[d]	* * * *	* * * *	* *	77	Bobcat predation, power line strikes, disseminated visceral coccidiosis(P)
Plovers					
Black-bellied Plover	*	*	—	5	Habitat alteration[f]
Lesser Golden-Plover	—	—	—	—	Unknown
Snowy Plover	*	*	—	6	Habitat alteration[f]
Wilson's Plover	*	*	—	2	Habitat alteration[f]
Semipalmated Plover	*	*	—	2	Habitat alteration[f]
Piping Plover	*	*	—	2	Habitat alteration[f]
Killdeer	* * * *	* *	—	24	Unknown
Mountain Plover	—	—	—	—	Unknown
Oystercatchers					
American Oystercatcher	*	*	—	5	Unknown
Stilts and Avocets					
Black-necked Stilt	* *	*	—	6	Botulism(P)
American Avocet	*	*	—	2	Unknown
Jacanas					
Northern Jacana	*	*	—	1	Unknown
Sandpipers and allies					
Greater Yellowlegs	—	—	—	—	Habitat alteration[f]
Lesser Yellowlegs	*	*	—	—	Habitat alteration[f]
Solitary Sandpiper	*	—	—	1	Habitat alteration[f]
Willet	* *	* *	—	43	Habitat alteration[f]
Spotted Sandpiper	*	*	—	3	Habitat alteration[f]
Upland Sandpiper	*	*	—	1	Habitat alteration[f]
Whimbrel	—	—	—	—	Habitat alteration[f]
Long-billed Curlew	—	—	—	—	Unknown
Black-tailed Godwit	—	—	—	—	Unknown
Hudsonian Godwit	—	—	—	—	Unknown

(continued)

Table 26.1. *(continued)*

| Group
Species | Numbers examined for[a] | | | Number
of diseases
or agents
identified | Major health problem/
factor of significance or
potential significance[b] |
	Total number examined	Parasitic or infectious agents	Toxins		
Bar-tailed Godwit	—	—	—	—	Unknown
Marbled Godwit	—	—	—	—	Unknown
Ruddy Turnstone	**	**	—	8	Habitat alteration[f]
Surfbird	—	—	—	—	Unknown
Red Knot	****	***	*	27	Habitat alteration[f], besnoitiosis
Sanderling	**	*	—	6	Habitat alteration[f], pox
Semipalmated Sandpiper	*	*	—	2	Habitat alteration[f]
Western Sandpiper	****	**	*	8	Habitat alteration[f], pesticides
Least Sandpiper	*	*	—	3	Habitat alteration[f]
White-rumped Sandpiper	—	—	—	—	Unknown
Baird's Sandpiper	—	—	—	—	Unknown
Pectoral Sandpiper	—	—	—	—	Unknown
Sharp-tailed Sandpiper	—	—	—	—	Unknown
Purple Sandpiper	—	—	—	—	Unknown
Dunlin	**	*	*	7	Habitat alteration[f]
Curlew Sandpiper	—	—	—	—	Unknown
Stilt Sandpiper	*	*	—	2	Habitat alteration[f]
Buff-breasted Sandpiper	—	—	—	—	Unknown
Ruff	—	—	—	—	Unknown
Short-billed Dowitcher	*	*	—	8	Habitat alteration[f]
Long-billed Dowitcher	*	—	—	—	Habitat alteration[f]
Common Snipe[e]	**	*	—	6	Unknown
American Woodcock[e]	**	*	*	15	Unknown
Phalaropes					
Wilson's Phalarope	—	—	—	—	Unknown
Red-necked Phalarope	**	—	—	1	Unknown
Red Phalarope	*	*	—	1	Unknown
Jaegers					
Pomarine Jaeger	—	—	—	—	Unknown
Parasitic Jaeger	—	—	—	—	Unknown
Long-tailed Jaeger	—	—	—	—	Unknown
Gulls					
Laughing Gull	****	****	**	37	Roadkills, habitat alteration[f], salmonellosis
Franklin's Gull	—	—	—	—	Unknown
Little Gull	—	—	—	—	Unknown
Common Black-headed Gull	—	—	—	—	Unknown
Bonaparte's Gull	*	*	*	5	Unknown
Band-tailed Gull	—	—	—	—	Unknown
Ring-billed Gull	***	**	*	15	Botulism, oiling
Herring Gull	**	**	—	14	Botulism, oiling
Thayer's Gull	—	—	—	—	Unknown

(continued)

Table 26.1. *(continued)*

Group Species	Total number examined	Numbers examined for[a]		Number of diseases or agents identified	Major health problem/ factor of significance or potential significance[b]
		Parasitic or infectious agents	Toxins		
Iceland Gull	—	—	—	—	Unknown
Lesser Black-backed Gull	—	—	—	—	Unknown
Glaucous Gull	*	*	—	1	Unknown
Great Black-backed Gull	*	*	—	1	Unknown
Black-legged Kittiwake	—	—	—	—	Unknown
Sabine's Gull	—	—	—	—	Unknown
Terns					
Gull-billed Tern	*	*	—	7	Habitat alteration[f]
Caspian Tern	*	*	*	6	Habitat alteration[f]
Royal Tern	* * * *	* *	* *	48	Roadkills, habitat alteration[f]
Sandwich Tern	* *	*	*	8	Habitat alteration[f]
Roseate Tern	* *	*	—	7	Habitat alteration[f]
Common Tern	* *	*	*	10	Habitat alteration[f]
Arctic Tern	*	*	—	1	Unknown
Forster's Tern	*	*	—	6	Habitat alteration[f]
Least Tern	* *	*	*	14	Habitat alteration[f], oiling
Bridled Tern	*	*	*	6	Unknown
Sooty Tern	* * * *	*	* *	19	Inclement weather, habitat alteration[f]
Black Tern	*	*	—	2	Habitat alteration[f]
Brown Noddy	* * *	*	—	8	Habitat alteration[f]
Black Noddy	—	—	—	—	Unknown
Skimmers					
Black Skimmer	* *	* *	*	29	Habitat alteration[f]
Auks, Murres, and Puffins					
Dovekie	* * *	*	—	4	Unknown
Razorbill	—	—	—	—	Unknown
Marbled Murrelet	—	—	—	—	Unknown
Atlantic Puffin	—	—	—	—	Unknown
Thick-billed Murre	—	—	—	—	Unknown
Pigeons and Doves					
Rock Dove	* * * *	* * * *	*	16	Trichomonosis, pox
Scaly-naped Pigeon	*	*	—	2	Unknown
White-crowned Pigeon	* * * *	* * * *	—	11	Trauma, trichomonosis
Band-tailed Pigeon	—	—	—	—	Unknown
European Turtle-Dove	—	—	—	—	Unknown
Eurasian Collared-Dove	* *	* *	—	6	Trichomonosis, paramyxovirus 1 infection
White-winged Dove[e]	* * * *	* * * *	—	35	Trichomonosis(P), brood parasitism(P)
Zenaida Dove	—	—	—	—	Unknown
Mourning Dove[e]	* * * *	* * * *	* *	49	Trichomonosis, pox
Common Ground-Dove	* * *	* * *	*	16	Trichomonosis, West Nile virus

(continued)

Table 26.1. *(continued)*

Group Species	Total number examined	Parasitic or infectious agents	Toxins	Number of diseases or agents identified	Major health problem/ factor of significance or potential significance[b]
Key West Quail-Dove	—	—	—	—	Unknown
Ruddy Quail-Dove	—	—	—	—	Unknown
Parakeets					
Budgerigar	—	—	—	4	Unknown
Monk Parakeet	**	**	—	3	Unknown
Canary-winged Parakeet	—	—	—	—	Unknown
New World Cuckoos					
Black-billed Cuckoo	***	*	*	6	Unknown
Yellow-billed Cuckoo	****	**	**	16	Unknown
Mangrove Cuckoo	*	—	—	1	Unknown
Anis					
Smooth-billed Ani	*	—	—	1	Unknown
Groove-billed Ani	—	—	—	—	Unknown
Owls					
Barn Owl	***	**	*	26	Trauma, poisoning, electrocution
Flammulated Owl	—	—	—	—	Unknown
Eastern Screech-Owl	****	****	**	41	Trauma, poisoning, pox, electrocution, trichomonosis, inclement weather
Great Horned Owl	****	****	*	53	Trauma, poisoning, electrocution
Burrowing Owl	***	**	—	19	Trauma
Barred Owl	****	***	**	55	Trauma, poisoning
Long-eared Owl	*	*	—	1	Unknown
Short-eared Owl	*	*	—	5	Unknown
Northern Saw-whet Owl	*	—	—	1	Unknown
Nighthawks					
Lesser Nighthawk	—	—	—	—	Unknown
Common Nighthawk	**	*	*	7	Trauma
Antillean Nighthawk	—	—	—	—	Unknown
Nightjars					
Chuck-will's-widow	***	*	—	6	Trauma
Whip-poor-will	*	*	—	3	Unknown
Swifts					
White-collared Swift	—	—	—	—	Unknown
Chimney Swift	**	*	—	4	Trauma
Antillean Palm Swift	—	—	—	—	Unknown
Hummingbirds					
Buff-bellied Hummingbird	—	—	—	—	Unknown
Bahama Woodstar	—	—	—	—	Unknown
Ruby-throated Hummingbird	*	*	—	2	Inclement weather

(continued)

Table 26.1. *(continued)*

Group Species	Total number examined	Parasitic or infectious agents	Toxins	Number of diseases or agents identified	Major health problem/ factor of significance or potential significance[b]
		Numbers examined for[a]			
Black-chinned					
Hummingbird	—	—	—	—	Unknown
Anna's Hummingbird	—	—	—	—	Unknown
Calliope Hummingbird	—	—	—	—	Unknown
Rufous Hummingbird	—	—	—	—	Unknown
Kingfishers					
Belted Kingfisher	* *	*	—	14	Unknown
Woodpeckers					
Red-headed Woodpecker	* *	* *	—	4	Unknown
Golden-fronted					
Woodpecker	—	—	—	—	Unknown
Red-bellied Woodpecker	* * *	* * *	* *	20	Unknown
Yellow-bellied Sapsucker	* * * *	*	—	4	Trauma
Downy Woodpecker	*	*	—	0	Unknown
Hairy Woodpecker	*	*	—	3	Unknown
Red-cockaded					
Woodpecker	* *	* *	—	10	Inclement weather
Northern Flicker	* *	* *	—	9	Unknown
Pileated Woodpecker	* *	*	—	10	Unknown
Tyrant Flycatchers					
Olive-sided Flycatcher	—	—	—		Unknown
Eastern Wood-Pewee	* * *	—	—	1	Trauma
Yellow-bellied Flycatcher	*	—	—	1	Unknown
Acadian Flycatcher	* * * *	*	—	1	Trauma
Alder Flycatcher	*	—	—	1	Trauma
Willow Flycatcher	* *	—	—	1	Trauma
Least Flycatcher	*	—	—	1	Unknown
Black Phoebe	—	—	—	—	Unknown
Eastern Phoebe	* *	*	*	3	Trauma
Say's Phoebe	—	—	—	—	Unknown
Vermilion Flycatcher	—	—	—	—	Unknown
Ash-throated Flycatcher	—	—	—	—	Unknown
Great Crested Flycatcher	* * *	* *	—	6	Trauma
Brown-crested Flycatcher	—	—	—	—	Unknown
La Sagra's Flycatcher	—	—	—	—	Unknown
Variegated Flycatcher	—	—	—	—	Unknown
Cassin's Kingbird	—	—	—	—	Unknown
Western Kingbird	—	—	—	—	Unknown
Eastern Kingbird	* *	*	—	4	Trauma
Gray Kingbird	—	—	—	—	Unknown
Loggerhead Kingbird	—	—	—	—	Unknown
Scissor-tailed Flycatcher	—	—	—	—	Unknown
Fork-tailed Flycatcher	—	—	—	—	Unknown

(continued)

Table 26.1. *(continued)*

Group Species	Total number examined	Parasitic or infectious agents	Toxins	Number of diseases or agents identified	Major health problem/ factor of significance or potential significance[b]
		Numbers examined for[a]			
Shrikes					
Loggerhead Shrike	✳ ✳ ✳ ✳	✳ ✳ ✳	✳	16	Unknown
Vireos					
White-eyed Vireo	✳ ✳ ✳ ✳	✳ ✳ ✳	✳	11	Trauma, brood parasitism
Thick-billed Vireo	—	—	—	—	Unknown
Bell's Vireo	—	—	—	—	Unknown
Solitary Vireo	✳ ✳ ✳	—	—	1	Trauma
Yellow-throated Vireo	✳ ✳	—	—	3	Trauma, brood parasitism
Warbling Vireo	✳ ✳	—	—	1	Unknown
Philadelphia Vireo	✳ ✳	—	—	1	Trauma
Red-eyed Vireo	✳ ✳ ✳ ✳	✳	✳	9	Trauma, brood parasitism
Yellow-green Vireo	—	—	—	—	Unknown
Black-whiskered Vireo	✳	✳	—	5	Brood parasitism
Jays and Crows					

Blue Jay	✳ ✳ ✳ ✳	✳ ✳ ✳ ✳	✳ ✳	43	West Nile virus,
					  salmonellosis, mycoplasmal
					  conjunctivitis(P)

Florida Scrub-Jay	✳ ✳ ✳ ✳	✳ ✳ ✳ ✳	—	31	EEE, starvation
American Crow[e]	✳ ✳ ✳ ✳	✳ ✳ ✳	✳	21	West Nile virus, trauma
Fish Crow[e]	✳ ✳	✳ ✳	—	15	West Nile virus
Larks					
Horned Lark	—	—	—	—	Unknown
Swallows					
Purple Martin	✳ ✳ ✳ ✳	✳ ✳	—	8	Inclement weather, cimicid infestations
Cuban Martin	—	—	—	—	Unknown
Southern Martin	—	—	—	—	Unknown
Tree Swallow	✳ ✳ ✳ ✳	✳ ✳	—	7	Trauma, inclement weather
Bahama Swallow	—	—	—	—	Unknown
Northern Rough-winged Swallow	—	—	—	—	Unknown
Bank Swallow	✳	✳	—	3	Unknown
Cliff Swallow	—	—	—	—	Unknown
Cave Swallow	—	—	—	—	Unknown
Barn Swallow	✳ ✳	✳	—	7	Unknown
Titmice					
Carolina Chickadee	✳ ✳	✳ ✳	✳ ✳	2	Unknown
Tufted Titmouse	✳	✳	—	1	Salmonellosis
Nuthatches					
Red-breasted Nuthatch	✳	—	—	1	Unknown
White-breasted Nuthatch	—	—	—	—	Unknown
Brown-headed Nuthatch	✳ ✳	✳ ✳	✳ ✳	4	Unknown
Creepers					
Brown Creeper	✳	—	—	1	Unknown

(continued)

Table 26.1. *(continued)*

Group Species	Total number examined	Parasitic or infectious agents	Toxins	Number of diseases or agents identified	Major health problem/ factor of significance or potential significance[b]
Wrens					
Rock Wren	—	—	—	—	Unknown
Carolina Wren	* * * *	* * *	—	10	Unknown
Bewick's Wren	*	—	—	1	Unknown
House Wren	* * * *	*	*	3	Trauma
Winter Wren	* *	—	—	1	Trauma
Sedge Wren	* * * *	*	—	3	Trauma
Marsh Wren	* * * *	—	—	2	Trauma
Bulbuls					
Red-whiskered Bulbul	—	—	—	—	Trauma
Kinglets					
Golden-crowned Kinglet	* * *	—	—	1	Trauma
Ruby-crowned Kinglet	* * * *	—	—	2	Trauma
Gnatcatchers					
Blue-gray Gnatcatcher	*	*	—	1	Brood parasitism
Thrushes and allies					
Northern Wheatear	—	—	—	—	Unknown
Eastern Bluebird	* * *	* *	* *	15	Inclement weather, brood parasitism
Veery	* * * *	*	*	4	Trauma
Grey-cheeked Thrush	* * * *	—	—	1	Trauma
Bicknell's Thrush	—	—	—	—	Unknown
Swainson's Thrush	* * * *	* *	*	8	Trauma
Hermit Thrush	* * * *	*	*	9	Trauma
Wood Thrush	* * * *	—	—	2	Trauma, brood parasitism
American Robin	* * * *	* * *	* * *	18	Inclement weather, pesticide poisoning, biotoxins, podoknemidokoptosis
Varied Thrush	—	—	—	—	Unknown
Mockingbirds, Thrashers, and Allies					
Gray Catbird	* * * *	* *	*	14	Trauma, inclement weather
Northern Mockingbird	* * * *	* * * *	* *	28	Trauma, pox
Bahama Mockingbird	—	—	—	—	Unknown
Sage Thrasher	—	—	—	—	Unknown
Brown Thrasher	* * * *	* *	*	16	Trauma, salmonellosis
Curved-billed Thrasher	—	—	—	—	Unknown
Starlings					
European Starling	* * * *	*	* * * *	22	Trauma
Pipits					
American Pipit (Water Pipit)	*	—	—	1	Trauma
Sprague's Pipit	*	*	—	1	Trauma
Waxwings					
Cedar Waxwing	* * * *	*	* *	4	Trauma, poisoning

(continued)

Table 26.1. *(continued)*

Group Species	Total number examined	Parasitic or infectious agents	Toxins	Number of diseases or agents identified	Major health problem/ factor of significance or potential significance[b]
Numbers examined for[a]					
Wood-Warblers					
Bachman's Warbler	*	—	—	1	Trauma
Blue-winged Warbler	* * *	*	—	2	Trauma
Golden-winged Warbler	* * *	—	—	1	Trauma
Tennessee Warbler	* * * *	—	—	1	Trauma
Orange-crowned Warbler	* * * *	—	—	1	Trauma
Nashville Warbler	*	*	—	2	Unknown
Northern Parula	* * * *	* *	—	7	Trauma, brood parasitism, inclement weather
Yellow Warbler	* * * *	*	—	4	Trauma
Chestnut-sided Warbler	* * * *	—	—	1	Trauma
Magnolia Warbler	* * * *	—	—	3	Trauma
Cape May Warbler	* * * *	*	—	3	Trauma
Black-throated Blue Warbler	* * * *	* *	*	6	Trauma, inclement weather
Yellow-rumped Warbler	* * * *	*	*	11	Trauma
Black-throated Gray Warbler	—	—	—	—	Unknown
Townsend's Warbler	—	—	—	—	Unknown
Black-throated Green Warbler	* * *	—	—	2	Trauma
Golden-cheeked Warbler	—	—	—	—	Unknown
Blackburnian Warbler	* * *	—	—	1	Trauma
Yellow-throated Warbler	* * * *	*	—	4	Trauma, inclement weather, brood parasitism
Pine Warbler	* * * *	*	* *	6	Trauma, brood parasitism
Kirtland's Warbler	—	—	—	—	Unknown
Prairie Warbler	* * * *	*	—	6	Trauma, brood parasitism
Palm Warbler	* * * *	—	*	7	Trauma, inclement weather
Bay-breasted Warbler	* * * *	—	—	1	Trauma
Blackpoll Warbler	* * * *	*	*	6	Trauma, inclement weather
Cerulean Warbler	* * * *	*	—	2	Trauma
Black-and-white Warbler	* * * *	* *	* *	7	Trauma
American Redstart	* * * *	* * *	* *	8	Trauma, inclement weather
Prothonotary Warbler	* * * *	*	—	1	Trauma
Worm-eating Warbler	* * * *	*	—	2	Trauma
Swainson's Warbler	* * * *	—	—	2	Trauma
Ovenbird	* * * *	* * *	* *	10	Trauma, inclement weather
Northern Waterthrush	* * * *	* *	*	7	Trauma
Louisiana Waterthrush	* * *	—	—	1	Trauma
Kentucky Warbler	* * * *	*	—	3	Trauma
Connecticut Warbler	* *	—	—	3	Trauma
Mourning Warbler	*	—	—	2	Unknown

(continued)

Table 26.1. *(continued)*

Group Species	Total number examined	Parasitic or infectious agents	Toxins	Number of diseases or agents identified	Major health problem/ factor of significance or potential significance[b]
		Numbers examined for[a]			
Common Yellowthroat	****	***	**	13	Trauma, inclement weather, brood parasitism
Hooded Warbler	****	*	—	3	Trauma
Wilson's Warbler	*	—	—	1	Unknown
Canada Warbler	**	*	—	3	Unknown
Yellow-breasted Chat	****	—	—	2	Trauma
Bananaquits					
Bananaquit	—	—	—	—	Unknown
Tanagers					
Stripe-headed Tanager	—	—	—	—	Unknown
Summer Tanager	****	*	—	5	Trauma, brood parasitism
Scarlet Tanager	****	—	—	1	Trauma, brood parasitism
Western Tanager	*	—	—	1	Unknown
Emberizids					
Green-tailed Towhee	—	—	—	—	Unknown
Eastern Towhee	****	***	*	21	Trauma, brood parasitism
Yellow-faced Grassquit	—	—	—	—	Unknown
Black-faced Grassquit	—	—	—	—	Unknown
Bachman's Sparrow	****	***	**	7	Trauma
Chipping Sparrow	****	*	*	10	Trauma, salmonellosis
Clay-colored Sparrow	*	—	—	1	Unknown
Field Sparrow	***	*	—	3	Trauma
Vesper Sparrow	****	*	—	3	Trauma
Lark Sparrow	—	—	—	—	Unknown
Black-throated Sparrow	—	—	—	—	Unknown
Lark Bunting	—	—	—	—	Unknown
Savannah Sparrow	****	**	*	15	Trauma
Grasshopper Sparrow	****	*	—	3	Trauma
(Eastern Grasshopper Sparrow)[g]	**	**	—	4	Trauma
(Florida Grasshopper Sparrow)[g]	****	****	—	11	Trauma, inclement weather
Henslow's Sparrow	***	**	—	3	Trauma
Le Conte's Sparrow	*	—	—	1	Unknown
Saltmarsh Sharp-tailed Sparrow	**	*	—	2	Trauma
Seaside Sparrow	***	*	—	5	Inclement weather
(Cape Sable Seaside Sparrow)[h]	*	—	—	1	Inclement weather
(MacGillivray's Seaside Sparrow)[h]	—	—	—	—	Unknown
(Scott's Seaside Sparrow)[h]	—	—	—	—	Unknown
Louisiana Seaside Sparrow	—	—	—	—	Unknown
Fox Sparrow	**	—	—	1	Trauma

(continued)

Table 26.1. *(continued)*

Group Species	Total number examined	Parasitic or infectious agents	Toxins	Number of diseases or agents identified	Major health problem/ factor of significance or potential significance[b]
		Numbers examined for[a]			
Song Sparrow	****	**	*	5	Trauma
Lincoln's Sparrow	*	—	—	1	Unknown
Swamp Sparrow	****	*	*	7	Trauma
White-throated Sparrow	****	—	*	5	Trauma, salmonellosis
Golden-crowned Sparrow	—	—	—	—	Unknown
White-crowned Sparrow	**	—	—	1	Trauma
Harris' Sparrow	—	—	—	—	Unknown
Dark-eyed Junco	**	—	—	1	Trauma
Lapland Longspur	—	—	—	—	Unknown
Chestnut-collared Longspur	—	—	—	—	Unknown
Snow Bunting	—	—	—	—	Unknown
Cardinals, Grossbeaks, and allies					
Northern Cardinal	****	****	*	29	Trauma, brood parasitism, salmonellosis
Rose-breasted Grosbeak	***	—	—	1	Trauma
Black-headed Grosbeak	—	—	—	—	Unknown
Blue Grosbeak	***	*	—	1	Trauma
Lazuli Bunting	—	—	—	—	Unknown
Indigo Bunting	****	*	—	4	Trauma
Painted Bunting	**	—	—	3	Brood parasitism
Dickcissel	**	—	—	1	Trauma
Blackbirds and Allies					
Bobolink	****	*	*	4	Trauma
Red-winged Blackbird	****	****	**	24	Trauma, brood parasitism, salmonellosis, starvation, podoknemidokoptosis
Tawny-shouldered Blackbird	—	—	—	—	Unknown
Eastern Meadowlark	****	***	*	24	Trauma
Western Meadowlark	—	—	—	—	Unknown
Yellow-headed Blackbird	—	—	—	—	Unknown
Rusty Blackbird	**	*	—	3	Trauma
Brewer's Blackbird	—	—	—	—	Unknown
Boat-tailed Grackle	****	**	*	22	Trauma, podoknemidokoptosis
Common Grackle	****	****	**	27	Salmonellosis, podoknemidokoptosis
Shiny Cowbird	—	—	—	—	Unknown
Bronzed Cowbird	—	—	—	—	Unknown
Brown-headed Cowbird	****	****	—	9	Trauma, salmonellosis, podoknemidokoptosis
Orchard Oriole	***	—	—	3	Trauma, brood parasitism
Spot-breasted Oriole	—	—	—	—	Unknown
Baltimore Oriole	**	*	—	2	Trauma

(continued)

Table 26.1. *(continued)*

Group Species	Numbers examined for[a]			Number of diseases or agents identified	Major health problem/ factor of significance or potential significance[b]
	Total number examined	Parasitic or infectious agents	Toxins		
Cardueline Finches					
Purple Finch	*	*	—	2	Unknown
House Finch	*	* *	—	3	Mycoplasmal conjunctivitis(P)
Red Crossbill	—	—	—	—	Unknown
Pine Siskin	* *	—	—	1	Unknown
American Goldfinch	* *	*	—	3	Trauma
Evening Grosbeak	*	—	—	1	Unknown
Old-World Sparrows					
House Sparrow	* * * *	* * * *	* *	11	Salmonellosis, EEE

a. Numbers of birds examined: — = 0, * = 1–9, ** = 10–49, *** = 50–99, **** = >100.
b. Listed in order of importance for each avian species. (P) = some evidence that this may be a significant factor, but no data to substantiate this for the avian species in Florida.
c. Our only health information on the rare White-faced Ibis was from Jasmin et al. (1972), who reported 11 found dead during an enterotoxemia outbreak; these birds were possibly misidentified.
d. The natural Whooping Crane population is extinct outside of Texas and Canada. All of our data came from an experimentally introduced population in central Florida.
e. For game species, the numbers reported do not include birds legally killed by hunters.
f. Destruction of beach nesting habitat.
g. Subspecies of the Grasshopper Sparrow.
h. Subspecies of the Seaside Sparrow.

The significance of the diseases of wild birds in Florida can be evaluated from three perspectives: (1) their effects on populations of wild birds, (2) their relationship to domesticated and exotic captive animals, and (3) their public health importance.

In table 26.1 the major health problems of significance or potential significance to each species of wild bird found in Florida are summarized. The judgment of importance or potential importance was based on the prevalence of the factor and/or known effects on the species in Florida or elsewhere. Assessments of the amount of effort expended on each avian species to delineate its health problems are reflected in the numbers of specimens examined and the numbers of disease agents identified. This table can be used also to determine areas where additional research is needed. For some species the numbers of diseases, parasites, and other factors are fairly extensive, but even so,

the full significance of these factors is only partially understood. There is a need for long-term studies and continued surveillance of avian populations in order to elucidate the effects of these diseases on populations.

The importance of environmental contaminants (pesticides, metals, radionuclides, etc.) to populations of wild birds in Florida is not fully understood. Although intensive studies on Bald Eagles and a few species of wading birds have been conducted, most other species have not been well examined. More needs to be done. The threats of carbamates and organophosphates to wild birds continue and should be monitored since these compounds are being used more and more in the state. In addition, traumas (road kills and tower strikes) take a sizeable toll on various birds, especially eagles and neotropical migrants, and will certainly increase with burgeoning population growth in Florida and the rapid proliferation of communi-

Table 26.2. Viruses of wild birds in Florida that are harmful or potentially harmful to domestic/exotic animals and/or humans

Virus	Wild bird hosts in Florida	Susceptible domestic/exotic animals	Infect humans?
Adenovirus	Terns, bobwhites	Poultry	Unknown
Avian pox[a]	27 species (including terns, wading birds, cranes, sandpipers, ducks, raptors, bobwhites, turkeys, doves, flickers, and songbirds)	Poultry, waterfowl, exotic birds	No
EEE[b]	46 species (including songbirds (24 spp.), wading birds, raptors, turkeys, bobwhites, cranes, piciforms, columbiforms, anhingas, cormorants, and killdeer)	Horses, emus, pheasants, poultry	Yes
West Nile	Blue Jays, crows, Northern Cardinals, many others	Horses, poultry	Yes
Everglades	Vultures, mockingbirds, and sparrows	None	Yes
Herpesvirus	Turkeys	Poultry	No
Highlands J	13 species (including songbirds (9 spp.), herons, ospreys, bobwhites, and doves)	Horses, partridges, poultry	No
Keystone	Jays	Horses, cattle, dogs	Yes
Newcastle Disease	Pelicans, cormorants, anhingas, and loons	Poultry, waterfowl	Yes
Rabies[c]	Hawks and vultures	All domestic animals	Yes
Retrovirus	Turkeys	Poultry	No
SLE[d]	28 species (including songbirds (10 spp.), wading birds, raptors, turkeys, bobwhites, cranes, columbiforms, anhingas, cormorants, pelicans, gulls, and ducks)	Poultry	Yes
Soldado	Noddies	Poultry	Yes

a. Avian poxviruses tend to be very host specific and therefore only closely related species should be susceptible.
b. Eastern equine encephalitis.
c. Based on serologic evidence only.
d. St. Louis encephalitis.

cation towers throughout the state. Ways to avoid these types of mortality need to be determined and applied.

At least 153 infectious and parasitic disease agents are potentially shared by or exchanged between wild birds, various exotic caged birds, poultry, and other domestic animals in Florida (tables 26.2–26.9). These include viruses (11), bacteria (26), fungi (5), protozoans (21), trematodes (27), cestodes (2), nematodes (27), acanthocephalans (2), ticks (4), mites (12), chewing lice (13), fleas (1), and louse flies (2). These numbers are minimal, since many were grouped in generic categories rather than listing total

Table 26.3. Bacteria of wild birds in Florida that are harmful or potentially harmful to domestic/exotic animals and/or humans

Bacteria	Wild bird hosts in Florida	Susceptible domestic/exotic animals[a]	Infect humans?
Acinetobacter caloaceticus	Loons, pelicans, storks, cranes, eagles, falcons	Ducks, parrots	Yes
Actinobacillus spp.	Gannets, eagles, ospreys	Waterfowl, parrots, domestic mammals	No
Aeromonas hydrophila	Loons, pelicans, gannets, egrets, herons, flamingos	Caged birds, waterfowl, raptors	No
Arizona hinshawii	Turkeys	Poultry	Yes
Campylobacter sp.	Loons, cranes	Poultry, ducks, caged birds, domestic mammals	Yes
Chlamydia psittaci	Spoonbills, perhaps Bald Eagles	Domestic turkeys, ducks, pigeons, caged psitticines, passeriforms	Yes
Citrobacter freundii	Herons, turkeys	Caged birds	No
Clostridium perfringens	Pelicans, terns, skimmers, gulls, egrets, herons, ibises, spoonbills, shorebirds, ducks	Poultry, caged birds, waterfowl, domestic mammals	Yes
Edwardsiella tarda	Loons, pelicans, cormorants, anhingas, gulls, herons, cranes, eagles, ospreys	Fish, amphibians, reptiles, swine, cattle, marine mammals, primates, camels, dogs, penguins	Yes
Enterobacter spp.	Loons, pelicans, terns, gulls, herons, cranes, ducks, raptors, turkeys, perching birds, woodpeckers	Poultry, caged birds	No
Enterococcus spp.	Pelicans, cormorants, gulls, storks, cranes, raptors, turkeys, woodpeckers	Caged birds	No
Erysipelothrix rhusiopathiae	Pelicans	Caged birds, emus, guinea fowl, raptors, ducks, poultry, swine, marine mammals	Yes
Escherichia coli[b]	Loons, pelicans, cormorants, terns, skimmers, gulls, shearwaters, gannets, herons, flamingos, storks, cranes, coots, shorebirds, ducks, raptors, bobwhites, turkeys, perching birds, woodpeckers	Caged birds, ducks, poultry, calves, swine	Yes
Haemophilus spp.	Eagles	Caged birds, poultry, domestic mammals	No
Klebsiella spp.	Grebes, loons, pelicans, cormorants, gulls, egrets,	Caged birds, poultry, domestic mammals	Yes

(continued)

Table 26.3. *(continued)*

Bacteria	Wild bird hosts in Florida	Susceptible domestic/exotic animals[a]	Infect humans?
	herons, storks, cranes, raptors, turkeys, perching birds, woodpeckers		
Mycobacterium spp.	Loons, cranes, turkeys	Caged birds, poultry, domestic mammals, zoo animals	Yes
Mycoplasma spp.	Cattle egrets, raptors, perching birds	Caged birds, poultry	No
Pasteurella multocida	Grebes, loons, cormorants, terns, herons, coots, flamingos, cranes, ducks, eagles, sparrows	Caged birds, poultry, waterfowl, domestic mammals, zoo animals	Yes
Proteus spp.	Loons, pelicans, cormorants, terns, gulls, gannets, herons, storks, cranes, coots, shorebirds, ducks, raptors, turkeys, perching birds	Caged birds, quail, domestic mammals	No
Pseudomonas aeruginosa	Herons, storks, raptors, bobwhites	Caged birds, domestic mammals, pen-reared mink and chinchillas	No
Salmonella spp.	32 serotypes identified in Florida birds, including loons, pelicans, cormorants, gulls, gannets, egrets, herons, cranes, shorebirds, ducks, raptors, turkeys, perching birds	Caged birds, poultry, domestic animals, zoo animals	Yes
Serratia spp.	Loons, gannets, storks, cranes, raptors, turkeys, perching birds, woodpeckers	Caged birds	No
Staphylococcus aureus	Pelicans, cormorants, raptors, turkeys	Poultry, raptors, waterfowl, domestic mammals	Yes
Streptococcus spp.	Grebes, loons, pelicans, shearwaters, herons, egrets, spoonbills, cranes, ducks, raptors, turkeys	Poultry, raptors, pigeons, waterfowl, domestic mammals	Yes
Vibrio spp.	Loons, pelicans, herons, raptors	Saltwater fish, shellfish	Yes
Weeksella virosa	Loons	Unknown	Yes
Yersinia enterocolitica	Pelicans, gulls	Waterfowl, caged birds, domestic mammals	Yes

a. Caged birds = Passeriformes (perching birds), Piciformes (woodpeckers and allies), Psittaciformes (parrots, parakeets, lories, macaws).
b. The avian strains of *E. coli* are not thought to be significant causes of infection in species other than birds (Friend 1999).

Table 26.4. Fungi of wild birds in Florida that are harmful or potentially harmful to domestic/exotic animals and/or humans

Fungus	Wild bird hosts in Florida	Susceptible domestic/exotic animals[a]	Infect humans?
Aspergillus spp.	Loons, pelicans, cormorants, gulls, terns, gannets, herons, egrets, cranes, raptors, bobwhites, perching birds	Poultry, caged birds, penguins, flamingos, marine mammals, other mammals	Yes
Candida spp.	Pelicans, herons, raptors, turkeys	Poultry, caged birds, marine mammals	Yes
Mucor spp.	Cranes, raptors, bobwhites	Caged birds	Yes
Rhinosporidium seeberi	Swans	Caged birds, waterfowl, domestic mammals	Yes
Rhizopus spp.	Eagles	Caged birds	Yes

a. Caged birds = Passeriformes (perching birds), Piciformes (woodpeckers and allies), Psittaciformes (parrots, parakeets, lories, macaws).

Table 26.5. Protozoans of wild birds in Florida that are harmful or potentially harmful to domestic/exotic animals and/or humans

Protozoan	Wild bird hosts in Florida	Susceptible animals[a]	Infect humans?
Atoxoplasma spp.	Cranes, gallinules, cowbirds	Caged passeriformes	No
Cryptosporidium spp.	Ducks	Poultry, caged birds, waterfowl	No[b]
Eimeria grus and *E. reichenowi*	Cranes	Exotic cranes	No
Eimeria labbeana	Doves	Exotic columbiforms	No
Eimeria lettyae	Bobwhites	Pen-reared bobwhites	No
Eimeria spp.	Ducks, mergansers, wild turkeys	Domestic and exotic waterfowl, domestic turkeys	No
Haemoproteus meleagridis	Wild turkeys	Domestic turkeys	No
Hexamita sp.	Cranes	Exotic cranes, caged birds, poultry	No
Histomonas meleagridis	Wild turkeys	Domestic turkeys	No
Isospora sp.	Jays, sparrows	Caged passeriforms	No
Leucocytozoon simondi	Ducks	Domestic and exotic waterfowl	No
Leucocytozoon smithi	Wild turkeys	Domestic turkeys	No
Plasmodium circumflexum	Jays, mockingbirds, owls	Caged passerines, waterfowl	No
Plasmodium elongatum and *P. relictum*	Raptors, herons	Exotic birds, especially penguins	No
Plasmodium hermani	Turkeys, bobwhites	Domestic turkeys	No
Sarcocystis falcatula	Cowbirds, grackles	Caged psittacines	No
Toxoplasma gondii	Gulls, egrets, mockingbirds, grackles	Domestic and exotic animals	Yes
Trichomonas gallinae	Pigeons, doves, owls	Caged and exotic birds, especially columbiforms	No

a. Caged birds = Passeriformes (perching birds), piciformes (woodpeckers and allies), Psittaciformes (parrots, parakeets, lories, macaws).
b. Immunocompromised humans are susceptible.

Table 26.6. Trematodes and cestodes of wild birds in Florida that are harmful or potentially harmful to domestic/exotic animals and/or humans

Helminth	Wild bird hosts in Florida	Susceptible domestic/exotic animals	Infect humans?
Trematodes			
Amphimerus spp.	Loons, kingfishers	Chickens, domestic and exotic ducks	Yes[a]
Athesmia heterolecithodes	Rails, gallinules	Domestic turkeys, exotic birds, monkeys, bats	No
Austrobilharzia spp.	Loons, doves	Exotic gaviids and columbiforms	Yes[b]
Brachylecithum spp.	Raptors, passeriforms	Caged psitticines	No
Cotylurus spp.	Loons, gulls, ducks	Chickens, domestic and exotic ducks	No
Cryptocotyle spp.	Gulls	Domestic and exotic ducks	No
Dendritobilharzia pulverulenta	Loons, pelicans, spoonbills, ducks	Domestic and exotic waterbirds	Yes[b]
Echinoparyphium recurvatum	Ducks, turkeys	Poultry, waterbirds, columbiforms	Yes[a]
Echinoparyphium spp.	Plovers, ducks, hawks	Poultry	No
Echinostoma trivolvis	Ducks, kestrels, turkeys, passeriforms	Turkeys, domestic ducks, caged passeriforms	Yes[a]
Echinostoma spp.	Gannets, gallinules, coots, kites, crows	Domestic and exotic waterfowl	No
Hypoderaeum conoideum	Ducks, geese	Poultry, pigeons, domestic and exotic waterfowl	Yes
Lutztrema microstomum	Blue Jays	Caged psittacines	No
Lyperosomum spp.	Ibises, rails, passeriforms	Caged psittacines	No
Notocotylus spp.	Rails, gallinules, coots, willets, ducks	Chickens, domestic and exotic ducks	No
Ornithobilharzia spp.	Cormorants, terns	Exotic waterbirds	Yes[b]
Philophthalmus gralli	Ducks	Domestic ducks, gulls, ostriches	No
Philophthalmus spp.	Rails, gulls, terns, willets	Domestic and exotic waterbirds	Unknown
Plagiorchis spp.	Owls	Poultry	Unknown
Platynosomum illiciens	Raptors	Caged psittacines	No
Prosthogomimus spp.	Loons, wading birds, cranes, rails, coots, ducks, raptors, turkeys	Poultry, domestic and exotic waterbirds	No
Renicola spp.	Loons, pelicans, cormorants, gulls, terns, skimmers, wading birds, rails, raptors	Domestic turkeys, exotic waterbirds, raptors	No
Ribeiroia ondatrae	Loons, pelicans, wading birds, cranes, ducks, raptors	Chickens, domestic ducks, exotic waterbirds, mammals	No
Tanaisia spp.	Ibises, cranes, rails, gallinules, coots, killdeer, turkeys, doves	Poultry, pigeons, caged passeriforms	No

(continued)

Table 26.6. *(continued)*

Helminth	Wild bird hosts in Florida	Susceptible domestic/exotic animals	Infect humans?
Trichobilharzia spp.	Ducks	Domestic and exotic ducks	Yes[b]
Typhlocoelum cucumerinum	Ducks	Domestic and exotic ducks	No
Zygocotyle lunata	Ducks, turkeys	Poultry, domestic and exotic waterfowl, cattle, sheep	No
Cestodes			
Raillietina tetragona	Bobwhites	Poultry	Unknown
Raillietina spp.	Owls, turkeys, doves, woodpeckers	Caged psittacines	Unknown

a. Occurs only rarely.

b. Causes "swimmer's itch" (dermatitis) in which larvae penetrate the skin; these larvae eventually die and no adult infections occur.

numbers of species or serotypes involved. For example, the number of bacteria was determined by considering *Salmonella* as 1 bacterium rather than listing 32 serotypes known from Florida wild birds. Of the 152 agents, 72 could be linked with poultry (chickens and domestic turkeys) and 112 with exotic or caged birds in zoological collections. The numbers could be further inflated if all 125 parasites and infectious disease agents from Wild Turkeys were included, since theoretically any agents that infect/infest Wild Turkeys should also be able to infect/infest domestic turkeys. Tables 26.2–26.9 were constructed by considering what has been found previously in Florida wild birds and what is known to occur in poultry, caged or aviary birds, and other exotic avian species in the United States. For this latter type of information, the following references were consulted: Hofstad et al. (1978) and Calnek et al. (1997) for poultry, and Ritchie et al. (1994), Rosskopf and Woerpel (1996), and Altman et al. (1997) for caged/exotic birds. When examining these lists of diseases and parasites that are common to both wild and domestic or exotic birds, it must be remembered that the flow of these diseases is not always in one direction. Some are disease agents primarily of wild birds and only secondarily infect domestic birds, and vice versa. An example of this is eastern equine encephalitis virus, an arbovirus that circulates

mainly in various species of wild birds, especially passeriforms, and spills over into horses, emus, and poultry during epizootic years when ecological conditions are optimal for transmission by mosquito vectors. Newcastle disease virus, on the other hand, is thought to be mainly a problem in poultry and secondarily in wild birds, although this has not been proven definitively. Much is yet to be learned about the complex relationships of disease agents shared by wild birds and their domestic or exotic counterparts.

At least 49 of the parasitic and infectious disease agents that occur in wild birds in Florida are potentially public health problems in that they have also been reported from humans (tables 26.2–26.9). These include viruses (7), bacteria (17), fungi (5), protozoans (1), trematodes (8), nematodes (3), ticks (4), mites (2), fleas (1), and louse flies (1). Of these, probably the most significant are eastern equine encephalitis virus, St. Louis encephalitis virus, West Nile virus, and the various serotypes of *Salmonella*. The other agents occur in low prevalences and result in only occasional human exposures.

It should be remembered that all the agents listed in the tables might not infect humans through direct contact with birds, but rather through the environment or various vectors and intermediate hosts. For example, it is not likely that a person will contract aspergillosis

Table 26.7. Nematodes and acanthocephalans of wild birds in Florida that are harmful or potentially harmful to domestic/exotic animals and/or humans

Helminth	Wild bird hosts in Florida	Susceptible domestic/exotic animals[a]	Infect humans?
Nematodes			
Amidostomum spp.	Cranes, rails, gallinules, coots, waterfowl	Caged birds, columbiforms, waterfowl	No
Aproctella stoddardi	Ducks, bobwhites, turkeys, doves, kingfishers	Poultry, game birds	No
Ascaridia spp.	Cranes, bobwhites, turkeys, doves	Exotic cranes, columbiformes, game birds, waterfowl	No
Capillaria contorta	Pelicans, cormorants, herons, ibises, raptors, ducks, bobwhites	Poultry, waterfowl, game birds	No
Chandlerella spp.	Cranes, owls, turkeys, grackles	Caged birds	No
Cheilospirura spinosa	Bobwhites, turkeys	Game birds	No
Contracaecum spp.	Loons, pelicans, cormorants, anhingas, shearwaters, gannets, frigatebirds, wading birds, waterfowl, raptors	Exotic birds	Yes[b]
Cosmocephalus obvelatus	Loons, pelicans, wading birds	Exotic waterbirds	No
Cyathostoma spp.	Loons, pelicans, ibises, cranes, hawks	Poultry, waterfowl, caged birds	No
Dispharynx nasuta	Cranes, killdeer, raptors, doves, passeriforms	Poultry, game birds, caged passeriforms	No
Echinuria uncinata	Ducks	Domestic and exotic waterfowl	No
Epomidiostomum spp.	Waterfowl	Domestic and exotic waterfowl	No
Eustrongylides ignotus	Wading birds	Exotic wading birds	Yes[b]
Eustrongylides tubifex	Common Loons	Domestic ducks, exotic waterbirds	No
Hamatospiculum sp.[c]	Eagles	Captive and exotic raptors	No
Heterakis spp.	Bobwhites, turkeys	Game birds, waterfowl	No
Hystrichis tricolor	Gallinules, soras, coots, snipe, ducks	Exotic waterbirds	No
Oxyspirura spp.	Kestrels, bobwhites, jays, woodpeckers	Poultry, waterfowl, game birds, caged birds	No
Ornithostrongylus spp.	Doves	Exotic columbiforms	No
Pelecitus spp.	Herons	Caged psittacines	No
Physocephalus sexalatus	Shrikes, owls, hawks	Pigs, horses, rabbits	No
Porrocaecum spp.	Herons, ducks, raptors, passeriforms	Caged birds	Yes[b]
Strongyloides spp.	Wading birds, cranes, soras, gallinules, coots, limpkins, killdeer, ducks, owls, bobwhites, turkeys, passeriforms	Poultry, waterfowl, caged birds	No
Syngamus trachea	Cormorants, ibises, cranes	Poultry, waterfowl, game birds	No
Synhimantus spp.	Pelicans, cormorants, wading birds, cranes, raptors, turkeys	Exotic birds	No

(continued)

Table 26.7. *(continued)*

Helminth	Wild bird hosts in Florida	Susceptible domestic/exotic animals[a]	Infect humans?
Tetrameres spp.	Grebes, pelicans, cormorants, anhingas, wading birds, flamingos, cranes, coots, rails, woodcock, waterfowl, raptors, columbiforms, kingfishers	Poultry, game birds, waterfowl, columbiforms, caged birds	No
Trichostrongylus cramae[d]	Cranes, geese, bobwhites, turkeys	Poultry, game birds, waterfowl, exotic cranes	No
Acanthocephalans			
Polymorphus spp.	Loons, cormorants, anhingas, wading birds, cranes, rails, gallinules, coots, ducks, raptors	Poultry	No
Prosthorhynchus sp.	Willets	Poultry	No

a. Caged birds = Passeriformes (perching birds), Piciformes (woodpeckers and allies), Psittaciformes (parrots, parakeets, lories, macaws).
b. Occurs only rarely.
c. In older literature, referred to as *Serratospiculum amaculata* or *Serratospiculoides amacutata*.
d. In older literature, referred to as *Trichostrongylus tenuis*.

Table 26.8. Ticks and mites of wild birds in Florida that are harmful or potentially harmful to domestic/exotic animals and/or humans

Arthropod	Wild bird hosts in Florida	Susceptible domestic/exotic animals[a]	Infest humans?
Ticks			
Amblyomma spp.	Raptors, bobwhites, passeriforms	Poultry, domestic mammals, caged psittacines	Yes
Argas spp.	Pelicans, cormorants	Poultry, caged psittacines	Rarely
Haemaphysalis spp.	Bobwhites, passeriforms	Poultry, domestic rabbits	Rarely
Ixodes spp.	Shearwaters, eagles, bobwhites, passeriforms	Caged psittacines, domestic mammals	Yes
Mites			
Analges spp.	Passeriforms	Exotic passeriforms and psittacines	No
Dermoglyphus spp.	Bobwhites, doves	Caged birds	No
Dermation spp.	Shearwaters, storm-petrels, wading birds, ducks	Caged birds	No
Dubininia spp.	Caracaras	Caged psittacines	No
Eutrombicula spp.	Passeriforms	Poultry, domestic mammals	Yes
Knemidokoptes jamaicensis	Passeriforms	Caged and exotic passeriforms	No
Megninia spp.	Wild Turkeys	Poultry	No
Myialges nudus	Cranes	Caged birds	No
Ornithonyssus spp.	Cormorants, storks, raptors, bobwhites, passeriforms	Poultry, psittacines, columbiforms	Yes
Colinolichus virginianus	Bobwhites	Pen-reared bobwhites, perhaps domestic poultry	No
Rivoltasia sp.	Bobwhites	Caged birds	No
Syringophilus columbae	White-crowned Pigeons	Columbiforms	No

a. Caged birds = Passeriformes (perching birds), Piciformes (woodpeckers and allies), Psittaciformes (parrots, parakeets, lories, macaws).

Table 26.9. Chewing lice, louse flies,, fleas of wild birds in Florida that are harmful or potentially harmful to domestic/exotic animals and/or humans

Arthropod	Wild bird hosts in Florida	Susceptible domestic/exotic animals[a]	Infest humans?
Chewing lice			
Anaticola crassicornis	Ducks, mergansers	Domestic waterfowl	No
Anatoecus icterodes and *A. dentatus*	Ducks, mergansers	Domestic waterfowl	No
Bonomiella columbae	Doves	Columbiforms	No
Campanulotes compar	Rock Doves	Domestic pigeons	No
Chelopistes meleagridis	Wild Turkeys	Domestic turkeys	No
Colpocephalum turbinatum	Raptors	Domestic pigeons, caged raptors	No
Columbicola columbae	Rock Doves	Domestic pigeons, caged psitticines	No
Menacanthus stramineus	Wild Turkeys	Poultry, guinea fowl	No
Oxylipeurus corpulentus and *O. polytrapezius*	Wild Turkeys	Domestic turkeys	No
Physconelloides zenaidurae	Mourning Doves	Domestic pigeons	No
Trinoton querquedulae	Ducks, mergansers	Domestic waterfowl	No
Louse flies			
Icosta spp.	Wading birds, cranes, owls	Caged birds	Unknown
Pseudolynchia spp.	Owls, doves, nighthawks, nightjars	Caged birds	Yes[b]
Fleas			
Echidnophaga gallinacea	Ducks, raptors, bobwhites, jays	Poultry, caged passeriforms, columbiforms, psittacines, domestic and exotic mammals	Yes

a. Caged birds = Passeriformes (perching birds), Piciformes (woodpeckers and allies), Psittaciformes (parrots, parakeets, lories, macaws).
b. Known to bite humans, but only as a temporary host.

through contact with infected birds, but rather through inhaling or ingesting the fungal spores from the contaminated environment. Most of the zoonotic diseases can be prevented by observing careful practices of sanitation and cleanliness. It must be remembered that not all birds in Florida are diseased and certainly not all have zoonotic infections or diseases. It is necessary to maintain a balanced and informed viewpoint in order to keep these aspects in proper perspective.

Surveillance of diseases, parasites, and environmental contaminants and of other morbid-ity and mortality factors should be an ongoing activity, not only to recognize trends and changes in environmental health in general, but also to evaluate the health of populations of free-ranging birds in Florida. Health and disease situations in wildlife populations are dynamic and change is often the rule, rather than the exception. Continuous monitoring will allow us to recognize new problems when they arise and also ensure that useful information will be obtained and made available for the improved management and conservation of Florida's unique and valuable avian resource.

Literature cited

Altman, R.B., S.L. Clubb, G.M. Dorrestein, and K. Quesenberry (eds.). 1997. *Avian medicine and surgery.* W.B. Saunders, Philadelphia. 1,070 pp.

Calnek, B.W., H.J. Barnes, C.W. Beard, L.R. McDougald, and Y.M. Saif (eds.). 1997. *Diseases of poultry.* 10th ed. Iowa State University Press, Ames. 1080 pp.

Friend, M. 1999. Miscellaneous bacterial diseases. In: *Field manual of wildlife diseases. General field procedures and diseases of birds.* M. Friend and J.C. Franson (eds.). U.S. Department of the Interior, U.S. Geological Survey, Biological Research Division, Information and Technology Report 1999-001. Washington, D.C. pp. 121–126.

Hofstad, M.S., B.W. Calnek, C.F. Helmboldt, W.M. Reid, and H.W. Yoder, Jr. (eds.). 1978. *Diseases of poultry.* 7th ed. Iowa State University Press, Ames. 949 pp.

Jasmin, A.M., D.E. Cooperrider, C.P. Powell, and J.N. Baucom. 1972. Enterotoxemia of wildfowl due to *Cl. perfringens* type C. *J. Wildl. Dis.* 8:79–84.

Ritchie, B.W., G.J. Harrison, and L.R. Harrison (eds.). 1994. *Avian medicine: principles and applications.* Wingers, Lake Worth, Fla. 1,384 pp.

Rosskopf, W.J., Jr., and R.W. Woerpel. (eds.) 1996. *Diseases of cage and aviary birds.* 3d ed. Williams and Wilkins, Baltimore. 1,088 pp.

Index

Index

About the Authors

Donald J. Forrester is professor of parasitology and wildlife diseases in the Department of Pathobiology, College of Veterinary Medicine, and affiliate professor of wildlife ecology in the Department of Wildlife Ecology and Conservation, College of Agriculture and Life Sciences, both at the University of Florida, Gainesville. From 1981 to 1986 he was editor of the *Journal of Wildlife Diseases*. He is the author of *Parasites and Diseases of Wild Mammals in Florida* (UPF, 1992) as well as more than 200 scientific publications, mainly on parasites and diseases of wildlife.

Marilyn G. Spalding is associate scientist in the Department of Pathobiology, College of Veterinary Medicine at the University of Florida, Gainesville, specializing in the diseases of wild birds, particularly water birds. She acts as the consulting veterinarian to the Florida Fish and Wildlife Conservation Commission in its efforts to reintroduce the whooping crane to Florida and has published 57 scientific publications, most dealing with diseases of wild birds.